STUDENT SOLUTIONS MANUAL

Lea Rosenberry

INTERMEDIATE ALGEBRA for College Students

Third Edition

Robert Blitzer

Upper Saddle River, NJ 07458

Executive Editor: Karin E. Wagner
Supplements Editor: Elizabeth Covello
Editorial Assistant: Rudy Leon
Assistant Managing Editor: John Matthews
Production Editor: Wendy A. Perez
Supplement Cover Manager: Paul Gourhan
Supplement Cover Designer: PM Workshop Inc.
Manufacturing Buyer: Ilene Kahn

Prentice-Hall, Inc.
Upper Saddle River, NJ 07458

Printed in the United States of America

10 9 8 7 6 5 4

ISBN 0-13-034289-0

Pearson Education Ltd., *London*
Pearson Education Australia Pty. Ltd., *Sydney*
Pearson Education Singapore, Pte. Ltd.
Pearson Education North Asia Ltd., *Hong Kong*
Pearson Education Canada, Inc., *Toronto*
Pearson Educacíon de Mexico, S.A. de C.V.
Pearson Education—Japan, *Tokyo*
Pearson Education Malaysia, Pte. Ltd.

Table of Contents

Chapter 1 **Algebra, Mathematical Models, and Problem Solving** **1**

1.1 Algebraic Expressions and Real Numbers 1

1.2 Operations with Real Numbers and Simplifying Algebraic Expressions 3

1.3 Graphing Equations 8

1.4 Solving Linear Equations 13

1.5 Problem Solving and Using Formulas 19

1.6 Properties of Integral Exponents 25

1.7 Scientific Notation 30

Chapter 1 Review 34

Chapter 1 Test 43

Chapter 2 **Functions and Linear Functions** **47**

2.1 Introduction to Functions 47

2.2 The Algebra of Functions 53

2.3 Linear Functions and Slope 57

2.4 The Point-Slope Form of the Equation of a Line 65

Chapter 2 Review 76

Chapter 2 Test 83

Cumulative Review Chapters 1-2 85

Chapter 3 **Systems of Linear Equations** **88**

3.1 Systems of Linear Equations in Two Variables 88

3.2 Problem Solving and Business Applications Using Systems of Equations 103

3.3 Systems of Linear Equations in Three Variables 113

3.4 Matrix Solutions to Linear Systems 128

3.5 Determinants and Cramer's Rule 140

Chapter 3 Review 153

Chapter 3 Test 169

Cumulative Review Chapters 1-3 174

Chapter 4 **Inequalities and Problem Solving ... 180**
4.1 Solving Linear Inequalities ... 180
4.2 Compound Inequalities ... 186
4.3 Equations and Inequalities Involving Absolute Value ... 195
4.4 Linear Inequalities in Two Variables ... 204
4.5 Linear Programming ... 217
Chapter 4 Review ... 227
Chapter 4 Test ... 243
Cumulative Review Chapters 1-4 ... 250

Chapter 5 **Polynomials, Polynomial Functions, and Factoring ... 255**
5.1 Introduction to Polynomials and Polynomial Functions ... 255
5.2 Multiplication of Polynomials ... 262
5.3 Greatest Common Factors and Factoring by Grouping ... 270
5.4 Factoring Trinomials ... 275
5.5 Factoring Special Forms ... 282
5.6 A General Factoring Strategy ... 289
5.7 Polynomial Equations and Their Applications ... 293
Chapter 5 Review ... 303
Chapter 5 Test ... 314
Cumulative Review Chapters 1-5 ... 318

Chapter 6 **Rational Expressions, Functions, and Equations ... 322**
6.1 Multiplying and Dividing Rational Expressions ... 322
6.2 Adding and Subtracting Rational Expressions ... 331
6.3 Complex Rational Expressions ... 342
6.4 Division of Polynomials ... 350
6.5 Synthetic Division and the Remainder Theorem ... 356
6.6 Rational Equations ... 361
6.7 Formulas and Applications of Rational Equations ... 373
6.8 Modeling Using Variation ... 383
Chapter 6 Review ... 390
Chapter 6 Test ... 407
Cumulative Review Chapters 1-6 ... 414

Chapter 7 **Radicals, Radical Functions, and Rational Exponents....................420**

7.1 Radical Expressions and Functions....................420

7.2 Rational Exponents....................425

7.3 Multiplying and Simplifying Radical Expressions....................432

7.4 Adding, Subtracting, and Dividing Radical Expressions....................437

7.5 Multiplying With More Than One Term and Rationalizing Denominators....................442

7.6 Radical Equations....................454

7.7 Complex Numbers....................461

Chapter 7 Review....................468

Chapter 7 Test....................476

Cumulative Review Chapters 1-7....................478

Chapter 8 **Quadratic Equations and Functions....................484**

8.1 The Square Root Property and Completing the Square....................484

8.2 The Quadratic Formula....................496

8.3 Quadratic Functions and Their Graphs....................508

8.4 Equations Quadratic in Form....................523

8.5 Quadratic and Rational Inequalities....................533

Chapter 8 Review....................555

Chapter 8 Test....................571

Cumulative Review Chapters 1-8....................577

Chapter 9 **Exponential and Logarithmic Functions....................584**

9.1 Exponential Functions....................584

9.2 Composite and Inverse Functions....................592

9.3 Logarithmic Functions....................601

9.4 Properties of Logarithms....................608

9.5 Exponential and Logarithmic Equations....................613

9.6 Exponential Growth and Decay; Modeling Data....................624

Chapter 9 Review....................628

Chapter 9 Test....................640

Cumulative Review Chapters 1-9....................644

Chapter 10 Conic Sections and Systems of Nonlinear Equations 650

10.1 Distance and Midpoint Formulas; Circles.......................................650
10.2 The Ellipse...658
10.3 The Hyperbola..667
10.4 The Parabola; Identifying Conic Sections.......................................675
10.5 Systems of Nonlinear Equations in Two Variables 688
Chapter 10 Review 703
Chapter 10 Test 720
Cumulative Review Chapters 1-10 726

Chapter 11 Arithmetic Sequences 732

11.1 Sequences and Summation Notation...732
11.2 Arithmetic Sequences..738
11.3 Geometric Sequences 744
11.4 The Binomial Theorem 751
Chapter 11 Review 763
Chapter 11 Test 772
Cumulative Review Chapters 1-11 775

Chapter 1

Check Points 1.1

1. a. $8x+5$

   b. $\frac{x}{7}-2x$

2. $81-0.6x=81-0.6(50)$
   $=81-30=51$
   In 1950, approximately 51% of American adults smoked cigarettes.

3. $8+6(x-3)^2=8+6(13-3)^2$
   $=8+6(10)^2$
   $=8+6(100)$
   $=8+600=608$

4. Since 2000 is 40 years after 1960, evaluate the formula for $x=40$.
   $P=0.72x^2+9.4x+783$
   $=0.72(40)^2+9.4(40)+783$
   $=0.72(1600)+376+783$
   $=1152+376+783$
   $=2311$
   In 2000, the gray wolf population in the United States was 2311. The model fits the data quite well.

5. a. True. 13 is an integer.

   b. True. 6 is not an element of the set.

6. a. −8 is less than −2. True.

   b. 7 is greater than −3. True.

   c. −1 is less than or equal to −4. False.

   d. 5 is greater than or equal to 5. True.

   e. 2 is greater than or equal to −14. True.

Problem Set 1.1

Practice Exercises

1. $x+5$

3. $x-4$

5. $4x$

7. $2x+10$

9. $6-\frac{1}{2}x$

11. $\frac{4}{x}-2$

13. $\frac{3}{5-x}$

15. $7+5(10)=7+50=57$

17. $6(3)-8=18-8=10$

19. $8^2+3(8)=64+24=88$

21. $7^2-6(7)+3=49-42+3=7+3=10$

23. $4+5(9-7)^3 = 4+5(2)^3 = 4+5(8)$
$= 4+40 = 44$

25. $8^2 - 3(8-2) = 64 - 3(6)$
$= 64 - 18 = 46$

27. $\{1, 2, 3, 4\}$

29. $\{-7, -6, -5, -4\}$

31. $\{8, 9, 10, \ldots\}$

33. $\{1, 3, 5, 7, 9\}$

35. True. Seven is an integer.

37. True. Seven is a rational number.

39. False. Seven is a rational number.

41. True. Three is not an irrational number.

43. False. $\frac{1}{2}$ is a rational number.

45. True. $\sqrt{2}$ is not a rational number.

47. False. $\sqrt{2}$ is a real number.

49. –6 is less than –2. True.

51. 5 is greater than –7. True.

53. 0 is less than –4. False. 0 is greater than –4.

55. –4 is less than or equal to 1. True.

57. –2 is less than or equal to –6. False.
–2 is greater than –6.

59. –2 is less than or equal to –2. True.

61. –2 is greater than or equal to –2. True.

63. 2 is less than or equal to $-\frac{1}{2}$. False.

2 is greater than $-\frac{1}{2}$.

Application Exercises

65. According to the graph, life expectancy for women born in the year 2000 is 80.0 years.
The formula predicts:
$E = 0.22t + 71 = 0.22(50) + 71$
$= 11 + 71 = 82.$
The formula models the data quite well.

67. According to the formula, in the year 2000, the number of infections per month for every 1000 computers is:
$N = 0.2x^2 - 1.2x + 2$
$= 0.2(10)^2 - 1.2(10) + 2$
$= 0.2(100) - 12 + 2$
$= 20 - 12 + 2 = 8 + 2 = 10.$
The bar graph shows 9.9 infections per month for every 1000 computers. The formula models the data quite well.

69. $C = \frac{5}{9}(50-32) = \frac{5}{9}(18) = 10$
10°C is equivalent to 50°F.

71. $h = 4 + 60t - 16t^2 = 4 + 60(2) - 16(2)^2$
$= 4 + 120 - 16(4) = 4 + 120 - 64$
$= 124 - 64 = 60$

Two seconds after it is kicked, the ball's height is 60 feet.

Writing in Mathematics

73. Answers will vary.

75. Answers will vary.

77. Answers will vary.

79. Answers will vary.

81. Answers will vary.

83. Answers will vary.

85. Answers will vary.

87. Answers will vary.

89. Answers will vary.

Critical Thinking

91. Statement **c.** is true. Some rational numbers are not positive. $-\frac{1}{2}$ is one example.

Statement **a.** is false. Not all rational numbers can be expressed as integers. By definition, a rational number is a number that can be expressed as an integer divided by a nonzero integer. For example, $\frac{1}{2}$ is a rational number, since it is an integer, 1, divided by a nonzero integer, 2. But, $\frac{1}{2}$ is not an integer. Therefore, statement **a.** is false.

Statement **b.** is false. All whole numbers are integers.

Statements **d.** is false. Irrational numbers can be negative. $-\sqrt{3}$ is one example.

Review Exercises

93. $(2\cdot 3+3)\cdot 5=45$

94. $(8+2)\cdot(4-3)=10$

95. 26 is not a perfect square and $\sqrt{26}$ cannot be simplified. Consider the numbers closest to 26, both smaller and larger, which are perfect squares. The first perfect square smaller than 26 is 25. The first perfect square larger than 26 is 36. We know that the square root of 26 will lie between these numbers. We have $-\sqrt{36}<-\sqrt{26}<-\sqrt{25}$. If we simplify, we have $-6<-\sqrt{26}<-5$. Therefore, $-\sqrt{26}$ lies between -6 and -5.

Check Points 1.2

1. **a.** $|-6|=6$

b. $|4.5|=4.5$

c. $|0|=0$

2. **a.** $-10+(-18)=-28$

b. $-0.2+0.9=0.7$

c. $-\frac{3}{5}+\frac{1}{2}=-\frac{3}{5}\cdot\frac{2}{2}+\frac{1}{2}\cdot\frac{5}{5}$
$=-\frac{6}{10}+\frac{5}{10}=-\frac{1}{10}$

3. **a.** $x=-8$
$-x=8$

b. $x=\frac{1}{3}$
$-x=-\frac{1}{3}$

4. **a.** $7-10=7+(-10)=-3$

b. $4.3-(-6.2)=4.3+6.2=10.5$

c. $-9-(-3)=-9+3=-6$

5. **a.** $(-5)^2=(-5)(-5)=25$

b. $-5^2=-5\cdot 5=-25$

c. $(-4)^3=(-4)(-4)(-4)=-64$

d. $(-3)^4=(-3)(-3)(-3)(-3)=81$

6. **a.** $\frac{32}{-4}=-8$

b. $-\frac{2}{3}\div\left(-\frac{5}{4}\right)=-\frac{2}{3}\cdot\left(-\frac{4}{5}\right)=\frac{8}{15}$

7. $3-5^2+12\div 2(-4)^2$
$=3-25+12\div 2(16)=3-25+6(16)$
$=3-25+96=-22+96=74$

8. $\frac{4+3(-2)^3}{2-(6-9)}=\frac{4+3(-8)}{2-(-3)}$
$=\frac{4+(-24)}{2+3}=\frac{-20}{5}=-4$

9. Commutative Property of Addition:
$4x+9=9+4x$
Commutative Property of Multiplication: $4x+9=x\cdot 4+9$

10. **a.** $6+(12+x)=(6+12)+x=18+x$

b. $-7(4x)=(-7\cdot 4)x=-28x$

11. $-4(7x+2)=-4\cdot 7x+(-4)(2)$
$=-28x+(-8)=-28x-8$

12. $3x+14x^2+11x+x^2$
$=3x+11x+14x^2+x^2$
$=(3+11)x+(14+1)x^2$
$=14x+15x^2$ or $15x^2+14x$

13. $8(2x-5)-4x$
$=16x-40-4x=16x-4x-40$
$=(16-4)x-40=12x-40$

14. $6+4[7-(x-2)]$
$=6+4[7-x+2]=6+4[9-x]$
$=6+36-4x=42-4x$

Problem Set 1.2

Practice Exercises

1. $|-7|=7$

3. $|4|=4$

5. $|-7.6| = 7.6$

7. $\left|\frac{\pi}{2}\right| = \frac{\pi}{2}$

9. $|-\sqrt{2}| = \sqrt{2}$

11. $-3+(-8) = -11$

13. $-14+10 = -4$

15. $-6.8+2.3 = -4.5$

17. $\frac{11}{15}+\left(-\frac{3}{5}\right) = \frac{11}{15}+\left(-\frac{9}{15}\right) = \frac{2}{15}$

19. $-3.7+(-4.5) = -8.2$

21. $0+(-12.4) = -12.4$

23. $12.4+(-12.4) = 0$

25. $x = 11$
$-x = -11$

27. $x = -5$
$-x = 5$

29. $x = 0$
$-x = 0$

31. $3-15 = 3+(-15) = -12$

33. $8-(-10) = 8+10 = 18$

35. $-20-(-5) = -20+5 = -15$

37. $\frac{1}{4}-\frac{1}{2} = \frac{1}{4}+\left(-\frac{1}{2}\right) = \frac{1}{4}+\left(-\frac{2}{4}\right) = -\frac{1}{4}$

39. $-2.3-(-7.8) = -2.3+7.8 = 5.5$

41. $0-(-\sqrt{2}) = 0+\sqrt{2} = \sqrt{2}$

43. $9(-10) = -90$

45. $(-3)(-11) = 33$

47. $\frac{15}{13}(-1) = -\frac{15}{13}$

49. $-\sqrt{2}\cdot 0 = 0$

51. $(-4)(-2)(-1) = (8)(-1) = -8$

53. $2(-3)(-1)(-2)(-4) = (-6)(-1)(-2)(-4)$
$= (6)(-2)(-4)$
$= (-12)(-4) = 48$

55. $(-10)^2 = (-10)(-10) = 100$

57. $-10^2 = -(10)(10) = -100$

59. $(-2)^3 = (-2)(-2)(-2) = -8$

61. $(-1)^4 = (-1)(-1)(-1)(-1) = 1$

63. Since a product with an odd number of negative factors is negative, $(-1)^{33} = -1$.

65. $\frac{12}{-4} = -3$

67. $\frac{-90}{-2} = 45$

69. $\frac{0}{-4.6} = 0$

71. $-\frac{4.6}{0}$ is undefined.

73. $-\frac{1}{2} \div \left(-\frac{7}{9}\right) = -\frac{1}{2} \cdot \left(-\frac{9}{7}\right) = \frac{9}{14}$

75. $6 \div \left(-\frac{2}{5}\right) = \frac{6}{1} \cdot \left(-\frac{5}{2}\right) = -\frac{30}{2} = -15$

77. $4(-5) - 6(-3) = -20 - (-18)$
$= -20 + 18 = -2$

79. $3(-2)^2 - 4(-3)^2 = 3(4) - 4(9)$
$= 12 - 36 = -24$

81. $8^2 - 16 \div 2^2 \cdot 4 - 3$
$= 64 - 16 \div 4 \cdot 4 - 3 = 64 - 4 \cdot 4 - 3$
$= 64 - 16 - 3 = 48 - 3 = 45$

83. $\frac{4^2 + 3^3}{5^2 - (-18)} = \frac{16 + 27}{25 - (-18)} = \frac{43}{25 + 18}$
$= \frac{43}{43} = 1$

85. $8 - 3[-2(2-5) - 4(8-6)]$
$= 8 - 3[-2(-3) - 4(2)] = 8 - 3[6 - 8]$
$= 8 - 3[-2] = 8 + 6 = 14$

87. $\frac{2(-2) - 4(-3)}{5 - 8} = \frac{-4 + 12}{-3} = \frac{8}{-3} = -\frac{8}{3}$

89. $15 - \sqrt{3 - (-1)} + 12 \div 2 \cdot 3$
$= 15 - \sqrt{4} + 12 \div 2 \cdot 3$
$= 15 - 2 + 12 \div 2 \cdot 3 = 15 - 2 + 6 \cdot 3$
$= 15 - 2 + 18 = 13 + 18 = 31$

91. Commutative Property of Addition:
$4x + 10 = 10 + 4x$
Commutative Property of Multiplication: $4x + 10 = x \cdot 4 + 10$

93. Commutative Property of Addition:
$7x - 5 = -5 + 7x$
Commutative Property of Multiplication: $7x - 5 = x \cdot 7 - 5$

95. $4 + (6 + x) = (4 + 6) + x = 10 + x$

97. $-7(3x) = (-7 \cdot 3)x = -21x$

99. $-\frac{1}{3}(-3y) = \left(-\frac{1}{3} \cdot -3\right)y = y$

101. $3(2x + 5) = 3 \cdot 2x + 3 \cdot 5 = 6x + 15$

103. $-7(2x + 3) = -7 \cdot 2x + (-7)3$
$= -14x - 21$

105. $-(3x - 6) = -1 \cdot 3x + (-1)6 = -3x - 6$

107. $7x + 5x = (7 + 5)x = 12x$

109. $6x^2 - x^2 = (6 - 1)x^2 = 5x^2$

111. $6x + 10x^2 + 4x + 2x^2$
$= 6x + 4x + 10x^2 + 2x^2$
$= (6 + 4)x + (10 + 2)x^2 = 10x + 12x^2$

113. $8(3x - 5) - 6x$
$= 8 \cdot 3x - 8 \cdot 5 - 6x = 24x - 40 - 6x$
$= 24x - 6x - 40 = (24 - 6)x - 40$
$= 18x - 40$

115. $5(3y-2)-(7y+2)$
$=5\cdot 3y-5\cdot 2-1\cdot 7y+(-1)2$
$=15y-10-7y-2$
$=15y-7y-10-2$
$=(15-7)y-12=8y-12$

117. $7-4[3-(4y-5)]$
$=7-4[3-4y+5]=7-12+16y-20$
$=16y-25$

119. $18x^2+4-[6(x^2-2)+5]$
$=18x^2+4-[6x^2-12+5]$
$=18x^2+4-[6x^2-7]$
$=18x^2+4-6x^2+7$
$=18x^2-6x^2+4+7$
$=(18-6)x^2+11=12x^2+11$

Application Exercises

121. $\$4.29-\$3.66=\$0.63$
The price difference is \$0.63 per gallon.

123. $\$4.29-\$2.55=\$1.74$
The average price is \$1.74 per gallon in the United States.

125. $\$233-(-\$244)=\$477$ billion
The difference is \$477 billion.

127. **a.** $0.6(220-a)=132-0.6a$

b. $132-0.6a=132-0.6(20)$
$=132-12=120$
The optimum heart rate for a 20-year-old runner is 120 beats per minute.

Writing in Mathematics

129. Answers will vary.

131. Answers will vary.

133. Answers will vary.

135. Answers will vary.

137. Answers will vary.

139. Answers will vary.

141. Answers will vary.

143. Answers will vary.

145. Answers will vary.

Critical Thinking Exercises

147. $(8-2)\cdot 3-4=14$

149. $$\frac{9[4-(1+6)]-(3-9)^2}{5+\frac{12}{5-\frac{6}{2+1}}}$$
$$=\frac{9[4-(7)]-(-6)^2}{5+\frac{12}{5-\frac{6}{3}}}$$
$$=\frac{9[-3]-36}{5+\frac{12}{5-2}}=\frac{-27-36}{5+\frac{12}{3}}$$
$$=\frac{-63}{5+4}=\frac{-63}{9}=-7$$

Review Exercises

150. $\dfrac{10}{x} - 4x$

151. $10 + 2(x-5)^4$
$= 10 + 2(7-5)^4 = 10 + 2(2)^4$
$= 10 + 2(16) = 10 + 32 = 42$

152. True. $\dfrac{1}{2}$ is not an irrational number.

Check Points 1.3

1.

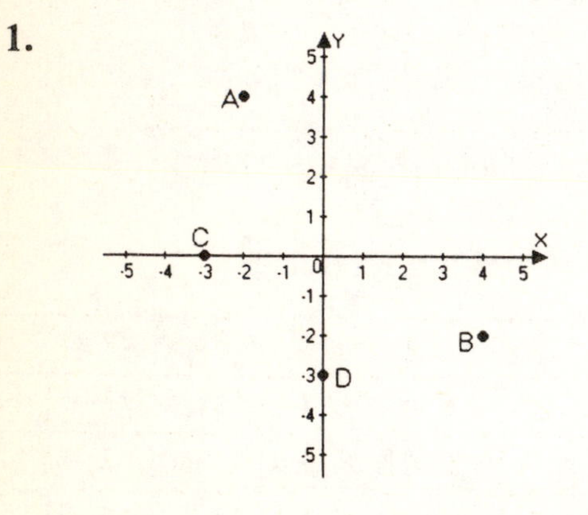

2.

x	$y = x^2 - 1$	(x, y)
-3	$y = (-3)^2 - 1 = 9 - 1 = 8$	$(-3, 8)$
-2	$y = (-2)^2 - 1 = 4 - 1 = 3$	$(-2, 3)$
-1	$y = (-1)^2 - 1 = 1 - 1 = 0$	$(-1, 0)$
0	$y = (0)^2 - 1 = 0 - 1 = -1$	$(0, -1)$
1	$y = (1)^2 - 1 = 1 - 1 = 0$	$(1, 0)$
2	$y = (2)^2 - 1 = 4 - 1 = 3$	$(2, 3)$
3	$y = (3)^2 - 1 = 9 - 1 = 8$	$(3, 8)$

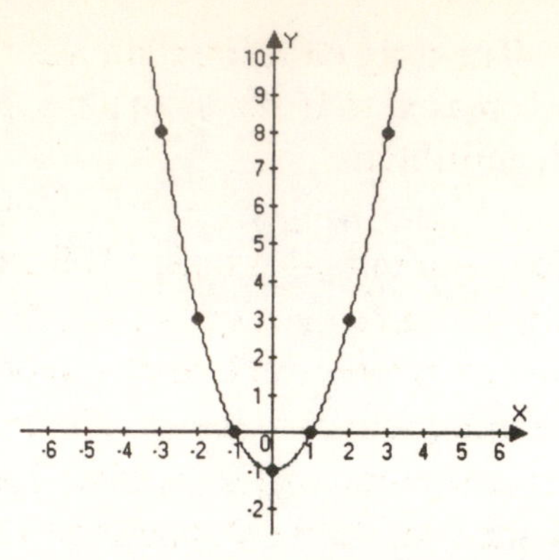

4.

x	$y = \lvert x+1 \rvert$	(x, y)
-4	$y = \lvert -4+1 \rvert = \lvert -3 \rvert = 3$	$(-4, 3)$
-3	$y = \lvert -3+1 \rvert = \lvert -2 \rvert = 2$	$(-3, 2)$
-2	$y = \lvert -2+1 \rvert = \lvert -1 \rvert = 1$	$(-2, 1)$
-1	$y = \lvert -1+1 \rvert = \lvert 0 \rvert = 0$	$(-1, 0)$
0	$y = \lvert 0+1 \rvert = \lvert 1 \rvert = 1$	$(0, 1)$
1	$y = \lvert 1+1 \rvert = \lvert 2 \rvert = 2$	$(1, 2)$
2	$y = \lvert 2+1 \rvert = \lvert 3 \rvert = 3$	$(2, 3)$

4. **a.** The drug concentration is increasing between 0 and 3 hours.

b. The drug concentration is decreasing between hours 3 and 13.

c. The maximum concentration is 0.05 milligrams per 100 milliliters at 3 hours.

d. The drug concentration has decreased to 0 milligrams per 100 milliliters.

5. The minimum x–value is –100 and the maximum x–value is 100. The distance between tick marks in 50. The minimum y–value is –100 and the maximum y–value is 100. The distance between tick marks in 10.

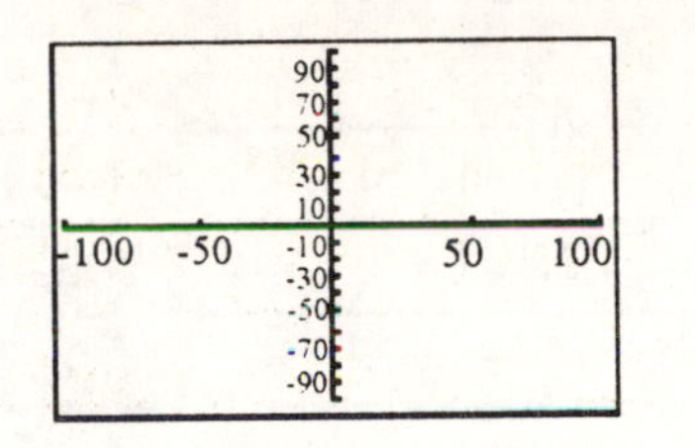

Problem Set 1.3

Practice Exercises

1. (1, 4)
See graphing answer section.

3. (-2, 3)
See graphing answer section.

5. (-3, -5)
See graphing answer section.

7. (4, -1)
See graphing answer section.

9. (-4, 0)
See graphing answer section.

11.

x	$y = x^2 - 2$	(x, y)
-3	$y = (-3)^2 - 2$ $= 9 - 2 = 7$	$(-3, 7)$
-2	$y = (-2)^2 - 2$ $= 4 - 2 = 2$	$(-2, 2)$
-1	$y = (-1)^2 - 2$ $= 1 - 2 = -1$	$(-1, -1)$
0	$y = (0)^2 - 2$ $= 0 - 2 = -2$	$(0, -2)$
1	$y = (1)^2 - 2$ $= 1 - 2 = -1$	$(1, -1)$
2	$y = (2)^2 - 2$ $= 4 - 2 = 2$	$(2, 2)$
3	$y = (3)^2 - 2$ $= 9 - 2 = 7$	$(3, 7)$

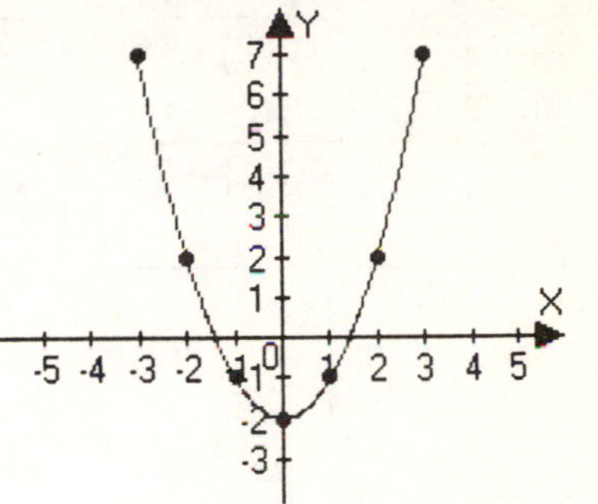

13.

x	$y = x - 2$	(x, y)
-3	$y = (-3) - 2 = -5$	$(-3, -5)$
-2	$y = (-2) - 2 = -4$	$(-2, -4)$
-1	$y = (-1) - 2 = -3$	$(-1, -3)$
0	$y = (0) - 2 = -2$	$(0, -2)$
1	$y = (1) - 2 = -1$	$(1, -1)$
2	$y = (2) - 2 = 0$	$(2, 0)$
3	$y = (3) - 2 = 1$	$(3, 1)$

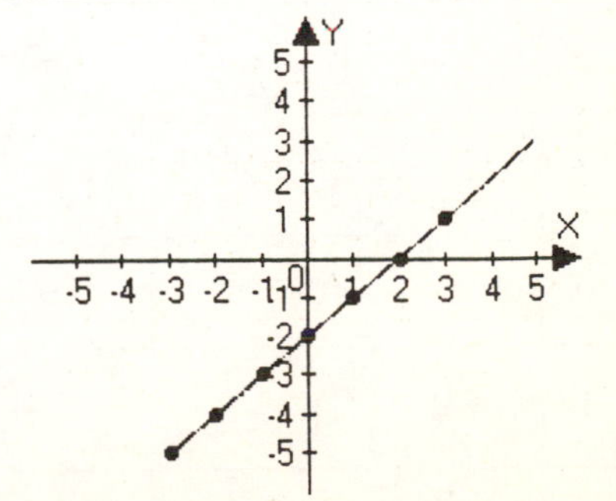

15.

x	$y = 2x+1$	(x,y)
-3	$y = 2(-3)+1$ $= -6+1 = -5$	$(-3,-5)$
-2	$y = 2(-2)+1$ $= -4+1 = -3$	$(-2,-3)$
-1	$y = 2(-1)+1$ $= -2+1 = -1$	$(-1,-1)$
0	$y = 2(0)+1$ $= 0+1 = 1$	$(0,1)$
1	$y = 2(1)+1$ $= 2+1 = 3$	$(1,3)$
2	$y = 2(2)+1$ $= 4+1 = 5$	$(2,5)$
3	$y = 2(3)+1$ $= 6+1 = 7$	$(3,7)$

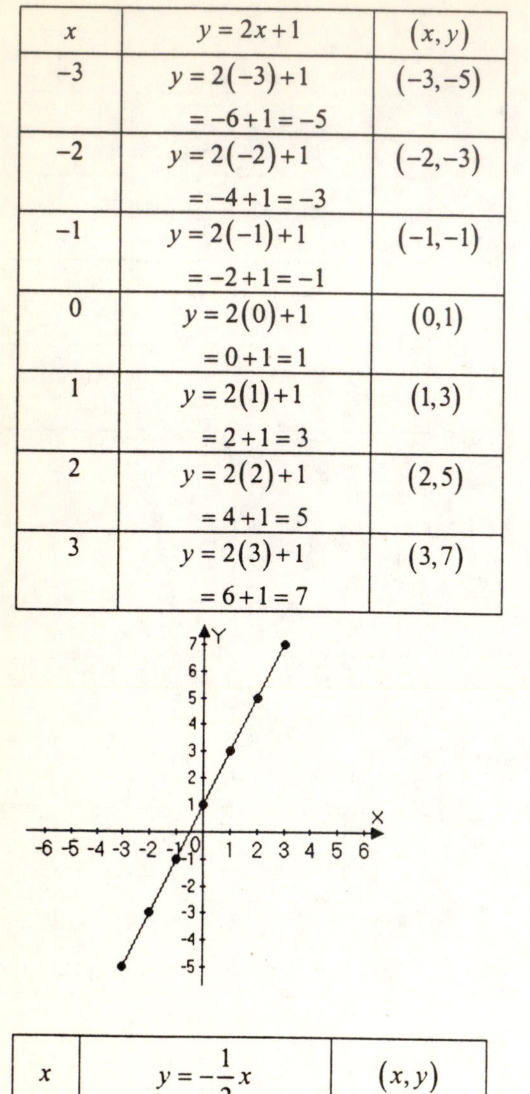

17.

x	$y = -\frac{1}{2}x$	(x,y)
-3	$y = -\frac{1}{2}(-3) = \frac{3}{2}$	$\left(-3,\frac{3}{2}\right)$
-2	$y = -\frac{1}{2}(-2) = 1$	$(-2,1)$
-1	$y = -\frac{1}{2}(-1) = \frac{1}{2}$	$\left(-1,\frac{1}{2}\right)$
0	$y = -\frac{1}{2}(0) = 0$	$(0,0)$
1	$y = -\frac{1}{2}(1) = -\frac{1}{2}$	$\left(1,-\frac{1}{2}\right)$
2	$y = -\frac{1}{2}(2) = -1$	$(2,-1)$
3	$y = -\frac{1}{2}(3) = -\frac{3}{2}$	$\left(3,-\frac{3}{2}\right)$

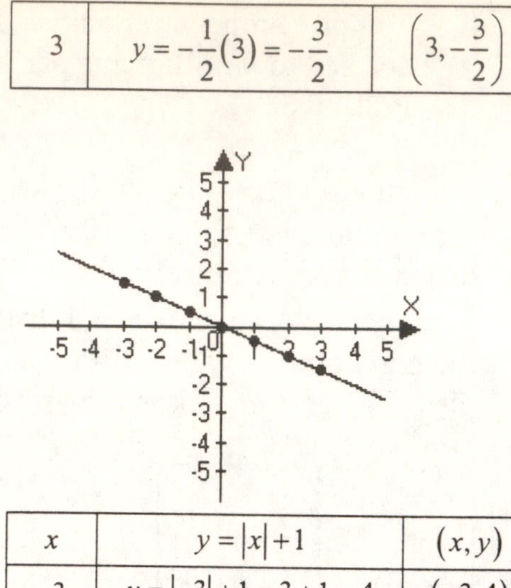

19.

x	$y = \lvert x\rvert+1$	(x,y)
-3	$y = \lvert -3\rvert+1 = 3+1 = 4$	$(-3,4)$
-2	$y = \lvert -2\rvert+1 = 2+1 = 3$	$(-2,3)$
-1	$y = \lvert -1\rvert+1 = 1+1 = 2$	$(-1,2)$
0	$y = \lvert 0\rvert+1 = 0+1 = 1$	$(0,1)$
1	$y = \lvert 1\rvert+1 = 1+1 = 2$	$(1,2)$
2	$y = \lvert 2\rvert+1 = 2+1 = 3$	$(2,3)$
3	$y = \lvert 3\rvert+1 = 3+1 = 4$	$(3,4)$

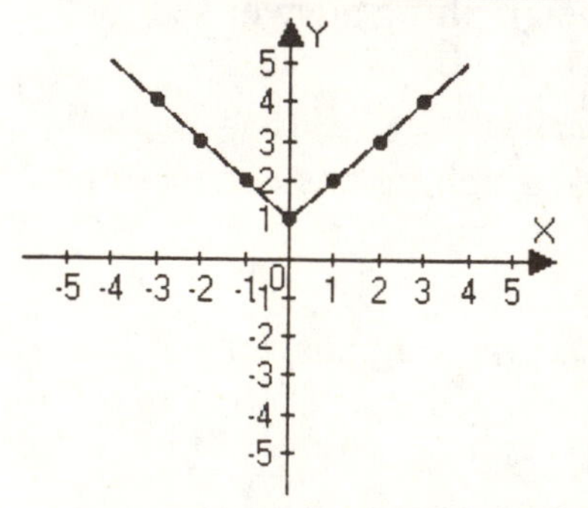

21.

x	$y = 2\lvert x\rvert$	(x,y)
-3	$y = 2\lvert -3\rvert = 2\cdot 3 = 6$	$(-3,6)$
-2	$y = 2\lvert -2\rvert = 2\cdot 2 = 4$	$(-2,4)$
-1	$y = 2\lvert -1\rvert = 2\cdot 1 = 2$	$(-1,2)$
0	$y = 2\lvert 0\rvert = 2\cdot 0 = 0$	$(0,0)$
1	$y = 2\lvert 1\rvert = 2\cdot 1 = 2$	$(1,2)$
2	$y = 2\lvert 2\rvert = 2\cdot 2 = 4$	$(2,4)$
3	$y = 2\lvert 3\rvert = 2\cdot 3 = 6$	$(3,6)$

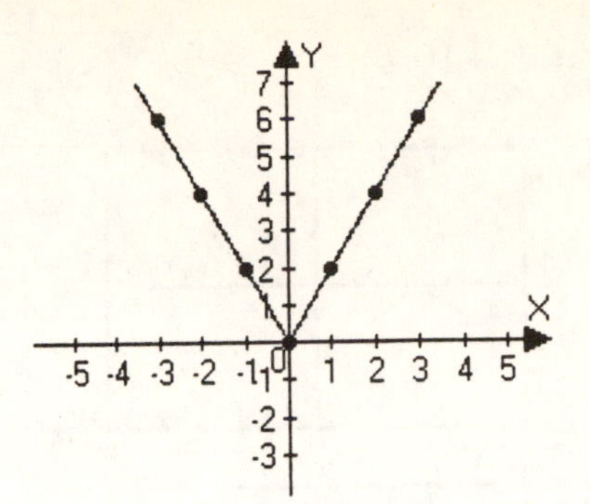

23.

x	$y = 4 - x^2$	(x, y)
-3	$y = 4-(-3)^2$ $= 4 - 9 = -5$	$(-3, -5)$
-2	$y = 4-(-2)^2$ $= 4 - 4 = 0$	$(-2, 0)$
-1	$y = 4-(-1)^2$ $= 4 - 1 = 3$	$(-1, 3)$
0	$y = 4-(0)^2$ $= 4 - 0 = 4$	$(0, 4)$
1	$y = 4-(1)^2$ $= 4 - 1 = 3$	$(1, 3)$
2	$y = 4-(2)^2$ $= 4 - 4 = 0$	$(2, 0)$
3	$y = 4-(3)^2$ $= 4 - 9 = -5$	$(3, -5)$

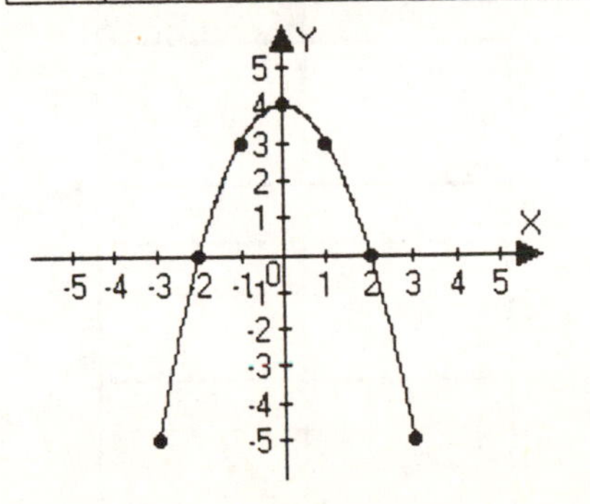

25.

x	$y = x^3$	(x, y)
-3	$y = (-3)^3 = -27$	$(-3, -27)$
-2	$y = (-2)^3 = -8$	$(-2, -8)$
-1	$y = (-1)^3 = -1$	$(-1, -1)$
0	$y = (0)^3 = 0$	$(0, 0)$
1	$y = (1)^3 = 1$	$(1, 1)$
2	$y = (2)^3 = 8$	$(2, 8)$
3	$y = (3)^3 = 27$	$(3, 27)$

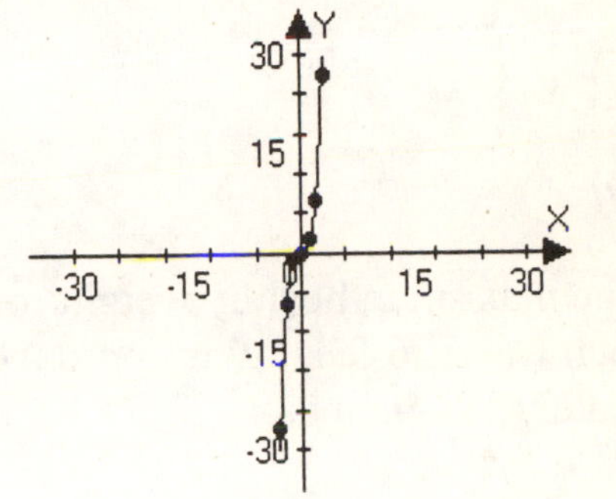

27. [-5, 5, 1] by [-5, 5, 1]
This matches figure (c).

29. [-20, 80, 10] by [-30, 70, 10]
This matches figure (b).

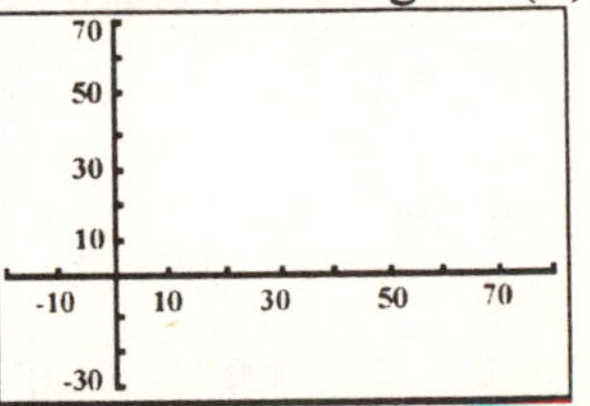

Application Exercises

31. The chance of divorce increases through year 4.

33. The chance of divorce is highest in the 4th year with chance of approximately 8.3%.

35. The percent of Americans voting was lowest in 1832 at 20%.

37. Approximately 52% of Americans voted in 2000.

39. (a)

41. (b)

43. (b)

45. The maximum healthy weight for a man who is 6 feet tall is approximately 181 pounds.

47. A man who is 70 inches tall and needs to gain 10 pounds weighs approximately 135 pounds.

Writing in Mathematics

49. Answers will vary.

51. Answers will vary.

53. Answers will vary.

55. Answers will vary.

Technology Exercises

57. Answers will vary. For example, consider Exercise 11.

$y = x^2 - 2$

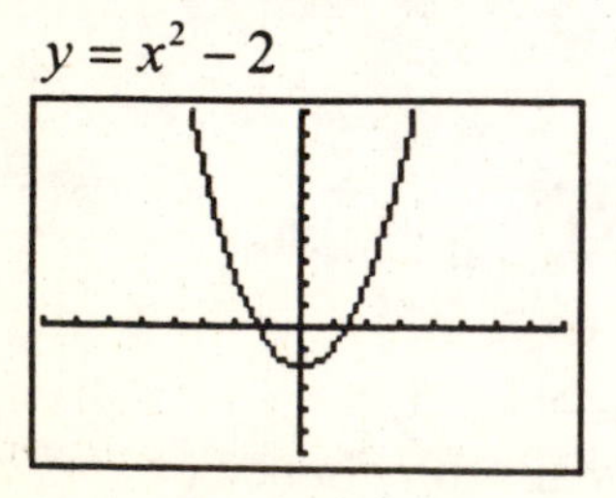

59. $y = x^2 + 10$

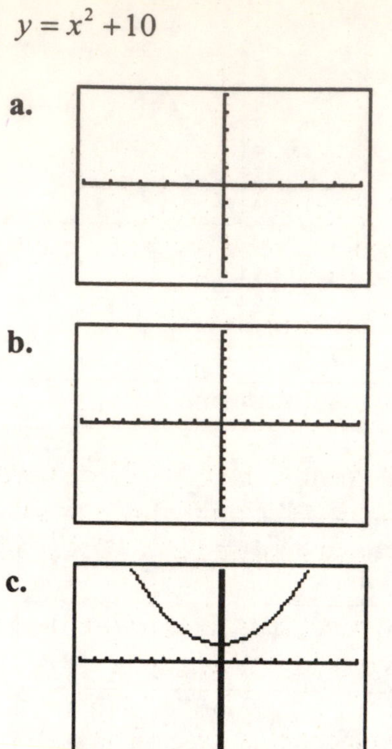

Graph **c.** gives a complete graph.

61. $y = \sqrt{x} + 18$

a.

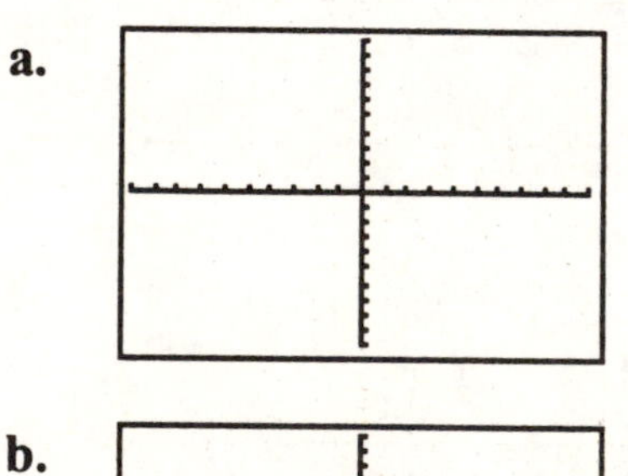

b.

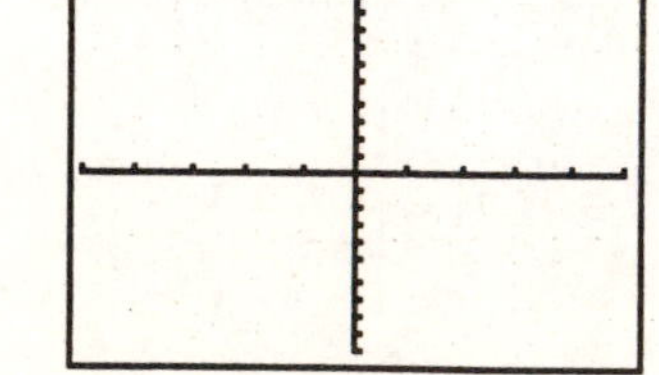

c. 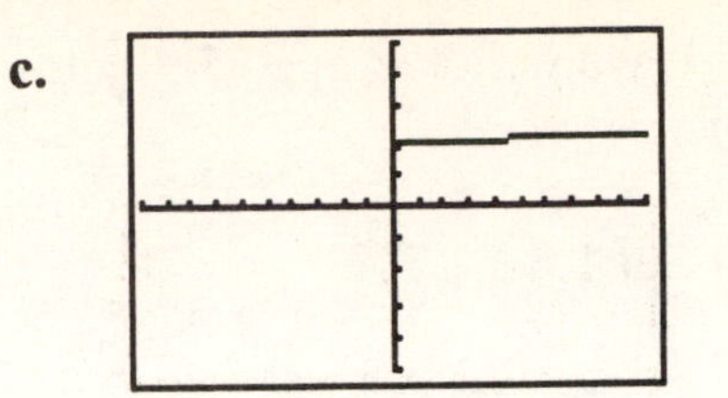

Graph **c.** gives a complete graph.

Critical Thinking Exercises

63. Statement **d.** is true.

Statement **a.** is false. If the product of a point's coordinates is positive, the point could be in quadrant I <u>or</u> III.

Statement **b.** is false. When a point lies on the x-axis, $y = 0$.

Statement **c.** is false. The majority of graphing utilities do not display numbers along the axes.

65. The ball will hit the ground approximately 8.8 seconds from when it is dropped.

Review Exercises

66. $|-14.3| = 14.3$

67. $[12-(13-17)]-[9-(6-10)]$
$= [12-(-4)]-[9-(-4)]$
$= [12+4]-[9+4] = 16-13 = 3$

68. $6x-5(4x+3)-10$
$= 6x-20x-15-10$
$= (6-20)x-(15+10)$
$= -14x-25$

Check Points 1.4

1.
$$4x+5=29$$
$$4x+5-5=29-5$$
$$4x=24$$
$$\frac{4x}{4}=\frac{24}{4}$$
$$x=6$$
The solution is 6 and the solution set is $\{6\}$.

2.
$$2x-12+x=6x-4+5x$$
$$3x-12=11x-4$$
$$3x-3x-12=11x-3x-4$$
$$-12=8x-4$$
$$-12+4=8x-4+4$$
$$-8=8x$$
$$\frac{-8}{8}=\frac{8x}{8}$$
$$-1=x$$
The solution is -1 and the solution set is $\{-1\}$.

3.
$$2(x-3)-17=13-3(x+2)$$
$$2x-6-17=13-3x-6$$
$$2x-23=7-3x$$
$$2x+3x-23=7-3x+3x$$
$$5x-23=7$$
$$5x-23+23=7+23$$
$$5x=30$$
$$\frac{5x}{5}=\frac{30}{5}$$
$$x=6$$
The solution is 6 and the solution set is $\{6\}$.

4.
$$\frac{x+5}{7}+\frac{x-3}{4}=\frac{5}{14}$$
$$28\left(\frac{x+5}{7}+\frac{x-3}{4}\right)=28\left(\frac{5}{14}\right)$$
$$4(x+5)+7(x-3)=2(5)$$
$$4x+20+7x-21=10$$
$$11x-1=10$$
$$11x-1+1=10+1$$
$$11x=11$$
$$\frac{11x}{11}=\frac{11}{11}$$
$$x=1$$
The solution is 1 and the solution set is $\{1\}$.

5.
$$4x-7=4(x-1)+3$$
$$4x-7=4x-4+3$$
$$4x-4x-7=4x-4x-1$$
$$-7=-1$$
Since this is false for all values of x, there is no solution and the equation is inconsistent.

6.
$$7x+9=9(x+1)-2x$$
$$7x+9=9x+9-2x$$
$$7x+9=7x+9$$
Since this is true for all values of x, the solution is $\{x|x \text{ is a real number}\}$ or $\mathbb{R}$ and the equation is an identity.

7.
$$W=0.3x+46.6$$
$$55.9=0.3x+46.6$$
$$55.9-46.6=0.3x+46.6-46.6$$
$$9.3=0.3x$$
$$\frac{9.3}{0.3}=\frac{0.3x}{0.3}$$
$$31=x$$
Since x is expressed as years after 1980, we use 1980 + 31 = 2001. In 2011, we will average 55.9 hours of work per week.

Problem Set 1.4

Practice Exercises

1.
$$5x+3=18$$
$$5x+3-3=18-3$$
$$5x=15$$
$$\frac{5x}{5}=\frac{15}{5}$$
$$x=3$$
The solution is 3 and the solution set is $\{3\}$.

3.
$$6x-3=63$$
$$6x-3+3=63+3$$
$$6x=66$$
$$\frac{6x}{6}=\frac{66}{6}$$
$$x=11$$
The solution is 11 and the solution set is $\{11\}$.

5.
$$14-5x=-41$$
$$14-5x-14=-41-14$$
$$-5x=-55$$
$$\frac{-5x}{-5}=\frac{-55}{-5}$$
$$x=11$$
The solution is 11 and the solution set is $\{11\}$.

7.
$$11x-(6x-5)=40$$
$$11x-6x+5=40$$
$$5x+5=40$$
$$5x+5-5=40-5$$
$$5x=35$$

$$\frac{5x}{5}=\frac{35}{5}$$
$$x=7$$

The solution is 7 and the solution set is $\{7\}$.

9.
$$2x-7=6+x$$
$$2x-x-7=6+x-x$$
$$x-7=6$$
$$x-7+7=6+7$$
$$x=13$$

The solution is 13 and the solution set is $\{13\}$.

11.
$$7x+4=x+16$$
$$7x-x+4=x-x+16$$
$$6x+4=16$$
$$6x+4-4=16-4$$
$$6x=12$$
$$\frac{6x}{6}=\frac{12}{6}$$
$$x=2$$

The solution is 2 and the solution set is $\{2\}$.

13.
$$8y-3=11y+9$$
$$8y-8y-3=11y-8y+9$$
$$-3=3y+9$$
$$-3-9=3y+9-9$$
$$-12=3y$$
$$\frac{-12}{3}=\frac{3y}{3}$$
$$-4=y$$

The solution is –4 and the solution set is $\{-4\}$.

15.
$$3(x-2)+7=2(x+5)$$
$$3x-6+7=2x+10$$
$$3x-2x-6+7=2x-2x+10$$
$$x-6+7=10$$
$$x+1=10$$
$$x+1-1=10-1$$
$$x=9$$

The solution is 9 and the solution set is $\{9\}$.

17.
$$3(x-4)-4(x-3)=x+3-(x-2)$$
$$3x-\cancel{12}-4x+\cancel{12}=\cancel{x}+3-\cancel{x}+2$$
$$-x=5$$
$$x=-5$$

The solution is –5 and the solution set is $\{-5\}$.

19.
$$16=3(x-1)-(x-7)$$
$$16=3x-3-x+7$$
$$16=2x+4$$
$$16-4=2x+4-4$$
$$12=2x$$
$$\frac{12}{2}=\frac{2x}{2}$$
$$6=x$$

The solution is 6 and the solution set is $\{6\}$.

21.
$$7(x+1)=4[x-(3-x)]$$
$$7x+7=4[x-3+x]$$
$$7x+7=4[2x-3]$$
$$7x+7=8x-12$$
$$7x-7x+7=8x-7x-12$$
$$7=x-12$$
$$7+12=x-12+12$$
$$19=x$$

The solution is 19 and the solution set is $\{19\}$.

23.
$$\frac{x}{3}=\frac{x}{2}-2$$
$$6\left(\frac{x}{3}\right)=6\left(\frac{x}{2}-2\right)$$
$$2x=3x-12$$
$$2x-3x=3x-3x-12$$
$$-x=-12$$
$$x=12$$
The solution is 12 and the solution set is $\{12\}$.

25.
$$20-\frac{x}{3}=\frac{x}{2}$$
$$6\left(20-\frac{x}{3}\right)=6\left(\frac{x}{2}\right)$$
$$120-2x=3x$$
$$120-2x+2x=3x+2x$$
$$120=5x$$
$$\frac{120}{5}=\frac{5x}{5}$$
$$24=x$$
The solution is 24 and the solution set is $\{24\}$.

27.
$$\frac{3x}{5}=\frac{2x}{3}+1$$
$$15\left(\frac{3x}{5}\right)=15\left(\frac{2x}{3}+1\right)$$
$$9x=10x+15$$
$$9x-10x=10x-10x+15$$
$$-x=15$$
$$x=-15$$
The solution is –15 and the solution set is $\{-15\}$.

29. $\frac{3x}{5}-x=\frac{x}{10}-\frac{5}{2}$
$$10\left(\frac{3x}{5}-x\right)=10\left(\frac{x}{10}-\frac{5}{2}\right)$$
$$6x-10x=x-25$$
$$-4x=x-25$$
$$-4x-x=x-x-25$$
$$-5x=-25$$
$$x=5$$
The solution is 5 and the solution set is $\{5\}$.

31.
$$\frac{x+3}{6}=\frac{2}{3}+\frac{x-5}{4}$$
$$12\left(\frac{x+3}{6}\right)=12\left(\frac{2}{3}\right)+12\left(\frac{x-5}{4}\right)$$
$$2(x+3)=4(2)+3(x-5)$$
$$2x+6=8+3x-15$$
$$2x+6=3x-7$$
$$-x+6=-7$$
$$-x=-13$$
$$x=13$$
The solution is 13 and the solution set is $\{13\}$.

33.
$$\frac{x}{4}=2+\frac{x-3}{3}$$
$$12\left(\frac{x}{4}\right)=12\left(2+\frac{x-3}{3}\right)$$
$$3x=24+4(x-3)$$
$$3x=24+4x-12$$
$$3x=12+4x$$
$$3x-4x=12+4x-4x$$
$$-x=12$$
$$x=-12$$
The solution is –12 and the solution set is $\{-12\}$.

35. $\frac{x+1}{3}=5-\frac{x+2}{7}$

$$21\left(\frac{x+1}{3}\right)=21\left(5-\frac{x+2}{7}\right)$$
$$7(x+1)=105-3(x+2)$$
$$7x+7=105-3x-6$$
$$7x+3x+7=105-3x+3x-6$$
$$10x+7=99$$
$$10x=92$$
$$x=\frac{92}{10}=\frac{46}{5}$$

The solution is $\frac{46}{5}$ and the solution set is $\left\{\frac{46}{5}\right\}$.

37. $5x+9=9(x+1)-4x$
$$5x+9=9x+9-4x$$
$$5x+9=5x+9$$
The solution set is $\{x|x$ is a real number$\}$ or $\mathbb{R}$. The equation is an identity.

39. $3(y+2)=7+3y$
$$3y+6=7+3y$$
$$3y-3y+6=7+3y-3y$$
$$6=7$$
There is no solution. The solution set is $\{\ \}$ or $\varnothing$. The equation is inconsistent.

41. $10x+3=8x+3$
$$10x-8x+3=8x-8x+3$$
$$2x=0$$
$$x=0$$
The solution set is $\{0\}$. The equation is conditional.

43. $4x+5x=8x$
$$9x=8x$$
$$9x-8x=8x-8x$$
$$x=0$$
The solution set is $\{0\}$. The equation is conditional.

45. $-4x-3(2-2x)=7+2x$
$$-4x-6+6x=7+2x$$
$$\cancel{2x}-6=7+\cancel{2x}$$
$$-6=7$$
There is no solution. The solution set is $\{\ \}$ or $\varnothing$. The equation is inconsistent.

47. $y+3(4y+2)=6(y+1)+5y$
$$y+12y+6=6y+6+5y$$
$$13y+6=11y+6$$
$$13y-11y+6=11y-11y+6$$
$$2y+6=6$$
$$2y=0$$
$$y=0$$
The solution set is $\{0\}$. The equation is conditional.

Application Exercises

49. If the desirable heart rate for a woman is 117 beats per minute, she is 40 years old. This solution is approximated by the point (40, 117) on the female graph.

51.
$$M = 0.19t + 54.91$$
$$73.91 = 0.19t + 54.91$$
$$19 = 0.19t$$
$$100 = t$$
U.S. men can expect to live to 73.91 years in the year 1900 + 100 = 2000.

53.
$$P = -0.22t + 9.6$$
$$0.5 = -0.22t + 9.6$$
$$-9.1 = -0.22t$$
$$41.4 = t$$
$$t \approx 41$$
The population will be 0.5 million in 1960 + 41 = 2001.

Writing in Mathematics

55. Answers will vary.

57. Answers will vary.

59. Answers will vary.

61. Answers will vary.

63. Answers will vary.

65. Answers will vary.

67. Answers will vary.

Technology Exercises

69. $9x + 3 - 3x = 2(3x + 1)$

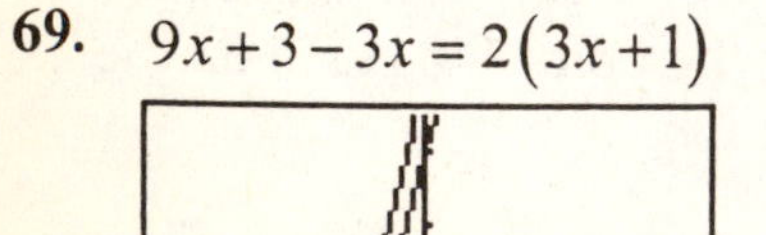

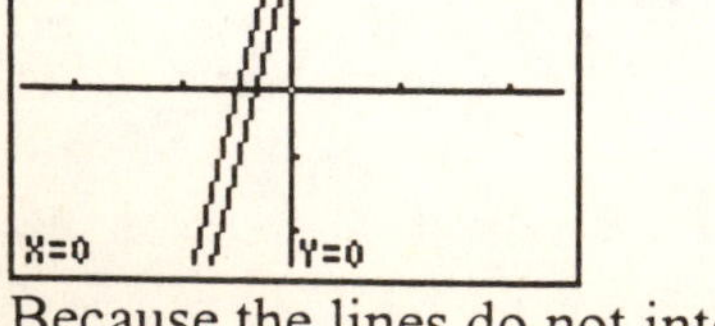

Because the lines do not intersect, there is no solution and the equation is inconsistent.

71. $\dfrac{2x-1}{3} - \dfrac{x-5}{6} = \dfrac{x-3}{4}$

The solution is –5 and the solution set is $\{-5\}$.

Verify by direct substitution.
$$\frac{2x-1}{3} - \frac{x-5}{6} = \frac{x-3}{4}$$
$$\frac{2(-5)-1}{3} - \frac{(-5)-5}{6} = \frac{(-5)-3}{4}$$
$$\frac{-10-1}{3} - \frac{-5-5}{6} = \frac{-5-3}{4}$$
$$\frac{-11}{3} - \frac{-10}{6} = \frac{-8}{4}$$
$$6\left(\frac{-11}{3} - \frac{-10}{6}\right) = 6(-2)$$
$$-22 - (-10) = -12$$
$$-22 + 10 = -12$$
$$-12 = -12$$

The equation is conditional and the solution set is $\{-5\}$.

Critical Thinking Exercises

73.
$$ax + b = c$$
$$ax + b - b = c - b$$
$$ax = c - b$$
$$x = \frac{c-b}{a}$$

75. Answers will vary.

Review Exercises

77.
$$-\frac{1}{5}-\left(-\frac{1}{2}\right)=-\frac{1}{5}+\frac{1}{2}$$
$$=-\frac{1}{5}\cdot\frac{2}{2}+\frac{1}{2}\cdot\frac{5}{5}$$
$$=-\frac{2}{10}+\frac{5}{10}=\frac{3}{10}$$

78. $4(-3)(-1)(-5)=(-12)(5)=-60$

79.

x	$y=4-x^2$	(x,y)
-3	$y=4-(-3)^2=4-9=-5$	$(-3,-5)$
-2	$y=4-(-2)^2=4-4=0$	$(-2,0)$
-1	$y=4-(-1)^2=4-1=3$	$(-1,3)$
0	$y=4-(0)^2=4-0=4$	$(0,4)$
1	$y=4-(1)^2=4-1=3$	$(1,3)$
2	$y=4-(2)^2=4-4=0$	$(2,0)$
3	$y=4-(3)^2=4-9=-5$	$(3,-5)$

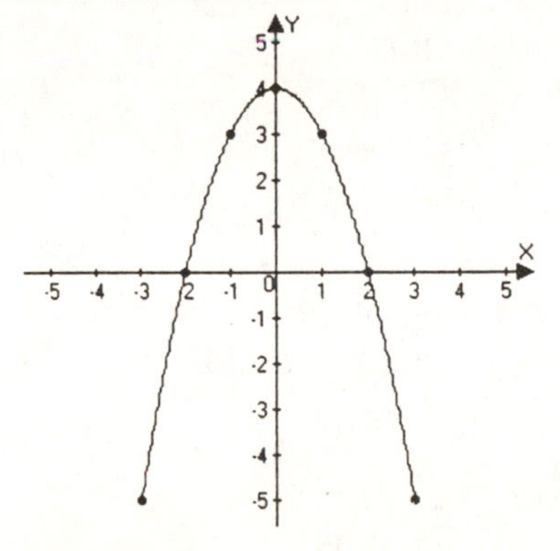

Check Points 1.5

1. Let x = the number of copies of *Purple Rain* sold (in millions)
Let $x + 2$ = the number of copies of *Saturday Night Fever* sold (in millions)
Let $x + 3$ = the number of copies of *The Bodyguard* sold (in millions)

$$x+(x+2)+(x+3)=44$$
$$x+x+2+x+3=44$$
$$3x+5=44$$
$$3x=39$$
$$x=13$$

Thus, 13 million copies of *Purple Rain* were sold, $x + 2 = 15$ million copies of *Saturday Night Fever* were sold, and $x + 3 = 16$ million copies of *The Bodyguard* were sold.

2. Let x = the number of years after 2000

$$298.5=275+2.35x$$
$$23.5=2.35x$$
$$10=x$$

The population will be 298.5 million in the year 2000 + 10 = 2010.

3. Let x = the number of minutes of long distance calls for the two plans to cost the same

$$15+0.08x=3+0.12x$$
$$15=3+0.04x$$
$$12=0.04x$$
$$300=x$$

After 300 minutes, the cost will be the same.

4. Let x = the price of the computer before the reduction

$$x-0.30x=980$$
$$0.70x=980$$
$$x=1400$$

Before the reduction, the computer was $1400.

5. Let x = the width of the basketball court
Let $x + 44$ = the length of the court
$$P = 2l + 2w$$
$$288 = 2(x+44) + 2x$$
$$288 = 2x + 88 + 2x$$
$$288 = 4x + 88$$
$$200 = 4x$$
$$50 = x$$
Find the length.
$x + 44 = 50 + 44 = 94$
The dimensions are 50 feet by 94 feet.

6.
$$2l + 2w = P$$
$$2l = P - 2w$$
$$\frac{2l}{2} = \frac{P-2w}{2}$$
$$l = \frac{P-2w}{2}$$

7.
$$V = lwh$$
$$\frac{V}{lw} = \frac{lwh}{lw}$$
$$h = \frac{V}{lw}$$

8.
$$\frac{W}{2} - 3H = 53$$
$$\frac{W}{2} = 53 + 3H$$
$$W = 2(53 + 3H)$$
$$W = 106 + 6H$$

9.
$$P = C + MC$$
$$P = C(1+M)$$
$$\frac{P}{1+M} = \frac{C(1+M)}{1+M}$$
$$C = \frac{P}{1+M}$$

Problem Set 1.5

Practice Exercises

1. Let x = a number
$$5x - 4 = 26$$
$$5x = 30$$
$$x = 6$$
The number is 6.

3. Let x = a number
$$x - 0.20x = 20$$
$$0.80x = 20$$
$$x = 25$$
The number is 25.

5. Let x = a number
$$0.60x + x = 192$$
$$1.6x = 192$$
$$x = 120$$
The number is 120.

7. Let x = a number
$$0.70x = 224$$
$$x = 320$$
The number is 320.

9. Let x = a number
$x + 26$ = another number
$$x + (x + 26) = 64$$
$$x + x + 26 = 64$$
$$2x + 26 = 64$$
$$2x = 38$$
$$x = 19$$
If $x = 19$, then $x + 26 = 45$.
The numbers are 19 and 45.

Application Exercises

11. Let x = the cost of Hurricane Hugo
$x + 5.5$ = the cost of the Northridge Earthquake
$x + 13$ = the cost of the Hurricane Andrew

$$x+(x+5.5)+(x+13)=39.5$$
$$x+x+5.5+x+13=39.5$$
$$3x+18.5=39.5$$
$$3x=21$$
$$x=7$$

The costs were: Hurricane Hugo \$7 billion, Northridge Earthquake \$12.5 billion, and Hurricane Andrew \$20 billion.

13. Let x = the measure of the 2nd angle
$2x$ = the measure of the 1st angle
$x - 20$ = the measure of the 3rd angle

$$x+2x+(x-20)=180$$
$$3x+x-20=180$$
$$4x-20=180$$
$$4x=200$$
$$x=50$$

The measure of the 1st angle is 100°, the 2nd angle is 50°, and the 3rd angle is 30°.

15. Let L = the life expectancy of an American man
Let y = the number of years after 1900

$$L=55+0.2y$$
$$85=55+0.2y$$
$$30=0.2y$$
$$150=y$$

The life expectancy will be 85 years in the year 1900 + 150 = 2050.

17. Let n = the percentage of babies born to unmarried parents
Let y = the number of years after 1990

$$L=28+0.6y$$
$$37=28+0.6y$$
$$9=0.6y$$
$$15=y$$

The percentage of babies born to unmarried parents will be 37% in the year 1990 + 15 = 2005.

19. Let w = the winning time for women
Let n = the number of years since 1980

$$w=41.78-0.19n$$
$$37.22=41.78-0.19n$$
$$-4.56=-0.19n$$
$$24=n$$

The winning time will be 37.22 seconds in the year 1980 + 24 = 2004.

21. Let p = the manufacturer's recommended list price
Plan A: $\text{Cost}=\$100+0.80p$
Plan B: $\text{Cost}=\$40+0.90p$

$$100+0.80p=40+0.90p$$
$$100=40+0.10p$$
$$60=0.10p$$
$$600=p$$

You would have to spend \$600 in one year for the costs under each plan to be equal.

$$\text{Cost}=\$100+0.80p$$
$$=100+0.80(600)$$
$$=100+480=580$$

The cost under each plan would be $580.

23. Cost per day: $1.25x
Cost with coupon book: $21 + $0.50x

$$1.25x = 21 + 0.50x$$
$$0.75x = 21$$
$$x = 28$$

The bus must be used 28 times in a month for the costs to be equal.

25. Let x = the cost of the television set

$$x - 0.20x = 336$$
$$0.80x = 336$$
$$x = 420$$

The television set's price is $420.

27. Let x = the nightly cost

$$x + 0.08x = 162$$
$$1.08x = 162$$
$$x = 150$$

The nightly cost is $150.

29. Let x = the annual salary with an Associate's degree

$$41,850 = x + 0.35x$$
$$41,850 = 1.35x$$
$$31,000 = x$$

The annual salary with an Associate's degree is $31,000.

31. Let c = the dealer's cost

$$584 = c + 0.25c$$
$$584 = 1.25c$$
$$467.20 = c$$

The dealer's cost is $467.20.

33. Let w = the width of the field
Let $2w$ = the length of the field

$$P = 2(\text{length}) + 2(\text{width})$$
$$300 = 2(2w) + 2(w)$$
$$300 = 4w + 2w$$
$$300 = 6w$$
$$50 = w$$

The dimensions are 50 yards by 100 yards.

35. Let w = the width of the field
Let $2w + 6$ = the length of the field

$$228 = 6w + 12$$
$$216 = 6w$$
$$36 = w$$

Use w to find the length of the field.

$$2w + 6 = 2(36) + 6$$
$$= 72 + 6 = 78$$

The dimensions are 36 feet by 78 feet.

37. Let s = the length of a shelf
Let $s + 3$ = the height of the bookcase

$$\text{Total length} = 4s + 2(s + 3)$$
$$18 = 4s + 2s + 6$$
$$18 = 6s + 6$$
$$12 = 6s$$
$$2 = s$$

Find the height.

$$s + 3 = 2 + 3$$
$$= 5$$

The length is 2 feet and the height is 5 feet.

39. Let i = the number of inches over 5 feet.

$$\text{Weight} = 100 + 5i$$
$$135 = 100 + 5i$$
$$35 = 5i$$
$$7 = i$$

The ideal weight of 135 pounds corresponds to a height of 5' 7".

41. $A = lw$

$l = \frac{A}{w}$

43. $A = \frac{1}{2}bh$

$2A = bh$

$b = \frac{2A}{h}$

45. $I = Prt$

$P = \frac{I}{rt}$

47. $T = D + pm$

$T - D = pm$

$p = \frac{T - D}{m}$

49. $A = \frac{1}{2}h(a+b)$

$2A = h(a+b)$

$\frac{2A}{h} = a + b$

$a = \frac{2A}{h} - b$

51. $V = \frac{1}{3}\pi r^2 h$

$3V = \pi r^2 h$

$h = \frac{3V}{\pi r^2}$

53. $y - y_1 = m(x - x_1)$

$m = \frac{y - y_1}{x - x_1}$

55. $V = \frac{d_1 - d_2}{t}$

$Vt = d_1 - d_2$

$d_1 = Vt + d_2$

57. $Ax + By = C$

$Ax = C - By$

$x = \frac{C - By}{A}$

59. $s = \frac{1}{2}at^2 + vt$

$2s = \not{2}\left(\frac{1}{\not{2}}at^2\right) + 2vt$

$2s = at^2 + 2vt$

$2s - at^2 = 2vt$

$\frac{2s - at^2}{2t} = \frac{2vt}{2t}$

$v = \frac{2s - at^2}{2t}$

61. $L = a + (n-1)d$

$L - a = (n-1)d$

$\frac{L - a}{d} = n - 1$

$n = \frac{L - a}{d} + 1$

63.
$$A = 2lw + 2lh + 2wh$$
$$A - 2wh = 2lw + 2lh$$
$$A - 2wh = l(2w + 2h)$$
$$l = \frac{A - 2wh}{2w + 2h}$$

65.
$$IR + Ir = E$$
$$I(R + r) = E$$
$$I = \frac{E}{R + r}$$

Technology Exercises

67. Answers will vary.

69. Answers will vary.

71. Answers will vary.

Technology Exercises

73. Exercise 17.

Y1=28+0.6X

X=15 Y=37

The percentage of babies born to unmarried parents will be 37% in the year 2005.

Exercise 18.

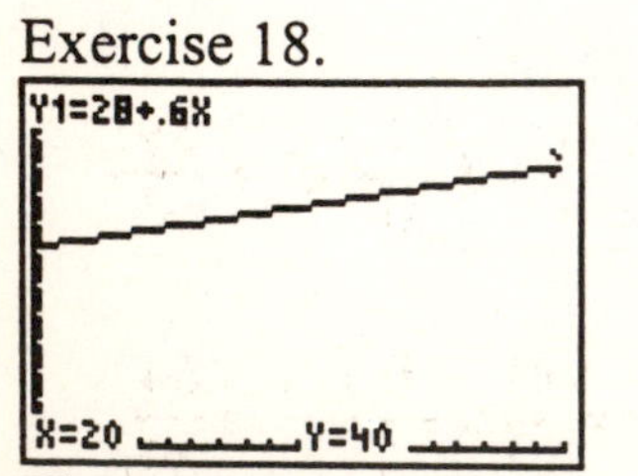

The percentage of babies born to unmarried parents will be 40% in the year 2010.

Critical Thinking Exercises

75. Let x = the original price of the dress
If the reduction in price is 40%, the price paid is 60%.
$$\text{price paid} = 0.60(0.60x)$$
$$72 = 0.60(0.60x)$$
$$72 = 0.36x$$
$$200 = x$$
The original price is \$200.

77. Let x = the amount a girl would receive
$2x$ = the amount Mrs. Ricardo would receive
$4x$ = the amount a boy would receive
$$\text{Total Savings} = x + 2x + 4x$$
$$14{,}000 = 7x$$
$$2{,}000 = x$$
Mrs. Ricardo received \$4000, the boy received \$8000, and the girl received \$2000.

79.
$$V = C - \frac{C - S}{L} N$$
$$V = C - \left(\frac{C - S}{L}\right)\frac{N}{1}$$
$$V = C - \frac{CN - SN}{L}$$
$$V = \frac{CL}{L} - \frac{CN - SN}{L}$$
$$V = \frac{CL - CN + SN}{L}$$
$$LV = CL - CN + SN$$
$$LV - SN = CL - CN$$
$$LV - SN = C(L - N)$$
$$C = \frac{LV - SN}{L - N}$$

Review Exercises

80. –6 is less than or equal to –6. True.

81.
$$\frac{(2+4)^2+(-1)^5}{12\div 2\cdot 3-3}=\frac{(6)^2+(-1)}{6\cdot 3-3}=\frac{36+(-1)}{18-3}=\frac{35}{15}=\frac{7}{3}$$

82.
$$\frac{2x}{3}-\frac{8}{3}=x$$
$$3\left(\frac{2x}{3}-\frac{8}{3}\right)=3(x)$$
$$3\left(\frac{2x}{3}\right)-3\left(\frac{8}{3}\right)=3x$$
$$\cancel{3}^1\left(\frac{2x}{\cancel{3}^1}\right)-\cancel{3}^1\left(\frac{8}{\cancel{3}^1}\right)=3x$$
$$2x-8=3x$$
$$-8=x$$
The solution is –8 and the solution set is $\{-8\}$.

Check Points 1.6

1. **a.** $b^6\cdot b^5=b^{6+5}=b^{11}$

b. $(4x^3y^4)(10x^2y^6)$
$$=4\cdot 10\cdot x^3\cdot x^2\cdot y^4\cdot y^6$$
$$=40x^{3+2}y^{4+6}$$
$$=40x^5y^{10}$$

2. **a.** $\frac{b^{15}}{b^3}=b^{15-3}=b^{12}$

b.
$$\frac{27x^{14}y^8}{3x^3y^5}=\frac{27}{3}\cdot\frac{x^{14}}{x^3}\cdot\frac{y^8}{y^5}$$
$$=9x^{14-3}y^{8-5}$$
$$=9x^{11}y^3$$

3. **a.** $7^0=1$

b. $(-5)^0=1$

c. $-5^0=-(5^0)=-1$

d. $10x^0=10\cdot 1=10$

e. $(10x)^0=1$

4. **a.** $5^{-2}=\frac{1}{5^2}=\frac{1}{25}$

b.
$$(-3)^{-4}=\frac{1}{(-3)^4}=\frac{1}{(-3)(-3)(-3)(-3)}=\frac{1}{81}$$

c. $\frac{1}{4^{-2}}=\frac{1}{\frac{1}{4^2}}=1\cdot\frac{4^2}{1}=4^2=16$

d. $3x^{-6}y^4=3\cdot\frac{1}{x^6}\cdot y^4=\frac{3y^4}{x^6}$

5. **a.** $\frac{7^{-2}}{4^{-3}}=\frac{4^3}{7^2}=\frac{4\cdot 4\cdot 4}{7\cdot 7}=\frac{64}{49}$

b. $\frac{1}{5x^{-2}} = \frac{x^2}{5}$

6. **a.** $(x^5)^3 = x^{5\cdot 3} = x^{15}$

b. $(y^7)^{-2} = y^{7\cdot(-2)} = y^{-14} = \frac{1}{y^{14}}$

c. $(b^{-3})^{-4} = b^{-3\cdot(-4)} = b^{12}$

7. **a.** $(2x)^4 = 2^4x^4 = 16x^4$

b. $(-3y^2)^3 = (-3)^3 y^{2\cdot 3} = -27y^6$

c. $(-4x^5y^{-1})^{-2} = (-4x^5y^{-1})^{-2}$

$= (-4)^{-2} x^{5(-2)}y^{-1(-2)}$

$= \frac{1}{(-4)^2} \cdot x^{-10}y^2$

$= \frac{1}{16} \cdot \frac{1}{x^{10}} \cdot y^2$

$= \frac{y^2}{16x^{10}}$

8. **a.** $\left(\frac{x^5}{4}\right)^3 = \frac{(x^5)^3}{4^3} = \frac{x^{15}}{64}$

b. $\left(\frac{2x^{-3}}{y^2}\right)^4 = \frac{2^4x^{-3(4)}}{y^{2(4)}}$

$= \frac{16x^{-12}}{y^8}$

$= \frac{16}{x^{12}y^8}$

9. **a.** $(-3x^{-6}y)(-2x^3y^4)^2$

$= (-3x^{-6}y)(-2)^2(x^3)^2(y^4)^2$

$= (-3)x^{-6}y(4)x^6y^8$

$= (-3)(4)x^{-6+6}y^{1+8}$

$= -12x^0y^9 = -12(1)y^9$

$= -12y^9$

b. $\left(\frac{10x^3y^5}{5x^6y^{-2}}\right)^2 = \left(2x^{3-6}y^{5-(-2)}\right)^2$

$= (2)^2(x^{-3})^2(y^7)^2$

$= 4x^{-6}y^{14} = \frac{4y^{14}}{x^6}$

Problem Set 1.6

Practice Exercises

1. $b^4 \cdot b^7 = b^{(4+7)} = b^{11}$

3. $x \cdot x^3 = x^{(1+3)} = x^4$

5. $2^3 \cdot 2^2 = 2^{(3+2)} = 2^5 = 32$

7. $3x^4 \cdot 2x^2 = 6x^{(4+2)} = 6x^6$

9. $(-2y^{10})(-10y^2) = 20y^{(10+2)} = 20y^{12}$

11. $(5x^3y^4)(20x^7y^8) = 100x^{(3+7)}y^{(4+8)}$

$= 100x^{10}y^{12}$

13. $\frac{b^{12}}{b^3} = b^{(12-3)} = b^9$

15. $\frac{15x^9}{3x^4} = 5x^{(9-4)} = 5x^5$

17. $\frac{x^9y^7}{x^4y^2} = x^{(9-4)}y^{(7-2)} = x^5y^5$

19. $\frac{50x^2y^7}{5xy^4} = 10x^{(2-1)}y^{(7-4)} = 10xy^3$

21. $\frac{-56a^{12}b^{10}c^8}{7ab^2c^4} = -8a^{(12-1)}b^{(10-2)}c^{(8-4)}$
$= -8a^{11}b^8c^4$

23. $6^0 = 1$

25. $(-4)^0 = 1$

27. $-4^0 = -1$

29. $13y^0 = 13(1) = 13$

31. $(13y)^0 = 1$

33. $3^{-2} = \frac{1}{3^2} = \frac{1}{9}$

35. $(-5)^{-2} = \frac{1}{(-5)^2} = \frac{1}{25}$

37. $-5^{-2} = -(5^{-2}) = -\frac{1}{5^2} = -\frac{1}{25}$

39. $x^2y^{-3} = \frac{x^2}{y^3}$

41. $8x^{-7}y^3 = \frac{8y^3}{x^7}$

43. $\frac{1}{5^{-3}} = 5^3 = 125$

45. $\frac{1}{(-3)^{-4}} = (-3)^4 = 81$

47. $\frac{x^{-2}}{y^{-5}} = \frac{y^5}{x^2}$

49. $\frac{a^{-4}b^7}{c^{-3}} = \frac{b^7c^3}{a^4}$

51. $(x^6)^{10} = x^{(6\cdot 10)} = x^{60}$

53. $(b^4)^{-3} = \frac{1}{(b^4)^3} = \frac{1}{b^{(4\cdot 3)}} = \frac{1}{b^{12}}$

55. $(7^{-4})^{-5} = 7^{-4\cdot(-5)} = 7^{20}$

57. $(4x)^3 = 4^3x^3 = 64x^3$

59. $(-3x^7)^2 = (-3)^2x^{7\cdot 2} = 9x^{14}$

61. $(2xy^2)^3 = 8x^{(1\cdot 3)}y^{(2\cdot 3)} = 8x^3y^6$

63. $(-3x^2y^5)^2 = (-3)^2x^{(2\cdot 2)}y^{(5\cdot 2)} = 9x^4y^{10}$

65. $(-3x^{-2})^{-3} = (-3)^{-3}(x^{-2})^{-3} = \frac{x^6}{(-3)^3}$
$= \frac{x^6}{-27} = -\frac{x^6}{27}$

67. $\left(5x^3y^{-4}\right)^{-2} = 5^{-2}\left(x^3\right)^{-2}\left(y^{-4}\right)^{-2}$
$= 5^{-2}x^{-6}y^8$
$= \dfrac{y^8}{25x^6}$

69. $\left(-2x^{-5}y^4z^2\right)^{-4} = (-2)^{-4}x^{20}y^{-16}z^{-8}$
$= \dfrac{x^{20}}{(-2)^4y^{16}z^8}$
$= \dfrac{x^{20}}{16y^{16}z^8}$

71. $\left(\dfrac{2}{x}\right)^4 = \dfrac{2^4}{x^4} = \dfrac{16}{x^4}$

73. $\left(\dfrac{x^3}{5}\right)^2 = \dfrac{x^{(3\cdot 2)}}{5^2} = \dfrac{x^6}{25}$

75. $\left(-\dfrac{3x}{y}\right)^4 = \dfrac{(-3)^4x^4}{y^4} = \dfrac{81x^4}{y^4}$

77. $\left(\dfrac{x^4}{y^2}\right)^6 = \dfrac{x^{(4\cdot 6)}}{y^{(2\cdot 6)}} = \dfrac{x^{24}}{y^{12}}$

79. $\left(\dfrac{x^3}{y^{-4}}\right)^3 = \dfrac{x^{(3\cdot 3)}}{y^{(-4\cdot 3)}} = \dfrac{x^9}{y^{-12}} = x^9y^{12}$

81. $\left(\dfrac{a^{-2}}{b^3}\right)^{-4} = \dfrac{a^{(-2\cdot(-4))}}{b^{(3\cdot(-4))}} = \dfrac{a^8}{b^{-12}} = a^8b^{12}$

83. $\dfrac{x^3}{x^9} = x^{3-9} = x^{-6} = \dfrac{1}{x^6}$

85. $\dfrac{20x^3}{-5x^4} = -4x^{3-4} = -4x^{-1} = -\dfrac{4}{x}$

87. $\dfrac{16x^3}{8x^{10}} = 2x^{3-10} = 2x^{-7} = \dfrac{2}{x^7}$

89. $\dfrac{20a^3b^8}{2ab^{13}} = 10a^{3-1}b^{8-13} = 10a^2b^{-5} = \dfrac{10a^2}{b^5}$

91. $x^3 \cdot x^{-12} = x^{3+(-12)} = x^{-9} = \dfrac{1}{x^9}$

93. $\left(2a^5\right)\left(-3a^{-7}\right) = -6a^{5+(-7)}$
$= -6a^{-2}$
$= -\dfrac{6}{a^2}$

95. $\dfrac{6x^2}{2x^{-8}} = 3x^{2-(-8)} = 3x^{2+8} = 3x^{10}$

97. $\dfrac{x^{-7}}{x^3} = x^{-7-3} = x^{-10} = \dfrac{1}{x^{10}}$

99. $\dfrac{30x^2y^5}{-6x^8y^{-3}} = -5x^{2-8}y^{5-(-3)}$
$= -5x^{-6}y^8$
$= -\dfrac{5y^8}{x^6}$

101. $\left(\dfrac{x^3}{x^{-5}}\right)^2 = \left(x^{3-(-5)}\right)^2 = \left(x^{3+5}\right)^2$
$= \left(x^8\right)^2 = x^{16}$

103. $\left(\dfrac{-15a^4b^2}{5a^{10}b^{-3}}\right)^3 = \left(-3a^{4-10}b^{2-(-3)}\right)^3$
$= \left(-3a^{-6}b^{2+3}\right)^3$
$= \left(-3a^{-6}b^5\right)^3$
$= (-3)^3\left(a^{-6}\right)^3\left(b^5\right)^3$
$= -27a^{-18}b^{15}$
$= -\dfrac{27b^{15}}{a^{18}}$

105. $\left(\dfrac{3a^{-5}b^2}{12a^3b^{-4}}\right)^0 = 1$

Recall the Zero Exponent Rule.

Application Exercises

107. If $x = 90$,

$y = 1000\left(\dfrac{1}{2}\right)^{\frac{90}{30}} = 1000\left(\dfrac{1}{2}\right)^3$
$= 1000\left(\dfrac{1}{8}\right) = 125$

The amount of cesium-137 in the atmosphere will be 125 kilograms. Chernobyl will be unsafe for human habitation.

109. If $n = 1$,

$d = \dfrac{3\left(2^{n-2}\right)+4}{10} = \dfrac{3\left(2^{1-2}\right)+4}{10}$
$= \dfrac{3\left(2^{-1}\right)+4}{10} = \dfrac{3\left(\frac{1}{2}\right)+4}{10}$
$= \dfrac{1.5+4}{10} = \dfrac{5.5}{10} = 0.55$

Mercury is 0.55 astronomical units from the sun.

111. If $n = 3$,

$d = \dfrac{3\left(2^{n-2}\right)+4}{10} = \dfrac{3\left(2^{3-2}\right)+4}{10}$
$= \dfrac{3(2)+4}{10} = \dfrac{6+4}{10}$
$= \dfrac{10}{10} = 1$

Earth is 1 astronomical unit from the sun.

Writing in Mathematics

113. Answers will vary.

115. Answers will vary.

117. Answers will vary.

119. Answers will vary.

Technology Exercises

121.

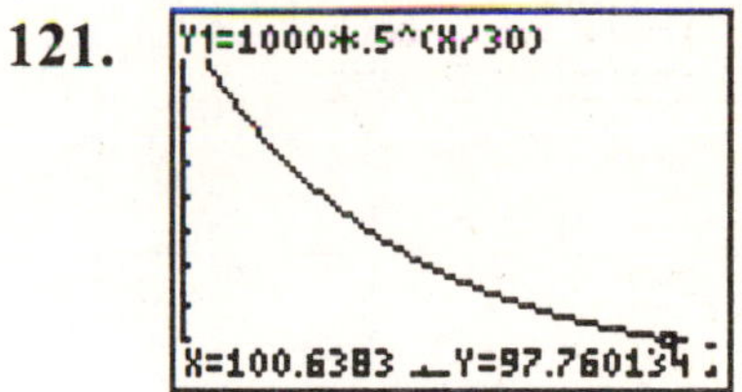

Approximately 100 years after 1986 in 2086, Chernobyl will be safe for human habitation.

123. Statement **d.** is true.

Statement **a.** is false.
$2^2 \cdot 2^4 = 2^{2+4} = 2^6$
Use the Product Rule.

Statement **b.** is false.
$5^6 \cdot 5^2 = 5^{6+2} = 5^8$
Use the Product Rule.

Statement **c.** is false.
$2^3 \cdot 3^2 = 8 \cdot 9 = 72$
Because the bases are not the same, we cannot apply the rules of exponents.

125. $x^{n-1} \cdot x^{3n+4} = x^{(n-1)+(3n+4)}$
$= x^{4n+3}$

127. $\left(\dfrac{x^{3-n}}{x^{6-n}}\right)^{-2} = \left(x^{(3-n)-(6-n)}\right)^{-2}$
$= \left(x^{3-n-6+n}\right)^{-2}$
$= \left(x^{-3}\right)^{-2}$
$= x^6$

Review Exercises

129.

x	$y = 2x - 1$	(x, y)
−3	$y = 2(-3) - 1$ $= -6 - 1 = -7$	$(-3, -7)$
−2	$y = 2(-2) - 1$ $= -4 - 1 = -5$	$(-2, -5)$
−1	$y = 2(-1) - 1$ $= -2 - 1 = -3$	$(-1, -3)$
0	$y = 2(0) - 1$ $= 0 - 1 = -1$	$(0, -1)$
1	$y = 2(1) - 1$ $= 2 - 1 = 1$	$(1, 1)$
2	$y = 2(2) - 1$ $= 4 - 1 = 3$	$(2, 3)$
3	$y = 2(3) - 1$ $= 6 - 1 = 5$	$(3, 5)$

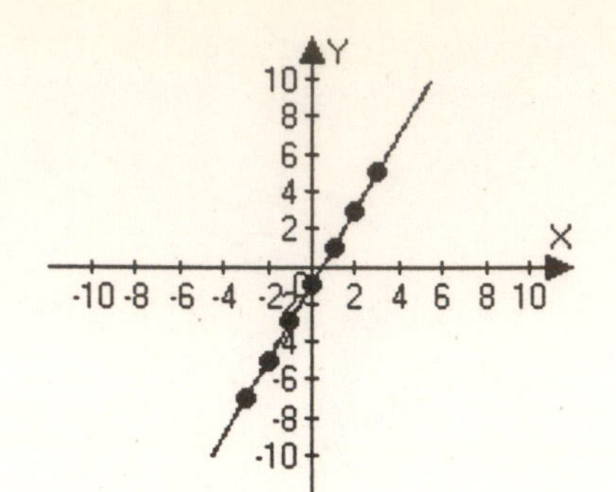

130. $Ax + By = C$
$By = C - Ax$
$\dfrac{\cancel{B}y}{\cancel{B}} = \dfrac{C - Ax}{B}$
$y = \dfrac{C - Ax}{B}$

131. Let w = the width of the playing field
Let $2w - 5$ = the length of the playing field
$P = 2(\text{length}) + 2(\text{width})$
$230 = 2(2w - 5) + 2w$
$230 = 4w - 10 + 2w$
$230 = 6w - 10$
$240 = 6w$
$40 = w$
Find the length.
$2w - 5 = 2(40) - 5 = 80 - 5 = 75$
The playing field is 40 meters by 75 meters.

Check Points 1.7

1. **a.** $2.6 \times 10^9 = 2{,}600{,}000{,}000$

b. $3.017 \times 10^{-6} = 0.000003017$

2. **a.** $7,410,000,000 = 7.41\times10^9$

b. $0.000000092 = 9.2\times10^{-8}$

3. $410\times10^7 = \left(4.1\times10^2\right)\times10^7$
$= 4.1\times10^{2+7}$
$= 4.1\times10^9$

4. **a.** $\left(7.1\times10^5\right)\left(5\times10^{-7}\right)$
$= (7.1\times5)\times\left(10^5\times10^{-7}\right)$
$= 35.5\times10^{5+(-7)}$
$= 35.5\times10^{-2}$
$= 3.55\times10^{-1}$

b. $\dfrac{\left(1.2\times10^6\right)}{\left(3\times10^{-3}\right)} = \dfrac{1.2}{3}\times\dfrac{10^6}{10^{-3}}$
$= 0.4\times10^{6-(-3)}$
$= 0.4\times10^9$
$= 4\times10^8$

5. $\dfrac{5.6\times10^{12}}{2.8\times10^8} = \dfrac{5.6}{2.8}\times\dfrac{10^{12}}{10^8}$
$= 2\times10^{12-8}$
$= 2\times10^4$
$= \$20,000$
Each citizen would have to pay \$20,000.

6. $d = rt$
$d = \left(1.55\times10^3\right)\times20000$
$d = \left(1.55\times10^3\right)\times\left(2\times10^4\right)$
$d = (1.55\times2)\times\left(10^3\times10^4\right)$
$d = 3.10\times10^{3+4}$
$d = 3.10\times10^7$
The distance from Venus to Mercury is 3.10×10^7 miles.

Problem Set 1.7

Practice Exercises

1. $3.8\times10^2 = 380$

3. $6\times10^{-4} = 0.0006$

5. $7.16\times10^6 = 7,160,000$

7. $1.4\times10^0 = 1.4\times1 = 1.4$

9. $7.9\times10^{-1} = 0.79$

11. $4.15\times10^{-3} = 0.00415$

13. $6.00001\times10^{10} = 60,000,100,000$

15. $32,000 = 3.2\times10^4$

17. $638,000,000,000,000,000$
$= 6.38\times10^{17}$

19. $317 = 3.17\times10^2$

21. $5716 = 5.716\times10^3$

23. $0.0027 = 2.7\times10^{-3}$

25. $0.00000000504 = 5.04\times10^{-9}$

27. $0.007 = 7\times10^{-3}$

29. $3.14159 = 3.14159\times10^0$

31. $(3\times10^4)(2.1\times10^3)$
$=(3\times2.1)(10^4\times10^3)$
$=6.3\times10^{4+3}$
$=6.3\times10^7$

33. $(1.6\times10^{15})(4\times10^{-11})$
$=(1.6\times4)(10^{15}\times10^{-11})$
$=6.4\times10^{15+(-11)}$
$=6.4\times10^4$

35. $(6.1\times10^{-8})(2\times10^{-4})$
$=(6.1\times2)(10^{-8}\times10^{-4})$
$=12.2\times10^{-8+(-4)}$
$=12.2\times10^{-12}$
$=1.22\times10^{-11}$

37. $(4.3\times10^8)(6.2\times10^4)$
$=(4.3\times6.2)(10^8\times10^4)$
$=26.66\times10^{8+4}$
$=26.66\times10^{12}$
$=2.666\times10^{13}$
$\approx2.67\times10^{13}$

39. $\frac{8.4\times10^8}{4\times10^5}=\frac{8.4}{4}\times\frac{10^8}{10^5}$
$=2.1\times10^{8-5}$
$=2.1\times10^3$

41. $\frac{3.6\times10^4}{9\times10^{-2}}=\frac{3.6}{9}\times\frac{10^4}{10^{-2}}$
$=0.4\times10^{4-(-2)}$
$=0.4\times10^6=4\times10^5$

43. $\frac{4.8\times10^{-2}}{2.4\times10^6}=\frac{4.8}{2.4}\times\frac{10^{-2}}{10^6}$
$=2\times10^{-2-6}=2\times10^{-8}$

45. $\frac{2.4\times10^{-2}}{4.8\times10^{-6}}=\frac{2.4}{4.8}\times\frac{10^{-2}}{10^{-6}}$
$=0.5\times10^{-2-(-6)}$
$=0.5\times10^4=5\times10^3$

47. $\frac{480{,}000{,}000{,}000}{0.00012}=\frac{4.8\times10^{11}}{1.2\times10^{-4}}$
$=\frac{4.8}{1.2}\times\frac{10^{11}}{10^{-4}}$
$=4\times10^{11-(-4)}$
$=4\times10^{15}$

49. $\frac{0.00072\times0.003}{0.00024}$
$=\frac{(7.2\times10^{-4})(3\times10^{-3})}{2.4\times10^{-4}}$
$=\frac{7.2\times3}{2.4}\times\frac{\cancel{10^{-4}}\cdot10^{-3}}{\cancel{10^{-4}}}$
$=9\times10^{-3}$

Application Exercises

51. 53.3 million = 53,300,000=5.33×10^7
5.33×10^7 people will be 65 and over in 2020.

53. $77.0-35.3=41.7$
41.7 million = 41,700,000=4.17×10^7
There will be 4.17×10^7 more people 65 and over in the 2040 than in the year 2000.

55. $\dfrac{1.9\times10^{12}}{2.8\times10^{8}}=\dfrac{1.9}{2.8}\times\dfrac{10^{12}}{10^{8}}$

$=.678571\times10^{12-8}$

$=.6786\times10^{4}$

$=6786$

≈ 6800

The per capita tax burden is approximately \$6800.

57. $4,000\left(2.8\times10^{8}\right)$

$=\left(4\times10^{3}\right)\left(2.8\times10^{8}\right)$

$=\left(4\times2.8\right)\left(10^{3}\times10^{8}\right)$

$=11.2\times10^{3+8}$

$=11.2\times10^{11}$

$=1.12\times10^{12}$

We spend $\$1.12\times10^{12}$ on health care nationwide each year.

59. $20,000\left(5.3\times10^{-23}\right)$

$=\left(2\times10^{4}\right)\left(5.3\times10^{-23}\right)$

$=\left(2\times5.3\right)\left(10^{4}\times10^{-23}\right)$

$=10.6\times10^{4+(-23)}$

$=10.6\times10^{-19}$

$=1.06\times10^{-18}$

The mass of 20,000 oxygen molecules is 1.06×10^{-18} grams.

Writing in Mathematics

61. Answers will vary.

63. Answers will vary.

65. Answers will vary.

Technology Exercises

67. Answers will vary. For example, consider Exercise 15.

$32,000=3.2\times10^{4}$

```
3.2*10^4
                32000
```

Critical Thinking Exercises

69. Statement **d.** is true.

$\left(4\times10^{3}\right)+\left(3\times10^{2}\right)=4000+300$

$=4300=43\times10^{2}$

Statement **a.** is false.

$534.7=5.347\times10^{2}$

Statement **b.** is false.

$\dfrac{8\times10^{30}}{4\times10^{-5}}=2\times10^{30-(-5)}=2\times10^{35}$

Statement **c.** is false.

$\left(7\times10^{5}\right)+\left(2\times10^{-3}\right)=700,000+0.002$

$=700,000.002$

71. Answers will vary.

Review Exercises

72. $9(10x-4)-(5x-10)$

$=90x-36-5x+10$

$=90x-5x-36+10$

$=85x-26$

73.
$$\frac{4x-1}{10}=\frac{5x+2}{4}-4$$
$$20\left(\frac{4x-1}{10}\right)=20\left(\frac{5x+2}{4}-4\right)$$
$$2(4x-1)=5(5x+2)-80$$
$$8x-2=25x+10-80$$
$$8x-2=25x-70$$
$$-2=17x-70$$
$$68=17x$$
$$4=x$$
The solution is 4 and the solution set is $\{4\}$.

74.
$$\left(8x^4y^{-3}\right)^{-2}=8^{-2}\left(x^4\right)^{-2}\left(y^{-3}\right)^{-2}$$
$$=8^{-2}x^{-8}y^6$$
$$=\frac{1}{64}\cdot\frac{1}{x^8}y^6$$
$$=\frac{y^6}{64x^8}$$

Chapter 1 Review

1.1

1. $2x-10$

2. $4+6x=6x+4$

3. $\frac{9}{x}+\frac{1}{2}x$

4.
$$x^2-7x+4=(10)^2-7(10)+4$$
$$=100-70+4$$
$$=30+4$$
$$=34$$

5.
$$6+2(x-8)^3=6+2(11-8)^3$$
$$=6+2(3)^3$$
$$=6+2(27)$$
$$=6+54$$
$$=60$$

6.
$$x^4-(x-y)=(2)^4-(2-1)$$
$$=16-1$$
$$=15$$

7. $\{1, 2\}$

8. $\{-3, -2, -1, 0, 1\}$

9. False. Zero is not a natural number.

10. True. –2 is a rational number.

11. True. $\frac{1}{3}$ is not an irrational number.

12. Negative five is less than two. True.

13. Negative seven is greater than or equal to negative three. False.

14. Negative seven is less than or equal to negative seven. True.

15.
$$S=0.015x^2+x+10$$
$$=0.015(60)^2+60+10$$
$$=0.015(3600)+70$$
$$=54+70$$
$$=124$$
The recommended safe distance is 124 feet.

1.2

16. $|-9.7| = 9.7$

17. $|5.003| = 5.003$

18. $|0| = 0$

19. $-2.4 + (-5.2) = -7.6$

20. $-6.8 + 2.4 = -4.4$

21. $-7 - (-20) = -7 + 20 = 13$

22. $(-3)(-20) = 60$

23. $$\begin{aligned} 4(-3)(-2)(-10) &= -12(-2)(-10) \\ &= 24(-10) \\ &= -240 \end{aligned}$$

24. $(-2)^4 = 16$

25. $-2^5 = -32$

26. $-\frac{2}{3} \div \frac{8}{5} = -\frac{2}{3} \cdot \frac{5}{8} = -\frac{\not{2} \cdot 5}{3 \cdot \not{2} \cdot 4} = -\frac{5}{12}$

27. $\frac{-35}{-5} = \frac{-\not{5} \cdot 7}{-\not{5}} = 7$

28. $\frac{54.6}{-6} = -9.1$

29. $$\begin{aligned} x &= -7 \\ -1(x) &= -1(-7) \\ -x &= 7 \end{aligned}$$

30. $$\begin{aligned} -11 - [-17 + (-3)] &= -11 - [-20] \\ &= -11 + 20 \\ &= 9 \end{aligned}$$

31. $\left(-\frac{1}{2}\right)^3 \cdot 2^4 = -\frac{1}{8} \cdot 16 = -\frac{16}{8} = -2$

32. $$\begin{aligned} -3[4 - (6 - 8)] &= -3[4 - (-2)] \\ &= -3[6] \\ &= -18 \end{aligned}$$

33. $$\begin{aligned} & 8^2 - 36 \div 3^2 \cdot 4 - (-7) \\ &= 64 - 36 \div 9 \cdot 4 - (-7) \\ &= 64 - 4 \cdot 4 - (-7) \\ &= 64 - 16 - (-7) \\ &= 48 - (-7) \\ &= 48 + 7 \\ &= 55 \end{aligned}$$

34. $$\begin{aligned} \frac{(-2)^4 + (-3)^2}{2^2 - (-21)} &= \frac{16 + 9}{4 - (-21)} \\ &= \frac{25}{4 + 21} \\ &= \frac{25}{25} \\ &= 1 \end{aligned}$$

35. $$\begin{aligned} \frac{(7-9)^3 - (-4)^2}{2 + 2(8) \div 4} &= \frac{(-2)^3 - 16}{2 + 16 \div 4} \\ &= \frac{-8 - 16}{2 + 4} \\ &= \frac{-24}{6} = -4 \end{aligned}$$

36. $4-(3-8)^2+3\div 6\cdot 4^2$
$=4-(-5)^2+3\div 6\cdot 4^2$
$=4-25+3\div 6\cdot 16$
$=4-25+\frac{1}{2}\cdot 16$
$=4-25+8$
$=-21+8=-13$

37. $5(2x-3)+7x=10x-15+7x$
$=17x-15$

38. $5x+7x^2-4x+2x^2=x+9x^2$
$=9x^2+x$

39. $3(4y-5)-(7y+2)$
$=12y-15-7y-2$
$=5y-17$

40. $8-2[3-(5x-1)]=8-2[3-5x+1]$
$=8-6+10x-2$
$=10x$

41. $6(2x-3)-5(3x-2)$
$=12x-18-15x+10$
$=-3x-8$

1.3

42. (−1, 3)

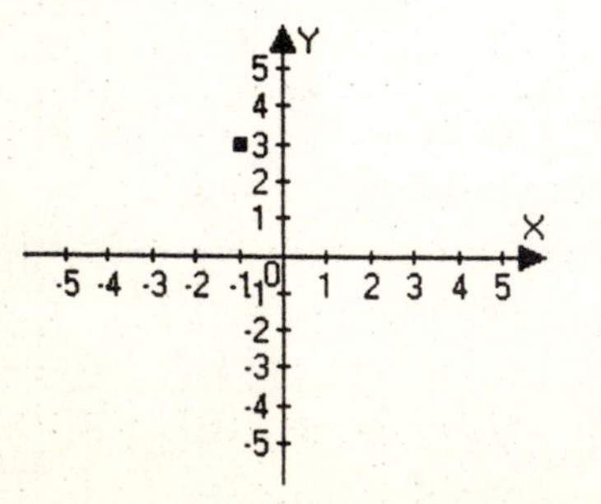

43. (2,−5)

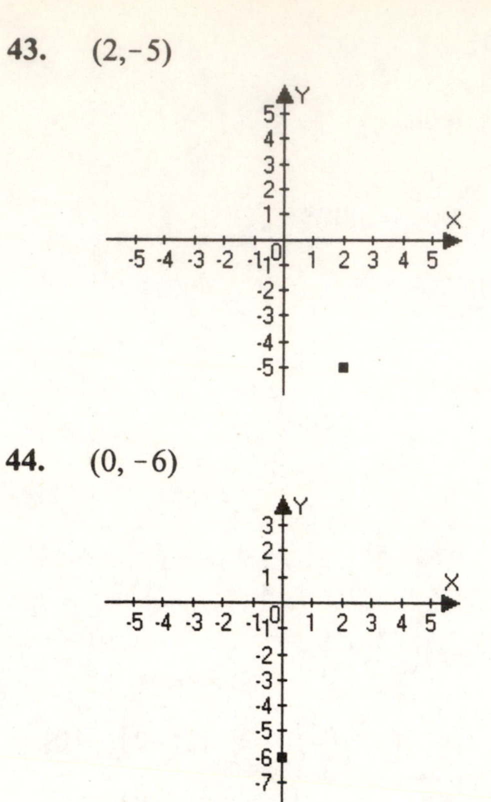

45.

x	$y=2x-2$	(x,y)
−3	$y=2(-3)-2$ $=-6-2=-8$	$(-3,-8)$
−2	$y=2(-2)-2$ $=-4-2=-6$	$(-2,-6)$
−1	$y=2(-1)-2$ $=-2-2=-4$	$(-1,-4)$
0	$y=2(0)-2$ $=0-2=-2$	$(0,-2)$
1	$y=2(1)-2$ $=2-2=0$	$(1,0)$
2	$y=2(2)-2$ $=4-2=2$	$(2,2)$
3	$y=2(3)-2$ $=6-2=4$	$(3,4)$

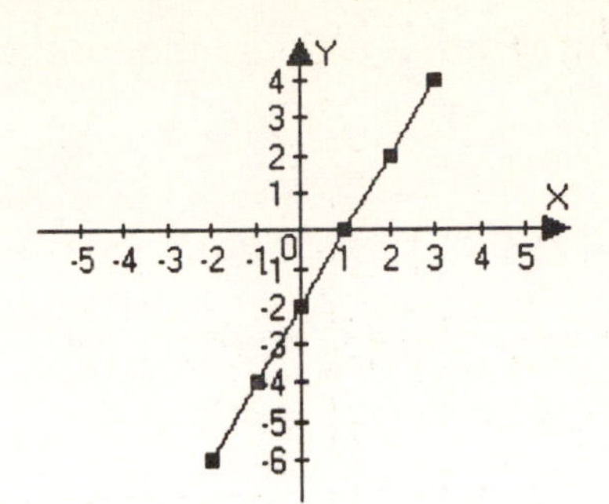

46.

x	$y = x^2 - 3$	(x, y)
-3	$y = (-3)^2 - 3$ $= 9 - 3 = 6$	$(-3, 6)$
-2	$y = (-2)^2 - 3$ $= 4 - 3 = 1$	$(-2, 1)$
-1	$y = (-1)^2 - 3$ $= 1 - 3 = -2$	$(-1, -2)$
0	$y = (0)^2 - 3$ $= 0 - 3 = -3$	$(0, -3)$
1	$y = (1)^2 - 3$ $= 1 - 3 = -2$	$(1, -2)$
2	$y = (2)^2 - 3$ $= 4 - 3 = 1$	$(2, 1)$
3	$y = (3)^2 - 3$ $= 9 - 3 = 6$	$(3, 6)$

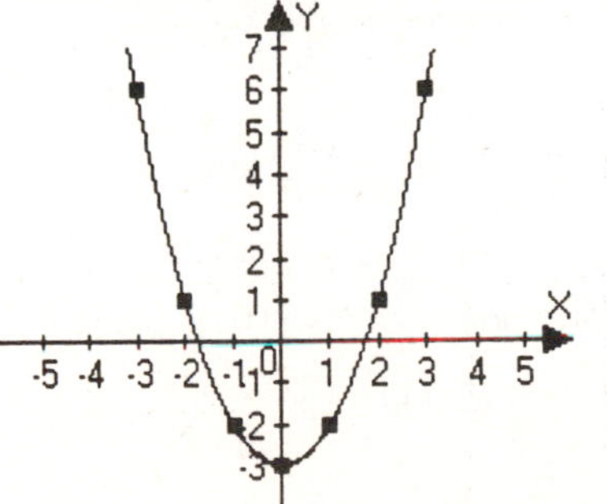

47.

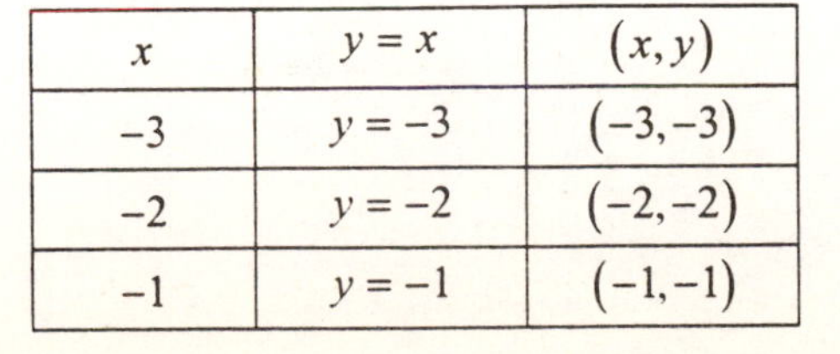

x	$y = x$	(x, y)
-3	$y = -3$	$(-3, -3)$
-2	$y = -2$	$(-2, -2)$
-1	$y = -1$	$(-1, -1)$

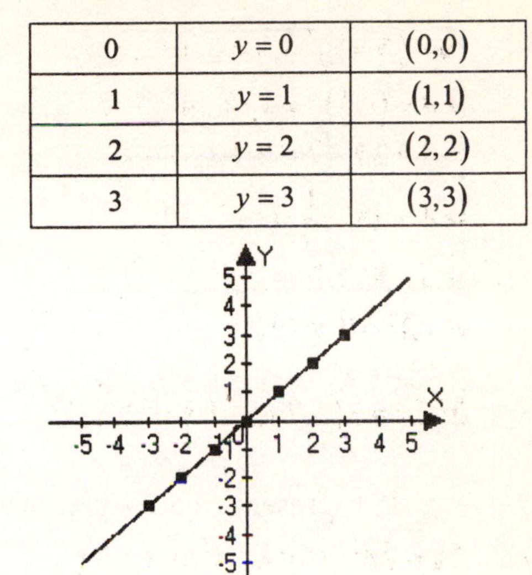

0	$y = 0$	$(0, 0)$
1	$y = 1$	$(1, 1)$
2	$y = 2$	$(2, 2)$
3	$y = 3$	$(3, 3)$

48.

x	$y = \lvert x \rvert - 2$	(x, y)
-3	$y = \lvert -3 \rvert - 2 = 3 - 2 = 1$	$(-3, 1)$
-2	$y = \lvert -2 \rvert - 2 = 2 - 2 = 0$	$(-2, 0)$
-1	$y = \lvert -1 \rvert - 2 = 1 - 2 = -1$	$(-1, -1)$
0	$y = \lvert 0 \rvert - 2 = 0 - 2 = -2$	$(0, -2)$
1	$y = \lvert 1 \rvert - 2 = 1 - 2 = -1$	$(1, -1)$
2	$y = \lvert 2 \rvert - 2 = 2 - 2 = 0$	$(2, 0)$
3	$y = \lvert 3 \rvert - 2 = 3 - 2 = 1$	$(3, 1)$

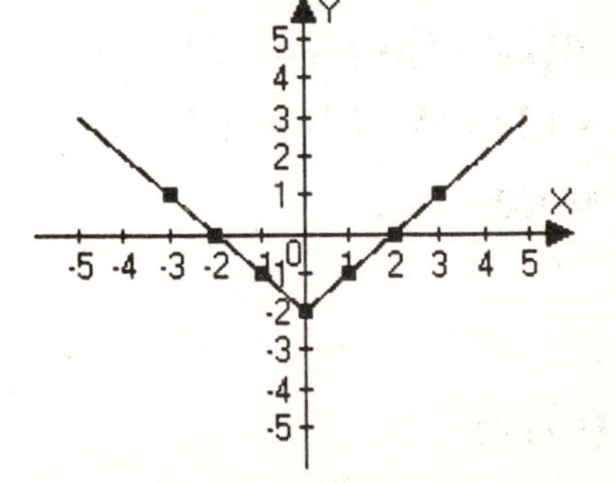

49. The minimum x-value is -20 and the maximum x-value is 40. The distance between tick marks is 10. The minimum y-value is -5 and the maximum y-value is 5. The distance between tick marks is 1.

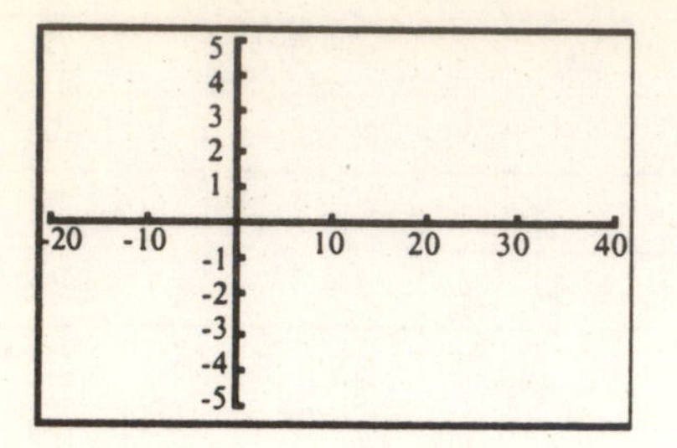

50. 20% of 75 year old Americans have Alzheimer's.

51. Age 85 represents a 50% prevalence.

52. Answers will vary.

53. Graph **c.** illustrates the description.

1.4

54.
$$2x-5=7$$
$$2x=12$$
$$x=6$$
The solution is 6 and the solution set is $\{6\}$.

55.
$$5x+20=3x$$
$$2x+20=0$$
$$2x=-20$$
$$x=-10$$
The solution is –10 and the solution set is $\{-10\}$.

56.
$$7(x-4)=x+2$$
$$7x-28=x+2$$
$$6x-28=2$$
$$6x=30$$
$$x=5$$
The solution is 5 and the solution set is $\{5\}$.

57.
$$1-2(6-x)=3x+2$$
$$1-12+2x=3x+2$$
$$-11+2x=3x+2$$
$$-11=x+2$$
$$-13=x$$
The solution is –13 and the solution set is $\{-13\}$.

58.
$$2(x-4)+3(x+5)=2x-2$$
$$2x-8+3x+15=2x-2$$
$$5x+7=2x-2$$
$$3x+7=-2$$
$$3x=-9$$
$$x=-3$$
The solution is –3 and the solution set is $\{-3\}$.

59.
$$2x-4(5x+1)=3x+17$$
$$2x-20x-4=3x+17$$
$$-18x-4=3x+17$$
$$-4=21x+17$$
$$-21=21x$$
$$-1=x$$
The solution is –1 and the solution set is $\{-1\}$.

60.
$$\frac{2x}{3}=\frac{x}{6}+1$$
$$6\left(\frac{2x}{3}\right)=6\left(\frac{x}{6}+1\right)$$

$$2(2x) = 6\left(\frac{x}{6}\right) + 6(1)$$
$$4x = x + 6$$
$$3x = 6$$
$$x = 2$$

The solution is 2 and the solution set is $\{2\}$.

61.
$$\frac{x}{2} - \frac{1}{10} = \frac{x}{5} + \frac{1}{2}$$
$$10\left(\frac{x}{2} - \frac{1}{10}\right) = 10\left(\frac{x}{5} + \frac{1}{2}\right)$$
$$10\left(\frac{x}{2}\right) - 10\left(\frac{1}{10}\right) = 10\left(\frac{x}{5}\right) + 10\left(\frac{1}{2}\right)$$
$$5x - 1 = 2x + 5$$
$$3x - 1 = 5$$
$$3x = 6$$
$$x = 2$$

The solution is 2 and the solution set is $\{2\}$.

62.
$$\frac{2x}{3} = 6 - \frac{x}{4}$$
$$12\left(\frac{2x}{3}\right) = 12\left(6 - \frac{x}{4}\right)$$
$$4(2x) = 12(6) - 12\left(\frac{x}{4}\right)$$
$$8x = 72 - 3x$$
$$11x = 72$$
$$x = \frac{72}{11}$$

The solution is $\frac{72}{11}$ and the solution set is $\left\{\frac{72}{11}\right\}$.

63.
$$\frac{x}{4} = 2 + \frac{x-3}{3}$$
$$12\left(\frac{x}{4}\right) = 12\left(2 + \frac{x-3}{3}\right)$$
$$3x = 12(2) + 12\left(\frac{x-3}{3}\right)$$
$$3x = 24 + 4(x-3)$$
$$3x = 24 + 4x - 12$$
$$3x = 12 + 4x$$
$$-x = 12$$
$$x = -12$$

The solution is –12 and the solution set is $\{-12\}$.

64.
$$\frac{3x+1}{3} - \frac{13}{2} = \frac{1-x}{4}$$
$$12\left(\frac{3x+1}{3} - \frac{13}{2}\right) = 12\left(\frac{1-x}{4}\right)$$
$$12\left(\frac{3x+1}{3}\right) - 12\left(\frac{13}{2}\right) = 3(1-x)$$
$$4(3x+1) - 6(13) = 3(1-x)$$
$$12x + 4 - 78 = 3 - 3x$$
$$12x - 74 = 3 - 3x$$
$$15x - 74 = 3$$
$$15x = 77$$
$$x = \frac{77}{15}$$

The solution is $\frac{77}{15}$ and the solution set is $\left\{\frac{77}{15}\right\}$.

65. $7x+5=5(x+3)+2x$

$7x+5=5x+15+2x$

$7x+5=7x+15$

$5=15$

There is no solution. The solution set is $\{\ \}$ or $\varnothing$. The equation is inconsistent.

66. $7x+13=4x-10+3x+23$

$7x+13=7x+13$

The solution set is $\{x|x \text{ is a real number}\}$ or $\mathbb{R}$. The equation is an identity.

67. $7x+13=3x-10+2x+23$

$7x+13=5x-10+23$

$7x+13=5x+13$

$2x+13=13$

$2x=0$

$x=0$

The solution set is $\{0\}$. The equation is conditional.

68. $M=420x+720$

$4080=420x+720$

$3360=420x$

$8=x$

The losses were 4080 million dollars in 1989 + 8 = 1997.

1.5

69. Let x = the cost of raising a child in a low income family
Let x + 63,000 = the cost of raising a child in a middle income family
Let $2x$ – 3000 = the cost of raising a child in a high income family

Adding the three costs, we have:

$4x+60,000=756,000$

$4x=696,000$

$x=174,000$

Solve for the other variables.

$x+63,000=174,000+63,000$

$=237,000$

$2x-3,000=2(174,000)-3,000$

$=348,000-3,000$

$=345,000$

The costs are $174,000 for a low income family, $237,000 for a middle income family, and $345,000 for a high income family.

70. Let x = the number of years after 1965

$W=1.2+0.4x$

$17.2=1.2+0.4x$

$16=0.4x$

$40=x$

Women will make up 17.2% of the military in the year 1965 + 40 = 2005.

71. Let x = the number of minutes of long distance
Plan A: $C=15+0.05x$
Plan B: $C=5+0.07x$
Set the costs equal.

$15+0.05x=5+0.07x$

$15=5+0.02x$

$10=0.02x$

$500=x$

The cost will be the same if 500 minutes of long distance are used.

72. Let x = the original price of the phone

$48 = x - 0.20x$

$48 = 0.80x$

$48 = 0.80x$

$60 = x$

The original price is \$60.

73. Let x = the amount sold to earn \$800 in one week

$800 = 300 + 0.05x$

$500 = 0.05x$

$500 = 0.05x$

$10{,}000 = x$

Sales must be \$10,000 in one week to earn \$800.

74. Let w = the width of the playing field

Let $3w - 6$ = the length of the playing field

$P = 2(\text{length}) + 2(\text{width})$

$340 = 2(3w - 6) + 2w$

$340 = 6w - 12 + 2w$

$340 = 8w - 12$

$352 = 8w$

$44 = w$

The dimensions are 44 yards by 126 yards.

75.

$V = \frac{1}{3}Bh$

$3V = Bh$

$h = \frac{3V}{B}$

76.

$y - y_1 = m(x - x_1)$

$\frac{y - y_1}{m} = \frac{\cancel{m}(x - x_1)}{\cancel{m}}$

$\frac{y - y_1}{m} = x - x_1$

$x = \frac{y - y_1}{m} + x_1$

77.

$E = I(R + r)$

$\frac{E}{I} = \frac{\cancel{I}(R + r)}{\cancel{I}}$

$\frac{E}{I} = R + r$

$R = \frac{E}{I} - r$

78.

$C = \frac{5F - 160}{9}$

$9C = 5F - 160$

$9C + 160 = 5F$

$F = \frac{9C + 160}{5}$

79.

$s = vt + gt^2$

$s - vt = gt^2$

$g = \frac{s - vt}{t^2}$

80.

$T = gr + gvt$

$T = g(r + vt)$

$g = \frac{T}{r + vt}$

1.6

81. $(-3x^7)(-5x^6) = 15x^{7+6} = 15x^{13}$

82. $x^2y^{-5} = \dfrac{x^2}{y^5}$

83. $\dfrac{3^{-2}x^4}{y^{-7}} = \dfrac{x^4y^7}{3^2} = \dfrac{x^4y^7}{9}$

84. $(x^3)^{-6} = x^{3\cdot(-6)} = x^{-18} = \dfrac{1}{x^{18}}$

85. $(7x^3y)^2 = 7^2x^{3\cdot2}y^{1\cdot2} = 49x^6y^2$

86. $\dfrac{16y^3}{-2y^{10}} = -8y^{3-10} = -8y^{-7} = -\dfrac{8}{y^7}$

87. $(-3x^4)(4x^{-11}) = -12x^{4+(-11)}$

$= -12x^{-7} = -\dfrac{12}{x^7}$

88. $\dfrac{12x^7}{4x^{-3}} = 3x^{7-(-3)} = 3x^{10}$

89. $\dfrac{-10a^5b^6}{20a^{-3}b^{11}} = \dfrac{-1}{2}a^{5-(-3)}b^{6-11}$

$= \dfrac{-1}{2}a^8b^{-5} = -\dfrac{a^8}{2b^5}$

90. $(-3xy^4)(2x^2)^3 = (-3xy^4)(2^3x^{2\cdot3})$

$= (-3xy^4)(8x^6)$

$= -24x^{1+6}y^4$

$= -24x^7y^4$

91. $2^{-2} + \dfrac{1}{2}x^0 = \dfrac{1}{2^2} + \dfrac{1}{2}\cdot1$

$= \dfrac{1}{4} + \dfrac{1}{2}$

$= \dfrac{1}{4} + \dfrac{2}{4} = \dfrac{3}{4}$

92. $(5x^2y^{-4})^{-3} = \left(\dfrac{5x^2}{y^4}\right)^{-3}$

$= \left(\dfrac{y^4}{5x^2}\right)^3$

$= \dfrac{y^{4\cdot3}}{5^3x^{2\cdot3}}$

$= \dfrac{y^{12}}{125x^6}$

93. $(3x^4y^{-2})(-2x^5y^{-3}) = \left(\dfrac{3x^4}{y^2}\right)\left(\dfrac{-2x^5}{y^3}\right)$

$= \left(\dfrac{-6x^{4+5}}{y^{2+3}}\right)$

$= -\dfrac{6x^9}{y^5}$

1.7

94. $7.16\times10^6 = 7{,}160{,}000$

95. $1.07\times10^{-4} = 0.000107$

96. $41{,}000{,}000{,}000{,}000 = 4.1\times10^{13}$

97. $0.00809 = 8.09\times10^{-3}$

98. $(4.2\times10^{13})(3\times10^{-6})$
$=(4.2\times3)(10^{13}\times10^{-6})$
$=(4.2\times3)(10^{13}\times10^{-6})$
$=12.6\times10^{13+(-6)}$
$=12.6\times10^{7}$
$=1.26\times10^{8}$

99. $\frac{5\times10^{-6}}{20\times10^{-8}}=\frac{5}{20}\times\frac{10^{-6}}{10^{-8}}$
$=0.25\times10^{-6-(-8)}$
$=0.25\times10^{-6+8}$
$=0.25\times10^{2}$
$=2.5\times10^{1}$

100. $150(2.8\times10^{8})=(1.5\times10^{2})(2.8\times10^{8})$
$=(1.5\times2.8)(10^{2}\times10^{8})$
$=4.2\times10^{2+8}$
$=4.2\times10^{10}$

Chapter 1 Test

1. $4x-5$

2. $8+2(x-7)^4=8+2(10-7)^4$
$=8+2(3)^4$
$=8+2(81)$
$=8+162$
$=170$

3. $\{-4,-3,-2,-1\}$

4. True. $\frac{1}{4}$ is not a natural number.

5. Negative three is greater than negative one. False.

6. $F=24t^2-260t+816$
$=24(10)^2-260(10)+816$
$=24(100)-2600+816$
$=2400-2600+816$
$=-200+816$
$=616$
There were 616 convictions in 2000.

7. $|-17.9|=17.9$

8. $-10.8+3.2=-7.6$

9. $-\frac{1}{4}-\left(-\frac{1}{2}\right)=-\frac{1}{4}+\frac{1}{2}=-\frac{1}{4}+\frac{2}{4}=\frac{1}{4}$

10. $2(-3)(-1)(-10)=-6(-1)(-10)$
$=6(-10)$
$=-60$

11. $-\frac{1}{4}\left(-\frac{1}{2}\right)=\frac{1}{8}$

12. $\frac{-27.9}{-9}=3.1$

13. $24-36\div4\cdot3=24-9\cdot3$
$=24-27=-3$

14. $(5^2-2^4)+[9\div(-3)]$
$=(25-16)+[-3]=(9)+[-3]=6$

15. $\dfrac{(8-10)^3-(-4)^2}{2+8(2)\div 4}$

$=\dfrac{(-2)^3-16}{2+16\div 4}=\dfrac{-8-16}{2+4}$

$=\dfrac{-24}{6}=-4$

16. $7x-4(3x+2)-10=7x-12x-8-10$

$=-5x-18$

17. $5(2y-6)-(4y-3)=10y-30-4y+3$

$=6y-27$

18. (-2, -4)

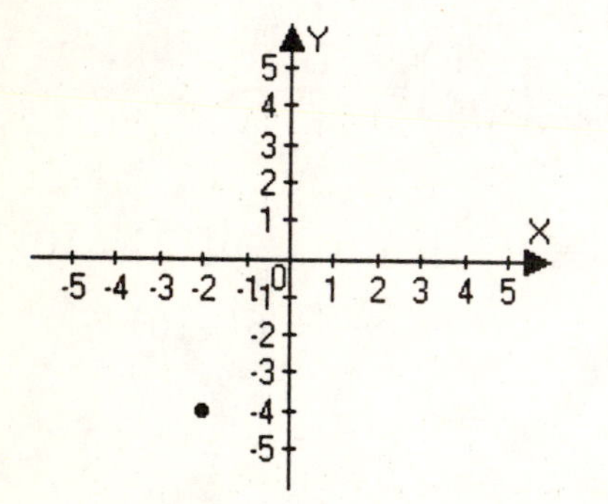

19.

x	$y=x^2-4$	(x,y)
–3	$y=(-3)^2-4=9-4=5$	$(-3,5)$
–2	$y=(-2)^2-4=4-4=0$	$(-2,0)$
–1	$y=(-1)^2-4=1-4=-3$	$(-1,-3)$
0	$y=(0)^2-4=0-4=-4$	$(0,-4)$
1	$y=(1)^2-4=1-4=-3$	$(1,-3)$
2	$y=(2)^2-4=4-4=0$	$(2,0)$
3	$y=(3)^2-4=9-4=5$	$(3,5)$

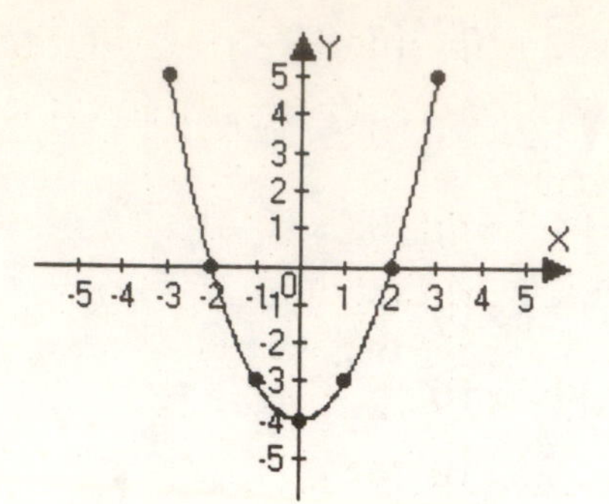

20. $3(2x-4)=9-3(x+1)$

$6x-12=9-3x-3$

$6x-12=6-3x$

$9x-12=6$

$9x=18$

$x=2$

The solution is 2 and the solution set is $\{2\}$.

21. $\dfrac{2x-3}{4}=\dfrac{x-4}{2}-\dfrac{x+1}{4}$

$2(2x-3)=8\left(\dfrac{x-4}{2}\right)-8\left(\dfrac{x+1}{4}\right)$

$4x-6=4(x-4)-2(x+1)$

$4x-6=4x-16-2x-2$

$4x-6=2x-18$

$2x-6=-18$

$2x=-12$

$x=-6$

The solution is –6 and the solution set is $\{-6\}$.

22. $3(x-4)+x=2(6+2x)$

$3x-12+x=12+4x$

$4x-12=12+4x$

$-12=12$

There is no solution. The solution set is $\{\ \}$ or $\varnothing$. The equation is inconsistent.

23. Let x = the first number
Let $2x + 3$ = the second number

$$x+2x+3=72$$
$$3x+3=72$$
$$3x=69$$
$$x=23$$

Find the second number.

$$2x+3=2(23)+3=46+3=49$$

The numbers are 23 and 49.

24. Let x = the number of years since the car was purchased.

$$\text{Value}=13{,}805-\$1820x$$
$$4705=13{,}805-\$1820x$$
$$-9100=-\$1820x$$
$$5=x$$

The car will have a value of $4705 in 5 years.

25. Let x = the original selling price

$$20=x-0.60x$$
$$20=0.40x$$
$$50=x$$

The original price is $50.

26. Let x = the width of the playing field
Let $x + 260$ = the length of the playing field

$$P=2(\text{length})+2(\text{width})$$
$$1000=2(x+260)+2x$$
$$1000=2x+520+2x$$
$$1000=4x+520$$
$$480=4x$$
$$x=120$$

The dimensions of the playing field are 120 yards by 380 yards.

27.

$$V=\frac{1}{3}lwh$$
$$3V=lwh$$
$$h=\frac{3V}{lw}$$

28.

$$Ax+By=C$$
$$By=C-Ax$$
$$y=\frac{C-Ax}{B}$$

29.

$$\left(-2x^{5}\right)\left(7x^{-10}\right)=-14x^{5+(-10)}$$
$$=-14x^{-5}$$
$$=-\frac{14}{x^{5}}$$

30.

$$\frac{-10x^{4}y^{3}}{-40x^{-2}y^{6}}=\frac{1}{4}x^{4-(-2)}y^{3-6}$$
$$=\frac{1}{4}x^{6}y^{-3}$$
$$=\frac{x^{6}}{4y^{3}}$$

31.

$$\left(4x^{-5}y^{2}\right)^{-3}=\left(\frac{4y^{2}}{x^{5}}\right)^{-3}=\left(\frac{x^{5}}{4y^{2}}\right)^{3}$$
$$=\frac{\left(x^{5}\right)^{3}}{(4)^{3}\left(y^{2}\right)^{3}}=\frac{x^{15}}{64y^{6}}$$

32. $3.8\times10^{-6}=0.0000038$

33. $407,000,000,000 = 4.07 \times 10^{11}$

34. $\dfrac{4\times 10^{-3}}{8\times 10^{-7}} = 0.5\times 10^{-3-(-7)}$

$= 0.5\times 10^{4}$

$= 5.0\times 10^{3}$

35. $2\left(6.08\times 10^{9}\right) = 12.16\times 10^{9}$

$= 1.216\times 10^{10}$

The population will be 1.216×10^{10}.

Chapter 2

Check Points 2.1

1. The domain is $\{5,10,15,20,25\}$.
The range is $\{12.8,16.2,18.9,20.7,$
$21.8\}$.

2. **a.** The relation is not a function. The domain element 5 is paired with more than one element of the range.

b. The relation is a function.

3. **a.** $f(6)=4(6)+5=24+5=29$

b. $g(-5)=3(-5)^2-10=3(25)-10$
$=75-10=65$

c. $h(-4)=(-4)^2-7(-4)+2$
$=16+28+2=46$

d. $F(a+h)=6(a+h)+9$
$=6a+6h+9$

4.

x	$f(x)=2x$	(x,y)
-2	$f(-2)=2(-2)=-4$	$(-2,-4)$
-1	$f(-1)=2(-1)=-2$	$(-1,-2)$
0	$f(0)=2(0)=0$	$(0,0)$
1	$f(1)=2(1)=2$	$(1,2)$
2	$f(2)=2(2)=4$	$(2,4)$

x	$g(x)=2x-3$	(x,y)
-2	$g(-2)=2(-2)-3$ $=-4-3=-7$	$(-2,-7)$
-1	$g(-1)=2(-1)-3$ $=-2-3=-5$	$(-1,-5)$
0	$g(0)=2(0)-3$ $=0-3=-3$	$(0,-3)$
1	$f(1)=2(1)-3$ $=2-3=-1$	$(1,-1)$
2	$g(2)=2(2)-3$ $=4-3=1$	$(2,1)$

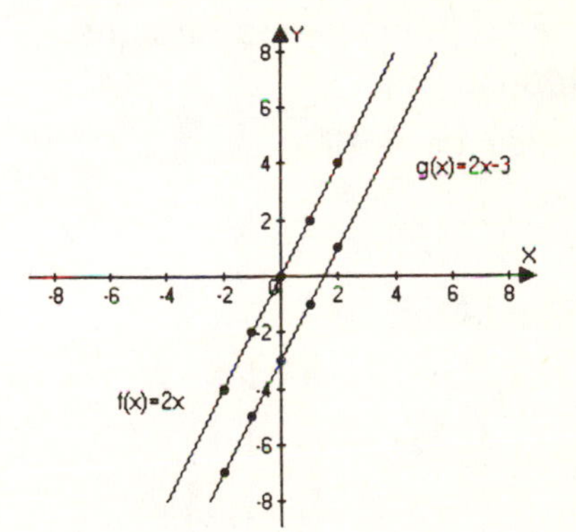

The graph of g shifts the graph of f down 3 units.

5. **a.** The vertical line test shows that this is the graph of a function.

b. The vertical line test shows that this is the graph of a function.

c. The vertical line test shows that this is not the graph of a function.

6. **a.** $f(10)\approx 16$

b. $x\approx 8$

Problem Set 2.1

Practice Exercises

1. The relation is a function.
Domain {1, 3, 5}
Range {2, 4, 5}

3. The relation is not a function.
Domain {3, 4}
Range {4, 5}

5. The relation is a function.
Domain {−3, −2, −1, 0}
Range {−3, −2, −1, 0}

7. The relation is not a function.
Domain {1}
Range {4, 5, 6}

9. a. $f(0)=0+1=1$
b. $f(5)=5+1=6$
c. $f(-8)=-8+1=-7$
d. $f(2a)=2a+1$
e. $f(a+2)=(a+2)+1$
$=a+2+1=a+3$

11. a. $g(0)=3(0)-2=0-2=-2$
b. $g(-5)=3(-5)-2$
$=-15-2=-17$
c. $g\left(\frac{2}{3}\right)=3\left(\frac{2}{3}\right)-2=2-2=0$
d. $g(4b)=3(4b)-2=12b-2$
e. $g(b+4)=3(b+4)-2$
$=3b+12-2=3b+10$

13. a. $h(0)=3(0)^2+5=3(0)+5$
$=0+5=5$
b. $h(-1)=3(-1)^2+5=3(1)+5$
$=3+5=8$
c. $h(4)=3(4)^2+5=3(16)+5$
$=48+5=53$
d. $h(-3)=3(-3)^2+5=3(9)+5$
$=27+5=32$
e. $h(4b)=3(4b)^2+5=3(16b^2)+5$
$=48b^2+5$

15. a. $f(0)=2(0)^2+3(0)-1$
$=0+0-1=-1$
b. $f(3)=2(3)^2+3(3)-1$
$=2(9)+9-1$
$=18+9-1=26$
c. $f(-4)=2(-4)^2+3(-4)-1$
$=2(16)-12-1$
$=32-12-1=19$
d. $f(b)=2(b)^2+3(b)-1$
$=2b^2+3b-1$
e. $f(5a)=2(5a)^2+3(5a)-1$
$=2(25a^2)+15a-1$
$=50a^2+15a-1$

17. a. $f(0)=\frac{2(0)-3}{(0)-4}=\frac{0-3}{0-4}$
$=\frac{-3}{-4}=\frac{3}{4}$

b.
$$f(3)=\frac{2(3)-3}{(3)-4}=\frac{6-3}{3-4}$$
$$=\frac{3}{-1}=-3$$

c.
$$f(-4)=\frac{2(-4)-3}{(-4)-4}=\frac{-8-3}{-8}$$
$$=\frac{-11}{-8}=\frac{11}{8}$$

d.
$$f(-5)=\frac{2(-5)-3}{(-5)-4}=\frac{-10-3}{-9}$$
$$=\frac{-13}{-9}=\frac{13}{9}$$

e.
$$f(a+h)=\frac{2(a+h)-3}{(a+h)-4}$$
$$=\frac{2a+2h-3}{a+h-4}$$

f. Four must be excluded from the domain, because 4 would make the denominator zero. Division by zero is undefined.

19.

x	$f(x)=x$	(x,y)
-2	-2	$(-2,-2)$
-1	-1	$(-1,-1)$
0	0	$(0,0)$
1	1	$(1,1)$
2	2	$(2,2)$

x	$g(x)=x+3$	(x,y)
-2	$g(-2)=-2+3=1$	$(-2,1)$
-1	$g(-1)=-1+3=2$	$(-1,2)$
0	$g(0)=0+3=3$	$(0,3)$
1	$g(1)=1+3=4$	$(1,4)$
2	$g(2)=2+3=5$	$(2,5)$

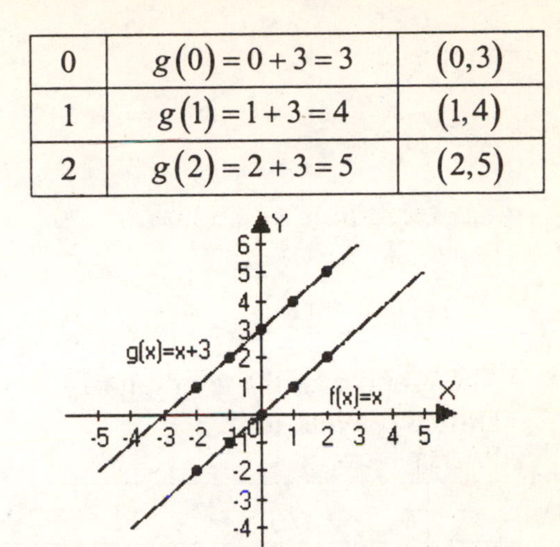

The graph of g is the graph of f shifted up 3 units.

21.

x	$f(x)=-2x$	(x,y)
-2	$f(-2)=-2(-2)=4$	$(-2,4)$
-1	$f(-1)=-2(-1)=2$	$(-1,2)$
0	$f(0)=-2(0)=0$	$(0,0)$
1	$f(1)=-2(1)=-2$	$(1,-2)$
2	$f(2)=-2(2)=-4$	$(2,-4)$

x	$g(x)=-2x-1$	(x,y)
-2	$g(-2)=-2(-2)-1$ $=4-1=3$	$(-2,3)$
-1	$g(-1)=-2(-1)-1$ $=2-1=1$	$(-1,1)$
0	$g(0)=-2(0)-1$ $=0-1=-1$	$(0,-1)$
1	$g(1)=-2(1)-1$ $=-2-1=-3$	$(1,-3)$
2	$g(2)=-2(2)-1$ $=-4-1=-5$	$(2,-5)$

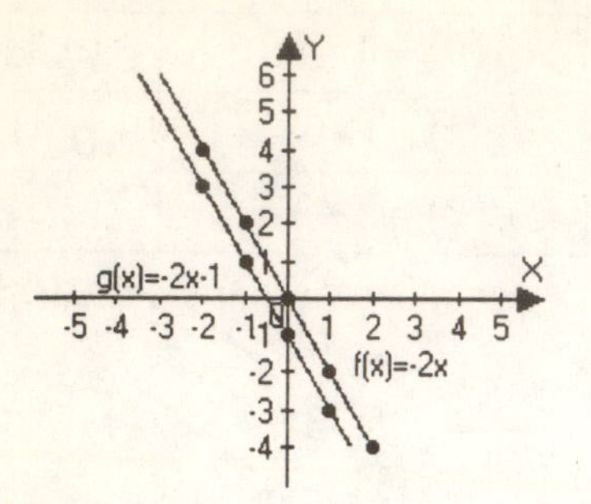

The graph of g is the graph of f shifted down 1 unit.

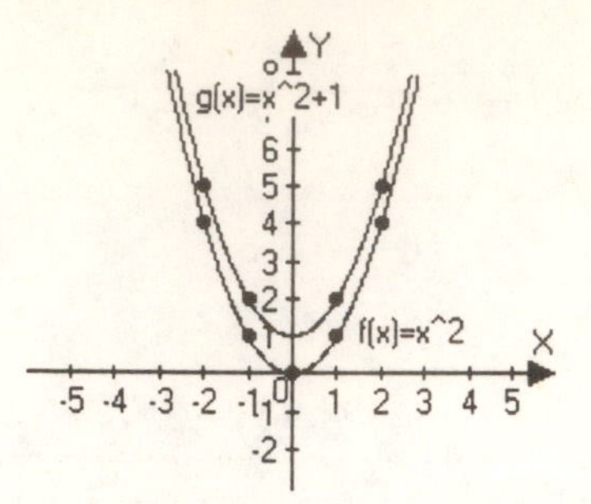

The graph of g is the graph of f shifted up 1 unit.

23.

x	$f(x)=x^2$	(x,y)
-2	$f(-2)=(-2)^2=4$	$(-2,4)$
-1	$f(-1)=(-1)^2=1$	$(-1,1)$
0	$f(0)=(0)^2=0$	$(0,0)$
1	$f(1)=(1)^2=1$	$(1,1)$
2	$f(2)=(2)^2=4$	$(2,4)$

x	$g(x)=x^2+1$	(x,y)
-2	$g(-2)=(-2)^2+1$ $=4+1=5$	$(-2,5)$
-1	$g(-1)=(-1)^2+1$ $=1+1=2$	$(-1,2)$
0	$g(0)=(0)^2+1$ $=0+1=1$	$(0,1)$
1	$g(1)=(1)^2+1$ $=1+1=2$	$(1,2)$
2	$g(2)=(2)^2+1$ $=4+1=5$	$(2,5)$

25.

x	$f(x)=\|x\|$	(x,y)
-2	$f(-2)=\|-2\|=2$	$(-2,2)$
-1	$f(-1)=\|-1\|=1$	$(-1,1)$
0	$f(0)=\|0\|=0$	$(0,0)$
1	$f(1)=\|1\|=1$	$(1,1)$
2	$f(2)=\|2\|=2$	$(2,2)$

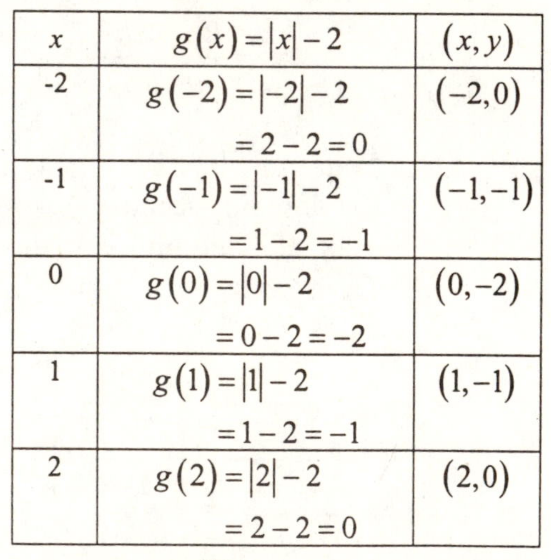

x	$g(x)=\|x\|-2$	(x,y)
-2	$g(-2)=\|-2\|-2$ $=2-2=0$	$(-2,0)$
-1	$g(-1)=\|-1\|-2$ $=1-2=-1$	$(-1,-1)$
0	$g(0)=\|0\|-2$ $=0-2=-2$	$(0,-2)$
1	$g(1)=\|1\|-2$ $=1-2=-1$	$(1,-1)$
2	$g(2)=\|2\|-2$ $=2-2=0$	$(2,0)$

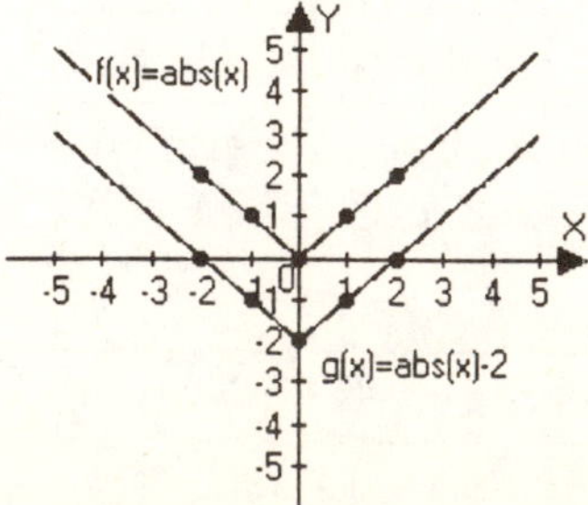

The graph of g is the graph of f shifted down 2 units.

27.

x	$f(x)=x^3$	(x,y)
-2	$f(-2)=(-2)^3=-8$	$(-2,-8)$
-1	$f(-1)=(-1)^3=-1$	$(-1,-1)$
0	$f(0)=(0)^3=0$	$(0,0)$
1	$f(1)=(1)^3=1$	$(1,1)$
2	$f(2)=(2)^3=8$	$(2,8)$

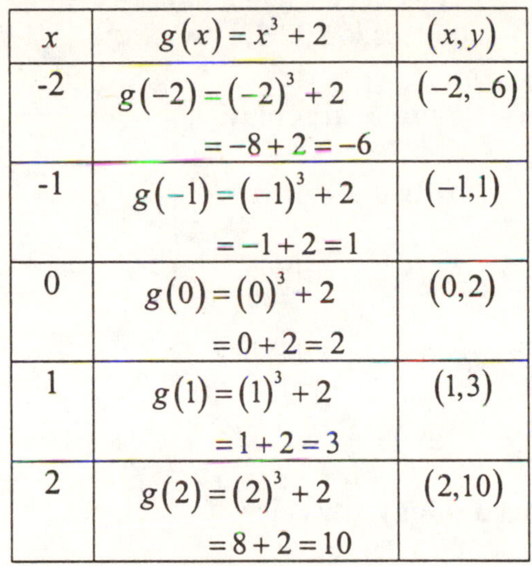

x	$g(x)=x^3+2$	(x,y)
-2	$g(-2)=(-2)^3+2$ $=-8+2=-6$	$(-2,-6)$
-1	$g(-1)=(-1)^3+2$ $=-1+2=1$	$(-1,1)$
0	$g(0)=(0)^3+2$ $=0+2=2$	$(0,2)$
1	$g(1)=(1)^3+2$ $=1+2=3$	$(1,3)$
2	$g(2)=(2)^3+2$ $=8+2=10$	$(2,10)$

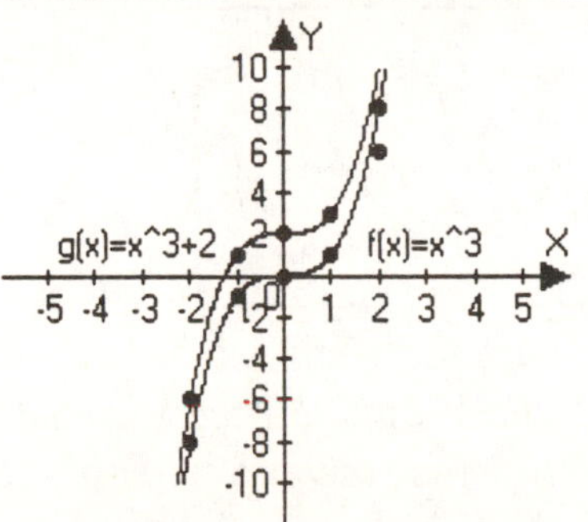

The graph of g is the graph of f shifted up 2 units.

29. The vertical line test shows that the graph represents a function.

31. The vertical line test shows that the graph represents a function.

33. The vertical line test shows that the graph does not represent a function.

35. The vertical line test shows that the graph represents a function.

37. $f(-2)=-4$

39. $f(4)=4$

41. $f(-3)=0$

43. $g(-4)=2$

45. $g(-10)=2$

47. When $x=-2$, $g(x)=1$.

Application Exercises

49. $\{(1, 31), (2, 53), (3, 70), (4, 86), (5, 86)\}$
Domain $\{1, 2, 3, 4, 5\}$
Range $\{31, 53, 70, 86\}$
The relation is a function.

51. $W(16)=0.07(16)+4.1$
$=1.12+4.1=5.22$
In 2000, there were 5.22 million women enrolled in U.S. colleges. This is represented by the point (16, 5.2) on the graph. (The coordinates are approximate since they are estimated from the graph.)

53. $W(20)=0.07(20)+4.1$
$=1.4+4.1=5.5$

$M(20) = 0.01(20) + 3.9$
$= 0.2 + 3.9 = 4.1$
$W(20) - M(20) = 5.5 - 4.1 = 1.4$
In 2004, there will be 1.4 million more women than men enrolled in U.S. colleges.

55. $f(20) = 0.4(20)^2 - 36(20) + 1000$
$= 0.4(400) - 720 + 1000$
$= 160 - 720 + 1000$
$= -560 + 1000 = 440$
Twenty-year-old drivers have 440 accidents per 50 million miles driven.

57. The graph reaches its lowest point at $x = 45$.
$f(45) = 0.4(45)^2 - 36(45) + 1000$
$= 0.4(2025) - 1620 + 1000$
$= 810 - 1620 + 1000$
$= -810 + 1000 = 190$
Drivers at age 45 have 190 accidents per 50 million miles driven. This is the least number of accidents for any driver between ages 16 and 74.

59. $f(60) \approx 3.1$
In 1960, 3.1% of the U.S. population was made up of Jewish Americans.

61. In 1919 and 1964, $f(x) = 3$. This means that in 1919 and 1964, 3% of the U.S. population was made up of Jewish Americans.

63. The percentage of Jewish Americans in the U.S. population reached a maximum in 1940. Using the graph to estimate, approximately 3.7% of the U.S. population were Jewish Americans.

65. Each year is paired with exactly one percentage. This means that each member of the domain is paired with one member of the range.

67. $f(3) = 0.76$
The cost of mailing a first-class letter weighing 3 ounces is $0.76.

69. The cost to mail a letter weighing 1.5 ounces is $0.55.

Writing in Mathematics

71. Answers will vary.

73. Answers will vary.

75. Answers will vary.

77. Answers will vary.

Technology Exercises

79.

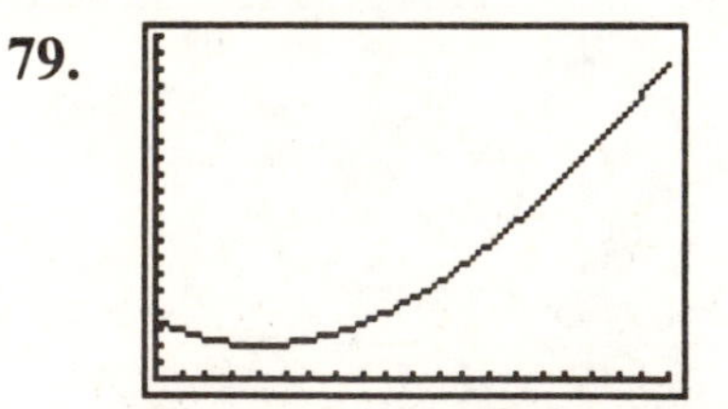

The number of physician's visits per year based on age first decreases and then increases over a person's life-time.

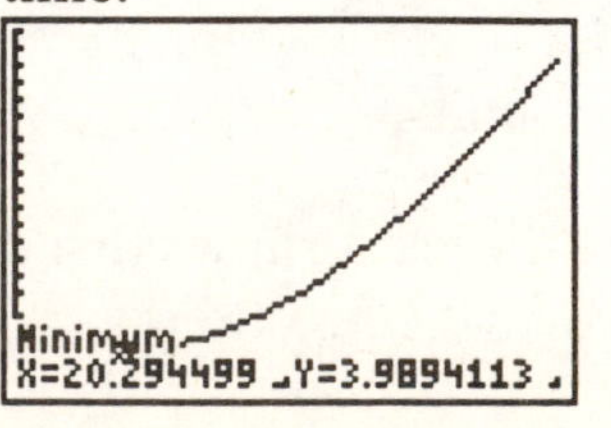

These are the approximate coordinates of the point (20.3, 4.0). The means that the minimum number of physician's visits per year is approximately 4. This occurs around age 20.

Critical Thinking Exercises

81. $f(a+h) = 3(a+h)+7 = 3a+3h+7$

$f(a) = 3a+7$

$$\frac{f(a+h)-f(a)}{h}$$
$$= \frac{(3a+3h+7)-(3a+7)}{h}$$
$$= \frac{3a+3h+7-3a-7}{h} = \frac{3h}{h} = 3$$

83. It is given that $f(x+y) = f(x)+f(y)$ and $f(1) = 3$. To find $f(2)$, rewrite 2 as 1 + 1.

$f(2) = f(1+1) = f(1)+f(1)$
$= 3+3 = 6$

Similarly:

$f(3) = f(2+1) = f(2)+f(1)$
$= 6+3 = 9$

$f(4) = f(3+1) = f(3)+f(1)$
$= 9+3 = 12$

While $f(x+y) = f(x)+f(y)$ is true for this function, it is not true for all functions. It is not true for $f(x) = x^2$, for example.

Review Exercises

84. $24 \div 4[2-(5-2)]^2 - 6$

$= 24 \div 4[2-(3)]^2 - 6$

$= 24 \div 4[-1]^2 - 6 = 24 \div 4[1] - 6$

$= 6[1] - 6 = 6 - 6 = 0$

85.
$$\left(\frac{3x^2y^{-2}}{y^3}\right)^{-2} = \left(\frac{3x^2}{y^2y^3}\right)^{-2} = \left(\frac{3x^2}{y^5}\right)^{-2}$$
$$= \left(\frac{y^5}{3x^2}\right)^2 = \frac{y^{5\cdot2}}{3^2x^{2\cdot2}} = \frac{y^{10}}{9x^4}$$

86.
$$\frac{x}{3} = \frac{3x}{5} + 4$$
$$15\left(\frac{x}{3}\right) = 15\left(\frac{3x}{5} + 4\right)$$
$$15\left(\frac{x}{3}\right) = 15\left(\frac{3x}{5}\right) + 15(4)$$
$$5x = 3(3x) + 60$$
$$5x = 9x + 60$$
$$5x - 9x = 9x - 9x + 60$$
$$-4x = 60$$
$$\frac{-4x}{-4} = \frac{60}{-4}$$
$$x = -15$$

The solution is –15 and the solution set is $\{-15\}$.

Check Points 2.2

1. a. Domain of $f = \{x | x \text{ is a real number}\}$.

b. Domain of $g = \{x | x \text{ is a real number and } x \neq -5\}$.

2. **a.**
$$(f+g)(x)$$
$$=f(x)+g(x)$$
$$=(3x^2+4x-1)+(2x+7)$$
$$=3x^2+4x-1+2x+7$$
$$=3x^2+6x+6$$

b.
$$(f+g)(4)=3(4)^2+6(4)+6$$
$$=3(16)+24+6$$
$$=48+24+6=78$$

3. **a.**
$$(f+g)(x)=f(x)+g(x)$$
$$=\frac{5}{x}+\frac{7}{x-8}$$

b. Domain of $f+g=\{x|x$ is a real number and, $x\neq 0$ and $x\neq 8\}$

4. **a.**
$$f(5)=(5)^2-2(5)=25-10=15$$
$$g(5)=5+3=8$$
$$(f+g)(5)=f(5)+g(5)$$
$$=15+8=23$$

b.
$$(f-g)(x)=f(x)-g(x)$$
$$=(x^2-2x)-(x+3)$$
$$=x^2-2x-x-3$$
$$=x^2-3x-3$$
$$(f-g)(-1)=(-1)^2-3(-1)-3$$
$$=1+3-3=1$$

c.
$$\left(\frac{f}{g}\right)(x)=\frac{f(x)}{g(x)}=\frac{x^2-2x}{x+3}$$
$$\left(\frac{f}{g}\right)(7)=\frac{7^2-2(7)}{7+3}$$
$$=\frac{49-14}{10}=\frac{35}{10}=\frac{7}{2}$$

d.
$$f(-4)=(-4)^2-2(-4)$$
$$=16+8=24$$
$$g(-4)=-4+3=-1$$
$$(fg)(-4)=f(-4)\cdot g(-4)$$
$$=24(-1)=-24$$

Problem Set 2.2

Practice Exercises

1. Domain of $f=\{x|x$ is a real number$\}$

3. Domain of $g=\{x|x$ is a real number and $x\neq -4\}$.

5. Domain of $f=\{x|x$ is a real number and $x\neq 3\}$.

7. Domain of $g=\{x|x$ is a real number and $x\neq 5\}$.

9. Domain of $f=\{x|x$ is a real number and $x\neq -7$ and $x\neq 9\}$

11. **a.**
$$(f+g)(x)=(3x+1)+(2x-6)$$
$$=3x+1+2x-6$$
$$=5x-5$$

b.
$$(f+g)(5)=5(5)-5$$
$$=25-5=20$$

13. **a.** $(f+g)(x) = (x-5)+(3x^2)$
$= x-5+3x^2$
$= 3x^2+x-5$

b. $(f+g)(5) = 3(5)^2+5-5$
$= 3(25) = 75$

15. **a.** $(f+g)(x)$
$= (2x^2-x-3)+(x+1)$
$= 2x^2-x-3+x+1 = 2x^2-2$

b. $(f+g)(5) = 2(5)^2-2$
$= 2(25)-2$
$= 50-2 = 48$

17. Domain of $f+g = \{x \mid x$ is a real number$\}$.

19. Domain of $f+g = \{x \mid x$ is a real number and $x \neq 5\}$.

21. Domain of $f+g = \{x \mid x$ is a real number and $x \neq 0$ and $x \neq 5\}$.

23. Domain of $f+g = \{x \mid x$ is a real number and $x \neq 2$ and $x \neq -3\}$.

25. Domain of $f+g = \{x \mid x$ is a real number and $x \neq 2\}$.

27. Domain of $f+g = \{x \mid x$ is a real number$\}$.

29. $(f+g)(x) = f(x)+g(x)$
$= x^2+4x+2-x$
$= x^2+3x+2$
$(f+g)(3) = (3)^2+3(3)+2$
$= 9+9+2 = 20$

31. $f(-2) = (-2)^2+4(-2)$
$= 4+(-8) = -4$
$g(-2) = 2-(-2) = 2+2 = 4$
$f(-2)+g(-2) = -4+4 = 0$

33. $(f-g)(x) = f(x)-g(x)$
$= (x^2+4x)-(2-x)$
$= x^2+4x-2+x$
$= x^2+5x-2$
$(f-g)(3) = (5)^2+5(5)-2$
$= 25+25-2 = 48$

35. From Exercise 31, we know $f(-2) = -4$, and $g(-2) = 4$.
$f(-2)-g(-2) = -4-4 = -8$

37. From Exercise 31, we know $f(-2) = -4$, and $g(-2) = 4$.
$(fg)(-2) = f(-2)\cdot g(-2)$
$= -4(4) = -16$

39. $f(5) = (5)^2+4(5) = 25+20 = 45$
$g(5) = 2-5 = -3$
$(fg)(5) = f(5)\cdot g(5)$
$= 45(-3) = -135$

41. $\left(\frac{f}{g}\right)(x) = \frac{f(x)}{g(x)} = \frac{x^2+4x}{2-x}$

$$\left(\frac{f}{g}\right)(1) = \frac{(1)^2 + 4(1)}{2-(1)} = \frac{1+4}{1} = \frac{5}{1} = 5$$

43. $$\left(\frac{f}{g}\right)(-1) = \frac{(-1)^2 + 4(-1)}{2-(-1)}$$
$$= \frac{1-4}{3} = \frac{-3}{3} = -1$$

45. Domain of $f + g = \{x | x$ is a real number$\}$.

47. $$\left(\frac{f}{g}\right)(x) = \frac{f(x)}{g(x)} = \frac{x^2+4x}{2-x}$$

Domain of $\frac{f}{g} = \{x | x$ is a real number and $x \neq 2\}$.

Application Exercises

49. $(D+C)(2000)$
$= (D)(2000) + (C)(2000)$
$= 14 + 6 = 20$
This means that the total veterinary costs for dogs and cats in the year 2000 was $20 billion.

51. Domain of $D + C$ = {1983, 1987, 1991,1996, 2000}.

53. $(f+g)(x)$ represents the total world population, $h(x)$.

55. $(f+g)(2000) = f(2000) + g(2000)$
$= h(2000) = 5.9$
This means that the total world population in the year 2000 was approximately 5.9 billion.

57. First, find $(R-C)(x)$.
$(R-C)(x) = 65x - (600,000 + 45x)$
$= 65x - 600,000 - 45x$
$= 20x - 600,000$
$(R-C)(20,000)$
$= 20(20,000) - 600,000$
$= 400,000 - 600,000 = -200,000$
This means that if the company produces and sells 20,000 radios, it will lose $200,000.
$(R-C)(30,000)$
$= 20(30,000) - 600,000$
$= 600,000 - 600,000 = 0$
If the company produces and sells 30,000 radios, it will break even with its costs equal to its revenue.
$(R-C)(40,000)$
$= 20(40,000) - 600,000$
$= 800,000 - 600,000 = 200,000$
This means that if the company produces and sells 40,000 radios, it will make a profit of $200,000.

Writing in Mathematics

59. Answers will vary.

61. Answers will vary.

63. Answers will vary.

Technology Exercises

65. $y_1 = 2x + 3$ $y_2 = 2 - 2x$
$y_3 = y_1 + y_2$

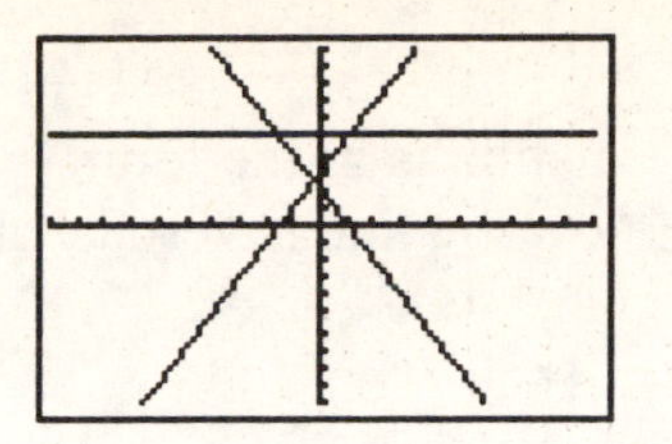

67. 

$y_1 = x \qquad y_2 = x - 4$

$y_3 = y_1 \cdot y_2$

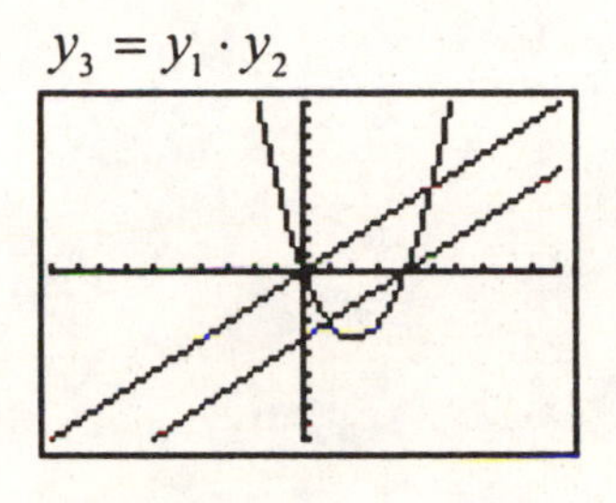

69. 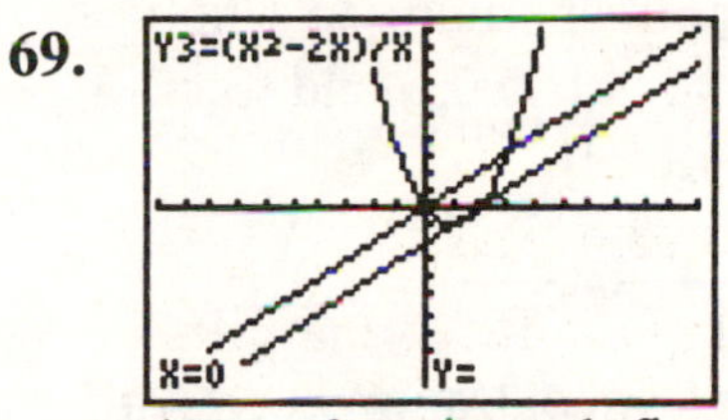

At $x = 0$, y is undefined. This is because at $x = 0$, the function, y_3 is undefined.

Critical Thinking Exercises

71. To create a graph that shows the population, in billions, of the world's less-developed regions from 1950 through 2050, at each year, x, subtract the population of the world's more-developed regions, $f(x)$, from the total world population, $h(x)$. This will yield the population of the world's less-developed regions, $g(x)$.

Review Exercises

72.

$$R = 3(a+b)$$
$$R = 3a + 3b$$
$$R - 3a = 3b$$
$$\frac{R-3a}{3} = \frac{3b}{3}$$
$$b = \frac{R-3a}{3} \text{ or } \frac{R}{3} - a$$

73.

$$3(6-x) = 3 - 2(x-4)$$
$$18 - 3x = 3 - 2x + 8$$
$$18 - 3x = 11 - 2x$$
$$18 = 11 + x$$
$$7 = x$$

The solution is 7 and the solution set is $\{7\}$.

74.

$$f(b+2) = 6(b+2) - 4$$
$$= 6b + 12 - 4 = 6b + 8$$

Check Points 2.3

1. $2x + 3y = 6$

Find the x–intercept by setting $y = 0$.

$$2x + 3(0) = 6$$
$$2x = 6$$
$$x = 3$$

Find the y–intercept by setting $x = 0$.

$$2(0) + 3y = 6$$
$$3y = 6$$
$$y = 2$$

2. **a.** $(-3,4)$ and $(-4,-2)$

$$m=\frac{4-(-2)}{-3-(-4)}=\frac{4+2}{-3+4}=\frac{6}{1}=6$$

b. $(4,-2)$ and $(-1,5)$

$$m=\frac{5-(-2)}{-1-4}=\frac{5+2}{-5}=-\frac{7}{5}$$

3. $y=3x-2$

The y–intercept is –2 and the slope is 3. We can write the slope as

$m=\frac{3}{1}=\frac{\text{rise}}{\text{run}}$.

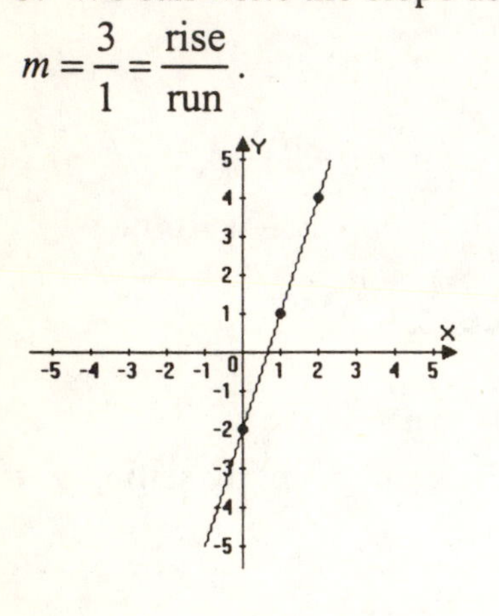

4. $y=\frac{3}{5}x+1$

The y–intercept is 1 and the slope is $\frac{3}{5}$. We can write the slope as

$m=\frac{3}{5}=\frac{\text{rise}}{\text{run}}$.

5. $x=-2$

Since the equation is of the form $x=a$, we know the line is vertical at $x=-2$.

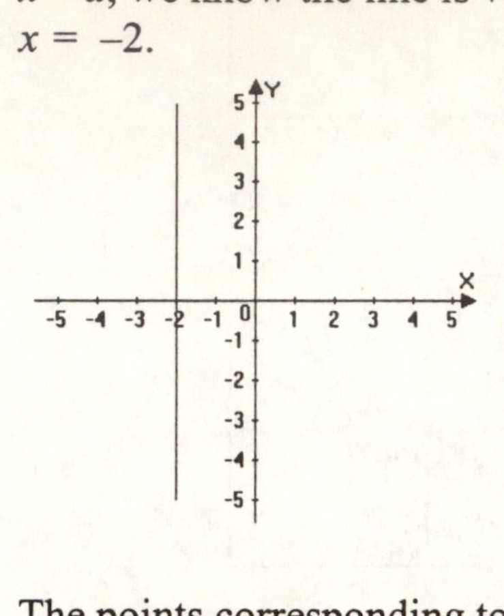

6. The points corresponding to 1 hour and 3 hours are (1, 0.03) and (3, 0.05), respectively.

$$m=\frac{0.05-0.03}{3-1}=\frac{0.02}{2}=0.01$$

The average concentration between 1 and 3 hours is 0.01 milligrams per 100 milliliters.

Problem Set 2.3

Practice Exercises

1. $x+y=5$

Find the x–intercept by setting $y=0$.

$$x+y=5$$
$$x+0=5$$
$$x=5$$

Find the y–intercept by setting $x=0$.

$$x+y=5$$
$$0+y=5$$
$$y=5$$

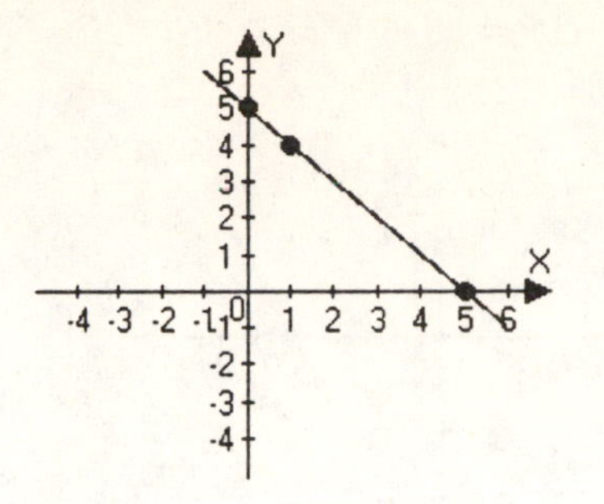

3. $3x + y = 6$

Find the x–intercept by setting $y = 0$.

$3x + 0 = 6$

$3x = 6$

$x = 2$

Find the y–intercept by setting $x = 0$.

$3(0) + y = 6$

$0 + y = 6$

$y = 6$

5. $2x - 6y = 12$

Find the x–intercept by setting $y = 0$.

$2x - \cancel{6(0)} = 12$

$2x = 12$

$x = 6$

Find the y–intercept by setting $x = 0$.

$2x - \cancel{6(0)} = 12$

$2x = 12$

$x = 6$

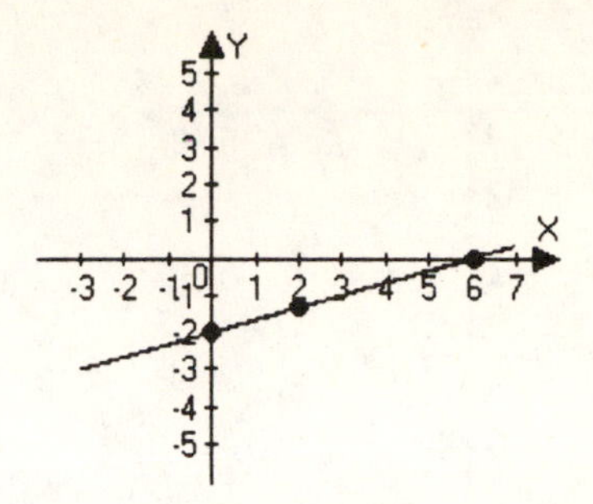

7. $-x + 3y = 6$

Find the x–intercept by setting $y = 0$.

$-x + \cancel{3(0)} = 6$

$-x = 6$

$x = -6$

Find the y–intercept by setting $x = 0$.

$-(0) + 3y = 6$

$3y = 6$

$y = 2$

9. $3x - y = 9$

Find the x–intercept by setting $y = 0$.

$3x - 0 = 9$

$3x = 9$

$x = 3$

Find the y–intercept by setting $x = 0$.

$\cancel{3(0)} - y = 9$

$-y = 9$

$y = -9$

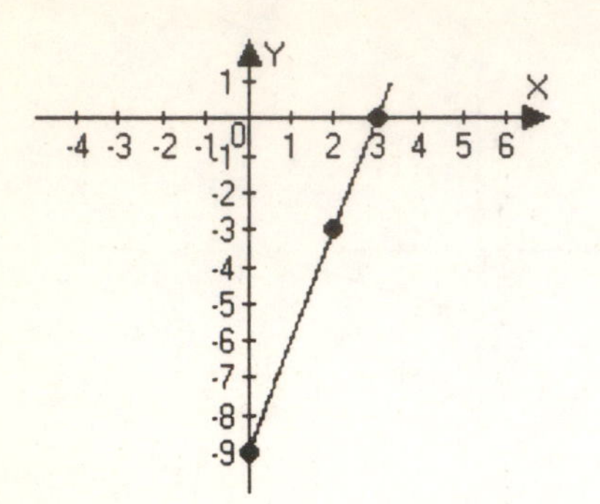

11. $3x = 2y + 6$

Find the x–intercept by setting $y = 0$.

$3x = 2(0) + 6$

$3x = 6$

$x = 2$

Find the y–intercept by setting $x = 0$.

$3(0) = 2y + 6$

$-6 = 2y$

$-3 = y$

13. $6x - 3y = 15$

Find the x–intercept by setting $y = 0$.

$6x - 3(0) = 15$

$6x = 15$

$x = \frac{15}{6} = \frac{5}{2}$

Find the y–intercept by setting $x = 0$.

$6(0) - 3y = 15$

$-3y = 15$

$y = -5$

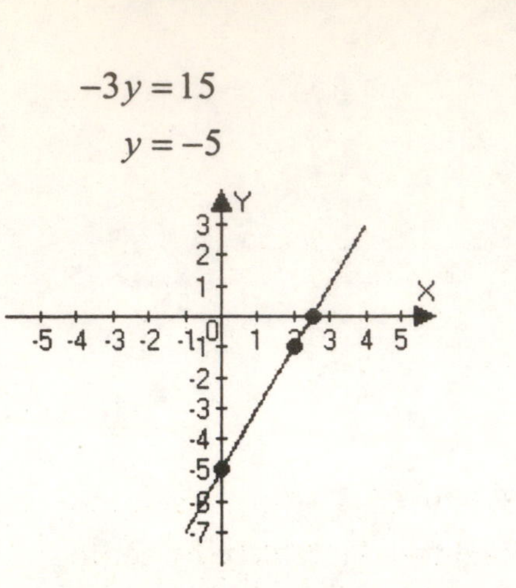

15. $m = \frac{10-7}{8-4} = \frac{3}{4}$

The line rises.

17. $m = \frac{2-1}{2-(-2)} = \frac{1}{2+2} = \frac{1}{4}$

The line rises.

19. $m = \frac{-2-(-2)}{4-3} = \frac{-2+2}{1} = \frac{0}{1} = 0$

The line is horizontal.

21. $m = \frac{4-(-1)}{-2-(-1)} = \frac{4+1}{-2+1} = \frac{5}{-1} = -5$

The line falls.

23. $m = \frac{3-(-2)}{5-5} = \frac{3+2}{0}$

m is undefined

The line is vertical.

25. $y = 2x + 1$

$m = 2 \qquad y\text{-intercept} = 1$

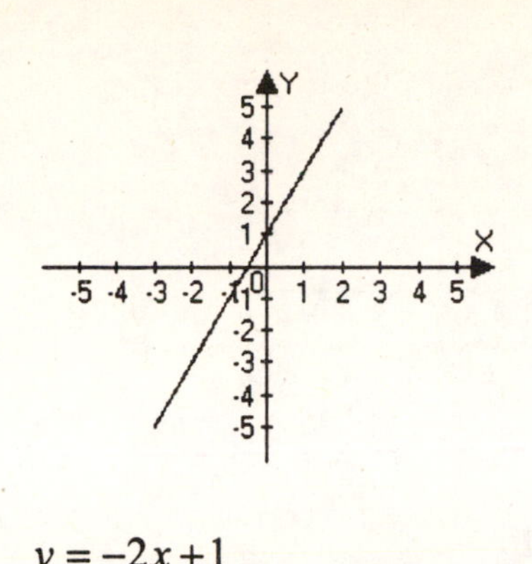

27. $y = -2x + 1$

$m = -2 \qquad y\text{-intercept} = 1$

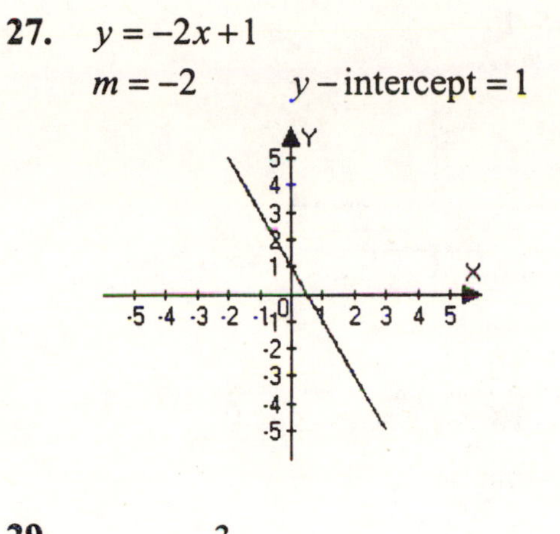

29. $f(x) = \dfrac{3}{4}x - 2$

$m = \dfrac{3}{4} \qquad y\text{-intercept} = -2$

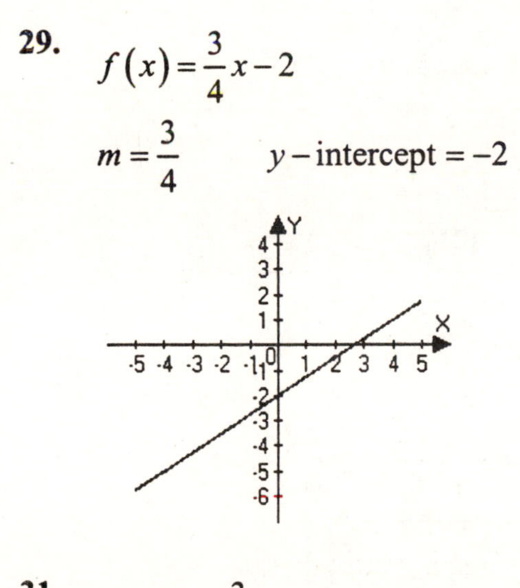

31. $f(x) = -\dfrac{3}{5}x + 7$

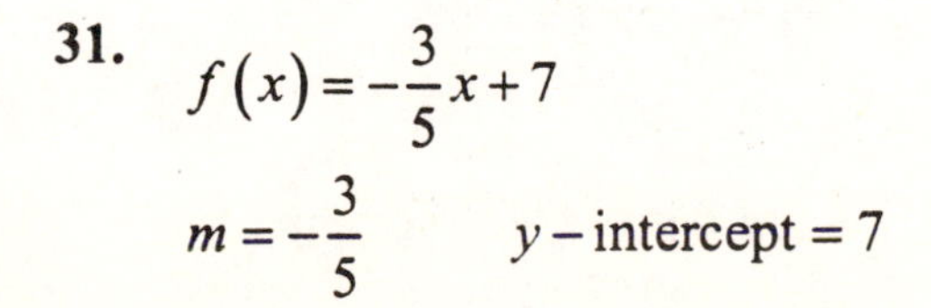

$m = -\dfrac{3}{5} \qquad y\text{-intercept} = 7$

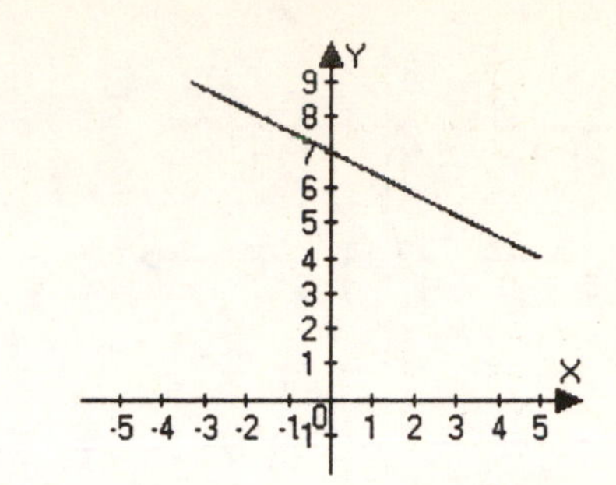

33. $y = -\dfrac{1}{2}x$

$m = -\dfrac{1}{2} \qquad y\text{-intercept} = 0$

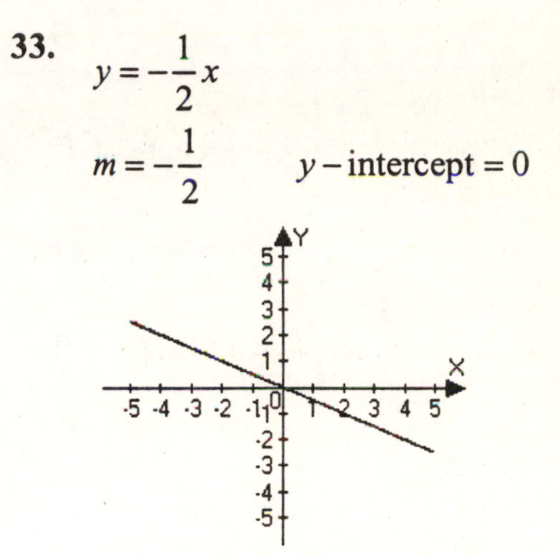

35. **a.** $2x + y = 0$

$y = -2x$

b. $m = -2 \qquad y\text{-intercept} = 0$

c.

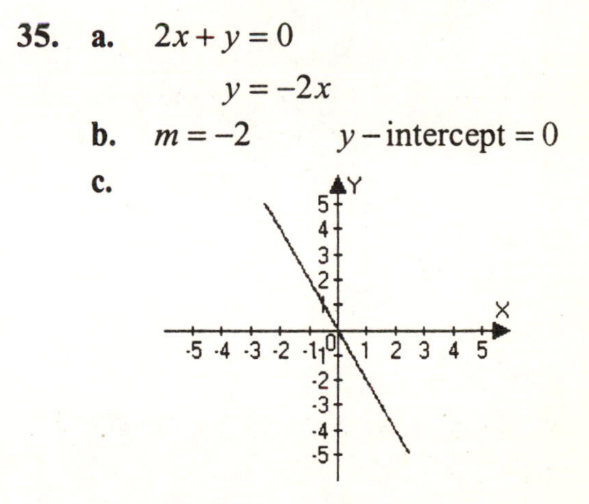

37. **a.** $4y = 5x$

$y = \dfrac{5}{4}x$

b. $m = \dfrac{5}{4} \qquad y\text{-intercept} = 0$

c.

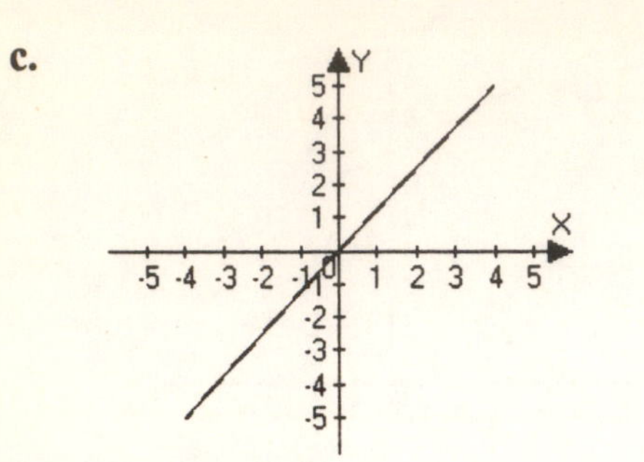

39. **a.** $3x + y = 4$

$$y = -3x + 4$$

b. $m = -3$ $\quad$ $y-$intercept = 4

c.

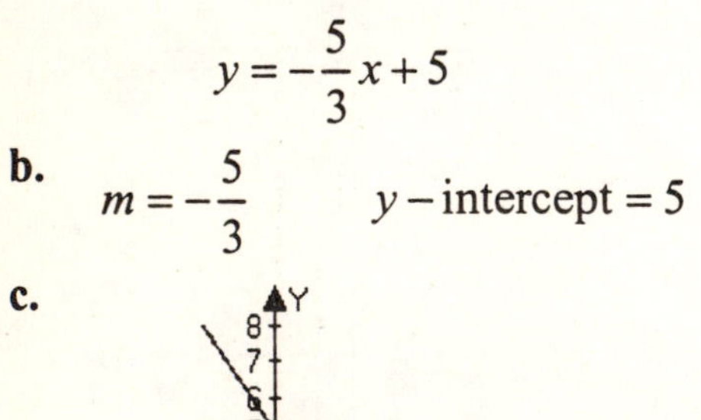

41. **a.** $5x + 3y = 15$

$$3y = -5x + 15$$

$$y = -\frac{5}{3}x + 5$$

b. $m = -\frac{5}{3}$ $\quad$ $y-$intercept = 5

c.

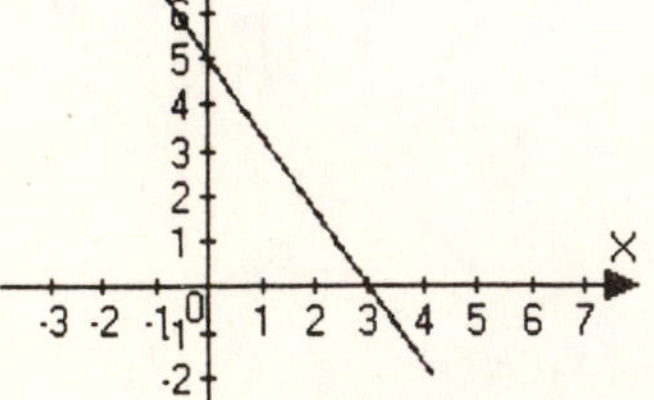

43. $y = 3$

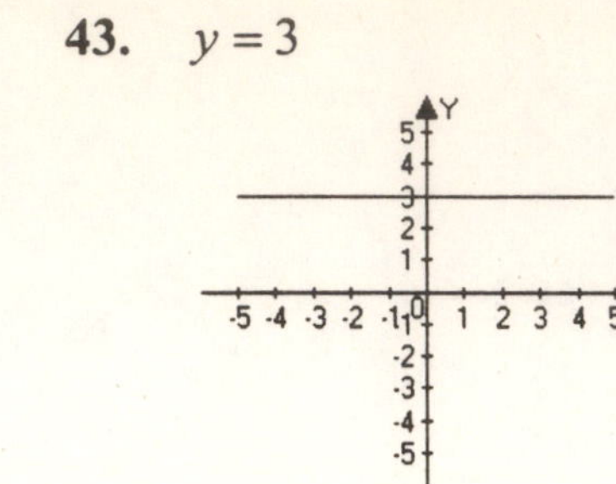

45. $y = -2$

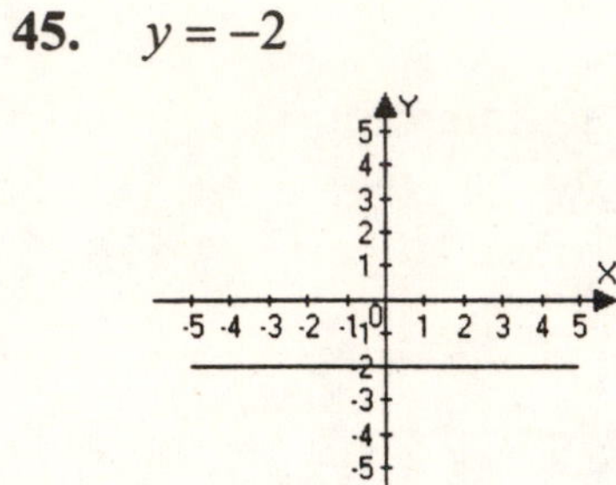

47. $3y = 18$

$$y = 6$$

49. $f(x) = 2$

$$y = 2$$

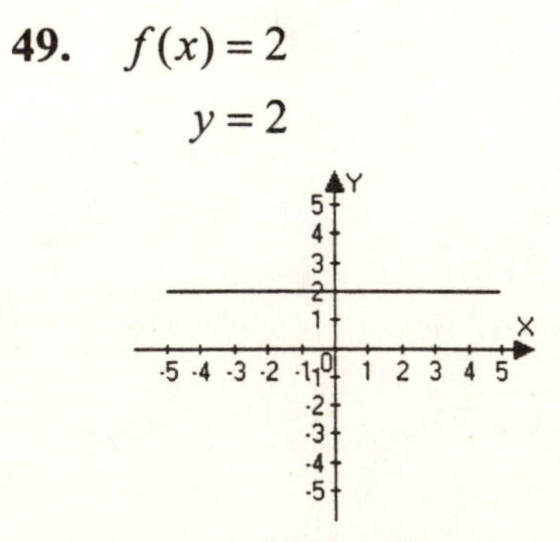

51. $x = 5$

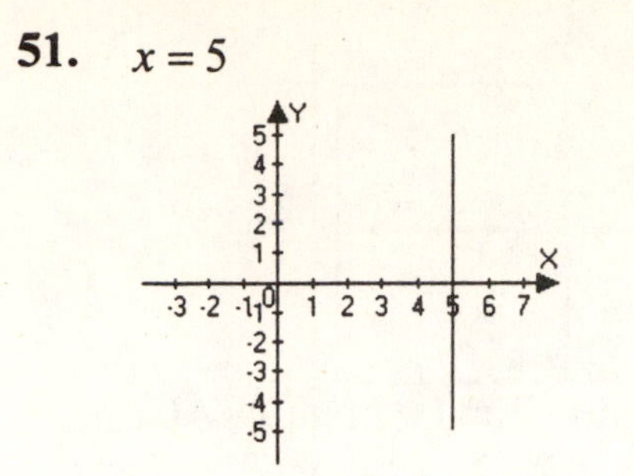

53. $3x = -12$

$x = -4$

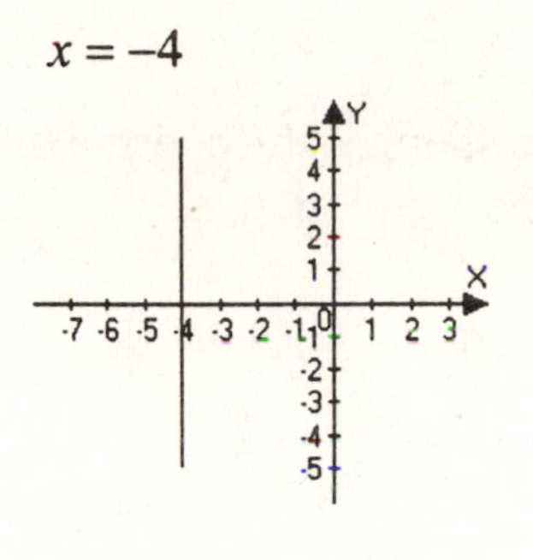

55. $x = 0$

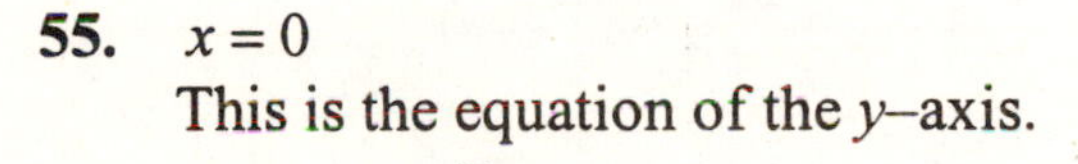
This is the equation of the y–axis.

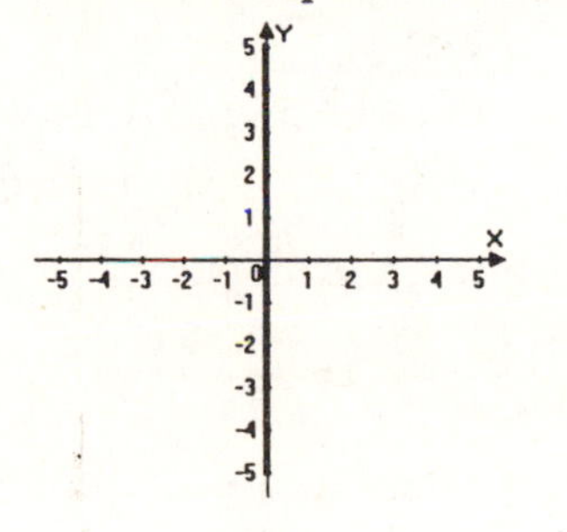

57. (2000, 6) (2050, 10.7)

$$m = \frac{10.7-6}{2050-2000} = \frac{4.7}{50} = 0.094$$

The average rate of change is 0.094. This means that the world population is expected to increase by 0.094 billion each year between 2000 and 2050.

59. (1975, 4100) (2000, 1400)

$$m = \frac{4100-1400}{1975-2000} = \frac{2700}{-25} = -108$$

The average rate of change is –108. This means that on average, cigarette consumption per U.S. adult is decreasing by 108 cigarettes per year between 1975 and 2000. (Answers will vary depending upon point estimation.)

61. (1940, 1950) (1996, 1950)

$$m = \frac{1950-1950}{1940-1996} = \frac{0}{-56} = 0$$

The average rate of change is 0. This means that on average, cigarette consumption per U.S. adult is showing no change between 1940 and 1996. (Answers will vary depending upon point estimation.)

63. (1940, 1950) (1945, 3400)

$$m = \frac{3400-1950}{1945-1940} = \frac{1450}{5} = 290$$

The average rate of change is 290. This means that on average, cigarette consumption per U.S. adult increased by 290 cigarettes per year between 1940 and 1945. (Answers will vary depending upon point estimation.)

65. Horizontal lines are of the form $y = b$. The equation that models the number of downhill skiers is: $y = 15$.

67. (1993, 16) (2000, 23)

$$m = \frac{2000-1993}{23-16} = \frac{7}{7} = 1$$

The rate of increase is 1. This means that the rate of increase in book sales was $1 billion per year between 1993 and 2000.

69. $y = 2x + 24$

$m = 2$

The average cost of a retail prescription drug in the United Stated has been increasing by \$2.00 per year since 1991.

71. $y = -0.7x + 60$

$m = -0.7$

The percentage of U.S. adults who read a newspaper has been decreasing by 0.7% per year since 1995.

Writing in Mathematics

73. Answers will vary.

75. Answers will vary.

77. Answers will vary.

79. Answers will vary.

81. Answers will vary.

83. Answers will vary.

85. Answers will vary.

87. Answers will vary.

Technology Exercises

89. Answers will vary. For example, consider Exercise 1.

$x + y = 5$

$y = -x + 5$

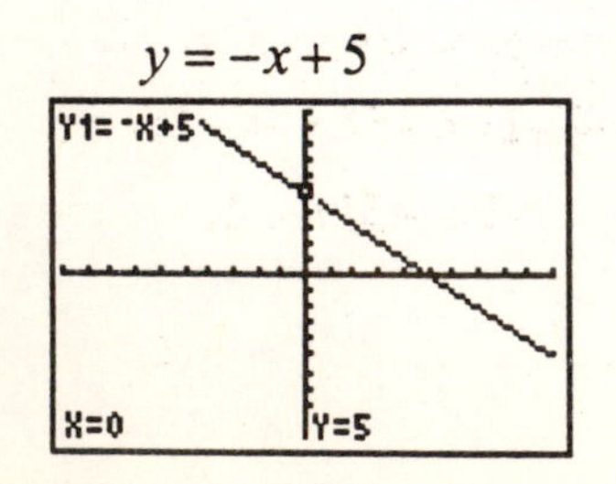

91. $y = -3x + 6$

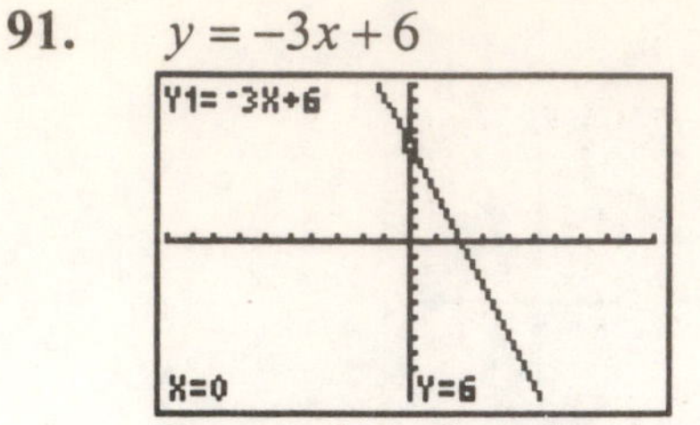

Two points found using [TRACE] are (0, 6) and (2, 0). Based on these points, the slope is:

$$m = \frac{6-0}{0-2} = \frac{6}{-2} = -3.$$

This is the same as the coefficient of x in the line's equation.

93. $y = \frac{3}{4}x - 2$

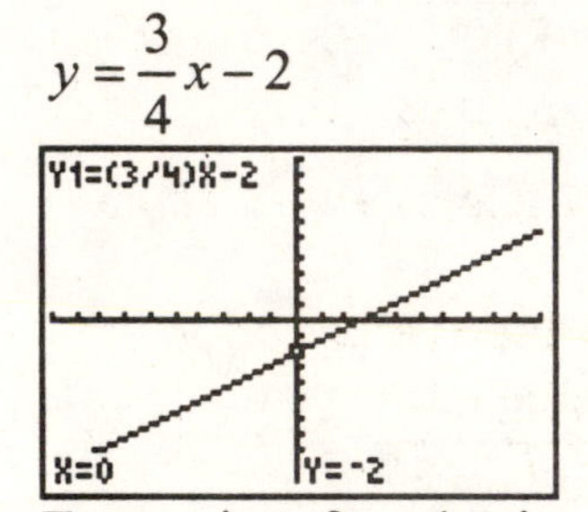

Two points found using [TRACE] are $(0, -2)$ and $(1, -1.25)$. Based on these points, the slope is:

$$m = \frac{-2-(-1.25)}{0-1} = \frac{-0.75}{-1} = 0.75.$$

This is equivalent to the coefficient of x in the line's equation.

Critical Thinking Exercises

95. We are given that the x-intercept is -2 and the y-intercept is 4. We can use the points $(-2, 0)$ and $(0, 4)$ to find the slope.

$$m = \frac{4-0}{0-(-2)} = \frac{4}{0+2} = \frac{4}{2} = 2$$

Using the slope and one of the intercepts, we can write the line in point-slope form.

$$y - y_1 = m(x - x_1)$$
$$y - 0 = 2(x - (-2))$$
$$y = 2(x + 2)$$
$$y = 2x + 4$$
$$-2x + y = 4$$

Find the x– and y–coefficients for the equation of the line with right-hand-side equal to 12. Multiply both sides of $-2x + y = 4$ by 3 to obtain 12 on the right-hand-side.

$$-2x + y = 4$$
$$3(-2x + y) = 3(4)$$
$$-6x + 3y = 12$$

The coefficients are –6 and 3.

97. In 1993, the average book price is

$$\frac{\$16 \text{ billion}}{2 \text{ billion books}} = \$8 \text{ per book.}$$

In 2000, the average book price is

$$\frac{\$23 \text{ billion}}{2 \text{ billion books}} = \$11.50 \text{ per book.}$$

The resulting ordered pairs are (1993, 8) and (2000, 11.50).

Use the pairs to find the slope.

$$m = \frac{11.50 - 8}{2000 - 1993} = \frac{3.50}{7} = 0.5.$$

The average book price increased \$0.50 per year from 1993 to 2000.

Review Exercises

98.
$$\left(\frac{4x^2}{y^{-3}}\right)^2 = \left(4x^2y^3\right)^2 = 4^2\left(x^2\right)^2\left(y^3\right)^2$$
$$= 16x^4y^6$$

99.
$$\left(8\times10^{-7}\right)\left(4\times10^3\right)$$
$$= (8\times4)\left(10^{-7}\times10^3\right) = 32\times10^{-4}$$
$$= \left(3.2\times10^1\right)\times10^{-4} = 3.2\times\left(10^1\times10^{-4}\right)$$
$$= 3.2\times10^{-3}$$

100.
$$5 - [3(x-4) - 6x] = 5 - [3x - 12 - 6x]$$
$$= 5 - 3x + 12 + 6x$$
$$= 3x + 17$$

Check Points 2.4

1. Slope = 6, passing through $(2, -5)$

First write the equation in point-slope form.

$$y - y_1 = m(x - x_1)$$
$$y - (-5) = 6(x - 2)$$
$$y + 5 = 6(x - 2)$$

Solve for y to obtain slope-intercept form.

$$y + 5 = 6(x - 2)$$
$$y + 5 = 6x - 12$$
$$y = 6x - 17$$

2. Passing through $(-2, -1)$ and $(-1, -6)$

First, find the slope:

$$m = \frac{-6 - (-1)}{-1 - (-2)} = \frac{-6 + 1}{-1 + 2} = \frac{-5}{1} = -5$$

a. Use the slope and one of the points to write the equation in point-slope form.

$$y - y_1 = m(x - x_1)$$
$$y - (-1) = -5(x - (-2))$$
$$y + 1 = -5(x + 2)$$

b. Solve for y to write the equation in slope-intercept form.

$$y+1=-5(x+2)$$
$$y+1=-5x-10$$
$$y=-5x-11$$

3. Write the equation of the line passing through (20, 226.5) and (30, 250). First, find the slope.

$$m=\frac{250-226.5}{30-20}=\frac{23.5}{10}=2.35$$

Now, use the slope and one of the points to write the equation in point-slope form.

$$y-y_1=m(x-x_1)$$
$$y-250=2.35(x-30)$$
$$y-250=2.35x-70.5$$
$$y=2.35x+179.5$$

To predict the U.S. population in 2010, substitute 2010 – 1960 = 50 for x.

$$y=2.35(50)+179.5$$
$$=117.5+179.5=297$$

If the current trend continues, the U.S. population will be 297 million in the year 2010.

4. Passing through (–2, 5) and parallel to the line $y=3x+1$.

The slope of the line $y=3x+1$ is 3. Since the line we are concerned with is parallel to this line, it will have the same slope. Use the slope and the point to write the equation of the line in point-slope form.

$$y-y_1=m(x-x_1)$$
$$y-5=3(x-(-2))$$
$$y-5=3(x+2)$$

Solve for y to rewrite the equation in slope-intercept form.

$$y-5=3(x+2)$$
$$y-5=3x+6$$
$$y=3x+11$$

Problem Set 2.4

Practice Exercises

1. Slope $=2$, passing through $(3,5)$

Point-Slope Form

$$y-y_1=m(x-x_1)$$
$$y-5=2(x-3)$$

Slope-Intercept Form

$$y-5=2(x-3)$$
$$y-5=2x-6$$
$$y=2x-1$$
$$f(x)=2x-1$$

3. Slope $=6$, passing through $(-2,5)$

Point-Slope Form

$$y-y_1=m(x-x_1)$$
$$y-5=6(x-(-2))$$
$$y-5=6(x+2)$$

Slope-Intercept Form

$$y-5=6(x+2)$$
$$y-5=6x+12$$
$$y=6x+17$$
$$f(x)=6x+17$$

5. Slope $=-3$, passing through $(-2,-3)$

Point-Slope Form

$$y-y_1=m(x-x_1)$$
$$y-(-3)=-3(x-(-2))$$
$$y+3=-3(x+2)$$

Slope-Intercept Form

$$y+3=-3(x+2)$$
$$y+3=-3x-6$$
$$y=-3x-9$$
$$f(x)=-3x-9$$

7. Slope $=-4$, passing through $(-4,0)$

Point-Slope Form

$$y-y_1=m(x-x_1)$$
$$y-0=-4(x-(-4))$$
$$y-0=-4(x+4)$$

Slope-Intercept Form

$$y-0=-4(x+4)$$
$$y=-4(x+4)$$
$$y=-4x-16$$
$$f(x)=-4x-16$$

9. Slope $=-1$, passing through $(-\frac{1}{2},-2)$

Point-Slope Form

$$y-y_1=m(x-x_1)$$
$$y-(-2)=-1\left(x-\left(-\frac{1}{2}\right)\right)$$
$$y+2=-\left(x+\frac{1}{2}\right)$$

Slope-Intercept Form

$$y+2=-\left(x+\frac{1}{2}\right)$$
$$y+2=-x-\frac{1}{2}$$
$$y=-x-\frac{5}{2}$$
$$f(x)=-x-\frac{5}{2}$$

11. Slope $=\frac{1}{2}$, passing through $(0,0)$

Point-Slope Form

$$y-y_1=m(x-x_1)$$
$$y-0=\frac{1}{2}(x-0)$$

Slope-Intercept Form

$$y-0=\frac{1}{2}(x-0)$$
$$y=\frac{1}{2}x$$
$$f(x)=\frac{1}{2}x$$

13. Slope $=-\frac{2}{3}$, passing through $(6,-2)$

Point-Slope Form

$$y-y_1=m(x-x_1)$$
$$y-(-2)=-\frac{2}{3}(x-6)$$
$$y+2=-\frac{2}{3}(x-6)$$

Slope-Intercept Form

$$y+2=-\frac{2}{3}(x-6)$$
$$y+2=-\frac{2}{3}x+4$$
$$y=-\frac{2}{3}x+2$$
$$f(x)=-\frac{2}{3}x+2$$

15. Passing through $(1,2)$ and $(5,10)$

First, find the slope:

$$m=\frac{10-2}{5-1}=\frac{8}{4}=2$$

Then use the slope and one of the points to write the equation in point-slope form.

$y - y_1 = m(x - x_1)$

$y - 10 = 2(x - 5)$

or

$y - 2 = 2(x - 1)$

Slope-Intercept Form

$y - 2 = 2(x - 1)$

$y - 2 = 2x - 2$

$y = 2x$

$f(x) = 2x$

17. Passing through (–3, 0) and (0, 3)
First, find the slope:

$m = \frac{3-0}{0-(-3)} = \frac{3}{3} = 1$

Then use the slope and one of the points to write the equation in point-slope form.

$y - y_1 = m(x - x_1)$

$y - 3 = 1(x - 0)$

or

$y - 0 = 1(x - (-3))$

$y - 0 = 1(x + 3)$

Slope-Intercept Form

$y - 0 = 1(x + 3)$

$y = x + 3$

$f(x) = x + 3$

19. Passing through (–3, –1) and (2, 4)
First, find the slope:

$m = \frac{4-(-1)}{2-(-3)} = \frac{5}{5} = 1$

Then use the slope and one of the points to write the equation in point-slope form.

$y - y_1 = m(x - x_1)$

$y - 4 = 1(x - 2)$

or

$y - (-1) = 1(x - (-3))$

$y + 1 = 1(x + 3)$

Slope-Intercept Form

$y + 1 = 1(x + 3)$

$y + 1 = x + 3$

$y = x + 2$

$f(x) = x + 2$

21. Passing through (–3, –2) and (3, 6)
First, find the slope:

$m = \frac{6-(-2)}{3-(-3)} = \frac{8}{6} = \frac{4}{3}$

Then use the slope and one of the points to write the equation in point-slope form.

$y - y_1 = m(x - x_1)$

$y - (-2) = \frac{4}{3}(x - (-3))$

$y + 2 = \frac{4}{3}(x + 3)$

or

$y - 6 = \frac{4}{3}(x - 3)$

Slope-Intercept Form

$y - 6 = \frac{4}{3}(x - 3)$

$y - 6 = \frac{4}{3}x - 4$

$y = \frac{4}{3}x + 2$

$f(x) = \frac{4}{3}x + 2$

23. Passing through $(-3,-1)$ and $(4,-1)$

First, find the slope:

$$m=\frac{-1-(-1)}{4-(-3)}=\frac{0}{7}=0$$

Then use the slope and one of the points to write the equation in point-slope form.

$$y-y_1=m(x-x_1)$$
$$y-(-1)=0(x-(-3))$$
$$y+1=0(x+3)$$

or

$$y-(-1)=0(x-4)$$
$$y+1=0(x-4)$$

Slope-Intercept Form

$$y+1=0(x-4)$$
$$y+1=0$$
$$y=-1$$
$$f(x)=-1$$

25. Passing through $(2,4)$ with $x-\text{intercept}=-2$

If the line has an $x-\text{intercept}=-2$, it passes through the point $(-2,0)$.

First, find the slope:

$$m=\frac{4-0}{2-(-2)}=\frac{4}{4}=1$$

Then use the slope and one of the points to write the equation in point-slope form.

$$y-y_1=m(x-x_1)$$
$$y-0=1(x-(-2))$$
$$y-0=1(x+2)$$

or

$$y-4=1(x-2)$$

Slope-Intercept Form

$$y-4=1(x-2)$$
$$y-4=x-2$$
$$y=x+2$$
$$f(x)=x+2$$

27. $x-\text{intercept}=-\frac{1}{2}$ and $y-\text{intercept}=4$

If the line has an $x-\text{intercept}=-\frac{1}{2}$, it passes through the point $\left(-\frac{1}{2},0\right)$. If the line has a $y-\text{intercept}=4$, it passes through $(0,4)$.

First, find the slope:

$$m=\frac{4-0}{0-\left(-\frac{1}{2}\right)}=\frac{4}{\frac{1}{2}}=4\cdot 2=8$$

Then use the slope and one of the points to write the equation in point-slope form.

$$y-y_1=m(x-x_1)$$
$$y-4=8(x-0)$$

or

$$y-0=8\left(x-\left(-\frac{1}{2}\right)\right)$$
$$y-0=8\left(x+\frac{1}{2}\right)$$

Slope-Intercept Form

$$y-4=8(x-0)$$
$$y-4=8x-0$$
$$y=8x+4$$
$$f(x)=8x+4$$

29. For $y = 5x$, $m = 5$. A line parallel to this line would have the same slope, $m = 5$. A line perpendicular to it would have slope $m = -\frac{1}{5}$.

31. For $y = -7x$, $m = -7$. A line parallel to this line would have the same slope, $m = -7$. A line perpendicular to it would have slope $m = \frac{1}{7}$.

33. For $y = \frac{1}{2}x + 3$, $m = \frac{1}{2}$. A line parallel to this line would have the same slope, $m = \frac{1}{2}$. A line perpendicular to it would have slope $m = -2$.

35. For $y = -\frac{2}{5}x - 1$, $m = -\frac{2}{5}$. A line parallel to this line would have the same slope, $m = -\frac{2}{5}$. A line perpendicular to it would have slope $m = \frac{5}{2}$.

37. To find the slope, we rewrite the equation in slope-intercept form.

$$4x + y = 7$$
$$y = -4x + 7$$

So, $m = -4$. A line parallel to this line would have the same slope, $m = -4$. A line perpendicular to it would have slope $m = \frac{1}{4}$.

39. To find the slope, we rewrite the equation in slope-intercept form.

$$2x + 4y - 8 = 0$$
$$4y = -2x + 8$$
$$y = -\frac{1}{2}x + 2$$

So, $m = -\frac{1}{2}$. A line parallel to this line would have the same slope, $m = -\frac{1}{2}$. A line perpendicular to it would have slope $m = 2$.

41. To find the slope, we rewrite the equation in slope-intercept form.

$$2x - 3y - 5 = 0$$
$$-3y = -2x + 5$$
$$y = \frac{2}{3}x - \frac{5}{3}$$

So, $m = \frac{2}{3}$. A line parallel to this line would have the same slope, $m = \frac{2}{3}$. A line perpendicular to it would have slope $m = -\frac{3}{2}$.

43. We know that $x = 6$ is a vertical line with undefined slope. A line parallel to it would also be vertical with undefined slope. A line perpendicular to it would be horizontal with slope $m = 0$.

45. Since L is parallel to $y = 2x$, we know it will have slope $m = 2$. We are given that it passes through (4, 2). We use the slope and point to write the equation in point-slope form.

$$y - y_1 = m(x - x_1)$$
$$y - 2 = 2(x - 4)$$

Solve for y to obtain slope-intercept form.

$$y-2=2(x-4)$$
$$y-2=2x-8$$
$$y=2x-6$$

In function notation, the equation of the line is $f(x)=2x-6$.

47. Since L is perpendicular to $y=2x$, we know it will have slope $m=-\frac{1}{2}$. We are given that it passes through (2, 4). We use the slope and point to write the equation in point-slope form.

$$y-y_1=m(x-x_1)$$
$$y-4=-\frac{1}{2}(x-2)$$

Solve for y to obtain slope-intercept form.

$$y-4=-\frac{1}{2}(x-2)$$
$$y-4=-\frac{1}{2}x+1$$
$$y=-\frac{1}{2}x+5$$

In function notation, the equation of the line is $f(x)=-\frac{1}{2}x+5$.

49. Since L is parallel to $y=-4x+3$, we know it will have slope $m=-4$. We are given that it passes through (–8, –10). We use the slope and point to write the equation in point-slope form.

$$y-y_1=m(x-x_1)$$
$$y-(-10)=-4(x-(-8))$$
$$y+10=-4(x+8)$$

Solve for y to obtain slope-intercept form.

$$y+10=-4(x+8)$$
$$y+10=-4x-32$$
$$y=-4x-42$$

In function notation, the equation of the line is $f(x)=-4x-42$.

51. Since L is perpendicular to $y=\frac{1}{5}x+6$, we know it will have slope $m=-5$. We are given that it passes through (2, –3). We use the slope and point to write the equation in point-slope form.

$$y-y_1=m(x-x_1)$$
$$y-(-3)=-5(x-2)$$
$$y+3=-5(x-2)$$

Solve for y to obtain slope-intercept form.

$$y+3=-5(x-2)$$
$$y+3=-5x+10$$
$$y=-5x+7$$

In function notation, the equation of the line is $f(x)=-5x+7$.

53. To find the slope, we rewrite the equation in slope-intercept form.

$$2x-3y-7=0$$
$$-3y=-2x+7$$
$$y=\frac{2}{3}x-\frac{7}{3}$$

Since the line is parallel to $y=\frac{2}{3}x-\frac{7}{3}$, we know it will have slope $m=\frac{2}{3}$. We are given that it passes through (–2, 2). We use the slope and point to write the equation in point-slope form.

$$y-y_1=m(x-x_1)$$
$$y-2=\frac{2}{3}(x-(-2))$$
$$y-2=\frac{2}{3}(x+2)$$

Solve for y to obtain slope-intercept form.

$$y-2=\frac{2}{3}(x+2)$$
$$y-2=\frac{2}{3}x+\frac{4}{3}$$
$$y=\frac{2}{3}x+\frac{10}{3}$$

In function notation, the equation of the line is $f(x)=\frac{2}{3}x+\frac{10}{3}$.

55. To find the slope, we rewrite the equation in slope-intercept form.

$$x-2y-3=0$$
$$-2y=-x+3$$
$$y=\frac{1}{2}x-\frac{3}{2}$$

Since the line is perpendicular to $y=\frac{1}{2}x-\frac{3}{2}$, we know it will have slope $m=-2$. We are given that it passes through (4, –7). We use the slope and point to write the equation in point-slope form.

$$y-y_1=m(x-x_1)$$
$$y-(-7)=-2(x-4)$$
$$y+7=-2(x-4)$$

Solve for y to obtain slope-intercept form.

$$y+7=-2(x-4)$$
$$y+7=-2x+8$$
$$y=-2x+1$$

In function notation, the equation of the line is $f(x)=-2x+1$.

Application Exercises

57. **a.** First, find the slope using $(2,34)$ and $(6,60)$.

$$m=\frac{60-34}{6-2}=\frac{26}{4}=\frac{13}{2}$$

Then use the slope and one of the points to write the equation in point-slope form.

$$y-y_1=m(x-x_1)$$
$$y-34=\frac{13}{2}(x-2)$$

b. Slope-Intercept Form

$$y-34=\frac{13}{2}(x-2)$$
$$y-34=\frac{13}{2}x-13$$
$$y=\frac{13}{2}x+21$$

c. To predict the number of accidents in 2010, let $x=2010-1994=16$.

$y = \frac{13}{2}x + 21$

$y = \frac{13}{2}(16) + 21$

$y = \frac{13}{2}(16) + 21$

$y = 104 + 21 = 125$

If the current trend continues, there will be 125 cell-phone related accidents in the year 2010.

59. We obtain the points $(0, 0.5)$ and $(15, 1.5)$ from the graph and use them to find the slope.

$m = \frac{1.5 - 0.5}{15 - 0} = \frac{1}{15}$

Then use the slope and one of the points to write the equation in point-slope form.

$y - y_1 = m(x - x_1)$

$y - 0.5 = \frac{1}{15}(x - 0)$

To predict the U.S. prison population in 2010, let $x = 2010 - 1980 = 30$.

$y - 0.5 = \frac{1}{15}(x - 0)$

$y - 0.5 = \frac{1}{15}(30 - 0)$

$y - 0.5 = \frac{1}{15}(30)$

$y - 0.5 = 2$

$y = 2.5$

If the current trend continues, there will be 2.5 million people in U.S. prisons and jails in the year 2010.

61. a. Scatter Plot for Percentage of Seniors who have ever used Marijuana

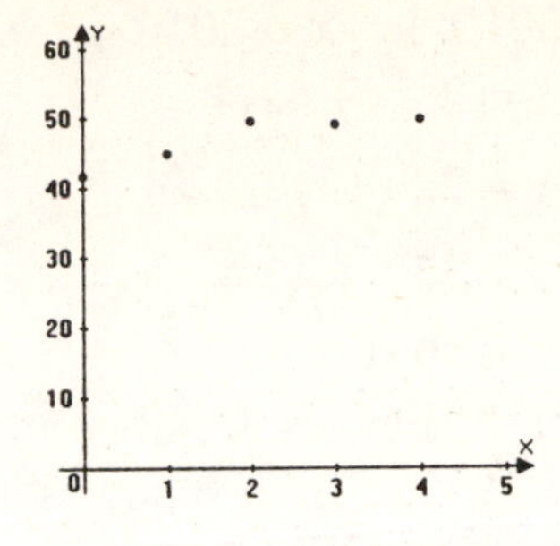

b. Scatter Plot with Line of Best Fit

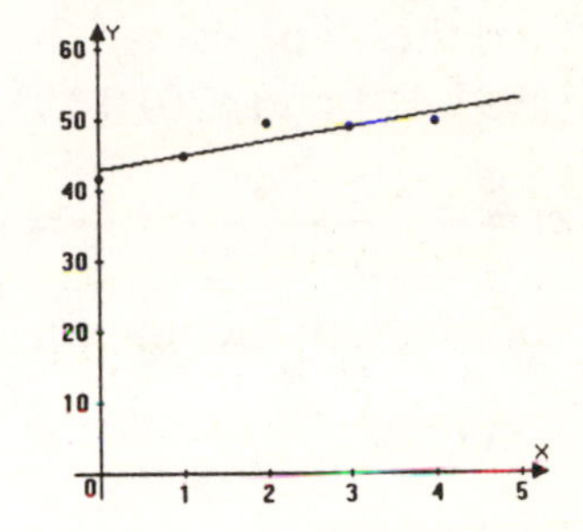

c. Answers will vary depending upon the data points chosen. One answer is as follows. First, find the slope using two points, $(1, 44.9)$ and $(3, 49.1)$.

$m = \frac{49.1 - 44.9}{3 - 1} = \frac{4.2}{2} = 2.1$

Then use the slope and one of the points to write the equation in point-slope form.

$y - y_1 = m(x - x_1)$

$y - 44.9 = 2.1(x - 1)$

To rewrite the equation in slope-intercept form, we solve for y.

$y - 44.9 = 2.1(x - 1)$

$y - 44.9 = 2.1x - 2.1$

$y = 2.1x + 42.8$

d. Answers will vary depending upon the equation obtained in part c. To predict the percentage of high school seniors who will try marijuana in the class of

2010, let $x = 2010 - 1995 = 15$.

$y = 2.5x + 42.4$

$y = 2.5(15) + 42.4$

$y = 37.5 + 42.4$

$y = 79.9$

If the current trend continues, 79.9% of high school seniors in 2010 will have tried marijuana.

63. **a.** Find the slope using (1995,14) and (2010,17).

$$m = \frac{17-14}{2010-1995} = \frac{3}{15} = \frac{1}{5}$$

The slope of the line for U.S. women is $\frac{1}{5}$.

b. Find the slope using (1995,10) and (2010,12).

$$m = \frac{12-10}{2010-1995} = \frac{2}{15}$$

The slope of the line for U.S. men is $\frac{2}{15}$.

c. Since the lines do not have the same slope, we know they are not parallel. When we compare the slopes, we see that the yearly increase in women living alone is greater than the yearly increase for men living alone.

Writing in Mathematics

65. Answers will vary.

67. Answers will vary.

69. Answers will vary.

71. Answers will vary.

Technology Exercises

73. **a.**

b.

c.

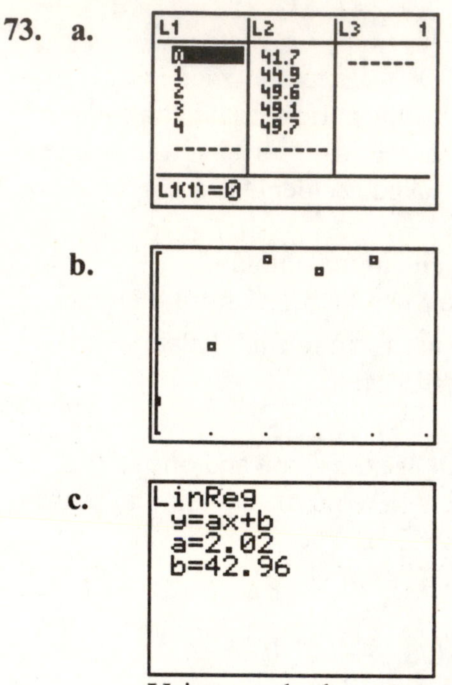

Using a calculator, we obtain $y = 2.02x + 42.96$. This is very close to the equation, $y = 2.5x + 42.4$, obtained in Exercise 61.

d.

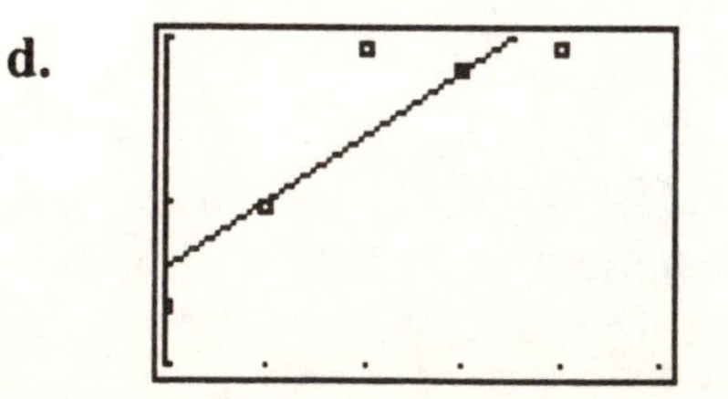

Critical Thinking Exercises

75. In order to obtain point-slope form of the line, we need one point on the line and the slope. Since the line has an x-intercept $= -3$, we know it passes through the point $(-3,0)$.

Next, find the slope. To find the slope, we use the slope of the line passing through $(0,0)$ and $(6,-2)$.

$$m=\frac{0-(-2)}{0-6}=\frac{2}{-6}=-\frac{1}{3}$$

The line we are concerned with is perpendicular to this line, so the slope will be $m=3$.

We use $m=3$ and the point $(-3,0)$ to write the equation in point-slope form.

$$y-y_1=m(x-x_1)$$
$$y-0=3(x-(-3))$$
$$y-0=3(x+3)$$

To rewrite the equation is slope-intercept form, we solve for y.

$$y-0=3(x+3)$$
$$y=3(x+3)$$
$$y=3x+9$$

In function notation, the equation of the line is $f(x)=3x+9$.

77. First, find the slope using $(a,0)$ and $(0,b)$.

$$m=\frac{b-0}{0-a}=-\frac{b}{a}$$

Then use the slope and one of the points to write the equation in point-slope form.

$$y-y_1=m(x-x_1)$$
$$y-0=-\frac{b}{a}(x-a)$$

Rewriting this equation, we obtain:

$$y-0=-\frac{b}{a}(x-a)$$
$$y=-\frac{b}{a}x+b$$
$$\frac{b}{a}x+y=b$$
$$\frac{1}{b}\left(\frac{b}{a}x+y\right)=\frac{1}{b}(b)$$
$$\frac{1}{a}x+\frac{1}{b}y=1$$
$$\frac{x}{a}+\frac{y}{b}=1$$

This is called the intercept form of a line, because the x- and y-intercepts are found in the denominators of the x- and y-terms, respectively.

Review Exercises

78. $f(-2)=3(-2)^2-8(-2)+5$
$=3(4)+16+5=12+16+5=33$

79. Since $(fg)(-1)=(f)(-1)\cdot(g)(-1)$, find $(f)(-1)$ and $(g)(-1)$.

$$f(-1)=(-1)^2-3(-1)+4$$
$$=1+3+4=8$$
$$g(-1)=2(-1)-5=-2-5=-7$$
$$(fg)(-1)=(f)(-1)\cdot(g)(-1)$$
$$=8(-7)=-56$$

80. Let x = the measure of the smallest angle
$x+20$ = the measure of the second angle
$2x$ = the measure of the third angle

$$x+(x+20)+2x=180$$
$$x+x+20+2x=180$$

$$4x+20=180$$
$$4x=160$$
$$x=40$$

Find the other angles.
$x + 20 = 40 + 20 = 60$
$2x = 2(40) = 80$
The angles are 40°, 60° and 80°.

Chapter 2 Review

2.1

1. The relation is a function.
Domain {3, 4, 5}
Range {10}

2. The relation is a function.
Domain {1, 2, 3, 4}
Range {-6, π, 12, 100}

3. The relation is not a function.
Domain {13, 15}
Range {14, 16, 17}

4. **a.** $f(0)=7(0)-5=0-5=-5$

b. $f(3)=7(3)-5=21-5=16$

c. $f(-10)=7(-10)-5$
$=-70-5=-75$

d. $f(2a)=7(2a)-5=14a-5$

e. $f(a+2)=7(a+2)-5$
$=7a+14-5=7a+9$

5. **a.** $g(0)=3(0)^2-5(0)+2$
$=0-0+2=2$

b. $g(5)=3(5)^2-5(5)+2$
$=3(25)-25+2$
$=75-25+2=52$

c. $g(-4)=3(-4)^2-5(-4)+2$
$=3(16)+20+2$
$=48+20+2=70$

d. $g(b)=3(b)^2-5(b)+2$
$=3b^2-5b+2$

e. $g(4a)=3(4a)^2-5(4a)+2$
$=3(16a^2)-20a+2$
$=48a^2-20a+2$

6.

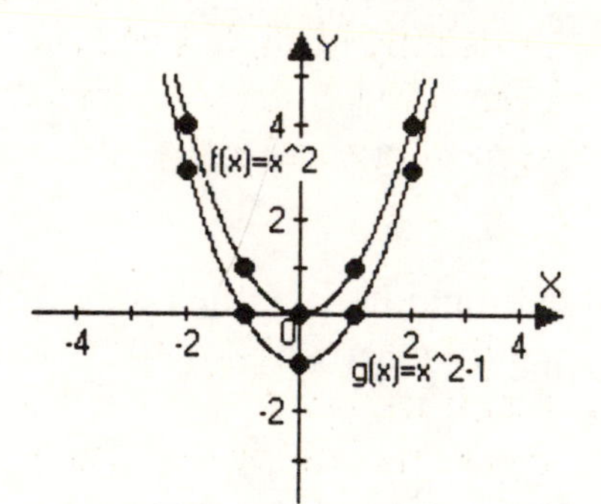

g shifts the graph of f down one unit

7.

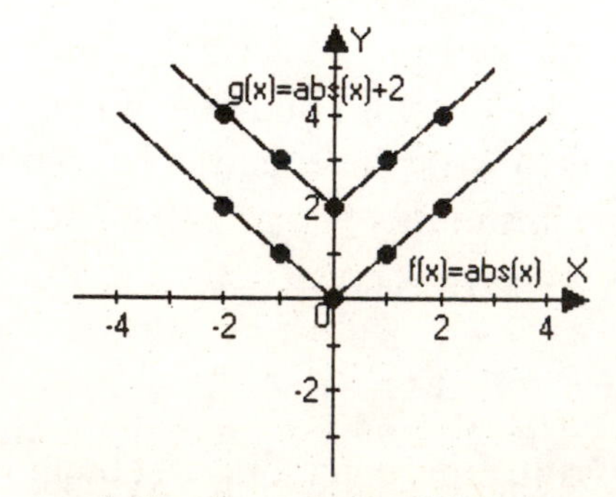

g shifts the graph of f up two units

8. The vertical line test shows that this is not the graph of a function.

9. The vertical line test shows that this is the graph of a function.

10. The vertical line test shows that this is the graph of a function.

11. The vertical line test shows that this is not the graph of a function.

12. $f(-2) = -3$

13. $f(0) = -2$

14. When $x = 3, f(x) = -5.$

15. **a.** The vulture's height is a function of its time in flight because every time, *t*, is associated with at most one height.

b. $f(15) = 0$
At time $t = 15$ seconds, the vulture is at height zero. This means that after 15 seconds, the vulture is on the ground.

c. The vulture's maximum height is 45 meters.

d. For $x = 7$ and 22, $f(x) = 20.$ This means that at times 7 seconds, and 20 seconds, the vulture is at a height of 22 meters.

e. The vulture began the flight at 45 meters and remained there for approximately 3 seconds. At that time, the vulture began to descend. It landed on the ground and stayed there from approximately 12 seconds to 17 seconds. The vulture then began to climb back up to a height of 44 meters.

2.2

16. Domain of $f = \{x | x \text{ is a real number}\}$

17. Domain of $f = \{x | x \text{ is a real number and } x \neq -8\}$

18. Domain of $f = \{x | x \text{ is a real number and } x \neq 5\}$

19. **a.**
$$\begin{aligned}(f+g)(x) &= (4x-5)+(2x+1) \\ &= 4x-5+2x+1 \\ &= 6x-4\end{aligned}$$

b.
$$\begin{aligned}(f+g)(3) &= 6(3)-4 \\ &= 18-4 = 14\end{aligned}$$

20. **a.**
$$\begin{aligned}&(f+g)(x) \\ &= (5x^2-x+4)+(x-3) \\ &= 5x^2-x+4+x-3 = 5x^2+1\end{aligned}$$

b.
$$\begin{aligned}(f+g)(3) &= 5(3)^2+1 = 5(9)+1 \\ &= 45+1 = 46\end{aligned}$$

21. Domain of $f + g = \{x | x \text{ is a real number and } x \neq 4\}$

22. Domain of $f+g=\{x|x$ is a real number and $x \neq -6$ and $x \neq -1\}$

23. $f(x)=x^2-2x, \quad g(x)=x-5$

$$\begin{aligned}(f+g)(x)&=(x^2-2x)+(x-5)\\&=x^2-2x+x-5\\&=x^2-x-5\end{aligned}$$

$$\begin{aligned}(f+g)(-2)&=(-2)^2-(-2)-5\\&=4+2-5=1\end{aligned}$$

24. From Exercise 23 we know $(f+g)(x)=x^2-x-5$. We can use this to find $f(3)+g(3)$.

$$\begin{aligned}f(3)+g(3)&=(f+g)(3)\\&=(3)^2-(3)-5\\&=9-3-5=1\end{aligned}$$

25. $f(x)=x^2-2x, \quad g(x)=x-5$

$$\begin{aligned}(f-g)(x)&=(x^2-2x)-(x-5)\\&=x^2-2x-x+5\\&=x^2-3x+5\end{aligned}$$

$$(f-g)(x)=x^2-3x+5$$

$$\begin{aligned}(f-g)(1)&=(1)^2-3(1)+5\\&=1-3+5=3\end{aligned}$$

26. From Exercise 25 we know $(f-g)(x)=x^2-3x+5$. We can use this to find $f(4)-g(4)$.

$$\begin{aligned}f(4)-g(4)&=(f-g)(4)\\&=(4)^2-3(4)+5\\&=16-12+5=9\end{aligned}$$

27. Since $(fg)(-3)=f(-3)\cdot g(-3)$, find $f(-3)$ and $g(-3)$ first.

$$\begin{aligned}f(-3)&=(-3)^2-2(-3)\\&=9+6=15\end{aligned}$$

$$g(-3)=-3-5=-8$$

$$\begin{aligned}(fg)(-3)&=f(-3)\cdot g(-3)\\&=15(-8)=-120\end{aligned}$$

28. $f(x)=x^2-2x, \quad g(x)=x-5$

$$\left(\frac{f}{g}\right)(x)=\frac{x^2-2x}{x-5}$$

$$\begin{aligned}\left(\frac{f}{g}\right)(4)&=\frac{(4)^2-2(4)}{4-5}=\frac{16-8}{-1}\\&=\frac{8}{-1}=-8\end{aligned}$$

29. $(f-g)(x)=x^2-3x+5$

Domain of $f-g=\{x|x$ is a real number$\}$

30. $\left(\frac{f}{g}\right)(x)=\frac{x^2-2x}{x-5}$

Domain of $\frac{f}{g}=\{x|x$ is a real number and $x \neq 5\}$

2.3

31. $x+2y=4$

Find the x–intercept by setting $y=0$ and the y–intercept by setting $x=0$.

$$\begin{aligned}x+2(0)&=4 &\quad 0+2y&=4\\x+0&=4 & 2y&=4\\x&=4 & y&=2\end{aligned}$$

Choose another point to use as a check. Let $x = 1$.

$1+2y=4$

$2y=3$

$y=\frac{3}{2}$

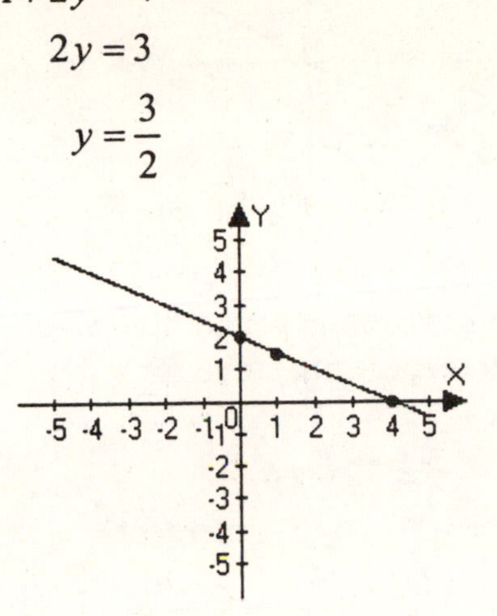

32. $2x-3y=12$

Find the x–intercept by setting $y = 0$ and the y–intercept by setting $x = 0$.

$2x-3(0)=12$ $\quad$ $2(0)-3y=12$

$2x+0=12$ $\quad$ $0-3y=12$

$2x=12$ $\quad$ $-3y=12$

$x=6$ $\quad$ $y=-4$

Choose another point to use as a check. Let $x = 1$.

$2(1)-3y=12$

$2-3y=12$

$-3y=10$

$y=-\frac{10}{3}$

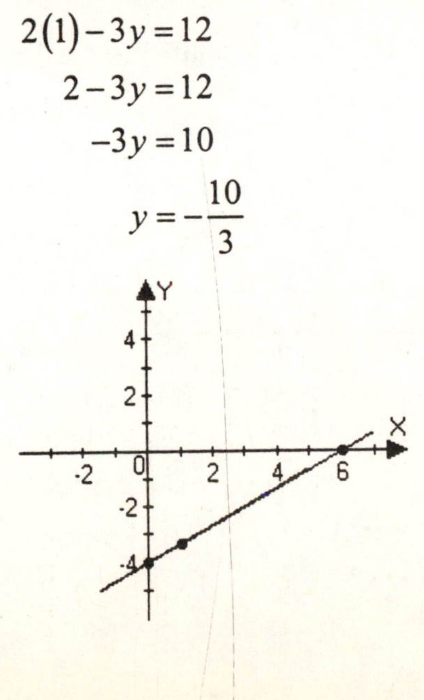

33. $4x=8-2y$

Find the x–intercept by setting $y = 0$ and the y–intercept by setting $x = 0$.

$4x=8-2(0)$ $\quad$ $4(0)=8-2y$

$4x=8-0$ $\quad$ $0=8-2y$

$4x=8$ $\quad$ $2y=8$

$x=2$ $\quad$ $y=4$

Choose another point to use as a check. Let $x = 1$.

$4(1)=8-2y$

$4=8-2y$

$-4=-2y$

$2=y$

34. $m=\frac{2-(-4)}{5-2}=\frac{6}{3}=2$

The line through the points rises.

35. $m=\frac{3-(-3)}{-2-7}=\frac{6}{-9}=-\frac{2}{3}$

The line through the points falls.

36. $m=\frac{2-(-1)}{3-3}=\frac{3}{0}$

m is undefined. The line through the points is vertical.

37. $m = \dfrac{4-4}{-3-(-1)} = \dfrac{0}{-2} = 0$

The line through the points is horizontal.

38. $y = 2x - 1$

$m = 2 \qquad y\text{-intercept} = -1$

39. $f(x) = -\dfrac{1}{2}x + 4$

$m = -\dfrac{1}{2} \qquad y\text{-intercept} = 4$

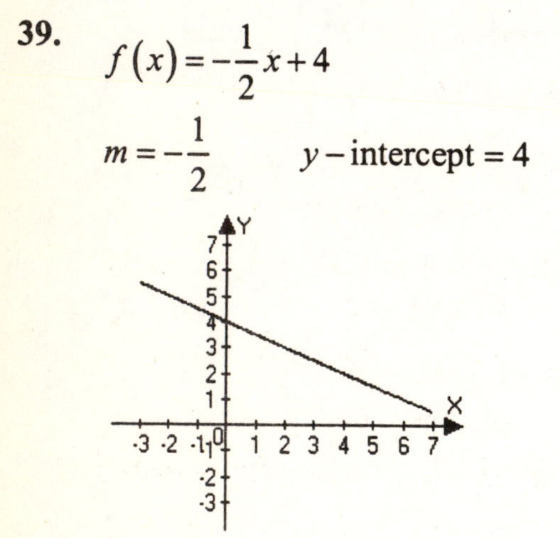

40. $y = \dfrac{2}{3}x$

$m = \dfrac{2}{3} \qquad y\text{-intercept} = 0$

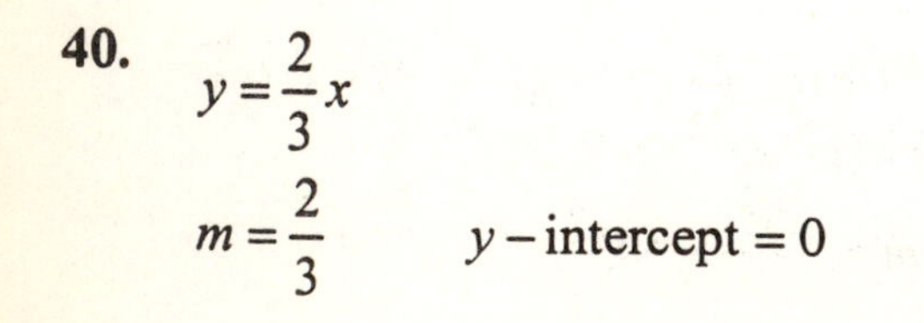

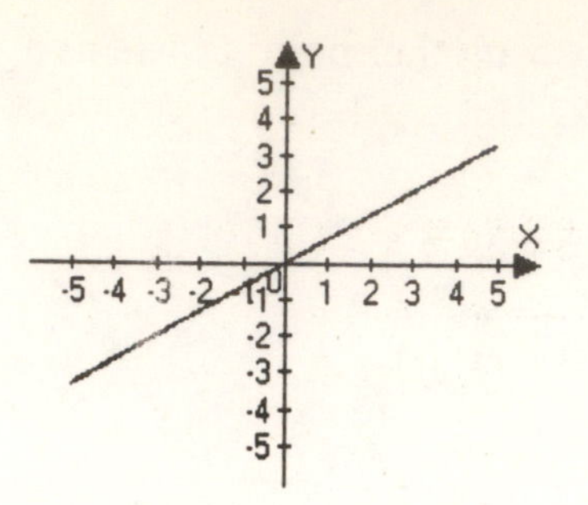

41. To rewrite the equation in slope-intercept form, solve for y.

$2x + y = 4$

$y = -2x + 4$

$m = -2 \qquad y\text{-intercept} = 4$

42. $-3y = 5x$

$y = -\dfrac{5}{3}x$

$m = -\dfrac{5}{3} \qquad y\text{-intercept} = 0$

43. $5x + 3y = 6$

$3y = -5x + 6$

$y = -\dfrac{5}{3}x + 2$

$m = -\dfrac{5}{3} \qquad y\text{-intercept} = 2$

44. $y = 2$

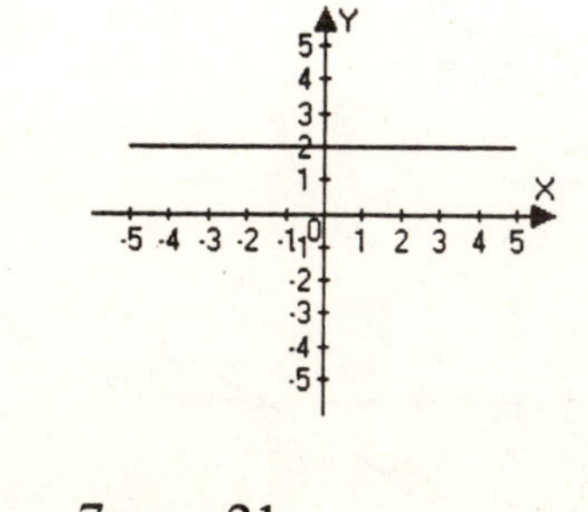

45. $7y = -21$

$y = -3$

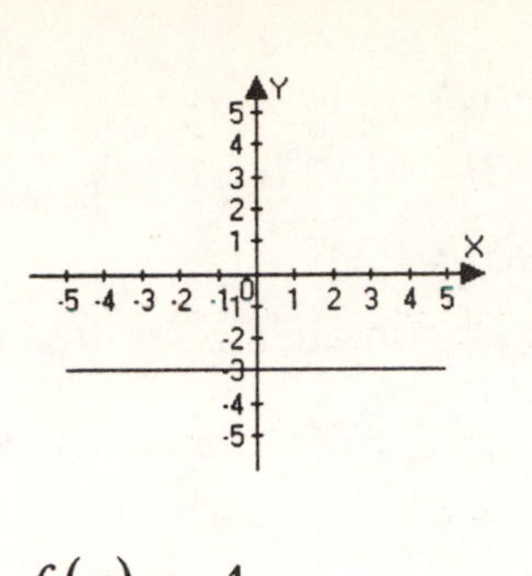

46. $f(x) = -4$

$y = -4$

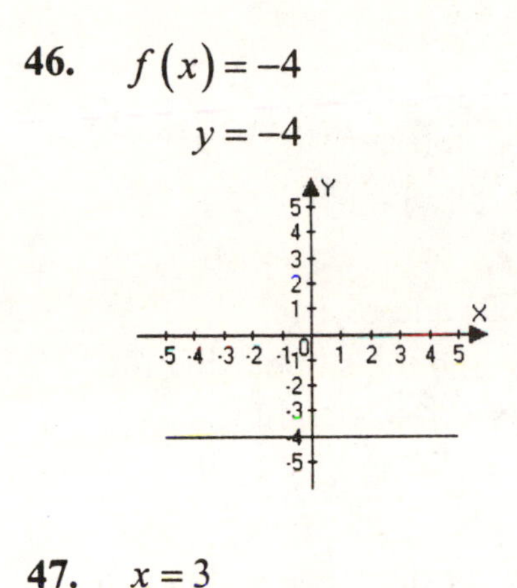

47. $x = 3$

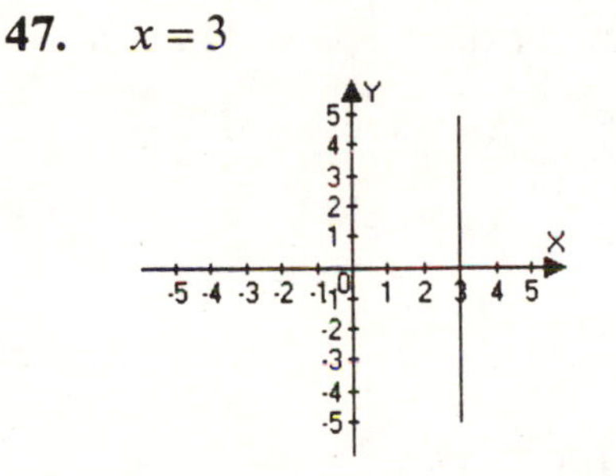

48. $2x = -10$

$x = -5$

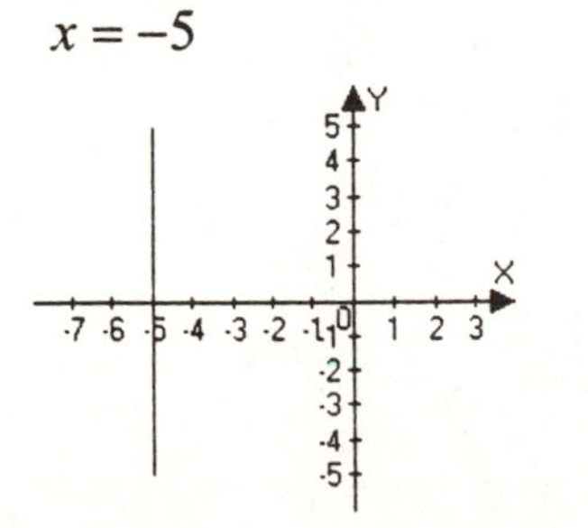

49. (1955, 21800) (2000, 74300)

$$m = \frac{74300 - 21800}{2000 - 1955} = \frac{52500}{45} = 1166\frac{2}{3}$$

The average rate of change is $1166\frac{2}{3}$. This means that on average, spending per uniformed member of the U.S. military has increased by approximately \$1167 between 1955 and 2000.

50. In $f(t) = -0.27 + 70.45$, the slope is -0.27. A slope of -0.27 indicates that the record time for the women's 400-meter has been decreasing by 0.27 seconds per year since 1900.

2.4

51. Slope $= -6$, passing through $(-3, 2)$

Point-Slope Form

$y - y_1 = m(x - x_1)$

$y - 2 = -6(x - (-3))$

$y - 2 = -6(x + 3)$

Slope-Intercept Form

$y - 2 = -6(x + 3)$

$y - 2 = -6x - 18$

$y = -6x - 16$

In function notation, the equation of the line is $f(x) = -6x - 16$.

52. Passing through $(1, 6)$ and $(-1, 2)$

First, find the slope:

$$m = \frac{6 - 2}{1 - (-1)} = \frac{4}{2} = 2$$

Then use the slope and one of the points to write the equation in point-slope form.

$y - y_1 = m(x - x_1)$

$y - 6 = 2(x - 1)$

or

$y-2=2(x-(-1))$
$y-2=2(x+1)$

Slope-Intercept Form

$y-6=2(x-1)$
$y-6=2x-2$
$y=2x+4$

In function notation, the equation of the line is $f(x)=2x+4$.

53. Rewrite $3x+y-9=0$ in slope-intercept form.

$3x+y-9=0$
$y=-3x+9$

Since the line we are concerned with is parallel to this line, we know it will have slope $m=-3$. We are given that it passes through (4, –7). We use the slope and point to write the equation in point-slope form.

$y-y_1=m(x-x_1)$
$y-(-7)=-3(x-4)$
$y+7=-3(x-4)$

Solve for y to obtain slope-intercept form.

$y+7=-3(x-4)$
$y+7=-3x+12$
$y=-3x+5$

In function notation, the equation of the line is $f(x)=-3x+5$.

54. The line is perpendicular to $y=\frac{1}{3}x+4$, so the slope is –3. We are given that it passes through (–2, 6). We use the slope and point to write the equation in point-slope form.

$y-y_1=m(x-x_1)$
$y-6=-3(x-(-2))$
$y-6=-3(x+2)$

Solve for y to obtain slope-intercept form.

$y-6=-3(x+2)$
$y-6=-3x-6$
$y=-3x$

In function notation, the equation of the line is $f(x)=-3x$.

55. **a.** (2,162) and (8,168)

First, find the slope using the points (2,162) and (8,168)

$$m=\frac{168-162}{8-2}=\frac{6}{6}=1$$

Then use the slope and one of the points to write the equation in point-slope form.

$y-y_1=m(x-x_1)$
$y-162=1(x-2)$

or

$y-168=1(x-8)$

b. Solve for y to obtain slope-intercept form.

$y-162=1(x-2)$
$y-162=x-2$
$y=x+160$
$f(x)=x+160$

c. To predict the average weight of Americans in 2005, let $x=2005-1990=15$.

$f(15)=15+160=175$

If the current trend continues, the average weight of Americans in 2005 will be 175 pounds.

Chapter 2 Test

1. The relation is a function.
   Domain {1, 3, 5, 6}
   Range {2, 4, 6}

2. The relation is not a function.
   Domain {2, 4, 6}
   Range {1, 3, 5, 6}

3. $f(a+4)=3(a+4)-2$
   $=3a+12-2=3a+10$

4. $f(-2)=4(-2)^2-3(-2)+6$
   $=4(4)+6+6=16+6+6=28$

5. 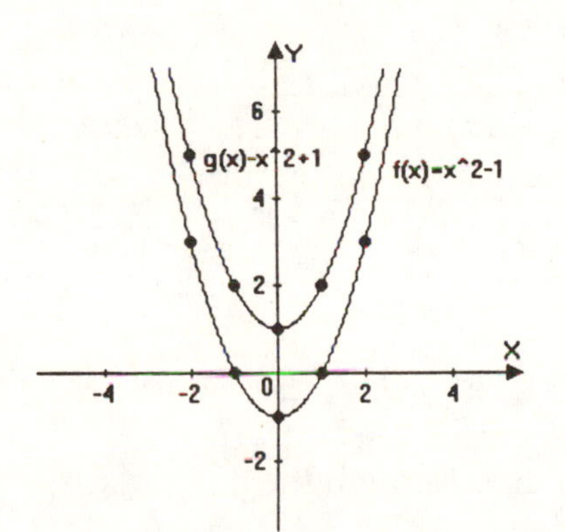

   g shifts the graph of f up 2 units

6. The vertical line test shows that this is the graph of a function.

7. The vertical line test shows that this is not the graph of a function.

8. $f(6)=-3$

9. When $x=-5$, $f(x)=-6$.

10. Domain of $f=\{x|x$ is a real number and $x\neq 10\}$

11. $f(x)=x^2+4x \qquad g(x)=x+2$
    $(f+g)(x)=f(x)+g(x)$
    $=(x^2+4x)+(x+2)$
    $=x^2+4x+x+2$
    $=x^2+5x+2$
    $(f+g)(3)=(3)^2+5(3)+2$
    $=9+15+2=26$

12. $f(x)=x^2+4x \qquad g(x)=x+2$
    $(f-g)(x)=f(x)-g(x)$
    $=(x^2+4x)-(x+2)$
    $=x^2+4x-x-2$
    $=x^2+3x-2$
    $(f-g)(-1)=(-1)^2+3(-1)-2$
    $=1-3-2=-4$

13. We know that $(fg)(x)=f(x)\cdot g(x)$. So, to find $(fg)(-5)$, we use $f(-5)$ and $g(-5)$.
    $f(-5)=(-5)^2+4(-5)=25-20=5$
    $g(-5)=-5+2=-3$
    $(fg)(-5)=f(-5)\cdot g(-5)$
    $=5(-3)=-15$

14. $f(x)=x^2+4x \qquad g(x)=x+2$

$$\left(\frac{f}{g}\right)(x)=\frac{x^2+4x}{x+2}$$

$$\left(\frac{f}{g}\right)(2)=\frac{(2)^2+4(2)}{2+2}=\frac{4+8}{4}=\frac{12}{4}=3$$

15. Domain of $\frac{f}{g}=\{x|x$ is a real number and $x\neq -2\}$

16. $4x-3y=12$

Find the x–intercept by setting $y=0$.

$$4x-3(0)=12$$
$$4x=12$$
$$x=3$$

Find the y–intercept by setting $x=0$.

$$4(0)-3y=12$$
$$-3y=12$$
$$y=-4$$

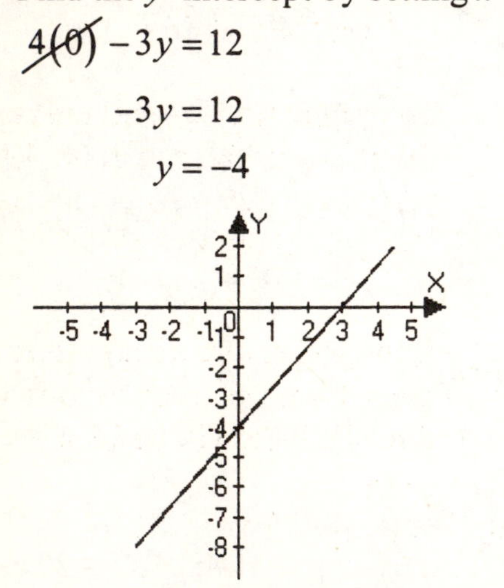

17. $f(x)=-\frac{1}{3}x+2$

$m=-\frac{1}{3} \qquad y-\text{intercept}=2$

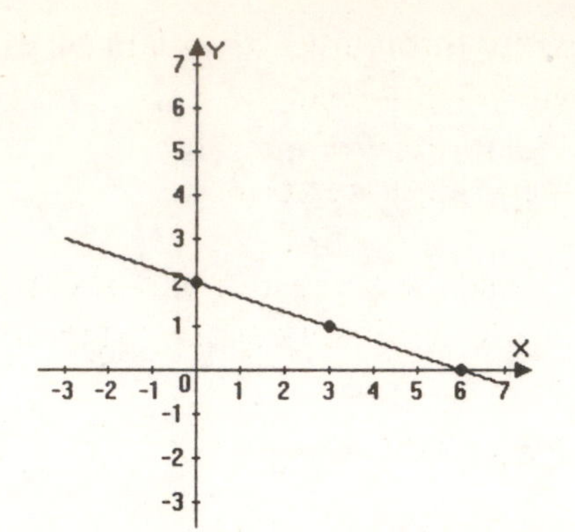

18. $f(x)=4$

$y=4$

An equation of the form $y=b$ is a horizontal line.

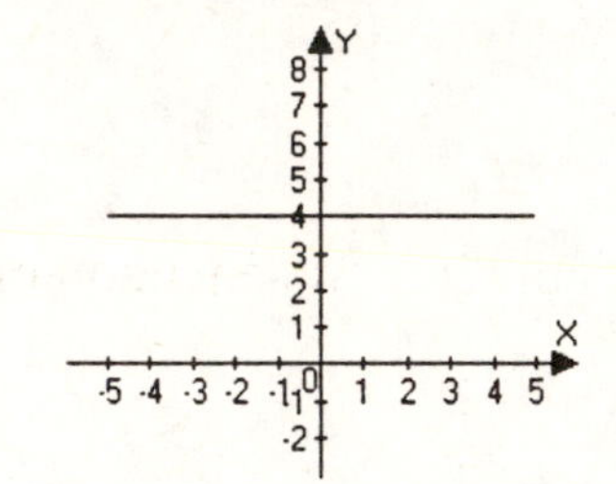

19. $m=\frac{4-2}{1-5}=\frac{2}{-4}=-\frac{1}{2}$

The line through the points falls.

20. $m=\frac{5-(-5)}{4-4}=\frac{10}{0}$

m is undefined

The line through the points is vertical.

21. $V(10)=3.6(10)+140$

$=36+140=176$

In the year 2005, there will be 176 million Super Bowl viewers.

22. The slope is 3.6. This means the number of Super Bowl viewers is increasing at a rate of 3.6 million per year.

23. Passing through $(-1,-3)$ and $(4,2)$

First, find the slope:

$$m=\frac{2-(-3)}{4-(-1)}=\frac{5}{5}=1$$

Then use the slope and one of the points to write the equation in point-slope form.

$$y-y_1=m(x-x_1)$$
$$y-(-3)=1(x-(-1))$$
$$y+3=1(x+1)$$

or

$$y-2=1(x-4)$$
$$y-2=x-4$$

Slope-Intercept Form

$$y-2=x-4$$
$$y=x-2$$

In function notation, the equation of the line is $f(x)=x-2$.

24. The line is perpendicular to $y=-\frac{1}{2}x-4$, so the slope is 2. We are given that it passes through $(-2,3)$. We use the slope and point to write the equation in point-slope form.

$$y-y_1=m(x-x_1)$$
$$y-3=2(x-(-2))$$
$$y-3=2(x+2)$$

Solve for y to obtain slope-intercept form.

$$y-3=2(x+2)$$
$$y-3=2x+4$$
$$y=2x+7$$

In function notation, the equation of the line is $f(x)=2x+7$.

25. **a.** First, find the slope using the points $(0,320)$ and $(5,530)$.

$$m=\frac{530-320}{5-0}=\frac{210}{5}=42$$

Then use the slope and one of the points to write the equation in point-slope form.

$$y-y_1=m(x-x_1)$$
$$y-320=42(x-0)$$

b. Solve for y to obtain slope-intercept form.

$$y-320=42x$$
$$y=42x+320$$
$$f(x)=42x+320$$

c. To predict the national average for one-way fares in 2008, let $x = 2008 - 1995 = 13$.

$$f(13)=42(13)+320$$
$$=546+320=866$$

If the current trend continues, the national average for one-way fares in 2008 will be $866.

Cumulative Review Exercises

Chapters 1 and 2

1. {0, 1, 2, 3}

2. False. π is an irrational number.

3. $$\frac{8-3^2\div 9}{|-5|-[5-(18\div 6)]^2}$$

$$=\frac{8-9\div 9}{5-[5-(3)]^2}=\frac{8-1}{5-[2]^2}$$

$$=\frac{7}{5-4}=\frac{7}{1}=7$$

4. $4-(2-9)^0+3^2\div 1+3$

$=4-(-7)^0+9\div 1+3=4-1+9\div 1+3$

$=4-1+9+3=3+9+3=15$

5. $3-[2(x-2)-5x]$

$=3-[2x-4-5x]=3-[-3x-4]$

$=3+3x+4=3x+7$

6. $2+3x-4=2(x-3)$

$3x-2=2x-6$

$x-2=-6$

$x=-4$

The solution is –4 and the solution set is $\{-4\}$.

7. $4x+12-8x=-6(x-2)+2x$

$12-4x=-6x+12+2x$

$12-4x=-4x+12$

$12=12$

$0=0$

The solution set is $\{x|x \text{ is a real number}\}$ or $\mathbb{R}$. The equation is an identity.

8. $$\frac{x-2}{4}=\frac{2x+6}{3}$$

$4(2x+6)=3(x-2)$

$8x+24=3x-6$

$5x+24=-6$

$5x=-30$

$x=-6$

The solution is –6 and the solution set is $\{-6\}$.

9. Let x = the price before reduction

$x-0.20x=1800$

$0.80x=1800$

$x=2250$

The price of the computer before the reduction was $2250.

10. $A=p+prt$

$A-p=prt$

$$t=\frac{A-p}{pr}$$

11. $$\left(3x^4y^{-5}\right)^{-2}=\left(\frac{3x^4}{y^5}\right)^{-2}=\left(\frac{y^5}{3x^4}\right)^2=\frac{y^{10}}{9x^8}$$

12. $$\left(\frac{3x^2y^{-4}}{x^{-3}y^2}\right)^2=\left(\frac{3x^2x^3}{y^2y^4}\right)^2$$

$$=\left(\frac{3x^5}{y^6}\right)^2=\frac{9x^{10}}{y^{12}}$$

13. $(7\times 10^{-8})(3\times 10^2)$

$=(7\times 3)(10^{-8}\times 10^2)=21\times 10^{-6}$

$=(2.1\times 10)\times 10^{-6}=2.1(10\times 10^{-6})$

$=2.1\times 10^{-5}$

14. The relation is a function.
Domain {1, 2, 3, 4, 6}
Range {5}

15.

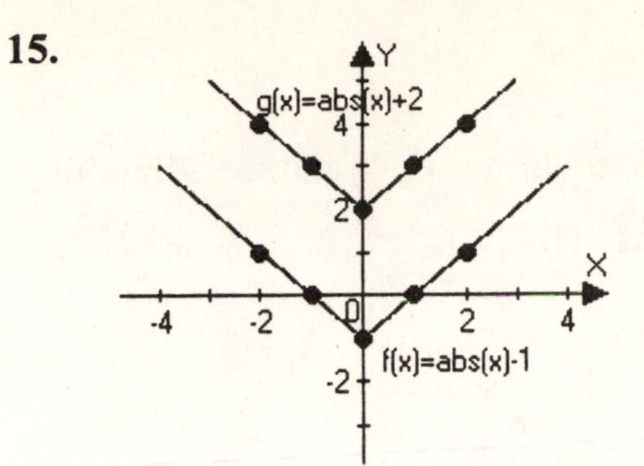

g shifts the graph of f up three units

16. Domain of $f = \{x | x$ is a real number and $x \neq 15\}$

17. $(f-g)(x)$

$= (3x^2 - 4x + 2) - (x^2 - 5x - 3)$

$= 3x^2 - 4x + 2 - x^2 + 5x + 3$

$= 2x^2 + x + 5$

$(f-g)(-1) = 2(-1)^2 + (-1) + 5$

$= 2(1) - 1 + 5 = 2 - 1 + 5 = 6$

18. $f(x) = -2x + 4$

$y = -2x + 4$

$m = -2 \qquad y\text{-intercept} = 4$

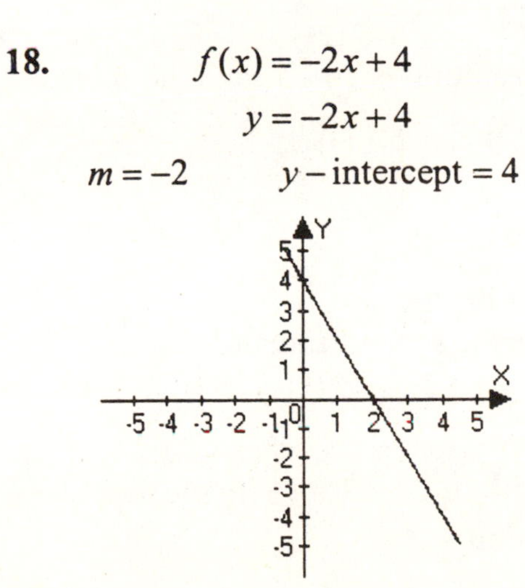

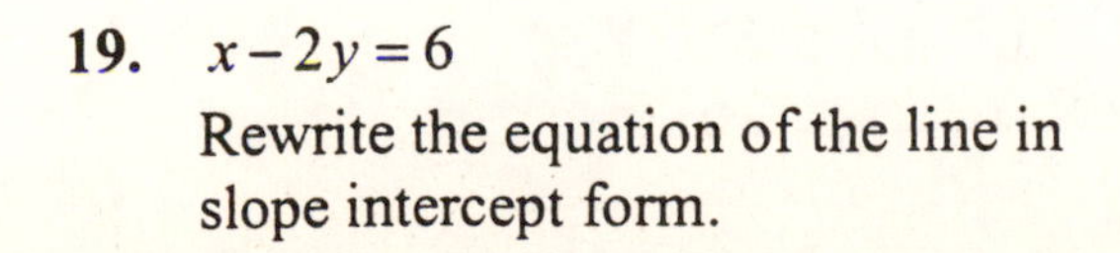

19. $x - 2y = 6$

Rewrite the equation of the line in slope intercept form.

$x - 2y = 6$

$-2y = -x + 6$

$y = \frac{1}{2}x - 3$

$m = \frac{1}{2} \qquad y\text{-intercept} = -3$

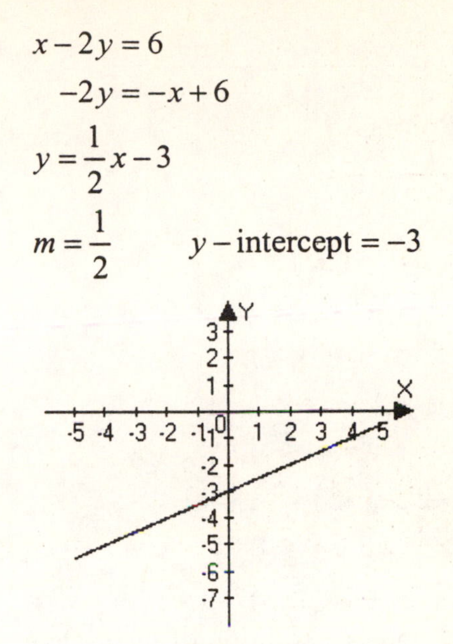

20. The line is parallel to $y = 4x + 7$, so the slope is 4. We are given that it passes through (3, –5). We use the slope and point to write the equation in point-slope form.

$y - y_1 = m(x - x_1)$

$y - (-5) = 4(x - 3)$

$y + 5 = 4(x - 3)$

Solve for y to obtain slope-intercept form.

$y + 5 = 4(x - 3)$

$y + 5 = 4x - 12$

$y = 4x - 17$

In function notation, the equation of the line is $f(x) = 4x - 17$.

Chapter 3

Check Points 3.1

1.

$$2x-4y=8$$
$$2(-2)-4(-3)=8$$
$$-4+12=8$$
$$8=8$$

$$2x-y=-1$$
$$2(-2)-(-3)=-1$$
$$-4+3=-1$$
$$-1=-1$$

The pair is a solution of the system.

2. Graph $y=-2x+6$ using the slope and y–intercept. The slope is –2 and the y–intercept is 6.

Graph $2x-y=-2$ using the intercepts. Find the x–intercept by setting $y=0$.

$$2x-0=-2$$
$$2x=-2$$
$$x=-1$$

Find the y–intercept by setting $x=0$.

$$2(0)-y=-2$$
$$-y=-2$$
$$y=2$$

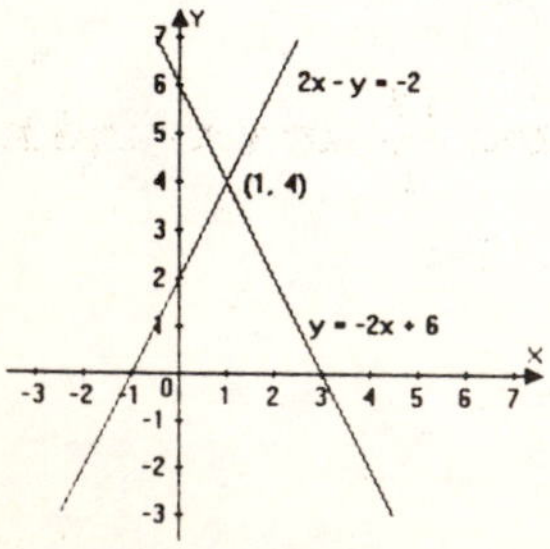

The solution is $(1,4)$ and the solution set is $\{(1,4)\}$.

3.

$$y=3x-7$$
$$5x-2y=8$$

Substitute $3x-7$ for y in the second equation to find x.

$$5x-2y=8$$
$$5x-2(3x-7)=8$$
$$5x-6x+14=8$$
$$-x+14=8$$
$$-x=-6$$
$$x=6$$

Back-substitute 6 for x to find y.

$$y=3x-7$$
$$y=3(6)-7$$
$$y=18-7$$
$$y=11$$

The solution is $(6,11)$ and the solution set is $\{(6,11)\}$.

4.

$$5x-4y=\ \ 9$$
$$x-2y=-3$$

Solve the second equation for x.

$$x-2y=-3$$
$$x=2y-3$$

Substitute $2y-3$ for x in the first equation.

$$5x-4y=9$$
$$5(2y-3)-4y=9$$
$$10y-15-4y=9$$

$6y-15=9$
$6y=24$
$y=4$
Back-substitute 4 for y to find x.
$x=2y-3$
$x=2(4)-3$
$x=8-3$
$x=5$
The solution is $(5,4)$ and the solution set is $\{(5,4)\}$.

5. $4x+5y=3$
$2x-3y=7$
Multiply the second equation by –2 and add to the first equation.
$$\begin{array}{r} 4x+5y=3 \\ -4x+6y=-14 \\ \hline 11y=-11 \\ y=-1 \end{array}$$
Back-substitute –1 for y to find x.
$4x+5y=3$
$4x+5(-1)=3$
$4x-5=3$
$4x=8$
$x=2$
The solution is $(2,-1)$ and the solution set is $\{(2,-1)\}$.

6. $3x=2-4y$
$5y=-1-2x$
Rewrite the equations in $Ax+By=C$ form.
$3x+4y=\ 2$
$2x+5y=-1$
Multiply the first equation by –2 and the second equation by 3.
$$\begin{array}{r} -6x-\ 8y=-4 \\ 6x+15y=-3 \\ \hline 7y=-7 \\ y=-1 \end{array}$$
Back-substitute –1 for y to find x.
$3x=2-4(-1)$
$3x=2+4$
$3x=6$
$x=2$
The solution is $(2,-1)$ and the solution set is $\{(2,-1)\}$.

7. $\frac{3x}{2}-2y=\frac{5}{2}$
$x-\frac{5y}{2}=-\frac{3}{2}$
Multiply both equations by 2 to clear fractions.
$3x-4y=\ 5$
$2x-5y=-3$
Multiply the first equation by –2 and the second equation by 3.
$$\begin{array}{r} -6x+\ 8y=-10 \\ 6x-15y=-9 \\ \hline -7y=-19 \\ y=\frac{19}{7} \end{array}$$
Back-substitute $\frac{19}{7}$ for y to find x.
$3x-4y=5$
$3x-4\left(\frac{19}{7}\right)=5$

$$3x-\frac{76}{7}=5$$
$$7(3x)-7\left(\frac{76}{7}\right)=7(5)$$
$$21x-76=35$$
$$21x=111$$
$$x=\frac{37}{7}$$

The solution is $\left(\frac{37}{7},\frac{19}{7}\right)$ and the solution set is $\left\{\left(\frac{37}{7},\frac{19}{7}\right)\right\}$.

8. $x+2y=4$
$3x+6y=13$
Multiply the first equation by –3.
$$-3x-6y=-12$$
$$3x+6y=13$$
$$0=1$$
There are no values of x and y for which $0=1$. There is no solution. The solution set is $\varnothing$ or $\{\ \}$.

9. $y=4x-4$
$8x-2y=8$
Substitute $4x-4$ for y to find x.
$$8x-2y=8$$
$$8x-2(4x-4)=8$$
$$8x-8x+8=8$$
$$8=8$$
Since the statement $8=8$ is true for all values of x and y, the solution set is $\{(x,y)|y=4x-4\}$ or $\{(x,y)|8x-2y=8\}$.

Problem Set 3.1

Practice Exercises

1.
$$x-y=12 \qquad x+y=2$$
$$7-(-5)=12 \qquad 7+(-5)=2$$
$$12=12 \qquad 2=2$$
The pair is a solution of the system.

3.
$$3x+4y=2 \qquad 2x+5y=1$$
$$3(2)+4(-1)=2 \qquad 2(2)+5(-1)=1$$
$$6-4=2 \qquad 4-5=1$$
$$2=2 \qquad -1=1$$
The pair is not a solution of the system.

5.
$$y=2x-13 \qquad 4x+9y=-7$$
$$-3=2(5)-13 \qquad 4(5)+9(-3)=-7$$
$$-3=10-13 \qquad 20-27=-7$$
$$-3=-3 \qquad -7=-7$$
The pair is a solution of the system.

7. $x+y=4$
Find the x–intercept by setting $y=0$.
$$x+0=4$$
$$x=4$$
Find the y–intercept by setting $x=0$.
$$0+y=4$$
$$y=4$$

$x-y=2$
Find the x–intercept by setting $y=0$.
$$x-0=2$$
$$x=2$$

Find the y–intercept by setting $x = 0$.

$$0 - y = 2$$
$$-y = 2$$
$$y = -2$$

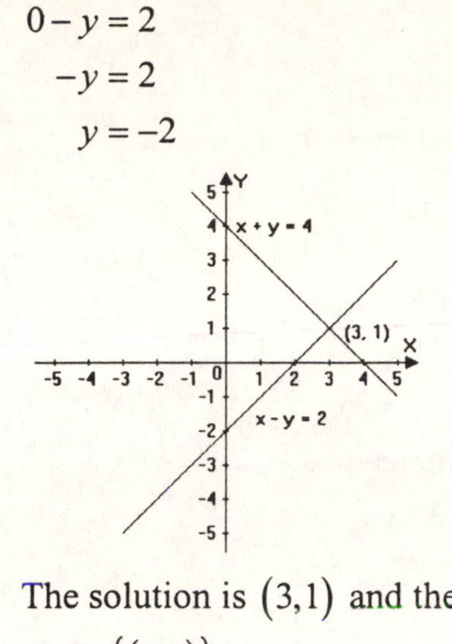

The solution is $(3,1)$ and the solution set is $\{(3,1)\}$.

9. $2x + y = 4$

Find the x–intercept by setting $y = 0$.

$$2x + 0 = 4$$
$$2x = 4$$
$$x = 2$$

Find the y–intercept by setting $x = 0$.

$$2(0) + y = 4$$
$$y = 4$$

$$y = 4x + 1$$

y – intercept $= 1$

slope $= 4$

The solution is $\left(\frac{1}{2}, 3\right)$ and the solution set is $\left\{\left(\frac{1}{2}, 3\right)\right\}$.

11. Solve both equations for y to obtain slopes and intercepts.

$$3x - 2y = 6 \qquad x - 4y = -8$$
$$-2y = -3x + 6 \qquad -4y = -x - 8$$
$$y = \frac{3}{2}x - 3 \qquad y = \frac{1}{4}x + 2$$

y – intercept $= -3$ $\qquad$ y – intercept $= 2$

slope $= \frac{3}{2}$ $\qquad$ slope $= \frac{1}{4}$

The solution is $(4,3)$ and the solution set is $\{(4,3)\}$.

13. Solve both equations for y to obtain slopes and intercepts.

$$2x + 3y = 6 \qquad 4x = -6y + 12$$
$$3y = -2x + 6 \qquad 6y = -4x + 12$$
$$y = -\frac{2}{3}x + 2 \qquad y = -\frac{4}{6}x + 2$$
$$y = -\frac{2}{3}x + 2$$

$y-\text{intercept}=2$ $\quad$ $y-\text{intercept}=2$

$\text{slope}=-\dfrac{2}{3}$ $\quad$ $\text{slope}=-\dfrac{2}{3}$

The lines coincide. The solution set is $\{(x,y)|2x+3y=6\}$ or $\{(x,y)|4x=-6y+12\}$.

15. Both equations are in slope–intercept form, so graph using slopes and intercepts.

$y=2x-2$ $\quad$ $y=-5x+5$

$y-\text{intercept}=-2$ $\quad$ $y-\text{intercept}=5$

$\text{slope}=2$ $\quad$ $\text{slope}=-5$

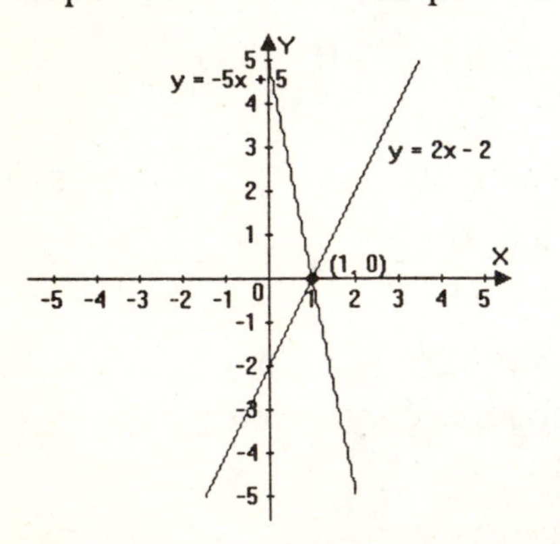

The solution is $(1,0)$ and the solution set is $\{(1,0)\}$.

17. Solve both equations for y to obtain slopes and intercepts.

$3x-y=4$

$-y=-3x+4$

$y=3x-4$

$y-\text{intercept}=-4$

$\text{slope}=3$

$6x-2y=4$

$-2y=-6x+4$

$y=3x-2$

$y-\text{intercept}=-2$

$\text{slope}=3$

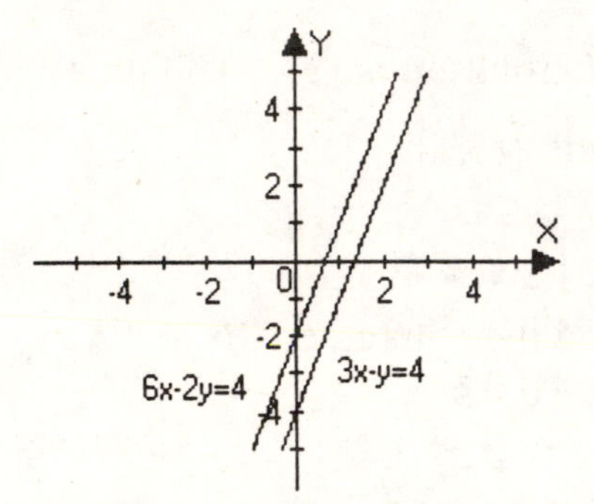

Since the lines do not intersect, there is no solution. The solution set is $\varnothing$ or $\{\ \}$.

19. Solve both equations for y to obtain slopes and intercepts.

$2x+y=4$

$y=-2x+4$

$y-\text{intercept}=4$

$\text{slope}=-2$

$4x+3y=10$

$3y=-4x+10$

$y=-\dfrac{4}{3}x+\dfrac{10}{3}$

$y-\text{intercept}=\frac{10}{3}=3\frac{1}{3}$

$\text{slope}=-\frac{4}{3}$

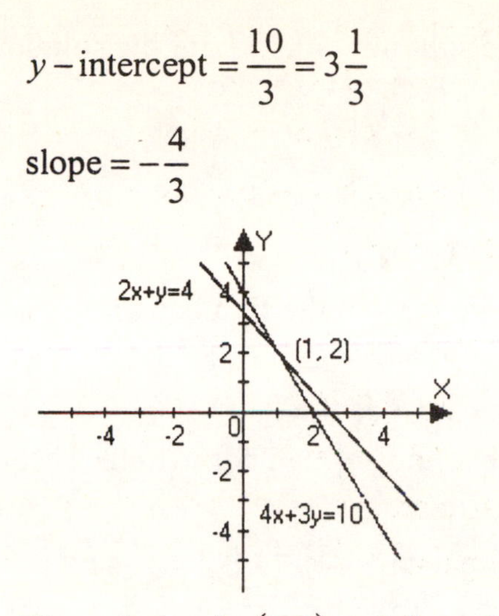

The solution is $(1,2)$ and the solution set is $\{(1,2)\}$.

21. Solve the first equation for y to obtain the slope and intercept.

$x-y=2$

$-y=-x+2$

$y=x-2$

$y-\text{intercept}=-2$

$\text{slope}=1$

$y=1$

This is a horizontal line passing through y–intercept, $(0,1)$.

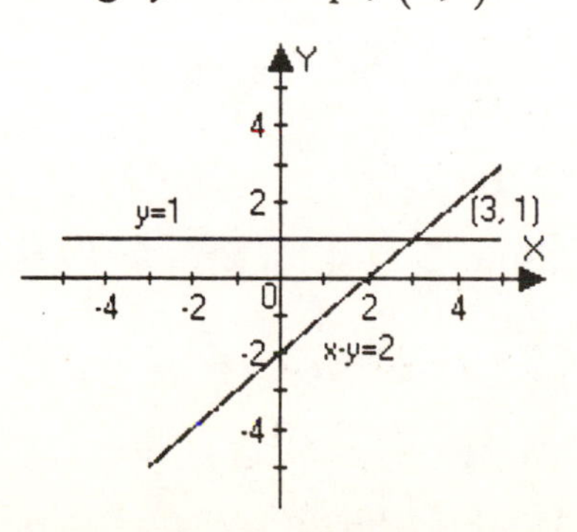

The solution is $(3,1)$ and the solution set is $\{(3,1)\}$.

23. Solve both equations for y to obtain slopes and intercepts.

$3x+y=3$

$y=-3x+3$

$y-\text{intercept}=3$

$\text{slope}=-1$

$6x+2y=12$

$2y=-6x+12$

$y=-3x+6$

$y-\text{intercept}=6$

$\text{slope}=-3$

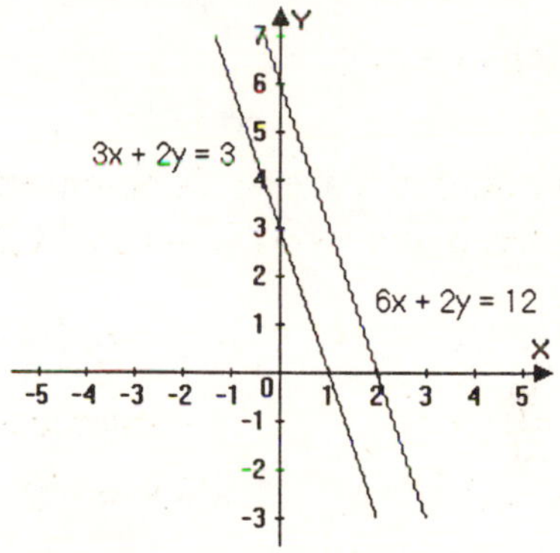

3x + 2y = 3

6x + 2y = 12

Since the lines do not intersect, there is no solution. The solution set is $\varnothing$ or $\{\ \}$.

25. $x+y=6$

$y=2x$

Substitute $2x$ for y in the first equation.

$x+y=6$

$x+2x=6$

$3x=6$

$x=2$

Back-substitute to find y.

$2+y=6$

$y=4$

The solution is $(2,4)$ and the solution set is $\{(2,4)\}$.

27. $2x+3y=9$

$x=y+2$

Substitute $y+2$ for x in the first equation.

$$2x+3y=9$$
$$2(y+2)+3y=9$$
$$2y+4+3y=9$$
$$5y+4=9$$
$$5y=5$$
$$y=1$$

Back-substitute to find x.

$$x=y+2$$
$$x=1+2$$
$$x=3$$

The solution is $(3,1)$ and the solution set is $\{(3,1)\}$.

29. $y=-3x+7$

$5x-2y=8$

Substitute $-3x+7$ for y in the second equation.

$$5x-2y=8$$
$$5x-2(-3x+7)=8$$
$$5x+6x-14=8$$
$$11x-14=8$$
$$11x=22$$
$$x=2$$

Back-substitute to find y.

$$y=-3(2)+7$$
$$y=-6+7$$
$$y=1$$

The solution is $(2,1)$ and the solution set is $\{(2,1)\}$.

31. $4x+y=5$

$2x-3y=13$

Solve for y in the first equation.

$$4x+y=5$$
$$y=-4x+5$$

Substitute $-4x+5$ for y in the second equation.

$$2x-3(-4x+5)=13$$
$$2x+12x-15=13$$
$$14x-15=13$$
$$14x=28$$
$$x=2$$

Back-substitute to find y.

$$4x+y=5$$
$$4(2)+y=5$$
$$8+y=5$$
$$y=-3$$

The solution is $(2,-3)$ and the solution set is $\{(2,-3)\}$.

33. $x-2y=4$

$2x-4y=5$

Solve for x in the first equation.

$$x-2y=4$$
$$x=2y+4$$

Substitute $2y+4$ for x in the second equation.

$$2x-4y=5$$
$$2(2y+4)-4y=5$$
$$4y+8-4y=5$$
$$4y+8-4y=5$$
$$8\neq 5$$

The system is inconsistent. There are no values of x and y for which 8 will equal 5. The solution set is $\varnothing$ or $\{\ \}$.

35. $2x+5y=-4$

$3x-y=11$

Solve for y in the second equation.

$$3x-y=11$$
$$-y=-3x+11$$
$$y=3x-11$$

Substitute $3x-11$ for y in the first equation.

$$2x+5y=-4$$
$$2x+5(3x-11)=-4$$
$$2x+15x-55=-4$$
$$17x-55=-4$$
$$17x=51$$
$$x=3$$

Back-substitute to find y.

$$3x-y=11$$
$$3(3)-y=11$$
$$9-y=11$$
$$-y=2$$
$$y=-2$$

The solution is $(3,-2)$ and the solution set is $\{(3,-2)\}$.

37. $2(x-1)-y=-3$

$y=2x+3$

Substitute $2x+3$ for y in the first equation.

$$2(x-1)-y=-3$$
$$2(x-1)-(2x+3)=-3$$
$$2x-2-2x-3=-3$$
$$-5\neq-3$$

Since there are no values of x and y for which –5 will equal –3, the system is inconsistent. The solution set is $\varnothing$ or $\{\ \}$.

39. $\frac{x}{4}-\frac{y}{4}=-1$

$x+4y=-9$

Solve for x in the second equation.

$$x+4y=-9$$
$$x=-4y-9$$

Substitute $-4y-9$ for x in the first equation.

$$\frac{x}{4}-\frac{y}{4}=-1$$
$$\frac{-4y-9}{4}-\frac{y}{4}=-1$$
$$4\left(\frac{-4y-9}{4}-\frac{y}{4}\right)=4(-1)$$
$$-4y-9-y=-4$$
$$-5y-9=-4$$
$$-5y=5$$
$$y=-1$$

Back-substitute to find x.

$$x+4y=-9$$
$$x+4(-1)=-9$$
$$x-4=-9$$
$$x=-5$$

The solution is $(-5,-1)$ and the solution set is $\{(-5,-1)\}$.

41. $y=\frac{2}{5}x-2$

$2x-5y=10$

Substitute $\frac{2}{5}x-2$ for y in the second equation.

$$2x-5y=10$$
$$2x-5\left(\frac{2}{5}x-2\right)=10$$
$$2x-2x+10=10$$
$$10=10$$

Since 10 = 10 for all values of x and y, the system is dependent. The solution set is $\left\{(x,y)\middle|\, y=\frac{2}{5}x-2\right\}$ or $\{(x,y)|\ 2x-5y=10\}$.

43. Solve by addition.

$$x+y=7$$
$$\underline{x-y=3}$$
$$2x=10$$
$$x=5$$

Back-substitute to find y.

$$x+y=7$$
$$5+y=7$$
$$y=2$$

The solution is $(5,2)$ and the solution set is $\{(5,2)\}$.

45. Solve by addition.

$$12x+3y=15$$
$$\underline{2x-3y=13}$$
$$14x=28$$
$$x=2$$

Back-substitute to find y.

$$12(2)+3y=15$$
$$24+3y=15$$
$$3y=-9$$
$$y=-3$$

The solution is $(2,-3)$ and the solution set is $\{(2,-3)\}$.

47.

$$x+3y=2$$
$$4x+5y=1$$

Multiply the first equation by –4.

$$-4x-12y=-8$$
$$\underline{4x+5y=1}$$
$$-7y=-7$$
$$y=1$$

Back-substitute to find x.

$$x+3y=2$$
$$x+3(1)=2$$
$$x+3=2$$
$$x=-1$$

The solution is $(-1,1)$ and the solution set is $\{(-1,1)\}$.

49.

$$6x-y=-5$$
$$4x-2y=6$$

Multiply the first equation by –2.

$$-12x+2y=10$$
$$\underline{4x-2y=6}$$
$$-8x=16$$
$$x=-2$$

Back-substitute to find y.

$$6(-2)-y=-5$$
$$-12-y=-5$$
$$-y=7$$
$$y=-7$$

The solution is $(-2,-7)$ and the solution set is $\{(-2,-7)\}$.

51. $3x-5y=11$

$2x-6y=2$

Multiply the first equation by –2 and the second equation by 3.

$$\begin{aligned} -6x+10y&=-22\\ 6x-18y&=\ \ 6\\ \hline -8y&=-16\\ y&=2 \end{aligned}$$

Back-substitute to find x.

$$\begin{aligned} 2x-6(2)&=2\\ 2x-12&=2\\ 2x&=14\\ x&=7 \end{aligned}$$

The solution is $(7,2)$ and the solution set is $\{(7,2)\}$.

53. $2x-5y=13$

$5x+3y=17$

Multiply the first equation by 3 and the second equation by 5.

$$\begin{aligned} 6x-15y&=39\\ 25x+15y&=85\\ \hline 31x&=124\\ x&=4 \end{aligned}$$

Back-substitute to find y.

$$\begin{aligned} 5(4)+3y&=17\\ 20+3y&=17\\ 3y&=-3\\ y&=-1 \end{aligned}$$

The solution is $(4,-1)$ and the solution set is $\{(4,-1)\}$.

55. $2x+6y=8$

$3x+9y=12$

Multiply the first equation by –3 and the second equation by 2.

$$\begin{aligned} -6x-18y&=-24\\ 6x+18y&=\ \ 24\\ \hline 0&=0 \end{aligned}$$

Since $0=0$ for all values of x and y, the system is dependent. The solution set is $\{(x,y)|\ 2x+6y=8\}$ or $\{(x,y)|\ 3x+9y=12\}$.

57. $2x-3y=4$

$4x+5y=3$

Multiply the first equation by –2.

$$\begin{aligned} -4x+6y&=-8\\ 4x+5y&=3\\ \hline 11y&=-5\\ y&=-\frac{5}{11} \end{aligned}$$

Back-substitute to find x.

$$\begin{aligned} 2x-3y&=4\\ 2x-3\left(-\frac{5}{11}\right)&=4\\ 2x+\frac{15}{11}&=4\\ 2x&=\frac{29}{11}\\ x&=\frac{29}{22} \end{aligned}$$

The solution is $\left(\frac{29}{22},-\frac{5}{11}\right)$ and the solution set is $\left\{\left(\frac{29}{22},-\frac{5}{11}\right)\right\}$.

59. $3x-7y=1$

$2x-3y=-1$

Multiply the first equation by –2 and the second equation by 3.

$$\begin{aligned} -6x+14y&=-2 \\ 6x-9y&=-3 \\ \hline 5y&=-5 \\ y&=-1 \end{aligned}$$

Back-substitute to find x.

$$\begin{aligned} 3x-7(-1)&=1 \\ 3x+7&=1 \\ 3x&=-6 \\ x&=-2 \end{aligned}$$

The solution is $(-2,-1)$ and the solution set is $\{(-2,-1)\}$.

61. $x=y+4$

$3x+7y=-18$

Substitute $y+4$ for x in the second equation.

$$\begin{aligned} 3x+7y&=-18 \\ 3(y+4)+7y&=-18 \\ 3y+12+7y&=-18 \\ 10y+12&=-18 \\ 10y&=-30 \\ y&=-3 \end{aligned}$$

Back-substitute to find x.

$$\begin{aligned} x&=y+4 \\ x&=-3+4 \\ x&=1 \end{aligned}$$

The solution is $(1,-3)$ and the solution set is $\{(1,-3)\}$.

63. $9x+\frac{4y}{3}=5$

$4x-\frac{y}{3}=5$

Multiply the second equation by 4.

$$\begin{aligned} 9x+\frac{4y}{3}&=5 \\ 16x-\frac{4y}{3}&=20 \\ \hline 25x&=25 \\ x&=1 \end{aligned}$$

Back-substitute to find y.

$$\begin{aligned} 4x-\frac{y}{3}&=5 \\ 4(1)-\frac{y}{3}&=5 \\ 4-\frac{y}{3}&=5 \\ -\frac{y}{3}&=1 \\ y&=-3 \end{aligned}$$

The solution is $(1,-3)$ and the solution set is $\{(1,-3)\}$.

65. $\frac{1}{4}x-\frac{1}{9}y=\frac{2}{3}$

$\frac{1}{2}x-\frac{1}{3}y=1$

Multiply the first equation by –2.

$$-2\left(\frac{1}{4}x\right)-(-2)\left(\frac{1}{9}y\right)=-2\left(\frac{2}{3}\right)$$

$$-\frac{1}{2}x+\frac{2}{9}y=-\frac{4}{3}$$

We now have a system of two equations in two variables.

$$\begin{aligned} -\frac{1}{2}x+\frac{2}{9}y&=-\frac{4}{3} \\ \frac{1}{2}x-\frac{1}{3}y&=1 \end{aligned}$$

Solve the addition.

$$\begin{aligned} -\frac{1}{2}x+\frac{2}{9}y &= -\frac{4}{3} \\ \frac{1}{2}x-\frac{1}{3}y &= 1 \\ \hline -\frac{1}{9}y &= -\frac{1}{3} \\ y &= -\frac{1}{3}(-9) \\ y &= 3 \end{aligned}$$

Back-substitute to find x.

$$\begin{aligned} \frac{1}{2}x-\frac{1}{3}y &= 1 \\ \frac{1}{2}x-\frac{1}{3}(3) &= 1 \\ \frac{1}{2}x-1 &= 1 \\ \frac{1}{2}x &= 2 \\ x &= 4 \end{aligned}$$

The solution is $(4,3)$ and the solution set is $\{(4,3)\}$.

67. $x=3y-1$

$2x-6y=-2$

Substitute $3y-1$ for x in the second equation.

$$\begin{aligned} 2x-6y &= -2 \\ 2(3y-1)-6y &= -2 \\ 6y-2-6y &= -2 \\ -2 &= -2 \end{aligned}$$

Since $-2=-2$ for all values of x and y, the system is dependent. The solution set is $\{(x,y)|\ x=3y-1\}$ or $\{(x,y)|\ 2x-6y=-2\}$.

69. $y=2x+1$

$y=2x-3$

Multiply the first equation by -1.

$$\begin{aligned} -y &= -2x-1 \\ y &= 2x-3 \\ \hline 0 &\neq -4 \end{aligned}$$

Since there are no values of x and y for which $0=-4$, the system is inconsistent. The solution set is $\varnothing$ or $\{\ \}$.

71. $0.4x+0.3y=2.3$

$0.2x-0.5y=0.5$

Multiply the second equation by -2.

$$\begin{aligned} 0.4x+0.3y &= 2.3 \\ -0.4x+1.0y &= -1.0 \\ \hline 1.3y &= 1.3 \\ y &= 1 \end{aligned}$$

Back-substitute to find x.

$$\begin{aligned} 0.2x-0.5y &= 0.5 \\ 0.2x-0.5(1) &= 0.5 \\ 0.2x-0.5 &= 0.5 \\ 0.2x &= 1.0 \\ x &= 5 \end{aligned}$$

The solution is $(5,1)$ and the solution set is $\{(5,1)\}$.

If you prefer to work with whole numbers and not decimals, you can multiple both of the original equations by 10. This will clear the equations of decimals. This is similar to the method we use to clear fractions.

73. $5x-40=6y$

$2y=8-3x$

Rewrite the equations in general form.

$5x-6y=40$

$3x+2y=8$

Multiply the second equation by 3.

$$\begin{array}{r} 5x-6y=40 \\ \underline{9x+6y=24} \\ 14x=64 \\ x=\frac{32}{7} \end{array}$$

Back-substitute to find y.

$2y=8-3x$

$2y=8-3\left(\frac{32}{7}\right)$

$2y=8-\frac{96}{7}$

$2y=-\frac{40}{7}$

$y=-\frac{40}{14}=-\frac{20}{7}$

The solution is $\left(\frac{32}{7},-\frac{20}{7}\right)$ and the solution set is $\left\{\left(\frac{32}{7},-\frac{20}{7}\right)\right\}$.

75. $3(x+y)=6$

$3(x-y)=-36$

Divide both equations by 3.

$x+y=\ \ 2$

$x-y=-12$

Solve the system by addition.

$$\begin{array}{r} x+y=\ \ 2 \\ \underline{x-y=-12} \\ 2x=-10 \\ x=-5 \end{array}$$

Back-substitute to find y.

$x+y=2$

$-5+y=2$

$y=7$

The solution is $(-5,7)$ and the solution set is $\{(-5,7)\}$.

77. $3(x-3)-2y=0$

$2(x-y)=-x-3$

Rewrite the equations in general form.

$3(x-3)-2y=0 \quad 2(x-y)=-x-3$

$3x-9-2y=0 \quad 2x-2y=-x-3$

$3x-2y=9 \quad 3x-2y=-3$

We now have a system of two equations in two variables.

$3x-2y=9$

$3x-2y=-3$

Multiply the second equation by -1.

$$\begin{array}{r} 3x-2y=9 \\ \underline{-3x+2y=3} \\ 0\neq 12 \end{array}$$

Since there are no values or x and y for which 0 = 12, the system is inconsistent. The solution set is $\varnothing$ or $\{\ \}$.

79. $x+2y-3=0$

$12=8y+4x$

Rewrite the equations in general form.

$x+2y-3=0$ $\quad$ $12=8y+4x$

$x+2y=3$ $\quad$ $8y+4x=12$

$4x+8y=12$

We now have a system of two equations in two variables.

$x+2y=3$

$4x+8y=12$

Multiply the first equation by -4.

$$-4x-8y=-12$$
$$4x+8y=12$$
$$0=0$$

Since $0 = 0$ for all values of x and y, the system is dependent. The solution set is $\{(x, y) | x+2y-3=0\}$ or $\{(x, y) | 12=8y+4x\}$.

81. $3x+4y=0$

$7x=3y$

After rewriting the second equation in general form, the system becomes:

$3x+4y=0$

$7x-3y=0$

Multiply the first equation by 3 and the second equation by 4.

$$9x+12y=0$$
$$28x-12y=0$$
$$37x=0$$
$$x=0$$

Back-substitute to find y.

$$7(0)=3y$$
$$0=3y$$
$$0=y$$

The solution is $(0,0)$ and the solution set is $\{(0,0)\}$.

Application Exercises

83. $y=-0.4x+28$

$0.07x+y=15$

Substitute $-0.4x+28$ for y in the second equation.

$$0.07x+y=15$$
$$0.07x+(-0.4x+28)=15$$
$$0.07x-0.4x+28=15$$
$$0.07x-0.4x+28=15$$
$$-0.33x+28=15$$
$$-0.33x=-13$$
$$x=39.\overline{39}$$
$$x\approx 39$$

Back-substitute to find y.

$$y=-0.4(39)+28$$
$$y=-15.6+28$$
$$y=12.4$$

The solution set is $\{(39,12.4)\}$. The number of deaths from car accidents and the number of deaths from gunfire will be the same in 1965 + 39 = 2004. The number of deaths caused by each in that year will be 12.4 per 100,000 persons. This result is demonstrated by the regression line in the figure. The lines intersect at the point year $(39,12.4)$.

85. $M=-0.41x+22$

$M=-0.18x+10$

Multiply the second equation by -1 and solve by addition.

$$\begin{aligned} M &= -0.41x + 22 \\ -M &= 0.18x - 10 \\ \hline 0 &= -0.23x + 12 \\ -12 &= -0.23x \end{aligned}$$

$52.1729 = x$
Back-substitute to find y.
$M = -0.41x + 22$

$M = -0.41(52.1729) + 22$

$M = -21.39 + 22$

$M \approx .61$
The solution is approximately $(52, 0.61)$. The infant mortality rate for blacks and whites will be the same in the year 1980 + 52 = 2032. The infant mortality rate at that time will be 0.61 deaths per 1000 live births.

87. **a.** Demand

$N = -13p + 760$

$N = -13(18) + 760$

$N = -234 + 760$

$N = 526$
Supply

$N = 2p + 430$

$N = 2(18) + 430$

$N = 36 + 430$

$N = 466$
526 videos are sold and 466 are sold at $18 per copy.

b. $N = -13p + 760$

$N = 2p + 430$
Multiply the second equation by –1 and solve by addition.

$$\begin{aligned} N &= -13p + 760 \\ -N &= -2p - 430 \\ \hline 0 &= -15p + 330 \\ -330 &= -15p \\ 22 &= p \end{aligned}$$

Back-substitute to find N.
$N = 2p + 430$

$N = 2(22) + 430$

$N = 44 + 430$

$N = 474$
At $22, the supply and demand will be equal. At this price, 474 copies can be supplied and sold each week.

Writing in Mathematics

89. Answers will vary.

91. Answers will vary.

93. Answers will vary.

95. Answers will vary.

97. Answers will vary.

99. Answers will vary.

Technology Exercises

101. Answers will vary. For example, consider Exercise 7.
$x + y = 4$
$x - y = 2$

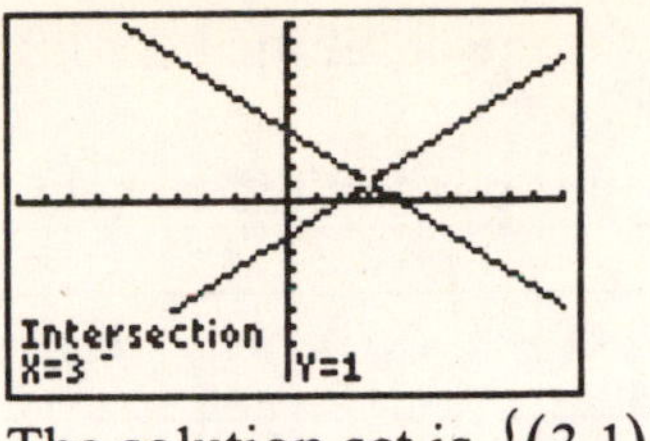

The solution set is $\{(3,1)\}$.

Critical Thinking Exercises

103. Answers will vary. One system of equations with solution set $\{(-2,7)\}$ is:

$$\begin{aligned} 3x+4y&=22 \\ -x-y&=-5. \end{aligned}$$

Review Exercises

105.
$$\begin{aligned} 6x&=10+5(3x-4) \\ 6x&=10+15x-20 \\ 6x&=15x-10 \\ -9x&=-10 \\ x&=\frac{10}{9} \end{aligned}$$

The solution is $\frac{10}{9}$ and the solution set is $\left\{\frac{10}{9}\right\}$.

106.
$$\begin{aligned} &\left(4x^2y^4\right)^2\left(-2x^5y^0\right)^3 \\ &=4^2x^{2\cdot 2}y^{4\cdot 2}(-2)^3x^{5\cdot 3}y^{0\cdot 3} \\ &=4^2x^4y^8(-2)^3x^{15}y^0 \\ &=16(-8)x^4x^{15}y^8y^0 \\ &=-128x^{19}y^8 \end{aligned}$$

107.
$$\begin{aligned} f(x)&=x^2-3x+7 \\ f(-1)&=(-1)^2-3(-1)+7 \\ &=1+3+7 \\ &=11 \end{aligned}$$

Check Points 3.2

1. Let x = the calorie content of a pan pizza
Let y = the calorie content of a beef burrito

$$\begin{aligned} x+2y&=1980 \\ 2x+\ \ y&=2670 \end{aligned}$$

Multiply the first equation by –2 and solve by addition.

$$\begin{aligned} -2x-4y&=-3960 \\ 2x+\ \ y&=\ \ 2670 \\ \hline -3y&=-1290 \\ y&=430 \end{aligned}$$

Back-substitute 430 for y to find x.

$$\begin{aligned} x+2y&=1980 \\ x+2(430)&=1980 \\ x+860&=1980 \\ x&=1120 \end{aligned}$$

There are 1120 calories in a pan pizza and 430 calories in a beef burrito.

2. Let x = the amount invested in a stock paying 6%
Let y = the amount invested in a stock paying 8%

	Principal	x Rate =	Interest
6%	x	0.06	$0.06x$
8%	y	0.08	$0.08y$

$$\begin{aligned} x+\qquad y&=16000 \\ 0.06x+0.08y&=\ 1180 \end{aligned}$$

Multiply the first equation by –0.06 and add to the second equation.

$$\begin{array}{r} -0.06x - 0.06y = -960 \\ 0.06x + 0.08y = 1180 \\ \hline 0.02y = 220 \\ y = 11000 \end{array}$$

Back-substitute 11000 for y to find x.

$$x + y = 16000$$
$$x + 11000 = 16000$$
$$x = 5000$$

You should invest \$5000 at 6% and \$11000 at 8%.

3. Let x = the amount of 18% acid solution

Let y = the amount of 45% acid solution

	# of ml's	× %	= Amount
18%	x	0.18	$0.18x$
45%	y	0.45	$0.45y$
36%	12	0.36	0.36(12)

$$x + y = 12$$
$$0.18x + 0.45y = 4.32$$

Multiply the first equation by –0.18 and add to the second equation.

$$\begin{array}{r} -0.18x - 0.18y = -2.16 \\ 0.18x + 0.45y = 4.32 \\ \hline 0.27y = 2.16 \\ y = 8 \end{array}$$

Back-substitute 8 for y to find x.

$$x + y = 12$$
$$x + 8 = 12$$
$$x = 4$$

The chemist must use 4 liters of the 18% solution and 8 liters of the 45% solution.

4. Let x = the speed of the boat in still water

Let y = the speed of the current

	r ×	t	= d
With Current	$x + y$	2	$2(x + y)$
Against the Current	$x - y$	3	$3(x - y)$

$$2(x + y) = 84$$
$$3(x - y) = 84$$

Rewrite the system in $Ax + By = C$ form.

$$2x + 2y = 84$$
$$3x - 3y = 84$$

Multiply the first equation by –3 and the second equation by 2.

$$\begin{array}{r} -6x - 6y = -252 \\ 6x - 6y = 168 \\ \hline -12y = -84 \\ y = 7 \end{array}$$

Back-substitute 7 for y to find x.

$$2x + 2y = 84$$
$$2x + 2(7) = 84$$
$$2x + 14 = 84$$
$$2x = 70$$
$$x = 35$$

The speed of the boat in still water is 35 miles per hour and the speed of the current is 7 miles per hour.

5.

$$\begin{aligned} P(x) &= R(x) - C(x) \\ &= 200x - (160{,}000 + 75x) \\ &= 200x - 160{,}000 - 75x \\ &= 125x - 160{,}000 \end{aligned}$$

The profit function is $P(x) = 125x - 160{,}000$.

6. **a.** $C(x) = 300,000 + 30x$

b. $R(x) = 80x$

c. $R(x) = C(x)$

$$80x = 300,000 + 30x$$
$$50x = 300,000$$
$$x = 6,000$$

Back-substitute 6000 for x in either the cost or the revenue function, we obtain:

$$R(x) = 80x$$
$$R(6000) = 80(6000)$$
$$= 480,000$$

The break-even point is (6,000, 480,000). The cost and revenue will be the same at $480,000, when 6,000 pairs of shoes are produced and sold.

Problem Set 3.2

Practice Exercises

1. Let x = the first number
Let y = the second number

$$x + y = 7$$
$$x - y = -1$$

Solve by addition:

$$x + y = 7$$
$$\underline{x - y = -1}$$
$$2x = 6$$
$$x = 3$$

Back-substitute to find y.

$$3 + y = 7$$
$$y = 4$$

The numbers are 3 and 4.

3. Let x = the first number
Let y = the second number

$$3x - y = 1$$
$$x + 2y = 12$$

Multiply the first equation by 2.

$$6x - 2y = 2$$
$$\underline{x + 2y = 12}$$
$$7x = 14$$
$$x = 2$$

Back-substitute to find y.

$$x + 2y = 12$$
$$2 + 2y = 12$$
$$2y = 10$$
$$y = 5$$

The numbers are 2 and 5.

5. **a.** $P(x) = R(x) - C(x)$

$$= (32x) - (25,500 + 15x)$$
$$= 32x - 25,500 - 15x$$
$$= 17x - 25,500$$

b. At the break-even point, $R(x) = C(x)$.

$$25500 + 15x = 32x$$
$$25500 = 17x$$
$$1500 = x$$
$$C(x) = 32x$$
$$C(1500) = 32(1500)$$
$$= 48000$$

Fifteen hundred units must be produced and sold to break-even. At this point, there will $48,000 in costs and revenue.

7. **a.** $P(x) = R(x) - C(x)$

$= (245x) - (105x + 70,000)$

$= 245x - 105x - 70,000$

$= 140x - 70,000$

b. At the break-even point, $R(x) = C(x)$.

$105x + 70000 = 245x$

$70000 = 140x$

$500 = x$

$C(x) = 245x$

$C(500) = 245(500)$

$= 122500$

Five hundred units must be produced and sold to break-even. At this point, there will $122,500 in costs and revenue.

Application Exercises

9. Let x = the number (in millions) of divorced Americans age 40-44
Let y = the number (in millions) of divorced Americans age 45-54

$x + y = 8.5$

$y - x = 2.3$

Rewrite the second equation and solve by addition:

$$\begin{aligned} x + y &= 8.5 \\ -x + y &= 2.3 \\ \hline 2y &= 10.8 \\ y &= 5.4 \end{aligned}$$

Back-substitute to find x.

$x + y = 8.5$

$x + 5.4 = 8.5$

$x = 3.1$

There are 3.1 million divorced Americans age 40-44 and 5.4 million divorced Americans age 45-54.

11. Let x = the cholesterol content of sponge cake
Let y = the cholesterol content of pound cake

$4x + 2y = 784$

$x + 3y = 366$

Multiply the second equation by –4 and solve by addition.

$$\begin{aligned} 4x + 2y &= 784 \\ -4x - 12y &= -1464 \\ \hline -10y &= -680 \\ y &= 68 \end{aligned}$$

Back-substitute to find x.

$x + 3y = 366$

$x + 3(68) = 366$

$x + 204 = 366$

$x + 204 = 366$

$x = 162$

The cholesterol content of sponge cake is 162 mg per slice. The cholesterol content of pound cake is 68 mg per slice.

13. Let x = the amount invested at 7%
Let y = the amount invested at 9%

$x + y = 20000$

$0.07x + 0.09y = 1550$

Solve the first equation for x.

$x + y = 20000$

$x = 20000 - y$

Substitute $20000 - y$ for x in the second equation.

$0.07(20000 - y) + 0.09y = 1550$

$1400 - 0.07y + 0.09y = 1550$

$$1400 + 0.02y = 1550$$
$$0.02y = 150$$
$$0.02y = 150$$
$$y = 7500$$

Back-substitute to find x.

$$x + y = 20000$$
$$x + 7500 = 20000$$
$$x = 12500$$

\$12,500 was invested at 7% and \$7,500 was invested at 9%.

15. Let x = the amount loaned at 8%
Let y = the amount loaned at 18%

$$x + y = 250000$$
$$0.08x + 0.18y = 23000$$

Solve the first equation for x.

$$x + y = 250000$$
$$x = 250000 - y$$

Substitute $250000 - y$ for x in the second equation.

$$0.08(250000 - y) + 0.18y = 23000$$
$$20000 - 0.08y + 0.18y = 23000$$
$$20000 + 0.10y = 23000$$
$$0.10y = 3000$$
$$y = 30000$$

Back-substitute to find x.

$$x + 30000 = 250000$$
$$x = 220000$$

\$220,000 was loaned at 8% annual mortgage interest and \$30,000 was loaned at 18% annual credit card interest.

17. Let x = the amount invested at 12%
Let y = the amount which lost 5%

$$x + y = 8000$$
$$0.12x - 0.05y = 620$$

Solve the first equation for x.

$$x + y = 8000$$
$$x = 8000 - y$$

Substitute $8000 - y$ for x in the second equation.

$$0.12(8000 - y) - 0.05y = 620$$
$$960 - 0.12y - 0.05y = 620$$
$$960 - 0.17y = 620$$
$$-0.17y = -340$$
$$y = 2000$$

Back-substitute to find x.

$$x + y = 8000$$
$$x + 2000 = 8000$$
$$x = 6000$$

\$6,000 was invested at 12% and \$2,000 was invested at a 5% loss.

19. Let x = the number of grams of 45% fat content cheese
Let y = the number of grams of 20% fat content cheese

$$x + \quad y = 30$$
$$0.45x + 0.20y = 0.30(30)$$

$$x + \quad y = 30$$
$$0.45x + 0.20y = 9$$

Solve the first equation for x.

$$x + y = 30$$
$$x = 30 - y$$

Substitute $30 - y$ for x in the second equation.

$$0.45x + 0.20y = 9$$
$$0.45(30 - y) + 0.20y = 9$$
$$13.5 - 0.25y = 9$$
$$-0.25y = -4.5$$
$$-0.25y = -4.5$$
$$y = 18$$

Back-substitute to find x.

$x+18=30$

$x=12$

Mix 12 grams of a 45% fat content cheese with 18 grams of a 20% fat content cheese.

21. Let x = the number of students at north campus before the merger
Let y = the number of students at south campus before the merger

$$\begin{aligned} x+\quad y&=1200\\ 0.10x+0.50y&=0.40(1200)\end{aligned}$$

$$\begin{aligned} x+\quad y&=1200\\ 0.10x+0.50y&=\ 480\end{aligned}$$

Solve the first equation for x.

$x+y=1200$

$x=1200-y$

Substitute $1200-y$ for x in the second equation.

$$0.10(1200-y)+0.50y=480$$
$$120-0.10y+0.50y=480$$
$$120+0.40y=480$$
$$0.40y=360$$
$$0.40y=360$$
$$y=900$$

Back-substitute to find x.

$x+y=1200$

$x+900=1200$

$x=300$

There were 300 students at north campus and 900 students at south campus before the merger.

23. Let x = the amount of \$6.00 per pound tea
Let y = the amount of \$8.00 per pound tea

$$\begin{aligned} x+\ y&=144\\ 6x+8y&=7.50(144)\end{aligned}$$

$$\begin{aligned} x+\ y&=144\\ 6x+8y&=1080\end{aligned}$$

Solve the first equation for x.

$x+y=144$

$x=144-y$

Substitute $144-y$ for x in the second equation.

$$6(144-y)+8y=1080$$
$$864-6y+8y=1080$$
$$864+2y=1080$$
$$2y=216$$
$$y=108$$

Back-substitute to find x.

$x+y=144$

$x+108=144$

$x=36$

Mix 36 pounds of the \$6.00 per pound tea with 108 pounds of the \$8.00 per pound tea.

25. Let x = the speed of the plane in still air
Let y = the speed of the wind

	r ×	t =	d
With Wind	$x+y$	5	800
Against Wind	$x-y$	8	800

$$\begin{aligned} 5(x+y)&=800\\ 8(x-y)&=800\end{aligned}$$

$$\begin{aligned} 5x+5y&=800\\ 8x-8y&=800\end{aligned}$$

Multiply the first equation by 8 and the second equation by 5.

$$40x + 40y = 6400$$
$$\underline{40x - 40y = 4000}$$
$$80x = 10400$$
$$x = 130$$

Back-substitute to find y.

$$5x + 5y = 800$$
$$5(130) + 5y = 800$$
$$650 + 5y = 800$$
$$5y = 150$$
$$y = 30$$

The speed of the plane in still air is 130 miles per hour and the speed of the wind is 30 miles per hour.

27. Let x = the crew's rowing rate
Let y = the rate of the current

	r ×	t =	d
With Current	$x + y$	2	16
Against Current	$x - y$	4	16

$$2(x + y) = 16$$
$$4(x - y) = 16$$

Rewrite the system is $Ax + By = C$ form.

$$2x + 2y = 16$$
$$4x - 4y = 16$$

Multiply the first equation by –2.

$$-4x - 4y = -32$$
$$\underline{4x - 4y = \ \ 16}$$
$$-8y = -16$$
$$y = 2$$

Back-substitute to find x.

$$2x + 2(2) = 16$$
$$2x + 4 = 16$$
$$2x = 12$$
$$x = 6$$

The crew's rowing rate is 6 kilometers per hour and the rate of the current is 2 kilometers per hour.

29. Let x = the speed in still water
Let y = the speed of the current

	r ·	t =	d
With Current	$x + y$	4	24
Against Current	$x - y$	6	$\frac{3}{4}(24)$

$$4(x + y) = 24$$
$$6(x - y) = \frac{3}{4}(24)$$

Rewrite the system is $Ax + By = C$ form.

$$4x + 4y = 24$$
$$6x - 6y = 18$$

Multiply the first equation by –3 and the second equation by 2.

$$-12x - 12y = -72$$
$$\underline{12x - 12y = 36}$$
$$-24y = -36$$
$$y = 1.5$$

Back-substitute to find x.

$$4x + 4y = 24$$
$$4x + 4(1.5) = 24$$
$$4x + 6 = 24$$
$$4x = 18$$
$$x = 4.5$$

The speed in still water is 4.5 miles per hour and the speed of the current is 1.5 miles per hour.

31. Let x = the length of the rectangular lot
Let y = the width of the rectangular lot

$$\begin{aligned} 2x + \ \ 2y &= 360 \\ 20x + 8(2y) &= 3280 \end{aligned}$$

$$\begin{aligned} 2x + \ \ 2y &= 360 \\ 20x + \ \ 16y &= 3280 \end{aligned}$$

Multiply the first equation by –8.

$$\begin{aligned} -16x - 16y &= -2880 \\ 20x + 16y &= 3280 \\ \hline 4x &= 400 \\ x &= 100 \end{aligned}$$

Back-substitute to find y.

$$\begin{aligned} 2x + 2y &= 360 \\ 2(100) + 2y &= 360 \\ 200 + 2y &= 360 \\ 2y &= 160 \\ y &= 80 \end{aligned}$$

The length is 100 feet and the width is 80 feet.

33. Let x = the number of rooms with kitchen
Let y = the number of rooms without kitchen

$$\begin{aligned} x + \ \ \ y &= 200 \\ 100x + 80y &= 17000 \end{aligned}$$

Multiply the first equation by –80.

$$\begin{aligned} -80x - 80y &= -16000 \\ 100x + 80y &= 17000 \\ \hline 20x &= 1000 \\ x &= 50 \end{aligned}$$

Back-substitute to find y.

$$\begin{aligned} x + y &= 200 \\ 50 + y &= 200 \\ y &= 150 \end{aligned}$$

There are 50 rooms with kitchen facilities and 150 rooms without kitchen facilities.

35. Let x = the number of apples
Let y = the number of avocados

$$\begin{aligned} 100x + 350y &= 1000 \\ 24x + \ \ 14y &= \ \ 100 \end{aligned}$$

Multiply the first equation by –24 and the second equation by 100.

$$\begin{aligned} -2400x - 8400y &= -24000 \\ 2400x + 1400y &= \ \ 10000 \\ \hline -7000y &= -14000 \\ y &= 2 \end{aligned}$$

Back-substitute to find x.

$$\begin{aligned} 24x + 14y &= 100 \\ 24x + 14(2) &= 100 \\ 24x + 28 &= 100 \\ 24x &= 72 \\ x &= 3 \end{aligned}$$

Three apples and 2 avocados should be eaten.

37. At the break-even point,
$R(x) = C(x)$.

$$\begin{aligned} 10000 + 30x &= 50x \\ 10000 &= 20x \\ 10000 &= 20x \\ 500 &= x \end{aligned}$$

Five hundred radios must be produced and sold to break-even.

39.
$$\begin{aligned} R(x) &= 50x \\ R(200) &= 50(200) \\ &= 10000 \\ C(x) &= 10000 + 30x \\ C(200) &= 10000 + 30(200) \\ &= 10000 + 6000 \\ &= 16000 \end{aligned}$$

$$R(200) - C(200) = 10000 - 16000$$
$$= -6000$$

This means that if 200 radios are produced and sold the company will lose \$6,000.

41. $P(x) = R(x) - C(x)$
$$= 50x - (10000 + 30x)$$
$$= 50x - 10000 - 30x$$
$$= 20x - 10000$$
$$P(x) = 20x - 10000$$
$$P(10000) = 20(10000) - 10000$$
$$= 200000 - 10000$$
$$= 190000$$

If 10,0000 radios are produced and sold the profit will be \$190,000.

43. **a.** The cost function is:
$$C(x) = 18,000 + 20x$$

b. The revenue function is:
$$R(x) = 80x$$

c. At the break-even point, $R(x) = C(x)$.
$$80x = 18000 + 20x$$
$$60x = 18000$$
$$x = 300$$
$$R(x) = 80x$$
$$R(300) = 80(300)$$
$$= 24,000$$

When approximately 300 canoes are produced the company will break-even with cost and revenue at \$24,000.

45. **a.** The cost function is:
$$C(x) = 30000 + 2500x$$

b. The revenue function is:
$$R(x) = 3125x$$

c. At the break-even point, $R(x) = C(x)$.
$$3125x = 30000 + 2500x$$
$$625x = 30000$$
$$x = 48$$

After 48 sold out performances, the investor will break-even.

Writing in Mathematics

47. Answers will vary.

49. Answers will vary.

51. Answers will vary.

53. Answers will vary.

Technology Exercises

55. $R(x) = 50x$
$$C(x) = 20x + 180$$

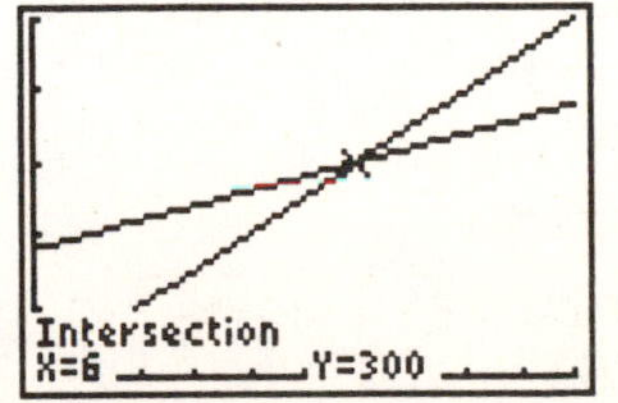

The intersection point, (6, 300), is the break-even point. This means that when 6 units are produced and sold, costs and revenue are the same at approximately \$300.

57. Answers will vary. For example, consider Exercise 43.

$R(x) = 80x$

$C(x) = 20000 + 20x$

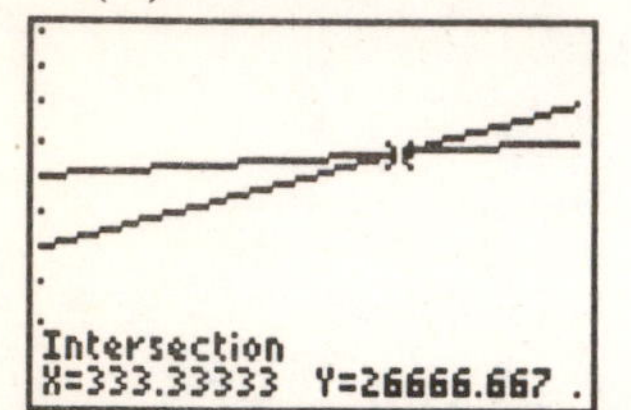

The break-even point is approximately 333 units. When 333 units are produced and sold, costs and revenue are the same at approximately \$26,666. This is the same result obtained in Exercise 43.

Critical Thinking Exercises

59.
$$x + y = 90$$
$$x + (3y + 20) = 180$$

Simplify the second equation.

$$x + (3y + 20) = 180$$
$$x + 3y + 20 = 180$$
$$x + 3y = 160$$

The system becomes:

$$x + y = 90$$
$$x + 3y = 160$$

Multiply the first equation by -1.

$$-x - y = -90$$
$$x + 3y = 160$$
$$2y = 70$$
$$y = 35$$

Back-substitute to find x.

$$x + 35 = 90$$
$$x = 55$$

The measure of $\angle x$ is 55° and the measure of $\angle y$ is 35°.

61. Let x = the cost of the mangos
Let y = the cost of the avocados

$$x + y = 67$$
$$0.20x - 0.02y = 8.56$$

Multiply the first equation by 0.02.

$$0.02x + 0.02y = 1.34$$
$$0.20x - 0.02y = 8.56$$
$$0.22x = 9.90$$
$$x = 45$$

Back-substitute to find y.

$$x + y = 67$$
$$45 + y = 67$$
$$y = 22$$

The dealer paid \$45 for the mangos and \$22 for the avocados.

Review Exercises

62. Passing through $(-2,5)$ and $(-6,13)$

First, find the slope:

$$m = \frac{y_2 - y_1}{x_2 - x_1}$$

$$m = \frac{13-5}{-6-(-2)} = \frac{8}{-6+2} = \frac{8}{-4} = -2$$

Use the slope and one of the points to write the equation in point-slope form.

$$y - y_1 = m(x - x_1)$$
$$y - 5 = -2(x - (-2))$$
$$y - 5 = -2(x + 2)$$

or

$$y - y_1 = m(x - x_1)$$
$$y - 13 = -2(x - (-6))$$
$$y - 13 = -2(x + 6)$$

Rewrite the equation in slope-intercept form by solving for y.

$$y-5=-2(x+2)$$
$$y-5=-2x-4$$
$$y=-2x+1$$

In function notation, the equation of the line is $f(x)=-2x+1$.

63. Since the line is parallel to $-x+y=7$, we can use it to obtain the slope. Rewriting the equation in slope-intercept form, we obtain $y=x+7$. The slope is $m=1$. We are given that it passes through (–3, 0). We use the slope and point to write the equation in point-slope form.

$$y-y_1=m(x-x_1)$$
$$y-0=1(x-(-3))$$
$$y-0=1(x+3)$$

Rewrite the equation in slope-intercept form by solving for *y*.

$$y-0=1(x+3)$$
$$y=x+3$$

In function notation, the equation of the line is $f(x)=x+3$.

64. Since the denominator of a fraction cannot be zero, the domain of $g=\{x|x \text{ is a real number and } x\neq 3\}$.

Check Points 3.3

1. Check by substituting the triple into each of the equations.

$(-1,-4,5)$

$$x-2y+3z=22$$
$$-1-2(-4)+3(5)=22$$
$$-1+8+15=22$$
$$22=22$$

True

$$2x-3y-z=5$$
$$2(-1)-3(-4)-5=5$$
$$-2+12-5=5$$
$$5=5$$

True

$$3x+y-5z=-32$$
$$3(-1)+(-4)-5(5)=-32$$
$$-3-4-25=-32$$
$$-32=-32$$

True

The ordered triple is a solution of the system.

2.

$$x+4y-z=20$$
$$3x+2y+z=8$$
$$2x-3y+2z=-16$$

Add the first and second equations to eliminate *z*.

$$x+4y-z=20$$
$$3x+2y+z=8$$
$$4x+6y=28$$

Multiply the first equation by 2 and add to the third equation.

$$2x+8y-2z=40$$
$$2x-3y+2z=-16$$
$$4x+5y=24$$

The system of two equations in two variables becomes:

$4x+6y=28$

$4x+5y=24$

Multiply the first equation by –1 and add.

$$\begin{aligned} -4x-6y&=-28 \\ 4x+5y&=24 \\ \hline -y&=-4 \\ y&=4 \end{aligned}$$

Back-substitute 4 for y to find x.

$$\begin{aligned} 4x+6y&=28 \\ 4x+6(4)&=28 \\ 4x+24&=28 \\ 4x&=4 \\ x&=1 \end{aligned}$$

Back-substitute 1 for x and 4 for y to find z.

$$\begin{aligned} x+4y-z&=20 \\ 1+4(4)-z&=20 \\ 1+16-z&=20 \\ 17-z&=20 \\ -z&=3 \\ z&=-3 \end{aligned}$$

The solution is $(1,4,-3)$ and the solution set is $\{(1,4,-3)\}$.

3. $$\begin{aligned} 2y-z&=7 \\ x+2y+z&=17 \\ 2x-3y+2z&=-1 \end{aligned}$$

Multiply the second equation by –2 and add to the third equation and solve by addition.

$$\begin{aligned} -2x-4y-2z&=-34 \\ 2x-3y+2z&=-1 \\ \hline -7y&=-35 \\ y&=5 \end{aligned}$$

Substitute 6 for y in the first equation to find z.

$$\begin{aligned} 2y-z&=7 \\ 2(5)-z&=7 \\ 10-z&=7 \\ -z&=-3 \\ z&=3 \end{aligned}$$

Back-substitute 6 for y and 5 for z to find x.

$$\begin{aligned} x+2y+z&=17 \\ x+2(5)+3&=17 \\ x+10+3&=17 \\ x+13&=17 \\ x&=4 \end{aligned}$$

The solution is $(4,5,3)$ and the solution set is $\{(4,5,3)\}$.

4. Using the function $y=ax^2+bx+c$ and the three points, we obtain three equations in three variables.

When $x=1$ and $y=4$,

$$\begin{aligned} y&=ax^2+bx+c \\ 4&=a(1)^2+b(1)+c \\ 4&=a(1)+b+c \\ 4&=a+b+c \end{aligned}$$

or $a+b+c=4$

When $x = 2$ and $y = 1$,

$y = ax^2 + bx + c$

$1 = a(2)^2 + b(2) + c$

$1 = a(4) + 2b + c$

$1 = 4a + 2b + c$

or $4a + 2b + c = 1$

When $x = 3$ and $y = 4$,

$y = ax^2 + bx + c$

$4 = a(3)^2 + b(3) + c$

$4 = a(9) + 3b + c$

$4 = 9a + 3b + c$

or $9a + 3b + c = 4$

The system of three equation in three variables becomes:

$a + b + c = 4$

$4a + 2b + c = 1$

$9a + 3b + c = 4$

Multiply the first equation by –1 and add to the second equation to eliminate c.

$-a - b - c = -4$

$4a + 2b + c = 1$

$3a + b = -3$

Multiply the first equation by –1 and add to the third equation to eliminate c.

$-a - b - c = -4$

$9a + 3b + c = 4$

$8a + 2b = 0$

The system of two equations in two variables becomes:

$3a + b = -3$

$8a + 2b = 0$

Solve the first equation for b.

$3a + b = -3$

$b = -3a - 3$

Substitute $-3a - 3$ for b in the second equation.

$8a + 2b = 0$

$8a + 2(-3a - 3) = 0$

$8a - 6a - 6 = 0$

$2a - 6 = 0$

$2a = 6$

$a = 3$

Back-substitute 3 for a to find b.

$b = -3a - 3$

$b = -3(3) - 3$

$b = -9 - 3$

$b = -12$

Back-substitute 3 for a and –12 for b to find c.

$a + b + c = 4$

$3 + (-12) + c = 4$

$-9 + c = 4$

$c = 13$

Now substitute the values for a, b, and c into $y = ax^2 + bx + c$. The function that models the given data is: $y = 3x^2 - 12x + 13$. In function notation, the equation is $f(x) = 3x^2 - 12x + 13$.

Problem Set 3.3

Practice Exercises

1. $x + y + z = 4$

$2 - 1 + 3 = 4$

$4 = 4$

True

$$x - 2y - z = 1$$
$$2 - 2(-1) - 3 = 1$$
$$2 + 2 - 3 = 1$$
$$1 = 1$$

True

$$2x - y - z = -1$$
$$2(2) - (-1) - 3 = -1$$
$$4 + 1 - 3 = -1$$
$$1 = -1$$

False

The ordered triple (2, –1, 3) does not make all three equations true, so it is not a solution.

3.

$$x - 2y = 2$$
$$4 - 2(1) = 2$$
$$4 - 2 = 2$$
$$2 = 2$$

True

$$2x + 3y = 11$$
$$2(4) + 3(1) = 11$$
$$8 + 3 = 11$$
$$11 = 11$$

True

$$y - 4z = -7$$
$$1 - 4(2) = -7$$
$$1 - 8 = -7$$
$$-7 = -7$$

True

The ordered triple (4, 1, 2) makes all three equations true, so it is a solution.

5.

$$x + y + 2z = 11$$
$$x + y + 3z = 14$$
$$x + 2y - z = 5$$

Multiply the second equation by –1 and add to the first equation..

$$x + y + 2z = 11$$
$$\underline{-x - y - 3z = -14}$$
$$-z = -3$$
$$z = 3$$

Back-substitute 3 for z in the first and third equations:

$$x + y + 2z = 11$$
$$x + y + 2(3) = 11$$
$$x + y + 6 = 11$$
$$x + y = 5$$

$$x + 2y - z = 5$$
$$x + 2y - 3 = 5$$
$$x + 2y = 8$$

We now have two equations in two variables.

$$x + y = 5$$
$$x + 2y = 8$$

Multiply the first equation by –1 and solve by addition.

$$-x - y = -5$$
$$\underline{x + 2y = 8}$$
$$y = 3$$

Back-substitute 3 for y into one of the equations in two variables.

$$x + y = 5$$
$$x + 3 = 5$$
$$x = 2$$

The solution is $(2, 3, 3)$ and the solution set is $\{(2, 3, 3)\}$.

7.

$$4x - y + 2z = 11$$
$$x + 2y - z = -1$$
$$2x + 2y - 3z = -1$$

Multiply the second equation by –4 and add to the first equation.

$$4x - y + 2z = 11$$
$$\underline{-4x - 8y + 4z = 4}$$
$$-9y + 6z = 15$$

Multiply the second equation by –2 and add it to the third equation.

$$\begin{array}{r} -2x-4y+2z=2 \\ 2x+2y-3z=-1 \\ \hline -2y-z=1 \end{array}$$

We now have two equations in two variables.

$-9y+6z=15$

$-2y-z=1$

Multiply the second equation by 6 and solve by addition.

$$\begin{array}{r} -9y+6z=15 \\ -12y-6z=6 \\ \hline -21y=21 \\ y=21 \end{array}$$

Back-substitute –1 for y in one of the equations in two variables.

$$\begin{aligned} -2y-z&=1 \\ -2(-1)-z&=1 \\ 2-z&=1 \\ -z&=-1 \\ z&=1 \end{aligned}$$

Back-substitute –1 for y and 1 for z in one of the original equations in three variables.

$$\begin{aligned} x+2y-z&=-1 \\ x+2(-1)-1&=-1 \\ x-2-1&=-1 \\ x-3&=-1 \\ x&=2 \end{aligned}$$

The solution is $(2,-1,1)$ and the solution set is $\{(2,-1,1)\}$.

9. $3x+2y-3z=-2$

$2x-5y+2z=-2$

$4x-3y+4z=10$

Multiply the second equation by –2 and add to the third equation.

$$\begin{array}{r} -4x+10y-4z=4 \\ 4x-3y+4z=10 \\ \hline 7y=14 \\ y=2 \end{array}$$

Back-substitute 2 for y in the first and third equations to obtain two equations in two unknowns.

$$\begin{aligned} 3x+2y-3z&=-2 \\ 3x+2(2)-3z&=-2 \\ 3x+4-3z&=-2 \\ 3x-3z&=-6 \end{aligned}$$

$$\begin{aligned} 4x-3y+4z&=10 \\ 4x-3(2)+4z&=10 \\ 4x-6+4z&=10 \\ 4x+4z&=16 \end{aligned}$$

The system of two equations in two variables becomes:

$3x-3z=-6$

$4x+4z=16$

Multiply the first equation by –4 and the second equation by 3.

$$\begin{array}{r} -12x+12z=24 \\ 12x+12z=48 \\ \hline 24z=72 \\ z=3 \end{array}$$

Back-substitute 3 for z to find x.

$$3x - 3z = -6$$
$$3x - 3(3) = -6$$
$$3x - 9 = -6$$
$$3x = 3$$
$$x = 1$$

The solution is $(1, 2, 3)$ and the solution set is $\{(1, 2, 3)\}$.

11. $2x - 4y + 3z = 17$

$x + 2y - z = 0$

$4x - y - z = 6$

Multiply the second equation by –1 and add it to the third equation.

$$-x - 2y + z = 0$$
$$\underline{4x - y - z = 6}$$
$$3x - 3y \quad = 6$$

Multiply the second equation by 3 and add it to the first equation.

$$2x - 4y + 3z = 17$$
$$\underline{3x + 6y - 3z = 0}$$
$$5x + 2y \quad = 17$$

The system in two variables becomes:

$$3x - 3y = 6$$
$$5x + 2y = 17$$

Multiply the first equation by 2 and the second equation by 3 and solve by addition.

$$6x - 6y = 12$$
$$\underline{15x + 6y = 51}$$
$$21x \quad = 63$$
$$x \quad = 3$$

Back-substitute 3 for x in one of the equations in two variables.

$$3x - 3y = 6$$
$$3(3) - 3y = 6$$
$$9 - 3y = 6$$
$$-3y = -3$$
$$y = 1$$

Back-substitute 3 for x and 1 for y in one of the original equations in three variables.

$$x + 2y - z = 0$$
$$3 + 2(1) - z = 0$$
$$3 + 2 - z = 0$$
$$5 - z = 0$$
$$5 = z$$

The solution is $(3, 1, 5)$ and the solution set is $\{(3, 1, 5)\}$.

13. $2x + y \quad = 2$

$x + y - z = 4$

$3x + 2y + z = 0$

Add the second and third equations together to obtain an equation in two variables.

$$x + y - z = 4$$
$$\underline{3x + 2y + z = 0}$$
$$4x + 3y \quad = 4$$

Use this equation and the first equation in the original system to write two equations in two variables.

$$2x + y = 2$$
$$4x + 3y = 4$$

Multiply the first equation by –2 and solve by addition.

$$-4x - 2y = -4$$
$$\underline{4x + 3y = 4}$$
$$y = 0$$

Back-substitute 0 for y in one of the equations in two unknowns.

$2x+y=2$

$2x+0=2$

$2x=2$

$x=1$

Back-substitute 1 for x and 0 for y in one of the equations in three unknowns.

$x+y-z=4$

$1+0-z=4$

$1-z=4$

$-z=3$

$z=-3$

The solution is $(1,0,-3)$ and the solution set is $\{(1,0,-3)\}$.

15. $x+y = -4$

$y-z= 1$

$2x+y+3z=-21$

Multiply the first equation by –1 and add to the second equation.

$-x-y = 4$

$\underline{y-z=1}$

$-x - z=5$

Multiply the second equation by –1 and add to the third equation.

$-y+z= -1$

$\underline{2x+y+3z=-21}$

$2x +4z=-22$

The system of two equations in two variables becomes.

$-x-z= 5$

$2x+4z=-22$

Multiply the first equation by 2 and add to the second equation.

$-2x-2z= 10$

$\underline{2x+4z=-22}$

$2z=-12$

$z=-6$

Back-substitute –6 for z in one of the equations in two variables.

$-x-z=5$

$-x-(-6)=5$

$-x+6=5$

$-x=-1$

$x=1$

Back-substitute 1 for x in the first equation of the original system.

$x+y=-4$

$1+y=-4$

$y=-5$

The solution is $(1,-5,-6)$ and the solution set is $\{(1,-5,-6)\}$.

17. $2x+ y+2z=1$

$3x- y+ z=2$

$x-2y- z=0$

Add the first and second equations to eliminate y.

$2x+y+2z=1$

$\underline{3x-y+ z=2}$

$5x +3z=3$

Multiply the second equation by –2 and add to the third equation.

$-6x+2y-2z=-4$

$\underline{x-2y- z= 0}$

$-5x -3z=-4$

We obtain two equations in two variables.

$5x+3z= 3$

$-5x-3z=-4$

Adding the two equations, we obtain:

$$\begin{aligned} 5x+3z &= 3 \\ -5x-3z &= -4 \\ \hline 0 &= -1 \end{aligned}$$

The system is inconsistent. There are no values of x, y, and z for which $0=-1$.

19. $5x-2y-5z=1$

$10x-4y-10z=2$

$15x-6y-15z=3$

Multiply the first equation by –2 and add to the second equation.

$$\begin{aligned} -10x+4y+10z &= -2 \\ 10x-4y-10z &= 2 \\ \hline 0 &= 0 \end{aligned}$$

The system is dependent and has infinitely many solutions.

21. $3(2x+y)+5z=-1$

$2(x-3y+4z)=-9$

$4(1+x)=-3(z-3y)$

Rewrite each equation and obtain the system of three equations in three variables.

$6x+3y+5z=-1$

$2x-6y+8z=-9$

$4x-9y+3z=-4$

Multiply the second equation by –3 and add to the first equation.

$$\begin{aligned} 6x+3y+5z &= -1 \\ -6x+18y-24z &= 27 \\ \hline 21y-19z &= 26 \end{aligned}$$

Multiply the second equation by –2 and add to the third equation.

$$\begin{aligned} -4x+12y-16z &= 18 \\ 4x-9y+3z &= -4 \\ \hline 3y-13z &= 14 \end{aligned}$$

The system of two variables in two equations is:

$21y-19z=26$

$3y-13z=14$

Multiply the second equation by –7 and add to the third equation.

$$\begin{aligned} 21y-19z &= 26 \\ -21y+91z &= -98 \\ \hline 72z &= -72 \\ z &= -1 \end{aligned}$$

Back-substitute –1 for z in one of the equations in two variables to find y.

$$\begin{aligned} 3y-13z &= 14 \\ 3y-13(-1) &= 14 \\ 3y+13 &= 14 \\ 3y &= 1 \\ y &= \frac{1}{3} \end{aligned}$$

Back-substitute –1 for z and $\frac{1}{3}$ for y in one of the original equations in three variables.

$$\begin{aligned} 6x+3y+5z &= -1 \\ 6x+1-5 &= -1 \\ 6x-4 &= -1 \\ 6x &= 3 \\ x &= \frac{1}{2} \end{aligned}$$

The solution is $\left(\frac{1}{2},\frac{1}{3},-1\right)$ and the solution set is $\left\{\left(\frac{1}{2},\frac{1}{3},-1\right)\right\}$.

23. Use each ordered pair to write an equation as follows:

$(x, y) = (-1, 6)$

$y = ax^2 + bx + c$

$6 = a(-1)^2 + b(-1) + c$

$6 = a - b + c$

$(x, y) = (1, 4)$

$y = ax^2 + bx + c$

$4 = a(1)^2 + b(1) + c$

$4 = a + b + c$

$(x, y) = (2, 9)$

$y = ax^2 + bx + c$

$9 = a(2)^2 + b(2) + c$

$9 = a(4) + 2b + c$

$9 = 4a + 2b + c$

The system of three equations in three variables is:

$a - b + c = 6$

$a + b + c = 4$

$4a + 2b + c = 9$

Add the first and second equations.

$a - b + c = 6$

$\underline{a + b + c = 4}$

$2a + 2c = 10$

Multiply the first equation by 2 and add to the third equation.

$2a - 2b + 2c = 12$

$\underline{4a + 2b + c = 9}$

$6a + 3c = 21$

The system of two equations in two variables becomes:

$2a + 2c = 10$

$6a + 3c = 21$

Multiply the first equation by –3 and add to the second equation.

$-6a - 6c = -30$

$\underline{6a + 3c = 21}$

$-3c = -9$

$c = 3$

Back-substitute 3 for c in one of the equations in two variables.

$2a + 2c = 10$

$2a + 2(3) = 10$

$2a + 6 = 10$

$2a = 4$

$a = 2$

Back-substitute 3 for c and 2 for a in one of the equations in three variables.

$a + b + c = 4$

$2 + b + 3 = 4$

$b + 5 = 4$

$b = -1$

The quadratic function is

$y = 2x^2 - x + 3.$

25. Use each ordered pair to write an equation.

$(x, y) = (-1, -4)$

$y = ax^2 + bx + c$

$-4 = a(-1)^2 + b(-1) + c$

$-4 = a - b + c$

$(x, y) = (1, -2)$

$y = ax^2 + bx + c$

$-2 = a(1)^2 + b(1) + c$

$-2 = a + b + c$

$(x, y) = (2, 5)$

$y = ax^2 + bx + c$

$5 = a(2)^2 + b(2) + c$

$5 = a(4) + 2b + c$

$5 = 4a + 2b + c$

The system of three equations in three variables is:

$a - b + c = -4$

$a + b + c = -2$

$4a + 2b + c = 5$

Multiply the second equation by –1 and add to the first equation.

$a - b + c = -4$

$\underline{-a - b - c = 2}$

$-2b = -2$

$b = 1$

Back-substitute 4 for b in first and third equations to obtain two equations in two variables.

$a - b + c = -4$

$a - 1 + c = -4$

$a + c = -3$

$4a + 2b + c = 5$

$4a + 2(1) + c = 5$

$4a + 2 + c = 5$

$4a + c = 3$

The system of two equations in two variables becomes:

$a + c = -3$

$4a + c = 3$

Multiply the first equation by –1 and add to the second equation.

$-a - c = 3$

$\underline{4a + c = 3}$

$3a = 6$

$a = 2$

Back-substitute 2 for a and 1 for b in one of the equations in three variables.

$a - b + c = -4$

$2 - 1 + c = -4$

$1 + c = -4$

$c = -5$

The quadratic function is

$y = 2x^2 + x - 5.$

27. Let x = the first number
Let y = the second number
Let z = the third number

$x + y + z = 16$

$2x + 3y + 4z = 46$

$5x - y = 31$

Multiply the first equation by –4 and add to the second equation.

$-4x - 4y - 4z = -64$

$\underline{2x + 3y + 4z = 46}$

$-2x - y = -18$

The system of two equations in two variables becomes:

$5x - y = 31$

$-2x - y = -18$

Multiply the first equation by –1 and add to the second equation.

$-5x + y = -31$

$\underline{-2x - y = -18}$

$-7x = -49$

$x = 7$

Back-substitute 7 for x in one of the equations in two variables.

$5x - y = 31$

$5(7) - y = 31$

$35 - y = 31$

$$-y = -4$$
$$y = 4$$

Back-substitute 7 for x and 4 for y in one of the equations in two variables.

$$x + y + z = 16$$
$$7 + 4 + z = 16$$
$$11 + z = 16$$
$$z = 5$$

The numbers are 7, 4 and 5.

Application Exercises

29. a.

1960	$(0,5.4)$
1970	$(10,4.7)$
1980	$(20,6.2)$

b. $y = ax^2 + bx + c$

$$5.4 = a(0)^2 + b(0) + c$$
$$5.4 = 0a + 0b + c$$

$$4.7 = a(10)^2 + b(10) + c$$
$$4.7 = a(100) + b(10) + c$$
$$4.7 = 100a + 10b + c$$

$$6.2 = a(20)^2 + b(20) + c$$
$$6.2 = a(400) + b(20) + c$$
$$6.2 = 400a + 20b + c$$

The system of three equations in three variables is:

$$0a + 0b + c = 5.4$$
$$100a + 10b + c = 4.7$$
$$400a + 20b + c = 6.2$$

31. $(1,224)$ $(3,176)$ $(4,104)$

a. $(x, y) = (1,224)$

$$y = ax^2 + bx + c$$
$$224 = a(1)^2 + b(1) + c$$
$$224 = a(1) + b + c$$
$$224 = a + b + c$$

$(x, y) = (3,176)$

$$y = ax^2 + bx + c$$
$$176 = a(3)^2 + b(3) + c$$
$$176 = a(9) + 3b + c$$
$$176 = 9a + 3b + c$$

$(x, y) = (4,104)$

$$y = ax^2 + bx + c$$
$$104 = a(4)^2 + b(4) + c$$
$$104 = a(16) + 4b + c$$
$$104 = 16a + 4b + c$$

The system of three equations in three variables becomes:

$$a + b + c = 224$$
$$9a + 3b + c = 176$$
$$16a + 4b + c = 104$$

Multiply the second equation by −1 and add to the first.

$$a + b + c = 224$$
$$\underline{-9a - 3b - c = -176}$$
$$-8a - 2b = 48$$

Multiply the second equation by −1 and add to the third.

$$-9a - 3b - c = -176$$
$$\underline{16a + 4b + c = 104}$$
$$7a + b = 72$$

The system of two equations in two variables becomes:

$$\begin{aligned} -8a - 2b &= 48 \\ 7a + b &= -72 \end{aligned}$$

Multiply the second equation by 2 and solve by addition.

$$\begin{aligned} -8a - 2b &= 48 \\ 14a + 2b &= -144 \\ \hline 6a &= -96 \\ a &= -16 \end{aligned}$$

Back-substitute to find b.

$$\begin{aligned} -8a - 2b &= 48 \\ -8(-16) - 2b &= 48 \\ 128 - 2b &= 48 \\ -2b &= -80 \\ b &= 40 \end{aligned}$$

Back-substitute -16 for a and 40 for b to find c.

$$\begin{aligned} a + b + c &= 224 \\ -16 + 40 + c &= 224 \\ 24 + c &= 224 \\ c &= 200 \end{aligned}$$

The quadratic function is $y = -16x^2 + 40x + 200.$

b. $y = -16x^2 + 40x + 200$ is the function used to describe the ball's height above the ground. When the ball hits the ground, its height is zero.

$$\begin{aligned} -16x^2 + 40x + 200 &= 0 \\ -8(2x^2 - 5x - 25) &= 0 \\ -8(2x^2 - 10x + 5x - 25) &= 0 \\ -8(2x(x-5) + 5(x-5)) &= 0 \\ -8(x-5)(2x+5) &= 0 \end{aligned}$$

Apply the zero product principle.

$$\begin{aligned} -8(x-5) &= 0 \quad \text{or} \quad & 2x + 5 &= 0 \\ x - 5 &= 0 & 2x &= -5 \\ x &= 5 & x &= -\frac{5}{2} \end{aligned}$$

We disregard $-\frac{5}{2}$ because we can't have a negative time measurement. The ball will hit the ground after 5 seconds.

33. Let x = the starting salary of Chemical Engineering majors
Let y = the starting salary of Mechanical Engineering majors
Let z = the starting salary of Electrical Engineering majors

$$\begin{aligned} x + y + z &= 121421 \\ x - y &= 2906 \\ y - z &= 1041 \end{aligned}$$

Multiply the second equation by -1 and add to the first equation.

$$\begin{aligned} x + y + z &= 121421 \\ -x + y &= -2906 \\ \hline 2y + z &= 118515 \end{aligned}$$

We obtain a system of two equations in two unknowns. Add the equations to eliminate z.

$$\begin{aligned} y - z &= 1041 \\ 2y + z &= 118515 \\ \hline 3y &= 119556 \\ y &= 39852 \end{aligned}$$

Back-substitute 39852 for y to find z.

$$\begin{aligned} y - z &= 1041 \\ 39852 - z &= 1041 \\ -z &= -38811 \\ z &= 38811 \end{aligned}$$

Back-substitute 39852 for y in one of the equations in two variables.

$$x - y = 2906$$
$$x - 39852 = 2906$$
$$x = 42758$$

The starting salaries are $42,758 for Chemical Engineers, $39,852 for Mechanical Engineers and $38,811for Electrical Engineers.

35. Let x = the amount invested at 8%
Let y = the amount invested at 10%
Let z = the amount invested at 12%

$$x + y + z = 6700$$
$$0.08x + 0.10y + 0.12z = 716$$
$$z - x - y = 300$$

Rewrite the system in $Ax + By + Cz = D$ form.

$$x + y + z = 6700$$
$$0.08x + 0.10y + 0.12z = 716$$
$$-x - y + z = 300$$

Add the first and third equations to find z.

$$x + y + z = 6700$$
$$\underline{-x - y + z = 300}$$
$$2z = 7000$$
$$z = 3500$$

Back-substitute 3500 for z to obtain two equations in two variables.

$$x + y + z = 6700$$
$$x + y + 3500 = 6700$$
$$x + y = 3200$$

$$0.08x + 0.10y + 0.12(3500) = 716$$
$$0.08x + 0.10y + 420 = 716$$
$$0.08x + 0.10y = 296$$

The system of two equations in two variables becomes:

$$x + y = 3200$$
$$0.08x + 0.10y = 296$$

Multiply the second equation by –10 and add it to the first equation.

$$x + y = 3200$$
$$\underline{-0.8x + -y = -2960}$$
$$0.2x = 240$$
$$x = 1200$$

Back-substitute 1200 for x in one of the equations in two variables.

$$x + y = 3200$$
$$1200 + y = 3200$$
$$y = 2000$$

$1200 was invested at 8%, $2000 was invested at 10%, and $3500 was invested at 12%.

37. Let x = the number of $8 tickets
Let y = the number of $10 tickets
Let z = the number of $12 tickets

$$x + y + z = 400$$
$$8x + 10y + 12z = 3700$$
$$x + y = 7z$$

Rewrite the system in $Ax + By + Cz = D$ form.

$$x + y + z = 400$$
$$8x + 10y + 12z = 3700$$
$$x + y - 7z = 0$$

Multiply the first equation by –1 and add to the third equation.

$$-x - y - z = -400$$
$$\underline{x + y - 7z = 0}$$
$$-8z = -400$$
$$z = 50$$

Back-substitute 50 for z in two of the original equations to obtain two of equations in two variables.

$$x+y+z=400$$
$$x+y+50=400$$
$$x+y=350$$

$$8x+10y+12z=3700$$
$$8x+10y+12(50)=3700$$
$$8x+10y+600=3700$$
$$8x+10y=3100$$

The system of two equations in two variables becomes:

$$x+\ \ y=\ 350$$
$$8x+10y=3100$$

Multiply the first equation by –8 and add to the second equation.

$$-8x-\ 8y=-2800$$
$$\underline{8x+10y=\ \ 3100}$$
$$2y=300$$
$$y=150$$

Back-substitute 50 for z and 150 for y in one of the original equations in three variables.

$$x+y+z=400$$
$$x+150+50=400$$
$$x+200=400$$
$$x=200$$

There were 200 \$8 tickets, 150 \$10 tickets, and 50 \$12 tickets sold.

39. Let A = the number of servings of A
Let B = the number of servings of B
Let C = the number of servings of C

$$40A+200B+400C=660$$
$$5A+\ \ 2B+\ \ 4C=\ 25$$
$$30A+\ 10B+300C=425$$

Multiply the second equation by –8 and add to the first equation to obtain an equation in two variables.

$$40A+200B+400C=\ 660$$
$$\underline{-40A-\ 16B-\ 32C=-200}$$
$$184B+368C=460$$

Multiply the second equation by –6 and add to the third equation to obtain an equation in two variables.

$$-30A-12B-\ 24C=-150$$
$$\underline{30A+10B+300C=\ \ 425}$$
$$-2B+276C=275$$

The system of two equations in two variables becomes:

$$184B+368C=460$$
$$-2B+276C=275$$

Multiply the second equation by 92 and eliminate B.

$$184B+\ \ 368C=\ \ 460$$
$$\underline{-184B+25392C=25300}$$
$$25760C=25760$$
$$C=1$$

Back-substitute 1 for C in one of the equations in two variables.

$$-2B+276C=275$$
$$-2B+276(1)=275$$
$$-2B+276=275$$
$$-2B=-1$$
$$B=\frac{1}{2}$$

Back-substitute 1 for C and $\frac{1}{2}$ for B in one of the original equations in three variables.

$$5A+2B+4C=25$$
$$5A+2\left(\frac{1}{2}\right)+4(1)=25$$
$$5A+1+4=25$$
$$5A+5=25$$
$$5A=20$$
$$A=4$$

To meet the requirements, 4 ounces of Food A, $\frac{1}{2}$ ounce of Food B, and 1 ounce of Food C should be used.

Writing in Mathematics

41. Answers will vary.

43. Answers will vary.

45. Answers will vary.

Technology Exercises

47. Answers will vary.

Critical Thinking Exercises

49. Statement **c.** is true. The variable terms of the second equation are multiples of the variable terms in the first equation, but the constants are not. If we multiply the first equation by 2 and add to the second equation, we obtain:

$$\begin{array}{r} -2x-2y+2z=-20 \\ 2x+2y-2z=7 \\ \hline 0=-13 \end{array}$$

This is a contradiction, so the system is inconsistent.

Statement **a.** is false. The ordered triple is one solution to the equation, but there are an infinite number of other ordered triples which satisfy the equation.

Statement **b.** is false.

$$x-y-z=-6$$
$$2-(-3)-5=-6$$
$$2+3-5=-6$$
$$0\neq-6$$

Statement **d.** is false. An equation with four variables can be satisfied by real numbers.

51. Let t = the number of triangles
Let r = the number of rectangles
Let p = the number of pentagons

$$t+r+p=40$$
$$3t+4r+5p=153$$
$$2r+5p=72$$

Multiply the first equation by –3 and add to the second equation.

$$\begin{array}{r} -3t-3r-3p=-120 \\ 3t+4r+5p=153 \\ \hline r+2p=33 \end{array}$$

We have two equations in two variables.

$$2r+5p=72$$
$$r+2p=33$$

Multiply the second equation by –2 and add to eliminate r.

$$\begin{array}{r} 2r+5p=72 \\ -2r-4p=-66 \\ \hline p=6 \end{array}$$

Back-substitute 6 for p in one of the equations in two variables.

$$r + 2p = 33$$
$$r + 2(6) = 33$$
$$r + 12 = 33$$
$$r = 21$$

Back-substitute 21 for r and 6 for p to find t.

$$t + r + p = 40$$
$$t + 21 + 6 = 40$$
$$t + 27 = 40$$
$$t = 13$$

There are 13 triangles, 21 rectangles, and 6 pentagons.

Review Exercises

53. $f(x) = -\frac{3}{4}x + 3$

Use the slope, $-\frac{3}{4}$, and the y–intercept, 3, to graph the line.

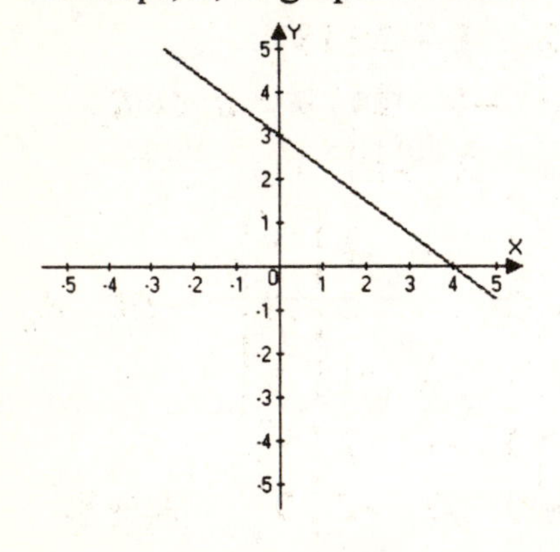

54. $-2x + y = 6$

Rewrite the equation in slope-intercept form.

$$-2x + y = 6$$
$$y = 2x + 6$$

Use the slope, 2, and the y–intercept, 6, to graph the line.

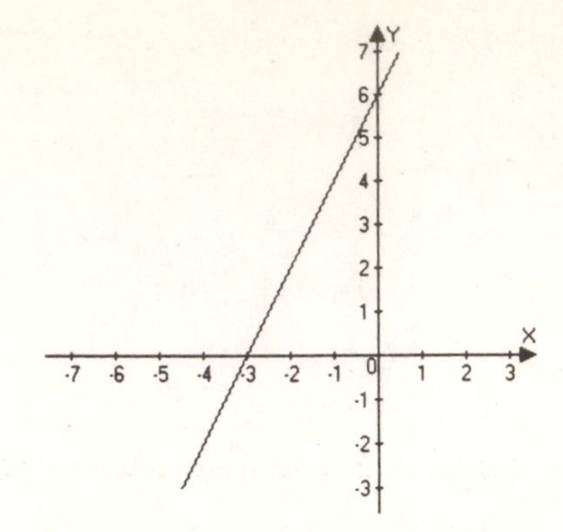

55. $f(x) = -5$

This is the horizontal line, $y = -5$.

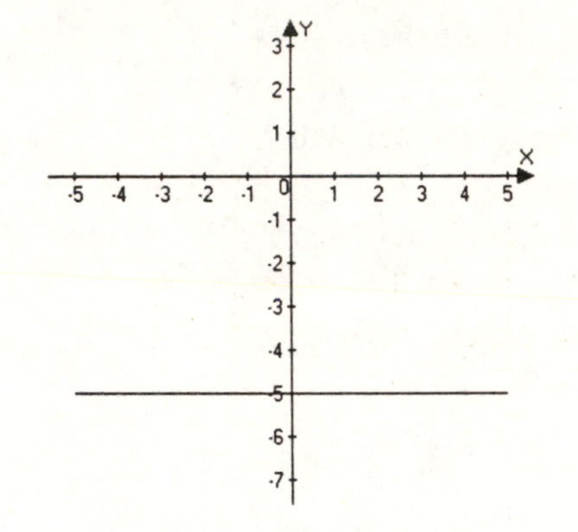

Check Points 3.4

1.
$$x - y + z = 8$$
$$0x + 1y - 12z = -15$$
$$0x + 0y + 1z = 1$$

Simplify the system.

$$x - y + z = 8$$
$$y - 12z = -15$$
$$z = 1$$

Since z is known, back-substitute 1 for z to find y.

$$y - 12z = -15$$
$$y - 12(1) = -15$$
$$y - 12 = -15$$
$$y = -3$$

Back-substitute –3 for y and 1 for z to find x.

$$\begin{aligned} x-y+z&=8 \\ x-(-3)+1&=8 \\ x+3+1&=8 \\ x+4&=8 \\ x&=4 \end{aligned}$$

The solution is $(4,-3,1)$ and the solution set is $\{(4,-3,1)\}$.

2. $\left[\begin{array}{ccc|c} 4 & 12 & -20 & 8 \\ 1 & 6 & -3 & 7 \\ -3 & -2 & 1 & -9 \end{array}\right]$

a. $\left[\begin{array}{ccc|c} 1 & 6 & -3 & 7 \\ 4 & 12 & -20 & 8 \\ -3 & -2 & 1 & -9 \end{array}\right]$

b. $\left[\begin{array}{ccc|c} 1 & 3 & -5 & 2 \\ 1 & 6 & -3 & 7 \\ -3 & -2 & 1 & -9 \end{array}\right]$

c. $\left[\begin{array}{ccc|c} 4 & 12 & -20 & 8 \\ 1 & 6 & -3 & 7 \\ 0 & 16 & -8 & 12 \end{array}\right]$

3.
$$\begin{aligned} 2x-\ \ y&=-4 \\ x+3y&=\ \ 5 \end{aligned}$$

$$\left[\begin{array}{cc|c} 2 & -1 & -4 \\ 1 & 3 & 5 \end{array}\right] \quad R_1 \leftrightarrow R_2$$

$$=\left[\begin{array}{cc|c} 1 & 3 & 5 \\ 2 & -1 & -4 \end{array}\right] \quad -2R_1+R_2$$

$$=\left[\begin{array}{cc|c} 1 & 3 & 5 \\ 0 & -7 & -14 \end{array}\right] \quad -\frac{1}{7}R_2$$

$$=\left[\begin{array}{cc|c} 1 & 3 & 5 \\ 0 & 1 & 2 \end{array}\right]$$

The resulting system is:

$$\begin{aligned} x+3y&=5 \\ y&=2 \end{aligned}$$

Back-substitute 2 for y to find x.

$$\begin{aligned} x+3y&=5 \\ x+3(2)&=5 \\ x+6&=5 \\ x&=-1 \end{aligned}$$

The solution is $(-1,2)$ and the solution set is $\{(-1,2)\}$.

4.
$$\begin{aligned} 2x+\ \ y+2z&=18 \\ x-\ \ y+2z&=\ \ 9 \\ x+2y-\ \ z&=\ \ 6 \end{aligned}$$

$$\left[\begin{array}{ccc|c} 2 & 1 & 2 & 18 \\ 1 & -1 & 2 & 9 \\ 1 & 2 & -1 & 6 \end{array}\right] \quad R_1 \leftrightarrow R_2$$

$$=\left[\begin{array}{ccc|c} 1 & -1 & 2 & 9 \\ 2 & 1 & 2 & 18 \\ 1 & 2 & -1 & 6 \end{array}\right] \quad -2R_1+R_2$$

$$=\left[\begin{array}{ccc|c} 1 & -1 & 2 & 9 \\ 0 & 3 & -2 & 0 \\ 1 & 2 & -1 & 6 \end{array}\right] \quad -R_1+R_3$$

$$=\left[\begin{array}{ccc|c} 1 & -1 & 2 & 9 \\ 0 & 3 & -2 & 0 \\ 0 & 3 & -3 & -3 \end{array}\right] \quad -R_2+R_3$$

$$=\left[\begin{array}{ccc|c} 1 & -1 & 2 & 9 \\ 0 & 3 & -2 & 0 \\ 0 & 0 & -1 & -3 \end{array}\right] \quad -\frac{1}{3}R_2$$

$$= \left[\begin{array}{ccc|c} 1 & -1 & 2 & 9 \\ 0 & 1 & -\frac{2}{3} & 0 \\ 0 & 0 & -1 & -3 \end{array}\right] \quad -R_3$$

$$= \left[\begin{array}{ccc|c} 1 & -1 & 2 & 9 \\ 0 & 1 & -\frac{2}{3} & 0 \\ 0 & 0 & 1 & 3 \end{array}\right]$$

The resulting system is:

$$x - y + 2z = 9$$
$$y - \frac{2}{3}z = 0$$
$$z = 3$$

Since z is known, back-substitute 3 for z to find y.

$$y - \frac{2}{3}z = 0$$
$$y - \frac{2}{3}(3) = 0$$
$$y - 2 = 0$$
$$y = 2$$

Back-substitute 2 for y and 3 for z to find x.

$$x - y + 2z = 9$$
$$x - 2 + 2(3) = 9$$
$$x - 2 + 6 = 9$$
$$x + 4 = 9$$
$$x = 5$$

The solution is $(5, 2, 3)$ and the solution set is $\{(5, 2, 3)\}$.

Problem Set 3.4

Practice Exercises

1. $x - 3y = 11$

$y = -3$

Substitute –3 for y in the first equation.

$$x - 3y = 11$$
$$x - 3(-3) = 11$$
$$x + 9 = 11$$
$$x = 2$$

The solution is $(2, -3)$ and the solution set is $\{(2, -3)\}$.

3. $x - 3y = 1$

$y = -1$

Substitute –1 for y in the first equation.

$$x - 3y = 1$$
$$x - 3(-1) = 1$$
$$x + 3 = 1$$
$$x = -2$$

The solution is $(-2, -1)$ and the solution set is $\{(-2, -1)\}$.

5. $x \quad - 4z = 5$

$y - 12z = 13$

$z = -\frac{1}{2}$

Substitute $-\frac{1}{2}$ for z in the second equation to find y.

$$y-12z=13$$
$$y-12\left(-\frac{1}{2}\right)=13$$
$$y+6=13$$
$$y=7$$

Substitute 7 for y in the first equation to find x.

$$x-4z=5$$
$$x-4\left(-\frac{1}{2}\right)=5$$
$$x+2=5$$
$$x=3$$

The solution is $\left(3,7,-\frac{1}{2}\right)$ and the solution set is $\left\{\left(3,7,-\frac{1}{2}\right)\right\}$.

7.
$$x+\frac{1}{2}y+\quad z=\frac{11}{2}$$
$$y+\frac{3}{2}z=7$$
$$z=4$$

Substitute 4 for z in the second equation to find y.

$$y+\frac{3}{2}z=7$$
$$y+\frac{3}{2}(4)=7$$
$$y+6=7$$
$$y=1$$

Substitute 1 for y and 4 for z in the first equation to find x.

$$x+\frac{1}{2}y+z=\frac{11}{2}$$
$$x+\frac{1}{2}(1)+4=\frac{11}{2}$$
$$x+\frac{9}{2}=\frac{11}{2}$$
$$x=\frac{2}{2}=1$$

The solution is $(1,1,4)$ and the solution set is $\{(1,1,4)\}$.

9. $\left[\begin{array}{cc|c} 2 & 2 & 5 \\ 1 & -\frac{3}{2} & 5 \end{array}\right]$ $R_1 \leftrightarrow R_2$

$=\left[\begin{array}{cc|c} 1 & -\frac{3}{2} & 5 \\ 2 & 2 & 5 \end{array}\right]$

11. $\left[\begin{array}{cc|c} -6 & 8 & -12 \\ 3 & 5 & -2 \end{array}\right]$ $-\frac{1}{6}R_1$

$=\left[\begin{array}{cc|c} 1 & -\frac{4}{3} & 2 \\ 3 & 5 & -2 \end{array}\right]$

13. $\left[\begin{array}{cc|c} 1 & -3 & 5 \\ 2 & 6 & 4 \end{array}\right]$ $-2R_1+R_2$

$=\left[\begin{array}{cc|c} 1 & -3 & 5 \\ 0 & 12 & -6 \end{array}\right]$

15. $\left[\begin{array}{cc|c} 1 & -\frac{3}{2} & \frac{7}{2} \\ 3 & 4 & 2 \end{array}\right]$ $-3R_1+R_2$

$=\left[\begin{array}{cc|c} 1 & -\frac{3}{2} & \frac{7}{2} \\ 0 & \frac{17}{2} & -\frac{17}{2} \end{array}\right]$

17. $\left[\begin{array}{ccc|c} 2 & -6 & 4 & 10 \\ 1 & 5 & -5 & 0 \\ 3 & 0 & 4 & 7 \end{array}\right] \quad \frac{1}{2}R_1$

$= \left[\begin{array}{ccc|c} 1 & -3 & 2 & 5 \\ 1 & 5 & -5 & 0 \\ 3 & 0 & 4 & 7 \end{array}\right]$

19. $\left[\begin{array}{ccc|c} 1 & -3 & 2 & 0 \\ 3 & 1 & -1 & 7 \\ 2 & -2 & 1 & 3 \end{array}\right] \quad -3R_1 + R_2$

$= \left[\begin{array}{ccc|c} 1 & -3 & 2 & 0 \\ 0 & 10 & -7 & 7 \\ 2 & -2 & 1 & 3 \end{array}\right]$

21. $\left[\begin{array}{ccc|c} 1 & 1 & -1 & 6 \\ 2 & -1 & 1 & -3 \\ 3 & -1 & -1 & 4 \end{array}\right] \quad \begin{array}{c} -2R_1 + R_2 \\ \text{and} \\ -3R_1 + R_3 \end{array}$

$= \left[\begin{array}{ccc|c} 1 & 1 & -1 & 6 \\ 0 & -3 & 3 & -15 \\ 0 & -4 & 2 & -14 \end{array}\right]$

23. $\left[\begin{array}{cc|c} 1 & 1 & 6 \\ 1 & -1 & 2 \end{array}\right] \quad -R_1 + R_2$

$= \left[\begin{array}{cc|c} 1 & 1 & 6 \\ 0 & -2 & -4 \end{array}\right] \quad -\frac{1}{2}R_2$

$= \left[\begin{array}{cc|c} 1 & 1 & 6 \\ 0 & 1 & 2 \end{array}\right]$

The resulting system is: $\begin{aligned} x + y &= 6 \\ y &= 2 \end{aligned}$

Back-substitute 2 for y in the first equation.

$x + y = 6$

$x + 2 = 6$

$x = 4$

The solution is $(4, 2)$ and the solution set is $\{(4, 2)\}$.

25. $\left[\begin{array}{cc|c} 2 & 1 & 3 \\ 1 & -3 & 12 \end{array}\right] \quad R_1 \leftrightarrow R_2$

$= \left[\begin{array}{cc|c} 1 & -3 & 12 \\ 2 & 1 & 3 \end{array}\right] \quad -2R_1 + R_2$

$= \left[\begin{array}{cc|c} 1 & -3 & 12 \\ 0 & 7 & -21 \end{array}\right] \quad \frac{1}{7}R_2$

$= \left[\begin{array}{cc|c} 1 & -3 & 12 \\ 0 & 1 & -3 \end{array}\right]$

The system is: $\begin{aligned} x - 3y &= 12 \\ y &= -3 \end{aligned}$

Back-substitute -3 for y in the first equation.

$x - 3y = 12$

$x - 3(-3) = 12$

$x + 9 = 12$

$x = 3$

The solution is $(3, -3)$ and the solution set is $\{(3, -3)\}$.

27. $\left[\begin{array}{cc|c} 5 & 7 & -25 \\ 11 & 6 & -8 \end{array}\right] \quad \frac{1}{5}R_1$

$= \left[\begin{array}{cc|c} 1 & \frac{7}{5} & -5 \\ 11 & 6 & -8 \end{array}\right] \quad -11R_1 + R_2$

$= \left[\begin{array}{cc|c} 1 & \frac{7}{5} & -5 \\ 0 & -\frac{47}{5} & 47 \end{array}\right] \quad -\frac{5}{47}R_2$

$$=\left[\begin{array}{cc|c} 1 & \frac{7}{5} & -5 \\ 0 & 1 & -5 \end{array}\right]$$

The resulting system is:

$$x+\frac{7}{5}y=-5$$
$$y=-5$$

Back-substitute -5 for y in the first equation.

$$x+\frac{7}{5}y=-5$$
$$x+\frac{7}{5}(-5)=-5$$
$$x-7=-5$$
$$x=2$$

The solution is $(2,-5)$ and the solution set is $\{(2,-5)\}$.

29. $\left[\begin{array}{cc|c} 4 & -2 & 5 \\ -2 & 1 & 6 \end{array}\right]$ $\frac{1}{4}R_1$

$$=\left[\begin{array}{cc|c} 1 & -\frac{1}{2} & \frac{5}{2} \\ -2 & 1 & 6 \end{array}\right] \quad 2R_1+R_2$$

$$=\left[\begin{array}{cc|c} 1 & -\frac{1}{2} & \frac{5}{2} \\ 0 & 0 & \frac{17}{2} \end{array}\right]$$

The resulting system is:

$$x-\frac{1}{2}y=\frac{5}{2}$$
$$0x+0y=\frac{17}{2}$$

This is a contradiction. The system is inconsistent.

31. $\left[\begin{array}{cc|c} 1 & -2 & 1 \\ -2 & 4 & -2 \end{array}\right]$ $2R_1+R_2$

$$=\left[\begin{array}{cc|c} 1 & -2 & 1 \\ 0 & 0 & 0 \end{array}\right]$$

The resulting system is:

$$x-2y=1$$
$$0x+0y=0$$

The system is dependent. There are infinitely many solutions.

33. $\left[\begin{array}{ccc|c} 1 & 1 & -1 & -2 \\ 2 & -1 & 1 & 5 \\ -1 & 2 & 2 & 1 \end{array}\right]$ $-2R_1+R_2$

$$=\left[\begin{array}{ccc|c} 1 & 1 & -1 & -2 \\ 0 & -3 & 3 & 9 \\ -1 & 2 & 2 & 1 \end{array}\right] \quad R_1+R_3$$

$$=\left[\begin{array}{ccc|c} 1 & 1 & -1 & -2 \\ 0 & -3 & 3 & 9 \\ 0 & 3 & 1 & -1 \end{array}\right] \quad R_2+R_3$$

$$=\left[\begin{array}{ccc|c} 1 & 1 & -1 & -2 \\ 0 & -3 & 3 & 9 \\ 0 & 0 & 4 & 8 \end{array}\right] \quad \frac{1}{4}R_3$$

$$=\left[\begin{array}{ccc|c} 1 & 1 & -1 & -2 \\ 0 & -3 & 3 & 9 \\ 0 & 0 & 1 & 2 \end{array}\right]$$

The resulting system is:

$$x+y-z=-2$$
$$y-z=-3$$
$$z=2$$

Back-substitute 2 for z to find y.

$$y-z=-3$$
$$y-2=-3$$
$$y=-1$$

Back-substitute 2 for z and -1 for y to find x.

$$x+y-z=-2$$
$$x-1-2=-2$$
$$x-3=-2$$
$$x=1$$

The solution is $(1,-1,2)$ and the solution set is $\{(1,-1,2)\}$.

35. $$\left[\begin{array}{ccc|c} 1 & 3 & 0 & 0 \\ 1 & 1 & 1 & 1 \\ 3 & -1 & -1 & 11 \end{array}\right] \quad -R_1+R_2$$

$$=\left[\begin{array}{ccc|c} 1 & 3 & 0 & 0 \\ 0 & -2 & 1 & 1 \\ 3 & -1 & -1 & 11 \end{array}\right] \quad -3R_1+R_3$$

$$=\left[\begin{array}{ccc|c} 1 & 3 & 0 & 0 \\ 0 & -2 & 1 & 1 \\ 0 & -10 & -1 & 11 \end{array}\right] \quad -\frac{1}{2}R_2$$

$$=\left[\begin{array}{ccc|c} 1 & 3 & 0 & 0 \\ 0 & 1 & -\frac{1}{2} & -\frac{1}{2} \\ 0 & -10 & -1 & 11 \end{array}\right] \quad -\frac{1}{10}R_3$$

$$=\left[\begin{array}{ccc|c} 1 & 3 & 0 & 0 \\ 0 & 1 & -\frac{1}{2} & -\frac{1}{2} \\ 0 & 1 & \frac{1}{10} & -\frac{11}{10} \end{array}\right] \quad -R_2+R_3$$

$$=\left[\begin{array}{ccc|c} 1 & 3 & 0 & 0 \\ 0 & 1 & -\frac{1}{2} & -\frac{1}{2} \\ 0 & 0 & \frac{3}{5} & -\frac{3}{5} \end{array}\right] \quad \frac{5}{3}R_3$$

$$=\left[\begin{array}{ccc|c} 1 & 3 & 0 & 0 \\ 0 & 1 & -\frac{1}{2} & -\frac{1}{2} \\ 0 & 0 & 1 & -1 \end{array}\right]$$

The resulting system is:

$$x+3y = 0$$
$$y-\frac{1}{2}z=-\frac{1}{2}$$
$$z=-1$$

Back-substitute -1 for z and solve for y.

$$y-\frac{1}{2}z=-\frac{1}{2}$$
$$y-\frac{1}{2}(-1)=-\frac{1}{2}$$
$$y+\frac{1}{2}=-\frac{1}{2}$$
$$y=-1$$

Back-substitute -1 for y to find x.

$$x+3y=0$$
$$x+3(-1)=0$$
$$x-3=0$$
$$x=3$$

The solution is $(3,-1,-1)$ and the solution set is $\{(3,-1,-1)\}$.

37. $$\left[\begin{array}{ccc|c} 2 & 2 & 7 & -1 \\ 2 & 1 & 2 & 2 \\ 4 & 6 & 1 & 15 \end{array}\right] \quad \frac{1}{2}R_1$$

$$=\left[\begin{array}{ccc|c} 1 & 1 & \frac{7}{2} & -\frac{1}{2} \\ 2 & 1 & 2 & 2 \\ 4 & 6 & 1 & 15 \end{array}\right] \quad -2R_1+R_2$$

$$= \left[\begin{array}{ccc|c} 1 & 1 & \frac{7}{2} & -\frac{1}{2} \\ 0 & -1 & -5 & 3 \\ 4 & 6 & 1 & 15 \end{array}\right] \quad -R_2$$

$$= \left[\begin{array}{ccc|c} 1 & 1 & \frac{7}{2} & -\frac{1}{2} \\ 0 & 1 & 5 & -3 \\ 4 & 6 & 1 & 15 \end{array}\right] \quad -4R_1 + R_3$$

$$= \left[\begin{array}{ccc|c} 1 & 1 & \frac{7}{2} & -\frac{1}{2} \\ 0 & 1 & 5 & -3 \\ 0 & 2 & -13 & 17 \end{array}\right] \quad -2R_2 + R_3$$

$$= \left[\begin{array}{ccc|c} 1 & 1 & \frac{7}{2} & -\frac{1}{2} \\ 0 & 1 & 5 & -3 \\ 0 & 0 & -23 & 23 \end{array}\right] \quad -R_3$$

$$= \left[\begin{array}{ccc|c} 1 & 1 & \frac{7}{2} & -\frac{1}{2} \\ 0 & 1 & 5 & -3 \\ 0 & 0 & 1 & -1 \end{array}\right]$$

The resulting system is:

$$x + y + \frac{7}{2}z = -\frac{1}{2}$$
$$y + 5z = -3$$
$$z = -1$$

Back-substitute -1 for z to find y.

$$y + 5z = -3$$
$$y + 5(-1) = -3$$
$$y - 5 = -3$$
$$y = 2$$

Back-substitute -1 for z and 2 for y to find x.

$$x + y + \frac{7}{2}z = -\frac{1}{2}$$
$$x + 2 + \frac{7}{2}(-1) = -\frac{1}{2}$$
$$x + 2 - \frac{7}{2} = -\frac{1}{2}$$
$$x - \frac{3}{2} = -\frac{1}{2}$$
$$x = 1$$

The solution is $(1, 2, -1)$ and the solution set is $\{(1, 2, -1)\}$.

39. $$\left[\begin{array}{ccc|c} 1 & 1 & 1 & 6 \\ 1 & 0 & -1 & -2 \\ 0 & 1 & 3 & 11 \end{array}\right] \quad R_2 \leftrightarrow R_3$$

$$= \left[\begin{array}{ccc|c} 1 & 1 & 1 & 6 \\ 0 & 1 & 3 & 11 \\ 1 & 0 & -1 & -2 \end{array}\right] \quad -R_1 + R_3$$

$$= \left[\begin{array}{ccc|c} 1 & 1 & 1 & 6 \\ 0 & 1 & 3 & 11 \\ 0 & -1 & -2 & -8 \end{array}\right] \quad R_2 + R_3$$

$$= \left[\begin{array}{ccc|c} 1 & 1 & 1 & 6 \\ 0 & 1 & 3 & 11 \\ 0 & 0 & 1 & 3 \end{array}\right]$$

The resulting system is:

$$x + y + z = 6$$
$$y + 3z = 11$$
$$z = 3$$

Back-substitute 3 for z to find y.

$$y + 3z = 11$$
$$y + 3(3) = 11$$
$$y + 9 = 11$$
$$y = 2$$

Back-substitute 3 for z and 2 for y to find x.

$$\begin{aligned} x+y+z&=6 \\ x+2+3&=6 \\ x+5&=6 \\ x&=1 \end{aligned}$$

The solution is $(1,2,3)$ and the solution set is $\{(1,2,3)\}$.

41. $\left[\begin{array}{ccc|c} 1 & -1 & 3 & 4 \\ 2 & -2 & 6 & 7 \\ 3 & -1 & 5 & 14 \end{array}\right]$ $-2R_1+R_2$ and $-3R_1+R_3$

$= \left[\begin{array}{ccc|c} 1 & -1 & 3 & 4 \\ 0 & 0 & 0 & -1 \\ 0 & 2 & -4 & 2 \end{array}\right]$

The resulting system is:

$$\begin{aligned} x-y+3z&=4 \\ 0x+0y+0z&=-1 \\ 2y-4z&=2 \end{aligned}$$

The second row is a contradiction, since $0x+0y+0z$ cannot equal -1. We conclude that the system is dependent and there are infinitely many solutions.

43. $\left[\begin{array}{ccc|c} 1 & -2 & 1 & 4 \\ 5 & -10 & 5 & 20 \\ -2 & 4 & -2 & -8 \end{array}\right]$ $\frac{1}{5}R_2$

$= \left[\begin{array}{ccc|c} 1 & -2 & 1 & 4 \\ 1 & -2 & 1 & 4 \\ -2 & 4 & -2 & -8 \end{array}\right]$

R_1 and R_2 are the same. The system is dependent and there are infinitely many solutions.

45. $\left[\begin{array}{ccc|c} 1 & 1 & 0 & 1 \\ 0 & 1 & 2 & -2 \\ 2 & 0 & -1 & 0 \end{array}\right]$ $-2R_1+R_3$

$= \left[\begin{array}{ccc|c} 1 & 1 & 0 & 1 \\ 0 & 1 & 2 & -2 \\ 0 & -2 & -1 & -2 \end{array}\right]$ $2R_2+R_3$

$= \left[\begin{array}{ccc|c} 1 & 1 & 0 & 1 \\ 0 & 1 & 2 & -2 \\ 0 & 0 & 3 & -6 \end{array}\right]$ $\frac{1}{3}R_3$

$= \left[\begin{array}{ccc|c} 1 & 1 & 0 & 1 \\ 0 & 1 & 2 & -2 \\ 0 & 0 & 1 & -2 \end{array}\right]$

The resulting system is:

$$\begin{aligned} x+y&=1 \\ y+2z&=-2 \\ z&=-2 \end{aligned}$$

Back-substitute –2 for z to find y.

$$\begin{aligned} y+2z&=-2 \\ y+2(-2)&=-2 \\ y-4&=-2 \\ y&=2 \end{aligned}$$

Back-substitute 2 for y to find x.

$$\begin{aligned} x+y&=1 \\ x+2&=1 \\ x&=-1 \end{aligned}$$

The solution is $(-1,2,-2)$ and the solution set is $\{(-1,2,-2)\}$.

Application Exercises

47. a. Use each ordered pair to write an equation as follows:

$$(x, y) = (1, 344)$$
$$y = ax^2 + bx + c$$
$$344 = a(1)^2 + b(1) + c$$
$$344 = a + b + c$$

$$(x, y) = (5, 480)$$
$$y = ax^2 + bx + c$$
$$480 = a(5)^2 + b(5) + c$$
$$480 = a(25) + 5b + c$$
$$480 = 25a + 5b + c$$

$$(x, y) = (10, 740)$$
$$y = ax^2 + bx + c$$
$$740 = a(10)^2 + b(10) + c$$
$$740 = a(100) + 10b + c$$
$$740 = 100a + 10b + c$$

The system of three equations in three variables is:

$$\begin{aligned} a + \quad b + c &= 344 \\ 25a + \ 5b + c &= 480 \\ 100a + 10b + c &= 740 \end{aligned}$$

$$\left[\begin{array}{ccc|c} 1 & 1 & 1 & 344 \\ 25 & 5 & 1 & 480 \\ 100 & 10 & 1 & 740 \end{array}\right] \begin{array}{c} -25R_1 + R_2 \\ \text{and} \\ -100R_1 + R_3 \end{array}$$

$$= \left[\begin{array}{ccc|c} 1 & 1 & 1 & 344 \\ 0 & -20 & -24 & -8120 \\ 0 & -90 & -99 & -33660 \end{array}\right] -\frac{1}{20}R_2$$

$$= \left[\begin{array}{ccc|c} 1 & 1 & 1 & 344 \\ 0 & 1 & \frac{6}{5} & 406 \\ 0 & -90 & -99 & -33660 \end{array}\right] -\frac{1}{90}R_3$$

$$= \left[\begin{array}{ccc|c} 1 & 1 & 1 & 344 \\ 0 & 1 & \frac{6}{5} & 406 \\ 0 & 1 & \frac{11}{10} & 374 \end{array}\right] -R_2 + R_3$$

$$= \left[\begin{array}{ccc|c} 1 & 1 & 1 & 344 \\ 0 & 1 & \frac{6}{5} & 406 \\ 0 & 0 & -\frac{1}{10} & -32 \end{array}\right] -10R_3$$

$$= \left[\begin{array}{ccc|c} 1 & 1 & 1 & 344 \\ 0 & 1 & \frac{6}{5} & 406 \\ 0 & 0 & 1 & 320 \end{array}\right]$$

The resulting system is:

$$\begin{aligned} x + y + \quad z &= 344 \\ y + \frac{6}{5}z &= 406 \\ z &= 320 \end{aligned}$$

Back-substitute 320 for z to find y.

$$y + \frac{6}{5}z = 406$$
$$y + \frac{6}{5}(320) = 406$$
$$y + 384 = 406$$
$$y = 22$$

Back-substitute 22 for y and 320 for z to find x.

$$x + y + z = 344$$
$$x + 22 + 320 = 344$$
$$x + 342 = 344$$
$$x = 2$$

The solution set is $\{(2, 22, 320)\}$.

The quadratic function is $y = 2x^2 + 22x + 320$.

b. $f(x)=2x^2+22x+320$

$$\begin{aligned} f(30) &= 2(30)^2+22(30)+320 \\ &= 2(900)+660+320 \\ &= 1800+660+320 \\ &= 2780 \end{aligned}$$

The model predicts that there will be 2,780,000 inmates in 2010.

c. Answers will vary. One possible answer is the number of prison cells available in the year 2010. If there isn't room to house 2,780,000 inmates, adjustments would have to be made.

49. Let x = the percentage of online users in the youngest group
Let y = the percentage of online users in the middle group
Let z = the percentage of online users in the oldest group

$$x+z=y+2$$
$$2z=x-3$$
$$x+y+z=100$$

Rewrite the system in standard form.

$$\begin{aligned} x-y+z &= 2 \\ -x \quad +2z &= -3 \\ x+y+z &= 100 \end{aligned}$$

$$\left[\begin{array}{ccc|c} 1 & -1 & 1 & 2 \\ -1 & 0 & 2 & -3 \\ 1 & 1 & 1 & 100 \end{array}\right] \begin{array}{c} R_1+R_2 \\ \text{and} \\ -R_1+R_3 \end{array}$$

$$=\left[\begin{array}{ccc|c} 1 & -1 & 1 & 2 \\ 0 & -1 & 3 & -1 \\ 0 & 2 & 0 & 98 \end{array}\right] R_2 \leftrightarrow R_3$$

$$=\left[\begin{array}{ccc|c} 1 & -1 & 1 & 2 \\ 0 & 2 & 0 & 98 \\ 0 & -1 & 3 & -1 \end{array}\right] \frac{1}{2}R_2$$

$$=\left[\begin{array}{ccc|c} 1 & -1 & 1 & 2 \\ 0 & 1 & 0 & 49 \\ 0 & -1 & 3 & -1 \end{array}\right] R_2+R_3$$

$$=\left[\begin{array}{ccc|c} 1 & -1 & 1 & 2 \\ 0 & 1 & 0 & 49 \\ 0 & 0 & 3 & 48 \end{array}\right] \frac{1}{3}R_3$$

$$=\left[\begin{array}{ccc|c} 1 & -1 & 1 & 2 \\ 0 & 1 & 0 & 49 \\ 0 & 0 & 1 & 16 \end{array}\right]$$

The resulting system is:

$$\begin{aligned} x-y+z &= 2 \\ y &= 49 \\ z &= 16 \end{aligned}$$

Back-substitute 49 for y and 16 for z to find x.

$$\begin{aligned} x-y+z &= 2 \\ x-49+16 &= 2 \\ x-33 &= 2 \\ x &= 35 \end{aligned}$$

35% of the online users are in the youngest group, 49% of the online users are in the middle group, and 16% of the online users are in the oldest group.

Writing in Mathematics

51. Answers will vary.

53. Answers will vary.

55. Answers will vary.

Technology Exercises

57. Answers will vary. For example, verify Exercise 11.

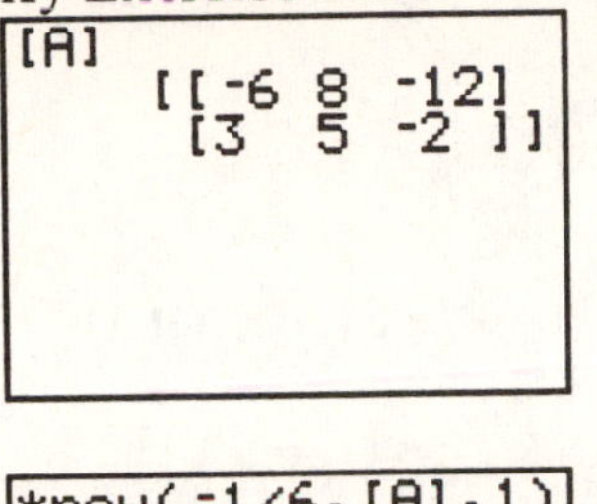


```
*row(-1/6,[A],1)
▸Frac
   [[1  -4/3 2 ]
    [3 5      -2]]
```

This matches the result obtained in Exercise 11.

Critical Thinking Exercises

59. Statement **d.** is true.

Statement **a.** is false. Multiplying a row by a negative fraction is permitted.

Statement **b.** is false because there are three variables in the system. The augmented matrix should be:

$$\left[\begin{array}{ccc|c} 1 & -3 & 0 & 5 \\ 0 & 1 & -2 & 7 \\ 2 & 0 & 1 & 4 \end{array}\right]$$

Statement **c.** is false. When solving a system of three equations in three variables, we use row operations to obtain ones along the diagonal and zeros below the ones.

61. a. $\left[\begin{array}{ccc|c} 3 & 1 & 3 & 14 \\ 7 & 5 & 8 & 32 \\ 1 & 3 & 2 & 9 \end{array}\right] \quad R_3 \leftrightarrow R_1$

$= \left[\begin{array}{ccc|c} 1 & 3 & 2 & 9 \\ 7 & 5 & 8 & 32 \\ 3 & 1 & 3 & 14 \end{array}\right] \quad -7R_1 + R_2$

$= \left[\begin{array}{ccc|c} 1 & 3 & 2 & 9 \\ 0 & -16 & -6 & -31 \\ 3 & 1 & 3 & 14 \end{array}\right] \quad -3R_1 + R_3$

$= \left[\begin{array}{ccc|c} 1 & 3 & 2 & 9 \\ 0 & -16 & -6 & -31 \\ 0 & -8 & -3 & -13 \end{array}\right] \quad -\frac{1}{2}R_2 + R_3$

$= \left[\begin{array}{ccc|c} 1 & 3 & 2 & 9 \\ 0 & -16 & -6 & -31 \\ 0 & 0 & 0 & -13 \end{array}\right]$

This is a contradiction. We have:

$0x + 0y + 0z = -13$

$0 = -13$

There is no solution to this system. No combination of these foods will meet the requirements.

b. $\left[\begin{array}{ccc|c} 3 & 1 & 3 & 14 \\ 7 & 5 & 8 & 37 \\ 1 & 3 & 2 & 9 \end{array}\right] \quad R_3 \leftrightarrow R_1$

$= \left[\begin{array}{ccc|c} 1 & 3 & 2 & 9 \\ 7 & 5 & 8 & 37 \\ 3 & 1 & 3 & 14 \end{array}\right] \quad -7R_1 + R_2$

$= \left[\begin{array}{ccc|c} 1 & 3 & 2 & 9 \\ 0 & -16 & -6 & -26 \\ 3 & 1 & 3 & 14 \end{array}\right] \quad -3R_1 + R_3$

$$= \left[\begin{array}{ccc|c} 1 & 3 & 2 & 9 \\ 0 & -16 & -6 & -26 \\ 0 & -8 & -3 & -13 \end{array}\right] \quad -\frac{1}{2}R_2 + R_3$$

$$= \left[\begin{array}{ccc|c} 1 & 3 & 2 & 9 \\ 0 & -16 & -6 & -26 \\ 0 & 0 & 0 & 0 \end{array}\right]$$

The system is dependent. There are infinitely many combinations of foods A, B, and C that will meet the requirements.

Review Exercises

62. $f(x) = -3x + 10$

$f(2a-1) = -3(2a-1) + 10$
$= -6a + 3 + 10$
$= -6a + 13$

63. $f(x) = 3x$ and $g(x) = 2x - 3$

$(fg)(x) = f(x) \cdot g(x)$
$= 3x(2x-3)$
$= 6x^2 - 9x$

$(fg)(-1) = 6(-1)^2 - 9(-1)$
$= 6(1) + 9$
$= 6 + 9 = 15$

64. $\dfrac{-4x^8 y^{-12}}{12x^{-3} y^{24}} = \dfrac{-4x^8 x^3}{12y^{24} y^{12}}$
$= \dfrac{-x^{11}}{3y^{36}}$

Check Points 3.5

1. a. $\begin{vmatrix} 10 & 9 \\ 6 & 5 \end{vmatrix} = 10(5) - 6(9)$
$= 50 - 54 = -4$

b. $\begin{vmatrix} 4 & 3 \\ -5 & -8 \end{vmatrix} = 4(-8) - (-5)(3)$
$= -32 + 15 = -17$

2. $5x + 4y = 12$
$3x - 6y = 24$

$D = \begin{vmatrix} 5 & 4 \\ 3 & -6 \end{vmatrix}$
$= 5(-6) - 3(4)$
$= -30 - 12 = -42$

$D_x = \begin{vmatrix} 12 & 4 \\ 24 & -6 \end{vmatrix}$
$= 12(-6) - 24(4)$
$= -72 - 96 = -168$

$D_y = \begin{vmatrix} 5 & 12 \\ 3 & 24 \end{vmatrix}$
$= 5(24) - 3(12)$
$= 120 - 36 = 84$

$x = \dfrac{D_x}{D} = \dfrac{-168}{-42} = 4$

$y = \dfrac{D_y}{D} = \dfrac{84}{-42} = -2$

The solution is $(4, -2)$ and the solution set is $\{(4, -2)\}$.

3. $$\begin{vmatrix} 2 & 1 & 7 \\ -5 & 6 & 0 \\ -4 & 3 & 1 \end{vmatrix} = 2\begin{vmatrix} 6 & 0 \\ 3 & 1 \end{vmatrix} - (-5)\begin{vmatrix} 1 & 7 \\ 3 & 1 \end{vmatrix} + (-4)\begin{vmatrix} 1 & 7 \\ 6 & 0 \end{vmatrix}$$

$$= 2(6(1)-3(0)) + 5(1(1)-3(7)) - 4(1(0)-6(7))$$
$$= 2(6-0) + 5(1-21) - 4(0-42)$$
$$= 2(6) + 5(-20) - 4(-42)$$
$$= 12 - 120 + 168 = 80$$

4. $$\begin{aligned} 3x - 2y + z &= 16 \\ 2x + 3y - z &= -9 \\ x + 4y + 3z &= 2 \end{aligned}$$

$$D = \begin{vmatrix} 3 & -2 & 1 \\ 2 & 3 & -1 \\ 1 & 4 & 3 \end{vmatrix} = 3\begin{vmatrix} 3 & -1 \\ 4 & 3 \end{vmatrix} - 2\begin{vmatrix} -2 & 1 \\ 4 & 3 \end{vmatrix} + 1\begin{vmatrix} -2 & 1 \\ 3 & -1 \end{vmatrix}$$

$$= 3(3(3)-4(-1)) - 2(-2(3)-4(1)) + 1(-2(-1)-3(1))$$
$$= 3(9+4) - 2(-6-4) + 1(2-3)$$
$$= 3(13) - 2(-10) + 1(-1)$$
$$= 39 + 20 - 1 = 58$$

$$D_x = \begin{vmatrix} 16 & -2 & 1 \\ -9 & 3 & -1 \\ 2 & 4 & 3 \end{vmatrix} = 16\begin{vmatrix} 3 & -1 \\ 4 & 3 \end{vmatrix} - (-9)\begin{vmatrix} -2 & 1 \\ 4 & 3 \end{vmatrix} + 2\begin{vmatrix} -2 & 1 \\ 3 & -1 \end{vmatrix}$$

$$= 16(3(3)-4(-1)) + 9(-2(3)-4(1)) + 2(-2(-1)-3(1))$$
$$= 16(9+4) + 9(-6-4) + 2(2-3)$$
$$= 16(13) + 9(-10) + 2(-1)$$
$$= 208 - 90 - 2 = 116$$

$$D_y = \begin{vmatrix} 3 & 16 & 1 \\ 2 & -9 & -1 \\ 1 & 2 & 3 \end{vmatrix} = 3\begin{vmatrix} -9 & -1 \\ 2 & 3 \end{vmatrix} - 2\begin{vmatrix} 16 & 1 \\ 2 & 3 \end{vmatrix} + 1\begin{vmatrix} 16 & 1 \\ -9 & -1 \end{vmatrix}$$

$$= 3(-9(3)-2(-1)) - 2(16(3)-2(1)) + 1(16(-1)-(-9)(1))$$
$$= 3(-27+2) - 2(48-2) + 1(-16+9)$$

$$= 3(-25) - 2(46) + 1(-7)$$
$$= -75 - 92 - 7 = -174$$

$$D_z = \begin{vmatrix} 3 & -2 & 16 \\ 2 & 3 & -9 \\ 1 & 4 & 2 \end{vmatrix} = 3\begin{vmatrix} 3 & -9 \\ 4 & 2 \end{vmatrix} - 2\begin{vmatrix} -2 & 16 \\ 4 & 2 \end{vmatrix} + 1\begin{vmatrix} -2 & 16 \\ 3 & -9 \end{vmatrix}$$
$$= 3(3(2) - 4(-9)) - 2(-2(2) - 4(16)) + 1(-2(-9) - 3(16))$$
$$= 3(6 + 36) - 2(-4 - 64) + 1(18 - 48)$$
$$= 3(42) - 2(-68) + 1(-30)$$
$$= 126 + 136 - 30 = 232$$

$$x = \frac{D_x}{D} = \frac{116}{58} = 2 \qquad y = \frac{D_y}{D} = \frac{-174}{58} = -3 \qquad z = \frac{D_z}{D} = \frac{232}{58} = 4$$

The solution is $(2, -3, 4)$ and the solution set is $\{(2, -3, 4)\}$.

Problem Set 3.5

Practice Exercises

1. $\begin{vmatrix} 5 & 7 \\ 2 & 3 \end{vmatrix} = 5(3) - 2(7) = 15 - 14 = 1$

3. $\begin{vmatrix} -4 & 1 \\ 5 & 6 \end{vmatrix} = -4(6) - 5(1)$

$= -24 - 5 = -29$

5. $\begin{vmatrix} -7 & 14 \\ 2 & -4 \end{vmatrix} = -7(-4) - 2(14)$

$= 28 - 28 = 0$

7. $\begin{vmatrix} -5 & -1 \\ -2 & -7 \end{vmatrix} = -5(-7) - (-2)(-1)$

$= 35 - 2 = 33$

9. $\begin{vmatrix} \frac{1}{2} & \frac{1}{2} \\ \frac{1}{8} & -\frac{3}{4} \end{vmatrix} = \frac{1}{2}\left(-\frac{3}{4}\right) - \frac{1}{8}\left(\frac{1}{2}\right)$

$= -\frac{3}{8} - \frac{1}{16} = -\frac{6}{16} - \frac{1}{16}$

$= -\frac{7}{16}$

11. $D = \begin{vmatrix} 1 & 1 \\ 1 & -1 \end{vmatrix} = 1(-1) - 1(1)$

$= -1 - 1 = -2$

$D_x = \begin{vmatrix} 7 & 1 \\ 3 & -1 \end{vmatrix} = 7(-1) - 3(1)$

$= -7 - 3 = -10$

$D_y = \begin{vmatrix} 1 & 7 \\ 1 & 3 \end{vmatrix} = 1(3) - 1(7)$

$= 3 - 7 = -4$

$x = \frac{D_x}{D} = \frac{-10}{-2} = 5$

$$y = \frac{D_y}{D} = \frac{-4}{-2} = 2$$

The solution is $(5,2)$ and the solution set is $\{(5,2)\}$.

13. $$D = \begin{vmatrix} 12 & 3 \\ 2 & -3 \end{vmatrix} = 12(-3) - 2(3)$$
$$= -36 - 6 = -42$$
$$D_x = \begin{vmatrix} 15 & 3 \\ 13 & -3 \end{vmatrix} = 15(-3) - 13(3)$$
$$= -45 - 39 = -84$$
$$D_y = \begin{vmatrix} 12 & 15 \\ 2 & 13 \end{vmatrix} = 12(13) - 2(15)$$
$$= 156 - 30 = 126$$
$$x = \frac{D_x}{D} = \frac{-84}{-42} = 2$$
$$y = \frac{D_y}{D} = \frac{126}{-42} = -3$$

The solution is $(2,-3)$ and the solution set is $\{(2,-3)\}$.

15. $$D = \begin{vmatrix} 4 & -5 \\ 2 & 3 \end{vmatrix} = 4(3) - 2(-5)$$
$$= 12 + 10 = 22$$
$$D_x = \begin{vmatrix} 17 & -5 \\ 3 & 3 \end{vmatrix} = 17(3) - 3(-5)$$
$$= 51 + 15 = 66$$
$$D_y = \begin{vmatrix} 4 & 17 \\ 2 & 3 \end{vmatrix} = 4(3) - 2(17)$$
$$= 12 - 34 = -22$$
$$x = \frac{D_x}{D} = \frac{66}{22} = 3$$
$$y = \frac{D_y}{D} = \frac{-22}{22} = -1$$

The solution is $(3,-1)$ and the solution set is $\{(3,-1)\}$.

17. $$D = \begin{vmatrix} 1 & -3 \\ 3 & -4 \end{vmatrix} = 1(-4) - 3(-3)$$
$$= -4 + 9 = 5$$
$$D_x = \begin{vmatrix} 4 & -3 \\ 12 & -4 \end{vmatrix} = 4(-4) - 12(-3)$$
$$= -16 + 36 = 20$$
$$D_y = \begin{vmatrix} 1 & 4 \\ 3 & 12 \end{vmatrix} = 1(12) - 3(4)$$
$$= 12 - 12 = 0$$
$$x = \frac{D_x}{D} = \frac{20}{5} = 4$$
$$y = \frac{D_y}{D} = \frac{0}{5} = 0$$

The solution is $(4,0)$ and the solution set is $\{(4,0)\}$.

19. $$D = \begin{vmatrix} 3 & -4 \\ 2 & 2 \end{vmatrix} = 3(2) - 2(-4)$$
$$= 6 + 8 = 14$$
$$D_x = \begin{vmatrix} 4 & -4 \\ 12 & 2 \end{vmatrix} = 4(2) - 12(-4)$$
$$= 8 + 48 = 56$$
$$D_y = \begin{vmatrix} 3 & 4 \\ 2 & 12 \end{vmatrix} = 3(12) - 2(4)$$
$$= 36 - 8 = 28$$
$$x = \frac{D_x}{D} = \frac{56}{14} = 4$$
$$y = \frac{D_y}{D} = \frac{28}{14} = 2$$

The solution is $(4,2)$ and the solution set is $\{(4,2)\}$.

21. First, rewrite the system in standard form.

$2x-3y=2$

$5x+4y=51$

$D=\begin{vmatrix}2 & -3\\5 & 4\end{vmatrix}=2(4)-5(-3)$

$=8+15=23$

$D_x=\begin{vmatrix}2 & -3\\51 & 4\end{vmatrix}=2(4)-51(-3)$

$=8+153=161$

$D_y=\begin{vmatrix}2 & 2\\5 & 51\end{vmatrix}=2(51)-5(2)$

$=102-10=92$

$x=\dfrac{D_x}{D}=\dfrac{161}{23}=7$

$y=\dfrac{D_y}{D}=\dfrac{92}{23}=4$

The solution is $(7,4)$ and the solution set is $\{(7,4)\}$.

23. $3x+3y=2$

$2x+2y=3$

First, rewrite the system in standard form.

$D=\begin{vmatrix}3 & 3\\2 & 2\end{vmatrix}=3(2)-2(3)$

$=6-6=0$

$D_x=\begin{vmatrix}2 & 3\\3 & 2\end{vmatrix}=2(2)-3(3)$

$=4-9=-5$

$D_y=\begin{vmatrix}2 & 3\\3 & 2\end{vmatrix}=2(2)-3(3)$

$=4-9=-5$

Because $D=0$ but neither D_x nor D_y is zero, Cramer's Rule cannot be used to solve the system. Instead, use matrices. $\left[\begin{array}{cc|c}3 & 3 & 2\\2 & 2 & 3\end{array}\right]$ $\frac{1}{3}R_1$

$=\left[\begin{array}{cc|c}1 & 1 & 2/3\\2 & 2 & 3\end{array}\right]$ $-2R_1+R_2$

$=\left[\begin{array}{cc|c}1 & 1 & 2/3\\0 & 0 & 5/3\end{array}\right]$

This is a contradiction. There are no values for x and y for which $0=\frac{5}{3}$.

The solution set is $\varnothing$ and the system is inconsistent.

25. Rewrite the system in standard form.

$3x+4y=16$

$6x+8y=32$

$D=\begin{vmatrix}3 & 4\\6 & 8\end{vmatrix}=3(8)-6(4)$

$=24-24=0$

$D_x=\begin{vmatrix}16 & 4\\32 & 8\end{vmatrix}=16(8)-32(4)$

$=128-128=128$

$D_y=\begin{vmatrix}3 & 16\\6 & 32\end{vmatrix}=3(32)-6(16)$

$=96-96=0$

Since $D=0$ and all determinants in the numerators are 0, the equations in the system are dependent and there are infinitely many solutions.

27. $\begin{vmatrix} 3 & 0 & 0 \\ 2 & 1 & -5 \\ 2 & 5 & -1 \end{vmatrix} = 3\begin{vmatrix} 1 & -5 \\ 5 & -1 \end{vmatrix} - 2\begin{vmatrix} 0 & 0 \\ 5 & -1 \end{vmatrix} + 2\begin{vmatrix} 0 & 0 \\ 1 & -5 \end{vmatrix}$

$= 3(1(-1)-5(-5)) - 2(0(-1)-5(0)) + 2(0(-5)-1(0))$

$= 3(-1+25) - 2(0-0) + 2(0-0)$

$= 3(24) - 2(0) + 2(0) = 72$

29. $\begin{vmatrix} 3 & 1 & 0 \\ -3 & 4 & 0 \\ -1 & 3 & -5 \end{vmatrix} = 3\begin{vmatrix} 4 & 0 \\ 3 & -5 \end{vmatrix} - (-3)\begin{vmatrix} 1 & 0 \\ 3 & -5 \end{vmatrix} + (-1)\begin{vmatrix} 1 & 0 \\ 4 & 0 \end{vmatrix}$

$= 3(4(-5)-3(0)) + 3(1(-5)-3(0)) - 1(1(0)-4(0))$

$= 3(-20-0) + 3(-5-0) - 1(0-0)$

$= 3(-20) + 3(-5) - 1(0) = -60 - 15 = -75$

31. $\begin{vmatrix} 1 & 1 & 1 \\ 2 & 2 & 2 \\ -3 & 4 & -5 \end{vmatrix} = 1\begin{vmatrix} 2 & 2 \\ 4 & -5 \end{vmatrix} - 2\begin{vmatrix} 1 & 1 \\ 4 & -5 \end{vmatrix} + (-3)\begin{vmatrix} 1 & 1 \\ 2 & 2 \end{vmatrix}$

$= 1(2(-5)-4(2)) - 2(1(-5)-4(1)) - 3(1(2)-2(1))$

$= 1(-10-8) - 2(-5-4) - 3(2-2)$

$= 1(-18) - 2(-9) - 3(0) = -18 + 18 = 0$

33. $x + y + z = 0$

$2x - y + z = -1$

$-x + 3y - z = -8$

$D = \begin{vmatrix} 1 & 1 & 1 \\ 2 & -1 & 1 \\ -1 & 3 & -1 \end{vmatrix} = 1\begin{vmatrix} -1 & 1 \\ 3 & -1 \end{vmatrix} - 2\begin{vmatrix} 1 & 1 \\ 3 & -1 \end{vmatrix} - 1\begin{vmatrix} 1 & 1 \\ -1 & 1 \end{vmatrix}$

$= 1(-1(-1)-3(1)) - 2(1(-1)-3(1)) - 1(1(1)-(-1)(1))$

$= 1(1-3) - 2(-1-3) - 1(1+1)$

$= 1(-2) - 2(-4) - 1(2) = -2 + 8 - 2 = 4$

$$D_x = \begin{vmatrix} 0 & 1 & 1 \\ -1 & -1 & 1 \\ -8 & 3 & -1 \end{vmatrix} = 0\begin{vmatrix} -1 & 1 \\ 3 & -1 \end{vmatrix} - (-1)\begin{vmatrix} 1 & 1 \\ 3 & -1 \end{vmatrix} - 8\begin{vmatrix} 1 & 1 \\ -1 & 1 \end{vmatrix}$$

$$= 1(1(-1)-3(1)) - 8(1(1)-(-1)(1))$$
$$= 1(-1-3) - 8(1+1)$$
$$= 1(-4) - 8(2) = -4 - 16 = -20$$

$$D_y = \begin{vmatrix} 1 & 0 & 1 \\ 2 & -1 & 1 \\ -1 & -8 & -1 \end{vmatrix} = 1\begin{vmatrix} -1 & 1 \\ -8 & -1 \end{vmatrix} - 2\begin{vmatrix} 0 & 1 \\ -8 & -1 \end{vmatrix} - 1\begin{vmatrix} 0 & 1 \\ -1 & 1 \end{vmatrix}$$

$$= 1(-1(-1)-(-8)1) - 2(0(-1)-(-8)1) - 1(0(1)-(-1)1)$$
$$= 1(1+8) - 2(0+8) - 1(0+1)$$
$$= 1(9) - 2(8) - 1(1) = 9 - 16 - 1 = -8$$

$$D_z = \begin{vmatrix} 1 & 1 & 0 \\ 2 & -1 & -1 \\ -1 & 3 & -8 \end{vmatrix} = 1\begin{vmatrix} -1 & -1 \\ 3 & -8 \end{vmatrix} - 2\begin{vmatrix} 1 & 0 \\ 3 & -8 \end{vmatrix} - 1\begin{vmatrix} 1 & 0 \\ -1 & -1 \end{vmatrix}$$

$$= 1(-1(-8)-3(-1)) - 2(1(-8)-3(0)) - 1(1(-1)-(-1)0)$$
$$= 1(8+3) - 2(-8-0) - 1(-1+0)$$
$$= 1(11) - 2(-8) - 1(-1) = 11 + 16 + 1 = 28$$

$$x = \frac{D_x}{D} = \frac{-20}{4} = -5 \qquad y = \frac{D_y}{D} = \frac{-8}{4} = -2 \qquad z = \frac{D_z}{D} = \frac{28}{4} = 7$$

The solution is $(-5, -2, 7)$ and the solution set is $\{(-5, -2, 7)\}$.

35. $4x - 5y - 6z = -1$

$x - 2y - 5z = -12$

$2x - y \quad = 7$

$$D = \begin{vmatrix} 4 & -5 & -6 \\ 1 & -2 & -5 \\ 2 & -1 & 0 \end{vmatrix} = 4\begin{vmatrix} -2 & -5 \\ -1 & 0 \end{vmatrix} - 1\begin{vmatrix} -5 & -6 \\ -1 & 0 \end{vmatrix} + 2\begin{vmatrix} -5 & -6 \\ -2 & -5 \end{vmatrix}$$

$$= 4(-2(0) - (-1)(-5)) - 1(-5(0) - (-1)(-6)) + 2(-5(-5) - (-2)(-6))$$

$$= 4(-5) - 1(-6) + 2(25 - 12)$$
$$= -20 + 6 + 2(13) = -20 + 6 + 26 = 12$$

$$D_x = \begin{vmatrix} -1 & -5 & -6 \\ -12 & -2 & -5 \\ 7 & -1 & 0 \end{vmatrix} = -1\begin{vmatrix} -2 & -5 \\ -1 & 0 \end{vmatrix} - (-12)\begin{vmatrix} -5 & -6 \\ -1 & 0 \end{vmatrix} + 7\begin{vmatrix} -5 & -6 \\ -2 & -5 \end{vmatrix}$$
$$= -1(\cancel{-2(0)} - (-1)(-5)) - (-12)(\cancel{-5(0)} - (-1)(-6)) + 7(-5(-5) - (-2)(-6))$$
$$= -1(-5) - (-12)(-6) + 7(25 - 12)$$
$$= 5 - 72 + 7(13) = 5 - 72 + 91 = 24$$

$$D_y = \begin{vmatrix} 4 & -1 & -6 \\ 1 & -12 & -5 \\ 2 & 7 & 0 \end{vmatrix} = 4\begin{vmatrix} -12 & -5 \\ 7 & 0 \end{vmatrix} - 1\begin{vmatrix} -1 & -6 \\ 7 & 0 \end{vmatrix} + 2\begin{vmatrix} -1 & -6 \\ -12 & -5 \end{vmatrix}$$
$$= 4(\cancel{-12(0)} - 7(-5)) - 1(\cancel{-1(0)} - 7(-6)) + 2(-1(-5) - (-12)(-6))$$
$$= 4(35) - 1(42) + 2(5 - 72)$$
$$= 140 - 42 + 2(-67) = 140 - 42 - 134 = -36$$

$$D_z = \begin{vmatrix} 4 & -5 & -1 \\ 1 & -2 & -12 \\ 2 & -1 & 7 \end{vmatrix} = 4\begin{vmatrix} -2 & -12 \\ -1 & 7 \end{vmatrix} - 1\begin{vmatrix} -5 & -1 \\ -1 & 7 \end{vmatrix} + 2\begin{vmatrix} -5 & -1 \\ -2 & -12 \end{vmatrix}$$
$$= 4(-2(7) - (-1)(-12)) - 1(-5(7) - (-1)(-1)) + 2(-5(-12) - (-2)(-1))$$
$$= 4(-14 - 12) - 1(-35 - 1) + 2(60 - 2)$$
$$= 4(-26) - 1(-36) + 2(58) = -104 + 36 + 116 = 48$$

$$x = \frac{D_x}{D} = \frac{24}{12} = 2 \qquad y = \frac{D_y}{D} = \frac{-36}{12} = -3 \qquad z = \frac{D_z}{D} = \frac{48}{12} = 4$$

The solution is $(2, -3, 4)$ and the solution set is $\{(2, -3, 4)\}$.

37. $x + y + z = 4$
$x - 2y + z = 7$
$x + 3y + 2z = 4$

$$D = \begin{vmatrix} 1 & 1 & 1 \\ 1 & -2 & 1 \\ 1 & 3 & 2 \end{vmatrix} = 1\begin{vmatrix} -2 & 1 \\ 3 & 2 \end{vmatrix} - 1\begin{vmatrix} 1 & 1 \\ 3 & 2 \end{vmatrix} + 1\begin{vmatrix} 1 & 1 \\ -2 & 1 \end{vmatrix}$$

$$=1(-2(2)-3(1))-1(1(2)-3(1))+1(1(1)-(-2)1)$$
$$=1(-4-3)-1(2-3)+1(1+2)$$
$$=1(-7)-1(-1)+1(3)=-7+1+3=-3$$

$$D_x=\begin{vmatrix}4 & 1 & 1\\ 7 & -2 & 1\\ 4 & 3 & 2\end{vmatrix}=4\begin{vmatrix}-2 & 1\\ 3 & 2\end{vmatrix}-7\begin{vmatrix}1 & 1\\ 3 & 2\end{vmatrix}+4\begin{vmatrix}1 & 1\\ -2 & 1\end{vmatrix}$$
$$=4(-2(2)-3(1))-7(1(2)-3(1))+4(1(1)-(-2)1)$$
$$=4(-4-3)-7(2-3)+4(1+2)$$
$$=4(-7)-7(-1)+4(3)=-28+7+12=-9$$

$$D_y=\begin{vmatrix}1 & 4 & 1\\ 1 & 7 & 1\\ 1 & 4 & 2\end{vmatrix}=1\begin{vmatrix}7 & 1\\ 4 & 2\end{vmatrix}-1\begin{vmatrix}4 & 1\\ 4 & 2\end{vmatrix}+1\begin{vmatrix}4 & 1\\ 7 & 1\end{vmatrix}$$
$$=1(7(2)-4(1))-1(4(2)-4(1))+1(4(1)-7(1))$$
$$=1(14-4)-1(8-4)+1(4-7)$$
$$=1(10)-1(4)+1(-3)=10-4-3=3$$

$$D_z=\begin{vmatrix}1 & 1 & 4\\ 1 & -2 & 7\\ 1 & 3 & 4\end{vmatrix}=1\begin{vmatrix}-2 & 7\\ 3 & 4\end{vmatrix}-1\begin{vmatrix}1 & 4\\ 3 & 4\end{vmatrix}+1\begin{vmatrix}1 & 4\\ -2 & 7\end{vmatrix}$$
$$=1(-2(4)-3(7))-1(1(4)-3(4))+1(1(7)-(-2)4)$$
$$=1(-8-21)-1(4-12)+1(7+8)$$
$$=1(-29)-1(-8)+1(15)=-29+8+15=-6$$

$$x=\frac{D_x}{D}=\frac{-9}{-3}=3 \qquad y=\frac{D_y}{D}=\frac{3}{-3}=-1 \qquad z=\frac{D_z}{D}=\frac{-6}{-3}=2$$

The solution is $(3,-1,2)$ and the solution set is $\{(3,-1,2)\}$.

39. $x \quad +2z=4$

$2y-z=5$

$2x+3y \quad =13$

$$D=\begin{vmatrix}1 & 0 & 2\\ 0 & 2 & -1\\ 2 & 3 & 0\end{vmatrix}=1\begin{vmatrix}2 & -1\\ 3 & 0\end{vmatrix}-0\begin{vmatrix}0 & 2\\ 3 & 0\end{vmatrix}+2\begin{vmatrix}0 & 2\\ 2 & -1\end{vmatrix}$$

$$= 1(2(0) - 3(-1)) + 2(0(-1) - 2(2))$$
$$= 1(3) + 2(-4) = 3 - 8 = -5$$

$$D_x = \begin{vmatrix} 4 & 0 & 2 \\ 5 & 2 & -1 \\ 13 & 3 & 0 \end{vmatrix} = 4\begin{vmatrix} 2 & -1 \\ 3 & 0 \end{vmatrix} - 5\begin{vmatrix} 0 & 2 \\ 3 & 0 \end{vmatrix} + 13\begin{vmatrix} 0 & 2 \\ 2 & -1 \end{vmatrix}$$
$$= 4(2(0) - 3(-1)) - 5(0(0) - 3(2)) + 13(0(-1) - 2(2))$$
$$= 4(3) - 5(-6) + 13(-4) = 12 + 30 - 52 = -10$$

$$D_y = \begin{vmatrix} 1 & 4 & 2 \\ 0 & 5 & -1 \\ 2 & 13 & 0 \end{vmatrix} = 1\begin{vmatrix} 5 & -1 \\ 13 & 0 \end{vmatrix} - 0\begin{vmatrix} 4 & 2 \\ 13 & 0 \end{vmatrix} + 2\begin{vmatrix} 4 & 2 \\ 5 & -1 \end{vmatrix}$$
$$= 1(5(0) - 13(-1)) + 2(4(-1) - 5(2))$$
$$= 1(13) + 2(-4 - 10)$$
$$= 1(13) + 2(-14) = 13 - 28 = -15$$

$$D_z = \begin{vmatrix} 1 & 0 & 4 \\ 0 & 2 & 5 \\ 2 & 3 & 13 \end{vmatrix} = 1\begin{vmatrix} 2 & 5 \\ 3 & 13 \end{vmatrix} - 0\begin{vmatrix} 0 & 4 \\ 3 & 13 \end{vmatrix} + 2\begin{vmatrix} 0 & 4 \\ 2 & 5 \end{vmatrix}$$
$$= 1(2(13) - 3(5)) + 2(0(5) - 2(4))$$
$$= 1(26 - 15) + 2(-8)$$
$$= 1(11) - 16 = 11 - 16 = -5$$

$$x = \frac{D_x}{D} = \frac{-10}{-5} = 2 \qquad y = \frac{D_y}{D} = \frac{-15}{-5} = 3 \qquad z = \frac{D_z}{D} = \frac{-5}{-5} = 1$$

The solution is $(2, 3, 1)$ and the solution set is $\{(2, 3, 1)\}$.

Application Exercises

41.

$$\text{Area} = \pm\frac{1}{2}\begin{vmatrix} 3 & -5 & 1 \\ 2 & 6 & 1 \\ -3 & 5 & 1 \end{vmatrix} = \pm\frac{1}{2}\left[3\begin{vmatrix} 6 & 1 \\ 5 & 1 \end{vmatrix} - 2\begin{vmatrix} -5 & 1 \\ 5 & 1 \end{vmatrix} - 3\begin{vmatrix} -5 & 1 \\ 6 & 1 \end{vmatrix}\right]$$
$$= \pm\frac{1}{2}\left[3(6(1) - 5(1)) - 2(-5(1) - 5(1)) - 3(-5(1) - 6(1))\right]$$

$$= \pm\frac{1}{2}\left[3(6-5)-2(-5-5)-3(-5-6)\right]$$
$$= \pm\frac{1}{2}\left[3(1)-2(-10)-3(-11)\right]$$
$$= \pm\frac{1}{2}[3+20+33]$$
$$= \pm\frac{1}{2}[56] = \pm 28$$

The area is 28 square units.

43. $$\begin{vmatrix} 3 & -1 & 1 \\ 0 & -3 & 1 \\ 12 & 5 & 1 \end{vmatrix} = 3\begin{vmatrix} -3 & 1 \\ 5 & 1 \end{vmatrix} - 0\begin{vmatrix} -1 & 1 \\ 5 & 1 \end{vmatrix} + 12\begin{vmatrix} -1 & 1 \\ -3 & 1 \end{vmatrix}$$
$$= 3(-3(1)-5(1))+12(-1(1)-(-3)1)$$
$$= 3(-3-5)+12(-1+3)$$
$$= 3(-8)+12(2) = -24+24 = 0$$

Because the determinant is equal to zero, the points are collinear.

45. $$\begin{vmatrix} x & y & 1 \\ 3 & -5 & 1 \\ -2 & 6 & 1 \end{vmatrix} = x\begin{vmatrix} -5 & 1 \\ 6 & 1 \end{vmatrix} - 3\begin{vmatrix} y & 1 \\ 6 & 1 \end{vmatrix} - 2\begin{vmatrix} y & 1 \\ -5 & 1 \end{vmatrix}$$
$$= x(-5(1)-6(1))-3(y(1)-6(1))-2(y(1)-(-5)1)$$
$$= x(-5-6)-3(y-6)-2(y+5)$$
$$= x(-11)-3y+18-2y-10$$
$$= -11x-5y+8$$

To find the equation of the line, set the determinant equal to zero.

$$-11x-5y+8=0$$

Solve for y to obtain slope-intercept form.

$$-11x-5y+8=0$$
$$-5y = 11x-8$$
$$y = -\frac{11}{5}x+\frac{8}{5}$$

Writing in Mathematics

47. Answers will vary.

49. Answers will vary.

51. Answers will vary.

53. Answers will vary.

Technology Exercises

55. Answers will vary. For example, verify Exercise 7.

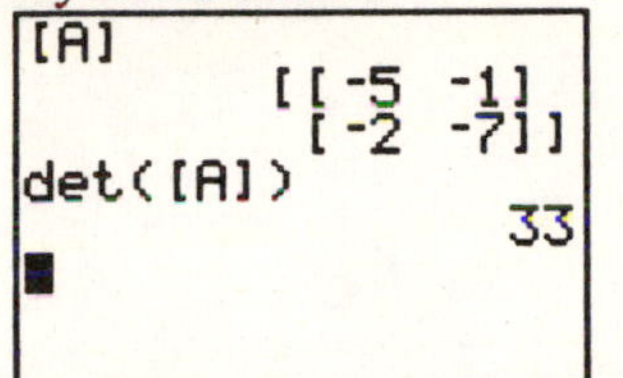

This is the same result obtained in Exercise 7.

Critical Thinking Exercises

57. Statement **d.** is true. Only one determinant is necessary to evaluate the determinant.

$$\begin{vmatrix} 2 & 3 & -2 \\ 0 & 1 & 3 \\ 0 & 4 & -1 \end{vmatrix}$$

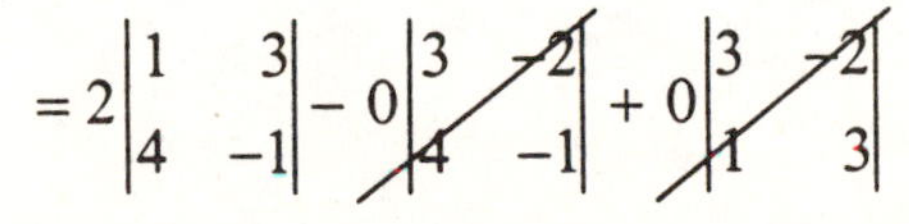

$$= 2\begin{vmatrix} 1 & 3 \\ 4 & -1 \end{vmatrix} - 0\begin{vmatrix} 3 & -2 \\ 4 & -1 \end{vmatrix} + 0\begin{vmatrix} 3 & -2 \\ 1 & 3 \end{vmatrix}$$

Two of the three 2×2 determinants are multiplied by zero so only one 2×2 is necessary to evaluate the expression.

Statement **a.** is false. For the determinant to equal zero, not every variable in the system must be zero.

Statement **b.** is false. Using Cramer's rule, we use $\frac{D_y}{D}$ to get the value of y.

Statement **c.** is false. Despite determinants being different (i.e., all entries are not identical), they can have the same value. This means that the numerators of the x and y when using Cramer's rule can have the same value, without being the same determinant. As a result, x and y can have the same value.

59. $\begin{vmatrix} 1 & 3 \\ 2 & 4 \end{vmatrix} = 1(4) - 2(3) = 4 - 6 = -2$

Switch the columns and re-evaluate.

$\begin{vmatrix} 3 & 1 \\ 4 & 2 \end{vmatrix} = 3(2) - 4(1) = 6 - 4 = 2$

When the columns of a second-order determinant are interchanged, the value of the determinant is multiplied by -1.

61. We are given two points, (x_1, y_1) and (x_2, y_2). To find the equation of a line using two points, we first find the slope, and then use the slope and one of the points to write the equation of the line in point-slope form. Here, the slope is: $m = \frac{y_2 - y_1}{x_2 - x_1}$. Using point slope form, we obtain: $y-y_1=\frac{y_2-y_1}{x_2-x_1}(x-x_1)$. To determine that this is equivalent to what is obtained when the determinant is set equal to zero, we multiply as follows:

$$y-y_1=\frac{y_2-y_1}{x_2-x_1}(x-x_1)$$
$$(x_2-x_1)(y-y_1)=(y_2-y_1)(x-x_1)$$
$$x_2y-x_2y_1-x_1y+x_1y_1=xy_2-x_1y_2-xy_1+x_1y_1$$

Now, evaluate the determinant to see if they are equivalent.

$$\begin{vmatrix} x & y & 1 \\ x_1 & y_1 & 1 \\ x_2 & y_2 & 1 \end{vmatrix}=0$$
$$x\begin{vmatrix} y_1 & 1 \\ y_2 & 1 \end{vmatrix}-x_1\begin{vmatrix} y & 1 \\ y_2 & 1 \end{vmatrix}+x_2\begin{vmatrix} y & 1 \\ y_1 & 1 \end{vmatrix}=0$$
$$x(y_1(1)-y_2(1))-x_1(y(1)-y_2(1))+x_2(y(1)-y_1(1))=0$$
$$x(y_1-y_2)-x_1(y-y_2)+x_2(y-y_1)=0$$
$$xy_1-xy_2-x_1y+x_1y_2+x_2y-x_2y_1=0$$
$$x_2y-x_2y_1-x_1y=xy_2-x_1y_2-xy_1$$

When we compare, we see that the equations not the same. Using the Addition Property of Equality, we add x_1y_1 to both sides of the equation and see that the expressions are the equivalent. This shows that the equation of a line through (x_1,y_1) and (x_2,y_2) is given by the determinant $\begin{vmatrix} x & y & 1 \\ x_1 & y_1 & 1 \\ x_2 & y_2 & 1 \end{vmatrix}=0$.

Review Exercises

62. $6x-4=2+6(x-1)$

$6x-4=2+6x-6$

$6x-4=6x-4$

$0=0$

Since $0=0$ for all x, the solution set is $\{x|x \text{ is a real number}\}$ or $\mathbb{R}$.

63. $-2x+3y=7$

$3y=2x+7$

$y=\frac{2x+7}{3}$ or $y=\frac{2}{3}x+\frac{7}{3}$

64. $\frac{4x+1}{3}=\frac{x-3}{6}+\frac{x+5}{6}$

$6\left(\frac{4x+1}{3}\right)=6\left(\frac{x-3}{6}\right)+6\left(\frac{x+5}{6}\right)$

$$2(4x+1)=x-3+x+5$$
$$8x+2=2x+2$$
$$6x+2=2$$
$$6x=0$$
$$x=\frac{0}{6}$$
$$x=0$$

The solution is 0 and the solution set is $\{0\}$.

Chapter 3 Review

3.1

1.
$$2x-5y=-2 \qquad 3x+4y=4$$
$$2(4)-5(2)=-2 \qquad 3(4)+4(2)=4$$
$$8-10=-2 \qquad 12+8=4$$
$$-2=-2 \qquad 20\neq 4$$

The pair is not a solution of the system.

2.
$$-x+2y=11 \qquad y=-\frac{x}{3}+\frac{4}{3}$$
$$-(-5)+2(3)=11 \qquad 3=-\frac{(-5)}{3}+\frac{4}{3}$$
$$5+6=11 \qquad 3=\frac{5}{3}+\frac{4}{3}$$
$$11=11 \qquad 3=\frac{9}{3}$$
$$3=3$$

The pair is a solution of the system.

3.
$$x+y=5$$
$$y=-x+5$$
$$m=-1$$
$$y\text{-intercept}=5$$

$$3x-y=3$$
$$-y=-3x+3$$
$$y=3x-3$$
$$m=3$$
$$y\text{-intercept}=-3$$

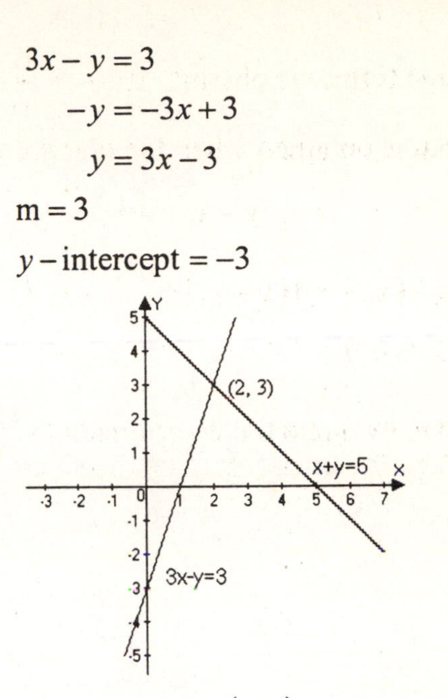

The solution is $(2,3)$ and the solution set is $\{(2,3)\}$.

4.
$$3x-2y=6$$
$$-2y=-3x+6$$
$$y=\frac{3}{2}x-3$$
$$m=\frac{3}{2}$$
$$y\text{-intercept}=-3$$

$$6x-4y=12$$
$$-4y=-6x+12$$
$$y=\frac{6}{4}x-3$$
$$y=\frac{3}{2}x-3$$
$$m=\frac{3}{2}$$
$$y\text{-intercept}=-3$$

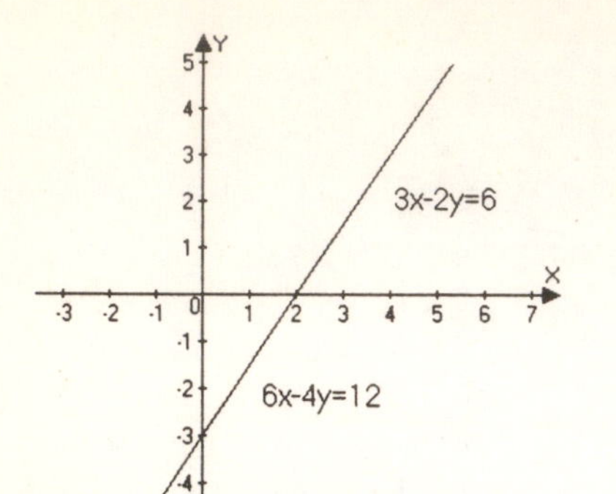

Since the lines coincide, the solution set is $\{(x, y)|3x-2y=6\}$ or $\{(x, y)|6x-4y=12\}$.

5.

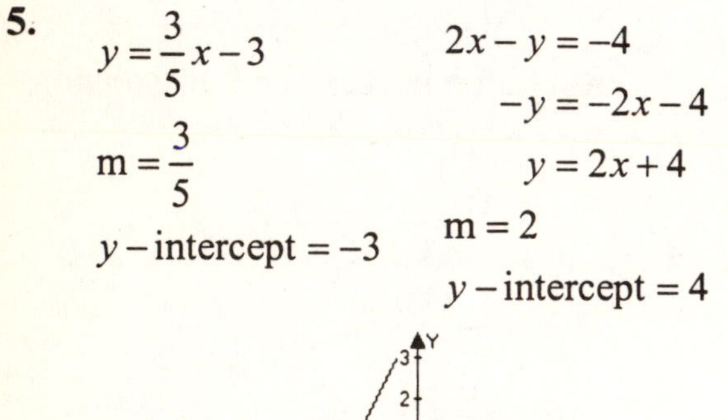

$y=\frac{3}{5}x-3$

$m=\frac{3}{5}$

$y-\text{intercept}=-3$

$2x-y=-4$

$-y=-2x-4$

$y=2x+4$

$m=2$

$y-\text{intercept}=4$

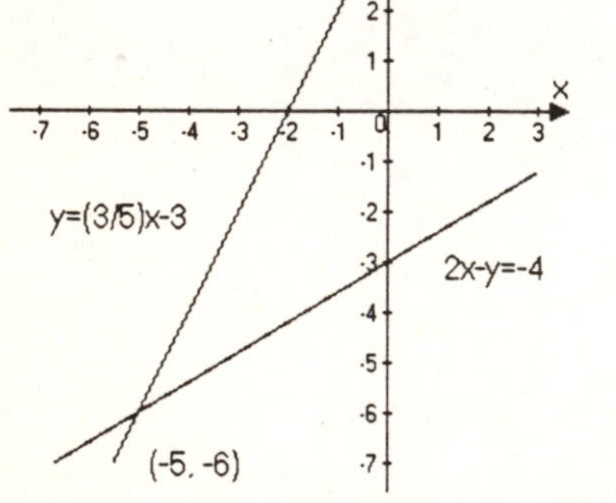

The solution is $(-5,-6)$ and the solution set is $\{(-5,-6)\}$.

6. $y=-x+4$

$m=-1$

$y-\text{intercept}=4$

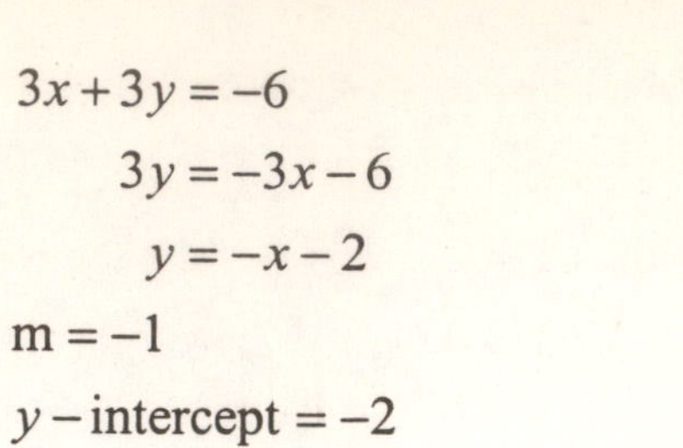

$3x+3y=-6$

$3y=-3x-6$

$y=-x-2$

$m=-1$

$y-\text{intercept}=-2$

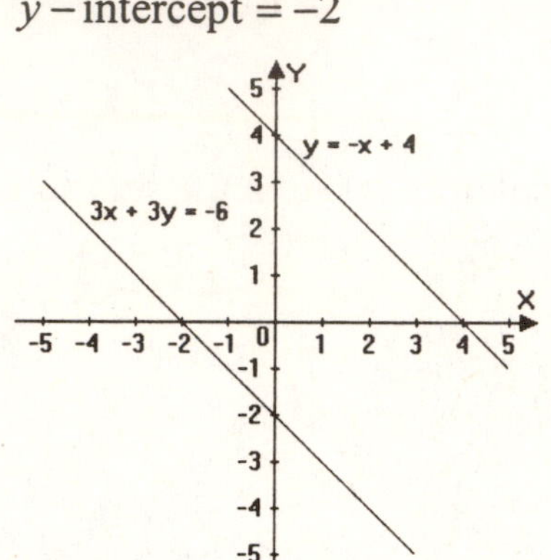

The lines do not intersect. The solution set is $\varnothing$ or $\{\ \}$.

7. $2x-\ y=\ 2$

$x+2y=11$

Solve the second equation for x.

$x+2y=11$

$x=11-2y$

Substitute for x in the first equation solve for y.

$2(11-2y)-y=2$

$22-4y-y=\ 2$

$22-5y=2$

$-5y=-20$

$y=4$

Back-substitute 4 for y to find x.

$2x-4=2$

$2x=6$

$x=3$

The solution is $(3,4)$ and the solution set is $\{(3,4)\}$.

8. $y = -2x + 3$

$3x + 2y = -17$

Substitute for y in the second equation and solve for x.

$$3x + 2y = -17$$
$$3x + 2(-2x + 3) = -17$$
$$3x - 4x + 6 = -17$$
$$-x + 6 = -17$$
$$-x = -23$$
$$x = 23$$

Back-substitute 23 for x to find y.

$$y = -2(23) + 3$$
$$y = -46 + 3$$
$$y = -43$$

The solution is $(23, -43)$ and the solution set is $\{(23, -43)\}$.

9. $3x + 2y = -8$

$2x + 5y = 2$

Multiply the first equation by -2 and the second equation by 3 and solve by addition.

$$\begin{aligned} -6x - 4y &= 16 \\ 6x + 15y &= 6 \\ \hline 11y &= 22 \\ y &= 2 \end{aligned}$$

Back-substitute 2 for y in one of the original equations.

$$3x + 2(2) = -8$$
$$3x + 4 = -8$$
$$3x = -12$$
$$x = -4$$

The solution is $(-4, 2)$ and the solution set is $\{(-4, 2)\}$.

10. $5x - 2y = 14$

$3x + 4y = 11$

Multiply the first equation by 2 and add to the second equation.

$$\begin{aligned} 10x - 4y &= 28 \\ 3x + 4y &= 11 \\ \hline 13x &= 39 \\ x &= 3 \end{aligned}$$

Back-substitute 3 for x in one of the original equations.

$$5x - 2y = 14$$
$$5(3) - 2y = 14$$
$$15 - 2y = 14$$
$$-2y = -1$$
$$y = \frac{1}{2}$$

The solution is $\left(3, \frac{1}{2}\right)$ and the solution set is $\left\{\left(3, \frac{1}{2}\right)\right\}$.

11. $y = 4 - x$

$3x + 3y = 12$

Substitute for x in the first equation solve for y.

$$3x + 3y = 12$$
$$3x + 3(4 - x) = 12$$
$$3x + 12 - 3x = 12$$
$$12 = 12$$

The system has infinitely many solutions. The solution set is $\{(x, y) | y = 4 - x\}$ or $\{(x, y) | 3x + 3y = 12\}$.

12. $\frac{x}{8}+\frac{3y}{4}=\frac{19}{8}$

$-\frac{x}{2}+\frac{3y}{4}=\frac{1}{2}$

To clear fractions, multiply the first equation by 8 and the second equation by 4.

$8\left(\frac{x}{8}\right)+8\left(\frac{3y}{4}\right)=8\left(\frac{19}{8}\right)$

$x+2(3y)=19$

$x+6y=19$

$4\left(-\frac{x}{2}\right)+4\left(\frac{3y}{4}\right)=4\left(\frac{1}{2}\right)$

$2(-x)+3y=2(1)$

$-2x+3y=2$

The system becomes:

$x+6y=19$

$-2x+3y=2$

Multiply the first equation by 2 and add to the second equation.

$2x+12y=38$

$\underline{-2x+3y=2}$

$15y=40$

$y=\frac{40}{15}$

$y=\frac{8}{3}$

Back-substitute $\frac{8}{3}$ for y in one of the equations to find x.

$x+6\left(\frac{8}{3}\right)=19$

$x+2(8)=19$

$x+16=19$

$x=3$

The solution is $\left(3,\frac{8}{3}\right)$ and the solution set is $\left\{\left(3,\frac{8}{3}\right)\right\}$.

13. $x-2y+3=0$

$2x-4y+7=0$

Rewrite the system is standard form.

$x-2y=-3$

$2x-4y=-7$

Multiply the first equation by –2 and add to the second equation.

$-2x+4y=6$

$\underline{2x-4y=-7}$

$0=-1$

There are no values for x and y that will make 0 = -1, so the solution set is $\varnothing$ or $\{\ \}$.

3.2

14. Let t = the cost of the televisions
Let s = the cost of the stereos

$3t+4s=2530$

$4t+3s=2510$

Multiply the first equation by –4 and the second by 3 and solve by addition.

$3t+4s=2530$

$-4(3t)+(-4)(4s)=-4(2530)$

$-12t-16s=-10120$

$4t+3s=2510$

$3(4t)+3(3s)=3(2510)$

$12t+9s=7530$

The system of two equations in two variables:

$$\begin{aligned} -12t-16s &= -10120 \\ 12t+\ \ 9s &= \ \ \ 7530 \\ \hline -7s &= -2590 \\ s &= 370 \end{aligned}$$

Back-substitute 370 for s in one of the original equations to find t.

$$\begin{aligned} 3t+4s &= 2530 \\ 3t+4(370) &= 2530 \\ 3t+1480 &= 2530 \\ 3t &= 1050 \\ t &= 350 \end{aligned}$$

The televisions cost \$350 each and the stereos costs \$370 each.

15. Let x = the amount invested at 8%
Let y = the amount invested at 10%

$$\begin{aligned} x+\qquad y &= 10000 \\ 0.08x+0.10y &= \ \ 940 \end{aligned}$$

Multiply the first equation by –0.08 and add.

$$\begin{aligned} -0.08x-0.08y &= -800 \\ 0.08x+0.10y &= \ \ 940 \\ \hline 0.02y &= 140 \\ y &= 7000 \end{aligned}$$

Back-substitute 7000 for y in one of the original equations to find x.

$$\begin{aligned} x+y &= 10000 \\ x+7000 &= 10000 \\ x &= 3000 \end{aligned}$$

There was \$3000 invested at 8% and \$7000 invested at 10%.

16. Let x = the amount of the 34% solution
Let y = the amount of the 4% solution

$$\begin{aligned} x+\qquad y &= 100 \\ 0.34x+0.04y &= 0.07(100) \end{aligned}$$

Simplified, the system becomes:

$$\begin{aligned} x+\qquad y &= 100 \\ 0.34x+0.04y &= \ \ 7 \end{aligned}$$

Multiply the first equation by –0.34 and add to the second equation.

$$\begin{aligned} -0.34x-0.34y &= -34 \\ 0.34x+0.04y &= \ \ 7 \\ \hline -0.30 &= -27 \\ y &= 90 \end{aligned}$$

Back-substitute 90 for y to find x.

$$\begin{aligned} x+y &= 100 \\ x+90 &= 100 \\ x &= 10 \end{aligned}$$

10 ml of the 34% solution and 90 ml of the 4% solution must be used.

17. Let r = the speed of the plane in still air
Let w = the speed of the wind

	r ·	t =	d
With Wind	**r + w**	3	3(*r* + *w*)
Against Wind	**r – w**	4	3(*r* - *w*)

$$\begin{aligned} 3(r+w) &= 2160 \\ 4(r-w) &= 2160 \end{aligned}$$

Simplified, the system becomes:

$$\begin{aligned} 3r+3w &= 2160 \\ 4r-4w &= 2160 \end{aligned}$$

Multiply the first equation by 4, the second equation by 3, and solve by addition.

$$\begin{aligned} 12r+12w &= 8640 \\ 12r-12w &= 6480 \\ \hline 24r &= 15120 \\ r &= 630 \end{aligned}$$

Back-substitute 630 for r to find w.

$$3r+3w=2160$$
$$3(630)+3w=2160$$
$$1890+3w=2160$$
$$3w=270$$
$$w=90$$

The speed of the plane in still air is 630 miles per hour and the speed of the wind is 90 miles per hour.

18. Let l = the length of the table
Let w = the width of the table

$$2l+2w=28$$
$$4l-3w=21$$

Multiply the first equation by –2 and solve by addition.

$$-4l-4w=-56$$
$$4l-3w=21$$
$$-7w=-35$$
$$w=5$$

Back-substitute 5 for w to find l.

$$2l+2w=28$$
$$2l+2(5)=28$$
$$2l+10=28$$
$$2l=18$$
$$l=9$$

The dimensions of the table are 9 feet by 5 feet.

19. $C(x)=22500+40x$

$$C(400)=22500+40(400)$$
$$=22500+16000$$
$$=38500$$
$$R(x)=85x$$
$$R(400)=85(400)$$
$$=34000$$
$$P(x)=R(x)-C(x)$$
$$P(400)=R(400)-C(400)$$
$$=34000-38500$$
$$=-4500$$

There is a $4500 profit loss when 400 calculators are sold.

20. $P(x)=R(x)-C(x)$

$$=85x-(22500+40x)$$
$$=85x-22500-40x$$
$$=45x-22,500$$

21. $R(x)=C(x)$

$$85x=22500+40x$$
$$45x=22500$$
$$x=500$$
$$R(x)=85x$$
$$R(500)=85(500)$$
$$=42500$$

The break-even point is (500, 42,500). This means that when 500 calculators are produced and sold, the cost is the same as the revenue at $42,500. At this point, there is no profit loss or gain.

22. **a.** $C(x)=60000+200x$

b. $R(x)=450x$

c. $R(x)=C(x)$

$$450x=60000+200x$$
$$250x=60000$$
$$x=240$$
$$R(x)=450x$$

$$R(240) = 450(240)$$
$$= 108,000$$

The break-even point is (240, 108,000). This means that when 240 desks are produced and sold, the cost is the same as the revenue at $108,000. At this point, there is no profit loss or gain.

3.3

23.

$$x + y + z = 0$$
$$-3 + (-2) + 5 = 0$$
$$-5 + 5 = 0$$
$$0 = 0$$

True

$$2x - 3y + z = 5$$
$$2(-3) - 3(-2) + 5 = 5$$
$$-6 + 6 + 5 = 5$$
$$5 = 5$$

True

$$4x + 2y + 4z = 3$$
$$4(-3) + 2(-2) + 4(5) = 3$$
$$-12 - 4 + 20 = 3$$
$$4 = 3$$

False

The ordered triple (–3, –2, 5) does not satisfy all three equations, so it is not a solution.

24.

$$2x - y + z = 1$$
$$3x - 3y + 4z = 5$$
$$4x - 2y + 3z = 4$$

Multiply the first equation by –2 and add to the third.

$$-4x + 2y - 2z = -2$$
$$\underline{4x - 2y + 3z = 4}$$
$$z = 2$$

Back-substitute 2 for z in two of the original equations to obtain a system of two equations in two variables.

$$2x - y + z = 1 \qquad 3x - 3y + 4z = 5$$
$$2x - y + 2 = 1 \qquad 3x - 3y + 4(2) = 5$$
$$2x - y = -1 \qquad 3x - 3y + 8 = 5$$
$$3x - 3y = -3$$

The system of two equations in two variables becomes:

$$2x - y = -1$$
$$3x - 3y = -3$$

Multiply the first equation by –3 and add solve by addition.

$$-6x + 3y = 3$$
$$\underline{3x - 3y = -3}$$
$$-3x = 0$$
$$x = 0$$

Back-substitute 0 for x to find y.

$$2x - y = -1$$
$$2(0) - y = -1$$
$$-y = -1$$
$$y = 1$$

The solution is $(0,1,2)$ and the solution set is $\{(0,1,2)\}$.

25.

$$x + 2y - z = 5$$
$$2x - y + 3z = 0$$
$$2y + z = 1$$

Multiply the first equation by –2 and add to the second equation.

$$\begin{aligned} -2x - 4y + 2z &= -10 \\ 2x - y + 3z &= 0 \\ \hline -5y + 5z &= -10 \end{aligned}$$

We now have two equations in two variables.

$$2y + z = 1$$
$$-5y + 5z = -10$$

Multiply the first equation by –5 and solve by addition.

$$\begin{aligned} -10y - 5z &= -5 \\ -5y + 5z &= -10 \\ \hline -15y &= -15 \\ y &= 1 \end{aligned}$$

Back-substitute 1 for y to find z.

$$2(1) + z = 1$$
$$2 + z = 1$$
$$z = -1$$

Back-substitute 1 for y and –1 for z to find x.

$$x + 2y - z = 5$$
$$x + 2(1) - (-1) = 5$$
$$x + 2 + 1 = 5$$
$$x + 3 = 5$$
$$x = 2$$

The solution is $(2,1,-1)$ and the solution set is $\{(2,1,-1)\}$.

26. $3x - 4y + 4z = 7$

$$x - y - 2z = 2$$
$$2x - 3y + 6z = 5$$

Multiply the second equation by –3 and add to the third equation.

$$\begin{aligned} -3x + 3y + 6z &= -6 \\ 2x - 3y + 6z &= 5 \\ \hline -x + 12z &= -1 \end{aligned}$$

Multiply the second equation by –4 and add to the first equation.

$$\begin{aligned} 3x - 4y + 4z &= 7 \\ -4x + 4y + 8z &= -8 \\ \hline -x + 12z &= -1 \end{aligned}$$

The system of two equations in two variables becomes:

$$-x + 12z = -1$$
$$-x + 12z = -1$$

The two equations in two variables are identical. The system is dependent. There are an infinite number of solutions to the system.

27. Use each ordered pair to write an equation as follows:

$$(x, y) = (1, 4)$$
$$y = ax^2 + bx + c$$
$$4 = a(1)^2 + b(1) + c$$
$$4 = a + b + c$$

$$(x, y) = (3, 20)$$
$$y = ax^2 + bx + c$$
$$20 = a(3)^2 + b(3) + c$$
$$20 = a(9) + 3b + c$$
$$20 = 9a + 3b + c$$

$$(x, y) = (-2, 25)$$
$$y = ax^2 + bx + c$$
$$25 = a(-2)^2 + b(-2) + c$$
$$25 = a(4) - 2b + c$$
$$25 = 4a - 2b + c$$

The system of three equations in three variables is:

$$a+b+c=4$$
$$9a+3b+c=20$$
$$4a-2b+c=25$$

Multiply the first equation by –1 and add to the second equation.

$$-a-b-c=-4$$
$$\underline{9a+3b+c=20}$$
$$8a+2b=16$$

Multiply the first equation by –1 and add to the third equation.

$$-a-b-c=-4$$
$$\underline{4a-2b+c=25}$$
$$3a-3b=21$$

The system of two equations in two variables becomes:

$$8a+2b=16$$
$$3a-3b=21$$

Multiply the first equation by 3, the second equation by 2 and solve by addition.

$$24a+6b=48$$
$$\underline{6a-6b=42}$$
$$30a=90$$
$$a=3$$

Back-substitute 3 for a to find b.

$$3(3)-3b=21$$
$$9-3b=21$$
$$-3b=12$$
$$b=-4$$

Back-substitute 3 for a and –4 for b to find c.

$$a+b+c=4$$
$$3+(-4)+c=4$$
$$-1+c=4$$
$$c=5$$

The quadratic function is $f(x)=3x^2-4x+5$.

28. $x+y+z=45$

$$x=y+1$$
$$2z-12=x$$

Rewrite the system in standard form.

$$x+y+z=45$$
$$x-y=1$$
$$-x+2z=12$$

Add the first and second equations to eliminate y.

$$x+y+z=45$$
$$\underline{x-y=1}$$
$$2x+z=46$$

We obtain a system of two equations in two variables.

$$-x+2z=12$$
$$2x+z=46$$

Multiply the first equation by 2 and solve by addition.

$$-2x+4z=24$$
$$\underline{2x+z=46}$$
$$5z=70$$
$$z=14$$

Back-substitute 14 for z to find x.

$$-x+2z=12$$
$$-x+2(14)=12$$
$$-x+28=12$$
$$-x=-16$$
$$x=16$$

Back-substitute 16 for x to find y.

$$x-y=1$$
$$16-y=1$$
$$-y=-15$$
$$y=15$$

The savings rate for Japan is 16%, the savings rate for Germany is 15%, and the savings rate for France is 14%.

3.4

29. $\left[\begin{array}{cc|c} 2 & 3 & -10 \\ 0 & 1 & -6 \end{array}\right]$

The system is:

$$\begin{aligned} 2x+3y &= -10 \\ y &= -6 \end{aligned}$$

Back-substitute –6 for y to find x.

$$\begin{aligned} 2x+3y &= -10 \\ 2x+3(-6) &= -10 \\ 2x-18 &= -10 \\ 2x &= 8 \\ x &= 4 \end{aligned}$$

The solution is $(4,-6)$ and the solution set is $\{(4,-6)\}$.

30. $\left[\begin{array}{ccc|c} 1 & 1 & 3 & 12 \\ 0 & 1 & -2 & -4 \\ 0 & 0 & 1 & 3 \end{array}\right]$

The system is:

$$\begin{aligned} x+y+3z &= 12 \\ y-2z &= -4 \\ z &= 3 \end{aligned}$$

Back-substitute 3 for z to find y.

$$\begin{aligned} y-2z &= -4 \\ y-2(3) &= -4 \\ y-6 &= -4 \\ y &= 2 \end{aligned}$$

Back-substitute 2 for y and 3 for z to find x.

$$\begin{aligned} x+2+3(3) &= 12 \\ x+2+9 &= 12 \\ x+11 &= 12 \\ x &= 1 \end{aligned}$$

The solution is $(1,2,3)$ and the solution set is $\{(1,2,3)\}$.

31. $\left[\begin{array}{cc|c} 1 & -8 & 3 \\ 0 & 7 & -14 \end{array}\right] \frac{1}{7}R_2$

$= \left[\begin{array}{cc|c} 1 & -8 & 3 \\ 0 & 1 & -2 \end{array}\right]$

32. $\left[\begin{array}{cc|c} 1 & -3 & 1 \\ 2 & 1 & -5 \end{array}\right] -2R_1 + R_2$

$= \left[\begin{array}{cc|c} 1 & -3 & 1 \\ 0 & 7 & -7 \end{array}\right]$

33. $\left[\begin{array}{ccc|c} 2 & -2 & 1 & -1 \\ 1 & 2 & -1 & 2 \\ 6 & 4 & 3 & 5 \end{array}\right] \frac{1}{2}R_1$

$= \left[\begin{array}{ccc|c} 1 & -1 & 1/2 & -1/2 \\ 1 & 2 & -1 & 2 \\ 6 & 4 & 3 & 5 \end{array}\right]$

34. $\left[\begin{array}{ccc|c} 1 & 2 & 2 & 2 \\ 0 & 1 & -1 & 2 \\ 0 & 5 & 4 & 1 \end{array}\right] -5R_2 + R_2$

$= \left[\begin{array}{ccc|c} 1 & 2 & 2 & 2 \\ 0 & 1 & -1 & 2 \\ 0 & 0 & 9 & -9 \end{array}\right]$

35. $\left[\begin{array}{cc|c} 1 & 4 & 7 \\ 3 & 5 & 0 \end{array}\right]$ $-3R_1+R_2$

$=\left[\begin{array}{cc|c} 1 & 4 & 7 \\ 0 & -7 & -21 \end{array}\right]$ $-\frac{1}{7}R_2$

$=\left[\begin{array}{cc|c} 1 & 4 & 7 \\ 0 & 1 & 3 \end{array}\right]$

The resulting system is:

$x+4y=7$

$y=3$

Back-substitute 3 for y to find x.

$x+4y=7$

$x+4(3)=7$

$x+12=7$

$x=-5$

The solution is $(-5,3)$ and the solution set is $\{(-5,3)\}$.

36. $\left[\begin{array}{cc|c} 2 & -3 & 8 \\ -6 & 9 & 4 \end{array}\right]$ $3R_1+R_2$

$=\left[\begin{array}{cc|c} 2 & -3 & 8 \\ 0 & 0 & 28 \end{array}\right]$

This is a contradiction. R_2 states that

$0x+0y=28$

$0=28$

There are no values of x and y for which $0=28$. The system is inconsistent and the solution set is $\varnothing$ or $\{\ \}$.

37. $\left[\begin{array}{ccc|c} 1 & 2 & 3 & -5 \\ 2 & 1 & 1 & 1 \\ 1 & 1 & -1 & 8 \end{array}\right]$ $-2R_1+R_2$

$=\left[\begin{array}{ccc|c} 1 & 2 & 3 & -5 \\ 0 & -3 & -5 & 11 \\ 1 & 1 & -1 & 8 \end{array}\right]$ $-R_1+R_3$

$=\left[\begin{array}{ccc|c} 1 & 2 & 3 & -5 \\ 0 & -3 & -5 & 11 \\ 0 & -1 & -4 & 13 \end{array}\right]$ $-\frac{1}{3}R_2$

$=\left[\begin{array}{ccc|c} 1 & 2 & 3 & -5 \\ 0 & 1 & \frac{5}{3} & -\frac{11}{3} \\ 0 & -1 & -4 & 13 \end{array}\right]$ R_2+R_3

$=\left[\begin{array}{ccc|c} 1 & 2 & 3 & -5 \\ 0 & 1 & \frac{5}{3} & -\frac{11}{3} \\ 0 & 0 & -\frac{7}{3} & \frac{28}{3} \end{array}\right]$ $-\frac{3}{7}R$

$=\left[\begin{array}{ccc|c} 1 & 2 & 3 & -5 \\ 0 & 1 & \frac{5}{3} & -\frac{11}{3} \\ 0 & 0 & 1 & -4 \end{array}\right]$

The resulting system is:

$x+2y+3z=-5$

$y+\frac{5}{3}z=-\frac{11}{3}$

$z=-4$

Back-substitute –4 for z to find y.

$y+\frac{5}{3}z=-\frac{11}{3}$

$y+\frac{5}{3}(-4)=-\frac{11}{3}$

$y-\frac{20}{3}=-\frac{11}{3}$

$y=\frac{9}{3}$

$y=3$

Back-substitute 3 for y and –4 for z to find x.

$$\begin{aligned} x+2y+3z &= -5 \\ x+2(3)+3(-4) &= -5 \\ x+6-12 &= -5 \\ x-6 &= -5 \\ x &= 1 \end{aligned}$$

The solution is $(1,3,-4)$ and the solution set is $\{(1,3,-4)\}$.

38. $\left[\begin{array}{ccc|c} 1 & -2 & 1 & 0 \\ 0 & 1 & -3 & -1 \\ 0 & 2 & 5 & -2 \end{array}\right]$ $-2R_2+R_3$

$=\left[\begin{array}{ccc|c} 1 & -2 & 1 & 0 \\ 0 & 1 & -3 & -1 \\ 0 & 0 & 11 & 0 \end{array}\right]$ $\frac{1}{11}R_3$

$=\left[\begin{array}{ccc|c} 1 & -2 & 1 & 0 \\ 0 & 1 & -3 & -1 \\ 0 & 0 & 1 & 0 \end{array}\right]$ $\frac{1}{11}R_3$

The resulting system is:

$$\begin{aligned} x-2y+z &= 0 \\ y-3z &= -1 \\ z &= 0 \end{aligned}$$

Back-substitute 0 for z to find y.

$$\begin{aligned} y-3z &= -1 \\ y-\cancel{3(0)} &= -1 \\ y &= -1 \end{aligned}$$

Back-substitute 1 for y and 0 for z to find x.

$$\begin{aligned} x-2y+z &= 0 \\ x-2(-1)+0 &= 0 \\ x+2 &= 0 \\ x &= -2 \end{aligned}$$

The solution is $(-2,-1,0)$ and the solution set is $\{(-2,-1,0)\}$.

3.5

39. $\begin{vmatrix} 3 & 2 \\ -1 & 5 \end{vmatrix} = 3(5)-(-1)2 = 15+2 = 17$

40. $\begin{vmatrix} -2 & -3 \\ -4 & -8 \end{vmatrix} = -2(-8)-(-4)(-3)$

$= 16-12 = 4$

41. $\begin{vmatrix} 2 & 4 & -3 \\ 1 & -1 & 5 \\ -2 & 4 & 0 \end{vmatrix} = 2\begin{vmatrix} -1 & 5 \\ 4 & 0 \end{vmatrix} - 1\begin{vmatrix} 4 & -3 \\ 4 & 0 \end{vmatrix} - 2\begin{vmatrix} 4 & -3 \\ -1 & 5 \end{vmatrix}$

$$\begin{aligned} &= 2(\cancel{-1(0)}-4(5))-1(\cancel{4(0)}-4(-3))-2(4(5)-(-1)(-3)) \\ &= 2(-20)-1(12)-2(20-3) \\ &= -40-12-2(17) \\ &= -40-12-34 = -86 \end{aligned}$$

42. $\begin{vmatrix} 4 & 7 & 0 \\ -5 & 6 & 0 \\ 3 & 2 & -4 \end{vmatrix} = 4\begin{vmatrix} 6 & 0 \\ 2 & -4 \end{vmatrix} - (-5)\begin{vmatrix} 7 & 0 \\ 2 & -4 \end{vmatrix} + 3\begin{vmatrix} 7 & 0 \\ 6 & 0 \end{vmatrix}$

$$= 4\left(6(-4) - \cancel{2(0)}\right) + 5\left(7(-4) - \cancel{2(0)}\right) + 3\left(\cancel{7(0)} - \cancel{6(0)}\right)$$

$$= 4(-24) + 5(-28) + \cancel{3(0)} = -96 - 140 = -236$$

43. $D = \begin{vmatrix} 1 & -2 \\ 3 & 2 \end{vmatrix} = 1(2) - 3(-2) = 2 + 6 = 8$

$D_x = \begin{vmatrix} 8 & -2 \\ -1 & 2 \end{vmatrix} = 8(2) - (-1)(-2) = 16 - 2 = 14$

$D_y = \begin{vmatrix} 1 & 8 \\ 3 & -1 \end{vmatrix} = 1(-1) - 3(8) = -1 - 24 = -25$

$x = \dfrac{D_x}{D} = \dfrac{14}{8} = \dfrac{7}{4}$ $y = \dfrac{D_y}{D} = \dfrac{-25}{8} = -\dfrac{25}{8}$

The solution is $\left(\dfrac{7}{4}, -\dfrac{25}{8}\right)$ and the solution set is $\left\{\left(\dfrac{7}{4}, -\dfrac{25}{8}\right)\right\}$.

44. $D = \begin{vmatrix} 7 & 2 \\ 2 & 1 \end{vmatrix} = 7(1) - 2(2) = 7 - 4 = 3$

$D_x = \begin{vmatrix} 0 & 2 \\ -3 & 1 \end{vmatrix} = \cancel{0(1)} - (-3)(2) = 6$ $D_y = \begin{vmatrix} 7 & 0 \\ 2 & -3 \end{vmatrix} = 7(-3) - \cancel{2(0)} = -21$

$x = \dfrac{D_x}{D} = \dfrac{6}{3} = 2$ $y = \dfrac{D_y}{D} = \dfrac{-21}{3} = -7$

The solution is $(2, -7)$ and the solution set is $\{(2, -7)\}$.

45. $D = \begin{vmatrix} 1 & 2 & 2 \\ 2 & 4 & 7 \\ -2 & -5 & -2 \end{vmatrix} = 1\begin{vmatrix} 4 & 7 \\ -5 & -2 \end{vmatrix} - 2\begin{vmatrix} 2 & 2 \\ -5 & -2 \end{vmatrix} - 2\begin{vmatrix} 2 & 2 \\ 4 & 7 \end{vmatrix}$

$$= 1(4(-2) - (-5)7) - 2(2(-2) - (-5)2) - 2(2(7) - 4(2))$$

$$= 1(-8 + 35) - 2(-4 + 10) - 2(14 - 8)$$

$$= 1(27) - 2(6) - 2(6) = 27 - 12 - 12 = 3$$

$$D_x = \begin{vmatrix} 5 & 2 & 2 \\ 19 & 4 & 7 \\ 8 & -5 & -2 \end{vmatrix} = 5\begin{vmatrix} 4 & 7 \\ -5 & -2 \end{vmatrix} - 19\begin{vmatrix} 2 & 2 \\ -5 & -2 \end{vmatrix} + 8\begin{vmatrix} 2 & 2 \\ 4 & 7 \end{vmatrix}$$

$$= 5(4(-2)-(-5)7) - 19(2(-2)-(-5)2) + 8(2(7)-4(2))$$

$$= 5(-8+35) - 19(-4+10) + 8(14-8)$$

$$= 5(27) - 19(6) + 8(6) = 135 - 114 + 48 = 69$$

$$D_y = \begin{vmatrix} 1 & 5 & 2 \\ 2 & 19 & 7 \\ -2 & 8 & -2 \end{vmatrix} = 1\begin{vmatrix} 19 & 7 \\ 8 & -2 \end{vmatrix} - 2\begin{vmatrix} 5 & 2 \\ 8 & -2 \end{vmatrix} - 2\begin{vmatrix} 5 & 2 \\ 19 & 7 \end{vmatrix}$$

$$= 1(19(-2)-(8)7) - 2(5(-2)-8(2)) - 2(5(7)-19(2))$$

$$= 1(-38-56) - 2(-10-16) - 2(35-38)$$

$$= 1(-94) - 2(-26) - 2(-3) = -94 + 52 + 6 = -36$$

$$D_z = \begin{vmatrix} 1 & 2 & 5 \\ 2 & 4 & 19 \\ -2 & -5 & 8 \end{vmatrix} = 1\begin{vmatrix} 4 & 19 \\ -5 & 8 \end{vmatrix} - 2\begin{vmatrix} 2 & 5 \\ -5 & 8 \end{vmatrix} - 2\begin{vmatrix} 2 & 5 \\ 4 & 19 \end{vmatrix}$$

$$= 1(4(8)-(-5)19) - 2(2(8)-(-5)5) - 2(2(19)-4(5))$$

$$= 1(32+95) - 2(16+25) - 2(38-20)$$

$$= 1(127) - 2(41) - 2(18) = 127 - 82 - 36 = 9$$

$$x = \frac{D_x}{D} = \frac{69}{3} = 23 \qquad y = \frac{D_y}{D} = \frac{-36}{3} = -12 \qquad z = \frac{D_z}{D} = \frac{9}{3} = 3$$

The solution is $(23, -12, 3)$ and the solution set is $\{(23, -12, 3)\}$.

46. Rewrite the system in standard form.

$$\begin{aligned} 2x + y \quad\quad &= -4 \\ y - 2z &= 0 \\ 3x \quad\quad - 2z &= -11 \end{aligned}$$

$$D = \begin{vmatrix} 2 & 1 & 0 \\ 0 & 1 & -2 \\ 3 & 0 & -2 \end{vmatrix} = 2\begin{vmatrix} 1 & -2 \\ 0 & -2 \end{vmatrix} - 0\begin{vmatrix} 1 & 0 \\ 0 & -2 \end{vmatrix} + 3\begin{vmatrix} 1 & 0 \\ 1 & -2 \end{vmatrix}$$

$$= 2(1(-2) - 0(-2)) + 3(1(-2) - 1(0))$$

$$= 2(-2) + 3(-2) = -4 - 6 = -10$$

$$D_x = \begin{vmatrix} -4 & 1 & 0 \\ 0 & 1 & -2 \\ -11 & 0 & -2 \end{vmatrix} = -4\begin{vmatrix} 1 & -2 \\ 0 & -2 \end{vmatrix} - 0\begin{vmatrix} 1 & 0 \\ 0 & -2 \end{vmatrix} - 11\begin{vmatrix} 1 & 0 \\ 1 & -2 \end{vmatrix}$$

$$= -4\left(1(-2) - 0(-2)\right) - 11\left(1(-2) - 1(0)\right)$$

$$= -4(-2) - 11(-2) = 8 + 22 = 30$$

$$D_y = \begin{vmatrix} 2 & -4 & 0 \\ 0 & 0 & -2 \\ 3 & -11 & -2 \end{vmatrix} = 2\begin{vmatrix} 0 & -2 \\ -11 & -2 \end{vmatrix} - 0\begin{vmatrix} -4 & 0 \\ -11 & -2 \end{vmatrix} + 3\begin{vmatrix} -4 & 0 \\ 0 & -2 \end{vmatrix}$$

$$= 2\left(0(-2) - (-11)(-2)\right) + 3\left(-4(-2) - 0(0)\right)$$

$$= 2(-22) + 3(8) = -44 + 24 = -20$$

$$D_z = \begin{vmatrix} 2 & 1 & -4 \\ 0 & 1 & 0 \\ 3 & 0 & -11 \end{vmatrix} = 2\begin{vmatrix} 1 & 0 \\ 0 & -11 \end{vmatrix} - 0\begin{vmatrix} 1 & -4 \\ 0 & -11 \end{vmatrix} + 3\begin{vmatrix} 1 & -4 \\ 1 & 0 \end{vmatrix}$$

$$= 2\left(1(-11) - 0(0)\right) + 3\left(1(0) - 1(-4)\right)$$

$$= 2(-11) + 3(4) = -22 + 12 = -10$$

$$x = \frac{D_x}{D} = \frac{30}{-10} = -3 \qquad y = \frac{D_y}{D} = \frac{-20}{-10} = 2 \qquad z = \frac{D_z}{D} = \frac{-10}{-10} = 1$$

The solution is $(-3, 2, 1)$ and the solution set is $\{(-3, 2, 1)\}$.

47. Use each ordered pair to write an equation as follows:

$(x, y) = (20, 400)$	$(x, y) = (40, 150)$	$(x, y) = (60, 400)$
$y = ax^2 + bx + c$	$y = ax^2 + bx + c$	$y = ax^2 + bx + c$
$400 = a(20)^2 + b(20) + c$	$150 = a(40)^2 + b(40) + c$	$400 = a(60)^2 + b(60) + c$
$400 = a(400) + 20b + c$	$150 = a(1600) + 40b + c$	$400 = a(3600) + 60b + c$
$400 = 400a + 20b + c$	$150 = 1600a + 40b + c$	$400 = 3600a + 60b + c$

The system of three equations in three variables is:

$$400a + 20b + c = 400$$
$$1600a + 40b + c = 150$$
$$3600a + 60b + c = 400$$

$$D=\begin{vmatrix}400 & 20 & 1\\1600 & 40 & 1\\3600 & 60 & 1\end{vmatrix}=400\begin{vmatrix}40 & 1\\60 & 1\end{vmatrix}-1600\begin{vmatrix}20 & 1\\60 & 1\end{vmatrix}+3600\begin{vmatrix}20 & 1\\40 & 1\end{vmatrix}$$

$$=400(40(1)-60(1))-1600(20(1)-60(1))+3600(20(1)-40(1))$$

$$=400(40-60)-1600(20-60)+3600(20-40)$$

$$=400(-20)-1600(-40)+3600(-20)$$

$$=-8000+64000-72000=-16000$$

$$D_x=\begin{vmatrix}400 & 20 & 1\\150 & 40 & 1\\400 & 60 & 1\end{vmatrix}=400\begin{vmatrix}40 & 1\\60 & 1\end{vmatrix}-150\begin{vmatrix}20 & 1\\60 & 1\end{vmatrix}+400\begin{vmatrix}20 & 1\\40 & 1\end{vmatrix}$$

$$=400(40(1)-60(1))-150(20(1)-60(1))+400(20(1)-40(1))$$

$$=400(40-60)-150(20-60)+400(20-40)$$

$$=400(-20)-150(-40)+400(-20)$$

$$=-8000+6000-8000=-10000$$

$$D_y=\begin{vmatrix}400 & 400 & 1\\1600 & 150 & 1\\3600 & 400 & 1\end{vmatrix}=400\begin{vmatrix}150 & 1\\400 & 1\end{vmatrix}-1600\begin{vmatrix}400 & 1\\400 & 1\end{vmatrix}+3600\begin{vmatrix}400 & 1\\150 & 1\end{vmatrix}$$

$$=400(150(1)-400(1))-1600(400(1)-400(1))+3600(400(1)-150(1))$$

$$=400(150-400)-1600(400-400)+3600(400-150)$$

$$=400(-250)-1600(0)+3600(250)$$

$$=400(-250)+3600(250)$$

$$=-100000+900000=800000$$

$$D_z=\begin{vmatrix}400 & 20 & 400\\1600 & 40 & 150\\3600 & 60 & 400\end{vmatrix}=400\begin{vmatrix}40 & 150\\60 & 400\end{vmatrix}-1600\begin{vmatrix}20 & 400\\60 & 400\end{vmatrix}+3600\begin{vmatrix}20 & 400\\40 & 150\end{vmatrix}$$

$$=400(40(400)-60(150))-1600(20(400)-60(400))+3600(20(150)-40(400))$$

$$=400(16000-9000)-1600(8000-24000)+3600(3000-16000)$$

$$=400(7000)-1600(-16000)+3600(-13000)$$

$$=2800000+25600000-46800000=-18400000$$

$$x=\frac{D_x}{D}=\frac{-10000}{-16000}=\frac{5}{8} \qquad y=\frac{D_y}{D}=\frac{800000}{-16000}=-50 \qquad z=\frac{D_z}{D}=\frac{-18400000}{-16000}=1150$$

The quadratic function is $f(x)=\frac{5}{8}x^2-50x+1150.$

To predict the average number of automobile accidents in which 30-year-olds and 50-year-olds are involved, find $f(30)$ and $f(50)$.

$$\begin{aligned} f(30) &= \frac{5}{8}(30)^2-50(30)+1150 \\ &= \frac{5}{8}(900)-1500+1150 \\ &= 562.5-1500+1150 \\ &= 212.5 \end{aligned}$$

$$\begin{aligned} f(50) &= \frac{5}{8}(50)^2-50(50)+1150 \\ &= \frac{5}{8}(2500)-2500+1150 \\ &= 1562.5-2500+1150 \\ &= 212.5 \end{aligned}$$

Both 30-year-olds and 50-year-olds are involved in approximately 212.5 accidents per day in the United States.

Chapter 3 Test

1. $x+y=6$

$y=-x+6$

$m=-1$

$y-\text{intercept}=6$

$4x-y=4$

$-y=-4x+4$

$y=4x-4$

$m=4$

$y-\text{intercept}=-4$

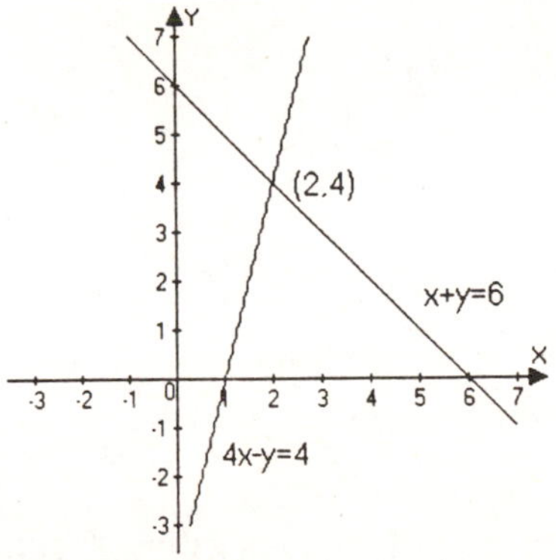

The solution is $(2,4)$ and the solution set is $\{(2,4)\}$.

2. $5x+4y=10$

$3x+5y=-7$

Multiply the first equation by –5, and the second equation by 4, and solve by addition.

$$\begin{aligned} -25x-20y &= -50 \\ 12x+20y &= -28 \\ \hline -13x &= -78 \\ x &= 6 \end{aligned}$$

Back-substitute 6 for x to find y.

$$\begin{aligned} 5x+4y &= 10 \\ 5(6)+4y &= 10 \\ 30+4y &= 10 \\ 4y &= -20 \\ y &= -5 \end{aligned}$$

The solution is $(6,-5)$ and the solution set is $\{(6,-5)\}$.

3. $x=y+4$

$3x+7y=-18$

Substitute $y+4$ for x to find y.

$$3x+7y=-18$$
$$3(y+4)+7y=-18$$
$$3y+12+7y=-18$$
$$10y+12=-18$$
$$10y=-30$$
$$y=-3$$

Back-substitute –3 for y to find x.

$$x=y+4$$
$$x=-3+4$$
$$x=1$$

The solution is $(1,-3)$ and the solution set is $\{(1,-3)\}$.

4. $4x=2y+6$

$y=2x-3$

Substitute $2x-3$ for y to find x.

$$4x=2y+6$$
$$4x=2(2x-3)+6$$
$$4x=4x-6+6$$
$$0=-6+6$$
$$0=0$$

The system is dependent. There are infinitely many solutions. The solution set is $\{(x,y)|4x=2y+6\}$ or $\{(x,y)\,y=2x-3\}$.

5. Let x = the amount of cholesterol in an ounce of shrimp

Let y = the amount of cholesterol in an ounce of scallops

$$3x+2y=156$$
$$5x+3y=300-45$$

Rewrite the system in standard form.

$$3x+2y=156$$
$$5x+3y=255$$

Multiply the first equation by –3 and the second equation by 2 and solve by addition.

$$-9x-6y=-468$$
$$10x+6y=510$$
$$x=42$$

Back-substitute 42 for x to find y.

$$3x+2y=156$$
$$3(42)+2y=156$$
$$126+2y=156$$
$$2y=30$$
$$y=15$$

There are 42 mg of cholesterol in an ounce of shrimp and 15 mg of cholesterol in an ounce of scallops.

6. Let x = the amount invested at 9%

Let y = the amount invested at 6%

$$x+y=6000$$
$$0.09x+0.06y=480$$

Multiply the first equation by –0.09 and add to the second equation.

$$-0.09x-0.09y=-540$$
$$0.09x+0.06y=480$$
$$-0.03y=-60$$
$$y=2000$$

Back-substitute 2000 for y to find x.

$$x+y=6000$$
$$x+2000=6000$$
$$x=4000$$

There is \$4000 invested at 9% and \$2000 invested at 6%.

7. Let x = the number of ounces of 20% solution

Let y = the number of ounces of 50% solution

$$x + y = 60$$
$$0.20x + 0.50y = 0.30(60)$$

Rewrite the system in standard form.

$$x + \quad y = 60$$
$$0.20x + 0.50y = 18$$

Multiply the first equation by –0.20 and add to the second equation.

$$-0.20x - 0.20y = -12$$
$$\underline{0.20x + 0.50y = \quad 18}$$
$$0.30y = 6$$
$$y = 20$$

Back-substitute 20 for y to find x.

$$x + y = 60$$
$$x + 20 = 60$$
$$x = 40$$

40 ounces of 20% solution and 20 ounces of 50% solution must be used.

8. Let r = the speed of the plane in still air
Let w = the speed of the wind

	r ·	t =	d
With Wind	$r+w$	2	1600
Against Wind	$r-w$	3	1950

$$2(r + w) = 1600$$
$$3(r - w) = 1950$$

Simplified, the system becomes:

$$2r + 2w = 1600$$
$$3r - 3w = 1950$$

Multiply the first equation by 3, the second equation by 2, and solve by addition.

$$6r + 6w = 4800$$
$$\underline{6r - 6w = 3900}$$
$$12r = 8700$$
$$r = 725$$

Back-substitute 725 for r to find w.

$$2r + 2w = 1600$$
$$2(725) + 2w = 1600$$
$$1450 + 2w = 1600$$
$$2w = 150$$
$$w = 75$$

The speed of the plane in still air is 725 kilometers per hour and the speed of the wind is 75 kilometers per hour.

9. $P(x) = R(x) - C(x)$
$$= 125x - (40x + 350,000)$$
$$= 125x - 40x - 350,000$$
$$= 85x - 350,000$$

10. Let x = the number of computers produced
$$C(x) = 360,000 + 850x$$

11. Let x = the number of computers sold
$$R(x) = 1150x$$

12. $R(x) = C(x)$
$$1150x = 360000 + 850x$$
$$300x = 360000$$
$$x = 1200$$
$$R(x) = 1150x$$
$$R(1200) = 1150(1200)$$
$$= 1,380,000$$

The break-even point is (1200, 1,380,000). When 1200 computers are produced and sold, the revenue will equal the cost at $1,380,000.

13.
$$x + \ y + \ z = 6$$
$$3x + 4y - 7z = 1$$
$$2x - \ y + 3z = 5$$

Multiply the first equation by 7 and add to the second equation.

$$\begin{array}{r} 7x+7y+7z=42 \\ \underline{3x+4y-7z=\ \ 1} \\ 10x+11y=43 \end{array}$$

Multiply the first equation by –3 and add to the third equation.

$$\begin{array}{r} -3x-3y-3z=-18 \\ \underline{2x-\ \ y+3z=\ \ \ 5} \\ -x-4y=-13 \end{array}$$

The system of two equations in two variables.

$$\begin{aligned} 10x+11y&=\ \ 43 \\ -x-\ 4y&=-13 \end{aligned}$$

Multiply the second equation by 10 and solve by addition.

$$\begin{array}{r} 10x+11y=\ \ \ \ 43 \\ \underline{-10x-40y=-130} \\ -29y=-87 \\ y=3 \end{array}$$

Back-substitute 3 for y to find x.

$$\begin{aligned} -x-4y&=-13 \\ -x-4(3)&=-13 \\ -x-12&=-13 \\ -x&=-1 \\ x&=1 \end{aligned}$$

Back-substitute 1 for x and 3 for y to find z.

$$\begin{aligned} x+y+z&=6 \\ 1+3+z&=6 \\ 4+z&=6 \\ z&=2 \end{aligned}$$

The solution is $(1,3,2)$ and the solution set is $\{(1,3,2)\}$.

14. $\left[\begin{array}{ccc|c} 1 & 0 & -4 & 5 \\ 6 & -1 & 2 & 10 \\ 2 & -1 & 4 & -3 \end{array}\right] \quad -6R_1+R_2$

$$=\left[\begin{array}{ccc|c} 1 & 0 & -4 & 5 \\ 0 & -1 & 26 & -20 \\ 2 & -1 & 4 & -3 \end{array}\right]$$

15. $\left[\begin{array}{cc|c} 2 & 1 & 6 \\ 3 & -2 & 16 \end{array}\right] \quad \frac{1}{2}R_1$

$$=\left[\begin{array}{cc|c} 1 & 1/2 & 3 \\ 3 & -2 & 16 \end{array}\right] \quad -3R_1+R_2$$

$$=\left[\begin{array}{cc|c} 1 & 1/2 & 3 \\ 0 & -7/2 & 7 \end{array}\right] \quad -\frac{2}{7}R_2$$

$$=\left[\begin{array}{cc|c} 1 & 1/2 & 3 \\ 0 & 1 & -2 \end{array}\right]$$

The resulting system is: $\begin{aligned} x+\frac{1}{2}y&=\ \ 3 \\ y&=-2 \end{aligned}$

Back-substitute –2 for y to find x.

$$\begin{aligned} x+\frac{1}{2}y&=3 \\ x+\frac{1}{2}(-2)&=3 \\ x-1&=3 \\ x&=4 \end{aligned}$$

The solution is $(4,-2)$ and the solution set is $\{(4,-2)\}$.

16. $\left[\begin{array}{ccc|c} 1 & -4 & 4 & -1 \\ 2 & -1 & 5 & 6 \\ -1 & 3 & -1 & 5 \end{array}\right] \quad -2R_1+R_2$

$$=\left[\begin{array}{ccc|c} 1 & -4 & 4 & -1 \\ 0 & 7 & -3 & 8 \\ -1 & 3 & -1 & 5 \end{array}\right] \quad \frac{1}{7}R_2$$

$$= \left[\begin{array}{rrr|r} 1 & -4 & 4 & -1 \\ 0 & 1 & -3/7 & 8/7 \\ -1 & 3 & -1 & 5 \end{array}\right] \quad R_1 + R_2$$

$$= \left[\begin{array}{rrr|r} 1 & -4 & 4 & -1 \\ 0 & 1 & -3/7 & 8/7 \\ 0 & -1 & 3 & 4 \end{array}\right] \quad R_2 + R_3$$

$$= \left[\begin{array}{rrr|r} 1 & -4 & 4 & -1 \\ 0 & 1 & -3/7 & 8/7 \\ 0 & 0 & 18/7 & 36/7 \end{array}\right] \quad \frac{7}{18}R_3$$

$$= \left[\begin{array}{rrr|r} 1 & -4 & 4 & -1 \\ 0 & 1 & -3/7 & 8/7 \\ 0 & 0 & 1 & 2 \end{array}\right]$$

The resulting system is:

$$x - 4y + 4z = -1$$
$$y - \frac{3}{7}z = \frac{8}{7}$$
$$z = 2$$

Back-substitute 2 for z to find y.

$$y - \frac{3}{7}(2) = \frac{8}{7}$$
$$y - \frac{6}{7} = \frac{8}{7}$$
$$y = \frac{14}{7}$$
$$y = 2$$

Back-substitute 2 for y and 2 for z to find x.

$$x - 4y + 4z = -1$$
$$x - 4(2) + 4(2) = -1$$
$$x - 8 + 8 = -1$$
$$x = -1$$

The solution is $(-1, 2, 2)$ and the solution set is $\{(-1, 2, 2)\}$.

17. $\begin{vmatrix} -1 & -3 \\ 7 & 4 \end{vmatrix} = -1(4) - 7(-3) = -4 + 21 = 17$

18. $\begin{vmatrix} 3 & 4 & 0 \\ -1 & 0 & -3 \\ 4 & 2 & 5 \end{vmatrix} = 3\begin{vmatrix} 0 & -3 \\ 2 & 5 \end{vmatrix} - (-1)\begin{vmatrix} 4 & 0 \\ 2 & 5 \end{vmatrix} + 4\begin{vmatrix} 4 & 0 \\ 0 & -3 \end{vmatrix}$

$$= 3\left(\cancel{0(5)} - 2(-3)\right) - (-1)\left(4(5) - \cancel{2(0)}\right) + 4\left(4(-3) - \cancel{0(0)}\right)$$
$$= 3(6) + 1(20) + 4(-12) = 18 + 20 - 48 = -10$$

19. $D = \begin{vmatrix} 4 & -3 \\ 3 & -1 \end{vmatrix} = 4(-1) - 3(-3) = -4 + 9 = 5$

$$D_x = \begin{vmatrix} 14 & -3 \\ 3 & -1 \end{vmatrix} = 14(-1) - 3(-3) = -14 + 9 = -5$$

$$D_y = \begin{vmatrix} 4 & 14 \\ 3 & 3 \end{vmatrix} = 4(3) - 3(14) = 12 - 42 = -30$$

$$x = \frac{D_x}{D} = \frac{-5}{5} = -1 \qquad y = \frac{D_y}{D} = \frac{-30}{5} = -6$$

The solution is $(-1,-6)$ and the solution set is $\{(-1,-6)\}$.

20.

$$D = \begin{vmatrix} 2 & 3 & 1 \\ 3 & 3 & -1 \\ 1 & -2 & -3 \end{vmatrix} = 2\begin{vmatrix} 3 & -1 \\ -2 & -3 \end{vmatrix} - 3\begin{vmatrix} 3 & 1 \\ -2 & -3 \end{vmatrix} + 1\begin{vmatrix} 3 & 1 \\ 3 & -1 \end{vmatrix}$$

$$= 2(3(-3)-(-2)(-1)) - 3(3(-3)-(-2)1) + 1(3(-1)-3(1))$$

$$= 2(-9-2) - 3(-9+2) + 1(-3-3)$$

$$= 2(-11) - 3(-7) + 1(-6) = -22 + 21 - 6 = -7$$

$$D_x = \begin{vmatrix} 2 & 3 & 1 \\ 0 & 3 & -1 \\ 1 & -2 & -3 \end{vmatrix} = 2\begin{vmatrix} 3 & -1 \\ -2 & -3 \end{vmatrix} - 0\cancel{\begin{vmatrix} 3 & 1 \\ -2 & -3 \end{vmatrix}} + 1\begin{vmatrix} 3 & 1 \\ 3 & -1 \end{vmatrix}$$

$$= 2(3(-3)-(-2)(-1)) + 1(3(-1)-3(1))$$

$$= 2(-9-2) + 1(-3-3) = 2(-11) + 1(-6) = -22 - 6 = -28$$

$$D_y = \begin{vmatrix} 2 & 2 & 1 \\ 3 & 0 & -1 \\ 1 & 1 & -3 \end{vmatrix} = 2\begin{vmatrix} 0 & -1 \\ 1 & -3 \end{vmatrix} - 3\begin{vmatrix} 2 & 1 \\ 1 & -3 \end{vmatrix} + 1\begin{vmatrix} 2 & 1 \\ 0 & -1 \end{vmatrix}$$

$$= 2(\cancel{0(-3)} - 1(-1)) - 3(2(-3)-1(1)) + 1(2(-1) - \cancel{0(1)})$$

$$= 2(1) - 3(-6-1) + 1(-2) = 2 - 3(-7) - 2 = 2 + 21 - 2 = 21$$

$$D_z = \begin{vmatrix} 2 & 3 & 2 \\ 3 & 3 & 0 \\ 1 & -2 & 1 \end{vmatrix} = 2\begin{vmatrix} 3 & 0 \\ -2 & 1 \end{vmatrix} - 3\begin{vmatrix} 3 & 2 \\ -2 & 1 \end{vmatrix} + 1\begin{vmatrix} 3 & 2 \\ 3 & 0 \end{vmatrix}$$

$$= 2(3(1) - \cancel{(-2)(0)}) - 3(3(1)-(-2)2) + 1(\cancel{3(0)} - 3(2))$$

$$= 2(3) - 3(3+4) + 1(-6)$$

$$= 6 - 3(7) - 6 = 6 - 21 - 6 = -21$$

$$x = \frac{D_x}{D} = \frac{-28}{-7} = 4 \qquad y = \frac{D_y}{D} = \frac{21}{-7} = -3 \qquad z = \frac{D_z}{D} = \frac{-21}{-7} = 3$$

The solution is $(4,-3,3)$ and the solution set is $\{(4,-3,3)\}$.

Cumulative Review Exercises

Chapters 1-3

1. $$\frac{6(8-10)^3+(-2)}{(-5)^2(-2)}=\frac{6(-2)^3-2}{(25)(-2)}$$
$$=\frac{6(-8)-2}{-50}$$
$$=\frac{-48-2}{-50}$$
$$=\frac{-50}{-50}=1$$

2. $$7x-[5-2(4x-1)]$$
$$=7x-[5-8x+2]$$
$$=7x-[7-8x]$$
$$=7x-7+8x$$
$$=15x-7$$

3. $$5-2(3-x)=2(2x+5)+1$$
$$5-6+2x=4x+10+1$$
$$-1+2x=4x+11$$
$$-1-2x=11$$
$$-2x=12$$
$$x=-6$$
The solution is –6 and the solution set is $\{-6\}$.

4. $$\frac{3x}{5}+4=\frac{x}{3}$$
$$15\left(\frac{3x}{5}\right)+15(4)=15\left(\frac{x}{3}\right)$$
$$3(3x)+60=5x$$
$$9x+60=5x$$
$$4x+60=0$$
$$4x=-60$$
$$x=-15$$
The solution is –15 and the solution set is $\{-15\}$.

5. $$3x-4=2(3x+2)-3x$$
$$3x-4=6x+4-3x$$
$$3x-4=3x+4$$
$$-4=4$$
This is a contradiction. There is no value of x for which –4 equals 4. The solution set is $\varnothing$ or $\{\ \}$.

6. Let x = the amount of the sales
$$200+0.05x=0.15x$$
$$200=0.10x$$
$$2000=x$$
For sales of \$2000, the earnings will be the same under either pay arrangement.

7. $$\frac{-5x^6y^{-10}}{20x^{-2}y^{20}}=\frac{-x^6x^2}{4y^{20}y^{10}}$$
$$=-\frac{x^8}{4y^{30}}$$

8. $$f(x)=-4x+5$$
$$f(a+2)=-4(a+2)+5$$
$$=-4a-8+5$$
$$=-4a-3$$

9. To find the denominator, set it equal to zero.

$$x+3=0$$
$$x=-3$$

The domain is $\{x|x \text{ is a real number and } x \neq -3\}$.

10. $(f-g)(x)$

$$=f(x)-g(x)$$
$$=2x^2-5x+2-(x^2-2x+3)$$
$$=2x^2-5x+2-x^2+2x-3$$
$$=x^2-3x-1$$

$$(f-g)(3)=(3)^2-3(3)-1$$
$$=9-9-1$$
$$=-1$$

11. $f(x)=-\frac{2}{3}x+2$

$$y=-\frac{2}{3}x+2$$

y − intercept = 2

$\text{m}=-\frac{2}{3}$

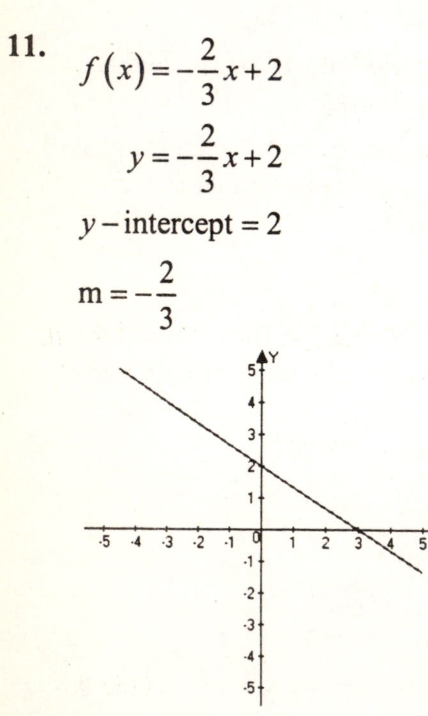

12. $2x-y=6$

$$-y=-2x+6$$
$$y=2x-6$$

y − intercept = −6

m = 2

13. First, find the slope.

$$m=\frac{4-(-2)}{2-4}=\frac{4+2}{-2}=\frac{6}{-2}=-3$$

Use the slope and one of the points to write the line in point-slope form.

$$y-y_1=m(x-x_1)$$
$$y-4=-3(x-2)$$

Solve for y to obtain slope-intercept form.

$$y-4=-3(x-2)$$
$$y-4=-3x+6$$
$$y=-3x+10$$

14. Parallel lines have the same slope. Find the slope of the given line.

$$3x+y=6$$
$$y=-3x+6$$

We obtain, $m=-3$.

Use the slope and the point to put the line in point-slope form.

$$y - y_1 = m(x - x_1)$$
$$y - 0 = -3(x - (-1))$$
$$y - 0 = -3(x + 1)$$

Solve for y to obtain slope-intercept form.

$$y - 0 = -3(x + 1)$$
$$y = -3x - 3$$

15. $3x + 12y = 25$

$2x - 6y = 12$

Multiply the second equation by 2 and solve by addition.

$$\begin{array}{r} 3x + 12y = 25 \\ \underline{4x - 12y = 24} \\ 7x = 49 \\ x = 7 \end{array}$$

Back-substitute 7 for x to find y.

$$2x - 6y = 12$$
$$2(7) - 6y = 12$$
$$14 - 6y = 12$$
$$-6y = -2$$
$$y = \frac{-2}{-6}$$
$$y = \frac{1}{3}$$

The solution is $\left(7, \frac{1}{3}\right)$ and the solution set is $\left\{\left(7, \frac{1}{3}\right)\right\}$.

16. $x + 3y - z = 5$

$-x + 2y + 3z = 13$

$2x - 5y - z = -8$

Add the first and second equations to eliminate x.

$$\begin{array}{r} x + 3y - z = 5 \\ \underline{-x + 2y + 3z = 13} \\ 5y + 2z = 18 \end{array}$$

Multiply the first equation by -2 and add to the third equation.

$$\begin{array}{r} -2x - 6y + 2z = -10 \\ \underline{2x - 5y - z = -8} \\ -11y + z = -18 \end{array}$$

The system of two equations in two variables becomes:

$$5y + 2z = 18$$
$$-11y + z = -18$$

Multiply the second equation by -2 and solve by addition.

$$\begin{array}{r} 5y + 2z = 18 \\ \underline{22y - 2z = 36} \\ 27y = 54 \\ y = 2 \end{array}$$

Back-substitute 2 for y to find z.

$$5y + 2z = 18$$
$$5(2) + 2z = 18$$
$$10 + 2z = 18$$
$$2z = 8$$
$$z = 4$$

Back-substitute 2 for y and 4 for z in one of the equations in three variables.

$$x + 3y - z = 5$$
$$x + 3(2) - 4 = 5$$
$$x + 6 - 4 = 5$$
$$x + 2 = 5$$
$$x = 3$$

The solution is $(3, 2, 4)$ and the solution set is $\{(3, 2, 4)\}$.

17. Let x = the number of pads
Let y = the number of pens

$$2x+19y=5.40$$
$$7x+4y=6.40$$

Multiply the first equation by 7 and the second equation by –2 and solve by addition.

$$\begin{aligned} 14x+133y &= 37.80 \\ -14x-8y &= -12.80 \\ \hline 125y &= 25.00 \\ y &= .20 \end{aligned}$$

Back-substitute .20 for y to find x.

$$\begin{aligned} 7x+4y &= 6.40 \\ 7x+4(.20) &= 6.40 \\ 7x+.80 &= 6.40 \\ 7x &= 5.60 \\ x &= .80 \end{aligned}$$

Pads cost \$.80 each and pens cost \$.20 each.

18.
$$\begin{vmatrix} 0 & 1 & -2 \\ -7 & 0 & -4 \\ 3 & 0 & 5 \end{vmatrix}$$
$$=0\begin{vmatrix} 0 & -4 \\ 0 & 5 \end{vmatrix}-(-7)\begin{vmatrix} 1 & -2 \\ 0 & 5 \end{vmatrix}+3\begin{vmatrix} 1 & -2 \\ 0 & -4 \end{vmatrix}$$
$$=7\left(1(5)-0(-2)\right)+3\left(1(-4)-0(-2)\right)$$
$$=7(5)+3(-4)=35-12=23$$

19.
$$\left[\begin{array}{ccc|c} 2 & 3 & -1 & -1 \\ 1 & 2 & 3 & 2 \\ 3 & 5 & -2 & -3 \end{array}\right] \quad R_1 \leftrightarrow R_2$$
$$=\left[\begin{array}{ccc|c} 1 & 2 & 3 & 2 \\ 2 & 3 & -1 & -1 \\ 3 & 5 & -2 & -3 \end{array}\right] \quad \begin{array}{c} -2R_1+R_2 \\ \text{and} \\ -3R_1+R_3 \end{array}$$
$$=\left[\begin{array}{ccc|c} 1 & 2 & 3 & 2 \\ 0 & -1 & -7 & -5 \\ 0 & -1 & -11 & -9 \end{array}\right] \quad -R_2$$
$$=\left[\begin{array}{ccc|c} 1 & 2 & 3 & 2 \\ 0 & 1 & 7 & 5 \\ 0 & -1 & -11 & -9 \end{array}\right] \quad R_2+R_3$$
$$=\left[\begin{array}{ccc|c} 1 & 2 & 3 & 2 \\ 0 & 1 & 7 & 5 \\ 0 & 0 & -4 & -4 \end{array}\right] \quad -\frac{1}{4}R_3$$
$$=\left[\begin{array}{ccc|c} 1 & 2 & 3 & 2 \\ 0 & 1 & 7 & 5 \\ 0 & 0 & 1 & 1 \end{array}\right]$$

The resulting system is:

$$\begin{aligned} x+2y+3z &= 2 \\ y+7z &= 5 \\ z &= 1 \end{aligned}$$

Back-substitute 1 for z to find y.

$$\begin{aligned} y+7z &= 5 \\ y+7(1) &= 5 \\ y+7 &= 5 \\ y &= -2 \end{aligned}$$

Back-substitute –2 for y and 1 for z to find x.

$$\begin{aligned} x+2y+3z &= 2 \\ x+2(-2)+3(1) &= 2 \\ x-4+3 &= 2 \\ x-1 &= 2 \\ x &= 3 \end{aligned}$$

The solution is $(3,-2,1)$ and the solution set is $\{(3,-2,1)\}$.

20.
$$D=\begin{vmatrix} 3 & 4 \\ -2 & 1 \end{vmatrix}=3(1)-(-2)4$$
$$=3+8=11$$

$$D_x = \begin{vmatrix} -1 & 4 \\ 8 & 1 \end{vmatrix} = -1(1) - 8(4)$$

$$= -1 - 32 = -33$$

$$D_y = \begin{vmatrix} 3 & -1 \\ -2 & 8 \end{vmatrix} = 3(8) - (-2)(-1)$$

$$= 24 - 2 = 22$$

$$x = \frac{D_x}{D} = \frac{-33}{11} = -3$$

$$y = \frac{D_y}{D} = \frac{22}{11} = 2$$

The solution is $(-3, 2)$ and the solution set is $\{(-3, 2)\}$.

Chapter 4

Check Points 4.1

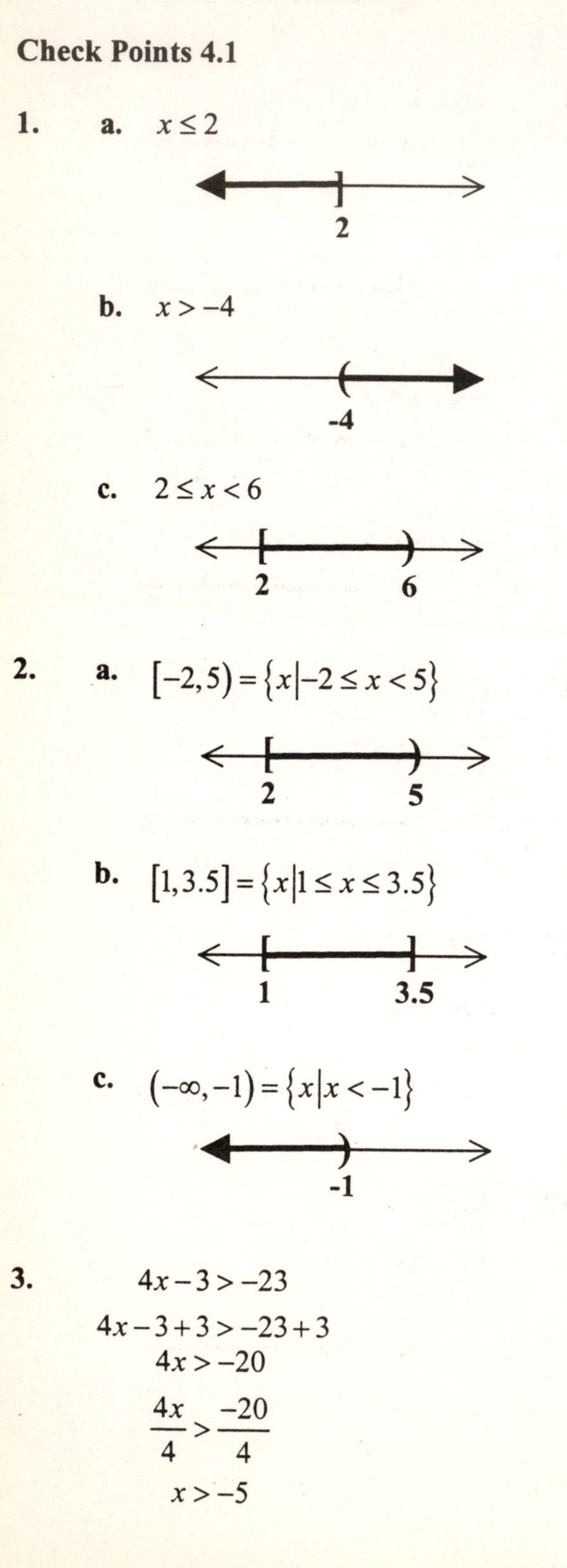

1. **a.** $x \le 2$

b. $x > -4$

c. $2 \le x < 6$

2. **a.** $[-2, 5) = \{x | -2 \le x < 5\}$

b. $[1, 3.5] = \{x | 1 \le x \le 3.5\}$

c. $(-\infty, -1) = \{x | x < -1\}$

3.

$$
\begin{aligned}
4x - 3 &> -23 \\
4x - 3 + 3 &> -23 + 3 \\
4x &> -20 \\
\frac{4x}{4} &> \frac{-20}{4} \\
x &> -5
\end{aligned}
$$

The solution set is $\{x | x > -5\}$ or $(-5, \infty)$.

4.

$$
\begin{aligned}
6 - 3x &\ge 5x - 2 \\
6 - 3x - 5x &\ge 5x - 5x - 2 \\
6 - 8x &\ge -2 \\
6 - 6 - 8x &\ge -2 - 6 \\
-8x &\ge -8 \\
\frac{-8x}{-8} &\le \frac{-8}{-8} \\
x &\le 1
\end{aligned}
$$

The solution set is $\{x | x \le 1\}$ or $(-\infty, 1]$.

5.

$$
\begin{aligned}
\frac{x-4}{2} &> \frac{x-2}{3} + \frac{5}{6} \\
6\left(\frac{x-4}{2}\right) &> 6\left(\frac{x-2}{3}\right) + 6\left(\frac{5}{6}\right) \\
3(x-4) &> 2(x-2) + 5 \\
3x - 12 &> 2x - 4 + 5 \\
3x - 12 &> 2x + 1 \\
3x - 2x - 12 &> 2x - 2x + 1 \\
x - 12 &> 1 \\
x - 12 + 12 &> 1 + 12 \\
x &> 13
\end{aligned}
$$

The solution set is $\{x | x > 13\}$ or $(13, \infty)$.

6. **a.**

$$3(x+1) > 3x+2$$
$$3x+3 > 3x+2$$
$$3x-3x+3 > 3x-3x+2$$
$$3 > 2$$

Since $3 > 2$ for all values of x, the solution set is $\{x | x \text{ is a real number}\}$, $\mathbb{R}$ or $(-\infty, \infty)$.

0

b.

$$x+1 \le x-1$$
$$x-x+1 \le x-x-1$$
$$1 \le -1$$

This is a contradiction. There are no values of x for which $1 \le -1$. The solution set is $\varnothing$ or $\{ \ \}$.

7. Let x = the number of miles driven
The cost from Basic Rental is:
$c_B = 260.$
The cost from Continental is:
$c_C = 80 + 25x.$

$$c_B = c_C$$
$$260 = 80 + 0.25x$$
$$180 = 0.25x$$
$$720 = x$$

When 720 miles are driven, the cost from both companies is the same at \$260.

Problem Set 4.1

Practice Exercises

1. $x > 6$

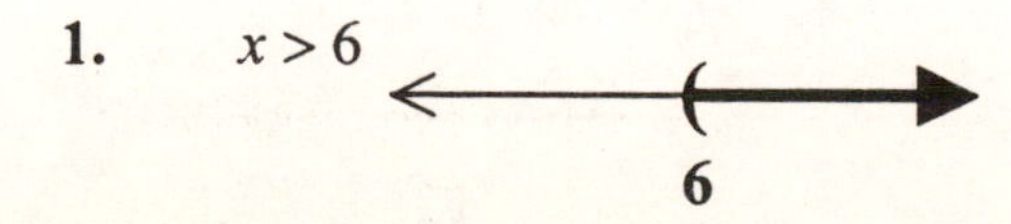

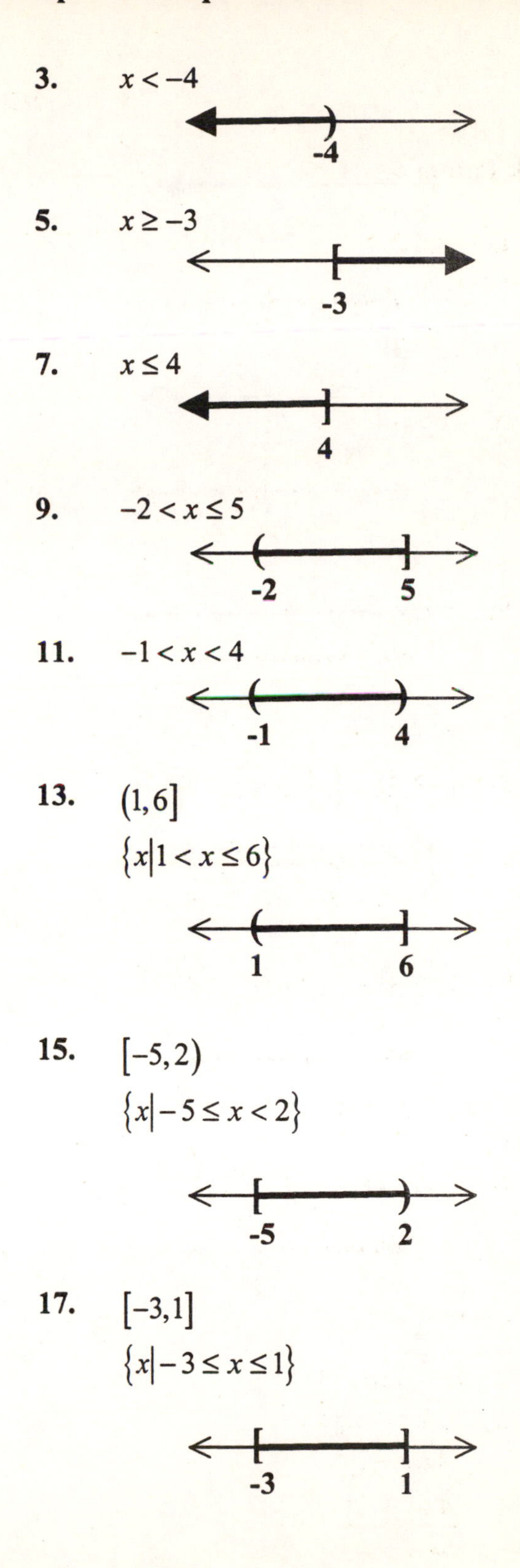

3. $x < -4$

-4

5. $x \ge -3$

-3

7. $x \le 4$

4

9. $-2 < x \le 5$

-2 5

11. $-1 < x < 4$

-1 4

13. $(1, 6]$

$\{x | 1 < x \le 6\}$

1 6

15. $[-5, 2)$

$\{x | -5 \le x < 2\}$

-5 2

17. $[-3, 1]$

$\{x | -3 \le x \le 1\}$

-3 1

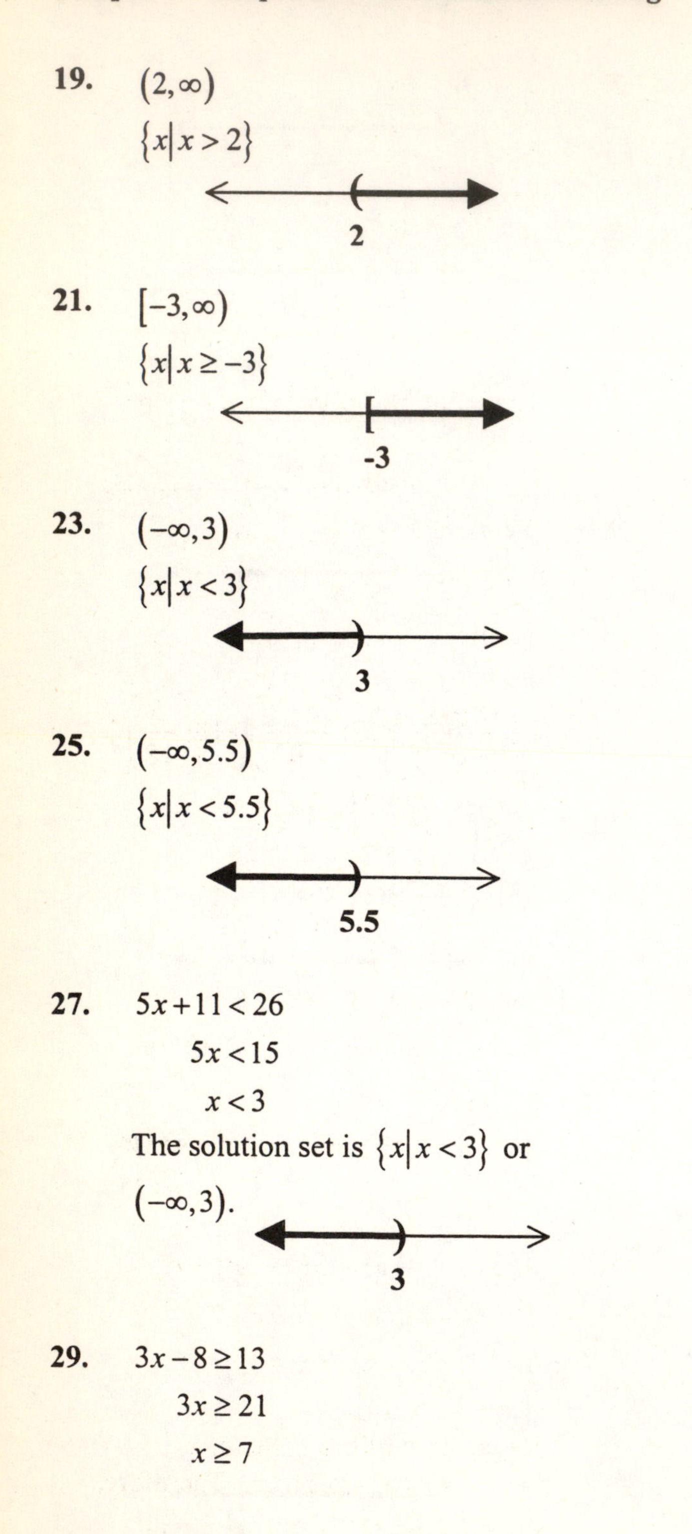

19. $(2,\infty)$

$\{x|x>2\}$

21. $[-3,\infty)$

$\{x|x\geq -3\}$

23. $(-\infty,3)$

$\{x|x<3\}$

25. $(-\infty,5.5)$

$\{x|x<5.5\}$

27.

$$5x+11<26$$
$$5x<15$$
$$x<3$$

The solution set is $\{x|x<3\}$ or $(-\infty,3)$.

29.

$$3x-8\geq 13$$
$$3x\geq 21$$
$$x\geq 7$$

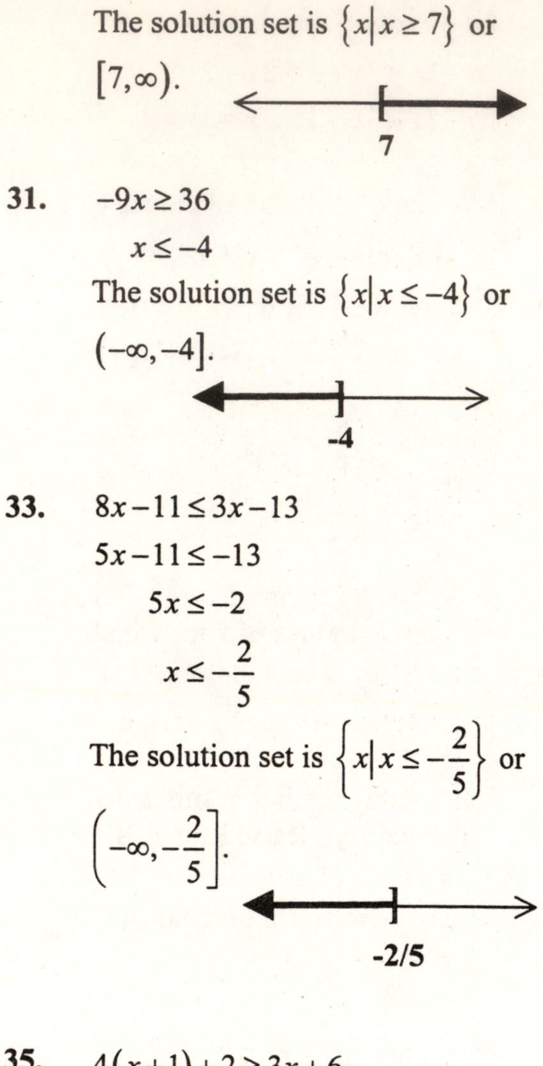

The solution set is $\{x|x\geq 7\}$ or $[7,\infty)$.

31.

$$-9x\geq 36$$
$$x\leq -4$$

The solution set is $\{x|x\leq -4\}$ or $(-\infty,-4]$.

33.

$$8x-11\leq 3x-13$$
$$5x-11\leq -13$$
$$5x\leq -2$$
$$x\leq -\frac{2}{5}$$

The solution set is $\left\{x\middle|x\leq -\frac{2}{5}\right\}$ or $\left(-\infty,-\frac{2}{5}\right]$.

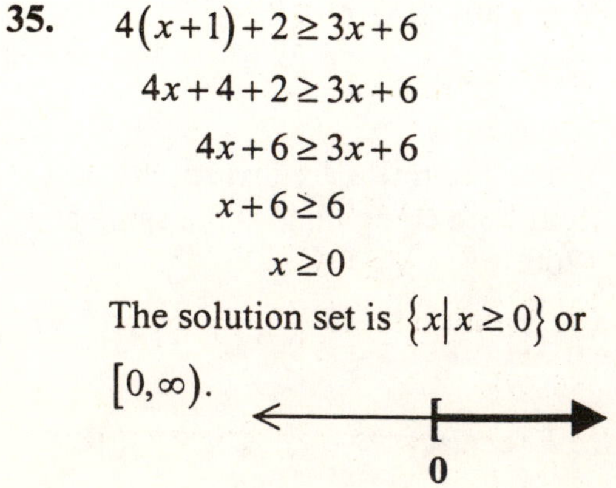

35.

$$4(x+1)+2\geq 3x+6$$
$$4x+4+2\geq 3x+6$$
$$4x+6\geq 3x+6$$
$$x+6\geq 6$$
$$x\geq 0$$

The solution set is $\{x|x\geq 0\}$ or $[0,\infty)$.

37.
$$2x-11<-3(x+2)$$
$$2x-11<-3x-6$$
$$5x-11<-6$$
$$5x<5$$
$$x<1$$
The solution set is $\{x|x<1\}$ or $(-\infty,1)$.

1

39.
$$1-(x+3)\geq 4-2x$$
$$1-x-3\geq 4-2x$$
$$-x-2\geq 4-2x$$
$$x-2\geq 4$$
$$x\geq 6$$
The solution set is $\{x|x\geq 6\}$ or $[6,\infty)$.

6

41.
$$\frac{x}{4}-\frac{1}{2}\leq\frac{x}{2}+1$$
$$4\left(\frac{x}{4}\right)-4\left(\frac{1}{2}\right)\leq 4\left(\frac{x}{2}\right)+4(1)$$
$$x-2\leq 2x+4$$
$$-x-2\leq 4$$
$$-x\leq 6$$
$$x\geq -6$$
The solution set is $\{x|x\geq -6\}$ or $[-6,\infty)$.

-6

43.
$$1-\frac{x}{2}>4$$
$$2(1)-2\left(\frac{x}{2}\right)>2(4)$$
$$2-x>8$$
$$-x>6$$
$$x<-6$$
The solution set is $\{x|x<-6\}$ or $(-\infty,-6)$.

-6

45.
$$\frac{x-4}{6}\geq\frac{x-2}{9}+\frac{5}{18}$$
$$18\left(\frac{x-4}{6}\right)\geq 18\left(\frac{x-2}{9}\right)+18\left(\frac{5}{18}\right)$$
$$3(x-4)\geq 2(x-2)+5$$
$$3x-12\geq 2x-4+5$$
$$3x-12\geq 2x+1$$
$$x-12\geq 1$$
$$x\geq 13$$
The solution set is $\{x|x\geq 13\}$ or $[13,\infty)$.

13

47.
$$4(3x-2)-3x<3(1+3x)-7$$
$$12x-8-3x<3+9x-7$$
$$9x-8<9x-4$$
$$-8<-4$$
Regardless of the value of x, $-8<-4$ is always true. Therefore, the solution set is $\{x|x$ is a real number$\}$, $\mathbb{R}$ or $(-\infty,\infty)$.

49. $8(x+1) \le 7(x+5)+x$

$8x+8 \le 7x+35+x$

$8x+8 \le 8x+35$

$8 \le 35$

Regardless of the value of x, $8 \le 35$ is always true. Therefore, the solution set is $\{x | x \text{ is a real number}\}$, $\mathbb{R}$ or $(-\infty, \infty)$.

51. $3x < 3(x-2)$

$3x < 3x-6$

$0 < -6$

There are no values of x for which $0 < -6$. The solution set is $\varnothing$ or $\{ \}$.

Application Exercises

53. $x < 40\%$

Playing Sports and Sports Events

55. $[51\%, 61\%]$

Exercise, Movies, Gardening, and Amusement Parks

57. $(51\%, 61\%)$

Movies and Gardening

59. $(39\%, 55\%]$

Gardening, Amusement Parks, and Home Improvement

61. $-0.7x+80 < 52$

$-0.7x < -28$

$x > 40$

This means that in all years after $1965 + 40 = 2005$, fewer than 52% of adults will read the daily newspaper.

63. $W < M$

$-0.19t+57 < -0.15t+50$

$-0.04t+57 < 50$

$-0.04t < -7$

$t > 175$

The women's winning time will be less than the men's winning time after $1900 + 175 = 2075$.

65. The daily rental cost to rent a truck from Basic Rental is $C_B = 50+0.20x$.
The daily cost to rent a truck from Continental is $C_C = 20+0.50x$.

$C_B < C_C$

$50+0.20x < 20+0.50x$

$50-0.30x < 20$

$-0.30x < -30$

$x > 100$

Basic Rental is a better deal when the truck is driven more than 100 miles in a day.

67. The tax bill assessed under the first tax bill is $T_1 = 1800+0.03x$.
The tax bill assessed under the second tax bill is $T_2 = 200+0.08x$.

$T_1 < T_2$

$1800+0.03x < 200+0.08x$

$1800-0.05x < 200$

$-0.05x < -1600$

$x > 32000$

The first tax bill is a better deal when the assessed value of the home is greater than \$32,000.

69. The cost is $C = 10,000 + 0.40x.$
The revenue is $R = 2x.$

$$C < R$$
$$10,000 + 0.40x < 2x$$
$$10,000 - 1.6x < 0$$
$$-1.6x < -10,000$$
$$x > 6250$$

The company will make a profit gain when more than 6,250 tapes are produced and sold each week.

71. Let x = the number of bags of cement that can be lifted safely in the elevator.

$$W_{operator} + W_{cement} \leq 2800$$
$$265 + 65x \leq 2800$$
$$65x \leq 2535$$
$$x \leq 39$$

39 bags of cement can be lifted safely on the elevator per trip.

Writing in Mathematics

73. Answers will vary.

75. Answers will vary.

77. Answers will vary.

79. Answers will vary.

Technology Exercises

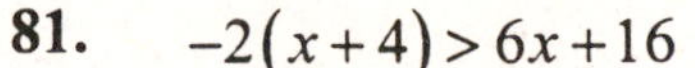

81. $-2(x+4) > 6x+16$

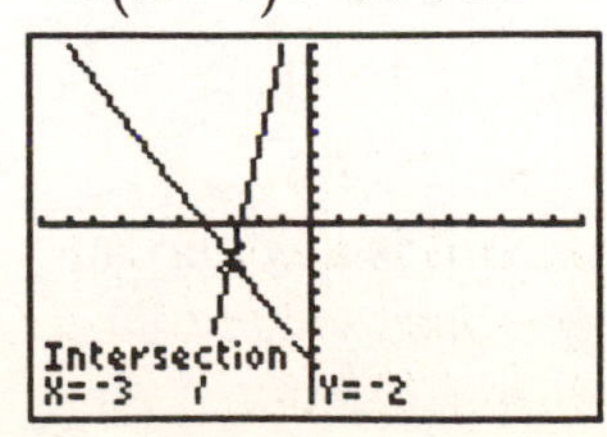

Moving from left to right on the graphing calculator screen, we see that the graph of $-2(x+4)$ is above the graph of $6x+16$ from $-\infty$ to -3. The solution set is $\{x | x < -3\}$ or $(-\infty, -3)$.

83. $2x+3 > 3(2x-4)-4x$

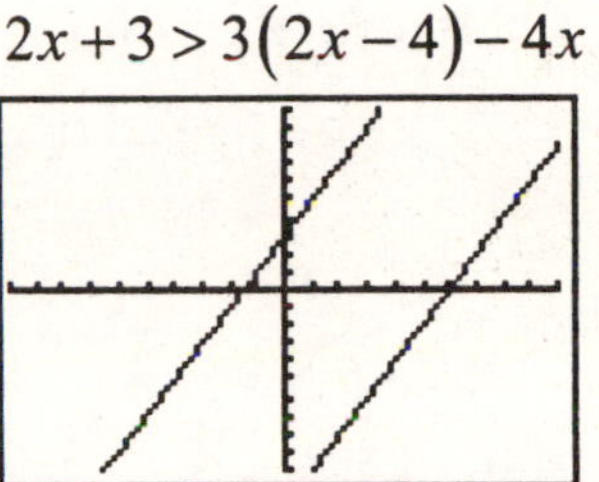

We see that the lines are parallel. Since the graph of $2x+3$ lies above the graph of $3(2x-4)-4x$ for all values of x, the solution set is $\{x | x \text{ is a real number}\}$, $\mathbb{R}$ or $(-\infty, \infty)$.

Critical Thinking Exercises

85. Statement **d.** is true.

$$-3x > 6$$
$$x < -2$$
$$-2 > x$$

Statement **a.** is false. The inequality $3x > 6$ is equivalent to $x > 2$.

Statement **b.** is false. $2x > 6$ is equivalent to $x > 3$. The smallest whole number or the smallest integer in the solution set is 4, not the smallest real number.

Statement **c.** is false. If x is at least 7, then $x \geq 7$.

87. We need to find the x-values for which the graph of $3(-x-5)-9$ is between 0 and 6. The x-coordinates of the intersection points of the lines are the endpoints of the region in which the inequality is true. The intersection points are $(-10,6)$ and $(-8,0)$. Therefore, the solution set is $\{x|-10<x<-8\}$ or $(-10,-8)$.

Review Exercises

88.
$$f(x)=x^2-2x+5$$
$$\begin{aligned}f(-4)&=(-4)^2-2(-4)+5\\&=16+8+5\\&=29\end{aligned}$$

89.
$$\begin{aligned}2x-\ y-\ z&=-3\\3x-2y-2z&=-5\\-x+\ y+2z&=\ 4\end{aligned}$$
Add the first and third equations to eliminate y.
$$\begin{aligned}2x-\ y-\ z&=-3\\-x+\ y+2z&=\ 4\\\hline x+z&=1\end{aligned}$$
Multiply the third equation by 2 and add to the second equation.
$$\begin{aligned}3x-2y-2z&=-5\\-2x+2y+4z&=\ 8\\\hline x+2z&=3\end{aligned}$$
The system of two equations in two variables becomes:
$$\begin{aligned}x+\ z&=1\\x+2z&=3\end{aligned}$$
Multiply the second equation by -1 and solve for z.
$$\begin{aligned}x+\ z&=\ 1\\-x-2z&=-3\\\hline -z&=-2\\z&=2\end{aligned}$$
Back-substitute 2 for z to find x.
$$\begin{aligned}x+z&=1\\x+2&=1\\x&=-1\end{aligned}$$
Back-substitute 2 for z and -1 and x in one of the original equations in three variables to find y.
$$\begin{aligned}2x-y-z&=-3\\2(-1)-y-2&=-3\\-2-y-2&=-3\\-y-4&=-3\\-y&=1\\y&=-1\end{aligned}$$
The solution is $(-1,-1,2)$ and the solution set is $\{(-1,-1,2)\}$.

91.
$$\begin{aligned}\left(\frac{2x^4y^{-2}}{4xy^3}\right)^3&=\frac{2^3x^{12}y^{-6}}{4^3x^3y^9}\\&=\frac{8}{64}x^{12-3}y^{-6-9}\\&=\frac{1}{8}x^9y^{-15}\\&=\frac{x^9}{8y^{15}}\end{aligned}$$

Check Points 4.2

1. $\{3,4,5,6,7\}\cap\{3,7,8,9\}=\{3,7\}$

2. $x+2<5$ and $2x-4<-2$

$x<3$ $\quad 2x<2$

$x<1$

$x<3$

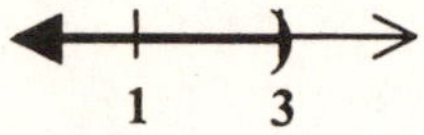

1 3

$x<1$

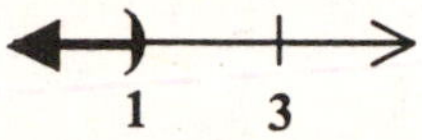

1 3

$x<3$ and $x<1$

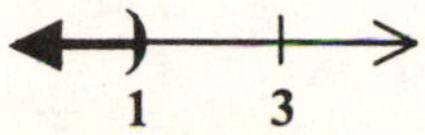

1 3

The solution set is $\{x|x<1\}$ or $(-\infty,1)$.

3. $4x-5>7$ and $5x-2<3$

$4x>12$ $\quad 5x<5$

$x>3$ $\quad x<1$

$x>3$

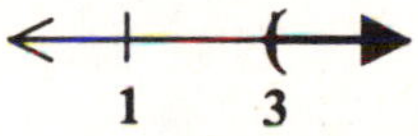

1 3

$x<1$

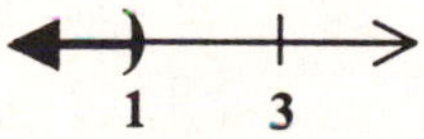

1 3

$x>3$ and $x<1$

1 3

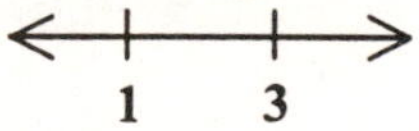

Since the two sets do not intersect, the solution set is $\varnothing$ or $\{\ \}$.

4.

$$1\le 2x+3<11$$
$$1-3\le 2x+3-3<11-3$$
$$-2\le 2x<8$$
$$-\frac{2}{2}\le\frac{2x}{2}<\frac{8}{2}$$
$$-1\le x<4$$

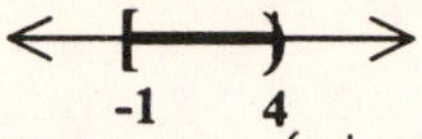

-1 4

The solution set is $\{x|-1\le x<4\}$ or $[-1,4)$.

5. $\{3,4,5,6,7\}\cup\{3,7,8,9\}$
$=\{3,4,5,6,7,8,9\}$

6. $3x-5\le -2$ or $10-2x<4$

$3x\le 3$ $\quad -2x<-6$

$x\le 1$ $\quad x>3$

$x\le 1$

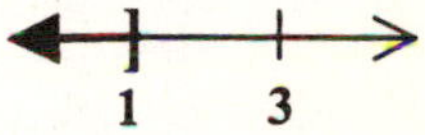

1 3

$x>3$

1 3

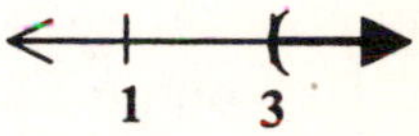

$x\le 1$ or $x>3$

1 3

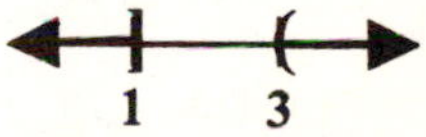

The solution set is $\{x|x\le 1 \text{ or } x>3\}$ or $(-\infty,1]\cup(3,\infty)$.

7. $2x+5\ge 3$ or $2x+3<3$

$2x\ge -2$ $\quad 2x<0$

$x\ge -1$ $\quad x<0$

$x\ge -1$

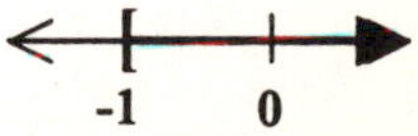

-1 0

$x<0$

-1 0

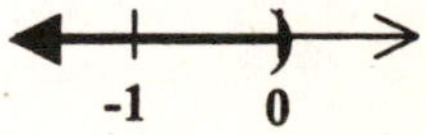

$x\ge -1$ or $x<0$

-1 0

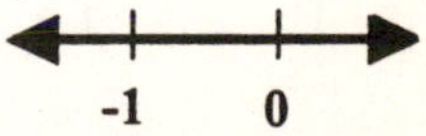

The solution set is $\mathbb{R}$, $(-\infty,\infty)$ or $\{x|x \text{ is a real number}\}$.

Problem Set 4.2

Practice Exercises

1. $\{1,2,3,4\} \cap \{2,4,5\} = \{2,4\}$

3. $\{1,3,5,7\} \cap \{2,4,6,8,10\} = \{\ \}$
The empty set is also denoted by $\varnothing$.

5. $x > 3$

$x > 6$

$x > 3$ and $x > 6$

The solution set is $\{x | x > 6\}$ or $(6, \infty)$.

7. $x \le 5$

$x \le 1$

$x \le 5$ and $x \le 1$

The solution set is $\{x | x \le 1\}$ or $(-\infty, 1]$.

9. $x < 2$

$x \ge -1$

$x < 2$ and $x \ge -1$

The solution set is $\{x | -1 \le x < 2\}$ or $[-1, 2)$.

11. $x > 2$

$x < -1$

$x > 2$ and $x < -1$

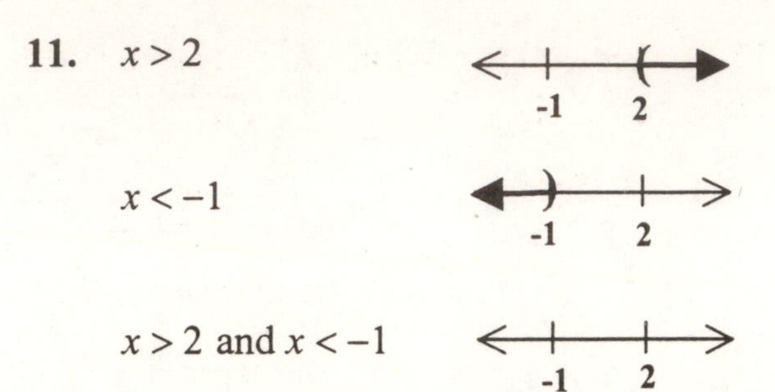

Since the two sets do not intersect, the solution set is $\varnothing$ or $\{\ \}$.

13. $5x < -20$
$x < -4$

$3x > -18$
$x > -6$

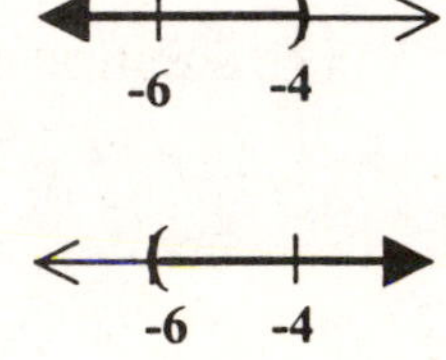

$x < -4$ and $x > -6$

The solution set is $\{x | -6 < x < -4\}$ or $(-6, -4)$.

15. $x - 4 \le 2$ and $3x + 1 > -8$
$x \le 6$ $\quad 3x > -9$
$x > -3$

$x \le 6$

$x > -3$

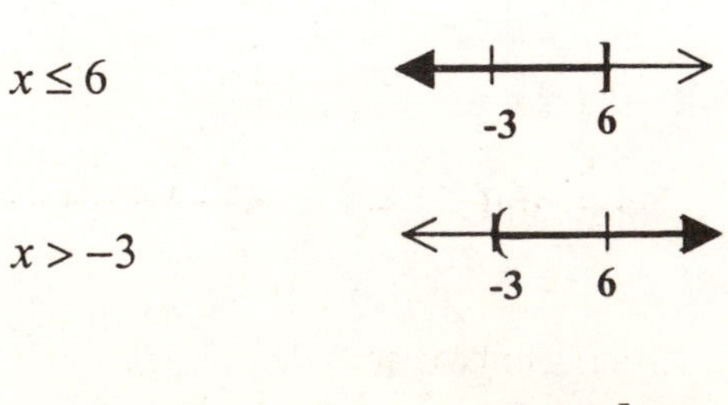

$x \le 6$ and $x > -3$

The solution set is $\{x|-3<x\le 6\}$ or $(-3,6]$.

17. $2x>5x-15$ and $7x>2x+10$

$-3x>-15$ $\quad$ $5x>10$

$x<5$ $\quad$ $x>2$

$x<5$

2 5

$x>2$

2 5

$x<5$ and $x>2$

2 5

The solution set is $\{x|2<x<5\}$ or $(2,5)$.

19. $4(1-x)<-6$ $\quad$ $\frac{x-7}{5}\le -2$

$4-4x<-6$ $\quad$ $5\left(\frac{x-7}{5}\right)\le 5(-2)$

$-4x<-10$ $\quad$ $x-7\le -10$

$x>\frac{5}{2}$ $\quad$ $x\le -3$

$x>\frac{5}{2}$

-3 5/2

$x\le -3$

-3 5/2

$x>\frac{5}{2}$ and $x\le -3$

-3 5/2

Since the two sets do not intersect, the solution set is $\varnothing$ or $\{\ \}$.

21. $x-1\le 7x-1$ and $4x-7<3-x$

$-1\le 6x-1$ $\quad$ $5x-7<3$

$0\le 6x$ $\quad$ $5x<10$

$0\le x$ $\quad$ $x<2$

$x\ge 0$

$x<2$

0 2

$x\ge 0$

0 2

$x<2$ and $x\ge 0$

0 2

The solution set is $\{x|0\le x<2\}$ or $[0,2)$.

23. $6<x+3<8$

$6-3<x+3-3<8-3$

$3<x<5$

3 5

The solution set is $\{x|3<x<5\}$ or $(3,5)$.

25. $-3\le x-2<1$

$-3+2\le x-2+2<1+2$

$-1\le x<3$

-1 3

The solution set is $\{x|-1\le x<3\}$ or $[-1,3)$.

27. $-11 < 2x - 1 \le -5$

$-11 + 1 < 2x - 1 + 1 \le -5 + 1$

$-10 < 2x \le -4$

$-5 < x \le -2$

-5 -2

The solution set is $\{x|-5 < x \le -2\}$ or $(-5, -2]$.

29. $-3 \le \frac{2x}{3} - 5 < -1$

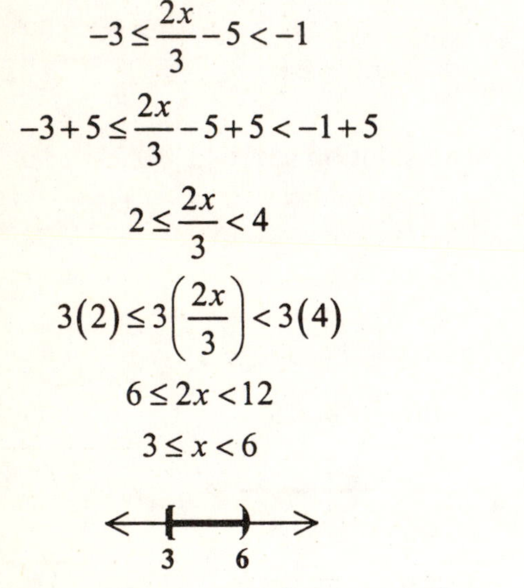

$-3 + 5 \le \frac{2x}{3} - 5 + 5 < -1 + 5$

$2 \le \frac{2x}{3} < 4$

$3(2) \le 3\left(\frac{2x}{3}\right) < 3(4)$

$6 \le 2x < 12$

$3 \le x < 6$

3 6

The solution set is $\{x|3 \le x < 6\}$ or $[3, 6)$.

31. $\{1,2,3,4\} \cup \{2,4,5\} = \{1,2,3,4,5\}$

33. $\{1,3,5,7\} \cup \{2,4,6,8,10\}$
$= \{1,2,3,4,5,6,7,8,10\}$

35. $x > 3$

3 6

$x > 6$

3 6

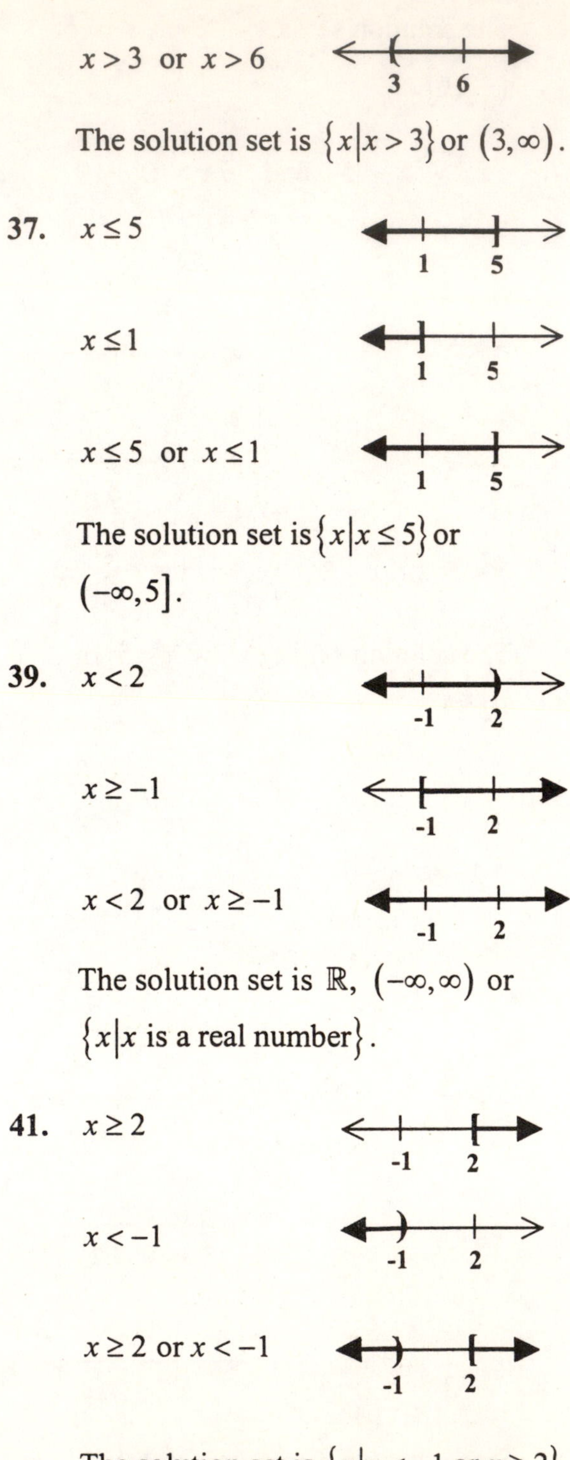

$x > 3$ or $x > 6$

3 6

The solution set is $\{x|x > 3\}$ or $(3, \infty)$.

37. $x \le 5$

1 5

$x \le 1$

1 5

$x \le 5$ or $x \le 1$

1 5

The solution set is $\{x|x \le 5\}$ or $(-\infty, 5]$.

39. $x < 2$

-1 2

$x \ge -1$

-1 2

$x < 2$ or $x \ge -1$

-1 2

The solution set is $\mathbb{R}$, $(-\infty, \infty)$ or $\{x|x \text{ is a real number}\}$.

41. $x \ge 2$

-1 2

$x < -1$

-1 2

$x \ge 2$ or $x < -1$

-1 2

The solution set is $\{x|x < -1 \text{ or } x \ge 2\}$ or $(-\infty, -1) \cup [2, \infty)$.

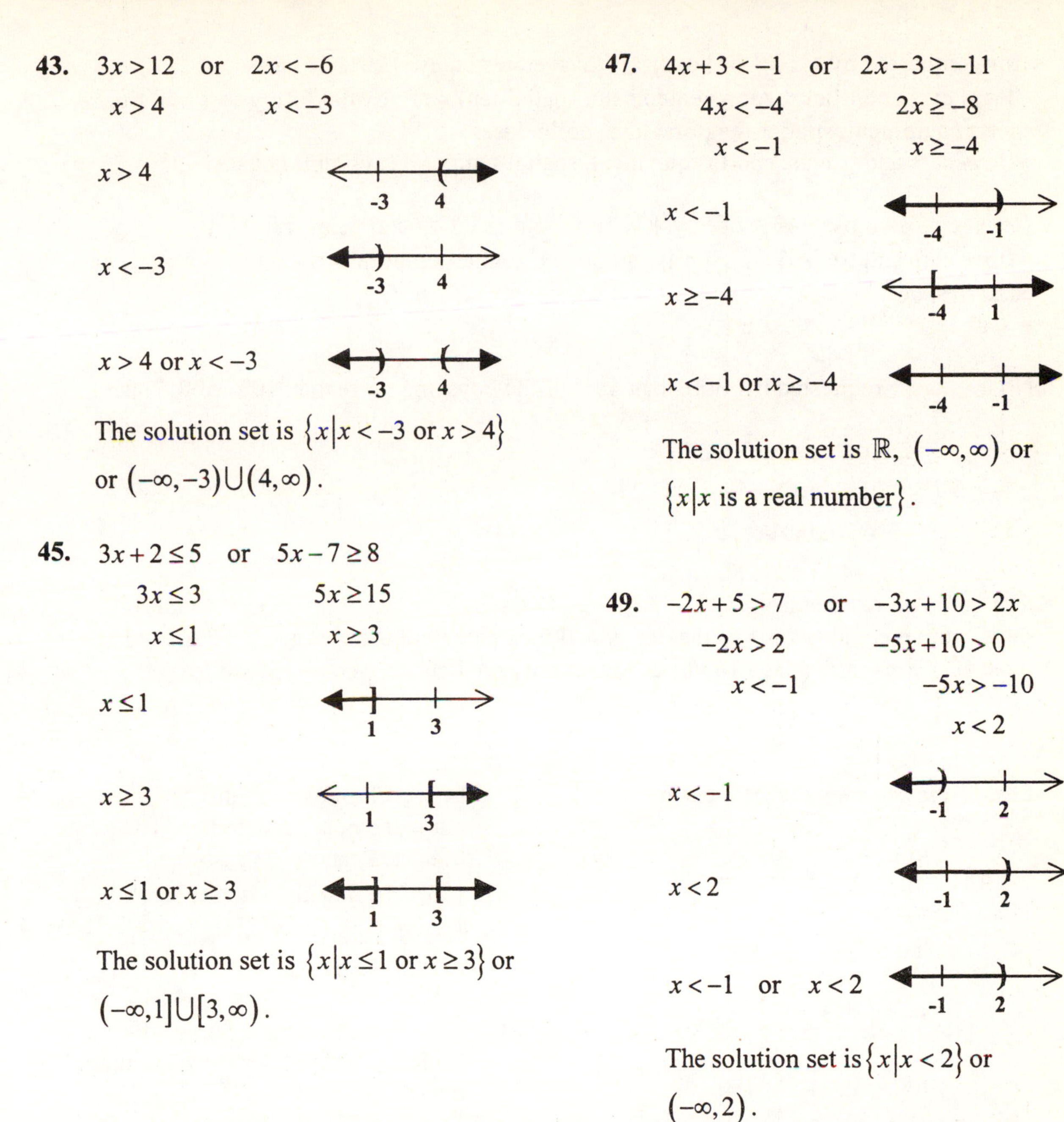

43. $3x > 12$ or $2x < -6$

$x > 4$ $\quad$ $x < -3$

$x > 4$

$x < -3$

$x > 4$ or $x < -3$

The solution set is $\{x \mid x < -3 \text{ or } x > 4\}$ or $(-\infty, -3) \cup (4, \infty)$.

45. $3x + 2 \le 5$ or $5x - 7 \ge 8$

$3x \le 3$ $\quad$ $5x \ge 15$

$x \le 1$ $\quad$ $x \ge 3$

$x \le 1$

$x \ge 3$

$x \le 1$ or $x \ge 3$

The solution set is $\{x \mid x \le 1 \text{ or } x \ge 3\}$ or $(-\infty, 1] \cup [3, \infty)$.

47. $4x + 3 < -1$ or $2x - 3 \ge -11$

$4x < -4$ $\quad$ $2x \ge -8$

$x < -1$ $\quad$ $x \ge -4$

$x < -1$

$x \ge -4$

$x < -1$ or $x \ge -4$

The solution set is $\mathbb{R}$, $(-\infty, \infty)$ or $\{x \mid x \text{ is a real number}\}$.

49. $-2x + 5 > 7$ or $-3x + 10 > 2x$

$-2x > 2$ $\quad$ $-5x + 10 > 0$

$x < -1$ $\quad$ $-5x > -10$

$x < 2$

$x < -1$

$x < 2$

$x < -1$ or $x < 2$

The solution set is $\{x \mid x < 2\}$ or $(-\infty, 2)$.

Application Exercises

51. {toys requested by > 10% of boys} ⋂ {toys requested by < 20% of girls}
={toys cars and trucks, sports equipment, spatial-temporal toys} ⋂ {toy cars and trucks, sports equipment, spatial-temporal toys, doll houses}
={toys cars and trucks, sports equipment, spatial-temporal toys}

53. {toys requested by > 10% of boys} $\cup$ {toys requested by < 20% of girls}
={toys cars and trucks, sports equipment, spatial-temporal toys} $\cup$ {toy cars and trucks, sports equipment, spatial-temporal toys, doll houses}
={toys cars and trucks, sports equipment, spatial-temporal toys, doll houses}

55. {toys requested by > 40% of boys} $\cap$ {toys requested by > 10% of girls}
={toys cars and trucks} $\cap$ {sports equipment, spatial-temporal toys, dolls, domestic accessories, doll houses}
= { }

No toys were requested by more than 40% of the boys and more than 10% of the girls.

57.
$$52 \le 0.01x + 56.7 \le 57.2$$
$$52 - 56.7 \le 0.01x + 56.7 - 56.7 \le 57.2 - 56.7$$
$$-4.7 \le 0.01x \le 0.5$$
$$-470 \le x \le 50$$
The range of years is from 1905 – 470 = 1435 through 1905 + 50 = 1955. Notice, although the inequality tells us that the global mean temperature was at least 52°F and at most 57.2°F as far back as 1435, our model only predicts temperature since 1905.

59. Let x = the score on the fifth exam

$$80 \le \frac{70+75+87+92+x}{5} < 90$$
$$80 \le \frac{324+x}{5} < 90$$
$$5(80) \le 5\left(\frac{324+x}{5}\right) < 5(90)$$
$$400 \le 324 + x < 450$$
$$400 - 324 \le 324 - 324 + x < 450 - 324$$
$$76 \le x < 126$$

A grade between 76 and 125 is needed on the fifth exam. (Because the inequality states the score must be less than 126, we say 125 is the highest possible score. In interval notation, we can use parentheses to exclude the maximum value. The range of scores can be expressed as $[76, 126)$.) If the highest grade is 100, the grade would need to be between 76 and 100.

61. Let x = the number of times the bridge is crossed per three month period
The cost with the 3-month pass is $C_3 = 7.50 + 0.50x$.
The cost with the 6-month pass is $C_6 = 30$.

Because we need to buy two 3-month passes per 6-month pass, we multiply the cost with the 3-month pass by 2.

$$2(7.50+0.50x)<30$$
$$15+x<30$$
$$x<15$$

We also must consider the cost without purchasing a pass. We need this cost to be less than the cost with a 3-month pass.

$$3x>7.50+0.50x$$
$$2.50x>7.50$$
$$x>3$$

The 3-month pass is the best deal when making more than 3 but less than 15 crossings per 3-month period.

Writing in Mathematics

63. Answers will vary.

65. Answers will vary.

67. Answer will vary.

69. Answers will vary.

Technology Exercises

71. $-1<\dfrac{x+4}{2}<3$

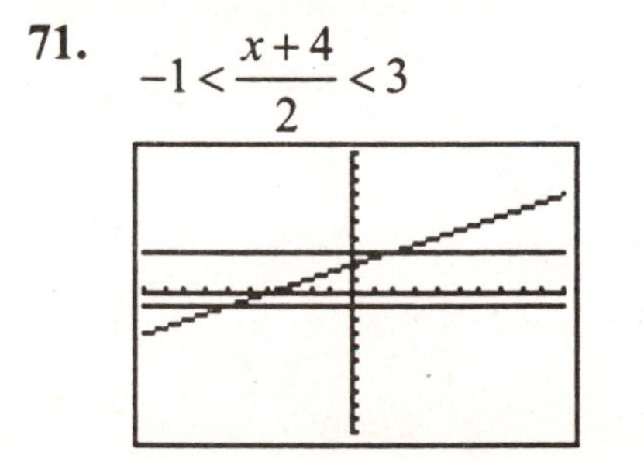

We need to find the range of the x-values of the points lying between the two constant functions. Using the intersection feature, we can determine the x-values of the endpoints of the range.

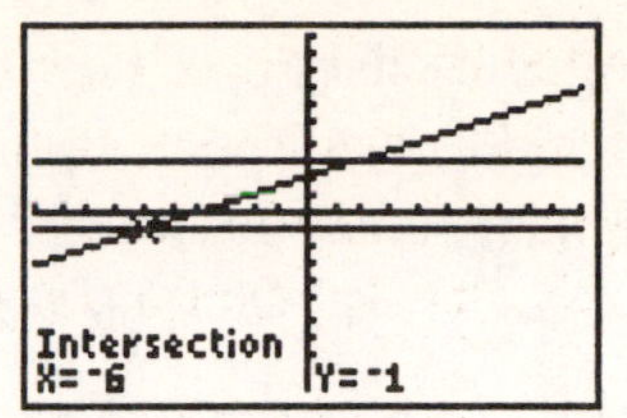

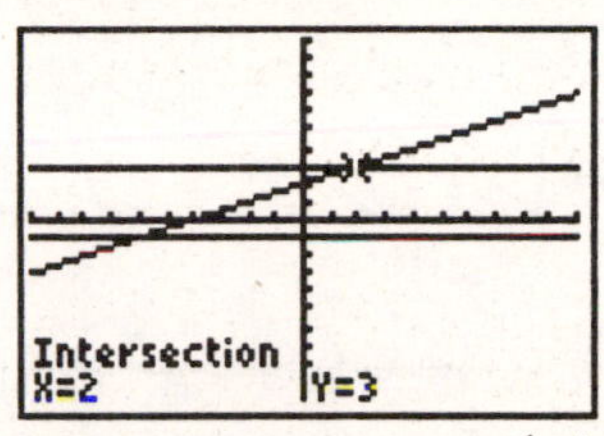

The solution set is $\{x|-6<x<2\}$ or $(-6,2)$.

73. $2\le 4-x\le 7$

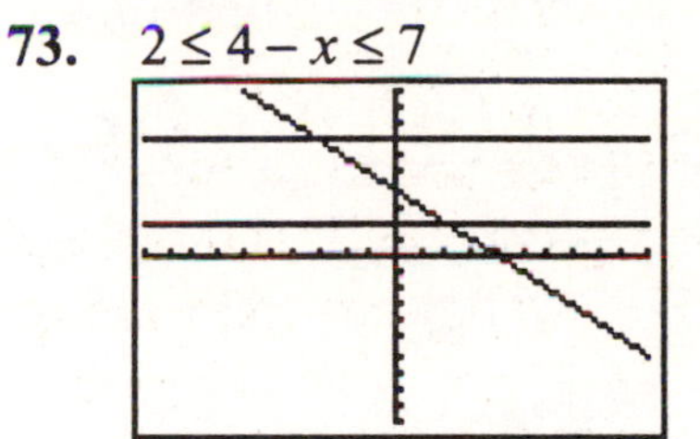

We need to find the range of the x-values of the points lying between the two constant functions. Using the intersection feature, we can determine the x-values of the endpoints of the range.

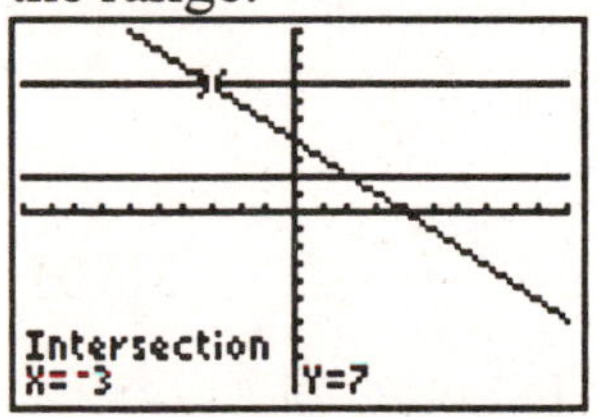

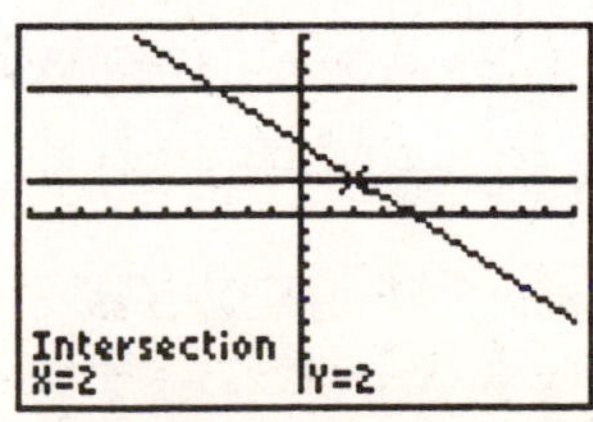

The solution set is $\{x|-3\le x\le 2\}$ or $[-3,2]$.

Critical Thinking Exercises

75. $-7 \le 8 - 3x \le 20$ and $-7 < 6x - 1 < 41$

$-15 \le -3x \le 12$ $\quad -6 < 6x < 42$

$5 \ge x \ge -4$ $\quad -1 < x < 7$

$-4 \le x \le 5$

$-4 \le x \le 5$

$-1 < x < 7$

$-4 \le x \le 5$ and $-1 < x < 7$

The solution set is $\{x|-1 < x \le 5\}$ or $(-1,5]$.

77. $f(x) = -x + 4$

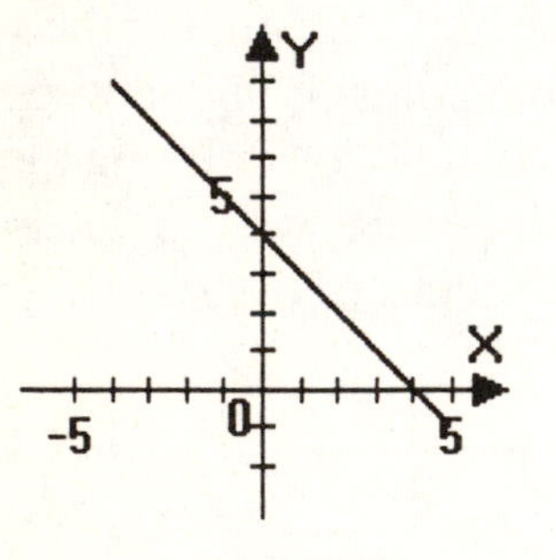

We need the values of $-x + 4 = 4 - x$ that fall between 2 and 7. Add the $y = 2$ and $y = 7$ lines to the graph.

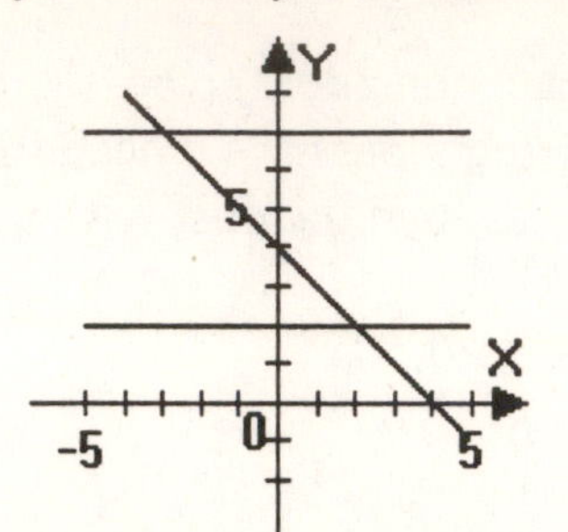

The function falls between 2 and 7 when x is between -3 and 2. The solution set is $\{x|-3 < x < 2\}$ or $(-3,2)$.

79. The domain of $f = \{x|x \le 4\}$ or $(-\infty,4]$.

81. The domain of $f + g =$ $\{x|-1 \le x \le 4\}$ or $[-1,4]$.

83. Let x = the number of nickels
Let $2x - 3$ = the number of dimes
$2x + 2$ = the number of quarters

$$3.20 \le 0.05x + 0.10(2x-3) + 0.25(2x+2) \le 5.45$$
$$3.20 \le 0.05x + 0.20x - 0.30 + 0.50x + 0.50 \le 5.45$$
$$3.20 \le 0.75x + 0.20 \le 5.45$$
$$3.00 \le 0.75x \le 5.25$$
$$4 \le x \le 7$$

The least possible number of nickels is 4 and the greatest possible number of nickels is 7.

Review Exercises

84. $f(x) = x^2 - 3x + 4 \qquad g(x) = 2x - 5$

$$\begin{aligned}(g-f)(x) &= g(x) - f(x) \\ &= (2x-5) - (x^2 - 3x + 4) \\ &= 2x - 5 - x^2 + 3x - 4 \\ &= -x^2 + 5x - 9\end{aligned}$$

$$\begin{aligned}(g-f)(x) &= -x^2 + 5x - 9 \\ (g-f)(-1) &= -(-1)^2 + 5(-1) - 9 \\ &= -1 - 5 - 9 \\ &= -15\end{aligned}$$

85. Passing through (4, 2) and perpendicular to the line $4x - 2y = 8$

The slope of the line $4x - 2y = 8$ can be found by rewriting the equation in slope-intercept form.

$$\begin{aligned}4x - 2y &= 8 \\ -2y &= -4x + 8 \\ y &= 2x - 4\end{aligned}$$

The slope of this line is 2. The slope of the line perpendicular to this line is $-\frac{1}{2}$.

Use the slope and the point to write the equation of the line in point-slope form. Then, solve for y and rewrite the equation using function notation.

$$\begin{aligned}y - y_1 &= m(x - x_1) \\ y - 2 &= -\frac{1}{2}(x - 4) \\ y - 2 &= -\frac{1}{2}x + 2 \\ y &= -\frac{1}{2}x + 4 \\ f(x) &= -\frac{1}{2}x + 4\end{aligned}$$

86. $$\begin{aligned}4 - [2(x-4) - 5] &= 4 - [2x - 8 - 5] \\ &= 4 - [2x - 13] \\ &= 4 - 2x + 13 \\ &= 17 - 2x\end{aligned}$$

Check Points 4.3

1. $|2x - 1| = 5$

$2x - 1 = 5$ or $2x - 1 = -5$

$2x = 6 \qquad 2x = -4$

$x = 3 \qquad x = -2$

The solutions are -2 and 3 and the solution set is $\{-2, 3\}$.

2. $|2x - 7| = |x + 3|$

$2x - 7 = x + 3$ or $2x - 7 = -(x + 3)$

$x - 7 = 3 \qquad 2x - 7 = -x - 3$

$x = 10 \qquad 3x - 7 = -3$

$3x = 4$

$x = \frac{4}{3}$

The solutions are $\frac{4}{3}$ and 10 and the solution set is $\left\{\frac{4}{3}, 10\right\}$.

3. $$\begin{aligned}|x - 2| &< 5 \\ -5 < x - 2 &< 5 \\ -5 + 2 < x - 2 + 2 &< 5 + 2 \\ -3 < x &< 7\end{aligned}$$

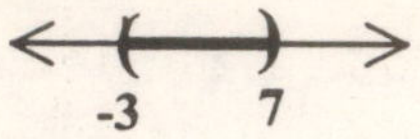

The solution set is $\{x|-3<x<7\}$ or $(-3,7)$.

4. $|2x-5|\geq 3$

$2x-5\leq -3$ or $2x-5\geq 3$

$2x\leq 2$ $\quad$ $2x\geq 8$

$x\leq 1$ $\quad$ $x\geq 4$

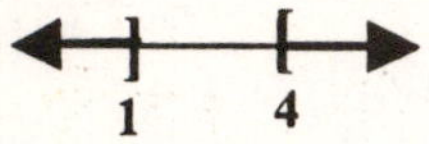

The solution set is $\{x|x\leq 1 \text{ or } x\geq 4\}$ or $(-\infty,1]\cup[4,\infty)$.

5.
$$|x-11|\leq 2.9$$
$$-2.9\leq x-11\leq 2.9$$
$$-2.9+11\leq x-11+11\leq 2.9+11$$
$$8.1\leq x\leq 13.9$$

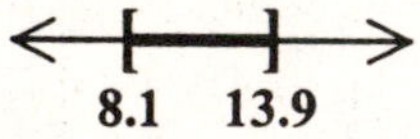

The solution set is $\{x|8.1\leq x\leq 13.9\}$ or $[8.1,13.9]$. The inequality describes the margin of error for children who think that not being able to do everything they want is a bad thing. The percentage of children in the population who think this is between 8.1% and 13.9%, inclusive.

Problem Set 4.3

Practice Exercises

1. $|x|=8$

$x=8$ or $x=-8$

The solutions are –8 and 8 and the solution set is $\{-8,8\}$.

3. $|x-2|=7$

$x-2=7$ or $x-2=-7$

$x=9$ $\quad$ $x=-5$

The solutions are –5 and 9 and the solution set is $\{-5,9\}$.

5. $|2x-1|=7$

$2x-1=7$ or $2x-1=-7$

$2x=8$ $\quad$ $2x=-6$

$x=4$ $\quad$ $x=-3$

The solutions are –3 and 4 and the solution set is $\{-3,4\}$.

7. $\left|\frac{4x-2}{3}\right|=2$

$\frac{4x-2}{3}=2$ or $\frac{4x-2}{3}=-2$

$4x-2=3(2)$ $\quad$ $4x-2=3(-2)$

$4x-2=6$ $\quad$ $4x-2=-6$

$4x=8$ $\quad$ $4x=-4$

$x=2$ $\quad$ $x=-1$

The solutions are –1 and 2 and the solution set is $\{-1,2\}$.

9. $|x|=-8$

The solution set is $\varnothing$ or $\{\ \}$. There are no values of x for which the absolute value of x is a negative number. By definition, absolute values are always positive.

11. $|x+3|=0$

Since the absolute value of the expression equals zero, we set the expression equal to zero and solve.

$x+3=0$

$x=-3$

The solution is –3 and the solution set is $\{-3\}$.

13. $2|y+6|=10$

$|y+6|=5$

$y+6=5$ or $y+6=-5$

$y=-1$ $y=-11$

The solutions are –11 and –1 and the solution set is $\{-11,-1\}$.

15. $3|2x-1|=21$

$|2x-1|=7$

$2x-1=7$ or $2x-1=-7$

$2x=8$ $2x=-6$

$x=4$ $x=-3$

The solutions are –3 and 4 and the solution set is $\{-3,4\}$.

17. $|6y-2|+4=32$

$|6y-2|=28$

$6y-2=28$ or $6y-2=-28$

$6y=30$ $6y=-26$

$y=5$ $y=-\frac{26}{6}$

$y=-\frac{13}{3}$

The solutions are $-\frac{13}{3}$ and 5 and the solution set is $\left\{-\frac{13}{3},5\right\}$.

19. $7|5x|+2=16$

$7|5x|=14$

$|5x|=2$

$5x=2$ or $5x=-2$

$x=\frac{2}{5}$ $x=-\frac{2}{5}$

The solutions are $-\frac{2}{5}$ and $\frac{2}{5}$ and the solution set is $\left\{-\frac{2}{5},\frac{2}{5}\right\}$.

21. $|x+1|+5=3$

$|x+1|=-2$

The solution set is $\varnothing$ or $\{\ \}$. By definition, absolute values are always positive.

23. $|4y+1|+10=4$

$|4y+1|=-6$

The solution set is $\varnothing$ or $\{\ \}$. By definition, absolute values are always positive.

25. $|2x-1|+3=3$

$|2x-1|=0$

Since the absolute value of the expression equals zero, we set the expression equal to zero and solve.

$$2x-1=0$$
$$2x=1$$
$$x=\frac{1}{2}$$

The solution is $\frac{1}{2}$ and the solution set is $\left\{\frac{1}{2}\right\}$.

27. $|5x-8|=|3x+2|$

$5x-8=3x+2$ or $5x-8=-3x-2$

$2x-8=2 \qquad 8x-8=-2$

$2x=10 \qquad 8x=6$

$x=5 \qquad x=\frac{6}{8}=\frac{3}{4}$

The solutions are $\frac{3}{4}$ and 5 and the solution set is $\left\{\frac{3}{4},5\right\}$.

29. $|2x-4|=|x-1|$

$2x-4=x-1$ or $2x-4=-x+1$

$x-4=-1 \qquad 3x-4=1$

$x=3 \qquad 3x=5$

$x=\frac{5}{3}$

The solutions are $\frac{5}{3}$ and 3 and the solution set is $\left\{\frac{5}{3},3\right\}$.

31. $|2x-5|=|2x+5|$

$2x-5=2x+5$ or $2x-5=-2x-5$

$-5\neq 5 \qquad 4x-5=-5$

$4x=0$

$x=0$

The solution is 0 and the solution set is $\{0\}$.

33. $|x-3|=|5-x|$

$x-3=5-x$ or $x-3=-(5-x)$

$2x-3=5 \qquad x-3=-5+x$

$2x=8 \qquad -3\neq -5$

$x=4$

The solution is 4 and the solution set is $\{4\}$.

35. $|2y-6|=|10-2y|$

$2y-6=10-2y$ or $2y-6=-10+2y$

$4y-6=10 \qquad -6\neq -10$

$4y=16$

$x=4$

The solution is 4 and the solution set is $\{4\}$.

37. $\left|\frac{2x}{3}-2\right|=\left|\frac{x}{3}+3\right|$

$$\frac{2x}{3}-2=\frac{x}{3}+3$$
$$3\left(\frac{2x}{3}\right)-3(2)=3\left(\frac{x}{3}\right)+3(3)$$
$$2x-6=x+9$$
$$x-6=9$$
$$x=15$$

or

$$\frac{2x}{3}-2=-\left(\frac{x}{3}+3\right)$$
$$\frac{2x}{3}-2=-\frac{x}{3}-3$$
$$3\left(\frac{2x}{3}\right)-3(2)=3\left(-\frac{x}{3}\right)-3(3)$$
$$2x-6=-x-9$$
$$3x-6=-9$$
$$3x=-3$$
$$x=-1$$

The solutions are -1 and 15 and the solution set is $\{-1,15\}$.

39. $|x|<3$

$-3<x<3$

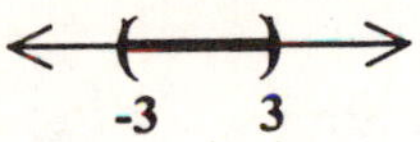

The solution set is $\{x|-3<x<3\}$ or $(-3,3)$.

41.
$$|x-2|<1$$
$$-1<x-2<1$$
$$-1+2<x-2+2<1+2$$
$$1<x<3$$

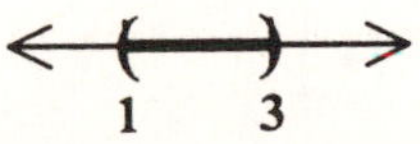

The solution set is $\{x|1<x<3\}$ or $(1,3)$.

43.
$$|x+2|\le 1$$
$$-1\le x+2\le 1$$
$$-1-2\le x+2-2\le 1-2$$
$$-3\le x\le -1$$

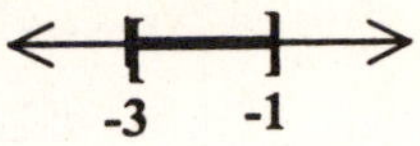

The solution set is $\{x|-3\le x\le -1\}$ or $[-3,-1]$.

45.
$$|2x-6|<8$$
$$-8<2x-6<8$$
$$-8+6<2x-6+6<8+6$$
$$-2<2x<14$$
$$-1<x<7$$

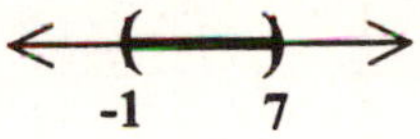

The solution set is $\{x|-1<x<7\}$ or $(-1,7)$.

47. $|x|>3$

$x<-3$ or $x>3$

The solution set is $\{x|x<-3$ and $x>3\}$ or $(-\infty,-3)\cup(3,\infty)$.

49. $|x+3|>1$

$x+3<-1$ or $x+3>1$

$x<-4$ $\qquad$ $x>-2$

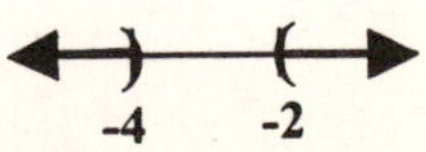

The solution set is $\{x | x < -4$ and $x > -2\}$ or $(-\infty, -4) \cup (-2, \infty)$.

51. $|x-4| \geq 2$

$x - 4 \leq -2$ or $x - 4 \geq 2$

$x \leq 2$ $\quad$ $x \geq 6$

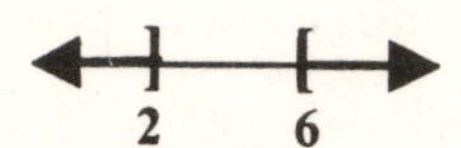

The solution set is $\{x | x \leq 2$ and $x \geq 6\}$ or $(-\infty, 2] \cup [6, \infty)$.

53. $|3x - 8| > 7$

$3x - 8 < -7$ or $3x - 8 > 7$

$3x < 1$ $\quad$ $3x > 15$

$x < \frac{1}{3}$ $\quad$ $x > 5$

The solution set is $\left\{x \middle| x < \frac{1}{3}\right.$ and $x > 5\}$ or $\left(-\infty, \frac{1}{3}\right) \cup (5, \infty)$.

55.

$$|2(x-1)+4| \leq 8$$
$$|2x - 2 + 4| \leq 8$$
$$|2x + 2| \leq 8$$
$$-8 \leq 2x + 2 \leq 8$$
$$-8 - 2 \leq 2x + 2 - 2 \leq 8 - 2$$
$$-10 \leq 2x \leq 6$$
$$-5 \leq x \leq 3$$

The solution set is $\{x | -5 \leq x \leq 3\}$ or $[-5, 3]$.

57.

$$\left|\frac{2y+6}{3}\right| < 2$$
$$-2 < \frac{2y+6}{3} < 2$$
$$3(-2) < 3\left(\frac{2y+6}{3}\right) < 3(2)$$
$$-6 < 2y + 6 < 6$$
$$-6 - 6 < 2y + 6 - 6 < 6 - 6$$
$$-12 < 2y < 0$$
$$-6 < y < 0$$

The solution set is $\{x | -6 < x < 0\}$ or $(-6, 0)$.

59. $\left|\frac{2x+2}{4}\right| \geq 2$

$\frac{2x+2}{4} \leq -2$ or $\frac{2x+2}{4} \geq 2$

$2x + 2 \leq -8$ $\quad$ $2x + 2 \geq 8$

$2x \leq -10$ $\quad$ $2x \geq 6$

$x \leq -5$ $\quad$ $x \geq 3$

The solution set is $\{x | x \leq -5$ and $x \geq 3\}$ or $(-\infty, -5] \cup [3, \infty)$.

61.

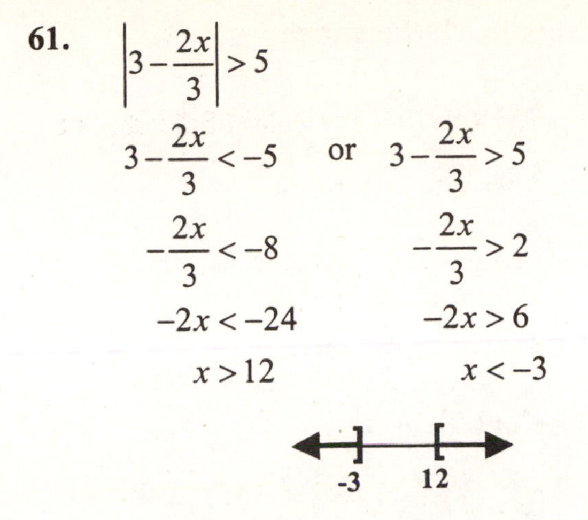

$$\left|3-\frac{2x}{3}\right|>5$$

$$3-\frac{2x}{3}<-5 \quad \text{or} \quad 3-\frac{2x}{3}>5$$

$$-\frac{2x}{3}<-8 \qquad -\frac{2x}{3}>2$$

$$-2x<-24 \qquad -2x>6$$

$$x>12 \qquad x<-3$$

The solution set is $\{x|x<-3$ and $x>12\}$ or $(-\infty,-3)\cup(12,\infty)$.

63. $|x-2|<-1$

The solution set is $\varnothing$ or $\{\ \}$. Since all absolute values are positive, there are no values of x that will make the absolute value of the expression less than -1.

65. $|x+6|>-10$

Since all absolute values are positive, we know that when simplified, the left hand side will be a positive number. We also know that any positive number is greater than any negative number. This means that regardless of the value of x, the left hand side will be greater than the right hand side of the inequality. The solution set is $\{x|x$ is a real number$\}$, $\mathbb{R}$ or $(-\infty,\infty)$.

67.

$$|x+2|+9\le 16$$

$$|x+2|\le 7$$

$$-7\le x+2\le 7$$

$$-7-2\le x+2-2\le 7-2$$

$$-9\le x\le 5$$

-9 5

The solution set is $\{x|-9\le x\le 5\}$ or $[-9,5]$.

69.

$$2|2x-3|+10>12$$

$$2|2x-3|>2$$

$$|2x-3|>1$$

$$2x-3<-1 \qquad 2x-3>1$$

$$2x<2 \quad \text{or} \quad 2x>4$$

$$x<1 \qquad x>2$$

1 2

The solution set is $\{x|x<1$ and $x>2\}$ or $(-\infty,1)\cup(2,\infty)$.

Application Exercises

71.

$$|x-60.2|\le 1.6$$

$$-1.6\le x-60.2\le 1.6$$

$$-1.6+60.2\le x-60.2+60.2\le 1.6+60.2$$

$$58.6\le x\le 61.8$$

The percentage of the U.S. population that watched M*A*S*H is between 58.6% and 61.8%, inclusive. The margin of error is 1.6%.

73.
$$|T-57| \le 7$$
$$-7 \le T-57 \le 7$$
$$-7+57 \le T-57+57 \le 7+57$$
$$50 \le T \le 64$$
The monthly average temperature for San Francisco, California ranges from 50°F to 64°F, inclusive.

75.
$$|x-8.6| \le 0.01$$
$$-0.01 \le x-8.6 \le 0.01$$
$$-0.01+8.6 \le x-8.6+8.6 \le 0.01+8.6$$
$$8.59 \le x \le 8.61$$
The length of the machine part must be between 8.59 and 8.61 centimeters, inclusive.

77.
$$\left|\frac{h-50}{5}\right| \ge 1.645$$

$$\frac{h-50}{5} \le -1.645$$
$$h-50 \le 5(-1.645)$$
$$h-50 \le -8.225$$
$$h \le 41.775$$

or

$$\frac{h-50}{5} \ge 1.645$$
$$h-50 \ge 5(1.645)$$
$$h-50 \ge 8.225$$
$$h \ge 58.225$$
The coin would be considered unfair if the tosses resulted in 41 or less heads, or 59 or more heads.

Writing in Mathematics

79. Answers will vary.

81. Answers will vary.

83. Answers will vary.

85. Answers will vary.

Technology Exercises

87. $|3(x+4)| = 12$

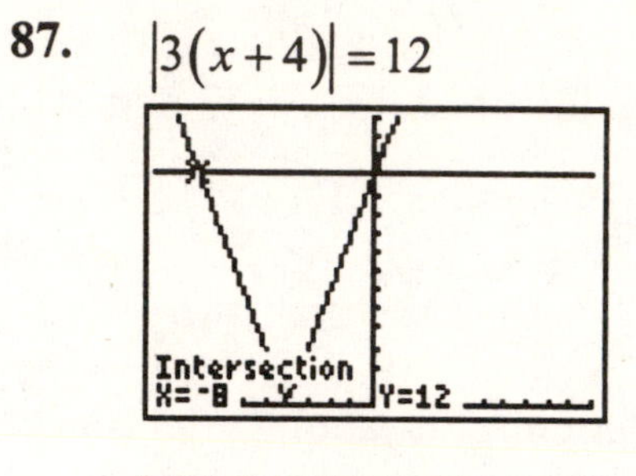

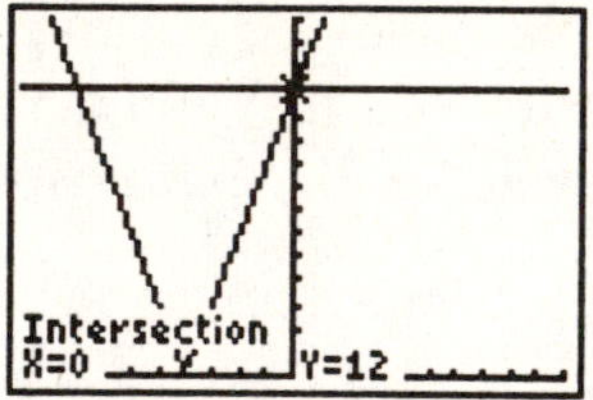

The solutions are –8 and 0 and the solution set is $\{-8, 0\}$.

89. 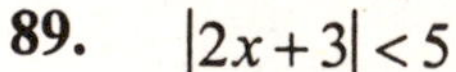 $|2x+3| < 5$

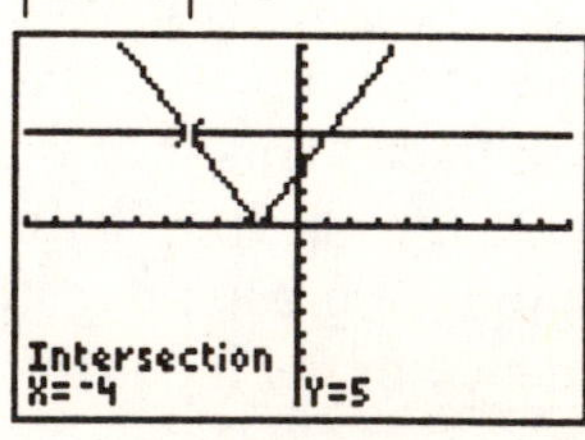

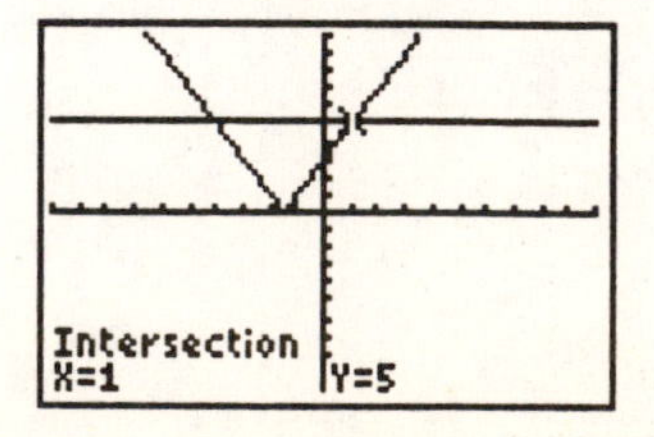

The solution set is $\{x|-4 < x < 1\}$ or $(-4,1)$.

91. $|x+4| < -1$

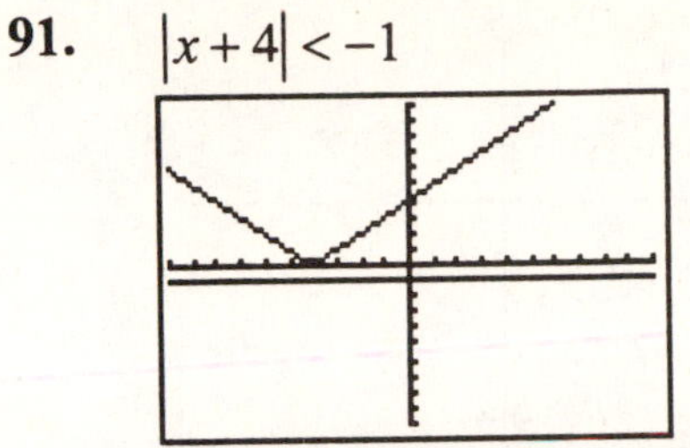

No part of the graph of the absolute value lies below the graph of the constant. The solution set is $\varnothing$ or $\{ \ \}$.

93. $|0.1x - 0.4| + 0.4 > 0.6$

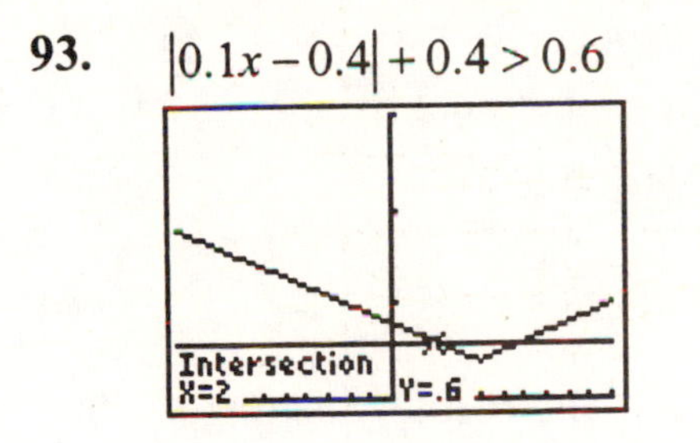

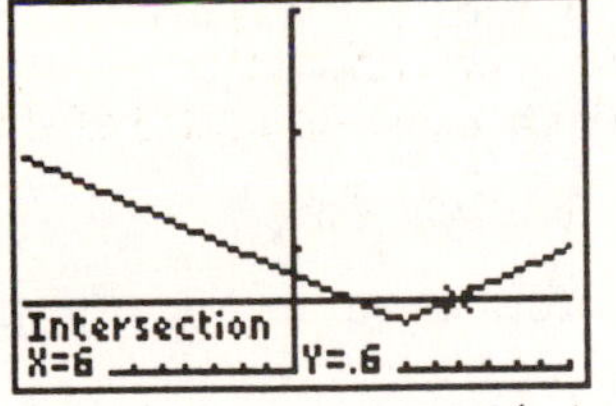

The solution set is $\{x|x < 2 \text{ or } x > 6\}$ or $(-\infty,2)\cup(6,\infty)$.

95. Answers will vary. For example, consider Exercise 5.

$|2x-1| = 5$

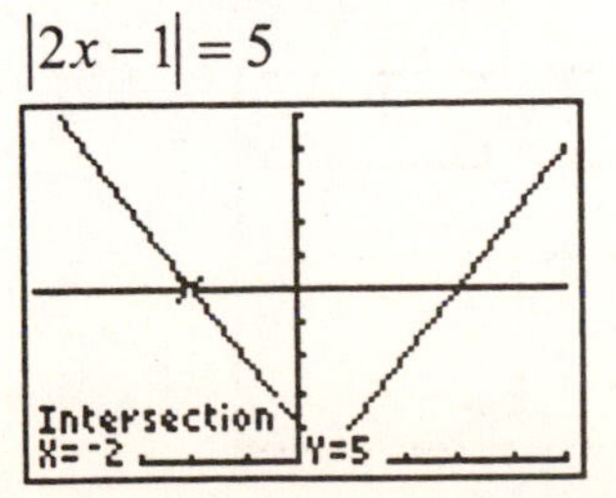

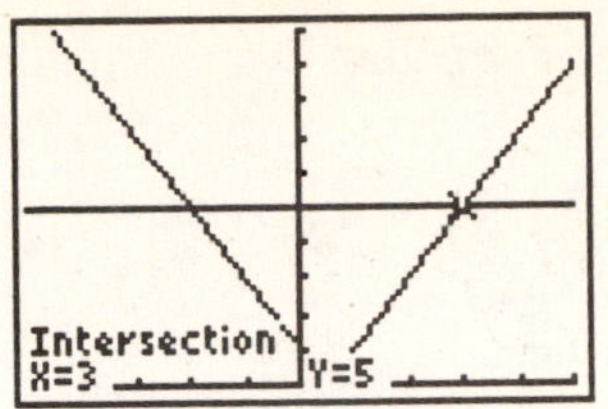

The solutions are -2 and 3 and the solution set is $\{-2,3\}$.

97. The solution of $|4-x| = 1$ are 3 and 5 and the solution set is $\{3,5\}$. These are the points of intersection of $y = |4-x|$ and $y = 1$.

99.

$$|p-0.3| \le 0.2$$
$$-0.2 \le p - 0.3 \le 0.2$$
$$-0.2 + 0.3 \le p - 0.3 + 0.3 \le 0.2 + 0.3$$
$$0.1 \le p \le 0.5$$

This means that at least 0.1% and at most 0.5% of the products are defective. First, multiply by 100,000 to determine the number of products that will be defective.

$100,000 \times 0.001 = 100$

$100,000 \times 0.005 = 500$

At least 100 and at most 500 products will be defective. At $5 per refund, the company will be paying out at least $500 and at most $2500.

Review Exercises

101. Solve for y to obtain slope-intercept form.

$$3x - 5y = 15$$
$$-5y = -3x + 15$$
$$y = \frac{3}{5}x - 3$$

The y-intercept is –3 and the slope is $\frac{3}{5}$.

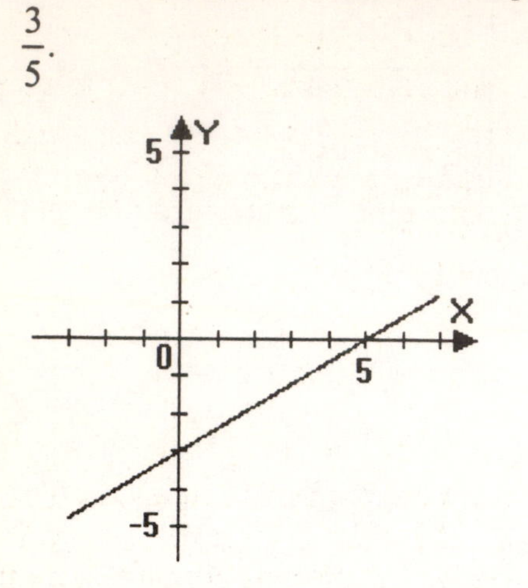

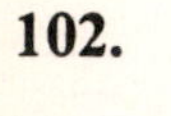

102. $f(x) = -\frac{2}{3}x$

$y = -\frac{2}{3}x$

The y-intercept is 0 and the slope is $-\frac{2}{3}$.

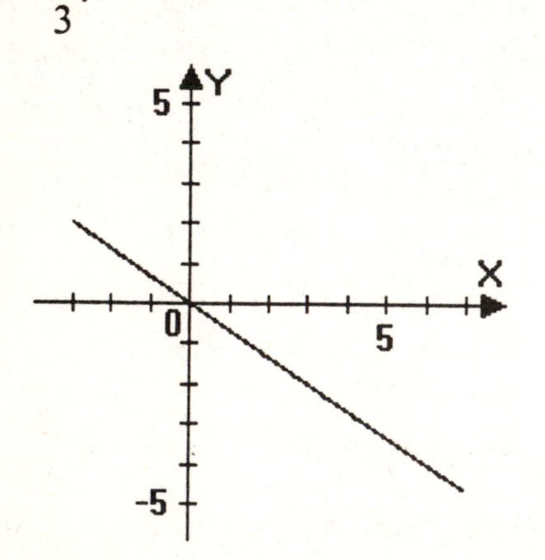

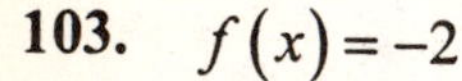

103. $f(x) = -2$

$y = -2$

When graphed, an equation of the form $y = b$ is a horizontal line.

$f(x) = -2$ is the horizontal line positioned at $y = -2$.

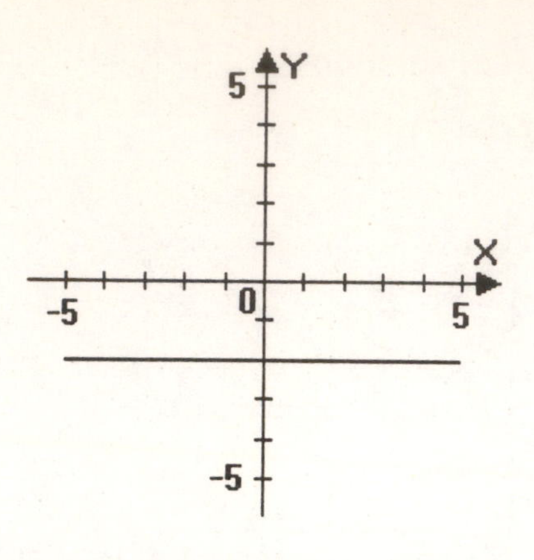

Check Points 4.4

1. $2x - 4y \ge 8$

First, find the intercepts to the equation $2x - 4y = 8$.

Find the x–intercept by setting $y = 0$.

$2x - 4(0) = 8$

$2x = 8$

$x = 4$

Find the y–intercept by setting $x = 0$.

$2(0) - 4y = 8$

$-4y = 8$

$y = -2$

Next, use the origin as a test point.

$2(0) - 4(0) \ge 8$

$0 - 0 \ge 8$

$0 \ge 8$

This is a false statement. This means that the origin will not fall in the shaded half-plane.

2. $y \geq -\frac{3}{4}x$

Replacing the inequality symbol with an equal sign, we have $y = -\frac{3}{4}x$.

Since the equation is in slope-intercept form, use the slope, $-\frac{3}{4}$, and the intercept, 0, to graph the equation. Next, we need to find a test point. We cannot use the origin, because it lies on the line. Use $(1,1)$ as a test point.

$1 \geq -\frac{3}{4}(1)$

$1 \geq -\frac{3}{4}$

This is a true statement, so we know the point $(1,1)$ lies in the shaded half-plane.

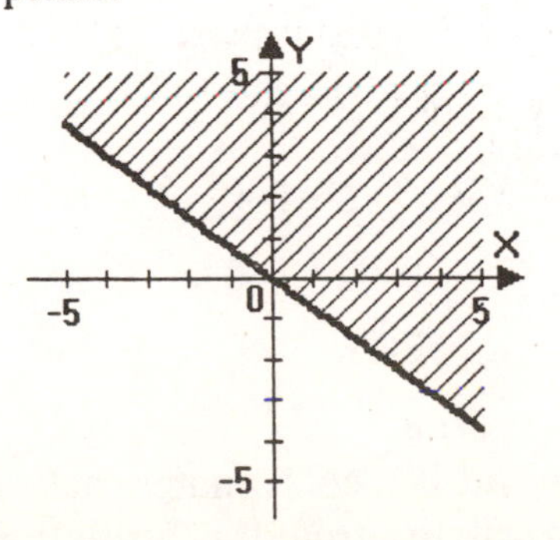

3. $x + 2y > 4$

$2x - 3y \leq -6$

First consider $x + 2y > 4$. If we solve for y in $x + 2y = 4$, we can graph the line using the slope and the y-intercept.

$x + 2y = 4$

$2y = -x + 4$

$y = -\frac{1}{2}x + 2$

y-intercept = 2 slope = $-\frac{1}{2}$

Now, use the origin as a test point.

$0 + 2(0) > 4$

$0 + 0 > 4$

$0 > 4$

This is a false statement. This means that the point, $(0,0)$, will not fall in the shaded half-plane.

Next consider $2x - 3y \leq -6$. If we solve for y in $2x - 3y = -6$, we can graph using the slope and the y-intercept.

$2x - 3y = -6$

$-3y = -2x - 6$

$y = \frac{2}{3}x + 2$

y-intercept = 2 slope = $\frac{2}{3}$

Now, use the origin as a test point.

$2(0) - 3(0) \leq -6$

$0 - 0 \leq -6$

$0 \leq -6$

This is a false statement. This means that the point $(0,0)$ will not fall in the shaded half-plane.

Next, graph each of the inequalities. The solution to the system is the intersection of the shaded half-planes.

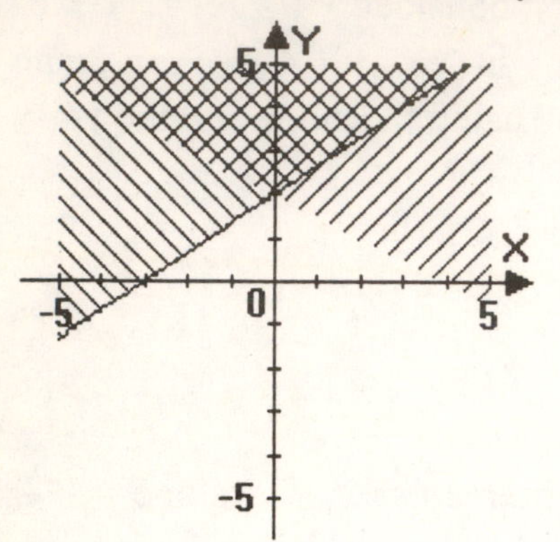

4. $x+y<2$

$-2 \le x < 1$

$y > -3$

First consider $x+y<2$. If we solve for y in $x+y=2$, we can graph using the slope and the y-intercept.

$x+y=2$

$y=-x+2$

y-intercept = 2

slope = - 1

Now, use the origin as a test point.

$x+y<2$

$0+0<2$

$0<2$

This is a true statement. This means that the point $(0,0)$ will fall in the shaded half-plane.

Next consider $-2 \le x < 1$. This is equivalent to $x \ge -2$ and $x < 1$. We need to graph the vertical lines, $x=-2$ and $x=1$. We know that $-2 \le x < 1$, so we shade the region between the two vertical lines.

Finally, consider $y > -3$. We need to graph the horizontal line, $y=-3$. Since $y > -3$, we shade the half-plane above the horizontal line.

Next, graph each of the inequalities. The solution to the system is the intersection of the graphs of the inequalities.

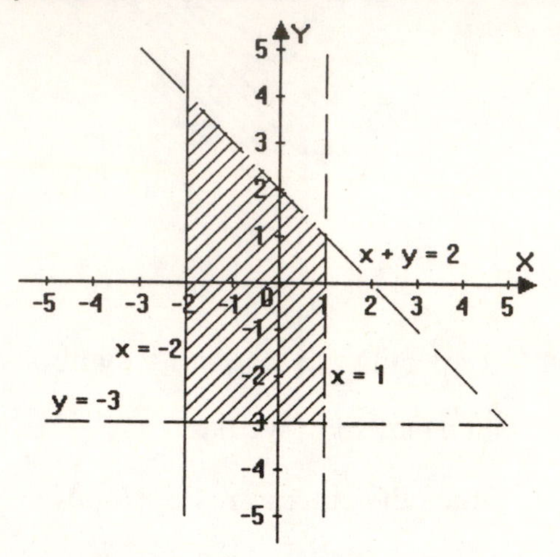

5. Point B = (60, 20)

Grasslands occur under the following conditions.

$T \ge 35$

$5T-7P \ge 70$

$3T-35P \le -140$

Show that point B satisfies each of the inequalities.

$T \ge 35$

$60 \ge 35$

True

$5T-7P \ge 70$

$5(60)-7(20) \ge 70$

$300-140 \ge 70$

$160 \ge 70$

True

$3T-35P \le -140$

$3(60)-35(20) \le -140$

$180-700 \le -140$

$-520 \le -140$

True

Since point B's coordinates makes each inequality true, it is a solution of the system that describes where grass-lands occur.

Problem Set 4.4

Practice Exercises

1. $x+y\ge 3$

First, graph the equation $x+y=3$.
Rewrite the equation in slope-intercept form by solving for y.

$x+y=3$

$y=-x+3$

y-intercept = 3
slope = −1
Next, use the origin as a test point.

$x+y\ge 3$

$0+0\ge 3$

$0\ge 3$

This is a false statement. This means that the point $(0,0)$ will not fall in the shaded half-plane.

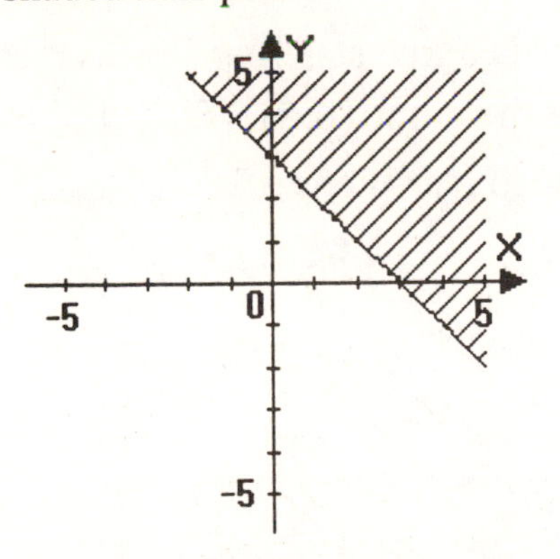

3. $x-y<5$

First, graph the equation $x-y=5$.
Rewrite the equation in slope-intercept form by solving for y.

$x-y=5$

$-y=-x+5$

$y=x-5$

y-intercept = −5
slope = 1
Next, use the origin as a test point.

$x-y<5$

$0-0<5$

$0<5$

This is a true statement. This means that the point $(0,0)$ will fall in the shaded half-plane.

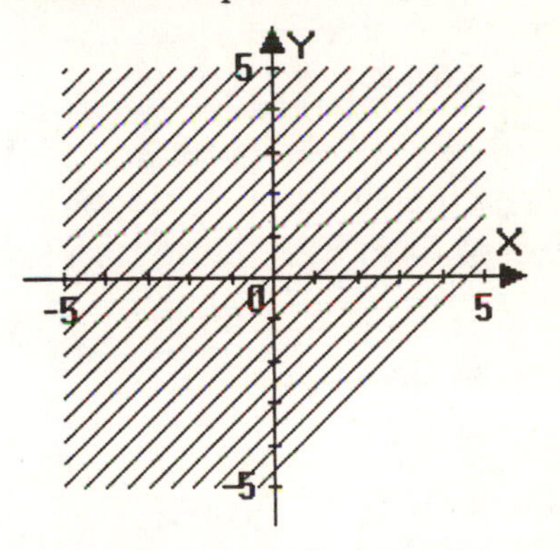

5. $x+2y>4$

First, graph the equation $x+2y=4$.
Rewrite the equation in slope-intercept form by solving for y.

$x+2y=4$

$2y=-x+4$

$y=-\frac{1}{2}x+2$

y-intercept = 2

slope = $-\frac{1}{2}$

Next, use the origin as a test point.

$0+2(0)>4$

$0+0>4$

$0>4$

This is a false statement. This means that the point $(0,0)$ will not fall in the shaded half-plane.

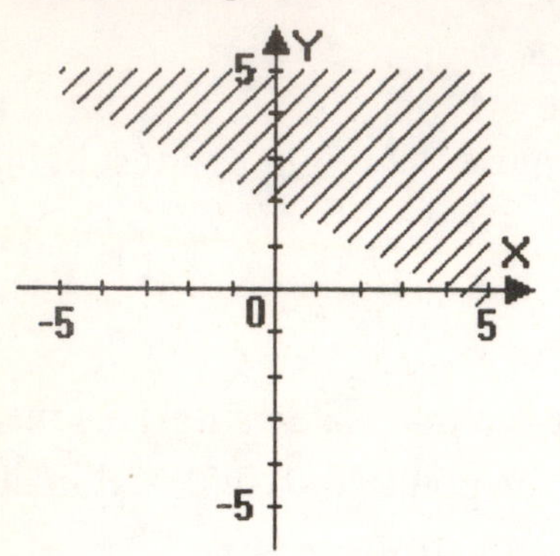

7. $3x - y \le 6$

First, graph the equation $3x - y = 6$.
Rewrite the equation in slope-intercept form by solving for y.

$3x - y = 6$

$-y = -3x + 6$

$y = 3x - 6$

y-intercept $= -6$ slope $= 3$

Next, use the origin as a test point.

$3(0) - 0 \le 6$

$0 - 0 \le 6$

$0 \le 6$

This is a true statement. This means that the point $(0,0)$ will fall in the shaded half-plane.

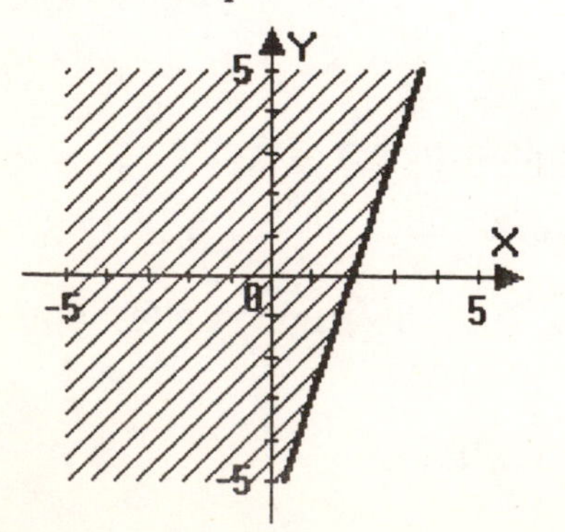

9. $\frac{x}{2} + \frac{y}{3} < 1$

First, graph the equation $\frac{x}{2} + \frac{y}{3} = 1$.
Rewrite the equation in slope-intercept form by solving for y.

$\frac{x}{2} + \frac{y}{3} = 1$

$6\left(\frac{x}{2}\right) + 6\left(\frac{y}{3}\right) = 6(1)$

$3x + 2y = 6$

$2y = -3x + 6$

$y = -\frac{3}{2}x + 3$

y-intercept $= 3$ slope $= -\frac{3}{2}$

Next, use the origin as a test point.

$\frac{0}{2} + \frac{0}{3} < 1$

$0 + 0 < 1$

$0 < 1$

This is a true statement. This means that the point $(0,0)$ will fall in the shaded half-plane.

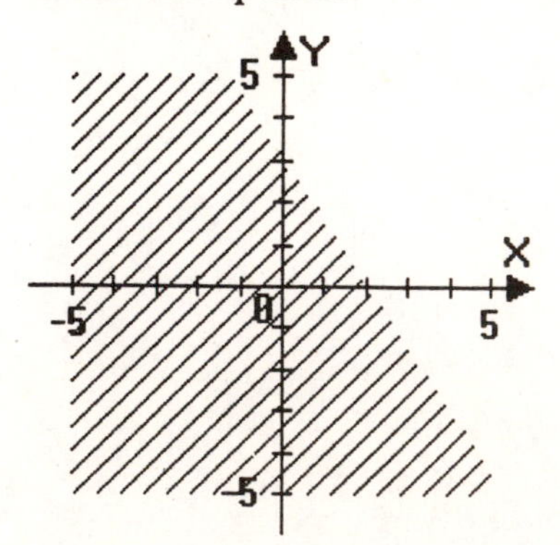

11. $y > \frac{1}{3}x$

Replacing the inequality symbol with

an equal sign, we have $y > \frac{1}{3}x$. Since the equation is in slope-intercept form, use the slope and the intercept to graph the equation. The y–intercept is 0 and the slope is $\frac{1}{3}$.

Next, we need to find a test point. We cannot use the origin this time, because it lies on the line. Use $(1,1)$ as a test point.

$1 > \frac{1}{3}(1)$

$1 > \frac{1}{3}$

This is a true statement, so we know the point $(1,1)$ lies in the shaded half-plane.

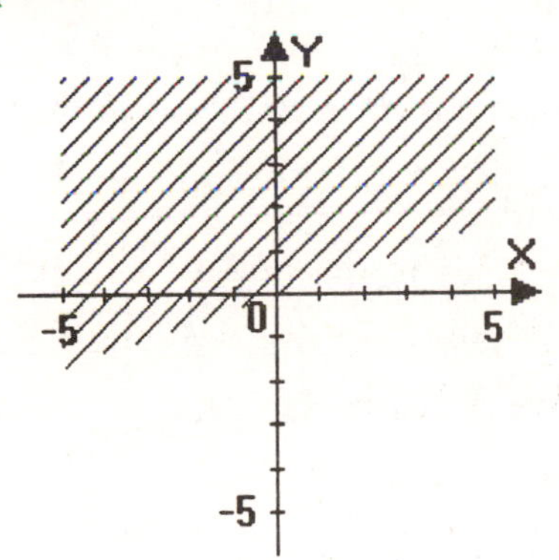

13. $y \le 3x + 2$

First, graph the equation $y = 3x + 2$. Since the equation is in slope-intercept form, use the slope and the intercept to graph the equation. The y–intercept is 2 and the slope is 3. Next, use the origin as a test point.

$0 \le 3(0) + 2$

$0 \le 2$

This is a true statement. This means that the point $(0,0)$ will fall in the shaded half-plane.

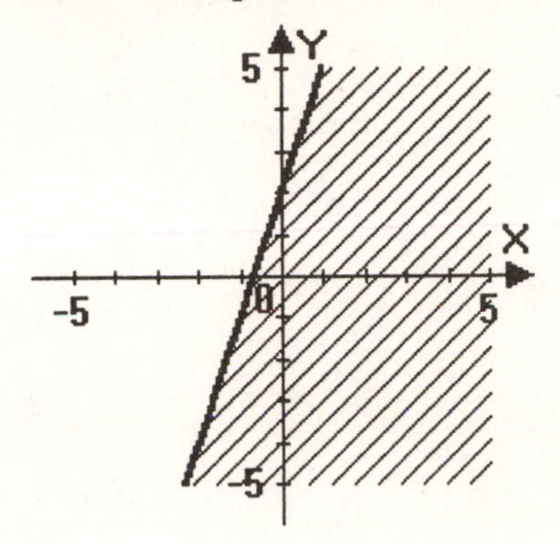

15. $y < -\frac{1}{4}x$

Replacing the inequality symbol with an equal sign, we have $y = -\frac{1}{4}x$. Since the equation is in slope-intercept form, use the slope and the intercept to graph the equation. The y–intercept is 0 and the slope is $-\frac{1}{4}$.

Next, we need to find a test point. We cannot use the origin this time, because it lies on the line. Use $(1,1)$ as a test point.

$1 < -\frac{1}{4}(1)$

$1 < -\frac{1}{4}$

This is a false statement, so we know the point $(1,1)$ does not lie in the shaded half-plane.

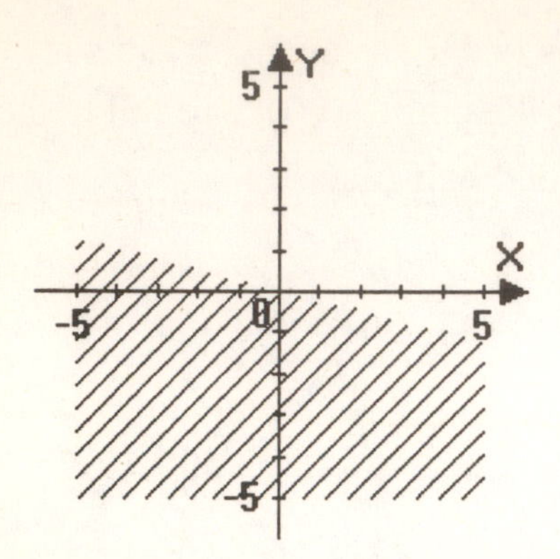

17. $x \le 2$
Replacing the inequality symbol with an equal sign, we have $x = 2$. We know that equations of the form $x = a$ are vertical lines with x-intercept = a.
Next, use the origin as a test point.
$x \le 2$
$0 \le 2$
This is a true statement, so we know the point $(0,0)$ lies in the shaded half-plane.

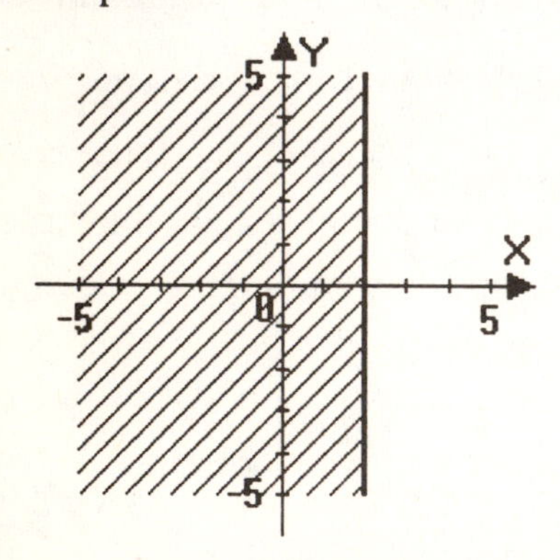

19. $y > -4$
Replacing the inequality symbol with an equal sign, we have $y = -4$. We know that equations of the form $y = b$ are horizontal lines with y-intercept = b.

Next, use the origin as a test point.
$y > -4$
$0 > -4$
This is a true statement, so we know the point $(0,0)$ lies in the shaded half-plane.

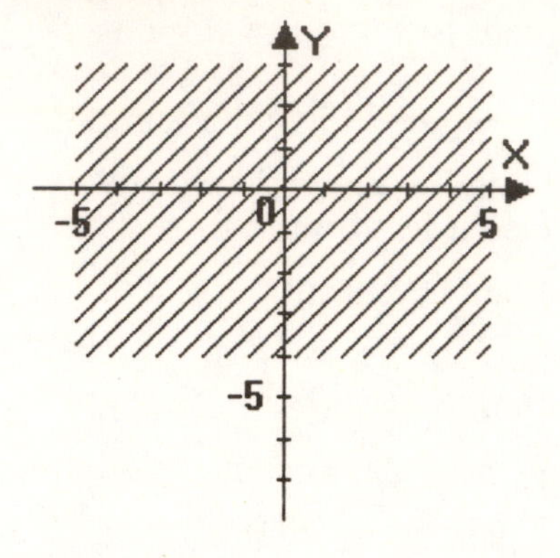

21. $y \ge 0$
Replacing the inequality symbol with an equal sign, we have $y = 0$. We know that equations of the form $y = b$ are vertical lines with y-intercept = b. In this case, we have $y = 0$, the equation of the x-axis.
Next, we need to find a test point. We cannot use the origin, because it lies on the line. Use $(1,1)$ as a test point.
$y \ge 0$
$1 \ge 0$
This is a true statement, so we know the point $(1,1)$ lies in the shaded half-plane.

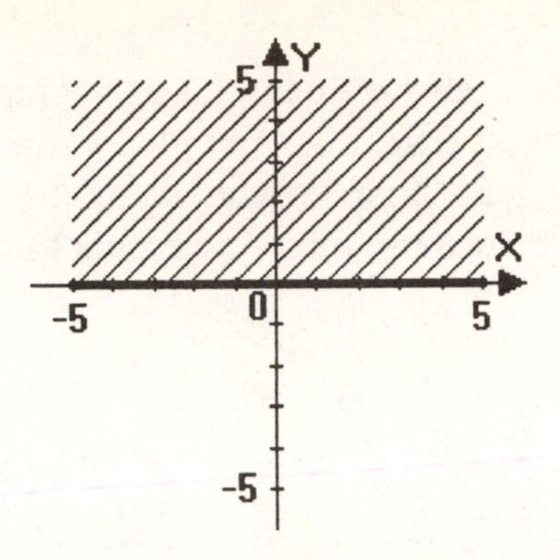

23. $3x + 6y \le 6$

$2x + y \le 8$

Graph the equations using the intercepts.

$3x + 6y = 6$ $\quad$ $2x + y = 8$

$x - \text{intercept} = 2$ $\quad$ $x - \text{intercept} = 4$

$y - \text{intercept} = 1$ $\quad$ $y - \text{intercept} = 8$

Use the origin as a test point to determine shading.

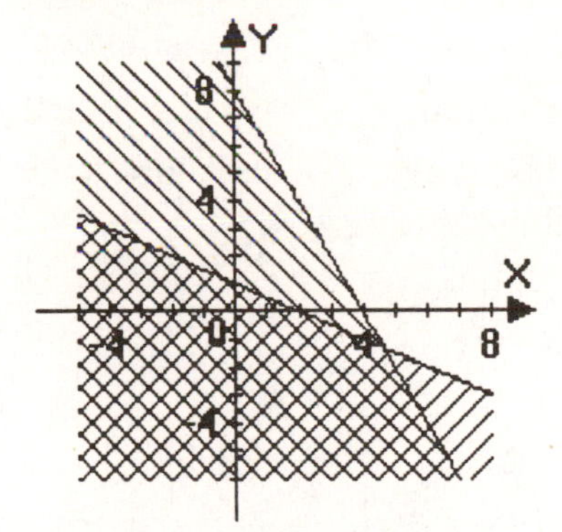

The solution set is the intersection of the shaded half-planes.

25. $2x - 5y \le 10$

$3x - 2y > 6$

Graph the equations using the intercepts.

$2x - 5y = 10$ $\quad$ $3x - 2y = 6$

$x - \text{intercept} = 5$ $\quad$ $x - \text{intercept} = 2$

$y - \text{intercept} = 2$ $\quad$ $y - \text{intercept} = -3$

Use the origin as a test point to determine shading.

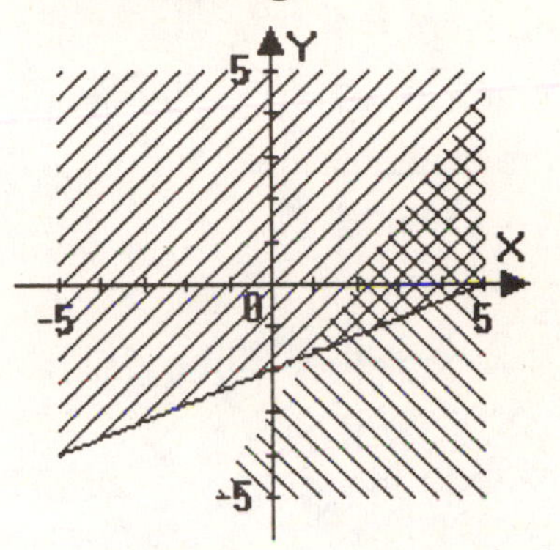

The solution set is the intersection of the shaded half-planes.

27. $y > 2x - 3$

$y < -x + 6$

Graph the equations using the intercepts.

$y = 2x - 3$ $\quad$ $y = -x + 6$

$x - \text{intercept} = \frac{3}{2}$ $\quad$ $x - \text{intercept} = 6$

$y - \text{intercept} = -3$ $\quad$ $y - \text{intercept} = 6$

Use the origin as a test point to determine shading.

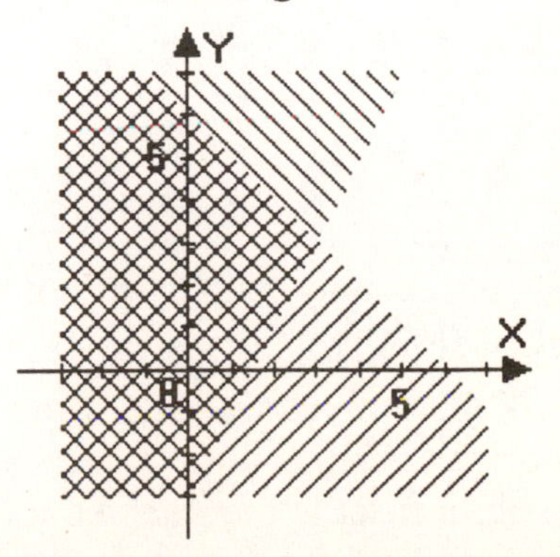

The solution set is the intersection of the shaded half-planes.

29. $x + 2y \le 4$

$y \ge x - 3$

Graph the equations using the intercepts.

$x + 2y = 4$	$y = x - 3$
$x - \text{intercept} = 4$	$x - \text{intercept} = 3$
$y - \text{intercept} = 2$	$y - \text{intercept} = -3$

Use the origin as a test point to determine shading.

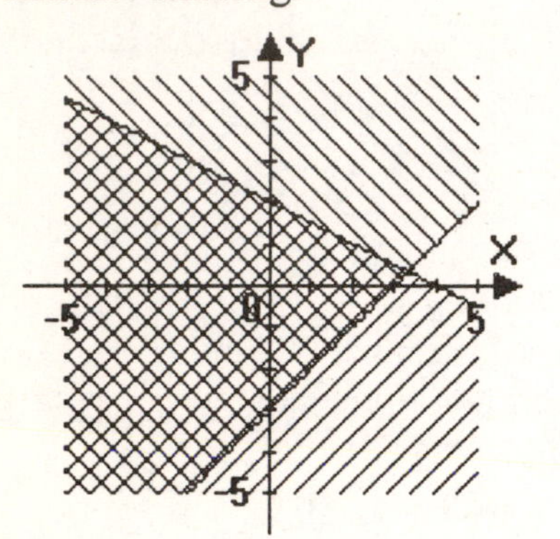

The solution set is the intersection of the shaded half-planes.

31. $x \le 2$

$y \ge -1$

Graph the vertical line, $x = 2$, and the horizontal line, $y = -1$. Use the origin as a test point to determine shading.

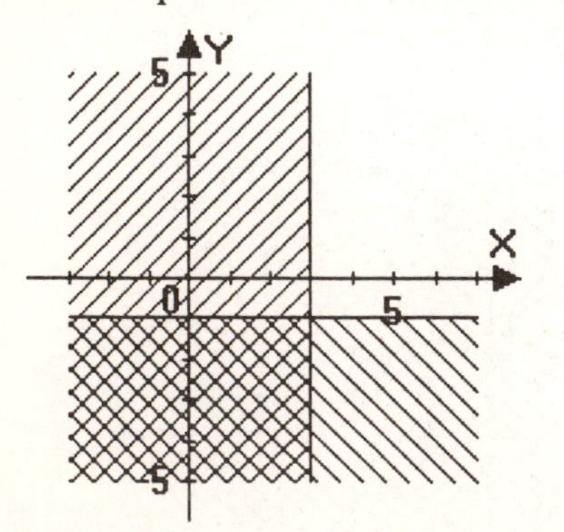

The solution set is the intersection of the shaded half-planes.

33. $-2 \le x < 5$

Since x lies between –2 and 5, graph the two vertical lines, $x = -2$ and $x = 5$. Since x lies between –2 and 5, shade between the two vertical lines.

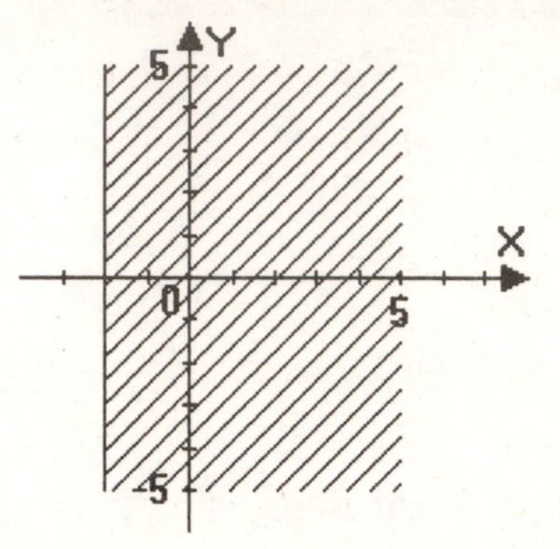

The solution is the shaded region.

35. $x - y \le 1$

$x \ge 2$

Graph the equations.

$x - y = 1$	$x = 2$
$x - \text{intercept} = 1$	$x - \text{intercept} = 2$
$y - \text{intercept} = -1$	vertical line

Use the origin as a test point to determine shading.

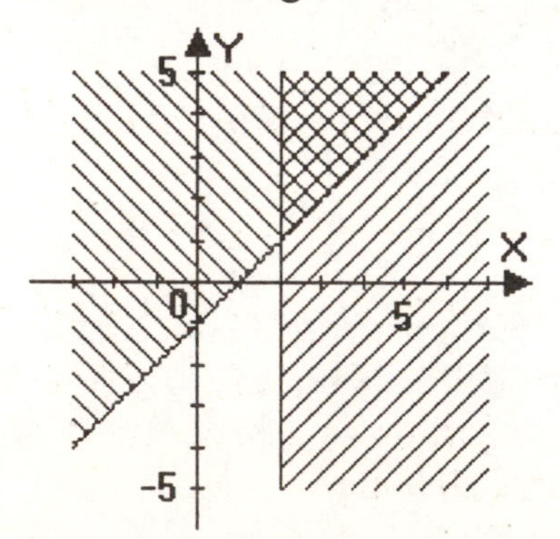

The solution set is the intersection of the shaded half-planes.

37. $x+y>4$

$x+y<-1$

Graph the equations using the intercepts.

$x+y=4$ $\quad$ $x+y=-1$

$x-\text{intercept}=4$ $\quad$ $x-\text{intercept}=-1$

$y-\text{intercept}=4$ $\quad$ $y-\text{intercept}=-1$

Use the origin as a test point to determine shading.

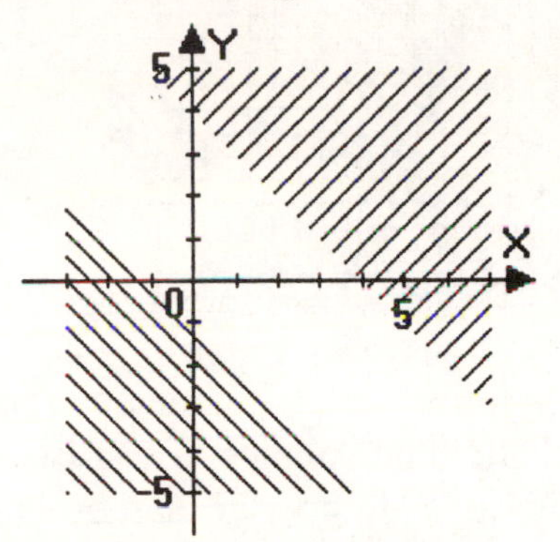

The solution set is the intersection of the shaded half-planes. Since the shaded half-planes do not intersect, there is no solution.

39. $x+y>4$

$x+y>-1$

Graph the equations using the intercepts.

$x+y=4$ $\quad$ $x+y=-1$

$x-\text{intercept}=4$ $\quad$ $x-\text{intercept}=-1$

$y-\text{intercept}=4$ $\quad$ $y-\text{intercept}=-1$

Use the origin as a test point to determine shading.

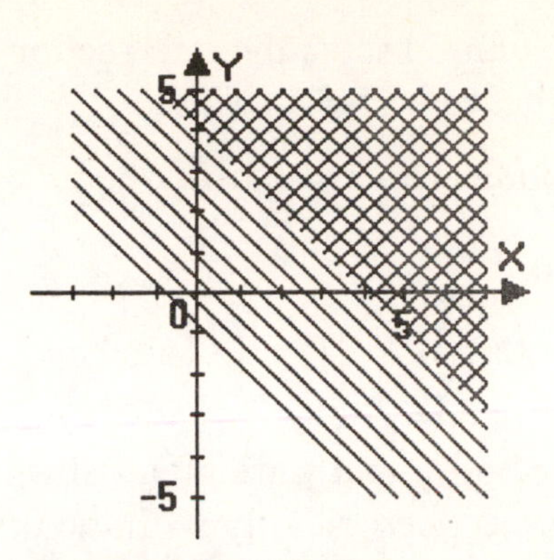

The solution set is the intersection of the shaded half-planes.

41. $x-y\le 2$

$x\ge -2$

$y\le 3$

Graph the equations using the intercepts.

$x-y=2$ $\quad$ $y=3$

$x-\text{intercept}=2$ $\quad$ $y-\text{intercept}=3$

$y-\text{intercept}=-2$ $\quad$ horizontal line

$x=-2$

$x-\text{intercept}=-2$

vertical line

Use the origin as a test point to determine shading.

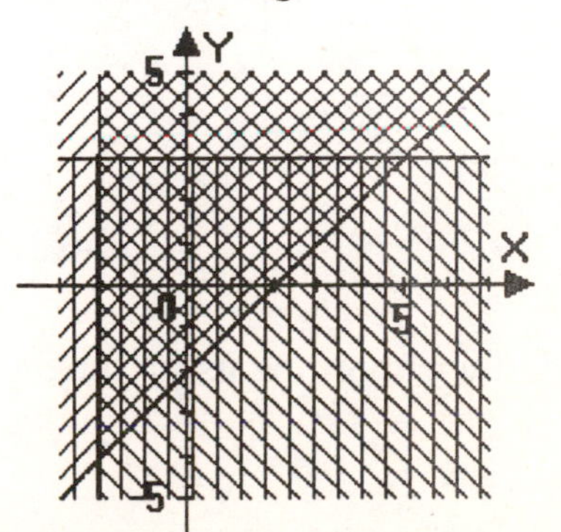

The solution set is the intersection of the shaded half-planes.

43. $x \geq 0$

$y \geq 0$

$2x+5y \leq 10$

$3x+4y \leq 12$

Since both x and y are greater than 0, we are concerned only with the first quadrant. Graph the other equations using the intercepts.

$2x+5y=10$	$3x+4y=12$
$x-\text{intercept}=5$	$x-\text{intercept}=4$
$y-\text{intercept}=2$	$y-\text{intercept}=3$

Use the origin as a test point to determine shading.

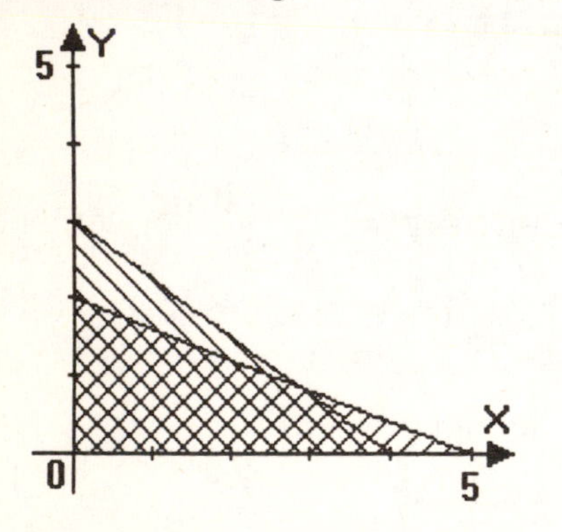

The solution set is the intersection of the shaded half-planes.

45. $3x+y \leq 6$

$2x-y \leq -1$

$x \geq -2$

$y \leq 4$

Graph the equations using the intercepts.

$3x+y=6$	$x=-2$
$x-\text{intercept}=2$	$x-\text{intercept}=-2$
$y-\text{intercept}=6$	vertical line

$2x-y=-1$	$y=4$
$x-\text{intercept}=-\frac{1}{2}$	$y-\text{intercept}=4$
$y-\text{intercept}=1$	horizontal line

Use the origin as a test point to determine shading.

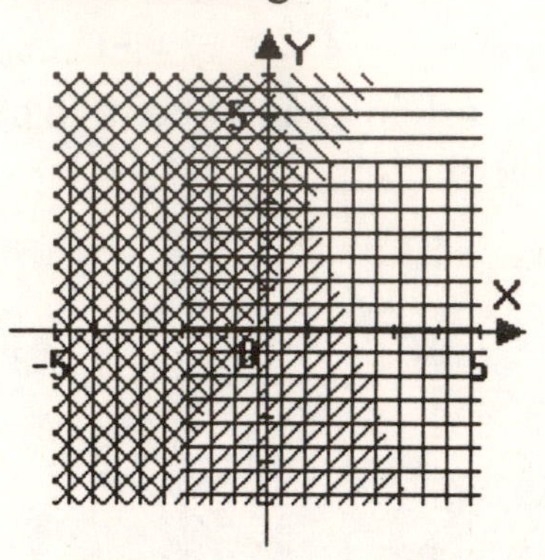

The solution set is the intersection of the shaded half-planes. Because all inequalities are greater than or equal to or less than or equal to, the boundaries of the shaded half-planes are also included in the solution set.

Application Exercises

47. **a.** The coordinates of point A are $(20,160)$. This means that a 20 year-old person with a pulse rate of 160 beats per minute falls within the target zone.

b. $10 \leq a \leq 70$

$10 \leq 20 \leq 70$

True

$$2a+3p \geq 450$$

$$2(20)+3(160) \geq 450$$

$$40+480 \geq 450$$

$$520 \geq 450$$

True

$40+480 \geq 450$

$520 \geq 450$

True

$a+p \leq 190$

$20+160 \leq 190$

$180 \leq 190$

True

Since point A makes all three inequalities true, it is a solution of the system.

49. Answers will vary. One example is the point (70, 40). Grasslands occur under the following conditions.

$T \geq 35$

$5T-7P \geq 70$

$3T-35P \leq -140$

Show that the point satisfies each of the inequalities.

$T \geq 35$

$70 \geq 35$

True

$5T-7P \geq 70$

$5(70)-7(40) \geq 70$

$350-280 \geq 70$

$70 \geq 70$

True

$3T-35P \leq -140$

$3(70)-35(40) \leq -140$

$210-1400 \leq -140$

$-1190 \leq -140$

True

Since the coordinates of the point (70, 40) make each inequality true, it is a solution of the system that describes where grasslands occur.

51. The maximum healthy weight for a person who is 6 feet tall is approximately 190 pounds.

53. The border of the underweight region is the line $7w-25h=-800$. Using a test point, we can determine what inequality symbol should be used. Consider the point (100, 70). Find the value of $7w-25h$.

$7w-25h=7(100)-25(70)$

$=700-1750$

$=-1050$

Since $-1050<-800$ is a true statement, the inequality should be written $7w-25h<-800$.

To verify, check the inequality using some other point in the region. One possible test point is $(75,75)$.

$7w-25h<-800$

$7(75)-25(75)<-800$

$525-1875<-800$

$-1350<-800$

The point makes the inequality true, so it lies in the underweight region and the inequality should be written $7w-25h<-800$.

55. The healthy weight region falls between the lines $7w-25h=-800$ and $w-5h=-170$. We need to replace the equal sign in each equation with an inequality sign. To determine which sign to use, we choose a point in the region and evaluate the expression of the left hand side of each equation.

One point within the region is (150, 70).

$7w-25h$

$=7(150)-25(70)$

$=1050-1750$

$=-700$

$w-5h$

$=-170-5(70)$

$=-170-350$

$=-520$

Since $-700 \geq -800$, we know that $7w - 25h \geq -800$. Also, since $-520 \leq -170$, we know that $w - 5h \leq -170$.
The system of inequalities that describes the healthy weight region is as follows.

$$7w - 25h \geq -800$$
$$w - 5h \leq -170$$

Writing in Mathematics

57. Answers will vary.

59. Answers will vary.

61. Answers will vary.

63. Answers will vary.

65. Answers will vary.

Technology Exercises

67. $y \leq 4x + 4$

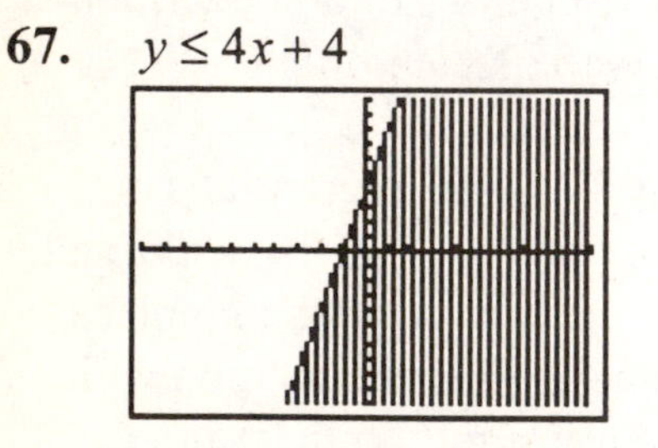

69. $2x + y \leq 6$

$y \leq -2x + 6$

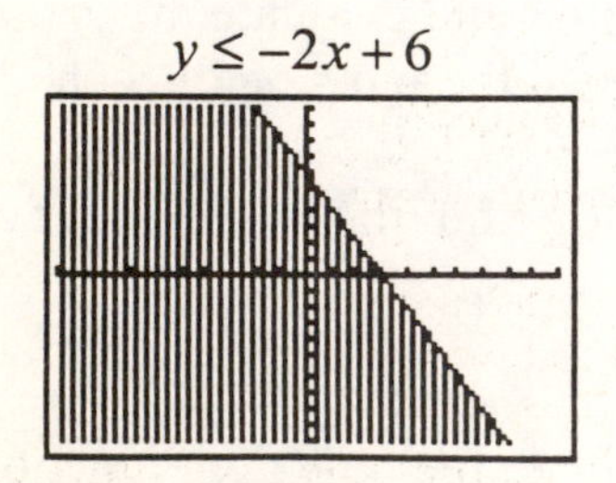

71. Answers will vary.

73. Answers will vary. For example, verify Exercise 23.

$3x + 6y \leq 6$

$2x + y \leq 8$

First solve both inequalities for y.

$3x + 6y \leq 6$	$2x + y \leq 8$
$6y \leq -3x + 6$	$y \leq -2x + 8$
$y \leq -\frac{1}{2}x + 1$	

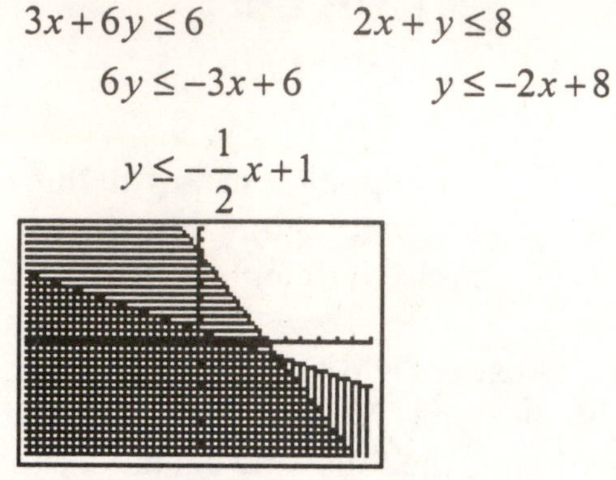

Critical Thinking Exercises

75. Answers will vary.

77.

$y \geq nx + b$	$y \leq mx + b$
y-intercept > 0	y-intercept > 0
slope < 0	slope > 0

The line $y = nx + b$ will have a y-intercept above the x-axis and a negative slope. As a result, the line will go downhill from left to right. The line $y = mx + b$ will have a y-intercept above the x-axis and a positive slope. As a result, the line will go uphill from left to right.

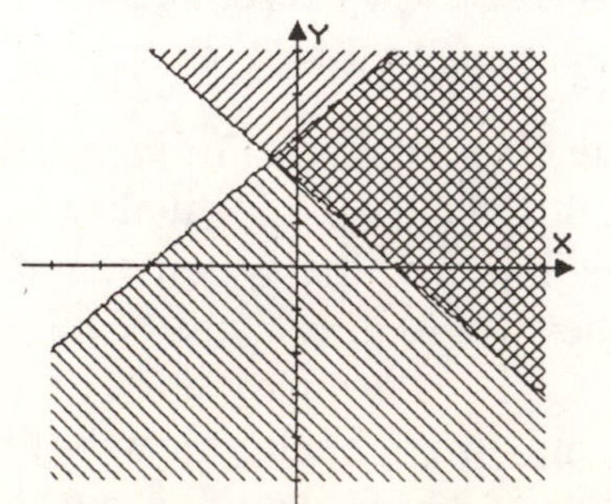

Review Exercises

78. $3x - y = 8$

$x - 5y = -2$

$$\left[\begin{array}{cc|c} 3 & -1 & 8 \\ 1 & -5 & -2 \end{array}\right] \quad R_1 \leftrightarrow R_2$$

$$= \left[\begin{array}{cc|c} 1 & -5 & -2 \\ 3 & -1 & 8 \end{array}\right] \quad -3R_1 + R_2$$

$$= \left[\begin{array}{cc|c} 1 & -5 & -2 \\ 0 & 14 & 14 \end{array}\right] \quad \frac{1}{14}R_2$$

$$= \left[\begin{array}{cc|c} 1 & -5 & -2 \\ 0 & 1 & 1 \end{array}\right]$$

The resulting system is:

$x - 5y = -2$

$y = 1.$

Since we know $y = 1$, we can use back-substitution to find x.

$x - 5(1) = -2$

$x - 5 = -2$

$x = 3$

The solution is $(3,1)$ and the solution set is $\{(3,1)\}$.

79. $y = 3x - 2$

$y = -2x + 8$

Both equations are in slope-intercept form, so use the slopes and y-intercepts to graph the lines.

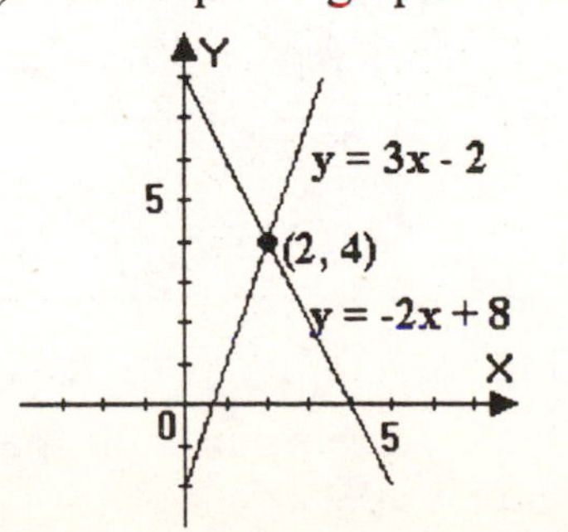

The solution is the intersection point $(2,4)$ and the solution set is $\{(2,4)\}$.

Check Points 4.5

1. The objective function is $z = 25x + 55y$.

2. Since not more than 80 bookshelves and desks can be manufactured per day, we have $x + y \leq 80$.

3. Since the number of bookshelves must be between 30 and 80 per day, we have $30 < x < 80$.
Since the number of desks must be between 10 and 30 per day inclusive, we have $10 \leq y \leq 30$.
Objective Function:
$z = 25x + 55y$
Constraints:
$x + y \leq 80$
$30 < x < 80$
$10 \leq y \leq 30$

4. First, graph the constraints.
$x + y \leq 80$
$30 < x < 80$
$10 \leq y \leq 30$

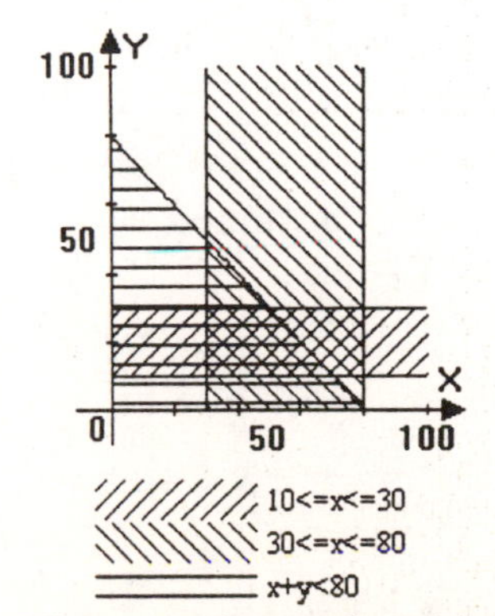

Next, evaluate the objective function at each corner of the region where the constraints intersect.

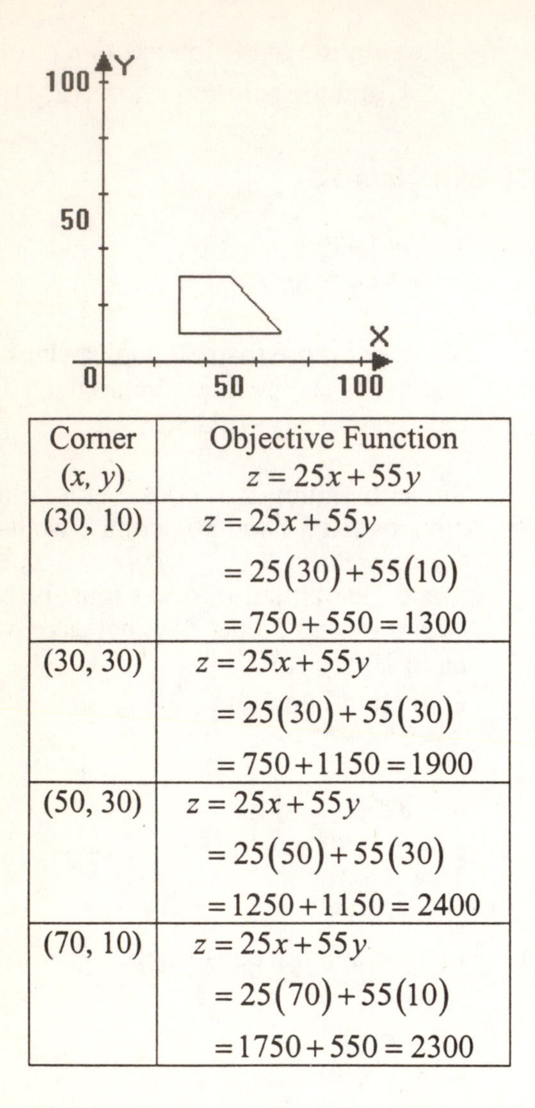

Corner (x, y)	Objective Function $z = 25x + 55y$
(30, 10)	$z = 25x + 55y$ $= 25(30) + 55(10)$ $= 750 + 550 = 1300$
(30, 30)	$z = 25x + 55y$ $= 25(30) + 55(30)$ $= 750 + 1150 = 1900$
(50, 30)	$z = 25x + 55y$ $= 25(50) + 55(30)$ $= 1250 + 1150 = 2400$
(70, 10)	$z = 25x + 55y$ $= 25(70) + 55(10)$ $= 1750 + 550 = 2300$

The maximum value of z is 2400 and occurs at the point (50, 30). In order to maximize profit, 50 bookshelves and 30 desks must be produced each day for a profit of \$2400.

5. First, graph the constraints.

$x \geq 0$

$y \geq 0$

$x + y \geq 1$

$x + y \leq 6$

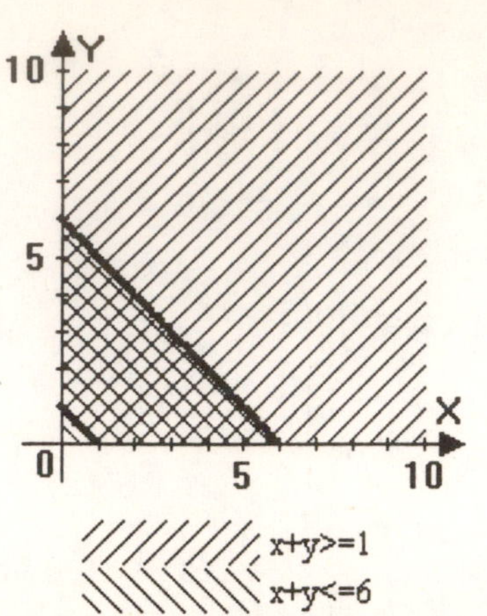

Next, evaluate the objective function at each corner of the region where the constraints intersect.

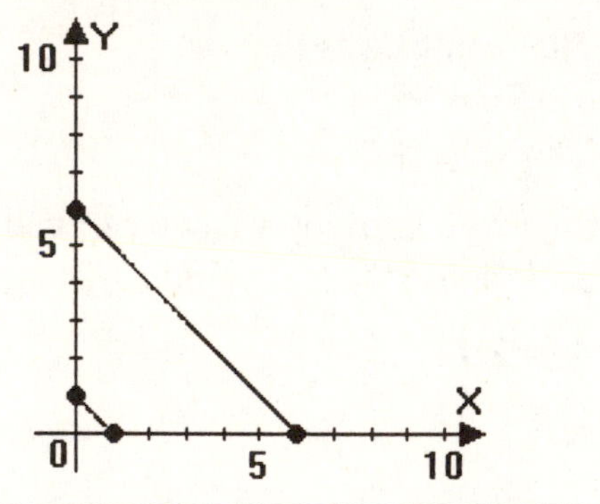

Corner (x, y)	Objective Function $z = 3x + 5y$
(0, 1)	$z = 3x + 5y$ $= 3(0) + 5(1) = 5$
(1, 0)	$z = 3x + 5y$ $= 3(1) + 5(0) = 3$
(0, 6)	$z = 3x + 5y$ $= 3(0) + 5(6) = 30$
(6, 0)	$z = 3x + 5y$ $= 3(6) + 5(0) = 18$

The maximum value of z occurs at the point (0, 6). At this point, the maximum value is 30.

Problem Set 4.5

Practice Exercises

1.

Corner (x, y)	Objective Function $z = 5x + 6y$
(1, 2)	$z = 5x + 6y$ $= 5(1) + 6(2)$ $= 5 + 12 = 17$
(8, 3)	$z = 5x + 6y$ $= 5(8) + 6(3)$ $= 40 + 18 = 58$
(7, 5)	$z = 5x + 6y$ $= 5(7) + 6(5)$ $= 35 + 30 = 65$
(2, 10)	$z = 5x + 6y$ $= 5(2) + 6(10)$ $= 10 + 60$ $= 70$

The maximum value is 70 and the minimum is 17.

3.

Corner (x, y)	Objective Function $z = 40x + 50y$
(0, 0)	$z = 40x + 50y$ $= 40(0) + 50(0)$ $= 0 + 0 = 0$
(8, 0)	$z = 40x + 50y$ $= 40(8) + \cancel{50(0)} = 320$
(4, 9)	$z = 40x + 50y$ $= 40(4) + 50(9)$ $= 160 + 450 = 610$
(0, 8)	$z = 40x + 50y$ $= \cancel{40(0)} + 50(8) = 400$

The maximum value is 610 and the minimum value is 0.

5. Objective Function: $z = 3x + 2y$

Constraints: $x \geq 0,\ y \geq 0$

$2x + y \leq 8$

$x + y \geq 4$

a.

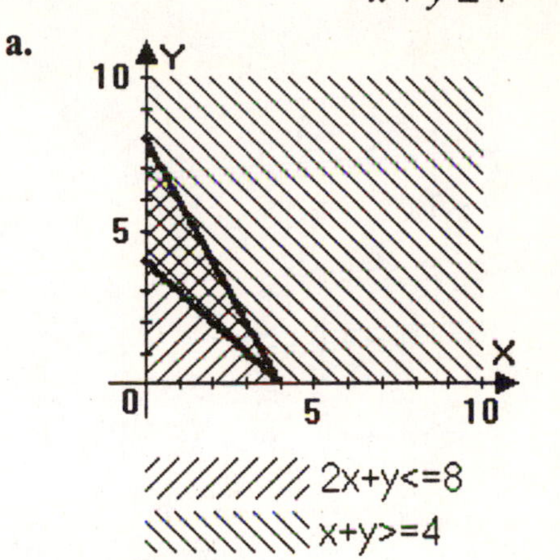

b.

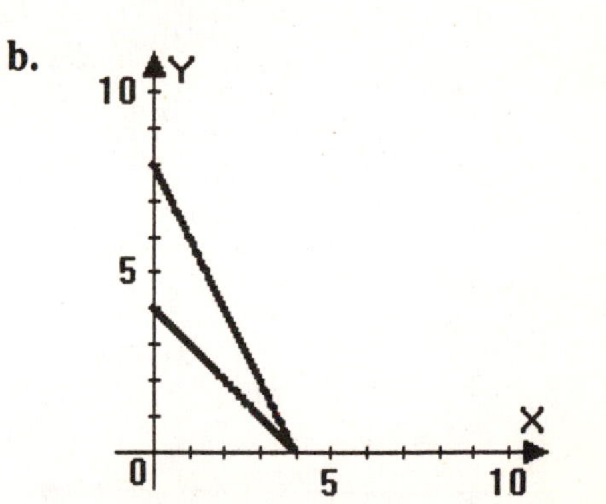

Corner (x, y)	Objective Function $z = 3x + 2y$
(4, 0)	$z = 3x + 2y$ $= 3(4) + \cancel{2(0)} = 12$
(0, 8)	$z = 3x + 2y$ $= \cancel{3(0)} + 2(8) = 16$

(0, 4)	$z = 3x + 2y$ $= 3(0) + 2(4) = 8$

c. The maximum value is 16. It occurs at the point (0, 8).

7. Objective Function: $z = 4x + y$

Constraints: $x \geq 0,\ y \geq 0$

$2x + 3y \leq 12$

$x + y \geq 3$

a.

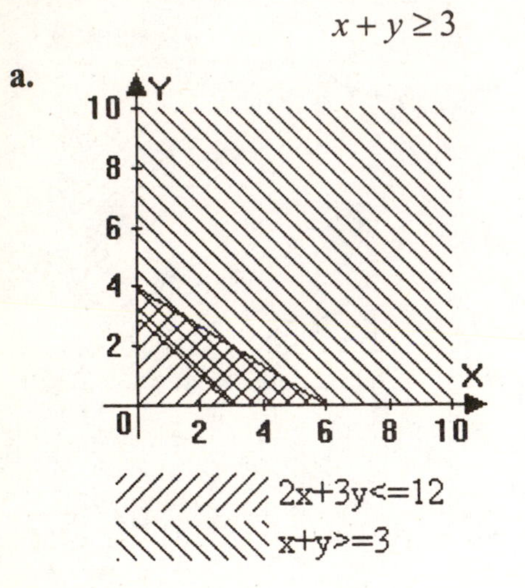

b.

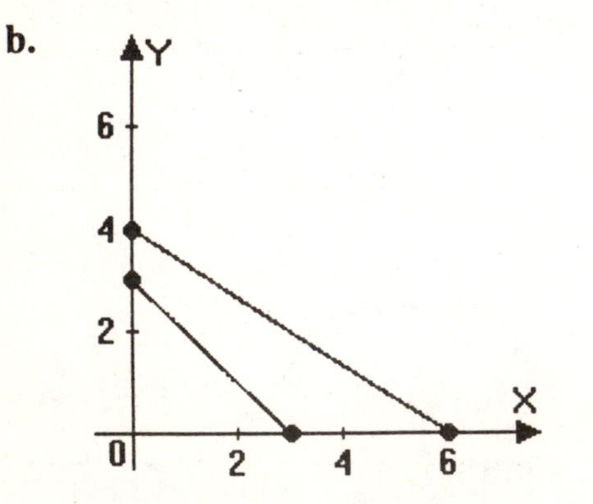

Corner (x, y)	Objective Function $z = 4x + y$
(3, 0)	$z = 4x + y$ $= 4(0) + 0 = 0$
(6, 0)	$z = 4x + y$ $= 4(6) + 0 = 24$
(0, 3)	$z = 4x + y$ $= 4(0) + 3 = 3$

c. The maximum value is 24. It occurs at the point (6, 0).

9. Objective Function: $z = 3x - 2y$

Constraints: $1 \leq x \leq 5$

$y \geq 2$

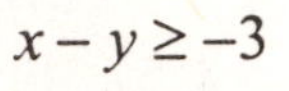

$x - y \geq -3$

a.

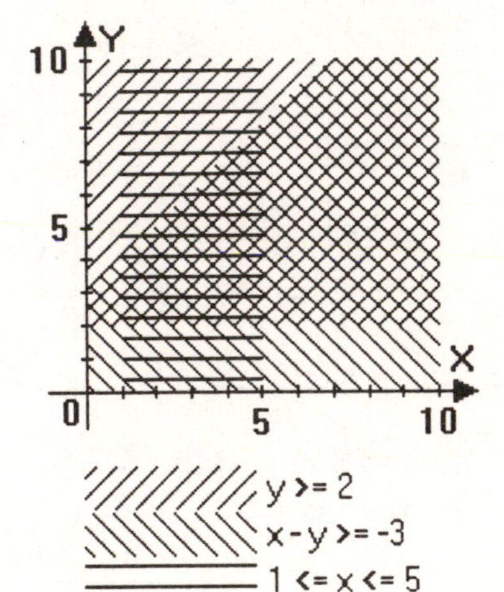

b.

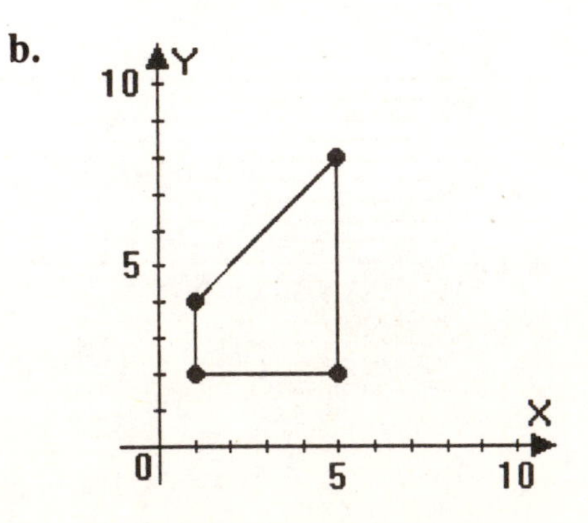

Corner (x, y)	Objective Function $z = 3x - 2y$
(1, 2)	$z = 3x - 2y$ $= 3(1) - 2(2)$ $= 3 - 4 = -1$
(5, 2)	$z = 3x - 2y$ $= 3(5) - 2(2)$ $= 15 - 4 = 11$
(5, 8)	$z = 3x - 2y$ $= 3(5) - 2(8)$ $= 15 - 16 = -1$
(1, 4)	$z = 3x - 2y$ $= 3(1) - 2(4)$ $= 3 - 8 = -5$

c. The maximum value is 11. It occurs at the point (5, 2).

11. Objective Function: $z = 4x + 2y$

Constraints: $x \geq 0,\ y \geq 0$

$2x + 3y \leq 12$

$3x + 2y \leq 12$

$x + y \geq 2$

a.

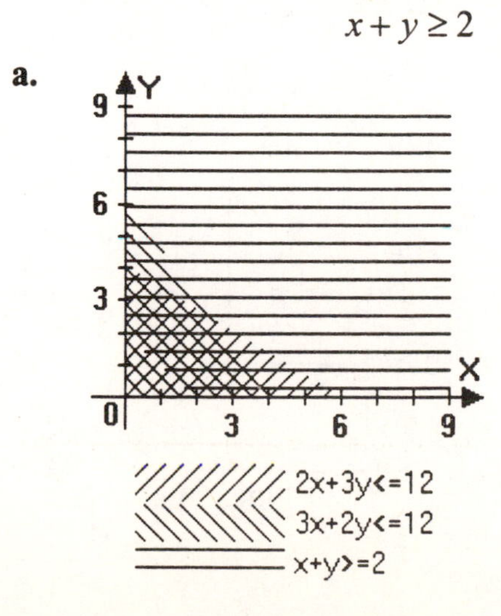

b.

Corner (x, y)	Objective Function $z = 4x + 2y$
(2, 0)	$z = 4x + 2y$ $= 4(2) + 2(0) = 8$
(4, 0)	$z = 4x + 2y$ $= 4(4) + 2(0) = 16$
(2.4,2.4)	$z = 4x + 2y$ $= 4(2.4) + 2(2.4)$ $= 9.6 + 4.8 = 14.4$
(0, 4)	$z = 4x + 2y$ $= 4(0) + 2(4) = 8$
(0, 2)	$z = 4x + 2y$ $= 4(0) + 2(2) = 4$

c. The maximum value is 16. It occurs at the point (4, 0).

13. Objective Function: $z = 10x + 12y$

Constraints: $x \geq 0,\ y \geq 0$

$x + y \leq 7$

$2x + y \leq 10$

$2x + 3y \leq 18$

a.

b.

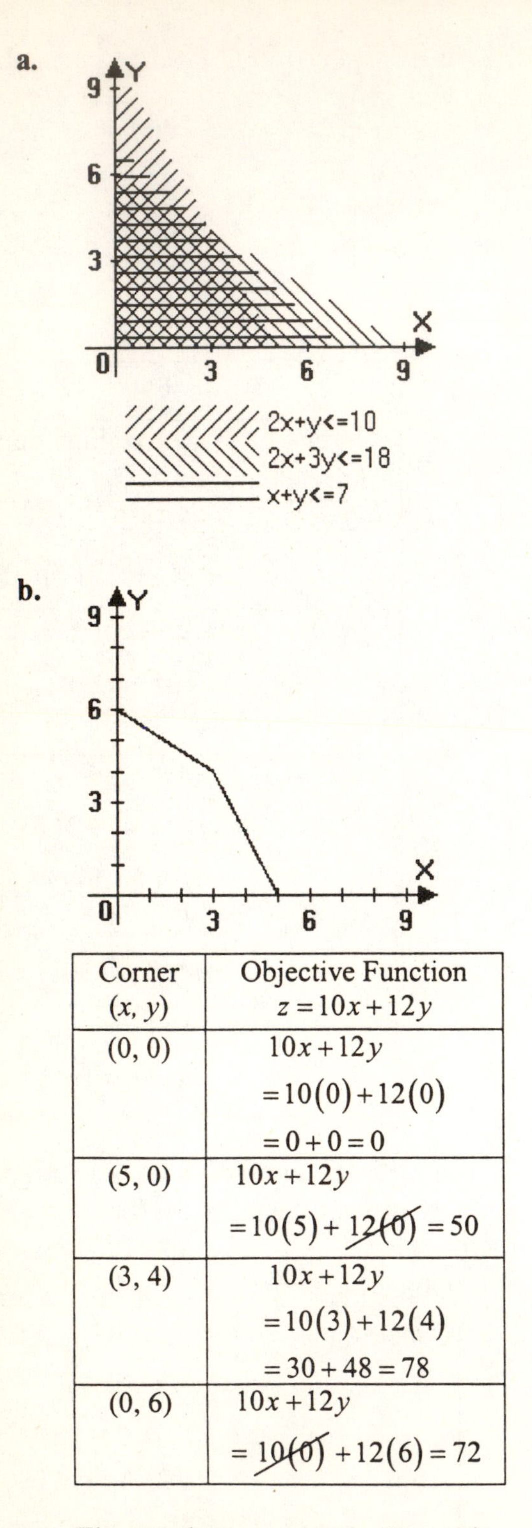

Corner (x, y)	Objective Function $z=10x+12y$
(0, 0)	$10x+12y$ $=10(0)+12(0)$ $=0+0=0$
(5, 0)	$10x+12y$ $=10(5)+12(0)=50$
(3, 4)	$10x+12y$ $=10(3)+12(4)$ $=30+48=78$
(0, 6)	$10x+12y$ $=10(0)+12(6)=72$

c. The maximum value is 78 and it occurs at the point (3, 4)

Application Exercises

15. a. The objective function is $z=125x+200y$.

b. Since we can make at most 450 console televisions, we have $x \leq 450$.
Since we can make at most 200 wide screen televisions, we have $y \leq 200$.
Since we can spend at most \$360,000 per month, we have $600x+900y \leq 360,000$.

c.

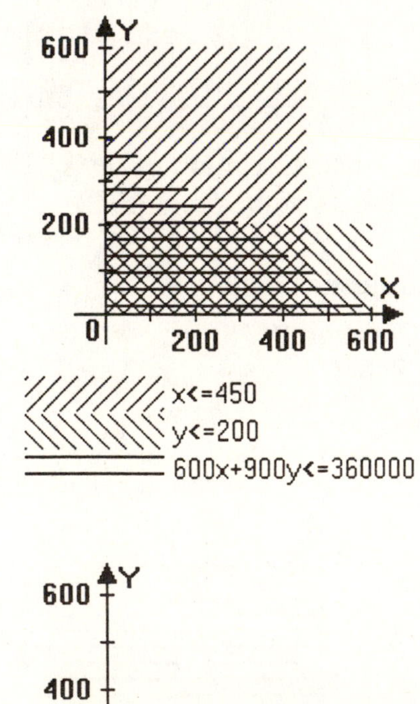

d.

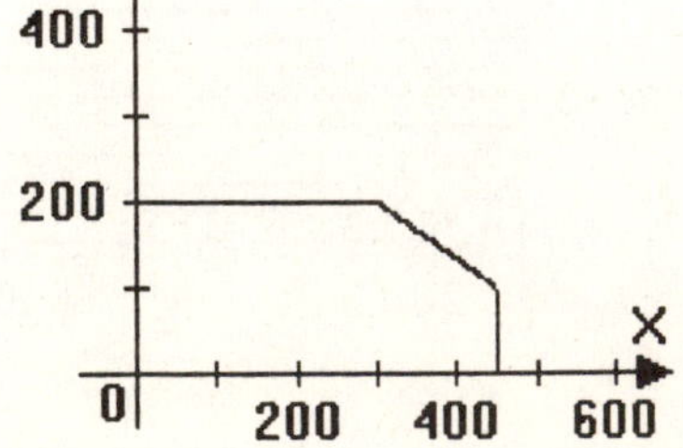

Corner (x, y)	Objective Function $z = 125x + 200y$
(0, 0)	$125x + 200y$ $= 125(0) + 200(0) = 0$
(0, 200)	$125x + 200y$ $= 125(0) + 200(200)$ $= 40,000$
(300, 200)	$125x + 200y$ $= 125(300) + 200(200)$ $= 37,500 + 40,000$ $= 77,500$
(450, 100)	$125x + 200y$ $= 125(450) + 200(100)$ $= 56,250 + 20,000$ $= 76,250$
(450, 0)	$125x + 200y$ $= 125(450) + 200(0)$ $= 56,250$

e. The television manufacturer will make the greatest profit by manufacturing 300 console televisions and 200 wide screen televisions each month. The maximum monthly profit is \$77,500.

17. Let x = the number of model A bicycles produced
Let y = the number of model B bicycles produced
The objective function is $z = 25x + 15y$.
The assembling constraint is $5x + 4y \leq 200$.
The painting constraint is $2x + 3y \leq 108$.

We also know that x and y must either be zero or a positive number. We cannot make a negative number of bicycles.
Next, graph the constraints.

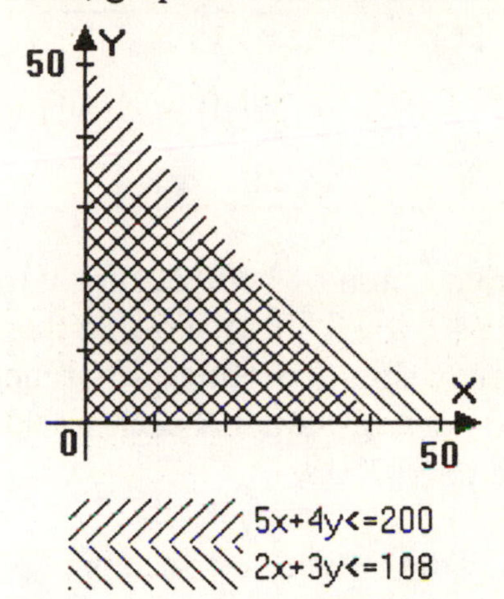

Using the graph, find the value of the objective function at each of the corner points.

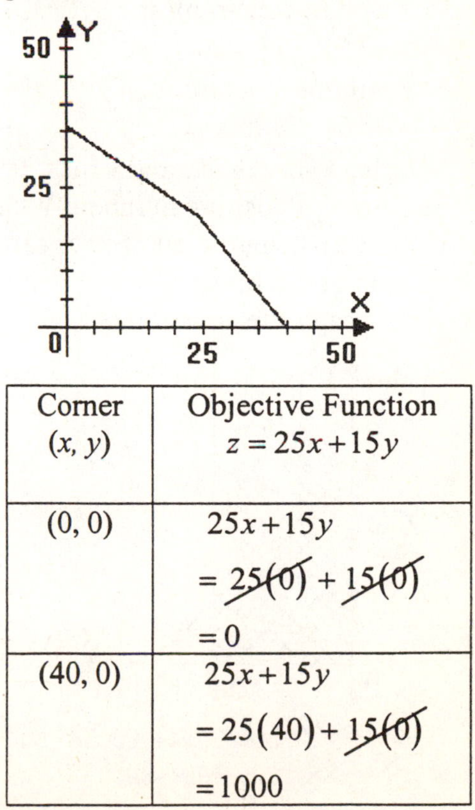

Corner (x, y)	Objective Function $z = 25x + 15y$
(0, 0)	$25x + 15y$ $= 25(0) + 15(0)$ $= 0$
(40, 0)	$25x + 15y$ $= 25(40) + 15(0)$ $= 1000$

(24, 20)	$25x+15y$ $=25(24)+15(20)$ $=600+300$ $=900$
(0, 36)	$25x+15y$ $=25(0)+15(36)$ $=540$

The maximum of 1000 occurs at the point (40, 0). This means that the company should produce 40 of model *A* and none of model *B* each week for a profit of $1000.

19. Let x = the number of cartons of food
Let y = the number of cartons of clothing
The objective function is $z=12x+5y$.
The weight constraint is
$50x+20y \le 19{,}000$.
The volume constraint is
$20x+10y \le 8000$.
We also know that x and y must either be zero or a positive number. We cannot have a negative number of cartons of food or clothing.
Next, graph the constraints.

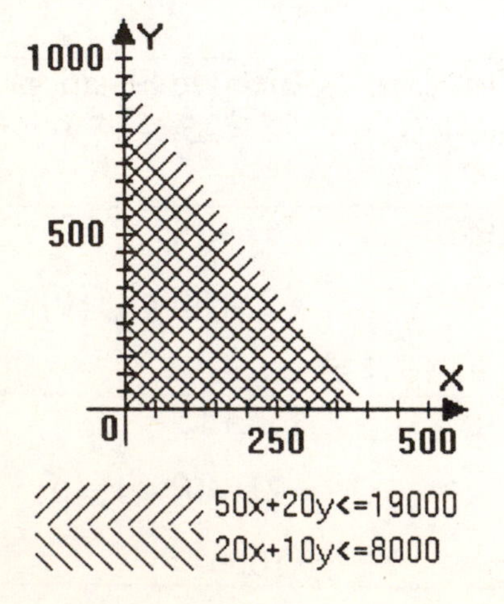

Using the graph, find the value of the objective function at each of the corner points.

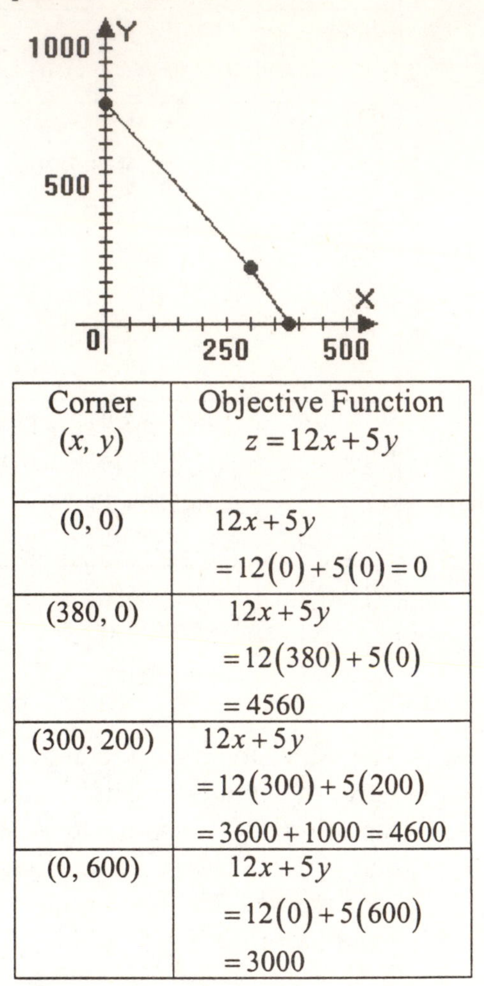

Corner (x, y)	Objective Function $z=12x+5y$
(0, 0)	$12x+5y$ $=12(0)+5(0)=0$
(380, 0)	$12x+5y$ $=12(380)+5(0)$ $=4560$
(300, 200)	$12x+5y$ $=12(300)+5(200)$ $=3600+1000=4600$
(0, 600)	$12x+5y$ $=12(0)+5(600)$ $=3000$

The maximum of 4600 occurs at the point (300, 200). This means that to maximize the number of people who are helped, 300 boxes of food and 200 boxes of clothing should be sent.

21. Let x = the number of parents
Let y = the number of students
The objective function is $z=2x+y$.
The seating constraint is $x+y \le 150$.
The two parents per student constraint is $2x \ge y$.

We also know that x and y must either be zero or a positive number. We cannot have a negative number of parents or students.
Next, graph the constraints.

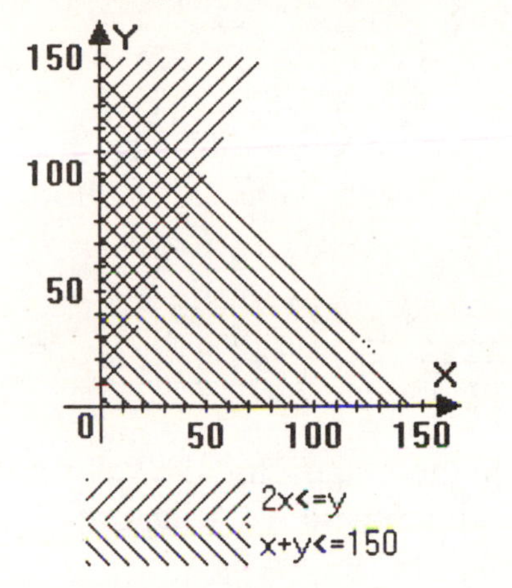

Using the graph, find the value of the objective function at each of the corner points.

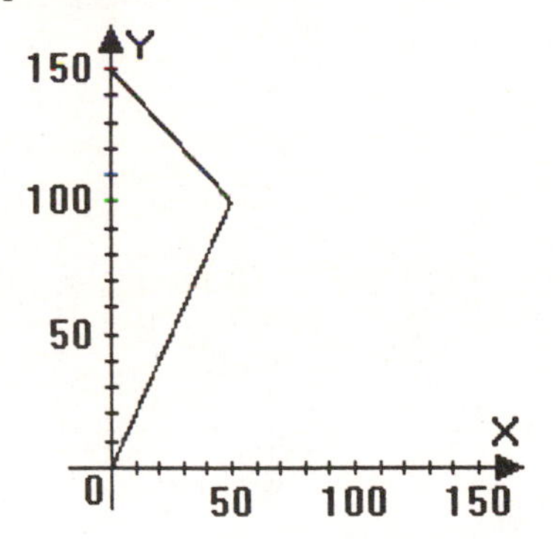

Corner (x, y)	Objective Function $z = 2x + y$
(0, 0)	$2x + y$ $= 2(0) + 0 = 0$
(50, 100)	$2x + y$ $= 2(50) + 100$ $= 100 + 100 = 200$
(0, 150)	$2x + y$ $= 2(0) + 100 = 100$

The maximum of 200 occurs at the point (50, 100). This means that to maximize the amount of money raised, 50 parents and 100 students should attend.

23. Let x = the number of Boeing 727s
Let y = the number of Falcon 20s
The objective function is $z = x + y$.
The hourly operating cost constraint is $1400x + 500y \le 35000$.
The total payload constraint is $42000x + 6000y \ge 672,000$.
The 727 constraint is $x \le 20$.
We also know that x and y must either be zero or a positive number. We cannot have a negative number of aircraft.
Next, graph the constraints.

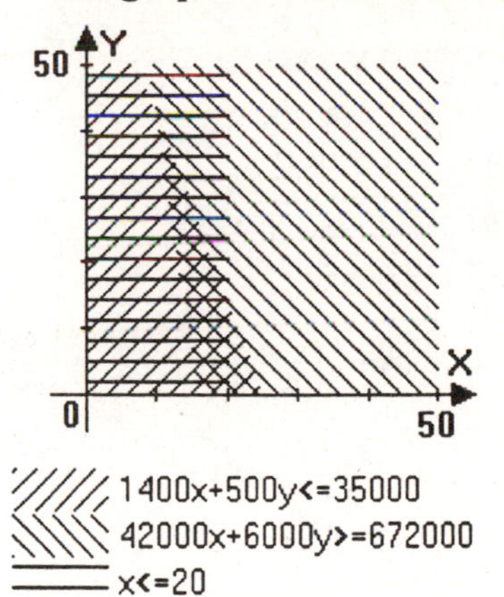

Using the graph, find the value of the objective function at each of the corner points.

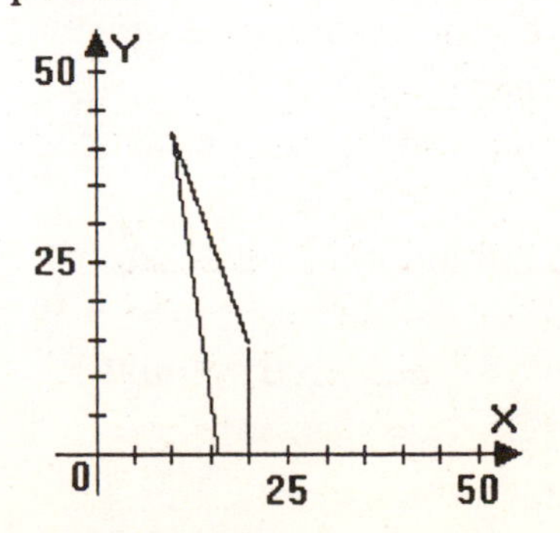

Corner (x, y)	Objective Function $z = x + y$
(16, 0)	$z = x + y$ $= 16 + 0 = 16$
(20, 0)	$z = x + y$ $= 20 + 0 = 20$
(20, 14)	$z = x + y$ $= 20 + 14 = 34$
(10, 42)	$z = x + y$ $= 10 + 42 = 52$

The maximum of 52 occurs at the point (10, 42). This means that to maximize the number of aircraft, 10 Boeing 727s and 42 Falcon 20s should be purchased.

Writing in Mathematics

25. Answers will vary.

27. Answers will vary.

Critical Thinking Exercises

29. Let x = the amount invested in stocks
Let y = the amount invested in bonds
The objective function is
$z = 0.12x + 0.08y$.
The total money constraint is
$x + y \le 10000$.
The minimum bond investment constraint is $y \ge 3000$.
The minimum stock investment constraint is $x \ge 2000$.
The stock versus bond constraint is
$y \ge x$.
We also know that x and y must either be zero or a positive number. We cannot invest a negative amount of money.

Next, graph the constraints.

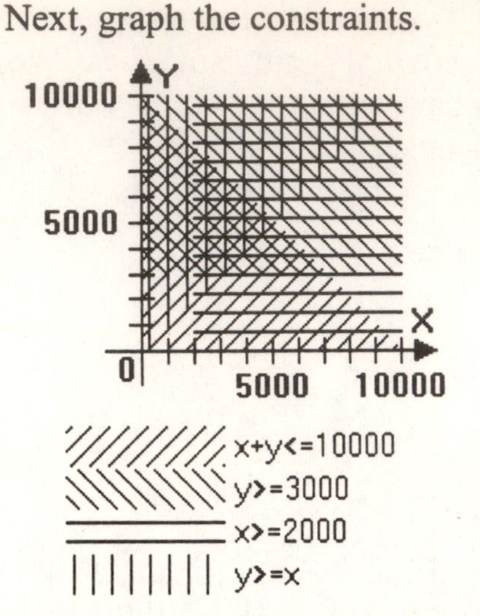

Using the graph, find the value of the objective function at each of the corner points.

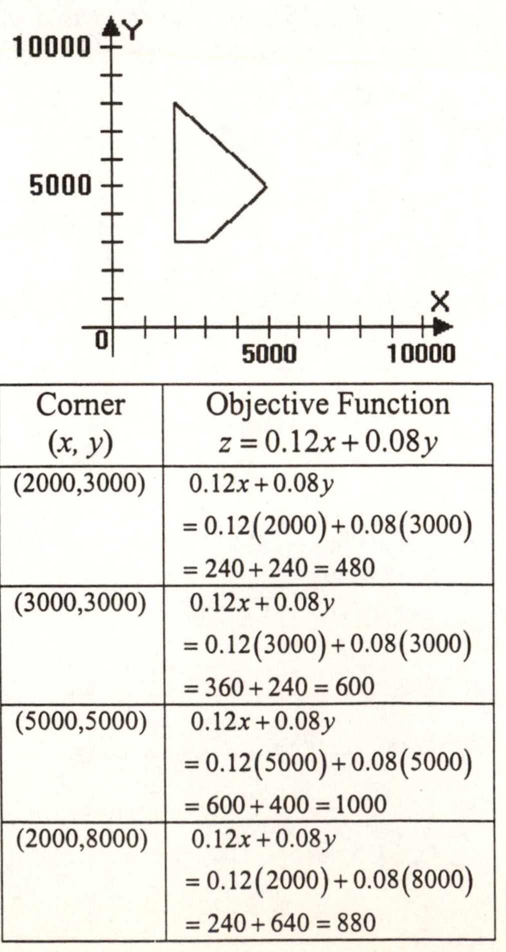

Corner (x, y)	Objective Function $z = 0.12x + 0.08y$
(2000,3000)	$0.12x + 0.08y$ $= 0.12(2000) + 0.08(3000)$ $= 240 + 240 = 480$
(3000,3000)	$0.12x + 0.08y$ $= 0.12(3000) + 0.08(3000)$ $= 360 + 240 = 600$
(5000,5000)	$0.12x + 0.08y$ $= 0.12(5000) + 0.08(5000)$ $= 600 + 400 = 1000$
(2000,8000)	$0.12x + 0.08y$ $= 0.12(2000) + 0.08(8000)$ $= 240 + 640 = 880$

The maximum of 1000 occurs at the point (5000, 5000). This means that to maximize the return on the investment, \$5000 should be invested in stocks and \$5000 should be invested in bonds.

Review Exercises

31. $(2x^4y^3)(3xy^4)^3 = 2x^4y^3 3^3 x^3 y^{12}$
$= 2 \cdot 3^3 x^4 x^3 y^3 y^{12}$
$= 2 \cdot 27 x^7 y^{15}$
$= 54x^7y^{15}$

32.
$$3P = \frac{2L - W}{4}$$
$$12P = 2L - W$$
$$12P + W = 2L$$
$$L = \frac{12P + W}{2}$$
$$L = \frac{12P}{2} + \frac{W}{2} \text{ or } 6P + \frac{W}{2}$$

33. $f(x) = x^3 + 2x^2 - 5x + 4$
$f(-1) = (-1)^3 + 2(-1)^2 - 5(-1) + 4$
$= -1 + 2(1) + 5 + 4$
$= -1 + 2 + 5 + 4 = 10$

Chapter 4 Review

4.1

1. $x > 5$

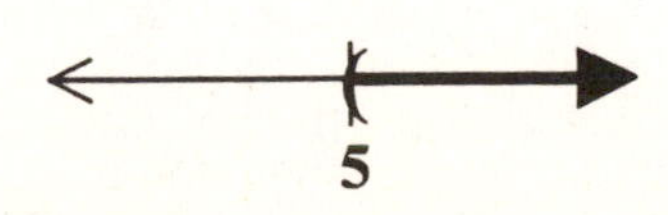

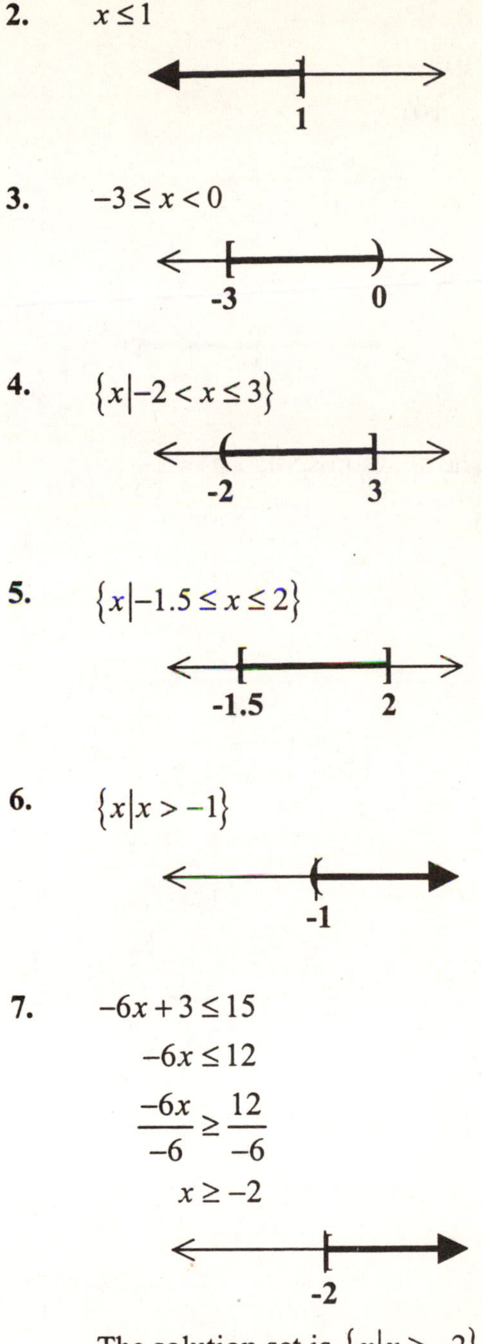

2. $x \le 1$

3. $-3 \le x < 0$

4. $\{x | -2 < x \le 3\}$

5. $\{x | -1.5 \le x \le 2\}$

6. $\{x | x > -1\}$

7. $-6x + 3 \le 15$
$-6x \le 12$
$\frac{-6x}{-6} \ge \frac{12}{-6}$
$x \ge -2$

The solution set is $\{x | x \ge -2\}$ or $[2, \infty)$.

8.
$$6x-9\geq -4x-3$$
$$10x-9\geq -3$$
$$10x\geq 6$$
$$x\geq \frac{6}{10}$$
$$x\geq \frac{3}{5}$$

3/5

The solution set is $\left\{x\middle|x\geq \frac{3}{5}\right\}$ or $\left[\frac{3}{5},\infty\right)$.

9.
$$\frac{x}{3}-\frac{3}{4}-1>\frac{x}{2}$$
$$12\left(\frac{x}{3}\right)-12\left(\frac{3}{4}\right)-12(1)>12\left(\frac{x}{2}\right)$$
$$4x-3(3)-12>6x$$
$$4x-9-12>6x$$
$$4x-21>6x$$
$$-2x-21>0$$
$$-2x>21$$
$$x<-\frac{21}{2}$$

-21/2

The solution set is $\left\{x\middle|x<-\frac{21}{2}\right\}$ or $\left(-\infty,-\frac{21}{2}\right)$.

10.
$$6x+5>-2(x-3)-25$$
$$6x+5>-2x+6-25$$
$$6x+5>-2x-19$$
$$8x+5>-19$$
$$8x>-24$$
$$x>-3$$

-3

The solution set is $\{x|x>-3\}$ or $(-3,\infty)$.

11.
$$3(2x-1)-2(x-4)\geq 7+2(3+4x)$$
$$6x-3-2x+8\geq 7+6+8x$$
$$4x+5\geq 13+8x$$
$$-4x+5\geq 13$$
$$-4x\geq 8$$
$$x\leq -2$$

-2

The solution set is $\{x|x\leq -2\}$ or $(-\infty,-2)$.

12.
$$2x+7\leq 5x-6-3x$$
$$2x+7\leq 2x-6$$
$$7\leq -6$$

This is a contradiction. Seven is not less than or equal to –6. There are no values of x for which $7\leq -6$. The solution set is $\varnothing$ or $\{\ \}$.

13. Let x = the number of checks written per month
The cost using the first method is $c_1=11+0.06x$.

The cost using the second method is $c_2 = 4 + 0.20x$.
The first method is a better deal if it costs less than the second method.

$$c_1 < c_2$$
$$11 + 0.06x < 4 + 0.20x$$
$$11 - 0.14x < 4$$
$$-0.14x < -7$$
$$\frac{-0.14x}{-0.14} > \frac{-7}{-0.14}$$
$$x > 50$$

The first method is a better deal when more than 50 checks per month are written.

14. Let x = the amount of sales per month in dollars
The salesperson's commission is $c = 500 + 0.20x$.
We are looking for the amount of sales, x, the salesman must make to receive more than \$3200 in income.

$$c > 3200$$
$$500 + 0.20x > 3200$$
$$0.20x > 2700$$
$$x > 13500$$

The salesman must sell at least \$13,500 to receive a total income that exceeds \$3200 per month.

4.2

15. $A \cap B = \{a, c\}$

16. $A \cap C = \{a\}$

17. $A \cup B = \{a, b, c, d, e\}$

18. $A \cup C = \{a, b, c, d, f, g\}$

19. $x \le 3$

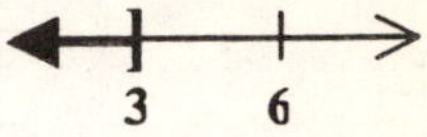

$x < 6$

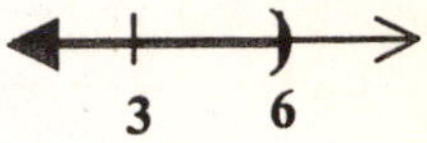

$x \le 3$ and $x < 6$

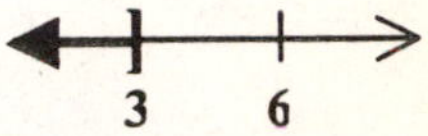

The solution set is $\{x | x \le 3\}$ or $(-\infty, 3]$.

20. $x \le 3$

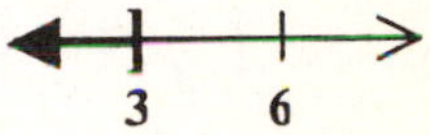

$x < 6$

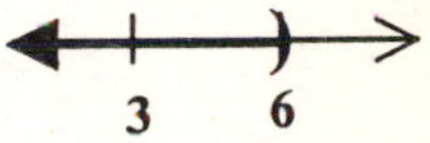

$x \le 3$ or $x < 6$

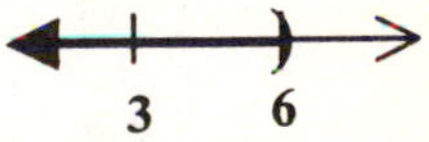

The solution set is $\{x | x < 6\}$ or $(-\infty, 6)$.

21. $-2x < -12$ and $x - 3 < 5$

$$\frac{-2x}{-2} > \frac{-12}{-2} \qquad x < 8$$
$$x > 6$$

$x < 8$

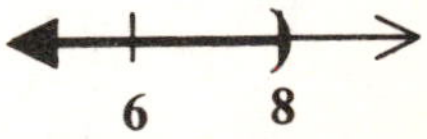

$x > 6$

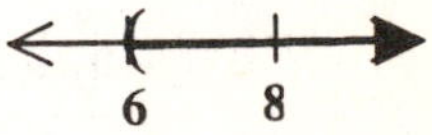

$x < 8$ and $x > 6$

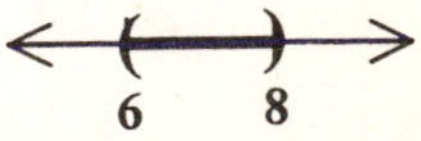

The solution set is $\{x | 6 < x < 8\}$ or $(6, 8)$.

22. $5x+3\le 18$ and $2x-7\le -5$

$5x\le 15$ $\quad$ $2x\le 2$

$x\le 3$ $\quad$ $x\le 1$

$x\le 3$

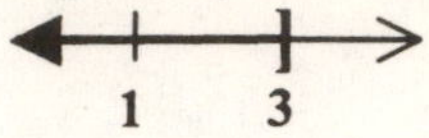

$x\le 1$

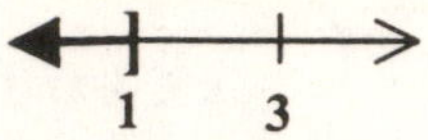

$x\le 3$ and $x\le 1$

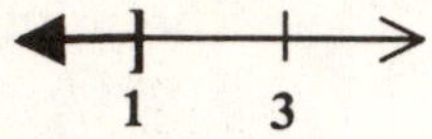

The solution set is $\{x|x\le 1\}$ or $(-\infty,1]$.

23. $2x-5>-1$ and $3x<3$

$2x>4$ $\quad$ $x<1$

$x>2$

$x>2$

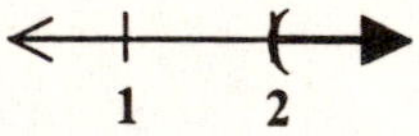

$x<1$

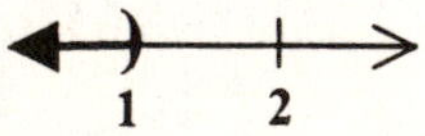

$x>2$ and $x<1$

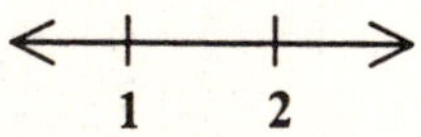

Since the two sets do not intersect, the solution set is $\varnothing$ or $\{\ \}$.

24. $2x-5>-1$ or $3x<3$

$2x>4$ $\quad$ $x<1$

$x>2$

$x>2$

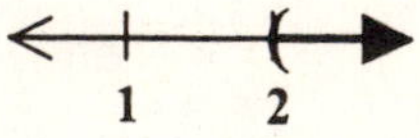

$x<1$

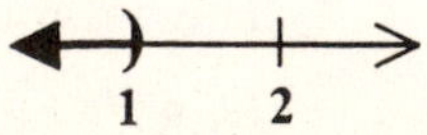

$x>2$ or $x<1$

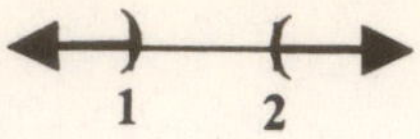

The solution set is $\{x|x<1 \text{ or } x>2\}$ or $(-\infty,1)\cup(2,\infty)$.

25. $x+1\le -3$ or $-4x+3<-5$

$x\le -4$ $\quad$ $-4x<-8$

$x>2$

$x\le -4$

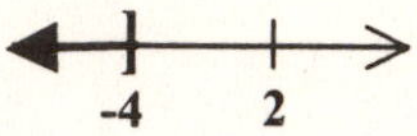

$x>2$

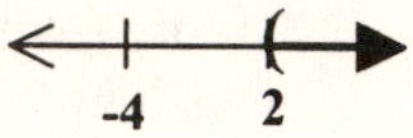

$x\le -4$ or $x>2$

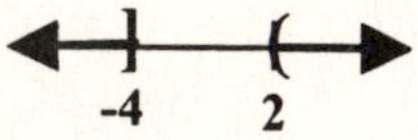

The solution set is $\{x|x\le -4$ and $x>2\}$ or $(-\infty,-4]\cup(2,\infty)$.

26. $5x-2\le -22$ or $-3x-2>4$

$5x\le -20$ $\quad$ $-3x>6$

$x\le -4$ $\quad$ $x<-2$

$x\le -4$

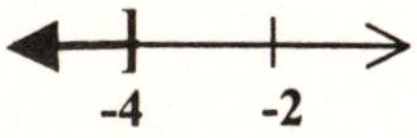

$x<-2$

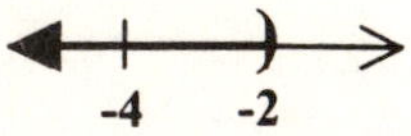

$x\le -4$ or $x<-2$

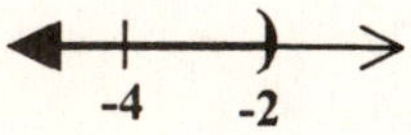

The solution set is $\{x|x<-2\}$ or $(-\infty,-2)$.

27. $5x+4\geq -11$ or $1-4x\geq 9$

$5x\geq -15$ $\quad -4x\geq 8$

$x\geq -3$ $\quad x\leq -2$

$x\geq -3$

-3 2

$x\leq -2$

-3 -2

$x\geq -3$ or $x\leq -2$

-3 -2

The solution set is $\mathbb{R}$, $(-\infty,\infty)$ or $\{x|x \text{ is a real number}\}$.

28.

$$-3<x+2\leq 4$$
$$-3-2<x+2-2\leq 4-2$$
$$-5<x\leq 2$$

-5 2

The solution set is $\{x|-5<x\leq 2\}$ or $(-5,2]$.

29.

$$-1\leq 4x+2\leq 6$$
$$-1-2\leq 4x+2-2\leq 6-2$$
$$-3\leq 4x\leq 4$$
$$-\frac{3}{4}\leq\frac{4x}{4}\leq\frac{4}{4}$$
$$-\frac{3}{4}\leq x\leq 1$$

-3/4 1

The solution set is $\left\{x\middle|-\frac{3}{4}\leq x\leq 1\right\}$ or $\left[-\frac{3}{4},1\right]$.

30. Let x = the grade on the fifth exam

$$80\leq\frac{72+73+94+80+x}{5}<90$$
$$80\leq\frac{319+x}{5}<90$$
$$5(80)\leq 5\left(\frac{319+x}{5}\right)<5(90)$$
$$400\leq 319+x<450$$
$$400-319\leq 319-319+x<450-319$$
$$81\leq x<131$$

You need to score at least 81 and less than 131 on the exam to receive a B. In interval notation, the range is $[81,131)$. If the highest score is 100, the range is $[81,100)$.

4.3

31. $|2x+1|=7$

$2x+1=7$ or $2x+1=-7$

$2x=6$ $\quad 2x=-8$

$x=3$ $\quad x=-4$

The solutions are –4 and 3 and the solution set is $\{-4,3\}$.

32. $|3x+2|=-5$

There are no values of x for which the absolute value of $3x + 2$ is a negative number. By definition, absolute values are always positive. The solution set is $\varnothing$ or $\{\ \}$.

33.

$$2|x-3|-7=10$$
$$2|x-3|=17$$
$$|x-3|=8.5$$

$$x-3=8.5 \quad \text{or} \quad x-3=-8.5$$
$$x=11.5 \qquad\qquad x=-5.5$$

The solutions are –5.5 and 11.5 and the solution set is $\{-5.5, 11.5\}$.

34. $|4x-3|=|7x+9|$

$$4x-3=7x+9$$
$$-3x-3=9$$
$$-3x=12$$
$$x=-4$$

or

$$4x-3=-7x-9$$
$$11x-3=-9$$
$$11x=-6$$
$$x=-\frac{6}{11}$$

The solutions are -4 and $-\frac{6}{11}$ and the solution set is $\left\{-4, -\frac{6}{11}\right\}$.

35. $|2x+3|\le 15$

$$-15\le 2x+3\le 15$$
$$-15-3\le 2x+3-3\le 15-3$$
$$-18\le 2x\le 12$$
$$-\frac{18}{2}\le\frac{2x}{2}\le\frac{12}{2}$$
$$-9\le x\le 6$$

-9 6

The solution set is $\{x|-9\le x\le 6\}$ or $[-9,6]$.

36. $\left|\frac{2x+6}{3}\right|>2$

$$\frac{2x+6}{3}<-2 \quad \text{or} \quad \frac{2x+6}{3}>2$$
$$2x+6<-6 \qquad 2x+6>6$$
$$2x<-12 \qquad 2x>0$$
$$x<-6 \qquad x>0$$

-6 0

The solution set is $\{x|x<-6 \text{ or } x>0\}$ or $(-\infty,-6)\cup(0,\infty)$.

37. $|2x+5|-7<-6$

$$|2x+5|<1$$
$$-1<2x+5<1$$
$$-1-5<2x+5-5<1-5$$
$$-6<2x<-4$$
$$-3<x<-2$$

-3 -2

The solution set is $\{x|-3<x<-2\}$ or $(-3,-2)$.

38. $|2x-3|+4=-10$

$$|2x-3|=-14$$

There are no values of x for which the absolute value of $2x-3$ is a negative number. By definition, absolute values are always positive. The solution set is $\varnothing$ or $\{\ \}$.

39. Answers will vary.

4.4

40. $3x-4y>12$

First, find the intercepts to the equation $3x-4y=12$.

Find the x–intercept by setting y =0.

$$3x-4y=12$$
$$3x-4(0)=12$$
$$3x=12$$
$$x=4$$

Find the y–intercept by setting x =0.

$$3x-4y=12$$
$$3(0)-4y=12$$
$$-4y=12$$
$$y=-3$$

Next, use the origin as a test point.

$$3x-4y>12$$
$$3(0)-4(0)>12$$
$$0>12$$

This is a false statement. This means that the point, $(0,0)$, will not fall in the shaded half-plane.

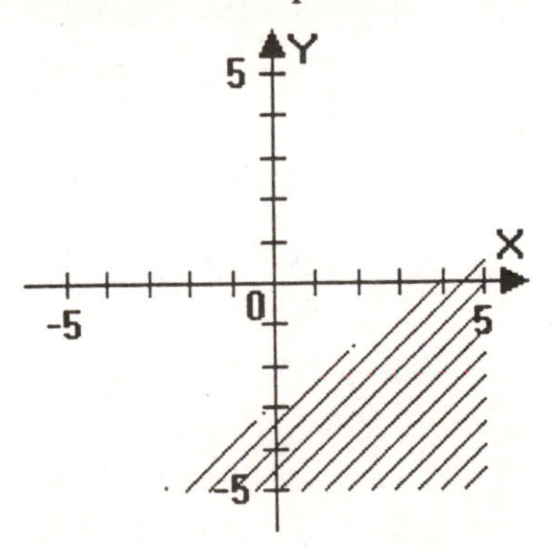

41. $x-3y\le 6$

First, find the intercepts to the equation $x-3y=6$.

Find the x–intercept by setting y =0, find the y–intercept by setting x =0.

$$x-3y=6 \qquad x-3y=6$$
$$x-3(0)=6 \qquad 0-3y=6$$
$$x=6 \qquad -3y=6$$
$$y=-2$$

Next, use the origin as a test point.

$$0-3(0)\le 6$$
$$0\le 6$$

This is a true statement. This means that the point, $(0,0)$, will fall in the shaded half-plane.

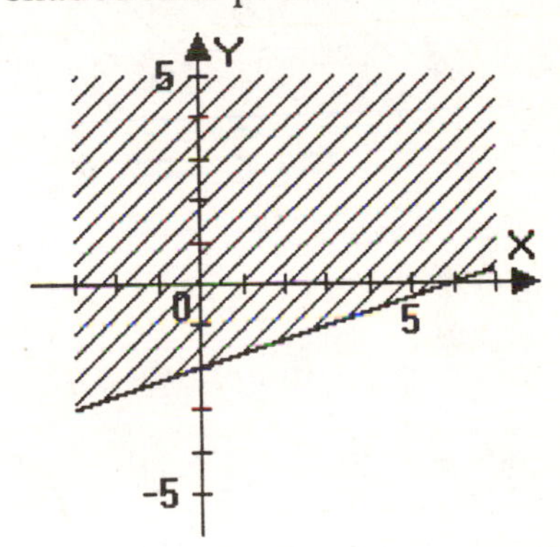

42. $y\le -\frac{1}{2}x+2$

Replacing the inequality symbol with an equal sign, we have $y=-\frac{1}{2}x+2$.

Since the equation is in slope-intercept form, use the slope and the intercept to graph the equation. The y–intercept is 2 and the slope is $-\frac{1}{2}$.

Next, use the origin as a test point.

$y \le -\frac{1}{2}x + 2$

$0 \le -\frac{1}{2}(0) + 2$

$0 \le 2$

This is a true statement. This means that the point $(0,0)$ will fall in the shaded half-plane.

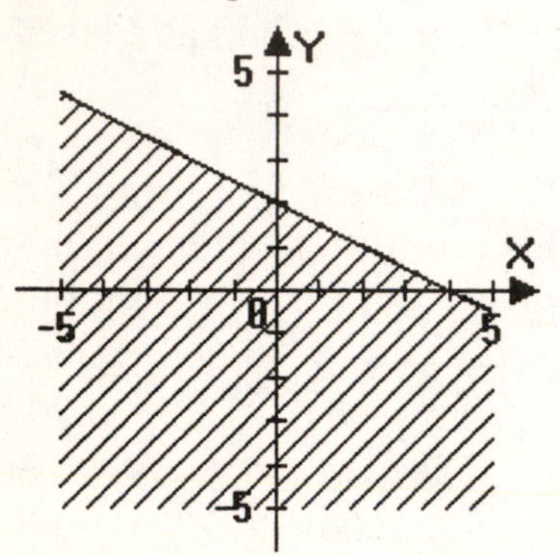

43. $y > \frac{3}{5}x$

Replacing the inequality symbol with an equal sign, we have $y = \frac{3}{5}x$.

Since the equation is in slope-intercept form, use the slope and the intercept to graph the equation. The y–intercept is 0 and the slope is $\frac{3}{5}$.

Next, we need to find a test point. We cannot use the origin this time, because it lies on the line.

Use $(1,1)$ as a test point.

$1 > \frac{3}{5}(1)$

$1 > \frac{3}{5}$

This is a true statement, so we know the point $(1,1)$ lies in the shaded half-plane.

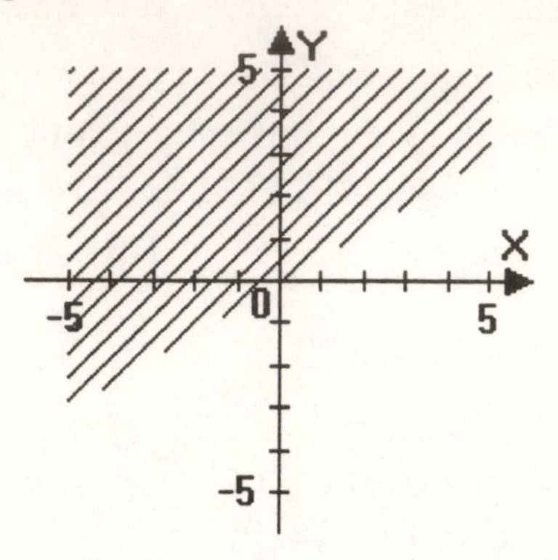

44. $x \le 2$

Replacing the inequality symbol with an equal sign, we have $x = 2$. We know that equations of the form $x = a$ are vertical lines with x–intercept $= a$.

Next, use the origin as a test point.

$x \le 2$

$0 \le 2$

This is a true statement, so we know the point $(0,0)$ lies in the shaded half-plane.

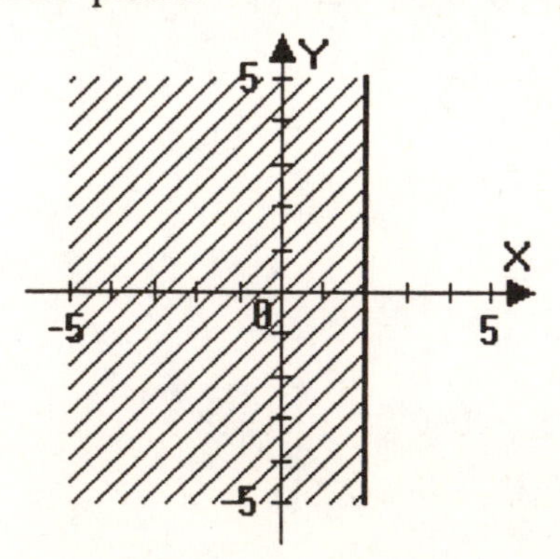

45. 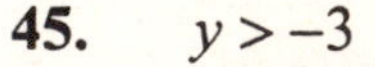 $y > -3$

Replacing the inequality symbol with an equal sign, we have $y = -3$. We know that equations of the form $y = b$ are horizontal lines with y-intercept $= b$.

Next, use the origin as a test point.

$y > -3$

$0 > -3$

This is a true statement, so we know the point $(0,0)$ lies in the shaded half-plane.

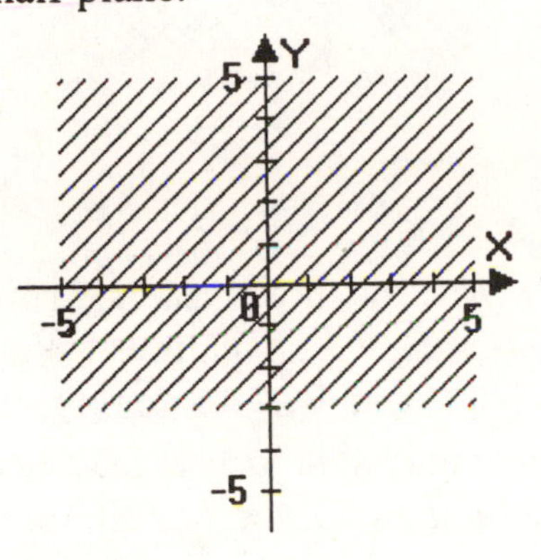

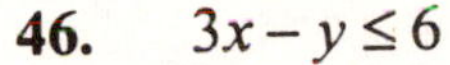

46. $3x - y \le 6$

$x + y \ge 2$

First consider $3x - y \le 6$. If we solve for y in $3x - y = 6$, we can graph the line using the slope and the y-intercept.

$3x - y = 6$

$-y = -3x + 6$

$y = 3x - 6$

y-intercept $= -6$

slope $= 3$

Now, use the origin as a test point.

$3(0) - 0 \le 6$

$0 \le 6$

This is a true statement. This means that the point $(0,0)$ will fall in the shaded half-plane.

Next consider $x + y \ge 2$. If we solve for y in $x + y = 2$, we can graph using the slope and the y-intercept.

$x + y = 2$

$y = -x + 2$

y-intercept $= 2$

slope $= -1$

Now, use the origin as a test point.

$0 + 0 \ge 2$

$0 \ge 2$

This is a false statement. This means that the point $(0,0)$ will not fall in the shaded half-plane.

Next, graph each of the inequalities. The solution to the system is the intersection of the shaded half-planes.

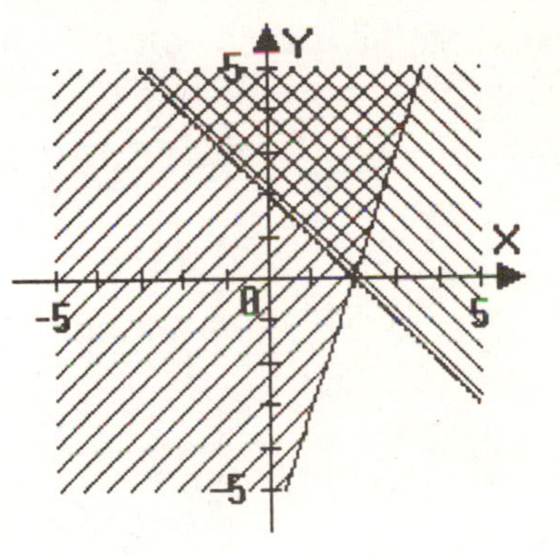

47. $y < -x + 4$

$y > x - 4$

First consider $y < -x + 4$. Change the inequality symbol to an equal sign. The line $y = -x + 4$ is in slope-intercept form and can be graphed using the slope and the y-intercept.

y-intercept $= 4$

slope $= -1$

Now, use the origin as a test point.

$0 < -0 + 4$

$0 < 4$

This is a true statement. This means that the point $(0,0)$ will fall in the shaded half-plane.
Next consider $y > x-4$. Change the inequality symbol to an equal sign. The line $y = x-4$ is in slope-intercept form and can be graphed using the slope and the y-intercept.
y-intercept $= -4$
slope $= 1$
Now, use the origin as a test point.
$0 > 0-4$
$0 > 4$
This is a false statement. This means that the point $(0,0)$ will not fall in the shaded half-plane.
Next, graph each of the inequalities. The solution to the system is the intersection of the shaded half-planes.

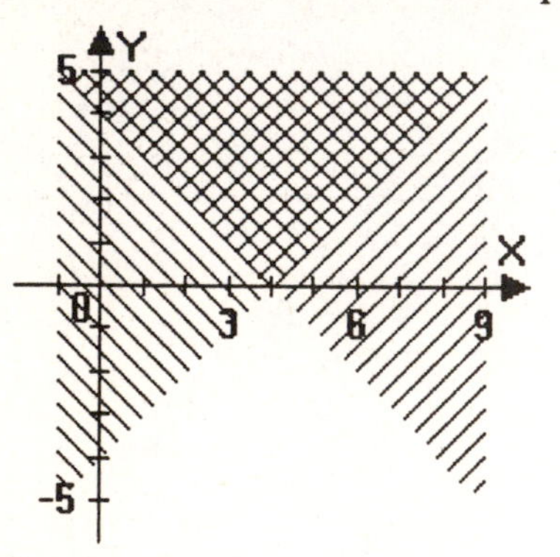

48. $-3 \le x < 5$
Rewrite the three part inequality as two separate inequalities. We have $-3 \le x$ and $x < 5$. We replace the inequality symbols with equal signs and obtain $-3 = x$ and $x = 5$. Equations of the form $x = a$ are vertical lines with x-intercept $= a$.
We know the shading in the graph will be between $x = -3$ and $x = 5$ because in the original inequality we see that x lies between -3 and 5.

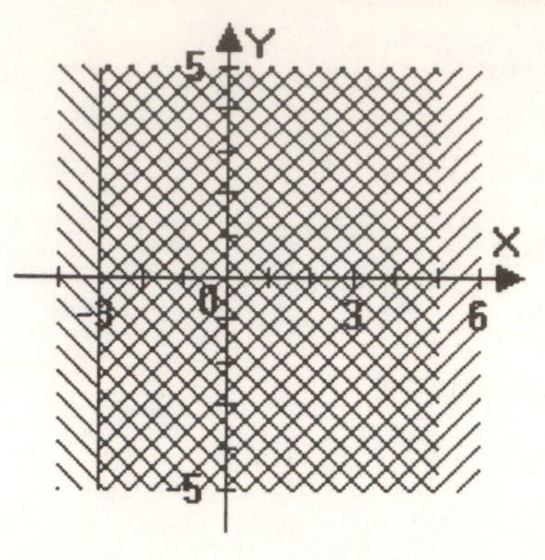

49. $-2 < y \le 6$
Rewrite the three part inequality as two separate inequalities. We have $-2 < y$ and $y \le 6$. We replace the inequality symbols with equal signs and obtain $-2 = y$ and $y = 6$. Equations of the form $y = b$ are vertical lines with y-intercept $= b$.
We know the shading in the graph will be between $y = -2$ and $y = 6$ because in the original inequality we see that y lies between -2 and 6.

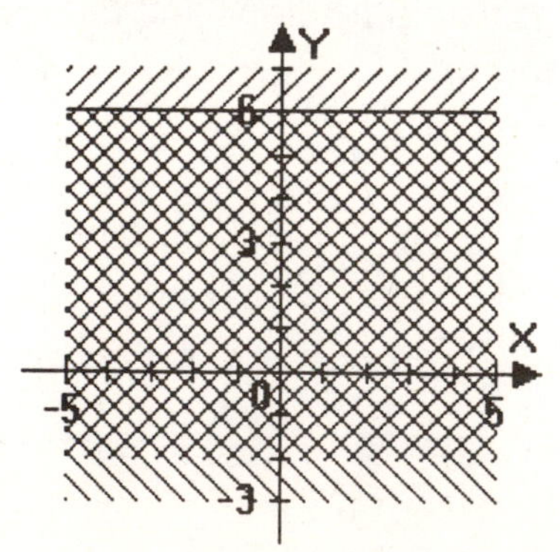

50. $x \ge 3$
$y \le 0$
First consider $x \ge 3$. Change the inequality symbol to an equal sign and

we obtain the vertical line $x = 3$. Because we have $x \geq 3$, we know the shading is to the right of the line $x = 3$.
Next consider $y \leq 0$. Change the inequality symbol to an equal sign and we obtain the horizontal line $y = 0$. (Recall that this is the equation of the x–axis.) Because we have $y \leq 0$, we know that the shading will be below the x–axis.
Next, graph each of the inequalities. The solution to the system is the intersection of the shaded half-planes.

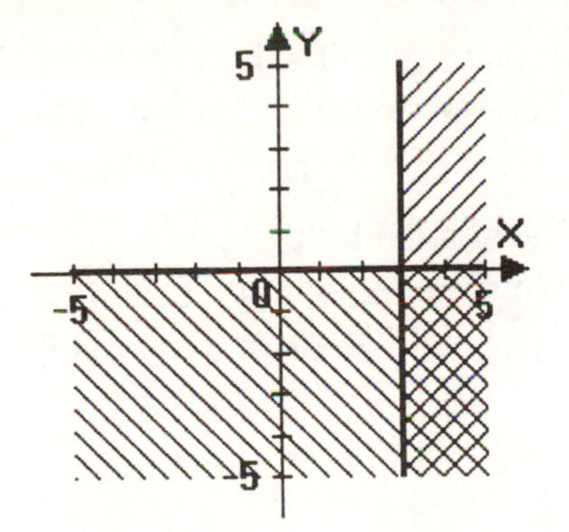

51. $2x - y > -4$

$x \geq 0$

First consider $2x - y > -4$. Replace the inequality symbol with an equal sign and we have $2x - y = -4$. Solve for y to obtain slope-intercept form.

$2x - y = -4$

$-y = -2x - 4$

$y = 2x + 4$

y-intercept = 4
slope = 2

Now, use the origin as a test point.

$2x - y > -4$

$2(0) - 0 > -4$

$0 > -4$

This is a false statement. This means that the point $(0,0)$ will not fall in the shaded half-plane.
Next consider $x \geq 0$. Change the inequality symbol to an equal sign and we obtain the horizontal line $x = 0$. (Recall that this is the equation of the y–axis.) Because we have $x \geq 0$, we know that the shading will be above the y–axis.
Next, graph each of the inequalities. The solution to the system is the intersection of the shaded half-planes.

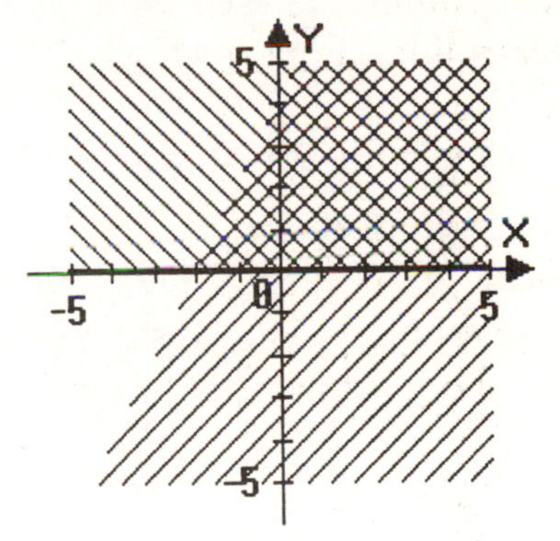

52. $x + y \leq 6$

$y \geq 2x - 3$

First consider $x + y \leq 6$. Replace the inequality symbol with an equal sign and we have $x + y = 6$. Solve for y to obtain slope-intercept form.

$x + y = 6$

$y = -x + 6$

y-intercept = 6
slope = – 1

Now, use the origin as a test point.

$0+0 \le 6$

$0 \le 6$

This is a true statement. This means that the point $(0,0)$ will fall in the shaded half-plane.

Next consider $y \ge 2x-3$. Replace the inequality symbol with an equal sign and we have $y = 2x-3$. The equation is in slope-intercept form, so we can use the slope and the y-intercept to graph the line.

y-intercept $= -3$

slope $= 2$

Now, use the origin as a test point.

$y \ge 2x-3$

$0 \ge 2(0)-3$

$0 \ge -3$

This is a true statement. This means that the point $(0,0)$ will fall in the shaded half-plane.

Next, graph each of the inequalities. The solution to the system is the intersection of the shaded half-planes.

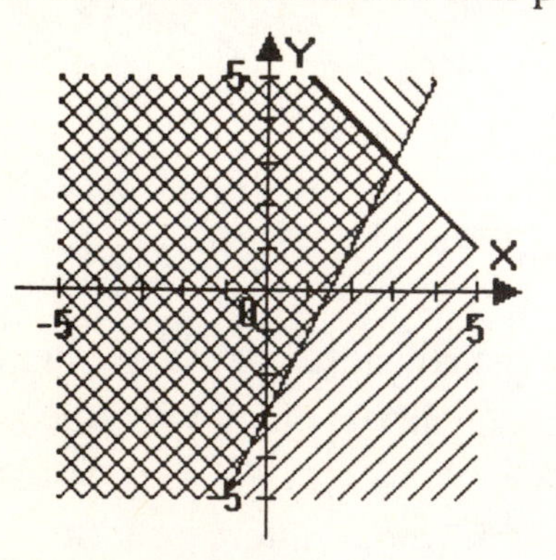

53. $3x+2y \ge 4$

$x-y \le 3$

$x \ge 0, \quad y \ge 0$

First consider $3x+2y \ge 4$. Replace the inequality symbol with an equal sign and we have $3x+2y=4$. Solve for y to obtain slope-intercept form.

$3x+2y=4$

$2y=-3x+4$

$y=-\frac{3}{2}x+2$

y-intercept $= 2$ slope $= -\frac{3}{2}$

Now, use the origin as a test point.

$3x+2y \ge 4$

$3(0)+2(0) \ge 4$

$0 \ge 4$

This is a false statement. This means that the point $(0,0)$ will not fall in the shaded half-plane.

Now consider $x-y \le 3$. Replace the inequality symbol with an equal sign and we have $x-y=3$. Solve for y to obtain slope-intercept form.

$x-y=3$

$-y=-x+3$

$y=x-3$

y-intercept $= -3$ slope $= 1$

Now, use the origin as a test point.

$x-y \le 3$

$0-0 \le 3$

$0 \le 3$

This is a true statement. This means that the point $(0,0)$ will fall in the shaded half-plane.

Now consider the inequalities $x \ge 0$ and $y \ge 0$. The inequalities mean that both x and y will be positive. This means that we only need to consider quadrant I.

Next, graph each of the inequalities. The solution to the system is the intersection of the shaded half-planes.

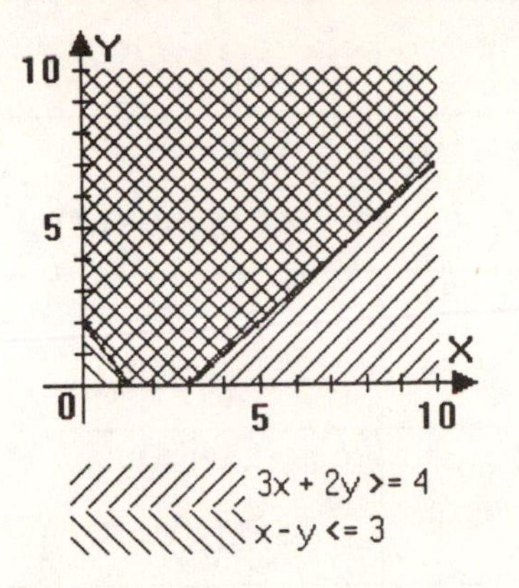

54. $2x - y > 2$

$2x - y < -2$

First consider $2x - y > 2$. Replace the inequality symbol with an equal sign and we have $2x - y = 2$. Solve for y to obtain slope-intercept form.

$2x - y = 2$

$-y = -2x + 2$

$y = 2x - 2$

y-intercept = -2 slope = 2

Now, use the origin as a test point.

$2x - y > 2$

$2(0) - 0 > 2$

$0 > 2$

This is a false statement. This means that the point $(0,0)$ will not fall in the shaded half-plane.

Now consider $2x - y < -2$. Replace the inequality symbol with an equal sign and we have $2x - y = -2$. Solve for y to obtain slope-intercept form.

$2x - y = -2$

$-y = -2x - 2$

$y = 2x + 2$

y-intercept = 2 slope = 2

Now, use the origin as a test point.

$2x - y < -2$

$2(0) - 0 < -2$

$0 < -2$

This is a false statement. This means that the point $(0,0)$ will not fall in the shaded half-plane.

Next, graph each of the inequalities. The solution to the system is the intersection of the shaded half-planes.

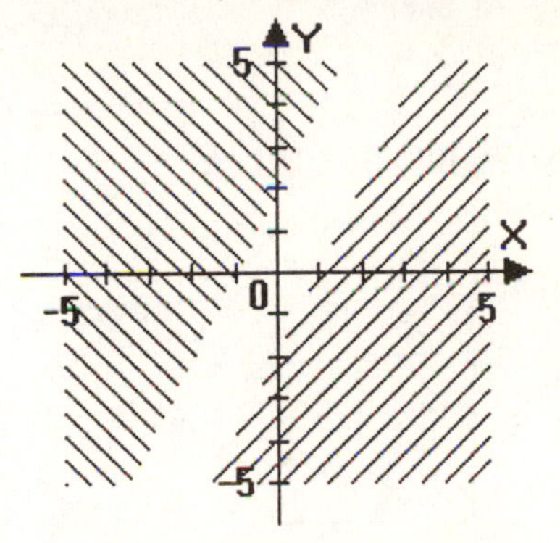

The graphs of the inequalities do not intersect, so there is no solution. The solution set is $\varnothing$ or $\{ \}$.

4.5

55.

Corner (x, y)	Objective Function $z = 2x + 3y$
(1, 0)	$z = 2x + 3y$ $= 2(1) + 3(0) = 2$
(4, 0)	$z = 2x + 3y$ $= 2(4) + 3(0) = 8$
(2, 2)	$z = 2x + 3y$ $= 2(2) + 3(2)$ $= 4 + 6 = 10$

$\left(\frac{1}{2}, \frac{1}{2}\right)$	$z = 2x + 3y$ $= 2\left(\frac{1}{2}\right) + 3\left(\frac{1}{2}\right)$ $= \frac{2}{2} + \frac{3}{2} = \frac{5}{2}$

The maximum value is 10 and the minimum is 2.

56. Objective Function: $z = 2x + 3y$

Constraints: $x \geq 0, \quad y \geq 0$

$x + y \leq 8$

$3x + 2y \geq 6$

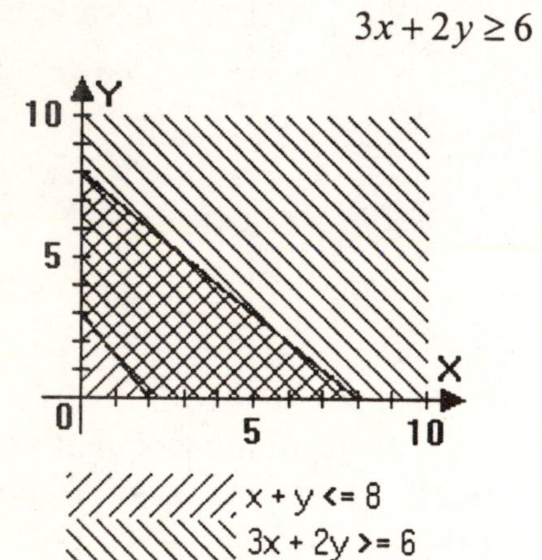

Using the graph, find the value of the objective function at each of the corner points.

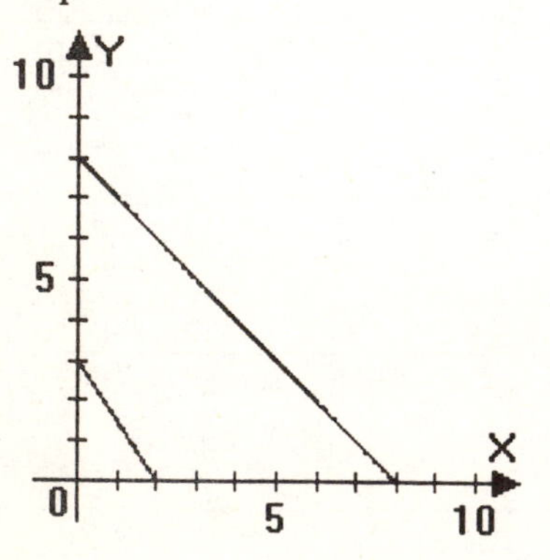

Corner (x, y)	Objective Function $z = 2x + 3y$
(2, 0)	$z = 2x + 3y$ $= 2(2) + 3(0) = 4$
(8, 0)	$z = 2x + 3y$ $= 2(8) + 3(0) = 16$
(0, 8)	$z = 2x + 3y$ $= 2(0) + 3(8) = 24$
(0, 3)	$z = 2x + 3y$ $= 2(0) + 3(3) = 9$

The maximum of 24 occurs at the point (0, 8).

57. Objective Function: $z = x + 4y$

Constraints: $0 \leq x \leq 5$

$0 \leq y \leq 7$

$x + y \geq 3$

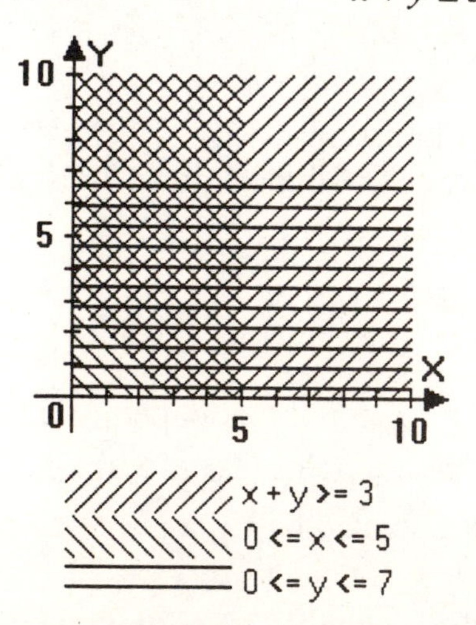

Using the graph, find the value of the objective function at each of the corner points.

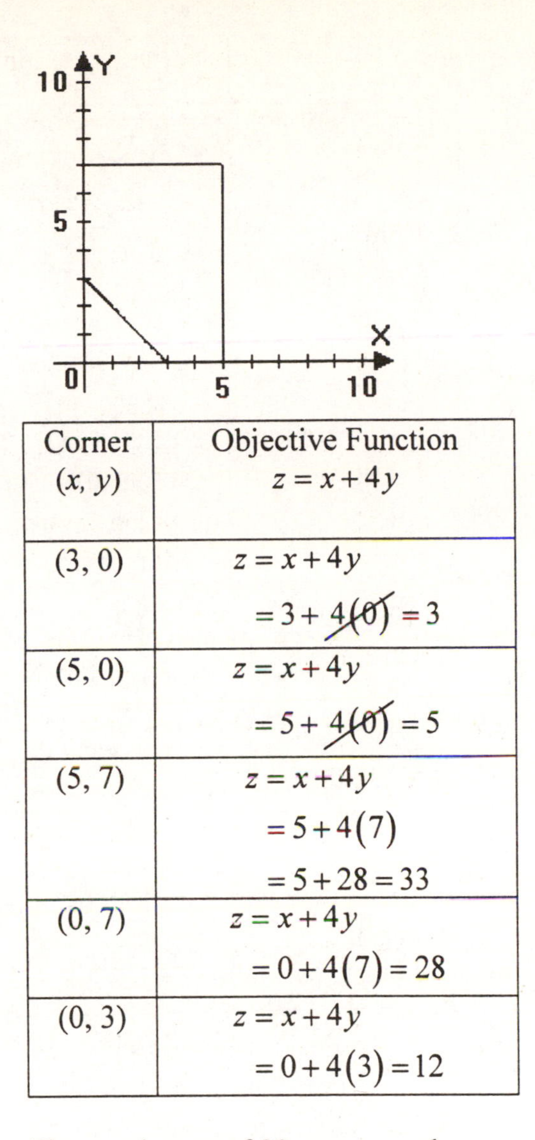

Corner (x, y)	Objective Function $z = x + 4y$
(3, 0)	$z = x + 4y$ $= 3 + 4(0) = 3$
(5, 0)	$z = x + 4y$ $= 5 + 4(0) = 5$
(5, 7)	$z = x + 4y$ $= 5 + 4(7)$ $= 5 + 28 = 33$
(0, 7)	$z = x + 4y$ $= 0 + 4(7) = 28$
(0, 3)	$z = x + 4y$ $= 0 + 4(3) = 12$

The maximum of 33 occurs at the point (5, 7).

58. Objective Function: $z = 5x + 6y$

Constraints: $x \geq 0, \ y \geq 0$

$y \leq x$

$2x + y \leq 12$

$2x + 3y \geq 6$

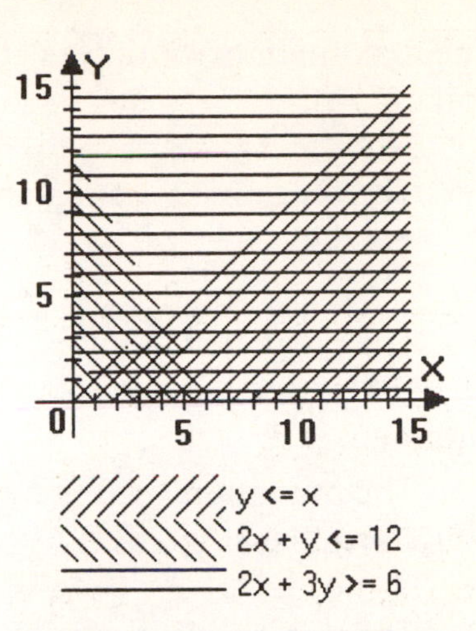

Using the graph, find the value of the objective function at each of the corner points.

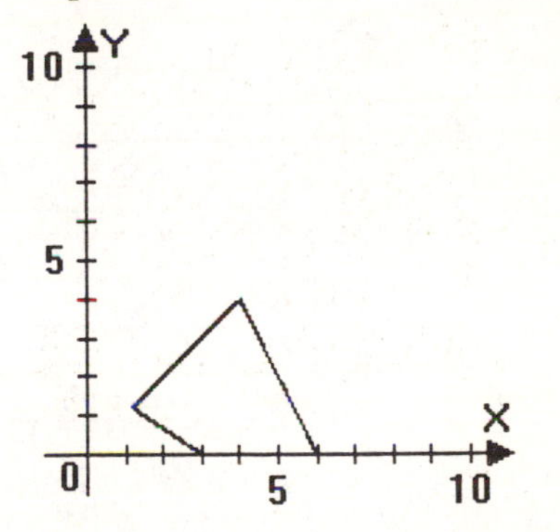

Corner (x, y)	Objective Function $z = 5x + 6y$
(3, 0)	$z = 5x + 6y$ $= 5(3) + 6(0) = 15$
(6, 0)	$z = 5x + 6y$ $= 5(6) + 6(0) = 30$
(4, 4)	$z = 5x + 6y$ $= 5(4) + 6(4)$ $= 20 + 24 = 44$
(1.2, 1.2)	$z = 5x + 6y$ $= 5(1.2) + 6(1.2)$ $= 6 + 7.2 = 13.2$

The maximum of 44 occurs at the point (4, 4).

59. **a.** The objective function is $z = 500x + 350y$.

b. The paper constraint is $x + y \le 200$.

The minimum writing paper constraint is $x \ge 10$.

The minimum newsprint constraint is $y \ge 80$.

c.

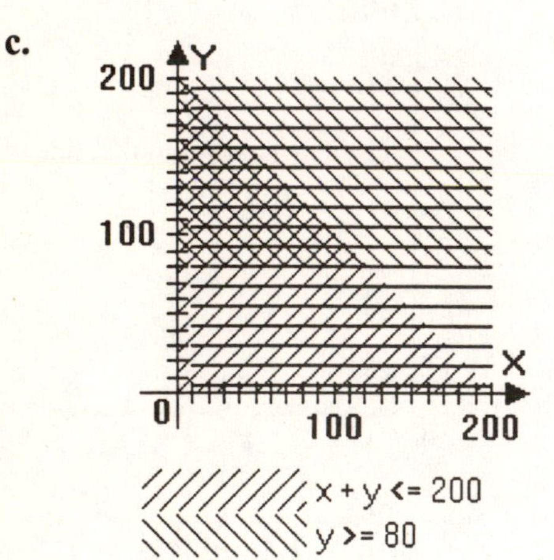

d.

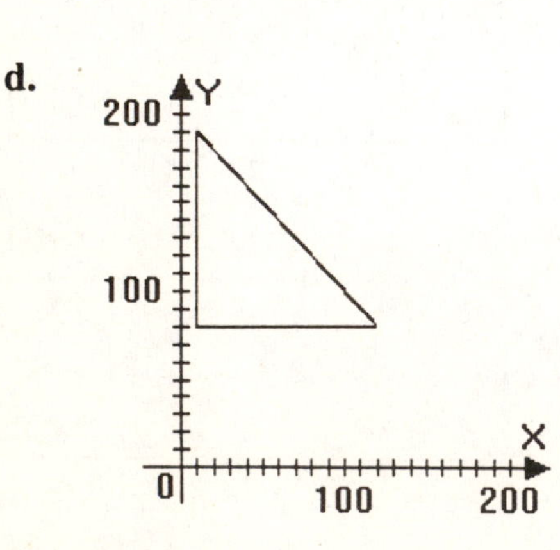

Corner (x, y)	Objective Function $z = 500x + 350y$
(10, 80)	$z = 500x + 350y$ $= 500(10) + 350(80)$ $= 5000 + 28000$ $= 33000$
(120, 80)	$z = 500x + 350y$ $= 500(120) + 350(80)$ $= 60000 + 28000$ $= 88000$
(10, 190)	$z = 500x + 350y$ $= 500(10) + 350(190)$ $= 5000 + 66500$ $= 71500$

e. The company will make the greatest profit by producing <u>120</u> units of writing paper and <u>80</u> units of newsprint each day. The maximum daily profit is <u>$88,000</u>.

60. Let x = the number of model A produced
Let y = the number of model B produced
The objective function is $z = 25x + 40y$.
The cutting department labor constraint is $0.9x + 1.8y \le 864$.
The assembly department labor constraint is $0.8x + 1.2y \le 672$.
We also know that x and y are either zero or a positive number. We cannot have a negative number of units produced.
Next, graph the constraints.

0.9x + 1.8y <= 864
0.8x + 1.2y <= 672

Using the graph, find the value of the objective function at each of the corner points.

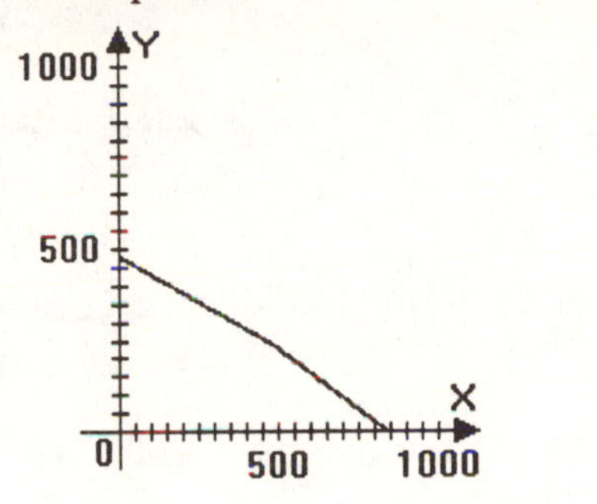

Corner (x, y)	Objective Function $z = 25x + 40y$
(0, 0)	$25x + 40y$ $= 25(0) + 40(0)$ $= 0$
(840, 0)	$25x + 40y$ $= 25(840) + 40(0)$ $= 21000$
(480, 240)	$25x + 40y$ $= 25(480) + 40(240)$ $= 12000 + 9600$ $= 21600$
(0, 480)	$25x + 40y$ $= 25(0) + 40(480)$ $= 19200$

The maximum of 21,600 occurs at the point (480, 240). This means that to maximize the profit, 480 of model A and 240 of model B should be manufactured monthly. This would result in a profit of \$21,600.

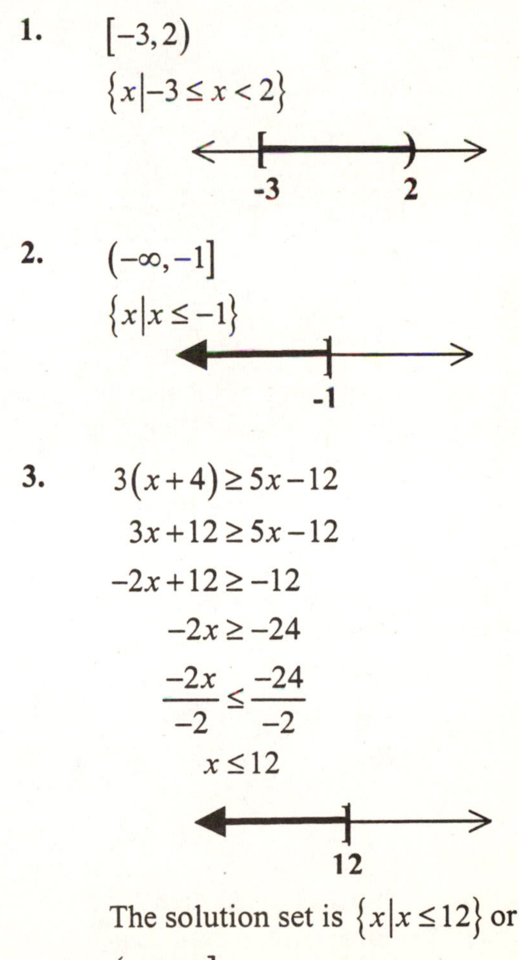

Chapter 4 Test

1. $[-3, 2)$

$\{x \mid -3 \le x < 2\}$

-3 2

2. $(-\infty, -1]$

$\{x \mid x \le -1\}$

-1

3.
$$3(x+4) \ge 5x - 12$$
$$3x + 12 \ge 5x - 12$$
$$-2x + 12 \ge -12$$
$$-2x \ge -24$$
$$\frac{-2x}{-2} \le \frac{-24}{-2}$$
$$x \le 12$$

12

The solution set is $\{x \mid x \le 12\}$ or $(-\infty, 12]$.

4.
$$\frac{x}{6}+\frac{1}{8}\le\frac{x}{2}-\frac{3}{4}$$
$$24\left(\frac{x}{6}\right)+24\left(\frac{1}{8}\right)\le 24\left(\frac{x}{2}\right)-24\left(\frac{3}{4}\right)$$
$$4x+3\le 12x-6(3)$$
$$4x+3\le 12x-18$$
$$-8x+3\le -18$$
$$-8x\le -21$$
$$\frac{-8x}{-8}\ge\frac{-21}{-8}$$
$$x\ge\frac{21}{8}$$

21/8

The solution set is $\left\{x \middle| x\ge\frac{21}{8}\right\}$ or $\left[\frac{21}{8},\infty\right)$.

5. Let x = the number of local calls
The monthly cost using Plan A is $C_A = 25$.
The monthly cost using Plan B is $C_B = 13+0.06x$.
For Plan A to be better deal, it must cost less than Plan B.
$$C_A < C_B$$
$$25 < 13+0.06x$$
$$12 < 0.06x$$
$$200 < x$$
$$x > 200$$
Plan A is a better deal when more than 200 local calls are made per month.

6. $\{2,4,6,8,10\}\cap\{4,6,12,14\}$
$=\{4,6\}$

7. $\{2,4,6,8,10\}\cup\{4,6,12,14\}$
$=\{2,4,6,8,10,12,14\}$

8. $2x+4<2$ and $x-3>-5$
$2x<-2$ $\quad x>-2$
$x<-1$

$x<-1$ (-2, -1)

$x>-2$ (-2, -1)

$x<-1$ and $x>-2$ (-2, -1)

The solution set is $\{x|-2<x<-1\}$ or $(-2,-1)$.

9. $x+6\ge 4$ and $2x+3\ge -2$
$x\ge -2$ $\quad 2x\ge -5$
$x\ge -\frac{5}{2}$

$x\ge -2$ (-5/2, -2)

$x\ge -\frac{5}{2}$ (-5/2, -2)

$x\ge -2$ and $x\ge -\frac{5}{2}$

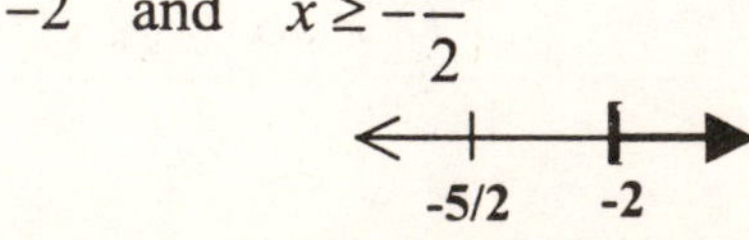

The solution set is $\{x | x \geq -2\}$ or $[-2, \infty)$.

10. $2x - 3 < 5$ or $3x - 6 \leq 4$

$2x < 8$ $\quad$ $3x \leq 10$

$x < 4$ $\quad$ $x \leq \frac{10}{3}$

$x < 4$

10/3 4

$x \leq \frac{10}{3}$

10/3 4

$x < 4$ or $x \leq \frac{10}{3}$

10/3 4

The solution set is $\{x | x < 4\}$ or $(-\infty, 4)$.

11. $x + 3 \leq -1$ or $-4x + 3 < -5$

$x \leq -4$ $\quad$ $-4x < -8$

$x > 2$

$x \leq -4$

-4 2

$x > 8$

-4 2

$x \leq -4$ or $x > 2$

-4 2

The solution set is $\{x | x \leq -4$ or $x > 2\}$ or $(-\infty, -4] \cup (2, \infty)$.

12.

$$-3 \leq \frac{2x+5}{3} < 6$$

$$3(-3) \leq 3\left(\frac{2x+5}{3}\right) < 3(6)$$

$$-9 \leq 2x + 5 < 18$$

$$-9 - 5 \leq 2x + 5 - 5 < 18 - 5$$

$$-14 \leq 2x < 13$$

$$-7 \leq x < \frac{13}{2}$$

-7 13/2

The solution set is $\left\{x \middle| -7 \leq x < \frac{13}{2}\right\}$ or $\left[-7, \frac{13}{2}\right)$.

13. $|5x + 3| = 7$

$5x + 3 = 7$ or $5x + 3 = -7$

$5x = 4$ $\quad$ $5x = -10$

$x = \frac{4}{5}$ $\quad$ $x = -2$

The solutions are -2 and $\frac{4}{5}$ and the solution set is $\left\{-2, \frac{4}{5}\right\}$.

14. $|6x + 1| = |4x + 15|$

$$6x + 1 = 4x + 15$$

$$2x + 1 = 15$$

$$2x = 14$$

$$x = 7$$

or

$$6x+1=-(4x+15)$$
$$6x+1=-4x-15$$
$$10x+1=-15$$
$$10x=-16$$
$$x=-\frac{16}{10}=-\frac{8}{5}$$

The solutions are $-\frac{8}{5}$ and 7 and the solution set is $\left\{-\frac{8}{5},7\right\}$.

15. $$|2x-1|<7$$
$$-7<2x-1<7$$
$$-7+1<2x-1+1<7+1$$
$$-6<2x<8$$
$$-3<x<4$$

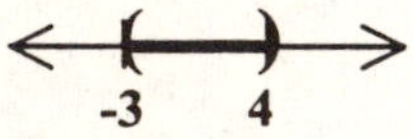

The solution set is $\{x|-3<x<4\}$ or $(-3,4)$.

16. $$|2x-3|\ge 5$$
$$2x-3\le -5 \quad \text{or} \quad 2x-3\ge 5$$
$$2x\le -2 \qquad\qquad 2x\ge 8$$
$$x\le -1 \qquad\qquad x\ge 4$$

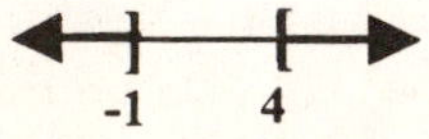

The solution set is $\{x|x\le -1$ or $x\ge 4\}$ or $(-\infty,-1]\cup[4,\infty)$.

17. $$|T-74|\le 8$$
$$-8\le T-74\le 8$$
$$-8+74\le T-74+74\le 8+74$$
$$66\le T\le 82$$

The monthly average temperature for Miami, Florida is between 66°F and 82°F, inclusive.

18. $3x-2y<6$

First, find the intercepts to the equation $3x-2y=6$.

Find the x–intercept by setting $y=0$.
$$3x-2y=6$$
$$3x-2(0)=6$$
$$3x=6$$
$$x=2$$

Find the y–intercept by setting $x=0$.
$$3x-2y=6$$
$$3(0)-2y=6$$
$$-2y=6$$
$$y=-3$$

Next, use the origin as a test point.
$$3x-2y<6$$
$$3(0)-2(0)<6$$
$$0<6$$

This is a true statement. This means that the point will fall in the shaded half-plane.

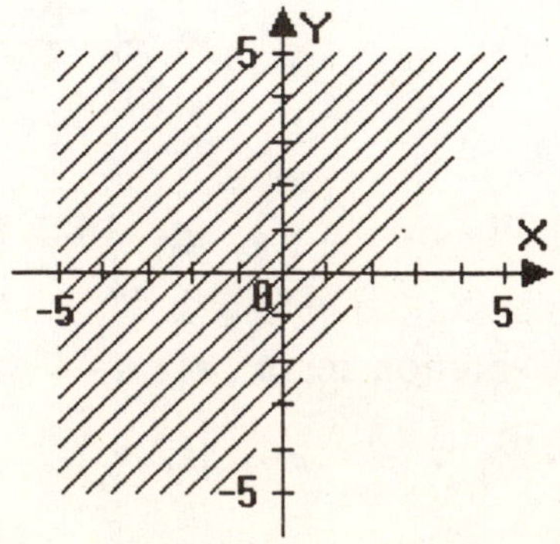

19. $y \geq \frac{1}{2}x - 1$

Replacing the inequality symbol with an equal sign, we have $y = \frac{1}{2}x - 1$. The equation is in slope-intercept form, so graph the line using the slope and the y-intercept.

y-intercept = −1 slope = $\frac{1}{2}$

Now, use the origin, $(0,0)$, as a test point.

$y \geq \frac{1}{2}x - 1$

$0 \geq \frac{1}{2}(0) - 1$

$0 \geq -1$

This is a true statement. This means that the point will fall in the shaded half-plane.

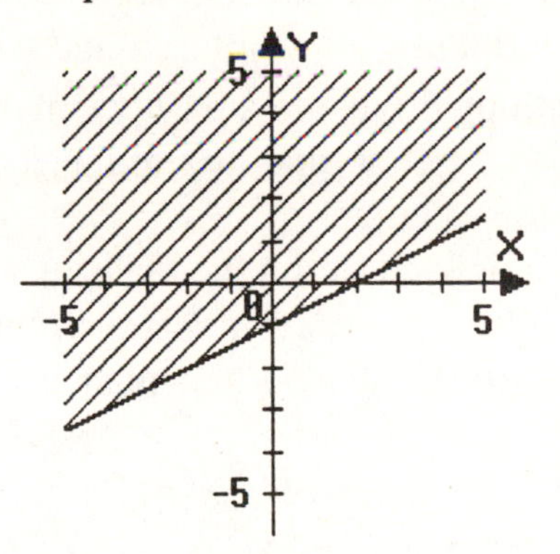

20. $y \leq -1$

Replacing the inequality symbol with an equal sign, we have $y = -1$. Equations of the form $y = b$ are horizontal lines with y-intercept = b, so this is a horizontal line at $y = -1$. Next, use the origin as a test point.

$y \leq -1$

$0 \leq -1$

This is a false statement, so we know the point $(0,0)$ does not lie in the shaded half-plane.

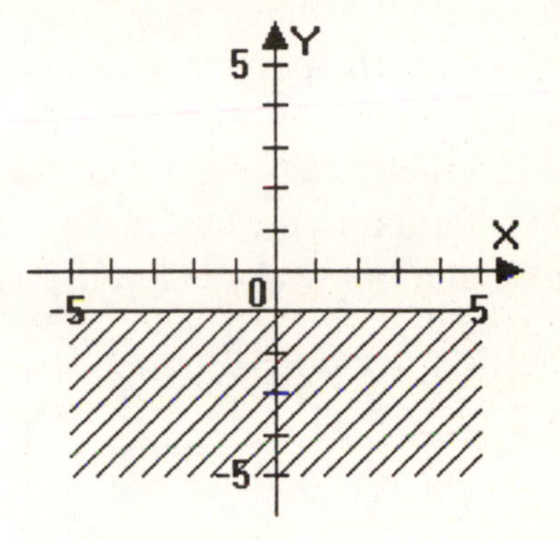

21. $x + y \geq 2$

$x - y \geq 4$

First consider $x + y \geq 2$. If we solve for y in $x + y = 2$, we can graph the line using the slope and the y-intercept.

$x + y = 2$

$y = -x + 2$

y-intercept = 2 slope = −1

Now, use the origin as a test point.

$x + y \geq 2$

$0 + 0 \geq 2$

$0 \geq 2$

This is a false statement. This means that the point will not fall in the shaded half-plane.

Next consider $x - y \geq 4$. If we solve for y in $x - y = 4$, we can graph using the slope and the y-intercept.

$x - y = 4$

$-y = -x + 4$

$y = x - 4$

y-intercept = −4 slope = 1

Now, use the origin as a test point.

$x - y \geq 4$

$0 - 0 \geq 4$

$0 \geq 4$

This is a false statement. This means that the point will not fall in the shaded half-plane.

Next, graph each of the inequalities. The solution to the system is the intersection of the shaded half-planes.

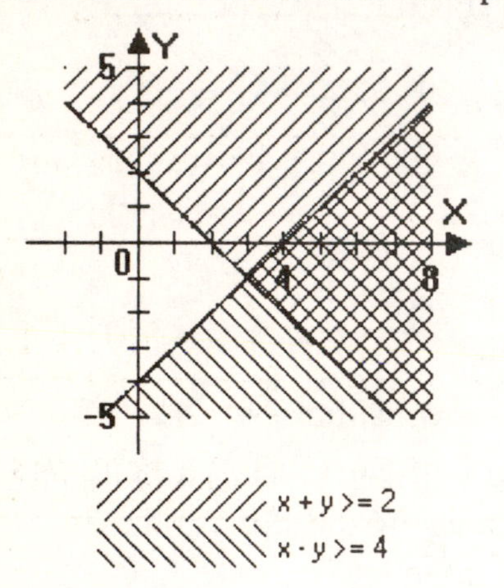

22. $3x + y \leq 9$

$2x + 3y \geq 6$

$x \geq 0, \quad y \geq 0$

First consider $3x + y \leq 9$. If we solve for y in $3x + y = 9$, we can graph the line using the slope and the y-intercept.

$3x + y = 9$

$y = -3x + 9$

y-intercept = 9 slope = −3

Now, use the origin as a test point.

$3x + y \leq 9$

$3(0) + 0 \leq 9$

$0 \leq 9$

This is a true statement. This means that the point will fall in the shaded half-plane.

Next consider $2x + 3y \geq 6$. If we solve for y in $2x + 3y = 6$, we can graph using the slope and the y-intercept.

$2x + 3y = 6$

$3y = -2x + 6$

$y = -\frac{2}{3}x + 2$

y-intercept = 2 slope = $-\frac{2}{3}$

Now, use the origin as a test point.

$2x + 3y \geq 6$

$2(0) + 3(0) \geq 6$

$0 \geq 6$

This is a false statement. This means that the point will not fall in the shaded half-plane.

Next consider the inequalities $x \geq 0$ and $y \geq 0$. When x and y are both positive, we are only concerned with the first quadrant of the coordinate system.

Graph each of the inequalities. The solution to the system is the intersection of the shaded half-planes.

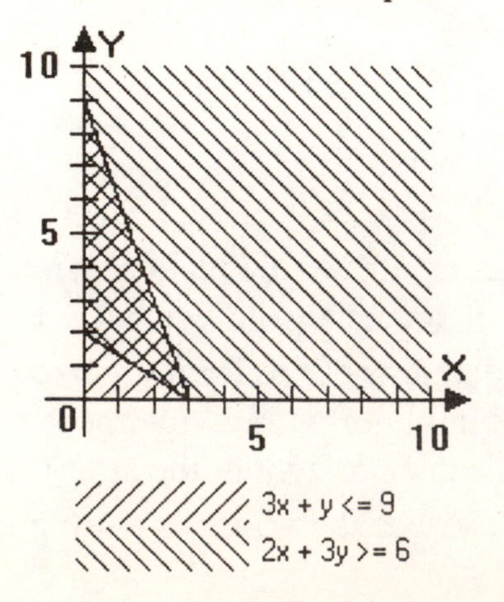

23. $-2 < x \le 4$
Rewrite the three part inequality as two separate inequalities. We have $-2 < x$ and $x \le 4$. We replace the inequality symbols with equal signs and obtain $-2 = x$ and $x = 4$. Equations of the form $x = a$ are vertical lines with x-intercept $= a$.
We know the shading will be between $x = -2$ and $x = 4$ because in the original inequality we see that x lies between –2 and 4.

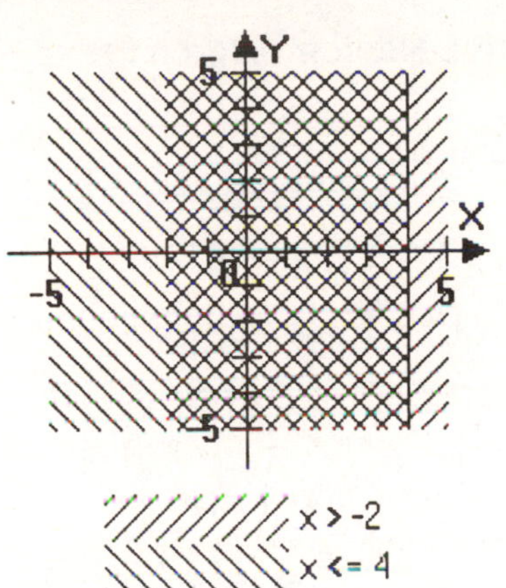

24. Objective Function: $z = 3x + 5y$
Constraints: $x \ge 0,\ y \ge 0$
$x + y \le 6$
$x \ge 2$

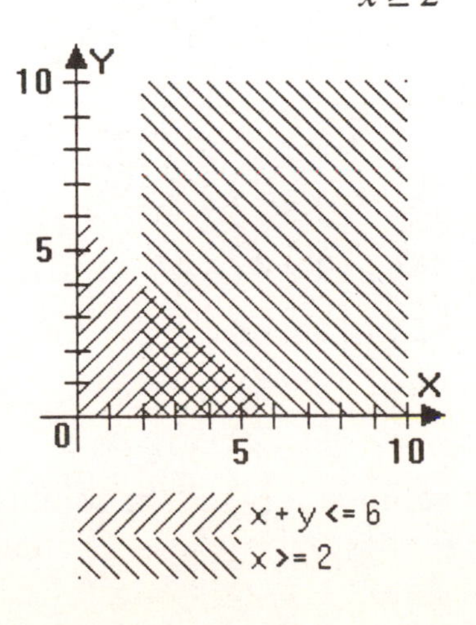

Using the graph, find the value of the objective function at each of the corner points.

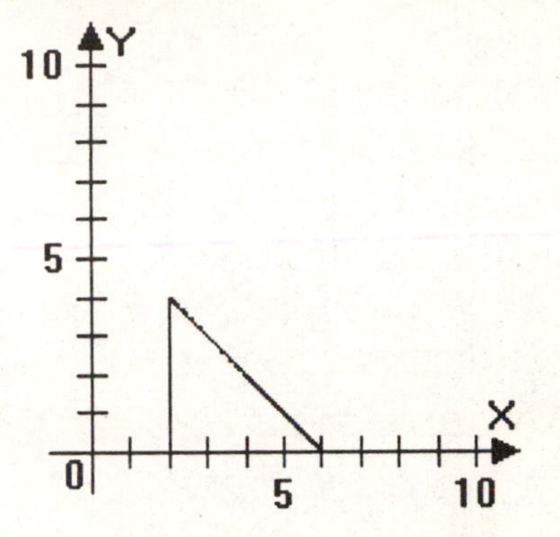

Corner (x, y)	Objective Function $z = 3x + 5y$
(2, 0)	$z = 3x + 5y$ $= 3(2) + 5(0) = 6$
(6, 0)	$z = 3x + 5y$ $= 3(6) + 5(0) = 18$
(2, 4)	$z = 3x + 5y$ $= 3(2) + 5(4)$ $= 6 + 20 = 26$

The maximum of 26 occurs at the point (2, 4).

25. Let x = the number of regular jet skis produced
Let y = the number of deluxe jet skis produced
The objective function is
$z = 200x + 250y$.
The regular jet ski demand constraint is $x \ge 50$.
The deluxe jet ski demand constraint is $y \ge 75$.
The quality constraint is $x + y \le 150$.

We also know that x and y are either zero or a positive number. We cannot have a negative number of units produced.
Next, graph the constraints.

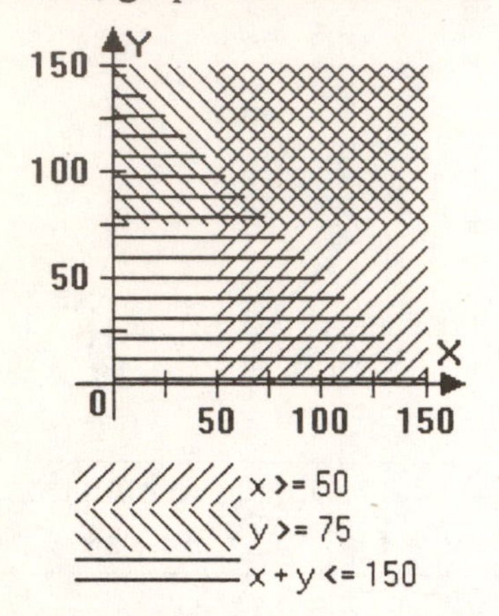

Using the graph, find the value of the objective function at each of the corner points.

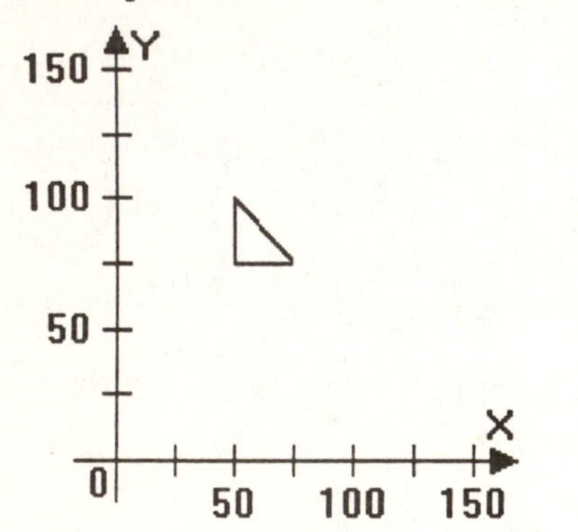

Corner (x, y)	Objective Function $z = 200x + 250y$
(50, 75)	$z = 200x + 250y$ $= 200(50) + 250(75)$ $= 10000 + 18750$ $= 28750$
(75, 75)	$z = 200x + 250y$ $= 200(75) + 250(75)$ $= 15000 + 18750$ $= 33750$
(50, 100)	$z = 200x + 250y$ $= 200(50) + 250(100)$ $= 10000 + 25000$ $= 35000$

The maximum of 35,000 occurs at the point (50, 100). This means that to maximize the profit, 50 regular jet skis and 100 deluxe jet skis should be manufactured weekly. This would result in a profit of $35,000.

Cumulative Review Exercises

Chapters 1-4

1.
$$\begin{aligned}5(x+1)+2 &= x-3(2x+1)\\ 5x+5+2 &= x-6x-3\\ 5x+7 &= -5x-3\\ 10x+7 &= -3\\ 10x &= -10\\ x &= -1\end{aligned}$$
The solution is –1 and the solution set is $\{-1\}$.

2.
$$\begin{aligned}\frac{2(x+6)}{3} &= 1+\frac{4x-7}{3}\\ 3\left(\frac{2(x+6)}{3}\right) &= 3(1)+3\left(\frac{4x-7}{3}\right)\\ 2(x+6) &= 3+4x-7\\ 2x+12 &= 4x-4\\ -2x+12 &= -4\\ -2x &= -16\\ x &= 8\end{aligned}$$
The solution is 8 and the solution set is $\{8\}$.

3. $\dfrac{-10x^2y^4}{15x^7y^{-3}} = \dfrac{-10}{15}x^{2-7}y^{4-(-3)}$

$= -\dfrac{2}{3}x^{-5}y^7 = -\dfrac{2y^7}{3x^5}$

4. $f(x) = x^2 - 3x + 4$

$f(-3) = (-3)^2 - 3(-3) + 4$

$= 9 + 9 + 4 = 22$

$f(2a) = (2a)^2 - 3(2a) + 4$

$= 4a^2 - 6a + 4$

5. $f(x) = 3x^2 - 4x + 1$

$g(x) = x^2 - 5x - 1$

$(f-g)(x) = f(x) - g(x)$

$= (3x^2 - 4x + 1) - (x^2 - 5x - 1)$

$= 3x^2 - 4x + 1 - x^2 + 5x + 1$

$= 2x^2 + x + 2$

$(f-g)(2) = 2(2)^2 + 2 + 2$

$= 2(4) + 2 + 2$

$= 8 + 2 + 2 = 12$

6. Since the line we are concerned with is perpendicular to the line, $y = 2x - 3$, we know the slopes are negative reciprocals. The slope of the line will be the negative reciprocal of 2 which is $-\dfrac{1}{2}$. Using the slope and the point, $(2,3)$, write the equation of the line in point-slope form.

$y - y_1 = m(x - x_1)$

$y - 3 = -\dfrac{1}{2}(x - 2)$

Solve for y to write the equation in function notation.

$y - 3 = -\dfrac{1}{2}(x - 2)$

$y - 3 = -\dfrac{1}{2}x + 1$

$y = -\dfrac{1}{2}x + 4$

$f(x) = -\dfrac{1}{2}x + 4$

7. $f(x) = 2x + 1$

$y = 2x + 1$

Find the x–intercept by setting $y = 0$, and the y–intercept by setting $x = 0$.

$y = 2x + 1$ $\quad$ $y = 2x + 1$

$0 = 2x + 1$ $\quad$ $y = 2(0) + 1$

$-1 = 2x$ $\quad$ $y = 1$

$-\dfrac{1}{2} = x$

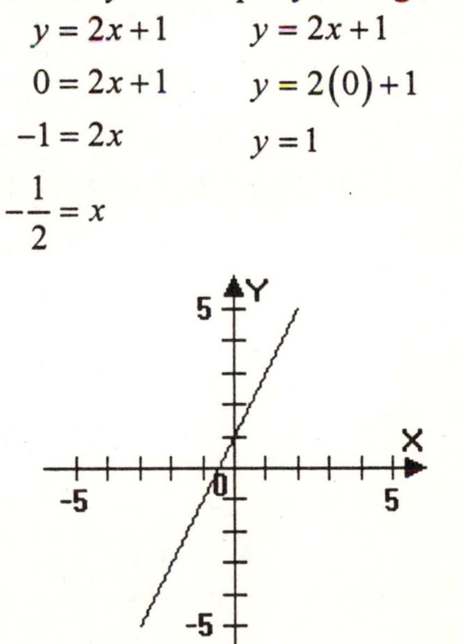

8. $y > 2x$

Consider the line $y = 2x$. Since the line is in slope-intercept form, we

know that the slope is 2 and the y-intercept is 0. Use this information to graph the line.
Since the origin, (0, 0), lies on the line, we cannot use it as a test point. Instead, use the point (1, 1).

$y > 2x$

$1 > 2(1)$

$1 > 2$

This is a false statement. This means that the point (1, 1) does not lie in the shaded region.

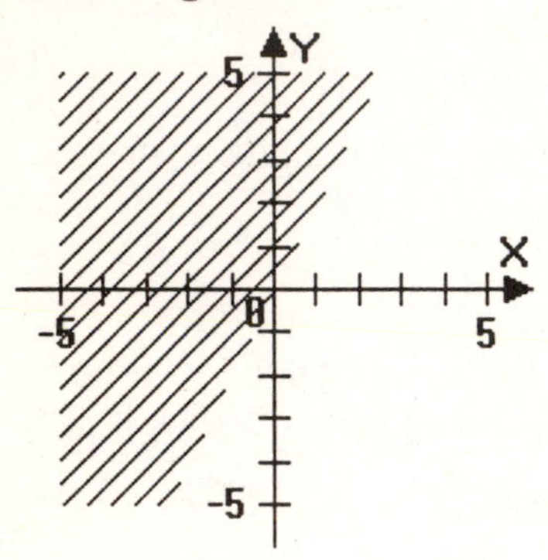

9. $2x - y \geq 6$

Graph the equation using the intercepts.

$2x - y = 6$

x – intercept = 3

y – intercept = –6

Use the origin as a test point to determine shading.

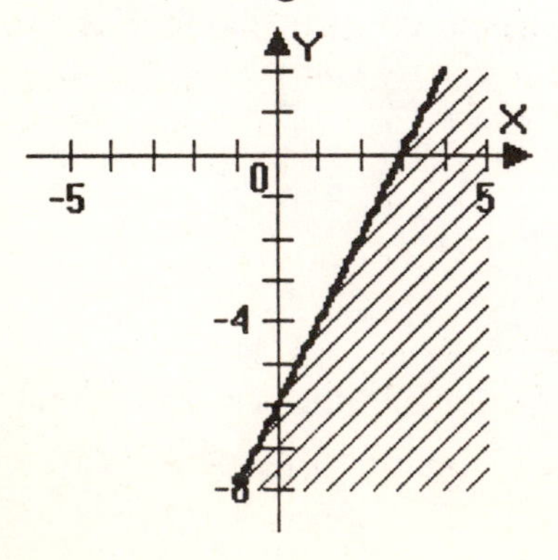

10. $f(x) = -1$

$y = -1$

Equations of the form $y = b$ are horizontal lines with y-intercept = b. This is the horizontal line at $y = -1$.

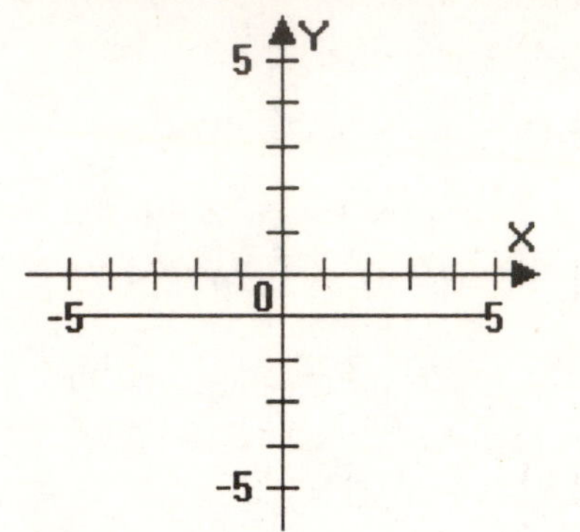

11.
$$3x - y + z = -15$$
$$x + 2y - z = 1$$
$$2x + 3y - 2z = 0$$

Add the first two equations to eliminate z.

$$3x - y + z = -15$$
$$x + 2y - z = 1$$
$$4x + y = -14$$

Multiply the first equation by 2 and add to the third equation.

$$6x - 2y + 2z = -30$$
$$2x + 3y - 2z = 0$$
$$8x + y = -30$$

The system of two equations in two variables becomes is as follows.

$$4x + y = -14$$
$$8x + y = -30$$

Multiply the first equation by –1 and add to the second equation.

$$\begin{aligned} -4x - y &= 14 \\ 8x + y &= -30 \\ \hline 4x &= -16 \\ x &= -4 \end{aligned}$$

Back-substitute –4 for x to find y.

$$\begin{aligned} 4(-4) + y &= -14 \\ -16 + y &= -14 \\ y &= 2 \end{aligned}$$

Back-substitute 2 for y and –4 for x to find z.

$$\begin{aligned} 3x - y + z &= -15 \\ 3(-4) - 2 + z &= -15 \\ -12 - 2 + z &= -15 \\ -14 + z &= -15 \\ z &= -1 \end{aligned}$$

The solution is $(-4, 2, -1)$ and the solution set is $\{(-4, 2, -1)\}$.

12. $2x - y = -4$

$x + 3y = 5$

$$\left[\begin{array}{cc|c} 2 & -1 & -4 \\ 1 & 3 & 5 \end{array}\right] \quad R_1 \leftrightarrow R_2$$

$$= \left[\begin{array}{cc|c} 1 & 3 & 5 \\ 2 & -1 & -4 \end{array}\right] \quad -2R_1 + R_2$$

$$= \left[\begin{array}{cc|c} 1 & 3 & 5 \\ 0 & -7 & -14 \end{array}\right] \quad -\frac{1}{7}R_2$$

$$= \left[\begin{array}{cc|c} 1 & 3 & 5 \\ 0 & 1 & 2 \end{array}\right]$$

The resulting system is:

$x + 3y = 5$

$y = 2.$

Back-substitute 2 for y to find x.

$$\begin{aligned} x + 3(2) &= 5 \\ x + 6 &= 5 \\ x &= -1 \end{aligned}$$

The solution is $(-1, 2)$ and the solution set is $\{(-1, 2)\}$.

13. $$\begin{vmatrix} 4 & 3 \\ -1 & -5 \end{vmatrix} = 4(-5) - (-1)3 = -20 + 3 = -17$$

14. Let x = the number of rooms with a kitchen

Let y = the number of rooms without a kitchen

$$\begin{aligned} x + y &= 60 \\ 90x + 80y &= 5260 \end{aligned}$$

Solve the first equation for y.

$$\begin{aligned} x + y &= 60 \\ y &= 60 - x \end{aligned}$$

Substitute $60 - x$ for y to find x.

$$\begin{aligned} 90x + 80y &= 5260 \\ 90x + 80(60 - x) &= 5260 \\ 90x + 4800 - 80x &= 5260 \\ 10x + 4800 &= 5260 \\ 10x &= 460 \\ x &= 46 \end{aligned}$$

Back-substitute 46 for x to find y.

$y = 60 - x = 60 - 46 = 14$

There are 46 rooms with kitchens and 14 rooms without kitchens.

15. Using the vertical line test, we see that graphs a. and b. are functions.

16.
$$\begin{aligned} \frac{x}{4} - \frac{3}{4} - 1 &\le \frac{x}{2} \\ 4\left(\frac{x}{4}\right) - 4\left(\frac{3}{4}\right) - 4(1) &\le 4\left(\frac{x}{2}\right) \\ x - 3 - 4 &\le 2x \\ x - 7 &\le 2x \end{aligned}$$

$$x \le 2x + 7$$
$$-x \le 7$$
$$x \ge -7$$

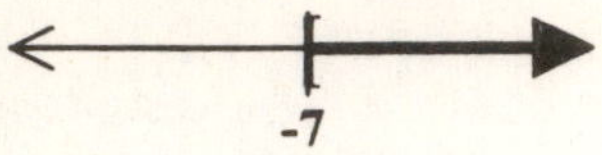

The solution set is $\{x|x \ge -7\}$ or $(-7, \infty)$.

17. $2x + 5 \le 11$ and $-3x > 18$

$2x \le 6$ $\quad x < -6$

$x \le 3$

$x \le 3$

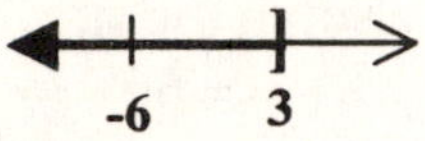

$x < -6$

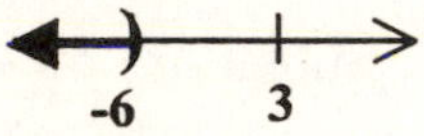

$x \le 3$ and $x < -6$

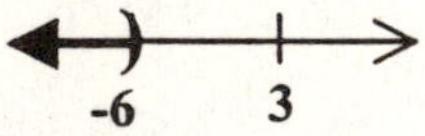

The solution set is $\{x|x < -6\}$ or $(-\infty, -6)$.

18. $x - 4 \ge 1$ or $-3x + 1 \ge -5 - x$

$x \ge 5$ $\quad -2x + 1 \ge -5$

$-2x \ge -6$

$x \le 3$

$x \ge 5$

$x \le 3$

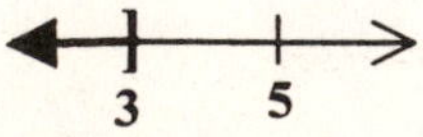

$x \ge 5$ or $x \le 3$

The solution set is $\{x|x \le 3 \text{ or } x \ge 5\}$ or $(-\infty, 3] \cup [5, \infty)$.

19. $|2x + 3| \le 17$

$$-17 \le 2x + 3 \le 17$$
$$-20 \le 2x \le 14$$
$$-10 \le x \le 7$$

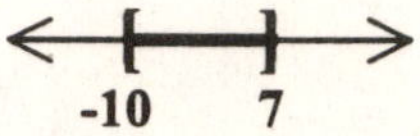

The solution set is $\{x|-10 \le x \le 7\}$ or $[-10, 7]$.

20. $|3x - 8| > 7$

$3x - 8 < -7$ or $3x - 8 > 7$

$3x < 1$ $\quad 3x > 15$

$x < \frac{1}{3}$ $\quad x > 5$

$x < \frac{1}{3}$

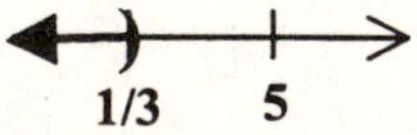

$x > 5$

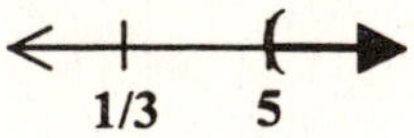

$x < \frac{1}{3}$ or $x > 5$

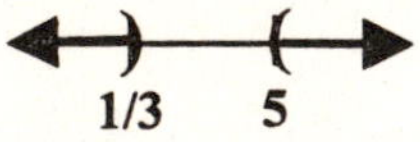

The solution set is $\left\{x \middle| x < \frac{1}{3} \text{ or } x > 5\right\}$ or $\left(-\infty, \frac{1}{3}\right) \cup (5, \infty)$.

Chapter 5

Check Points 5.1

1. The coefficient of $8x^4y^5$ is 8 and the degree is 9.
   The coefficient of $-7x^3y^2$ is -7 and the degree is 5.
   The coefficient of $-x^2y$ is -1 and the degree is 3.
   The coefficient of $-5x$ is -5 and the degree is 1.
   The coefficient of 11 is 11 and the degree is 0.
   The degree of the polynomial is 9.
   The leading term is $8x^4y^5$ and the leading coefficient is 8.

2. $f(x)=4x^3-3x^2-5x+6$
   $$\begin{aligned} f(2) &= 4(2)^3-3(2)^2-5(2)+6 \\ &= 4(8)-3(4)-10+6 \\ &= 32-12-10+6 \\ &= 16 \end{aligned}$$

3. In the polynomial, $f(x)=x^4-4x^2$, the leading term is x^4 with leading coefficient, 1. The degree of the polynomial is 4. This is a polynomial of even-degree with a positive leading coefficient. Applying the Leading Coefficient Test, we know that the graph will rise to the left and to the right.

4. Use the Leading Coefficient Test to determine end behavior. In the polynomial,
   $f(x)=-0.27x^3+9.2x^2-102.9x+400$
   the leading term is $-0.27x^3$ with leading coefficient, -0.27. The degree of the polynomial is 3. This is a polynomial of odd-degree with a negative leading coefficient. Applying the Leading Coefficient Test, we know that the graph will rise to the left and fall to the right. Because the graph falls to the right, at some point the ratio of students to computers will be negative. This is impossible. If a function has a graph that decreases without bound over time, it cannot model non-negative phenomena over long time periods.

5. In the polynomial,
   $f(x)=x^3+13x^2+10x-4$, the leading term is x^3 with leading coefficient, 1. The degree of the polynomial is 3. This is a polynomial of odd-degree with a positive leading coefficient. Applying the Leading Coefficient Test, we know that the graph will fall to the left and rise to the right. The graph in Figure 5.8 rises to the left and right. This means that the graph does not show the end behavior of the polynomial.

6. $$\begin{aligned} &(-7x^3+4x^2+3)+(4x^3+6x^2-13) \\ &= -7x^3+4x^2+3+4x^3+6x^2-13 \\ &= -7x^3+4x^3+4x^2+6x^2+3-13 \\ &= -3x^3+10x^2-10 \end{aligned}$$

7. In addition to the method is Check Point 7, polynomials can be added vertically as shown below. The methods are comparable and the choice of

method is a matter of personal preference.

$$\begin{array}{r} 7xy^3 - 5xy^2 - 3y \\ \underline{2xy^3 + 8xy^2 - 12y - 9} \\ 9xy^3 + 3xy^2 - 15y - 9 \end{array}$$

8. $(14x^3 - 5x^2 + x - 9) - (4x^3 - 3x^2 - 7x + 1)$
$= 14x^3 - 5x^2 + x - 9 - 4x^3 + 3x^2 + 7x - 1$
$= 14x^3 - 4x^3 - 5x^2 + 3x^2 + x + 7x - 9 - 1$
$= 10x^3 - 2x^2 + 8x - 10$

9. $(6x^2y^5 - 2xy^3 - 8) - (-7x^2y^5 - 4xy^3 + 2)$
$= 6x^2y^5 - 2xy^3 - 8 + 7x^2y^5 + 4xy^3 - 2$
$= 6x^2y^5 + 7x^2y^5 - 2xy^3 + 4xy^3 - 8 - 2$
$= 13x^2y^5 + 2xy^3 - 10$

Problem Set 5.1

Practice Exercises

1. The coefficient of $-x^4$ is –1 and the degree is 4.
The coefficient of x^2 is 1 and the degree is 2.
The degree of the polynomial is 4.
The leading term is $-x^4$ and the leading coefficient is –1.

3. The coefficient of $5x^3$ is 5 and the degree is 3.
The coefficient of $7x^2$ is 7 and the degree is 2.
The coefficient of $-x$ is –1 and the degree is 1.
The coefficient of 9 is 9 and the degree is 0.
The degree of the polynomial is 3.
The leading term is $5x^3$ and the leading coefficient is 5.

5. The coefficient of $3x^2$ is 3 and the degree is 2.
The coefficient of $-7x^4$ is –7 and the degree is 4.
The coefficient of $-x$ is –1 and the degree is 1.
The coefficient of 6 is 6 and the degree is 0.
The degree of the polynomial is 4.
The leading term is $-7x^4$ and the leading coefficient is –7.

7. The coefficient of x^3y^2 is 1 and the degree is 5.
The coefficient of $-5x^2y^7$ is –5 and the degree is 9.
The coefficient of $6y^2$ is 6 and the degree is 2.
The coefficient of –3 is –3 and the degree is 0.
The degree of the polynomial is 9.
The leading term is $-5x^2y^7$ and the leading coefficient is –5.

9. The coefficient of x^5 is 1 and the degree is 5.
The coefficient of $3x^2y^4$ is 3 and the degree is 6.
The coefficient of $7xy$ is 7 and the degree is 2.
The coefficient of $9x$ is 9 and the degree is 1.
The coefficient of –2 is –2 and the degree is 0.
The degree of the polynomial is 6.

The leading term is $3x^2y^4$ and the leading coefficient is 3.

11. $f(x) = x^2 - 5x + 6$

$f(3) = (3)^2 - 5(3) + 6$

$= 9 - 15 + 6 = 0$

13. $f(x) = x^2 - 5x + 6$

$f(-1) = (-1)^2 - 5(-1) + 6$

$= 1 + 5 + 6 = 12$

15. $g(x) = 2x^3 - x^2 + 4x - 1$

$g(3) = 2(3)^3 - (3)^2 + 4(3) - 1$

$= 2(27) - 9 + 12 - 1$

$= 54 - 9 + 12 - 1 = 56$

17. $g(x) = 2x^3 - x^2 + 4x - 1$

$g(-2) = 2(-2)^3 - (-2)^2 + 4(-2) - 1$

$= 2(-8) - 4 - 8 - 1$

$= -16 - 4 - 8 - 1$

$= -29$

19. $g(x) = 2x^3 - x^2 + 4x - 1$

$g(0) = 2(0)^3 - (0)^2 + 4(0) - 1$

$= 2(0) - 0 - 0 - 1$

$= 0 - 0 - 0 - 1$

$= -1$

21. Polynomial function.

23. Not a polynomial function.

25. Since the degree of the polynomial is 4, an even number, and the leading coefficient is -1, the graph will fall to the left and to the right. The graph of the polynomial is graph (b).

27. Since the degree of the polynomial is 2, an even number, and the leading coefficient is 1, the graph will rise to the left and to the right. The graph of the polynomial is graph (a).

29.

$$\begin{aligned}(-6x^3 + 5x^2 - 8x + 9) + (17x^3 + 2x^2 - 4x - 13) &= -6x^3 + 5x^2 - 8x + 9 + 17x^3 + 2x^2 - 4x - 13 \\ &= -6x^3 + 17x^3 + 5x^2 + 2x^2 - 8x - 4x + 9 - 13 \\ &= 11x^3 + 7x^2 - 12x - 4\end{aligned}$$

31.

$$\begin{aligned}\left(\frac{2}{5}x^4 + \frac{2}{3}x^3 + \frac{5}{8}x^2 + 7\right) + \left(-\frac{4}{5}x^4 + \frac{1}{3}x^3 - \frac{1}{4}x^2 - 7\right) &= \frac{2}{5}x^4 + \frac{2}{3}x^3 + \frac{5}{8}x^2 + 7 - \frac{4}{5}x^4 + \frac{1}{3}x^3 - \frac{1}{4}x^2 - 7 \\ &= \frac{2}{5}x^4 - \frac{4}{5}x^4 + \frac{2}{3}x^3 + \frac{1}{3}x^3 + \frac{5}{8}x^2 - \frac{1}{4}x^2 + 7 - 7 \\ &= -\frac{2}{5}x^4 + \frac{3}{3}x^3 + \left(\frac{5}{8} - \frac{2}{8}\right)x^2 \\ &= -\frac{2}{5}x^4 + \frac{3}{3}x^3 + \frac{3}{8}x^2 = -\frac{2}{5}x^4 + x^3 + \frac{3}{8}x^2\end{aligned}$$

33. $\left(7x^2y-5xy\right)+\left(2x^2y-xy\right)=7x^2y-5xy+2x^2y-xy$
$=7x^2y+2x^2y-5xy-xy$
$=9x^2y-6xy$

35. $\left(5x^2y+9xy+12\right)+\left(-3x^2y+6xy+3\right)=5x^2y+9xy+12-3x^2y+6xy+3$
$=5x^2y-3x^2y+9xy+6xy+12+3$
$=2x^2y+15xy+15$

37. $\left(9x^4y^2-6x^2y^2+3xy\right)+\left(-18x^4y^2-5x^2y-xy\right)=9x^4y^2-6x^2y^2+3xy-18x^4y^2-5x^2y-xy$
$=9x^4y^2-18x^4y^2-6x^2y^2-5x^2y+3xy-xy$
$=-9x^4y^2-6x^2y^2-5x^2y+2xy$

39. $\left(17x^3-5x^2+4x-3\right)-\left(5x^3-9x^2-8x+11\right)=17x^3-5x^2+4x-3-5x^3+9x^2+8x-11$
$=17x^3-5x^3-5x^2+9x^2+4x+8x-3-11$
$=12x^3+4x^2+12x-14$

41. $\left(13y^5+9y^4-5y^2+3y+6\right)-\left(-9y^5-7y^3+8y^2+11\right)$
$=13y^5+9y^4-5y^2+3y+6+9y^5+7y^3-8y^2-11$
$=13y^5+9y^5+9y^4+7y^3-5y^2-8y^2+3y+6-11$
$=22y^5+9y^4+7y^3-13y^2+3y-5$

43. $\left(x^3+7xy-5y^2\right)-\left(6x^3-xy+4y^2\right)=x^3+7xy-5y^2-6x^3+xy-4y^2$
$=x^3-6x^3+7xy+xy-5y^2-4y^2$
$=-5x^3+8xy-9y^2$

45. $\left(3x^4y^2+5x^3y-3y\right)-\left(2x^4y^2-3x^3y-4y+6x\right)=3x^4y^2+5x^3y-3y-2x^4y^2+3x^3y+4y-6x$
$=3x^4y^2-2x^4y^2+5x^3y+3x^3y-3y+4y-6x$
$=x^4y^2+8x^3y+y-6x$

47. $\left(3a^2b^4 - 5ab^2 + 7ab\right) - \left(-5a^2b^4 - 8ab^2 - ab\right) = 3a^2b^4 - 5ab^2 + 7ab + 5a^2b^4 + 8ab^2 + ab$
$= 3a^2b^4 + 5a^2b^4 - 5ab^2 + 8ab^2 + 7ab + ab$
$= 8a^2b^4 + 3ab^2 + 8ab$

49. $\left(x^3 + 2x^2y - y^3\right) - \left(-4x^3 - x^2y + xy^2 + 3y^3\right) = x^3 + 2x^2y - y^3 + 4x^3 + x^2y - xy^2 - 3y^3$
$= x^3 + 4x^3 + 2x^2y + x^2y - xy^2 - y^3 - 3y^3$
$= 5x^3 + 3x^2y - xy^2 - 4y^3$

Application Exercises

51. $f(x) = 0.036x^2 - 2.8x + 58.14$
$f(50) = 0.036(50)^2 - 2.8(50) + 58.14$
$= 0.036(2500) - 140 + 58.14$
$= 90 - 140 + 58.14 = 8.14$
There are 8.14 deaths per year per thousand people who are 50 years old.

53. The point on the graph is approximately (50, 8.1).

55. At around age 57, approximately 16 people per thousand die each year.

57. $f(x) = -0.001183x^4 + 0.05495x^3 - 0.8523x^2 + 9.054x + 6.748$
$f(10) = -0.001183(10)^4 + 0.05495(10)^3 - 0.8523(10)^2 + 9.054(10) + 6.748$
$= -0.001183(10000) + 0.05495(1000) - 0.8523(100) + 9.054(10) + 6.748$
$= -11.83 + 54.95 - 85.23 + 90.54 + 6.748$
$= 55.178 \approx 55$
If your dog is 10 years old, the equivalent age in human years is approximately 55.

59. The equivalent age for dogs is 2 years.

61. The leading coefficient is –0.75 and the degree is 4. This means that the graph will fall to the right. Over time, the number of particles in the body will go to zero.

63. In the polynomial, $f(x) = -x^4 + 21x^2 + 100$, the leading coefficient is –1 and the degree is 4. Applying the Leading Coefficient Test, we know that even-degree polynomials with negative leading coefficient will fall to the left and to the right. Since the graph falls to the right, we know that the elk population will die out over time.

Writing in Mathematics

65. Answers vary.

67. Answers vary.

69. Answers vary.

71. Answers vary.

73. Answers vary.

75. Answers vary.

Technology Exercises

77. The graph falls to the right and left, so the polynomial must be of even degree with a negative leading coefficient. One example is $-x^4 - 2x^3$.

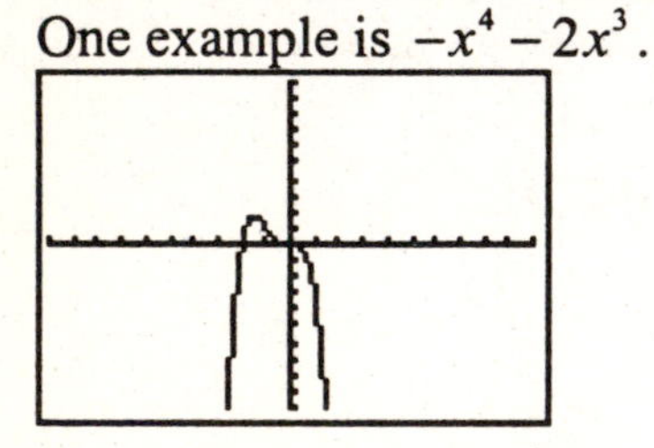

79. The graph falls to the left and rises to the right, so the polynomial must be of odd degree with a positive leading coefficient. One example is $x^3 + 3x^2$.

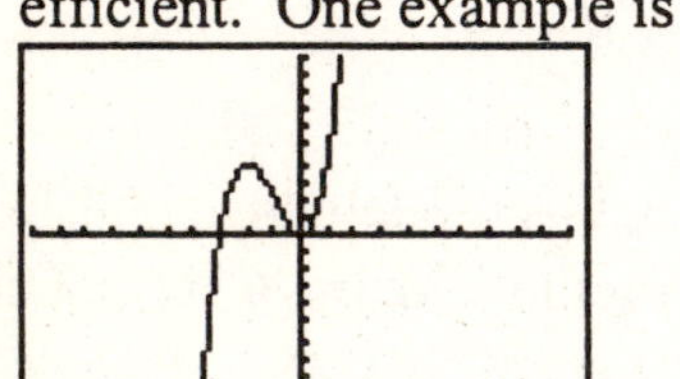

81. $f(x) = -2x^3 + 6x^2 + 3x - 1$

Since we have an odd-degree polynomial with a negative coefficient, the Leading Coefficient Test predicts that the graph will rise to the left and fall to the right.

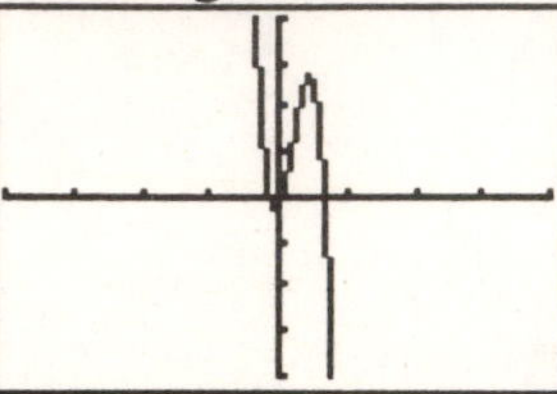

The viewing rectangle large enough to show the end behavior is shown.

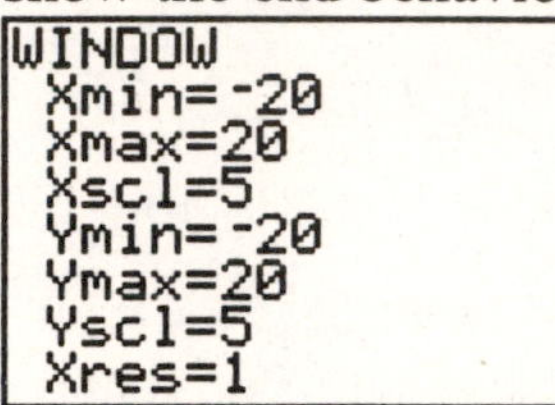

WINDOW
Xmin=-20
Xmax=20
Xscl=5
Ymin=-20
Ymax=20
Yscl=5
Xres=1

83. $f(x) = -x^5 + 5x^4 - 6x^3 + 2x + 20$

Since we have an odd-degree polynomial with a negative coefficient, the Leading Coefficient Test predicts that the graph will rise to the left and fall to the right.

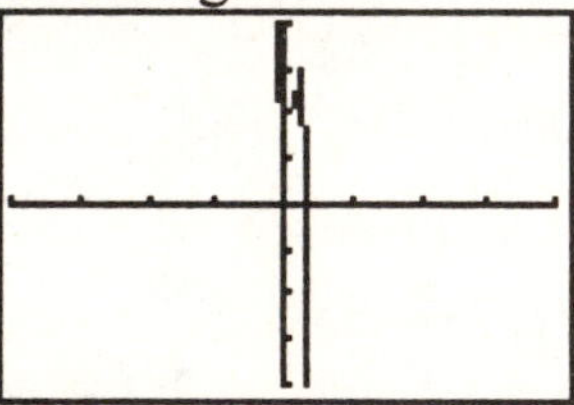

The viewing rectangle large enough to show the end behavior is shown.

WINDOW
Xmin=-40
Xmax=40
Xscl=10
Ymin=-40
Ymax=40
Yscl=10
Xres=1

85. $f(x) = -x^4 + 2x^3 - 6x$

$g(x) = -x^4$

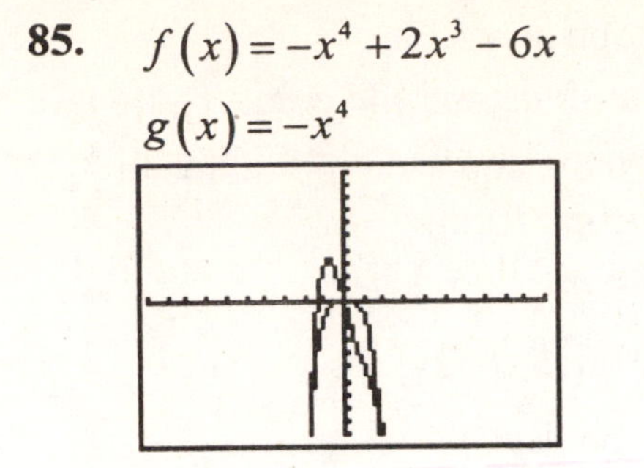

Critical Thinking Exercises

87. $(x^{2n} - 3x^n + 5) + (4x^{2n} - 3x^n - 4) - (2x^{2n} - 5x^n - 3)$

$= x^{2n} - 3x^n + 5 + 4x^{2n} - 3x^n - 4 - 2x^{2n} + 5x^n + 3$

$= x^{2n} + 4x^{2n} - 2x^{2n} - 3x^n - 3x^n + 5x^n + 5 - 4 + 3$

$= 3x^{2n} - x^n - 4$

89. $(4x^2 + 2x - 3) + (5x^2 - 5x + 8)$

$= 4x^2 + 2x - 3 + 5x^2 - 5x + 8$

$= 9x^2 - 3x + 5$

Review Exercises

90. $9(x-1) = 1 + 3(x-2)$

$9x - 9 = 1 + 3x - 6$

$9x - 9 = 3x - 5$

$6x - 9 = -5$

$6x = 4$

$x = \frac{4}{6} = \frac{2}{3}$

The solution is $\frac{2}{3}$ and the solution set is $\left\{\frac{2}{3}\right\}$.

91. $2x - 3y < -6$

Replacing the inequality symbol with an equal sign, we have $2x - 3y = -6$. Solve for y to obtain slope-intercept form.

$2x - 3y = -6$

$-3y = -2x - 6$

$y = \frac{2}{3}x + 2$

The slope is $\frac{2}{3}$ and the y–intercept is 2. Next, use the origin as a test point.

$2(0) - 3(0) < -6$

$0 < -6$

This is a false statement. This means that the point, $(0,0)$, will not fall in the shaded half-plane.

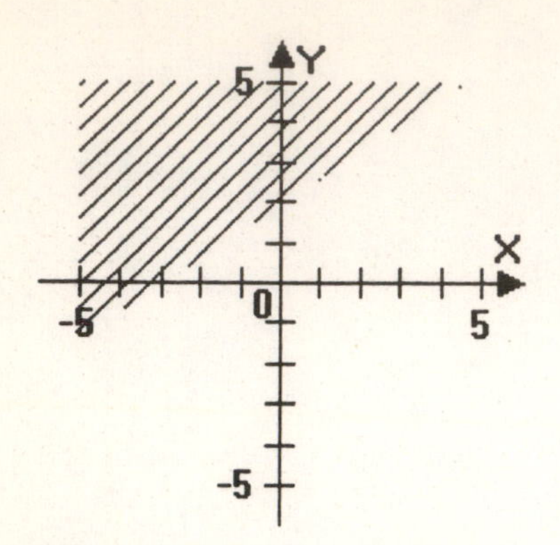

92. Since the line is parallel to $3x - y = 9$, it will have the same slope. To find the slope, put the equation is slope intercept form.

$$3x - y = 9$$
$$-y = -3x + 9$$
$$y = 3x - 9$$

The slope is $m = 3$.

Use the slope and the point, $(-2, 5)$, to write the equation of the line in point-slope form.

$$y - y_1 = m(x - x_1)$$
$$y - 5 = 3(x - (-2))$$
$$y - 5 = 3(x + 2)$$

Solve for y to obtain slope-intercept form.

$$y - 5 = 3x + 6$$
$$y = 3x + 11$$

Check Points 5.2

1. **a.** $(6x^5y^7)(-3x^2y^4)$
$$= 6(-3)x^5x^2y^7y^4$$
$$= -18x^{5+2}y^{7+4}$$
$$= -18x^7y^{11}$$

b. $(10x^4y^3z^6)(3x^6y^3z^2)$
$$= 10(3)x^4x^6y^3y^3z^6z^2$$
$$= 30x^{4+6}y^{3+3}z^{6+2}$$
$$= 30x^{10}y^6z^8$$

2. **a.** $6x^4(2x^5 - 3x^2 + 4)$
$$= 6x^4 \cdot 2x^5 - 6x^4 \cdot 3x^2 + 6x^4 \cdot 4$$
$$= 12x^9 - 18x^6 + 24x^4$$

b. $2x^4y^3(5xy^6 - 4x^3y^4 - 5)$
$$= 2x^4y^3 \cdot 5xy^6 - 2x^4y^3 \cdot 4x^3y^4 - 2x^4y^3 \cdot 5$$
$$= 10x^5y^9 - 8x^7y^7 - 10x^4y^3$$

3.
$$\begin{array}{r} x^2 - 4x + 3 \\ 5x + 2 \\ \hline 5x^3 - 20x^2 + 15x \\ 2x^2 - \ \ 8x + 6 \\ \hline 5x^3 - 18x^2 + \ \ 7x + 6 \end{array}$$

4.
$$
\begin{array}{r}
3xy^4 - 2xy^2 + \quad y \\
4xy^2 + 2y \\
\hline
12x^2y^6 - 8x^2y^4 + 4xy^3 \qquad\qquad\qquad \\
-4xy^3 + 6xy^5 + 2y^2 \\
\hline
12x^2y^6 - 8x^2y^4 \qquad\qquad + 6xy^5 + 2y^2
\end{array}
$$

5. **a.** $(x+5)(x+3) = x^2 + 3x + 5x + 15$
$= x^2 + 8x + 15$

b. $(7x+4y)(2x-y) = 7x \cdot 2x + 7x(-y) + 4y \cdot 2x + 4y(-y)$
$= 14x^2 - 7xy + 8xy - 4y^2 = 14x^2 + xy - 4y^2$

c. $(4x^3 - 5)(x^3 - 3x)$
$= 4x^3 \cdot x^3 + 4x^3(-3x) + (-5)x^3 + (-5)(-3x)$
$= 4x^6 - 12x^4 - 5x^3 + 15x$

6. **a.** $(x+10)^2 = x^2 + 2 \cdot x \cdot 10 + 10^2$
$= x^2 + 20x + 100$

b. $(4x+5y)^2 = (4x)^2 + 2 \cdot 4x \cdot 5y + (5y)^2$
$= 16x^2 + 40xy + 25y^2$

7. **a.** $(x-5)^2 = x^2 + 2 \cdot x \cdot (-5) + (-5)^2$
$= x^2 - 10x + 25$

b. $(2x - 6y^4)^2 = (2x)^2 + 2 \cdot 2x \cdot (-6y^4) + (-6y^4)^2$
$= 4x^2 - 24xy^4 + 36y^8$

8. **a.** $(x+3)(x-3) = x^2 - 3^2$
$= x^2 - 9$

b. $(5x+7y)(5x-7y) = (5x)^2 - (7y)^2$
$= 25x^2 - 49y^2$

c. $(5ab^2 - 4a)(5ab^2 + 4a) = (5ab^2)^2 - (4a)^2$
$= 25a^2b^4 - 16a^2$

9. **a.** $(3x+2+5y)(3x+2-5y)$
$= ((3x+2)+5y)((3x+2)-5y)$
$= (3x+2)^2 - (5y)^2$
$= (3x)^2 + 2 \cdot 3x \cdot 2 + 2^2 - 25y^2$
$= 9x^2 + 12x + 4 - 25y^2$

b. $(2x+y+3)^2$
$= ((2x+y)+3)^2$
$= (2x+y)^2 + 2 \cdot (2x+y) \cdot 3 + 3^2$
$= (2x)^2 + 2 \cdot 2x \cdot y + y^2 + 6(2x+y) + 9$
$= 4x^2 + 4xy + y^2 + 12x + 6y + 9$

10. $f(x) = x - 3$ and $g(x) = x - 7$

a. $(fg)(x) = f(x) \cdot g(x)$
$= (x-3)(x-7)$
$= x^2 - 7x - 3x + 21$
$= x^2 - 10x + 21$

b. $(fg)(x) = x^2 - 10x + 21$
$(fg)(2) = (2)^2 - 10(2) + 21$
$= 4 - 20 + 21$
$= 5$

11. a. $f(x) = x^2 - 5x + 4$
$f(a+3) = (a+3)^2 - 5(a+3) + 4$
$= a^2 + 6a + 9 - 5a - 15 + 4$
$= a^2 + a - 2$

b. $f(a+h) - f(a) = \left[(a+h)^2 - 5(a+h) + 4\right] - \left[a^2 - 5a + 4\right]$
$= \left[a^2 + 2ah + h^2 - 5a - 5h + 4\right] - a^2 + 5a - 4$
$= \cancel{a^2} + 2ah + h^2 \cancel{-5a} - 5h + \cancel{4} - \cancel{a^2} + \cancel{5a} - \cancel{4}$
$= 2ah + h^2 - 5h$

Problem Set 5.2

Practice Exercises

1. $(3x^2)(5x^4)$
$= 3(5)x^2 \cdot x^4$
$= 15x^{2+4}$
$= 15x^6$

3. $(3x^2y^4)(5xy^7)$
$= 3(5)x^2 \cdot x \cdot y^4 \cdot y^7$
$= 15x^{2+1}y^{4+7}$
$= 15x^3y^{11}$

5. $(-3xy^2z^5)(2xy^7z^4)$
$= -3(2)x \cdot x \cdot y^2 \cdot y^7 \cdot z^5 \cdot z^4$
$= -6x^{1+1}y^{2+7}z^{5+4}$
$= -6x^2y^9z^9$

7. $4x^2(3x+2) = 4x^2 \cdot 3x + 4x^2 \cdot 2$
$= 12x^3 + 8x^2$

9. $2y(y^2 - 5y) = 2y \cdot y^2 - 2y \cdot 5y$
$= 2y^3 - 10y^2$

11. $5x^3(2x^5 - 4x^2 + 9)$
$= 5x^3 \cdot 2x^5 - 5x^3 \cdot 4x^2 + 5x^3 \cdot 9$
$= 10x^8 - 20x^5 + 45x^3$

13. $4xy(7x+3y) = 4xy \cdot 7x + 4xy \cdot 3y$
$= 28x^2y + 12xy^2$

15. $3ab^2(6a^2b^3 + 5ab)$
$= 3ab^2 \cdot 6a^2b^3 + 3ab^2 \cdot 5ab$
$= 18a^3b^5 + 15a^2b^3$

17. $-4x^2y(3x^4y^2 - 7xy^3 + 6)$
$= -4x^2y \cdot 3x^4y^2 + 4x^2y \cdot 7xy^3 - 4x^2y \cdot 6$
$= -12x^6y^3 + 28x^3y^4 - 24x^2y$

19.
$$\begin{array}{r} x^2 + 2x + 5 \\ \underline{x - 3} \\ x^3 + 2x^2 + 5x \\ \underline{-3x^2 - 6x - 15} \\ x^3 - x^2 - x - 15 \end{array}$$

21.
$$\begin{array}{r} x^2 + x + 1 \\ \underline{x - 1} \\ x^3 + x^2 + x \\ \underline{-x^2 - x - 1} \\ x^3 \qquad -1 \end{array}$$

23.
$$\begin{array}{r} a^2 + ab + b^2 \\ \underline{a - b} \\ a^3 + a^2b + ab^2 \\ \underline{-a^2b - ab^2 - b^3} \\ a^3 \qquad -b^3 \end{array}$$

25.
$$\begin{array}{r} x^2 + 2x - 1 \\ \underline{x^2 + 3x - 4} \\ x^4 + 2x^3 - x^2 \\ 3x^3 + 6x^2 - 3x \\ \underline{-4x^2 - 8x + 4} \\ x^4 + 5x^3 + x^2 - 11x + 4 \end{array}$$

27.
$$\begin{array}{r} x^2 - 3xy + y^2 \\ \underline{x - y} \\ x^3 - 3x^2y + xy^2 \\ \underline{-x^2y + 3xy^2 - y^3} \\ x^3 - 4x^2y + 4xy^2 - y^3 \end{array}$$

29.
$$\begin{array}{r} x^2y^2 - 2xy + 4 \\ \underline{xy + 2} \\ x^3y^3 - 2x^2y^2 + 4xy \\ \underline{2x^2y^2 - 4xy + 8} \\ x^3y^3 \qquad +8 \end{array}$$

31. $(x+4)(x+7) = x^2 + 7x + 4x + 28$
$= x^2 + 11x + 28$

33. $(y+5)(y-6) = y^2 - 6y + 5y - 30$
$= y^2 - y - 30$

35. $(5x+3)(2x+1) = 10x^2 + 5x + 6x + 3$
$= 10x^2 + 11x + 3$

37. $(3y-4)(2y-1) = 6y^2 - 3y - 8y + 4$
$= 6y^2 - 11y + 4$

39. $(3x-2)(5x-4)$
$=15x^2-12x-10x+8$
$=15x^2-22x+8$

41. $(x-3y)(2x+7y)$
$=2x^2+7xy-6xy-21y^2$
$=2x^2+xy-21y^2$

43. $(7xy+1)(2xy-3)$
$=14x^2y^2-21xy+2xy-3$
$=14x^2y^2-19xy-3$

45. $(x-4)(x^2-5)=x^3-5x-4x^2+20$
$=x^3-4x^2-5x+20$

47. $(8x^3+3)(x^2-5)$
$=8x^5-40x^3+3x^2-15$

49. $(x+3)^2=x^2+2(3x)+9$
$=x^2+6x+9$

51. $(y-5)^2=y^2+2(-5y)+25$
$=y^2-10y+25$

53. $(2x+y)^2=4x^2+2(2xy)+y^2$
$=4x^2+4xy+y^2$

55. $(5x-3y)^2=25x^2+2(-15xy)+9y^2$
$=25x^2-30xy+9y^2$

57. $(2x^2+3y)^2=4x^4+2(6x^2y)+9y^2$
$=4x^4+12x^2y+9y^2$

59. $(4xy^2-xy)^2$
$=16x^2y^4+2(4x^2y^3)+x^2y^2$
$=16x^2y^4+8x^2y^3+x^2y^2$

61. $(x+4)(x-4)=x^2-16$

63. $(5x+3)(5x-3)=25x^2-9$

65. $(4x+7y)(4x-7y)=16x^2-49y^2$

67. $(y^3+2)(y^3-2)=y^6-4$

69. $(1-y^5)(1+y^5)=1-y^{10}$

71. $(7xy^2-10y)(7xy^2+10y)$
$=49x^2y^4-100y^2$

73. $[(2x+3)+4y][(2x+3)-4y]$
$=(2x+3)^2-16y^2$
$=4x^2+2(6x)+9-16y^2$
$=4x^2+12x+9-16y^2$

75. $(x+y+3)(x+y-3)$
$=((x+y)+3)((x+y)-3)$
$=(x+y)^2-9$
$=x^2+2xy+y^2-9$

77. $(5x+7y-2)(5x+7y+2)$
$=((5x+7y)-2)((5x+7y)+2)$
$=(5x+7y)^2-4$

$= 25x^2 + 2(35xy) + 49y^2 - 4$
$= 25x^2 + 70xy + 49y^2 - 4$

79. $[5y + (2x+3)][5y - (2x+3)]$
$= 25y^2 - (2x+3)^2$
$= 25y^2 - (4x^2 + 2(6x) + 9)$
$= 25y^2 - (4x^2 + 12x + 9)$
$= 25y^2 - 4x^2 - 12x - 9$

81.
$$\begin{array}{l} x+y+1 \\ \underline{x+y+1} \\ x^2+xy \quad +x \\ \quad xy \qquad +y^2+y \\ \underline{\qquad\quad x \qquad +y \ +1} \\ x^2+2xy+2x+y^2+2y+1 \end{array}$$
or
$x^2 + 2xy + y^2 + 2x + 2y + 1$

83. $(x+1)(x-1)(x^2+1)$
$= (x^2-1)(x^2+1) = x^4 - 1$

85. **a.** $(fg)(x) = f(x) \cdot g(x)$
$= (x-2)(x+6)$
$= x^2 + 6x - 2x - 12$
$= x^2 + 4x - 12$

b. $(fg)(-1) = (-1)^2 + 4(-1) - 12$
$= 1 - 4 - 12 = -15$

c. $(fg)(0) = (0)^2 + 4(0) - 12$
$= 0 + 0 - 12 = -12$

87. **a.** $(fg)(x)$
$= f(x) \cdot g(x)$
$= (x-3)(x^2+3x+9)$
$= x(x^2+3x+9) - 3(x^2+3x+9)$
$= x^3 + 3x^2 + 9x - 3x^2 - 9x - 27$
$= x^3 - 27$

b. $(fg)(-2) = (-2)^3 - 27$
$= -8 - 27 = -35$

c. $(fg)(0) = (0)^3 - 27 = -27$

89. **a.** $f(a+2) = (a+2)^2 - 3(a+2) + 7$
$= a^2 + 4a + 4 - 3a - 6 + 7 = a^2 + a + 5$

b. $f(a+h) - f(a) = (a+h)^2 - 3(a+h) + 7 - (a^2 - 3a + 7)$
$= \cancel{a^2} + 2ah + h^2 - \cancel{3a} - 3h + \cancel{7} - \cancel{a^2} + \cancel{3a} - \cancel{7}$
$= 2ah + h^2 - 3h$

91. **a.**
$$\begin{aligned} f(a+2) &= 3(a+2)^2 + 2(a+2) - 1 \\ &= 3(a^2+4a+4) + 2a + 4 - 1 \\ &= 3a^2 + 12a + 12 + 2a + 4 - 1 \\ &= 3a^2 + 14a + 15 \end{aligned}$$

b.
$$\begin{aligned} f(a+h) - f(a) &= 3(a+h)^2 + 2(a+h) - 1 - (3a^2 + 2a - 1) \\ &= 3(a^2 + 2ah + h^2) + 2a + 2h - 1 - 3a^2 - 2a + 1 \\ &= \cancel{3a^2} + 6ah + 3h^2 + \cancel{2a} + 2h - \cancel{1} - \cancel{3a^2} - \cancel{2a} + \cancel{1} \\ &= 6ah + 3h^2 + 2h \end{aligned}$$

Application Exercises

93. **a.** $x^2 + 6x + 4x + 24 = x^2 + 10x + 24$

b.
$$\begin{aligned} (x+6)(x+4) &= x^2 + 4x + 6x + 24 \\ &= x^2 + 10x + 24 \end{aligned}$$

95. **a.**
$$\begin{aligned} (x+9)(x+3) &= x^2 + 3x + 9x + 27 \\ &= x^2 + 12x + 27 \end{aligned}$$

b.
$$\begin{aligned} (x+5)(x+1) &= x^2 + x + 5x + 5 \\ &= x^2 + 6x + 5 \end{aligned}$$

c.
$$\begin{aligned} &(x^2 + 12x + 27) - (x^2 + 6x + 5) \\ &= x^2 + 12x + 27 - x^2 - 6x - 5 \\ &= 6x + 22 \end{aligned}$$

97. **a.**
$$\begin{aligned} (8-2x)(10-2x) &= 80 - 16x - 20x + 4x^2 \\ &= 80 - 36x + 4x^2 \\ &= 4x^2 - 36x + 80 \end{aligned}$$

b.
$$\begin{aligned} &x(80 - 36x + 2x^2) \\ &= 80x - 36x^2 + 4x^3 \\ &= 4x^3 - 36x^2 + 80x \end{aligned}$$

Writing in Mathematics

99. Answers will vary.

101. Answers will vary.

103. Answers will vary.

105. Answers will vary.

107. Answers will vary.

Technology Exercises

109.

The functions y_1 and y_2 are the same. Verify by multiplying.

$$\begin{array}{r} x^2-3x+2 \\ x-4 \\ \hline x^3-3x^2+2x \\ -4x^2+12x-8 \\ \hline x^3-7x^2+14x-8 \end{array}$$

111.

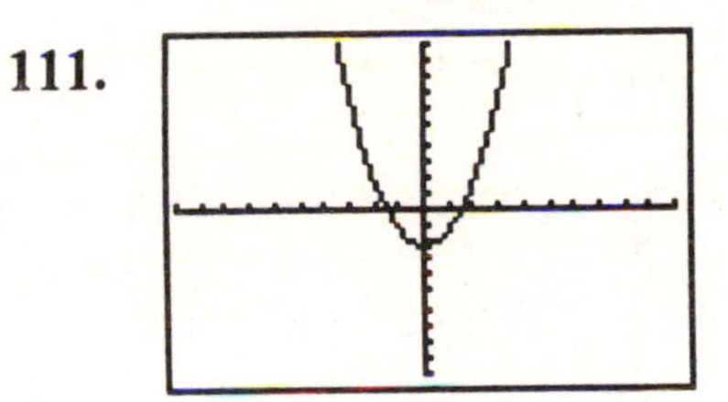

The functions y_1 and y_2 are the same. Verify by multiplying.

$(x+1.5)(x-1.5)=x^2-2.25$

Critical Thinking Exercises

113. Statement **d.** is true.

$$\begin{aligned}&(3x+7)(3x-2)\\&=9x^2-6x+21x-14\\&=9x^2+15x-14\end{aligned}$$

Statement **a.** is false. Do not confuse $(f+g)(x)=f(x)+g(x)$ with $(f)(a+h)=f(a)+f(h)$.

Statement **b.** is false.

$$\begin{aligned}(x-5)^2&=x^2-2\cdot x\cdot 5+(-5)^2\\&=x^2-10x+25\end{aligned}$$

Statement **c.** is false.

$$\begin{aligned}(x+1)^2&=x^2+2\cdot x\cdot 1+1^2\\&=x^2+2x+1\end{aligned}$$

115. First divide the figure into two rectangular solids, one tall and one short. To find the total volume, we use $V_{\text{total}}=V_{\text{tall}}+V_{\text{short}}$. The only measurement that is missing is the length of the taller rectangular solid. Since the distance across the entire figure is $x+3$, and the distance across the shorter rectangular solid is 3, the distance across the taller rectangular solid is $x+3-3=x$.

$$\begin{aligned}V_{\text{total}}&=V_{\text{tall}}+V_{\text{short}}\\&=l_t w_t h_t+l_s w_s h_s\\&=3\cdot x\cdot(2x-1)+x\cdot x\cdot(x+1)\\&=3x(2x-1)+x^2(x+1)\\&=6x^2-3x+x^3+x^2\\&=x^3+7x^2-3x\end{aligned}$$

117.

$$\begin{aligned}&(y^n+2)(y^n-2)-(y^n-3)^2\\&=y^{2n}-4-\left(y^{2n}-2(3y^n)+9\right)\\&=y^{2n}-4-\left(y^{2n}-6y^n+9\right)\\&=y^{2n}-4-y^{2n}+6y^n-9\\&=6y^n-13\end{aligned}$$

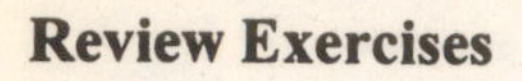

Review Exercises

118. $|3x+4| \geq 10$

$$3x+4 \leq -10 \qquad 3x+4 \geq 10$$
$$3x \leq -14 \quad \text{or} \quad 3x \geq 6$$
$$x \leq -\frac{14}{3} \qquad x \geq 2$$

The solution set is $\left\{x \middle| x \leq -\frac{14}{3} \text{ or } x \geq 2\right\}$ or $\left(-\infty, -\frac{14}{3}\right] \cup [2, \infty)$.

119. $2-6x \leq 20$

$$-6x \leq 18$$
$$x \geq -3$$

The solution set is $\{x | x \geq -3\}$ or $[-3, \infty)$.

120. $8,034,000,000 = 8.034 \times 10^9$

Check Points 5.3

1. $20x^2 + 30x = 10x \cdot 2x + 10x \cdot 3$
$= 10x(2x+3)$

2. a. $9x^4 + 21x^2 = 3x^2 \cdot 3x^2 + 3x^2 \cdot 7$
$= 3x^2(3x^2+7)$

b. $15x^3y^2 - 25x^4y^3$
$= 5x^3y^2 \cdot 3 - 5x^3y^2 \cdot 5xy$
$= 5x^3y^2(3-5xy)$

c. $16x^4y^5 - 8x^3y^4 + 4x^2y^3$
$= 4x^2y^3 \cdot 4x^2y^2 - 4x^2y^3 \cdot 2xy + 4x^2y^3 \cdot 1$
$= 4x^2y^3(4x^2y^2 - 2xy + 1)$

3. $-2x^3 + 10x^2 - 6x$
$= -2x \cdot x^2 - 2x \cdot (-5x) - 2x \cdot 3$
$= -2x(x^2 - 5x + 3)$

4. a. $3(x-4) + 7a(x-4)$
$= (x-4)(3+7a)$

b. $7x(a+b) - (a+b)$
$= (a+b)(7x-1)$

5. $x^3 - 4x^2 + 5x - 20$
$= (x^3 - 4x^2) + (5x - 20)$
$= x^2(x-4) + 5(x-4)$
$= (x-4)(x^2+5)$

6. $4x^2 + 20x - 3xy - 15y$
$= (4x^2 + 20x) + (-3xy - 15y)$
$= 4x(x+5) + (-3y)(x+5)$
$= (x+5)(4x + (-3y))$
$= (x+5)(4x - 3y)$

Problem Set 5.3

Practice Exercises

1. $10x^2+4x=2x\cdot 5x+2x\cdot 2$
$=2x(5x+2)$

3. $y^2-4y=y\cdot y-y\cdot 4=y(y-4)$

5. $x^3+5x^2=x^2\cdot x+x^2\cdot 5=x^2(x+5)$

7. $12x^4-8x^2=4x^2\cdot 3x^2-4x^2\cdot 2$
$=4x^2(3x^2-2)$

9. $32x^4+2x^3+8x^2$
$=2x^2\cdot 16x^2+2x^2\cdot x+2x^2\cdot 4$
$=2x^2(16x^2+x+4)$

11. $4x^2y^3+6xy=2xy\cdot 2xy^2+2xy\cdot 3$
$=2xy(2xy^2+3)$

13. $30x^2y^3-10xy^2$
$=10xy^2\cdot 3xy-10xy^2\cdot 1$
$=10xy^2(3xy-1)$

15. $12xy-6xz+4xw$
$=2x\cdot 6y-2x\cdot 3z+2x\cdot 2w$
$=2x(6y-3z+2w)$

17. $15x^3y^6-9x^4y^4+12x^2y^5$
$=3x^2y^4\cdot 5xy^2-3x^2y^4\cdot 3x^2+3x^2y^4\cdot 4y$
$=3x^2y^4(5xy^2-3x^2+4y)$

19. $-4x+12=-4\cdot x+(-4)(-3)$
$=-4(x-3)$

21. $-8x-48=-8\cdot x+(-8)6$
$=-8(x+6)$

23. $-2x^2+6x-14$
$=-2\cdot x^2+(-2)(-3x)+(-2)(7)$
$=-2(x^2+(-3x)+7)$
$=-2(x^2-3x+7)$

25. $-5y^2+40x=-5\cdot y^2+(-5)(-8x)$
$=-5(y^2+(-8x))$
$=-5(y^2-8x)$

27. $-4x^3+32x^2-20x$
$=-4x\cdot x^2+(-4x)(-8x)+(-4x)(5)$
$=-4x(x^2+(-8x)+5)$
$=-4x(x^2-8x+5)$

29. $-x^2-7x+5$
$=-1\cdot x^2+(-1)(7x)+(-1)(-5)$
$=-1(x^2+7x+(-5))$
$=-1(x^2+7x-5)$

31. $4(x+3)+a(x+3)$
$=(x+3)(4+a)$

33. $x(y-6)-7(y-6)$
$=(y-6)(x-7)$

35. $3x(x+y)-(x+y)$
$=3x(x+y)-1(x+y)$
$=(x+y)(3x-1)$

37. $4x^2(3x-1)+3x-1$
$=4x^2(3x-1)+1(3x-1)$
$=(3x-1)(4x^2+1)$

39. $(x+2)(x+3)+(x-1)(x+3)$
$=(x+3)(x+2+x-1)$
$=(x+3)(2x+1)$

41. $x^2+3x+5x+15$
$=x(x+3)+5(x+3)$
$=(x+3)(x+5)$

43. $x^2+7x-4x-28$
$=x(x+7)-4(x+7)$
$=(x+7)(x-4)$

45. $x^3-3x^2+4x-12$
$=x^2(x-3)+4(x-3)$
$=(x-3)(x^2+4)$

47. $xy-6x+2y-12$
$=x(y-6)+2(y-6)$
$=(y-6)(x+2)$

49. $xy+x-7y-7$
$=x(y+1)-7(y+1)$
$=(y+1)(x-7)$

51. $10x^2-12xy+35xy-42y^2$
$=2x(5x-6y)+7y(5x-6y)$
$=(5x-6y)(2x+7y)$

53. $4x^3-x^2-12x+3$
$=x^2(4x-1)-3(4x-1)$
$=(4x-1)(x^2-3)$

55. $x^2-ax-bx+ab$
$=x(x-a)-b(x-a)$
$=(x-a)(x-b)$

57. $x^3-12-3x^2+4x$
$=x^3-3x^2+4x-12$
$=x^2(x-3)+4(x-3)$
$=(x-3)(x^2+4)$

Application Exercises

59. $ay-by+bx-ax$
$=y(a-b)+x(b-a)$
$=y(a-b)+x(-1)(a-b)$
$=y(a-b)-x(a-b)$
$=(a-b)(y-x)$

61. $f(t)=-16t^2+40t$

a. $f(2)=-16(2)^2+40(2)$
$=-16(4)+80$
$=-64+80$
$=16$

After 2 seconds, the ball will be at a height of 16 feet.

b. $f(2.5) = -16(2.5)^2 + 40(2.5)$
$= -16(6.25) + 100$
$= -100 + 100 = 0$
After 2.5 seconds, the ball will hit the ground.

c. $-16t^2 + 40t = -4t \cdot 4t - 4t(-10)$
$= -8t(2t - 5)$

d. $f(t) = -4t(4t - 10)$
$f(2) = -4(2)(4(2) - 10)$
$= -8(8 - 10)$
$= -8(-2) = 16$
$f(2.5) = -4(2.5)(4(2.5) - 10)$
$= -10(10 - 10)$
$= -10(0) = 0$
These are the same answers as parts **a.** and **b.** This shows that the factorization is equivalent to the original polynomial and that the factorization is correct.

63. **a.** $(x - 0.4x) - 0.4(x - 0.4x)$
$= (x - 0.4x)(1 - 0.4)$
$= (0.6x)(0.6)$
$= 0.36x$

b. The computer is selling at 36% of its original price.

65. $A = P + Pr + (P + Pr)r$
$= (P + Pr)1 + (P + Pr)r$
$= (P + Pr)(1 + r)$
$= (P \cdot 1 + P \cdot r)(1 + r)$
$= P(1 + r)(1 + r)$
$= P(1 + r)^2$

67. $A = \pi r^2 + 2rl$
$= r \cdot \pi r + r \cdot 2l$
$= r(\pi r + 2l)$

Writing in Mathematics

69. Answers will vary.

71. Answers will vary.

73. Answers will vary.

Technology Exercises

75. $x^2 - 4x = x(x - 4)$

The graphs coincide. The polynomial is factored correctly.

77. $x^2 + 2x + x + 2 = x(x + 2) + 1$

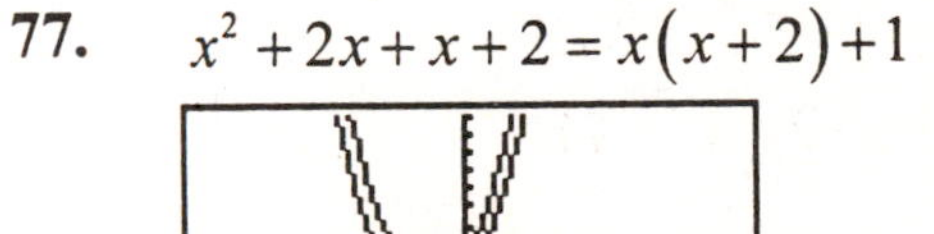

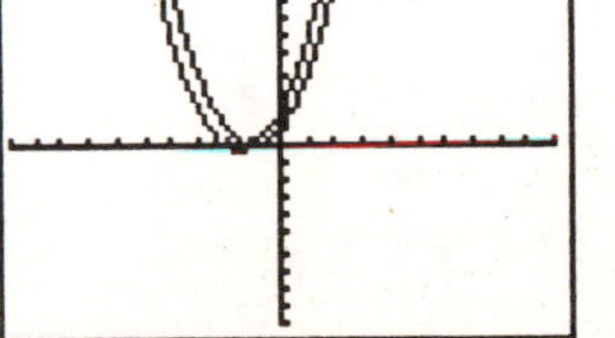

The graphs do not coincide. Factor the polynomial correctly.
$x^2 + 2x + x + 2 = x(x + 2) + 1(x + 2)$
$= (x + 2)(x + 1)$

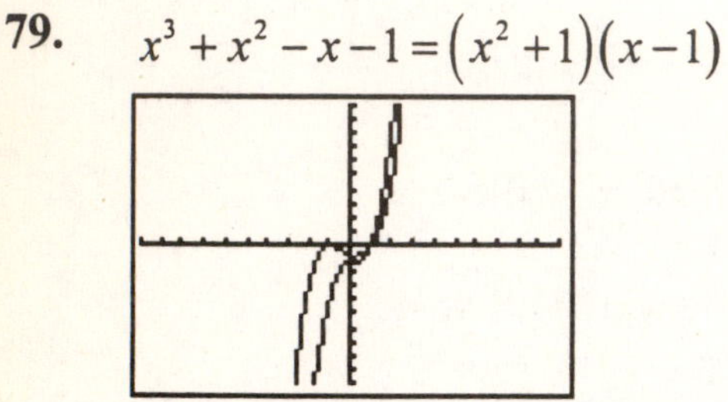

The new factorization is correct.
The graphs coincide.

79. $x^3 + x^2 - x - 1 = (x^2 + 1)(x - 1)$

The graphs do not coincide. Factor the polynomial correctly.

$$x^3 + x^2 - x - 1 = x^2(x+1) - 1(x+1)$$
$$= (x+1)(x^2 - 1)$$
$$= (x+1)(x+1)(x-1)$$
$$= (x+1)^2(x-1)$$

The new factorization is correct.
The graphs coincide.

Critical Thinking Exercises

81. $x^{4n} + x^{2n} + x^{3n}$
$$= x^{2n} \cdot x^{2n} + x^{2n} \cdot 1 + x^{2n} \cdot x^n$$
$$= x^{2n}(x^{2n} + 1 + x^n)$$

83. $8y^{2n+4} + 16^{2n+3} - 12y^{2n}$
$$= 4y^{2n} \cdot 2y^4 + 4y^{2n} \cdot 4y^3 - 4y^{2n} \cdot 3$$
$$= 4y^{2n}(2y^4 + 4y^3 - 3)$$

85. Answers will vary. One example is $2a^2b + 2ab^2 - 3abc - 3b^2c$.

Review Exercises

86. $3x - 2y = 8$
$2x - 5y = 10$

Because $x = \frac{D_x}{D}$ and $y = \frac{D_y}{D}$, we need to set up and evaluate D, D_x, and D_y.

$$D = \begin{vmatrix} 3 & -2 \\ 2 & -5 \end{vmatrix} = (3)(-5) - (2)(-2)$$
$$= -15 + 4 = -11$$
$$D_x = \begin{vmatrix} 8 & -2 \\ 10 & -5 \end{vmatrix} = (8)(-5) - (10)(-2)$$
$$= -40 + 20 = -20$$
$$D_y = \begin{vmatrix} 3 & 8 \\ 2 & 10 \end{vmatrix} = (3)(10) - (2)(8)$$
$$= 30 - 16 = 14$$
$$x = \frac{D_x}{D} = \frac{-20}{-11} = \frac{20}{11}$$
$$y = \frac{D_y}{D} = \frac{14}{-11} = -\frac{14}{11}$$

The solution is $\left(\frac{20}{11}, -\frac{14}{11}\right)$ and the solution set is $\left\{\left(\frac{20}{11}, -\frac{14}{11}\right)\right\}$.

87. **a.** The relation is a function.

b. The relation is not a function.

88. Let w = the width of the rectangle
Let $2w + 2$ = the length of the rectangle

$$P = 2l + 2w$$
$$22 = 2(2w+2) + 2w$$
$$22 = 4w + 4 + 2w$$
$$22 = 6w + 4$$
$$18 = 6w$$
$$3 = w$$

To find the length:

$$2w + 2 = 2(3) + 2 = 6 + 2 = 8$$

The length is 8 feet and the width is 3 feet.

Check Points 5.4

1. $x^2 + 6x + 8 = (x+4)(x+2)$

or

$= (x+2)(x+4)$

product: $4(2) = 8$
sum: $4 + 2 = 6$

2. $x^2 - 9x + 20 = (x-5)(x-4)$
product: $-5(-4) = 20$
sum: $-5 + (-4) = -9$

3. $y^2 + 19y - 66 = (y-3)(y+22)$
product: $-3(22) = -66$
sum: $-3 + 22 = 19$

4. $x^2 - 5xy + 6y^2 = (x-3y)(x-2y)$
product: $-3(-2) = 6$
sum: $-3 + (-2) = -5$

5. $3x^3 - 15x^2 - 42x$
First, factor out the common monomial. Then factor the remaining trinomial (if possible).

$$3x^3 - 15x^2 - 42x = 3x(x^2 - 5x - 14)$$
$$= 3x(x-7)(x+2)$$

Check the factoring of the polynomial.
product: $-7(2) = -14$
sum: $-7 + 2 = -5$

6. Let $t = x^3$

$$x^6 - 7x^3 + 10$$
$$= (x^3)^2 - 7x^3 + 10$$
$$= t^2 - 7t + 10$$
$$= (t-5)(t-2)$$

Substitute x^3 for t.

$$= (x^3 - 5)(x^3 - 2)$$

7. $3x^2 - 20x + 28 = (3x-14)(x-2)$
Check:

$$(3x)(x) = 3x^2$$
$$(3x)(-2) + (-14)(x) = -6x - 14x$$
$$= -20x$$
$$(-14)(-2) = 28$$

8. $6x^6 + 19x^5 - 7x^4 = x^4(6x^2 + 19x - 7)$
$= x^4(3x-1)(2x+7)$

Check:

$$(3x)(2x) = 6x^2$$
$$(3x)(7) + (-1)(2x) = 21x - 2x = 19x$$
$$(-1)(7) = -7$$

9. $2x^2 - 7xy + 3y^2 = (2x - y)(x - 3y)$

Check:

$(2x)(x) = 2x^2$

$(2x)(-3y) + (-y)(x) = -6xy - xy$

$= -7xy$

$(-y)(-3y) = 3y^2$

10. Let $t = y^2$

$3y^4 + 10y^2 - 8$

$= 3(y^2)^2 + 10y^2 - 8$

$= 3t^2 + 10t - 8$

$= (3t - 2)(t + 4)$ Substitute y^2 for t.

$= (3y^2 - 2)(y^2 + 4)$

Check:

$(3y^2 - 2)(y^2 + 4)$

$(3y^2)(y^2) = 3y^4$

$(3y^2)(4) + (-2)(y^2) = 12y^2 - 2y^2$

$= 10y^2$

$(-2)(4) = -8$

11. Factor $8x^2 - 22x + 5$ by grouping. Since, $ac = 8(5) = 40$, find factors of 40 with a difference of 22. The factors are –20 and –2. Rewrite the middle term and factor by grouping.

$8x^2 - 22x + 5$

$= 8x^2 - 20x - 2x + 5$

$= 4x(2x - 5) - 1(2x - 5)$

$= (2x - 5)(4x - 1)$

Problem Set 5.4

Practice Exercises

1. $x^2 + 5x + 6 = (x + 3)(x + 2)$

product: 3(2) = 6
sum: 3 + 2 = 5

3. $x^2 + 8x + 12 = (x + 6)(x + 2)$

product: 6(2) = 12
sum: 6 + 2 = 8

5. $x^2 + 9x + 20 = (x + 5)(x + 4)$

product: 5(4) = 20
sum: 5 + 4 = 9

7. $y^2 + 10y + 16 = (y + 8)(y + 2)$

product: 8(2) = 16
sum: 8 + 2 = 10

9. $x^2 - 8x + 15 = (x - 5)(x - 3)$

product: –5(–3)=15
sum: –5 + (–3) = –8

11. $y^2 - 12y + 20 = (y - 10)(y - 2)$

product: –10(–2) = 20
sum: –10 + (–2) = –12

13. $a^2 + 5a - 14 = (a + 7)(a - 2)$

product: 7(–2) = –14
sum: 7 + (–2) = 5

15. $x^2 + x - 30 = (x + 6)(x - 5)$

product: 6(–5)= –30
sum: 6 + (–5) = 1

17. $x^2 - 3x - 28 = (x - 7)(x + 4)$

product: –7(4)= –28
sum: –7 + 4 = –3

19. $y^2 - 5y - 36 = (y-9)(y+4)$
product: $-9(4) = -36$
sum: $-9 + 4 = -5$

21. $x^2 - x + 7$
The trinomial is not factorable. There are no factors of 7 that add up to –1. The polynomial is prime.

23. $x^2 - 9xy + 14y^2 = (x - 7y)(x - 2y)$
product: $-7y(-2y) = 14y^2$
sum: $-7y + (-2y) = -9y$

25. $x^2 - xy - 30y^2 = (x - 6y)(x + 5y)$
product: $-6(5) = -30$
sum: $-6 + 5 = -1$

27. $x^2 + xy + y^2$
The trinomial is not factorable. There are no factors of 1 that add up to 1. The polynomial is prime.

29. $a^2 - 18ab + 80b^2 = (a - 10b)(a - 8b)$
Product: $-10(-8) = 80$
Sum: $-10 + (-8) = -18$

31. $3x^2 + 3x - 18 = 3(x^2 + x - 6)$
$= 3(x+3)(x-2)$

33. $2x^3 - 14x^2 + 24x = 2x(x^2 - 7x + 12)$
$= 2x(x-4)(x-3)$

35. $3y^3 - 15y^2 + 18y$
$= 3y(y^2 - 5y + 6)$
$= 3y(y-3)(y-2)$

37. $2x^4 - 26x^3 - 96x^2$
$= 2x^2(x^2 - 13x - 48)$
$= 2x^2(x-16)(x+3)$

39. Let $t = x^3$
$x^6 - x^3 - 6 = (x^3)^2 - x^3 - 6$
$= t^2 - t - 6$
$= (t-3)(t+2)$
Substitute x^3 for t.
$= (x^3 - 3)(x^3 + 2)$

41. Let $t = x^2$
$x^4 - 5x^2 - 6 = (x^2)^2 - 5x^2 - 6$
$= t^2 - 5t - 6$
$= (t-6)(t+1)$
Substitute x^2 for t.
$= (x^2 - 6)(x^2 + 1)$

43. Let $t = x + 1$
$(x+1)^2 + 6(x+1) + 5$
$= t^2 + 6t + 5$
$= (t+5)(t+1)$
Substitute $x+1$ for t.
$= ((x+1)+5)((x+1)+1)$
$= (x+1+5)(x+1+1)$
$= (x+6)(x+2)$

45. $3x^2 + 8x + 5 = (3x+5)(x+1)$

47. $5x^2 + 56x + 11 = (5x+1)(x+11)$

49. $3y^2 + 22y - 16 = (3y-2)(y+8)$

51. $4y^2+9y+2=(y+2)(4y+1)$

53. $10x^2+19x+6=(5x+2)(2x+3)$

55. $8x^2-18x+9=(4x-3)(2x-3)$

57. $6y^2-23y+15=(6y-5)(y-3)$

59. $6y^2+14y+3$
The trinomial is not factorable. The polynomial is prime.

61. $3x^2+4xy+y^2=(3x+y)(x+y)$

63. $6x^2-7xy-5y^2=(2x+y)(3x-5y)$

65. $15x^2-31xy+10y^2$
$=(3x-5y)(5x-2y)$

67. $3a^2-ab-14b^2=(3a-7b)(a+2b)$

69. $15x^3-25x^2+10x$
$=5x(3x^2-5x+2)$
$=5x(3x-2)(x-1)$

71. $24x^4+10x^3-4x^2$
$=2x^2(12x^2+5x-2)$
$=2x^2(3x+2)(4x-1)$

73. $15y^5-2y^4-y^3=y^3(15y^2-2y-1)$
$=y^3(3y-1)(5y+1)$

75. $24x^2+3xy-27y^2$
$=3(8x^2+xy-9y^2)$
$=3(8x+9y)(x-y)$

77. $6a^2b-2ab-60b$
$=2b(3a^2-a-30)$
$=2b(3a-10)(a+3)$

79. $12x^2y-34xy^2+14y^3$
$=2y(6x^2-17xy+7y^2)$
$=2y(3x-7y)(2x-y)$

81. $13x^3y^3+39x^3y^2-52x^3y$
$=13x^3y(y^2+3y-4)$
$=13x^3y(y+4)(y-1)$

83. Let $t=x^2$
$2x^4-x^2-3$
$=2(x^2)^2-x^2-3$
$=2t^2-t-3$
$=(2t-3)(t+1)$
Substitute x^2 for t.
$=(2x^2-3)(x^2+1)$

85. Let $t=x^3$
$2x^6+11x^3+15$
$=2(x^3)^2+11x^3+15$
$=2t^2+11t+15$
$=(2t+5)(t+3)$
Substitute x^3 for t.
$=(2x^3+5)(x^3+3)$

87. Let $t = y^5$

$2y^{10} + 5y^5 + 3$
$= 2(y^5)^2 + 5y^5 + 3$
$= 2t^2 + 5t + 3$
$= (2t + 3)(t + 1)$

Substitute y^5 for t.

$= (2y^5 + 3)(y^5 + 1)$

89. Let $t = x + 1$

$5(x+1)^2 + 12(x+1) + 7$
$= 5t^2 + 12t + 7$
$= (5t + 7)(t + 1)$

Substitute $x + 1$ for t.

$= (5(x+1) + 7)((x+1) + 1)$
$= (5x + 5 + 7)(x + 1 + 1)$
$= (5x + 12)(x + 2)$

91. Let $t = x - 3$

$2(x-3)^2 - 5(x-3) - 7$
$= 2t^2 - 5t - 7$
$= (2t - 7)(t + 1)$

Substitute $x - 3$ for t.

$= (2(x-3) - 7)((x-3) + 1)$
$= (2x - 6 - 7)(x - 3 + 1)$
$= (2x - 13)(x - 2)$

Application Exercises

93. **a.** $f(1) = -16(1)^2 + 16(1) + 32$
$= -16(1) + 16 + 32$
$= -16 + 16 + 32$
$= 32$

After 1 second, the diver will be 32 feet above the water.

b. $f(2) = -16(2)^2 + 16(2) + 32$
$= -16(4) + 32 + 32$
$= -64 + 32 + 32 = 0$

After 2 seconds, the diver will hit the water.

c. $-16t^2 + 16t + 32$
$= -16(t^2 - t - 2)$
$= -16(t - 2)(t + 1)$

d. $f(t) = -16(t - 2)(t + 1)$
$f(1) = -16(1 - 2)(1 + 1)$
$= -16(-1)(2) = 32$
$f(2) = -16(2 - 2)(2 + 1)$
$= -16(0)(3) = 0$

95. **a.** $x \cdot x + 1 \cdot x + 1 \cdot x + 1 \cdot 1 + 1 \cdot x + 1 \cdot 1$
$= x^2 + x + x + 1 + x + 1$
$= x^2 + 3x + 2$

b. $(x + 1)(x + 2) = x^2 + 2x + x + 2$
$= x^2 + 3x + 2$

c. Answers will vary.

Writing in Mathematics

97. Answers will vary.

99. Answers will vary.

101. Answers will vary.

103. Answers will vary.

Technology Exercises

105. $x^2+7x+12=(x+4)(x+3)$

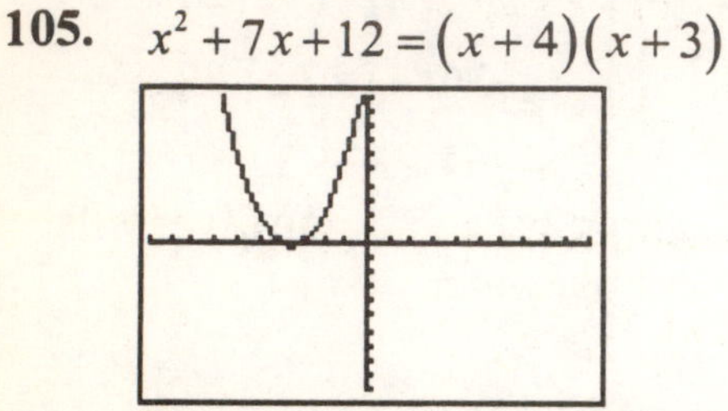

The graphs coincide. The polynomial is factored correctly.

107. $6x^3+5x^2-4x=x(3x+4)(2x-1)$

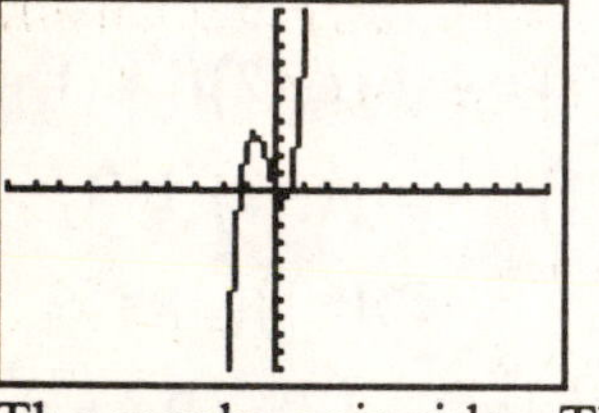

The graphs coincide. The polynomial is factored correctly.

Critical Thinking Exercises

109. Statement **b.** is true.

$8y^2-51y+18=(8y-3)(y-6)$

Statement **a.** is false.

$6y^6-19y^5+10y^4$

$=y^4(6y^2-19y+10)$

$=y^4(3y-2)(2y-5)$

Statement **c.** is false. The coefficient of the middle term can be a prime number, regardless of the number of variables in the problem.

Statement **d.** is false. The correct factorization is

$12x^2-19xy+5y=(4x-5y)(3x+y)$.

111. Use the factors of 3 and 5 to find b. The factors of 3 are 1 and 3 and the factors of 5 are 1 and 5. Multiply the combinations of the factors as follows.

$(3x+5)(x+1)=3x^2+3x+5x+5$

$=3x^2+8x+5$

or

$(3x+1)(x+5)=3x^2+15x+x+5$

$=3x^2+16x+5$

In addition to 8 and 16, –8 and –16 can also be obtained.

$(3x-5)(x-1)$

$=3x^2-3x-5x+5$

$=3x^2-8x+5$

or

$(3x-1)(x-5)$

$=3x^2-15x-x+5$

$=3x^2-16x+5$

The integers are –16, –8, 8 and 16.

113. Let $t=x^n$

$4x^{2n}-9x^n+5$

$=4(x^n)^2-9x^n+5$

$=4t^2-9t+5$

$=(4t-5)(t-1)$

Substitute x^n for t.

$=(4x^n-5)(x^n-1)$

Review Exercises

114. $-2x \le 6$ and $-2x + 3 < -7$

$x \ge -3$ $\quad -2x < -10$

$x > 5$

$x \ge -3$

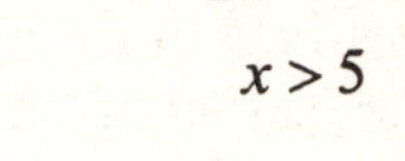

$x > 5$

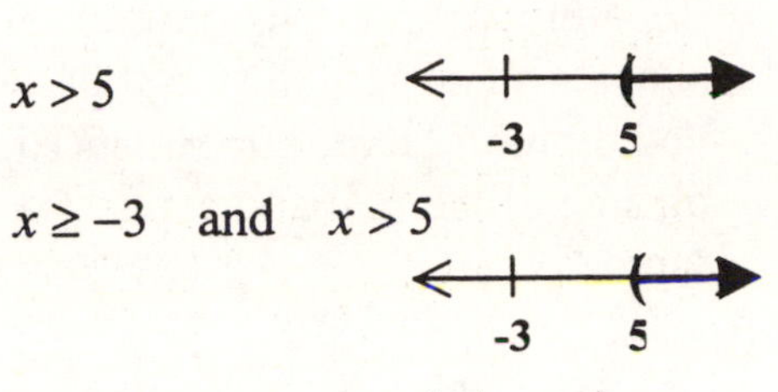

$x \ge -3$ and $x > 5$

The solution set is $\{x | x > 5\}$ or $(5, \infty)$.

115. $2x - y - 2z = -1$

$x - 2y - z = 1$

$x + y + z = 4$

Multiply the second equation by –2 and add to the first equation.

$$\begin{aligned} 2x - y - 2z &= -1 \\ -2x + 4y + 2z &= -2 \\ \hline 3y &= -3 \\ y &= -1 \end{aligned}$$

Add the second and third equations to eliminate z.

$$\begin{aligned} x - 2y - z &= 1 \\ x + y + z &= 4 \\ \hline 2x - y &= 5 \end{aligned}$$

Back-substitute –1 for y to find x.

$$\begin{aligned} 2x - (-1) &= 5 \\ 2x + 1 &= 5 \\ 2x &= 4 \\ x &= 2 \end{aligned}$$

Back-substitute 2 for x and –1 for y to find z.

$$\begin{aligned} x + y + z &= 4 \\ 2 + (-1) + z &= 4 \\ 1 + z &= 4 \\ z &= 3 \end{aligned}$$

The solution is $(2, -1, 3)$ or $\{(2, -1, 3)\}$.

116. $x - y \ge -4$

$x + 2y \ge 2$

First consider $x - y \ge -4$. Solve for y in $x - y = -4$, and graph using the slope and y-intercept.

$$\begin{aligned} x - y &= -4 \\ -y &= -x - 4 \\ y &= x + 4 \end{aligned}$$

y-intercept = 4 slope = 1

Now, use the origin, $(0, 0)$, as a test point.

$$\begin{aligned} 0 - 0 &\ge -4 \\ 0 &\ge -4 \end{aligned}$$

This is a true statement and the point $(0, 0)$ will fall in the shaded half-plane.

Next consider $x + 2y \ge 2$. Solve for y in $x + 2y = 2$, and graph using the slope and the y-intercept.

$$\begin{aligned} x + 2y &= 2 \\ 2y &= -x + 2 \\ y &= -\frac{1}{2}x + 1 \end{aligned}$$

y-intercept = 1 slope = $-\frac{1}{2}$

Now, use the origin, $(0,0)$, as a test point.

$0 + 2(0) \geq 2$

$0 \geq 2$

This is a false statement, so the point $(0,0)$ will not fall in the shaded half-plane.

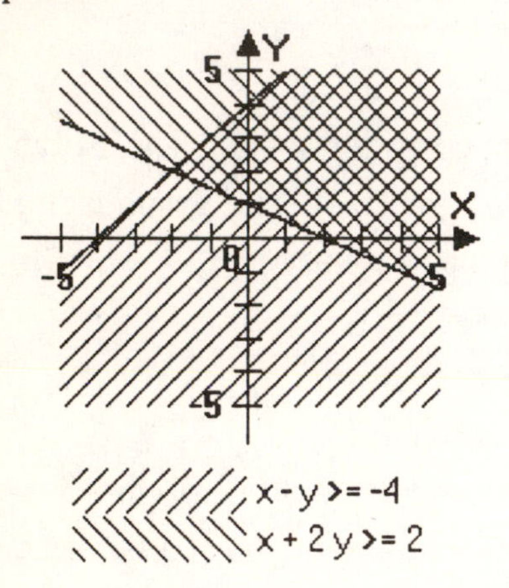

Check Points 5.5

1. **a.** $16x^2 - 25 = (4x)^2 - 5^2$
$= (4x+5)(4x-5)$

b. $100y^6 - 9x^4$
$= (10y^3)^2 - (3x^2)^2$
$= (10y^3 + 3x^2)(10y^3 - 3x^2)$

2. $6y - 6x^2y^7 = 6y(1 - x^2y^6)$
$= 6y(1^2 - (xy^3)^2)$
$= 6y(1 + xy^3)(1 - xy^3)$

3. $16x^4 - 81$
$= (4x^2)^2 - 9^2$
$= (4x^2 + 9)(4x^2 - 9)$
$= (4x^2 + 9)((2x)^2 - 3^2)$
$= (4x^2 + 9)(2x + 3)(2x - 3)$

Recall that $4x^2 + 9$ is the sum of two squares. The sum of two squares with no common factor other than 1 is a prime polynomial and cannot be factored.

4. $x^3 + 7x^2 - 4x - 28$
$= x^2(x+7) - 4(x+7)$
$= (x+7)(x^2 - 4)$
$= (x+7)(x^2 - 2^2)$
$= (x+7)(x+2)(x+2)$

5. **a.** $x^2 + 6x + 9 = x^2 + 2 \cdot x \cdot 3 + 3^2$
$= (x+3)^2$

b. $16x^2 + 40xy + 25y$
$= (4x)^2 + 2 \cdot 4x \cdot 5y + (5y)^2$
$= (4x + 5y)^2$

c. $4y^4 - 20y^2 + 25$
$= (2y^2)^2 - 2 \cdot 2y^2 \cdot 5 + 5^2$
$= (2y^2 - 5)^2$

6. $x^2+10x+25-y^2$
$=(x^2+10x+25)-y^2$
$=(x^2+2\cdot x\cdot 5+5^2)-y^2$
$=(x+5)^2-y^2$
$=((x+5)+y)((x+5)-y)$
$=(x+5+y)(x+5-y)$

7. a^2-b^2+4b-4
$=a^2-(b^2-4b+4)$
$=a^2-(b^2-2\cdot b\cdot 2+2^2)$
$=a^2-(b-2)^2$
$=(a+(b-2))(a-(b-2))$
$=(a+b-2)(a-b+2)$

8. **a.** x^3+27
$=x^3+3^3$
$=(x+3)(x^2-x\cdot 3+3^2)$
$=(x+3)(x^2-3x+9)$

b. $x^6+1000y^3$
$=(x^2)^3+(10y)^3$
$=(x^2+10y)((x^2)^2-x^2 10y+(10y)^2)$
$=(x^2+10y)(x^4-10x^2y+100y^2)$

9. **a.** x^3-8
$=x^3-2^3$
$=(x-2)(x^2+x\cdot 2+2^2)$
$=(x-2)(x^2+2x+4)$

b. $1-27x^3y^3$
$=1^3-(3xy)^3$
$=(1-3xy)(1^2+1\cdot 3xy+(3xy)^2)$
$=(1-3xy)(1+3xy+9x^2y^2)$

Problem Set 5.5

Practice Exercises

1. $x^2-4=x^2-2^2$
$=(x+2)(x-2)$

3. $9x^2-25=(3x)^2-5^2$
$=(3x+5)(3x-5)$

5. $9-25y^2=3^2-(5y)^2$
$=(3+5y)(3-5y)$

7. $36x^2-49y^2=(6x)^2-(7y)^2$
$=(6x+7y)(6x-7y)$

9. $x^2y^2-1=(xy)^2-1^2$
$=(xy+1)(xy-1)$

11. $9x^4-25y^6=(3x^2)^2-(5y^3)^2$
$=(3x^2+5y^3)(3x^2-5y^3)$

13. $x^{14}-y^4=(x^7)^2-(y^2)^2$
$=(x^7+y^2)(x^7-y^2)$

15. $(x-3)^2 - y^2$
$= ((x-3)+y)((x-3)-y)$
$= (x-3+y)(x-3-y)$

17. $a^2 - (b-2)^2$
$= (a+(b-2))(a-(b-2))$
$= (a+b-2)(a-b+2)$

19. $2x^3 - 8x = 2x(x^2 - 4)$
$= 2x(x^2 - 2^2)$
$= 2x(x+2)(x-2)$

21. $50 - 2y^2 = 2(25 - y^2)$
$= 2(5^2 - y^2)$
$= 2(5+y)(5-y)$

23. $8x^2 - 8y^2 = 8(x^2 - y^2)$
$= 8(x+y)(x-y)$

25. $2x^3y - 18xy = 2xy(x^2 - 9)$
$= 2xy(x^2 - 3^2)$
$= 2xy(x+3)(x-3)$

27. $a^3b^2 - 49ac^2 = a(a^2b^2 - 49c^2)$
$= a((ab)^2 - (7c)^2)$
$= a(ab+7c)(ab-7c)$

29. $5y - 5x^2y^7 = 5y(1 - x^2y^6)$
$= 5y(1^2 - (xy^3)^2)$
$= 5y(1+xy^3)(1-xy^3)$

31. $8x^2 + 8y^2 = 8(x^2 + y^2)$

33. $x^2 + 25y^2$
Prime
The sum of two squares with no common factor other than 1 is a prime polynomial.

35. $x^4 - 16 = (x^2)^2 - 4^2$
$= (x^2+4)(x^2-4)$
$= (x^2+4)(x^2-2^2)$
$= (x^2+4)(x+2)(x-2)$

37. $81x^4 - 1 = (9x^2)^2 - 1^2$
$= (9x^2+1)(9x^2-1)$
$= (9x^2+1)((3x)^2 - 1^2)$
$= (9x^2+1)(3x+1)(3x-1)$

39. $2x^5 - 2xy^4$
$= 2x(x^4 - y^4)$
$= 2x((x^2)^2 - (y^2)^2)$
$= 2x(x^2+y^2)(x^2-y^2)$
$= 2x(x^2+y^2)(x+y)(x-y)$

41. $x^3+3x^2-4x-12$
$=(x^3+3x^2)+(-4x-12)$
$=x^2(x+3)+(-4)(x+3)$
$=(x+3)(x^2-4)$
$=(x+3)(x^2-2^2)$
$=(x+3)(x+2)(x-2)$

43. x^3-7x^2-x+7
$=(x^3-7x^2)+(-x+7)$
$=x^2(x-7)+(-1)(x-7)$
$=(x-7)(x^2-1)$
$=(x-7)(x^2-1^2)$
$=(x-7)(x+1)(x-1)$

45. $x^2+4x+4=x^2+2\cdot x\cdot 2+2^2$
$=(x+2)^2$

47. $x^2-10x+25=x^2-2\cdot x\cdot 5+5^2$
$=(x-5)^2$

49. $x^4-4x^2+4=(x^2)^2-2\cdot x^2\cdot 2+2^2$
$=(x^2-2)^2$

51. $9y^2+6y+1=(3y)^2+2\cdot y\cdot 3+1^2$
$=(3y+1)^2$

53. $64y^2-16y+1=(8y)^2-2\cdot y\cdot 8+1^2$
$=(8y-1)^2$

55. $x^2-12xy+36y^2$
$=x^2-2\cdot x\cdot 6y+(6y)^2$
$=(x-6y)^2$

57. $x^2-8xy+64y^2$
Prime
Because the first and third terms of the polynomial are perfect squares, check to see if the trinomial is a perfect square. The middle term would have to be $-16xy$ for the polynomial to be a perfect square. Since this is not the case and the polynomial cannot be factored in any other way, we conclude that the polynomial is prime.

59. $9x^2+48xy+64y^2$
$=(3x)^2+2\cdot 3x\cdot 8y+(8y)^2$
$=(3x+8y)^2$

61. $x^2-6x+9-y^2$
$=(x^2-6x+9)-y^2$
$=(x^2-2\cdot x\cdot 3+3^2)-y^2$
$=(x-3)^2-y^2$
$=((x-3)+y)((x-3)-y)$
$=(x-3+y)(x-3-y)$

63. $x^2+20x+100-x^4$
$=(x^2+20x+100)-x^4$
$=(x^2+2\cdot x\cdot 10+10^2)-x^4$
$=(x+10)^2-(x^2)^2$

$=\left((x+10)+x^2\right)\left((x+10)-x^2\right)$
$=\left(x+10+x^2\right)\left(x+10-x^2\right)$

65. $9x^2-30x+25-36y^2$
$=\left(9x^2-30x+25\right)-36y^2$
$=\left((3x)^2-2\cdot 3x\cdot 5+5^2\right)-36y^2$
$=(3x-5)^2-(6y)^2$
$=\left((3x-5)+6y\right)\left((3x-5)-6y\right)$
$=(3x-5+6y)(3x-5-6y)$

67. x^4-x^2-2x-1
$=x^4-\left(x^2+2x+1\right)$
$=x^4-\left(x^2+2\cdot x\cdot 1+1^2\right)$
$=\left(x^2\right)^2-(x+1)^2$
$=\left(x^2+(x+1)\right)\left(x^2-(x+1)\right)$
$=\left(x^2+x+1\right)\left(x^2-x-1\right)$

69. $z^2-x^2+4xy-4y^2$
$=z^2-\left(x^2-4xy+4y^2\right)$
$=z^2-\left(x^2-2\cdot x\cdot 2y+(2y)^2\right)$
$=z^2-(x-2y)^2$
$=\left(z+(x-2y)\right)\left(z-(x-2y)\right)$
$=(z+x-2y)(z-x+2y)$

71. $x^3+64=x^3+4^3$
$=(x+4)\left(x^2-x\cdot 4+4^2\right)$
$=(x+4)\left(x^2-4x+16\right)$

73. $x^3-27=x^3-3^3$
$=(x-3)\left(x^2+3\cdot x+3^2\right)$
$=(x-3)\left(x^2+3x+9\right)$

75. $8y^3+1$
$=(2y)^3+1^3$
$=(2y+1)\left((2y)^2-2y\cdot 1+1^2\right)$
$=(2y+1)\left(4y^2-2y+1\right)$

77. $125x^3-8$
$=(5x)^3-2^3$
$=(5x-2)\left(25x^2+5x\cdot 2+2^2\right)$
$=(5x-2)\left(25x^2+10x+4\right)$

79. x^3y^3+27
$=(xy)^3+3^3$
$=(xy+3)\left((xy)^2-xy\cdot 3+3^2\right)$
$=(xy+3)\left(x^2y^2-3xy+9\right)$

81. $64x-x^4$
$=x\left(64-x^3\right)$
$=x\left(4^3-x^3\right)$
$=x(4-x)\left(4^2+4\cdot x+x^2\right)$
$=x(4-x)\left(16+4x+x^2\right)$

83. x^6+27y^3
$=\left(x^2\right)^3+(3y)^3$

$$=(x^2+3y)\left((x^2)^2-x^2\cdot 3y+(3y)^2\right)$$
$$=(x^2+3y)(x^4-3x^2y+9y^2)$$

85. $125x^6-64y^6$
$$=(5x^2)^3-(4y^2)^3$$
$$=(5x^2-4y^2)\left((5x^2)^2+5x^2 4y^2+(4y^2)^2\right)$$
$$=(5x^2-4y^2)(25x^4+20x^2y^2+16y^4)$$

87. x^9+1
$$=(x^3)^3+1^3$$
$$=(x^3+1)\left((x^3)^2-x^3\cdot 1+1^2\right)$$
$$=(x^3+1)(x^6-x^3+1)$$
$$=(x^3+1^3)(x^6-x^3+1)$$
$$=(x+1)(x^2-x\cdot 1+1^2)(x^6-x^3+1)$$
$$=(x+1)(x^2-x+1)(x^6-x^3+1)$$

Application Exercises

89. $x^2-4^2=(x+4)(x-4)$

91. $(3x)^2-4\cdot 2^2=(3x)^2-4\cdot 4$
$$=(3x)^2-4^2$$
$$=(3x+4)(3x-4)$$

93. $V_{\text{shaded}}=V_{\text{outside}}-V_{\text{inside}}$
$$=a\cdot a\cdot 4a-b\cdot b\cdot 4a$$
$$=4a^3-4ab^2$$
$$=4a(a^2-b^2)$$
$$=4a(a+b)(a-b)$$

Writing in Mathematics

95. Answers vary.

97. Answers vary.

Technology Exercises

99. $9x^2-4=(3x+2)(3x-2)$

The graphs coincide. The polynomial is factored correctly.

101. $9x^2+12x+4=(3x+2)^2$

The graphs coincide. The polynomial is factored correctly.

103. $(2x+3)^2-9=4x(x+3)$

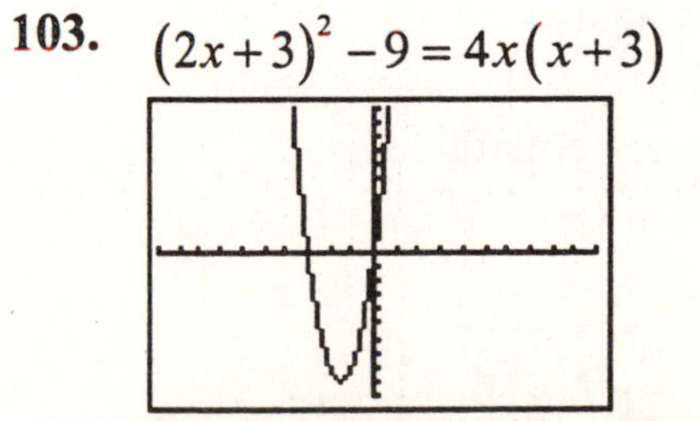

The graphs coincide. The polynomial is factored correctly.

105. $x^3 - 1 = (x-1)(x^2 - x + 1)$

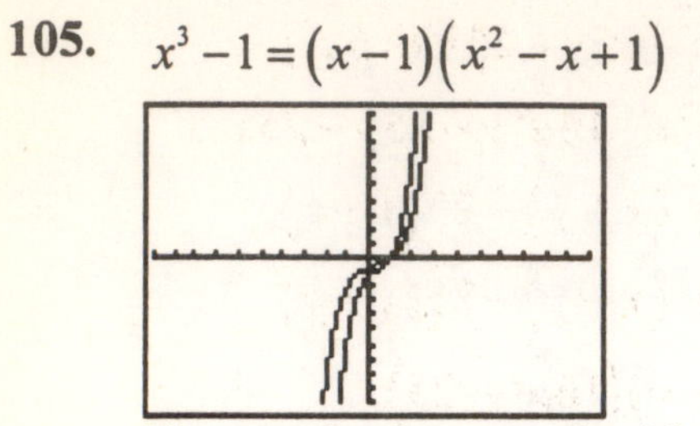

The graphs do not coincide. The factorization is not correct.

$x^3 - 1 = (x-1)(x^2 + x + 1)$

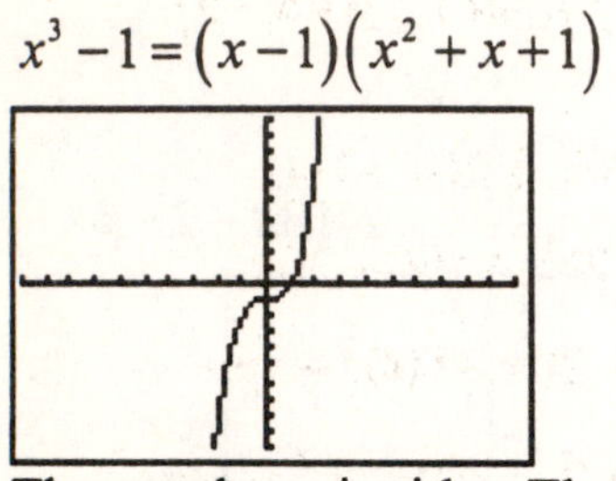

The graphs coincide. The new factorization is correct.

Critical Thinking Exercises

107. Statement **d.** is true.

Statement **a.** is false.
$9x^2 + 15x + 25$ is not a perfect square trinomial. The middle term would have to be $2 \cdot 3x \cdot 5 = 30x$.

Statement **b.** is false.
$x^3 - 27$ is factored incorrectly. The middle term of the second factor should be $3x$.

Statement **c.** is false.
$x^3 - 64 = (x-4)(x^2 + 4x + 16)$

109.
$$36x^{2n} - y^{2n} = (6x^n)^2 - (y^n)^2 = (6x^n + y^n)(6x^n - y^n)$$

111.
$$4x^{2n} + 20x^n y^m + 25y^{2m} = (2x^n)^2 + 2 \cdot 2x^n \cdot 5y^m + (5y^m)^2 = (2x^n + 5y^m)^2$$

113. In a perfect square trinomial, the middle term is $2AB$. Rewrite the middle term in $2AB$ form.
$kx^2 + 8xy + y^2$
$= kx^2 + 2 \cdot 4x \cdot y + y^2$
This means that $A = 4x$. The first term is $A^2 = (4x)^2 = 16x^2$. From this, we see that $k = 16$.

Review Problems

115. $2x + 2 \geq 12$ and $\dfrac{2x-1}{3} \leq 7$

$2x \geq 10$ $\quad$ $2x - 1 \leq 21$

$x \geq 5$ $\quad$ $2x \leq 22$

$x \leq 11$

$x \geq 5$

$x \leq 11$

$x \geq 5$ and $x \leq 11$

The solution set is $\{x | 5 \leq x \leq 11\}$ or $[5, 11]$.

116. $3x - 2y = -8$
$x + 6y = 4$
Write the augmented matrix and solve by Gaussian elimination.

$$\left[\begin{array}{cc|c} 3 & -2 & -8 \\ 1 & 6 & 4 \end{array}\right] \quad R_1 \leftrightarrow R_2$$

$$= \left[\begin{array}{cc|c} 1 & 6 & 4 \\ 3 & -2 & -8 \end{array}\right] \quad -3R_1 + R_2$$

$$= \left[\begin{array}{cc|c} 1 & 6 & 4 \\ 0 & -20 & -20 \end{array}\right] \quad -\frac{1}{20}R_2$$

$$= \left[\begin{array}{cc|c} 1 & 6 & 4 \\ 0 & 1 & 1 \end{array}\right]$$

The system becomes

$$x + 6y = 4$$
$$y = 1.$$

Back-substitute 1 for y, to find x.

$$x + 6(1) = 4$$
$$x + 6 = 4$$
$$x = -2$$

The solution is set is $(-2,1)$ or $\{(-2,1)\}$.

117. $\dfrac{4^{-2}x^{-3}y^{11}}{x^2y^{-5}} = \dfrac{y^5y^{11}}{4^2x^3x^2} = \dfrac{y^{16}}{16x^5}$

Check Points 5.6

1.
$$3x^3 - 30x^2 + 75x$$
$$= 3x\left(x^2 - 10y + 25\right)$$
$$= 3x\left(x^2 - 2\cdot x\cdot 5 + 5^2\right)$$
$$= 3x(x-5)^2$$

2.
$$3x^2y - 12xy - 36y = 3y\left(x^2 - 4x - 12\right)$$
$$= 3y(x-6)(x+2)$$

3.
$$16a^2x - 25y - 25x + 16a^2y$$
$$= 16a^2x + 16a^2y - 25y - 25x$$
$$= 16a^2(x+y) - 25(y+x)$$
$$= 16a^2(x+y) - 25(x+y)$$
$$= (x+y)\left(16a^2 - 25\right)$$
$$= (x+y)\left((4a)^2 - 5^2\right)$$
$$= (x+y)(4a+5)(4a-5)$$

4.
$$x^2 - 36a^2 + 20x + 100$$
$$= x^2 + 20x + 100 - 36a^2$$
$$= \left(x^2 + 20x + 100\right) - 36a^2$$
$$= \left(x^2 + 2\cdot x\cdot 10 + 10^2\right) - 36a^2$$
$$= (x+10)^2 - 36a^2$$
$$= (x+10)^2 - (6a)^2$$
$$= ((x+10)+6a)((x+10)-6a)$$
$$= (x+10+6a)(x+10-6a)$$

5.
$$x^{10} + 512x$$
$$= x\left(x^9 + 512\right)$$
$$= x\left(x^9 + 8^3\right)$$
$$= x\left(\left(x^3\right)^3 + 8^3\right)$$
$$= x\left(x^3 + 8\right)\left(\left(x^3\right)^2 - x^3\cdot 8 + 8^2\right)$$
$$= x\left(x^3 + 8\right)\left(x^6 - 8x^3 + 64\right)$$
$$= x\left(x^3 + 2^3\right)\left(x^6 - 8x^3 + 64\right)$$
$$= x(x+2)\left(x^2 - x\cdot 2 + 2^2\right)\left(x^6 - 8x^3 + 64\right)$$
$$= x(x+2)\left(x^2 - 2x + 4\right)\left(x^6 - 8x^3 + 64\right)$$

Problem Set 5.6

Practice Exercises

1. $x^3-16x=x(x^2-16)$
$=x(x+4)(x-4)$

3. $3x^2+18x+27=3(x^2+6x+9)$
$=3(x+3)^2$

5. $81x^3-3=3(27x^3-1)$
$=3(3x-1)(9x^2+3x+1)$

7. $x^2y-16y+32-2x^2$
$=(x^2y-16y)+(-2x^2+32)$
$=y(x^2-16)+(-2)(x^2-16)$
$=(x^2-16)(y+(-2))$
$=(x+4)(x-4)(y-2)$

9. $4a^2b-2ab-30b$
$=2b(2a^2-a-15)$
$=2b(2a+5)(a-3)$

11. $ay^2-4a-4y^2+16$
$=a(y^2-4)+(-4)(y^2-4)$
$=(y^2-4)(a+(-4))$
$=(y+2)(y-2)(a-4)$

13. $11x^5-11xy^2=11x(x^4-y^2)$
$=11x(x^2+y)(x^2-y)$

15. $4x^5-64x$
$=4x(x^4-16)$
$=4x(x^2+4)(x^2-4)$
$=4x(x^2+4)(x+2)(x-2)$

17. $x^3-4x^2-9x+36$
$=x^2(x-4)+(-9)(x-4)$
$=(x-4)(x^2+(-9))$
$=(x-4)(x^2-9)$
$=(x-4)(x+3)(x-3)$

19. $2x^5+54x^2=2x^2(x^3+27)$
$=2x^2(x+3)(x^2-3x+9)$

21. $3x^4y-48y^5$
$=3y(x^4-16y^4)$
$=3y(x^2+4y^2)(x^2-4y^2)$
$=3y(x^2+4y^2)(x+2y)(x-2y)$

23. $12x^3+36x^2y+27xy^2$
$=3x(4x^2+12xy+9y^2)$
$=3x(2x+3y)^2$

25. $x^2-12x+36-49y^2$
$=(x^2-12x+36)-49y^2$
$=(x-6)^2-(7y)^2$
$=((x-6)+7y)((x-6)-7y)$
$=(x-6+7y)(x-6-7y)$

27. $4x^2 + 25y^2$
Prime
The sum of two squares with no common factor other than 1 is a prime polynomial.

29. $12x^3y - 12xy^3 = 12xy(x^2 - y^2)$
$= 12xy(x + y)(x - y)$

31. $6bx^2 + 6by^2 = 6b(x^2 + y^2)$

33. $x^4 - xy^3 + x^3y - y^4$
$= x(x^3 - y^3) + y(x^3 - y^3)$
$= (x^3 - y^3)(x + y)$
$= (x - y)(x^2 + xy - y^2)(x + y)$

35. $x^2 - 4a^2 + 12x + 36$
$= x^2 + 12x + 36 - 4a^2$
$= (x^2 + 12x + 36) - 4a^2$
$= (x + 6)^2 - 4a^2$
$= ((x + 6) + 2a)((x + 6) - 2a)$
$= (x + 6 + 2a)(x + 6 - 2a)$

37. $5x^3 + x^6 - 14$
$= x^6 + 5x^3 - 14$
$= (x^3 + 7)(x^3 - 2)$

39. $4x - 14 + 2x^3 - 7x^2$
$= 2(2x - 7) + x^2(2x - 7)$
$= (2x - 7)(2 + x^2)$

41. $54x^3 - 16y^3$
$= 2(27x^3 - 8y^3)$
$= 2(3x - 2y)(9x^2 + 6xy + 4y^2)$

43. $x^2 + 10x - y^2 + 25$
$= x^2 + 10x + 25 - y^2$
$= (x + 5)^2 - y^2$
$= ((x + 5) + y)((x + 5) - y)$
$= (x + 5 + y)(x + 5 - y)$

45. $x^8 - y^8$
$= (x^4 + y^4)(x^4 - y^4)$
$= (x^4 + y^4)(x^2 + y^2)(x^2 - y^2)$
$= (x^4 + y^4)(x^2 + y^2)(x + y)(x - y)$

47. $x^3y - 16xy^3$
$= xy(x^2 - 16y^2)$
$= xy(x + 4y)(x - 4y)$

49. $x + 8x^4$
$= x(1 + 8x^3)$
$= x(1 + 2x)(1 - 2x + 4x^2)$

51. $16y^2 - 4y - 2$
$= 2(8y^2 - 2y - 1)$
$= 2(4y + 1)(2y - 1)$

53. $14y^3 + 7y^2 - 10y$
$= y(14y^2 + 7y - 10)$

55. $27x^2 + 36xy + 12y^2$
$= 3(9x^2 + 12xy + 4y^2)$
$= 3(3x + 2y)^2$

57. $12x^3 + 3xy^2 = 3x(4x^2 + y^2)$

59. $x^6y^6 - x^3y^3$
$= x^3y^3(x^3y^3 - 1)$
$= x^3y^3(xy - 1)(x^2y^2 + xy + 1)$

Application Exercises

61. **a.** $x(x + y) - y(x + y)$

b. $x(x + y) - y(x + y)$
$= (x + y)(x - y)$

63. **a.** $xy + xy + xy + 3x(x) = 3xy + 3x^2$

b. $3xy + 3x^2 = 3x(y + x)$

Writing in Mathematics

65. Answers will vary.

Technology Exercises

67. $4x^2 - 12x + 9 = (4x - 3)^2$

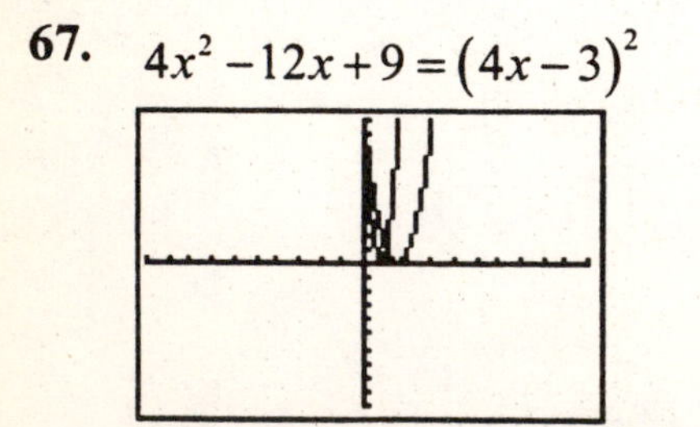

The graphs do not coincide. The factorization is not correct.

$4x^2 - 12x + 9 = (2x - 3)^2$

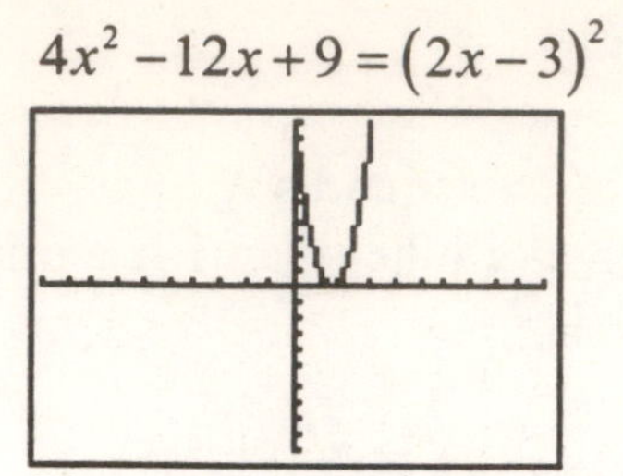

The graphs coincide. The new factorization is correct.

69. $x^4 - 16 = (x^2 + 4)(x + 2)(x - 2)$

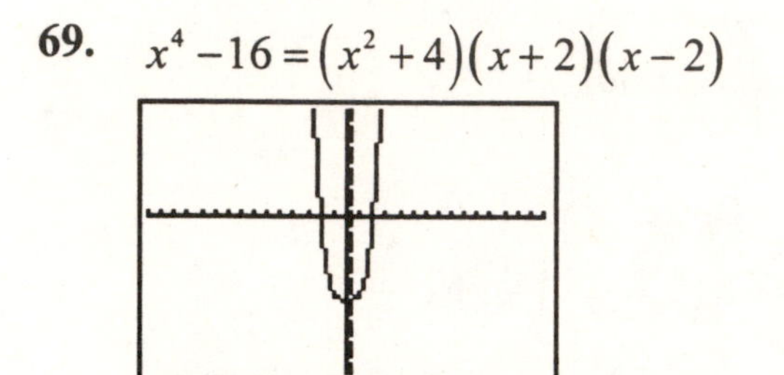

The graphs coincide so the factorization is correct.

Critical Thinking Exercises

71. Statement **b.** is true. The trinomial is prime because there are no factors of –4 that sum to –4.

Statement **a.** is false. $x^4 - 16$ is not factored completely.
$(x^2 + 4)(x^2 - 4)$
$= (x^2 + 4)(x + 2)(x - 2)$

Statement **c.** is false. $x^2 + 36$ is the sum of two squares. The sum of two squares with no common factor other than 1 is a prime polynomial.

Statement **d.** is false. The correct factorization is:
$x^3 - 64 = (x - 4)(x^2 + 4x + 16)$.

73. Let $t = x^2 - 8$

$$\begin{aligned}
&(x^2-8)^2-4(x^2-8)-32\\
&=t^2-4t-32\\
&=(t-8)(t+4)\\
&=((x^2-8)-8)((x^2-8)+4)\\
&=(x^2-8-8)(x^2-8+4)\\
&=(x^2-16)(x^2-4)\\
&=(x+4)(x-4)(x+2)(x-2)
\end{aligned}$$

75.

$$\begin{aligned}
&x^{4n+1}-xy^{4n}\\
&=x^{4n}x-xy^{4n}\\
&=x(x^{4n}-y^{4n})\\
&=x(x^{2n}+y^{2n})(x^{2n}-y^{2n})\\
&=x(x^{2n}+y^{2n})(x^n+y^n)(x^n-y^n)
\end{aligned}$$

77.

$$\begin{aligned}
&x^4+4\\
&=x^4+4x^2+4-4x^2\\
&=(x^4+4x^2+4)-4x^2\\
&=(x^2+2)^2-4x^2\\
&=((x^2+2)+2x)((x^2+2)-2x)\\
&=(x^2+2+2x)(x^2+2-2x)\\
&=(x^2+2x+2)(x^2-2x+2)
\end{aligned}$$

Review Exercises

79.

$$\frac{3x-1}{5}+\frac{x+2}{2}=-\frac{3}{10}$$

$$10\left(\frac{3x-1}{5}\right)+10\left(\frac{x+2}{2}\right)=10\left(-\frac{3}{10}\right)$$

$$\begin{aligned}
2(3x-1)+5(x+2)&=-3\\
6x-2+5x+10&=-3\\
11x+8&=-3\\
11x&=-11\\
x&=-1
\end{aligned}$$

The solution is -1 and the solution set is $\{-1\}$.

80.

$$\begin{aligned}
&(4x^3y^{-1})^2(2x^{-3}y)^{-1}\\
&=4^2x^6y^{-2}2^{-1}x^3y^{-1}\\
&=4^2 2^{-1}x^6x^3y^{-2}y^{-1}\\
&=16\cdot\frac{1}{2}x^9y^{-3}\\
&=\frac{8x^9}{y^3}
\end{aligned}$$

81.

$$\begin{vmatrix}0&-3&2\\1&5&3\\-2&1&4\end{vmatrix}$$

$$=0\begin{vmatrix}5&3\\1&4\end{vmatrix}-1\begin{vmatrix}-3&2\\1&4\end{vmatrix}-2\begin{vmatrix}-3&2\\5&3\end{vmatrix}$$

$$\begin{aligned}
&=-1(-3\cdot4-1\cdot2)-2(-3\cdot3-5\cdot2)\\
&=-1(-12-2)-2(-9-10)\\
&=-1(-14)-2(-19)\\
&=14+38=52
\end{aligned}$$

Check Points 5.7

1.

$$\begin{aligned}
2x^2-9x&=5\\
2x^2-9x-5&=0\\
(2x+1)(x-5)&=0
\end{aligned}$$

Apply the zero product principle.

$2x+1=0 \qquad x-5=0$

$2x=-1 \qquad x=5$

$x=-\frac{1}{2}$

The solutions are $-\frac{1}{2}$ and 5 and the solution set is $\left\{-\frac{1}{2},5\right\}$.

2. **a.**

$$3x^2=2x$$
$$3x^2-2x=0$$
$$x(3x-2)=0$$

Apply the zero product principle.

$x=0 \qquad 3x-2=0$

$3x=2$

$x=\frac{2}{3}$

The solutions are 0 and $\frac{2}{3}$ and the solution set is $\left\{0,\frac{2}{3}\right\}$.

b.

$$x^2+7=10x-18$$
$$x^2-10x+25=0$$
$$(x-5)^2=0$$

Apply the zero product principle to obtain the double-root.

$x-5=0$

$x=5$

The solution is 5 and the solution set is $\{5\}$.

c.

$$(x-2)(x+3)=6$$
$$x^2+3x-2x-6=6$$
$$x^2+x-6=6$$
$$x^2+x-12=0$$
$$(x+4)(x-3)=0$$

Apply the zero product principle.

$x+4=0 \qquad x-3=0$

$x=-4 \qquad x=3$

The solutions are –4 and 3 and the solution set is $\{-4,3\}$.

3.

$$2x^3+3x^2=8x+12$$
$$2x^3+3x^2-8x-12=0$$
$$x^2(2x+3)-4(2x+3)=0$$
$$(2x+3)(x^2-4)=0$$
$$(2x+3)(x+2)(x-2)=0$$

Apply the zero product principle.

$2x+3=0 \qquad x+2=0 \qquad x-2=0$

$2x=-3 \qquad x=-2 \qquad x=2$

$x=-\frac{3}{2}$

The solutions are $-2,-\frac{3}{2}$ and 2, and the solution set is $\left\{-2,-\frac{3}{2},2\right\}$.

4.

$$-16t^2+48t+160=192$$
$$-16t^2+48t-32=0$$
$$-16(t^2-3t+2)=0$$
$$-16(t-2)(t-1)=0$$

Apply the zero product principle.

$-16(t-2)=0 \qquad t-1=0$

$t-2=0 \qquad t=1$

$t=2$

The solution set is $\{1,2\}$. The ball will be at a height of 192 feet at 1 second and 2 seconds after it is thrown. This can be shown on the graph as the points $(1,192)$ and $(2,192)$.

5. The total area is the area of the garden and the area of the path.

$$A_{total} = A_{garden} + A_{path}$$
$$320 = (16 \cdot 12) + (2 \cdot 16x + 2 \cdot 12x + 4x \cdot x)$$
$$320 = 192 + (32x + 24x + 4x^2)$$
$$320 = 192 + 56x + 4x^2$$
$$0 = -128 + 56x + 4x^2$$
$$0 = 4x^2 + 56x - 128$$
$$0 = 4(x^2 + 14x - 32)$$
$$0 = 4(x+16)(x-2)$$

Apply the zero product principle.

$4(x+16) = 0 \qquad x - 2 = 0$

$x + 16 = 0 \qquad x = 2$

$x = -16$

The solution set is $\{-16,2\}$. We disregard -16 because it is not possible to have a negative length measurement. The path should be 2 feet wide.

6.
$$x^2 + (x+1)^2 = (x+2)^2$$
$$x^2 + x^2 + 2x + 1 = x^2 + 4x + 4$$
$$2x^2 + 2x + 1 = x^2 + 4x + 4$$
$$x^2 - 2x - 3 = 0$$
$$(x-3)(x+1) = 0$$

Apply the zero product principle.

$x - 3 = 0 \qquad x + 1 = 0$

$x = 3 \qquad x = -1$

The solution set is $\{-1,3\}$. We disregard -1 because it is not possible to have a negative length measurement. The guy wire is $x + 2 = 3 + 2 = 5$ feet long.

Problem Set 5.7

Practice Exercises

1.
$$x^2 + x - 12 = 0$$
$$(x+4)(x-3) = 0$$

Apply the zero product principle.

$x + 4 = 0 \qquad x - 3 = 0$

$x = -4 \qquad x = 3$

The solutions are -4 and 3 and the solution set is $\{-4,3\}$.

3.
$$x^2 + 6x = 7$$
$$x^2 + 6x - 7 = 0$$
$$(x+7)(x-1) = 0$$

Apply the zero product principle.

$x + 7 = 0 \qquad x - 1 = 0$

$x = -7 \qquad x = 1$

The solutions are -7 and 1 and the solution set is $\{-7,1\}$.

5.
$$3x^2 + 10x - 8 = 0$$
$$(3x-2)(x+4) = 0$$

Apply the zero product principle.

$3x - 2 = 0 \qquad x + 4 = 0$

$3x = 2 \qquad x = -4$

$x = \frac{2}{3}$

The solutions are -4 and $\frac{2}{3}$ and the solution set is $\left\{-4, \frac{2}{3}\right\}$.

7. $$5x^2 = 8x - 3$$
$$5x^2 - 8x + 3 = 0$$
$$(5x-3)(x-1) = 0$$
Apply the zero product principle.
$$5x - 3 = 0 \qquad x - 1 = 0$$
$$5x = 3 \qquad x = 1$$
$$x = \frac{3}{5}$$
The solutions are $\frac{3}{5}$ and 1 and the solution set is $\left\{\frac{3}{5}, 1\right\}$.

9. $$3x^2 = 2 - 5x$$
$$3x^2 + 5x - 2 = 0$$
$$(3x-1)(x+2) = 0$$
Apply the zero product principle.
$$3x - 1 = 0 \qquad x + 2 = 0$$
$$3x = 1 \qquad x = -2$$
$$x = \frac{1}{3}$$
The solutions are -2 and $\frac{1}{3}$ and the solution set is $\left\{-2, \frac{1}{3}\right\}$.

11. $$x^2 = 8x$$
$$x^2 - 8x = 0$$
$$x(x-8) = 0$$
Apply the zero product principle.
$$x = 0 \qquad x - 8 = 0$$
$$x = 8$$
The solutions are 0 and 8 and the solution set is $\{0, 8\}$.

13. $$3x^2 = 5x$$
$$3x^2 - 5x = 0$$
$$x(3x-5) = 0$$
Apply the zero product principle.
$$x = 0 \qquad 3x - 5 = 0$$
$$3x = 5$$
$$x = \frac{5}{3}$$
The solutions are 0 and $\frac{5}{3}$ and the solution set is $\left\{0, \frac{5}{3}\right\}$.

15. $$x^2 + 4x + 4 = 0$$
$$(x+2)^2 = 0$$
There are two identical factors. Apply the zero product principle and solve the equation one time to obtain the double-root.
$$x + 2 = 0$$
$$x = -2$$
The solution is -2 and the solution set is $\{-2\}$.

17. $$x^2 = 14x - 49$$
$$x^2 - 14x + 49 = 0$$
$$(x-7)^2 = 0$$
Apply the zero product principle to obtain the double-root.

$$x-7=0$$
$$x=7$$
The solution is 7 and the solution set is $\{7\}$.

19. $$9x^2 = 30x-25$$
$$9x^2-30x+25=0$$
$$(3x-5)^2=0$$
Apply the zero product principle to obtain the double-root.
$$3x-5=0$$
$$3x=5$$
$$x=\frac{5}{3}$$
The solution is $\frac{5}{3}$ and the solution set is $\left\{\frac{5}{3}\right\}$.

21. $$x^2-25=0$$
$$(x+5)(x-5)=0$$
Apply the zero product principle.
$x+5=0 \qquad x-5=0$
$x=-5 \qquad x=5$
The solutions are –5 and 5 and the solution set is $\{-5,5\}$.

23. $$9x^2=100$$
$$9x^2-100=0$$
$$(3x+10)(3x-10)=0$$
Apply the zero product principle.
$3x+10=0 \qquad 3x-10=0$
$3x=-10 \qquad 3x=10$
$x=-\frac{10}{3} \qquad x=\frac{10}{3}$
The solutions are $-\frac{10}{3}$ and $\frac{10}{3}$ and the solution set is $\left\{-\frac{10}{3},\frac{10}{3}\right\}$.

25. $$x(x-3)=18$$
$$x^2-3x=18$$
$$x^2-3x-18=0$$
$$(x-6)(x+3)=0$$
Apply the zero product principle.
$x-6=0 \qquad x+3=0$
$x=6 \qquad x=-3$
The solutions are –3 and 6 and the solution set is $\{-3,6\}$.

27. $$(x-3)(x+8)=-30$$
$$x^2+8x-3x-24=-30$$
$$x^2+5x-24=-30$$
$$x^2+5x+6=0$$
$$(x+3)(x+2)=0$$
Apply the zero product principle.
$x+3=0 \qquad x+2=0$
$x=-3 \qquad x=-2$
The solutions are –3 and –2 and the solution set is $\{-3,-2\}$.

29. $$x(x+8)=16(x-1)$$
$$x^2+8x=16x-16$$
$$x^2-8x+16=0$$
$$(x-4)^2=0$$
Apply the zero product principle to obtain the double-root.
$$x-4=0$$
$$x=4$$

The solution is 4 and the solution set is $\{4\}$.

31.
$$(x+1)^2-5(x+2)=3x+7$$
$$(x^2+2x+1)-5x-10=3x+7$$
$$x^2+2x+1-5x-10=3x+7$$
$$x^2-3x-9=3x+7$$
$$x^2-6x-16=0$$
$$(x-8)(x+2)=0$$
Apply the zero product principle.
$x-8=0 \quad x+2=0$
$x=8 \quad x=-2$
The solutions are –2 and 8 and the solution set is $\{-2,8\}$.

33.
$$x(8x+1)=3x^2-2x+2$$
$$8x^2+x=3x^2-2x+2$$
$$5x^2+3x-2=0$$
$$(5x-2)(x+1)=0$$
Apply the zero product principle.
$5x-2=0 \quad x+1=0$
$5x=2 \quad x=-1$
$x=\frac{2}{5}$
The solutions are -1 and $\frac{2}{5}$ and the solution set is $\left\{-1,\frac{2}{5}\right\}$.

35.
$$\frac{x^2}{18}+\frac{x}{2}+1=0$$
$$18\left(\frac{x^2}{18}\right)+18\left(\frac{x}{2}\right)+18(1)=18(0)$$
$$x^2+9x+18=0$$
$$(x+3)(x+6)=0$$
Apply the zero product principle.
$x+3=0 \quad x+6=0$
$x=-3 \quad x=-6$
The solutions are –6 and –3 and the solution set is $\{-6,-3\}$.

37.
$$x^3+4x^2-25x-100=0$$
$$x^2(x+4)-25(x+4)=0$$
$$(x+4)(x^2-25)=0$$
$$(x+4)(x+5)(x-5)=0$$
Apply the zero product principle.
$x+4=0 \quad x+5=0 \quad x-5=0$
$x=-4 \quad x=-5 \quad x=5$
The solutions are –5, –4, and 5 and the solution set is $\{-5,-4,5\}$.

39.
$$x^3-x^2=25x-25$$
$$x^3-x^2-25x+25=0$$
$$x^2(x-1)-25(x-1)=0$$
$$(x-1)(x^2-25)=0$$
$$(x-1)(x+5)(x-5)=0$$
Apply the zero product principle.
$x-1=0 \quad x+5=0 \quad x-5=0$
$x=1 \quad x=-5 \quad x=5$
The solutions are –5, 1 and 5 and the solution set is $\{-5,1,5\}$.

41.
$$3x^4-48x^2=0$$
$$3x^2(x^2-16)=0$$
$$3x^2(x+4)(x-4)=0$$

Apply the zero product principle.

$3x^2 = 0 \quad x+4=0 \quad x-4=0$

$x=0 \quad x=-4 \quad x=4$

The solutions are –4, 0 and 4 and the solution set is $\{-4,0,4\}$.

43. $$2x^4 = 16x$$

$$2x^4 - 16x = 0$$

$$2x(x^3-8)=0$$

$$2x(x-2)(x^2+2x+4)=0$$

Apply the zero product principle.

$2x = 0 \quad x-2=0$

$x=0 \quad x=2$

$x^2+2x+4=0$

Since x^2+2x+4 will always be greater than zero, there is no solution.

The solutions are 0 and 2 and the solution set is $\{0,2\}$.

45. $$2x^3+16x^2+30x=0$$

$$2x(x^2+8x+15)=0$$

$$2x(x+5)(x+3)=0$$

Apply the zero product principle.

$2x=0 \quad x+5=0 \quad x+3=0$

$x=0 \quad x=-5 \quad x=-3$

The solutions are –5, –3 and 0 and the solution set is $\{-5,-3,0\}$.

47. $$x^2-6x+8=0$$

$$(x-4)(x-2)=0$$

Apply the zero product principle.

$x-4=0 \quad x-2=0$

$x=4 \quad x=2$

The x-intercepts are 2 and 4. This corresponds to graph (d).

49. $$x^2+6x+8=0$$

$$(x+4)(x+2)=0$$

Apply the zero product principle.

$x+4=0 \quad x+2=0$

$x=-4 \quad x=-2$

The x-intercepts are -2 and -4. This corresponds to graph (c).

Application Exercises

51. $$-16t^2+20t+300=0$$

$$-4(4t^2-5t-75)=0$$

$$-4(t-5)(4t+15)=0$$

Apply the zero product principle.

$-4(t-5)=0 \quad 4t+15=0$

$t-5=0 \quad 4t=-15$

$t=5 \quad t=-\frac{15}{4}$

The solution set is $\left(-\frac{15}{4},5\right)$. Disregard $-\frac{15}{4}$ because we can't have a negative time measurement. The ball will hit the ground at t = 5 seconds.

53. $$-16t^2+20t+300=276$$

$$-16t^2+20t+24=0$$

$$-4(4t^2+5t+6)=0$$

$$-4(4t+3)(t-2)=0$$

Apply the zero product principle.

$$-4(4t+3)=0 \qquad t-2=0$$
$$4t+3=0 \qquad t=2$$
$$4t=-3$$
$$t=-\frac{3}{4}$$

The solution set is $\left(-\frac{3}{4},2\right)$. In this application, we disregard $-\frac{3}{4}$, because we can't have a negative time measurement and conclude that the ball will be at a height of 276 feet at t = 2 seconds.

55.
$$-\frac{1}{2}x^2+4x+19=27$$
$$-\frac{1}{2}x^2+4x-8=0$$
$$-2\left(-\frac{1}{2}x^2+4x-8\right)=-2(0)$$
$$x^2-8x+16=0$$
$$(x-4)^2=0$$

Apply the zero product principle by solving the equation one time to obtain the double-root.

$$x-4=0$$
$$x=4$$

The solution set is $\{4\}$. This means that 27 million people will receive food stamps in 1994.

57. In Exercise 55, we found that 4 years after 1994, 27 million people will receive food stamps. This corresponds to the point $(4,27)$ on the graph.

59. Let w = the width
Let $w+3$ = the length

Area = lw

$$54=(w+3)w$$
$$54=w^2+3w$$
$$0=w^2+3w-54$$
$$0=(w+9)(w-6)$$

Apply the zero product principle.

$$w+9=0 \qquad w-6=0$$
$$w=-9 \qquad w=6$$

The solution set is $\{-9,6\}$. Disregard –9 because we can't have a negative length measurement. The width is 6 feet and the length is 6 feet + 3 feet = 9 feet.

61. Let x = the length of the side of the original square
Let $x + 3$ = the length of the side of the new, larger square

$$(x+3)^2=64$$
$$x^2+6x+9=64$$
$$x^2+6x-55=0$$
$$(x+11)(x-5)=0$$

Apply the zero product principle.

$$x+11=0 \qquad x-5=0$$
$$x=-11 \qquad x=5$$

The solution set is $\{-11,5\}$. Disregard -11 because we can't have a negative length measurement. This means that x, the length of the side of the original square, is 5 inches.

63. Let x = the width of the path

$$(20+2x)(10+2x)=600$$
$$200+40x+20x+4x^2=600$$
$$200+60x+4x^2=600$$

$$4x^2+60x+200=600$$
$$4x^2+60x-400=0$$
$$4\left(x^2+15x-100\right)=0$$
$$4(x+20)(x-5)=0$$

Apply the zero product principle.

$4(x+20)=0$ $\quad x-5=0$

$x+20=0$ $\quad x=5$

$x=-20$

The solution set is $\{-20,5\}$. Disregard -20 because we can't have a negative width measurement. The width of the path is 5 meters.

65. a. $(2x+12)(2x+10)-10(12)$

$=4x^2+20x+24x+\cancel{120}-\cancel{120}$

$=4x^2+44x$

b.
$$4x^2+44x=168$$
$$4x^2+44x-168=0$$
$$4\left(x^2+11x-42\right)=0$$
$$4(x+14)(x-3)=0$$

Apply the zero product principle.

$4(x+14)=0$ $\quad x-3=0$

$x+14=0$ $\quad x=3$

$x=-14$

The solution set is $\{-14,3\}$. In this application, we disregard -14 because we can't have a negative width measurement. The width of the border is 3 feet.

67. Volume $= lwh$

$$200=x\cdot x\cdot 2$$
$$200=2x^2$$
$$0=2x^2-200$$
$$0=2\left(x^2-100\right)$$
$$0=2(x+10)(x-10)$$

Apply the zero product principle.

$2(x+10)=0$ $\quad x-10=0$

$x+10=0$ $\quad x=10$

$x=-10$

The solution set is $\{-10,10\}$. Disregard -10 because we can't have a negative width measurement. The length and width of the open box are 10 inches.

69.
$$\text{leg}^2+\text{leg}^2=\text{hypotenuse}^2$$
$$(2x+2)^2+x^2=13^2$$
$$4x^2+8x+4+x^2=169$$
$$5x^2+8x+4=169$$
$$5x^2+8x-165=0$$
$$(5x+33)(x-5)=0$$

Apply the zero product principle.

$5x+33=0$ $\quad x-5=0$

$5x=-33$ $\quad x=5$

$x=-\dfrac{33}{5}$

The solution set is $\left\{-\dfrac{33}{5},5\right\}$. In this application, we disregard $-\dfrac{33}{5}$ because we can't have a negative width measurement. The width of the closet is 5 feet. The length is $2x+2=2(5)+2=10+2=12$ feet.

71. Let x = the length of the wire

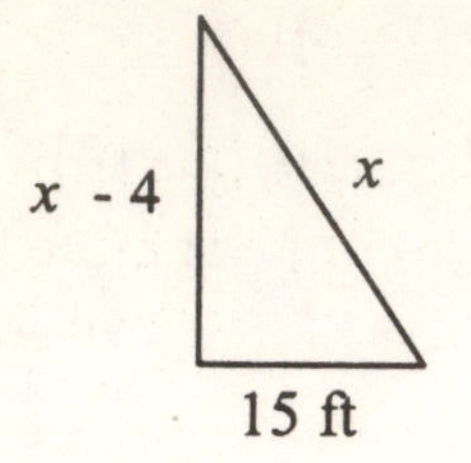

$$\text{leg}^2 + \text{leg}^2 = \text{hypotenuse}^2$$
$$(x-4)^2 + 15^2 = x^2$$
$$\cancel{x^2} - 8x + 16 + 225 = \cancel{x^2}$$
$$-8x + 241 = 0$$
$$-8x = -241$$
$$-8x = -241$$
$$x = \frac{241}{8} = 30\frac{1}{8}$$

The length of the wire is $30\frac{1}{8}$ or 30.125 feet.

Writing in Mathematics

73. Answers will vary.

75. Answers will vary.

77. Answers will vary.

79. Answers will vary.

81. Answers will vary.

83. Answers will vary.

Technology Exercises

85. $y = x^3 + 3x^2 - x - 3$

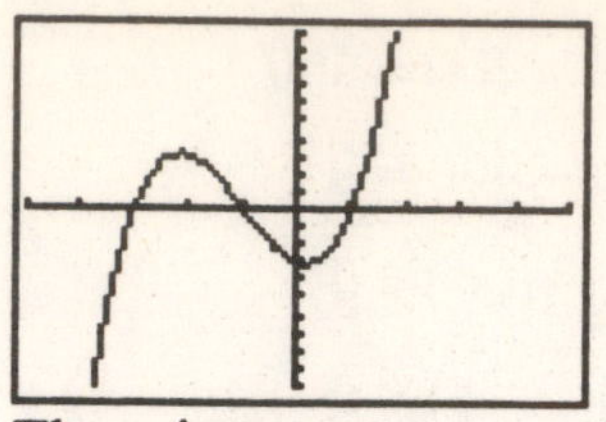

The x-intercepts are -3, -1, and 1. Use the intercepts to solve the equation.

$$x^3 + 3x^2 - x - 3 = 0$$
$$x^2(x+3) - 1(x+3) = 0$$
$$(x+3)(x^2-1) = 0$$
$$(x+3)(x+1)(x-1) = 0$$

Apply the zero product principle.

$x+3=0 \quad x+1=0 \quad x-1=0$

$x=-3 \quad x=-1 \quad x=1$

The solutions are -3, -1 and 1 and the solution set is $\{-3,-1,1\}$.

87. $y = -x^4 + 4x^3 - 4x^2$

The x-intercepts are 0 and 2. Use the intercepts to solve the equation.

$$-x^4 + 4x^3 - 4x^2 = 0$$
$$-x^2(x^2 - 4x + 4) = 0$$
$$-x^2(x-2)^2 = 0$$

Apply the zero product principle.

$-x^2 = 0 \quad (x-2)^2 = 0$

$x^2 = 0 \quad x-2=0$

$x=0 \quad x=2$

The solutions are 0 and 2 and the solution set is $\{0,2\}$.

Critical Thinking Exercises

89. Set the denominator equal to zero and solve.

$$x^2+4x-45=0$$
$$(x+9)(x-5)=0$$

Apply the zero product principle.

$x+9=0$ $\quad$ $x-5=0$

$x=-9$ $\quad$ $x=5$

The domain is $\{x|x$ is a real number and $x\neq -9$ and $x\neq 5\}$ or $(-\infty,-9)\cup(-9,5)\cup(5,\infty)$.

91. Use the solutions, –3 and 7, to find the factors of the polynomial.

$x=-3$ $\quad$ $x=7$

$x+3=0$ $\quad$ $x-7=0$

Next, multiply the factors.

$$(x+3)(x-7)=0$$
$$x^2-7x+3x-21=0$$
$$x^2-4x-21=0$$

The quadratic equation in standard form is $x^2-4x-21=0$.

Review Exercises

93. $|3x-2|=8$

$3x-2=-8$ or $3x-2=8$

$3x=-6$ $\quad$ $3x=10$

$x=-2$ $\quad$ $x=\frac{10}{3}$

The solutions are -2 and $\frac{10}{3}$ and the solution set is $\left\{-2,\frac{10}{3}\right\}$.

94.
$$3(5-7)^2+\sqrt{16}+12\div(-3)$$
$$=3(-2)^2+4+(-4)$$
$$=3(4)+4+(-4)$$
$$=12+4+(-4)$$
$$=12$$

95. Let x = the amount invested at 5%

Let y = the amount invested at 8%

$$x+y=3000$$
$$0.05x+0.08y=189$$

Multiply the first equation by –0.05 and add to the second equation.

$$-0.05x-0.05y=-0.05(3000)$$
$$-0.05x-0.05y=-150$$

$$-0.05x-0.05y=-150$$
$$0.05x+0.08y=189$$
$$0.03y=39$$
$$y=1300$$

Back-substitute 1300 for y in one of the original equations.

$$x+y=3000$$
$$x+1300=3000$$
$$x=1700$$

\$1700 was invested at 5% and \$1300 was invested at 8%.

Chapter 5 Review

5.1

1. The coefficient of $-5x^3$ is –5 and the degree is 3.
The coefficient of $7x^2$ is 7 and the degree is 2.
The coefficient of $-x$ is –1 and the degree is 1.

The coefficient of 2 is 2 and the degree is 0.
The degree of the polynomial is 3.
The leading term is $-5x^3$ and the leading coefficient is –5.

2. The coefficient of $8x^4y^2$ is 8 and the degree is 6.
The coefficient of $-7xy^6$ is –7 and the degree is 7.
The coefficient of $-x^3y$ is –1 and the degree is 4.
The degree of the polynomial is 7.
The leading term is $-7xy^6$ and the leading coefficient is –7.

3.
$$\begin{aligned} f(x) &= x^3 - 4x^2 + 3x - 1 \\ f(-2) &= (-2)^3 - 4(-2)^2 + 3(-2) - 1 \\ &= -8 - 4(4) - 6 - 1 \\ &= -8 - 16 - 6 - 1 = -31 \end{aligned}$$

4.
$$\begin{aligned} f(x) &= 0.26x^2 + 2.09x + 16.4 \\ f(10) &= 0.26(10)^2 + 2.09(10) + 16.4 \\ &= 0.26(100) + 20.9 + 16.4 \\ &= 26 + 20.9 + 16.4 = 63.3 \end{aligned}$$

There are 63,300 people waiting for organ transplants in the United States in the year $1990 + 10 = 2000$.

5. Since the degree of the polynomial is 3, an odd number, and the leading coefficient is –1, the graph will rise to the left and fall to the right. The graph of the polynomial is graph (**c**).

6. Since the degree of the polynomial is 6, an even number, and the leading coefficient is 1, the graph will rise to the left and to the right. The graph of the polynomial is graph (**b**).

7. Since the degree of the polynomial is 5, an odd number, and the leading coefficient is 1, the graph will fall to the left and rise to the right. The graph of the polynomial is graph (**a**).

8. Since the degree of the polynomial is 4, an even number, and the leading coefficient is –1, the graph will fall to the left and fall to the right. The graph of the polynomial is graph (**d**).

9. Since the degree of the polynomial is 3, an odd number, and the leading coefficient is negative, the graph will fall to the right. This means that over time, the model predicts the number of families living below the poverty level will go to zero. This is probably not realistic and the model is valid only for a limited period of time.

10. $$\begin{aligned}\left(-8x^3+5x^2-7x+4\right)+\left(9x^3-11x^2+6x-13\right) &= -8x^3+5x^2-7x+4+9x^3-11x^2+6x-13\\ &= -8x^3+9x^3+5x^2-11x^2-7x+6x+4-13\\ &= x^3-6x^2-x-9\end{aligned}$$

11. $$\begin{aligned}\left(7x^3y-13x^2y-6y\right)+\left(5x^3y+11x^2y-8y-17\right) &= 7x^3y-13x^2y-6y+5x^3y+11x^2y-8y-17\\ &= 7x^3y+5x^3y-13x^2y+11x^2y-6y-8y-17\\ &= 12x^3y-2x^2y-14y-17\end{aligned}$$

12. $$\begin{aligned}\left(7x^3-6x^2+5x-11\right)-\left(-8x^3+4x^2-6x-3\right) &= 7x^3-6x^2+5x-11+8x^3-4x^2+6x+3\\ &= 7x^3+8x^3-6x^2-4x^2+5x+6x-11+3\\ &= 15x^3-10x^2+11x-8\end{aligned}$$

13. $$\begin{aligned}\left(4x^3y^2-7x^3y-4\right)-\left(6x^3y^2-3x^3y+4\right) &= 4x^3y^2-7x^3y-4-6x^3y^2+3x^3y-4\\ &= 4x^3y^2-6x^3y^2-7x^3y+3x^3y-4-4\\ &= -2x^3y^2-4x^3y-8\end{aligned}$$

14. $$\begin{aligned}\left(x^3+4x^2y-y^3\right)-\left(-2x^3-x^2y+xy^2+7y^3\right) &= x^3+4x^2y-y^3+2x^3+x^2y-xy^2-7y^3\\ &= x^3+2x^3+4x^2y+x^2y-xy^2-y^3-7y^3\\ &= 3x^3+5x^2y-xy^2-8y^3\end{aligned}$$

5.2

15. $$\begin{aligned}&\left(4x^2yz^5\right)\left(-3x^4yz^2\right)\\ &= 4x^2yz^5(-3)x^4yz^2\\ &= 4(-3)x^2x^4yyz^5z^2\\ &= -12x^6y^2z^7\end{aligned}$$

16. $$\begin{aligned}&6x^3\left(\frac{1}{3}x^5-4x^2-2\right)\\ &= 6x^3\left(\frac{1}{3}x^5\right)-6x^3\left(4x^2\right)-6x^3(2)\\ &= 6\left(\frac{1}{3}\right)x^3x^5-6\cdot 4x^3x^2-6\cdot 2x^3\\ &= 2x^8-24x^5-12x^3\end{aligned}$$

17. $7xy^2(3x^4y^2 - 5xy - 1)$
$= 7xy^2(3x^4y^2) - 7xy^2(5xy) - 7xy^2(1)$
$= 7 \cdot 3xx^4y^2y^2 - 7 \cdot 5xxy^2y - 7xy^2$
$= 21x^5y^4 - 35x^2y^3 - 7xy^2$

18.
$$\begin{array}{r} 3x^2 + 7x - 4 \\ \underline{2x + 5} \\ 6x^3 + 14x^2 - 8x \\ \underline{15x^2 + 35x - 20} \\ 6x^3 + 29x^2 + 27x - 20 \end{array}$$

19.
$$\begin{array}{r} x^2 + x - 1 \\ \underline{x^2 + 3x + 2} \\ x^4 + x^3 - x^2 \\ 3x^3 + 3x^2 - 3x \\ \underline{2x^2 + 2x - 2} \\ x^4 + 4x^3 + 4x^2 - x - 2 \end{array}$$

20. $(4x - 1)(3x - 5)$
$= 12x^2 - 20x - 3x + 5$
$= 12x^2 - 23x + 5$

21. $(3xy - 2)(5xy + 4)$
$= 15x^2y^2 + 12xy - 10xy - 8$
$= 15x^2y^2 + 2xy - 8$

22. Two methods can be used to multiply the binomials.
Using the FOIL Method:
$(3x + 7y)^2$
$= (3x + 7y)(3x + 7y)$
$= 9x^2 + 21xy + 21xy + 49y^2$
$= 9x^2 + 42xy + 49y^2$
Recognizing a Perfect Square Trinomial:
$(3x + 7y)^2$
$= (3x)^2 + (7y)^2$
$= 9x^2 + 2 \cdot 3x \cdot 7y + 49y^2$
$= 9x^2 + 42xy + 49y^2$

23. Two methods can be used to multiply the binomials.
Using the FOIL Method:
$(x^2 - 5y)^2$
$= (x^2 - 5y)(x^2 - 5y)$
$= x^4 - 5x^2y - 5x^2y + 25y^2$
$= x^4 - 10x^2y + 25y^2$
Recognizing a Perfect Square Trinomial:
$(x^2 - 5y)^2$
$= (x^2)^2 + (-5y)^2$
$= x^4 + 2 \cdot x^2(-5y) + 25y^2$
$= x^4 - 10x^2y + 25y^2$

24. Two methods can be used to multiply the binomials.
Using the FOIL Method:
$(2x + 7y)(2x - 7y)$
$= 4x^2 - 14xy + 14xy - 49y^2$
$= 4x^2 - 49y^2$
Recognizing the Difference of Two Squares:

$(2x+7y)(2x-7y)$
$=(2x)^2-(7y)^2$
$=4x^2-49y^2$

25. Two methods can be used to multiply the binomials.
Using the FOIL Method:
$(3xy^2-4x)(3xy^2+4x)$
$=9x^2y^4+12x^2y^2-12x^2y^2-16x^2$
$=9x^2y^4-16x^2$
Recognizing the Difference of Two Squares:
$(3xy^2-4x)(3xy^2+4x)$
$=(3xy^2)^2-(4x)^2=9x^2y^4-16x^2$

26. Two methods can be used to multiply the binomials.
Using the FOIL Method:
$[(x+3)+5y][(x+3)-5y]$
$=(x+3)^2-5(x+3)y+5(x+3)y-25y^2$
$=(x+3)^2-25y^2$
$=x^2+6x+9-25y^2$
Recognizing the Difference of Two Squares:
$[(x+3)+5y][(x+3)-5y]$
$=(x+3)^2-(5y)^2=x^2+6x+9-25y^2$

27.
$$\begin{array}{l} x+y+4 \\ \underline{x+y+4} \\ x^2+xy+4x \\ \quad xy \qquad +y^2+4y \\ \underline{\qquad\quad 4x \qquad +4y+16} \\ x^2+2xy+8x+y^2+8y+16 \end{array}$$
or
$x^2+2xy+y^2+8x+8y+16$

28. $f(x)=x-3$ and $g(x)=2x+5$
$(fg)(x)=f(x)\cdot g(x)$
$=(x-3)(2x+5)$
$=2x^2+5x-6x-15$
$=2x^2-x-15$
$(fg)(-4)=2(-4)^2-(-4)-15$
$=2(16)+4-15$
$=32+4-15$
$=21$

29. $f(x)=x^2-7x+2$

a. $f(a-1)$
$=(a-1)^2-7(a-1)+2$
$=a^2-2a+1-7a+7+2$
$=a^2-9a+10$

b. $f(a+h)-f(a)$
$=(a+h)^2-7(a+h)+2-(a^2-7a+2)$
$=\cancel{a^2}+2ah+h^2\cancel{-7a}-7h\cancel{+2}\cancel{-a^2}\cancel{+7a}\cancel{-2}$
$=2ah+h^2-7h$

5.3

30. $16x^3+24x^2=8x^2(2x+3)$

31. $2x-36x^2=2x(1-18x)$

32. $21x^2y^2-14xy^2+7xy$
$=7xy(3xy-2y+1)$

33. $18x^3y^2-27x^2y=9x^2y(2xy-3)$

34. $-12x^2+8x-48=-4(3x^2-2x+12)$

35. $-x^2-11x+14=-1(x^2+11x-14)$
$=-(x^2+11x-14)$

36. $x^3-x^2-2x+2=x^2(x-1)-2(x-1)$
$=(x-1)(x^2-2)$

37. $xy-3x-5y+15=x(y-3)-5(y-3)$
$=(y-3)(x-5)$

38. $5ax-15ay+2bx-6by$
$=5a(x-3y)+2b(x-3y)$
$=(x-3y)(5a+2b)$

5.4

39. $x^2+8x+15=(x+5)(x+3)$

40. $x^2+16x-80=(x+20)(x-4)$

41. $x^2+16xy-17y^2=(x+17y)(x-y)$

42. $3x^3-36x^2+33x=3x(x^2-12x+11)$
$=3x(x-11)(x-1)$

43. $3x^2+22x+7=(3x+1)(x+7)$

44. $6x^2-13x+6=(2x-3)(3x-2)$

45. $5x^2-6xy-8y^2=(5x+4y)(x-2y)$

46. $6x^3+5x^2-4x=x(6x^2+5x-4)$
$=x(2x-1)(3x+4)$

47. $2x^2+11x+15=(2x+5)(x+3)$

48. Let $t=x^3$
$x^6+x^3-30=(x^3)^2+x^3-30$
$=t^2+t-30$
$=(t+6)(t-5)$
$=(x^3+6)(x^3-5)$

49. Let $t=x^2$
$x^4-10x^2-39=(x^2)^2-10x^2-39$
$=t^2-10t-39$
$=(t-13)(t+3)$
$=(x^2-13)(x^2+3)$

50. Let $t=x+5$
$(x+5)^2+10(x+5)+24$
$=t^2+10t+24$
$=(t+6)(t+4)$

$=((x+5)+6)((x+5)+4)$
$=(x+5+6)(x+5+4)$
$=(x+11)(x+9)$

51. $5x^6+17x^3+6=5(x^3)^2+17x^3+6$
$=5t^2+17t+6$
$=(5t+2)(t+3)$
$=(5x^3+2)(x^3+3)$

5.5

52. $4x^2-25=(2x+5)(2x-5)$

53. $1-81x^2y^2=(1+9xy)(1-9xy)$

54. $x^8-y^6=(x^4+y^3)(x^4-y^3)$

55. $(x-1)^2-y^2$
$=((x-1)+y)((x-1)-y)$
$=(x-1+y)(x-1-y)$

56. $x^2+16x+64$
$=x^2+2\cdot x\cdot 8+8^2$
$=(x+8)^2$

57. $9x^2-6x+1=(3x)^2-2\cdot 3x\cdot 1+1^2$
$=(3x-1)^2$

58. $25x^2+20xy+4y^2$
$=(5x)^2+2\cdot 5x\cdot 2y+(2y)^2$
$=(5x+2y)^2$

59. $49x^2+7x+1$
Prime

60. $25x^2-40xy+16y^2$
$=(5x)^2-2\cdot 5x\cdot 4y+(4y)^2$
$=(5x-4y)^2$

61. $x^2+18x+81-y^2$
$=(x^2+18x+81)-y^2$
$=(x+9)^2-y^2$
$=((x+9)+y)((x+9)-y)$
$=(x+9+y)(x+9-y)$

62. $z^2-25x^2+10x-1$
$=z^2-(25x^2-10x+1)$
$=z^2-((5x)^2-2\cdot 5x\cdot 1+1^2)$
$=z^2-(5x-1)^2$
$=(z+(5x-1))(z-(5x-1))$
$=(z+5x-1)(z-5x+1)$

63. $64x^3+27$
$=(4x)^3+(3)^3$
$=(4x+3)((4x)^2-4x\cdot 3+3^2)$
$=(4x+3)(16x^2-12x+9)$

64. $125x^3-8$
$=(5x)^3-(2)^3$
$=(5x-2)((5x)^2+5x\cdot 2+2^2)$
$=(5x-2)(25x^2+10x+4)$

65. x^3y^3+1
$=(xy)^3+1^3$
$=(xy+1)\left((xy)^2-xy\cdot 1+1^2\right)$
$=(xy+1)\left(x^2y^2-xy+1\right)$

5.6

66. $15x^2+3x=3x(5x+1)$

67. $12x^4-3x^2$
$=3x^2\left(4x^2-1\right)$
$=3x^2\left((2x)^2-1^2\right)$
$=3x^2(2x+1)(2x-1)$

68. $20x^4-24x^3+28x^2-12x$
$=4x\left(5x^3-6x^2+7x-3\right)$

69. x^3-15x^2+26x
$=x\left(x^2-15x+26\right)$
$=x(x-2)(x-13)$

70. $-2y^4+24y^3-54y^2$
$=-2y^2\left(y^2-12y+27\right)$
$=-2y^2(y-9)(y-3)$

71. $9x^2-30x+25$
$=(3x)^2-2\cdot 3x\cdot 5+5^2$
$=(3x-5)^2$

72. $5x^2-45=5\left(x^2-9\right)$
$=5(x+3)(x-3)$

73. $2x^3-x^2-18x+9$
$=x^2(2x-1)-9(2x-1)$
$=(2x-1)\left(x^2-9\right)$
$=(2x-1)(x+3)(x-3)$

74. $6x^2-23xy+7y^2$
$=(3x-y)(2x-7y)$

75. $2y^3+12y^2+18y$
$=2y\left(y^2+6y+9\right)$
$=2y\left(y^2+2\cdot y\cdot 3+3^2\right)$
$=2y(y+3)^2$

76. $x^2+6x+9-4a^2$
$=\left(x^2+6x+9\right)-4a^2$
$=\left(x^2+2\cdot x\cdot 3+3^2\right)-(2a)^2$
$=(x+3)^2-(2a)^2$
$=((x+3)+2a)((x+3)-2a)$
$=(x+3+2a)(x+3-2a)$

77. $8x^3-27$
$=(2x)^3-3^3$
$=(2x-3)\left((2x)^2+2x\cdot 3+3^2\right)$
$=(2x-3)\left(4x^2+6x+9\right)$

78. $x^5 - x$
$= x(x^4 - 1)$
$= x((x^2)^2 - 1^2)$
$= x(x^2 + 1)(x^2 - 1)$
$= x(x^2 + 1)(x^2 - 1^2)$
$= x(x^2 + 1)(x + 1)(x - 1)$

79. Let $t = x^2$
$x^4 - 6x^2 + 9 = (x^2)^2 - 6x^2 + 9$
$= t^2 - 6t + 9$
$= t^2 - 2 \cdot t \cdot 3 + 3^2$
$= (t - 3)^2$
$= (x^2 - 3)^2$

80. $x^2 + xy + y^2$
Prime

81. $4a^3 + 32$
$= 4(a^3 + 8)$
$= 4(a^3 + 2^3)$
$= 4(a + 2)(a^2 - a \cdot 2 + 2^2)$
$= 4(a + 2)(a^2 - 2a + 4)$

82. $x^4 - 81$
$= (x^2)^2 - 9^2$
$= (x^2 + 9)(x^2 - 9)$
$= (x^2 + 9)(x^2 - 3^2)$
$= (x^2 + 9)(x + 3)(x - 3)$

83. $ax + 3bx - ay - 3by$
$= x(a + 3b) - y(a + 3b)$
$= (a + 3b)(x - y)$

84. $27x^3 - 125y^3$
$= (3x)^3 - (5y)^3$
$= (3x - 5y)((3x)^2 + 3x \cdot 5y + (5y)^2)$
$= (3x - 5y)(9x^2 + 15xy + 25y^2)$

85. $10x^3y + 22x^2y - 24xy$
$= 2xy(5x^2 + 11x - 12)$
$= 2xy(5x - 4)(x + 3)$

86. Let $t = x^3$
$6x^6 + 13x^3 - 5$
$= 6(x^3)^2 + 13x^3 - 5$
$= 6t^2 + 13t - 5$
$= (2x^3 + 5)(3x^3 - 1)$

87. **a.** $2xy + 2y^2$
b. $2xy + 2y^2 = 2y(x + y)$

88. **a.** $x^2 - 4y^2$
b. $x \;\; - 4y^2 = x^2 - (2y)^2$
$= (x + 2y)(x - 2y)$

5.7

89. $x^2 + 6x + 5 = 0$
$(x + 5)(x + 1) = 0$
Apply the zero product principle.

$x+5=0 \qquad x+1=0$

$x=-5 \qquad x=-1$

The solutions are –5 and –1 and the solution set is $\{-5,-1\}$.

90. $3x^2 = 22x-7$

$3x^2-22x+7=0$

$(3x-1)(x-7)=0$

Apply the zero product principle.

$3x-1=0 \qquad x-7=0$

$3x=1 \qquad x=7$

$x=\frac{1}{3}$

The solutions are $\frac{1}{3}$ and 7 and the solution set is $\left\{\frac{1}{3},7\right\}$.

91. $(x+3)(x-2)=50$

$x^2-2x+3x-6=50$

$x^2+x-6=50$

$x^2+x-56=0$

$(x+8)(x-7)=0$

Apply the zero product principle.

$x+8=0 \qquad x-7=0$

$x=-8 \qquad x=7$

The solutions are –8 and 7 and the solution set is $\{-8,7\}$.

92. $3x^2=12x$

$3x^2-12x=0$

$3x(x-4)=0$

Apply the zero product principle.

$3x=0 \qquad x-4=0$

$x=0 \qquad x=4$

The solutions are 0 and 4 and the solution set is $\{0,4\}$.

93. $x^3+5x^2=9x+45$

$x^3+5x^2-9x-45=0$

$(x^3+5x^2)+(-9x-45)=0$

$x^2(x+5)+(-9)(x+5)=0$

$(x+5)(x^2-9)=0$

$(x+5)(x+3)(x-3)=0$

Apply the zero product principle.

$x+5=0 \qquad x+3=0 \qquad x-3=0$

$x=-5 \qquad x=-3 \qquad x=3$

The solutions are –5, –3 and 3 and the solution set is $\{-5,-3,3\}$.

94. $-16t^2+128t+144=0$

$-16(t^2-8t+9)=0$

$-16(t-9)(t+1)=0$

Apply the zero product principle.

$-16(t-9)=0 \qquad t+1=0$

$t-9=0 \qquad t=-1$

$t=9$

The solution set is $\{-1,9\}$. Disregard –1 because we can't have a negative time measurement. The rocket will hit the water at $t=9$ seconds.

95. $2x^2+22x+320=740$

$2x^2+22x-420=0$

$2(x^2+11x-210)=0$

$2(x+21)(x-10)=0$

Apply the zero product principle.

$2(x+21)=0 \qquad x-10=0$

$x+21=0 \qquad x=10$

$x=-21$

The solution set is $\{-21,10\}$. Disregard -21 because we can't have a negative number of prisoners. There will be 740,000 inmates in prisons in the year 1990.

96. Let w = the width of the sign
Let $w+3$ = the length of the sign
Area = lw

$54=(w+3)w$

$54=w^2+3w$

$0=w^2+3w-54$

$0=(w+9)(w-6)$

Apply the zero product principle.

$w+9=0 \qquad w-6=0$

$w=-9 \qquad w=6$

The solution set is $\{-9,6\}$. Disregard -9 because we can't have a negative length measurement. The width is 6 feet and the length is 6 + 3 = 9 feet.

97. Let x = the width of the border

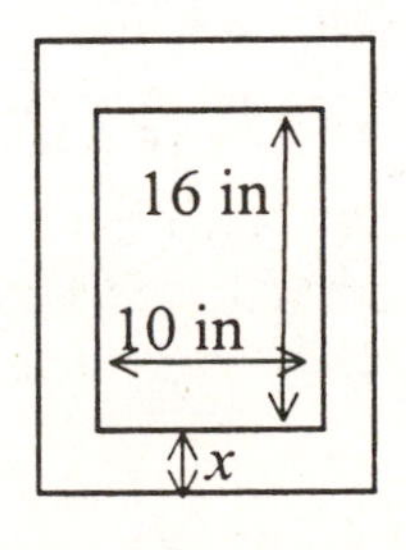

Area = lw

$280=(2x+16)(2x+10)$

$280=4x^2+20x+32x+160$

$0=4x^2+52x-120$

$0=4(x^2+13x-30)$

$0=4(x+15)(x-2)$

Apply the zero product principle.

$4(x+15)=0 \qquad x-2=0$

$x+15=0 \qquad x=2$

$x=-15$

The solution set is $\{-15,2\}$. Disregard -15 because we can't have a negative length measurement. The width of the frame is 2 inches.

98.

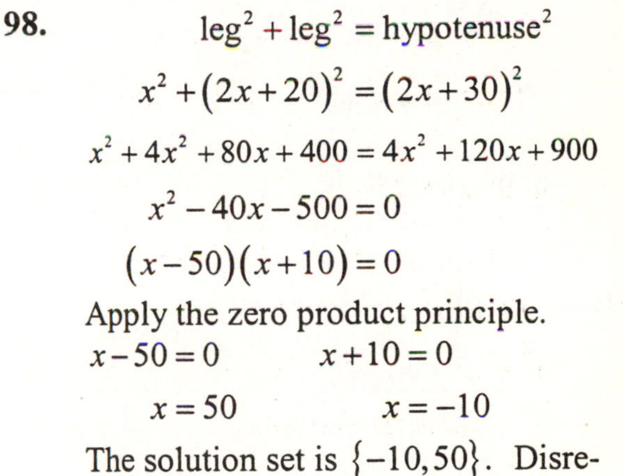

$\text{leg}^2+\text{leg}^2=\text{hypotenuse}^2$

$x^2+(2x+20)^2=(2x+30)^2$

$x^2+4x^2+80x+400=4x^2+120x+900$

$x^2-40x-500=0$

$(x-50)(x+10)=0$

Apply the zero product principle.

$x-50=0 \qquad x+10=0$

$x=50 \qquad x=-10$

The solution set is $\{-10,50\}$. Disregard -10 because we can't have a negative length measurement. We have x = 50 yards. Find the lengths of the other sides.

$2x+20$	$2x+30$
$=2(50)+20$	$=2(50)+30$
$=100+20$	$=100+30$
$=120$	$=130$

The three sides are 50 yards, 120 yards, and 130 yards.

Chapter 5 Test

1. The degree of the polynomial is 3 and the leading coefficient is –6.

2. The degree of the polynomial is 9 and the leading coefficient is 7.

3.
$$f(x) = 3x^3 + 5x^2 - x + 6$$
$$f(0) = 3(0)^3 + 5(0)^2 - 0 + 6$$
$$= 3(0) + 5(0) + 6$$
$$= 0 + 0 + 6 = 6$$
$$f(-2) = 3(-2)^3 + 5(-2)^2 - (-2) + 6$$
$$= 3(-8) + 5(4) + 2 + 6$$
$$= -24 + 20 + 2 + 6 = 4$$

4. Since the degree of the polynomial is 2, an even number, and the leading coefficient is negative, the graph will fall to the left and to the right.

5. Since the degree of the polynomial is 3, an odd number, and the leading coefficient is positive, the graph will fall to the left and rise to the right.

6.
$$(4x^3y - 19x^2y - 7y) + (3x^3y + x^2y + 6y - 9)$$
$$= 4x^3y - 19x^2y - 7y + 3x^3y + x^2y + 6y - 9$$
$$= 4x^3y + 3x^3y - 19x^2y + x^2y - 7y + 6y - 9$$
$$= 7x^3y - 18x^2y - y - 9$$

7.
$$(6x^2 - 7x - 9) - (-5x^2 + 6x - 3)$$
$$= 6x^2 - 7x - 9 + 5x^2 - 6x + 3$$
$$= 6x^2 + 5x^2 - 7x - 6x - 9 + 3$$
$$= 11x^2 - 13x - 6$$

8.
$$(-7x^3y)(-5x^4y^2)$$
$$= -7(-5)x^3x^4yy^2$$
$$= 35x^7y^3$$

9.
$$\begin{array}{r} x^2 - 3xy - y^2 \\ x - y \\ \hline x^3 - 3x^2y - xy^2 \\ -x^2y + 3xy^2 + y^3 \\ \hline x^3 - 4x^2y + 2xy^2 + y^3 \end{array}$$

10.
$$(7x - 9y)(3x + y)$$
$$= 21x^2 + 7xy - 27xy - 9y^2$$
$$= 21x^2 - 20xy - 9y^2$$

11.
$$(2x - 5y)(2x + 5y)$$
$$= 4x^2 + \cancel{10xy} - \cancel{10xy} - 25y^2$$
$$= 4x^2 - 25y^2$$

12.
$$(4y - 7)^2$$
$$= (4y)^2 + 2 \cdot 4y \cdot (-7) + (-7)^2$$
$$= 16y^2 - 56y + 49$$

13.
$$[(x + 2) + 3y][(x + 2) - 3y]$$
$$= (x + 2)^2 - 3(x + 2)y + 3(x + 2)y - 9y^2$$
$$= (x + 2)^2 - 9y^2$$
$$= x^2 + 4x + 4 - 9y^2$$

14. $f(x) = x + 2$ and $g(x) = 3x - 5$
$$(fg)(x) = f(x) \cdot g(x)$$
$$= (x + 2)(3x - 5)$$

$$= 3x^2 - 5x + 6x - 10$$
$$= 3x^2 + x - 10$$
$$(fg)(-5) = 3(-5)^2 + (-5) - 10$$
$$= 3(25) - 5 - 10$$
$$= 75 - 5 - 10$$
$$= 60$$

15. $f(x) = x^2 - 5x + 3$
$$f(a+h) - f(a)$$
$$= (a+h)^2 - 5(a+h) + 3 - (a^2 - 5a + 3)$$
$$= a^2 + 2ah + h^2 - 5a - 5h + 3 - a^2 + 5a - 3$$
$$= 2ah + h^2 - 5h$$

16. $14x^3 - 15x^2 = x^2(14x - 15)$

17. $81y^2 - 25 = (9y)^2 - 5^2$
$$= (9y+5)(9y-5)$$

18. $x^3 + 3x^2 - 25x - 75$
$$= x^2(x+3) - 25(x+3)$$
$$= (x+3)(x^2 - 25)$$
$$= (x+3)(x^2 - 5^2)$$
$$= (x+3)(x+5)(x-5)$$

19. $25x^2 - 30x + 9 = (5x)^2 - 2 \cdot 5x \cdot 3 + 3^2$
$$= (5x-3)^2$$

20. $x^2 + 10x + 25 - 9y^2$
$$= (x^2 + 10x + 25) - 9y^2$$
$$= (x^2 + 2 \cdot x \cdot 5 + 5^2) - 9y^2$$
$$= (x+5)^2 - (3y)^2$$
$$= ((x+5) + 3y)((x+5) - 3y)$$
$$= (x + 5 + 3y)(x + 5 - 3y)$$

21. $x^4 + 1$
Prime

22. $y^2 - 16y - 36$
$$= (y-18)(y+2)$$

23. $14x^2 + 41x + 15$
$$= (2x+5)(7x+3)$$

24. $5x^3 - 5 = 5(x^3 - 1)$
$$= 5(x^3 - 1^3)$$
$$= 5(x-1)(x^2 + x \cdot 1 + 1^2)$$
$$= 5(x-1)(x^2 + x + 1)$$

25. $12x^2 - 3y^2 = 3(4x^2 - y^2)$
$$= 3((2x)^2 - y^2)$$
$$= 3(2x+y)(2x-y)$$

26. $12x^2 - 34x + 10$
$$= 2(6x^2 - 17x + 5)$$
$$= 2(3x-1)(2x-5)$$

27. $3x^4 - 3$
$$= 3(x^4 - 1)$$
$$= 3((x^2)^2 - 1^2)$$

$= 3(x^2+1)(x^2-1)$
$= 3(x^2+1)(x^2-1^2)$
$= 3(x^2+1)(x+1)(x-1)$

28. $x^8 - y^8$
$= (x^4)^2 - (y^4)^2$
$= (x^4+y^4)(x^4-y^4)$
$= (x^4+y^4)((x^2)^2-(y^2)^2)$
$= (x^4+y^4)(x^2+y^2)(x^2-y^2)$
$= (x^4+y^4)(x^2+y^2)(x+y)(x-y)$

29. $12x^2y^4 + 8x^3y^2 - 36x^2y$
$= 4x^2y(3y^3+2xy-9)$

30. Let $t = x^3$
$x^6 - 12x^3 - 28$
$= (x^3)^2 - 12x^3 - 28$
$= t^2 - 12t - 28$
$= (t-14)(t+2)$ Substitute x^3 for t.
$= (x^3-14)(x^3+2)$

31. Let $t = x^2$
$x^4 - 2x^2 - 24$
$= (x^2)^2 - 2x^2 - 24$
$= t^2 - 2t - 24$
$= (t-6)(t+4)$ Substitute x^3 for t.
$= (x^2-6)(x^2+4)$

32. $12x^2y - 27xy + 6y$
$= 3y(4x^2-9x+2)$
$= 3y(x-2)(4x-1)$

33. $y^4 - 3y^3 + 2y^2 - 6y$
$= y(y^3-3y^2+2y-6)$
$= y(y^2(y-3)+2(y-3))$
$= y(y-3)(y^2+2)$

34. $3x^2 = 5x+2$
$3x^2 - 5x - 2 = 0$
$(3x+1)(x-2) = 0$
Apply the zero product principle.
$3x+1=0 \qquad x-2=0$
$3x=-1 \qquad x=2$
$x = -\frac{1}{3}$

The solutions are $-\frac{1}{3}$ and 2 and the solution set is $\left\{-\frac{1}{3}, 2\right\}$.

35. $(5x+4)(x-1) = 2$
$5x^2 - 5x + 4x - 4 = 2$
$5x^2 - x - 4 = 2$
$5x^2 - x - 6 = 0$
$(5x-6)(x+1) = 0$
Apply the zero product principle.
$5x-6=0 \qquad x+1=0$
$5x=6 \qquad x=-1$
$x = \frac{6}{5}$

The solutions are -1 and $\frac{6}{5}$ and the solution set is $\left\{-1, \frac{6}{5}\right\}$.

36. $15x^2 - 5x = 0$

$5x(3x-1) = 0$

Apply the zero product principle.

$5x = 0 \qquad 3x - 1 = 0$

$x = 0 \qquad 3x = 1$

$x = \frac{1}{3}$

The solutions are 0 and $\frac{1}{3}$ and the solution set is $\left\{0, \frac{1}{3}\right\}$.

37. $x^3 - 4x^2 - x + 4 = 0$

$x^2(x-4) - 1(x-4) = 0$

$(x-4)(x^2-1) = 0$

$(x-4)(x+1)(x-1) = 0$

Apply the zero product principle.

$x - 4 = 0 \qquad x + 1 = 0 \qquad x - 1 = 0$

$x = 4 \qquad x = -1 \qquad x = 1$

The solutions are -1, 1 and 4 and the solution set is $\{-1, 1, 4\}$.

38. $-16t^2 + 48t + 448 = 0$

$-16(t^2 - 3t - 28) = 0$

$-16(t-7)(t+4) = 0$

Apply the zero product principle.

$-16(t-7) = 0 \qquad t + 4 = 0$

$t - 7 = 0 \qquad t = -4$

$t = 7$

The solution set is $\{-4, 7\}$. Disregard -4 because we can't have a negative time measurement. The baseball will hit the water at $t = 7$ seconds.

39. Let l = the length of the room

Let $2l - 7$ = the width of the room

Area = lw

$15 = l(2l - 7)$

$15 = 2l^2 - 7l$

$0 = 2l^2 - 7l - 15$

$0 = (2l + 3)(l - 5)$

Apply the zero product principle.

$2l + 3 = 0 \qquad l - 5 = 0$

$2l = -3 \qquad l = 5$

$l = -\frac{3}{2}$

The solution set is $\left\{-\frac{3}{2}, 5\right\}$. Disregard $-\frac{3}{2}$ because we can't have a negative length measurement. The length is 5 yards and the width is $2w - 7 = 2(5) - 7 = 10 - 7 = 3$ yards.

40. $\text{leg}^2 + \text{leg}^2 = \text{hypotenuse}^2$

$x^2 + 12^2 = (2x-3)^2$

$x^2 + 144 = 4x^2 - 12x + 9$

$0 = 3x^2 - 12x - 135$

$0 = 3(x^2 - 4x - 45)$

$0 = 3(x-9)(x+5)$

Apply the zero product principle.

$3(x-9) = 0 \qquad x + 5 = 0$

$x - 9 = 0 \qquad x = -5$

$x = 9$

The solution set is $\{-5, 9\}$. Disregard –5 because we can't have a negative length measurement. The lengths of the sides are 9 and 12 units and the length of the hypotenuse is $2x - 3 = 2(9) - 3 = 18 - 3 = 15$ units.

Cumulative Review Exercises

Chapters 1-5

1. $8(x+2)-3(2-x)=4(2x+6)-2$

$$8x+16-6+3x=8x+24-2$$
$$11x+10=8x+22$$
$$3x=12$$
$$x=4$$

The solution is 4 and the solution set is $\{4\}$.

2. $2x+4y=-6$

$x=2y-5$

Rewrite the system in standard form.

$$2x+4y=-6$$
$$x-2y=-5$$

Multiply the second equation by 2 and add to the first equation.

$$2x+4y=-6$$
$$2x-4y=-10$$
$$4x=-16$$
$$x=-4$$

Back-substitute –4 for x in one of the original equations.

$$2x+4y=-6$$
$$2(-4)+4y=-6$$
$$-8+4y=-6$$
$$4y=2$$
$$y=\frac{2}{4}=\frac{1}{2}$$

The solution is $\left(-4, \frac{1}{2}\right)$ and the solution set is $\left\{\left(-4, \frac{1}{2}\right)\right\}$.

3. $2x-y+3z=0$

$2y+z=1$

$x+2y-z=5$

Multiply the third equation by –2 and add to the first equation.

$$2x-y+3z=0$$
$$-2x-4y+2z=-10$$
$$-5y+5z=-10$$

We now have 2 equations in 2 variables.

$$2y+z=1$$
$$-5y+5z=-10$$

Multiply the first equation by –5 and add to the second equation.

$$-10y-5z=-5$$
$$-5y+5z=-10$$
$$-15y=-15$$
$$y=1$$

Back-substitute 1 for y in one of the equations in 2 variables.

$$2y+z=1$$
$$2(1)+z=1$$
$$2+z=1$$
$$z=-1$$

Back-substitute 1 for y and –1 for z in one of the original equations in 3 variables.

$$2x - y + 3z = 0$$
$$2x - 1 + 3(-1) = 0$$
$$2x - 1 - 3 = 0$$
$$2x - 4 = 0$$
$$2x = 4$$
$$x = 2$$

The solution is $(2,1,-1)$ and the solution set is $\{(2,1,-1)\}$.

4. $2x + 4 < 10$ and $3x - 1 > 5$

$2x < 6$ $\quad$ $3x > 6$

$x < 3$ $\quad$ $x > 2$

$x < 3$

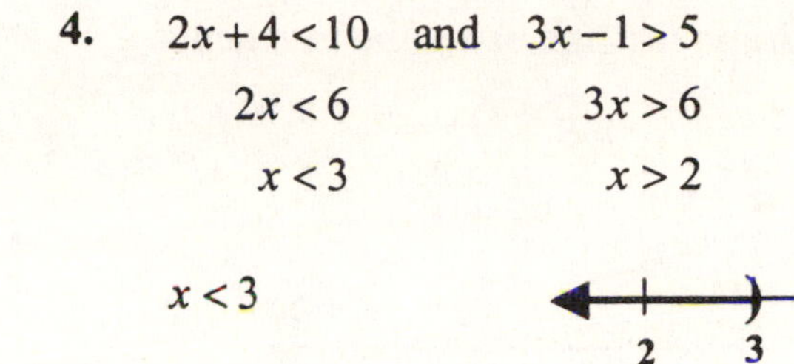

$x > 2$

2 3

$x < 3$ and $x > 2$

2 3

The solution set is $\{x \mid 2 < x < 3\}$ or $(2,3)$.

5. $|2x - 5| \geq 9$

$2x - 5 \leq -9$ or $2x - 5 \geq 9$

$2x \leq -4$ $\quad$ $2x \geq 14$

$x \leq -2$ $\quad$ $x \geq 7$

-2 7

The solution set is $\{x \mid x \leq -2 \text{ or } x \geq 7\}$ or $(-\infty, -2] \cup [7, \infty)$.

6. $$2x^2 = 7x - 5$$
$$2x^2 - 7x + 5 = 0$$
$$(2x - 5)(x - 1) = 0$$

Apply the zero product principle.

$2x - 5 = 0$ $\quad$ $x - 1 = 0$

$2x = 5$ $\quad$ $x = 1$

$x = \frac{5}{2}$

The solutions are 1 and $\frac{5}{2}$ or the solution set is $\left\{1, \frac{5}{2}\right\}$.

7. $$2x^3 + 6x^2 = 20x$$
$$2x^3 + 6x^2 - 20x = 0$$
$$2x(x^2 + 3x - 10) = 0$$
$$2x(x + 5)(x - 2) = 0$$

Apply the zero product principle.

$2x = 0$ $\quad$ $x + 5 = 0$ $\quad$ $x - 2 = 0$

$x = 0$ $\quad$ $x = -5$ $\quad$ $x = 2$

The solutions are –5, 0, and 2 and the solution set is $\{-5, 0, 2\}$.

8. $$x = \frac{ax + b}{c}$$
$$cx = ax + b$$
$$cx - ax = b$$
$$x(c - a) = b$$
$$x = \frac{b}{c - a}$$

9. First, find the slope.

$$m = \frac{5 - (-3)}{2 - (-2)} = \frac{5 + 3}{2 + 2} = \frac{8}{4} = 2$$

Write the equation in point-slope form.

$$y - y_1 = m(x - x_1)$$
$$y - 5 = 2(x - 2)$$
$$y - 5 = 2x - 4$$
$$y = 2x + 1$$

Write the equation of the line using function notation.

$$f(x) = 2x + 1$$

10. Let x = the number of votes for the loser
Let y = the number of votes for the winner

$$x + y = 2800$$
$$x + 160 = y$$

Rewrite the system in standard form.

$$x + y = 2800$$
$$x - y = -160$$

Add the equations to eliminate y.

$$x + y = 2800$$
$$\underline{x - y = \ \ 160}$$
$$2x = 2640$$
$$x = 1320$$

Back-substitute 1320 for x to find y.

$$x + 160 = y$$
$$1320 + 160 = y$$
$$1480 = y$$

The solution set is $\{(1320, 1480)\}$.
The loser received 1320 votes and the winner received 1480 votes.

11. $f(x) = -\frac{1}{3}x + 1$

First, find the intercepts. To find the y–intercept, set $x = 0$.

$$y = -\frac{1}{3}(0) + 1$$
$$y = 0 + 1$$
$$y = 1$$

To find the x–intercept, set $y = 0$.

$$0 = -\frac{1}{3}x + 1$$
$$-3(-1) = -3\left(-\frac{1}{3}x\right)$$
$$3 = x$$

Graph the line using the intercepts.

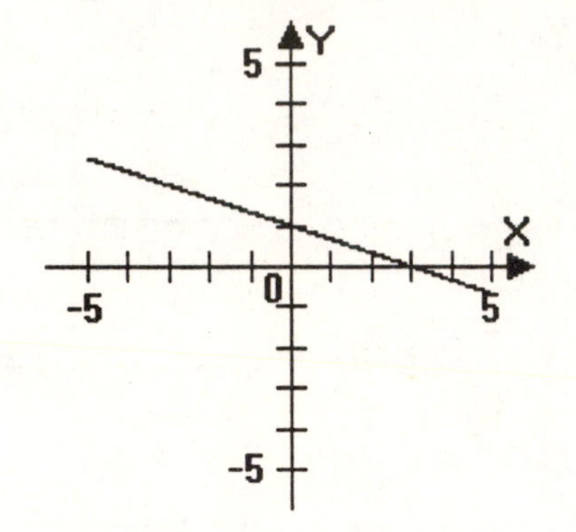

12. $4x - 5y < 20$

First, find the intercepts. To find the y–intercept, set $x = 0$.

$$4(0) - 5y = 20$$
$$0 - 5y = 20$$
$$-5y = 20$$
$$y = -4$$

To find the x–intercept, set $y = 0$.

$$4x - 5(0) = 20$$
$$4x - 0 = 20$$
$$4x = 20$$
$$x = 5$$

Graph the inequality using the intercepts.

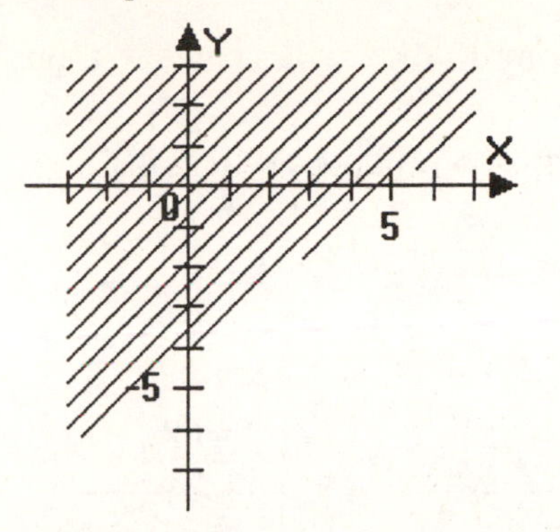

13. $y \le -1$
We know that $y = -1$ is a horizontal line and the y–intercept is -1. This is sufficient information to graph the inequality.

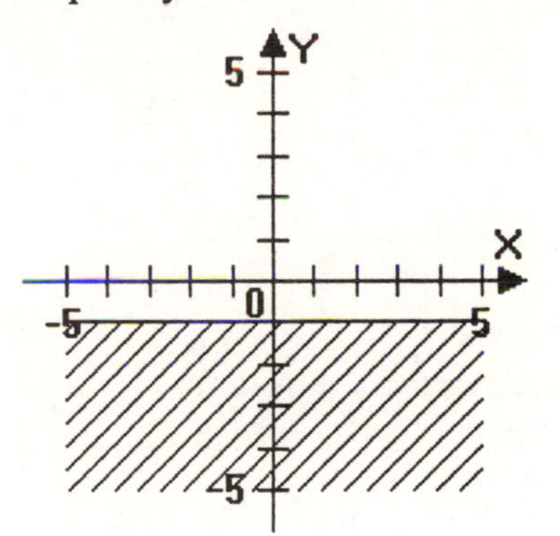

14. $$\frac{-8x^3y^6}{16x^9y^{-4}} = \frac{-8}{16}x^{3-9}y^{6-(-4)}$$
$$= \frac{-8}{16}x^{3-9}y^{6+4}$$
$$= \frac{-1}{2}x^{-6}y^{10}$$
$$= -\frac{y^{10}}{2x^6}$$

15. $0.0000706 = 7.06 \times 10^{-5}$

16. $$(3x^2 - y)^2 = (3x^2)^2 - 2 \cdot 3x^2 \cdot y + y^2$$
$$= 9x^4 - 6x^2y + y^2$$

17. $$(3x^2 - y)(3x^2 + y)$$
$$= (3x^2)^2 + \cancel{3x^2y} - \cancel{3x^2y} - y^2$$
$$= 9x^4 - y^2$$

18. $$x^3 - 3x^2 - 9x + 27$$
$$= x^2(x-3) - 9(x-3)$$
$$= (x-3)(x^2-9)$$
$$= (x-3)(x^2-3^2)$$
$$= (x-3)(x+3)(x-3)$$
$$= (x-3)^2(x+3)$$

19. $$x^6 - x^2$$
$$= x^2(x^4 - 1)$$
$$= x^2\left((x^2)^2 - 1^2\right)$$
$$= x^2(x^2+1)(x^2-1)$$
$$= x^2(x^2+1)(x^2-1^2)$$
$$= x^2(x^2+1)(x+1)(x-1)$$

20. $$14x^3y^2 - 28x^4y^2$$
$$= 14x^3y^2(1-2x)$$

Chapter 6

Check Points 6.1

1. **a.**
$$f(x)=\frac{120x}{100-x}$$
$$f(40)=\frac{120(40)}{100-40}$$
$$=\frac{4800}{60}=80$$
This corresponds to the point $(40,80)$. The cost to remove 40% of the lake's pollutants is $80,000.

b.
$$f(x)=\frac{120x}{100-x}$$
$$f(60)=\frac{120(60)}{100-60}$$
$$=\frac{7200}{40}=180$$
This corresponds to the point $(60,180)$. The cost to remove 60% of the lake's pollutants is $180,000.

2. The domain of *f* is the set of all real numbers except those for which the denominator is zero. Set the denominator equal to zero and solve.
$$2x^2+5x-3=0$$
$$(2x-1)(x+3)=0$$
Apply the zero product principle.
$$2x-1=0 \qquad x+3=0$$
$$2x=1 \qquad x=-3$$
$$x=\frac{1}{2}$$
Domain of $f=\left\{x \mid x \text{ is a real number and } x\neq-3 \text{ and } x\neq\frac{1}{2}\right\}$ or $(-\infty,-3)\cup\left(-3,\frac{1}{2}\right)\cup\left(\frac{1}{2},\infty\right)$.

3.
$$\frac{x^2+7x+10}{x+2}=\frac{\cancel{(x+2)}(x+5)}{1\cancel{(x+2)}}$$
$$=x+5$$

4. **a.**
$$\frac{x^2-2x-15}{3x^2+8x-3}=\frac{(x-5)\cancel{(x+3)}}{(3x-1)\cancel{(x+3)}}$$
$$=\frac{x-5}{3x-1}$$

b.
$$\frac{3x^2+9xy-12y^2}{9x^3-9xy^2}$$
$$=\frac{3\left(x^2+3xy-4y^2\right)}{9x\left(x^2-y^2\right)}$$
$$=\frac{\cancel{3}\cancel{(x-y)}(x+4y)}{\cancel{3}\cdot 3x(x+y)\cancel{(x-y)}}$$
$$=\frac{x+4y}{3x(x+y)}$$

5.
$$\frac{x+4}{x-7}\cdot\frac{x^2-4x-21}{x^2-16}$$
$$=\frac{1\cancel{(x+4)}}{1\cancel{(x-7)}}\cdot\frac{\cancel{(x-7)}(x+3)}{\cancel{(x+4)}(x-4)}$$
$$=\frac{x+3}{x-4}$$

6. $\frac{4x+8}{6x-3x^2}\cdot\frac{3x^2-4x-4}{9x^2-4}$

$=\frac{4(x+2)}{3x(2-x)}\cdot\frac{(3x+2)(x-2)}{(3x+2)(3x-2)}$

$=\frac{-4(x+2)}{3x(3x-2)}$ or $-\frac{4(x+2)}{3x(3x-2)}$

7. a. $(9x^2-49)\div\frac{3x-7}{9}$

$=\frac{9x^2-49}{1}\div\frac{3x-7}{9}$

$=\frac{9x^2-49}{1}\cdot\frac{9}{3x-7}$

$=\frac{(3x+7)(3x-7)}{1}\cdot\frac{9}{3x-7}$

$=9(3x+7)$

b. $\frac{x^2-x-12}{5x}\div\frac{x^2-10x+24}{x^2-6x}$

$=\frac{x^2-x-12}{5x}\cdot\frac{x^2-6x}{x^2-10x+24}$

$=\frac{(x-4)(x+3)}{5x}\cdot\frac{x(x-6)}{(x-6)(x-4)}$

$=\frac{x+3}{5}$

Problem Set 6.1

Practice Exercises

1. $f(-2)=\frac{(-2)^2-9}{-2+3}=\frac{4-9}{1}=\frac{-5}{1}=-5$

$f(0)=\frac{0^2-9}{0+3}=-3$

$f(5)=\frac{5^2-9}{5+3}=\frac{25-9}{8}=\frac{16}{8}=2$

3. $f(x)=\frac{x^2-2x-3}{4-x}$

$f(-1)=\frac{(-1)^2-2(-1)-3}{4-(-1)}$

$=\frac{1+2-3}{4+1}=\frac{0}{5}=0$

$f(2)=\frac{2^2-2(2)-3}{4-2}$

$=\frac{4-4-3}{2}=-\frac{3}{2}$

$f(6)=\frac{6^2-2(6)-3}{4-6}$

$=\frac{36-12-3}{-2}=-\frac{21}{2}$

5. $g(t)=\frac{2t^3-5}{t^2+1}$

$g(-1)=\frac{2(-1)^3-5}{(-1)^2+1}=\frac{2(-1)-5}{1+1}$

$=\frac{-2-5}{2}=-\frac{7}{2}$

$g(0)=\frac{2(0)^3-5}{0^2+1}=\frac{-5}{1}=-5$

$g(2)=\frac{2(2)^3-5}{(2)^2+1}=\frac{2(8)-5}{4+1}$

$=\frac{16-5}{5}=\frac{11}{5}$

7. The domain is all real numbers except those which make the denominator zero. Set the denominator equal to zero and solve.

$x-5=0$

$x=5$

Domain of $f=\{x|x$ is a real number and $x\neq 5\}$ or $(-\infty,5)\cup(5,\infty)$.

9. The domain is all real numbers except those which make the denominator equal to zero. Set the denominator equal to zero and solve.

$x-1=0 \qquad x+3=0$

$x=1 \qquad x=-3$

Domain of $f=\{x|x$ is a real number and $x\neq -3$ and $x\neq 1\}$ or $(-\infty,-3)\cup(-3,1)\cup(1,\infty)$.

11. The domain is all real numbers except those which make the denominator equal to zero. Set the denominator equal to zero and solve. Since $x+5$ is a double root, we only need solve the equation once.

$x+5=0$

$x=-5$

Domain of $f=\{x|x$ is a real number and $x\neq -5\}$ or $(-\infty,-5)\cup(-5,\infty)$.

13. The domain is all real numbers except those which make the denominator equal to zero. Set the denominator equal to zero and solve.

$x^2-8x+15=0$

$(x-5)(x-3)=0$

Apply the zero product principle.

$x-5=0 \qquad x-3=0$

$x=5 \qquad x=3$

Domain of $f=\{x|x$ is a real number and $x\neq 3$ and $x\neq 5\}$ or $(-\infty,3)\cup(3,5)\cup(5,\infty)$.

15. The domain is all real numbers except those which make the denominator equal to zero. Set the denominator equal to zero and solve.

$3x^2-2x-8=0$

$(3x+4)(x-2)=0$

Apply the zero product principle.

$3x+4=0 \qquad x-2=0$

$3x=-4 \qquad x=2$

$x=-\frac{4}{3}$

Domain of $f=\left\{x|x \text{ is a real number and } x\neq -\frac{4}{3} \text{ and } x\neq 2\right\}$ or $\left(-\infty,-\frac{4}{3}\right)\cup\left(-\frac{4}{3},2\right)\cup(2,\infty)$.

17. $f(4)=4$

19. Domain of $f=\{x|x$ is a real number and $x\neq -2$ and $x\neq 2\}$ or $(-\infty,-2)\cup(-2,2)\cup(2,\infty)$.

21. As x decreases, the value of the function approaches 3. The equation of the horizontal asymptote is $y=3$.

23. There is no point on the graph with an x–coordinate of -2.

25. The graph is not continuous. Furthermore, it neither rises nor falls to the left or the right.

27.
$$\frac{x^2-4}{x-2}=\frac{(x+2)\cancel{(x-2)}}{1\cancel{(x-2)}}$$
$$=x+2$$

29.
$$\frac{x+2}{x^2-x-6}=\frac{1\cancel{(x+2)}}{(x-3)\cancel{(x+2)}}$$
$$=\frac{1}{x-3}$$

31.
$$\frac{4x+20}{x^2+5x}=\frac{4\cancel{(x+5)}}{x\cancel{(x+5)}}$$
$$=\frac{4}{x}$$

33.
$$\frac{4y-20}{y^2-25}=\frac{4\cancel{(y-5)}}{(y+5)\cancel{(y-5)}}$$
$$=\frac{4}{y+5}$$

35.
$$\frac{3x-5}{25-9x^2}=\frac{1\overset{-1}{\cancel{(3x-5)}}}{(5+3x)\cancel{(5-3x)}}$$
$$=\frac{-1}{5+3x}\text{ or }-\frac{1}{5+3x}$$

Or, by the Commutative Property of Addition, we have $\frac{-1}{3x+5}$ or $-\frac{1}{3x+5}$.

37.
$$\frac{y^2-49}{y^2-14y+49}=\frac{(y+7)\cancel{(y-7)}}{(y-7)\cancel{(y-7)}}$$
$$=\frac{y+7}{y-7}$$

39.
$$\frac{x^2+7x-18}{x^2-3x+2}=\frac{(x+9)\cancel{(x-2)}}{\cancel{(x-2)}(x-1)}$$
$$=\frac{x+9}{x-1}$$

41. $\frac{3x+7}{3x+10}$

The rational expression cannot be simplified.

43.
$$\frac{x^2-x-12}{16-x^2}=\frac{\overset{-1}{\cancel{(x-4)}}(x+3)}{\cancel{(4-x)}(4+x)}$$
$$=-\frac{x+3}{4+x}\text{ or }-\frac{x+3}{x+4}$$

45.
$$\frac{x^2+3xy-10y^2}{3x^2-7xy+2y^2}=\frac{(x+5y)\cancel{(x-2y)}}{(3x-y)\cancel{(x-2y)}}$$
$$=\frac{x+5y}{3x-y}$$

47.
$$\frac{x^3-8}{x^2-4}=\frac{\cancel{(x-2)}(x^2+2x+4)}{(x+2)\cancel{(x-2)}}$$
$$=\frac{x^2+2x+4}{x+2}$$

49.
$$\frac{x^3+4x^2-3x-12}{x+4}$$

$$= \frac{x^2(x+4)-3(x+4)}{x+4}$$

$$= \frac{(x+4)(x^2-3)}{1(x+4)}$$

$$= x^2 - 3$$

51. $$\frac{x-3}{x+7} \cdot \frac{3x+21}{2x-6} = \frac{1(x-3)}{1(x+7)} \cdot \frac{3(x+7)}{2(x-3)}$$

$$= \frac{3}{2}$$

53. $$\frac{x^2-49}{x^2-4x-21} \cdot \frac{x+3}{x}$$

$$= \frac{(x+7)(x-7)}{(x-7)(x+3)} \cdot \frac{1(x+3)}{x}$$

$$= \frac{x+7}{x}$$

55. $$\frac{x^2-9}{x^2-x-6} \cdot \frac{x^2+5x+6}{x^2+x-6}$$

$$= \frac{(x+3)(x-3)}{(x-3)(x+2)} \cdot \frac{(x+3)(x+2)}{(x+3)(x-2)}$$

$$= \frac{x+3}{x-2}$$

57. $$\frac{x^2+4x+4}{x^2+8x+16} \cdot \frac{(x+4)^3}{(x+2)^3}$$

$$= \frac{(x+2)^2}{(x+4)^2} \cdot \frac{(x+4)^3}{(x+2)^3}$$

$$= \frac{(x+2)^2}{(x+4)^2} \cdot \frac{(x+4)^3}{(x+2)^3} \quad \text{(reduces to } \frac{x+4}{x+2}\text{)}$$

$$= \frac{x+4}{x+2}$$

59. $$\frac{8y+2}{y^2-9} \cdot \frac{3-y}{4y^2+y}$$

$$= \frac{2(4y+1)}{(y+3)(y-3)} \cdot \frac{1(3-y)}{y(4y+1)}$$

$$= \frac{-2}{y(y+3)} \text{ or } -\frac{2}{y(y+3)}$$

61. $$\frac{y^3-8}{y^2-4} \cdot \frac{y+2}{2y}$$

$$= \frac{(y-2)(y^2+2y+4)}{(y+2)(y-2)} \cdot \frac{1(y+2)}{2y}$$

$$= \frac{y^2+2y+4}{2y}$$

63. $$(x-3) \cdot \frac{x^2+x+1}{x^2-5x+6}$$

$$= \frac{1(x-3)}{1} \cdot \frac{x^2+x+1}{(x-3)(x-2)}$$

$$= \frac{x^2+x+1}{x-2}$$

65. $$\frac{x^2+xy}{x^2-y^2} \cdot \frac{4x-4y}{x}$$

$$= \frac{x(x+y)}{(x+y)(x-y)} \cdot \frac{4(x-y)}{x}$$

$$= \frac{x(x+y)}{(x+y)(x-y)} \cdot \frac{4(x-y)}{x} = 4$$

67. $\frac{x^2+2xy+y^2}{x^2-2xy+y^2} \cdot \frac{4x-4y}{3x+3y}$

$$= \frac{(x+y)^2}{(x-y)^2} \cdot \frac{4(x-y)}{3(x+y)}$$

$$= \frac{4(x+y)}{3(x-y)}$$

69. $\frac{x+5}{7} \div \frac{4x+20}{9}$

$$= \frac{x+5}{7} \cdot \frac{9}{4x+20}$$

$$= \frac{1(x+5)}{7} \cdot \frac{9}{4(x+5)} = \frac{9}{28}$$

71. $\frac{4}{y-6} \div \frac{40}{7y-42}$

$$= \frac{4}{y-6} \cdot \frac{7y-42}{40}$$

$$= \frac{4}{1(y-6)} \cdot \frac{7(y-6)}{4 \cdot 10} = \frac{7}{10}$$

73. $\frac{x^2-2x}{15} \div \frac{x-2}{5}$

$$= \frac{x^2-2x}{15} \cdot \frac{5}{x-2}$$

$$= \frac{x(x-2)}{3 \cdot 5} \cdot \frac{5}{1(x-2)} = \frac{x}{3} \text{ or } \frac{1}{3}x$$

75. $\frac{y^2-25}{2y-2} \div \frac{y^2+10y+25}{y^2+4y-5}$

$$= \frac{y^2-25}{2y-2} \cdot \frac{y^2+4y-5}{y^2+10y+25}$$

$$= \frac{(y+5)(y-5)}{2(y-1)} \cdot \frac{(y+5)(y-1)}{(y+5)^2}$$

$$= \frac{y-5}{2}$$

77. $(x^2-16) \div \frac{x^2+3x-4}{x^2+4}$

$$= \frac{x^2-16}{1} \div \frac{x^2+3x-4}{x^2+4}$$

$$= \frac{x^2-16}{1} \cdot \frac{x^2+4}{x^2+3x-4}$$

$$= \frac{(x+4)(x-4)}{1} \cdot \frac{x^2+4}{(x+4)(x-1)}$$

$$= \frac{(x-4)(x^2+4)}{x-1}$$

79. $\frac{y^2-4y-21}{y^2-10y+25} \div \frac{y^2+2y-3}{y^2-6y+5}$

$$= \frac{y^2-4y-21}{y^2-10y+25} \cdot \frac{y^2-6y+5}{y^2+2y-3}$$

$$= \frac{(y-7)(y+3)}{(y-5)^2} \cdot \frac{(y-5)(y-1)}{(y+3)(y-1)}$$

$$= \frac{y-7}{y-5}$$

81. $\frac{8x^3-1}{4x^2+2x+1} \div \frac{x-1}{(x-1)^2}$

$$=\frac{8x^3-1}{4x^2+2x+1}\cdot\frac{(x-1)^2}{x-1}$$

$$=\frac{(2x-1)\left(4x^2+2x+1\right)}{1\left(4x^2+2x+1\right)}\cdot\frac{(x-1)^2}{1(x-1)}$$

$$=(2x-1)(x-1)$$

83.

$$\frac{x^2-4y^2}{x^2+3xy+2y^2}\div\frac{x^2-4xy+4y^2}{x+y}$$

$$=\frac{x^2-4y^2}{x^2+3xy+2y^2}\cdot\frac{x+y}{x^2-4xy+4y^2}$$

$$=\frac{(x+2y)(x-2y)}{(x+2y)(x+y)}\cdot\frac{1(x+y)}{(x-2y)^2}$$

$$=\frac{1}{x-2y}$$

85.

$$\frac{x^4-y^8}{x^2+y^4}\div\frac{x^2-y^4}{3x^2}$$

$$=\frac{x^4-y^8}{x^2+y^4}\cdot\frac{3x^2}{x^2-y^4}$$

$$=\frac{\left(x^2+y^4\right)\left(x^2-y^4\right)}{1\left(x^2+y^4\right)}\cdot\frac{3x^2}{\left(x+y^2\right)\left(x-y^2\right)}$$

$$=\frac{3x^2\left(x^2-y^4\right)}{\left(x+y^2\right)\left(x-y^2\right)}$$

$$=\frac{3x^2\left(x+y^2\right)\left(x-y^2\right)}{1\left(x+y^2\right)\left(x-y^2\right)}=3x^2$$

Application Exercises

87. From the graph, we see that $f(60)=195$. This corresponds to the point $(60,195)$. The cost to inoculate 60% of the population against a particular strain of flu is $195,000,000.

89. The value 100 must be excluded from the domain. We cannot inoculate 100% of the population.

91. From the graph, we see that $P(10)=7$. This is represented by the point $(10,7)$. 7% of the people with 10 years of education are unemployed.

93. The function's value is approaching 0 to the far right of the graph. There is no education level that leads to guaranteed employment. This is indicated by the horizontal asymptote $y=0$, showing that unemployment will never be 0%.

95. From the graph, we see that $P(10)=90$. This is represented by the point $(10,90)$ and refers to an incidence ratio of 10. From the chart, we know that smokers between the ages of 55 and 64 are 10 times more likely than nonsmokers to die from lung cancer. Also, 90% of the deaths from lung cancer in this group are smoking-related.

97. The horizontal asymptote of the graph is $y = 100$. This means that as incidence ratio increases the percentage of smoking-related deaths increases. The percentage will never reach 100.

Writing in Mathematics

99. Answers will vary.

101. Answers will vary.

103. Answers will vary.

105. Answers will vary.

107. Answers will vary.

109. Answers will vary.

Technology Exercises

111. $\frac{x^2+x}{3x} \cdot \frac{6x}{x+1} = 2x$

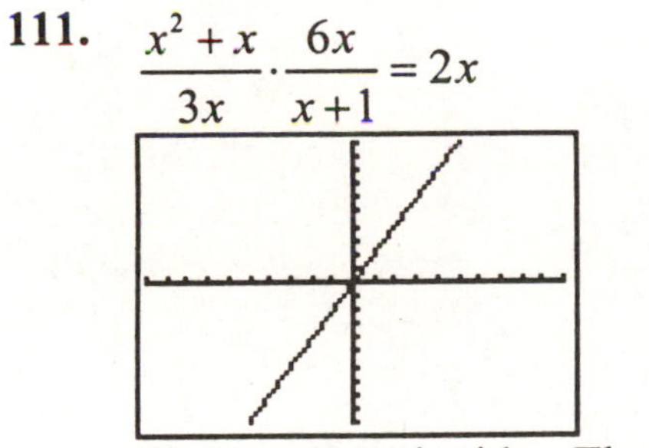

The graphs coincide. The multiplication is correct.

113. $\frac{x^2-9}{x+4} \div \frac{x-3}{x+4} = x-3$

The graphs do not coincide. The division is incorrect.

$$\frac{x^2-9}{x+4} \div \frac{x-3}{x+4}$$
$$= \frac{x^2-9}{x+4} \cdot \frac{x+4}{x-3}$$
$$= \frac{(x+3)\cancel{(x-3)}}{\cancel{x+4}} \cdot \frac{\cancel{x+4}}{\cancel{x-3}}$$
$$= x+3$$

The graphs coincide. The division is now correct.

115. a. $f(x) = \frac{x^2-x-2}{x-2}$

$g(x) = x+1$

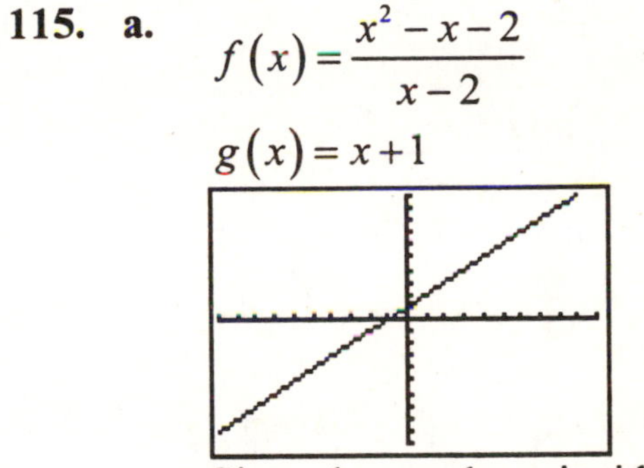

Since the graphs coincide, we know that $f(x) = g(x)$.

b. $f(x) = \frac{x^2-x-2}{x-2}$

$$= \frac{\cancel{(x-2)}(x+1)}{\cancel{x-2}}$$
$$= x+1$$

Although when simplified the expressions are the same, f and g do not represent the same function. The domains of the two functions are not the same since the domain of f excludes

$x = 2$. Specifically, $g(2) = 3$, while $f(2)$ is undefined.

c. $f(x) = \dfrac{x^2 - x - 2}{x - 2}$

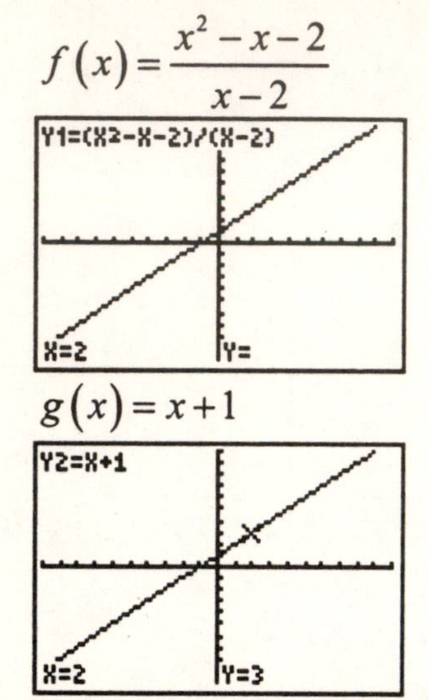

$g(x) = x + 1$

Since $f(2)$ is not defined, we know that 2 is not in the domain of f. Since $g(2) = 3$, we know that 2 is in the domain of g.

Critical Thinking Exercises

117. Statement **d.** is true.

Statement **a.** is false.

$$\frac{x^2 - 25}{x - 5} = \frac{(x+5)\cancel{(x-5)}}{\cancel{x-5}} = x + 5$$

Statement **b.** is false. $\dfrac{x^2 + 7}{7}$ cannot be simplified.

Statement **c.** is false. The domain of $f = \{x \mid x \text{ is a real number}\}$. Normally we would exclude any value of x that would make the denominator equal to zero. However, there are no values of x for which $x^2 + 1$ will equal zero.

119.
$$\frac{f(a+h) - f(a)}{h} = \frac{\left((a+h)^2 - 5(a+h) + 3\right) - \left(a^2 - 5a + 3\right)}{h}$$
$$= \frac{\left(a^2 + 2ah + h^2 - 5a - 5h + 3\right) - \left(a^2 - 5a + 3\right)}{h}$$
$$= \frac{\cancel{a^2} + 2ah + h^2 - \cancel{5a} - 5h + \cancel{3} - \cancel{a^2} + \cancel{5a} - \cancel{3}}{h}$$
$$= \frac{2ah + h^2 - 5h}{h} = \frac{h(2a + h - 5)}{h} = 2a + h - 5$$

120.
$$\frac{y^{2n} - 1}{y^{2n} + 3y^n + 2} \div \frac{y^{2n} + y^n - 12}{y^{2n} - y^n - 6} = \frac{y^{2n} - 1}{y^{2n} + 3y^n + 2} \cdot \frac{y^{2n} - y^n - 6}{y^{2n} + y^n - 12}$$
$$= \frac{\cancel{(y^n + 1)}(y^n - 1)}{\cancel{(y^n + 2)}\cancel{(y^n + 1)}} \cdot \frac{\cancel{(y^n - 3)}\cancel{(y^n + 2)}}{(y^n + 4)\cancel{(y^n - 3)}} = \frac{y^n - 1}{y^n + 4}$$

Review Exercises

122. $4x-5y\geq 20$

First, find the intercepts to the equation $4x-5y=20$.

Find the x–intercept by setting $y=0$.

$$4x-5(0)=20$$
$$4x=20$$
$$x=5$$

Find the y–intercept by setting $x=0$.

$$4(0)-5y=20$$
$$-5y=20$$
$$y=-4$$

Next, use the origin, $(0,0)$, as a test point.

$$4(0)-5(0)\geq 20$$
$$0\geq 20$$

This is a false statement. This means that the point, $(0,0)$, will not fall in the shaded half-plane.

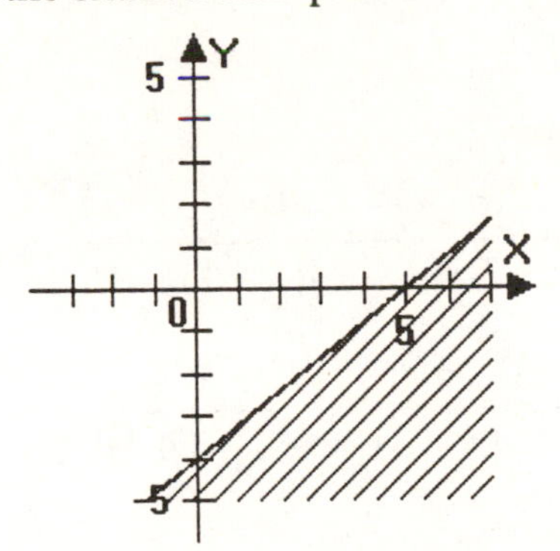

123.

$$\begin{array}{r} x^2-3x-6 \\ 2x-5 \\ \hline 2x^3-6x^2-12x \\ -5x^2+15x+30 \\ \hline 2x^3-11x^2+3x+30 \end{array}$$

124.

$$\begin{aligned} &x^4-3x^3+2x^2-6x \\ &=x\left(x^3-3x^2+2x-6\right) \\ &=x\left(\left(x^3-3x^2\right)+\left(2x-6\right)\right) \\ &=x\left(x^2\left(x-3\right)+2\left(x-3\right)\right) \\ &=x\left(x-3\right)\left(x^2+2\right) \end{aligned}$$

Check Points 6.2

1.

$$\begin{aligned} &\frac{x^2-5x-15}{x^2+5x+6}+\frac{2x+5}{x^2+5x+6} \\ &=\frac{x^2-5x-15+2x+5}{x^2+5x+6} \\ &=\frac{x^2-3x-10}{x^2+5x+6} \\ &=\frac{(x-5)(x+2)}{(x+3)(x+2)} \\ &=\frac{x-5}{x+3} \end{aligned}$$

2.

$$\begin{aligned} &\frac{5x-y}{x^2-y^2}-\frac{4x-2y}{x^2-y^2} \\ &=\frac{5x-y-(4x-2y)}{(x+y)(x-y)} \\ &=\frac{5x-y-4x+2y}{(x+y)(x-y)} \\ &=\frac{x+y}{(x+y)(x-y)} \\ &=\frac{1}{x-y} \end{aligned}$$

3.

$$6x^2=2\cdot 3\ x^2$$
$$9x=\ 3^2x$$
$$\text{LCD}=2\cdot 3^2x^2=18x^2$$

4. $5x^2+15x = 5x(x+3)$

$x^2+6x+9 = (x+3)^2$

$\text{LCD} = 5x(x+3)^2$

5. $\frac{7}{6x^2}+\frac{2}{9x}$

In Check Point 3, we determined the LCD of $6x^2$ and $9x$ is $18x^2$.

$$\frac{7}{6x^2}+\frac{2}{9x}=\frac{7\cdot 3}{6x^2\cdot 3}+\frac{2\cdot 2x}{9x\cdot 2x}$$

$$=\frac{21}{18x^2}+\frac{4x}{18x^2}$$

$$=\frac{21+4x}{18x^2} \text{ or } \frac{4x+21}{18x^2}$$

6. $\frac{x}{x-4}+\frac{x-2}{x+4}$

Because the denominators have no common factors, the LCD is $(x+4)(x-4)$.

$$=\frac{x(x+4)}{(x+4)(x-4)}+\frac{(x-2)(x-4)}{(x+4)(x-4)}$$

$$=\frac{x(x+4)+(x-2)(x-4)}{(x+4)(x-4)}$$

$$=\frac{x^2+4x+x^2-6x+8}{(x+4)(x-4)}$$

$$=\frac{2x^2-2x+8}{(x+4)(x-4)}$$

$$=\frac{2(x^2-x+4)}{(x+4)(x-4)}$$

7. $$\frac{2x-3}{x^2-5x+6}-\frac{x+4}{x^2-2x-3}=\frac{2x-3}{(x-3)(x-2)}-\frac{x+4}{(x-3)(x+1)}$$

The LCD is $(x-3)(x-2)(x+1)$.

$$=\frac{(2x-3)(x+1)}{(x-3)(x-2)(x+1)}-\frac{(x+4)(x-2)}{(x-3)(x-2)(x+1)}=\frac{(2x-3)(x+1)-(x+4)(x-2)}{(x-3)(x-2)(x+1)}$$

$$=\frac{2x^2-x-3-(x^2+2x-8)}{(x-3)(x-2)(x+1)}=\frac{2x^2-x-3-x^2-2x+8}{(x-3)(x-2)(x+1)}=\frac{x^2-3x+5}{(x-3)(x-2)(x+1)}$$

8. $$\frac{y-1}{y-2}+\frac{y-6}{y^2-4}-\frac{y+1}{y+2}=\frac{y-1}{y-2}+\frac{y-6}{(y+2)(y-2)}-\frac{y+1}{y+2}$$

The LCD is $(y+2)(y-2)$.

$$\frac{(y-1)(y+2)}{(y+2)(y-2)}+\frac{y-6}{(y+2)(y-2)}-\frac{(y+1)(y-2)}{(y+2)(y-2)}=\frac{(y-1)(y+2)+(y-6)-(y+1)(y-2)}{(y+2)(y-2)}$$

$$=\frac{y^2+y-2+y-6-\left(y^2-y-2\right)}{(y+2)(y-2)}=\frac{y^2+y-2+y-6-y^2+y+2}{(y+2)(y-2)}=\frac{3y-6}{(y+2)(y-2)}$$

$$=\frac{3\cancel{(y-2)}}{(y+2)\cancel{(y-2)}}=\frac{3}{y+2}$$

9. $\frac{4x-7y}{x-3y}+\frac{x-2y}{3y-x}$

Since the denominators are opposites, multiply one of the fractions by $\frac{-1}{-1}$.

$$=\frac{4x-7y}{x-3y}+\frac{(-1)}{(-1)}\cdot\frac{x-2y}{3y-x}$$

$$=\frac{4x-7y}{x-3y}+\frac{-x+2y}{-3y+x}$$

$$=\frac{4x-7y}{x-3y}+\frac{-x+2y}{x-3y}$$

$$=\frac{4x-7y-x+2y}{x-3y}=\frac{3x-5y}{x-3y}$$

Problem Set 6.2

Practice Exercises

1. $\frac{7}{9x}+\frac{5}{9x}=\frac{7+5}{9x}=\frac{12}{9x}=\frac{4\cdot\cancel{3}}{3\cdot\cancel{3}x}=\frac{4}{3x}$

3. $\frac{x}{x-4}+\frac{9x+7}{x-4}=\frac{x+9x+7}{x-4}=\frac{10x+7}{x-4}$

5. $\frac{x^2-2x}{x^2+3x}+\frac{x^2+x}{x^2+3x}=\frac{x^2-2x+x^2+x}{x^2+3x}$

$$=\frac{2x^2-x}{x^2+3x}$$

$$=\frac{\cancel{x}(2x-1)}{\cancel{x}(x+3)}$$

$$=\frac{2x-1}{x+3}$$

7. $\frac{y^2}{y^2-9}+\frac{9-6y}{y^2-9}=\frac{y^2+9-6y}{y^2-9}$

$$=\frac{\cancel{(y-3)}^2}{(y+3)\cancel{(y-3)}}\quad (y-3)$$

$$=\frac{y-3}{y+3}$$

9. $\frac{3x}{4x-3}-\frac{2x-1}{4x-3}=\frac{3x-(2x-1)}{4x-3}$

$$=\frac{3x-2x+1}{4x-3}$$

$$=\frac{x+1}{4x-3}$$

11. $\frac{x^2-2}{x^2+6x-7}-\frac{19-4x}{x^2+6x-7}$

$$=\frac{x^2-2-(19-4x)}{x^2+6x-7}=\frac{x^2-2-19+4x}{x^2+6x-7}$$

$$=\frac{x^2+4x-21}{x^2+6x-7}=\frac{\cancel{(x+7)}(x-3)}{\cancel{(x+7)}(x-1)}$$

$$=\frac{x-3}{x-1}$$

$$=\frac{x-3}{x-1}$$

13. $$\frac{20y^2+5y+1}{6y^2+y-2}-\frac{8y^2-12y-5}{6y^2+y-2}$$
$$=\frac{20y^2+5y+1-(8y^2-12y-5)}{6y^2+y-2}$$
$$=\frac{20y^2+5y+1-8y^2+12y+5}{6y^2+y-2}$$
$$=\frac{12y^2+17y+6}{6y^2+y-2}$$
$$=\frac{(4y+3)\cancel{(3y+2)}}{\cancel{(3y+2)}(2y-1)}$$
$$=\frac{4y+3}{2y-1}$$

15. $$\frac{2x^3-3y^3}{x^2-y^2}-\frac{x^3-2y^3}{x^2-y^2}$$
$$=\frac{2x^3-3y^3-(x^3-2y^3)}{x^2-y^2}$$
$$=\frac{2x^3-3y^3-x^3+2y^3}{x^2-y^2}$$
$$=\frac{x^3-y^3}{x^2-y^2}$$
$$=\frac{\cancel{(x-y)}(x^2+xy+y^2)}{(x+y)\cancel{(x-y)}}$$
$$=\frac{x^2+xy+y^2}{x+y}$$

17. $25x^2=5^2\quad x^2$
$35x=5\cdot 7x$
$\text{LCD}=5^2\cdot 7x^2=175x^2$

19. $x-5 = x-5$
$x^2-25=(x+5)(x-5)$
$\text{LCD}=(x+5)(x-5)$

21. $y^2-100=(y+10)(y-10)$
$y(y-10)=y\quad(y-10)$
$\text{LCD}=y(y+10)(y-10)$

23. $x^2-16=(x+4)(x-4)$
$x^2-8x+16=(x-4)^2$
$\text{LCD}=(x+4)(x-4)^2$

25. $y^2-5y-6=(y-6)\quad(y+1)$
$y^2-4y-5=(y-5)(y+1)$
$\text{LCD}=(y-6)(y-5)(y+1)$

27. $$\frac{3}{5x^2}+\frac{10}{x}$$
The LCD is $5x^2$
$$=\frac{3}{5x^2}+\frac{10\cdot 5x}{x\cdot 5x}$$
$$=\frac{3}{5x^2}+\frac{50x}{5x^2}=\frac{3+50x}{5x^2}$$

29. $$\frac{3}{x-2}+\frac{4}{x+3}$$
The LCD is $(x-2)(x+3)$.
$$=\frac{3}{(x-2)}\cdot\frac{(x+3)}{(x+3)}+\frac{4}{(x+3)}\cdot\frac{(x-2)}{(x-2)}$$
$$=\frac{3(x+3)+4(x-2)}{(x-2)(x+3)}$$
$$=\frac{3x+9+4x-8}{(x-2)(x+3)}=\frac{7x+1}{(x-2)(x+3)}$$

31. $\frac{x}{x^2+5x+6}+\frac{4}{x^2-x-6}=\frac{x}{(x+3)(x+2)}+\frac{4}{(x+2)(x-3)}$

The LCD is $(x+3)(x+2)(x-3)$.

$=\frac{x(x-3)}{(x+3)(x+2)(x-3)}+\frac{4(x+3)}{(x+3)(x+2)(x-3)}=\frac{x(x-3)+4(x+3)}{(x+3)(x+2)(x-3)}$

$=\frac{x^2-3x+4x+12}{(x+3)(x+2)(x-3)}=\frac{x^2+x+12}{(x+3)(x+2)(x-3)}=\frac{x^2+x+12}{(x+3)(x+2)(x-3)}$

33. $\frac{y-7}{y+4}+\frac{y+4}{y-7}$

Since the denominators have no common factors, the LCD is $(y+4)(y-7)$.

$=\frac{(y-7)(y-7)}{(y+4)(y-7)}+\frac{(y+4)(y+4)}{(y+4)(y-7)}$

$=\frac{(y-7)(y-7)+(y+4)(y+4)}{(y+4)(y-7)}$

$=\frac{y^2-14y+49+y^2+8y+16}{(y+4)(y-7)}$

$=\frac{2y^2-6y+65}{(y+4)(y-7)}$

35. $\frac{8x}{x^2-16}-\frac{5}{x+4}$

$=\frac{8x}{(x+4)(x-4)}-\frac{5}{x+4}$

The LCD is $(x+4)(x-4)$.

$=\frac{8x}{(x+4)(x-4)}-\frac{5(x-4)}{(x+4)(x-4)}$

$=\frac{8x-5(x-4)}{(x+4)(x-4)}=\frac{8x-5x+20}{(x+4)(x-4)}$

$=\frac{3x+20}{(x+4)(x-4)}$

37. $\frac{3y+7}{y^2-5y+6}-\frac{3}{y-3}$

$=\frac{3y+7}{(y-3)(y-2)}-\frac{3}{y-3}$

The LCD is $(y-3)(y-2)$.

$=\frac{3y+7}{(y-3)(y-2)}-\frac{3(y-2)}{(y-3)(y-2)}$

$=\frac{3y+7-3(y-2)}{(y-3)(y-2)}$

$=\frac{3y+7-3y+6}{(y-3)(y-2)}=\frac{13}{(y-3)(y-2)}$

39. $\frac{x^2-6}{x^2+9x+18}-\frac{x-4}{x+6}$

$=\frac{x^2-6}{(x+3)(x+6)}-\frac{x-4}{x+6}$

The LCD is $(x+3)(x+6)$.

$=\frac{x^2-6}{(x+3)(x+6)}-\frac{(x-4)(x+3)}{(x+3)(x+6)}$

$=\frac{x^2-6-(x-4)(x+3)}{(x+3)(x+6)}$

$=\frac{x^2-6-(x^2-x-12)}{(x+3)(x+6)}$

$$=\frac{x^2-6-x^2+x+12}{(x+3)(x+6)}=\frac{x+6}{(x+3)(x+6)}=\frac{1\cancel{(x+6)}}{(x+3)\cancel{(x+6)}}=\frac{1}{x+3}$$

41. $$\frac{4x+1}{x^2+7x+12}+\frac{2x+3}{x^2+5x+4}=\frac{4x+1}{(x+3)(x+4)}+\frac{2x+3}{(x+4)(x+1)}$$

The LCD is $(x+3)(x+4)(x+1)$.

$$=\frac{(4x+1)(x+1)}{(x+3)(x+4)(x+1)}+\frac{(2x+3)(x+3)}{(x+3)(x+4)(x+1)}=\frac{(4x+1)(x+1)+(2x+3)(x+3)}{(x+3)(x+4)(x+1)}$$

$$=\frac{4x^2+5x+1+2x^2+9x+9}{(x+3)(x+4)(x+1)}=\frac{6x^2+14x+10}{(x+3)(x+4)(x+1)}=\frac{2(3x^2+7x+5)}{(x+3)(x+4)(x+1)}$$

43. $$\frac{x+4}{x^2-x-2}-\frac{2x+3}{x^2+2x-8}=\frac{x+4}{(x-2)(x+1)}-\frac{2x+3}{(x+4)(x-2)}$$

The LCD is $(x-2)(x+1)(x+4)$.

$$=\frac{(x+4)(x+4)}{(x-2)(x+1)(x+4)}-\frac{(2x+3)(x+1)}{(x+4)(x+1)(x-2)}=\frac{(x+4)(x+4)-(2x+3)(x+1)}{(x-2)(x+1)(x+4)}$$

$$=\frac{x^2+8x+16-(2x^2+5x+3)}{(x-2)(x+1)(x+4)}=\frac{x^2+8x+16-2x^2-5x-3}{(x-2)(x+1)(x+4)}=\frac{-x^2+3x+13}{(x-2)(x+1)(x+4)}$$

$$=\frac{-(x^2-3x-13)}{(x-2)(x+1)(x+4)}=-\frac{x^2-3x-13}{(x-2)(x+1)(x+4)}$$

45. $$4+\frac{1}{x-3}=\frac{4}{1}+\frac{1}{x-3}$$

The LCD is $(x-3)$.

$$=\frac{4(x-3)}{x-3}+\frac{1}{x-3}=\frac{4(x-3)+1}{x-3}$$

$$=\frac{4x-12+1}{x-3}$$

$$=\frac{4x-11}{x-3}$$

47. $$\frac{y-7}{y^2-16}+\frac{7-y}{16-y^2}$$

The LCD is y^2-16.

$$=\frac{y-7}{y^2-16}+\frac{(-1)}{(-1)}\cdot\frac{7-y}{16-y^2}$$

$$=\frac{y-7}{y^2-16}+\frac{-7+y}{-16+y^2}$$

$$=\frac{y-7}{y^2-16}+\frac{y-7}{y^2-16}$$

$$=\frac{y-7+y-7}{y^2-16}=\frac{2y-14}{(y+4)(y-4)}$$

49. $\frac{x+3}{3x+6}+\frac{x}{4-x^2}=\frac{x+3}{3(x+2)}+\frac{x}{(2+x)(2-x)}=\frac{x+3}{3(x+2)}+\frac{x}{(x+2)(2-x)}$

The LCD is $3(x+2)(2-x)$.

$=\frac{(x+3)(2-x)}{3(x+2)(2-x)}+\frac{3x}{3(x+2)(2-x)}=\frac{(x+3)(2-x)+3x}{3(x+2)(2-x)}=\frac{2x-x^2+6-3x+3x}{3(x+2)(2-x)}$

$=\frac{-x^2+2x+6}{3(x+2)(2-x)}=\frac{-1}{-1}\cdot\frac{-x^2+2x+6}{3(x+2)(2-x)}=\frac{x^2-2x-6}{3(x+2)(x-2)}$

51. $\frac{2x}{x-4}+\frac{64}{x^2-16}-\frac{2x}{x+4}=\frac{2x}{x-4}+\frac{64}{(x+4)(x-4)}-\frac{2x}{x+4}$

The LCD is $(x+4)(x-4)$.

$=\frac{2x(x+4)}{(x+4)(x-4)}+\frac{64}{(x+4)(x-4)}-\frac{2x(x-4)}{(x+4)(x-4)}=\frac{2x(x+4)+64-2x(x-4)}{(x+4)(x-4)}$

$=\frac{2x^2+8x+64-2x^2+8x}{(x+4)(x-4)}=\frac{16x+64}{(x+4)(x-4)}=\frac{16\cancel{(x+4)}}{\cancel{(x+4)}(x-4)}=\frac{16}{x-4}$

53. $\frac{7x}{x^2-y^2}-\frac{3}{y-x}=\frac{7x}{(x+y)(x-y)}-\frac{(-1)3}{(-1)(y-x)}$

$=\frac{7x}{(x+y)(x-y)}-\frac{-3}{(-y+x)}=\frac{7x}{(x+y)(x-y)}-\frac{-3}{(x-y)}$

The LCD is $(x+y)(x-y)$.

$=\frac{7x}{(x+y)(x-y)}-\frac{-3(x+y)}{(x+y)(x-y)}=\frac{7x-(-3)(x+y)}{(x+y)(x-y)}=\frac{7x+3(x+y)}{(x+y)(x-y)}$

$=\frac{7x+3x+3y}{(x+y)(x-y)}=\frac{10x+3y}{(x+y)(x-y)}$

55. $\frac{3}{5x+6}-\frac{4}{x-2}+\frac{x^2-x}{5x^2-4x-12}=\frac{3}{5x+6}-\frac{4}{x-2}+\frac{x^2-x}{(5x+6)(x-2)}$

The LCD is $(5x+6)(x-2)$.

$=\frac{3(x-2)}{(5x+6)(x-2)}-\frac{4(5x+6)}{(5x+6)(x-2)}+\frac{x^2-x}{(5x+6)(x-2)}=\frac{3(x-2)-4(5x+6)+x^2-x}{(5x+6)(x-2)}$

$$=\frac{3x-6-20x-24+x^2-x}{(5x+6)(x-2)}=\frac{x^2-18x-30}{(5x+6)(x-2)}$$

57. $\frac{3x-y}{x^2-9xy+20y^2}+\frac{2y}{x^2-25y^2}=\frac{3x-y}{(x-5y)(x-4y)}+\frac{2y}{(x+5y)(x-5y)}$

The LCD is $(x+5y)(x-5y)(x-4y)$.

$$=\frac{(3x-y)(x+5y)}{(x+5y)(x-5y)(x-4y)}+\frac{2y(x-4y)}{(x+5y)(x-5y)(x-4y)}=\frac{(3x-y)(x+5y)+2y(x-4y)}{(x+5y)(x-5y)(x-4y)}$$

$$=\frac{3x^2+14xy-5y^2+2xy-8y^2}{(x+5y)(x-5y)(x-4y)}=\frac{3x^2+16xy-13y^2}{(x+5y)(x-5y)(x-4y)}$$

59. $\frac{3x}{x^2-4}+\frac{5x}{x^2+x-2}-\frac{3}{x^2-4x+4}=\frac{3x}{(x+2)(x-2)}+\frac{5x}{(x+2)(x-1)}-\frac{3}{(x-2)^2}$

The LCD is $(x+2)(x-2)^2(x-1)$.

$$=\frac{3x(x-2)(x-1)}{(x+2)(x-2)^2(x-1)}+\frac{5x(x-2)^2}{(x+2)(x-2)^2(x-1)}-\frac{3(x+2)(x-1)}{(x+2)(x-2)^2(x-1)}$$

$$=\frac{3x(x-2)(x-1)+5x(x-2)^2-3(x+2)(x-1)}{(x+2)(x-2)^2(x-1)}$$

$$=\frac{3x\left(x^2-3x+2\right)+5x\left(x^2-4x+4\right)-3\left(x^2+x-2\right)}{(x+2)(x-2)^2(x-1)}$$

$$=\frac{3x^3-9x^2+6x+5x^3-20x^2+20x-3x^2-3x+6}{(x+2)(x-2)^2(x-1)}$$

$$=\frac{8x^3-32x^2+23x+6}{(x+2)(x-2)^2(x-1)}$$

61. $\frac{6a+5b}{6a^2+5ab-4b^2}-\frac{a+2b}{9a^2-16b^2}=\frac{6a+5b}{(3a+4b)(2a-b)}-\frac{a+2b}{(3a+4b)(3a-4b)}$

The LCD is $(3a+4b)(2a-b)(3a-4b)$.

$$=\frac{(6a+5b)(3a-4b)}{(3a+4b)(2a-b)(3a-4b)}-\frac{(a+2b)(2a-b)}{(3a+4b)(2a-b)(3a-4b)}$$

$$= \frac{(6a+5b)(3a-4b)-(a+2b)(2a-b)}{(3a+4b)(2a-b)(3a-4b)} = \frac{18a^2-9ab-20b^2-(2a^2+3ab-2b^2)}{(3a+4b)(2a-b)(3a-4b)}$$

$$= \frac{18a^2-9ab-20b^2-2a^2-3ab+2b^2}{(3a+4b)(2a-b)(3a-4b)} = \frac{16a^2-12ab-18b^2}{(3a+4b)(2a-b)(3a-4b)}$$

$$= \frac{2(8a^2-6ab-9b^2)}{(3a+4b)(2a-b)(3a-4b)} = \frac{2(4a+3b)(2a-3b)}{(3a+4b)(2a-b)(3a-4b)}$$

Application Exercises

63.
$$T(0) = \frac{470}{0+70} + \frac{250}{0+65} = \frac{470}{70} + \frac{250}{65} = 6.7 + 3.8 = 10.5$$

This corresponds to the point $(0, 10.5)$ on the graph. If you drive zero miles per hour over the speed limit, total driving time is approximately 11 hours.

65.
$$T(x) = \frac{470}{x+70} + \frac{250}{x+65}$$

$$= \frac{470(x+65)}{(x+70)(x+65)} + \frac{250(x+70)}{(x+70)(x+65)}$$

$$= \frac{470(x+65)+250(x+70)}{(x+70)(x+65)}$$

$$= \frac{470x+30550+250x+17500}{(x+70)(x+65)}$$

$$= \frac{720x+48050}{(x+70)(x+65)}$$

$$T(0) = \frac{720(0)+48050}{(0+70)(0+65)}$$

$$= \frac{48050}{(70)(65)} \approx 11$$

67. Answers will vary. In order to make the trip in 9 hours, you need to drive approximately 12 miles per hour over the speed limit.

69.
$$P = 2\left(\frac{x}{x+4}\right) + 2\left(\frac{x}{x+5}\right)$$

$$= \frac{2x}{x+4} + \frac{2x}{x+5}$$

$$= \frac{2x(x+5)}{(x+4)(x+5)} + \frac{2x(x+4)}{(x+4)(x+5)}$$

$$= \frac{2x(x+5)+2x(x+4)}{(x+4)(x+5)}$$

$$= \frac{2x^2+10x+2x^2+8x}{(x+4)(x+5)}$$

$$= \frac{4x^2+18x}{(x+4)(x+5)}$$

$$= \frac{2x(2x+9)}{(x+4)(x+5)}$$

71.
$$f(x) = \frac{27,725(x-14)}{x^2+9} - 5x$$

$$= \frac{27,725(x-14)}{x^2+9} - \frac{5x}{1}$$

$$= \frac{27,725(x-14)}{x^2+9} - \frac{5x(x^2+9)}{(x^2+9)}$$

$$=\frac{27,725(x-14)-5x(x^2+9)}{x^2+9}$$

$$=\frac{27,725x-388,150-5x^3-45x}{x^2+9}$$

$$=\frac{-5x^3+27,680x-388,150}{x^2+9}$$

Writing in Mathematics

73. Answers will vary.

75. Answers will vary.

77. Answers will vary.

Technology Exercises

79. $f(x)=\frac{27,725(x-14)}{x^2+9}-5x$

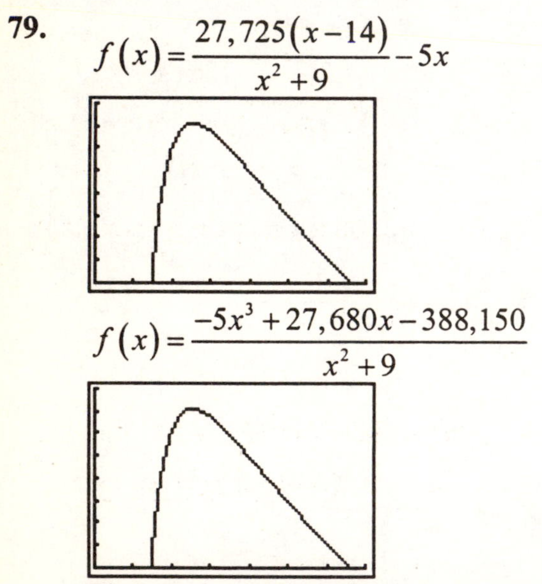

$f(x)=\frac{-5x^3+27,680x-388,150}{x^2+9}$

The graphs are identical.

As age increases, the number of arrests per 100,000 drivers increases until about age 25 and then decreases.

Critical Thinking Exercises

81. $\left(\frac{2x+3}{x+1}\cdot\frac{x^2+4x-5}{2x^2+x-3}\right)-\frac{2}{x+2}$

$$=\left(\frac{\cancel{2x+3}}{x+1}\cdot\frac{(x+5)\cancel{(x-1)}}{\cancel{(2x+3)}\cancel{(x-1)}}\right)-\frac{2}{x+2}$$

$$=\frac{(x+5)}{(x+1)}-\frac{2}{x+2}$$

$$=\frac{(x+5)(x+2)}{(x+1)(x+2)}-\frac{2(x+1)}{(x+1)(x+2)}$$

$$=\frac{(x+5)(x+2)-2(x+1)}{(x+1)(x+2)}$$

$$=\frac{x^2+7x+10-2x-2}{(x+1)(x+2)}$$

$$=\frac{x^2+5x+8}{(x+1)(x+2)}$$

83. $(x-y)^{-1}+(x-y)^{-2}$

$$=\frac{1}{(x-y)}+\frac{1}{(x-y)^2}$$

$$=\frac{(x-y)}{(x-y)(x-y)}+\frac{1}{(x-y)^2}$$

$$=\frac{x-y+1}{(x-y)^2}$$

85.

$$(f-g)(x)=f(x)-g(x)=\frac{2x}{x^2-9}-\frac{6x^2}{x^2+3x-18}=\frac{2x}{(x+3)(x-3)}-\frac{6x^2}{(x+6)(x-3)}$$

$$=\frac{2x(x+6)}{(x+3)(x-3)(x+6)}-\frac{6x^2(x+3)}{(x+3)(x-3)(x+6)}=\frac{2x(x+6)-6x^2(x+3)}{(x+3)(x-3)(x+6)}$$

$$=\frac{2x^2+12x-6x^3-18x^2}{(x+3)(x-3)(x+6)}=\frac{-6x^3-16x^2+12x}{(x+3)(x-3)(x+6)}=\frac{-2x(3x^2+8x-6)}{(x+3)(x-3)(x+6)}$$

$$=-\frac{2x(3x^2+8x-6)}{(x+3)(x-3)(x+6)}$$

87.

$$\left(\frac{g}{f}\right)(x)$$

$$=g(x)\div f(x)$$

$$=\frac{6x^2}{x^2+3x-18}\div\frac{2x}{x^2-9}$$

$$=\frac{6x^2}{x^2+3x-18}\cdot\frac{x^2-9}{2x}$$

$$=\frac{3\cdot\cancel{2}\cdot x\cdot\cancel{x}}{(x+6)\cancel{(x-3)}}\cdot\frac{(x+3)\cancel{(x-3)}}{\cancel{2}\cancel{x}}$$

$$=\frac{3x(x+3)}{(x+6)}$$

89.

$$|3x-1|\le 14$$

$$-14\le 3x-1\le 14$$

$$-14+1\le 3x-1+1\le 14+1$$

$$-13\le 3x\le 15$$

$$-\frac{13}{3}\le x\le 5$$

The solution set is $\left\{x\,\middle|\,-\frac{13}{3}\le x\le 5\right\}$ or $\left[-\frac{13}{3},5\right]$.

Review Exercises

88.

$$\left(\frac{3x^2y^{-2}}{y^3}\right)^{-2}=\left(\frac{3x^2}{y^2y^3}\right)^{-2}=\left(\frac{3x^2}{y^5}\right)^{-2}$$

$$=\left(\frac{y^5}{3x^2}\right)^2=\frac{y^{10}}{9x^4}$$

90.

$$50x^3-18x=2x(25x^2-9)$$

$$=2x(5x+3)(5x-3)$$

Check Points 6.3

1. $$\frac{\frac{x}{y}-1}{\frac{x^2}{y^2}-1}=\frac{y^2}{y^2}\cdot\frac{\frac{x}{y}-1}{\frac{x^2}{y^2}-1}=\frac{y^2\cdot\frac{x}{y}-y^2\cdot 1}{y^2\cdot\frac{x^2}{y^2}-y^2\cdot 1}=\frac{xy-y^2}{x^2-y^2}=\frac{y(x-y)}{(x+y)(x-y)}=\frac{y}{x+y}$$

2. $$\frac{\frac{1}{x+7}-\frac{1}{x}}{7}=\frac{x(x+7)}{x(x+7)}\cdot\frac{\frac{1}{x+7}-\frac{1}{x}}{7}=\frac{x(x+7)\cdot\frac{1}{x+7}-x(x+7)\cdot\frac{1}{x}}{7x(x+7)}$$
$$=\frac{x-(x+7)}{7x(x+7)}=\frac{x-x-7}{7x(x+7)}=\frac{-7}{7x(x+7)}=-\frac{1}{x(x+7)}$$

3. $$\frac{\frac{x+1}{x-1}-\frac{x-1}{x+1}}{\frac{x-1}{x+1}+\frac{x+1}{x-1}}=\frac{\frac{(x+1)(x+1)}{(x-1)(x+1)}-\frac{(x-1)(x-1)}{(x-1)(x+1)}}{\frac{(x-1)(x-1)}{(x+1)(x-1)}+\frac{(x+1)(x+1)}{(x+1)(x-1)}}=\frac{\frac{(x+1)(x+1)-(x-1)(x-1)}{(x-1)(x+1)}}{\frac{(x-1)(x-1)+(x+1)(x+1)}{(x+1)(x-1)}}$$
$$=\frac{\frac{x^2+2x+1-(x^2-2x+1)}{(x-1)(x+1)}}{\frac{x^2-2x+1+x^2+2x+1}{(x+1)(x-1)}}=\frac{\frac{x^2+2x+1-x^2+2x-1}{(x-1)(x+1)}}{\frac{2x^2+2}{(x+1)(x-1)}}$$
$$=\frac{\frac{4x}{(x-1)(x+1)}}{\frac{2(x^2+1)}{(x+1)(x-1)}}=\frac{\frac{4x}{(x-1)(x+1)}}{\frac{2(x^2+1)}{(x+1)(x-1)}}=\frac{2\cdot 2x}{(x-1)(x+1)}\cdot\frac{(x+1)(x-1)}{2(x^2+1)}=\frac{2x}{x^2+1}$$

4. $$\frac{1-4x^{-2}}{1-7x^{-1}+10x^{-2}}=\frac{1-\frac{4}{x^2}}{1-\frac{7}{x}+\frac{10}{x^2}}=\frac{x^2}{x^2}\cdot\frac{1-\frac{4}{x^2}}{1-\frac{7}{x}+\frac{10}{x^2}}=\frac{x^2\cdot 1-x^2\cdot\frac{4}{x^2}}{x^2\cdot 1-x^2\cdot\frac{7}{x}+x^2\cdot\frac{10}{x^2}}$$
$$=\frac{x^2-4}{x^2-7x+10}=\frac{(x+2)(x-2)}{(x-5)(x-2)}=\frac{x+2}{x-5}$$

Problem Set 6.3

Practice Exercises

1.
$$\frac{4+\frac{2}{x}}{1-\frac{3}{x}}=\frac{x}{x}\cdot\frac{4+\frac{2}{x}}{1-\frac{3}{x}}=\frac{x\cdot 4+\cancel{x}\cdot\frac{2}{\cancel{x}}}{x\cdot 1-\cancel{x}\cdot\frac{3}{\cancel{x}}}=\frac{4x+2}{x-3}$$

3.
$$\frac{\frac{3}{x}+\frac{x}{3}}{\frac{x}{3}-\frac{3}{x}}=\frac{3x}{3x}\cdot\frac{\frac{3}{x}+\frac{x}{3}}{\frac{x}{3}-\frac{3}{x}}=\frac{3\cancel{x}\cdot\frac{3}{\cancel{x}}+\cancel{3}x\cdot\frac{x}{\cancel{3}}}{\cancel{3}x\cdot\frac{x}{\cancel{3}}-3\cancel{x}\cdot\frac{3}{\cancel{x}}}=\frac{3\cdot 3+x\cdot x}{x\cdot x-3\cdot 3}=\frac{9+x^2}{x^2-9}=\frac{x^2+9}{(x+3)(x-3)}$$

5.
$$\frac{\frac{1}{x}+\frac{1}{y}}{\frac{1}{x}-\frac{1}{y}}=\frac{xy}{xy}\cdot\frac{\frac{1}{x}+\frac{1}{y}}{\frac{1}{x}-\frac{1}{y}}=\frac{\cancel{x}y\cdot\frac{1}{\cancel{x}}+x\cancel{y}\cdot\frac{1}{\cancel{y}}}{\cancel{x}y\cdot\frac{1}{\cancel{x}}-x\cancel{y}\cdot\frac{1}{\cancel{y}}}=\frac{y+x}{y-x}$$

7.
$$\frac{8x^{-2}-2x^{-1}}{10x^{-1}-6x^{-2}}=\frac{\frac{8}{x^2}-\frac{2}{x}}{\frac{10}{x}-\frac{6}{x^2}}=\frac{x^2}{x^2}\cdot\frac{\frac{8}{x^2}-\frac{2}{x}}{\frac{10}{x}-\frac{6}{x^2}}=\frac{\cancel{x^2}\cdot\frac{8}{\cancel{x^2}}-x^{\cancel{2}}\cdot\frac{2}{\cancel{x}}}{x^{\cancel{2}}\cdot\frac{10}{\cancel{x}}-\cancel{x^2}\cdot\frac{6}{\cancel{x^2}}}=\frac{8-x\cdot 2}{x\cdot 10-6}$$
$$=\frac{8-2x}{10x-6}=\frac{\cancel{2}(4-x)}{\cancel{2}(5x-3)}=\frac{4-x}{5x-3}$$

9.
$$\frac{\frac{1}{x-2}}{1-\frac{1}{x-2}}=\frac{x-2}{x-2}\cdot\frac{\frac{1}{x-2}}{1-\frac{1}{x-2}}=\frac{\cancel{(x-2)}\cdot\frac{1}{\cancel{x-2}}}{(x-2)\cdot 1-\cancel{(x-2)}\cdot\frac{1}{\cancel{x-2}}}=\frac{1}{x-2-1}=\frac{1}{x-3}$$

11.
$$\frac{\frac{1}{x+5}-\frac{1}{x}}{5}=\frac{x(x+5)}{x(x+5)}\cdot\frac{\frac{1}{x+5}-\frac{1}{x}}{5}=\frac{x\cancel{(x+5)}\cdot\frac{1}{\cancel{x+5}}-\cancel{x}(x+5)\cdot\frac{1}{\cancel{x}}}{5x(x+5)}=\frac{x-(x+5)}{5x(x+5)}$$

$$= \frac{\cancel{x} - x - 5}{5x(x+5)} = -\frac{\cancel{5}}{\cancel{5}x(x+5)} = -\frac{1}{x(x+5)}$$

13.

$$\frac{\frac{4}{x+4}}{\frac{1}{x+4} - \frac{1}{x}} = \frac{x(x+4)}{x(x+4)} \cdot \frac{\frac{4}{x+4}}{\frac{1}{x+4} - \frac{1}{x}} = \frac{x\cancel{(x+4)} \cdot \frac{4}{\cancel{x+4}}}{x\cancel{(x+4)} \cdot \frac{1}{\cancel{x+4}} - \cancel{x}(x+4) \cdot \frac{1}{\cancel{x}}}$$

$$= \frac{x \cdot 4}{x - (x+4)} = \frac{4x}{\cancel{x} - x - 4} = -\frac{\cancel{4}x}{\cancel{4}} = -x$$

15.

$$\frac{\frac{1}{x-1} + 1}{\frac{1}{x+1} - 1} = \frac{(x+1)(x-1)}{(x+1)(x-1)} \cdot \frac{\frac{1}{x-1} + 1}{\frac{1}{x+1} - 1} = \frac{(x+1)\cancel{(x-1)} \cdot \frac{1}{\cancel{x-1}} + (x+1)(x-1) \cdot 1}{\cancel{(x+1)}(x-1) \cdot \frac{1}{\cancel{x+1}} - (x+1)(x-1) \cdot 1} = \frac{(x+1) + (x^2 - 1)}{(x-1) - (x^2 - 1)}$$

$$= \frac{x + 1 + x^2 - 1}{x - 1 - x^2 + 1} = \frac{x^2 + x}{x - x^2} = \frac{\cancel{x}(x+1)}{\cancel{x}(1-x)} = \frac{x+1}{-1(x-1)} = -\frac{x+1}{x-1}$$

17.

$$\frac{x^{-1} + y^{-1}}{(x+y)^{-1}} = \frac{\frac{1}{x} + \frac{1}{y}}{\frac{1}{(x+y)}} = \frac{xy(x+y)}{xy(x+y)} \cdot \frac{\frac{1}{x} + \frac{1}{y}}{\frac{1}{(x+y)}} = \frac{\cancel{x}y(x+y) \cdot \frac{1}{\cancel{x}} + x\cancel{y}(x+y) \cdot \frac{1}{\cancel{y}}}{xy\cancel{(x+y)} \cdot \frac{1}{\cancel{(x+y)}}}$$

$$= \frac{y(x+y) + x(x+y)}{xy} = \frac{(x+y)(y+x)}{xy} = \frac{(x+y)(x+y)}{xy}$$

19.

$$\frac{\frac{x+2}{x-2} - \frac{x-2}{x+2}}{\frac{x-2}{x+2} + \frac{x+2}{x-2}} = \frac{(x+2)(x-2)}{(x+2)(x-2)} \cdot \frac{\frac{x+2}{x-2} - \frac{x-2}{x+2}}{\frac{x-2}{x+2} + \frac{x+2}{x-2}}$$

$$= \frac{(x+2)\cancel{(x-2)} \cdot \frac{x+2}{\cancel{x-2}} - \cancel{(x+2)}(x-2) \cdot \frac{x-2}{\cancel{x+2}}}{\cancel{(x+2)}(x-2) \cdot \frac{x-2}{\cancel{x+2}} + (x+2)\cancel{(x-2)} \cdot \frac{x+2}{\cancel{x-2}}}$$

$$=\frac{(x+2)(x+2)-(x-2)(x-2)}{(x-2)(x-2)+(x+2)(x+2)}=\frac{x^2+4x+4-(x^2-4x+4)}{x^2-4x+4+x^2+4x+4}$$

$$=\frac{x^2+4x+4-x^2+4x-4}{2x^2+8}=\frac{8x}{2(x^2+4)}=\frac{4x}{x^2+4}$$

21.
$$\frac{\frac{2}{x^3y}+\frac{5}{xy^4}}{\frac{5}{x^3y}-\frac{3}{xy}}=\frac{x^3y^4}{x^3y^4}\cdot\frac{\frac{2}{x^3y}+\frac{5}{xy^4}}{\frac{5}{x^3y}-\frac{3}{xy}}=\frac{x^3y^4\cdot\frac{2}{x^3y}+x^3y^4\cdot\frac{5}{xy^4}}{x^3y^4\cdot\frac{5}{x^3y}-x^3y^4\cdot\frac{3}{xy}}$$

$$=\frac{y^3\cdot 2+x^2\cdot 5}{y^3\cdot 5-x^2y^3\cdot 3}=\frac{2y^3+5x^2}{5y^3-3x^2y^3}=\frac{2y^3+5x^2}{y^3(5-3x^2)}$$

23.
$$\frac{\frac{3}{x+2}-\frac{3}{x-2}}{\frac{5}{x^2-4}}=\frac{\frac{3}{x+2}-\frac{3}{x-2}}{\frac{5}{(x+2)(x-2)}}=\frac{(x+2)(x-2)}{(x+2)(x-2)}\cdot\frac{\frac{3}{x+2}-\frac{3}{x-2}}{\frac{5}{(x+2)(x-2)}}$$

$$=\frac{(x+2)(x-2)\cdot\frac{3}{x+2}-(x+2)(x-2)\cdot\frac{3}{x-2}}{(x+2)(x-2)\cdot\frac{5}{(x+2)(x-2)}}$$

$$=\frac{(x-2)\cdot 3-(x+2)\cdot 3}{5}=\frac{3x-6-(3x+6)}{5}=\frac{3x-6-3x-6}{5}=-\frac{12}{5}$$

25.
$$\frac{3a^{-1}+3b^{-1}}{4a^{-2}-9b^{-2}}=\frac{\frac{3}{a}+\frac{3}{b}}{\frac{4}{a^2}-\frac{9}{b^2}}=\frac{a^2b^2}{a^2b^2}\cdot\frac{\frac{3}{a}+\frac{3}{b}}{\frac{4}{a^2}-\frac{9}{b^2}}=\frac{a^2b^2\cdot\frac{3}{a}+a^2b^2\cdot\frac{3}{b}}{a^2b^2\cdot\frac{4}{a^2}-a^2b^2\cdot\frac{9}{b^2}}$$

$$=\frac{ab^2\cdot 3+a^2b\cdot 3}{b^2\cdot 4-a^2\cdot 9}=\frac{3ab^2+3a^2b}{4b^2-9a^2}=\frac{3ab(b+a)}{(2b+3a)(2b-3a)}$$

27.
$$\frac{\frac{4x}{x^2-4}-\frac{5}{x-2}}{\frac{2}{x-2}+\frac{3}{x+2}}=\frac{\frac{4x}{(x+2)(x-2)}-\frac{5}{x-2}}{\frac{2}{x-2}+\frac{3}{x+2}}=\frac{(x+2)(x-2)}{(x+2)(x-2)}\cdot\frac{\frac{4x}{(x+2)(x-2)}-\frac{5}{x-2}}{\frac{2}{x-2}+\frac{3}{x+2}}$$

$$=\frac{(x+2)(x-2)\cdot\frac{4x}{(x+2)(x-2)}-(x+2)(x-2)\cdot\frac{5}{x-2}}{(x+2)(x-2)\cdot\frac{2}{x-2}+(x+2)(x-2)\cdot\frac{3}{x+2}}$$

$$=\frac{4x-(x+2)\cdot 5}{(x+2)\cdot 2+(x-2)\cdot 3}=\frac{4x-5x-10}{2x+4+3x-6}=\frac{-x-10}{5x-2}=-\frac{x+10}{5x-2}$$

29.
$$\frac{\frac{2y}{y^2+4y+3}}{\frac{1}{y+3}+\frac{2}{y+1}}=\frac{\frac{2y}{(y+3)(y+1)}}{\frac{1}{y+3}+\frac{2}{y+1}}=\frac{(y+3)(y+1)}{(y+3)(y+1)}\cdot\frac{\frac{2y}{(y+3)(y+1)}}{\frac{1}{y+3}+\frac{2}{y+1}}$$

$$=\frac{(y+3)(y+1)\cdot\frac{2y}{(y+3)(y+1)}}{(y+3)(y+1)\cdot\frac{1}{y+3}+(y+3)(y+1)\cdot\frac{2}{y+1}}$$

$$=\frac{2y}{(y+1)+(y+3)\cdot 2}=\frac{2y}{y+1+2y+6}=\frac{2y}{3y+7}$$

31.
$$\frac{\frac{2}{a^2}-\frac{1}{ab}-\frac{1}{b^2}}{\frac{1}{a^2}-\frac{3}{ab}+\frac{2}{b^2}}=\frac{a^2b^2}{a^2b^2}\cdot\frac{\frac{2}{a^2}-\frac{1}{ab}-\frac{1}{b^2}}{\frac{1}{a^2}-\frac{3}{ab}+\frac{2}{b^2}}=\frac{a^2b^2\cdot\frac{2}{a^2}-a^2b^2\cdot\frac{1}{ab}-a^2b^2\cdot\frac{1}{b^2}}{a^2b^2\cdot\frac{1}{a^2}-a^2b^2\cdot\frac{3}{ab}+a^2b^2\cdot\frac{2}{b^2}}$$

$$=\frac{b^2\cdot 2-ab-a^2}{b^2-ab\cdot 3+a^2\cdot 2}=\frac{2b^2-ab-a^2}{b^2-3ab+2a^2}=\frac{(2b+a)(b-a)}{(b-2a)(b-a)}=\frac{2b+a}{b-2a}$$

33.
$$\frac{\frac{2x}{x^2-25}+\frac{1}{3x-15}}{\frac{5}{x-5}+\frac{3}{4x-20}}=\frac{\frac{2x}{(x+5)(x-5)}+\frac{1}{3(x-5)}}{\frac{5}{x-5}+\frac{3}{4(x-5)}}=\frac{12(x+5)(x-5)}{12(x+5)(x-5)}\cdot\frac{\frac{2x}{(x+5)(x-5)}+\frac{1}{3(x-5)}}{\frac{5}{x-5}+\frac{3}{4(x-5)}}$$

$$= \frac{12(x+5)(x-5) \cdot \frac{2x}{(x+5)(x-5)} + 12(x+5)(x-5) \cdot \frac{1}{3(x-5)}}{12(x+5)(x-5) \cdot \frac{5}{x-5} + 12(x+5)(x-5) \cdot \frac{3}{4(x-5)}}$$

$$= \frac{12 \cdot 2x + 4(x+5)}{12(x+5) \cdot 5 + 3(x+5) \cdot 3} = \frac{24x + 4x + 20}{60(x+5) + 9(x+5)}$$

$$= \frac{28x + 20}{(x+5)(60+9)} = \frac{4(7x+5)}{69(x+5)}$$

35. $$\frac{\frac{3}{x+2y} - \frac{2y}{x^2+2xy}}{\frac{3y}{x^2+2xy} + \frac{5}{x}} = \frac{\frac{3}{x+2y} - \frac{2y}{x(x+2y)}}{\frac{3y}{x(x+2y)} + \frac{5}{x}} = \frac{x(x+2y)}{x(x+2y)} \cdot \frac{\frac{3}{x+2y} - \frac{2y}{x(x+2y)}}{\frac{3y}{x(x+2y)} + \frac{5}{x}}$$

$$= \frac{x(x+2y) \cdot \frac{3}{x+2y} - x(x+2y) \cdot \frac{2y}{x(x+2y)}}{x(x+2y) \cdot \frac{3y}{x(x+2y)} + x(x+2y) \cdot \frac{5}{x}}$$

$$= \frac{x \cdot 3 - 2y}{3y + (x+2y) \cdot 5} = \frac{3x - 2y}{3y + 5x + 10y} = \frac{3x - 2y}{5x + 13y}$$

37. $$\frac{x}{1 - \frac{1}{1+\frac{1}{x}}} = \frac{x}{1 - \frac{x}{x} \cdot \frac{1}{1+\frac{1}{x}}} = \frac{x}{1 - \frac{x \cdot 1}{x \cdot 1 + x \cdot \frac{1}{x}}} = \frac{x}{1 - \frac{x}{x+1}} = \frac{x+1}{x+1} \cdot \frac{x}{1 - \frac{x}{x+1}}$$

$$= \frac{(x+1)x}{(x+1)1 - (x+1)\frac{x}{x+1}} = \frac{x(x+1)}{(x+1) - (x+1)\frac{x}{x+1}} = \frac{x(x+1)}{x+1-x} = \frac{x(x+1)}{1} = x(x+1)$$

Application Exercises

39. a.

$$A=\frac{Pi}{1-\frac{1}{(1+i)^n}}=\frac{(1+i)^n}{(1+i)^n}\cdot\frac{Pi}{1-\frac{1}{(1+i)^n}}=\frac{Pi(1+i)^n}{(1+i)^n\cdot 1-\cancel{(1+i)^n}\cdot\frac{1}{\cancel{(1+i)^n}}}=\frac{Pi(1+i)^n}{(1+i)^n-1}$$

b.

$$A=\frac{Pi(1+i)^n}{(1+i)^n-1}=\frac{(20000)(0.01)(1+0.01)^{48}}{(1+0.01)^{48}-1}=\frac{(20000)(0.01)(1.01)^{48}}{(1.01)^{48}-1}$$

$$=\frac{(20000)(0.01)(1.612)}{(1.612)-1}=\frac{322.4}{0.612}=526.80$$

You will pay approximately $527 each month.

41.

$$R=\frac{1}{\frac{1}{R_1}+\frac{1}{R_2}+\frac{1}{R_3}}=\frac{R_1R_2R_3}{R_1R_2R_3}\cdot\frac{1}{\frac{1}{R_1}+\frac{1}{R_2}+\frac{1}{R_3}}$$

$$=\frac{R_1R_2R_3\cdot 1}{\cancel{R_1}R_2R_3\cdot\frac{1}{\cancel{R_1}}+R_1\cancel{R_2}R_3\cdot\frac{1}{\cancel{R_2}}+R_1R_2\cancel{R_3}\cdot\frac{1}{\cancel{R_3}}}=\frac{R_1R_2R_3}{R_2R_3+R_1R_3+R_1R_2}$$

$$R=\frac{R_1R_2R_3}{R_2R_3+R_1R_3+R_1R_2}=\frac{4\cdot 8\cdot 12}{8\cdot 12+4\cdot 12+4\cdot 8}=\frac{384}{96+48+32}=\frac{384}{176}\approx 2.18$$

The combined resistance is approximately 2.18 ohms.

Writing in Mathematics

43. Answers will vary.

45. Answers will vary.

Technology Exercises

47.

$$\frac{x-\frac{1}{2x+1}}{1-\frac{x}{2x+1}}=2x-1$$

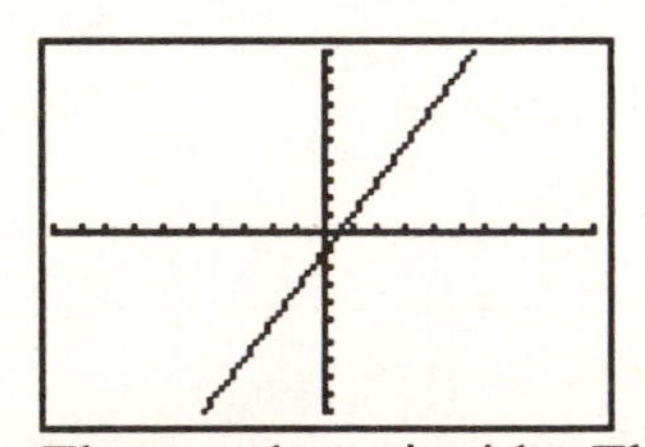

The graphs coincide. The simplification is correct.

49. $$\frac{\frac{1}{x}+\frac{1}{3}}{\frac{1}{3x}}=x+\frac{1}{3}$$

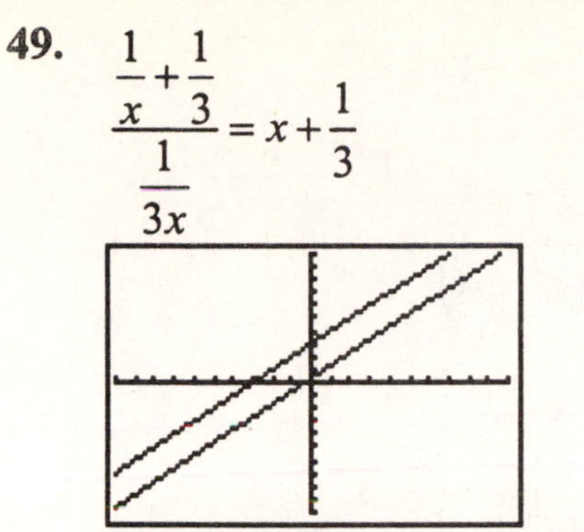

The graphs do not coincide. Simplify the complex fraction.

$$\frac{\frac{1}{x}+\frac{1}{3}}{\frac{1}{3x}}=\frac{3x}{3x}\cdot\frac{\frac{1}{x}+\frac{1}{3}}{\frac{1}{3x}}$$

$$=\frac{3\cancel{x}\cdot\frac{1}{\cancel{x}}+\cancel{3}x\cdot\frac{1}{\cancel{3}}}{\cancel{3x}\cdot\frac{1}{\cancel{3x}}}$$

$$=\frac{3+x}{1}=x+3$$

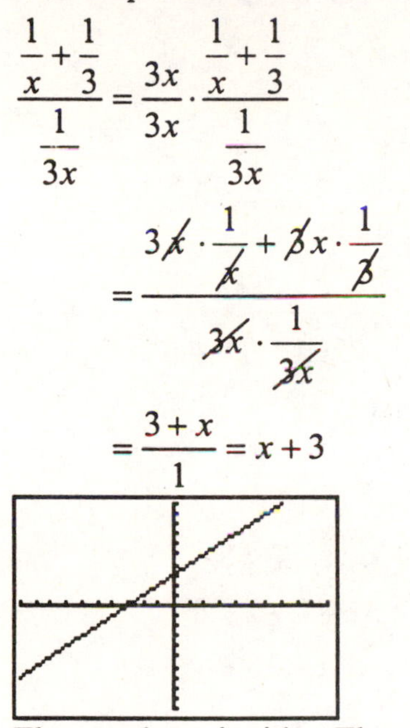

The graphs coincide. The revised simplification is correct.

Critical Thinking Exercises

51. $$\frac{\frac{x+h}{x+h+1}-\frac{x}{x+1}}{h}$$

$$=\frac{\frac{(x+h)(x+1)}{(x+h+1)(x+1)}-\frac{x(x+h+1)}{(x+h+1)(x+1)}}{h}$$

$$=\frac{\frac{(x+h)(x+1)-x(x+h+1)}{(x+h+1)(x+1)}}{h}$$

$$=\frac{\frac{\cancel{x^2}+\cancel{x}+\cancel{hx}+h-\cancel{x^2}-\cancel{hx}-\cancel{x}}{(x+h+1)(x+1)}}{h}$$

$$=\frac{\frac{h}{(x+h+1)(x+1)}}{h}$$

$$=\frac{\frac{h}{(x+h+1)(x+1)}}{\frac{h}{1}}$$

$$=\frac{\cancel{h}}{(x+h+1)(x+1)}\cdot\frac{1}{\cancel{h}}$$

$$=\frac{1}{(x+h+1)(x+1)}$$

53. $$x+\frac{1}{x+\frac{1}{x+\frac{1}{x}}}=x+\frac{1}{x+\frac{x}{x}\cdot\frac{1}{x+\frac{1}{x}}}=x+\frac{1}{x+\frac{x\cdot 1}{x\cdot x+\cancel{x}\cdot\frac{1}{\cancel{x}}}}=x+\frac{1}{x+\frac{x}{x^2+1}}$$

$$=x+\frac{x^2+1}{x^2+1}\cdot\frac{1}{x+\frac{x}{x^2+1}}=x+\frac{(x^2+1)1}{(x^2+1)x+\cancel{(x^2+1)}\frac{x}{\cancel{x^2+1}}}$$

$$= x + \frac{x^2+1}{x^3+x+x} = \frac{x}{1} + \frac{x^2+1}{x^3+2x} = \frac{x\left(x^3+2x\right)}{x^3+2x} + \frac{x^2+1}{x^3+2x}$$

$$= \frac{x\left(x^3+2x\right)+x^2+1}{x^3+2x} = \frac{x^4+2x^2+x^2+1}{x\left(x^2+2\right)} = \frac{x^4+3x^2+1}{x\left(x^2+2\right)}$$

55. $f(x) = \frac{1}{x^2}$

$$\frac{f(a+h)-f(a)}{h} = \frac{\frac{1}{(a+h)^2} - \frac{1}{a^2}}{h} = \frac{a^2(a+h)^2}{a^2(a+h)^2} \cdot \frac{\frac{1}{(a+h)^2} - \frac{1}{a^2}}{h}$$

$$= \frac{a^2\cancel{(a+h)^2} \cdot \frac{1}{\cancel{(a+h)^2}} - \cancel{a^2}(a+h)^2 \cdot \frac{1}{\cancel{a^2}}}{a^2(a+h)^2 h}$$

$$= \frac{a^2-(a+h)^2}{a^2(a+h)^2 h} = \frac{a^2-\left(a^2+2ah+h^2\right)}{a^2(a+h)^2 h} = \frac{a^2-a^2-2ah-h^2}{a^2(a+h)^2 h}$$

$$= \frac{-2ah-h^2}{a^2(a+h)^2 h} = \frac{-\cancel{h}(2a+h)}{a^2(a+h)^2 \cancel{h}} = -\frac{2a+h}{a^2(a+h)^2}$$

Review Exercises

56.

$$x^2+27 = 12x$$
$$x^2-12x+27 = 0$$
$$(x-9)(x-3) = 0$$

Apply the zero product principle.

$x-9=0$ $\quad$ $x-3=0$

$x=9$ $\quad$ $x=3$

The solutions are 3 and 9 and the solution set is $\{3,9\}$.

57. $\left(4x^2-y\right)^2$

$$= \left(4x^2\right)^2 + 2\cdot 4x^2(-y) + (-y)^2$$
$$= 16x^4 - 8x^2y + y^2$$

58.

$$-4 < 3x-7 < 8$$
$$-4+7 < 3x-7+7 < 8+7$$
$$3 < 3x < 15$$
$$1 < x < 5$$

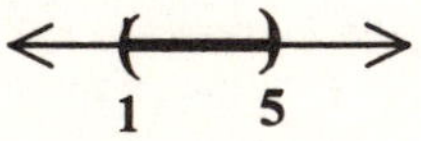

The solution set is $\{x \mid 1 < x < 5\}$ or $(1,5)$.

Check Points 6.4

1. $\left(16x^3-32x^2+2x+4\right) \div 4x$

$$= \frac{16x^3-32x^2+2x+4}{4x}$$

$$=\frac{16x^3}{4x}-\frac{32x^2}{4x}+\frac{2x}{4x}+\frac{4}{4x}$$
$$=4x^2-8x+\frac{1}{2}+\frac{1}{x}$$

2. $$\frac{15x^4y^5-5x^3y^4+10x^2y^2}{5x^2y^3}$$
$$=\frac{15x^4y^5}{5x^2y^3}-\frac{5x^3y^4}{5x^2y^3}+\frac{10x^2y^2}{5x^2y^3}$$
$$=3x^2y^2-xy+\frac{2}{y}$$

3.
$$\begin{array}{r|l} & x+5 \\ x+9 & x^2+14x+45 \\ & \underline{x^2+9x} \\ & \quad 5x+45 \\ & \quad \underline{5x+45} \\ & \qquad 0 \end{array}$$
$$\frac{x^2+14x+45}{x+9}=x+5$$

4. $(-9+7x-4x^2+4x^3)\div(2x-1)$

Rewrite the polynomials in descending order and divide.
$$\begin{array}{r|l} & 2x^2-x+3 \\ 2x-1 & 4x^3-4x^2+7x-9 \\ & \underline{4x^3-2x^2} \\ & \quad -2x^2+7x \\ & \quad \underline{-2x^2+x} \\ & \qquad 6x-9 \\ & \qquad \underline{6x-3} \\ & \qquad\quad -6 \end{array}$$
$$\frac{4x^3-4x^2+7x-9}{2x-1}$$
$$=2x^2-x+3-\frac{6}{2x-1}$$

5. $(2x^4+3x^3-7x-10)\div(x^2-2x)$

Rewrite the dividend with the missing power of x and divide.
$$\begin{array}{r|l} & 2x^2+7x+14 \\ x^2-2x & 2x^4+3x^3+0x^2-7x-10 \\ & \underline{2x^4-4x^3} \\ & \quad 7x^3+0x^2 \\ & \quad \underline{7x^3-14x^2} \\ & \qquad 14x^2-7x \\ & \qquad \underline{14x^2-28x} \\ & \qquad\quad 21x-10 \end{array}$$
$$\frac{2x^4+3x^3-7x-10}{x^2-2x}$$
$$=2x^2+7x+14+\frac{21x-10}{x^2-2x}$$

Problem Set 6.4

Practice Exercises

1. $$\frac{25x^7-15x^5+10x^3}{5x^3}$$
$$=\frac{25x^7}{5x^3}-\frac{15x^5}{5x^3}+\frac{10x^3}{5x^3}$$
$$=5x^4-3x^2+2$$

3. $$\frac{18x^3+6x^2-9x-6}{3x}$$
$$=\frac{18x^3}{3x}+\frac{6x^2}{3x}-\frac{9x}{3x}-\frac{6}{3x}$$
$$=6x^2+2x-3-\frac{2}{x}$$

5. $\dfrac{28x^3-7x^2-16x}{4x^2}=\dfrac{28x^3}{4x^2}-\dfrac{7x^2}{4x^2}-\dfrac{16x}{4x^2}$

$=7x-\dfrac{7}{4}-\dfrac{4}{x}$

7. $\dfrac{25x^8-50x^7+3x^6-40x^5}{-5x^5}$

$=\dfrac{25x^8}{-5x^5}-\dfrac{50x^7}{-5x^5}+\dfrac{3x^6}{-5x^5}-\dfrac{40x^5}{-5x^5}$

$=-5x^3+10x^2-\dfrac{3}{5}x+8$

9. $\dfrac{18a^3b^2-9a^2b-27ab^2}{9ab}$

$=\dfrac{18a^3b^2}{9ab}-\dfrac{9a^2b}{9ab}-\dfrac{27ab^2}{9ab}$

$=2a^2b-a-3b$

11. $\dfrac{36x^4y^3-18x^3y^2-12x^2y}{6x^3y^3}$

$=\dfrac{36x^4y^3}{6x^3y^3}-\dfrac{18x^3y^2}{6x^3y^3}-\dfrac{12x^2y}{6x^3y^3}$

$=6x-\dfrac{3}{y}-\dfrac{2}{xy^2}$

13.

$$\begin{array}{r} x+3 \\ x+5\overline{)x^2+8x+15} \\ \underline{x^2+5x} \\ 3x+15 \\ \underline{3x+15} \\ 0 \end{array}$$

$\dfrac{x^2+8x+15}{x+5}=x+3$

15.

$$\begin{array}{r} x^2+x-2 \\ x-3\overline{)x^3-2x^2-5x+6} \\ \underline{x^3-3x^2} \\ x^2-5x \\ \underline{x^2-3x} \\ -2x+6 \\ \underline{-2x+6} \\ 0 \end{array}$$

$\dfrac{x^3-2x^2-5x+6}{x-3}=x^2+x-2$

17.

$$\begin{array}{r} x-2 \\ x-5\overline{)x^2-7x+12} \\ \underline{x^2-5x} \\ -2x+12 \\ \underline{-2x+10} \\ 2 \end{array}$$

$\dfrac{x^2-7x+12}{x-5}=x-2+\dfrac{2}{x-5}$

19.

$$\begin{array}{r} x+5 \\ 2x+3\overline{)2x^2+13x+5} \\ \underline{2x^2+3x} \\ 10x+5 \\ \underline{10x+15} \\ -10 \end{array}$$

$\dfrac{2x^2+13x+5}{2x+3}=x+5-\dfrac{10}{2x+3}$

21.

$$\begin{array}{r} x^2+2x+3 \\ x+1\overline{\smash{)}x^3+3x^2+5x+4} \\ \underline{x^3+x^2} \\ 2x^2+5x \\ \underline{2x^2+2x} \\ 3x+4 \\ \underline{3x+3} \\ 1 \end{array}$$

$$\frac{x^3+3x^2+5x+4}{x+1} = x^2+2x+3+\frac{1}{x+1}$$

23.

$$\begin{array}{r} 2y^2+3y-1 \\ 2y+3\overline{\smash{)}4y^3+12y^2+7y-3} \\ \underline{4y^3+6y^2} \\ 6y^2+7y \\ \underline{6y^2+9y} \\ -2y-3 \\ \underline{-2y-3} \\ 0 \end{array}$$

$$\frac{4y^3+12y^2+7y-3}{2y+3} = 2y^2+3y-1$$

25.

$$\begin{array}{r} 3x^2-3x+1 \\ 3x+2\overline{\smash{)}9x^3-3x^2-3x+4} \\ \underline{9x^3+6x^2} \\ -9x^2-3x \\ \underline{-9x^2-6x} \\ 3x+4 \\ \underline{3x+2} \\ 2 \end{array}$$

$$\frac{9x^3-3x^2-3x+4}{3x+2} = 3x^2-3x+1+\frac{2}{3x+2}$$

27. $(4x^3-6x-11)\div(2x-4)$

Rewrite the dividend with the missing power of x and divide.

$$\begin{array}{r} 2x^2+4x+5 \\ 2x-4\overline{\smash{)}4x^3+0x^2-6x-11} \\ \underline{4x^3-8x^2} \\ 8x^2-6x \\ \underline{8x^2-16x} \\ 10x-11 \\ \underline{10x-20} \\ 9 \end{array}$$

$$\frac{4x^3-6x-11}{2x-4} = 2x^2+4x+5+\frac{9}{2x-4}$$

29. $(4y^3-5y)\div(2y-1)$

Rewrite the dividend with the missing powers of y and divide. Notice: If there is no constant term in the polynomial, it should also be added for the purposes of long division.

$$\begin{array}{r} 2y^2+y-2 \\ 2y-1\overline{\smash{)}4y^3+0y^2-5y+0} \\ \underline{4y^3-2y^2} \\ 2y^2-5y \\ \underline{2y^2-y} \\ -4y+0 \\ \underline{-4y+2} \\ -2 \end{array}$$

$$\frac{4y^3-5y}{2y-1}=2y^2+y-2-\frac{2}{2y-1}$$

31. $\left(4y^4-17y^2+14y-3\right)\div(2y-3)$

Rewrite the dividend with the missing power of y and divide.

$$\begin{array}{r} 2y^3+3y^2-4y+1 \\ 2y-3\overline{\big)4y^4+0y^3-17y^2+14y-3} \\ \underline{4y^3-6y^3} \\ 6y^3-17y^2 \\ \underline{6y^3-\ 9y^2} \\ -8y^2+14y \\ \underline{-8y^2+12y} \\ 2y-3 \\ \underline{2y-3} \\ 0 \end{array}$$

$$\frac{4y^4-17y^2+14y-3}{2y-3}$$
$$=2y^3+3y^2-4y+1$$

33. $\left(4x^4+3x^3+4x^2+9x-6\right)\div\left(x^2+3\right)$

Rewrite the divisor with the missing power of x and divide.

$$\begin{array}{r} 4x^2+3x-8 \\ x^2+0x+3\overline{\big)4x^4+3x^3+\ 4x^2+9x-6} \\ \underline{4x^4+0x^3+12x^2} \\ 3x^3-\ 8x^2+9x \\ \underline{3x^3+\ 0x^2+9x} \\ -8x^2+0x-\ 6 \\ \underline{-8x^2+0x-24} \\ 18 \end{array}$$

$$\frac{4x^4+3x^3+4x^2+9x-6}{x^2+3}$$
$$=4x^2+3x-8+\frac{18}{x^2+3}$$

35. $\left(15x^4+3x^3+4x^2+4\right)\div\left(3x^2-1\right)$

Rewrite the dividend and the divisor with the missing powers of x and divide.

$$\begin{array}{r} 5x^2+\ x+3 \\ 3x^2+0x-1\overline{\big)15x^4+3x^3+4x^2+0x+4} \\ \underline{15x^4+0x^3-5x^2} \\ 3x^3+9x^2+0x \\ \underline{3x^3+0x^2-\ x} \\ 9x^2+\ x+4 \\ \underline{9x^2+0x-3} \\ x+7 \end{array}$$

$$\frac{15x^4+3x^3+4x^2+4}{3x^2-1}$$
$$=5x^2+x+3+\frac{7+x}{3x^2-1}$$
$$\text{or } 5x^2+x+3+\frac{x+7}{3x^2-1}$$

Application Exercises

37.
$$f(30)=\frac{80(30)-8000}{30-110}$$
$$=\frac{2400-8000}{-80}=\frac{-5600}{-80}=70$$

With a tax rate percentage of 30, the government tax revenue will be \$70 ten billions or \$700 billion. This is shown on the graph as the point $(30,70)$.

39. $(80x-8000)\div(x-110)$

$$\begin{array}{r} 80 \\ x-110\overline{)80x-8000} \\ \underline{80x-8800} \\ 800 \end{array}$$

$$\frac{80x-8000}{x-110}=80+\frac{800}{x-110}$$

$$f(x)=80+\frac{800}{x-110}$$

$$f(30)=80+\frac{800}{30-110}$$

$$=80+\frac{800}{-80}$$

$$=80-10$$

$$=70$$

This is the same answer obtained in Exercise 37.

Writing in Mathematics

41. Answers will vary.

43. Answers will vary.

45. Answers will vary.

Technology Exercises

47. $(6x^2+16x+8)\div(3x+2)=2x+4$

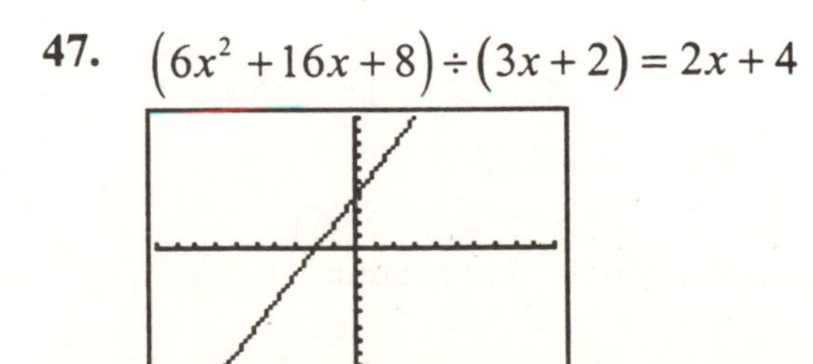

The graphs coincide. The division is correct.

49. $(3x^2-4x+6)\div(x-2)$

$$=3x+2+\frac{10}{x-2}$$

The graphs coincide. The division is correct.

Critical Thinking Exercises

51. Statement **d.** is true. If long division results in a zero remainder, the divisor is a factor of the dividend.

Statement **a.** is false. Since some polynomials cannot be factored this method cannot always be used.

Statement **b.** is false. In many cases, long-division results in a remainder with a variable term in the denominator. Polynomials do not have variable terms in any denominator.

Statement **c.** is false. There is no such restriction in long division.

53. $(x^{3n}+1)\div(x^n+1)$

Rewrite the dividend with the missing power of x and divide.

$$\begin{array}{r} x^{2n} - x^n + 1 \\ x^n + 1 \overline{) x^{3n} + 0x^{2n} + 0x^n + 1} \\ \underline{x^{3n} + x^{2n}} \\ -x^{2n} + 0x^n \\ \underline{-x^{2n} - x^n} \\ x^n + 1 \\ \underline{x^n + 1} \\ 0 \end{array}$$

$$\frac{x^{3n}+1}{x^n+1} = x^{2n} - x^n + 1$$

55. $(20x^3 + 23x^2 - 10x + k) \div (4x+3)$

$$\begin{array}{r} 5x^2 + 2x - 4 \\ 4x+3 \overline{) 20x^3 + 23x^2 - 10x + k} \\ \underline{20x^3 + 15x^2} \\ 8x^2 - 10x \\ \underline{8x^2 + 6x} \\ -16x + k \\ \underline{-16x - 12} \\ k - (-12) \end{array}$$

Set the remainder equal to zero to find k.

$k - (-12) = 0$

$k + 12 = 0$

$k = -12$

When $k = -12$, the remainder will be zero.

Review Exercises

56. $|2x - 3| > 4$

$2x - 3 < -4$ $\quad$ $2x - 3 > 4$

$x < -\frac{1}{2}$ $\quad$ $2x > 7$

$x > \frac{7}{2}$

-1/2 7/2

The solution set is $\left\{ x \middle| x < -\frac{1}{2} \text{ or } x > \frac{7}{2} \right\}$ or $\left(-\infty, -\frac{1}{2}\right) \cup \left(\frac{7}{2}, \infty\right)$.

57. $40{,}610{,}000 = 4.061 \times 10^7$

58. $2x - 4[x - 3(2x+1)]$

$= 2x - 4[x - 6x - 3]$

$= 2x - 4[-5x - 3]$

$= 2x + 20x + 12$

$= 22x + 12$

Check Points 6.5

1. $(x^3 - 7x - 6) \div (x+2)$

$$\begin{array}{r|rrrr} \underline{-2} & 1 & 0 & -7 & -6 \\ & & -2 & 4 & 6 \\ \hline & 1 & -2 & -3 & 0 \end{array}$$

$(x^3 - 7x - 6) \div (x+2) = x^2 - 2x - 3$

2. Divide $f(x)$ by $x - (-4) = x + 4$.

$$\begin{array}{r|rrrr} \underline{-4} & 3 & 4 & -5 & 3 \\ & & -12 & 32 & -108 \\ \hline & 3 & -8 & 27 & -105 \end{array}$$

$f(-4) = -105$

3. To show that –1 is a solution to the equation, show that when the polynomial is divided by $x + 1$ the remainder is zero.

$$\begin{array}{r|rrrr} -1 & 15 & 14 & -3 & -2 \\ & & -15 & 1 & 2 \\ \hline & 15 & -1 & -2 & 0 \end{array}$$

The remainder is zero and –1 is a solution to the equation.

$15x^3 + 14x^2 - 3x - 2$

$= (x+1)(15x^2 - x - 2)$

To solve the equation, we set it equal to zero and factor.

$$15x^3 + 14x^2 - 3x - 2 = 0$$
$$(x+1)(15x^2 - x - 2) = 0$$
$$(x+1)(3x+1)(5x-2) = 0$$

Apply the zero product principle.

$x+1=0 \qquad 3x+1=0 \qquad 5x-2=0$

$x=-1 \qquad 3x=-1 \qquad 5x=2$

$x=-\frac{1}{3} \qquad x=\frac{2}{5}$

The solutions are $-1, -\frac{1}{3}$ and $\frac{2}{5}$ and the solution set is $\left\{-1, -\frac{1}{3}, \frac{2}{5}\right\}$.

Problem Set 6.5

Practice Problems

1. $(2x^2 + x - 10) \div (x - 2)$

$$\begin{array}{r|rrr} 2 & 2 & 1 & -10 \\ & & 4 & 10 \\ \hline & 2 & 5 & 0 \end{array}$$

$(2x^2 + x - 10) \div (x - 2) = 2x + 5$

3. $(3x^2 + 7x - 20) \div (x + 5)$

$$\begin{array}{r|rrr} -5 & 3 & 7 & -20 \\ & & -15 & 40 \\ \hline & 3 & -8 & 20 \end{array}$$

$(3x^2 + 7x - 20) \div (x + 5) = 3x - 8 + \frac{20}{x+5}$

5. $(4x^3 - 3x^2 + 3x - 1) \div (x - 1)$

$$\begin{array}{r|rrrr} 1 & 4 & -3 & 3 & -1 \\ & & 4 & 1 & 4 \\ \hline & 4 & 1 & 4 & 3 \end{array}$$

$(4x^3 - 3x^2 + 3x - 1) \div (x - 1)$

$= 4x^2 + x + 4 + \frac{3}{x-1}$

7. $(6x^5 - 2x^3 + 4x^2 - 3x + 1) \div (x - 2)$

$$\begin{array}{r|rrrrrr} 2 & 6 & 0 & -2 & 4 & -3 & 1 \\ & & 12 & 24 & 44 & 96 & 186 \\ \hline & 6 & 12 & 22 & 48 & 93 & 187 \end{array}$$

$(6x^5 - 2x^3 + 4x^3 - 3x + 1) \div (x - 2)$

$= 6x^4 + 12x^3 + 22x^2 + 48x + 93 + \frac{187}{x-2}$

9. $(x^2 - 5x - 5x^3 + x^4) \div (5 + x)$

Rewrite the polynomials in descending order.

$(x^4 - 5x^3 + x^2 - 5x) \div (x + 5)$

$$\begin{array}{r|rrrrr} -5 & 1 & -5 & 1 & -5 & 0 \\ & & -5 & 50 & -255 & 1300 \\ \hline & 1 & -10 & 51 & -260 & 1300 \end{array}$$

$(x^2 - 5x - 5x^3 + x^4) \div (5 + x)$

$$= x^3 - 10x^2 + 51x - 260 + \frac{1300}{5+x}$$

11. $\left(3x^3 + 2x^2 - 4x + 1\right) \div \left(x - \frac{1}{3}\right)$

$$\begin{array}{r|rrrr} \frac{1}{3} & 3 & 2 & -4 & 1 \\ & & 1 & 1 & -1 \\ \hline & 3 & 3 & -3 & 0 \end{array}$$

$\left(3x^3 + 2x^2 - 4x - 1\right) \div \left(x - \frac{1}{3}\right)$

$= 3x^2 + 3x - 3$

13. $\frac{x^5 + x^3 - 2}{x - 1}$

$$\begin{array}{r|rrrrrr} 1 & 1 & 0 & 1 & 0 & 0 & -2 \\ & & 1 & 1 & 2 & 2 & 2 \\ \hline & 1 & 1 & 2 & 2 & 2 & 0 \end{array}$$

$$\frac{x^5 + x^3 - 2}{x - 1} = x^4 + x^3 + 2x^2 + 2x + 2$$

15. $\frac{x^4 - 256}{x - 4}$

$$\begin{array}{r|rrrrr} 4 & 1 & 0 & 0 & 0 & -256 \\ & & 4 & 16 & 64 & 256 \\ \hline & 1 & 4 & 16 & 64 & 0 \end{array}$$

$$\frac{x^4 - 256}{x - 4} = x^3 + 4x^2 + 16x + 64$$

17. $\frac{2x^5 - 3x^4 + x^3 - x^2 + 2x - 1}{x + 2}$

$$\begin{array}{r|rrrrrr} -2 & 2 & -3 & 1 & -1 & 2 & -1 \\ & & -4 & 14 & -30 & 62 & -128 \\ \hline & 2 & -7 & 15 & -31 & 64 & -129 \end{array}$$

$$\frac{2x^5 - 3x^4 + x^3 - x^2 + 2x - 1}{x + 2}$$

$$= 2x^4 - 7x^3 + 15x^2 - 31x + 64 - \frac{129}{x+2}$$

19. $f(x) = 2x^3 - 11x^2 + 7x - 5$

$$\begin{array}{r|rrrr} 4 & 2 & -11 & 7 & -5 \\ & & 8 & -12 & -20 \\ \hline & 2 & -3 & -5 & -25 \end{array}$$

$f(4) = -25$

21. $f(x) = 3x^3 - 7x^2 - 2x + 5$

$$\begin{array}{r|rrrr} -3 & 3 & -7 & -2 & 5 \\ & & -9 & 48 & -138 \\ \hline & 3 & -16 & 46 & -133 \end{array}$$

$f(-3) = -133$

23. $f(x) = x^4 + 5x^3 + 5x^2 - 5x - 6$

$$\begin{array}{r|rrrrr} 3 & 1 & 5 & 5 & -5 & -6 \\ & & 3 & 24 & 87 & 246 \\ \hline & 1 & 8 & 29 & 82 & 240 \end{array}$$

$f(3) = 240$

25. $f(x) = 7x^4 - 3x^3 + 6x + 9$

$$\begin{array}{r|rrrrr} -5 & 7 & -3 & 0 & 6 & 9 \\ & & -35 & 190 & -950 & 4720 \\ \hline & 7 & -38 & 190 & -944 & 4729 \end{array}$$

$f(-5) = 4729$

27. $x^3 - 4x^2 + x + 6 = 0$

$$\begin{array}{r|rrrr} -1 & 1 & -4 & 1 & 6 \\ & & -1 & 5 & -6 \\ \hline & 1 & -5 & 6 & 0 \end{array}$$

The remainder is zero and -1 is a solution to the equation.

$x^3-4x^2+x+6=(x+1)(x^2-5x+6)$

To solve the equation, we set it equal to zero and factor.

$$x^3-4x^2+x+6=0$$
$$(x+1)(x^2-5x+6)=0$$
$$(x+1)(x-3)(x-2)=0$$

Apply the zero product principle.

$x+1=0 \quad x-3=0 \quad x-2=0$

$x=-1 \quad x=3 \quad x=2$

The solutions are -1, 2 and 3 and the solution set is $\{-1,2,3\}$.

29. $2x^3-5x^2+x+2=0$

$$\begin{array}{r|rrrr} 2 & 2 & -5 & 1 & 2 \\ & & 4 & -2 & -2 \\ \hline & 2 & -1 & -1 & 0 \end{array}$$

The remainder is zero and 2 is a solution to the equation.

$2x^3-5x^2+x+2=(x-2)(2x^2-x-1)$

To solve the equation, we set it equal to zero and factor.

$$2x^3-5x^2+x+2=0$$
$$(x-2)(2x^2-x-1)=0$$
$$(x-2)(2x+1)(x-1)=0$$

Apply the zero product principle.

$x-2=0 \quad 2x+1=0 \quad x-1=0$

$x=2 \quad 2x=-1 \quad x=1$

$x=-\frac{1}{2}$

The solutions are $-\frac{1}{2}$, 1 and 2 and the solution set is $\left\{-\frac{1}{2},1,2\right\}$.

31. $6x^3+25x^2-24x+5=0$

$$\begin{array}{r|rrrr} -5 & 6 & 25 & -24 & 5 \\ & & -30 & 25 & -5 \\ \hline & 6 & -5 & 1 & 0 \end{array}$$

The remainder is zero and 3 is a solution to the equation.

$6x^3+25x^2-24x+5$

$=(x+5)(6x^2-5x+1)$

To solve the equation, we set it equal to zero and factor.

$$6x^3+25x^2-24x+5=0$$
$$(x+5)(6x^2-5x+1)=0$$
$$(x+5)(3x-1)(2x-1)=0$$

Apply the zero product principle.

$x+5=0 \quad 3x-1=0 \quad 2x-1=0$

$x=-5 \quad 3x=1 \quad 2x=1$

$x=\frac{1}{3} \quad x=\frac{1}{2}$

The solutions are -5, $\frac{1}{3}$ and $\frac{1}{2}$ and the solution set is $\left\{-5,\frac{1}{3},\frac{1}{2}\right\}$.

Application Exercises

33. **a.** $14x^3-17x^2-16x-177=0$

$$\begin{array}{r|rrrr} 3 & 14 & -17 & -16 & -177 \\ & & 42 & 75 & 177 \\ \hline & 14 & 25 & 59 & 0 \end{array}$$

$14x^3-17x^2-16x-177$

$=(x-3)(14x^2+25x+59)$

b. $f(x)=14x^3-17x^2-16x+34$

We need to find x when

$f(x) = 211.$

$f(x) = 14x^3 - 17x^2 - 16x + 34$

$211 = 14x^3 - 17x^2 - 16x + 34$

$0 = 14x^3 - 17x^2 - 16x - 177$

This is the equation obtained in part **a.** One solution is 3. It can be used to find other solutions (if they exist).

$14x^3 - 17x^2 - 16x - 177 = 0$

$(x-3)(14x^2 + 25x + 59) = 0$

The polynomial $14x^2 + 25x + 59$ cannot be factored, so the only solution is $x = 3$. The female moth's abdominal width is 3 millimeters.

Writing in Mathematics

35. Answers will vary.

37. Answers will vary.

Technology Exercise

39. Answers will vary. For example, consider Exercise 27.

$x^3 - 4x^2 + x + 6 = 0$

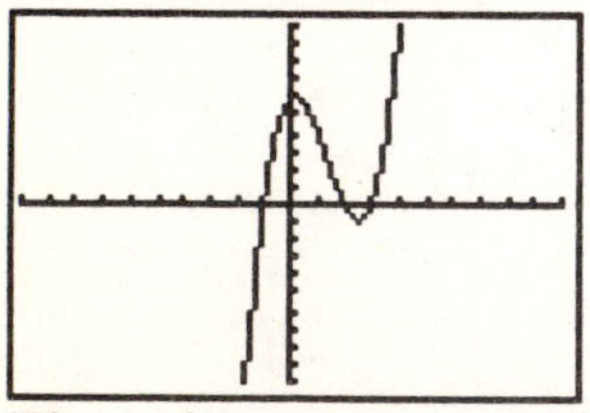

The x–intercepts are –1, 2, and 3. This is the same solution set obtained in Exercise 27.

Critical Thinking Exercises

41. $x^4 - 4x^3 - 9x^2 + 16x + 20 = 0$

5	1	–4	–9	16	20
		5	5	–20	–20
	1	1	–4	–4	0

The remainder is zero and 5 is a solution to the equation.

$x^4 - 4x^3 - 9x^2 + 16x + 20$

$= (x-5)(x^3 + x^2 - 4x - 4)$

To solve the equation, we set it equal to zero and factor.

$$(x-5)(x^3 + x^2 - 4x - 4) = 0$$

$$(x-5)(x^2(x+1) - 4(x+1)) = 0$$

$$(x-5)(x+1)(x^2 - 4) = 0$$

$$(x-5)(x+1)(x+2)(x-2) = 0$$

Apply the zero product principle.

$x - 5 = 0 \qquad x + 1 = 0$

$x = 5 \qquad x = -1$

$x + 2 = 0 \qquad x - 2 = 0$

$x = -2 \qquad x = 2$

The solutions are –2, –1, 2 and 5 and the solution set is $\{-2, -1, 2, 5\}$.

Review Exercises

42. $4x + 3 - 13x - 7 < 2(3 - 4x)$

$-9x - 4 < 6 - 8x$

$-x - 4 < 6$

$-x < 10$

$x > -10$

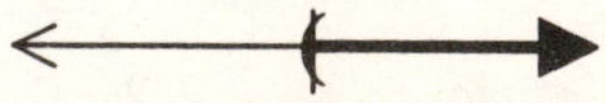

-10

The solution set is $\{x|x>-10\}$ or $(-10,\infty)$.

43. $2x(x+3)+6(x-3)=-28$

$$2x^2+6x+6x-18=-28$$
$$2x^2+12x-18=-28$$
$$2x^2+12x+10=0$$
$$2(x^2+6x+5)=0$$
$$2(x+5)(x+1)=0$$

Apply the zero product principle

$2(x+5)=0 \qquad x+1=0$

$x+5=0 \qquad x=-1$

$x=-5$

The solutions are –5 and –1 and the solution set is $\{-5,-1\}$.

44. $7x-6y=17$

$3x+\ y=18$

$$D=\begin{vmatrix}7 & -6\\ 3 & 1\end{vmatrix}=7(1)-3(-6)$$
$$=7+18=25$$
$$D_x=\begin{vmatrix}17 & -6\\ 18 & 1\end{vmatrix}=17(1)-18(-6)$$
$$=17+108=125$$
$$D_y=\begin{vmatrix}7 & 17\\ 3 & 18\end{vmatrix}=7(18)-3(17)$$
$$=126-51=75$$
$$x=\frac{D_x}{D}=\frac{125}{25}=5$$
$$y=\frac{D_y}{D}=\frac{75}{25}=3$$

The solution is $(5,3)$ and the solution set is $\{(5,3)\}$.

Check Points 6.6

1. $\dfrac{x+6}{2x}+\dfrac{x+24}{5x}=2$

So that denominators will not equal zero, x cannot equal zero. To eliminate fractions, multiply by the LCD, $10x$.

$$10x\left(\frac{x+6}{2x}+\frac{x+24}{5x}\right)=10x(2)$$
$$\frac{10x}{1}\cdot\frac{x+6}{2x}+\frac{10x}{1}\cdot\frac{x+24}{5x}=20x$$
$$5(x+6)+2(x+24)=20x$$
$$5x+30+2x+48=20x$$
$$7x+78=20x$$
$$78=13x$$
$$6=x$$

The solution is 6 and the solution set is $\{6\}$.

2. $\frac{x-3}{x+1}=\frac{x-2}{x+6}$

So that denominators will not equal zero, x cannot equal –1 or –6. To eliminate fractions, multiply by the LCD, $(x+1)(x+6)$.

$$(x+1)(x+6)\left(\frac{x-3}{x+1}\right)=(x+1)(x+6)\left(\frac{x-2}{x+6}\right)$$
$$(x+6)(x-3)=(x+1)(x-2)$$
$$\cancel{x^2}+3x-18=\cancel{x^2}-x-2$$
$$4x-18=-2$$
$$4x=16$$
$$x=4$$

The solution is 4 and the solution set is $\{4\}$.

3. $\frac{8x}{x+1}=4-\frac{8}{x+1}$

So that denominators will not equal zero, x cannot equal –1. To eliminate fractions, multiply by the LCD, $x+1$.

$$(x+1)\left(\frac{8x}{x+1}\right)=(x+1)\left(4-\frac{8}{x+1}\right)$$
$$8x=(x+1)4-(x+1)\frac{8}{x+1}$$
$$8x=4x+4-8$$
$$8x=4x-4$$
$$4x=-4$$
$$x=-1$$

Since –1 will make one or more of the denominators equal to zero, we disregard the solution. The solution set is $\varnothing$ or $\{\ \}$.

4. $\frac{x}{2}+\frac{12}{x}=5$

So that the denominator will not equal zero, x cannot equal 0. To eliminate fractions, multiply by the LCD, $2x$.

$$2x\cdot\left(\frac{x}{2}+\frac{12}{x}\right)=2x\cdot 5$$
$$2x\cdot\frac{x}{2}+2x\cdot\frac{12}{x}=10x$$

$$x \cdot x + 2 \cdot 12 = 10x$$
$$x^2 + 24 = 10x$$
$$x^2 - 10x + 24 = 0$$
$$(x-6)(x-4) = 0$$

Apply the zero product principle.

$x - 6 = 0 \qquad x - 4 = 0$

$x = 6 \qquad x = 4$

The solutions are 4 and 6 and the solution set is $\{4,6\}$.

5. $$\frac{3}{x-3} + \frac{5}{x-4} = \frac{x^2 - 20}{(x-4)(x-3)}$$

So that the denominator will not equal zero, x cannot equal 3 or 4. To eliminate fractions, multiply by the LCD, $(x-4)(x-3)$.

$$(x-4)(x-3) \cdot \left(\frac{3}{x-3} + \frac{5}{x-4}\right) = (x-4)(x-3) \cdot \frac{x^2 - 20}{(x-4)(x-3)}$$
$$(x-4) \cdot 3 + (x-3) \cdot 5 = x^2 - 20$$
$$3x - 12 + 5x - 15 = x^2 - 20$$
$$8x - 27 = x^2 - 20$$
$$0 = x^2 - 8x + 7$$
$$0 = (x-7)(x-1)$$

Apply the zero product principle.

$x - 7 = 0 \qquad x - 1 = 0$

$x = 7 \qquad x = 1$

The solutions are 1 and 7 and the solution set is $\{1,7\}$.

6. $$120 = \frac{120x}{100 - x}$$
$$(100 - x)120 = (100 - x)\frac{120x}{100 - x}$$
$$12000 - 120x = 120x$$
$$12000 = 240x$$
$$50 = x$$

If voter's commit \$120,000, 50% of the lake's pollutants can be removed.

Problem Set 6.6

Practice Exercises

1. $\frac{1}{x}+2=\frac{3}{x}$

So that the denominator will not equal zero, x cannot be zero. To eliminate fractions, multiply by the LCD, x.

$$x\left(\frac{1}{x}+2\right)=x\left(\frac{3}{x}\right)$$
$$x\cdot\frac{1}{x}+x\cdot 2=3$$
$$1+2x=3$$
$$2x=2$$
$$x=1$$

The solution is 1 and the solution set is $\{1\}$.

3. $\frac{5}{x}+\frac{1}{3}=\frac{6}{x}$

So that the denominator will not equal zero, x cannot equal 0. To eliminate fractions, multiply by the LCD, $3x$.

$$3x\left(\frac{5}{x}+\frac{1}{3}\right)=3x\left(\frac{6}{x}\right)$$
$$3x\cdot\frac{5}{x}+3x\cdot\frac{1}{3}=3\cdot 6$$
$$3\cdot 5+x\cdot 1=18$$
$$15+x=18$$
$$x=3$$

The solution is 3 and the solution set is $\{3\}$.

5. $\frac{x-2}{2x}+1=\frac{x+1}{x}$

So that the denominator will not equal zero, x cannot equal 0. To eliminate fractions, multiply by the LCD, $2x$.

$$2x\left(\frac{x-2}{2x}+1\right)=2x\left(\frac{x+1}{x}\right)$$
$$2x\cdot\frac{x-2}{2x}+2x\cdot 1=2(x+1)$$
$$x-2+2x=2x+2$$
$$-2+3x=2x+2$$
$$-2+x=2$$
$$x=4$$

The solution is 4 and the solution set is $\{4\}$.

7. $\frac{3}{x+1}=\frac{5}{x-1}$

So that the denominator will not equal zero, x cannot equal 1 or –1. To eliminate fractions, multiply by the LCD, $(x+1)(x-1)$.

$$(x+1)(x-1)\frac{3}{x+1}=(x+1)(x-1)\frac{5}{x-1}$$
$$(x-1)\cdot 3=(x+1)\cdot 5$$
$$3x-3=5x+5$$
$$-3=2x+5$$
$$-8=2x$$
$$-4=x$$

The solution is –4 and the solution set is $\{-4\}$.

9. $\frac{x-6}{x+5}=\frac{x-3}{x+1}$

So that the denominator will not equal zero, x cannot equal –5 or –1.

To eliminate fractions, multiply by the LCD, $(x+5)(x+1)$.

$$(x+5)(x+1)\frac{x-6}{x+5}=(x+5)(x+1)\frac{x-3}{x+1}$$

$$(x+1)(x-6)=(x+5)(x-3)$$

$$\cancel{x^2}-5x-6=\cancel{x^2}+2x-15$$

$$-5x-6=2x-15$$

$$-7x-6=-15$$

$$-7x=-9$$

$$x=\frac{9}{7}$$

The solution is $\frac{9}{7}$ and the solution set is $\left\{\frac{9}{7}\right\}$.

11. $\frac{x+6}{x+3}=\frac{3}{x+3}+2$

So that the denominator will not equal zero, x cannot equal -3. To eliminate fractions, multiply by the LCD, $x+3$.

$$(x+3)\left(\frac{x+6}{x+3}\right)=(x+3)\left(\frac{3}{x+3}+2\right)$$

$$x+6=3+2x+6$$

$$x+6=2x+9$$

$$6=x+9$$

$$-3=x$$

Since -3 would result in a zero denominator, we disregard it and conclude that the solution set is $\varnothing$ or $\{\ \}$.

13. $1-\frac{4}{x+7}=\frac{5}{x+7}$

So that the denominator will not equal zero, x cannot equal -7. To eliminate fractions, multiply by the LCD, $x+7$.

$$(x+7)\left(1-\frac{4}{x+7}\right)=(x+7)\left(\frac{5}{x+7}\right)$$

$$(x+7)\cdot 1-(x+7)\cdot\frac{4}{x+7}=5$$

$$x+7-4=5$$

$$x+3=5$$

$$x=2$$

The solution is 2 and the solution set is $\{2\}$.

15. $\frac{4x}{x+2}+\frac{2}{x-1}=4$

So that the denominator will not equal zero, x cannot equal -2 or 1. To eliminate fractions, multiply by the LCD, $(x+2)(x-1)$.

$$(x+2)(x-1)\cdot\left(\frac{4x}{x+2}+\frac{2}{x-1}\right)=(x+2)(x-1)\cdot 4$$

$$(x-1)4x+(x+2)2=(x^2+x-2)4$$

$$4x^2-4x+2x+4=4x^2+4x-8$$

$$\cancel{4x^2}-2x+4=\cancel{4x^2}+4x-8$$

$$-2x+4=4x-8$$
$$-6x+4=-8$$
$$-6x=-12$$
$$x=2$$

The solution is 2 and the solution set is $\{2\}$.

17. $\dfrac{8}{x^2-9}+\dfrac{4}{x+3}=\dfrac{2}{x-3}$

$\dfrac{8}{(x+3)(x-3)}+\dfrac{4}{x+3}=\dfrac{2}{x-3}$

So that the denominator will not equal zero, x cannot equal –3 or 3. To eliminate fractions, multiply by the LCD, $(x+3)(x-3)$.

$$(x+3)(x-3)\cdot\left(\frac{8}{(x+3)(x-3)}+\frac{4}{x+3}\right)=(x+3)(x-3)\cdot\frac{2}{x-3}$$
$$8+4(x-3)=2(x+3)$$
$$8+4x-12=2x+6$$
$$4x-4=2x+6$$
$$2x-4=6$$
$$2x=10$$
$$x=5$$

The solution is 5 and the solution set is $\{5\}$.

19. $x+\dfrac{7}{x}=-8$

So that the denominator will not equal zero, x cannot equal 0. To eliminate fractions, multiply by the LCD, x.

$$x\left(x+\frac{7}{x}\right)=x(-8)$$
$$x^2+7=-8x$$
$$x^2+8x+7=0$$
$$(x+7)(x+1)=0$$

Apply the zero product principle.

$x+7=0 \qquad x+1=0$

$x=-7 \qquad x=-1$

The solutions are –7 and –1 and the solution set is $\{-7,-1\}$.

21. $\frac{6}{x}-\frac{x}{3}=1$

So that the denominator will not equal zero, x cannot equal 0. To eliminate fractions, multiply by the LCD, $3x$.

$$3x\left(\frac{6}{x}-\frac{x}{3}\right)=3x(1)$$

$$3x\left(\frac{6}{x}\right)-3x\left(\frac{x}{3}\right)=3x$$

$$3(6)-x(x)=3x$$

$$18-x^2=3x$$

$$0=x^2+3x-18$$

$$0=(x+6)(x-3)$$

Apply the zero product principle.

$x+6=0 \qquad x-3=0$

$x=-6 \qquad x=3$

The solutions are –6 and 3 and the solution set is $\{-6,3\}$.

23. $\frac{x+6}{3x-12}=\frac{5}{x-4}+\frac{2}{3}$

$$\frac{x+6}{3(x-4)}=\frac{5}{x-4}+\frac{2}{3}$$

So that the denominator will not equal zero, x cannot equal 4. To eliminate fractions, multiply by the LCD, $3(x-4)$.

$$3(x-4)\left(\frac{x+6}{3(x-4)}\right)=3(x-4)\left(\frac{5}{x-4}+\frac{2}{3}\right)$$

$$x+6=3(x-4)\cdot\frac{5}{x-4}+3(x-4)\cdot\frac{2}{3}$$

$$x+6=3\cdot 5+2(x-4)$$

$$x+6=15+2x-8$$

$$x+6=7+2x$$
$$6=7+x$$
$$-1=x$$

The solution is –1 and the solution set is $\{-1\}$.

25. $$\frac{1}{x-1}+\frac{1}{x+1}=\frac{2}{x^2-1}$$
$$\frac{1}{x-1}+\frac{1}{x+1}=\frac{2}{(x+1)(x-1)}$$

So that the denominator will not equal zero, x cannot equal –1 or 1. To eliminate fractions, multiply by the LCD, $(x+1)(x-1)$.

$$(x+1)(x-1)\left(\frac{1}{x-1}+\frac{1}{x+1}\right)=(x+1)(x-1)\cdot\frac{2}{(x+1)(x-1)}$$
$$(x+1)+(x-1)=2$$
$$x+1+x-1=2$$
$$x+x=2$$
$$2x=2$$
$$x=1$$

Since 1 would result in a zero denominator, we disregard it and conclude that the solution set is $\varnothing$ or $\{\ \}$.

27. $$\frac{5}{x+4}+\frac{3}{x+3}=\frac{12x+19}{x^2+7x+12}$$
$$\frac{5}{x+4}+\frac{3}{x+3}=\frac{12x+19}{(x+4)(x+3)}$$

So that the denominator will not equal zero, x cannot equal –4 or –3. To eliminate fractions, multiply by the LCD, $(x+4)(x+3)$.

$$(x+4)(x+3)\left(\frac{5}{x+4}+\frac{3}{x+3}\right)=(x+4)(x+3)\left(\frac{12x+19}{(x+4)(x+3)}\right)$$
$$(x+4)(x+3)\cdot\left(\frac{5}{x+4}\right)+(x+4)(x+3)\cdot\left(\frac{3}{x+3}\right)=12x+19$$
$$(x+3)\cdot 5+(x+4)\cdot 3=12x+19$$
$$5x+15+3x+12=12x+19$$
$$8x+27=12x+19$$

$$-4x+27=19$$
$$-4x=-8$$
$$x=2$$

The solution is 2 and the solution set is $\{2\}$.

29. $$\frac{4x}{x+3}-\frac{12}{x-3}=\frac{4x^2+36}{x^2-9}$$
$$\frac{4x}{x+3}-\frac{12}{x-3}=\frac{4x^2+36}{(x+3)(x-3)}$$

So that the denominator will not equal zero, x cannot equal –3 or 3. To eliminate fractions, multiply by the LCD, $(x+3)(x-3)$.

$$(x+3)(x-3)\left(\frac{4x}{x+3}-\frac{12}{x-3}\right)=(x+3)(x-3)\left(\frac{4x^2+36}{(x+3)(x-3)}\right)$$
$$(x+3)(x-3)\left(\frac{4x}{x+3}\right)-(x+3)(x-3)\left(\frac{12}{x-3}\right)=4x^2+36$$
$$(x-3)(4x)-(x+3)(12)=4x^2+36$$
$$4x^2-12x-(12x+36)=4x^2+36$$
$$\cancel{4x^2}-12x-12x-36=\cancel{4x^2}+36$$
$$-24x-36=36$$
$$-24x=72$$
$$x=-3$$

Since –3 would result in a zero denominator, we disregard it and conclude that the solution set is $\varnothing$ or $\{\ \}$.

31. $$\frac{4}{x^2+3x-10}+\frac{1}{x^2+9x+20}=\frac{2}{x^2+2x-8}$$
$$\frac{4}{(x+5)(x-2)}+\frac{1}{(x+5)(x+4)}=\frac{2}{(x+4)(x-2)}$$

So that the denominator will not equal zero, x cannot equal –5, –4, or 2. To eliminate fractions, multiply by the LCD, $(x+5)(x+4)(x-2)$.

$$(x+5)(x+4)(x-2)\left(\frac{4}{(x+5)(x-2)}+\frac{1}{(x+5)(x+4)}\right)=(x+5)(x+4)(x-2)\left(\frac{2}{(x+4)(x-2)}\right)$$
$$(x+4)\cdot 4+(x-2)\cdot 1=(x+5)\cdot 2$$

$$4x+16+x-2=2x+10$$
$$5x+14=2x+10$$
$$3x+14=10$$
$$3x=-4$$
$$x=-\frac{4}{3}$$

The solution is $-\frac{4}{3}$ and the solution set is $\left\{-\frac{4}{3}\right\}$.

Application Exercises

33.

$$f(x)=\frac{250x}{100-x}$$
$$375=\frac{250x}{100-x}$$
$$(100-x)(375)=(100-x)\left(\frac{250x}{100-x}\right)$$
$$37500-375x=250x$$
$$37500=625x$$
$$60=x$$

If the government commits \$375 million for the project, 60% of pollutants can be removed.

35.

$$f(x)=\frac{5x+30}{x}$$
$$8=\frac{5x+30}{x}$$
$$x(8)=x\left(\frac{5x+30}{x}\right)$$
$$8x=5x+30$$
$$3x=30$$
$$x=10$$

The students will remember an average of 8 words after 10 days. This is shown on the graph as the point (10, 8).

37. The horizontal asymptote is $y = 5$. The means that on average, the students will remember more than 5 words over an extended period of time.

39.

$$f(x)=\frac{0.9x-0.4}{0.9x+0.1}$$
$$0.95=\frac{0.9x-0.4}{0.9x+0.1}$$
$$\frac{0.95}{1}=\frac{0.9x-0.4}{0.9x+0.1}$$
$$0.95(0.9x+0.1)=0.9x-0.4$$
$$0.855x+0.095=0.9x-0.4$$
$$0.095=0.045x-0.4$$
$$0.495=0.045x$$
$$11=x$$

It will take 11 learning trials for 0.95 or 95% of the responses to be correct. This is shown on the graph as the point (11, 0.95).

41. As the number of learning trials increases, the proportion of correct responses increases. Initially the proportion of correct responses increases rapidly, but slows down over time.

43.

$$f(x)=\frac{10{,}000}{x}+3x$$
$$350=\frac{10{,}000}{x}+3x$$
$$x(350)=x\left(\frac{10{,}000}{x}+3x\right)$$
$$350x=x\left(\frac{10{,}000}{x}\right)+x(3x)$$

$$350x = 10,000 + 3x^2$$
$$0 = 3x^2 - 350x + 10,000$$
$$0 = (3x - 200)(x - 50)$$

Apply the zero product principle.

$3x - 200 = 0$ or $x - 50 = 0$

$3x = 200$ $\quad x = 50$

$x = \dfrac{200}{3}$

$x \approx 67$

The solutions are 50 and 67. This means that if the grocery store orders 50 or 67 cases of soup at a time, the yearly inventory cost is \$350. The points (50, 350) and (67, 350) on the graph correspond to these solutions.

Writing in Mathematics

45. Answers will vary.

47. Answers will vary.

49. Answers will vary.

51. Answers will vary.

Technology Exercises

53. $\dfrac{50}{x} = 2x$

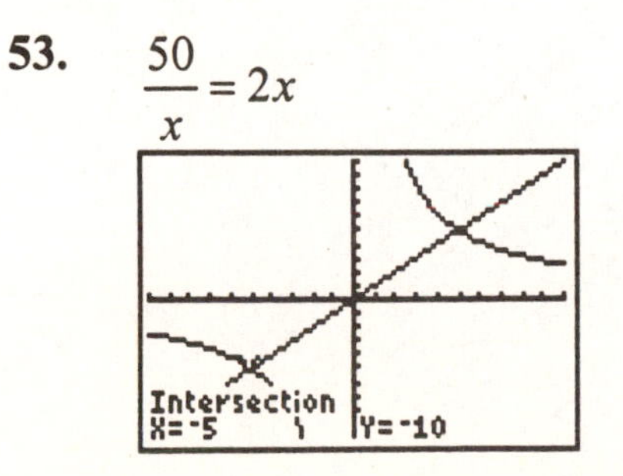

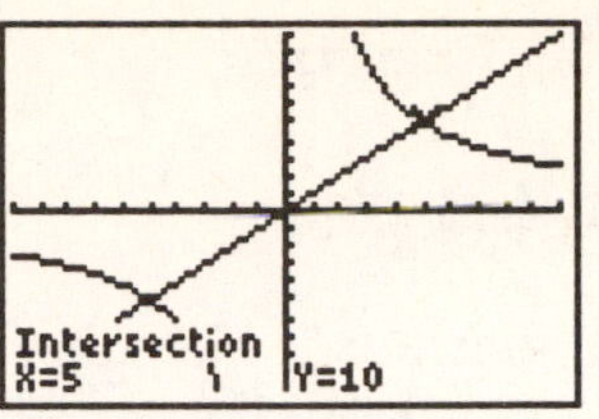

Check the solutions by substituting –5 and 5 in the original equation.

$x = 5$ $\quad x = -5$

$\dfrac{50}{5} = 2(5)$ $\quad \dfrac{50}{-5} = 2(-5)$

$10 = 10$ $\quad -10 = -10$

The solutions check. The solutions are –5 and 5 and the solution set is $\{-5, 5\}$.

55. $\dfrac{2}{x} = x + 1$

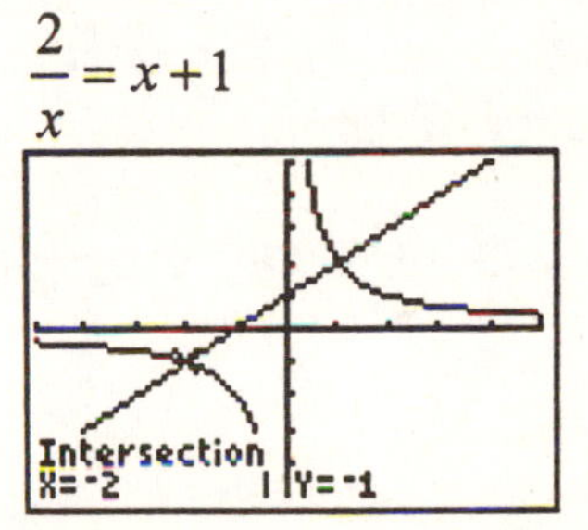

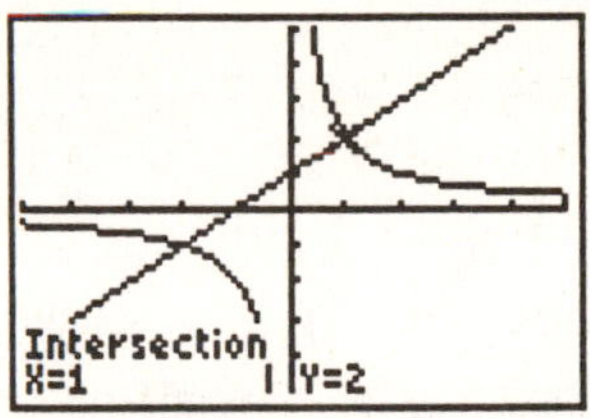

Check the solutions by substituting –2 and 1 in the original equation.

$x = -2$ $\quad x = 1$

$\dfrac{2}{-2} = -2 + 1$ $\quad \dfrac{2}{1} = 1 + 1$

$-1 = -1$ $\quad 2 = 2$

The solutions check. The solutions are –2 and 1 and the solution set is $\{-2, 1\}$.

Critical Thinking Exercises

57. Statement **d.** is true.

Statement **a.** is false. Not all of the solutions of the resulting equation are solutions of the rational equation. Solutions which will result in a denominator of zero must be discarded.

Statement **b.** is false. The only time you can multiply by the LCD to clear fractions is in an equation. The expression given here does not have an equal sign and is not an equation.

Statement **c.** is false. Zero does not satisfy the equation. Zero results in a denominator of zero.

59. Answers will vary. One example is
$$\frac{1}{x}+3=\frac{1}{x}+5.$$

Review Exercises

61. $x+2y\geq 2$
$x-y\geq -4$
First consider $x+2y\geq 2$. If we solve for y in $x+2y=2$, we can graph using the slope and the y-intercept.
$$x+2y=2$$
$$2y=-x+2$$
$$y=-\frac{1}{2}x+1$$
y-intercept = 1 slope = $-\frac{1}{2}$
Now, use the origin, $(0,0)$, as a test point.
$$x+2y\geq 2$$
$$0+2(0)\geq 2$$
$$0\geq 2$$
This is a false statement. This means that the point $(0,0)$ will not fall in the shaded half-plane.
Next consider $x-y\geq -4$. If we solve for y in $x-y=-4$, we can graph using the slope and the y-intercept.
$$x-y=-4$$
$$-y=-x-4$$
$$y=x+4$$
y-intercept = 4 slope = 1
Now, use the origin, $(0,0)$, as a test point.
$$x-y\geq -4$$
$$0-0\geq -4$$
$$0\geq -4$$
This is a true statement. This means that the point $(0,0)$ will fall in the shaded half-plane.
Next, graph each of the inequalities. The solution to the system is the intersection of the shaded half-planes.

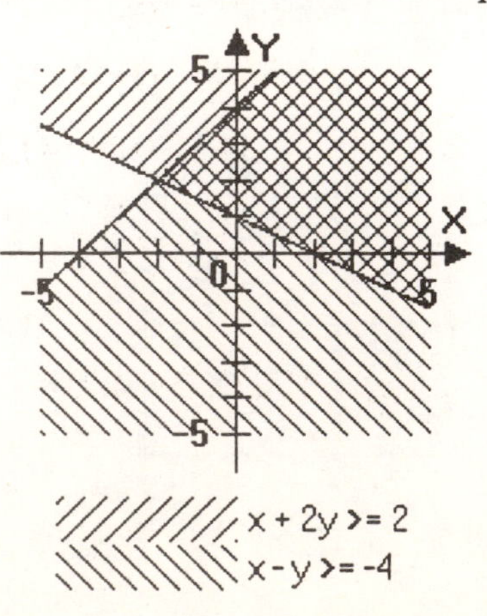

62.

$$\frac{x-4}{2}-\frac{1}{5}=\frac{7x+1}{20}$$
$$20\left(\frac{x-4}{2}-\frac{1}{5}\right)=20\left(\frac{7x+1}{20}\right)$$
$$20\left(\frac{x-4}{2}\right)-20\left(\frac{1}{5}\right)=7x+1$$
$$10(x-4)-4(1)=7x+1$$
$$10x-40-4=7x+1$$
$$10x-44=7x+1$$
$$3x-44=1$$
$$3x=45$$
$$x=15$$

The solution is 15 and the solution set is $\{15\}$.

Check Points 6.7

1.

$$\frac{1}{x}+\frac{1}{y}=\frac{1}{z}$$
$$xyz\left(\frac{1}{x}+\frac{1}{y}\right)=xyz\left(\frac{1}{z}\right)$$
$$xyz\left(\frac{1}{x}\right)+xyz\left(\frac{1}{y}\right)=xy$$
$$yz+xz=xy$$
$$yz=xy-xz$$
$$yz=x(y-z)$$
$$x=\frac{yz}{y-z}$$

2. **a.** $C(x)=500,000+400x$

b. $\overline{C}(x)=\frac{500,000+400x}{x}$

c.

$$450=\frac{500,000+400x}{x}$$
$$x(450)=x\left(\frac{500,000+400x}{x}\right)$$
$$450x=500,000+400x$$
$$50x=500,000$$
$$x=10,000$$

10,000 wheelchairs must be produced each month for the average cost to be $450.

3.

	d	r	$t=\frac{d}{r}$
Riding	40	$4x$	$\frac{40}{4x}$
Walking	5	x	$\frac{5}{x}$

$$\frac{40}{4x}+\frac{5}{x}=5$$
$$4x\left(\frac{40}{4x}+\frac{5}{x}\right)=4x(5)$$
$$4x\left(\frac{40}{4x}\right)+4x\left(\frac{5}{x}\right)=20x$$
$$40+4(5)=20x$$
$$40+20=20x$$
$$60=20x$$
$$3=x$$

The cyclist was riding at a rate of $4x=4(3)=12$ miles per hour.

4.

	Part Done in 1 Hour	Time Working Together	Part Done in x Hours
First Person	$\frac{1}{8}$	x	$\frac{x}{8}$
Second Person	$\frac{1}{4}$	x	$\frac{x}{4}$

$$\frac{x}{8}+\frac{x}{4}=1$$
$$8\left(\frac{x}{8}+\frac{x}{4}\right)=8(1)$$
$$8\left(\frac{x}{8}\right)+8\left(\frac{x}{4}\right)=8$$
$$x+2x=8$$
$$3x=8$$
$$x=\frac{8}{3}=2\frac{2}{3}$$

If they work together, the job will take $2\frac{2}{3}$ hours or 2 hours and 40 minutes.

5.

	Part Done in 1 Hour	Time Working Together	Part Done in 6 Hours
carpenter	$\frac{1}{3x}$	6	$\frac{6}{3x}$
apprentice	$\frac{1}{x}$	6	$\frac{6}{x}$

$$\frac{6}{3x}+\frac{6}{x}=1$$
$$3x\left(\frac{6}{3x}+\frac{6}{x}\right)=3x(1)$$
$$3x\left(\frac{6}{3x}\right)+3x\left(\frac{6}{x}\right)=3x$$
$$6+3(6)=3x$$
$$6+18=3x$$
$$24=3x$$
$$8=x$$

Working alone, the carpenter can do the job in 8 hours and the apprentice can do the job in 24 hours.

Problem Set 6.7

Practice Exercises

1.

$$\frac{V_1}{V_2}=\frac{P_2}{P_1}$$
$$P_1V_2\left(\frac{V_1}{V_2}\right)=P_1V_2\left(\frac{P_2}{P_1}\right)$$
$$P_1(V_1)=V_2(P_2)$$
$$P_1V_1=P_2V_2$$
$$\frac{P_1V_1}{V_1}=\frac{P_2V_2}{V_1}$$
$$P_1=\frac{P_2V_2}{V_1}$$

3.

$$\frac{1}{p}+\frac{1}{q}=\frac{1}{f}$$
$$fpq\left(\frac{1}{p}+\frac{1}{q}\right)=fpq\left(\frac{1}{f}\right)$$
$$fpq\left(\frac{1}{p}\right)+fpq\left(\frac{1}{q}\right)=pq$$
$$fq+fp=pq$$
$$f(q+p)=pq$$
$$\frac{f(q+p)}{q+p}=\frac{pq}{q+p}$$
$$f=\frac{pq}{q+p}$$

5.

$$P = \frac{A}{1+r}$$
$$(1+r)(P) = (1+r)\left(\frac{A}{1+r}\right)$$
$$P + Pr = A$$
$$Pr = A - P$$
$$\frac{Pr}{P} = \frac{A-P}{P}$$
$$r = \frac{A-P}{P}$$

7.

$$F = \frac{Gm_1m_2}{d^2}$$
$$d^2(F) = d^2\left(\frac{Gm_1m_2}{d^2}\right)$$
$$d^2F = Gm_1m_2$$
$$\frac{d^2F}{Gm_2} = \frac{Gm_1m_2}{Gm_2}$$
$$m_1 = \frac{d^2F}{Gm_2}$$

9.

$$z = \frac{x - \bar{x}}{s}$$
$$s(z) = s\left(\frac{x - \bar{x}}{s}\right)$$
$$zs = x - \bar{x}$$
$$x = \bar{x} + zs$$

11.

$$I = \frac{E}{R+r}$$
$$(R+r)(I) = (R+r)\left(\frac{E}{R+r}\right)$$
$$IR + Ir = E$$
$$IR = E - Ir$$
$$\frac{IR}{I} = \frac{E - Ir}{I}$$
$$R = \frac{E - Ir}{I}$$

13.

$$f = \frac{f_1f_2}{f_1 + f_2}$$
$$(f_1 + f_2)(f) = (f_1 + f_2)\left(\frac{f_1f_2}{f_1 + f_2}\right)$$
$$ff_1 + ff_2 = f_1f_2$$
$$ff_1 = f_1f_2 - ff_2$$
$$ff_1 + ff_2 = f_1f_2$$
$$ff_2 = f_1f_2 - ff_1$$
$$ff_2 = f_1(f_2 - f)$$
$$\frac{ff_2}{f_2 - f} = \frac{f_1(f_2 - f)}{f_2 - f}$$
$$f_1 = \frac{ff_2}{f_2 - f}$$

Application Exercises

15. Approximately 50,000 wheelchairs must be produced each month for the average cost to be $410 per chair.

17. The horizontal asymptote is $y = 400$. This means that despite the number of wheelchairs produced, the cost will approach but never reach $400.

19. **a.** $C(x) = 100{,}000 + 100x$

b. $\overline{C}(x) = \frac{100{,}000 + 100x}{x}$

c.
$$300 = \frac{100,000 + 100x}{x}$$
$$x(300) = x\left(\frac{100,000 + 100x}{x}\right)$$
$$300x = 100,000 + 100x$$
$$200x = 100,000$$
$$x = 500$$

500 mountain bikes must be produced each month for the average cost to be \$300.

21. The walking rate is 5 miles per hour.

23. The time spent driving and hiking increases as the walking rate decreases.

25.

	d	r	$t = \frac{d}{r}$
Motorcycle	40	$x + 20$	$\frac{40}{x+20}$
Bicycle	15	x	$\frac{15}{x}$

$$\frac{40}{x+20} = \frac{15}{x}$$
$$x(x+20)\left(\frac{40}{x+20}\right) = x(x+20)\left(\frac{15}{x}\right)$$
$$x(40) = (x+20)(15)$$
$$40x = 15x + 300$$
$$25x = 300$$
$$x = 12$$

The average rate for the bicycle is 12 miles per hour and the average rate for the motorcycle is $x + 20 =$ $12 + 20 = 32$ miles per hour.

27.

	d	r	$t = \frac{d}{r}$
To Campus	5	$x + 9$	$\frac{5}{x+9}$
From Campus	5	x	$\frac{5}{x}$

$$\frac{5}{x+9} + \frac{5}{x} = \frac{7}{6}$$
$$6x(x+9)\left(\frac{5}{x+9} + \frac{5}{x}\right) = 6x(x+9)\left(\frac{7}{6}\right)$$
$$6x(x+9)\left(\frac{5}{x+9}\right) + 6x(x+9)\left(\frac{5}{x}\right) = x(x+9)(7)$$
$$6x(5) + 6(x+9)(5) = 7x(x+9)$$
$$30x + 30(x+9) = 7x^2 + 63x$$
$$30x + 30x + 270 = 7x^2 + 63x$$

$$60x+270=7x^2+63x$$
$$0=7x^2+3x-270$$
$$0=(7x+45)(x-6)$$

Apply the zero product principle.

$7x+45=0$ $\quad$ $x-6=0$

$7x=-45$ $\quad$ $x=6$

$x=-\frac{45}{7}$

The solution set is $\left\{-\frac{45}{7},6\right\}$. We disregard $-\frac{45}{7}$ because we can't have a negative time measurement. The average rate on the trip home is 6 miles per hour.

29. The time with the current is $\frac{15}{8+x}$. The time against the current is $\frac{9}{8-x}$.

$$\frac{15}{8+x}=\frac{9}{8-x}$$
$$(8+x)(8-x)\left(\frac{15}{8+x}\right)=(8+x)(8-x)\left(\frac{9}{8-x}\right)$$
$$(8-x)(15)=(8+x)(9)$$
$$120-15x=72+9x$$
$$120=72+24x$$
$$48=24x$$
$$2=x$$

The rate of the current is 2 miles per hour.

31.

	d	r	$t=\frac{d}{r}$
Down Stream	6	$x+2$	$\frac{6}{x+2}$
Up Stream	2	$x-2$	$\frac{2}{x-2}$

$$\frac{6}{x+2}=\frac{2}{x-2}$$
$$(x+2)(x-2)\left(\frac{6}{x+2}\right)=(x+2)(x-2)\left(\frac{2}{x-2}\right)$$

$$(x-2)(6)=(x+2)(2)$$
$$6x-12=2x+4$$
$$4x-12=4$$
$$4x=16$$
$$x=4$$

The canoe's rate in still water is 4 miles per hour.

33. Think of the speed of the sidewalk as a "current." Walking with or against the movement of the sidewalk is the same as paddling with or against a current.

	d	r	$t=\frac{d}{r}$
With Sidewalk	100	$x+1.8$	$\frac{100}{x+1.8}$
Against Sidewalk	40	$x-1.8$	$\frac{40}{x-1.8}$

$$\frac{100}{x+1.8}=\frac{40}{x-1.8}$$
$$(x+1.8)(x-1.8)\left(\frac{100}{x+1.8}\right)=(x+1.8)(x-1.8)\left(\frac{40}{x-1.8}\right)$$
$$(x-1.8)(100)=(x+1.8)(40)$$
$$100x-180=40x+72$$
$$60x=252$$
$$x=4.2$$

The walking speed on a nonmoving sidewalk is 4.2 miles per hour.

35.

	d	r	$t=\frac{d}{r}$
Fast Runner	x	8	$\frac{x}{8}$
Slow Runner	x	6	$\frac{x}{6}$

$$\frac{x}{6}-\frac{x}{8}=\frac{1}{2}$$
$$24\left(\frac{x}{6}-\frac{x}{8}\right)=24\left(\frac{1}{2}\right)$$
$$24\left(\frac{x}{6}\right)-24\left(\frac{x}{8}\right)=12$$
$$4x-3x=12$$
$$x=12$$

Each person ran 12 miles.

37.

	Part Done in 1 Minute	Time Working Together	Part Done in x Minutes
You	$\frac{1}{40}$	x	$\frac{x}{40}$
Your Sister	$\frac{1}{30}$	x	$\frac{x}{30}$

$$\frac{x}{40}+\frac{x}{30}=1$$
$$120\left(\frac{x}{40}+\frac{x}{30}\right)=120(1)$$
$$120\left(\frac{x}{40}\right)+120\left(\frac{x}{30}\right)=120$$
$$3x+4x=120$$
$$7x=120$$
$$x\approx 17.1$$

If they work together, it will take about 17 minutes to wash the car. There is not enough time to finish the job before the parents return home.

39.

	Part Done in 1 Hour	Time Working Together	Part Done in x Hours
First Pipe	$\frac{1}{3}$	x	$\frac{x}{3}$
Second Pipe	$\frac{1}{6}$	x	$\frac{x}{6}$

$$\frac{x}{3}+\frac{x}{6}=1$$
$$6\left(\frac{x}{3}+\frac{x}{6}\right)=6(1)$$
$$6\left(\frac{x}{3}\right)+6\left(\frac{x}{6}\right)=6$$
$$2x+x=6$$
$$3x=6$$
$$x=2$$

If both pipes are used, the pool will be filled in 2 hours.

41.

	Part Done in 1 Hour	Time Working Together	Part Done in 3 Hours
You	$\frac{1}{x}$	3	$\frac{3}{x}$
Your Cousin	$\frac{1}{4}$	3	$\frac{3}{4}$

$$\frac{3}{x}+\frac{3}{4}=1$$
$$4x\left(\frac{3}{x}\right)+4x\left(\frac{3}{4}\right)=4x(1)$$
$$4(3)+x(3)=4x$$
$$12+3x=4x$$
$$12=x$$

Working alone, it would take you 12 hours to finish the job.

43.

	Part Done in 1 Hour	Time Working Together	Part Done in x Hours
Crew 1	$\frac{1}{20}$	x	$\frac{x}{20}$
Crew 2	$\frac{1}{30}$	x	$\frac{x}{30}$
Crew 3	$\frac{1}{60}$	x	$\frac{x}{60}$

$$\frac{x}{20}+\frac{x}{30}+\frac{x}{60}=1$$
$$60\left(\frac{x}{20}+\frac{x}{30}+\frac{x}{60}\right)=60(1)$$

$$60\left(\frac{x}{20}\right)+60\left(\frac{x}{30}\right)+60\left(\frac{x}{60}\right)=60$$
$$3x+2x+x=60$$
$$6x=60$$
$$x=10$$

If the three crews work together, it will take 10 hours to dispense the food and water.

45.

	Part Done in 1 Hour	Time Working Together	Part Done in 6 Hours
Lou	$\frac{1}{x+5}$	6	$\frac{6}{x+5}$
Bud	$\frac{1}{x}$	6	$\frac{6}{x}$

$$\frac{6}{x+5}+\frac{6}{x}=1$$
$$x(x+5)\left(\frac{6}{x+5}+\frac{6}{x}\right)=x(x+5)(1)$$
$$x(x+5)\left(\frac{6}{x+5}\right)+x(x+5)\left(\frac{6}{x}\right)=x(x+5)$$
$$x(6)+(x+5)(6)=x^2+5x$$
$$6x+6x+30=x^2+5x$$
$$12x+30=x^2+5x$$
$$0=x^2-7x-30$$
$$0=(x-10)(x+3)$$

Apply the zero product principle.

$x-10=0$ $\quad$ $x+3=0$

$x=10$ $\quad$ $x=-3$

We disregard –3 because we can't have a negative time measurement. Working alone, it would take Bud 10 hours to paint the room.

47.

	Part Done in 1 Minute	Time Working Together	Part Done in x Minutes
Faucet	$\frac{1}{5}$	x	$\frac{x}{5}$
Drain	$\frac{1}{10}$	x	$\frac{x}{10}$

$$\frac{x}{5}-\frac{x}{10}=1$$
$$10\left(\frac{x}{5}-\frac{x}{10}\right)=10(1)$$
$$10\left(\frac{x}{5}\right)-10\left(\frac{x}{10}\right)=10$$
$$2x-x=10$$
$$x=10$$

It will take 10 minutes for the sink to fill if the drain is left open.

Writing in Mathematics

49. Answers will vary.

51. Answers will vary.

53. Answers will vary.

55. Answers will vary.

Technology Exercises

57. Exercise 45.

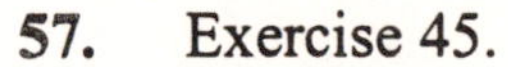

$$\frac{6}{x+5}+\frac{6}{x}=1$$

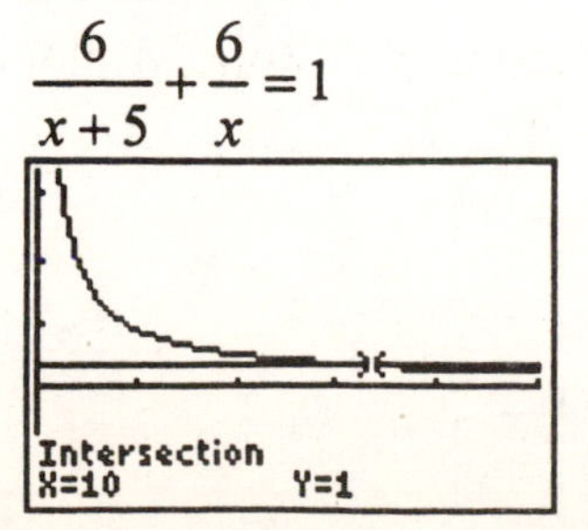

The solution is 10 and the solution set is $\{10\}$.

Exercise 46.

$$\frac{12}{x+10}+\frac{12}{x}=1$$

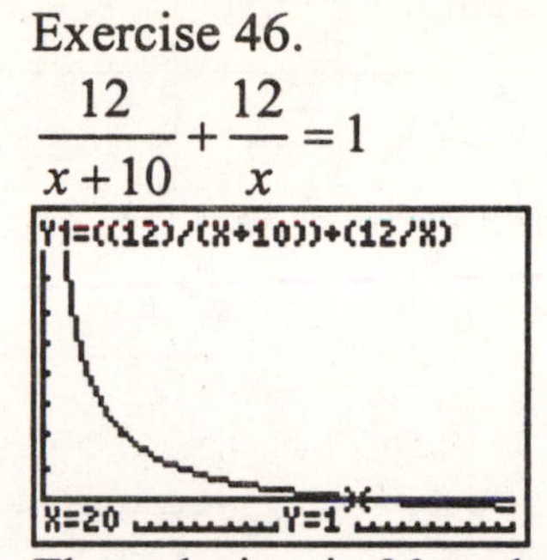

The solution is 20 and the solution set is $\{20\}$.

Critical Thinking Exercises

59. Statement **c.** is true. You can account for anything that "takes away" from the job getting done by subtracting it instead of adding it to the right hand side of the equation.

Statement **a.** is false. As production level increases, the average cost for a company to produce each unit of its product decreases.

Statement **b.** is false. The equation contains two terms involving p. If you subtract qf from both sides and then divide by f, there will still be a p term on the other side of the equation.

Statement **d.** is false. Statement **c.** is true.

61.

	d	r	$t=\frac{d}{r}$
Old Schedule	351	$x-2$	$\frac{351}{x-2}$
New Schedule	351	x	$\frac{351}{x}$

$$\frac{351}{x-2}-\frac{351}{x}=\frac{1}{4}$$

$$4x(x-2)\left(\frac{351}{x-2}-\frac{351}{x}\right)=4x(x-2)\left(\frac{1}{4}\right)$$

$$4x(x-2)\left(\frac{351}{x-2}\right)-4x(x-2)\left(\frac{351}{x}\right)=x(x-2)$$

$$4x(351)-4(x-2)(351)=x^2-2x$$

$$1404x-1404(x-2)=x^2-2x$$

$$\cancel{1404x}-\cancel{1404x}+2808=x^2-2x$$

$$2808=x^2-2x$$

$$0=x^2-2x-2808$$

$$0=(x-54)(x+52)$$

Apply the zero product principle.

$x-54=0$ $\quad$ $x+52=0$

$x=54$ $\quad$ $x=-52$

We disregard –52 because we can't have a negative average rate. The average rate of the train on the new schedule is 54 miles per hour.

Review Exercises

63. $x^2+4x+4-9y^2$

$=\left(x^2+4x+4\right)-9y^2$

$=(x+2)^2-(3y)^2$

$=((x+2)+3y)((x+2)-3y)$

$=(x+2+3y)(x+2-3y)$

64. $2x+5y=-5$

$x+2y=-1$

$$\left[\begin{array}{cc|c}2&5&-5\\1&2&-1\end{array}\right]\quad R_1\leftrightarrow R_2$$

$$=\left[\begin{array}{cc|c}1&2&-1\\2&5&-5\end{array}\right]\quad -2R_1+R_2$$

$$=\left[\begin{array}{cc|c}1&2&-1\\0&1&-3\end{array}\right]$$

The resulting system is:

$$x+2y=-1$$
$$y=-3.$$

Since we know $y=-3$, we can use back-substitution to find x.

$$x+2y=-1$$
$$x+2(-3)=-1$$
$$x-6=-1$$
$$x=5$$

The solution is $(5,-3)$ and the solution set is $\{(5,-3)\}$.

65.
$$x+y+z=4$$
$$2x+5y=1$$
$$x-y-2z=0$$

Multiply the first equation by 2 and add to the third equation.

$$2x+2y+2z=8$$
$$x-y-2z=0$$
$$3x+y=8$$

We can use this equation and the second equation in the original system to form a system of two equations in two variables.

$$2x+5y=1$$
$$3x+y=8$$

Multiply the second equation by –5 and solve by addition.

$$2x+5y=1$$
$$-15x-5y=-40$$
$$-13x=-39$$
$$x=3$$

Back-substitute 3 for x to find y.

$$2x+5y=1$$
$$2(3)+5y=1$$
$$6+5y=1$$
$$5y=-5$$
$$y=-1$$

Back-substitute –1 for y and 3 for x to find z.

$$x+y+z=4$$
$$3+(-1)+z=4$$
$$2+z=4$$
$$z=2$$

The solution is $(3,-1,2)$ and the solution set is $\{(3,-1,2)\}$.

Check Points 6.8

1. Since pressure varies directly as distance, we have $P=kd$.
Use the given values to find k.

$$P=kd$$
$$25=k60$$
$$\frac{25}{60}=\frac{k60}{60}$$
$$k=\frac{5}{12}$$

The equation becomes $P=\frac{5}{12}d$.
When $d=330$ feet,
$P=\frac{5}{12}d=\frac{5}{12}(330)=137.5$.
A submarine 330 feet below the surface will experience a pressure of 137.5 pounds per square inch.

2. Since distance varies directly as the square of the speed, we have $d=ks^2$.
Use the given values to find k.

$$d=ks^2$$
$$200=k(60)^2$$
$$200=k(3600)$$

$$\frac{200}{3600}=\frac{k(3600)}{3600}$$
$$\frac{1}{18}=k$$

The equation becomes $d=\frac{1}{18}s^2$.

When $s = 100$, we have

$$d=\frac{1}{18}s^2=\frac{1}{18}(10000)\approx 556.$$

An automobile traveling at 100 miles per hour will require approximately 556 feet to stop.

3. Since price, P, varies inversely as the supply, we have $P=\frac{k}{s}$.

Use the given values to find k.

$$P=\frac{k}{s}$$
$$19.50=\frac{k}{4}$$
$$4(19.50)=4\left(\frac{k}{4}\right)$$
$$78=k$$

The equation becomes $P=\frac{78}{s}$.

When $s = 3$ million barrels,

$$P=\frac{78}{s}=\frac{78}{3}=26.$$

When the daily production level is 3 million barrels, the price will be \$26 per barrel.

4. Since the number of minutes varies directly with the number of problems and inversely with the number of people, we have $t=\frac{kn}{p}$, where t is the number of minutes, n is the number of problems and p is the number of people.

Use the given values to find k.

$$t=\frac{kn}{p}$$
$$32=\frac{k(16)}{4}$$
$$32=k(4)$$
$$\frac{32}{4}=\frac{k(4)}{4}$$
$$8=k$$

The equation becomes $t=\frac{8n}{p}$.

When $p = 8$ and $n = 24$,

$$t=\frac{8n}{p}=\frac{\cancel{8}(24)}{\cancel{8}}=24.$$

It will take 24 minutes for 8 people to solve 24 problems.

5. Since the volume of a cone varies jointly with its height and the square of its radius, we have $V = khr^2$.

Use the given values to find k.

$$V=khr^2$$
$$120\pi=k(10)(6)^2$$
$$120\pi=k(10)(36)$$
$$120\pi=k(360)$$
$$\frac{120\pi}{360}=\frac{k(360)}{360}$$
$$\frac{1}{3}\pi=k$$

The equation becomes $V=\frac{1}{3}\pi hr^2$.

When $r = 12$ feet and $h = 2$ feet, we have the following.

$$V = \frac{1}{3}\pi(2)(12)^2$$
$$= \frac{1}{3}\pi(2)(144)$$
$$= \frac{1}{3}\pi(288)$$
$$= 96\pi$$

A cone with a height of 2 feet and a radius of 12 feet has a volume of 96π cubic feet.

Problem Set 6.8

Practice Exercises

1. Since y varies directly with x, we have $y = kx$.
Use the given values to find k.
$$y = kx$$
$$35 = k \cdot 5$$
$$\frac{35}{5} = \frac{k \cdot 5}{5}$$
$$7 = k$$
The equation becomes $y = 7x$.
When $x = 12$, $y = 7x = 7 \cdot 12 = 84$.

3. Since y varies inversely with x, we have $y = \frac{k}{x}$.
Use the given values to find k.
$$y = \frac{k}{x}$$
$$10 = \frac{k}{5}$$
$$5 \cdot 10 = 5 \cdot \frac{k}{5}$$
$$50 = k$$
The equation becomes $y = \frac{50}{x}$.
When $x = 2$, $y = \frac{50}{2} = 25$.

5. Since y varies inversely as x and inversely as the square of z, we have $y = \frac{kx}{z^2}$.
Use the given values to find k.
$$y = \frac{kx}{z^2}$$
$$20 = \frac{k(50)}{5^2}$$
$$20 = \frac{k(50)}{25}$$
$$20 = 2k$$
$$10 = k$$
The equation becomes $y = \frac{10x}{z^2}$.
When $x = 3$ and $z = 6$,
$$y = \frac{10x}{z^2} = \frac{10(3)}{6^2} = \frac{10(3)}{36} = \frac{30}{36} = \frac{5}{6}.$$

7. Since y varies jointly as x and y, we have $y = kxy$.
Use the given values to find k.
$$y = kxy$$
$$25 = k(2)(5)$$
$$25 = k(10)$$
$$\frac{25}{10} = \frac{k(10)}{10}$$
$$\frac{5}{2} = k$$
The equation becomes $y = \frac{5}{2}xy$.
When $x = 8$ and $z = 12$,

$$y=\frac{5}{2}(8)(12)=\frac{5}{\not{2}}(\overset{4}{\not{8}})(12)=240.$$

Application Exercises

9. **a.** $G=kW$

b. $G=0.02W$

c. $G=0.02(52)=1.04$

At the end of the year, the fingernails would be 1.04 inches long.

11. Since C varies directly as M, we have $C=kM$.
Use the given values to find k.

$$C=kM$$
$$400=k(3000)$$
$$\frac{400}{3000}=\frac{k(3000)}{3000}$$
$$k=\frac{400}{3000}=\frac{2}{15}$$

The equation becomes $C=\frac{2}{15}M$.

When $M=450$,

$$C=\frac{2}{15}M=\frac{2}{\not{15}}(\overset{30}{\not{450}})=60.$$

The ticket for a 450 mile trip will cost $60.

13. Since the speed of an aircraft varies directly as it Mach number, we have $s=km$.
Use the given values to find k.

$$s=km$$
$$1502.2=k(2.03)$$
$$\frac{1502.2}{2.03}=\frac{k(2.03)}{2.03}$$
$$740=k$$

The equation becomes $s=740m$.
When $m=3.3$, $s=740(3.3)=2442$.
The Blackbird's speed is 2442 miles per hour.

15. Since a man's weight varies directly as the cube of his height, we have $w=kh^3$.
Use the given values to find k.

$$w=kh^3$$
$$170=k(70)^3$$
$$170=k(343{,}000)$$
$$\frac{170}{343{,}000}=\frac{k(343{,}000)}{343{,}000}$$
$$0.000496=k$$

The equation becomes $w=0.000496h^3$.
When $h=107$,

$$w=0.000496(107)^3$$
$$=0.000496(1{,}225{,}043)\approx 607.$$

Robert Wadlow's weight was approximately 607 pounds.

17. Since time varies inversely as driving rate, we have $t=\frac{k}{d}$.
Use the given values to find k.

$$t=\frac{k}{d}$$
$$1.5=\frac{k}{20}$$
$$20(1.5)=20\left(\frac{k}{20}\right)$$
$$30=k$$

The equation becomes $t = \frac{30}{d}$.

When $d = 60$, $t = \frac{30}{60} = 0.5$.
When the average rate is 60 miles per hour, it will take $\frac{1}{2}$ hour to get to campus.

19. Since volume varies inversely as pressure, we have $v = \frac{k}{p}$.
Use the given values to find k.

$$v = \frac{k}{p}$$
$$32 = \frac{k}{8}$$
$$8(32) = 8\left(\frac{k}{8}\right)$$
$$256 = k$$

The equation becomes $v = \frac{256}{p}$.

When $d = 40$, $v = \frac{256}{40} = 6.4$.
When the pressure is 40 cubic centimeters, the volume is 6.4 pounds.

21. Since index varies directly as weight and inversely as height, we have
$I = \frac{kw}{h}$.
Use the given values to find k.

$$I = \frac{kw}{h}$$
$$21 = \frac{k(150)}{70}$$
$$\frac{70}{150}(21) = \frac{\cancel{70}}{\cancel{150}}\left(\frac{k(\cancel{150})}{\cancel{70}}\right)$$
$$9.8 = k$$

The equation becomes $I = \frac{9.8w}{h}$.

When $w = 240$ and $h = 74$,
$I = \frac{9.8(240)}{74} \approx 31.8$.
The body mass index is approximately 32. This is not in the desirable range.

23. Since intensity varies inversely as the square of the pressure, we have
$I = \frac{k}{d^2}$.
Use the given values to find k.

$$I = \frac{k}{d^2}$$
$$25 = \frac{k}{4^2}$$
$$25 = \frac{k}{16}$$
$$16(25) = 16\left(\frac{k}{16}\right)$$
$$400 = k$$

The equation becomes $I = \frac{400}{d^2}$.

When $d = 6$, $I = \frac{400}{6^2} = \frac{400}{36} \approx 11.1$.
At a distance of 6 feet, the intensity of illumination is 11.1 foot-candles.

25. Since kinetic energy varies jointly as the mass and the square of the velocity, we have $E = kmv^2$.
Use the given values to find k.

$$E = kmv^2$$
$$36 = k(8)(3)^2$$
$$36 = k(72)$$
$$\frac{36}{72} = \frac{k(72)}{72}$$
$$0.5 = k$$

The equation becomes $E = 0.5mv^2$.
When $m = 4$ and $v = 6$,

$$E = 0.5(4)(6)^2 = 0.5(4)(36) = 72.$$

A mass of 4 grams and velocity of 6 centimeters per second has a kinetic energy of 72 ergs.

27. Since the average number of phone calls varies jointly as the product of the populations and inversely as the square of the distance, we have

$$n = \frac{kp_1p_2}{d^2}.$$

Use the given values to find k.

$$n = \frac{kp_1p_2}{d^2}$$
$$158,233 = \frac{k(2538)(1818)}{(608)^2}$$
$$158,233 = \frac{k(4,614,084)}{369,664}$$
$$158,233 = 12.4818k$$
$$12677.1 \approx k$$

The equation becomes

$$n = \frac{12677.1p_1p_2}{d^2}.$$

When $p_1 = 1225$, $p_2 = 2970$ and $d = 3403$,

$$n = \frac{12677.1(1225)(2970)}{(3403)^2}$$
$$= \frac{46,122,459,090}{11,580,409} \approx 3982.8.$$

The average number of phone calls between Orlando and Seattle is about 3983 per day.

29. Since the force of wind varies jointly with the area of the window and the square of the wind's speed, we have $f = kas^2$.
Use the given values to find k.

$$f = kas^2$$
$$150 = k(4 \cdot 5)(30)^2$$
$$150 = k(20)(900)$$
$$150 = k(18000)$$
$$\frac{150}{18000} = \frac{k(18000)}{18000}$$
$$\frac{1}{120} = k$$

The equation becomes $f = \frac{1}{120}as^2$.
When $s = 60$, $a = 3 \cdot 4 = 12$,

$$f = \frac{1}{120}(12)(60)^2 = \frac{1}{10}(3600) = 360.$$

The force is 360 pounds. Since the windows are only made to withstand a 300 pound force, the hurricane shutters should be placed on the windows.

Writing in Mathematics

31. Answers will vary.

33. Answers will vary.

35. Answers will vary.

37. Answers will vary.

Technology Exercises

39. Answers will vary. For example, consider Exercise 11.

$C=\frac{2}{15}M$

Graph the equation $y=\frac{2}{15}x$.

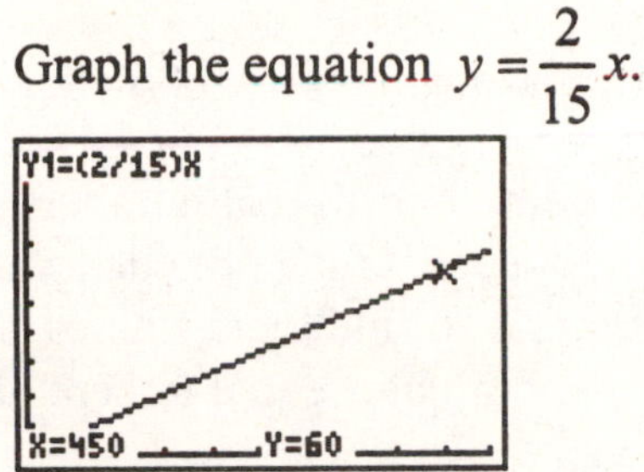

The ticket for a 450 mile trip will cost \$60.

Critical Thinking Exercises

41. Since illumination varies inversely as the square of the distance from the light source, we have $I=\frac{k}{d^2}$. If the lamp is raised from 15 inches to 30 inches, the distance has been multiplied by two. In the formula, the distance is squared so the denominator will be multiplied by a factor of 4. Since illumination varies inversely as the square of the distance, it will be multiplied by a factor of $\frac{1}{4}$.

The illumination for a 15-inch and a 30-inch distance is:

$d=15$

$I=\frac{k}{d^2}=\frac{k}{15^2}=\frac{k}{225}$

$d=30$

$I=\frac{k}{d^2}=\frac{k}{30^2}=\frac{k}{900}$

We can rewrite $\frac{k}{900}$ to show that it is indeed multiplied by a factor of $\frac{1}{4}\cdot\frac{k}{900}=\frac{k}{4\cdot 225}=\frac{1}{4}\left(\frac{k}{225}\right)$.

If the distance is changed from 15 inches to 30 inches the illumination is decreased to $\frac{1}{4}$ of the former intensity.

43. Since the brightness of a source point varies inversely as the square of its distance from an observer, we have $B=\frac{k}{d^2}$. We can now see things that are only $\frac{1}{50}$ as bright.

$B=\frac{1}{50}\cdot\frac{k}{d^2}=\frac{k}{50d^2}$

$=\frac{k}{(7.07)^2 d^2}=\frac{k}{(7.07d)^2}$

The distance that can be seen is about 7 times farther with the space telescope.

Review Exercises

44. $\begin{vmatrix}-1 & 2\\ 3 & -4\end{vmatrix}=-1(-4)-3(2)=4-6=-2$

45. $x^2y-9y-3x^2+27$

$=y(x^2-9)-3(x^2-9)$

$=(x^2-9)(y-3)$

$=(x+3)(x-3)(y-3)$

46. $7xy + x^2y^2 - 5x^3 - 7$
The degree of the polynomial is 4.

Chapter 6 Review

6.1

1. $f(x) = \frac{x^2 + 2x - 3}{x^2 - 4}$

a. $f(4) = \frac{(4)^2 + 2(4) - 3}{(4)^2 - 4}$
$= \frac{16 + 8 - 3}{16 - 4} = \frac{21}{12} = \frac{7}{4}$

b. $f(0) = \frac{0^2 + 2(0) - 3}{0^2 - 4}$
$= \frac{0 + 0 - 3}{0 - 4} = \frac{-3}{-4} = \frac{3}{4}$

c. $f(-2) = \frac{(-2)^2 + 2(-2) - 3}{(-2)^2 - 4}$
$= \frac{4 - 4 - 3}{4 - 4} = \frac{-3}{0}$
Division by zero is undefined.
$f(-2)$ does not exist.

d. $f(-3) = \frac{(-3)^2 + 2(-3) - 3}{(-3)^2 - 4}$
$= \frac{9 - 6 - 3}{9 - 4} = \frac{0}{5} = 0$

2. The domain is all real numbers except those that make the denominator zero. To find these values, set the denominator equal to zero and solve.
$(x - 3)(x + 4) = 0$
Apply the zero product principle.
$x - 3 = 0 \qquad x + 4 = 0$
$x = 3 \qquad x = -4$
The domain of $f = \{x | x$ is a real number and $x \neq -4$ and $x \neq 3\}$ or $(-\infty, -4) \cup (-4, 3) \cup (3, \infty)$.

3. The domain is all real numbers except those that make the denominator zero. To find these values, set the denominator equal to zero and solve.
$x^2 + x - 2 = 0$
$(x + 2)(x - 1) = 0$
Apply the zero product principle.
$x + 2 = 0 \qquad x - 1 = 0$
$x = -2 \qquad x = 1$
The domain of $f = \{x | x$ is a real number and $x \neq -2$ and $x \neq 1\}$ or $(-\infty, -2) \cup (-2, 1) \cup (1, \infty)$.

4. $\frac{5x^3 - 35x}{15x^2} = \frac{\cancel{5x}(x^2 - 7)}{3x \cdot \cancel{5x}} = \frac{x^2 - 7}{3x}$

5. $\frac{x^2 + 6x - 7}{x^2 - 49} = \frac{(x + 7)(x - 1)}{(x + 7)(x - 7)}$
$= \frac{\cancel{(x+7)}(x - 1)}{\cancel{(x+7)}(x - 7)}$
$= \frac{x - 1}{x - 7}$

6. $\frac{6x^2 + 7x + 2}{2x^2 - 9x - 5} = \frac{(3x + 2)(2x + 1)}{(2x + 1)(x - 5)}$

$$= \frac{(3x+2)\cancel{(2x+1)}}{\cancel{(2x+1)}(x-5)}$$

$$= \frac{3x+2}{x-5}$$

7. $$\frac{x^2+4}{x^2-4} = \frac{x^2+4}{(x+2)(x-2)}$$

8. $$\frac{x^3-8}{x^2-4} = \frac{\cancel{(x-2)}(x^2+2x+4)}{(x+2)\cancel{(x-2)}}$$

$$= \frac{x^2+2x+4}{x+2}$$

9. $$\frac{5x^2-5}{3x+12} \cdot \frac{x+4}{x-1} = \frac{5(x^2-1)}{3\cancel{(x+4)}} \cdot \frac{\cancel{x+4}}{x-1}$$

$$= \frac{5(x^2-1)}{3(x-1)}$$

$$= \frac{5(x+1)\cancel{(x-1)}}{3\cancel{(x-1)}}$$

$$= \frac{5(x+1)}{3}$$

10. $$\frac{2x+5}{4x^2+8x-5} \cdot \frac{4x^2-4x+1}{x+1}$$

$$= \frac{\cancel{2x+5}}{\cancel{(2x+5)}\cancel{(2x-1)}} \cdot \frac{\cancel{(2x-1)}(2x-1)}{x+1}$$

$$= \frac{2x-1}{x+1}$$

11. $$\frac{x^2-9x+14}{x^3+2x^2} \cdot \frac{x^2-4}{x^2-4x+4}$$

$$= \frac{(x-7)\cancel{(x-2)}}{x^2\cancel{(x+2)}} \cdot \frac{\cancel{(x+2)}\cancel{(x-2)}}{\cancel{(x-2)}\cancel{(x-2)}}$$

$$= \frac{x-7}{x^2}$$

12. $$\frac{1}{x^2+8x+15} \div \frac{3}{x+5}$$

$$= \frac{1}{x^2+8x+15} \cdot \frac{x+5}{3}$$

$$= \frac{1}{\cancel{(x+5)}(x+3)} \cdot \frac{\cancel{x+5}}{3}$$

$$= \frac{1}{3(x+3)}$$

13. $$\frac{x^2+16x+64}{2x^2-128} \div \frac{x^2+10x+16}{x^2-6x-16}$$

$$= \frac{x^2+16x+64}{2x^2-128} \cdot \frac{x^2-6x-16}{x^2+10x+16}$$

$$= \frac{\cancel{(x+8)}(x+8)}{2(x^2-64)} \cdot \frac{(x-8)\cancel{(x+2)}}{\cancel{(x+8)}\cancel{(x+2)}}$$

$$= \frac{(x+8)(x-8)}{2(x^2-64)}$$

$$= \frac{\cancel{(x+8)}\cancel{(x-8)}}{2\cancel{(x+8)}\cancel{(x-8)}} = \frac{1}{2}$$

14. $$\frac{y^2-16}{y^3-64} \div \frac{y^2-3y-18}{y^2+5y+6}$$

$$= \frac{y^2 - 16}{y^3 - 64} \cdot \frac{y^2 + 5y + 6}{y^2 - 3y - 18}$$
$$= \frac{(y+4)\cancel{(y-4)}}{\cancel{(y-4)}(y^2 + 4x + 16)} \cdot \frac{\cancel{(y+3)}(y+2)}{(y-6)\cancel{(y+3)}}$$
$$= \frac{(y+4)(y+2)}{(y-6)(y^2 + 4x + 16)}$$

15. **a.** 50 deer were introduced into the habitat.

b. After 10 years, the population is 150 deer.

c. The equation of the horizontal asymptote is $y = 225$. This means that the deer population will increase over time to 225, but will never reach it.

6.2

16.
$$\frac{4x+1}{3x-1} + \frac{8x-5}{3x-1} = \frac{4x+1+8x-5}{3x-1}$$
$$= \frac{12x-4}{3x-1} = \frac{4(3x-1)}{3x-1}$$
$$= 4$$

17.
$$\frac{2x-7}{x^2-9} - \frac{x-4}{x^2-9} = \frac{2x-7-(x-4)}{x^2-9}$$
$$= \frac{2x-7-x+4}{x^2-9}$$
$$= \frac{\cancel{x-3}}{(x+3)\cancel{(x-3)}}$$
$$= \frac{1}{x+3}$$

18.
$$\frac{4x^2-11x+4}{x-3} - \frac{x^2-4x+10}{x-3}$$
$$= \frac{4x^2-11x+4-(x^2-4x+10)}{x-3}$$
$$= \frac{4x^2-11x+4-x^2+4x-10}{x-3}$$
$$= \frac{3x^2-7x-6}{x-3}$$
$$= \frac{\cancel{(x-3)}(3x+2)}{\cancel{x-3}}$$
$$= 3x+2$$

19.
$$9x^3 = \quad 3^2 x^2$$
$$12x = 2^2 \cdot 3\, x$$
$$\text{LCD} = 2^2 \cdot 3^2 x^3 = 4 \cdot 9x^3 = 36x^3$$

20.
$$x^2 + 2x - 35 = (x+7)(x-5)$$
$$x^2 + 9x + 14 = (x+7) \qquad (x+2)$$
$$\text{LCD} = (x+7)(x-5)(x+2)$$

21.
$$\frac{1}{x} + \frac{2}{x-5} = \frac{1(x-5)}{x(x-5)} + \frac{2x}{x(x-5)}$$
$$= \frac{1(x-5)+2x}{x(x-5)}$$
$$= \frac{x-5+2x}{x(x-5)}$$
$$= \frac{3x-5}{x(x-5)}$$

22. $$\frac{2}{x^2-5x+6}+\frac{3}{x^2-x-6}=\frac{2}{(x-3)(x-2)}+\frac{3}{(x-3)(x+2)}$$
$$=\frac{2(x+2)}{(x-3)(x-2)(x+2)}+\frac{3(x-2)}{(x-3)(x-2)(x+2)}$$
$$=\frac{2(x+2)+3(x-2)}{(x-3)(x-2)(x+2)}=\frac{2x+4+3x-6}{(x-3)(x-2)(x+2)}$$
$$=\frac{5x-2}{(x-3)(x-2)(x+2)}$$

23. $$\frac{x-3}{x^2-8x+15}+\frac{x+2}{x^2-x-6}=\frac{x-3}{(x-3)(x-5)}+\frac{\cancel{x+2}}{(x-3)(\cancel{x+2})}=\frac{x-3}{(x-3)(x-5)}+\frac{1}{x-3}$$
$$=\frac{x-3}{(x-3)(x-5)}+\frac{1(x-5)}{(x-3)(x-5)}=\frac{x-3+1(x-5)}{(x-3)(x-5)}$$
$$=\frac{x-3+x-5}{(x-3)(x-5)}=\frac{2x-8}{(x-3)(x-5)}\text{ or }\frac{2(x-4)}{(x-3)(x-5)}$$

24. $$\frac{3x^2}{9x^2-16}-\frac{x}{3x+4}=\frac{3x^2}{(3x+4)(3x-4)}-\frac{x}{3x+4}=\frac{3x^2}{(3x+4)(3x-4)}-\frac{x(3x-4)}{(3x+4)(3x-4)}$$
$$=\frac{3x^2-x(3x-4)}{(3x+4)(3x-4)}=\frac{3x^2-3x^2+4x}{(3x+4)(3x-4)}=\frac{4x}{(3x+4)(3x-4)}$$

25. $$\frac{y}{y^2+5y+6}-\frac{2}{y^2+3y+2}=\frac{y}{(y+3)(y+2)}-\frac{2}{(y+2)(y+1)}$$
$$=\frac{y(y+1)}{(y+3)(y+2)(y+1)}-\frac{2(y+3)}{(y+3)(y+2)(y+1)}$$
$$=\frac{y(y+1)-2(y+3)}{(y+3)(y+2)(y+1)}=\frac{y^2+y-2y-6}{(y+3)(y+2)(y+1)}$$
$$=\frac{y^2-y-6}{(y+3)(y+2)(y+1)}=\frac{(y-3)\cancel{(y+2)}}{(y+3)\cancel{(y+2)}(y+1)}$$
$$=\frac{y-3}{(y+3)(y+1)}$$

26.
$$\frac{x}{x+3}+\frac{x}{x-3}-\frac{9}{x^2-9}=\frac{x}{x+3}+\frac{x}{x-3}-\frac{9}{(x+3)(x-3)}$$
$$=\frac{x(x-3)}{(x+3)(x-3)}+\frac{x(x+3)}{(x-3)(x+3)}-\frac{9}{(x+3)(x-3)}$$
$$=\frac{x(x-3)+x(x+3)-9}{(x+3)(x-3)}=\frac{x^2-3x+x^2+3x-9}{(x+3)(x-3)}=\frac{2x^2-9}{(x+3)(x-3)}$$

27.
$$\frac{3x^2}{x-y}+\frac{3y^2}{y-x}=\frac{3x^2}{x-y}+\frac{-1(3y^2)}{-1(y-x)}=\frac{3x^2}{x-y}+\frac{-3y^2}{x-y}=\frac{3x^2-3y^2}{x-y}=\frac{3(x^2-y^2)}{x-y}$$
$$=\frac{3(x+y)\cancel{(x-y)}}{\cancel{x-y}}=3(x+y) \text{ or } 3x+3y$$

6.3

28.
$$\frac{\frac{3}{x}-3}{\frac{8}{x}-8}=\frac{x}{x}\cdot\frac{\frac{3}{x}-3}{\frac{8}{x}-8}=\frac{x\cdot\frac{3}{x}-x\cdot 3}{x\cdot\frac{8}{x}-x\cdot 8}=\frac{3-3x}{8-8x}=\frac{3\cancel{(1-x)}}{8\cancel{(1-x)}}=\frac{3}{8}$$

29.
$$\frac{\frac{5}{x}+1}{1-\frac{25}{x^2}}=\frac{x^2}{x^2}\cdot\frac{\frac{5}{x}+1}{1-\frac{25}{x^2}}=\frac{x^2\cdot\frac{5}{x}+x^2\cdot 1}{x^2\cdot 1-x^2\cdot\frac{25}{x^2}}=\frac{5x+x^2}{x^2-25}=\frac{x\cancel{(5+x)}}{\cancel{(x+5)}(x-5)}=\frac{x}{x-5}$$

30.
$$\frac{3-\frac{1}{x+3}}{3+\frac{1}{x+3}}=\frac{x+3}{x+3}\cdot\frac{3-\frac{1}{x+3}}{3+\frac{1}{x+3}}=\frac{(x+3)\cdot 3-(x+3)\cdot\frac{1}{x+3}}{(x+3)\cdot 3+(x+3)\cdot\frac{1}{x+3}}=\frac{3x+9-1}{3x+9+1}=\frac{3x+8}{3x+10}$$

31.
$$\frac{\frac{4}{x+3}}{\frac{2}{x-2}-\frac{1}{x^2+x-6}}=\frac{\frac{4}{x+3}}{\frac{2}{x-2}-\frac{1}{(x+3)(x-2)}}=\frac{(x+3)(x-2)}{(x+3)(x-2)}\cdot\frac{\frac{4}{x+3}}{\frac{2}{x-2}-\frac{1}{(x+3)(x-2)}}$$
$$=\frac{(x-2)4}{(x+3)2-1}=\frac{4x-8}{2x+6-1}=\frac{4x-8}{2x+5}=\frac{4(x-2)}{2x+5}$$

32.
$$\frac{x^{-2}+x^{-1}}{x^{-2}-x^{-1}}=\frac{\frac{1}{x^2}+\frac{1}{x}}{\frac{1}{x^2}-\frac{1}{x}}=\frac{x^2}{x^2}\cdot\frac{\frac{1}{x^2}+\frac{1}{x}}{\frac{1}{x^2}-\frac{1}{x}}=\frac{x^2\cdot\frac{1}{x^2}+x^2\cdot\frac{1}{x}}{x^2\cdot\frac{1}{x^2}-x^2\cdot\frac{1}{x}}=\frac{1+x}{1-x}$$

6.4

33.
$$\frac{15x^3-30x^2+10x-2}{5x^2}=\frac{15x^3}{5x^2}-\frac{30x^2}{5x^2}+\frac{10x}{5x^2}-\frac{2}{5x^2}=3x-6+\frac{2}{x}-\frac{2}{5x^2}$$

34.
$$\frac{36x^4y^3+12x^2y^3-60x^2y^2}{6xy^2}=\frac{36x^4y^3}{6xy^2}+\frac{12x^2y^3}{6xy^2}-\frac{60x^2y^2}{6xy^2}=6x^3y+2xy-10x$$

35.
$$\begin{array}{r} 3x-7 \\ 2x+3\overline{)6x^2-5x+5} \\ \underline{6x^2+9x} \\ -14x+5 \\ \underline{-14x-21} \\ 26 \end{array}$$

$$\frac{6x^2-5x+5}{2x+3}=3x-7+\frac{26}{2x+3}$$

36.
$$\begin{array}{r} 2x^2-4x+1 \\ 5x-3\overline{)10x^3-26x^2+17x-13} \\ \underline{10x^3-6x^2} \\ -20x^2+17x \\ \underline{-20x^2+12x} \\ 5x-13 \\ \underline{5x-3} \\ -10 \end{array}$$

$$\frac{10x^3-26x^2+17x-13}{5x-3}=2x^2-4x+1-\frac{10}{5x-3}$$

37.

$$\begin{array}{r} x^5+5x^4+8x^3+16x^2+33x+63 \\ x-2\overline{)x^6+3x^5-2x^4+0x^3+x^2-3x+2} \\ \underline{x^6-2x^5} \\ 5x^5-2x^4 \\ \underline{5x^5-10x^4} \\ 8x^4+0x^3 \\ \underline{8x^4-16x^3} \\ 16x^3+x^2 \\ \underline{16x^3-32x^2} \\ 33x^2-3x \\ \underline{33x^2-66x} \\ 63x+2 \\ \underline{63x-126} \\ 128 \end{array}$$

$$\frac{x^6+3x^5-2x^4+x^2-3x+2}{x-2}=x^5+5x^4+8x^3+16x^2+33x+63+\frac{128}{x-2}$$

38.

$$\begin{array}{r} 2x^2+3x-1 \\ 2x^2+0x+1\overline{)4x^4+6x^3+0x^2+3x-1} \\ \underline{4x^4+0x^3+2x^2} \\ 6x^3-2x^2+3x \\ \underline{6x^3+0x^2+3x} \\ -2x^2+0x-1 \\ \underline{-2x^2+0x-1} \\ 0 \end{array}$$

$$\frac{4x^4+6x^3+3x-1}{2x^2+1}=2x^2+3x-1$$

6.5

39. $(4x^3-3x^2-2x+1)\div(x+1)$

$$\begin{array}{r|rrrr} -1 & 4 & -3 & -2 & 1 \\ & & -4 & 7 & -5 \\ \hline & 4 & -7 & 5 & -4 \end{array}$$

$(4x^3-3x^2-2x+1)\div(x+1)$

$=4x^2-7x+5-\dfrac{4}{x+1}$

40. $(3x^4-2x^2-10x-20)\div(x-2)$

$$\begin{array}{r|rrrrr} 2 & 3 & 0 & -2 & -10 & -20 \\ & & 6 & 12 & 20 & 20 \\ \hline & 3 & 6 & 10 & 10 & 0 \end{array}$$

$(3x^4-2x^2-10x-20)\div(x-2)$

$=3x^3+6x^2+10x+10$

41. $(x^4+16)\div(x+4)$

$$\begin{array}{r|rrrrr} -4 & 1 & 0 & 0 & 0 & 16 \\ & & -4 & 16 & -64 & 256 \\ \hline & 1 & -4 & 16 & -64 & 272 \end{array}$$

$(x^4+16)\div(x+4)$

$=x^3-4x^2+16x-64+\dfrac{272}{x+4}$

42. Divide $f(x)$ by $x-2$.

$$\begin{array}{r|rrrr} 2 & 2 & -5 & 4 & -1 \\ & & 4 & -2 & 4 \\ \hline & 2 & -1 & 2 & 3 \end{array}$$

$f(2)=3$

43. $f(x)=3x^4+6x^3-2x+4$

$$\begin{array}{r|rrrrr} -4 & 3 & 6 & 0 & -2 & 4 \\ & & -12 & 24 & -96 & 392 \\ \hline & 3 & -6 & 24 & -98 & 394 \end{array}$$

$f(-4)=396$

44. To show that –2 is a solution to the equation, show that when the polynomial is divided by $x+2$ the remainder is zero.

$$\begin{array}{r|rrrr} -2 & 2 & -1 & -8 & 4 \\ & & -4 & 10 & -4 \\ \hline & 2 & -5 & 2 & 0 \end{array}$$

Since the remainder is zero, 2 is a solution to the equation.

45. $x^4-x^3-7x^2+x+6=0$

$$\begin{array}{r|rrrrr} 4 & 1 & -1 & -7 & 1 & 6 \\ & & 4 & 12 & 20 & 84 \\ \hline & 1 & 3 & 5 & 21 & 90 \end{array}$$

Since the remainder is not zero, 4 is not a solution to the equation.

46. To show that $\frac{1}{2}$ is a solution to the equation, show that when the polynomial is divided by $x-\frac{1}{2}$ the remainder is zero.

$$\begin{array}{r|rrrr} \frac{1}{2} & 6 & 1 & -4 & 1 \\ & & 3 & 2 & -1 \\ \hline & 6 & 4 & -2 & 0 \end{array}$$

$6x^3+x^2-4x+1$

$=\left(x-\dfrac{1}{2}\right)(6x^2+4x-2)$

To solve the equation, we set it equal to zero and factor.

$$\left(x-\frac{1}{2}\right)\left(6x^2+4x-2\right)=0$$

$$\left(x-\frac{1}{2}\right)\left(2\left(3x^2+2x-1\right)\right)=0$$

$$\left(x-\frac{1}{2}\right)\left(2\left(3x-1\right)\left(x+1\right)\right)=0$$

$$2\left(x-\frac{1}{2}\right)\left(3x-1\right)\left(x+1\right)=0$$

6.6

47. $\frac{3}{x}+\frac{1}{3}=\frac{5}{x}$

So that denominators will not equal zero, x cannot equal zero. To eliminate fractions, multiply by the LCD, $3x$.

$$\frac{3}{x}+\frac{1}{3}=\frac{5}{x}$$

$$3x\left(\frac{3}{x}+\frac{1}{3}\right)=3x\left(\frac{5}{x}\right)$$

$$3x\left(\frac{3}{x}\right)+3x\left(\frac{1}{3}\right)=3(5)$$

$$3(3)+x(1)=15$$

$$9+x=15$$

$$x=6$$

The solution set is $\{6\}$.

48. $\frac{5}{3x+4}=\frac{3}{2x-8}$

To find the restrictions on x, set the denominators equal to zero and solve.

$$3x+4=0 \qquad 2x-8=0$$

$$3x=-4 \qquad 2x=8$$

$$x=-\frac{4}{3} \qquad x=4$$

To eliminate fractions, multiply by the LCD, $3x$.

$$\frac{5}{3x+4}=\frac{3}{2x-8}$$

$$(3x+4)(2x-8)\left(\frac{5}{3x+4}\right)=(3x+4)(2x-8)\left(\frac{3}{2x-8}\right)$$

$$(2x-8)(5)=(3x+4)(3)$$

$$10x-40=9x+12$$

$$x-40=12$$

$$x=52$$

The solution set is $\{52\}$.

49. $\frac{1}{x-5}-\frac{3}{x+5}=\frac{6}{x^2-25}$

$\frac{1}{x-5}-\frac{3}{x+5}=\frac{6}{(x+5)(x-5)}$

So that denominators will not equal zero, x cannot equal 5 or –5. To eliminate fractions, multiply by the LCD, $(x+5)(x-5)$.

$$\frac{1}{x-5}-\frac{3}{x+5}=\frac{6}{(x+5)(x-5)}$$

$$(x+5)(x-5)\left(\frac{1}{x-5}-\frac{3}{x+5}\right)=(x+5)(x-5)\left(\frac{6}{(x+5)(x-5)}\right)$$

$$(x+5)(x-5)\left(\frac{1}{x-5}\right)-(x+5)(x-5)\left(\frac{3}{x+5}\right)=6$$

$$(x+5)(1)-(x-5)(3)=6$$

$$x+5-(3x-15)=6$$

$$x+5-3x+15=6$$

$$-2x+20=6$$

$$-2x=-14$$

$$x=7$$

The solution set is $\{7\}$.

50. $\frac{x+5}{x+1}-\frac{x}{x+2}=\frac{4x+1}{x^2+3x+2}$

$\frac{x+5}{x+1}-\frac{x}{x+2}=\frac{4x+1}{(x+2)(x+1)}$

So that denominators will not equal zero, x cannot equal –1 or –2. To eliminate fractions, multiply by the LCD, $(x+2)(x+1)$.

$$\frac{x+5}{x+1}-\frac{x}{x+2}=\frac{4x+1}{(x+2)(x+1)}$$

$$(x+2)(x+1)\left(\frac{x+5}{x+1}-\frac{x}{x+2}\right)=(x+2)(x+1)\left(\frac{4x+1}{(x+2)(x+1)}\right)$$

$$(x+2)(x+1)\left(\frac{x+5}{x+1}\right)-(x+2)(x+1)\left(\frac{x}{x+2}\right)=4x+1$$

$$(x+2)(x+5)-(x+1)(x)=4x+1$$
$$x^2+7x+10-(x^2+x)=4x+1$$
$$x^2+7x+10-x^2-x=4x+1$$
$$6x+10=4x+1$$
$$2x+10=1$$
$$2x=-9$$
$$x=-\frac{9}{2}$$

The solution set is $\left\{-\frac{9}{2}\right\}$.

51. $\frac{2}{3}-\frac{5}{3x}=\frac{1}{x^2}$

So that denominators will not equal zero, x cannot equal zero. To eliminate fractions, multiply by the LCD, $3x^2$.

$$3x^2\left(\frac{2}{3}-\frac{5}{3x}\right)=3x^2\left(\frac{1}{x^2}\right)$$
$$3x^2\left(\frac{2}{3}\right)-3x^2\left(\frac{5}{3x}\right)=3(1)$$
$$x^2(2)-x(5)=3$$
$$2x^2-5x=3$$
$$2x^2-5x-3=0$$
$$(2x+1)(x-3)=0$$

Apply the zero product principle.

$2x+1=0$ $\qquad$ $x-3=0$

$2x=-1$ $\qquad$ $x=3$

$x=-\frac{1}{2}$

The solution set is $\left\{-\frac{1}{2},3\right\}$.

52. $\frac{2}{x-1}=\frac{1}{4}+\frac{7}{x+2}$

So that denominators will not equal zero, x cannot equal 1 or –2. To eliminate fractions, multiply by the LCD, $4(x-1)(x+2)$.

$$4(x-1)(x+2)\left(\frac{2}{x-1}\right)=4(x-1)(x+2)\left(\frac{1}{4}+\frac{7}{x+2}\right)$$

$$4(x+2)(2)=4(x-1)(x+2)\left(\frac{1}{4}\right)+4(x-1)(x+2)\left(\frac{7}{x+2}\right)$$

$$8x+16=(x-1)(x+2)+4(x-1)(7)$$

$$8x+16=x^2+x-2+28x-28$$

$$8x+16=x^2+29x-30$$

$$0=x^2+21x-46$$

$$0=(x+23)(x-2)$$

Apply the zero product principle.

$x+23=0$ $\qquad$ $x-2=0$

$x=-23$ $\qquad$ $x=2$

The solution set is $\{-23,2\}$.

53. $$\frac{2x+7}{x+5}-\frac{x-8}{x-4}=\frac{x+18}{x^2+x-20}$$

$$\frac{2x+7}{x+5}-\frac{x-8}{x-4}=\frac{x+18}{(x+5)(x-4)}$$

So that denominators will not equal zero, x cannot equal –5 or 4. To eliminate fractions, multiply by the LCD, $(x+5)(x-4)$.

$$\frac{2x+7}{x+5}-\frac{x-8}{x-4}=\frac{x+18}{(x+5)(x-4)}$$

$$(x+5)(x-4)\left(\frac{2x+7}{x+5}-\frac{x-8}{x-4}\right)=(x+5)(x-4)\left(\frac{x+18}{(x+5)(x-4)}\right)$$

$$(x+5)(x-4)\left(\frac{2x+7}{x+5}\right)-(x+5)(x-4)\left(\frac{x-8}{x-4}\right)=x+18$$

$$(x-4)(2x+7)-(x+5)(x-8)=x+18$$

$$2x^2-x-28-(x^2-3x-40)=x+18$$

$$2x^2-x-28-x^2+3x+40=x+18$$

$$x^2+2x+12=x+18$$

$$x^2+x-6=0$$
$$(x+3)(x-2)=0$$

Apply the zero product principle.

$$x+3=0 \qquad x-2=0$$
$$x=-3 \qquad x=2$$

The solution set is $\{-3,2\}$.

54.

$$f(x)=\frac{4x}{100-x}$$
$$16=\frac{4x}{100-x}$$
$$(100-x)(16)=(100-x)\left(\frac{4x}{100-x}\right)$$
$$1600-16x=4x$$
$$1600=20x$$
$$80=x$$

80% of the pollutants can be removed for $16 million.

6.7

55.

$$P=\frac{R-C}{n}$$
$$n(P)=n\left(\frac{R-C}{n}\right)$$
$$nP=R-C$$
$$nP+C=R$$
$$C=R-nP$$

56.

$$\frac{P_1V_1}{T_1}=\frac{P_2V_2}{T_2}$$
$$T_1T_2\left(\frac{P_1V_1}{T_1}\right)=T_1T_2\left(\frac{P_2V_2}{T_2}\right)$$
$$T_2(P_1V_1)=T_1(P_2V_2)$$
$$P_1T_2V_1=P_2T_1V_2$$
$$\frac{P_1T_2V_1}{P_2V_2}=\frac{P_2T_1V_2}{P_2V_2}$$
$$T_1=\frac{P_1T_2V_1}{P_2V_2}$$

57.

$$T=\frac{A-P}{Pr}$$
$$Pr(T)=Pr\left(\frac{A-P}{Pr}\right)$$
$$PrT=A-P$$
$$PrT+P=A$$
$$P(rT+1)=A$$
$$\frac{P(rT+1)}{rT+1}=\frac{A}{rT+1}$$
$$P=\frac{A}{rT+1}$$

58.

$$\frac{1}{R}=\frac{1}{R_1}+\frac{1}{R_2}$$
$$RR_1R_2\left(\frac{1}{R}\right)=RR_1R_2\left(\frac{1}{R_1}+\frac{1}{R_2}\right)$$
$$R_1R_2=RR_1R_2\left(\frac{1}{R_1}\right)+RR_1R_2\left(\frac{1}{R_2}\right)$$
$$R_1R_2=RR_2+RR_1$$
$$R_1R_2=R(R_2+R_1)$$
$$\frac{R_1R_2}{R_2+R_1}=\frac{R(R_2+R_1)}{R_2+R_1}$$
$$R=\frac{R_1R_2}{R_2+R_1}$$

59.

$$I=\frac{nE}{R+nr}$$

$$(R+nr)(I)=(R+nr)\left(\frac{nE}{R+nr}\right)$$
$$IR+Inr=nE$$
$$IR=nE-Inr$$
$$IR=n(E-Ir)$$
$$\frac{IR}{E-Ir}=\frac{n(E-Ir)}{E-Ir}$$
$$n=\frac{IR}{E-Ir}$$

60. **a.** $C(x)=50{,}000+25x$

b. $\overline{C}(x)=\frac{50{,}000+25x}{x}$

c.
$$35=\frac{50{,}000+25x}{x}$$
$$x(35)=x\left(\frac{50{,}000+25x}{x}\right)$$
$$35x=50{,}000+25x$$
$$10x=50{,}000$$
$$x=5000$$

5000 graphing calculators must be produced each month to have an average cost of $35.

61.

	d	r	$t=\frac{d}{r}$
Riding	60	$3x$	$\frac{60}{3x}$
Walking	8	x	$\frac{8}{x}$

$$\frac{60}{3x}+\frac{8}{x}=7$$
$$3x\left(\frac{60}{3x}+\frac{8}{x}\right)=3x(7)$$
$$3x\left(\frac{60}{3x}\right)+3x\left(\frac{8}{x}\right)=21x$$
$$60+3(8)=21x$$
$$60+24=21x$$
$$84=21x$$
$$4=x$$

The cyclist was riding at a rate of $3x=3(4)=12$ miles per hour.

62.

	d	r	$t=\frac{d}{r}$
Down Stream	12	$x+3$	$\frac{12}{x+3}$
Up Stream	12	$x-3$	$\frac{12}{x-3}$

$$\frac{12}{x+3}+\frac{12}{x-3}=3$$

$$(x+3)(x-3)\left(\frac{12}{x+3}+\frac{12}{x-3}\right)=(x+3)(x-3)(3)$$

$$(x+3)(x-3)\left(\frac{12}{x+3}\right)+(x+3)(x-3)\left(\frac{12}{x-3}\right)=3\left(x^2-9\right)$$

$$(x-3)(12)+(x+3)(12)=3x^2-27$$

$$12x-36+12x+36=3x^2-27$$

$$24x=3x^2-27$$

$$0=3x^2-24x-27$$

$$0=3\left(x^2-8x-9\right)$$

$$0=3(x-9)(x+1)$$

Apply the zero product principle.

$3(x-9)=0 \qquad x+1=0$

$x-9=0 \qquad x=-1$

$x=9$

The solutions are –1 and 9. We disregard –1 because we cannot have a negative rate. The boat's rate in still water is 9 miles per hour.

63.

	Part Done in 1 Hour	Time Working Together	Part Done in x Hours
First Person	$\frac{1}{3}$	x	$\frac{x}{3}$
Second Person	$\frac{1}{6}$	x	$\frac{x}{6}$

$$\frac{x}{3}+\frac{x}{6}=1$$

$$6\left(\frac{x}{3}+\frac{x}{6}\right)=6(1)$$

$$6\left(\frac{x}{3}\right)+6\left(\frac{x}{6}\right)=6$$

$$2x+x=6$$

$$3x=6$$

$$x=2$$

If they work together, it will take 2 hours to clean the house. There is not enough time to finish the job before the TV program starts.

64.

	Part Done in 1 Hour	Time Working Together	Part Done in 20 Hours
Fast Crew	$\frac{1}{x-9}$	20	$\frac{20}{x-9}$
Slow Crew	$\frac{1}{x}$	20	$\frac{20}{x}$

$$\frac{20}{x-9}+\frac{20}{x}=1$$

$$x(x-9)\left(\frac{20}{x-9}+\frac{20}{x}\right)=x(x-9)(1)$$

$$x(x-9)\left(\frac{20}{x-9}\right)+x(x-9)\left(\frac{20}{x}\right)=x^2-9x$$

$$x(20)+(x-9)(20)=x^2-9x$$

$$20x+20x-180=x^2-9x$$

$$40x-180=x^2-9x$$

$$0=x^2-49x+180$$

$$0=x^2-49x+180$$

$$0=(x-45)(x-4)$$

Apply the zero product principle.

$x-45=0 \qquad x-4=0$

$x=45 \qquad x=4$

The solutions are 45 and 4. We disregard 4, because the fast crew's rate would be $4-9=-5$. No crew can do the job in a negative number of hours. It would take the slow crew 45 hours and the fast crew $45-9=36$ hours to complete the job working alone.

65.

	Part Done in 1 Minute	Time Working Together	Part Done in x Minutes
Faucet	$\frac{1}{60}$	x	$\frac{x}{60}$
Drain	$\frac{1}{80}$	x	$\frac{x}{80}$

$$\frac{x}{60}-\frac{x}{80}=1$$
$$240\left(\frac{x}{60}-\frac{x}{80}\right)=240(1)$$
$$240\left(\frac{x}{60}\right)-240\left(\frac{x}{80}\right)=240$$
$$4x-3x=240$$
$$x=240$$

It will take 240 minutes or 4 hours to fill the pond.

6.8

66. Since the bill varies directly as electricity used, we have $b=ke$. Use the given values to find k.

$$b=ke$$
$$98=k(1400)$$
$$\frac{98}{1400}=\frac{k(1400)}{1400}$$
$$0.07=k$$

The equation becomes $b=0.07e$. When $e=2200$ kilowatts, $b=0.07(2200)=154$.

If 2200 kilowatts are used, the electric bill will be \$154.

67. Since distance varies directly as the square of the time, we have $d=kt^2$. Use the given values to find k.

$$d=kt^2$$
$$144=k(3)^2$$
$$144=k(9)$$
$$\frac{144}{9}=\frac{k(9)}{9}$$
$$16=k$$

The equation becomes $d=16t^2$. When $t=10$,

$d=16(10)^2=16(100)=1600$.

A skydiver will fall 1600 feet in 10 seconds.

68. Since driving time varies inversely as the rate of travel, we have $t=\frac{k}{r}$. Use the given values to find k.

$$t=\frac{k}{r}$$
$$4=\frac{k}{50}$$
$$50(4)=50\left(\frac{k}{50}\right)$$
$$200=k$$

The equation becomes $t=\frac{200}{r}$.

When $r=40$, $t=\frac{200}{40}=5$.

It will take 5 hours at 40 miles per hour.

69. Since loudness varies inversely as the square of the distance, we have

$l = \frac{k}{d^2}$.

Use the given values to find k.

$$l = \frac{k}{d^2}$$
$$28 = \frac{k}{8^2}$$
$$28 = \frac{k}{64}$$
$$64(28) = 64\left(\frac{k}{64}\right)$$
$$1792 = k$$

The equation becomes $l = \frac{1792}{d^2}$.

When $d = 4$, $l = \frac{1792}{(4)^2} = \frac{1792}{16} = 112$.

At a distance of 4 feet, the loudness of the stereo is 112 decibels.

70. Since time varies directly as the number of computers and inversely as the number of workers, we have $t = \frac{kn}{w}$.

Use the given values to find k.

$$t = \frac{kn}{w}$$
$$10 = \frac{k(30)}{6}$$
$$10 = 5k$$
$$\frac{10}{5} = \frac{5k}{5}$$
$$2 = k$$

The equation becomes $t = \frac{2n}{w}$.

When $n = 40$ and $w = 5$,

$$t = \frac{2(40)}{5} = \frac{80}{5} = 16.$$

It will take 16 hours for 5 workers to assemble 40 computers.

71. Since the volume varies jointly as height and the area of the base, we have $v = kha$.

Use the given values to find k.

$$175 = k(15)(35)$$
$$175 = k(525)$$
$$\frac{175}{525} = \frac{k(525)}{525}$$
$$\frac{1}{3} = k$$

The equation becomes $v = \frac{1}{3}ha$.

When $h = 20$ feet and $a = 120$ square feet, $v = \frac{1}{3}(20)(120) = 800$.

If the height is 20 feet and the area is 120 square feet, the volume will be 800 cubic feet.

Chapter 6 Test

1. The domain is all real numbers except those that make the denominator zero. To find these values, set the denominator equal to zero and solve.

$$x^2 - 7x + 10 = 0$$
$$(x-5)(x-2) = 0$$

Apply the zero product principle.

$$x - 5 = 0 \qquad x - 2 = 0$$
$$x = 5 \qquad x = 2$$

The domain of $f = \{x \mid x \text{ is a real number and } x \neq 2 \text{ and } x \neq 5\}$ or

$(-\infty, 2)\cup(2,5)\cup(5,\infty)$.

$$f(x)=\frac{x^2-2x}{x^2-7x+10}$$
$$=\frac{x\cancel{(x-2)}}{(x-5)\cancel{(x-2)}}=\frac{x}{x-5}$$

2. $$\frac{x^2}{x^2-16}\cdot\frac{x^2+7x+12}{x^2+3x}$$
$$=\frac{x^2}{(x+4)(x-4)}\cdot\frac{(x+4)(x+3)}{x(x+3)}$$
$$=\frac{x^2}{\cancel{(x+4)}(x-4)}\cdot\frac{\cancel{(x+4)}\cancel{(x+3)}}{x\cancel{(x+3)}}$$
$$=\frac{x}{x-4}$$

3. $$\frac{x^3+27}{x^2-1}\div\frac{x^2-3x+9}{x^2-2x+1}$$
$$=\frac{x^3+27}{x^2-1}\cdot\frac{x^2-2x+1}{x^2-3x+9}$$
$$=\frac{(x+3)\cancel{(x^2-3x+9)}}{(x+1)\cancel{(x-1)}}\cdot\frac{(x-1)^{\cancel{2}}}{\cancel{x^2-3x+9}}$$
$$=\frac{(x+3)(x-1)}{x+1}$$

4. $$\frac{x^2-5x-2}{6x^2-11x-35}-\frac{x^2-7x+5}{6x^2-11x-35}$$
$$=\frac{x^2-5x-2-(x^2-7x+5)}{6x^2-11x-35}$$
$$=\frac{x^2-5x-2-x^2+7x-5}{6x^2-11x-35}$$
$$=\frac{2x-7}{6x^2-11x-35}$$
$$=\frac{\cancel{2x-7}}{\cancel{(2x-7)}(3x+5)}=\frac{1}{3x+5}$$

5. $$\frac{x}{x+3}+\frac{5}{x-3}$$
$$=\frac{x(x-3)}{(x+3)(x-3)}+\frac{5(x+3)}{(x+3)(x-3)}$$
$$=\frac{x(x-3)+5(x+3)}{(x+3)(x-3)}$$
$$=\frac{x^2-3x+5x+15}{(x+3)(x-3)}$$
$$=\frac{x^2+2x+15}{(x+3)(x-3)}$$

6. $$\frac{x^2-5x-2}{6x^2-11x-35}-\frac{x^2-7x+5}{6x^2-11x-35}=\frac{x^2-5x-2-(x^2-7x+5)}{6x^2-11x-35}$$
$$=\frac{3x^2-7x+4}{(x-3)(x-1)(x+2)}=\frac{(3x-4)\cancel{(x-1)}}{(x-3)\cancel{(x-1)}(x+2)}$$
$$=\frac{3x-4}{(x-3)(x+2)}$$

7. $$\frac{5x}{x^2-4}-\frac{2}{x^2+x-2}=\frac{5x}{(x+2)(x-2)}-\frac{2}{(x+2)(x-1)}=$$
$$\frac{5x(x-1)}{(x+2)(x-2)(x-1)}-\frac{2(x-2)}{(x+2)(x-2)(x-1)}$$
$$=\frac{5x(x-1)-2(x-2)}{(x+2)(x-2)(x-1)}=\frac{5x^2-5x-(2x-4)}{(x+2)(x-2)(x-1)}$$
$$=\frac{5x^2-5x-2x+4}{(x+2)(x-2)(x-1)}=\frac{5x^2-7x+4}{(x+2)(x-2)(x-1)}$$

8. $$\frac{x-4}{x-5}-\frac{3}{x+5}-\frac{10}{x^2-25}=\frac{(x-4)(x+5)}{(x+5)(x-5)}-\frac{3(x-5)}{(x+5)(x-5)}-\frac{10}{(x+5)(x-5)}$$
$$=\frac{(x-4)(x+5)-3(x-5)-10}{(x+5)(x-5)}=\frac{x^2+x-20-3x+15-10}{(x+5)(x-5)}$$
$$=\frac{x^2-2x-15}{(x+5)(x-5)}=\frac{\cancel{(x-5)}(x+3)}{(x+5)\cancel{(x-5)}}=\frac{x+3}{x+5}$$

9. $$\frac{1}{10-x}+\frac{x-1}{x-10}=\frac{-1(1)}{-1(10-x)}+\frac{x-1}{x-10}=\frac{-1}{x-10}+\frac{x-1}{x-10}=\frac{-1+x-1}{x-10}=\frac{x-2}{x-10}$$

10. $$\frac{\frac{x}{4}-\frac{1}{x}}{1+\frac{x+4}{x}}=\frac{4x}{4x}\cdot\frac{\frac{x}{4}-\frac{1}{x}}{1+\frac{x+4}{x}}=\frac{\cancel{4}x\cdot\frac{x}{\cancel{4}}-4\cancel{x}\cdot\frac{1}{\cancel{x}}}{4x\cdot 1+4\cancel{x}\cdot\frac{x+4}{\cancel{x}}}=\frac{x^2-4}{4x+4(x+4)}=\frac{(x+2)(x-2)}{4x+4x+16}$$
$$=\frac{(x+2)(x-2)}{8x+16}=\frac{\cancel{(x+2)}(x-2)}{8\cancel{(x+2)}}=\frac{x-2}{8}$$

11. $$\frac{\frac{1}{x}-\frac{3}{x+2}}{\frac{2}{x^2+2x}}=\frac{\frac{1}{x}-\frac{3}{x+2}}{\frac{2}{x(x+2)}}=\frac{x(x+2)}{x(x+2)}\cdot\frac{\frac{1}{x}-\frac{3}{x+2}}{\frac{2}{x(x+2)}}=\frac{x(x+2)\cdot\frac{1}{x}-x(x+2)\cdot\frac{3}{x+2}}{x(x+2)\cdot\frac{2}{x(x+2)}}$$
$$=\frac{x+2-3x}{2}=\frac{-2x+2}{2}=\frac{-\cancel{2}(x-1)}{\cancel{2}}=-(x-1)=1-x$$

12. $\dfrac{12x^4y^3+16x^2y^3-10x^2y^2}{4x^2y}=\dfrac{12x^4y^3}{4x^2y}+\dfrac{16x^2y^3}{4x^2y}-\dfrac{10x^2y^2}{4x^2y}=3x^2y^2+4y^2-\dfrac{5y}{2}$

13.

$$\begin{array}{r} 3x^2-3x+1 \\ 3x+2\overline{)9x^3-3x^2-3x+4} \\ \underline{9x^3+6x^2} \\ -9x^2-3x \\ \underline{-9x^2-6x} \\ 3x+4 \\ \underline{3x+2} \\ 2 \end{array}$$

$$\frac{9x^3-3x^2-3x+4}{3x+2}$$

$$=3x^2-3x+1+\frac{2}{3x+2}$$

14.

$$\begin{array}{r} 3x^2+2x+3 \\ x^2+0x-1\overline{)3x^4+2x^3+0x^2-8x+6} \\ \underline{3x^4+0x^3-3x^2} \\ 2x^3+3x^2-8x \\ \underline{2x^3+0x^2-2x} \\ 3x^2-6x+6 \\ \underline{3x^2+0x-3} \\ -6x+9 \end{array}$$

$$\frac{3x^4+2x^3-8x+6}{x^2-1}$$

$$=3x^2+2x+3+\frac{-6x+9}{x^2-1}$$

$$=3x^2+2x+3-\frac{6x-9}{x^2-1}$$

15. $(3x^4+11x^3-20x^2+7x+35)\div(x+5)$

$$\begin{array}{r|rrrrr} \underline{-5} & 3 & 11 & -20 & 7 & 35 \\ & & -15 & 20 & 0 & -35 \\ \hline & 3 & -4 & 0 & 7 & 0 \end{array}$$

$(3x^4-2x^2-10x)\div(x-2)$

$=3x^3-4x^2+7$

16. Divide $f(x)$ by $x-(-2)=x+2$.

$$\begin{array}{r|rrrrr} \underline{-2} & 1 & -2 & -11 & 5 & 34 \\ & & -2 & 8 & 6 & -22 \\ \hline & 1 & -4 & -3 & 11 & 12 \end{array}$$

$f(-2)=12$

17. $2x^3-3x^2-11x+6$

$$\begin{array}{r|rrrr} \underline{-2} & 2 & -3 & -11 & 6 \\ & & -4 & 14 & -6 \\ \hline & 2 & -7 & 3 & 0 \end{array}$$

Since the remainder is 0, -2 is a solution.

18. $\dfrac{4}{x+1}=\dfrac{5}{2x+1}$

To find the restrictions on x, set the denominators equal to zero and solve.

$x+1=0$ $\qquad 2x+1=0$

$x=-1$ $\qquad 2x=-1$

$x=-\dfrac{1}{2}$

To eliminate fractions, multiply by the LCD, $(x+1)(2x+1)$.

$$(x+1)(2x+1)\left(\frac{4}{x+1}\right)=(x+1)(2x+1)\left(\frac{5}{2x+1}\right)$$
$$(2x+1)(4)=(x+1)(5)$$
$$8x+4=5x+5$$
$$3x+4=5$$
$$3x=1$$
$$x=\frac{1}{3}$$

The solution set is $\left\{\frac{1}{3}\right\}$.

19.
$$\frac{x+1}{x^2+2x-3}-\frac{1}{x+3}=\frac{1}{x-1}$$
$$\frac{x+1}{(x+3)(x-1)}-\frac{1}{x+3}=\frac{1}{x-1}$$

So that denominators will not equal zero, x cannot equal 1 or -3. To eliminate fractions, multiply by the LCD, $(x+3)(x-1)$.

$$(x+3)(x-1)\left(\frac{x+1}{(x+3)(x-1)}-\frac{1}{x+3}\right)=(x+3)(x-1)\left(\frac{1}{x-1}\right)$$
$$(x+3)(x-1)\left(\frac{x+1}{(x+3)(x-1)}\right)-(x+3)(x-1)\left(\frac{1}{x+3}\right)=(x+3)(1)$$
$$x+1-(x-1)=x+3$$
$$x+1-x+1=x+3$$
$$2=x+3$$
$$-1=x$$

The solution set is $\{-1\}$.

20.

$$f(t)=\frac{250(3t+5)}{t+25}$$
$$125=\frac{250(3t+5)}{t+25}$$
$$(t+25)(125)=(t+25)\left(\frac{250(3t+5)}{t+25}\right)$$
$$125t+3125=250(3t+5)$$
$$125t+3125=750t+1250$$
$$-625t+3125=1250$$
$$-625t=-1875$$
$$t=3$$

It will take 3 years for the elk population to reach 125.

21.

$$R=\frac{as}{a+s}$$
$$(a+s)R=(a+s)\left(\frac{as}{a+s}\right)$$
$$aR+Rs=as$$
$$aR=as-Rs$$
$$aR-as=-Rs$$
$$a(R-s)=-Rs$$
$$\frac{a(R-s)}{R-s}=-\frac{Rs}{R-s}$$
$$a=-\frac{Rs}{R-s} \text{ or } \frac{Rs}{s-R}$$

22. a. $C(x)=300{,}000+10x$

b. $\overline{C}(x)=\frac{300{,}000+10x}{x}$

c.

$$25=\frac{300{,}000+10x}{x}$$
$$x(25)=x\left(\frac{300{,}000+10x}{x}\right)$$
$$25x=300{,}000+10x$$
$$15x=300{,}000$$
$$x=20{,}000$$

20,000 televisions must be produced for the average cost to be $25.

23.

	Part Done in 1 Hour	Time Working Together	Part Done in x Hours
Fill Pipe	$\frac{1}{3}$	x	$\frac{x}{3}$
Drain Pipe	$\frac{1}{4}$	x	$\frac{x}{4}$

$$\frac{x}{3}-\frac{x}{4}=1$$
$$12\left(\frac{x}{3}-\frac{x}{4}\right)=12(1)$$
$$12\left(\frac{x}{3}\right)-12\left(\frac{x}{4}\right)=12$$
$$4x-3x=12$$
$$x=12$$

It will take 12 hours to fill the pool.

24.

	d	r	$t = \frac{d}{r}$
Down Stream	3	$20 + x$	$\frac{3}{20+x}$
Up Stream	2	$20 - x$	$\frac{2}{20-x}$

$$\frac{3}{20+x} = \frac{2}{20-x}$$
$$(20+x)(20-x)\left(\frac{3}{20+x}\right) = (20+x)(20-x)\left(\frac{2}{20-x}\right)$$
$$(20-x)(3) = (20+x)(2)$$
$$60 - 3x = 40 + 2x$$
$$60 - 5x = 40$$
$$-5x = -20$$
$$x = 4$$

The current's rate in still water is 4 miles per hour.

25. Since intensity varies inversely as the square of the distance, we have $I = \frac{k}{d^2}$.

Use the given values to find k.

$$I = \frac{k}{d^2}$$
$$20 = \frac{k}{15^2}$$
$$20 = \frac{k}{225}$$
$$225(20) = 225\left(\frac{k}{225}\right)$$
$$4500 = k$$

The equation becomes $I = \frac{4500}{d^2}$. When $d = 10$, $I = \frac{4500}{10^2} = \frac{4500}{100} = 45$.

At a distance of 10 feet, the light's intensity if 45 foot-candles.

Cumulative Review Exercises

Chapters 1-6

1. $2x+5 \le 11$ and $-3x > 18$

$2x \le 6 \qquad \frac{-3x}{-3} < \frac{18}{-3}$

$x \le 3 \qquad x < -6$

$x \le 3$

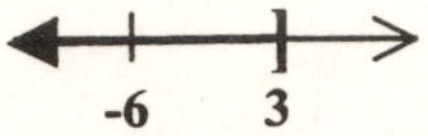

$x < -6$

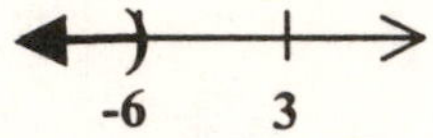

$x \le 3$ and $x < -6$

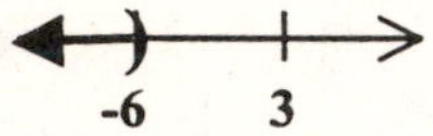

The solution set is $\{x|x < -6\}$ or $(-\infty, -6)$.

2.
$$2x^2 = 7x+4$$
$$2x^2 - 7x - 4 = 0$$
$$(2x+1)(x-4) = 0$$
Apply the zero product principle.

$2x+1=0 \qquad x-4=0$

$2x=-1 \qquad x=4$

$x = -\frac{1}{2}$

The solutions are $-\frac{1}{2}$ and 4 and the solution set is $\left\{-\frac{1}{2}, 4\right\}$.

3.
$$4x+3y+3z = 4$$
$$3x \qquad +2z = 2$$
$$2x-5y \qquad = -4$$

Multiply the first equation by 5 and the third equation by 3.

$$20x+15y+15z = 20$$
$$6x-15y \qquad = -12$$
$$26x \qquad +15z = 8$$

We now have a system of two equations in two variables.

$$3x+ 2z = 2$$
$$26x+15z = 8$$

Multiply the first equation by –15 and the second equation by 2.

$$-45x-30z = -30$$
$$52x+30z = 16$$
$$7x \qquad = -14$$
$$x \qquad = -2$$

Back-substitute –2 for x to find z.

$$3x+2z = 2$$
$$3(-2)+2z = 2$$
$$-6+2z = 2$$
$$2z = 8$$
$$z = 4$$

Back-substitute–2 for x to find y.

$$2x-5y = -4$$
$$2(-2)-5y = -4$$
$$-4-5y = -4$$
$$-5y = 0$$
$$y = 0$$

The solution is $(-2,0,4)$ and the solution set is $\{(-2,0,4)\}$.

4.
$$|3x-4| \le 10$$
$$-10 \le 3x-4 \le 10$$
$$-10+4 \le 3x-4+4 \le 10+4$$
$$-6 \le 3x \le 14$$

$$\frac{-6}{3} \le \frac{3x}{3} \le \frac{14}{3}$$

$$-2 \le x \le \frac{14}{3}$$

-2 14/3

The solution set is $\left\{x \middle| -2 \le x \le \frac{14}{3}\right\}$ or $\left[-2, \frac{14}{3}\right]$.

5.

$$\frac{x}{x-8} + \frac{6}{x-2} = \frac{x^2}{x^2 - 10x + 16}$$

$$\frac{x}{x-8} + \frac{6}{x-2} = \frac{x^2}{(x-8)(x-2)}$$

So that denominators will not equal zero, x cannot equal 2 or 8. To eliminate fractions, multiply by the LCD, $(x-8)(x-2)$.

$$(x-8)(x-2)\left(\frac{x}{x-8} + \frac{6}{x-2}\right) = (x-8)(x-2)\left(\frac{x^2}{(x-8)(x-2)}\right)$$

$$(x-8)(x-2)\left(\frac{x}{x-8}\right) + (x-8)(x-2)\left(\frac{6}{x-2}\right) = x^2$$

$$(x-2)(x) + (x-8)(6) = x^2$$

$$\cancel{x^2} - 2x + 6x - 48 = \cancel{x^2}$$

$$4x - 48 = 0$$

$$4x = 48$$

$$x = 12$$

The solution set is $\{12\}$.

6.

$$I = \frac{2R}{w + 2s}$$

$$(w+2s)(I) = (w+2s)\left(\frac{2R}{w+2s}\right)$$

$$Iw + 2Is = 2R$$

$$2Is = 2R - Iw$$

$$\frac{2Is}{2I} = \frac{2R - Iw}{2I}$$

$$s = \frac{2R - Iw}{2I}$$

7. $2x-y=4$
$x+y=5$

$$2x-y=4$$
$$-y=-2x+4$$
$$y=2x-4$$

m = 2

y – intercept = –4

$$x+y=5$$
$$y=-x+5$$

m = –1

y – intercept = 5

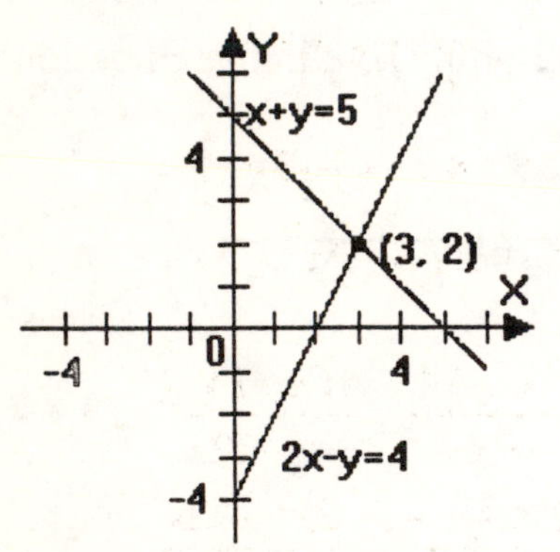

The solution set is $\{(3,2)\}$.

8. Slope = –3, passing through (1, –5)

Point-Slope Form

$$y-y_1=m(x-x_1)$$
$$y-(-5)=-3(x-1)$$
$$y+5=-3(x-1)$$

Slope-Intercept Form

$$y+5=-3(x-1)$$
$$y+5=-3x+3$$
$$y=-3x-2$$

In function notation, the equation of the line is $f(x)=-3x-2$.

9. $y=|x|+2$

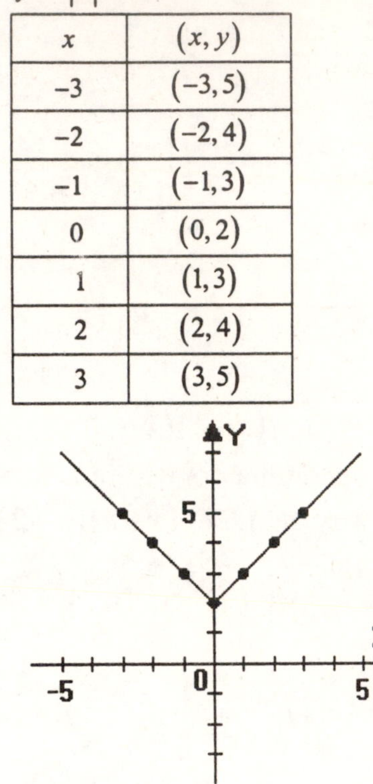

x	(x,y)
–3	(–3, 5)
–2	(–2, 4)
–1	(–1, 3)
0	(0, 2)
1	(1, 3)
2	(2, 4)
3	(3, 5)

10. $y\geq 2x-1$
$x\geq 1$

First consider $y\geq 2x-1$. Replace the inequality symbol with an equal sign and we have $y=2x-1$. Since the equation is in slope-intercept form, we know that the y-intercept is –1 and the slope is 2.

Now, use the origin, $(0,0)$, as a test point.

$$y\geq 2x-1$$
$$0\geq 2(0)-1$$
$$0\geq -1$$

This is a true statement. This means that the point $(0,0)$ will fall in the

shaded half-plane.
Next consider $x \geq 1$. Replace the inequality symbol with an equal sign and we have $x = 1$. We know that equations of the form $x = b$ are vertical lines through the point $(b, 0)$. Since the inequality is greater than or equal to, we know that the shading will extend from $x = 1$ toward ∞.
Next, graph each of the inequalities. The solution to the system is the intersection of the shaded half-planes.

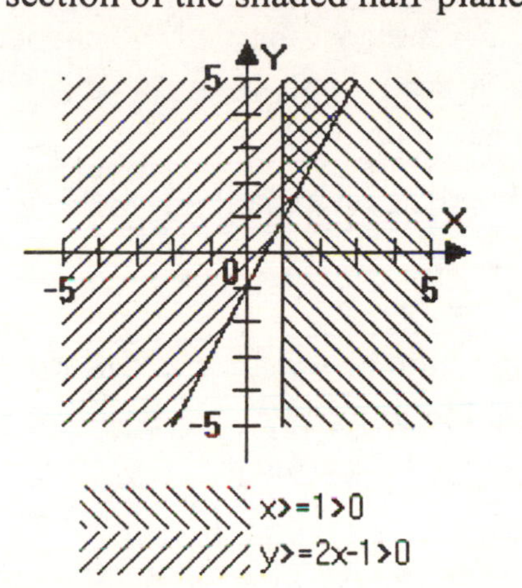

11. $2x - y < 4$

First graph the line, $2x - y = 4$. Solve for y to obtain slope-intercept form.

$$2x - y = 4$$
$$-y = -2x + 4$$
$$y = 2x - 4$$

slope = 2
y–intercept = –4
Now, use the origin, $(0,0)$, as a test point.

$$2x - y < 4$$
$$2(0) - 0 < 4$$
$$0 - 0 < 4$$
$$0 < 4$$

This is a true statement. This means that the point $(0,0)$ will fall in the shaded half-plane.
Next, graph the inequality.

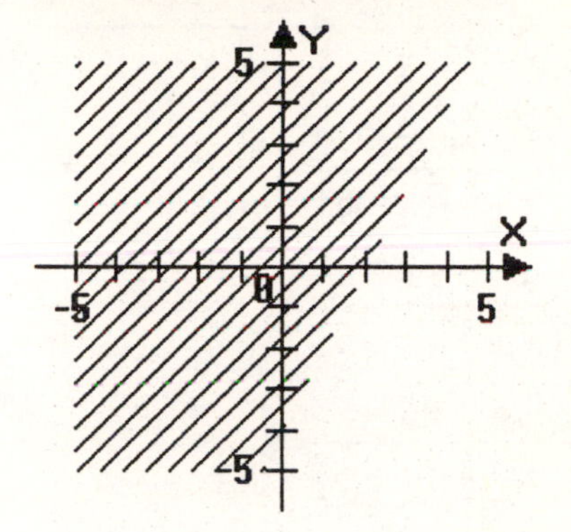

12. $\left[(x+2)+3y\right]\left[(x+2)-3y\right]$

$$= \left[(x+2)^2 - (3y)^2\right]$$
$$= \left[(x^2+4x+4) - 9y^2\right]$$
$$= x^2 + 4x + 4 - 9y^2$$

13. $\dfrac{2x^2+x-1}{2x^2-9x+4} \div \dfrac{6x+15}{3x^2-12x}$

$$= \frac{2x^2+x-1}{2x^2-9x+4} \cdot \frac{3x^2-12x}{6x+15}$$
$$= \frac{\cancel{(2x-1)}(x+1)}{\cancel{(2x-1)}\cancel{(x-4)}} \cdot \frac{\cancel{3}x\cancel{(x-4)}}{\cancel{3}(2x+5)}$$
$$= \frac{x(x+1)}{2x+5}$$

14. $\dfrac{3x}{x^2-9x+20} - \dfrac{5}{2x-8}$

$$= \frac{3x}{(x-4)(x-5)} - \frac{5}{2(x-4)}$$

The LCD is $2(x-4)(x-5)$.

$$\frac{3x}{(x-4)(x-5)}-\frac{5}{2(x-4)}$$
$$=\frac{2\cdot 3x}{2(x-4)(x-5)}-\frac{5(x-5)}{2(x-4)(x-5)}$$
$$=\frac{6x-5(x-5)}{2(x-4)(x-5)}$$
$$=\frac{6x-5x+25}{2(x-4)(x-5)}$$
$$=\frac{x+25}{2(x-4)(x-5)}$$

15.
$$\begin{array}{r} 3x+4 \\ x+2\overline{\big)3x^2+10x+10} \\ \underline{3x^2+\ 6x} \\ 4x+10 \\ \underline{4x+\ 8} \\ 2 \end{array}$$

$$\frac{3x^2+10x+10}{x+2}=3x+4+\frac{2}{x+2}$$

16. $xy-6x+2y-12$
$=x(y-6)+2(y-6)$
$=(y-6)(x+2)$

17. $24x^3y+16x^2y-30xy$
$=2xy(12x^2+8x-15)$
$=2xy(2x+3)(6x-5)$

18. $s(t)=-16t^2+48t+64$
$0=-16t^2+48t+64$
$0=-16(t^2-3t-4)$
$0=-16(t-4)(t+1)$

Apply the zero product principle.
$-16(t-4)=0 \qquad t+1=0$
$t-4=0 \qquad t=-1$
$t=4$

The solutions are –1 and 4. Disregard –1 because we can't have a negative time measurement. The ball will hit the ground in 4 seconds.

19. Let x = the cost for the basic cable service
Let y = the cost for a movie channel
$x+\ y=35$
$x+2y=45$
Solve the first equation for x.
$x+y=35$
$x=35-y$
Substitute $35-y$ for x in the second equation to find y.
$x+2y=45$
$(35-y)+2y=45$
$35-y+2y=45$
$35+y=45$
$y=10$
Back-substitute 10 for y to find x.
$x=35-10$
$x=25$
The cost of basic cable is \$25 and the cost for each movie channel is \$10.

20.

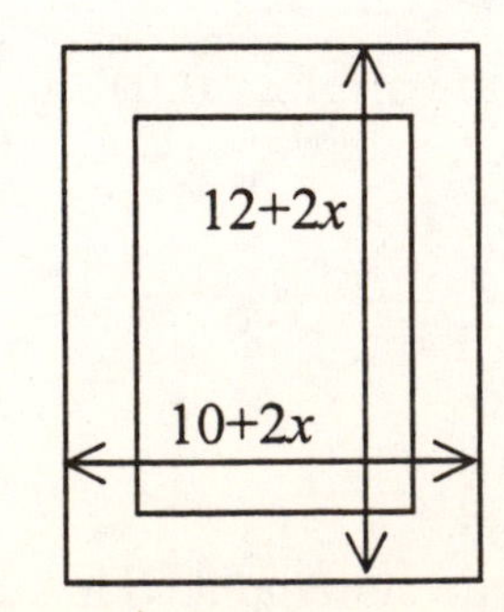

$$A = lw$$
$$168 = (12 + 2x)(10 + 2x)$$
$$168 = 120 + 44x + 4x^2$$
$$0 = -48 + 44x + 4x^2$$
$$0 = 4x^2 + 44x - 48$$
$$0 = 4(x^2 + 11x - 12)$$
$$0 = 4(x + 12)(x - 1)$$

Apply the zero product principle.

$4(x+12) = 0$ $\quad$ $x - 1 = 0$

$x + 12 = 0$ $\quad$ $x = 1$

$x = -12$

The solutions are –12 and 1. We disregard –12 because we cannot have a negative length measurement. The width of the rock border is 1 foot.

Chapter 7

Check Points 7.1

1. **a.** $\sqrt{64} = 8$

b. $-\sqrt{49} = -7$

c. $\sqrt{\frac{16}{25}} = \frac{4}{5}$

d. $\sqrt{0.0081} = 0.09$

e. $\sqrt{9+16} = \sqrt{25} = 5$

f. $\sqrt{9} + \sqrt{16} = 3 + 4 = 7$

2. **a.** $f(3) = \sqrt{12(3) - 20}$
$= \sqrt{36 - 20}$
$= \sqrt{16} = 4$

b. $g(-5) = -\sqrt{9 - 3(-5)}$
$= -\sqrt{9 + 15}$
$= -\sqrt{24} \approx -4.90$

3. $f(x) = \sqrt{9x - 27}$
The domain of a square root function is the set of real numbers for which the radicand is nonnegative. Set the radicand greater than or equal to zero and solve.
$9x - 27 \geq 0$
$9x \geq 27$
$x \geq 3$
The domain of f is $\{x | x \geq 3\}$ or $[3, \infty)$.

4. Since 1997 is 4 years after 1994, find $f(4)$.
$f(4) = 6.75\sqrt{4} + 12$
$= 6.75(2) + 12$
$= 13.5 + 12 = 25.5$
The amount of new student loans is \$25.5 billion. The model describes the data very well.

5. **a.** $\sqrt{(-7)^2} = |-7| = 7$

b. $\sqrt{(x+8)^2} = |x + 8|$

c. $\sqrt{49x^{10}} = \sqrt{(7x^5)^2} = |7x^5|$

d. $\sqrt{x^2 - 6x + 9} = \sqrt{(x-3)^2}$
$= |x - 3|$

6. **a.** $f(33) = \sqrt[3]{33 - 6} = \sqrt[3]{27} = 3$

b. $g(-5) = \sqrt[3]{2(-5) + 2}$
$= \sqrt[3]{-10 + 2}$
$= \sqrt[3]{-8} = -2$

7. $\sqrt[3]{-27x^3} = \sqrt[3]{(-3x)^3} = -3x$

8. **a.** $\sqrt[4]{16} = \sqrt[4]{2^4} = 2$

b. $-\sqrt[4]{16} = -\sqrt[4]{2^4} = -2$

c. $\sqrt[4]{-16}$ is not a real number. The index is even and the radicand is negative.

d. $\sqrt[5]{-1} = \sqrt[5]{(-1)^5} = -1$

9 **a.** $\sqrt[4]{(x+6)^4} = \sqrt[4]{(x+6)^4} = |x+6|$

b. $\sqrt[5]{(3x-2)^5} = \sqrt[5]{(3x-2)^5} = 3x-2$

c. $\sqrt[6]{(-8)^6} = |-8| = 8$

Problem Set 7.1

Practice Exercises

1. $\sqrt{36} = 6$ because $6^2 = 36$

3. $-\sqrt{36} = -6$ because $6^2 = 36$

5. $\sqrt{-36}$
Not a real number

7. $\sqrt{\frac{1}{25}} = \frac{1}{5}$ because $\left(\frac{1}{5}\right)^2 = \frac{1}{25}$

9. $-\sqrt{0.04} = -0.2$ because $(0.2)^2 = 0.04$

11. $\sqrt{25-16} = \sqrt{9} = 3$

13. $\sqrt{25} - \sqrt{16} = 5 - 4 = 1$

15. $\sqrt{16-25} = \sqrt{-9}$
Not a real number

17. $f(18) = \sqrt{18-2} = \sqrt{16} = 4$
$f(3) = \sqrt{3-2} = \sqrt{1} = 1$
$f(2) = \sqrt{2-2} = \sqrt{0} = 0$
$f(-2) = \sqrt{-2-2} = \sqrt{-4}$
Not a real number

19.
$$g(11) = -\sqrt{2(11)+3} = -\sqrt{22+3} = -\sqrt{25} = -5$$
$$g(1) = -\sqrt{2(1)+3} = -\sqrt{2+3} = -\sqrt{5} \approx -2.24$$
$$g(-1) = -\sqrt{2(-1)+3} = -\sqrt{-2+3} = -\sqrt{1} = -1$$
$$g(-2) = -\sqrt{2(-2)+3} = -\sqrt{-4+3} = -\sqrt{-1}$$
Not a real number

21.
$h(5) = \sqrt{(5-1)^2} = \sqrt{(4)^2} = |4| = 4$
$h(3) = \sqrt{(3-1)^2} = \sqrt{(2)^2} = |2| = 2$
$h(0) = \sqrt{(0-1)^2} = \sqrt{(-1)^2} = |-1| = 1$
$h(-5) = \sqrt{(-5-1)^2} = \sqrt{(-6)^2} = |-6| = 6$

23. To find the domain, set the radicand greater than or equal to zero and solve.

$x-3\geq 0$

$x\geq 3$

The domain of f is $\{x|x\geq 3\}$ or $[3,\infty)$. This corresponds to graph (c).

25. To find the domain, set the radicand greater than or equal to zero and solve.

$3x+15\geq 0$

$3x\geq -15$

$x\geq -5$

The domain of f is $\{x|x\geq -5\}$ or $[-5,\infty)$. This corresponds to graph (d).

27. To find the domain, set the radicand greater than or equal to zero and solve.

$6-2x\geq 0$

$-2x\geq -6$

$x\leq 3$

The domain of f is $\{x|x\leq 3\}$ or $(-\infty,3]$. This corresponds to graph (e).

29. $\sqrt{5^2}=5$

31. $\sqrt{(-4)^2}=|-4|=4$

33. $\sqrt{(x-1)^2}=|x-1|$

35. $\sqrt{36x^4}=\sqrt{(6x^2)^2}=|6x^2|=6x^2$

37. $-\sqrt{100x^6}=-\sqrt{(10x^3)^2}$

$=-|10x^3|=-10|x^3|$

39. $\sqrt{x^2+12x+36}=\sqrt{(x+6)^2}=|x+6|$

41. $-\sqrt{x^2-8x+16}=-\sqrt{(x-4)^2}$

$=-|x-4|$

43. $\sqrt[3]{27}=3$ because $3^3=27$

45. $\sqrt[3]{-27}=-3$ because $(-3)^3=-27$

47. $\sqrt[3]{\frac{1}{125}}=\frac{1}{5}$ because $\left(\frac{1}{5}\right)^3=\frac{1}{125}$

49. $f(28)=\sqrt[3]{28-1}=\sqrt[3]{27}=3$

$f(9)=\sqrt[3]{9-1}=\sqrt[3]{8}=2$

$f(0)=\sqrt[3]{0-1}=\sqrt[3]{-1}=-1$

$f(-63)=\sqrt[3]{-63-1}=\sqrt[3]{-64}=-4$

51. $g(2)=-\sqrt[3]{8(2)-8}=-\sqrt[3]{16-8}$

$=-\sqrt[3]{8}=-2$

$g(1)=-\sqrt[3]{8(1)-8}=-\sqrt[3]{8-8}$

$=-\sqrt[3]{0}=-0=0$

$g(0)=-\sqrt[3]{8(0)-8}=-\sqrt[3]{-8}$

$=-(-2)=2$

53. $\sqrt[4]{1} = 1$ because $1^4 = 1$

55. $\sqrt[4]{16} = 2$ because $2^4 = 16$

57. $-\sqrt[4]{16} = -2$ because $2^4 = 16$

59. $\sqrt[4]{-16}$
Not a real number

61. $\sqrt[5]{-1} = -1$ because $(-1)^5 = -1$

63. $\sqrt[6]{-1}$
Not a real number

65. $-\sqrt[4]{256} = -4$ because $4^4 = 256$

67. $\sqrt[6]{64} = 2$ because $2^6 = 64$

69. $-\sqrt[5]{32} = -2$ because $2^5 = 32$

71. $\sqrt[3]{x^3} = x$

73. $\sqrt[4]{y^4} = |y|$

75. $\sqrt[3]{-8x^3} = -2x$

77. $\sqrt[3]{(-5)^3} = -5$

79. $\sqrt[4]{(-5)^4} = |-5| = 5$

81. $\sqrt[4]{(x+3)^4} = |x+3|$

83. $\sqrt[5]{-32(x-1)^5} = -2(x-1)$

Application Exercises

85. $f(48) = 2.9\sqrt{48} + 20.1$
$= 2.9(6.9) + 20.1$
$= 20.1 + 20.1 \approx 40.2$
The model predicts the median height of boys who are 48 months old to be 40.1 inches. The model predicts the median height very well. According to the table, the median height is 40.8.

87. $f(245) = \sqrt{20(245)} = \sqrt{4900} = 70$
The officer should not believe the motorist. The model predicts that the motorist's speed was 70 miles per hour. This is well above the 50 miles per hour speed limit.

Writing in Mathematics

89. Answers will vary.

91. Answers will vary.

93. Answers will vary.

95. Answers will vary.

97. Answers will vary.

99. Answers will vary.

Technology Exercises

101. $y = \sqrt{x}$
$y = \sqrt{x} + 4$
$y = \sqrt{x} - 3$

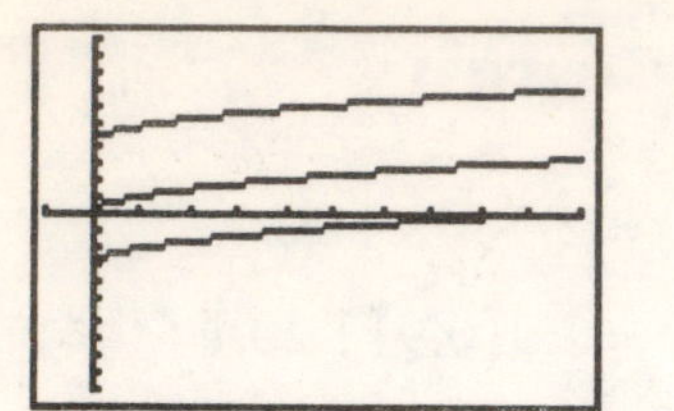

The graphs have the same shape but differ in their orientation on the y-axis.

103. $y_1 = \sqrt{x^2}$

$y_2 = -x$

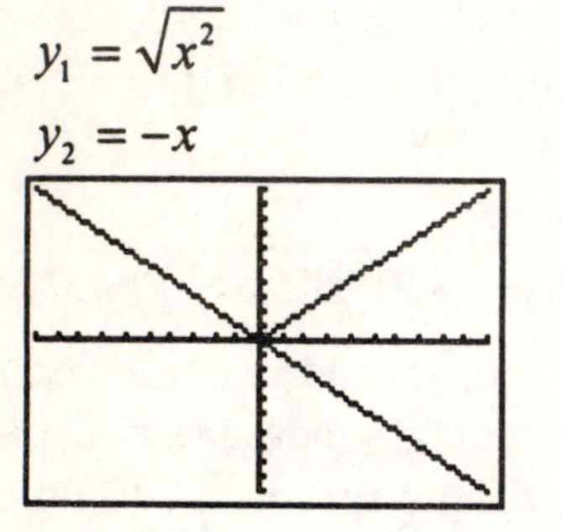

a. $\sqrt{x^2} = -x$ for $\{x|x \le 0\}$.

b. $\sqrt{x^2} \ne -x$ for $\{x|x > 0\}$.

Critical Thinking Exercises

105. Answers will vary. One example is $f(x) = \sqrt{5-x}$.

107. $\sqrt{(2x+3)^{10}} = \sqrt{\left((2x+3)^5\right)^2}$

$= \left|(2x+3)^5\right|$

109. $h(x) = \sqrt{x+3}$

x	$h(x)$
−3	0
−2	1
1	2
6	3

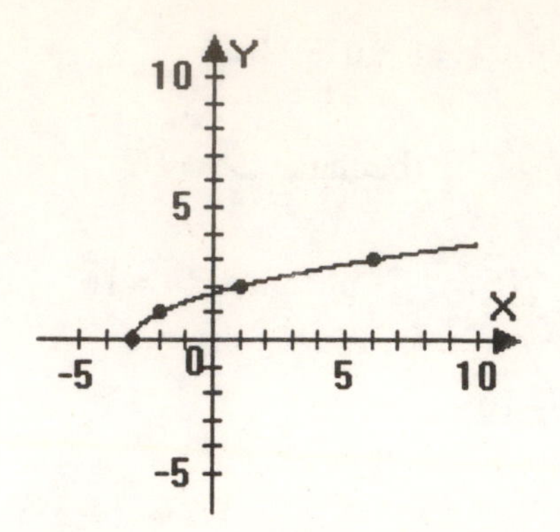

The graph of h is the graph of f shifted three units to the left.

Review Exercises

110. $3x - 2[x - 3(x+5)]$

$= 3x - 2[x - 3x - 15]$

$= 3x - 2[-2x - 15]$

$= 7x + 30$

111. $(-3x^{-4}y^3)^{-2} = (-3)^{-2}x^8y^{-6}$

$= \dfrac{x^8}{(-3)^2 y^6} = \dfrac{x^8}{9y^6}$

112. $|3x-4| > 11$

$3x - 4 < -11$ or $3x - 4 > 11$

$3x < -7$ $\quad$ $3x > 15$

$x < -\dfrac{7}{3}$ $\quad$ $x > 5$

The solution set is $\left\{x \middle| x < -\dfrac{7}{3}\right.$ or $x > 5\}$ or $\left(-\infty, -\dfrac{7}{3}\right) \cup (5, \infty)$.

Check Points 7.2

1. a. $25^{\frac{1}{2}} = \sqrt{25} = 5$

   b. $(-8)^{\frac{1}{3}} = \sqrt[3]{-8} = -2$

   c. $(5xy^2)^{\frac{1}{4}} = \sqrt[4]{5xy^2}$

2. a. $\sqrt[4]{5xy} = (5xy)^{\frac{1}{4}}$

   b. $\sqrt[5]{\frac{a^3b}{2}} = \left(\frac{a^3b}{2}\right)^{\frac{1}{5}}$

3. a. $27^{\frac{4}{3}} = \left(\sqrt[3]{27}\right)^4 = (3)^4 = 81$

   b. $4^{\frac{3}{2}} = \left(\sqrt{4}\right)^3 = (2)^3 = 8$

   c. $-16^{\frac{3}{4}} = -\left(\sqrt[4]{16}\right)^3 = -(2)^3 = -8$

4. a. $\sqrt[3]{6^4} = 6^{\frac{4}{3}}$

   b. $\left(\sqrt[5]{2xy}\right)^7 = (2xy)^{\frac{7}{5}}$

5. a. $25^{-\frac{1}{2}} = \frac{1}{25^{\frac{1}{2}}} = \frac{1}{\sqrt{25}} = \frac{1}{5}$

   b. $64^{-\frac{1}{3}} = \frac{1}{64^{\frac{1}{3}}} = \frac{1}{\sqrt[3]{64}} = \frac{1}{4}$

   c. $32^{-\frac{4}{5}} = \frac{1}{32^{\frac{4}{5}}} = \frac{1}{\left(\sqrt[5]{32}\right)^4}$
   $= \frac{1}{2^4} = \frac{1}{16}$

   d. $(3xy)^{-\frac{5}{9}} = \frac{1}{(3xy)^{\frac{5}{9}}}$

6. Since 1997 is 16 years after 1981, find $S(16)$.

   $S(16) = 63.25(16)^{\frac{1}{4}}$
   $= 63.25\sqrt[4]{16}$
   $= 63.25(2)$
   $= 126.5$

   In 1997, the average sale price was \$126.5 thousand or \$126,500.

7. a. $7^{\frac{1}{2}} \cdot 7^{\frac{1}{3}} = 7^{\frac{1}{2}+\frac{1}{3}} = 7^{\frac{3}{6}+\frac{2}{6}} = 7^{\frac{5}{6}}$

   b. $\frac{50x^{\frac{1}{3}}}{10x^{\frac{4}{3}}} = \frac{50}{10}x^{\frac{1}{3}-\frac{4}{3}} = 5x^{-\frac{3}{3}}$
   $= 5x^{-1} = \frac{5}{x}$

   c. $\left(9.1^{\frac{2}{5}}\right)^{\frac{3}{4}} = 9.1^{\frac{2}{5}\cdot\frac{3}{4}} = 9.1^{\frac{6}{20}} = 9.1^{\frac{3}{10}}$

d. $\left(x^{-\frac{3}{5}}y^{\frac{1}{4}}\right)^{\frac{1}{3}} = x^{-\frac{3}{5}\cdot\frac{1}{3}}y^{\frac{1}{4}\cdot\frac{1}{3}}$

$= x^{-\frac{1}{5}}y^{\frac{1}{12}} = \dfrac{y^{\frac{1}{12}}}{x^{\frac{1}{5}}}$

8. a. $\sqrt[6]{x^3} = x^{\frac{3}{6}} = x^{\frac{1}{2}} = \sqrt{x}$

b. $\sqrt[3]{8a^{12}} = \left(8a^{12}\right)^{\frac{1}{3}} = (8)^{\frac{1}{3}}\left(a^{12}\right)^{\frac{1}{3}}$

$= 2a^{\frac{12}{3}} = 2a^4$

c. $\sqrt[8]{x^4y^2} = \left(x^4y^2\right)^{\frac{1}{8}}$

$= \left(x^4\right)^{\frac{1}{8}}\left(y^2\right)^{\frac{1}{8}} = x^{\frac{4}{8}}y^{\frac{2}{8}}$

$= x^{\frac{2}{4}}y^{\frac{1}{4}} = \left(x^2y\right)^{\frac{1}{4}}$

$= \sqrt[4]{x^2y}$

d. $\dfrac{\sqrt{x}}{\sqrt[3]{x}} = \dfrac{x^{\frac{1}{2}}}{x^{\frac{1}{3}}} = x^{\frac{1}{2}-\frac{1}{3}} = x^{\frac{3}{6}-\frac{2}{6}}$

$= x^{\frac{1}{6}} = \sqrt[6]{x}$

e. $\sqrt{\sqrt[3]{x}} = \sqrt{x^{\frac{1}{3}}} = \left(x^{\frac{1}{3}}\right)^{\frac{1}{2}}$

$= x^{\frac{1}{3}\cdot\frac{1}{2}} = x^{\frac{1}{6}}$

$= \sqrt[6]{x}$

Problem Set 7.2

Practice Exercises

1. $49^{1/2} = \sqrt{49} = 7$

3. $(-27)^{1/3} = \sqrt[3]{-27} = -3$

5. $-16^{1/4} = -\sqrt[4]{16} = -2$

7. $(xy)^{1/3} = \sqrt[3]{xy}$

9. $\left(2xy^3\right)^{1/5} = \sqrt[5]{2xy^3}$

11. $81^{3/2} = \left(\sqrt{81}\right)^3 = 9^3 = 729$

13. $125^{2/3} = \left(\sqrt[3]{125}\right)^2 = 5^2 = 25$

15. $(-32)^{3/5} = \left(\sqrt[5]{-32}\right)^3 = (-2)^3 = -8$

17. $27^{2/3} + 16^{3/4} = \left(\sqrt[3]{27}\right)^2 + \left(\sqrt[4]{16}\right)^3$

$= 3^2 + 2^3$

$= 9 + 8 = 17$

19. $(xy)^{4/7} = \left(\sqrt[7]{xy}\right)^4$ or $\sqrt[7]{(xy)^4}$

21. $\sqrt{7} = 7^{1/2}$

23. $\sqrt[3]{5} = 5^{1/3}$

25. $\sqrt[5]{11x} = (11x)^{1/5}$

27. $\sqrt{x^3} = x^{3/2}$

29. $\sqrt[5]{x^3} = x^{3/5}$

31. $\sqrt[5]{x^2 y} = \left(x^2 y\right)^{1/5}$

33. $\left(\sqrt{19xy}\right)^3 = (19xy)^{3/2}$

35. $\left(\sqrt[6]{7xy^2}\right)^5 = \left(7xy^2\right)^{5/6}$

37. $2x\sqrt[3]{y^2} = 2xy^{2/3}$

39. $49^{-1/2} = \frac{1}{49^{1/2}} = \frac{1}{\sqrt{49}} = \frac{1}{7}$

41. $27^{-1/3} = \frac{1}{27^{1/3}} = \frac{1}{\sqrt[3]{27}} = \frac{1}{3}$

43. $16^{-3/4} = \frac{1}{16^{3/4}} = \frac{1}{\left(\sqrt[4]{16}\right)^3} = \frac{1}{2^3} = \frac{1}{8}$

45. $8^{-2/3} = \frac{1}{8^{2/3}} = \frac{1}{\left(\sqrt[3]{8}\right)^2} = \frac{1}{2^2} = \frac{1}{4}$

47. $\left(\frac{8}{27}\right)^{-1/3} = \left(\frac{27}{8}\right)^{1/3} = \sqrt[3]{\frac{27}{8}} = \frac{3}{2}$

49. $(-64)^{-2/3} = \frac{1}{(-64)^{2/3}} = \frac{1}{\left(\sqrt[3]{-64}\right)^2}$

$= \frac{1}{(-4)^2} = \frac{1}{16}$

51. $(2xy)^{-7/10} = \frac{1}{(2xy)^{7/10}}$

$= \frac{1}{\sqrt[10]{(2xy)^7}}$ or $\frac{1}{\left(\sqrt[10]{2xy}\right)^7}$

53. $5xz^{-1/3} = \frac{5xz^{-1/3}}{1} = \frac{5x}{z^{1/3}}$

55. $3^{3/4} \cdot 3^{1/4} = 3^{(3/4)+(1/4)}$

$= 3^{4/4} = 3^1 = 3$

57. $\frac{16^{3/4}}{16^{1/4}} = 16^{(3/4)-(1/4)} = 16^{2/4}$

$= 16^{1/2} = \sqrt{16} = 4$

59. $x^{1/2} \cdot x^{1/3} = x^{(1/2)+(1/3)}$

$= x^{(3/6)+(2/6)} = x^{5/6}$

61. $\frac{x^{4/5}}{x^{1/5}} = x^{(4/5)-(1/5)} = x^{3/5}$

63. $\frac{x^{1/3}}{x^{3/4}} = x^{(1/3)-(3/4)} = x^{(4/12)-(9/12)}$

$= x^{-5/12} = \frac{1}{x^{5/12}}$

65. $\left(5^{\frac{2}{3}}\right)^3 = 5^{\frac{2}{3}\cdot 3} = 5^2 = 25$

67. $\left(y^{-2/3}\right)^{1/4} = y^{(-2/3)\cdot(1/4)} = y^{-2/12}$

$= y^{-1/6} = \frac{1}{y^{1/6}}$

69. $\left(2x^{1/5}\right)^5 = 2^5 x^{(1/5)\cdot 5} = 32x^1 = 32x$

71. $\left(25x^4y^6\right)^{1/2} = 25^{1/2}\left(x^4\right)^{1/2}\left(y^6\right)^{1/2}$
$= \sqrt{25}x^{4(1/2)}y^{6(1/2)}$
$= 5x^2y^3$

73. $\left(x^{1/2}y^{-3/5}\right)^{1/2} = \left(\dfrac{x^{1/2}y^{-3/5}}{1}\right)^{1/2}$
$= \left(\dfrac{x^{1/2}}{y^{3/5}}\right)^{1/2}$
$= \dfrac{x^{(1/2)\cdot(1/2)}}{y^{(3/5)\cdot(1/2)}} = \dfrac{x^{1/4}}{y^{3/10}}$

75. $\dfrac{3^{1/2}\cdot 3^{3/4}}{3^{1/4}} = 3^{(1/2)+(3/4)-(1/4)}$
$= 3^{(2/4)+(3/4)-(1/4)}$
$= 3^{4/4} = 3^1 = 3$

77. $\dfrac{\left(3y^{1/4}\right)^3}{y^{1/12}} = \dfrac{3^3y^{(1/4)\cdot 3}}{y^{1/12}} = \dfrac{27y^{3/4}}{y^{1/12}}$
$= 27y^{(3/4)-(1/12)}$
$= 27y^{(9/12)-(1/12)}$
$= 27y^{8/12} = 27y^{2/3}$

79. $\sqrt[8]{x^2} = x^{2/8} = x^{1/4} = \sqrt[4]{x}$

81. $\sqrt[3]{8a^6} = 8^{1/3}a^{6/3} = 2a^2$

83. $\sqrt[5]{x^{10}y^{15}} = x^{10/5}y^{15/5} = x^2y^3$

85. $\left(\sqrt[3]{xy}\right)^{18} = (xy)^{18/3} = (xy)^6 = x^6y^6$

87. $\sqrt[10]{(3y)^2} = (3y)^{2/10} = (3y)^{1/5} = \sqrt[5]{3y}$

89. $\left(\sqrt[6]{2a}\right)^4 = (2a)^{4/6} = (2a)^{2/3}$
$= \left(4a^2\right)^{1/3} = \sqrt[3]{4a^2}$

91. $\sqrt[9]{x^6y^3} = x^{6/9}y^{3/9}$
$= x^{2/3}y^{1/3} = \sqrt[3]{x^2y}$

93. $\sqrt{2}\cdot\sqrt[3]{2} = 2^{1/2}\cdot 2^{1/3} = 2^{(1/2)+(1/3)}$
$= 2^{(3/6)+(2/6)} = 2^{5/6}$
$= \sqrt[6]{2^5}$ or $\sqrt[6]{32}$

95. $\dfrac{\sqrt[4]{x}}{\sqrt[5]{x}} = \dfrac{x^{1/4}}{x^{1/5}} = x^{(1/4)-(1/5)}$
$= x^{(5/20)-(4/20)}$
$= x^{1/20} = \sqrt[20]{x}$

97. $\dfrac{\sqrt[3]{y^2}}{\sqrt[6]{y}} = \dfrac{y^{2/3}}{y^{1/6}} = y^{(2/3)-(1/6)}$
$= y^{(4/6)-(1/6)} = y^{3/6}$
$= y^{1/2} = \sqrt{y}$

99. $\sqrt[4]{\sqrt{x}} = \sqrt[4]{x^{1/2}} = \left(x^{1/2}\right)^{1/4}$
$= x^{(1/2)\cdot(1/4)} = x^{1/8}$
$= \sqrt[8]{x}$

101. $\sqrt{\sqrt{x^2 y}} = \sqrt{\left(x^2 y\right)^{1/2}} = \left(\left(x^2 y\right)^{1/2}\right)^{1/2}$
$= \left(x^2 y\right)^{(1/2)\cdot(1/2)} = \left(x^2 y\right)^{1/4}$
$= \sqrt[4]{x^2 y}$

103. $\sqrt[4]{\sqrt[3]{2x}} = \sqrt[4]{(2x)^{1/3}} = \left((2x)^{1/3}\right)^{1/4}$
$= (2x)^{(1/3)\cdot(1/4)} = (2x)^{1/12}$
$= \sqrt[12]{2x}$

105. $\left(\sqrt[4]{x^3 y^5}\right)^{12} = \left(\left(x^3 y^5\right)^{1/4}\right)^{12}$
$= \left(x^3 y^5\right)^{(1/4)\cdot 12}$
$= \left(x^3 y^5\right)^{12/4} = \left(x^3 y^5\right)^3$
$= x^{3\cdot 3} y^{5\cdot 3} = x^9 y^{15}$

Application Exercises

107. $f(8) = 29(8)^{1/3}$
$= 29\sqrt[3]{8}$
$= 29(2) = 58$
There are 58 plant species on an 8 square mile island.

109. $f(9) = 0.07(9)^{3/2}$
$= 0.07\left(\sqrt{9}\right)^3$
$= 0.07(3)^3$
$= 0.07(27) \approx 1.9$
The duration is approximately 1.9 hours.

111. $P(8) = \dfrac{73(8)^{1/3} - 28(8)^{2/3}}{8}$
$= \dfrac{73\sqrt[3]{8} - 28\left(\sqrt[3]{8}\right)^2}{8}$
$= \dfrac{73(2) - 28(2)^2}{8}$
$= \dfrac{146 - 28(4)}{8} = \dfrac{146 - 112}{8}$
$= \dfrac{34}{8} = 4.25$
4.25% of employees tested positive.

113. $f(80) = 70(80)^{3/4}$
$= 70(26.75)$
$= 1872.5$
A person weighing 80 kilograms needs approximately 1872 calories to maintain life.

115. $L + 1.25\sqrt{S} - 9.8\sqrt[3]{D} \le 16.296$

a. $L + 1.25S^{1/2} - 9.8D^{1/3} \le 16.296$

b.

$$L+1.25S^{1/2}-9.8D^{1/3}\le 16.296$$
$$20.85+1.25(276.4)^{1/2}-9.8(18.55)^{1/3}\le 16.296$$
$$20.85+1.25\sqrt{276.4}-9.8\sqrt[3]{18.55}\le 16.296$$
$$20.85+1.25(16.625)-9.8(2.647)\le 16.296$$
$$20.85+20.781-25.941\le 16.296$$
$$15.69\le 16.296$$

The yacht is eligible to enter the America's Cup.

Writing in Mathematics

117. Answers will vary.

119. Answers will vary.

121. Answers will vary.

123. Answers will vary.

125. Answers will vary.

Technology Exercises

127. For example, consider Exercise 45.

```
8^(-2/3)▸Frac
            1/4
```

129. a. $y=x^{2/6}$

b. $y=x^{1/3}$

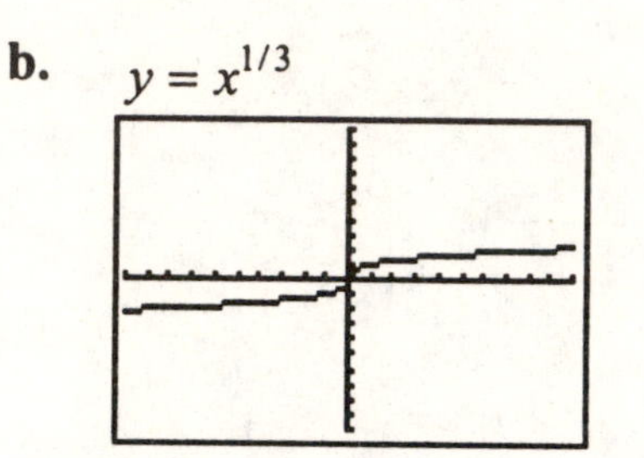

c. The graphs are not the same. In part **a.**, the domain is $\{x|x\ge 0\}$. In part **b.**, the domain is $\{x|x$ is a real number$\}$. Note that we are taking a sixth root in part **a.** We cannot take the sixth root of a negative number; thus, $\left(x^{1/6}\right)^2$ is undefined for negative numbers.

Critical Thinking Exercises

131.

$$2^{5/2}\cdot 2^{3/4}\div 2^{1/4}=2^{(5/2)+(3/4)-(1/4)}$$
$$=2^{(10/4)+(3/4)-(1/4)}$$
$$=2^{12/4}=2^3=8$$

The son is 8 years old.

133.
$$\left[3+\left(27^{2/3}+32^{2/5}\right)\right]^{3/2}-9^{1/2}=\left[3+\left(\left(\sqrt[3]{27}\right)^2+\left(\sqrt[5]{32}\right)^2\right)\right]^{3/2}-\sqrt{9}$$
$$=\left[3+\left(3^2+2^2\right)\right]^{3/2}-3=\left[3+(9+4)\right]^{3/2}-3$$
$$=\left[3+(13)\right]^{3/2}-3=\left[16\right]^{3/2}-3$$
$$=\left(\sqrt{16}\right)^3-3=(4)^3-3$$
$$=64-3=61$$

Review Exercises

135. First, find the slope.

$$m=\frac{y_2-y_1}{x_2-x_1}=\frac{3-1}{4-5}=\frac{2}{-1}=-2$$

Use the slope and one of the points to write the equation of the line in point-slope form.

$$y-y_1=m(x-x_1)$$
$$y-3=-2(x-4)$$

Solve for y to write the equation in slope–intercept form.

$$y-3=-2(x-4)$$
$$y-3=-2x+8$$
$$y=-2x+11 \text{ or } f(x)=-2x+11$$

136.
$$y\le-\frac{3}{2}x+3$$

First graph the line, $y=-\frac{3}{2}x+3$.

Since the line is in slope–intercept form we can identify the slope and y–intercept.

slope $=-\frac{3}{2}$ $\quad$ y–intercept $=3$

Now, use the origin, $(0,0)$, as a test point.

$$0\le-\frac{3}{2}(0)+3$$
$$0\le3$$

This is a true statement. This means that the point $(0,0)$ will fall in the shaded half-plane.

Next, graph the inequality.

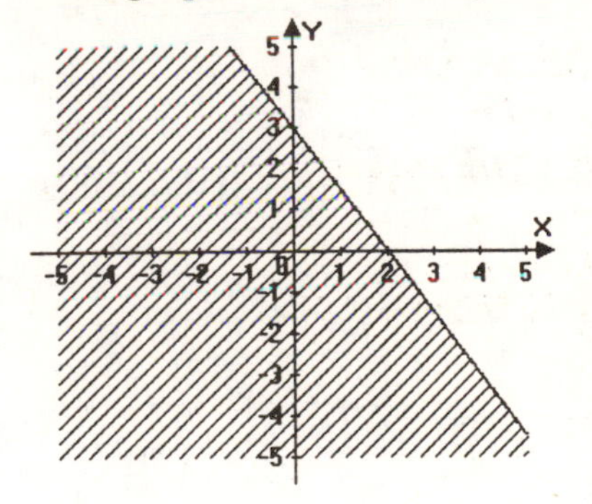

137.
$$5x-3y=3$$
$$7x+y=25$$
$$D=\begin{vmatrix}5 & -3\\7 & 1\end{vmatrix}=5(1)-7(-3)$$
$$=5+21=26$$
$$D_x=\begin{vmatrix}3 & -3\\25 & 1\end{vmatrix}=3(1)-25(-3)$$
$$=3+75=78$$

$$D_y = \begin{vmatrix} 5 & 3 \\ 7 & 25 \end{vmatrix} = 5(25) - 7(3)$$
$$= 125 - 21 = 104$$
$$x = \frac{D_x}{D} = \frac{78}{26} = 3$$
$$y = \frac{D_y}{D} = \frac{104}{26} = 4$$

The solution is $(3,4)$ and the solution set is $\{(3,4)\}$.

Check Points 7.3

1. **a.** $\sqrt{5} \cdot \sqrt{11} = \sqrt{5 \cdot 11} = \sqrt{55}$

b. $\sqrt{x+4} \cdot \sqrt{x-4}$
$$= \sqrt{(x+4)(x-4)}$$
$$= \sqrt{x^2 - 16}$$

c. $\sqrt[3]{6} \cdot \sqrt[3]{10} = \sqrt[3]{6 \cdot 10} = \sqrt[3]{60}$

d. $\sqrt[7]{2x} \cdot \sqrt[7]{6x^3} = \sqrt[7]{2x \cdot 6x^3} = \sqrt[7]{12x^4}$

2. **a.** $\sqrt{80} = \sqrt{16 \cdot 5} = \sqrt{16} \cdot \sqrt{5} = 4\sqrt{5}$

b. $\sqrt[3]{40} = \sqrt[3]{8 \cdot 5} = \sqrt[3]{8} \cdot \sqrt[3]{5} = 2\sqrt[3]{5}$

c. $\sqrt[4]{32} = \sqrt[4]{16 \cdot 2} = \sqrt[4]{16} \cdot \sqrt[4]{2} = 2\sqrt[4]{2}$

d. $\sqrt{200x^2y} = \sqrt{100x^2 \cdot 2y}$
$$= \sqrt{100x^2} \cdot \sqrt{2y}$$
$$= 10|x|\sqrt{2y}$$

3. $f(x) = \sqrt{3x^2 - 12x + 12}$
$$= \sqrt{3(x^2 - 4x + 4)}$$
$$= \sqrt{3(x-2)^2}$$
$$= |x-2|\sqrt{3}$$

4. $\sqrt{x^9y^{11}z^3} = \sqrt{x^8 \cdot x \cdot y^{10} \cdot y \cdot z^2 \cdot z}$
$$= \sqrt{x^8y^{10}z^2} \cdot \sqrt{xyz}$$
$$= x^4y^5z\sqrt{xyz}$$

5. $\sqrt[3]{40x^{10}y^{14}} = \sqrt[3]{8 \cdot 5 \cdot x^9 \cdot x \cdot y^{12} \cdot y^2}$
$$= \sqrt[3]{8x^9y^{12}} \cdot \sqrt[3]{5xy^2}$$
$$= 2x^3y^4\sqrt[3]{5xy^2}$$

6. $\sqrt[5]{32x^{12}y^2z^8} = \sqrt[5]{32 \cdot x^{10} \cdot x^2 \cdot y^2 \cdot z^5 \cdot z^3}$
$$= \sqrt[5]{32x^{10}z^5} \cdot \sqrt[5]{x^2y^2z^3}$$
$$= 2x^2z\sqrt[5]{x^2y^2z^3}$$

7. **a.** $\sqrt{6} \cdot \sqrt{2} = \sqrt{6 \cdot 2}$
$$= \sqrt{12} = \sqrt{4 \cdot 3}$$
$$= \sqrt{4} \cdot \sqrt{3} = 2\sqrt{3}$$

b. $10\sqrt[3]{16} \cdot 5\sqrt[3]{2}$
$$= 50\sqrt[3]{16 \cdot 2}$$
$$= 50\sqrt[3]{32} = 50\sqrt[3]{8 \cdot 4}$$
$$= 50\sqrt[3]{8} \cdot \sqrt[3]{4} = 50 \cdot 2 \cdot \sqrt[3]{4}$$
$$= 100\sqrt[3]{4}$$

c. $\sqrt[4]{4x^2y}\cdot\sqrt[4]{8x^6y^3}=\sqrt[4]{4x^2y\cdot 8x^6y^3}$
$=\sqrt[4]{32x^8y^4}$
$=\sqrt[4]{16\cdot 2\cdot x^8y^4}$
$=\sqrt[4]{16x^8y^4}\cdot\sqrt[4]{2}$
$=2x^2y\sqrt[4]{2}$

Problem Set 7.3

Practice Exercises

1. $\sqrt{3}\cdot\sqrt{5}=\sqrt{3\cdot 5}=\sqrt{15}$

3. $\sqrt[3]{2}\cdot\sqrt[3]{9}=\sqrt[3]{2\cdot 9}=\sqrt[3]{18}$

5. $\sqrt[4]{11}\cdot\sqrt[4]{3}=\sqrt[4]{11\cdot 3}=\sqrt[4]{33}$

7. $\sqrt{3x}\cdot\sqrt{11y}=\sqrt{3x\cdot 11y}=\sqrt{33xy}$

9. $\sqrt[5]{6x^3}\cdot\sqrt[5]{4x}=\sqrt[5]{6x^3\cdot 4x}=\sqrt[5]{24x^4}$

11. $\sqrt{x+3}\cdot\sqrt{x-3}=\sqrt{(x+3)(x-3)}$
$=\sqrt{x^2-9}$

13. $\sqrt[6]{x-4}\cdot\sqrt[6]{(x-4)^4}$
$=\sqrt[6]{(x-4)(x-4)^4}=\sqrt[6]{(x-4)^5}$

15. $\sqrt{\frac{2x}{3}}\cdot\sqrt{\frac{3}{2}}=\sqrt{\frac{2x}{3}\cdot\frac{3}{2}}$
$=\sqrt{\frac{\cancel{2}x}{\cancel{3}}\cdot\frac{\cancel{3}}{\cancel{2}}}$
$=\sqrt{x}$

17. $\sqrt[4]{\frac{x}{7}}\cdot\sqrt[4]{\frac{3}{y}}=\sqrt[4]{\frac{x}{7}\cdot\frac{3}{y}}=\sqrt[4]{\frac{3x}{7y}}$

19. $\sqrt[7]{7x^2y}\cdot\sqrt[7]{11x^3y^2}=\sqrt[7]{7x^2y\cdot 11x^3y^2}$
$=\sqrt[7]{7\cdot 11x^2x^3yy^2}$
$=\sqrt[7]{77x^5y^3}$

21. $\sqrt{50}=\sqrt{25\cdot 2}=\sqrt{25}\cdot\sqrt{2}=5\sqrt{2}$

23. $\sqrt{45}=\sqrt{9\cdot 5}=\sqrt{9}\cdot\sqrt{5}=3\sqrt{5}$

25. $\sqrt{75x}=\sqrt{25\cdot 3x}$
$=\sqrt{25}\cdot\sqrt{3x}=5\sqrt{3x}$

27. $\sqrt[3]{16}=\sqrt[3]{8\cdot 2}=\sqrt[3]{8}\cdot\sqrt[3]{2}=2\sqrt[3]{2}$

29. $\sqrt[3]{27x^3}=\sqrt[3]{27\cdot x^3}=\sqrt[3]{27}\cdot\sqrt[3]{x^3}=3x$

31. $\sqrt[3]{-16x^2y^3}=\sqrt[3]{-8\cdot 2x^2y^3}$
$=\sqrt[3]{-8y^3}\cdot\sqrt[3]{2x^2}$
$=-2y\sqrt[3]{2x^2}$

33. $f(x)=\sqrt{36(x+2)^2}=6|x+2|$

35. $f(x)=\sqrt[3]{32(x+2)^3}$
$=\sqrt[3]{8\cdot 4(x+2)^3}$
$=\sqrt[3]{8(x+2)^3}\cdot\sqrt[3]{4}$
$=2(x+2)\sqrt[3]{4}$

37. $f(x)=\sqrt{3x^2-6x+3}$

$$= \sqrt{3(x^2-2x+1)}$$
$$= \sqrt{3(x-1)^2}$$
$$= |x-1|\sqrt{3}$$

39. $\sqrt{x^7} = \sqrt{x^6 \cdot x} = \sqrt{x^6} \cdot \sqrt{x} = x^3\sqrt{x}$

41. $\sqrt{x^8y^9} = \sqrt{x^8y^8y} = \sqrt{x^8y^8}\sqrt{y}$
$= x^4y^4\sqrt{y}$

43. $\sqrt{48x^3} = \sqrt{16 \cdot 3x^2x} = \sqrt{16x^2} \cdot \sqrt{3x}$
$= 4x\sqrt{3x}$

45. $\sqrt[3]{y^8} = \sqrt[3]{y^6 \cdot y^2} = \sqrt[3]{y^6} \cdot \sqrt[3]{y^2}$
$= y^2\sqrt[3]{y^2}$

47. $\sqrt[3]{x^{14}y^3z} = \sqrt[3]{x^{12}x^2y^3z}$
$= \sqrt[3]{x^{12}y^3} \cdot \sqrt[3]{x^2z}$
$= x^4y\sqrt[3]{x^2z}$

49. $\sqrt[3]{81x^8y^6} = \sqrt[3]{27 \cdot 3x^6x^2y^6}$
$= \sqrt[3]{27x^6y^6} \cdot \sqrt[3]{3x^2}$
$= 3x^2y^2\sqrt[3]{3x^2}$

51. $\sqrt[3]{(x+y)^5} = \sqrt[3]{(x+y)^3 \cdot (x+y)^2}$
$= \sqrt[3]{(x+y)^3} \cdot \sqrt[3]{(x+y)^2}$
$= (x+y)\sqrt[3]{(x+y)^2}$

53. $\sqrt[5]{y^{17}} = \sqrt[5]{y^{15} \cdot y^2} = \sqrt[5]{y^{15}} \cdot \sqrt[5]{y^2}$
$= y^3\sqrt[5]{y^2}$

55. $\sqrt[5]{64x^6y^{17}} = \sqrt[5]{32 \cdot 2x^5xy^{15}y^2}$
$= \sqrt[5]{32x^5y^{15}} \cdot \sqrt[5]{2xy^2}$
$= 2xy^3\sqrt[5]{2xy^2}$

57. $\sqrt[4]{80x^{10}} = \sqrt[4]{16 \cdot 5x^8x^2}$
$= \sqrt[4]{16x^8} \cdot \sqrt[4]{5x^2}$
$= 2x^2\sqrt[4]{5x^2}$

59. $\sqrt[4]{(x-3)^{10}} = \sqrt[4]{(x-3)^8(x-3)^2}$
$= \sqrt[4]{(x-3)^8} \cdot \sqrt[4]{(x-3)^2}$
$= (x-3)^2\sqrt[4]{(x-3)^2}$

61. $\sqrt{12} \cdot \sqrt{2} = \sqrt{12 \cdot 2} = \sqrt{24}$
$= \sqrt{4 \cdot 6} = \sqrt{4} \cdot \sqrt{6}$
$= 2\sqrt{6}$

63. $\sqrt{5x} \cdot \sqrt{10y} = \sqrt{5x \cdot 10y} = \sqrt{50xy}$
$= \sqrt{25 \cdot 2xy} = 5\sqrt{2xy}$

65. $\sqrt{12x} \cdot \sqrt{3x} = \sqrt{12x \cdot 3x}$
$= \sqrt{36x^2} = 6x$

67. $\sqrt{50xy} \cdot \sqrt{4xy^2} = \sqrt{50xy \cdot 4xy^2}$
$= \sqrt{200x^2y^3}$
$= \sqrt{100 \cdot 2x^2y^2y}$
$= \sqrt{100x^2y^2} \cdot \sqrt{2y}$
$= 10xy\sqrt{2y}$

69. $2\sqrt{5} \cdot 3\sqrt{40} = 2 \cdot 3\sqrt{5 \cdot 40} = 6\sqrt{200}$

$= 6\sqrt{100 \cdot 2} = 6\sqrt{100} \cdot \sqrt{2}$
$= 6 \cdot 10\sqrt{2} = 60\sqrt{2}$

71. $\sqrt[3]{12} \cdot \sqrt[3]{4} = \sqrt[3]{12 \cdot 4} = \sqrt[3]{48} = \sqrt[3]{8 \cdot 6}$
$= \sqrt[3]{8} \cdot \sqrt[3]{6} = 2\sqrt[3]{6}$

73. $\sqrt{5x^3} \cdot \sqrt{8x^2} = \sqrt{5x^3 \cdot 8x^2} = \sqrt{40x^5}$
$= \sqrt{4 \cdot 10x^4 x}$
$= \sqrt{4x^4} \cdot \sqrt{10x}$
$= 2x^2\sqrt{10x}$

75. $\sqrt[3]{25x^4y^2} \cdot \sqrt[3]{5xy^{12}}$
$= \sqrt[3]{25x^4y^2 \cdot 5xy^{12}}$
$= \sqrt[3]{125x^5y^{14}}$
$= \sqrt[3]{125x^3x^2y^{12}y^2}$
$= \sqrt[3]{125x^3y^{12}} \cdot \sqrt[3]{x^2y^2}$
$= 5xy^4\sqrt[3]{x^2y^2}$

77. $\sqrt[4]{8x^2y^3z^6} \cdot \sqrt[4]{2x^4yz}$
$= \sqrt[4]{8x^2y^3z^6 \cdot 2x^4yz}$
$= \sqrt[4]{16x^6y^4z^7}$
$= \sqrt[4]{16x^4x^2y^4z^4z^3}$
$= \sqrt[4]{16x^4y^4z^4} \cdot \sqrt[4]{x^2z^3}$
$= 2xyz\sqrt[4]{x^2z^3}$

79. $\sqrt[5]{8x^4y^6} \cdot \sqrt[5]{8xy^7}$
$= \sqrt[5]{8x^4y^6 \cdot 8xy^7}$
$= \sqrt[5]{64x^5y^{13}}$
$= \sqrt[5]{32 \cdot 2x^5y^{10}y^3}$
$= \sqrt[5]{32x^5y^{10}} \cdot \sqrt[5]{2y^3}$
$= 2xy^2\sqrt[5]{2y^3}$

81. $\sqrt[3]{x-y} \cdot \sqrt[3]{(x-y)^7}$
$= \sqrt[3]{(x-y) \cdot (x-y)^7}$
$= \sqrt[3]{(x-y)^8}$
$= \sqrt[3]{(x-y)^6(x-y)^2}$
$= \sqrt[3]{(x-y)^6} \cdot \sqrt[3]{(x-y)^2}$
$= (x-y)^2\sqrt[3]{(x-y)^2}$

Application Exercises

83. $r(8) = 4\sqrt{8} = 4\sqrt{4 \cdot 2} = 4\sqrt{4} \cdot \sqrt{2}$
$= 4 \cdot 2\sqrt{2} = 8\sqrt{2}$

The maximum rate a cyclist should travel around a corner of radius 8 feet is $8\sqrt{2} \approx 8(1.414) \approx 11.3$ miles per hour.

85. $T(320) = \sqrt{\frac{320}{16}} = \sqrt{20} = \sqrt{4 \cdot 5}$
$= \sqrt{4} \cdot \sqrt{5} = 2\sqrt{5}$

It will take $2\sqrt{5} \approx 4.5$ seconds for the ball to hit the ground.

87. **a.** $C(32) = \frac{7.644}{\sqrt[4]{32}} = \frac{7.644}{\sqrt[4]{16 \cdot 2}}$
$= \frac{7.644}{2\sqrt[4]{2}} = \frac{3.822}{\sqrt[4]{2}}$

The cardiac index of a 32-year-old is $\frac{3.822}{\sqrt[4]{2}}$.

b. $\frac{3.822}{\sqrt[4]{2}} = \frac{3.822}{1.189} = \frac{3.822}{1.189} \approx 3.21$

The cardiac index of a 32-year-old is 3.21 liters per minute per square meter. This is shown on the graph as the point (32, 3.21).

Writing in Mathematics

89. Answers will vary.

91. Answers will vary.

93. Answers will vary.

Technology Exercises

95. $\sqrt{x^4} = x^2$

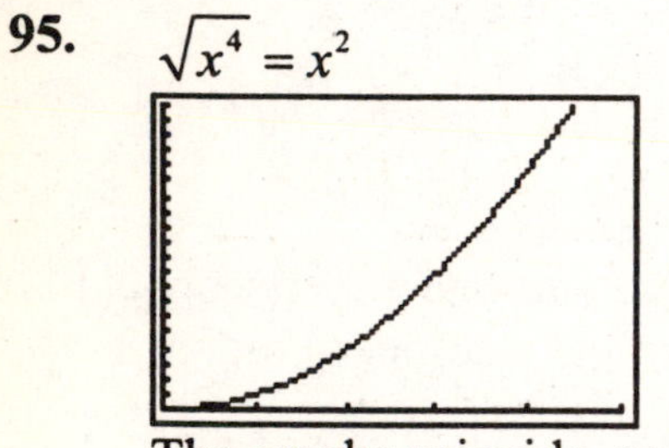

The graphs coincide, so the simplification is correct.

97. $\sqrt{3x^2 - 6x + 3} = (x-1)\sqrt{3}$

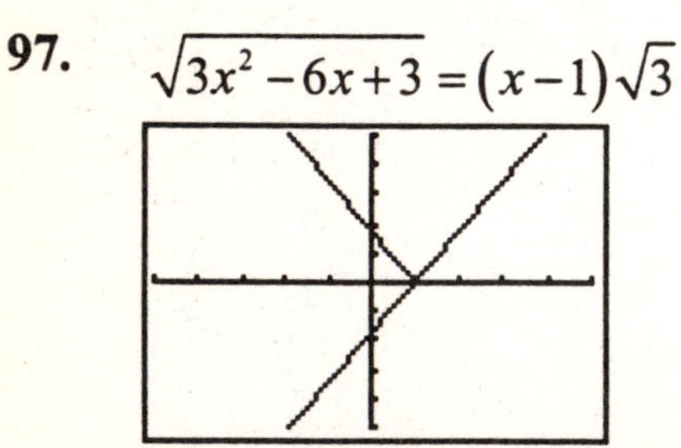

The graphs do not coincide.
To correct the simplification:

$$\sqrt{3x^2 - 6x + 3} = \sqrt{3(x^2 - 2x + 1)}$$
$$= \sqrt{3(x-1)^2}$$
$$= |x-1|\sqrt{3}$$

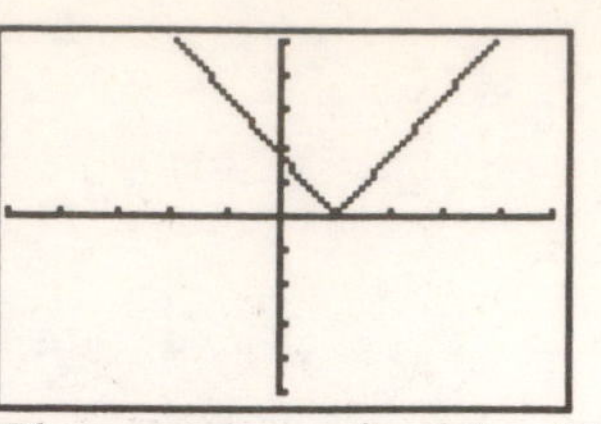

The graphs coincide. The simplification is correct.

Critical Thinking Exercises

99. Statement **d.** is true.

$\sqrt[5]{3^{25}} = 3^5 = 243$

Statement **a.** is false.

$2\sqrt{5} \cdot 6\sqrt{5} = 12\sqrt{5 \cdot 5} = 12 \cdot 5 = 60$

Statement **b.** is false.

$\sqrt[3]{4} \cdot \sqrt[3]{4} = \sqrt[3]{4 \cdot 4} = \sqrt[3]{16}$

Statement **c.** is false.

$\sqrt{12} = \sqrt{4 \cdot 3} = 2\sqrt{3}$

101. Assume the cube root of a number is x. Triple the cube root and work in reverse to find the number.

$3x = \sqrt[3]{(3x)^3} = \sqrt[3]{27x^3}$

The number must be multiplied by 27 for the cube root to be tripled.

103. $f(x) = \sqrt{(x-1)^2}$

x	$f(x)$
-3	4
-2	3
-1	2
0	1
1	0
2	1
3	2

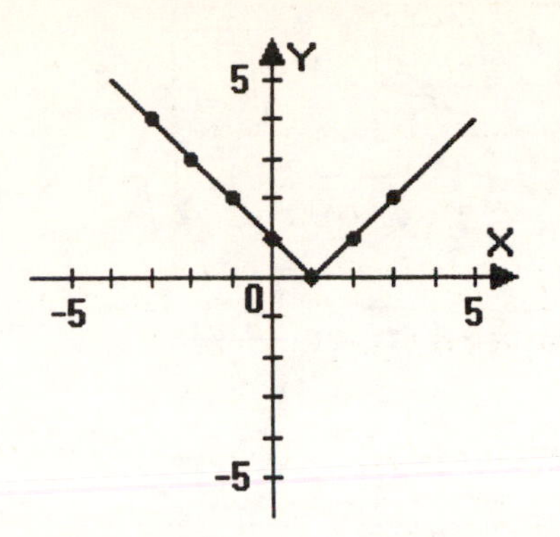

Review Exercises

104. $2x - 1 \le 21$ and $2x + 2 \ge 12$

$2x \le 22$ $\quad$ $2x \ge 10$

$x \le 11$ $\quad$ $x \ge 5$

The solution set is $\{x | 5 \le x \le 11\}$ or $[5, 11]$.

105. $5x + 2y = 2$

$4x + 3y = -4$

Multiply the first equation by -3, the second equation by 2 and solve by addition.

$$-15x - 6y = -6$$
$$8x + 6y = -8$$
$$-7x = -14$$
$$x = 2$$

Back-substitute 2 for x to find y.

$$5x + 2y = 2$$
$$5(2) + 2y = 2$$
$$10 + 2y = 2$$
$$2y = -8$$
$$y = -4$$

The solution is $(2, -4)$ and the solution set is $\{(2, -4)\}$.

106. $64x^3 - 27$

$= (4x)^3 - (3)^3$

$= (4x - 3)\left((4x)^2 + 4x \cdot 3 + 3^2\right)$

$= (4x - 3)\left(16x^2 + 12x + 9\right)$

Check Points 7.4

1. **a.** $8\sqrt{13} + 2\sqrt{13} = (8 + 2)\sqrt{13}$

$= 10\sqrt{13}$

b. $9\sqrt[3]{7} - 6x\sqrt[3]{7} + 12\sqrt[3]{7}$

$= (9 - 6x + 12)\sqrt[3]{7}$

$= (21 - 6x)\sqrt[3]{7}$

c. $7\sqrt[4]{3x} - 2\sqrt[4]{3x} + 2\sqrt[3]{3x}$

$= \left(7\sqrt[4]{3x} - 2\sqrt[4]{3x}\right) + 2\sqrt[3]{3x}$

$= \left((7 - 2)\sqrt[4]{3x}\right) + 2\sqrt[3]{3x}$

$= 5\sqrt[4]{3x} + 2\sqrt[3]{3x}$

2. **a.** $3\sqrt{20} + 5\sqrt{45} = 3\sqrt{4 \cdot 5} + 5\sqrt{9 \cdot 5}$

$= 3 \cdot 2\sqrt{5} + 5 \cdot 3\sqrt{5}$

$= 6\sqrt{5} + 15\sqrt{5}$

$= (6 + 15)\sqrt{5}$

$= 21\sqrt{5}$

b. $3\sqrt{12x} - 6\sqrt{27x}$

$= 3\sqrt{4 \cdot 3x} - 6\sqrt{9 \cdot 3x}$

$= 3 \cdot 2\sqrt{3x} - 6 \cdot 3\sqrt{3x}$

$= 6\sqrt{3x} - 18\sqrt{3x}$

$= (6 - 18)\sqrt{3x} = -12\sqrt{3x}$

c. $8\sqrt{5}-6\sqrt{2}$
Cannot be simplified.

3. a. $3\sqrt[3]{24}-5\sqrt[3]{81}=3\sqrt[3]{8\cdot 3}-5\sqrt[3]{27\cdot 3}$
$=3\cdot 2\sqrt[3]{3}-5\cdot 3\sqrt[3]{3}$
$=6\sqrt[3]{3}-15\sqrt[3]{3}$
$=(6-15)\sqrt[3]{3}$
$=-9\sqrt[3]{3}$

b. $5\sqrt[3]{x^2y}+\sqrt[3]{27x^5y^4}$
$=5\sqrt[3]{x^2y}+\sqrt[3]{27x^3x^2y^3y}$
$=5\sqrt[3]{x^2y}+\sqrt[3]{27x^3y^3\cdot x^2y}$
$=5\sqrt[3]{x^2y}+3xy\sqrt[3]{x^2y}$
$=(5+3xy)\sqrt[3]{x^2y}$

4. a. $\sqrt[3]{\frac{24}{125}}=\frac{\sqrt[3]{24}}{\sqrt[3]{125}}=\frac{\sqrt[3]{8\cdot 3}}{5}=\frac{2\sqrt[3]{3}}{5}$

b. $\sqrt{\frac{9x^3}{y^{10}}}=\frac{\sqrt{9x^3}}{\sqrt{y^{10}}}=\frac{\sqrt{9x^2\cdot x}}{y^5}=\frac{3x\sqrt{x}}{y^5}$

c. $\sqrt[3]{\frac{8y^7}{x^{12}}}=\frac{\sqrt[3]{8y^7}}{\sqrt[3]{x^{12}}}=\frac{\sqrt[3]{8y^6y}}{x^4}=\frac{2x^2\sqrt[3]{y}}{x^4}$

5. a. $\frac{\sqrt{40x^5}}{\sqrt{2x}}=\sqrt{\frac{40x^5}{2x}}=\sqrt{20x^4}$
$=\sqrt{4\cdot 5x^4}=\sqrt{4x^4\cdot 5}$
$=2x^2\sqrt{5}$

b. $\frac{\sqrt{50xy}}{2\sqrt{2}}=\frac{1}{2}\cdot\sqrt{\frac{50xy}{2}}$
$=\frac{1}{2}\cdot\sqrt{25xy}=\frac{1}{2}\cdot 5\sqrt{xy}$
$=\frac{5}{2}\sqrt{xy}$ or $\frac{5\sqrt{xy}}{2}$

c. $\frac{\sqrt[3]{48x^7y}}{\sqrt[3]{6xy^{-2}}}=\sqrt[3]{\frac{48x^7y}{6xy^{-2}}}$
$=\sqrt[3]{8x^6y^3}=2x^2y$

Problem Set 7.4

Practice Exercises

1. $8\sqrt{5}+3\sqrt{5}=(8+3)\sqrt{5}=11\sqrt{5}$

3. $9\sqrt[3]{6}-2\sqrt[3]{6}=(9-2)\sqrt[3]{6}=7\sqrt[3]{6}$

5. $4\sqrt[5]{2}+3\sqrt[5]{2}-5\sqrt[5]{2}=(4+3-5)\sqrt[5]{2}$
$=2\sqrt[5]{2}$

7. $3\sqrt{13}-2\sqrt{5}-2\sqrt{13}+4\sqrt{5}$
$=3\sqrt{13}-2\sqrt{13}-2\sqrt{5}+4\sqrt{5}$
$=(3-2)\sqrt{13}+(-2+4)\sqrt{5}$
$=\sqrt{13}+2\sqrt{5}$

9. $3\sqrt{5}-\sqrt[3]{x}+4\sqrt{5}+3\sqrt[3]{x}$
$=3\sqrt{5}+4\sqrt{5}-\sqrt[3]{x}+3\sqrt[3]{x}$
$=(3+4)\sqrt{5}+(-1+3)\sqrt[3]{x}$
$=7\sqrt{5}+2\sqrt[3]{x}$

11. $\sqrt{3}+\sqrt{27}=\sqrt{3}+\sqrt{9\cdot 3}=\sqrt{3}+3\sqrt{3}$
$=(1+3)\sqrt{3}=4\sqrt{3}$

13. $7\sqrt{12}+\sqrt{75}=7\sqrt{4\cdot 3}+\sqrt{25\cdot 3}$
$=7\cdot 2\sqrt{3}+5\sqrt{3}$
$=14\sqrt{3}+5\sqrt{3}$
$=(14+5)\sqrt{3}=19\sqrt{3}$

15. $3\sqrt{32x}-2\sqrt{18x}$
$=3\sqrt{16\cdot 2x}-2\sqrt{9\cdot 2x}$
$=3\cdot 4\sqrt{2x}-2\cdot 3\sqrt{2x}$
$=12\sqrt{2x}-6\sqrt{2x}=6\sqrt{2x}$

17. $5\sqrt[3]{16}+\sqrt[3]{54}=5\sqrt[3]{8\cdot 2}+\sqrt[3]{27\cdot 2}$
$=5\cdot 2\sqrt[3]{2}+3\sqrt[3]{2}$
$=10\sqrt[3]{2}+3\sqrt[3]{2}$
$=(10+3)\sqrt[3]{2}=13\sqrt[3]{2}$

19. $3\sqrt{45x^3}+\sqrt{5x}=3\sqrt{9\cdot 5x^2x}+\sqrt{5x}$
$=3\cdot 3x\sqrt{5x}+\sqrt{5x}$
$=9x\sqrt{5x}+\sqrt{5x}$
$=(9x+1)\sqrt{5x}$

21. $\sqrt[3]{54xy^3}+y\sqrt[3]{128x}$
$=\sqrt[3]{27\cdot 2xy^3}+y\sqrt[3]{64\cdot 2x}$
$=3y\sqrt[3]{2x}+4y\sqrt[3]{2x}$
$=(3y+4y)\sqrt[3]{2x}=7y\sqrt[3]{2x}$

23. $\sqrt[3]{54x^4}-\sqrt[3]{16x}=\sqrt[3]{27\cdot 2x^3x}-\sqrt[3]{8\cdot 2x}$
$=3x\sqrt[3]{2x}-2\sqrt[3]{2x}$
$=(3x-2)\sqrt[3]{2x}$

25. $\sqrt{9x-18}+\sqrt{x-2}$
$=\sqrt{9(x-2)}+\sqrt{x-2}$
$=3\sqrt{x-2}+\sqrt{x-2}$
$=(3+1)\sqrt{x-2}$
$=4\sqrt{x-2}$

27. $2\sqrt[3]{x^4y^2}+3x\sqrt[3]{xy^2}$
$=2\sqrt[3]{x^3xy^2}+3x\sqrt[3]{xy^2}$
$=2x\sqrt[3]{xy^2}+3x\sqrt[3]{xy^2}$
$=(2x+3x)\sqrt[3]{xy^2}$
$=5x\sqrt[3]{xy^2}$

29. $\sqrt{\dfrac{11}{4}}=\dfrac{\sqrt{11}}{\sqrt{4}}=\dfrac{\sqrt{11}}{2}$

31. $\sqrt[3]{\dfrac{19}{27}}=\dfrac{\sqrt[3]{19}}{\sqrt[3]{27}}=\dfrac{\sqrt[3]{19}}{3}$

33. $\sqrt{\dfrac{x^2}{36y^8}}=\dfrac{\sqrt{x^2}}{\sqrt{36y^8}}=\dfrac{x}{6y^4}$

35. $\sqrt{\dfrac{8x^3}{25y^6}}=\dfrac{\sqrt{8x^3}}{\sqrt{25y^6}}=\dfrac{\sqrt{4\cdot 2x^2x}}{5y^3}$
$=\dfrac{2x\sqrt{2x}}{5y^3}$

37. $\sqrt[3]{\dfrac{x^4}{8y^3}}=\dfrac{\sqrt[3]{x^4}}{\sqrt[3]{8y^3}}=\dfrac{\sqrt[3]{x^3x}}{2y}=\dfrac{x\sqrt[3]{x}}{2y}$

39. $\sqrt[3]{\frac{50x^8}{27y^{12}}} = \frac{\sqrt[3]{50x^8}}{\sqrt[3]{27y^{12}}} = \frac{\sqrt[3]{50x^6x^2}}{3y^4}$

$= \frac{x^2\sqrt[3]{50x^2}}{3y^4}$

39. $\sqrt[3]{\frac{50x^8}{27y^{12}}} = \frac{\sqrt[3]{50x^8}}{\sqrt[3]{27y^{12}}}$

$= \frac{\sqrt[3]{50x^6x^2}}{3y^4} = \frac{x^2\sqrt[3]{50x^2}}{3y^4}$

41. $\sqrt[4]{\frac{9y^6}{x^8}} = \frac{\sqrt[4]{9y^6}}{\sqrt[4]{x^8}} = \frac{\sqrt[4]{9y^4y^2}}{x^2} = \frac{y\sqrt[4]{9y^2}}{x^2}$

43. $\sqrt[5]{\frac{64x^{13}}{y^{20}}} = \frac{\sqrt[5]{64x^{13}}}{\sqrt[5]{y^{20}}} = \frac{\sqrt[5]{32\cdot 2x^{10}x^3}}{y^4}$

$= \frac{2x^2\sqrt[5]{2x^3}}{y^4}$

45. $\frac{\sqrt{40}}{\sqrt{5}} = \sqrt{\frac{40}{5}} = \sqrt{8} = \sqrt{4\cdot 2} = 2\sqrt{2}$

47. $\frac{\sqrt[3]{48}}{\sqrt[3]{6}} = \sqrt[3]{\frac{48}{6}} = \sqrt[3]{8} = 2$

49. $\frac{\sqrt{54x^3}}{\sqrt{6x}} = \sqrt{\frac{54x^3}{6x}} = \sqrt{9x^2} = 3x$

51. $\frac{\sqrt{x^5y^3}}{\sqrt{xy}} = \sqrt{\frac{x^5y^3}{xy}} = \sqrt{x^4y^2} = x^2y$

53. $\frac{\sqrt{200x^3}}{\sqrt{10x^{-1}}} = \sqrt{\frac{200x^3}{10x^{-1}}} = \sqrt{20x^{3-(-1)}}$

$= \sqrt{20x^4} = \sqrt{4\cdot 5x^4} = 2x^2\sqrt{5}$

55. $\frac{\sqrt{72xy}}{2\sqrt{2}} = \frac{1}{2}\sqrt{\frac{72xy}{2}} = \frac{1}{2}\sqrt{36xy}$

$= \frac{1}{2}\cdot 6\sqrt{xy} = 3\sqrt{xy}$

57. $\frac{\sqrt[3]{24x^3y^5}}{\sqrt[3]{3y^2}} = \sqrt[3]{\frac{24x^3y^5}{3y^2}} = \sqrt[3]{8x^3y^3} = 2xy$

59. $\frac{\sqrt[4]{32x^{10}y^8}}{\sqrt[4]{2x^2y^{-2}}} = \sqrt[4]{\frac{32x^{10}y^8}{2x^2y^{-2}}} = \sqrt[4]{16x^8y^{8-(-2)}}$

$= \sqrt[4]{16x^8y^{10}} = \sqrt[4]{16x^8y^8y^2}$

$= 2x^2y^2\sqrt[4]{y^2}$

61. $\frac{\sqrt[3]{x^2+5x+6}}{\sqrt[3]{x+2}} = \sqrt[3]{\frac{x^2+5x+6}{x+2}}$

$= \sqrt[3]{\frac{\cancel{(x+2)}(x+3)}{\cancel{x+2}}}$

$= \sqrt[3]{x+3}$

Application Exercises

63. a. $R_f\frac{\sqrt{c^2-v^2}}{\sqrt{c^2}} = R_f\sqrt{\frac{c^2-v^2}{c^2}}$

$= R_f\sqrt{\frac{c^2}{c^2}-\frac{v^2}{c^2}}$

$= R_f\sqrt{1-\left(\frac{v}{c}\right)^2}$

b.
$$R_f\sqrt{1-\left(\frac{v}{c}\right)^2} = R_f\sqrt{1-\left(\frac{c}{c}\right)^2}$$
$$= R_f\sqrt{1-(1)^2}$$
$$= R_f\sqrt{1-1}$$
$$= R_f\sqrt{0}$$
$$= R_f \cdot 0 = 0$$

Your aging rate is zero. This means that a person moving close to the speed of light does not age relative to a friend on Earth.

65. $P = 2l + 2w$
$$= 2\left(2\sqrt{20}\right) + 2\left(\sqrt{125}\right)$$
$$= 4\sqrt{20} + 2\sqrt{125}$$
$$= 4\sqrt{4\cdot 5} + 2\sqrt{25\cdot 5}$$
$$= 4\cdot 2\sqrt{5} + 2\cdot 5\sqrt{5}$$
$$= 8\sqrt{5} + 10\sqrt{5}$$
$$= (8+10)\sqrt{5} = 18\sqrt{5}$$

The perimeter is $18\sqrt{5}$ feet.

$A = lw = 2\sqrt{20}\cdot\sqrt{125}$
$$= 2\sqrt{20\cdot 125} = 2\sqrt{2500}$$
$$= 2\cdot 50 = 100$$

The area is 100 square feet.

Writing in Mathematics

67. Answers will vary.

69. Answers will vary.

71. Answers will vary.

Technology Exercises

73. Answers will vary. For example, consider Exercise 11.

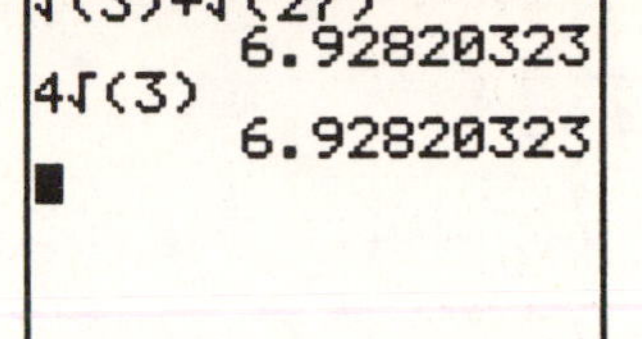

75. $\sqrt{16x} - \sqrt{9x} = \sqrt{7x}$

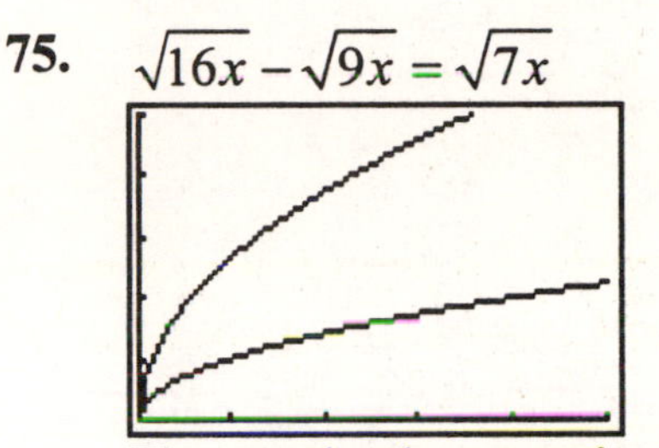

The graphs do not coincide.
Correct the simplification.

$$\sqrt{16x} - \sqrt{9x} = 4\sqrt{x} - 3\sqrt{x}$$
$$= (4-3)\sqrt{x} = \sqrt{x}$$

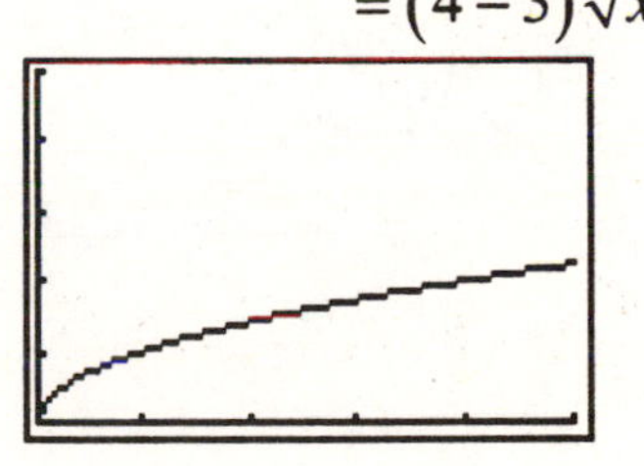

Critical Thinking Exercises

77. Statement **d.** is true.

Statement **a.** is false.
$\sqrt{5} + \sqrt{5} = 2\sqrt{5}$

Statement **b.** is false.
$4\sqrt{3} + 5\sqrt{3} = 9\sqrt{3}$

Statement **c.** is false. In order for two radical expressions to be combined, both the index and the radicand must be the same. Just because two radical expressions are completely simplified does guarantee that the index and radicands match.

79. $\frac{\sqrt{20}}{3}+\frac{\sqrt{45}}{4}-\sqrt{80}$

$=\frac{\sqrt{4\cdot5}}{3}+\frac{\sqrt{9\cdot5}}{4}-\sqrt{16\cdot5}$

$=\frac{2\sqrt{5}}{3}+\frac{3\sqrt{5}}{4}-4\sqrt{5}$

$=\frac{2}{3}\sqrt{5}+\frac{3}{4}\sqrt{5}-4\sqrt{5}$

$=\left(\frac{2}{3}+\frac{3}{4}-4\right)\sqrt{5}$

$=\left(\frac{2}{3}\cdot\frac{4}{4}+\frac{3}{4}\cdot\frac{3}{3}-\frac{4}{1}\cdot\frac{12}{12}\right)\sqrt{5}$

$=\left(\frac{8}{12}+\frac{9}{12}-\frac{48}{12}\right)\sqrt{5}$

$=-\frac{31}{12}\sqrt{5}$ or $-\frac{31\sqrt{5}}{12}$

Review Exercises

81. $2(3x-1)-4=2x-(6-x)$

$6x-2-4=2x-6+x$

$6x-6=x-6$

$5x=0$

$x=0$

The solution is 0 and the solution set is $\{0\}$.

82. $x^2-8xy+12y^2=(x-6y)(x-2y)$

83. $\frac{2}{x^2+5x+6}+\frac{3x}{x^2+6x+9}$

$=\frac{2}{(x+3)(x+2)}+\frac{3x}{(x+3)^2}$

$=\frac{2(x+3)}{(x+3)^2(x+2)}+\frac{3x(x+2)}{(x+3)^2(x+2)}$

$=\frac{2(x+3)+3x(x+2)}{(x+3)^2(x+2)}$

$=\frac{2x+6+3x^2+6x}{(x+3)^2(x+2)}$

$=\frac{3x^2+8x+6}{(x+3)^2(x+2)}$

Check Points 7.5

1. **a.** $\sqrt{6}\left(x+\sqrt{10}\right)=\sqrt{6}\cdot x+\sqrt{6}\cdot\sqrt{10}$

$=x\sqrt{6}+\sqrt{60}$

$=x\sqrt{6}+\sqrt{4\cdot15}$

$=x\sqrt{6}+2\sqrt{15}$

b. $\sqrt[3]{y}\left(\sqrt[3]{y^2}-\sqrt[3]{7}\right)=\sqrt[3]{y}\cdot\sqrt[3]{y^2}-\sqrt[3]{y}\cdot\sqrt[3]{7}$

$=\sqrt[3]{y^3}-\sqrt[3]{7y}$

$=y-\sqrt[3]{7y}$

c. $\left(6\sqrt{5}+3\sqrt{2}\right)\left(2\sqrt{5}-4\sqrt{2}\right)=\left(6\sqrt{5}\right)\left(2\sqrt{5}\right)+\left(6\sqrt{5}\right)\left(-4\sqrt{2}\right)+\left(3\sqrt{2}\right)\left(2\sqrt{5}\right)+\left(3\sqrt{2}\right)\left(-4\sqrt{2}\right)$

$$=12\sqrt{25}+\left(-24\sqrt{10}\right)+6\sqrt{10}+\left(-12\sqrt{4}\right)$$
$$=12\cdot 5+(-24+6)\sqrt{10}+(-12\cdot 2)$$
$$=60-18\sqrt{10}-24=36-18\sqrt{10}$$

2. a. $\left(\sqrt{5}+\sqrt{6}\right)^2$

$$=\left(\sqrt{5}\right)^2+2\cdot\sqrt{5}\cdot\sqrt{6}+\left(\sqrt{6}\right)^2$$
$$=5+2\sqrt{30}+6=11+2\sqrt{30}$$

b. $\left(\sqrt{6}+\sqrt{5}\right)\left(\sqrt{6}-\sqrt{5}\right)$

$$=\left(\sqrt{6}\right)^2-\left(\sqrt{5}\right)^2$$
$$=6-5=1$$

c. $\left(\sqrt{a}-\sqrt{7}\right)\left(\sqrt{a}+\sqrt{7}\right)$

$$=\left(\sqrt{a}\right)^2-\left(\sqrt{7}\right)^2$$
$$=a-7$$

3. a. $\dfrac{\sqrt{3}}{\sqrt{7}}=\dfrac{\sqrt{3}}{\sqrt{7}}\cdot\dfrac{\sqrt{7}}{\sqrt{7}}=\dfrac{\sqrt{21}}{\sqrt{49}}=\dfrac{\sqrt{21}}{7}$

b. $\sqrt[3]{\dfrac{2}{9}}=\dfrac{\sqrt[3]{2}}{\sqrt[3]{9}}=\dfrac{\sqrt[3]{2}}{\sqrt[3]{3^2}}=\dfrac{\sqrt[3]{2}}{\sqrt[3]{3^2}}\cdot\dfrac{\sqrt[3]{3}}{\sqrt[3]{3}}$

$$=\frac{\sqrt[3]{6}}{\sqrt[3]{3^3}}=\frac{\sqrt[3]{6}}{3}$$

4. a. $\sqrt{\dfrac{2x}{7y}}=\dfrac{\sqrt{2x}}{\sqrt{7y}}=\dfrac{\sqrt{2x}}{\sqrt{7y}}\cdot\dfrac{\sqrt{7y}}{\sqrt{7y}}$

$$=\frac{\sqrt{14xy}}{7y}$$

b. $\dfrac{\sqrt[3]{x}}{\sqrt[3]{9y}}=\dfrac{\sqrt[3]{x}}{\sqrt[3]{3^2 y}}\cdot\dfrac{\sqrt[3]{3y^2}}{\sqrt[3]{3y^2}}=\dfrac{\sqrt[3]{3xy^2}}{\sqrt[3]{3^3 y^3}}$

$$=\frac{\sqrt[3]{3xy^2}}{3y}$$

c. $\dfrac{6x}{\sqrt[5]{8x^2y^4}}=\dfrac{6x}{\sqrt[5]{2^3x^2y^4}}\cdot\dfrac{\sqrt[5]{2^2x^3y}}{\sqrt[5]{2^2x^3y}}$

$$=\frac{6x\sqrt[5]{2^2x^3y}}{\sqrt[5]{2^5x^5y^5}}$$
$$=\frac{6x\sqrt[5]{4x^3y}}{2xy}=\frac{3\sqrt[5]{4x^3y}}{y}$$

5. $\dfrac{18}{2\sqrt{3}+3}=\dfrac{18}{2\sqrt{3}+3}\cdot\dfrac{2\sqrt{3}-3}{2\sqrt{3}-3}$

$$=\frac{18\left(2\sqrt{3}-3\right)}{\left(2\sqrt{3}+3\right)\left(2\sqrt{3}-3\right)}$$
$$=\frac{18\left(2\sqrt{3}-3\right)}{\left(2\sqrt{3}\right)^2-3^2}=\frac{18\left(2\sqrt{3}-3\right)}{4\cdot 3-9}$$
$$=\frac{18\left(2\sqrt{3}-3\right)}{12-9}=\frac{\overset{6}{\cancel{18}}\left(2\sqrt{3}-3\right)}{\underset{1}{\cancel{3}}}$$
$$=6\left(2\sqrt{3}-3\right)=12\sqrt{3}-18$$

6. $\dfrac{3+\sqrt{7}}{\sqrt{5}-\sqrt{2}}$

$=\dfrac{3+\sqrt{7}}{\sqrt{5}-\sqrt{2}}\cdot\dfrac{\sqrt{5}+\sqrt{2}}{\sqrt{5}+\sqrt{2}}$

$=\dfrac{3\sqrt{5}+3\sqrt{2}+\sqrt{7}\sqrt{5}+\sqrt{7}\sqrt{2}}{\left(\sqrt{5}\right)^2-\left(\sqrt{2}\right)^2}$

$=\dfrac{3\sqrt{5}+3\sqrt{2}+\sqrt{35}+\sqrt{14}}{5-2}$

$=\dfrac{3\sqrt{5}+3\sqrt{2}+\sqrt{35}+\sqrt{14}}{3}$

7. $\dfrac{\sqrt{x+3}-\sqrt{x}}{3}$

$=\dfrac{\sqrt{x+3}-\sqrt{x}}{3}\cdot\dfrac{\sqrt{x+3}+\sqrt{x}}{\sqrt{x+3}+\sqrt{x}}$

$=\dfrac{\left(\sqrt{x+3}\right)^2-\left(\sqrt{x}\right)^2}{3\left(\sqrt{x+3}+\sqrt{x}\right)}$

$=\dfrac{x+3-x}{3\left(\sqrt{x+3}+\sqrt{x}\right)}$

$=\dfrac{3}{3\left(\sqrt{x+3}+\sqrt{x}\right)}=\dfrac{1}{\sqrt{x+3}+\sqrt{x}}$

Problem Set 7.5

Practice Exercises

1. $\sqrt{2}\left(x+\sqrt{7}\right)=\sqrt{2}\cdot x+\sqrt{2}\sqrt{7}$

$=x\sqrt{2}+\sqrt{14}$

3. $\sqrt{6}\left(7-\sqrt{6}\right)=\sqrt{6}\cdot 7-\sqrt{6}\sqrt{6}$

$=7\sqrt{6}-\sqrt{36}=7\sqrt{6}-6$

5. $\sqrt{3}\left(4\sqrt{6}-2\sqrt{3}\right)$

$=\sqrt{3}\cdot 4\sqrt{6}-\sqrt{3}\cdot 2\sqrt{3}$

$=4\sqrt{18}-2\sqrt{9}$

$=4\sqrt{9\cdot 2}-2\cdot 3$

$=4\cdot 3\sqrt{2}-6=12\sqrt{2}-6$

7. $\sqrt[3]{2}\left(\sqrt[3]{6}+4\sqrt[3]{5}\right)=\sqrt[3]{2}\cdot\sqrt[3]{6}+\sqrt[3]{2}\cdot 4\sqrt[3]{5}$

$=\sqrt[3]{12}+4\sqrt[3]{10}$

9. $\sqrt[3]{x}\left(\sqrt[3]{16x^2}-\sqrt[3]{x}\right)$

$=\sqrt[3]{x}\cdot\sqrt[3]{16x^2}-\sqrt[3]{x}\cdot\sqrt[3]{x}$

$=\sqrt[3]{x}\cdot\sqrt[3]{8\cdot 2x^2}-\sqrt[3]{x^2}$

$=\sqrt[3]{8\cdot 2x^3}-\sqrt[3]{x^2}$

$=2x\sqrt[3]{2}-\sqrt[3]{x^2}$

11. $\left(5+\sqrt{2}\right)\left(6+\sqrt{2}\right)$

$=5\cdot 6+5\sqrt{2}+6\sqrt{2}+\sqrt{2}\sqrt{2}$

$=30+(5+6)\sqrt{2}+2$

$=32+11\sqrt{2}$

13. $\left(6+\sqrt{5}\right)\left(9-4\sqrt{5}\right)$

$=6\cdot 9-6\cdot 4\sqrt{5}+9\sqrt{5}-4\sqrt{5}\sqrt{5}$

$=54-24\sqrt{5}+9\sqrt{5}-4\cdot 5$

$=54+(-24+9)\sqrt{5}-20$

$=34+(-15)\sqrt{5}$

$=34-15\sqrt{5}$

15. $\left(6-3\sqrt{7}\right)\left(2-5\sqrt{7}\right)$

$=6\cdot 2-6\cdot 5\sqrt{7}-2\cdot 3\sqrt{7}+3\sqrt{7}\cdot 5\sqrt{7}$

$= 12 - 30\sqrt{7} - 6\sqrt{7} + 15 \cdot 7 = 12 + (-30 - 6)\sqrt{7} + 105 = 117 + (-36)\sqrt{7} = 117 - 36\sqrt{7}$

17. $\left(\sqrt{2} + \sqrt{7}\right)\left(\sqrt{3} + \sqrt{5}\right) = \sqrt{2}\sqrt{3} + \sqrt{2}\sqrt{5} + \sqrt{7}\sqrt{3} + \sqrt{7}\sqrt{5} = \sqrt{6} + \sqrt{10} + \sqrt{21} + \sqrt{35}$

19. $\left(\sqrt{2} - \sqrt{7}\right)\left(\sqrt{3} - \sqrt{5}\right) = \sqrt{2}\sqrt{3} - \sqrt{2}\sqrt{5} - \sqrt{7}\sqrt{3} + \sqrt{7}\sqrt{5} = \sqrt{6} - \sqrt{10} - \sqrt{21} + \sqrt{35}$

21. $\left(3\sqrt{2} - 4\sqrt{3}\right)\left(2\sqrt{2} + 5\sqrt{3}\right) = 3\sqrt{2}\left(2\sqrt{2}\right) + 3\sqrt{2}\left(5\sqrt{3}\right) - 4\sqrt{3}\left(2\sqrt{2}\right) - 4\sqrt{3}\left(5\sqrt{3}\right)$

$= 6 \cdot 2 + 15\sqrt{6} - 8\sqrt{6} - 20 \cdot 3 = 12 + 7\sqrt{6} - 60$

$= 7\sqrt{6} - 48$ or $-48 + 7\sqrt{6}$

23. $\left(\sqrt{3} + \sqrt{5}\right)^2 = \left(\sqrt{3}\right)^2 + 2\sqrt{3}\sqrt{5} + \left(\sqrt{5}\right)^2 = 3 + 2\sqrt{15} + 5 = 8 + 2\sqrt{15}$

25. $\left(\sqrt{3x} - \sqrt{y}\right)^2 = \left(\sqrt{3x}\right)^2 - 2\sqrt{3x}\sqrt{y} + \left(\sqrt{y}\right)^2 = 3x - 2\sqrt{3xy} + y$

27. $\left(\sqrt{5} + 7\right)\left(\sqrt{5} - 7\right) = \sqrt{5}\sqrt{5} - 7\sqrt{5} + 7\sqrt{5} - 7 \cdot 7$

$= 5 - \cancel{7\sqrt{5}} + \cancel{7\sqrt{5}} - 49 = 5 - 49 = -44$

29. $\left(2 - 5\sqrt{3}\right)\left(2 + 5\sqrt{3}\right) = 2 \cdot 2 + 2 \cdot 5\sqrt{3} - 2 \cdot 5\sqrt{3} - 5\sqrt{3} \cdot 5\sqrt{3}$

$= 4 + \cancel{10\sqrt{3}} - \cancel{10\sqrt{3}} - 25 \cdot 3 = 4 - 75 = -71$

31. $\left(3\sqrt{2} + 2\sqrt{3}\right)\left(3\sqrt{2} - 2\sqrt{3}\right) = 3\sqrt{2} \cdot 3\sqrt{2} - 3\sqrt{2} \cdot 2\sqrt{3} + 3\sqrt{2} \cdot 2\sqrt{3} - 2\sqrt{3} \cdot 2\sqrt{3}$

$= 9 \cdot 2 - \cancel{6\sqrt{6}} + \cancel{6\sqrt{6}} - 4 \cdot 3 = 18 - 12 = 6$

33. $\left(3 - \sqrt{x}\right)\left(2 - \sqrt{x}\right)$

$= 3 \cdot 2 - 3\sqrt{x} - 2\sqrt{x} + \sqrt{x}\sqrt{x}$

$= 6 + (-3 - 2)\sqrt{x} + x$

$= 6 + (-5)\sqrt{x} + x$

$= 6 - 5\sqrt{x} + x$

35. $\left(\sqrt[3]{x} - 4\right)\left(\sqrt[3]{x} + 5\right)$

$= \sqrt[3]{x}\sqrt[3]{x} + 5\sqrt[3]{x} - 4\sqrt[3]{x} - 4 \cdot 5$

$= \sqrt[3]{x^2} + (5 - 4)\sqrt[3]{x} - 20$

$= \sqrt[3]{x^2} + \sqrt[3]{x} - 20$

37. $\left(x+\sqrt[3]{y^2}\right)\left(2x-\sqrt[3]{y^2}\right)$

$= x \cdot 2x - x\sqrt[3]{y^2} + 2x\sqrt[3]{y^2} - \sqrt[3]{y^2}\sqrt[3]{y^2}$

$= 2x^2 + (-x+2x)\sqrt[3]{y^2} - \sqrt[3]{y^4}$

$= 2x^2 + x\sqrt[3]{y^2} - \sqrt[3]{y^3 y}$

$= 2x^2 + x\sqrt[3]{y^2} - y\sqrt[3]{y}$

39. $\dfrac{\sqrt{2}}{\sqrt{5}} = \dfrac{\sqrt{2}}{\sqrt{5}} \cdot \dfrac{\sqrt{5}}{\sqrt{5}} = \dfrac{\sqrt{2\cdot 5}}{\sqrt{5\cdot 5}} = \dfrac{\sqrt{10}}{5}$

41. $\sqrt{\dfrac{11}{x}} = \dfrac{\sqrt{11}}{\sqrt{x}} = \dfrac{\sqrt{11}}{\sqrt{x}} \cdot \dfrac{\sqrt{x}}{\sqrt{x}}$

$= \dfrac{\sqrt{11x}}{\sqrt{x^2}} = \dfrac{\sqrt{11x}}{x}$

43. $\dfrac{9}{\sqrt{3y}} = \dfrac{9}{\sqrt{3y}} \cdot \dfrac{\sqrt{3y}}{\sqrt{3y}} = \dfrac{9\sqrt{3y}}{\sqrt{3y\cdot 3y}}$

$= \dfrac{\overset{3}{\cancel{9}}\sqrt{3y}}{\underset{1}{\cancel{3}}y} = \dfrac{3\sqrt{3y}}{y}$

45. $\dfrac{1}{\sqrt[3]{2}} = \dfrac{1}{\sqrt[3]{2}} \cdot \dfrac{\sqrt[3]{2^2}}{\sqrt[3]{2^2}} = \dfrac{\sqrt[3]{2^2}}{\sqrt[3]{2^3}} = \dfrac{\sqrt[3]{4}}{2}$

47. $\dfrac{6}{\sqrt[3]{4}} = \dfrac{6}{\sqrt[3]{4}} \cdot \dfrac{\sqrt[3]{4^2}}{\sqrt[3]{4^2}} = \dfrac{6\sqrt[3]{4^2}}{\sqrt[3]{4}\sqrt[3]{4^2}}$

$= \dfrac{6\sqrt[3]{16}}{\sqrt[3]{4^3}} = \dfrac{6\sqrt[3]{8\cdot 2}}{4} = \dfrac{6\cdot 2\sqrt[3]{2}}{4}$

$= \dfrac{\overset{3}{\cancel{12}}\sqrt[3]{2}}{\underset{1}{\cancel{4}}} = 3\sqrt[3]{2}$

49. $\sqrt[3]{\dfrac{2}{3}} = \dfrac{\sqrt[3]{2}}{\sqrt[3]{3}} = \dfrac{\sqrt[3]{2}}{\sqrt[3]{3}} \cdot \dfrac{\sqrt[3]{3^2}}{\sqrt[3]{3^2}} = \dfrac{\sqrt[3]{2\cdot 3^2}}{\sqrt[3]{3^3}}$

$= \dfrac{\sqrt[3]{2\cdot 9}}{3} = \dfrac{\sqrt[3]{18}}{3}$

51. $\dfrac{4}{\sqrt[3]{x}} = \dfrac{4}{\sqrt[3]{x}} \cdot \dfrac{\sqrt[3]{x^2}}{\sqrt[3]{x^2}} = \dfrac{4\sqrt[3]{x^2}}{\sqrt[3]{x}\sqrt[3]{x^2}}$

$= \dfrac{4\sqrt[3]{x^2}}{\sqrt[3]{x^3}} = \dfrac{4\sqrt[3]{x^2}}{x}$

53. $\sqrt[3]{\dfrac{2}{y^2}} = \dfrac{\sqrt[3]{2}}{\sqrt[3]{y^2}} = \dfrac{\sqrt[3]{2}}{\sqrt[3]{y^2}} \cdot \dfrac{\sqrt[3]{y}}{\sqrt[3]{y}}$

$= \dfrac{\sqrt[3]{2y}}{\sqrt[3]{y^3}} = \dfrac{\sqrt[3]{2y}}{y}$

55. $\dfrac{7}{\sqrt[3]{2x^2}} = \dfrac{7}{\sqrt[3]{2x^2}} \cdot \dfrac{\sqrt[3]{2^2x}}{\sqrt[3]{2^2x}} = \dfrac{7\sqrt[3]{2^2x}}{\sqrt[3]{2x^2}\sqrt[3]{2^2x}}$

$= \dfrac{7\sqrt[3]{4x}}{\sqrt[3]{2^3x^3}} = \dfrac{7\sqrt[3]{4x}}{2x}$

57. $\sqrt[3]{\dfrac{2}{xy^2}} = \dfrac{\sqrt[3]{2}}{\sqrt[3]{xy^2}} = \dfrac{\sqrt[3]{2}}{\sqrt[3]{xy^2}} \cdot \dfrac{\sqrt[3]{x^2y}}{\sqrt[3]{x^2y}}$

$= \dfrac{\sqrt[3]{2}\sqrt[3]{x^2y}}{\sqrt[3]{xy^2}\sqrt[3]{x^2y}} = \dfrac{\sqrt[3]{2x^2y}}{\sqrt[3]{x^3y^3}}$

$= \dfrac{\sqrt[3]{2x^2y}}{xy}$

59. $\dfrac{3}{\sqrt[4]{x}} = \dfrac{3}{\sqrt[4]{x}} \cdot \dfrac{\sqrt[4]{x^3}}{\sqrt[4]{x^3}} = \dfrac{3\sqrt[4]{x^3}}{\sqrt[4]{xx^3}}$

$= \dfrac{3\sqrt[4]{x^3}}{\sqrt[4]{x^4}} = \dfrac{3\sqrt[4]{x^3}}{x}$

61. $\frac{6}{\sqrt[5]{8x^3}} = \frac{6}{\sqrt[5]{2^3x^3}} \cdot \frac{\sqrt[5]{2^2x^2}}{\sqrt[5]{2^2x^2}} = \frac{6\sqrt[5]{4x^2}}{\sqrt[5]{2^5x^5}}$

$= \frac{6\sqrt[5]{4x^2}}{2x} = \frac{3\sqrt[5]{4x^2}}{x}$

63. $\frac{2x^2y}{\sqrt[5]{4x^2y^4}} = \frac{2x^2y}{\sqrt[5]{2^2x^2y^4}} \cdot \frac{\sqrt[5]{2^3x^3y}}{\sqrt[5]{2^3x^3y}}$

$= \frac{2x^2y\sqrt[5]{8x^3y}}{\sqrt[5]{2^5x^5y^5}}$

$= \frac{2x^2y\sqrt[5]{8x^3y}}{2xy}$

$= x\sqrt[5]{8x^3y}$

65. $\frac{8}{\sqrt{5}+2} = \frac{8}{\sqrt{5}+2} \cdot \frac{\sqrt{5}-2}{\sqrt{5}-2}$

$= \frac{8\sqrt{5}-8\cdot 2}{\sqrt{5}\sqrt{5}-2\sqrt{5}+2\sqrt{5}-2\cdot 2}$

$= \frac{8\sqrt{5}-16}{5-2\sqrt{5}+2\sqrt{5}-4}$

$= \frac{8\sqrt{5}-16}{5-4} = \frac{8\sqrt{5}-16}{1}$

$= 8\sqrt{5}-16$

67. $\frac{13}{\sqrt{11}-3} = \frac{13}{\sqrt{11}-3} \cdot \frac{\sqrt{11}+3}{\sqrt{11}+3}$

$= \frac{13\left(\sqrt{11}+3\right)}{\left(\sqrt{11}-3\right)\left(\sqrt{11}+3\right)}$

$= \frac{13\sqrt{11}+13\cdot 3}{\sqrt{11}\cdot\sqrt{11}+3\sqrt{11}-3\sqrt{11}-3\cdot 3}$

$= \frac{13\sqrt{11}+39}{11+3\sqrt{11}-3\sqrt{11}-9}$

$= \frac{13\sqrt{11}+39}{11-9}$

$= \frac{13\sqrt{11}+39}{2}$

69. $\frac{6}{\sqrt{5}+\sqrt{3}}$

$= \frac{6}{\sqrt{5}+\sqrt{3}} \cdot \frac{\sqrt{5}-\sqrt{3}}{\sqrt{5}-\sqrt{3}}$

$= \frac{6\left(\sqrt{5}-\sqrt{3}\right)}{\left(\sqrt{5}+\sqrt{3}\right)\left(\sqrt{5}-\sqrt{3}\right)}$

$= \frac{6\sqrt{5}-6\sqrt{3}}{\sqrt{5}\sqrt{5}-\sqrt{3}\sqrt{5}+\sqrt{3}\sqrt{5}-\sqrt{3}\sqrt{3}}$

$= \frac{6\sqrt{5}-6\sqrt{3}}{5-\sqrt{15}+\sqrt{15}-3}$

$= \frac{6\sqrt{5}-6\sqrt{3}}{5-3}$

$= \frac{6\sqrt{5}-6\sqrt{3}}{2}$

$= \frac{2\left(3\sqrt{5}-3\sqrt{3}\right)}{2}$

$= 3\sqrt{5}-3\sqrt{3}$

71. $\frac{\sqrt{a}}{\sqrt{a}-\sqrt{b}} = \frac{\sqrt{a}}{\sqrt{a}-\sqrt{b}} \cdot \frac{\sqrt{a}+\sqrt{b}}{\sqrt{a}+\sqrt{b}}$

$= \frac{\sqrt{a}\left(\sqrt{a}+\sqrt{b}\right)}{\left(\sqrt{a}-\sqrt{b}\right)\left(\sqrt{a}+\sqrt{b}\right)}$

$$=\frac{\sqrt{a}\sqrt{a}+\sqrt{a}\sqrt{b}}{\sqrt{a}\sqrt{a}+\sqrt{a}\sqrt{b}-\sqrt{a}\sqrt{b}-\sqrt{b}\sqrt{b}}=\frac{a+\sqrt{ab}}{a-b}$$

73.

$$\frac{25}{5\sqrt{2}-3\sqrt{5}}=\frac{25}{5\sqrt{2}-3\sqrt{5}}\cdot\frac{5\sqrt{2}+3\sqrt{5}}{5\sqrt{2}+3\sqrt{5}}=\frac{25\left(5\sqrt{2}+3\sqrt{5}\right)}{\left(5\sqrt{2}-3\sqrt{5}\right)\left(5\sqrt{2}+3\sqrt{5}\right)}$$

$$=\frac{125\sqrt{2}+75\sqrt{5}}{5\cdot5\sqrt{2\cdot2}+\cancel{5\cdot3\sqrt{2\cdot5}}-\cancel{5\cdot3\sqrt{2\cdot5}}-3\cdot3\sqrt{5\cdot5}}=\frac{125\sqrt{2}+75\sqrt{5}}{25\cdot2-9\cdot5}$$

$$=\frac{125\sqrt{2}+75\sqrt{5}}{50-45}=\frac{125\sqrt{2}+75\sqrt{5}}{5}=\frac{\cancel{5}\left(25\sqrt{2}+15\sqrt{5}\right)}{\cancel{5}}=25\sqrt{2}+15\sqrt{5}$$

75.

$$\frac{\sqrt{5}+\sqrt{3}}{\sqrt{5}-\sqrt{3}}=\frac{\sqrt{5}+\sqrt{3}}{\sqrt{5}-\sqrt{3}}\cdot\frac{\sqrt{5}+\sqrt{3}}{\sqrt{5}+\sqrt{3}}=\frac{\left(\sqrt{5}+\sqrt{3}\right)^2}{\left(\sqrt{5}-\sqrt{3}\right)\left(\sqrt{5}+\sqrt{3}\right)}$$

$$=\frac{\left(\sqrt{5}\right)^2+2\sqrt{5}\sqrt{3}+\left(\sqrt{3}\right)^2}{\sqrt{5}\cdot\sqrt{5}+\sqrt{5}\cdot\sqrt{3}-\sqrt{5}\cdot\sqrt{3}-\sqrt{3}\sqrt{3}}=\frac{5+2\sqrt{15}+3}{5+\cancel{\sqrt{15}}-\cancel{\sqrt{15}}-3}$$

$$=\frac{8+2\sqrt{15}}{5-3}=\frac{\cancel{2}\left(4+\sqrt{15}\right)}{\cancel{2}}=4+\sqrt{15}$$

77.

$$\frac{\sqrt{x}+1}{\sqrt{x}+3}=\frac{\sqrt{x}+1}{\sqrt{x}+3}\cdot\frac{\sqrt{x}-3}{\sqrt{x}-3}=\frac{\sqrt{x}\cdot\sqrt{x}-3\sqrt{x}+1\sqrt{x}-3\cdot1}{\sqrt{x}\cdot\sqrt{x}-\cancel{3\sqrt{x}}+\cancel{3\sqrt{x}}-3\cdot3}$$

$$=\frac{\sqrt{x^2}+(-3+1)\sqrt{x}-3}{\sqrt{x^2}-9}=\frac{x+(-2)\sqrt{x}-3}{x-9}=\frac{x-2\sqrt{x}-3}{x-9}$$

79.

$$\frac{5\sqrt{3}-3\sqrt{2}}{3\sqrt{2}-2\sqrt{3}}=\frac{5\sqrt{3}-3\sqrt{2}}{3\sqrt{2}-2\sqrt{3}}\cdot\frac{3\sqrt{2}+2\sqrt{3}}{3\sqrt{2}+2\sqrt{3}}=\frac{5\sqrt{3}\cdot3\sqrt{2}+5\sqrt{3}\cdot2\sqrt{3}-3\sqrt{2}\cdot3\sqrt{2}-3\sqrt{2}\cdot2\sqrt{3}}{3\sqrt{2}\cdot3\sqrt{2}+3\sqrt{2}\cdot2\sqrt{3}-3\sqrt{2}\cdot2\sqrt{3}-2\sqrt{3}\cdot2\sqrt{3}}$$

$$=\frac{15\sqrt{6}+10\cdot3-9\cdot2-6\sqrt{6}}{9\cdot2+\cancel{6\sqrt{6}}-\cancel{6\sqrt{6}}-4\cdot3}=\frac{15\sqrt{6}+30-18-6\sqrt{6}}{18-12}$$

$$=\frac{9\sqrt{6}+12}{6}=\frac{\cancel{3}\left(3\sqrt{6}+4\right)}{\cancel{3}\cdot2}=\frac{3\sqrt{6}+4}{2}$$

81.

$$\frac{2\sqrt{x}+\sqrt{y}}{\sqrt{y}-2\sqrt{x}} = \frac{2\sqrt{x}+\sqrt{y}}{\sqrt{y}-2\sqrt{x}} \cdot \frac{\sqrt{y}+2\sqrt{x}}{\sqrt{y}+2\sqrt{x}} = \frac{2\sqrt{x}\sqrt{y}+2\sqrt{x}\cdot 2\sqrt{x}+\sqrt{y}\sqrt{y}+2\sqrt{x}\sqrt{y}}{\sqrt{y}\sqrt{y}+2\sqrt{x}\sqrt{y}-2\sqrt{x}\sqrt{y}-2\sqrt{x}\cdot 2\sqrt{x}}$$

$$= \frac{2\sqrt{xy}+4\sqrt{x^2}+\sqrt{y^2}+2\sqrt{xy}}{\sqrt{y^2}+\cancel{2\sqrt{xy}}-\cancel{2\sqrt{xy}}-4\sqrt{x^2}} = \frac{2\sqrt{xy}+4x+y+2\sqrt{xy}}{y-4x} = \frac{4\sqrt{xy}+4x+y}{y-4x}$$

83.

$$\sqrt{\frac{3}{2}} = \frac{\sqrt{3}}{\sqrt{2}} \cdot \frac{\sqrt{3}}{\sqrt{3}} = \frac{\sqrt{3}\sqrt{3}}{\sqrt{2}\sqrt{3}} = \frac{3}{\sqrt{6}}$$

85.

$$\frac{\sqrt[3]{4x}}{\sqrt[3]{y}} = \frac{\sqrt[3]{4x}}{\sqrt[3]{y}} \cdot \frac{\sqrt[3]{4^2x^2}}{\sqrt[3]{4^2x^2}} = \frac{\sqrt[3]{4^3x^3}}{\sqrt[3]{4^2x^2y}} = \frac{4x}{\sqrt[3]{16x^2y}} = \frac{4x}{\sqrt[3]{8\cdot 2x^2y}} = \frac{4x}{2\sqrt[3]{2x^2y}} = \frac{2x}{\sqrt[3]{2x^2y}}$$

87.

$$\frac{\sqrt{x}+3}{\sqrt{x}} = \frac{\sqrt{x}+3}{\sqrt{x}} \cdot \frac{\sqrt{x}-3}{\sqrt{x}-3} = \frac{\sqrt{x}\cdot\sqrt{x}-\cancel{3\sqrt{x}}+\cancel{3\sqrt{x}}-3\cdot 3}{\sqrt{x}\cdot\sqrt{x}-3\sqrt{x}} = \frac{\sqrt{x^2}-9}{\sqrt{x^2}-3\sqrt{x}} = \frac{x-9}{x-3\sqrt{x}}$$

89.

$$\frac{\sqrt{a}+\sqrt{b}}{\sqrt{a}-\sqrt{b}} = \frac{\sqrt{a}+\sqrt{b}}{\sqrt{a}-\sqrt{b}} \cdot \frac{\sqrt{a}-\sqrt{b}}{\sqrt{a}-\sqrt{b}} = \frac{\sqrt{a}\cdot\sqrt{a}-\sqrt{a}\sqrt{b}+\sqrt{a}\sqrt{b}-\sqrt{b}\sqrt{b}}{\sqrt{a}\cdot\sqrt{a}-\sqrt{a}\sqrt{b}-\sqrt{a}\sqrt{b}+\sqrt{b}\sqrt{b}}$$

$$= \frac{\sqrt{a^2}-\cancel{\sqrt{ab}}+\cancel{\sqrt{ab}}-\sqrt{b^2}}{\sqrt{a^2}-\sqrt{ab}-\sqrt{ab}+\sqrt{b^2}} = \frac{a-b}{a-2\sqrt{ab}+b}$$

91.

$$\frac{\sqrt{x+5}-\sqrt{x}}{5} = \frac{\sqrt{x+5}-\sqrt{x}}{5} \cdot \frac{\sqrt{x+5}+\sqrt{x}}{\sqrt{x+5}+\sqrt{x}} = \frac{\left(\sqrt{x+5}\right)^2+\sqrt{x+5}\cdot\sqrt{x}-\sqrt{x+5}\cdot\sqrt{x}-\left(\sqrt{x}\right)^2}{5\left(\sqrt{x+5}+\sqrt{x}\right)}$$

$$= \frac{x+5+\sqrt{x(x+5)}-\sqrt{x(x+5)}-x}{5\left(\sqrt{x+5}+\sqrt{x}\right)} = \frac{5}{5\left(\sqrt{x+5}+\sqrt{x}\right)} = \frac{1}{\sqrt{x+5}+\sqrt{x}}$$

93.

$$\frac{\sqrt{x}+\sqrt{y}}{x^2-y^2} = \frac{\sqrt{x}+\sqrt{y}}{x^2-y^2} \cdot \frac{\sqrt{x}-\sqrt{y}}{\sqrt{x}-\sqrt{y}} = \frac{\left(\sqrt{x}\right)^2-\cancel{\sqrt{xy}}+\cancel{\sqrt{xy}}-\left(\sqrt{y}\right)^2}{x^2\sqrt{x}-x^2\sqrt{y}-y^2\sqrt{x}+y^2\sqrt{y}}$$

$$= \frac{x-y}{x^2\left(\sqrt{x}-\sqrt{y}\right)-y^2\left(\sqrt{x}-\sqrt{y}\right)} = \frac{x-y}{\left(\sqrt{x}-\sqrt{y}\right)\left(x^2-y^2\right)}$$

$$= \frac{\cancel{x-y}}{\left(\sqrt{x}-\sqrt{y}\right)(x+y)\cancel{(x-y)}} = \frac{1}{\left(\sqrt{x}-\sqrt{y}\right)(x+y)}$$

Application Exercises

95. $P(4) = 6.85\sqrt{4} + 19 = 6.85 \cdot 2 + 19$

$= 13.7 + 19 = 32.7$

Approximately 33% of U.S. households are online. This model predicts the number of households extremely well. The actual percentage shown on the graph is 33%.

97. $\frac{\text{change in percent}}{\text{change in time}} = \frac{33-19}{2001-1997}$

$= \frac{14}{4} = 3.5$

The average yearly increase from 1997 to 2001 is 3.5%

99. $6.85\left(\frac{\sqrt{t+h}-\sqrt{t}}{h}\right)$

$= 6.85\left(\frac{\sqrt{0+4}-\sqrt{0}}{4}\right) = 6.85\left(\frac{\sqrt{4}}{4}\right)$

$= 6.85\left(\frac{2}{4}\right) = 6.85\left(\frac{1}{2}\right) = 6.85\left(\frac{1}{2}\right) \approx 3.4$

The model predicts the yearly increase in percentage very well. The average yearly increase predicted in Exercise 97 is 3.5%. This is very close to 3.4% found here.

101. a. $6.85\left(\frac{\sqrt{t+h}-\sqrt{t}}{h}\right)$

$= 6.85\left(\frac{\sqrt{t+h}-\sqrt{t}}{h} \cdot \frac{\sqrt{t+h}+\sqrt{t}}{\sqrt{t+h}+\sqrt{t}}\right)$

$= 6.85\left(\frac{\left(\sqrt{t+h}\right)^2-\left(\sqrt{t}\right)^2}{h\sqrt{t+h}+h\sqrt{t}}\right)$

$= 6.85\left(\frac{t+h-t}{h\sqrt{t+h}+h\sqrt{t}}\right)$

$= 6.85\left(\frac{h}{h\left(\sqrt{t+h}+\sqrt{t}\right)}\right)$

$= 6.85\left(\frac{1}{\sqrt{t+h}+\sqrt{t}}\right)$

$= \frac{6.85}{\sqrt{t+h}+\sqrt{t}}$

b. $\frac{6.85}{\sqrt{t+h}+\sqrt{t}} = \frac{6.85}{\sqrt{t+0}+\sqrt{t}}$

$= \frac{6.85}{\sqrt{t}+\sqrt{t}} = \frac{6.85}{2\sqrt{t}} = \frac{3.425}{\sqrt{t}}$

c. $\frac{3.425}{\sqrt{t}} = \frac{3.425}{\sqrt{4}} = \frac{3.425}{2} = 1.7125$

The rate of change in percentage in 2001 was approximately 1.7%.

103. a. $P(25) = \frac{25\left(13+\sqrt{25}\right)}{5\sqrt{25}}$

$= \frac{25(13+5)}{5 \cdot 5}$

$= \frac{25(18)}{25} = 18$

18% of 25-year-olds must pay more taxes.

b. $p(x) = \frac{x\left(13+\sqrt{x}\right)}{5\sqrt{x}}$

$= \frac{13x + x\sqrt{x}}{5\sqrt{x}} \cdot \frac{\sqrt{x}}{\sqrt{x}}$

$$= \frac{\sqrt{x}\left(13x + x\sqrt{x}\right)}{5\sqrt{x}\cdot\sqrt{x}}$$

$$= \frac{13x\sqrt{x} + x\left(\sqrt{x}\right)^2}{5x}$$

$$= \frac{13x\sqrt{x} + x\cdot x}{5x}$$

$$= \frac{x\left(13\sqrt{x} + x\right)}{5x}$$

$$= \frac{13\sqrt{x} + x}{5}$$

105. Perimeter $= 2l + 2w$

$$= 2\left(\sqrt{8}+1\right) + 2\left(\sqrt{8}-1\right)$$
$$= 2\sqrt{8} + \not{2} + 2\sqrt{8} - \not{2}$$
$$= (2+2)\sqrt{8} = 4\sqrt{8}$$
$$= 4\sqrt{4\cdot 2} = 4\cdot 2\sqrt{2} = 8\sqrt{2}$$

The perimeter is $8\sqrt{2}$ inches.

Area $= lw = \left(\sqrt{8}+1\right)\left(\sqrt{8}-1\right)$

$$= \left(\sqrt{8}\right)^2 - \not{\sqrt{8}} + \not{\sqrt{8}} - 1$$
$$= 8 - 1 = 7$$

The area is 7 square inches.

107. $\frac{7\sqrt{2\cdot 2\cdot 3}}{6} = \frac{7\cdot 2\sqrt{3}}{6}$

$$= \frac{7\cdot \not{2}\sqrt{3}}{\not{2}\cdot 3} = \frac{7}{3}\sqrt{3}$$

Writing in Mathematics

109. Answers will vary.

111. Answers will vary.

113. Answers will vary.

115. Answers will vary.

117. Answers will vary.

Technology Exercises

119. $\left(\sqrt{x}-1\right)\left(\sqrt{x}-1\right) = x+1$

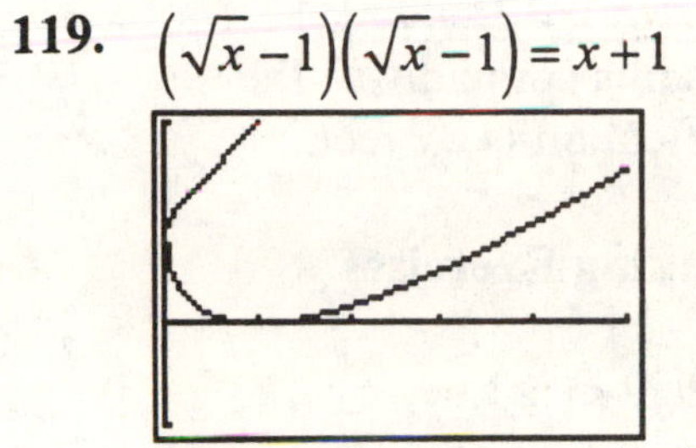

The graphs do not coincide. Correct the simplification.

$$\left(\sqrt{x}-1\right)\left(\sqrt{x}-1\right)$$
$$= \left(\sqrt{x}\right)^2 - \sqrt{x} - \sqrt{x} + 1$$
$$= x - 2\sqrt{x} + 1$$

The graphs coincide. The new simplification is correct.

121. $\left(\sqrt{x}+1\right)^2 = x+1$

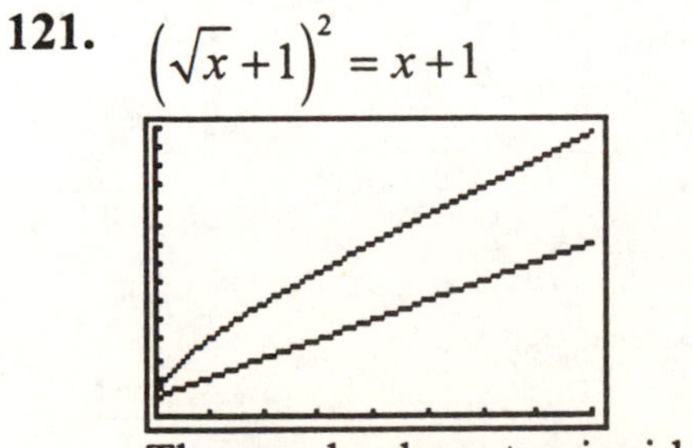

The graphs do not coincide. Correct the simplification.

$$\left(\sqrt{x}+1\right)^2 = \left(\sqrt{x}\right)^2 + 2\sqrt{x}\cdot 1 + 1^2$$
$$= x + 2\sqrt{x} + 1$$

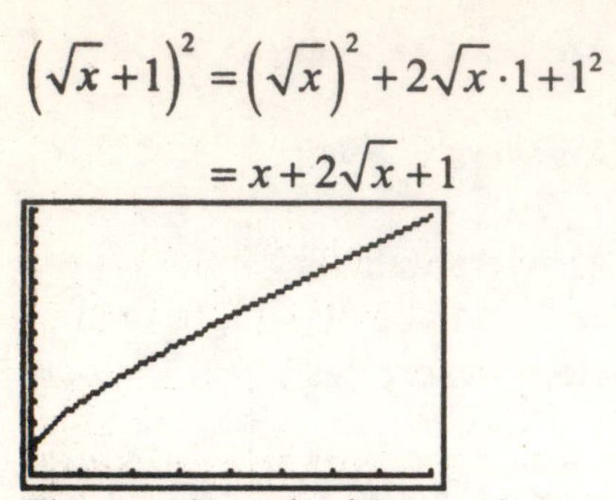

The graphs coincide, so the new simplification is correct.

Critical Thinking Exercises

123. Statement **c.** is true.

$$\frac{4\sqrt{x}}{\sqrt{x}-y} = \frac{4\sqrt{x}}{\sqrt{x}-y}\cdot\frac{\sqrt{x}+y}{\sqrt{x}+y}$$
$$= \frac{4\sqrt{x}\sqrt{x}+4\sqrt{x}\cdot y}{\left(\sqrt{x}\right)^2 - \cancel{y\sqrt{x}} + \cancel{y\sqrt{x}} - y^2}$$
$$= \frac{4x+4y\sqrt{x}}{x-y^2} = \frac{4\left(x+y\sqrt{x}\right)}{x-y^2}$$

Statement **a.** is false.

$$\frac{\sqrt{3}+7}{\sqrt{3}-2} = \frac{\sqrt{3}+7}{\sqrt{3}-2}\cdot\frac{\sqrt{3}+2}{\sqrt{3}+2}$$
$$= \frac{\left(\sqrt{3}\right)^2 + 2\sqrt{3} + 7\sqrt{3} + 14}{\left(\sqrt{3}\right)^2 - 2^2}$$
$$= \frac{3+9\sqrt{3}+14}{3-4} = \frac{17+9\sqrt{3}}{-1}$$
$$= -\left(17+9\sqrt{3}\right) = -17-9\sqrt{3}$$

Statement **b.** is false.

$$\frac{4}{\sqrt{x+y}} = \frac{4}{\sqrt{x+y}}\cdot\frac{\sqrt{x+y}}{\sqrt{x+y}} = \frac{4\sqrt{x+y}}{x+y}$$

Statement **d.** is false.

$$\left(\sqrt{x}-7\right)^2$$
$$= \left(\sqrt{x}\right)^2 + 2\cdot\sqrt{x}\cdot(-7) + (-7)^2$$
$$= x - 14\sqrt{x} + 49$$

125.

$$\left(\sqrt{2+\sqrt{3}}+\sqrt{2-\sqrt{3}}\right)^2 = \left(\sqrt{2+\sqrt{3}}\right)^2 + 2\sqrt{2+\sqrt{3}}\sqrt{2-\sqrt{3}} + \left(\sqrt{2-\sqrt{3}}\right)^2$$
$$= 2+\sqrt{3}+2\sqrt{\left(2+\sqrt{3}\right)\left(2-\sqrt{3}\right)}+2-\sqrt{3}$$
$$= 2+\cancel{\sqrt{3}}+2\sqrt{4-\cancel{2\sqrt{3}}+\cancel{2\sqrt{3}}-3}+2-\cancel{\sqrt{3}}$$
$$= 4+2\sqrt{4-3}$$
$$= 4+2\sqrt{1}$$
$$= 4+2 = 6$$

127.
$$3x^2+2=6x$$
$$3\left(\frac{3+\sqrt{3}}{3}\right)^2+2=6\left(\frac{3+\sqrt{3}}{3}\right)$$
$$3\left(\frac{3^2+2\cdot 3\sqrt{3}+\left(\sqrt{3}\right)^2}{9}\right)+2=2\left(3+\sqrt{3}\right)$$
$$\not{3}\left(\frac{12+6\sqrt{3}}{3\cdot\not{3}}\right)+2=6+2\sqrt{3}$$
$$\frac{12+6\sqrt{3}}{3}+2=6+2\sqrt{3}$$
$$\frac{\not{3}\left(4+2\sqrt{3}\right)}{\not{3}}+2=6+2\sqrt{3}$$
$$4+2\sqrt{3}+2=6+2\sqrt{3}$$
$$6+2\sqrt{3}=6+2\sqrt{3}$$

$\frac{3+\sqrt{3}}{3}$ is a solution.

Review Exercises

128. $\frac{2}{x-2}+\frac{3}{x^2-4}$
$$=\frac{2}{x-2}+\frac{3}{(x+2)(x-2)}$$
$$=\frac{2}{x-2}\cdot\frac{(x+2)}{(x+2)}+\frac{3}{(x+2)(x-2)}$$
$$=\frac{2(x+2)}{(x-2)(x+2)}+\frac{3}{(x+2)(x-2)}$$
$$=\frac{2(x+2)+3}{(x-2)(x+2)}$$
$$=\frac{2x+4+3}{(x-2)(x+2)}$$
$$=\frac{2x+7}{(x-2)(x+2)} \text{ or } \frac{2x+7}{x^2-4}$$

129. Using the results from Exercise 128, we know that the left side of the equation simplifies to $\frac{2x+7}{(x-2)(x+2)}$. Setting this equal to zero, we have $\frac{2x+7}{(x-2)(x+2)}=0$.

To solve, set each factor equal to zero.

$$2x+7=0 \qquad x-2=0$$
$$2x=-7 \qquad x=2$$
$$x=-\frac{7}{2}$$

$$x+2=0$$
$$x=-2$$

The solutions are $-\frac{7}{2}$, -2 and 2. We disregard -2 and 2 because these values result in a zero denominator.

The solution set is $\left\{-\frac{7}{2}\right\}$.

130. $f(x)=x^4-3x^2-2x+5$

To find $f(-2)$, we use synthetic division to divide by $-(-2)=2$.

-2	1	0	-3	-2	5
		-2	4	-2	8
	1	-2	1	-4	13

The remainder is 13. This means that $f(-2)=13$.

Check Points 7.6

1.
$$\sqrt{3x+4}=8$$
$$\left(\sqrt{3x+4}\right)^2=8^2$$
$$3x+4=64$$
$$3x=60$$
$$x=20$$

Check:
$$\sqrt{3x+4}=8$$
$$\sqrt{3(20)+4}=8$$
$$\sqrt{60+4}=8$$
$$\sqrt{64}=8$$
$$8=8 \quad \text{True}$$

The solution is 20 and the solution set is $\{20\}$.

2.
$$\sqrt{x-1}+7=2$$
$$\sqrt{x-1}=-5$$
$$\left(\sqrt{x-1}\right)^2=(-5)^2$$
$$x-1=25$$
$$x=26$$

Check:
$$\sqrt{x-1}+7=2$$
$$\sqrt{26-1}+7=2$$
$$\sqrt{25}+7=2$$
$$5+7=2$$
$$12\neq 2 \quad \text{False}$$

26 is an extraneous solution. There is no solution. The solution set is $\varnothing$ or $\{\ \}$.

3.
$$\sqrt{6x+7}-x=2$$
$$\sqrt{6x+7}=x+2$$
$$\left(\sqrt{6x+7}\right)^2=(x+2)^2$$
$$6x+7=x^2+4x+4$$
$$0=x^2-2x-3$$
$$0=(x-3)(x+1)$$

Apply the zero product principle.

$x-3=0$ or $x+1=0$

$x=3$ $\qquad$ $x=-1$

Check:
$$x=3$$
$$\sqrt{6x+7}-x=2$$
$$\sqrt{6(3)+7}-3=2$$
$$\sqrt{18+7}-3=2$$
$$\sqrt{25}-3=2$$
$$5-3=2$$
$$2=2 \quad \text{True}$$

Check:
$$x=-1$$
$$\sqrt{6x+7}-x=2$$
$$\sqrt{6(-1)+7}-(-1)=2$$
$$\sqrt{-6+7}+1=2$$
$$\sqrt{1}+1=2$$
$$1+1=2$$
$$2=2 \quad \text{True}$$

The solutions are –1 and 3 and the solution set is $\{-1,3\}$.

4.
$$\sqrt{x+5}-\sqrt{x-3}=2$$
$$\sqrt{x+5}=\sqrt{x-3}+2$$
$$\left(\sqrt{x+5}\right)^2=\left(\sqrt{x-3}+2\right)^2$$
$$\cancel{x}+5=\cancel{x}-3+4\sqrt{x-3}+4$$

$$4 = 4\sqrt{x-3}$$
$$\frac{4}{4} = \frac{4\sqrt{x-3}}{4}$$
$$1 = \sqrt{x-3}$$
$$1^2 = \left(\sqrt{x-3}\right)^2$$
$$1 = x-3$$
$$4 = x$$

Check:

$$\sqrt{x+5} - \sqrt{x-3} = 2$$
$$\sqrt{4+5} - \sqrt{4-3} = 2$$
$$\sqrt{9} - \sqrt{1} = 2$$
$$3-1 = 2$$
$$2 = 2 \quad \text{True}$$

The solution is 4 and the solution set is $\{4\}$.

5.

$$(2x-3)^{\frac{1}{3}} + 3 = 0$$
$$\sqrt[3]{2x-3} + 3 = 0$$
$$\sqrt[3]{2x-3} = -3$$
$$\left(\sqrt[3]{2x-3}\right)^3 = (-3)^3$$
$$2x-3 = -27$$
$$2x = -24$$
$$x = -12$$

Check:

$$(2x-3)^{\frac{1}{3}} + 3 = 0$$
$$(2(-12)-3)^{\frac{1}{3}} + 3 = 0$$
$$(-24-3)^{\frac{1}{3}} + 3 = 0$$
$$(-27)^{\frac{1}{3}} + 3 = 0$$
$$\sqrt[3]{-27} + 3 = 0$$
$$-3+3 = 0$$
$$0 = 0 \quad \text{True}$$

The solution is -12 and the solution set is $\{-12\}$.

6. We are interested in $\frac{1}{10}$ of the total population or $\frac{1}{10} \cdot 280 = 28$ million Americans.

$$28 = 2.6\sqrt{x} + 11$$
$$17 = 2.6\sqrt{x}$$
$$\frac{17}{2.6} = \sqrt{x}$$
$$\left(\frac{17}{2.6}\right)^2 = \left(\sqrt{x}\right)^2$$
$$42.75 \approx x$$

This means that $\frac{1}{10}$ of the total population or 28 million Americans will live alone approximately 43 years after 1970 in the year 1970 + 43 = 2013.

Problem Set 7.6

Practice Exercises

1.

$$\sqrt{3x-2} = 4$$
$$\left(\sqrt{3x-2}\right)^2 = 4^2$$
$$3x-2 = 16$$
$$3x = 18$$
$$x = 6$$

Check:

$$\sqrt{3(6)-2} = 4$$
$$\sqrt{18-2} = 4$$

$$\sqrt{16} = 4$$
$$4 = 4$$

The solution is 6 and the solution set is $\{6\}$.

3. $\sqrt{5x-4} - 9 = 0$

$$\sqrt{5x-4} = 9$$
$$\left(\sqrt{5x-4}\right)^2 = 9^2$$
$$5x - 4 = 81$$
$$5x = 85$$
$$x = 17$$

Check:

$$\sqrt{5(17)-4} - 9 = 0$$
$$\sqrt{85-4} - 9 = 0$$
$$\sqrt{81} - 9 = 0$$
$$9 - 9 = 0$$
$$0 = 0$$

The solution is 17 and the solution set is $\{17\}$.

5. $\sqrt{3x+7} + 10 = 4$

$$\sqrt{3x+7} = -6$$

Since the square root of a number is always positive, the solution set is $\{\ \}$ or $\varnothing$.

7. $x = \sqrt{7x+8}$

$$x^2 = \left(\sqrt{7x+8}\right)^2$$
$$x^2 = 7x + 8$$
$$x^2 - 7x - 8 = 0$$
$$(x-8)(x+1) = 0$$

Apply the zero product principle.

$x - 8 = 0 \qquad x + 1 = 0$

$x = 8 \qquad x = -1$

Check:

$8 = \sqrt{7(8)+8} \qquad -1 = \sqrt{7(-1)+8}$

$$8 = \sqrt{56+8}$$
$$8 = \sqrt{64}$$
$$8 = 8$$

We disregard -1 because square roots are always positive. The solution is 8 and the solution set is $\{8\}$.

9. $\sqrt{5x+1} = x + 1$

$$\left(\sqrt{5x+1}\right)^2 = (x+1)^2$$
$$5x + \cancel{1} = x^2 + 2x + \cancel{1}$$
$$0 = x^2 - 3x$$
$$0 = x(x-3)$$

Apply the zero product principle.

$x = 0 \qquad x - 3 = 0$

$x = 3$

Both solutions check. The solutions

are 0 and 3 and the solution set is $\{0, 3\}$.

11. $x = \sqrt{2x-2} + 1$

$$x - 1 = \sqrt{2x-2}$$
$$(x-1)^2 = \left(\sqrt{2x-2}\right)^2$$
$$x^2 - 2x + 1 = 2x - 2$$
$$x^2 - 4x + 3 = 0$$
$$(x-3)(x-1) = 0$$

Apply the zero product principle.

$x-3=0 \qquad x-1=0$

$x=3 \qquad x=1$

Both solutions check. The solutions are 1 and 3 and the solution set is $\{1,3\}$.

13. $x-2\sqrt{x-3}=3$

$x-3=2\sqrt{x-3}$

$(x-3)^2=\left(2\sqrt{x-3}\right)^2$

$x^2-6x+9=4(x-3)$

$x^2-6x+9=4x-12$

$x^2-10x+21=0$

$(x-7)(x-3)=0$

Apply the zero product principle.

$x-7=0 \qquad x-3=0$

$x=7 \qquad x=3$

Both solutions check. The solutions are 3 and 7 and the solution set is $\{3,7\}$.

15. $\sqrt{2x-5}=\sqrt{x+4}$

$\left(\sqrt{2x-5}\right)^2=\left(\sqrt{x+4}\right)^2$

$2x-5=x+4$

$x-5=4$

$x=9$

The solution checks. The solution is 9 and the solution set is $\{9\}$.

17. $\sqrt[3]{2x+11}=3$

$\left(\sqrt[3]{2x+11}\right)^3=3^3$

$2x+11=27$

$2x=16$

$x=8$

The solution checks. The solution is 8 and the solution set is $\{8\}$.

19. $\sqrt[3]{2x-6}-4=0$

$\sqrt[3]{2x-6}=4$

$\left(\sqrt[3]{2x-6}\right)^3=4^3$

$2x-6=64$

$2x=70$

$x=35$

The solution checks. The solution is 35 and the solution set is $\{35\}$.

21. $\sqrt{x-7}=7-\sqrt{x}$

$\left(\sqrt{x-7}\right)^2=\left(7-\sqrt{x}\right)^2$

$\not{x}-7=49-14\sqrt{x}+\not{x}$

$-7=49-14\sqrt{x}$

$-56=-14\sqrt{x}$

$\frac{-56}{-14}=\frac{-14\sqrt{x}}{-14}$

$4=\sqrt{x}$

$4^2=\left(\sqrt{x}\right)^2$

$16=x$

The solution checks. The solution is 16 and the solution set is $\{16\}$.

23. $\sqrt{x+2}+\sqrt{x-1}=3$

$\sqrt{x+2}=3-\sqrt{x-1}$

$\left(\sqrt{x+2}\right)^2=\left(3-\sqrt{x-1}\right)^2$

$$\not{x}+2=9-6\sqrt{x-1}+\not{x}-1$$
$$2=8-6\sqrt{x-1}$$
$$-6=-6\sqrt{x-1}$$
$$\frac{-6}{-6}=\frac{-6\sqrt{x-1}}{-6}$$
$$1=\sqrt{x-1}$$
$$1^2=\left(\sqrt{x-1}\right)^2$$
$$1=x-1$$
$$2=x$$

The solution checks. The solution is 2 and the solution set is $\{2\}$.

25. $$(2x+3)^{1/3}+4=6$$
$$(2x+3)^{1/3}=2$$
$$\left((2x+3)^{1/3}\right)^3=2^3$$
$$2x+3=8$$
$$2x=5$$
$$x=\frac{5}{2}$$

The solution checks. The solution is $\frac{5}{2}$ and the solution set is $\left\{\frac{5}{2}\right\}$.

27. $$(3x+1)^{1/4}+7=9$$
$$(3x+1)^{1/4}=2$$
$$\left((3x+1)^{1/4}\right)^4=2^4$$
$$3x+1=16$$
$$3x=15$$
$$x=5$$

The solution checks. The solution is 5 and the solution set is $\{5\}$.

29. $$(x+2)^{1/2}+8=4$$
$$(x+2)^{1/2}=-4$$
$$\sqrt{x+2}=-4$$

The square root of a number must be positive. The solution set is $\varnothing$.

31. $$\sqrt{2x-3}-\sqrt{x-2}=1$$
$$\sqrt{2x-3}=\sqrt{x-2}+1$$
$$\left(\sqrt{2x-3}\right)^2=\left(\sqrt{x-2}+1\right)^2$$
$$2x-3=x-2+2\sqrt{x-2}+1$$
$$2x-3=x-1+2\sqrt{x-2}$$
$$x-2=2\sqrt{x-2}$$
$$(x-2)^2=\left(2\sqrt{x-2}\right)^2$$
$$x^2-4x+4=4(x-2)$$
$$x^2-4x+4=4x-8$$
$$x^2-8x+12=0$$
$$(x-6)(x-2)=0$$

Apply the zero product principle.

$x-6=0$ $\quad$ $x-2=0$

$x=6$ $\quad$ $x=2$

Both solutions check. The solutions are 2 and 6 and the solution set is $\{2,6\}$.

33. $$3x^{1/3}=\left(x^2+17x\right)^{1/3}$$
$$\left(3x^{1/3}\right)^3=\left(\left(x^2+17x\right)^{1/3}\right)^3$$
$$3^3x=x^2+17x$$
$$27x=x^2+17x$$
$$0=x^2-10x$$
$$0=x(x-10)$$

Apply the zero product principle.

$x = 0 \qquad x - 10 = 0$

$x = 10$

Both solutions check. The solutions are 0 and 10 and the solution set is $\{0, 10\}$.

35. $(x+8)^{1/4} = (2x)^{1/4}$

$\left((x+8)^{1/4}\right)^4 = \left((2x)^{1/4}\right)^4$

$x + 8 = 2x$

$8 = x$

The solution checks. The solution is 8 and the solution set is $\{8\}$.

Application Exercises

37. $304 = 4\sqrt{x} + 280$

$24 = 4\sqrt{x}$

$\frac{24}{4} = \frac{4\sqrt{x}}{4}$

$6 = \sqrt{x}$

$6^2 = \left(\sqrt{x}\right)^2$

$36 = x$

The average score will return to 304 thirty-six years after 1982 in 2018.

39. $32.25 = 6.75\sqrt{x} + 12$

$20.25 = 6.75\sqrt{x}$

$\frac{20.25}{6.75} = \frac{6.75\sqrt{x}}{6.75}$

$3 = \sqrt{x}$

$3^2 = \left(\sqrt{x}\right)^2$

$9 = x$

The loan amount will reach \$32.25 billion 9 years after 1993 in the year 2002.

41. $40000 = 5000\sqrt{100 - x}$

$\frac{40000}{5000} = \frac{5000\sqrt{100 - x}}{5000}$

$8 = \sqrt{100 - x}$

$8^2 = \left(\sqrt{100 - x}\right)^2$

$64 = 100 - x$

$-36 = -x$

$36 = x$

40,000 people in the group will survive to age 36. This is shown on the graph as the point $(36,\ 40{,}000)$.

43. $87 = 29x^{1/3}$

$\frac{87}{29} = \frac{29x^{1/3}}{29}$

$3 = x^{1/3}$

$3^3 = \left(x^{1/3}\right)^3$

$27 = x$

A Galápagos island with an area of 27 square miles will have 87 plant species.

45. $365 = 0.2x^{1/3}$

$\frac{365}{0.2} = \frac{0.2x^{1/3}}{0.2}$

$1825 = x^{1/3}$

$1825^3 = \left(x^{1/3}\right)^3$

$6{,}078{,}390{,}625 = x$

The average distance of the Earth from the sun is approximately 6078 million or 6,078,000,000 kilometers.

Writing in Mathematics

47. Answers will vary.

49. Answers will vary.

51. Answers will vary.

53. Answers will vary.

Technology Exercises

55. $\sqrt{x+3}=5$

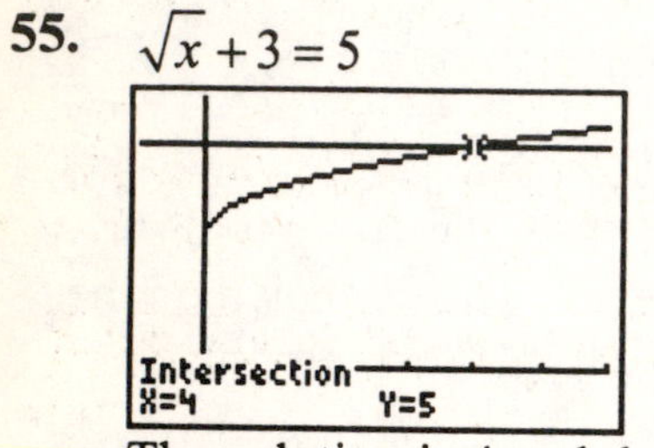

The solution is 4 and the solution set is $\{4\}$.

57. $4\sqrt{x}=x+3$

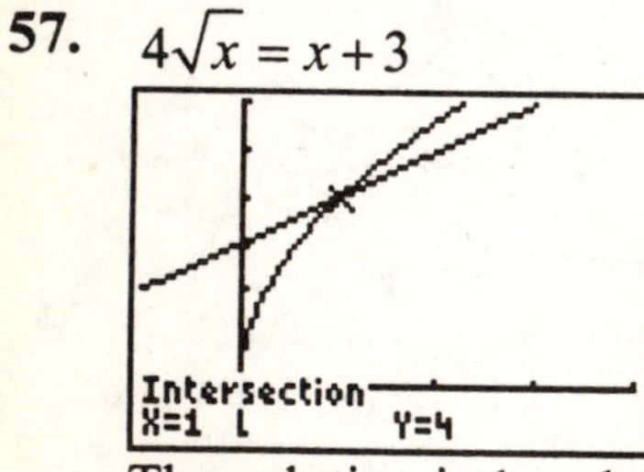

The solution is 1 and the solution set is $\{1\}$.

Critical Thinking Exercises

59. Statement **c.** is true. To show this, substitute for T in the equation for L and simplify.

$$L=\frac{8T^2}{\pi^2}=\frac{8\left(2\pi\sqrt{\frac{L}{32}}\right)^2}{\pi^2}=\frac{8\left(4\pi^2\frac{L}{32}\right)}{\pi^2}$$

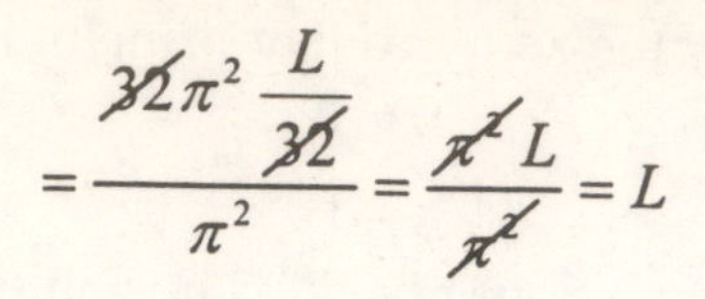

$$=\frac{32\pi^2\frac{L}{32}}{\pi^2}=\frac{\pi^2 L}{\pi^2}=L$$

Statement **a.** is false. The first step is to square both sides, obtaining $x+6=x^2+4x+4$.

Statement **b.** is false. The equation $\sqrt{x+4}=-5$ has no solution. By definition, absolute values are positive.

Statement **d.** is false. We know that an equation with an absolute value equal to a negative number has no solution. In this case, however, we do not know that $-x$ represents a negative number. If x is negative, then $-x$ is positive.

61.
$$\sqrt[3]{x\sqrt{x}}=9$$
$$\left(\sqrt[3]{x\sqrt{x}}\right)^3=9^3$$
$$x\sqrt{x}=729$$
$$\left(x\sqrt{x}\right)^2=729^2$$
$$x^2x=531441$$
$$x^3=531441$$
$$x=81$$

The solution checks. The solution is 81 and the solution set is $\{81\}$.

63.
$$(x-4)^{2/3}=25$$
$$\left((x-4)^{2/3}\right)^{3/2}=25^{3/2}$$

$$x-4=\left(\sqrt{25}\right)^3$$
$$x-4=5^3$$
$$x-4=125$$
$$x=129$$

The solution checks. The solution is 129 and the solution set is $\{129\}$.

Review Exercises

64. $\dfrac{4x^4-3x^3+2x^2-x-1}{x+3}$

$$=\frac{4x^4-3x^3+2x^2-x-1}{x-(-3)}$$

$$\begin{array}{r|rrrrr} -3 & 4 & -3 & 2 & -1 & -1 \\ & & -12 & 45 & -141 & 426 \\ \hline & 4 & -15 & 47 & -142 & 425 \end{array}$$

$$\frac{4x^4-3x^3+2x^2-x-1}{x+3}$$
$$=4x^3-15x^2+47x-142+\frac{425}{x+3}$$

65. $\dfrac{3x^2-12}{x^2+2x-8}\div\dfrac{6x+18}{x+4}$

$$=\frac{3x^2-12}{x^2+2x-8}\cdot\frac{x+4}{6x+18}$$
$$=\frac{3\left(x^2-4\right)}{\cancel{(x+4)}(x-2)}\cdot\frac{\cancel{x+4}}{6(x+3)}$$
$$=\frac{3(x+2)\cancel{(x-2)}}{\cancel{(x-2)}}\cdot\frac{1}{6(x+3)}$$
$$=\frac{3(x+2)}{1}\cdot\frac{1}{6(x+3)}$$
$$=\frac{3(x+2)}{6(x+3)}=\frac{x+2}{2(x+3)}$$

66. $y^2-6y+9-25x^2$

$$=\left(y^2-6y+9\right)-25x^2$$
$$=(y-3)^2-(5x)^2$$
$$=\left((y-3)+5x\right)\left((y-3)-5x\right)$$
$$=(y-3+5x)(y-3-5x)$$

Check Points 7.7

1. **a.** $\sqrt{-16}=\sqrt{16(-1)}=\sqrt{16}\sqrt{-1}=4i$

b. $\sqrt{-5}=\sqrt{5(-1)}=\sqrt{5}\sqrt{-1}=\sqrt{5}i$

c. $\sqrt{-50}=\sqrt{50(-1)}=\sqrt{25\cdot 2}\sqrt{-1}$
$$=5\sqrt{2}i$$

2. **a.** $(5-2i)+(3+3i)$

$$=5-2i+3+3i$$
$$=5+3-2i+3i$$
$$=(5+3)+(-2i+3i)$$
$$=8+(-2+3)i$$
$$=8+1i=8+i$$

b. $(2+6i)-(12-4i)$

$$=2+6i-12+4i$$
$$=2-12+6i+4i$$
$$=(2-12)+(6i+4i)$$
$$=-10+(6+4)i$$
$$=-10+10i$$

3. **a.** $7i(2-9i)$

$$=7i\cdot 2-7i\cdot 9i=14i-63i^2$$
$$=14i-63(-1)=63+14i$$

b. $(5+4i)(6-7i)$
$= 30-35i+24i-28i^2$
$= 30-11i-28(-1)$
$= 30-11i+28$
$= 58-11i$

4. $\sqrt{-5}\cdot\sqrt{-7} = \sqrt{5}\sqrt{-1}\cdot\sqrt{7}\sqrt{-1}$
$= \sqrt{5}i\cdot\sqrt{7}i = \sqrt{35}i^2$
$= \sqrt{35}(-1) = -\sqrt{35}$

5. $\frac{6+2i}{4-3i} = \frac{6+2i}{4-3i}\cdot\frac{4+3i}{4+3i}$
$= \frac{24+18i+8i+6i^2}{4^2-(3i)^2}$
$= \frac{24+26i+6(-1)}{16-9i^2}$
$= \frac{24+26i-6}{16-9(-1)}$
$= \frac{18+26i}{16+9}$
$= \frac{18+26i}{25} = \frac{18}{25}+\frac{26}{25}i$

6. $\frac{3-2i}{4i} = \frac{3-2i}{4i}\cdot\frac{i}{i} = \frac{3i-2i^2}{4i^2}$
$= \frac{3i-2(-1)}{4(-1)} = \frac{2+3i}{-4}$
$= \frac{2}{-4}+\frac{3i}{-4} = -\frac{1}{2}-\frac{3}{4}i$

7. **a.** $i^{16} = (1^2)^8 = (-1)^8 = 1$

b. $i^{25} = i^{24}\cdot i = (i^2)^{12} i$
$= (-1)^{12} i = 1i = i$

c. $i^{35} = i^{34}i = (i^2)^{17}\cdot i = (-1)^{17}\cdot i$
$= (-1)\cdot i = -i$

Problem Set 7.7

Practice Exercises

1. $\sqrt{-49} = \sqrt{49\cdot -1} = \sqrt{49}\cdot\sqrt{-1} = 7i$

3. $\sqrt{-17} = \sqrt{17\cdot -1} = \sqrt{17}\cdot\sqrt{-1} = \sqrt{17}\ i$

5. $\sqrt{-75} = \sqrt{25\cdot 3\cdot -1}$
$= \sqrt{25}\cdot\sqrt{3}\cdot\sqrt{-1} = 5\sqrt{3}\ i$

7. $\sqrt{-28} = \sqrt{4\cdot 7\cdot -1}$
$= \sqrt{4}\cdot\sqrt{7}\cdot\sqrt{-1} = 2\sqrt{7}\ i$

9. $-\sqrt{-150} = -\sqrt{25\cdot 6\cdot -1}$
$= -\sqrt{25}\cdot\sqrt{6}\cdot\sqrt{-1} = -5\sqrt{6}\ i$

11. $7+\sqrt{-16} = 7+\sqrt{16\cdot -1}$
$= 7+\sqrt{16}\cdot\sqrt{-1} = 7+4i$

13. $5+\sqrt{-5} = 5+\sqrt{5\cdot -1}$
$= 5+\sqrt{5}\cdot\sqrt{-1}$
$= 5+\sqrt{5}\ i$

15. $6-\sqrt{-18} = 6-\sqrt{9\cdot 2\cdot -1}$
$= 6-\sqrt{9}\cdot\sqrt{2}\cdot\sqrt{-1}$
$= 6-3\sqrt{2}\ i$

17. $(3+2i)+(5+i)$
$=3+2i+5+i=3+5+2i+i$
$=(3+5)+(2+1)i=8+3i$

19. $(7+2i)+(1-4i)$
$=7+2i+1-4i=7+1+2i-4i$
$=(7+1)+(2-4)i=8-2i$

21. $(10+7i)-(5+4i)$
$=10+7i-5-4i=10-5+7i-4i$
$=(10-5)+(7-4)i=5+3i$

23. $(9-4i)-(10+3i)$
$=9-4i-10-3i=9-10-4i-3i$
$=(9-10)+(-4-3)i$
$=-1+(-7)i=-1-7i$

25. $(3+2i)-(5-7i)$
$=3+2i-5+7i=3-5+2i+7i$
$=(3-5)+(2+7)i=-2+9i$

27. $(-5+4i)-(-13-11i)$
$=-5+4i+13+11i$
$=-5+13+4i+11i$
$=(-5+13)+(4+11)i=8+15i$

29. $8i-(14-9i)$
$=8i-14+9i=-14+8i+9i$
$=-14+(8+9)i=-14+17i$

31. $\left(2+\sqrt{3}\,i\right)+\left(7+4\sqrt{3}\,i\right)$
$=2+\sqrt{3}\,i+7+4\sqrt{3}\,i$
$=2+7+\sqrt{3}\,i+4\sqrt{3}\,i$
$=(2+7)+\left(\sqrt{3}+4\sqrt{3}\right)i=9+5\sqrt{3}\,i$

33. $2i(5+3i)$
$=2i\cdot 5+2i\cdot 3i=10i+6i^2$
$=10i+6(-1)=-6+10i$

35. $3i(7i-5)$
$=3i\cdot 7i-3i\cdot 5=21i^2-15i$
$=21(-1)-15i=-21-15i$

37. $-7i(2-5i)$
$=-7i\cdot 2-(-7i)5i=-14i+35i^2$
$=-14i+35(-1)=-35-14i$

39. $(3+i)(4+5i)=12+15i+4i+5i^2$
$=12+15i+4i+5(-1)$
$=12-5+15i+4i$
$=7+19i$

41. $(7-5i)(2-3i)$
$=14-21i-10i+15i^2$
$=14-21i-10i+15(-1)$
$=14-15-21i-10i=-1-31i$

43. $(6-3i)(-2+5i)$
$=-12+30i+6i-15i^2$
$=-12+30i+6i-15(-1)$
$=-12+15+30i+6i=3+36i$

45. $(3+5i)(3-5i)$
$=9-\cancel{15i}+\cancel{15i}-25i^2$

$$= 9 - 25(-1) = 9 + 25$$
$$= 34 = 34 + 0i$$

47. $(-5+3i)(-5-3i)$
$$= 25 + 15i - 15i - 9i^2$$
$$= 25 - 9(-1) = 25 + 9$$
$$= 34 = 34 + 0i$$

49. $(3-\sqrt{2}\,i)(3+\sqrt{2}\,i)$
$$= 9 + 3\sqrt{2}\,i - 3\sqrt{2}\,i - 2i^2$$
$$= 9 - 2(-1) = 9 + 2$$
$$= 11 = 11 + 0i$$

51. $(2+3i)^2$
$$= 4 + 2\cdot 6i + 9i^2 = 4 + 12i + 9(-1)$$
$$= 4 - 9 + 12i = -5 + 12i$$

53. $(5-2i)^2 = 25 - 2\cdot 10i + 4i^2$
$$= 25 - 20i + 4(-1)$$
$$= 25 - 4 - 20i = 21 - 20i$$

55. $\sqrt{-7}\cdot\sqrt{-2} = \sqrt{7}\sqrt{-1}\cdot\sqrt{2}\sqrt{-1}$
$$= \sqrt{7}\,i\cdot\sqrt{2}\,i = \sqrt{14}\,i^2$$
$$= \sqrt{14}(-1) = -\sqrt{14}$$
$$= -\sqrt{14} + 0i$$

57. $\sqrt{-9}\cdot\sqrt{-4}$
$$= \sqrt{9}\sqrt{-1}\cdot\sqrt{4}\sqrt{-1} = 3i\cdot 2i = 6i^2$$
$$= 6(-1) = -6 = -6 + 0i$$

59. $\sqrt{-7}\cdot\sqrt{-25} = \sqrt{7}\sqrt{-1}\cdot\sqrt{25}\sqrt{-1}$
$$= \sqrt{7}\,i\cdot 5i = 5\sqrt{7}\,i^2$$
$$= 5\sqrt{7}(-1) = -5\sqrt{7}$$
$$= -5\sqrt{7} + 0i$$

61. $\sqrt{-8}\cdot\sqrt{-3} = \sqrt{4\cdot 2}\sqrt{-1}\cdot\sqrt{3}\sqrt{-1}$
$$= 2\sqrt{2}\,i\cdot\sqrt{3}\,i = 2\sqrt{6}\,i^2$$
$$= 2\sqrt{6}(-1) = -2\sqrt{6}$$
$$= -2\sqrt{6} + 0i$$

63. $\dfrac{2}{3+i} = \dfrac{2}{3+i}\cdot\dfrac{3-i}{3-i} = \dfrac{6-2i}{3^2-i^2}$
$$= \frac{6-2i}{9-(-1)} = \frac{6-2i}{9+1}$$
$$= \frac{6-2i}{10} = \frac{6}{10} - \frac{2i}{10}$$
$$= \frac{3}{5} - \frac{1}{5}i$$

65. $\dfrac{2i}{1+i} = \dfrac{2i}{1+i}\cdot\dfrac{1-i}{1-i} = \dfrac{2i-2i^2}{1^2-i^2}$
$$= \frac{2i-2(-1)}{1-(-1)} = \frac{2+2i}{1+1}$$
$$= \frac{2+2i}{2} = \frac{2}{2} + \frac{2i}{2} = 1 + i$$

67. $\dfrac{7}{4-3i} = \dfrac{7}{4-3i}\cdot\dfrac{4+3i}{4+3i} = \dfrac{28+21i}{4^2-(3i)^2}$
$$= \frac{28+21i}{16-9i^2} = \frac{28+21i}{16-9(-1)}$$
$$= \frac{28+21i}{16+9} = \frac{28+21i}{25}$$
$$= \frac{28}{25} + \frac{21}{25}i$$

69. $$\frac{6i}{3-2i}=\frac{6i}{3-2i}\cdot\frac{3+2i}{3+2i}=\frac{18i+12i^2}{3^2-(2i)^2}$$
$$=\frac{18i+12(-1)}{9-4i^2}=\frac{-12+18i}{9-4(-1)}$$
$$=\frac{-12+18i}{9+4}=\frac{-12+18i}{13}$$
$$=-\frac{12}{13}+\frac{18}{13}i$$

71. $$\frac{1+i}{1-i}=\frac{1+i}{1-i}\cdot\frac{1+i}{1+i}=\frac{1+2i+i^2}{1^2-i^2}$$
$$=\frac{1+2i+(-1)}{1-(-1)}=\frac{2i}{2}$$
$$=i \text{ or } 0+i$$

73. $$\frac{2-3i}{3+i}=\frac{2-3i}{3+i}\cdot\frac{3-i}{3-i}$$
$$=\frac{6-2i-9i+3i^2}{3^2-i^2}$$
$$=\frac{6-11i+3(-1)}{9-(-1)}$$
$$=\frac{6-3-11i}{9+1}$$
$$=\frac{3-11i}{10}=\frac{3}{10}-\frac{11}{10}i$$

75. $$\frac{5-2i}{3+2i}=\frac{5-2i}{3+2i}\cdot\frac{3-2i}{3-2i}$$
$$=\frac{15-10i-6i+4i^2}{3^2-(2i)^2}$$
$$=\frac{15-10i-6i+4i^2}{3^2-(2i)^2}$$
$$=\frac{15-16i+4(-1)}{9-4i^2}$$
$$=\frac{15-4-16i}{9-4(-1)}$$
$$=\frac{11-16i}{9+4}$$
$$=\frac{11-16i}{13}=\frac{11}{13}-\frac{16}{13}i$$

77. $$\frac{4+5i}{3-7i}=\frac{4+5i}{3-7i}\cdot\frac{3+7i}{3+7i}$$
$$=\frac{12+28i+15i+35i^2}{3^2-(7i)^2}$$
$$=\frac{12+43i+35(-1)}{9-49i^2}$$
$$=\frac{12-35+43i}{9-49(-1)}$$
$$=\frac{-23+43i}{9+49}=\frac{-23+43i}{58}$$
$$=-\frac{23}{58}+\frac{43}{58}i$$

79. $$\frac{7}{3i}=\frac{7}{3i}\cdot\frac{-3i}{-3i}=\frac{-21i}{-9i^2}=\frac{-21i}{-9(-1)}$$
$$=\frac{-21i}{9}=-\frac{7}{3}i \text{ or } 0-\frac{7}{3}i$$

81. $$\frac{8-5i}{2i}=\frac{8-5i}{2i}\cdot\frac{-2i}{-2i}=\frac{-16i+10i^2}{-4i^2}$$
$$=\frac{-16i+10(-1)}{-4(-1)}=\frac{-10-16i}{4}$$
$$=-\frac{10}{4}-\frac{16}{4}i=-\frac{5}{2}-4i$$

83. $$\frac{4+7i}{-3i}=\frac{4+7i}{-3i}\cdot\frac{3i}{3i}=\frac{12i+21i^2}{-9i^2}$$

$$= \frac{12i + 21(-1)}{-9(-1)} = \frac{-21 + 12i}{9}$$
$$= -\frac{21}{9} + \frac{12}{9}i = -\frac{7}{3} + \frac{4}{3}i$$

85. $i^{10} = (i^2)^5 = (-1)^5 = -1$

87. $i^{11} = (i^2)^5 i = (-1)^5 i = -i$

89. $i^{22} = (i^2)^{11} = (-1)^{11} = -1$

91. $i^{200} = (i^2)^{100} = (-1)^{100} = 1$

93. $i^{17} = (i^2)^8 i = (-1)^8 i = i$

95. $(-i)^4 = (-1)^4 i^4 = i^4 = (i^2)^2$
$= (-1)^2 = 1$

97. $(-i)^9 = (-1)^9 i^9 = (-1)(i^2)^4 i$
$= (-1)(-1)^4 i = (-1)i$
$= -i$

99. $i^{24} + i^2 = (i^2)^{12} + (-1)$
$= (-1)^{12} + (-1)$
$= 1 + (-1) = 0$

Application Exercises

101. $E = IR = (4 - 5i)(3 + 7i)$
$= 12 + 28i - 15i - 35i^2$
$= 12 + 13i - 35(-1)$
$= 12 + 35 + 13i = 47 + 13i$

The voltage of the circuit is $(47 + 13i)$ volts.

103. Sum:
$(5 + \sqrt{15}\, i) + (5 - \sqrt{15}\, i)$
$= 5 + \cancel{\sqrt{15}\, i} + 5 - \cancel{\sqrt{15}\, i}$
$= 5 + 5 = 10$
Product:
$(5 + \sqrt{15}\, i)(5 - \sqrt{15}\, i)$
$= 25 - \cancel{5\sqrt{15}\, i} + \cancel{5\sqrt{15}\, i} - 15i^2$
$= 25 - 15(-1) = 25 + 15 = 40$

Writing in Mathematics

105. Answers will vary.

107. Answers will vary.

109. Answers will vary.

111. Answers will vary.

113. Answers will vary.

115. Answers will vary.

117. Answers will vary.

Critical Thinking Exercises

119. Statement **d.** is true.
$(x + yi)(x - yi)$
$= x^2 - \cancel{xyi} + \cancel{xyi} - y^2 i^2$
$= x^2 - y^2(-1)$
$= x^2 + y^2$

Statement **a.** is false. All irrational numbers are complex numbers.

Statement **b.** is false.

$$(3+7i)(3-7i)$$
$$=3^2-(7i)^2=9-49i^2$$
$$=9-49(-1)$$
$$=9+49=58$$

Statement **c.** is false.

$$\frac{7+3i}{5+3i}=\frac{7+3i}{5+3i}\cdot\frac{5-3i}{5-3i}$$
$$=\frac{35-21i+15i-9i^2}{5^2-(3i)^2}$$
$$=\frac{35-6i-9(-1)}{25-9i^2}$$
$$=\frac{35-6i+9}{25-9(-1)}$$
$$=\frac{44-6i}{25+9}=\frac{44-6i}{34}$$
$$=\frac{44}{34}-\frac{6}{34}i=\frac{22}{17}-\frac{3}{17}i$$

121.
$$\frac{4}{(2+i)(3-i)}$$
$$=\frac{4}{6-2i+3i-i^2}$$
$$=\frac{4}{6+i-(-1)}=\frac{4}{6+1+i}$$
$$=\frac{4}{7+i}\cdot\frac{7-i}{7-i}=\frac{28-4i}{7^2-i^2}$$
$$=\frac{28-4i}{49-(-1)}=\frac{28-4i}{50}$$
$$=\frac{28}{50}-\frac{4}{50}i=\frac{14}{25}-\frac{2}{25}i$$

123.
$$x^2-2x+2$$
$$=(1+i)^2-2(1+i)+2$$
$$=1+\cancel{2i}+i^2-\cancel{2}-\cancel{2i}+\cancel{2}$$
$$=1+(-1)=0$$

Review Exercises

124.
$$\frac{\frac{x}{y^2}+\frac{1}{y}}{\frac{y}{x^2}+\frac{1}{x}}=\frac{\frac{x}{y^2}+\frac{1}{y}}{\frac{y}{x^2}+\frac{1}{x}}\cdot\frac{x^2y^2}{x^2y^2}$$
$$=\frac{\frac{x}{\cancel{y^2}}\cdot x^2\cancel{y^2}+\frac{1}{\cancel{y}}\cdot x^2y^{\cancel{2}}}{\frac{y}{\cancel{x^2}}\cdot\cancel{x^2}y^2+\frac{1}{\cancel{x}}\cdot x^{\cancel{2}}y^2}$$
$$=\frac{x^3+x^2y}{y^3+xy^2}=\frac{x^2\cancel{(x+y)}}{y^2\cancel{(y+x)}}$$
$$=\frac{x^2}{y^2}$$

125.
$$\frac{1}{x}+\frac{1}{y}=\frac{1}{z}$$
$$\frac{1}{\cancel{x}}\cdot\cancel{x}yz+\frac{1}{\cancel{y}}\cdot x\cancel{y}z=\frac{1}{\cancel{z}}\cdot xy\cancel{z}$$
$$yz+xz=xy$$
$$yz=xy-xz$$
$$yz=x(y-z)$$
$$x=\frac{yz}{y-z}$$

126.
$$2x-\frac{x-3}{8}=\frac{1}{2}+\frac{x+5}{2}$$
$$2x\cdot 8-\frac{x-3}{\cancel{8}}\cdot\cancel{8}=\frac{1}{2}\cdot 8+\frac{x+5}{2}\cdot 8$$

$$16x - x + 3 = 4 + 4(x+5)$$
$$16x - x + 3 = 4 + 4x + 20$$
$$15x + 3 = 4x + 24$$
$$11x + 3 = 24$$
$$11x = 21$$
$$x = \frac{21}{11}$$

The solution is $\frac{21}{11}$ and the solution set is $\left\{\frac{21}{11}\right\}$.

Chapter 7 Review

7.1

1. $\sqrt{81} = 9$ because $9^2 = 81$

2. $-\sqrt{\frac{1}{100}} = -\frac{1}{10}$ because $\left(\frac{1}{10}\right)^2 = \frac{1}{100}$

3. $\sqrt[3]{-27} = -3$ because $(-3)^3 = -27$

4. $\sqrt[4]{-16}$
Not a real number
The index is even and the radicand is negative.

5. $\sqrt[5]{-32} = -2$ because $(-2)^5 = -32$

6. $f(15) = \sqrt{2(15)-5} = \sqrt{30-5}$
$= \sqrt{25} = 5$
$f(4) = \sqrt{2(4)-5} = \sqrt{8-5} = \sqrt{3}$
$f\left(\frac{5}{2}\right) = \sqrt{2\left(\frac{5}{2}\right)-5} = \sqrt{5-5}$
$= \sqrt{0} = 0$
$f(1) = \sqrt{2(1)-5} = \sqrt{2-5} = \sqrt{-3}$
Not a real number

7. $g(4) = \sqrt[3]{4(4)-8} = \sqrt[3]{16-8} = \sqrt[3]{8} = 2$
$g(0) = \sqrt[3]{4(0)-8} = \sqrt[3]{-8} = -2$
$g(-14) = \sqrt[3]{4(-14)-8} = \sqrt[3]{-56-8}$
$= \sqrt[3]{-64} = -4$

8. To find the domain, set the radicand greater than or equal to zero and solve the resulting inequality.
$x - 2 \geq 0$
$x \geq 2$
The domain of f is $\{x | x \geq 2\}$ or $[2, \infty)$.

9. To find the domain, set the radicand greater than or equal to zero and solve the resulting inequality.
$100 - 4x \geq 0$
$-4x \geq -100$
$\frac{-4x}{-4} \leq \frac{-100}{-4}$
$x \leq 25$
The domain of g is $\{x | x \leq 25\}$ or $(-\infty, 25]$.

10. $\sqrt{25x^2} = 5|x|$

11. $\sqrt{(x+14)^2} = |x+14|$

12. $\sqrt{x^2-8x+16}=\sqrt{(x-4)^2}$
$=|x-4|$

13. $\sqrt[3]{64x^3}=4x$

14. $\sqrt[4]{16x^4}=2|x|$

15. $\sqrt[5]{-32(x+7)^5}=-2(x+7)$

7.2

16. $(5xy)^{\frac{1}{3}}=\sqrt[3]{5xy}$

17. $16^{\frac{3}{2}}=\left(\sqrt{16}\right)^3=(4)^3=64$

18. $32^{\frac{4}{5}}=\left(\sqrt[5]{32}\right)^4=(2)^4=16$

19. $\sqrt{7x}=(7x)^{\frac{1}{2}}$

20. $\left(\sqrt[3]{19xy}\right)^5=(19xy)^{\frac{5}{3}}$

21. $8^{-\frac{2}{3}}=\dfrac{1}{8^{\frac{2}{3}}}=\dfrac{1}{\left(\sqrt[3]{8}\right)^2}=\dfrac{1}{(2)^2}=\dfrac{1}{4}$

22. $3x(ab)^{-\frac{4}{5}}=\dfrac{3x}{(ab)^{\frac{4}{5}}}$
$=\dfrac{3x}{\left(\sqrt[5]{ab}\right)^4}$ or $\dfrac{3x}{\sqrt[5]{(ab)^4}}$

23. $x^{\frac{1}{3}}\cdot x^{\frac{1}{4}}=x^{\frac{1}{3}+\frac{1}{4}}=x^{\frac{4}{12}+\frac{3}{12}}=x^{\frac{7}{12}}$

24. $\dfrac{5^{\frac{1}{2}}}{5^{\frac{1}{3}}}=5^{\frac{1}{2}-\frac{1}{3}}=5^{\frac{3}{6}-\frac{2}{6}}=5^{\frac{1}{6}}$

25. $\left(8x^6y^3\right)^{\frac{1}{3}}=8^{\frac{1}{3}}x^{6\cdot\frac{1}{3}}y^{3\cdot\frac{1}{3}}=2x^2y$

26. $\left(x^{-\frac{2}{3}}y^{\frac{1}{4}}\right)^{\frac{1}{2}}=x^{-\frac{2}{3}\cdot\frac{1}{2}}y^{\frac{1}{4}\cdot\frac{1}{2}}$
$=x^{-\frac{1}{3}}y^{\frac{1}{8}}=\dfrac{y^{\frac{1}{8}}}{x^{\frac{1}{3}}}$

27. $\sqrt[3]{x^9y^{12}}=\left(x^9y^{12}\right)^{\frac{1}{3}}$
$=x^{9\cdot\frac{1}{3}}y^{12\cdot\frac{1}{3}}=x^3y^4$

28. $\sqrt[9]{x^3y^9}=\left(x^3y^9\right)^{\frac{1}{9}}=x^{3\cdot\frac{1}{9}}y^{9\cdot\frac{1}{9}}$
$=x^{\frac{1}{3}}y=y\sqrt[3]{x}$

29. $\sqrt{x}\cdot\sqrt[3]{x}=x^{\frac{1}{2}}x^{\frac{1}{3}}=x^{\frac{1}{2}+\frac{1}{3}}=x^{\frac{3}{6}+\frac{2}{6}}$
$=x^{\frac{5}{6}}=\sqrt[6]{x^5}$

30. $\dfrac{\sqrt[3]{x^2}}{\sqrt[4]{x^2}}=\dfrac{x^{\frac{2}{3}}}{x^{\frac{2}{4}}}=x^{\frac{2}{3}-\frac{1}{2}}$
$=x^{\frac{4}{6}-\frac{3}{6}}=x^{\frac{1}{6}}=\sqrt[6]{x}$

31.
$$\sqrt[5]{\sqrt[3]{x}} = \sqrt[5]{x^{\frac{1}{3}}} = \left(x^{\frac{1}{3}}\right)^{\frac{1}{5}} = x^{\frac{1}{3}\cdot\frac{1}{5}}$$
$$= x^{\frac{1}{15}} = \sqrt[15]{x}$$

32. Since 2012 is 27 years after 1985, find $f(27)$.
$$f(27) = 350(27)^{\frac{2}{3}} = 350\left(\sqrt[3]{27}\right)^2$$
$$= 350(3)^2 = 350(9) = 3150$$
Expenditures will be \$3150 million or \$3,150,000,000 in the year 2012.

7.3

33. $\sqrt{3x}\cdot\sqrt{7y} = \sqrt{21xy}$

34. $\sqrt[5]{7x^2}\cdot\sqrt[5]{11x} = \sqrt[5]{77x^3}$

35. $\sqrt[6]{x-5}\cdot\sqrt[6]{(x-5)^4} = \sqrt[6]{(x-5)^5}$

36.
$$f(x) = \sqrt{7x^2 - 14x + 7}$$
$$= \sqrt{7(x^2 - 2x + 1)}$$
$$= \sqrt{7(x-1)^2} = \sqrt{7}|x-1|$$

37.
$$\sqrt{20x^3} = \sqrt{4\cdot 5\cdot x^2\cdot x} = \sqrt{4x^2\cdot 5x}$$
$$= 2x\sqrt{5x}$$

38.
$$\sqrt[3]{54x^8y^6} = \sqrt[3]{27\cdot 2\cdot x^6\cdot x^2y^6}$$
$$= \sqrt[3]{27x^6y^6\cdot 2x^2}$$
$$= 3x^2y^2\sqrt[3]{2x^2}$$

39.
$$\sqrt[4]{32x^3y^{11}} = \sqrt[4]{16\cdot 2\cdot x^3y^8\cdot y^3}$$
$$= \sqrt[4]{16y^8\cdot 2x^3y^3}$$
$$= 2y^2\sqrt[4]{2x^3y^3}$$

40.
$$\sqrt{6x^3}\cdot\sqrt{4x^2} = \sqrt{24x^5} = \sqrt{4\cdot 6\cdot x^4\cdot x}$$
$$= \sqrt{4x^4\cdot 6x} = 2x^2\sqrt{6x}$$

41.
$$\sqrt[3]{4x^2y}\cdot\sqrt[3]{4xy^4} = \sqrt[3]{16x^3y^5}$$
$$= \sqrt[3]{8\cdot 2\cdot x^3\cdot y^3\cdot y^2}$$
$$= \sqrt[3]{8x^3y^3\cdot 2y^2}$$
$$= 2xy\sqrt[3]{2y^2}$$

42.
$$\sqrt[5]{2x^4y^3}\cdot\sqrt[5]{8xy^6} = \sqrt[5]{16x^5y^9}$$
$$= \sqrt[5]{16\cdot x^5\cdot y^5\cdot y^4}$$
$$= \sqrt[5]{x^5y^5\cdot 16y^4}$$
$$= xy\sqrt[5]{16y^4}$$

43.
$$\sqrt{x+1}\cdot\sqrt{x-1} = \sqrt{(x+1)(x-1)}$$
$$= \sqrt{x^2-1}$$

7.4

44. $6\sqrt[3]{3} + 2\sqrt[3]{3} = (6+2)\sqrt[3]{3} = 8\sqrt[3]{3}$

45.
$$5\sqrt{18} - 3\sqrt{8} = 5\sqrt{9\cdot 2} - 3\sqrt{4\cdot 2}$$
$$= 5\cdot 3\sqrt{2} - 3\cdot 2\sqrt{2}$$
$$= 15\sqrt{2} - 6\sqrt{2}$$
$$= (15-6)\sqrt{2} = 9\sqrt{2}$$

46.
$$\sqrt[3]{27x^4} + \sqrt[3]{xy^6}$$
$$= \sqrt[3]{27x^3x} + \sqrt[3]{xy^6}$$

$$= 3x\sqrt[3]{x} + y^2\sqrt[3]{x}$$
$$= (3x + y^2)\sqrt[3]{x}$$

47. $2\sqrt[3]{6} - 5\sqrt[3]{48} = 2\sqrt[3]{6} - 5\sqrt[3]{8\cdot 6}$
$$= 2\sqrt[3]{6} - 5\cdot 2\sqrt[3]{6}$$
$$= 2\sqrt[3]{6} - 10\sqrt[3]{6}$$
$$= (2-10)\sqrt[3]{6} = -8\sqrt[3]{6}$$

48. $\sqrt[3]{\dfrac{16}{125}} = \sqrt[3]{\dfrac{8\cdot 2}{125}} = \dfrac{2}{5}\sqrt[3]{2}$

49. $\sqrt{\dfrac{x^3}{100y^4}} = \sqrt{\dfrac{x^2\cdot x}{100y^4}}$
$$= \frac{x}{10y^2}\sqrt{x} \text{ or } \frac{x\sqrt{x}}{10y^2}$$

50. $\sqrt[4]{\dfrac{3y^5}{16x^{20}}} = \sqrt[4]{\dfrac{y^4\cdot 3y}{16x^{20}}}$
$$= \frac{y}{2x^5}\sqrt[4]{3y} \text{ or } \frac{y\sqrt[4]{3y}}{2x^5}$$

51. $\dfrac{\sqrt{48}}{\sqrt{2}} = \sqrt{\dfrac{48}{2}} = \sqrt{24} = \sqrt{4\cdot 6} = 2\sqrt{6}$

52. $\dfrac{\sqrt[3]{32}}{\sqrt[3]{2}} = \sqrt[3]{\dfrac{32}{2}} = \sqrt[3]{16} = \sqrt[3]{8\cdot 2} = 2\sqrt[3]{2}$

53. $\dfrac{\sqrt[4]{64x^7}}{\sqrt[4]{2x^2}} = \sqrt[4]{\dfrac{64x^7}{2x^2}} = \sqrt[4]{32x^5}$
$$= \sqrt[4]{16\cdot 2\cdot x^4\cdot x}$$
$$= \sqrt[4]{16x^4\cdot 2x} = 2x\sqrt[4]{2x}$$

54. $\dfrac{\sqrt{200x^3y^2}}{\sqrt{2x^{-2}y}} = \sqrt{\dfrac{200x^3y^2}{2x^{-2}y}} = \sqrt{100x^5y}$
$$= \sqrt{100x^4xy} = 10x^2\sqrt{xy}$$

7.5

55. $\sqrt{3}\left(2\sqrt{6} + 4\sqrt{15}\right) = 2\sqrt{18} + 4\sqrt{45}$
$$= 2\sqrt{9\cdot 2} + 4\sqrt{9\cdot 5}$$
$$= 2\cdot 3\sqrt{2} + 4\cdot 3\sqrt{5}$$
$$= 6\sqrt{2} + 12\sqrt{5}$$

56. $\sqrt[3]{5}\left(\sqrt[3]{50} - \sqrt[3]{2}\right) = \sqrt[3]{250} - \sqrt[3]{10}$
$$= \sqrt[3]{125\cdot 2} - \sqrt[3]{10}$$
$$= 5\sqrt[3]{2} - \sqrt[3]{10}$$

57. $\left(\sqrt{7} - 3\sqrt{5}\right)\left(\sqrt{7} + 6\sqrt{5}\right)$
$$= 7 + 6\sqrt{35} - 3\sqrt{35} - 18\cdot 5$$
$$= 7 + 3\sqrt{35} - 90$$
$$= 3\sqrt{35} - 83 \text{ or } -83 + 3\sqrt{35}$$

58. $\left(\sqrt{x} - \sqrt{11}\right)\left(\sqrt{y} - \sqrt{11}\right)$
$$= \sqrt{xy} - \sqrt{11x} - \sqrt{11y} + 11$$

59. $\left(\sqrt{5} + \sqrt{8}\right)^2 = 5 + 2\cdot\sqrt{5}\cdot\sqrt{8} + 8$
$$= 13 + 2\sqrt{40}$$
$$= 13 + 2\sqrt{4\cdot 10}$$
$$= 13 + 2\cdot 2\sqrt{10}$$
$$= 13 + 4\sqrt{10}$$

60. $\left(2\sqrt{3}-\sqrt{10}\right)^2$
$= 4\cdot 3 - 2\cdot 2\sqrt{3}\cdot\sqrt{10} + 10$
$= 12 - 4\sqrt{30} + 10 = 22 - 4\sqrt{30}$

61. $\left(\sqrt{7}+\sqrt{13}\right)\left(\sqrt{7}-\sqrt{13}\right)$
$= \left(\sqrt{7}\right)^2 - \left(\sqrt{13}\right)^2 = 7 - 13 = -6$

62. $\left(7-3\sqrt{5}\right)\left(7+3\sqrt{5}\right) = 7^2 - \left(3\sqrt{5}\right)^2$
$= 49 - 9\cdot 5$
$= 49 - 45 = 4$

63. $\frac{4}{\sqrt{6}} = \frac{4}{\sqrt{6}}\cdot\frac{\sqrt{6}}{\sqrt{6}} = \frac{4\sqrt{6}}{6} = \frac{2\sqrt{6}}{3}$

64. $\sqrt{\frac{2}{7}} = \frac{\sqrt{2}}{\sqrt{7}} = \frac{\sqrt{2}}{\sqrt{7}}\cdot\frac{\sqrt{7}}{\sqrt{7}} = \frac{\sqrt{14}}{7}$

65. $\frac{12}{\sqrt[3]{9}} = \frac{12}{\sqrt[3]{3^2}}\cdot\frac{\sqrt[3]{3}}{\sqrt[3]{3}} = \frac{12\sqrt[3]{3}}{\sqrt[3]{3^3}}$
$= \frac{12\sqrt[3]{3}}{3} = 4\sqrt[3]{3}$

66. $\sqrt{\frac{2x}{5y}} = \frac{\sqrt{2x}}{\sqrt{5y}}\cdot\frac{\sqrt{5y}}{\sqrt{5y}} = \frac{\sqrt{10xy}}{\sqrt{5^2y^2}} = \frac{\sqrt{10xy}}{5y}$

67. $\frac{14}{\sqrt[3]{2x^2}} = \frac{14}{\sqrt[3]{2x^2}}\cdot\frac{\sqrt[3]{2^2x}}{\sqrt[3]{2^2x}} = \frac{14\sqrt[3]{2^2x}}{\sqrt[3]{2^3x^3}}$
$= \frac{14\sqrt[3]{4x}}{2x} = \frac{7\sqrt[3]{4x}}{x}$

68. $\sqrt[4]{\frac{7}{3x}} = \frac{\sqrt[4]{7}}{\sqrt[4]{3x}} = \frac{\sqrt[4]{7}}{\sqrt[4]{3x}}\cdot\frac{\sqrt[4]{3^3x^3}}{\sqrt[4]{3^3x^3}}$
$= \frac{\sqrt[4]{7\cdot 3^3x^3}}{\sqrt[4]{3^4x^4}} = \frac{\sqrt[4]{7\cdot 27x^3}}{3x}$
$= \frac{\sqrt[4]{189x^3}}{3x}$

69. $\frac{5}{\sqrt[5]{32x^4y}} = \frac{5}{\sqrt[5]{2^5x^4y}}\cdot\frac{\sqrt[5]{xy^4}}{\sqrt[5]{xy^4}}$
$= \frac{5\sqrt[5]{xy^4}}{\sqrt[5]{2^5x^5y^5}} = \frac{5\sqrt[5]{xy^4}}{2xy}$

70. $\frac{6}{\sqrt{3}-1} = \frac{6}{\sqrt{3}-1}\cdot\frac{\sqrt{3}+1}{\sqrt{3}+1}$
$= \frac{6\left(\sqrt{3}+1\right)}{\left(\sqrt{3}\right)^2-1^2} = \frac{6\left(\sqrt{3}+1\right)}{3-1}$
$= \frac{6\left(\sqrt{3}+1\right)}{2} = 3\left(\sqrt{3}+1\right)$
$= 3\sqrt{3}+3$

71. $\frac{\sqrt{7}}{\sqrt{5}+\sqrt{3}} = \frac{\sqrt{7}}{\sqrt{5}+\sqrt{3}}\cdot\frac{\sqrt{5}-\sqrt{3}}{\sqrt{5}-\sqrt{3}}$
$= \frac{\sqrt{35}-\sqrt{21}}{\left(\sqrt{5}\right)^2-\left(\sqrt{3}\right)^2}$
$= \frac{\sqrt{35}-\sqrt{21}}{5-3} = \frac{\sqrt{35}-\sqrt{21}}{2}$

72. $\frac{10}{2\sqrt{5}-3\sqrt{2}}$
$= \frac{10}{2\sqrt{5}-3\sqrt{2}}\cdot\frac{2\sqrt{5}+3\sqrt{2}}{2\sqrt{5}+3\sqrt{2}}$
$= \frac{10\left(2\sqrt{5}+3\sqrt{2}\right)}{\left(2\sqrt{5}\right)^2-\left(3\sqrt{2}\right)^2} = \frac{10\left(2\sqrt{5}+3\sqrt{2}\right)}{4\cdot 5-9\cdot 2}$

$$=\frac{10(2\sqrt{5}+3\sqrt{2})}{20-18}=\frac{10(2\sqrt{5}+3\sqrt{2})}{2}$$
$$=5(2\sqrt{5}+3\sqrt{2})=10\sqrt{5}+15\sqrt{2}$$

73. $$\frac{\sqrt{x}+5}{\sqrt{x}-3}=\frac{\sqrt{x}+5}{\sqrt{x}-3}\cdot\frac{\sqrt{x}+3}{\sqrt{x}+3}$$
$$=\frac{x+3\sqrt{x}+5\sqrt{x}+15}{(\sqrt{x})^2-3^2}$$
$$=\frac{x+8\sqrt{x}+15}{x-9}$$

74. $$\frac{\sqrt{7}+\sqrt{3}}{\sqrt{7}-\sqrt{3}}=\frac{\sqrt{7}+\sqrt{3}}{\sqrt{7}-\sqrt{3}}\cdot\frac{\sqrt{7}+\sqrt{3}}{\sqrt{7}+\sqrt{3}}$$
$$=\frac{7+2\cdot\sqrt{7}\cdot\sqrt{3}+3}{(\sqrt{7})^2-(\sqrt{3})^2}$$
$$=\frac{10+2\sqrt{21}}{7-3}=\frac{10+2\sqrt{21}}{4}$$
$$=\frac{2(5+\sqrt{21})}{4}=\frac{5+\sqrt{21}}{2}$$

75. $$\frac{2\sqrt{3}+\sqrt{6}}{2\sqrt{6}+\sqrt{3}}$$
$$=\frac{2\sqrt{3}+\sqrt{6}}{2\sqrt{6}+\sqrt{3}}\cdot\frac{2\sqrt{6}-\sqrt{3}}{2\sqrt{6}-\sqrt{3}}$$
$$=\frac{4\sqrt{18}-2\cdot3+2\cdot6-\sqrt{18}}{(2\sqrt{6})^2-(\sqrt{3})^2}$$
$$=\frac{3\sqrt{18}-6+12}{4\cdot6-3}=\frac{3\sqrt{9\cdot2}+6}{24-3}$$
$$=\frac{3\cdot3\sqrt{2}+6}{21}=\frac{9\sqrt{2}+6}{21}$$
$$=\frac{3(3\sqrt{2}+2)}{21}=\frac{3\sqrt{2}+2}{7}$$

76. $$\sqrt{\frac{2}{7}}=\frac{\sqrt{2}}{\sqrt{7}}=\frac{\sqrt{2}}{\sqrt{7}}\cdot\frac{\sqrt{2}}{\sqrt{2}}=\frac{2}{\sqrt{14}}$$

77. $$\frac{\sqrt[3]{3x}}{\sqrt[3]{y}}=\frac{\sqrt[3]{3x}}{\sqrt[3]{y}}\cdot\frac{\sqrt[3]{3^2x^2}}{\sqrt[3]{3^2x^2}}$$
$$=\frac{\sqrt[3]{3^3x^3}}{\sqrt[3]{3^2x^2y}}=\frac{3x}{\sqrt[3]{9x^2y}}$$

78. $$\frac{\sqrt{7}}{\sqrt{5}+\sqrt{3}}=\frac{\sqrt{7}}{\sqrt{5}+\sqrt{3}}\cdot\frac{\sqrt{7}}{\sqrt{7}}$$
$$=\frac{7}{\sqrt{35}+\sqrt{21}}$$

79. $$\frac{\sqrt{7}+\sqrt{3}}{\sqrt{7}-\sqrt{3}}$$
$$=\frac{\sqrt{7}+\sqrt{3}}{\sqrt{7}-\sqrt{3}}\cdot\frac{\sqrt{7}-\sqrt{3}}{\sqrt{7}-\sqrt{3}}$$
$$=\frac{(\sqrt{7})^2-(\sqrt{3})^2}{7-2\sqrt{7}\sqrt{3}+3}=\frac{7-3}{10-2\sqrt{21}}$$
$$=\frac{4}{10-2\sqrt{21}}=\frac{4}{2(5-\sqrt{21})}$$
$$=\frac{2}{5-\sqrt{21}}$$

7.6

80. $$\sqrt{2x+4}=6$$
$$(\sqrt{2x+4})^2=6^2$$
$$2x+4=36$$

$$2x = 32$$
$$x = 16$$

The solution checks. The solution is 16 and the solution set is $\{16\}$.

81. $\sqrt{x-5}+9=4$

$$\sqrt{x-5}=-5$$

The square root of a number is always positive. The solution set is $\varnothing$ or $\{\ \}$.

82. $\sqrt{2x-3}+x=3$

$$\sqrt{2x-3}=3-x$$
$$\left(\sqrt{2x-3}\right)^2=(3-x)^2$$
$$2x-3=9-6x+x^2$$
$$0=12-8x+x^2$$
$$0=x^2-8x+12$$
$$0=(x-6)(x-2)$$

Apply the zero product principle.

$x-6=0 \qquad x-2=0$

$x=6 \qquad x=2$

6 is an extraneous solution. The solution is 2 and the solution set is $\{2\}$.

83. $\sqrt{x-4}+\sqrt{x+1}=5$

$$\sqrt{x-4}=5-\sqrt{x+1}$$
$$\left(\sqrt{x-4}\right)^2=\left(5-\sqrt{x+1}\right)^2$$
$$x-4=25-10\sqrt{x+1}+x+1$$
$$-30=-10\sqrt{x+1}$$
$$\frac{-30}{-10}=\frac{-10\sqrt{x+1}}{-10}$$
$$3=\sqrt{x+1}$$
$$3^2=\left(\sqrt{x+1}\right)^2$$
$$9=x+1$$
$$8=x$$

The solution checks. The solution is 8 and the solution set is $\{8\}$.

84. $\left(x^2+6x\right)^{\frac{1}{3}}+2=0$

$$\left(x^2+6x\right)^{\frac{1}{3}}=-2$$
$$\sqrt[3]{x^2+6x}=-2$$
$$\left(\sqrt[3]{x^2+6x}\right)^3=(-2)^3$$
$$x^2+6x=-8$$
$$x^2+6x+8=0$$
$$(x+4)(x+2)=0$$

Apply the zero product principle.

$x+4=0 \qquad x+2=0$

$x=-4 \qquad x=-2$

Both solutions check. The solutions are –4 and –2, and the solution set is $\{-4,-2\}$.

85.

$$4=\sqrt{\frac{x}{16}}$$
$$4^2=\left(\sqrt{\frac{x}{16}}\right)^2$$
$$16=\frac{x}{16}$$
$$256=x$$

The hammer was dropped from a height of 256 feet.

86.
$$20,000 = 5000\sqrt{100-x}$$
$$\frac{20,000}{5000} = \frac{5000\sqrt{100-x}}{5000}$$
$$4 = \sqrt{100-x}$$
$$4^2 = \left(\sqrt{100-x}\right)^2$$
$$16 = 100 - x$$
$$-84 = -x$$
$$84 = x$$
20,000 people in the group will survive to 84 years old.

7.7

87. $\sqrt{-81} = \sqrt{81 \cdot -1} = \sqrt{81}\sqrt{-1} = 9i$

88.
$$\sqrt{-63} = \sqrt{9 \cdot 7 \cdot -1}$$
$$= \sqrt{9}\sqrt{7}\sqrt{-1} = 3\sqrt{7}i$$

89.
$$-\sqrt{-8} = -\sqrt{4 \cdot 2 \cdot -1}$$
$$= -\sqrt{4}\sqrt{2}\sqrt{-1} = -2\sqrt{2}i$$

90.
$$(7+12i)+(5-10i)$$
$$= 7 + 12i + 5 - 10i = 12 + 2i$$

91.
$$(8-3i)-(17-7i) = 8 - 3i - 17 + 7i$$
$$= -9 + 4i$$

92.
$$4i(3i-2) = 4i \cdot 3i - 4i \cdot 2$$
$$= 12i^2 - 8i$$
$$= 12(-1) - 8i$$
$$= -12 - 8i$$

93.
$$(7-5i)(2+3i) = 14 + 21i - 10i - 15i^2$$
$$= 14 + 11i - 15(-1)$$
$$= 14 + 11i + 15$$
$$= 29 + 11i$$

94.
$$(3-4i)^2 = 3^2 - 2 \cdot 3 \cdot 4i + (4i)^2$$
$$= 9 - 24i + 16i^2$$
$$= 9 - 24i + 16(-1)$$
$$= 9 - 24i - 16$$
$$= -7 - 24i$$

95.
$$(7+8i)(7-8i)$$
$$= 7^2 - (8i)^2 = 49 - 64i^2$$
$$= 49 - 64(-1) = 49 + 64$$
$$= 113 = 113 + 0i$$

96.
$$\sqrt{-8} \cdot \sqrt{-3} = \sqrt{4 \cdot 2 \cdot -1} \cdot \sqrt{3 \cdot -1}$$
$$= 2\sqrt{2}i \cdot \sqrt{3}i = 2\sqrt{6}i^2$$
$$= 2\sqrt{6}(-1) = -2\sqrt{6}$$
$$= -2\sqrt{6} + 0i$$

97.
$$\frac{6}{5+i} = \frac{6}{5+i} \cdot \frac{5-i}{5-i} = \frac{30-6i}{25-i^2}$$
$$= \frac{30-6i}{25-(-1)} = \frac{30-6i}{25+1}$$
$$= \frac{30-6i}{26} = \frac{30}{26} - \frac{6}{26}i$$
$$= \frac{15}{13} - \frac{3}{13}i$$

98.
$$\frac{3+4i}{4-2i} = \frac{3+4i}{4-2i} \cdot \frac{4+2i}{4+2i}$$
$$= \frac{12+6i+16i+8i^2}{16-4i^2}$$
$$= \frac{12+22i+8(-1)}{16-4(-1)}$$

$$= \frac{12+22i-8}{16+4} = \frac{4+22i}{20}$$
$$= \frac{4}{20} + \frac{22}{20}i = \frac{1}{5} + \frac{11}{10}i$$

99. $\frac{5+i}{3i} = \frac{5+i}{3i} \cdot \frac{i}{i} = \frac{5i+i^2}{3i^2}$
$$= \frac{5i+(-1)}{3(-1)} = \frac{5i-1}{-3}$$
$$= \frac{-1}{-3} + \frac{5}{-3}i = \frac{1}{3} - \frac{5}{3}i$$

100. $i^{16} = \left(i^2\right)^8 = (-1)^8 = 1$

101. $i^{23} = i^{22} \cdot i = \left(i^2\right)^{11} i = (-1)^{11} i$
$$= (-1)i = -i$$

Chapter 7 Test

1. a. $f(-14) = \sqrt{8-2(-14)}$
$$= \sqrt{8+28} = \sqrt{36} = 6$$

b. To find the domain, set the radicand greater than or equal to zero and solve the resulting inequality.
$8 - 2x \geq 0$
$-2x \geq -8$
$x \leq 4$
The domain of f is $\{x | x \leq 4\}$ or $(-\infty, 4]$.

2. $27^{-\frac{4}{3}} = \frac{1}{27^{\frac{4}{3}}} = \frac{1}{\left(\sqrt[3]{27}\right)^4} = \frac{1}{(3)^4} = \frac{1}{81}$

3. $\left(25x^{-\frac{1}{2}}y^{\frac{1}{4}}\right)^{\frac{1}{2}} = 25^{\frac{1}{2}}x^{-\frac{1}{4}}y^{\frac{1}{8}} = 5x^{-\frac{1}{4}}y^{\frac{1}{8}}$
$$= \frac{5y^{\frac{1}{8}}}{x^{\frac{1}{4}}} = \frac{5\sqrt[8]{y}}{\sqrt[4]{x}}$$

4. $\sqrt[8]{x^4} = \left(x^4\right)^{\frac{1}{8}} = x^{4 \cdot \frac{1}{8}} = x^{\frac{1}{2}} = \sqrt{x}$

5. $\sqrt[4]{x} \cdot \sqrt[5]{x} = x^{\frac{1}{4}} \cdot x^{\frac{1}{5}} = x^{\frac{1}{4}+\frac{1}{5}} = x^{\frac{5}{20}+\frac{4}{20}}$
$$= x^{\frac{9}{20}} = \sqrt[20]{x^9}$$

6. $\sqrt{75x^2} = \sqrt{25 \cdot 3x^2} = 5|x|\sqrt{3}$

7. $\sqrt{x^2 - 10x + 25} = \sqrt{(x-5)^2}$
$$= |x-5|$$

8. $\sqrt[3]{16x^4y^8} = \sqrt[3]{8 \cdot 2 \cdot x^3 \cdot x \cdot y^6 \cdot y^2}$
$$= \sqrt[3]{8x^3y^6 \cdot 2xy^2}$$
$$= 2xy^2\sqrt[3]{2xy^2}$$

9. $\sqrt[5]{-\frac{32}{x^{10}}} = \sqrt[5]{-\frac{2^5}{\left(x^2\right)^5}} = -\frac{2}{x^2}$

10. $\sqrt[3]{5x^2} \cdot \sqrt[3]{10y} = \sqrt[3]{50x^2y}$

11. $\sqrt[4]{8x^3y} \cdot \sqrt[4]{4xy^2} = \sqrt[4]{32x^4y^3}$
$$= \sqrt[4]{16 \cdot 2 \cdot x^4 \cdot y^3}$$
$$= \sqrt[4]{16x^4 \cdot 2y^3}$$
$$= 2x\sqrt[4]{2y^3}$$

12. $3\sqrt{18}-4\sqrt{32}=3\sqrt{9\cdot 2}-4\sqrt{16\cdot 2}$
$=3\cdot 3\sqrt{2}-4\cdot 4\sqrt{2}$
$=9\sqrt{2}-16\sqrt{2}=-7\sqrt{2}$

13. $\sqrt[3]{8x^4}+\sqrt[3]{xy^6}=\sqrt[3]{8x^3\cdot x}+\sqrt[3]{xy^6}$
$=2x\sqrt[3]{x}+y^2\sqrt[3]{x}$
$=\left(2x+y^2\right)\sqrt[3]{x}$

14. $\dfrac{\sqrt[3]{16x^8}}{\sqrt[3]{2x^4}}=\sqrt[3]{\dfrac{16x^8}{2x^4}}=\sqrt[3]{8x^4}$
$=\sqrt[3]{8x^3\cdot 2}=2x\sqrt[3]{2}$

15. $\sqrt{3}\left(4\sqrt{6}-\sqrt{5}\right)=\sqrt{3}\cdot 4\sqrt{6}-\sqrt{3}\cdot\sqrt{5}$
$=4\sqrt{18}-\sqrt{15}$
$=4\sqrt{9\cdot 2}-\sqrt{15}$
$=4\cdot 3\sqrt{2}-\sqrt{15}$
$=12\sqrt{2}-\sqrt{15}$

16. $\left(5\sqrt{6}-2\sqrt{2}\right)\left(\sqrt{6}+\sqrt{2}\right)$
$=5\cdot 6+5\sqrt{12}-2\sqrt{12}-2\cdot 2$
$=30+3\sqrt{12}-4=26+3\sqrt{4\cdot 3}$
$=26+3\cdot 2\sqrt{3}=26+6\sqrt{3}$

17. $\left(7-\sqrt{3}\right)^2=49-2\cdot 7\cdot\sqrt{3}+3$
$=52-14\sqrt{3}$

18. $\sqrt{\dfrac{5}{x}}=\dfrac{\sqrt{5}}{\sqrt{x}}\cdot\dfrac{\sqrt{x}}{\sqrt{x}}=\dfrac{\sqrt{5x}}{x}$

19. $\dfrac{5}{\sqrt[3]{5x^2}}=\dfrac{5}{\sqrt[3]{5x^2}}\cdot\dfrac{\sqrt[3]{5^2x}}{\sqrt[3]{5^2x}}=\dfrac{5\sqrt[3]{5^2x}}{\sqrt[3]{5^3x^3}}$
$=\dfrac{5\sqrt[3]{25x}}{5x}=\dfrac{\sqrt[3]{25x}}{x}$

20. $\dfrac{\sqrt{2}-\sqrt{3}}{\sqrt{2}+\sqrt{3}}=\dfrac{\sqrt{2}-\sqrt{3}}{\sqrt{2}+\sqrt{3}}\cdot\dfrac{\sqrt{2}-\sqrt{3}}{\sqrt{2}-\sqrt{3}}$
$=\dfrac{2-2\sqrt{2}\sqrt{3}+3}{2-3}$
$=\dfrac{5-2\sqrt{6}}{-1}=-5+2\sqrt{6}$

21. $3+\sqrt{2x-3}=x$
$\sqrt{2x-3}=x-3$
$\left(\sqrt{2x-3}\right)^2=(x-3)^2$
$2x-3=x^2-6x+9$
$0=x^2-8x+12$
$0=(x-6)(x-2)$

Apply the zero product rule.

$x-6=0 \qquad x-2=0$
$x=6 \qquad x=2$

2 is an extraneous solution. The solution is 6 and the solution set is $\{6\}$.

22. $\sqrt{x+9}-\sqrt{x-7}=2$
$\sqrt{x+9}=2+\sqrt{x-7}$
$\left(\sqrt{x+9}\right)^2=\left(2+\sqrt{x-7}\right)^2$
$x+9=4+2\cdot 2\cdot\sqrt{x-7}+x-7$
$x+9=4\sqrt{x-7}+x-3$
$12=4\sqrt{x-7}$
$3=\sqrt{x-7}$
$3^2=\left(\sqrt{x-7}\right)^2$
$9=x-7$
$16=x$

The solution is 16 and the solution set is $\{16\}$.

23. $(11x+6)^{\frac{1}{3}}+3=0$

$$(11x+6)^{\frac{1}{3}}=-3$$
$$\sqrt[3]{11x+6}=-3$$
$$\left(\sqrt[3]{11x+6}\right)^3=(-3)^3$$
$$11x+6=-27$$
$$11x=-33$$
$$x=-3$$

The solution is –3 and the solution set is $\{-3\}$.

24. $40.4=2.9\sqrt{x}+20.1$

$$20.3=2.9\sqrt{x}$$
$$7=\sqrt{x}$$
$$7^2=\left(\sqrt{x}\right)^2$$
$$49=x$$

Boys who are 49 months of age have an average height of 40.4 inches.

25. $\sqrt{-75}=\sqrt{25\cdot3\cdot-1}$

$$=\sqrt{25}\cdot\sqrt{3}\cdot\sqrt{-1}=5\sqrt{3}i$$

26. $(5-3i)-(6-9i)=5-3i-6+9i$

$$=5-6-3i+9i$$
$$=-1+6i$$

27. $(3-4i)(2+5i)=6+15i-8i-20i^2$

$$=6+7i-20(-1)$$
$$=6+7i+20$$
$$=26+7i$$

28. $\sqrt{-9}\cdot\sqrt{-4}=\sqrt{9\cdot-1}\cdot\sqrt{4\cdot-1}$

$$=\sqrt{9}\cdot\sqrt{-1}\cdot\sqrt{4}\cdot\sqrt{-1}$$
$$=3\cdot i\cdot2\cdot i=6i^2=6(-1)$$
$$=-6 \text{ or } -6+0i$$

29. $\dfrac{3+i}{1-2i}=\dfrac{3+i}{1-2i}\cdot\dfrac{1+2i}{1+2i}=\dfrac{3+6i+i+2i^2}{1-4i^2}$

$$=\frac{3+7i+2(-1)}{1-4(-1)}=\frac{3+7i-2}{1+4}$$
$$=\frac{1+7i}{5}=\frac{1}{5}+\frac{7}{5}i$$

30. $i^{35}=i^{34}\cdot i=\left(i^2\right)^{17}\cdot i=(-1)^{17}\cdot i$

$$=(-1)i=-i$$

Cumulative Review Exercises

Chapters 1-7

1.
$$2x-y+z=-5$$
$$x-2y-3z=6$$
$$x+y-2z=1$$

Add the first and third equations to eliminate y.

$$2x-y+z=-5$$
$$x+y-2z=1$$
$$3x-z=-4$$

Multiply the third equation by 2 and add to the second equation.

$$x-2y-3z=6$$
$$2x+2y-4z=2$$
$$3x-7z=8$$

We now have a system of two equations in two variables.

$$3x-z=-4$$
$$3x-7z=8$$

Multiply the first equation by -1 and add to the second equation.

$$\begin{aligned} -3x + z &= 4 \\ 3x - 7z &= 8 \\ \hline -6z &= 12 \\ z &= -2 \end{aligned}$$

Back-substitute -2 for z to find x.

$$\begin{aligned} -3x + z &= 4 \\ -3x - 2 &= 4 \\ -3x &= 6 \\ x &= -2 \end{aligned}$$

Back-substitute -2 for x and z in one of the original equations to find y.

$$\begin{aligned} 2x - y + z &= -5 \\ 2(-2) - y - 2 &= -5 \\ -4 - y - 2 &= -5 \\ -y - 6 &= -5 \\ -y &= 1 \\ y &= -1 \end{aligned}$$

The solution is $(-2,-1,-2)$ or the solution set is $\{(-2,-1,-2)\}$.

2.

$$\begin{aligned} 3x^2 - 11x &= 4 \\ 3x^2 - 11x - 4 &= 0 \\ (3x+1)(x-4) &= 0 \end{aligned}$$

Apply the zero product principle.

$$\begin{aligned} 3x + 1 &= 0 & x - 4 &= 0 \\ 3x &= -1 & x &= 4 \\ x &= -\frac{1}{3} \end{aligned}$$

The solutions are $-\frac{1}{3}$ and 4 and the solution set is $\left\{-\frac{1}{3}, 4\right\}$.

3.

$$\begin{aligned} 2(x+4) &< 5x + 3(x+2) \\ 2x + 8 &< 5x + 3x + 6 \\ 2x + 8 &< 8x + 6 \\ -6x + 8 &< 6 \\ -6x &< -2 \\ \frac{-6x}{-6} &> \frac{-2}{-6} \\ x &> \frac{1}{3} \end{aligned}$$

The solution set is $\left\{x \middle| x > \frac{1}{3}\right\}$ or $\left(\frac{1}{3}, \infty\right)$.

4.

$$\frac{1}{x+2} + \frac{15}{x^2-4} = \frac{5}{x-2}$$

$$\frac{1}{x+2} + \frac{15}{(x+2)(x-2)} = \frac{5}{x-2}$$

So that denominators will not equal zero, x cannot equal 2 or -2. To eliminate fractions, multiply by the LCD, $(x+2)(x-2)$.

$$(x+2)(x-2)\left(\frac{1}{x+2}+\frac{15}{(x+2)(x-2)}\right)=(x+2)(x-2)\left(\frac{5}{x-2}\right)$$

$$(x+2)(x-2)\left(\frac{1}{x+2}\right)+(x+2)(x-2)\left(\frac{15}{(x+2)(x-2)}\right)=(x+2)(5)$$

$$x-2+15=5x+10$$
$$x+13=5x+10$$
$$-4x+13=10$$
$$-4x=-3$$
$$x=\frac{3}{4}$$

The solution is $\frac{3}{4}$ and the solution set is $\left\{\frac{3}{4}\right\}$.

5. $\sqrt{x+2}-\sqrt{x+1}=1$

$$\sqrt{x+2}=1+\sqrt{x+1}$$
$$\left(\sqrt{x+2}\right)^2=\left(1+\sqrt{x+1}\right)^2$$
$$x+2=1+2\sqrt{x+1}+x+1$$
$$x+2=2+2\sqrt{x+1}+x$$
$$0^2=\left(2\sqrt{x+1}\right)^2$$
$$0=4(x+1)$$
$$0=4x+4$$
$$-4=4x$$
$$-1=x$$

The solution checks. The solution is -1 and the solution set is $\{-1\}$.

6. $x+2y<2$

$2y-x>4$

First consider $x+2y<2$. Replace the inequality symbol with an equal sign and we have $x+2y=2$. Solve for y to put the equation in slope-intercept form.

$$x+2y=2$$
$$2y=-x+2$$
$$y=-\frac{1}{2}x+1$$

slope $=-\frac{1}{2}$ y-intercept $=1$

Now, use the origin as a test point.

$$0+2(0)<2$$
$$0<2$$

This is a true statement. This means that the point $(0,0)$ will fall in the shaded half-plane.

Next consider $2y-x>4$. Replace the inequality symbol with an equal sign and we have $2y-x=4$. Solve for y to put the equation in slope-intercept form.

$$2y-x=4$$
$$2y=x+4$$
$$y=\frac{1}{2}x+2$$

slope $= \frac{1}{2}$ $\quad$ y–intercept $= 2$

Now, use the origin as a test point.

$$2(0) - 0 > 4$$
$$0 > 4$$

This is a false statement. This means that the point $(0, 0)$ will not fall in the shaded half-plane.

Next, graph each of the inequalities. The solution to the system is the intersection of the shaded half-planes.

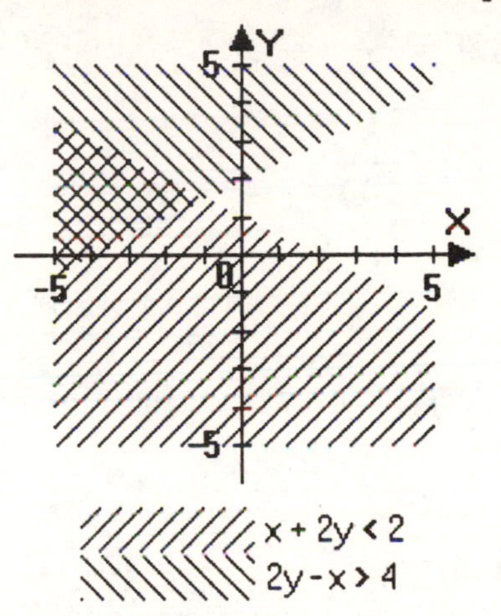

7.

$$\frac{8x^2}{3x^2-12} \div \frac{40}{x-2}$$
$$= \frac{8x^2}{3x^2-12} \cdot \frac{x-2}{40}$$
$$= \frac{8x^2}{3(x^2-4)} \cdot \frac{x-2}{40}$$
$$= \frac{\overset{1}{\cancel{8}}x^2}{3(x+2)\cancel{(x-2)}} \cdot \frac{\cancel{x-2}}{\underset{5}{\cancel{40}}}$$
$$= \frac{x^2}{3 \cdot 5(x+2)} = \frac{x^2}{15(x+2)}$$

8.

$$\frac{x+\frac{1}{y}}{y+\frac{1}{x}} = \frac{xy}{xy} \cdot \frac{x+\frac{1}{y}}{y+\frac{1}{x}} = \frac{xy \cdot x + xy \cdot \frac{1}{y}}{xy \cdot y + xy \cdot \frac{1}{x}}$$
$$= \frac{x^2y+x}{xy^2+y} = \frac{x(xy+1)}{y(xy+1)} = \frac{x}{y}$$

9.

$$\begin{array}{r} 4x^2 - 5x - 2 \\ 2x - 3 \\ \hline 8x^3 - 10x^2 - 4x \\ -12x^2 + 15x + 6 \\ \hline 8x^3 - 22x^2 + 11x + 6 \end{array}$$

10.

$$\frac{7x}{x^2-2x-15} - \frac{2}{x-5}$$
$$= \frac{7x}{(x-5)(x+3)} - \frac{2}{x-5}$$
$$= \frac{7x}{(x-5)(x+3)} - \frac{2(x+3)}{(x-5)(x+3)}$$
$$= \frac{7x-2(x+3)}{(x-5)(x+3)} = \frac{7x-2x-6}{(x-5)(x+3)}$$
$$= \frac{5x-6}{(x-5)(x+3)}$$

11.

$$7(8-10)^3 - 7 + 3 \div (-3)$$
$$= 7(-2)^3 - 7 + 3 \div (-3)$$
$$= 7(-8) - 7 + 3 \div (-3)$$
$$= -56 - 7 + (-1) = -64$$

12.

$$\sqrt{80x} - 5\sqrt{20x} + 2\sqrt{45x}$$
$$= \sqrt{16 \cdot 5x} - 5\sqrt{4 \cdot 5x} + 2\sqrt{9 \cdot 5x}$$

$= 4\sqrt{5x} - 5 \cdot 2\sqrt{5x} + 2 \cdot 3\sqrt{5x}$
$= 4\sqrt{5x} - 10\sqrt{5x} + 6\sqrt{5x} = 0$

13. $\frac{\sqrt{3}-2}{2\sqrt{3}+5} = \frac{\sqrt{3}-2}{2\sqrt{3}+5} \cdot \frac{2\sqrt{3}-5}{2\sqrt{3}-5}$
$= \frac{2 \cdot 3 - 5\sqrt{3} - 4\sqrt{3} + 10}{4 \cdot 3 - 25}$
$= \frac{6 - 9\sqrt{3} + 10}{12 - 25}$
$= \frac{16 - 9\sqrt{3}}{-13}$
$= -\frac{16 - 9\sqrt{3}}{13}$

14.
$$\begin{array}{r|l} & 2x^2 + x + 5 \\ x-2 & 2x^3 - 3x^2 + 3x - 4 \\ & \underline{2x^3 - 4x^2} \\ & \quad x^2 + 3x \\ & \quad \underline{x^2 - 2x} \\ & \qquad 5x - 4 \\ & \qquad \underline{5x - 10} \\ & \qquad\quad 6 \end{array}$$

$\frac{2x^3 - 3x^2 + 3x - 4}{x - 2}$
$= 2x^2 + x + 5 + \frac{6}{x-2}$

15. $\left(2\sqrt{3} + 5\sqrt{2}\right)\left(\sqrt{3} - 4\sqrt{2}\right)$
$= 2 \cdot 3 - 8\sqrt{6} + 5\sqrt{6} - 20 \cdot 2$
$= 6 - 3\sqrt{6} - 40 = -34 - 3\sqrt{6}$

16. $24x^2 + 10x - 4 = 2\left(12x^2 + 5x - 2\right)$
$= 2(3x + 2)(4x - 1)$

17. $16x^4 - 1 = \left(4x^2 + 1\right)\left(4x^2 - 1\right)$
$= \left(4x^2 + 1\right)(2x + 1)(2x - 1)$

18. Since light varies inversely as the square of the distance, we have
$l = \frac{k}{d^2}$.
Use the given values to find k.
$l = \frac{k}{d^2}$
$120 = \frac{k}{10^2}$
$120 = \frac{k}{100}$
$12{,}000 = k$
The equation becomes $l = \frac{12{,}000}{d^2}$.
When $d = 15$,
$l = \frac{12{,}000}{15^2} = \frac{12{,}000}{225} \approx 53.3$.
At a distance of 15 feet, approximately 53 lumens are provided.

19. Let x = the amount invested at 7%
Let y = the amount invested at 9%
$x + \quad y = 6000$
$0.07x + 0.09y = 510$
Solve the first equation for y.
$x + y = 6000$
$y = 6000 - x$
Substitute and solve.
$0.07x + 0.09(6000 - x) = 510$
$0.07x + 540 - 0.09x = 510$
$540 - 0.02x = 510$
$-0.02x = -30$
$x = 1500$

Back-substitute 1500 for x to find y.

$y = 6000 - x$

$y = 6000 - 1500$

$y = 4500$

\$1500 was invested at 7% and \$4500 was invested at 9%.

20. Let x = the number of students enrolled last year

$x - 0.12x = 2332$

$0.88x = 2332$

$x = 2650$

There were 2650 students enrolled last year.

Chapter 8

Check Points 8.1

1. $5x^2 = 15$

$\frac{5x^2}{5} = \frac{15}{5}$

$x^2 = 3$

Apply the square root property.

$x = \sqrt{3}$ or $x = -\sqrt{3}$

Check:

$x = \sqrt{3}$	$x = -\sqrt{3}$
$5x^2 = 15$	$5x^2 = 15$
$5\left(\sqrt{3}\right)^2 = 15$	$5\left(-\sqrt{3}\right)^2 = 15$
$5(3) = 15$	$5(3) = 15$
$15 = 15$	$15 = 15$
True	True

The solutions are $\sqrt{3}$ and $-\sqrt{3}$ and the solution set is $\left\{-\sqrt{3}, \sqrt{3}\right\}$ or $\left\{\pm\sqrt{3}\right\}$.

2. $2x^2 - 5 = 0$

$2x^2 = 5$

$\frac{2x^2}{2} = \frac{5}{2}$

$x^2 = \frac{5}{2}$

Apply the square root property.

$x = \sqrt{\frac{5}{2}}$ or $x = -\sqrt{\frac{5}{2}}$

Because the proposed solutions are opposites, rationalize both denominators at once.

$$x = \pm\sqrt{\frac{5}{2}} = \pm\frac{\sqrt{5}}{\sqrt{2}} \cdot \frac{\sqrt{2}}{\sqrt{2}} = \pm\frac{\sqrt{10}}{2}$$

The solutions are $\frac{\sqrt{10}}{2}$ and $-\frac{\sqrt{10}}{2}$ and the solution set is $\left\{-\frac{\sqrt{10}}{2}, \frac{\sqrt{10}}{2}\right\}$ or $\left\{\pm\frac{\sqrt{10}}{2}\right\}$.

3. $4x^2 + 9 = 0$

$4x^2 = -9$

$x^2 = -\frac{9}{4}$

Apply the square root property.

$x = \sqrt{-\frac{9}{4}}$ or $x = -\sqrt{-\frac{9}{4}}$

$x = \sqrt{\frac{9}{4}}\sqrt{-1}$ $\quad$ $x = -\sqrt{\frac{9}{4}}\sqrt{-1}$

$x = \frac{3}{2}i$ $\quad$ $x = -\frac{3}{2}i$

The solutions are $\pm\frac{3}{2}i$ and the solution set is $\left\{\pm\frac{3}{2}i\right\}$.

4. $(x-2)^2 = 7$

Apply the square root property.

$x - 2 = \sqrt{7}$ or $x - 2 = -\sqrt{7}$

$x = 2 + \sqrt{7}$ $\quad$ $x = 2 - \sqrt{7}$

The solutions are $2 \pm \sqrt{7}$ and the solution set is $\left\{2 \pm \sqrt{7}\right\}$.

5. **a.** $x^2 + 10x + \_\_\_\_$

Since $b = 10$, we add

$$\left(\frac{b}{2}\right)^2 = \left(\frac{10}{2}\right)^2 = (5)^2 = 25.$$

$$x^2 + 10x + 25 = (x+5)^2$$

b. $x^2 - 3x + \_\_\_\_$

Since $b = -3$, we add

$$\left(\frac{b}{2}\right)^2 = \left(\frac{-3}{2}\right)^2 = \frac{9}{4}.$$

$$x^2 - 3x + \frac{9}{4} = \left(x - \frac{3}{2}\right)^2$$

c. $x^2 + \frac{3}{4}x + \_\_\_\_$

Since $b = \frac{3}{4}$, we add

$$\left(\frac{b}{2}\right)^2 = \left(\frac{3}{4} \div 2\right)^2 = \left(\frac{3}{4} \cdot \frac{1}{2}\right)^2 = \left(\frac{3}{8}\right)^2 = \frac{9}{64}.$$

$$x^2 + \frac{3}{4}x + \frac{9}{64} = \left(x + \frac{3}{8}\right)^2$$

6. $x^2 - 10x + 18 = 0$

$x^2 - 10x \quad = -18$

Since $b = -10$, we add

$$\left(\frac{b}{2}\right)^2 = \left(\frac{-10}{2}\right)^2 = (-5)^2 = 25.$$

$$x^2 - 10x + 25 = -18 + 25$$

$$(x-5)^2 = 7$$

Apply the square root property.

$$x - 5 = \pm\sqrt{7}$$

$$x = 5 \pm \sqrt{7}$$

The solutions are $5 \pm \sqrt{7}$ and the solution set is $\{5 \pm \sqrt{7}\}$.

7. $2x^2 - 10x - 1 = 0$

$$x^2 - 5x - \frac{1}{2} = 0$$

$$x^2 - 5x \quad = \frac{1}{2}$$

Since $b = -5$, we add

$$\left(\frac{b}{2}\right)^2 = \left(\frac{-5}{2}\right)^2 = \frac{25}{4}.$$

$$x^2 - 5x + \frac{25}{4} = \frac{1}{2} + \frac{25}{4}$$

$$\left(x - \frac{5}{2}\right)^2 = \frac{27}{4}$$

Apply the square root property.

$$x - \frac{5}{2} = \pm\sqrt{\frac{27}{4}}$$

$$x - \frac{5}{2} = \pm\frac{\sqrt{27}}{2}$$

$$x - \frac{5}{2} = \pm\frac{\sqrt{9 \cdot 3}}{2}$$

$$x - \frac{5}{2} = \pm\frac{3\sqrt{3}}{2}$$

$$x = \frac{5}{2} \pm \frac{3\sqrt{3}}{2}$$

$$x = \frac{5 \pm 3\sqrt{3}}{2}$$

The solutions are $\frac{5 \pm 3\sqrt{3}}{2}$ and the solution set is $\left\{\frac{5 \pm 3\sqrt{3}}{2}\right\}$.

8. $A = P(1+r)^t$

$4320 = 3000(1+r)^2$

$\frac{4320}{3000} = (1+r)^2$

$1.44 = (1+r)^2$

Apply the square root property.

$1+r = \sqrt{1.44}$ or $1+r = -\sqrt{1.44}$

$1+r = 1.2$ $\quad$ $1+r = -1.2$

$r = 0.2$ $\quad$ $r = -2.2$

We reject –2.2 because we cannot have a negative interest rate. The solution is 0.2 and we conclude that the annual interest rate is 20%.

9. $x^2 + 20^2 = 50^2$

$x^2 + 400 = 2500$

$x^2 = 2100$

Apply the square root property.

$x = \pm\sqrt{2100}$

$x = \pm\sqrt{21 \cdot 100}$

$x = \pm 10\sqrt{21}$

We reject $-10\sqrt{21}$ because we cannot have a negative length measurement. The solution is $10\sqrt{21}$ feet or $10\sqrt{21} \approx 45.8$ feet. The wire is attached Approximately 45.8 feet from the base of the antenna.

Problem Set 8.1

Practice Exercises

1. $3x^2 = 75$

$x^2 = 25$

Apply the square root property.

$x = \pm\sqrt{25}$

$x = \pm 5$

The solutions are ± 5 and the solution set is $\{\pm 5\}$.

3. $7x^2 = 42$

$x^2 = 6$

Apply the square root property.

$x = \pm\sqrt{6}$

The solutions are $\pm\sqrt{6}$ and the solution set is $\{\pm\sqrt{6}\}$.

5. $16x^2 = 25$

$x^2 = \frac{25}{16}$

Apply the square root property.

$x = \pm\sqrt{\frac{25}{16}}$

$x = \pm\frac{5}{4}$

The solutions are $\pm\frac{5}{4}$ and the solution set is $\left\{\pm\frac{5}{4}\right\}$.

7. $3x^2 - 2 = 0$

$3x^2 = 2$

$x^2 = \frac{2}{3}$

Apply the square root property.

$x = \pm\sqrt{\frac{2}{3}}$

Because the proposed solutions are opposites, rationalize both denominators at once.

$x = \pm\sqrt{\frac{2}{3}} = \pm\frac{\sqrt{2}}{\sqrt{3}}\cdot\frac{\sqrt{3}}{\sqrt{3}} = \pm\frac{\sqrt{6}}{3}$

The solutions are $\pm\frac{\sqrt{6}}{3}$ and the solution set is $\left\{\pm\frac{\sqrt{6}}{3}\right\}$.

9. $25x^2 + 16 = 0$

$25x^2 = -16$

$x^2 = -\frac{16}{25}$

Apply the square root property.

$x = \pm\sqrt{-\frac{16}{25}}$

$x = \pm\sqrt{\frac{16}{25}}\sqrt{-1}$

$x = \pm\frac{4}{5}i$

$x = 0 \pm \frac{4}{5}i$

The solutions are $0 \pm \frac{4}{5}i$ and the solution set is $\left\{0 \pm \frac{4}{5}i\right\}$.

11. $(x+7)^2 = 9$

Apply the square root property.

$x + 7 = \sqrt{9}$ or $x + 7 = -\sqrt{9}$

$x + 7 = 3$ $\quad$ $x + 7 = -3$

$x = -4$ $\quad$ $x = -10$

The solutions are -4 and -10 and the solution set is $\{-10, -4\}$.

13. $(x-3)^2 = 5$

Apply the square root property.

$x - 3 = \pm\sqrt{5}$

$x = 3 \pm \sqrt{5}$

The solutions are $3 \pm \sqrt{5}$ and the solution set is $\{3 \pm \sqrt{5}\}$.

15. $(x+2)^2 = 8$

Apply the square root property.

$x + 2 = \pm\sqrt{8}$

$x + 2 = \pm\sqrt{4\cdot 2}$

$x + 2 = \pm 2\sqrt{2}$

$x = -2 \pm 2\sqrt{2}$

The solutions are $-2 \pm 2\sqrt{2}$ and the solution set is $\{-2 \pm 2\sqrt{2}\}$.

17. $(x-5)^2 = -9$

Apply the square root property.

$x - 5 = \pm\sqrt{-9}$

$x - 5 = \pm 3i$

$x = 5 \pm 3i$

The solutions are $5 \pm 3i$ and the solution set is $\{5 \pm 3i\}$.

19. $\left(x+\frac{3}{4}\right)^2=\frac{11}{16}$

Apply the square root property.

$$x+\frac{3}{4}=\pm\sqrt{\frac{11}{16}}$$

$$x+\frac{3}{4}=\pm\frac{\sqrt{11}}{4}$$

$$x=-\frac{3}{4}\pm\frac{\sqrt{11}}{4}$$

$$x=\frac{-3\pm\sqrt{11}}{4}$$

The solutions are $\frac{-3\pm\sqrt{11}}{4}$ and the solution set is $\left\{\frac{-3\pm\sqrt{11}}{4}\right\}$.

21. $x^2-6x+9=36$

$$(x-3)^2=36$$

Apply the square root property.

$x-3=\sqrt{36}$ or $x-3=-\sqrt{36}$

$x-3=6$ $\quad$ $x-3=-6$

$x=9$ $\quad$ $x=-3$

The solutions are 9 and -3 and the solution set is $\{-3,9\}$.

23. x^2+2x+____

Since $b=2$, we add

$$\left(\frac{b}{2}\right)^2=\left(\frac{2}{2}\right)^2=(1)^2=1.$$

$$x^2+2x+1=(x+1)^2$$

25. x^2-14x+____

Since $b=-14$, we add

$$\left(\frac{b}{2}\right)^2=\left(\frac{-14}{2}\right)^2=(-7)^2=49.$$

$$x^2-14x+49=(x-7)^2$$

27. x^2+7x+____

Since $b=7$, we add

$$\left(\frac{b}{2}\right)^2=\left(\frac{7}{2}\right)^2=\frac{49}{4}.$$

$$x^2+7x+\frac{49}{4}=\left(x+\frac{7}{2}\right)^2$$

29. $x^2-\frac{1}{2}x+$____

Since $b=-\frac{1}{2}$, we add

$$\left(\frac{b}{2}\right)^2=\left(\frac{-1}{2}\div 2\right)^2=\left(\frac{-1}{2}\cdot\frac{1}{2}\right)^2$$

$$=\left(\frac{-1}{4}\right)^2=\frac{1}{16}.$$

$$x^2-\frac{1}{2}x+\frac{1}{16}=\left(x-\frac{1}{4}\right)^2$$

31. $x^2+\frac{4}{3}x+$____

Since $b=\frac{4}{3}$, we add

$$\left(\frac{b}{2}\right)^2=\left(\frac{4}{3}\div 2\right)^2=\left(\frac{4}{3}\cdot\frac{1}{2}\right)^2=\left(\frac{2}{3}\right)^2=\frac{4}{9}.$$

$$x^2+\frac{4}{3}x+\frac{4}{9}=\left(x+\frac{2}{3}\right)^2$$

33. $x^2-\frac{9}{4}x+$____

Since $b = -\frac{9}{4}$, we add

$$\left(\frac{b}{2}\right)^2 = \left(-\frac{9}{4} \div 2\right)^2 = \left(-\frac{9}{4} \cdot \frac{1}{2}\right)^2$$

$$= \left(-\frac{9}{8}\right)^2 = \frac{81}{64}.$$

$$x^2 - \frac{9}{4}x + \frac{81}{64} = \left(x - \frac{9}{8}\right)^2$$

35. $x^2 + 6x = -8$

$x^2 + 6x \qquad = -8$

Since $b = 6$, we add

$$\left(\frac{b}{2}\right)^2 = \left(\frac{6}{2}\right)^2 = (3)^2 = 9.$$

$$x^2 + 6x + 9 = -8 + 9$$

$$(x+3)^2 = 1$$

Apply the square root property.

$x + 3 = \sqrt{1}$ or $x + 3 = -\sqrt{1}$

$x + 3 = 1 \qquad x + 3 = -1$

$x = -2 \qquad x = -4$

The solutions are -2 and -4 and the solution set is $\{-4, -2\}$.

37. $x^2 + 6x = -2$

$x^2 + 6x \qquad = -2$

Since $b = 6$, we add

$$\left(\frac{b}{2}\right)^2 = \left(\frac{6}{2}\right)^2 = (3)^2 = 9.$$

$$x^2 + 6x + 9 = -2 + 9$$

$$(x+3)^2 = 7$$

Apply the square root property.

$x + 3 = \pm\sqrt{7}$

$x = -3 \pm \sqrt{7}$

The solutions are $-3 \pm \sqrt{7}$ and the solution set is $\{-3 \pm \sqrt{7}\}$.

39. $x^2 + 4x + 1 = 0$

$x^2 + 4x = -1$

$x^2 + 4x \qquad = -1$

Since $b = 4$, we add

$$\left(\frac{b}{2}\right)^2 = \left(\frac{4}{2}\right)^2 = (2)^2 = 4.$$

$$x^2 + 4x + 4 = -1 + 4$$

$$(x+2)^2 = 3$$

Apply the square root property.

$x + 2 = \pm\sqrt{3}$

$x = -2 \pm \sqrt{3}$

The solutions are $-2 \pm \sqrt{3}$ and the solution set is $\{-2 \pm \sqrt{3}\}$.

41. $x^2 + 2x + 2 = 0$

$x^2 + 2x = -2$

$x^2 + 2x \qquad = -2$

Since $b = 2$, we add

$$\left(\frac{b}{2}\right)^2 = \left(\frac{2}{2}\right)^2 = (1)^2 = 1.$$

$$x^2 + 2x + 1 = -2 + 1$$

$$(x+1)^2 = -1$$

Apply the square root property.

$x + 1 = \pm\sqrt{-1}$

$x + 1 = \pm i$

$x = -1 \pm i$

The solutions are $-1 \pm i$ and the solution set is $\{-1 \pm i\}$.

43. $x^2 + 3x - 1 = 0$

$x^2 + 3x = 1$

$x^2 + 3x \quad = 1$

Since $b = 3$, we add

$\left(\frac{b}{2}\right)^2 = \left(\frac{3}{2}\right)^2 = \frac{9}{4}.$

$x^2 + 3x + \frac{9}{4} = 1 + \frac{9}{4}$

$\left(x + \frac{3}{2}\right)^2 = \frac{13}{4}$

Apply the square root property.

$x + \frac{3}{2} = \pm\sqrt{\frac{13}{4}}$

$x + \frac{3}{2} = \pm\frac{\sqrt{13}}{2}$

$x = -\frac{3}{2} \pm \frac{\sqrt{13}}{2}$

$x = \frac{-3 \pm \sqrt{13}}{2}$

The solutions are $\frac{-3 \pm \sqrt{13}}{2}$ and the solution set is $\left\{\frac{-3 \pm \sqrt{13}}{2}\right\}$.

45. $x^2 = 7x - 3$

$x^2 - 7x = -3$

$x^2 - 7x \quad = -3$

Since $b = -7$, we add

$\left(\frac{b}{2}\right)^2 = \left(\frac{-7}{2}\right)^2 = \frac{49}{4}.$

$x^2 - 7x + \frac{49}{4} = -3 + \frac{49}{4}$

$\left(x - \frac{7}{2}\right)^2 = -\frac{12}{4} + \frac{49}{4}$

$\left(x - \frac{7}{2}\right)^2 = \frac{37}{4}$

Apply the square root property.

$x - \frac{7}{2} = \pm\sqrt{\frac{37}{4}}$

$x - \frac{7}{2} = \pm\frac{\sqrt{37}}{2}$

$x = \frac{7}{2} \pm \frac{\sqrt{37}}{2}$

$x = \frac{7 \pm \sqrt{37}}{2}$

The solutions are $\frac{7 \pm \sqrt{37}}{2}$ and the solution set is $\left\{\frac{7 \pm \sqrt{37}}{2}\right\}$.

47. $x^2 + x - 1 = 0$

$x^2 + x = 1$

$x^2 + x \quad = 1$

Since $b = 1$, we add

$\left(\frac{b}{2}\right)^2 = \left(\frac{1}{2}\right)^2 = \frac{1}{4}.$

$x^2 + x + \frac{1}{4} = 1 + \frac{1}{4}$

$\left(x + \frac{1}{2}\right)^2 = \frac{5}{4}$

Apply the square root property.

$x + \frac{1}{2} = \pm\sqrt{\frac{5}{4}}$

$x + \frac{1}{2} = \pm\frac{\sqrt{5}}{2}$

$$x = -\frac{1}{2} \pm \frac{\sqrt{5}}{2}$$

$$x = \frac{-1 \pm \sqrt{5}}{2}$$

The solutions are $\frac{-1 \pm \sqrt{5}}{2}$ and the solution set is $\left\{\frac{-1 \pm \sqrt{5}}{2}\right\}$.

49. $2x^2 - 3x + 1 = 0$

$$x^2 - \frac{3}{2}x + \frac{1}{2} = 0$$

$$x^2 - \frac{3}{2}x = -\frac{1}{2}$$

$$x^2 - \frac{3}{2}x \qquad = -\frac{1}{2}$$

Since $b = -\frac{3}{2}$, we add

$$\left(\frac{b}{2}\right)^2 = \left(-\frac{3}{2} \div 2\right)^2 = \left(-\frac{3}{2} \cdot \frac{1}{2}\right)^2$$

$$= \left(-\frac{3}{4}\right)^2 = \frac{9}{16}.$$

$$x^2 - \frac{3}{2}x + \frac{9}{16} = -\frac{1}{2} + \frac{9}{16}$$

$$\left(x - \frac{3}{4}\right)^2 = -\frac{8}{16} + \frac{9}{16}$$

$$\left(x - \frac{3}{4}\right)^2 = \frac{1}{16}$$

Apply the square root property.

$$x - \frac{3}{4} = \pm\sqrt{\frac{1}{16}}$$

$$x - \frac{3}{4} = \pm\frac{1}{4}$$

$$x = \frac{3}{4} \pm \frac{1}{4}$$

$$x = \frac{3}{4} + \frac{1}{4} \text{ or } \frac{3}{4} - \frac{1}{4}$$

$$x = \frac{4}{4} \text{ or } \frac{2}{4}$$

$$x = 1 \text{ or } \frac{1}{2}$$

The solutions are $\frac{1}{2}$ and 1 and the solution set is $\left\{\frac{1}{2}, 1\right\}$.

51. $2x^2 + 10x + 11 = 0$

$$x^2 + 5x + \frac{11}{2} = 0$$

$$x^2 + 5x = -\frac{11}{2}$$

$$x^2 + 5x \qquad = -\frac{11}{2}$$

Since $b = 5$, we add

$$\left(\frac{b}{2}\right)^2 = \left(\frac{5}{2}\right)^2 = \frac{25}{4}.$$

$$x^2 + 5x + \frac{25}{4} = -\frac{11}{2} + \frac{25}{4}$$

$$\left(x + \frac{5}{2}\right)^2 = -\frac{22}{4} + \frac{25}{4}$$

$$\left(x + \frac{5}{2}\right)^2 = \frac{3}{4}$$

Apply the square root property.

$$x + \frac{5}{2} = \pm\sqrt{\frac{3}{4}}$$

$$x + \frac{5}{2} = \pm\frac{\sqrt{3}}{2}$$

$$x = -\frac{5}{2} \pm \frac{\sqrt{3}}{2}$$

$$x = \frac{-5 \pm \sqrt{3}}{2}$$

The solutions are $\frac{-5 \pm \sqrt{3}}{2}$ and the solution set is $\left\{\frac{-5 \pm \sqrt{3}}{2}\right\}$.

53. $3x^2 - 2x - 4 = 0$

$$x^2 - \frac{2}{3}x - \frac{4}{3} = 0$$

$$x^2 - \frac{2}{3}x = \frac{4}{3}$$

$$x^2 - \frac{2}{3}x \quad = \frac{4}{3}$$

Since $b = -\frac{2}{3}$, we add

$$\left(\frac{b}{2}\right)^2 = \left(-\frac{2}{3} \div 2\right)^2 = \left(-\frac{2}{3} \cdot \frac{1}{2}\right)^2$$

$$= \left(-\frac{1}{3}\right)^2 = \frac{1}{9}.$$

$$x^2 - \frac{2}{3}x + \frac{1}{9} = \frac{4}{3} + \frac{1}{9}$$

$$\left(x - \frac{1}{3}\right)^2 = \frac{12}{9} + \frac{1}{9}$$

$$\left(x - \frac{1}{3}\right)^2 = \frac{13}{9}$$

Apply the square root property.

$$x - \frac{1}{3} = \pm\sqrt{\frac{13}{9}}$$

$$x - \frac{1}{3} = \pm\frac{\sqrt{13}}{3}$$

$$x = \frac{1}{3} \pm \frac{\sqrt{13}}{3}$$

$$x = \frac{1 \pm \sqrt{13}}{3}$$

The solutions are $\frac{1 \pm \sqrt{13}}{3}$ and the solution set is $\left\{\frac{1 \pm \sqrt{13}}{3}\right\}$.

55. $8x^2 - 4x + 1 = 0$

$$x^2 - \frac{1}{2}x + \frac{1}{8} = 0$$

$$x^2 - \frac{1}{2}x = -\frac{1}{8}$$

$$x^2 - \frac{1}{2}x \quad = -\frac{1}{8}$$

Since $b = -\frac{1}{2}$, we add

$$\left(\frac{b}{2}\right)^2 = \left(-\frac{1}{2} \div 2\right)^2 = \left(-\frac{1}{2} \cdot \frac{1}{2}\right)^2$$

$$= \left(-\frac{1}{4}\right)^2 = \frac{1}{16}.$$

$$x^2 - \frac{1}{2}x + \frac{1}{16} = -\frac{1}{8} + \frac{1}{16}$$

$$\left(x - \frac{1}{4}\right)^2 = -\frac{2}{16} + \frac{1}{16}$$

$$\left(x - \frac{1}{4}\right)^2 = -\frac{1}{16}$$

Apply the square root property.

$$x - \frac{1}{4} = \pm\sqrt{-\frac{1}{16}}$$

$$x - \frac{1}{4} = \pm\frac{1}{4}i$$

$$x = \frac{1}{4} \pm \frac{1}{4}i$$

The solutions are $\frac{1}{4} \pm \frac{1}{4}i$ and the solution set is $\left\{\frac{1}{4} \pm \frac{1}{4}i\right\}$.

Application Exercises

57. $2880 = 2000(1+r)^2$

$\frac{2880}{2000} = (1+r)^2$

$1.44 = (1+r)^2$

Apply the square root property.

$1+r = \pm\sqrt{1.44}$

$1+r = \pm 1.2$

$r = -1 \pm 1.2$

$r = -1+1.2 \text{ or } -1-1.2$

$r = 0.2 \text{ or } -2.2$

We reject –2.2 because we cannot have a negative interest rate. The solution is 0.2 and we conclude that the annual interest rate is 20%.

59. $3360 = 3125(1+r)^2$

$\frac{3360}{3125} = (1+r)^2$

$1.0752 = (1+r)^2$

Apply the square root property.

$1+r = \pm\sqrt{1.0752}$

$1+r = \pm 1.0369$

$r = -1 \pm 1.0369$

$r = -1+1.0369 \text{ or } -1-1.0369$

$r = 0.0369 \text{ or } -2.0369$

We reject –2.0369 because we cannot have a negative interest rate. The solution is 0.0369 and we conclude that the annual interest rate is 3.69%.

61. $20 = 0.4x^2 + 0.5$

$19.5 = 0.4x^2$

$\frac{19.5}{0.4} = x^2$

$48.75 = x^2$

Apply the square root property.

$x = \pm\sqrt{48.75}$

$x = \pm 6.98$

$x \approx \pm 7$

We disregard –7 because we can't have a negative number of years. The solution is 7 and we conclude that there will be 20 million cable-TV modem users in the year 1996 + 7 = 2003.

63. $4800 = 16t^2$

$\frac{4800}{16} = t^2$

$300 = t^2$

Apply the square root property.

$t = \pm\sqrt{300}$

$t = \pm\sqrt{100 \cdot 3}$

$t = \pm 10\sqrt{3}$

$t \approx \pm 17.3$

We disregard –17.3 because we can't have a negative time measurement. The solution is 17.3 and we conclude that the sky diver was in a free fall for $10\sqrt{3}$ or approximately 17.3 seconds.

65.

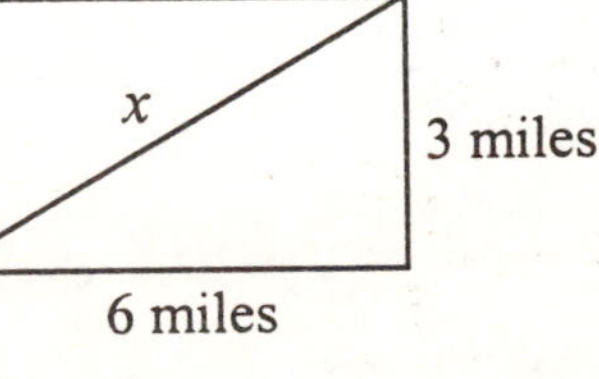

$6^2 + 3^2 = x^2$

$36 + 9 = x^2$

$45 = x^2$

Apply the square root property.

$x = \pm\sqrt{45}$

$x = \pm\sqrt{9 \cdot 5}$

$x = \pm 3\sqrt{5}$

We disregard $-3\sqrt{5}$ because we can't have a negative length measurement. The solution is $3\sqrt{5}$ and we conclude that the pedestrian route is $3\sqrt{5}$ or approximately 6.7 miles long.

67. $x^2 + 15^2 = 20^2$

$x^2 + 225 = 400$

$x^2 = 175$

Apply the square root property.

$x = \pm\sqrt{175}$

$x = \pm\sqrt{25 \cdot 7}$

$x = \pm 5\sqrt{7}$

We disregard $-5\sqrt{7}$ because we can't have a negative length measurement. The solution is $5\sqrt{7}$ and we conclude that the ladder reaches $5\sqrt{7}$ or approximately 13.2 feet up the house.

69.

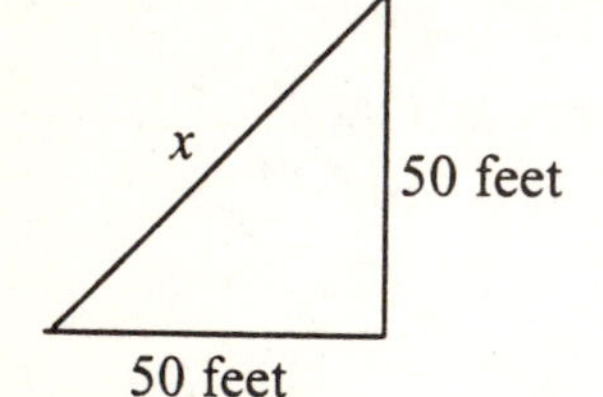

$50^2 + 50^2 = x^2$

$2500 + 2500 = x^2$

$5000 = x^2$

Apply the square root property.

$x = \pm\sqrt{5000}$

$x = \pm\sqrt{2500 \cdot 2}$

$x = \pm 50\sqrt{2} \approx \pm 70.7$

We disregard -70.7 because we can't have a negative length measurement. The solution is 70.7 and we conclude that a supporting wire of $50\sqrt{2}$ or approximately 70.7 feet is required.

71. $E = mc^2$

$\frac{E}{m} = c^2$

$\pm\sqrt{\frac{E}{m}} = c$

Rationalize the denominator.

$c = \pm\frac{\sqrt{E}}{\sqrt{m}} \cdot \frac{\sqrt{m}}{\sqrt{m}}$

$c = \pm\frac{\sqrt{Em}}{m}$

Because the speed of light is positive, we disregard the negative solution.

$c = \sqrt{\frac{E}{m}}$ or $\frac{\sqrt{Em}}{m}$

73. $A = lw$

$144 = (x+2+2)(x+2+2)$

$144 = (x+4)(x+4)$

$144 = x^2 + 8x + 16$

$0 = x^2 + 8x - 128$

$0 = (x+16)(x-8)$

Apply the zero product principle.

$x + 16 = 0 \qquad x - 8 = 0$

$x = -16 \qquad x = 8$

We disregard -16 because we can't have a negative length measurement. The solution is 8 and we conclude that the length of the original square is 8 meters.

Writing in Mathematics

75. Answers will vary.

77. Answers will vary.

79. Answers will vary.

81. Answers will vary.

83. Answers will vary.

Technology Exercises

85. $(x-1)^2-9=0$

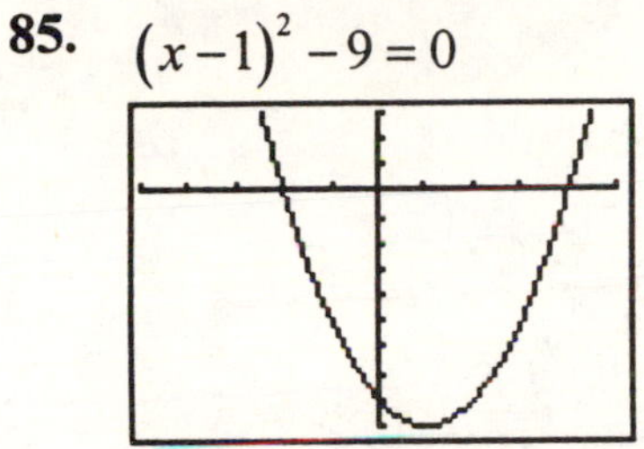

The solutions are –2 and 4 and the solution set is $\{-2,4\}$.

Check:

$x=-2$	$x=4$
$(-2-1)^2-9=0$	$(4-1)^2-9=0$
$(-3)^2-9=0$	$(3)^2-9=0$
$9-9=0$	$9-9=0$
$0=0$	$0=0$
True	True

Critical Thinking Exercises

87. Statement **d.** is true. To find x–intercepts, set y equal to 0. In this case, setting y equal to zero will result in complex solutions.

Statement **a.** is false. $(x-5)^2=12$ is equivalent to $x-5=\pm 2\sqrt{3}$.

Statement **b.** is false. Divide both sides of the equation by 2 before completing the square. Add $\left(\frac{b}{2}\right)^2=\left(\frac{-3}{2}\right)^2=\frac{9}{4}$.

Statement **c.** is false. Not all quadratic equations can be solved by factoring.

89. $x^2+x+c=0$

$x^2+x \quad = -c$

Since $b = 1$, we add

$\left(\frac{b}{2}\right)^2=\left(\frac{1}{2}\right)^2=\frac{1}{4}$.

$$x^2+x+\frac{1}{4}=-c+\frac{1}{4}$$

$$\left(x+\frac{1}{2}\right)^2=-c+\frac{1}{4}$$

Apply the square root property.

$$x+\frac{1}{2}=\pm\sqrt{-c+\frac{1}{4}}$$

$$x=-\frac{1}{2}\pm\sqrt{-c+\frac{1}{4}}$$

Simplify the solutions.

$$x=-\frac{1}{2}\pm\sqrt{\frac{-c}{1}\cdot\frac{4}{4}+\frac{1}{4}}$$

$$=-\frac{1}{2}\pm\sqrt{\frac{-4c}{4}+\frac{1}{4}}$$

$$=-\frac{1}{2}\pm\sqrt{\frac{-4c+1}{4}}$$

$$= -\frac{1}{2} \pm \frac{\sqrt{1-4c}}{2} = \frac{-1 \pm \sqrt{1-4c}}{2}$$

The solutions are $\frac{-1 \pm \sqrt{1-4c}}{2}$ and the solution set is $\left\{\frac{-1 \pm \sqrt{1-4c}}{2}\right\}$.

91. $x^4 - 8x^2 + 15 = 0$

Let $t = x^2$

$$x^4 - 8x^2 + 15 = 0$$
$$(x^2)^2 - 8x^2 + 15 = 0$$
$$t^2 - 8t + 15 = 0$$
$$(t-5)(t-3) = 0$$

Apply the zero product principle.

$t - 5 = 0 \qquad t - 3 = 0$

$t = 5 \qquad t = 3$

Substitute x^2 for t.

$t = 5 \qquad t = 3$

$x^2 = 5 \qquad x^2 = 3$

Apply the square root property.

$x = \pm\sqrt{5} \qquad x = \pm\sqrt{3}$

The solutions are $\pm\sqrt{5}$ and $\pm\sqrt{3}$, and the solution set is $\{\pm\sqrt{5}, \pm\sqrt{3}\}$.

Review Exercises

92. $4x - 2 - 3[4 - 2(3 - x)]$

$= 4x - 2 - 3[4 - 6 + 2x]$

$= 4x - 2 - 3[-2 + 2x]$

$= 4x - 2 + 6 - 6x$

$= 4 - 2x$

93. $1 - 8x^3 = 1^3 - (2x)^3$

$= (1 - 2x)(1 + 2x + 4x^2)$

94. $(x^4 - 5x^3 + 2x^2 - 6) \div (x - 3)$

3	1	−5	2	0	−6
		3	−6	−12	−36
	1	−2	−4	−12	−42

$(x^4 - 5x^3 + 2x^2 - 6) \div (x - 3)$

$= x^3 - 2x^2 - 4x - 12 - \frac{42}{x-3}$

Check Points 8.2

1. $2x^2 + 9x - 5 = 0$

$a = 2 \quad b = 9 \quad c = -5$

$$x = \frac{-9 \pm \sqrt{9^2 - 4(2)(-5)}}{2(2)}$$
$$= \frac{-9 \pm \sqrt{81 - (-40)}}{4}$$
$$= \frac{-9 \pm \sqrt{121}}{4} = \frac{-9 \pm 11}{4}$$

Evaluate the expression to obtain two solutions.

$x = \frac{-9+11}{4}$ or $x = \frac{-9-11}{4}$

$x = \frac{2}{4} \qquad x = \frac{-20}{4}$

$x = \frac{1}{2} \qquad x = -5$

The solutions are $\frac{1}{2}$ and -5 and the solution set is $\left\{-5, \frac{1}{2}\right\}$.

2. $2x^2 = 6x - 1$

$2x^2 - 6x + 1 = 0$

$a = 2 \quad b = -6 \quad c = 1$

$$x = \frac{-(-6) \pm \sqrt{(-6)^2 - 4(2)(1)}}{2(2)}$$

$$= \frac{6 \pm \sqrt{36 - 8}}{4} = \frac{6 \pm \sqrt{28}}{4}$$

$$= \frac{6 \pm \sqrt{4 \cdot 7}}{4} = \frac{6 \pm 2\sqrt{7}}{4}$$

$$= \frac{2(3 \pm \sqrt{7})}{4} = \frac{3 \pm \sqrt{7}}{2}$$

The solutions are $\frac{3 \pm \sqrt{7}}{2}$ and the solution set is $\left\{\frac{3 \pm \sqrt{7}}{2}\right\}$.

3. $3x^2 + 5 = -6x$

$3x^2 + 6x + 5 = 0$

$a = 3 \quad b = 6 \quad c = 5$

$$x = \frac{-6 \pm \sqrt{6^2 - 4(3)(5)}}{2(3)}$$

$$= \frac{-6 \pm \sqrt{36 - 60}}{6} = \frac{-6 \pm \sqrt{-24}}{6}$$

$$= \frac{-6 \pm \sqrt{4 \cdot 6(-1)}}{6} = \frac{-6 \pm 2\sqrt{6}i}{6}$$

$$= \frac{-6}{6} \pm \frac{2\sqrt{6}i}{6} = -1 \pm \frac{\sqrt{6}}{3}i$$

The solutions are $-1 \pm \frac{\sqrt{6}}{3}i$ and the solution set is $\left\{-1 \pm \frac{\sqrt{6}}{3}i\right\}$.

4. **a.** $x^2 + 6x + 9 = 0$

$a = 1 \quad b = 6 \quad c = 9$

$b^2 - 4ac = 6^2 - 4(1)(9)$

$= 36 - 36 = 0$

Since the discriminant is zero, there is one real solution.

b. $2x^2 - 7x - 4 = 0$

$a = 2 \quad b = -7 \quad c = -4$

$b^2 - 4ac = (-7)^2 - 4(2)(-4)$

$= 49 - (-32) = 81$

Since the discriminant is greater than zero, there are two unequal real solutions. Also, since the discriminant is a perfect square, the solutions are rational.

c. $3x^2 - 2x + 4 = 0$

$a = 3 \quad b = -2 \quad c = 4$

$b^2 - 4ac = (-2)^2 - 4(3)(4)$

$= 4 - 48 = -44$

Since the discriminant is less than zero, there is no solution. There are two imaginary solutions which are complex conjugates.

5. **a.** Because the solution set is $\left\{-\frac{3}{5}, \frac{1}{4}\right\}$, we have

$x = -\frac{3}{5}$ or $x = \frac{1}{4}$

$x + \frac{3}{5} = 0 \qquad x - \frac{1}{4} = 0$

$5x + 3 = 0 \qquad 4x - 1 = 0.$

Use the zero-product principle in reverse.

$$(5x+3)(4x-1)=0$$
$$20x^2-5x+12x-3=0$$
$$20x^2+7x-3=0$$

b. Because the solution set is $\{-7i, 7i\}$ we have

$x=7i$ or $x=-7i$

$x-7i=0$ $\quad$ $x+7i=0$

Use the zero-product principle in reverse.

$$(x-7i)(x+7i)=0$$
$$x^2+7i-7i-49i^2=0$$
$$x^2-49(-1)=0$$
$$x^2+49=0$$

6. $f(x)=-1.65x^2+51.8x+111.44$

$$330=-1.65x^2+51.8x+111.44$$
$$0=-1.65x^2+51.8x-218.56$$

$a=-1.65 \quad b=51.8 \quad c=-218.56$

$$x=\frac{-51.8\pm\sqrt{(51.8)^2-4(-1.65)(-218.56)}}{2(-1.65)}$$
$$=\frac{-51.8\pm\sqrt{2683.24-1442.496}}{-3.3}$$
$$=\frac{-51.8\pm\sqrt{1240.744}}{-3.3}$$
$$=\frac{-51.8\pm 35.224}{-3.3}$$
$$=\frac{-51.8+35.224}{-3.3} \text{ or } \frac{-51.8-35.224}{-3.3}$$
$$\approx -26.37 \text{ or } 5.02$$

We disregard –26.37 because we can't have a negative number of years. The solution is 5.02. This means that approximately 5 years after 1990 in 1995, there were 330,000 U.S. AIDS deaths. The actual number of deaths was approximately 340,000. The function describes the data very well for that year.

Problem Set 8.2

Practice Exercises

1. $x^2+8x+12=0$

$a=1 \quad b=8 \quad c=12$

$$x=\frac{-8\pm\sqrt{8^2-4(1)(12)}}{2(1)}$$
$$=\frac{-8\pm\sqrt{64-48}}{2}$$
$$=\frac{-8\pm\sqrt{16}}{2}=\frac{-8\pm 4}{2}$$

Evaluate the expression to obtain two solutions.

$x=\frac{-8-4}{2}$ or $x=\frac{-8+4}{2}$

$x=\frac{-12}{2}$ $\quad$ $x=\frac{-4}{2}$

$x=-6$ $\quad$ $x=-2$

The solutions are –2 and –6 and the solution set is $\{-6, -2\}$.

3. $2x^2-7x=-5$

$$2x^2-7x+5=0$$

$a=2 \quad b=-7 \quad c=5$

$$x=\frac{-(-7)\pm\sqrt{(-7)^2-4(2)(5)}}{2(2)}$$
$$=\frac{7\pm\sqrt{49-40}}{4}$$
$$=\frac{7\pm\sqrt{9}}{4}=\frac{7\pm 3}{4}$$

Evaluate the expression to obtain two solutions.

$$x = \frac{7-3}{4} \quad \text{or} \quad x = \frac{7+3}{4}$$

$$x = \frac{4}{4} \qquad x = \frac{10}{4}$$

$$x = 1 \qquad x = \frac{5}{2}$$

The solutions are 1 and $\frac{5}{2}$ and the solution set is $\left\{1, \frac{5}{2}\right\}$.

5. $$x^2 + 3x - 20 = 0$$

$a = 1 \quad b = 3 \quad c = -20$

$$x = \frac{-3 \pm \sqrt{3^2 - 4(1)(-20)}}{2(1)}$$

$$= \frac{-3 \pm \sqrt{9-(-80)}}{2} = \frac{-3 \pm \sqrt{89}}{2}$$

The solutions are $\frac{-3 \pm \sqrt{89}}{2}$ and the solution set is $\left\{\frac{-3 \pm \sqrt{89}}{2}\right\}$.

7. $$3x^2 - 7x = 3$$

$$3x^2 - 7x - 3 = 0$$

$a = 3 \quad b = -7 \quad c = -3$

$$x = \frac{-(-7) \pm \sqrt{(-7)^2 - 4(3)(-3)}}{2(3)}$$

$$= \frac{7 \pm \sqrt{49-(-36)}}{6} = \frac{7 \pm \sqrt{85}}{6}$$

The solutions are $\frac{7 \pm \sqrt{85}}{6}$ and the solution set is $\left\{\frac{7 \pm \sqrt{85}}{6}\right\}$.

9. $$6x^2 = 2x + 1$$

$$6x^2 - 2x - 1 = 0$$

$a = 6 \quad b = -2 \quad c = -1$

$$x = \frac{-(-2) \pm \sqrt{(-2)^2 - 4(6)(-1)}}{2(6)}$$

$$= \frac{2 \pm \sqrt{4-(-24)}}{12} = \frac{2 \pm \sqrt{28}}{12}$$

$$= \frac{2 \pm \sqrt{4 \cdot 7}}{12} = \frac{2 \pm 2\sqrt{7}}{12}$$

$$= \frac{2\left(1 \pm \sqrt{7}\right)}{12} = \frac{1 \pm \sqrt{7}}{6}$$

The solutions are $\frac{1 \pm \sqrt{7}}{6}$ and the solution set is $\left\{\frac{1 \pm \sqrt{7}}{6}\right\}$.

11. $$4x^2 - 3x = -6$$

$$4x^2 - 3x + 6 = 0$$

$a = 4 \quad b = -3 \quad c = 6$

$$x = \frac{-(-3) \pm \sqrt{(-3)^2 - 4(4)(6)}}{2(4)}$$

$$= \frac{3 \pm \sqrt{9-96}}{8} = \frac{3 \pm \sqrt{-87}}{8}$$

$$= \frac{3 \pm \sqrt{87(-1)}}{8} = \frac{3 \pm \sqrt{87}i}{8}$$

$$= \frac{3}{8} \pm \frac{\sqrt{87}}{8}i$$

The solutions are $\frac{3}{8} \pm \frac{\sqrt{87}}{8}i$ and the solution set is $\left\{\frac{3}{8} \pm \frac{\sqrt{87}}{8}i\right\}$.

13. $x^2 - 4x + 8 = 0$

$a = 1 \quad b = -4 \quad c = 8$

$$x = \frac{-(-4) \pm \sqrt{(-4)^2 - 4(1)(8)}}{2(1)}$$
$$= \frac{4 \pm \sqrt{16 - 32}}{2} = \frac{4 \pm \sqrt{-16}}{2}$$
$$= \frac{4 \pm 4i}{2} = \frac{4}{2} \pm \frac{4}{2}i = 2 \pm 2i$$

The solutions are $2 \pm 2i$ and the solution set is $\{2 \pm 2i\}$.

15. $3x^2 = 8x - 7$

$3x^2 - 8x + 7 = 0$

$a = 3 \quad b = -8 \quad c = 7$

$$x = \frac{-(-8) \pm \sqrt{(-8)^2 - 4(3)(7)}}{2(3)}$$
$$= \frac{8 \pm \sqrt{64 - 84}}{6} = \frac{8 \pm \sqrt{-20}}{6}$$
$$= \frac{8 \pm \sqrt{4 \cdot 5(-1)}}{6} = \frac{8 \pm 2\sqrt{5}i}{6}$$
$$= \frac{8}{6} \pm \frac{2}{6}\sqrt{5}i = \frac{4}{3} \pm \frac{\sqrt{5}}{3}i$$

The solutions are $\frac{4}{3} \pm \frac{\sqrt{5}}{3}i$ and the solution set is $\left\{\frac{4}{3} \pm \frac{\sqrt{5}}{3}i\right\}$.

17. $x^2 + 8x + 3 = 0$

$a = 1 \quad b = 8 \quad c = 3$

$$b^2 - 4ac = 8^2 - 4(1)(3)$$
$$= 64 - 12 = 50$$

Since the discriminant is positive and not a perfect square, there are two irrational solutions.

19. $x^2 + 6x + 8 = 0$

$a = 1 \quad b = 6 \quad c = 8$

$$b^2 - 4ac = (6)^2 - 4(1)(8)$$
$$= 36 - 32 = 4$$

Since the discriminant is greater than zero, there are two unequal real solutions. Also, since the discriminant is a perfect square, the solutions are rational.

21. $2x^2 + x + 3 = 0$

$a = 2 \quad b = 1 \quad c = 3$

$$b^2 - 4ac = 1^2 - 4(2)(3)$$
$$= 1 - 24 = -23$$

Since the discriminant is negative, there are no real solutions. There are two imaginary solutions that are complex conjugates.

23. $2x^2 + 6x = 0$

$a = 2 \quad b = 6 \quad c = 0$

$$b^2 - 4ac = (6)^2 - 4(1)(0)$$
$$= 36 - 0 = 36$$

Since the discriminant is greater than zero, there are two unequal real solutions. Also, since the discriminant is a perfect square, the solutions are rational.

25. $5x^2+3=0$

$a=5 \quad b=0 \quad c=3$

$b^2-4ac=0^2-4(5)(3)$

$=0-60=-60$

Since the discriminant is negative, there are no real solutions. There are two imaginary solutions that are complex conjugates.

27. $9x^2=12x-4$

$9x^2-12x+4=0$

$a=9 \quad b=-12 \quad c=4$

$b^2-4ac=(-12)^2-4(9)(4)$

$=144-144=0$

Since the discriminant is zero, there is one repeated rational solution.

29. $3x^2-4x=4$

$3x^2-4x-4=0$

$(3x+2)(x-2)=0$

Apply the zero product principle.

$3x+2=0$ or $x-2=0$

$3x=-2 \qquad x=2$

$x=-\frac{2}{3}$

The solutions are $-\frac{2}{3}$ and 2 and the solution set is $\left\{-\frac{2}{3},2\right\}$.

31. $x^2-2x=1$

Since $b=-2$, we add

$\left(\frac{b}{2}\right)^2=\left(\frac{-2}{2}\right)^2=(-1)^2=1.$

$x^2-2x+1=1+1$

$(x-1)^2=2$

Apply the square root principle.

$x-1=\pm\sqrt{2}$

$x=1\pm\sqrt{2}$

The solutions are $1\pm\sqrt{2}$ and the solution set is $\left\{1\pm\sqrt{2}\right\}$.

33. $(2x-5)(x+1)=2$

$2x^2+2x-5x-5=2$

$2x^2-3x-5=2$

$2x^2-3x-7=0$

$a=2 \quad b=-3 \quad c=-7$

$$x=\frac{-(-3)\pm\sqrt{(-3)^2-4(2)(-7)}}{2(2)}$$

$$=\frac{3\pm\sqrt{9-(-56)}}{4}=\frac{3\pm\sqrt{65}}{4}$$

The solutions are $\frac{3\pm\sqrt{65}}{4}$ and the solution set is $\left\{\frac{3\pm\sqrt{65}}{4}\right\}$.

35. $(3x-4)^2=16$

Apply the square root property.

$3x-4=\sqrt{16}$ or $3x-4=-\sqrt{16}$

$3x-4=4 \qquad 3x-4=-4$

$3x=8 \qquad 3x=0$

$x=\frac{8}{3} \qquad x=0$

The solutions are $\frac{8}{3}$ and 0 and the solution set is $\left\{0, \frac{8}{3}\right\}$.

37. $\frac{x^2}{2}+2x+\frac{2}{3}=0$

Multiply both sides of the equation by 6 to clear fractions.

$3x^2+12x+4=0$

$a=3 \quad b=12 \quad c=4$

$$x=\frac{-12\pm\sqrt{12^2-4(3)(4)}}{2(3)}$$
$$=\frac{-12\pm\sqrt{144-48}}{6}=\frac{-12\pm\sqrt{96}}{6}$$
$$=\frac{-12\pm\sqrt{16\cdot 6}}{6}=\frac{-12\pm 4\sqrt{6}}{6}$$
$$=\frac{2\left(-6\pm 2\sqrt{6}\right)}{6}=\frac{-6\pm 2\sqrt{6}}{3}$$

The solutions are $\frac{-6\pm 2\sqrt{6}}{3}$ and the solution set is $\left\{\frac{-6\pm 2\sqrt{6}}{3}\right\}$.

39. $(3x-2)^2=10$

Apply the square root property.

$3x-2=\pm\sqrt{10}$

$3x=2\pm\sqrt{10}$

$x=\frac{2\pm\sqrt{10}}{3}$

The solutions are $\frac{2\pm\sqrt{10}}{3}$ and the solution set is $\left\{\frac{2\pm\sqrt{10}}{3}\right\}$.

41. $\frac{1}{x}+\frac{1}{x+2}=\frac{1}{3}$

The LCD is $3x(x+2)$.

$$3x(x+2)\left(\frac{1}{x}+\frac{1}{x+2}\right)=3x(x+2)\left(\frac{1}{3}\right)$$
$$3(x+2)+3x=x(x+2)$$
$$3x+6+3x=x^2+2x$$
$$6+6x=x^2+2x$$
$$0=x^2-4x-6$$

$a=1 \quad b=-4 \quad c=-6$

$$x=\frac{-(-4)\pm\sqrt{(-4)^2-4(1)(-6)}}{2(1)}$$
$$=\frac{4\pm\sqrt{16-(-24)}}{2}=\frac{4\pm\sqrt{40}}{2}$$
$$=\frac{4\pm\sqrt{4\cdot 10}}{2}=\frac{4\pm 2\sqrt{10}}{2}$$
$$=\frac{2\left(2\pm 1\sqrt{10}\right)}{2}=2\pm\sqrt{10}$$

The solutions are $2\pm\sqrt{10}$ and the solution set is $\left\{2\pm\sqrt{10}\right\}$.

43. $(2x-6)(x+2)=5(x-1)-12$

$$2x^2+4x-6x-12=5x-5-12$$
$$2x^2-2x-12=5x-17$$
$$2x^2-7x+5=0$$
$$(2x-5)(x-1)=0$$

Apply the zero product principle.

$2x-5=0 \quad \text{or} \quad x-1=0$

$2x=5 \qquad x=1$

$x=\frac{5}{2}$

The solutions are 1 and $\frac{5}{2}$ and the solution set is $\left\{1, \frac{5}{2}\right\}$.

45. Because the solution set is $\{-3, 5\}$, we have

$$x = -3 \quad \text{or} \quad x = 5$$
$$x + 3 = 0 \qquad x - 5 = 0.$$

Use the zero-product principle in reverse.

$$(x+3)(x-5) = 0$$
$$x^2 - 5x + 3x - 15 = 0$$
$$x^2 - 2x - 15 = 0$$

47. Because the solution set is $\left\{-\frac{2}{3}, \frac{1}{4}\right\}$, we have

$$x = -\frac{2}{3} \quad \text{or} \quad x = \frac{1}{4}$$
$$x + \frac{2}{3} = 0 \qquad 4x - 1 = 0.$$
$$3x + 2 = 0$$

Use the zero-product principle in reverse.

$$(3x+2)(4x-1) = 0$$
$$12x^2 - 3x + 8x - 2 = 0$$
$$12x^2 + 5x - 2 = 0$$

49. Because the solution set is $\{-6i, 6i\}$, we have

$$x = 6i \quad \text{or} \quad x = -6i$$
$$x - 6i = 0 \qquad x + 6i = 0.$$

Use the zero-product principle in reverse.

$$(x-6i)(x+6i) = 0$$
$$x^2 + \cancel{6i} - \cancel{6i} - 36i^2 = 0$$
$$x^2 - 36(-1) = 0$$
$$x^2 + 36 = 0$$

51. Because the solution set is $\{-\sqrt{2}, \sqrt{2}\}$, we have

$$x = \sqrt{2} \quad \text{or} \quad x = -\sqrt{2}$$
$$x - \sqrt{2} = 0 \qquad x + \sqrt{2} = 0.$$

Use the zero-product principle in reverse.

$$(x-\sqrt{2})(x+\sqrt{2}) = 0$$
$$x^2 + \cancel{x\sqrt{2}} - \cancel{x\sqrt{2}} - 2 = 0$$
$$x^2 - 2 = 0$$

53. Because the solution set is $\{-2\sqrt{5}, 2\sqrt{5}\}$ we have

$$x = 2\sqrt{5} \quad \text{or} \quad x = -2\sqrt{5}$$
$$x - 2\sqrt{5} = 0 \qquad x + 2\sqrt{5} = 0.$$

Use the zero-product principle in reverse.

$$(x-2\sqrt{5})(x+2\sqrt{5}) = 0$$
$$x^2 + \cancel{2x\sqrt{5}} - \cancel{2x\sqrt{5}} - 4 \cdot 5 = 0$$
$$x^2 - 20 = 0$$

55. Because the solution set is $\{1+i, 1-i\}$, we have

$$x = 1 + i \quad \text{or} \quad x = 1 - i$$
$$x - (1+i) = 0 \qquad x - (1-i) = 0.$$

Use the zero-product principle in reverse.

$$(x-(1+i))(x-(1-i))=0$$
$$x^2-x(1-i)-x(1+i)+(1+i)(1-i)=0$$
$$x^2-x+\cancel{xi}-x-\cancel{xi}+1-i^2=0$$
$$x^2-x-x+1-(-1)=0$$
$$x^2-2x+2=0$$

57. Because the solution set is $\{1+\sqrt{2}, 1-\sqrt{2}\}$, we have

$x=1+\sqrt{2}$ or $x=1-\sqrt{2}$

$x-(1+\sqrt{2})=0$ $\quad x-(1-\sqrt{2})=0.$

Use the zero-product principle in reverse.

$$(x-(1+\sqrt{2}))(x-(1-\sqrt{2}))=0$$
$$x^2-x(1-\sqrt{2})-x(1+\sqrt{2})+(1+\sqrt{2})(1-\sqrt{2})=0$$
$$x^2-x+\cancel{x\sqrt{2}}-x-\cancel{x\sqrt{2}}+1-2=0$$
$$x^2-2x-1=0$$

Application Exercises

59.
$$f(x)=23.4x^2-259.1x+815.8$$
$$1000=23.4x^2-259.1x+815.8$$
$$0=23.4x^2-259.1x-184.2$$

$a=23.4 \quad b=-259.1 \quad c=-184.2$

$$x=\frac{-(-259.1)\pm\sqrt{(-259.1)^2-4(23.4)(-184.2)}}{2(23.4)}$$
$$=\frac{259.1\pm\sqrt{67132.81-(-17241.12)}}{46.8}$$
$$=\frac{259.1\pm\sqrt{84373.93}}{46.8}\approx\frac{259.1\pm 290.47}{46.8}$$
$$\approx 11.7 \text{ or } -0.67$$

We disregard –0.67 because can't have a negative number of policemen. The solution is approximately 12 and we conclude that in the year 1990 + 12 = 2002, there will be 1000 police officers convicted of felonies.

61. $f(x) = 0.013x^2 - 1.19x + 28.24$

$$10 = 0.013x^2 - 1.19x + 28.24$$
$$0 = 0.013x^2 - 1.19x + 18.24$$

$a = 0.013 \quad b = -1.19 \quad c = 18.24$

$$x = \frac{-(-1.19) \pm \sqrt{(-1.19)^2 - 4(0.013)(18.24)}}{2(0.013)} = \frac{1.19 \pm \sqrt{1.4161 - 0.94848}}{0.026}$$

$$= \frac{1.19 \pm \sqrt{0.46762}}{0.026} \approx \frac{1.19 \pm 0.68383}{0.026}$$

Evaluate the expression to obtain two solutions.

$x = \frac{1.19 + 0.68383}{0.026}$ or $x = \frac{1.19 - 0.68383}{0.026}$

$x = \frac{1.87383}{0.026}$ $\qquad$ $x = \frac{0.50617}{0.026}$

$x \approx 72.1$ $\qquad$ $x \approx 19$

Drivers of approximately age 19 and age 72 are expected to be involved in 10 fatal crashes per 100 million miles driven. The model doesn't seem to predict the number of accidents very well. The model overestimates the number of fatal accidents.

63. $f(t) = -16t^2 + 60t + 4$

$$0 = -16t^2 + 60t + 4$$

$a = -16 \quad b = 60 \quad c = 4$

$$x = \frac{-60 \pm \sqrt{60^2 - 4(-16)(4)}}{2(-16)}$$

$$= \frac{-60 \pm \sqrt{3600 - (-256)}}{-32}$$

$$= \frac{-60 \pm \sqrt{3856}}{-32} \approx \frac{-60 \pm 62.1}{-32}$$

≈ -0.07 or 3.8

We disregard –0.07 because we can't have a negative time measurement. The solution is 3.8. The ball will hit the ground in approximately 3.8 seconds.

65. Let x = the width of the rectangle
Let $x + 3$ = the length of the rectangle

$$A = lw$$
$$36 = x(x+3)$$
$$36 = x^2 + 3x$$
$$0 = x^2 + 3x - 36$$

$a = 1 \quad b = 3 \quad c = -36$

$$x = \frac{-3 \pm \sqrt{3^2 - 4(1)(-36)}}{2(1)}$$

$$= \frac{-3 \pm \sqrt{9 - (-144)}}{2}$$

$$= \frac{-3 \pm \sqrt{153}}{2} \approx \frac{-3 \pm 12.4}{2}$$

Evaluate the expression to obtain two solutions.

$x = \frac{-3+12.4}{2}$ or $x = \frac{-3-12.4}{2}$

$x = \frac{9.4}{2}$ $\quad x = \frac{-15.4}{2}$

$x = 4.7$ $\quad x = -7.7$

We disregard –7.7 because we can't have a negative length measurement. The solution is 4.7 and we conclude that the rectangle's dimensions are 4.7 meters by 4.7 + 3 = 7.7 meters.

67. Let x = the length of the one leg
Let $x + 1$ = the length of the other leg

$$x^2 + (x+1)^2 = 4^2$$
$$x^2 + x^2 + 2x + 1 = 16$$
$$2x^2 + 2x + 1 = 16$$
$$2x^2 + 2x - 15 = 0$$

$a = 2 \quad b = 2 \quad c = -15$

$$x = \frac{-2 \pm \sqrt{2^2 - 4(2)(-15)}}{2(2)}$$
$$= \frac{-2 \pm \sqrt{4-(-120)}}{4}$$
$$= \frac{-2 \pm \sqrt{124}}{4} \approx \frac{-2 \pm 11.1}{4}$$

Evaluate the expression to obtain two solutions.

$x = \frac{-2+11.1}{4}$ or $x = \frac{-2-11.1}{4}$

$x = \frac{9.1}{4}$ $\quad x = \frac{-13.1}{4}$

$x = 2.275$ $\quad x = -3.275$

We disregard –3.275 because we can't have a negative length measurement. The solution is 2.275 and we conclude that the triangle's legs are approximately 2.3 feet and 2.3 + 1 = 3.3 feet.

69.
$$x(20-2x) = 13$$
$$20x - 2x^2 = 13$$
$$0 = 2x^2 - 20x + 13$$

$a = 2 \quad b = -20 \quad c = 13$

$$x = \frac{-(-20) \pm \sqrt{(-20)^2 - 4(2)(13)}}{2(2)}$$
$$= \frac{20 \pm \sqrt{400-104}}{4}$$
$$= \frac{20 \pm \sqrt{296}}{4}$$
$$\approx \frac{20+17.2}{4} \text{ or } \frac{20-17.2}{4}$$
$$\approx \frac{37.2}{4} \text{ or } \frac{2.8}{4}$$
$$\approx 9.3 \text{ or } 0.7$$

A gutter with depth 9.3 or 0.7 inches will have a cross-sectional area of 13 square inches.

Writing in Mathematics

71. Answers will vary.

73. Answers will vary.

75. Answers will vary.

77. Answers will vary.

Technology Exercises

79. Answers will vary. For example, consider Exercise 17.

$x^2 - 5x + 13 = 0$

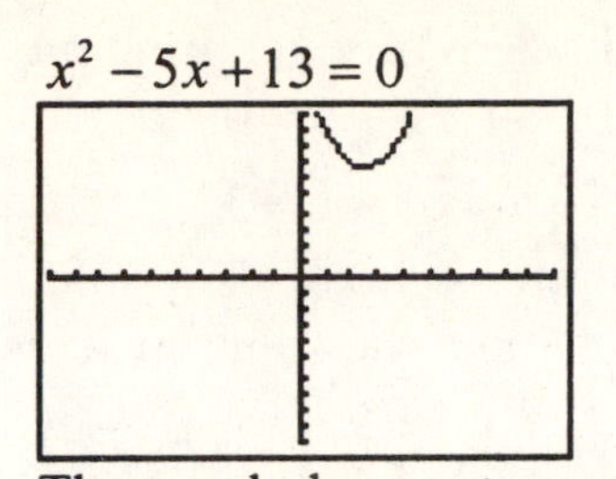

The graph does not cross the x–axis and we conclude that there are 2 imaginary solutions which are complex conjugates.

81. $f(x) = x(20 - 2x)$

Y1=X(20-2X)

X=5 Y=50

The depth of the gutter that will maximize water flow is 5 inches. This will allow for a water flow of 50 square inches. The situation described in Exercise 69 does not take full advantage of the sheets of aluminum.

Critical Thinking Exercises

83.

$$x^2 - 3\sqrt{2}x = -2$$

$$x^2 - 3\sqrt{2}x + 2 = 0$$

$$a = 1 \quad b = -3\sqrt{2} \quad c = 2$$

$$x = \frac{-\left(-3\sqrt{2}\right) \pm \sqrt{\left(-3\sqrt{2}\right)^2 - 4(1)(2)}}{2(1)}$$

$$= \frac{3\sqrt{2} \pm \sqrt{9 \cdot 2 - 8}}{2}$$

$$= \frac{3\sqrt{2} \pm \sqrt{18 - 8}}{2} = \frac{3\sqrt{2} \pm \sqrt{10}}{2}$$

The solutions are $\frac{3\sqrt{2} \pm \sqrt{10}}{2}$ and the solution set is $\left\{\frac{3\sqrt{2} \pm \sqrt{10}}{2}\right\}$.

85. The dimensions of the pool are 12 meters by 8 meters. With the tile, the dimensions will be $12 + 2x$ meters by $8 + 2x$ meters. If we take the area of the pool with the tile and subtract the area of the pool without the tile, we are left with the area of the tile only.

$$(12 + 2x)(8 + 2x) - 12(8) = 120$$

$$\cancel{96} + 24x + 16x + 4x^2 - \cancel{96} = 120$$

$$4x^2 + 40x = 120$$

$$4x^2 + 40x - 120 = 0$$

$$x^2 + 10x - 30 = 0$$

$$a = 1 \quad b = 10 \quad c = -30$$

$$x = \frac{-10 \pm \sqrt{10^2 - 4(1)(-30)}}{2(1)}$$

$$= \frac{-10 \pm \sqrt{100 + 120}}{2}$$

$$= \frac{-10 \pm \sqrt{220}}{2} \approx \frac{-10 \pm 14.8}{2}$$

Evaluate the expression to obtain two solutions.

$$x = \frac{-10 + 14.8}{2} \quad \text{or} \quad x = \frac{-10 - 14.8}{2}$$

$$x = \frac{4.8}{2} \qquad x = \frac{-24.8}{2}$$

$$x = 2.4 \qquad x = -12.4$$

We disregard -12.4 because we can't have a negative width measurement. The solution is 2.4 and we conclude that the width of the uniform tile border is 2.4 meters. This is more than the 2-meter requirement, so the tile meets the zoning laws.

Review Exercises

87. $|5x+2| = |4-3x|$

$5x+2 = 4-3x$ or $5x+2 = -(4-3x)$

$8x+2 = 4$ $\quad$ $5x+2 = -4+3x$

$8x = 2$ $\quad$ $2x+2 = -4$

$x = \frac{1}{4}$ $\quad$ $2x = -6$

$x = -3$

The solutions are $\frac{1}{4}$ and -3 and the solution set is $\left\{-3, \frac{1}{4}\right\}$.

88. $\sqrt{2x-5} - \sqrt{x-3} = 1$

$$\sqrt{2x-5} = \sqrt{x-3} + 1$$
$$\left(\sqrt{2x-5}\right)^2 = \left(\sqrt{x-3}+1\right)^2$$
$$2x-5 = x-3+2\sqrt{x-3}+1$$
$$2x-5 = x-2+2\sqrt{x-3}$$
$$x-3 = 2\sqrt{x-3}$$
$$(x-3)^2 = \left(2\sqrt{x-3}\right)^2$$
$$x^2-6x+9 = 4(x-3)$$
$$x^2-6x+9 = 4x-12$$
$$x^2-10x+21 = 0$$
$$(x-7)(x-3) = 0$$

Apply the zero product principle.

$x-7=0$ or $x-3=0$

$x=7$ $\quad$ $x=3$

Both solutions check. The solutions are 3 and 7 and the solution set is $\{3,7\}$.

89. $$\frac{5}{\sqrt{3}+x} = \frac{5}{\sqrt{3}+x} \cdot \frac{\sqrt{3}-x}{\sqrt{3}-x}$$
$$= \frac{5\left(\sqrt{3}-x\right)}{3-x^2} = \frac{5\sqrt{3}-5x}{3-x^2}$$

Check Points 8.3

1. $f(x) = -(x-1)^2 + 4$

Since $a=-1$ is negative, the parabola opens downward. The vertex of the parabola is $(h,k) = (1,4)$. Replace $f(x)$ with 0 to find x–intercepts.

$$0 = -(x-1)^2 + 4$$
$$-4 = -(x-1)^2$$
$$4 = (x-1)^2$$

Apply the square root property.

$x-1 = \sqrt{4}$ or $x-1 = -\sqrt{4}$

$x-1 = 2$ $\quad$ $x-1 = -2$

$x = 3$ $\quad$ $x = -1$

The x–intercepts are -1 and 3. Set $x=0$ and solve for y to obtain the y–intercept.

$$y = -(0-1)^2 + 4 = -(-1)^2 + 4$$
$$= -(1) + 4 = 3$$

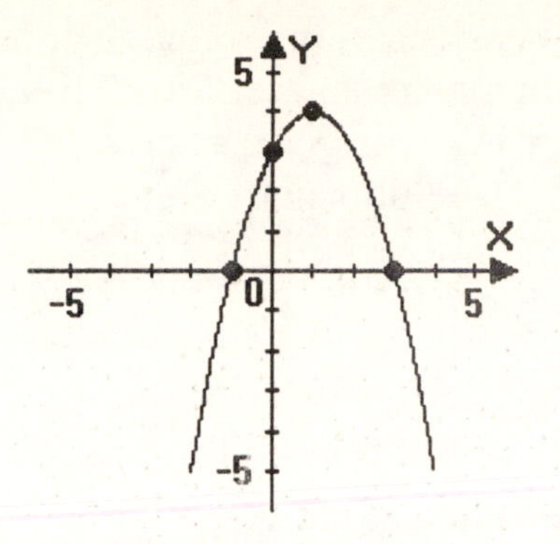

2. $f(x)=(x-2)^2+1$

Since $a=1$ is positive, the parabola opens upward. The vertex of the parabola is $(h,k)=(2,1)$. Replace $f(x)$ with 0 to find x–intercepts.

$0=(x-2)^2+1$

$-1=(x-2)^2$

Apply the square root property.

$x-2=\sqrt{-1}$ or $x-2=-\sqrt{-1}$

$x-2=i$ 　　 $x-2=-i$

$x=2+i$ 　　 $x=2-i$

Because this solution has no real solutions, the parabola has no x–intercepts. Set $x=0$ and solve for y to obtain the y–intercept.

$y=(0-2)^2+1=(-2)^2+1$

$=4+1=5$

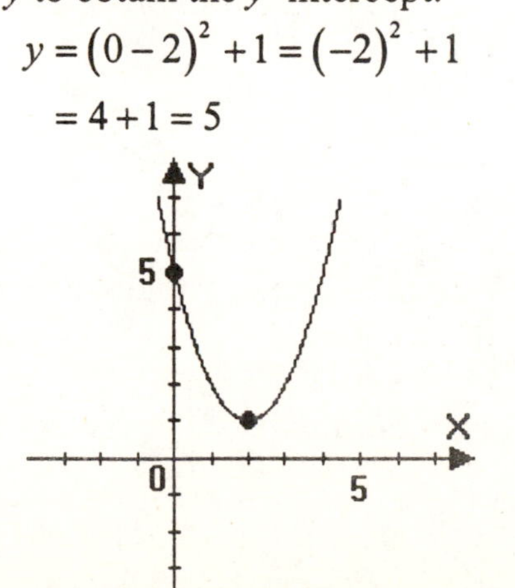

3. $f(x)=x^2-2x-3$

Since $a=1$ is positive, the parabola opens upward. The x–coordinate of the vertex of the parabola is

$-\frac{b}{2a}=-\frac{-2}{2(1)}=-\frac{-2}{2}=1$ and the

y–coordinate of the vertex of the parabola is

$f\left(-\frac{b}{2a}\right)=f(1)=(1)^2-2(1)-3$

$=1-2-3=-4.$

The vertex is $(1,-4)$. Replace $f(x)$ with 0 to find x–intercepts.

$0=x^2-2x-3$

$0=(x-3)(x+1)$

Apply the zero product principle.

$x-3=0$ or $x+1=0$

$x=3$ 　　 $x=-1$

The x–intercepts are –1 and 3.
Set $x=0$ and solve for y to obtain the y–intercept.

$y=(0)^2-2(0)-3=-3$

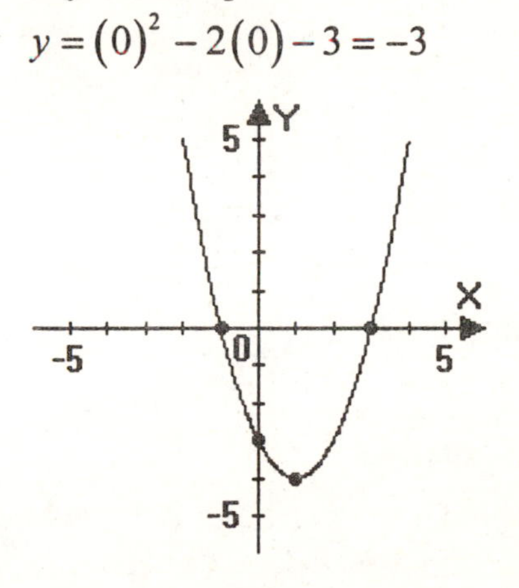

4. $f(x)=0.4x^2-36x+1000$

Since $a=0.4$ is positive, we know the function opens upward and has a minimum at

$x = -\frac{b}{2a} = -\frac{-36}{2(0.4)} = -\frac{-36}{0.8} = 45.$

Drivers of age 45 have the least number of accidents per 50 million miles driven. At this age, the number of accidents drivers have is

$$\begin{aligned} f(45) &= 0.4(45)^2 - 36(45) + 1000 \\ &= 0.4(2025) - 1620 + 1000 \\ &= 810 - 1620 + 1000 \\ &= 190. \end{aligned}$$

The minimum number of accidents is 190 per 50 million miles driven.

5. Maximize the area of a rectangle constructed with 120 feet of fencing.
Let x = the length of the rectangle
Let y = the width of the rectangle
Since we need an equation in one variable, use the perimeter to express y in terms of x.

$$\begin{aligned} 2x + 2y &= 120 \\ 2y &= 120 - 2x \\ y &= \frac{120 - 2x}{2} \\ y &= 60 - x \end{aligned}$$

We need to maximize $A = xy = x(60 - x)$. Rewrite A as a function of x.

$$A(x) = x(60 - x)$$
$$A(x) = -x^2 + 60x$$

Since $a = -1$ is negative, we know the function opens downward and has a maximum at

$x = -\frac{b}{2a} = -\frac{60}{2(-1)} = -\frac{60}{-2} = 30.$

When the length x is 30, the width y is $y = 60 - x = 60 - 30 = 30$. The dimensions of the rectangular region with maximum area are 30 feet by 30 feet. This gives an area of $30 \cdot 30 = 900$ square feet.

Problem Set 8.3

Practice Exercises

1. The vertex of the graph is the point $(-1, -1)$. This means that the equation is $f(x) = (x - (-1))^2 - 1$
$= (x + 1)^2 - 1$.

3. The vertex of the graph is the point $(-1, 1)$. This means that the equation is $g(x) = (x - (-1))^2 + 1 = (x + 1)^2 + 1$.

5. The vertex of the graph is the point $(0, -1)$. This means that the equation is $f(x) = (x - 0)^2 - 1 = x^2 - 1$.

7. The vertex of the graph is the point $(1, 0)$. This means that the equation is $f(x) = (x - 1)^2 + 0 = (x - 1)^2$.

9. $f(x) = 2(x - 3)^2 + 1$
The vertex is $(3, 1)$.

11. $f(x) = -2(x + 1)^2 + 5$
The vertex is $(-1, 5)$.

13. $f(x) = 2x^2 - 8x + 3$
The x–coordinate of the vertex of the parabola is $-\frac{b}{2a} = -\frac{-8}{2(2)} = -\frac{-8}{4} = 2$ and the y–coordinate of the vertex of

the parabola is

$$f\left(-\frac{b}{2a}\right) = f(2) = 2(2)^2 - 8(2) + 3$$
$$= 2(4) - 16 + 3$$
$$= 8 - 16 + 3 = -5.$$

The vertex is $(2, -5)$.

15. $f(x) = -x^2 - 2x + 8$

The x–coordinate of the vertex of the parabola is

$-\frac{b}{2a} = -\frac{-2}{2(-1)} = -\frac{-2}{-2} = -1$ and the y–coordinate of the vertex of the parabola is

$$f\left(-\frac{b}{2a}\right) = f(-1)$$
$$= -(-1)^2 - 2(-1) + 8$$
$$= -1 + 2 + 8 = 9.$$

The vertex is $(-1, 9)$.

17. $f(x) = (x-4)^2 - 1$

Since $a = 1$ is positive, the parabola opens upward. The vertex of the parabola is $(h,k) = (4,-1)$. Replace $f(x)$ with 0 to find x–intercepts.

$0 = (x-4)^2 - 1$

$1 = (x-4)^2$

Apply the square root property.

$x - 4 = \sqrt{1}$ or $x - 4 = -\sqrt{1}$

$x - 4 = 1$ $\quad$ $x - 4 = -1$

$x = 5$ $\quad$ $x = 3$

The x–intercepts are 5 and 3.
Set $x = 0$ and solve for y to obtain the y–intercept.

$y = (0-4)^2 - 1 = (-4)^2 - 1$
$= 16 - 1 = 15$

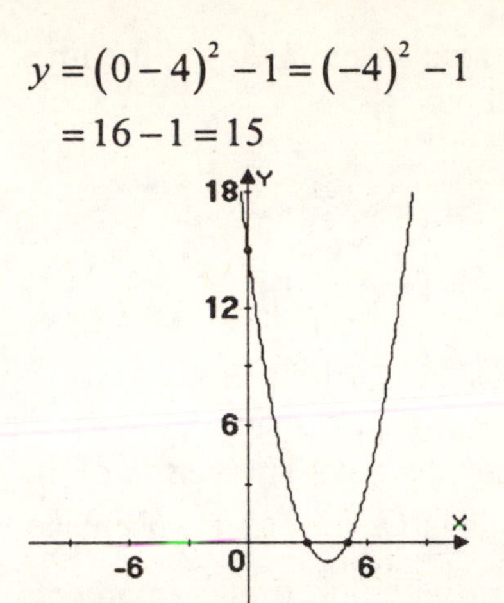

The axis of symmetry is the line $x = 4$.

19. $f(x) = (x-1)^2 + 2$

Since $a = 1$ is positive, the parabola opens upward. The vertex of the parabola is $(h,k) = (1,2)$. Replace $f(x)$ with 0 to find x–intercepts.

$0 = (x-1)^2 + 2$

$-2 = (x-1)^2$

Because the solutions to the equation are imaginary, we know that there are no x–intercepts. Set $x = 0$ and solve for y to obtain the y–intercept.

$y = (0-1)^2 + 2 = (-1)^2 + 2$
$= 1 + 2 = 3$

The y–intercept is 3.

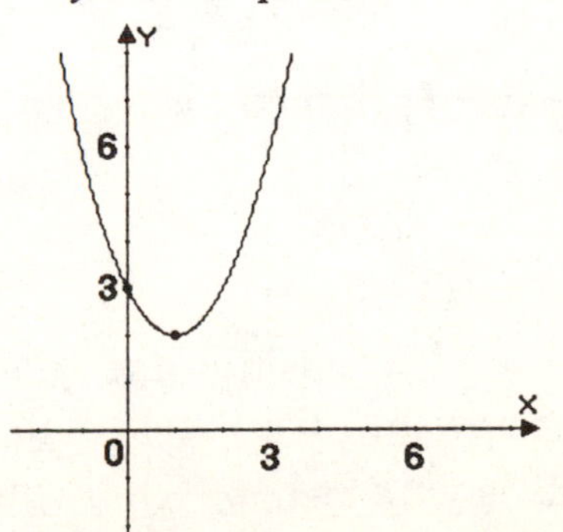

The axis of symmetry is the line $x = 1$.

21. $y - 1 = (x-3)^2$

$y = (x-3)^2 + 1$

$f(x) = (x-3)^2 + 1$

Since $a = 1$ is positive, the parabola opens upward. The vertex of the parabola is $(h,k) = (3,1)$. Replace $f(x)$ with 0 to find x–intercepts.

$0 = (x-3)^2 + 1$

$-1 = (x-3)^2$

Because the solutions to the equation are imaginary, we know that there are no x–intercepts. Set $x = 0$ and solve for y to obtain the y–intercept.

$y = (0-3)^2 + 1 = (-3)^2 + 1$

$= 9 + 1 = 10$

The y–intercept is 10.

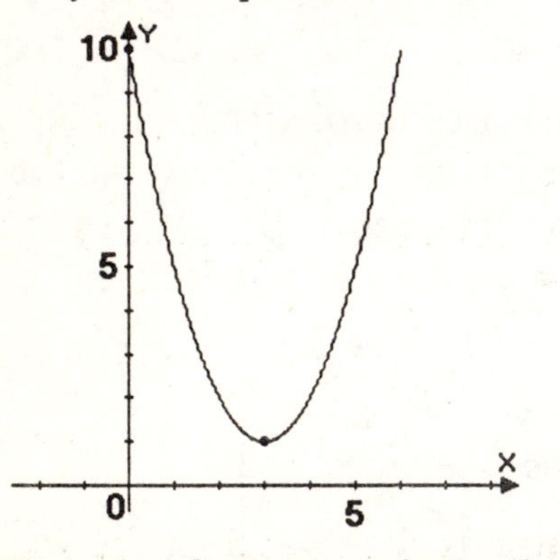

The axis of symmetry is the line $x = 3$.

23. $f(x) = 2(x+2)^2 - 1$

Since $a = 2$ is positive, the parabola opens upward. The vertex of the parabola is $(h,k) = (-2,-1)$. Replace $f(x)$ with 0 to find x–intercepts.

$0 = 2(x+2)^2 - 1$

$1 = 2(x+2)^2$

$\frac{1}{2} = (x+2)^2$

Apply the square root property.

$x + 2 = \pm\sqrt{\frac{1}{2}}$

$x = -2 \pm \sqrt{\frac{1}{2}}$

$x \approx -2 - \sqrt{\frac{1}{2}}$ or $-2 + \sqrt{\frac{1}{2}}$

$x \approx -2.7$ or -1.3

The x–intercepts are -1.3 and -2.7. Set $x = 0$ and solve for y to obtain the y–intercept.

$y = 2(0+2)^2 - 1 = 2(2)^2 - 1$

$= 2(4) - 1 = 8 - 1 = 7$

The y–intercept is 7.

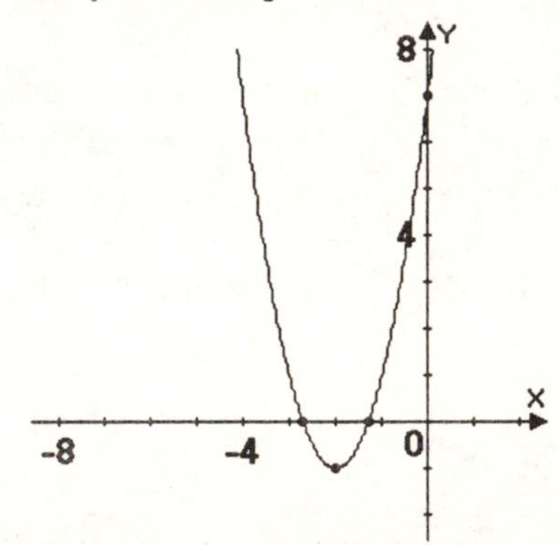

The axis of symmetry is the line $x = -2$.

25. $f(x) = 4 - (x-1)^2$

$f(x) = -(x-1)^2 + 4$

Since $a=-1$ is negative, the parabola opens downward. The vertex of the parabola is $(h,k)=(1,4)$. Replace $f(x)$ with 0 to find x–intercepts.

$$0=-(x-1)^2+4$$
$$-4=-(x-1)^2$$
$$4=(x-1)^2$$

Apply the square root property.

$\sqrt{4}=x-1$ or $-\sqrt{4}=x-1$

$2=x-1$ $\quad -2=x-1$

$3=x$ $\quad -1=x$

The x–intercepts are –1 and 3.
Set $x=0$ and solve for y to obtain the y–intercept.

$$y=-(0-1)^2+4$$
$$=-(-1)^2+4=-1+4=3$$

The y–intercept is 3.

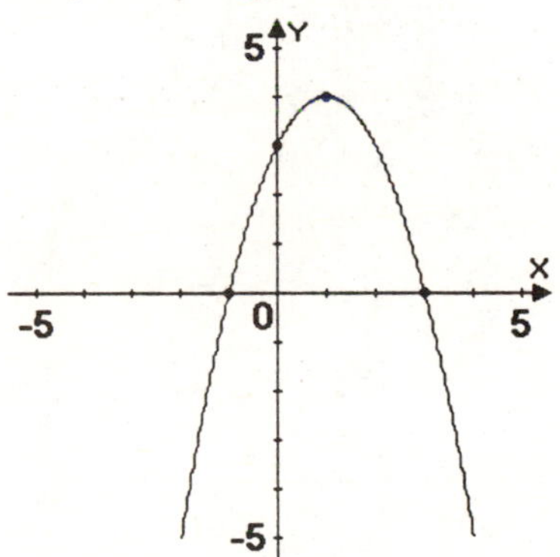

The axis of symmetry is the line $x=1$.

27. $f(x)=x^2+2x-3$

Since $a=1$ is positive, the parabola opens upward. The x–coordinate of the vertex of the parabola is

$$-\frac{b}{2a}=-\frac{2}{2(1)}=-\frac{2}{2}=-1$$ and the y–coordinate of the vertex of the parabola is

$$f\left(-\frac{b}{2a}\right)=f(-1)=(-1)^2+2(-1)-3$$
$$=1-2-3=-4.$$

The vertex is $(-1, -4)$. Replace $f(x)$ with 0 to find x–intercepts.

$$0=x^2+2x-3$$
$$0=(x+3)(x-1)$$

Apply the zero product principle.

$x+3=0$ or $x-1=0$

$x=-3$ $\quad x=1$

The x–intercepts are –3 and 1. Set $x=0$ and solve for y to obtain the y–intercept.

$$y=(0)^2+2(0)-3$$
$$y=-3$$

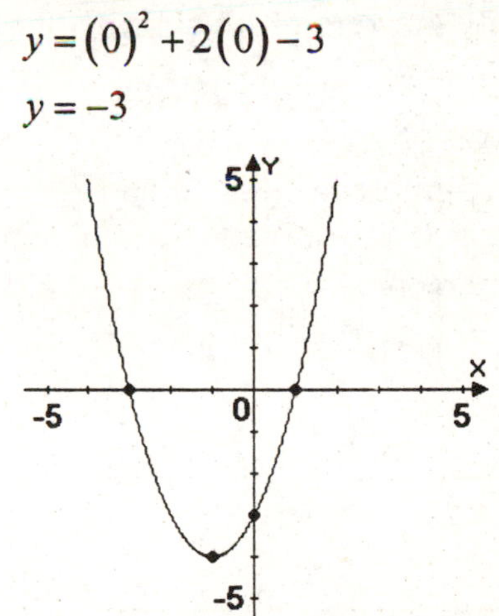

The axis of symmetry is the line $x=-1$.

29. $f(x)=x^2+3x-10$

Since $a=1$ is positive, the parabola opens upward. The x–coordinate of the vertex of the parabola is

$-\frac{b}{2a}=-\frac{3}{2(1)}=-\frac{3}{2}$ and the y–coordinate of the vertex of the parabola is

$f\left(-\frac{b}{2a}\right)$

$=f\left(-\frac{3}{2}\right)=\left(-\frac{3}{2}\right)^2+3\left(-\frac{3}{2}\right)-10$

$=\frac{9}{4}-\frac{9}{2}-10=\frac{9}{4}-\frac{18}{4}-\frac{40}{4}=-\frac{49}{4}.$

The vertex is $\left(-\frac{3}{2},-\frac{49}{4}\right)$.

Replace $f(x)$ with 0 to find x–intercepts.

$0=x^2+3x-10$

$0=(x+5)(x-2)$

Apply the zero product principle.

$x+5=0$ or $x-2=0$

$x=-5$ $\quad x=2$

The x–intercepts are –5 and 2.

Set $x = 0$ and solve for y to obtain the y–intercept.

$y=0^2+3(0)-10=-10$

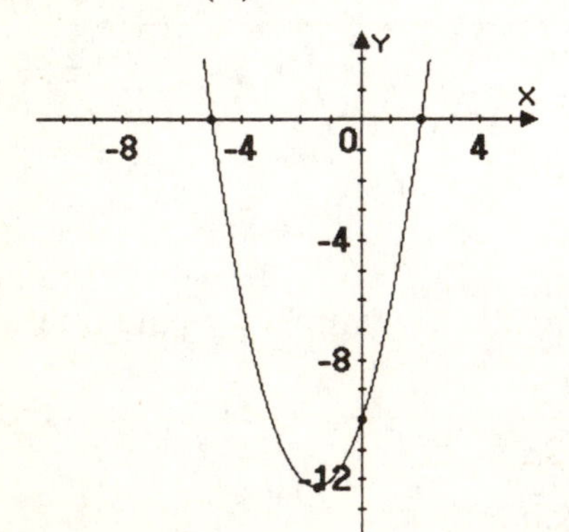

The axis of symmetry is the line $x=-2$.

31. $f(x)=2x-x^2+3$

$f(x)=-x^2+2x+3$

Since $a=-1$ is negative, the parabola opens downward. The x–coordinate of the vertex of the parabola is $-\frac{b}{2a}=-\frac{2}{2(-1)}=-\frac{2}{-2}=1$ and the y–coordinate of the vertex of the parabola is

$f\left(-\frac{b}{2a}\right)=f(1)=-(1)^2+2(1)+3$

$=-1+2+3=4.$

The vertex is $(1,4)$. Replace $f(x)$ with 0 to find x–intercepts.

$0=-x^2+2x+3$

$0=x^2-2x-3$

$0=(x-3)(x+1)$

Apply the zero product principle.

$x-3=0$ or $x+1=0$

$x=3$ $\quad x=-1$

The x–intercepts are 3 and –1. Set $x = 0$ and solve for y to obtain the y–intercept.

$y=-(0)^2+2(0)+3=3$

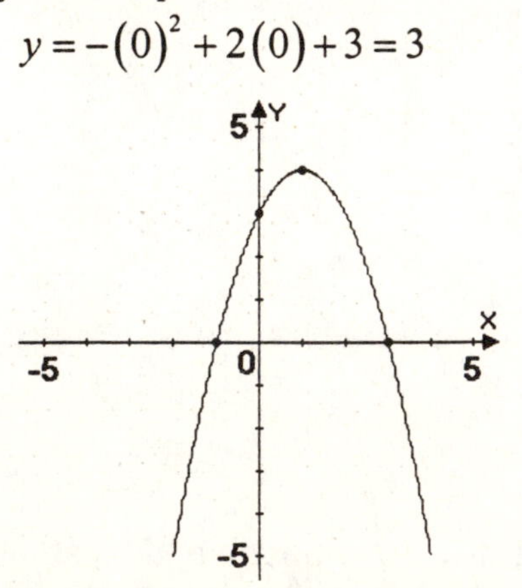

The axis of symmetry is the line $x = 1$.

33. $f(x) = 2x - x^2 - 2$

$f(x) = -x^2 + 2x - 2$

Since $a = -1$ is negative, the parabola opens downward. The x–coordinate of the vertex is

$-\frac{b}{2a} = -\frac{2}{2(-1)} = -\frac{2}{-2} = 1$ and the y–coordinate of the vertex is

$f\left(-\frac{b}{2a}\right) = f(1) = -(1)^2 + 2(1) - 2$

$= -1 + 2 - 2 = -1.$

The vertex is $(1, -1)$. Replace $f(x)$ with 0 to find x–intercepts.

$0 = -x^2 + 2x - 2$

$x^2 - 2x = -2$

Since $b = -2$, we add

$\left(\frac{b}{2}\right)^2 = \left(\frac{-2}{2}\right)^2 = (-1)^2 = 1$

$x^2 - 2x + 1 = -2 + 1$

$(x-1)^2 = -1$

Because the solutions to the equation are imaginary, we know that there are no x–intercepts. Set $x = 0$ and solve for y to obtain the y–intercept.

$y = 2(0) - 0^2 - 2 = -2$

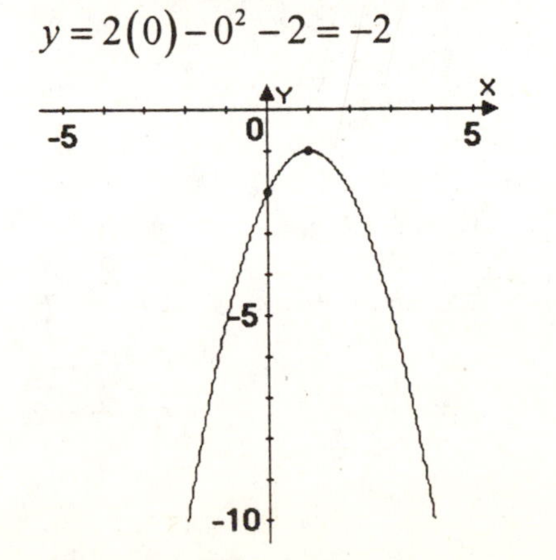

The axis of symmetry is the line $x = 1$.

35. $f(x) = 3x^2 - 12x - 1$

Since $a = 3$, the parabola opens upward and has a minimum. The x–coordinate of the minimum is

$-\frac{b}{2a} = -\frac{-12}{2(3)} = -\frac{-12}{6} = 2$ and the y–coordinate of the minimum is

$f\left(-\frac{b}{2a}\right) = f(2) = 3(2)^2 - 12(2) - 1$

$= 3(4) - 24 - 1 = 12 - 24 - 1$

$= -13.$

The minimum is $(2, -13)$.

37. $f(x) = -4x^2 + 8x - 3$

Since $a = -4$, the parabola opens downward and has a maximum. The x–coordinate of the maximum is

$-\frac{b}{2a} = -\frac{8}{2(-4)} = -\frac{8}{-8} = 1$ and the y–coordinate of the maximum is

$f\left(-\frac{b}{2a}\right) = f(1) = -4(1)^2 + 8(1) - 3$

$= -4(1) + 8 - 3$

$= -4 + 8 - 3 = 1.$

The maximum is $(1, 1)$.

39. $f(x) = 5x^2 - 5x$

Since $a = 5$, the parabola opens upward and has a minimum. The x–coordinate of the minimum is

$-\frac{b}{2a} = -\frac{-5}{2(5)} = -\frac{-5}{10} = \frac{1}{2}$ and the y–coordinate of the minimum is

$$f\left(-\frac{b}{2a}\right)=f\left(\frac{1}{2}\right)=5\left(\frac{1}{2}\right)^2-5\left(\frac{1}{2}\right)$$
$$=5\left(\frac{1}{4}\right)-\frac{5}{2}=\frac{5}{4}-\frac{10}{4}=-\frac{5}{4}.$$

The minimum is $\left(\frac{1}{2},-\frac{5}{4}\right)$.

Application Exercises

41. $f(x)=-3.1x^2+51.4x+4024.5$

The x-coordinate of the maximum is

$$x=-\frac{b}{2a}=-\frac{51.5}{2(-3.1)}=-\frac{51.5}{-6.2}\approx 8.3.$$

The maximum average annual per capita cigarette consumption occurred in about 1960 + 8 = 1968.

$$f(8)=-3.1(8)^2+51.4(8)+4024.5$$
$$=-3.1(64)+411.2+4024.5$$
$$=-198.4+411.2+4024.5$$
$$=4237.3.$$

The maximum consumption was approximately 4237 cigarettes.

43. $f(x)=104.5x^2-1501.5x+6016$

The x-coordinate of the minimum is

$$x=-\frac{b}{2a}=-\frac{-1501.5}{2(104.5)}$$
$$=-\frac{-1501.5}{209}\approx 7.2.$$

$f(7.2)$
$$=104.5(7.2)^2-1501.5(7.2)+6016$$
$$=104.5(51.84)-10810.8+6016$$
$$=622.48$$

The minimum death rate is approximately 622 per year per 100,000 males among U.S. men who average 7.2 hours of sleep per night.

45. $s(t)=-16t^2+64t+160$

a. The t-coordinate of the minimum is

$$t=-\frac{b}{2a}=-\frac{64}{2(-16)}=-\frac{64}{-32}=2.$$

The s-coordinate of the minimum is

$$s(2)=-16(2)^2+64(2)+160$$
$$=-16(4)+128+160$$
$$=-64+128+160=224$$

The ball reaches a maximum height of 244 feet 2 seconds after it is thrown.

b. $$0=-16t^2+64t+160$$
$$0=t^2-4t-10$$

$a=1 \quad b=-4 \quad c=-10$

$$t=\frac{-(-4)\pm\sqrt{(-4)^2-4(1)(-10)}}{2(1)}$$
$$=\frac{4\pm\sqrt{16+40}}{2}=\frac{4\pm\sqrt{56}}{2}$$
$$\approx\frac{4\pm 7.48}{2}$$

Evaluate the expression to obtain two solutions.

$x=\frac{4+7.48}{2}$ or $x=\frac{4-7.48}{2}$

$x=\frac{11.48}{2}$ $\quad$ $x=\frac{-3.48}{2}$

$x=5.74$ $\quad$ $x=-1.74$

We disregard -1.74 because we can't have a negative time

measurement. The solution is 5.74 and we conclude that the ball will hit the ground in approximately 5.7 seconds.

c. $s(0) = -16(0)^2 + 64(0) + 160$
$= -16(0) + 0 + 160 = 160$
At $t = 0$, the ball has not yet been thrown and is at a height of 160 feet. This is the height of the building.

d.

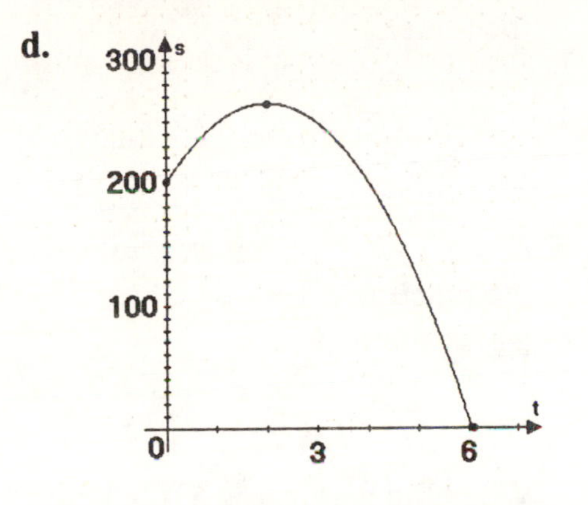

47. The vertex is (15, 41.1). The year is defined as the number of years after 1920, so the first coordinate represents 1920 + 15 = 1935. This means that the Consumer Price Index was 41.1 in the year 1935. What cost \$100 in 1967 cost \$41.10 in 1935.

49. Maximize the area of a rectangle constructed along a river with 120 feet of fencing.
Let x = the length of the rectangle
Let y = the width of the rectangle
Since we need an equation in one variable, use the perimeter to express y in terms of x. Only one length should be accounted for because the side along the river will not be fenced.

$$x + 2y = 120$$
$$2y = 120 - x$$
$$y = \frac{120 - x}{2}$$
$$y = 60 - \frac{1}{2}x$$

We need to maximize
$A = xy = x\left(60 - \frac{1}{2}x\right)$. Rewrite A as a function of x.

$$A(x) = x\left(60 - \frac{1}{2}x\right) = 60x - \frac{1}{2}x^2$$
$$= -\frac{1}{2}x^2 + 60x$$

Since $a = -\frac{1}{2}$ is negative, we know the function opens downward and has a maximum at

$$x = -\frac{b}{2a} = -\frac{60}{2\left(-\frac{1}{2}\right)} = -\frac{60}{-1} = 60.$$

When the length, x, is 60, the width, y, is $y = 60 - \frac{1}{2}(60) = 60 - 30 = 30$.
The dimensions of the rectangular plot with maximum area are 60 feet by 30 feet. This gives an area of $60 \cdot 30 = 1800$ square feet.

51. Let x = one of the numbers
Let $16 - x$ = the other number
We need to maximize the function

$$P(x) = x(16 - x)$$
$$= 16x - x^2$$
$$= -x^2 + 16x.$$

The maximum is

$$x = -\frac{b}{2a} = -\frac{16}{2(-1)} = -\frac{16}{-2} = 8.$$

The other number is
$16-x=16-8=8.$
The numbers which maximize the product are 8 and 8. The maximum product is $8 \cdot 8 = 64$.

53. Let x = one of the numbers
Let $x - 10$ = the other number
We need to minimize the function
$$P(x)=x(x-10)$$
$$=x^2-10x.$$
The minimum is
$$x=-\frac{b}{2a}=-\frac{-10}{2(1)}=-\frac{-10}{2}=5.$$
The other number is
$x-10=5-10=-5.$
The numbers which minimize the product are 5 and –5. The minimum product is $5(-5)=-25$.

55. Maximize the area of a rectangle constructed with 50 yards of fencing.
Let x = the length of the rectangle
Let y = the width of the rectangle

Since we need an equation in one variable, use the perimeter to express y in terms of x.
$$2x+2y=50$$
$$2y=50-2x$$
$$y=\frac{50-2x}{2}$$
$$y=25-x$$
We need to maximize
$A=xy=x(25-x)$. Rewrite A as a function of x.
$$A(x)=x(25-x)=-x^2+25x$$
Since $a = -1$ is negative, we know the function opens downward and has a maximum at
$$x=-\frac{b}{2a}=-\frac{25}{2(-1)}=-\frac{25}{-2}=12.5.$$
When the length x is 12.5, the width y is $y=25-x=25-12.5=12.5.$
The dimensions of the rectangular region with maximum area are 12.5 yards by 12.5 yards. This gives an area of $12.5 \cdot 12.5=156.25$ square feet.

57. Maximize the cross-sectional area of the gutter:
$$A(x)=x(20-2x)$$
$$=20x-2x^2$$
$$=-2x^2+20x.$$
Since $a = -2$ is negative, we know the function opens downward and has a maximum at
$$x=-\frac{b}{2a}=-\frac{20}{2(-2)}=-\frac{20}{-4}=5.$$
When the height x is 5, the width is
$20-2x=20-2(5)=20-10=10.$
$$A(5)=-2(5)^2+20(5)=-2(25)+100$$
$$=-50+100=50$$
The maximum cross-sectional area is 50 square inches. This occurs when the gutter is 5 inches deep and 10 inches wide.

59. **a.** $C(x)=525+0.55x$

b. $P(x)=R(x)-C(x)$
$$=(-0.001x^2+3x)-(525+0.55x)$$
$$=-0.001x^2+3x-525-0.55x$$
$$=-0.001x^2+2.45x-525$$

c. $P(x) = R(x) - C(x)$

$= (-0.001x^2 + 3x) - (525 + 0.55x)$

$= -0.001x^2 + 3x - 525 - 0.55x$

$= -0.001x^2 + 2.45x - 525$

Since $a = -0.001$ is negative, we know the function opens downward and has a maximum at

$$x = -\frac{b}{2a} = -\frac{2.45}{2(-0.001)}$$

$$= -\frac{2.45}{-0.002} = 1225.$$

When the number of units x is 1225, the profit is

$P(1225) = -0.001(1225)^2 + 2.45(1225) - 525$

$= -0.001(1500625) + 3001.25 - 525$

$= -1500.625 + 3001.25 - 525$

$= 975.625.$

The store maximizes its weekly profit when 1225 roast beef sandwiches are produced and sold. This results in a profit of \$975.63.

Writing in Mathematics

61. Answers will vary.

63. Answers will vary.

65. Answers will vary.

67. Answers will vary.

Technology Exercises

69. a. $y = 2x^2 - 82x + 720$

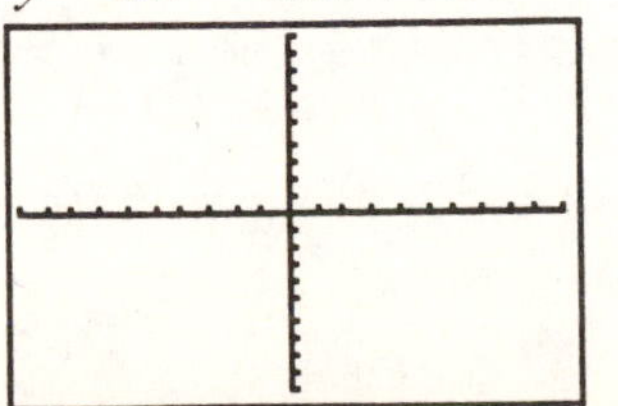

The function has no values that fall within the window.

b. $y = 2x^2 - 82x + 720$

The x–coordinate of the vertex of the parabola is

$$-\frac{b}{2a} = -\frac{-82}{2(2)} = -\frac{-82}{4} = 20.5$$

and the y–coordinate of the vertex of the parabola is

$$f\left(-\frac{b}{2a}\right) = f(20.5)$$

$= 2(20.5)^2 - 82(20.5) + 720$

$= 2(420.25) - 1681 + 720$

$= 840.5 - 1681 + 720 = -120.5.$

The vertex is $(20.5, -120.5)$.

c. Using the viewing window $[0,30,10]$ by $[-130,10,20]$, we have the following.

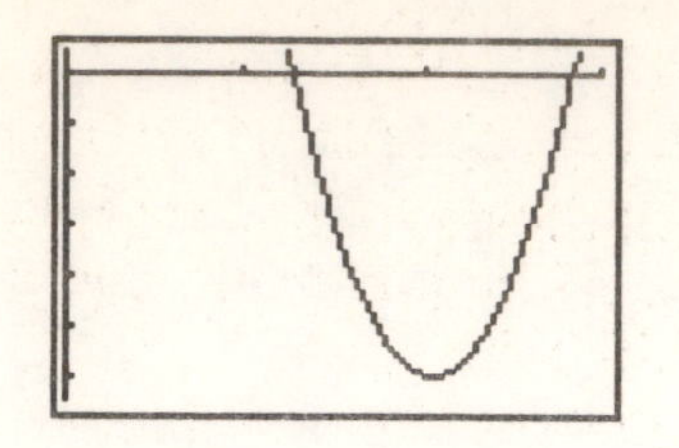

d. Answers will vary.

71. $y = -4x^2 + 20x + 160$

The x–coordinate of the vertex of the parabola is

$-\frac{b}{2a} = -\frac{20}{2(-4)} = -\frac{20}{-8} = 2.5$ and the

y–coordinate of the vertex of the parabola is

$$f\left(-\frac{b}{2a}\right) = f(2.5)$$
$$= -4(2.5)^2 + 20(2.5) + 160$$
$$= -4(6.25) + 50 + 160$$
$$= -25 + 50 + 160$$
$$= 185.$$

The vertex is (2.5, 185).

Using the viewing window $[-20, 20, 10]$ by $[0, 200, 20]$, we have the following.

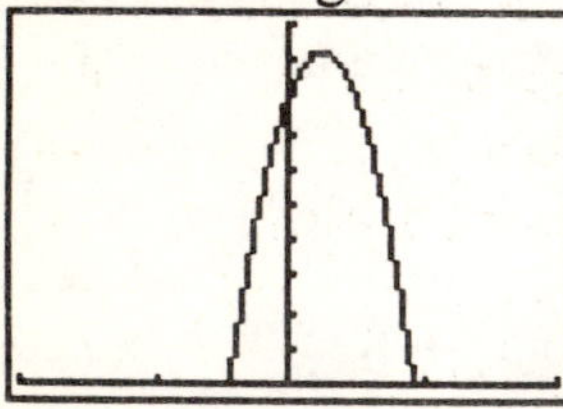

73. $y = 0.01x^2 + 0.6x + 100$

The x–coordinate of the vertex of the parabola is

$-\frac{b}{2a} = -\frac{0.6}{2(0.01)} = -\frac{0.6}{0.02} = -30$ and

the y–coordinate of the vertex of the parabola is

$$f\left(-\frac{b}{2a}\right) = f(-30)$$
$$= 0.01(-30)^2 + 0.6(-30) + 100$$
$$= 0.01(900) - 18 + 100$$
$$= 9 - 18 + 100 = 91.$$

The vertex is $(-30, 91)$.

Using the viewing window $[-150, 150, 20]$ by $[0, 200, 20]$, we have the following.

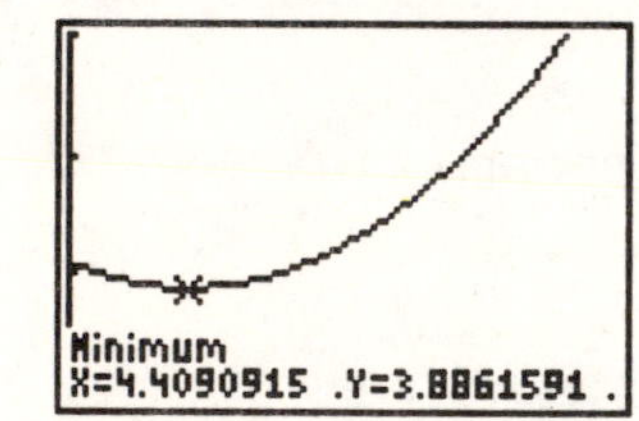

75.

Minimum
X=4.4090915 .Y=3.8861591

The vertex is approximately (4.4, 3.9). This means that approximately 4 years after 1970, in 1974, 3.9 million people held more than one job. This was a minimum.

Critical Thinking Exercises

77. Statement **a.** is true. Since quadratic functions represent parabolas, we know that the function has a maximum or a minimum. This means that the range cannot be $(-\infty, \infty)$.

Statement **b.** is false. The vertex is (5, −1).

Statement **c.** is false. The graph has no x–intercepts. To find x–intercepts, set $y = 0$ and solve for x.

$$0 = -2(x+4)^2 - 8$$
$$8 = -2(x+4)^2$$
$$-4 = (x+4)^2$$

Because the solutions to the equation are imaginary, we know that there are no x–intercepts.

Statement **d.** is false. The x–coordinate of the maximum is

$-\frac{b}{2a} = -\frac{1}{2(-1)} = -\frac{1}{-2} = \frac{1}{2}$ and the

y–coordinate of the vertex of the parabola is

$$f\left(-\frac{b}{2a}\right) = f\left(\frac{1}{2}\right) = -\left(\frac{1}{2}\right)^2 + \frac{1}{2} + 1$$
$$= -\frac{1}{4} + \frac{1}{2} + 1$$
$$= -\frac{1}{4} + \frac{2}{4} + \frac{4}{4} = \frac{5}{4}.$$

The maximum y–value is $\frac{5}{4}$.

79. $f(x) = (x-3)^2 + 2$

Since the vertex is $(3,2)$, we know that the axis of symmetry is the line $x = 3$. The point (6, 11) is on the parabola and lies three units to the right of the axis of symmetry. This means that the point (0, 11) will also lie on the parabola since it lies 3 units to the right of the axis of symmetry.

81. Let x = the number of trees over 50 that will be planted

The function describing the annual yield per lemon tree when $x + 50$ trees are planted per acre is

$$f(x) = (x+50)(320-4x)$$
$$= 320x - 4x^2 + 16000 - 200x$$
$$= -4x^2 + 120x + 16000.$$

This represents the number of lemon trees planted per acre multiplied by yield per tree.

The x–coordinate of the maximum is

$-\frac{b}{2a} = -\frac{120}{2(-4)} = -\frac{120}{-8} = 15$ and the

y–coordinate of the vertex of the parabola is

$$f\left(-\frac{b}{2a}\right) = f(15)$$
$$= -4(15)^2 + 120(15) + 16000$$
$$= -4(225) + 1800 + 16000$$
$$= -900 + 1800 + 16000 = 16900.$$

The maximum lemon yield is 16,900 pounds when $50 + 15 = 65$ lemon trees are planted per acre.

Review Exercises

82. $$\frac{2}{x+5} + \frac{1}{x-5} = \frac{16}{x^2-25}$$
$$\frac{2}{x+5} + \frac{1}{x-5} = \frac{16}{(x+5)(x-5)}$$

The LCD is $(x+5)(x-5)$, so multiply each side of the equation by the LCD and clear fractions.

$$(x+5)(x-5)\left(\frac{2}{x+5}+\frac{1}{x-5}\right)=(x+5)(x-5)\left(\frac{16}{(x+5)(x-5)}\right)$$

$$(x+5)(x-5)\left(\frac{2}{x+5}\right)+(x+5)(x-5)\left(\frac{1}{x-5}\right)=16$$

$$(x-5)(2)+(x+5)(1)=16$$

$$2x-10+x+5=16$$

$$3x-5=16$$

$$3x=21$$

$$x=7$$

Since $x = 7$ will not make any of the denominators zero, the solution is 7 and the solution set is $\{7\}$.

83. $$\frac{1+\frac{2}{x}}{1-\frac{4}{x^2}}=\frac{x^2}{x^2}\cdot\frac{1+\frac{2}{x}}{1-\frac{4}{x^2}}=\frac{x^2\cdot 1+x^2\cdot\frac{2}{x}}{x^2\cdot 1-x^2\cdot\frac{4}{x^2}}=\frac{x^2+x\cdot 2}{x^2-4}=\frac{x^2+2x}{(x+2)(x-2)}=\frac{x\cancel{(x+2)}}{\cancel{(x+2)}(x-2)}=\frac{x}{x-2}$$

84. $2x+3y=6$

$x-4y=14$

$$D=\begin{vmatrix}2 & 3\\ 1 & -4\end{vmatrix}=2(-4)-1(3)=-8-3=-11$$

$$D_x=\begin{vmatrix}6 & 3\\ 14 & -4\end{vmatrix}=6(-4)-14(3)=-24-42=-66$$

$$D_y=\begin{vmatrix}2 & 6\\ 1 & 14\end{vmatrix}=2(14)-1(6)=28-6=22$$

$$x=\frac{D_x}{D}=\frac{-66}{-11}=6$$

$$y=\frac{D_y}{D}=\frac{22}{-11}=-2$$

The solution is $(6,-2)$ and the solution set is $\{(6,-2)\}$.

Check Points 8.4

1. Let $t = x^2$.

$$x^4 - 17x^2 + 16 = 0$$
$$\left(x^2\right)^2 - 17x^2 + 16 = 0$$
$$t^2 - 17t + 16 = 0$$
$$(t-16)(t-1) = 0$$

Apply the zero product principle.

$t - 16 = 0$ or $t - 1 = 0$

$t = 16$ $\quad$ $t = 1$

Replace t by x^2.

$x^2 = 16$ or $x^2 = 1$

$x = \pm 4$ $\quad$ $x = \pm 1$

The solutions are ± 1 and ± 4 and the solution set is $\{-4, -1, 1, 4\}$.

2. Let $t = \sqrt{x}$.

$$x - 2\sqrt{x} - 8 = 0$$
$$\left(\sqrt{x}\right)^2 - 2\sqrt{x} - 8 = 0$$
$$t^2 - 2t - 8 = 0$$
$$(t-4)(t+2) = 0$$

Apply the zero product principle.

$t - 4 = 0$ or $t + 2 = 0$

$t = 4$ $\quad$ $t = -2$

Replace t by $\sqrt{x}$.

$\sqrt{x} = 4$ or ~~$\sqrt{x} = -2$~~

$x = 4^2$

$x = 16$

We disregard -2 because the square root of x cannot be a negative number. We need to check the solution, 16, because both sides of the equation were raised to an even power.

Check:

$$x - 2\sqrt{x} - 8 = 0$$
$$16 - 2\sqrt{16} - 8 = 0$$
$$16 - 2 \cdot 4 - 8 = 0$$
$$16 - 8 - 8 = 0$$
$$0 = 0$$

The solution is 16 and the solution set is $\{16\}$.

3. Let $t = x^2 - 4$.

$$\left(x^2 - 4\right)^2 + \left(x^2 - 4\right) - 6 = 0$$
$$t^2 + t - 6 = 0$$
$$t^2 + t - 6 = 0$$
$$(t+3)(t-2) = 0$$

Apply the zero product principle.

$t + 3 = 0$ or $t - 2 = 0$

$t = -3$ $\quad$ $t = 2$

Replace t by $x^2 - 4$.

$x^2 - 4 = -3$ or $x^2 - 4 = 2$

$x^2 = 1$ $\quad$ $x^2 = 6$

$x = \pm 1$ $\quad$ $x = \pm\sqrt{6}$

The solutions are ± 1 and $\pm\sqrt{6}$ and the solution set is $\left\{-\sqrt{6}, -1, 1, \sqrt{6}\right\}$.

4. Let $t = x^{-1}$.

$$2x^{-2} + x^{-1} - 1 = 0$$
$$2\left(x^{-1}\right)^2 + x^{-1} - 1 = 0$$
$$2t^2 + t - 1 = 0$$
$$(2t-1)(t+1) = 0$$

Apply the zero product principle.

$2t-1=0$ or $t+1=0$

$2t=1 \qquad t=-1$

$t=\frac{1}{2}$

Replace t by x^{-1}.

$x^{-1}=\frac{1}{2}$ or $x^{-1}=-1$

$\frac{1}{x}=\frac{1}{2} \qquad \frac{1}{x}=-1$

$x=2 \qquad x=-1$

The solutions are –1 and 2, and the solution set is $\{-1,2\}$.

5. Let $t=x^{\frac{1}{3}}$.

$$3x^{\frac{2}{3}}-11x^{\frac{1}{3}}-4=0$$

$$3\left(x^{\frac{1}{3}}\right)^2-11x^{\frac{1}{3}}-4=0$$

$$3t^2-11t-4=0$$

$$(3t+1)(t-4)=0$$

Apply the zero product principle.

$3t+1=0$ or $t-4=0$

$3t=-1 \qquad t=4$

$t=-\frac{1}{3}$

Replace t by $x^{\frac{1}{3}}$.

$x^{\frac{1}{3}}=-\frac{1}{3}$ or $x^{\frac{1}{3}}=4$

$\left(x^{\frac{1}{3}}\right)^3=\left(-\frac{1}{3}\right)^3 \qquad \left(x^{\frac{1}{3}}\right)^3=(4)^3$

$x=-\frac{1}{27} \qquad x=64$

The solutions are $-\frac{1}{27}$ and 64, and the solution set is $\left\{-\frac{1}{27},64\right\}$.

Problem Set 8.4

Practice Exercises

1. Let $t=x^2$.

$$x^4-5x^2+4=0$$

$$\left(x^2\right)^2-5x^2+4=0$$

$$t^2-5t+4=0$$

$$(t-4)(t-1)=0$$

Apply the zero product principle.

$t-4=0$ or $t-1=0$

$t=4 \qquad t=1$

Replace t by x^2.

$x^2=4$ or $x^2=1$

$x=\pm 2 \qquad x=\pm 1$

The solutions are ± 1 and ± 2 and the solution set is $\{-2,-1,1,2\}$.

3. Let $t=x^2$.

$$x^4-11x^2+18=0$$

$$\left(x^2\right)^2-11x^2+18=0$$

$$t^2-11t+18=0$$

$$(t-9)(t-2)=0$$

Apply the zero product principle.

$t-9=0$ or $t-2=0$

$t=9 \qquad t=2$

Replace t by x^2.

$x^2=9$ or $x^2=2$

$x=\pm 3 \qquad x=\pm\sqrt{2}$

The solutions are $\pm\sqrt{2}$ and ± 3 and the solution set is $\{-3,-\sqrt{2},\sqrt{2},3\}$.

5. Let $t = x^2$.

$$x^4 + 2x^2 = 8$$
$$x^4 + 2x^2 - 8 = 0$$
$$\left(x^2\right)^2 + 2x^2 - 8 = 0$$
$$t^2 + 2t - 8 = 0$$
$$(t+4)(t-2) = 0$$

Apply the zero product principle.

$t + 4 = 0$ or $t - 2 = 0$

$t = -4$ $\quad$ $t = 2$

Replace t by x^2.

$x^2 = -4$ or $x^2 = 2$

$x = \pm\sqrt{-4}$ $\quad$ $x = \pm\sqrt{2}$

$x = \pm 2i$

The solutions are $\pm 2i$ and $\pm\sqrt{2}$ and the solution set is $\left\{-\sqrt{2}i, -\sqrt{2}, \sqrt{2}, \sqrt{2}i\right\}$.

7. Let $t = \sqrt{x}$.

$$x + \sqrt{x} - 2 = 0$$
$$\left(\sqrt{x}\right)^2 + \sqrt{x} - 2 = 0$$
$$t^2 + t - 2 = 0$$
$$(t+2)(t-1) = 0$$

Apply the zero product principle.

$t + 2 = 0$ or $t - 1 = 0$

$t = -2$ $\quad$ $t = 1$

Replace t by $\sqrt{x}$.

~~$\sqrt{x} = -2$~~ or $\sqrt{x} = 1$

$x = 1$

We disregard –2 because the square root of x cannot be a negative number. We need to check the solution, 1, because both sides of the equation were raised to an even power.

Check:

$$1 + \sqrt{1} - 2 = 0$$
$$1 + 1 - 2 = 0$$
$$2 - 2 = 0$$
$$0 = 0$$

The solution is 1 and the solution set is $\{1\}$.

9. Let $t = x^{\frac{1}{2}}$.

$$x - 4x^{\frac{1}{2}} - 21 = 0$$
$$\left(x^{\frac{1}{2}}\right)^2 - 4x^{\frac{1}{2}} - 21 = 0$$
$$t^2 - 4t - 21 = 0$$
$$(t-7)(t+3) = 0$$

Apply the zero product principle.

$t - 7 = 0$ or $t + 3 = 0$

$t = 7$ $\quad$ $t = -3$

Replace t by $x^{\frac{1}{2}}$.

$x^{\frac{1}{2}} = 7$ or $x^{\frac{1}{2}} = -3$

$\sqrt{x} = 7$ $\quad$ ~~$\sqrt{x} = -3$~~

$x = 49$

We disregard –2 because the square root of x cannot be a negative number. We need to check the solution, 49, because both sides of the equation were raised to an even power.

Check:

$$49 - 4(49)^{\frac{1}{2}} - 21 = 0$$
$$49 - 4(7) - 21 = 0$$
$$49 - 28 - 21 = 0$$

$$49-49=0$$
$$0=0$$
The solution is 49 and the solution set is $\{49\}$.

11. Let $t=\sqrt{x}$.

$$x-13\sqrt{x}+40=0$$
$$\left(\sqrt{x}\right)^2-13\sqrt{x}+40=0$$
$$t^2-13t+40=0$$
$$(t-5)(t-8)=0$$

Apply the zero product principle.

$t-5=0$ or $t-8=0$

$t=5$ $\quad t=8$

Replace t by $\sqrt{x}$.

$\sqrt{x}=5$ or $\sqrt{x}=8$

$x=25$ $\quad x=64$

Both solutions must be checked since both sides of the equation were raised to an even power.

$x=25$

$$25-13\sqrt{25}+40=0$$
$$25-13(5)+40=0$$
$$25-65+40=0$$
$$65-65=0$$
$$0=0$$

$x=64$

$$64-13\sqrt{64}+40=0$$
$$64-13(8)+40=0$$
$$64-104+40=0$$
$$104-104=0$$
$$0=0$$

Both solutions check. The solutions are 25 and 64 and the solution set is $\{25,64\}$.

13. Let $t=x-5$.

$$(x-5)^2-4(x-5)-21=0$$
$$t^2-4t-21=0$$
$$(t-7)(t+3)=0$$

Apply the zero product principle.

$t-7=0$ or $t+3=0$

$t=7$ $\quad t=-3$

Replace t by $x-5$.

$x-5=7$ or $x-5=-3$

$x=12$ $\quad x=2$

The solutions are 2 and 12 and the solution set is $\{2,12\}$.

15. Let $t=x^2-1$.

$$\left(x^2-1\right)^2-\left(x^2-1\right)=2$$
$$\left(x^2-1\right)^2-\left(x^2-1\right)-2=0$$
$$t^2-t-2=0$$
$$(t-2)(t+1)=0$$

Apply the zero product principle.

$t-2=0$ or $t+1=0$

$t=2$ $\quad t=-1$

Replace t by x^2-1.

$x^2-1=2$ or $x^2-1=-1$

$x^2=3$ $\quad x^2=0$

$x=\pm\sqrt{3}$ $\quad x=0$

The solutions are $\pm\sqrt{3}$ and 0 and the solution set is $\left\{-\sqrt{3},0,\sqrt{3}\right\}$.

17. Let $t=x^2+3x$.

$$\left(x^2+3x\right)^2-8\left(x^2+3x\right)-20=0$$
$$t^2-8t-20=0$$
$$(t-10)(t+2)=0$$

Apply the zero product principle.

$t-10=0$ or $t+2=0$

$t=10$ $t=-2$

Replace t by x^2+3x.

First, consider $t = 10$.

$$x^2+3x=10$$
$$x^2+3x-10=0$$
$$(x+5)(x-2)=0$$

Apply the zero product principle.

$x+5=0$ or $x-2=0$

$x=-5$ $x=2$

Next, consider $t = -2$.

$$x^2+3x=-2$$
$$x^2+3x+2=0$$
$$(x+2)(x+1)=0$$

Apply the zero product principle.

$x+2=0$ or $x+1=0$

$x=-2$ $x=-1$

The solutions are $-5, -2, -1$, and 2 and the solution set is $\{-5,-2, -1,2\}$.

19. Let $t=x^{-1}$.

$$x^{-2}-x^{-1}-20=0$$
$$(x^{-1})^2-x^{-1}-20=0$$
$$t^2-t-20=0$$
$$(t-5)(t+4)=0$$

Apply the zero product principle.

$t-5=0$ or $t+4=0$

$t=5$ $t=-4$

Replace t by x^{-1}.

$x^{-1}=5$ or $x^{-1}=-4$

$\frac{1}{x}=5$ $\frac{1}{x}=-4$

$5x=1$ $-4x=1$

$x=\frac{1}{5}$ $x=-\frac{1}{4}$

The solutions are $-\frac{1}{4}$ and $\frac{1}{5}$, and the solution set is $\left\{-\frac{1}{4},\frac{1}{5}\right\}$.

21. Let $t=x^{-1}$.

$$2x^{-2}-7x^{-1}+3=0$$
$$2(x^{-1})^2-7x^{-1}+3=0$$
$$2t^2-7t+3=0$$
$$(2t-1)(t-3)=0$$

Apply the zero product principle.

$2t-1=0$ or $t-3=0$

$2t=1$ $t=3$

$t=\frac{1}{2}$

Replace t by x^{-1}.

$x^{-1}=\frac{1}{2}$ or $x^{-1}=3$

$\frac{1}{x}=\frac{1}{2}$ $\frac{1}{x}=3$

$x=2$ $3x=1$

$x=\frac{1}{3}$

The solutions are $\frac{1}{3}$ and 2, and the solution set is $\left\{\frac{1}{3},2\right\}$.

23. Let $t = x^{-1}$.

$$x^{-2} - 4x^{-1} = 3$$
$$x^{-2} - 4x^{-1} - 3 = 0$$
$$\left(x^{-1}\right)^2 - 4x^{-1} - 3 = 0$$
$$t^2 - 4t - 3 = 0$$

$a = 1 \quad b = -4 \quad c = -3$

Use the quadratic formula.

$$t = \frac{-(-4) \pm \sqrt{(-4)^2 - 4(1)(-3)}}{2(1)}$$
$$= \frac{4 \pm \sqrt{16+12}}{2} = \frac{4 \pm \sqrt{28}}{2}$$
$$= \frac{4 \pm 2\sqrt{7}}{2} = \frac{2\left(2 \pm \sqrt{7}\right)}{2} = 2 \pm \sqrt{7}$$

Replace t by x^{-1}.

$$x^{-1} = 2 \pm \sqrt{7}$$
$$\frac{1}{x} = 2 \pm \sqrt{7}$$
$$\left(2 \pm \sqrt{7}\right)x = 1$$
$$x = \frac{1}{2 \pm \sqrt{7}}$$

Rationalize the denominator.

$$x = \frac{1}{2 \pm \sqrt{7}} \cdot \frac{2 \mp \sqrt{7}}{2 \mp \sqrt{7}} = \frac{2 \mp \sqrt{7}}{2^2 - \left(\sqrt{7}\right)^2}$$
$$= \frac{2 \mp \sqrt{7}}{4-7} = \frac{2 \mp \sqrt{7}}{-3} = \frac{-2 \pm \sqrt{7}}{3}$$

The solutions are $\frac{-2 \pm \sqrt{7}}{3}$, and the solution set is $\left\{\frac{-2-\sqrt{7}}{3}, \frac{-2+\sqrt{7}}{3}\right\}$.

25. Let $t = x^{\frac{1}{3}}$.

$$x^{\frac{2}{3}} - x^{\frac{1}{3}} - 6 = 0$$
$$\left(x^{\frac{1}{3}}\right)^2 - x^{\frac{1}{3}} - 6 = 0$$
$$t^2 - t - 6 = 0$$
$$(t-3)(t+2) = 0$$

Apply the zero product principle.

$t - 3 = 0 \quad$ or $\quad t + 2 = 0$

$t = 3 \qquad\qquad t = -2$

Replace t by $x^{\frac{1}{3}}$.

$x^{\frac{1}{3}} = 3 \quad$ or $\quad x^{\frac{1}{3}} = -2$

$\left(x^{\frac{1}{3}}\right)^3 = 3^3 \qquad \left(x^{\frac{1}{3}}\right)^3 = (-2)^3$

$x = 27 \qquad\qquad x = -8$

The solutions are -8 and 27, and the solution set is $\{-8, 27\}$.

27. Let $t = x^{\frac{1}{5}}$.

$$x^{\frac{2}{5}} + x^{\frac{1}{5}} - 6 = 0$$
$$\left(x^{\frac{1}{5}}\right)^2 + x^{\frac{1}{5}} - 6 = 0$$
$$t^2 + t - 6 = 0$$
$$(t+3)(t-2) = 0$$

Apply the zero product principle.

$t + 3 = 0 \quad$ or $\quad t - 2 = 0$

$t = -3 \qquad\qquad t = 2$

Replace t by $x^{\frac{1}{5}}$.

$x^{\frac{1}{5}} = -3 \quad$ or $\quad x^{\frac{1}{5}} = 2$

$\left(x^{\frac{1}{5}}\right)^5 = (-3)^5 \qquad \left(x^{\frac{1}{5}}\right)^5 = (2)^5$

$x = -243 \qquad\qquad x = 32$

The solutions are -243 and 32, and the solution set is $\{-243, 32\}$.

29. Let $t = x^{\frac{1}{4}}$.

$$2x^{\frac{1}{2}} - x^{\frac{1}{4}} = 1$$
$$2\left(x^{\frac{1}{4}}\right)^2 - x^{\frac{1}{4}} - 1 = 0$$
$$2t^2 - t - 1 = 0$$
$$(2t+1)(t-1) = 0$$

Apply the zero product principle.

$2t+1=0$ or $t-1=0$

$2t = -1$ $\quad t = 1$

$t = -\frac{1}{2}$

Replace t by $x^{\frac{1}{4}}$.

$x^{\frac{1}{4}} = -\frac{1}{2}$ or $x^{\frac{1}{4}} = 1$

$\left(x^{\frac{1}{4}}\right)^4 = \left(-\frac{1}{2}\right)^4$ $\quad \left(x^{\frac{1}{4}}\right)^4 = 1^4$

$x = \frac{1}{16}$ $\quad x = 1$

Since both sides of the equations were raised to an even power, the solutions must be checked.

First, check $x = \frac{1}{16}$.

$$2\left(\frac{1}{16}\right)^{\frac{1}{2}} - \left(\frac{1}{16}\right)^{\frac{1}{4}} = 1$$
$$2\left(\frac{1}{4}\right) - \frac{1}{2} = 1$$
$$\frac{1}{2} - \frac{1}{2} = 1$$
$$0 \neq 1$$

The solution does not check, so disregard $x = \frac{1}{16}$.

Next, check $x = 1$.

$$2(1)^{\frac{1}{2}} - (1)^{\frac{1}{4}} = 1$$
$$2(1) - 1 = 1$$
$$2 - 1 = 1$$
$$1 = 1$$

The solution checks. The solution is 1 and the solution set is $\{1\}$.

31. Let $t = x - \frac{8}{x}$.

$$\left(x - \frac{8}{x}\right)^2 + 5\left(x - \frac{8}{x}\right) - 14 = 0$$
$$t^2 + 5t - 14 = 0$$
$$(t+7)(t-2) = 0$$

Apply the zero product principle.

$t+7=0$ or $t-2=0$

$t = -7$ $\quad t = 2$

Replace t by $x - \frac{8}{x}$.

First, consider $t = -7$.

$$x - \frac{8}{x} = -7$$
$$x\left(x - \frac{8}{x}\right) = x(-7)$$
$$x^2 - 8 = -7x$$
$$x^2 + 7x - 8 = 0$$
$$(x+8)(x-1) = 0$$

Apply the zero product principle.

$x+8=0$ or $x-1=0$

$x = -8$ $\quad x = 1$

Next, consider $t = 2$.

$$x-\frac{8}{x}=2$$

$$x\left(x-\frac{8}{x}\right)=x(2)$$

$$x^2-8=2x$$

$$x^2-2x-8=0$$

$$(x-4)(x+2)=0$$

Apply the zero product principle.

$x-4=0$ or $x+2=0$

$x=4$ $\quad x=-2$

The solutions are $-8, -2, 1,$ and 4, and the solution set is $\{-8,-2,1,4\}$.

33. $f(x)=x^4-5x^2+4$

$$y=x^4-5x^2+4$$

Set $y=0$ to find the x–intercept(s).

$$0=x^4-5x^2+4$$

Let $t=x^2$.

$$x^4-5x^2+4=0$$

$$\left(x^2\right)^2-5x^2+4=0$$

$$t^2-5t+4=0$$

$$(t-4)(t-1)=0$$

Apply the zero product principle.

$t-1=0$ or $t-4=0$

$t=1$ $\quad t=4$

Substitute x^2 for t.

$x^2=1$ or $x^2=4$

$x=\pm1$ $\quad x=\pm2$

The intercepts are ±1 and ±2. The corresponding graph is graph **c**.

35. $f(x)=x^{\frac{1}{3}}+2x^{\frac{1}{6}}-3$

$$y=x^{\frac{1}{3}}+2x^{\frac{1}{6}}-3$$

Set $y=0$ to find the x–intercept(s).

$$0=x^{\frac{1}{3}}+2x^{\frac{1}{6}}-3$$

Let $t=x^{\frac{1}{6}}$.

$$x^{\frac{1}{3}}+2x^{\frac{1}{6}}-3=0$$

$$\left(x^{\frac{1}{6}}\right)^2+2x^{\frac{1}{6}}-3=0$$

$$t^2+2t-3=0$$

$$(t+3)(t-1)=0$$

Apply the zero product principle.

$t+3=0$ or $t-1=0$

$t=-3$ $\quad t=1$

Substitute $x^{\frac{1}{6}}$ for t.

$x^{\frac{1}{6}}=-3$ or $x^{\frac{1}{6}}=1$

$\left(x^{\frac{1}{6}}\right)^6=(-3)^6$ $\quad \left(x^{\frac{1}{6}}\right)^6=(1)^6$

$x=729$ $\quad x=1$

Since both sides of the equations were raised to an even power, the solutions must be checked.

First check $x=729$.

$$(729)^{\frac{1}{3}}+2(729)^{\frac{1}{6}}-3=0$$

$$9+2(3)-3=0$$

$$9+6-3=0$$

$$15-3=0$$

$$12\neq0$$

Next check $x=1$.

$$(1)^{\frac{1}{3}}+2(1)^{\frac{1}{6}}-3=0$$

$$1+2(1)-3=0$$

$$1+2-3=0$$

$$3-3=0$$

$$0=0$$

Since 729 does not check, we disregard it. The intercept is 1. The corresponding graph is graph **e.**

37. $f(x) = (x+2)^2 - 9(x+2) + 20$

$y = (x+2)^2 - 9(x+2) + 20$

Set $y = 0$ to find the x–intercept(s).

$(x+2)^2 - 9(x+2) + 20 = 0$

Let $t = x+2$.

$(x+2)^2 - 9(x+2) + 20 = 0$

$t^2 - 9t + 20 = 0$

$(t-5)(t-4) = 0$

Apply the zero product principle.

$t - 5 = 0$ or $t - 4 = 0$

$t = 5$ $\quad$ $t = 4$

Substitute $x+2$ for t.

$x+2 = 5$ or $x+2 = 4$

$x = 3$ $\quad$ $x = 2$

The intercepts are 2 and 3. The corresponding graph is graph **f.**

39. $P(x) = 0.04(x+40)^2 - 3(x+40) + 104$

$60 = 0.04(x+40)^2 - 3(x+40) + 104$

$0 = 0.04(x+40)^2 - 3(x+40) + 44$

Let $t = x + 40$.

$0.04(x+40)^2 - 3(x+40) + 44 = 0$

$0.04t^2 - 3t + 44 = 0$

Solve using the quadratic formula.

$a = 0.04 \quad b = -3 \quad c = 44$

$$t = \frac{-(-3) \pm \sqrt{(-3)^2 - 4(0.04)(44)}}{2(0.04)}$$

$$= \frac{3 \pm \sqrt{9 - 7.04}}{0.08} = \frac{3 \pm \sqrt{1.96}}{0.08}$$

$$= \frac{3+1.4}{0.08} \text{ or } \frac{3-1.4}{0.08}$$

$= 55$ or 20

Since x represents the number of years a person's age is above or below 40, t = x + 40 is the percentage we are looking for.

The ages at which 60% of us feel that having a clean house is very important are 20 and 55. From the graph, we see that at 20, 58%, and at 55, 52% feel that a clean house if very important. The function models the data fairly well.

Writing in Mathematics

41. Answers will vary.

43. Answers will vary.

Technology Exercises

45. $x^6 - 7x^3 - 8 = 0$

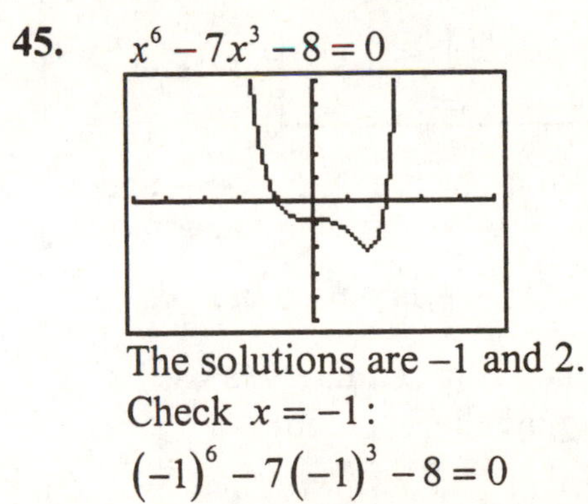

The solutions are –1 and 2.

Check $x = -1$:

$(-1)^6 - 7(-1)^3 - 8 = 0$

$1 - 7(-1) - 8 = 0$

$1 + 7 - 8 = 0$

$8 - 8 = 0$

$0 = 0$

Check $x = 2$ using the same method.

47. $x^4 - 10x^2 + 9 = 0$

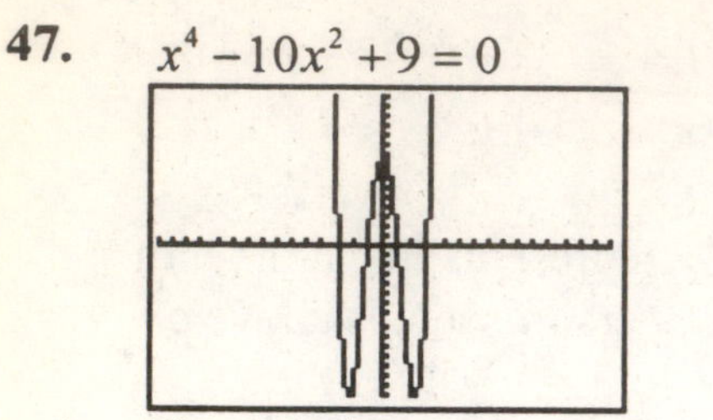

The solutions are –3, –1, 1 and 3.
Check:

$$(-3)^4 - 10(-3)^2 + 9 = 0$$
$$81 - 10(9) + 9 = 0$$
$$81 - 90 + 9 = 0$$
$$90 - 90 = 0$$
$$0 = 0$$

Check $x = -1$, 1, and 3 using the same method.

49.
$$2(x+1)^2 = 5(x+1) + 3$$
$$2(x+1)^2 - 5(x+1) - 3 = 0$$

The solutions are $-1.5 = -\frac{3}{2}$ and 2.

Because $-\frac{3}{2}$ is not an integer, the calculate zero feature was used to determine the intercept.
Check:
$x = -1.5$

$$2(-1.5+1)^2 = 5(-1.5+1) + 3$$
$$2(-0.5)^2 = 5(-0.5) + 3$$
$$2(0.25) = -2.5 + 3$$
$$0.5 = 0.5$$

Check $x = 2$ using the same method.

51.
$$x^{\frac{1}{2}} + 4x^{\frac{1}{4}} = 5$$
$$x^{\frac{1}{2}} + 4x^{\frac{1}{4}} - 5 = 0$$

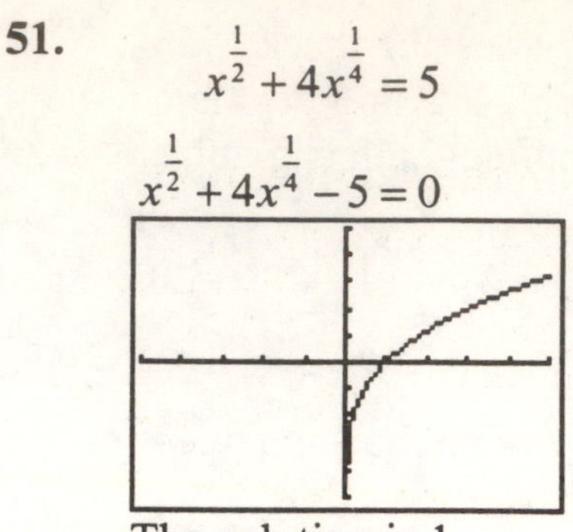

The solution is 1.
Check:
$x = 1$

$$1^{\frac{1}{2}} + 4(1)^{\frac{1}{4}} = 5$$
$$1 + 4(1) = 5$$
$$1 + 4 = 5$$
$$5 = 5$$

Critical Thinking Exercises

53. Statement **b.** is true.

Statement **a.** is false. Any method that can be used to solve a quadratic equation can also be used to solve a quadratic-in-form equation.

Statement **c.** is false. The equation would be quadratic in form if the third term were a constant.

Statement **d.** is false. To solve the equation, let $t = \sqrt{x}$.

55.
$$5x^6 + x^3 = 18$$
$$5x^6 + x^3 - 18 = 0$$

Let $t = x^3$.

$$5x^6 + x^3 - 18 = 0$$
$$5(x^3)^2 + x^3 - 18 = 0$$

$$5t^2 + t - 18 = 0$$
$$(5t - 9)(t + 2) = 0$$

Apply the zero product principle.

$5t - 9 = 0$ or $t + 2 = 0$

$5t = 9$ $\quad t = -2$

$t = \frac{9}{5}$

Substitute x^3 for t.

$x^3 = \frac{9}{5}$ or $x^3 = -2$

$x = \sqrt[3]{\frac{9}{5}}$ $\quad x = \sqrt[3]{-2}$

Rationalize the denominator.

$$\sqrt[3]{\frac{9}{5}} = \frac{\sqrt[3]{9}}{\sqrt[3]{5}} \cdot \frac{\sqrt[3]{5^2}}{\sqrt[3]{5^2}} = \frac{\sqrt[3]{9 \cdot 5^2}}{\sqrt[3]{5^3}} = \frac{\sqrt[3]{225}}{5}$$

The solutions are $\sqrt[3]{-2}$ and $\frac{\sqrt[3]{225}}{5}$ and the solution set is $\left\{\sqrt[3]{-2}, \frac{\sqrt[3]{225}}{5}\right\}$.

Review Exercises

57. $\frac{2x^2}{10x^3 - 2x^2} = \frac{2x^2}{2x^2(5x-1)} = \frac{1}{5x-1}$

58. $\frac{2+i}{1-i} = \frac{2+i}{1-i} \cdot \frac{1+i}{1+i} = \frac{2+2i+i+i^2}{1^2 - i^2}$

$= \frac{2+3i-1}{1-(-1)} = \frac{1+3i}{2} = \frac{1}{2} + \frac{3}{2}i$

59. $2x + y = 6$

$x - 2y = 8$

$$\left[\begin{array}{cc|c} 2 & 1 & 6 \\ 1 & -2 & 8 \end{array}\right] R_1 \leftrightarrow R_2$$

$$= \left[\begin{array}{cc|c} 1 & -2 & 8 \\ 2 & 1 & 6 \end{array}\right] -2R_1 + R_2$$

$$= \left[\begin{array}{cc|c} 1 & -2 & 8 \\ 0 & 5 & -10 \end{array}\right] \frac{1}{5}R_2$$

$$= \left[\begin{array}{cc|c} 1 & -2 & 8 \\ 0 & 1 & -2 \end{array}\right]$$

The resulting system is:

$x - 2y = 8$

$y = -2.$

Back-substitute –2 for y to find x.

$x - 2(-2) = 8$

$x + 4 = 8$

$x = 4$

The solution is $(4, -2)$ and the solution set is $\{(4, -2)\}$.

Check Points 8.5

1. $x^2 + 2x - 3 < 0$

Solve the related quadratic equation.

$x^2 + 2x - 3 = 0$

$(x + 3)(x - 1) = 0$

Apply the zero product principle.

$x+3=0$ or $x-1=0$

$x=-3$ $\quad x=1$

The boundary points are –3 and 1.

Test Interval	Test Number	Test	Conclusion
$(-\infty,-3)$	-4	$(-4)^2+2(-4)-3<0$ $5<0$, false	$(-\infty,-3)$ does not belong to the solution set.
$(-3,1)$	0	$0^2+2(0)-3<0$ $-3<0$, true	$(-3,1)$ belongs to the solution set.
$(1,\infty)$	2	$2^2+2(2)-3<0$ $5<0$, false	$(1,\infty)$ does not belong to the solution set.

The solution set is $(1,\infty)$ or $\{x \mid -3<x<1\}$.

2. $x^2-x\ge 20$

$x^2-x-20\ge 0$

Solve the related quadratic equation.

$x^2-x-20=0$

$(x-5)(x+4)=0$

Apply the zero product principle.

$x-5=0$ or $x+4=0$

$x=5$ $\quad x=-4$

The boundary points are –4 and 5.

Test Interval	Test Number	Test	Conclusion
$(-\infty,-4]$	-5	$(-5)^2-(-5)\ge 20$ $30\ge 20$, true	$(-\infty,-4]$ belongs to the solution set.
$[-4,5]$	0	$0^2-0\ge 20$ $0\ge 20$, false	$[-4,5]$ does not belong to the solution set.
$[5,\infty)$	6	$6^2-6\ge 20$ $30\ge 20$, true	$[5,\infty)$ belongs to the solution set.

The solution set is $(-\infty,-4]\cup[5,\infty)$ or $\{x \mid x\le -4 \text{ or } x\ge 5\}$.

3. $\dfrac{x-5}{x+2}>0$

Find the values of x that make the numerator and denominator zero.

$x-5=0$ $\quad x+2=0$

$x=5$ $\quad x=-2$

The boundary points are -2 and 5.

Test Interval	Test Number	Test	Conclusion
$(-\infty,-2)$	-3	$\frac{-3-5}{-3+2}>0$ $8>0$, true	$(-\infty,-2)$ belongs to the solution set.
$(-2,5)$	0	$\frac{0-5}{0+2}>0$ $-\frac{5}{2}>0$, false	$(-2,5)$ does not belong to the solution set.
$(5,\infty)$	6	$\frac{6-5}{6+2}>0$ $\frac{1}{8}>0$, true	$(5,\infty)$ belongs to the solution set.

The solution set is $(-\infty,-2)\cup(5,\infty)$ or $\{x|x<-2 \text{ or } x>5\}$.

-2 5

4. $\frac{2x}{x+1}\le 1$

Express the inequality so that one side is zero.

$$\frac{2x}{x+1}-1\le 0$$

$$\frac{2x}{x+1}-\frac{x+1}{x+1}\le 0$$

$$\frac{2x-(x+1)}{x+1}\le 0$$

$$\frac{2x-x-1}{x+1}\le 0$$

$$\frac{x-1}{x+1}\le 0$$

Find the values of x that make the numerator and denominator zero.

$x-1=0 \qquad x+1=0$

$x=1 \qquad\quad x=-1$

The boundary points are -1 and 1. We exclude -1 from the solution set, since this would make the denominator zero.

Test Interval	Test Number	Test	Conclusion
$(-\infty,-1)$	-2	$\frac{2(-2)}{-2+1}\le 1$ $4\le 1$, true	$(-\infty,-1)$ does not belong to the solution set.

$(-1,1]$	0	$\frac{2(0)}{0+1} \le 1$ $0 \le 1$, true	$(-1,1]$ belongs to the solution set.
$[1,\infty)$	2	$\frac{2(2)}{2+1} \le 1$ $\frac{4}{3} \le 1$, false	$[1,\infty)$ does not belong to the solution set.

The solution set is $(-1,1]$ or $\{x \mid -1 < x \le 1\}$.

-1 1

5. $s(t) = -16t^2 + 80t$

To find when the height is more than 64 feet above the ground, solve the inequality $-16t^2 + 80t > 64$.

Solve the related quadratic equation.

$$-16t^2 + 80t = 64$$
$$-16t^2 + 80t - 64 = 0$$
$$t^2 - 5t + 4 = 0$$
$$(t-4)(t-1) = 0$$

Apply the zero product principle.

$t - 4 = 0$ or $t - 1 = 0$

$t = 4$ $\quad$ $t = 1$

The boundary points are 1 and 4.

Test Interval	Test Number	Test	Conclusion
$(1,4)$	2	$-16(2)^2 + 80(2) > 64$ $96 > 64$, true	$(1,4)$ belongs to the solution set.
$(4,\infty)$	5	$-16(5)^2 + 80(5) > 64$ $0 > 64$, false	$(4,\infty)$ does not belong to the solution set.

The solution set is $(1,4)$. This means that the ball will be more than 64 feet above the ground between 1 and 4 seconds.

Problem Set 8.5

Practice Exercises

1. $(x-4)(x+2) > 0$

Solve the related quadratic equation.

$(x-4)(x+2) = 0$

Apply the zero product principle.

$x-4=0$ or $x+2=0$

$x=4$ $x=-2$

The boundary points are –2 and 4.

Test Interval	Test Number	Test	Conclusion
$(-\infty,-2)$	-3	$(-3-4)(-3+2)>0$ $7>0$, true	$(-\infty,-2)$ belongs to the solution set.
$(-2,4)$	0	$(0-4)(0+2)>0$ $-8>0$, false	$(-2,4)$ does not belong to the solution set.
$(4,\infty)$	5	$(5-4)(5+2)>0$ $7>0$, true	$(4,\infty)$ belongs to the solution set.

The solution set is $(-\infty,-2)\cup(4,\infty)$ or $\{x|x<-2 \text{ or } x>4\}$.

-2 4

3. $(x-7)(x+3)\le 0$

Solve the related quadratic equation.

$(x-7)(x+3)=0$

Apply the zero product principle.

$x-7=0$ or $x+3=0$

$x=7$ $x=-3$

The boundary points are –3 and 7.

Test Interval	Test Number	Test	Conclusion
$(-\infty,-3]$	-4	$(-4-7)(-4+3)\le 0$ $11\le 0$, false	$(-\infty,-3]$ does not belong to the solution set.
$[-3,7]$	0	$(0-7)(0+3)\le 0$ $-21\le 0$, true	$[-3,7]$ belongs to the solution set.
$[7,\infty)$	8	$(8-7)(8+3)\le 0$ $11\le 0$, false	$[7,\infty)$ does not belong to the solution set.

The solution set is $[-3,7]$ or $\{x|-3\le x\le 7\}$.

-3 7

5. $x^2-5x+4>0$

Solve the related quadratic equation.

$x^2-5x+4=0$

$(x-4)(x-1)=0$

Apply the zero product principle.

$x-4=0$ or $x-1=0$

$x=4$ $\quad$ $x=1$

The boundary points are 1 and 4.

Test Interval	Test Number	Test	Conclusion
$(-\infty,1)$	0	$0^2-5(0)+4>0$ $4>0$, true	$(-\infty,1)$ belongs to the solution set.
$(1,4)$	2	$2^2-5(2)+4>0$ $-2>0$, false	$(1,4)$ does not belong to the solution set.
$(4,\infty)$	5	$5^2-5(5)+4>0$ $4>0$, true	$(4,\infty)$ belongs to the solution set.

The solution set is $(-\infty,1)\cup(4,\infty)$ or $\{x|x<1 \text{ or } x>4\}$.

1 4

7. $x^2+5x+4>0$

Solve the related quadratic equation.

$x^2+5x+4=0$

$(x+4)(x+1)=0$

Apply the zero product principle.

$x+4=0$ or $x+1=0$

$x=-4$ $\quad$ $x=-1$

The boundary points are –1 and –4.

Test Interval	Test Number	Test	Conclusion
$(-\infty,-4)$	-5	$(-5)^2+5(-5)+4>0$ $4>0$, true	$(-\infty,-4)$ belongs to the solution set.
$(-4,-1)$	-2	$(-2)^2+5(-2)+4>0$ $-2>0$, false	$(-4,-1)$ does not belong to the solution set.
$(-1,\infty)$	0	$0^2+5(0)+4>0$ $4>0$, true	$(-1,\infty)$ belongs to the solution set.

The solution set is $(-\infty,-4)\cup(-1,\infty)$ or $\{x|x<-4 \text{ or } x>-1\}$.

-4 -1

9. $x^2 - 6x + 8 \le 0$

Solve the related quadratic equation.

$$x^2 - 6x + 8 = 0$$
$$(x-4)(x-2) = 0$$

Apply the zero product principle.

$x - 4 = 0$ or $x - 2 = 0$

$x = 4$ $\quad$ $x = 2$

The boundary points are 2 and 4.

Test Interval	Test Number	Test	Conclusion
$(-\infty, 2]$	0	$0^2 - 6(0) + 8 \le 0$ $8 \le 0$, false	$(-\infty, 2]$ does not belong to the solution set.
$[2, 4]$	3	$3^2 - 6(3) + 8 \le 0$ $-1 \le 0$, true	$[2, 4]$ belongs to the solution set.
$[4, \infty)$	5	$5^2 - 6(5) + 8 \le 0$ $3 \le 0$, false	$[4, \infty)$ does not belong to the solution set.

The solution set is $[2, 4]$ or $\{x \mid 2 \le x \le 4\}$.

2 4

11. $3x^2 + 10x - 8 \le 0$

Solve the related quadratic equation.

$$3x^2 + 10x - 8 = 0$$
$$(3x-2)(x+4) = 0$$

Apply the zero product principle.

$3x - 2 = 0$ or $x + 4 = 0$

$3x = 2$ $\quad$ $x = -4$

$x = \frac{2}{3}$

The boundary points are -4 and $\frac{2}{3}$.

Test Interval	Test Number	Test	Conclusion
$(-\infty, -4]$	-5	$3(-5)^2 + 10(-5) - 8 \le 0$ $17 \le 0$, false	$(-\infty, -4]$ does not belong to the solution set.
$\left[-4, \frac{2}{3}\right]$	0	$3(0)^2 + 10(0) - 8 \le 0$ $-8 \le 0$, true	$\left[-4, \frac{2}{3}\right]$ belongs to the solution set.

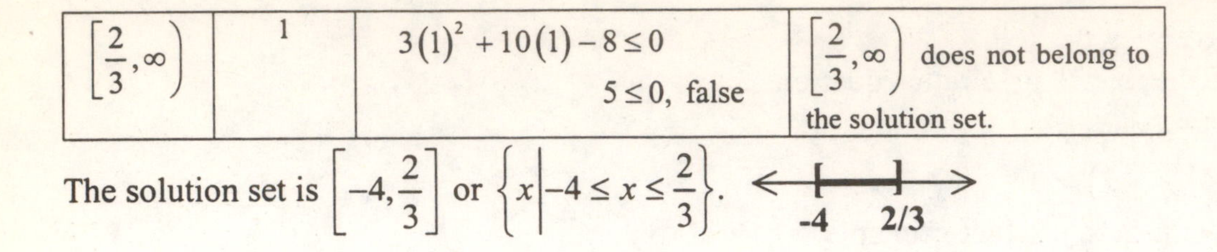

$\left[\frac{2}{3},\infty\right)$	1	$3(1)^2+10(1)-8\le 0$ $5\le 0$, false	$\left[\frac{2}{3},\infty\right)$ does not belong to the solution set.

The solution set is $\left[-4,\frac{2}{3}\right]$ or $\left\{x\middle|-4\le x\le\frac{2}{3}\right\}$.

13. $2x^2+x<15$

$2x^2+x-15<0$

Solve the related quadratic equation.

$2x^2+x-15=0$

$(2x-5)(x+3)=0$

Apply the zero product principle.

$2x-5=0$ or $x+3=0$

$2x=5$ $\quad x=-3$

$x=\frac{5}{2}$

The boundary points are -3 and $\frac{5}{2}$.

Test Interval	Test Number	Test	Conclusion
$(-\infty,-3)$	-4	$2(-4)^2+(-4)<15$ $28<15$, false	$(-\infty,-3)$ does not belong to the solution set.
$\left(-3,\frac{5}{2}\right)$	0	$2(0)^2+0<15$ $0<15$, true	$\left(-3,\frac{5}{2}\right)$ belongs to the solution set.
$\left(\frac{5}{2},\infty\right)$	3	$2(3)^2+3<15$ $21<15$, false	$\left(\frac{5}{2},\infty\right)$ does not belong to the solution set.

The solution set is $\left(-3,\frac{5}{2}\right)$ or $\left\{x\middle|-3<x<\frac{5}{2}\right\}$.

-3 5/2

15. $4x^2+7x<-3$

$4x^2+7x+3<0$

Solve the related quadratic equation.

$4x^2+7x+3=0$

$(4x+3)(x+1)=0$

Apply the zero product principle.

$4x+3=0$ or $x+1=0$

$4x=-3$ $\quad x=-1$

$x=-\frac{3}{4}$

The boundary points are -1 and $-\frac{3}{4}$.

Test Interval	Test Number	Test	Conclusion
$(-\infty,-1)$	-2	$4(-2)^2+7(-2)<-3$ $2<-3$, false	$(-\infty,-1)$ does not belong to the solution set.
$\left(-1,-\frac{3}{4}\right)$	$-\frac{7}{8}$	$4\left(-\frac{7}{8}\right)^2+7\left(-\frac{7}{8}\right)<-3$ $-3\frac{1}{16}<-3$, true	$\left(-1,-\frac{3}{4}\right)$ belongs to the solution set.
$\left(-\frac{3}{4},\infty\right)$	0	$4(0)^2+7(0)<-3$ $0<-3$, false	$\left(-\frac{3}{4},\infty\right)$ does not belong to the solution set.

The solution set is $\left(-1,-\frac{3}{4}\right)$ or $\left\{x \middle| -1<x<-\frac{3}{4}\right\}$.

-1 -3/4

17. $x^2-4x\geq 0$

Solve the related quadratic equation.

$x^2-4x=0$

$x(x-4)=0$

Apply the zero product principle.

$x=0$ or $x-4=0$

$x=4$

The boundary points are 0 and 4.

Test Interval	Test Number	Test	Conclusion
$(-\infty,0]$	-1	$(-1)^2-4(-1)\geq 0$ $5\geq 0$, true	$(-\infty,0]$ belongs to the solution set.
$[0,4]$	1	$(1)^2-4(1)\geq 0$ $-3\geq 0$, false	$[0,4]$ does not belong to the solution set.

$[4,\infty)$	5	$(5)^2-4(5)\geq 0$ $5\geq 0$, true	$[4,\infty)$ belongs to the solution set.

The solution set is $(-\infty,0]\cup[4,\infty)$ or $\{x|x\leq 0 \text{ or } x\geq 4\}$.

0 4

19. $2x^2+3x>0$

Solve the related quadratic equation.

$$2x^2+3x=0$$
$$x(2x+3)=0$$

Apply the zero product principle.

$x=0$ or $2x+3=0$

$$2x=-3$$
$$x=-\frac{3}{2}$$

The boundary points are $-\frac{3}{2}$ and 0.

Test Interval	Test Number	Test	Conclusion
$\left(-\infty,-\frac{3}{2}\right)$	-2	$2(-2)^2+3(-2)>0$ $2>0$, true	$\left(-\infty,-\frac{3}{2}\right)$ belongs to the solution set.
$\left(-\frac{3}{2},0\right)$	-1	$2(-1)^2+3(-1)>0$ $-1>0$, false	$\left(-\frac{3}{2},0\right)$ does not belong to the solution set.
$(0,\infty)$	1	$2(1)^2+3(1)>0$ $5>0$, true	$(0,\infty)$ belongs to the solution set.

The solution set is $\left(-\infty,-\frac{3}{2}\right)\cup(0,\infty)$ or $\left\{x\middle|x<-\frac{3}{2} \text{ or } x>0\right\}$.

-3/2 0

21. $-x^2+x\geq 0$

Solve the related quadratic equation.

$$-x^2+x=0$$
$$-x(x-1)=0$$

Apply the zero product principle.

$-x=0$ or $x-1=0$

$x=0$ $\quad$ $x=1$

The boundary points are 0 and 1.

Test Interval	Test Number	Test	Conclusion
$(-\infty,0]$	-1	$-(-1)^2+(-1)\geq 0$ $-2\geq 0$, false	$(-\infty,0]$ does not belong to the solution set.
$[0,1]$	$\frac{1}{2}$	$-\left(\frac{1}{2}\right)^2+\frac{1}{2}\geq 0$ $\frac{1}{4}\geq 0$, true	$[0,1]$ belongs to the solution set.
$[1,\infty)$	2	$-(2)^2+2\geq 0$ $-2\geq 0$, false	$[1,\infty)$ does not belong to the solution set.

The solution set is $[0,1]$ or $\{x|0\leq x\leq 1\}$.

0 1

23.

$$x^2\leq 4x-2$$
$$x^2-4x+2\leq 0$$

Solve the related quadratic equation, using the quadratic formula.

$x^2-4x+2=0$

$a=1 \quad b=-4 \quad c=2$

$$x=\frac{-(-4)\pm\sqrt{(-4)^2-4(1)(2)}}{2(1)}=\frac{4\pm\sqrt{16-8}}{2}=\frac{4\pm\sqrt{8}}{2}=\frac{4\pm\sqrt{4\cdot 2}}{2}$$

$$=\frac{4\pm 2\sqrt{2}}{2}=\frac{2\left(2\pm\sqrt{2}\right)}{2}=2\pm\sqrt{2}$$

The boundary points are $2-\sqrt{2}$ and $2+\sqrt{2}$.

Test Interval	Test Number	Test	Conclusion
$\left(-\infty,2-\sqrt{2}\right]$	0	$0^2\leq 4(0)-2$ $0\leq -2$, false	$\left(-\infty,2-\sqrt{2}\right]$ does not belong to the solution set.
$\left[2-\sqrt{2},2+\sqrt{2}\right]$	2	$2^2\leq 4(2)-2$ $4\leq 6$, true	$\left[2-\sqrt{2},2+\sqrt{2}\right]$ belongs to the solution set.
$\left[2+\sqrt{2},\infty\right)$	4	$4^2\leq 4(4)-2$ $16\leq 14$, false	$\left[2+\sqrt{2},\infty\right)$ does not belong to the solution set.

The solution set is $\left[2-\sqrt{2},2+\sqrt{2}\right]$ or $\left\{x\middle|2-\sqrt{2}\leq x\leq 2+\sqrt{2}\right\}$.

$2-\sqrt{2}$ $\quad$ $2+\sqrt{2}$

25. $x^2 - 6x + 9 < 0$

Solve the related quadratic equation.

$x^2 - 6x + 9 = 0$

$(x-3)^2 = 0$

Apply the zero product principle to obtain the double root.

$x - 3 = 0$

$x = 3$

The boundary point is 3.

Test Interval	Test Number	Test	Conclusion
$(-\infty, 3)$	0	$0^2 - 6(0) + 9 < 0$ $9 < 0$, false	$(-\infty, 3)$ does not belong to the solution set.
$(3, \infty)$	4	$4^2 - 6(4) + 9 < 0$ $1 < 0$, false	$(3, \infty)$ does not belong to the solution set.

There is no solution. The solution set is $\varnothing$ or $\{ \quad \}$.

27. $\dfrac{x-4}{x+3} > 0$

Find the values of x that make the numerator and denominator zero.

$x - 4 = 0 \qquad x + 3 = 0$

$x = 4 \qquad x = -3$

The boundary points are -3 and 4. We exclude -3 from the solution set, since this would make the denominator zero.

Test Interval	Test Number	Test	Conclusion
$(-\infty, -3)$	-4	$\dfrac{-4-4}{-4+3} > 0$ $8 > 0$, true	$(-\infty, -3)$ belongs to the solution set.
$(-3, 4)$	0	$\dfrac{0-4}{0+3} > 0$ $\dfrac{-4}{3} > 0$, false	$(-3, 4)$ does not belong to the solution set.
$(4, \infty)$	5	$\dfrac{5-4}{5+3} > 0$ $\dfrac{1}{8} > 0$, true	$(4, \infty)$ belongs to the solution set.

The solution set is $(-\infty, -3) \cup (4, \infty)$ or $\{x \mid x < -3 \text{ or } x > 4\}$.

-3 4

29. $\frac{x+3}{x+4} < 0$

Find the values of x that make the numerator and denominator zero.

$x+3=0 \qquad x+4=0$

$x=-3 \qquad x=-4$

The boundary points are –4 and –3.

Test Interval	Test Number	Test	Conclusion
$(-\infty,-4)$	-5	$\frac{-5+3}{-5+4} < 0$ $2 < 0$, false	$(-\infty,-4)$ does not belong to the solution set.
$(-4,-3)$	-3.5	$\frac{-3.5+3}{-3.5+4} < 0$ $-1 < 0$, true	$(-4,-3)$ belongs to the solution set.
$(-3,\infty)$	0	$\frac{0+3}{0+4} < 0$ $\frac{3}{4} < 0$, false	$(-3,\infty)$ does not belong to the solution set.

The solution set is $(-4,-3)$ or $\{x \mid -4 < x < -3\}$.

-4 -3

31. $\frac{-x+2}{x-4} \geq 0$

Find the values of x that make the numerator and denominator zero.

$-x+2=0$ and $x-4=0$

$-x=-2 \qquad x=4$

$x=2$

The boundary points are 2 and 4. We exclude 4 from the solution set because 4 would make the denominator zero.

Test Interval	Test Number	Test	Conclusion
$(-\infty,2]$	0	$\frac{-0+2}{0-4} \geq 0$ $-\frac{1}{2} \geq 0$, false	$(-\infty,2]$ does not belong to the solution set.
$[2,4)$	3	$\frac{-3+2}{3-4} \geq 0$ $1 \geq 0$, true	$[2,4)$ belongs to the solution set.

$(4,\infty)$	5	$\frac{-5+2}{5-4}\ge 0$ $-3\ge 0$, false	$(4,\infty)$ does not belong to the solution set.

The solution set is $[2,4)$ or $\{x\|2\le x<4\}$.

2 4

33. $\frac{4-2x}{3x+4}\le 0$

Find the values of x that make the numerator and denominator zero.

$4-2x=0$ and $3x+4=0$

$-2x=-4$ $\quad 3x=-4$

$x=2$ $\quad x=-\frac{4}{3}$

The boundary points are $-\frac{4}{3}$ and 2. We exclude $-\frac{4}{3}$ from the solution set because $-\frac{4}{3}$ would make the denominator zero.

Test Interval	Test Number	Test	Conclusion
$\left(-\infty,-\frac{4}{3}\right)$	-2	$\frac{4-2(-2)}{3(-2)+4}\le 0$ $-4\le 0$, true	$\left(-\infty,-\frac{4}{3}\right)$ belongs to the solution set.
$\left(-\frac{4}{3},2\right]$	0	$\frac{4-2(0)}{3(0)+4}\le 0$ $1\le 0$, false	$\left(-\frac{4}{3},2\right]$ does not belong to the solution set.
$[2,\infty)$	3	$\frac{4-2(3)}{3(3)+4}\le 0$ $-\frac{2}{13}\le 0$, true	$[2,\infty)$ belongs to the solution set.

The solution set is $\left(-\infty,-\frac{4}{3}\right)\cup[2,\infty)$ or $\left\{x\middle|x<-\frac{4}{3}\text{ and }x\ge 2\right\}$.

-4/3 2

35. $\frac{x}{x-3}>0$

Find the values of x that make the numerator and denominator zero.

$x=0$ and $x-3=0$

$x=3$

The boundary points are 0 and 3.

Test Interval	Test Number	Test	Conclusion
$(-\infty, 0)$	-1	$\frac{-1}{-1-3} > 0$ $\frac{1}{4} > 0$, true	$\left(-\infty, -\frac{4}{3}\right)$ belongs to the solution set.
$(0, 3)$	1	$\frac{1}{1-3} > 0$ $-\frac{1}{2} > 0$, false	$(0, 3)$ does not belong to the solution set.
$(3, \infty)$	4	$\frac{4}{4-3} > 0$ $4 > 0$, true	$(3, \infty)$ belongs to the solution set.

The solution set is $(-\infty, 0) \cup (3, \infty)$ or $\{x | x < 0 \text{ and } x > 3\}$.

37. $\frac{x+1}{x+3} < 2$

Express the inequality so that one side is zero.

$$\frac{x+1}{x+3} - 2 < 0$$

$$\frac{x+1}{x+3} - \frac{2(x+3)}{x+3} < 0$$

$$\frac{x+1-2(x+3)}{x+3} < 0$$

$$\frac{x+1-2x-6}{x+3} < 0$$

$$\frac{-x-5}{x+3} < 0$$

Find the values of x that make the numerator and denominator zero.

$-x-5=0$ $\quad$ $x+3=0$

$-x=5$ $\quad$ $x=-3$

$x=-5$

The boundary points are –5 and –3.

Test Interval	Test Number	Test	Conclusion
$(-\infty, -5)$	-6	$\frac{-6+1}{-6+3} < 2$ $\frac{5}{3} < 2$, true	$(-\infty, -5)$ belongs to the solution set.
$(-5, -3)$	-4	$\frac{-4+1}{-4+3} < 2$ $3 < 2$, false	$(-5, -3)$ does not belong to the solution set.
$(-3, \infty)$	0	$\frac{0+1}{0+3} < 2$ $\frac{1}{3} < 2$, true	$(-3, \infty)$ belongs to the solution set.

The solution set is $(-\infty, -5) \cup (-3, \infty)$ or $\{x | x < -5 \text{ and } x > -3\}$.

-5 -3

39. $\frac{x+4}{2x-1} \le 3$

Express the inequality so that one side is zero.

$$\frac{x+4}{2x-1} - 3 \le 0$$

$$\frac{x+4}{2x-1} - \frac{3(2x-1)}{2x-1} \le 0$$

$$\frac{x+4-3(2x-1)}{2x-1} \le 0$$

$$\frac{x+4-6x+3}{2x-1} \le 0$$

$$\frac{-5x+7}{2x-1} \le 0$$

Find the values of x that make the numerator and denominator zero.

$-5x + 7 = 0$ $\quad$ $2x - 1 = 0$

$-5x = -7$ $\quad$ $2x = 1$

$x = \frac{7}{5}$ $\quad$ $x = \frac{1}{2}$

The boundary points are $\frac{1}{2}$ and $\frac{7}{5}$. We exclude $\frac{1}{2}$ from the solution set because $\frac{1}{2}$ would make the denominator zero.

Test Interval	Test Number	Test	Conclusion
$\left(-\infty, \frac{1}{2}\right)$	0	$\frac{0+4}{2(0)-1} \leq 3$ $-4 \leq 3$, true	$\left(-\infty, \frac{1}{2}\right)$ belongs to the solution set.
$\left(\frac{1}{2}, \frac{7}{5}\right]$	1	$\frac{1+4}{2(1)-1} \leq 3$ $5 \leq 3$, false	$\left(\frac{1}{2}, \frac{7}{5}\right]$ does not belong to the solution set.
$\left[\frac{7}{5}, \infty\right)$	2	$\frac{2+4}{2(2)-1} \leq 3$ $2 \leq 3$, true	$\left[\frac{7}{5}, \infty\right)$ belongs to the solution set.

The solution set is $\left(-\infty, \frac{1}{2}\right) \cup \left[\frac{7}{5}, \infty\right)$ or $\left\{x \middle| x < \frac{1}{2} \text{ and } x \geq \frac{7}{5}\right\}$.

1/2 7/5

41. $\frac{x-2}{x+2} \leq 2$

Express the inequality so that one side is zero.

$$\frac{x-2}{x+2} - 2 \leq 0$$
$$\frac{x-2}{x+2} - \frac{2(x+2)}{x+2} \leq 0$$
$$\frac{x-2-2(x+2)}{x+2} \leq 0$$
$$\frac{x-2-2x-4}{x+2} \leq 0$$
$$\frac{-x-6}{x+2} \leq 0$$

Find the values of x that make the numerator and denominator zero.

$-x-6=0$ $\quad$ $x+2=0$

$-x=6$ $\quad$ $x=-2$

$x=-6$

The boundary points are -6 and -2. We exclude -2 from the solution set because -2 would make the denominator zero.

Test Interval	Test Number	Test	Conclusion
$(-\infty, -6]$	-7	$\frac{-7-2}{-7+2} \leq 2$ $\frac{9}{5} \leq 2$, true	$(-\infty, -6]$ belongs to the solution set.

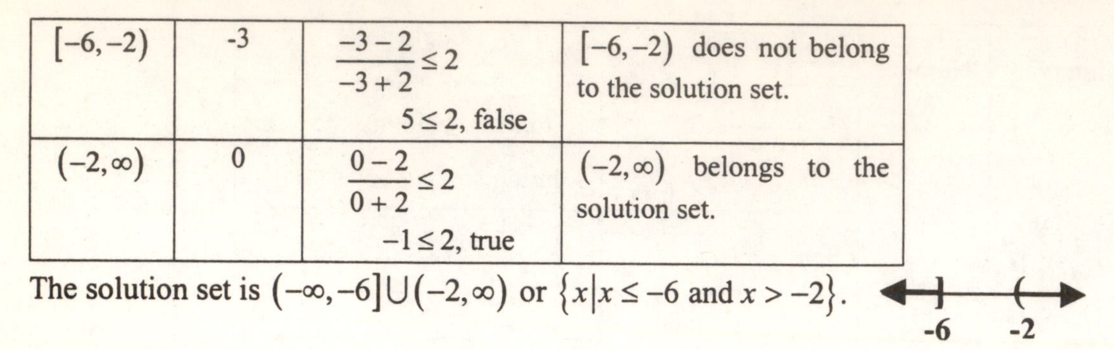

$[-6,-2)$	-3	$\frac{-3-2}{-3+2} \leq 2$ $5 \leq 2$, false	$[-6,-2)$ does not belong to the solution set.
$(-2,\infty)$	0	$\frac{0-2}{0+2} \leq 2$ $-1 \leq 2$, true	$(-2,\infty)$ belongs to the solution set.

The solution set is $(-\infty,-6] \cup (-2,\infty)$ or $\{x \mid x \leq -6 \text{ and } x > -2\}$.

Application Exercises

43. $s(t) = -16t^2 + 48t + 160$

To find when the height exceeds the height of the building, solve the inequality $-16t^2 + 48t + 160 > 160$.

Solve the related quadratic equation.

$$-16t^2 + 48t + 160 = 160$$
$$-16t^2 + 48t = 0$$
$$t^2 - 3t = 0$$
$$t(t-3) = 0$$

Apply the zero product principle.

$t = 0$ or $t - 3 = 0$

$t = 3$

The boundary points are 0 and 3.

Test Interval	Test Number	Test	Conclusion
$(0,3)$	1	$-16(1)^2 + 48(1) + 160 > 160$ $192 > 160$, true	$(0,3)$ belongs to the solution set.
$(3,\infty)$	4	$-16(4)^2 + 48(4) + 160 > 160$ $96 > 160$, true	$(3,\infty)$ does not belong to the solution set.

The solution set is $(0,3)$. This means that the ball exceeds the height of the building between 0 and 3 seconds.

45. $f(8) = 27(8) + 163 = 216 + 163 = 379$

$g(8) = 1.2(8)^2 + 15.2(8) + 181.4 = 1.2(64) + 121.6 + 181.4$

$= 76.8 + 121.6 + 181.4 = 379.8$

Since the graph indicates that Medicare spending will reach $379 billion, we conclude that both functions model the data quite well.

47. $g(x) = 1.2x^2 + 15.2x + 181.4$

To find when spending exceeds $536.6 billion, solve the inequality

$1.2x^2 + 15.2x + 181.4 > 536.6.$

Solve the related quadratic equation using the quadratic formula.

$1.2x^2 + 15.2x + 181.4 = 536.6$

$1.2x^2 + 15.2x - 355.2 = 0$

$a = 1.2 \quad b = 15.2 \quad c = -355.2$

$$x = \frac{-15.2 \pm \sqrt{15.2^2 - 4(1.2)(-355.2)}}{2(1.2)} = \frac{-15.2 \pm \sqrt{231.04 + 1704.96}}{2.4}$$

$$= \frac{-15.2 \pm \sqrt{1936}}{2.4} = \frac{-15.2 \pm 44}{2.4}$$

$$= \frac{-15.2 - 44}{2.4} \text{ or } \frac{-15.2 + 44}{2.4} = -24\frac{2}{3} \text{ or } 12$$

We disregard $-24\frac{2}{3}$ since x represents the number of years after 1995 and cannot be negative. The boundary point is 12.

Test Interval	Test Number	Test	Conclusion
$(0,12)$	1	$1.2(1)^2 + 15.2(1) + 181.4 > 536.6$ $197.8 > 536.6$, false	$(0,12)$ does not belong to the solution set.
$(12,\infty)$	13	$1.2(13)^2 + 15.2(13) + 181.4 > 536.6$ $581.8 > 536.6$, true	$(12,\infty)$ belongs to the solution set.

The solution set is $(12,\infty)$. This means that spending will exceed $536.6 billion after $1995 + 12 = 2007$.

49. $f(18) = 1.3(18)^2 + 32(18) + 303 = 1.3(324) + 576 + 303 = 421.2 + 576 + 303 = 1300.2$

The function predicts that there will be 1300.2 thousand or 1,300,200 inmates. The function models the actual number for 1998 very well.

51. $f(x) = 1.3x^2 + 32x + 303$

To find when the number of inmates exceeds 2433, solve the inequality $1.3x^2 + 32x + 303 > 2433$.

Solve the related quadratic equation using the quadratic formula.

$$1.3x^2 + 32x + 303 = 2433$$

$$1.3x^2 + 32x - 2130 = 0$$

$a = 1.3 \quad b = 32 \quad c = -2130$

$$x = \frac{-32 \pm \sqrt{32^2 - 4(1.3)(-2130)}}{2(1.3)} = \frac{-32 \pm \sqrt{1024 + 11076}}{2.6} = \frac{-32 \pm \sqrt{12100}}{2.6} = \frac{-32 \pm 110}{2.6}$$

$$= \frac{-32 - 110}{2.6} \text{ or } \frac{-32 + 110}{2.6}$$

$$= -54\frac{8}{13} \text{ or } 30$$

We disregard $-54\frac{8}{13}$ since x represents the number of years after 1980 and cannot be negative. The boundary point is 30.

Test Interval	Test Number	Test	Conclusion
$(0,30)$	1	$1.3(1)^2 + 32(1) + 303 > 2433$ $336.3 > 2433$, false	$(0,30)$ does not belong to the solution set.
$(30,\infty)$	31	$1.3(31)^2 + 32(31) + 303 > 2433$ $2544.3 > 2433$, true	$(30,\infty)$ belongs to the solution set.

The solution set is $(30,\infty)$. This means the number of inmates will exceed 2433 thousand or 2,433,000 after the year 1980 + 30 = 2010.

53. $\overline{C}(x) = \frac{500,000 + 400x}{x}$

To find when the cost of producing each wheelchair does not exceed \$425, solve the inequality $\frac{500,000 + 400x}{x} \le 425$.

Express the inequality so that one side is zero.

$$\frac{500,000+400x}{x}-425\le 0$$
$$\frac{500,000+400x}{x}-\frac{425x}{x}\le 0$$
$$\frac{500,000+400x-425x}{x}\le 0$$
$$\frac{500,000-25x}{x}\le 0$$

Find the values of x that make the numerator and denominator zero.

$500,000-25x=0$ $\qquad x=0$
$500,000=25x$
$20,000=x$

The boundary points are 0 and 20,000.

Test Interval	Test Number	Test	Conclusion
$[0,20000]$	1	$\frac{500,000+400(1)}{1}\le 425$ $500,400\le 425$, false	$[0,20000]$ does not belong to the solution set.
$[20000,\infty)$	25,000	$\frac{500,000+400(25,000)}{25,000}\le 425$ $420\le 425$, true	$[20000,\infty)$ belongs to the solution set.

The solution set is $[20000,\infty)$. This means that the company's production level will have to be at least 20,000 wheelchairs per week. The boundary corresponds to the point (20,000, 425) on the graph. When production is 20,000 or more per month, the average cost is \$425 or less.

Writing in Mathematics

55. Answers will vary.

57. Answers will vary.

Technology Exercises

59. $x^2+3x-10>0$

The solution set is $\{x|x<-5$ or $x>2\}$ or $(-\infty,-5)\cup(2,\infty)$.

61. $\frac{x-4}{x-1}\le 0$

The solution set is $\{x|1<x\le 4\}$ or $(1,4]$.

63. $\dfrac{1}{x+1} \le \dfrac{2}{x+4}$

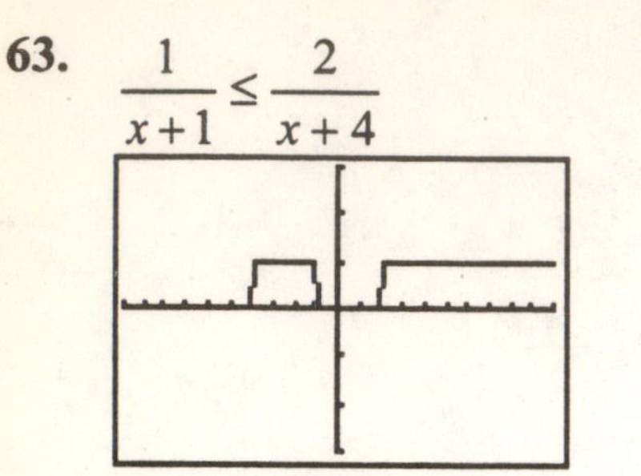

The solution set is $\{x | -4 < x < -1$ or $x \ge 2\}$ or $(-4,-1) \cup [2,\infty)$.

Critical Thinking Exercises

65. Statement **d.** is true.

Statement **a.** is false.

$x^2 > 25$

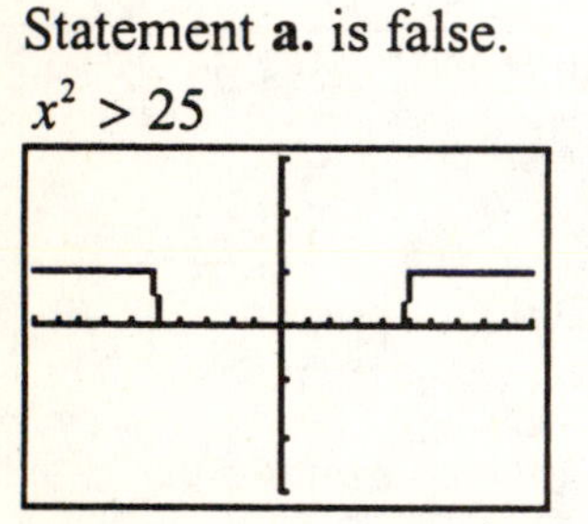

The solution set is $\{x | x < -5 \text{ or } x > 5\}$ or $(-\infty,-5) \cup (5,\infty)$.

Statement **b.** is false. The inequality cannot be solved by multiplying both sides by $x + 3$. We do not know if $x + 3$ is positive or negative. Thus, we do not know whether or not to reverse the sense of the inequality.

Statement **c.** is false. The solution sets are not the same.

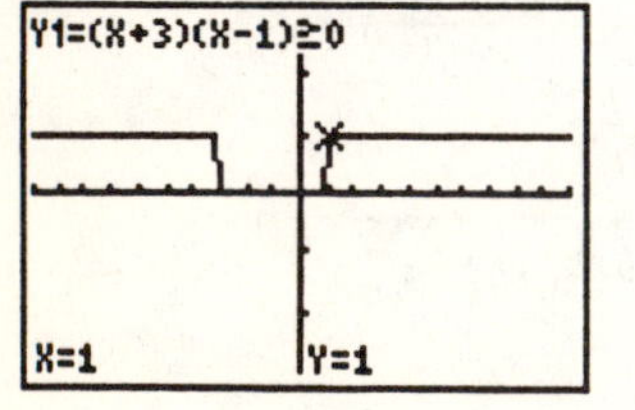

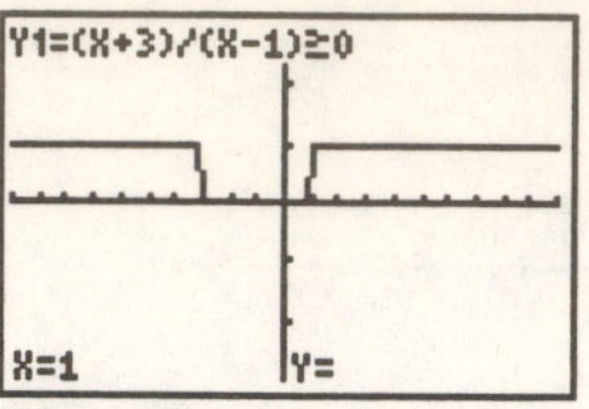

The value, 1, is included in the domain of the first inequality, but not included in the domain of the second inequality.

67. Answers will vary. An example is $\dfrac{x-3}{x+4} \ge 0$.

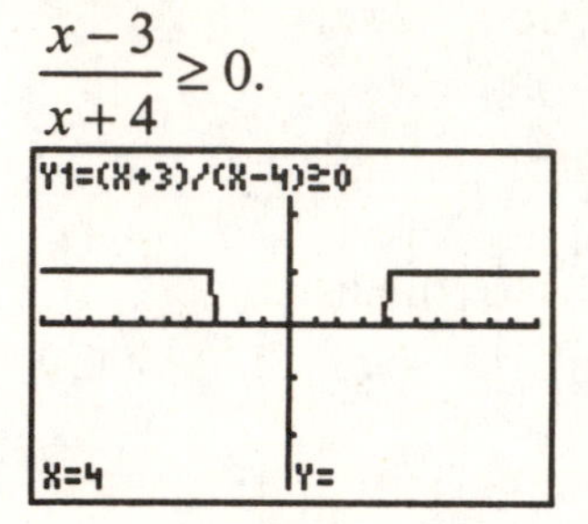

69. $(x-2)^2 \le 0$

Since the left hand side of the inequality is a square, we know it cannot be negative. In addition, the inequality calls for a number that is less than or equal to zero. The only possible solution is for the left hand side to equal zero. The left hand side of the inequality is zero when x is 2. Hence, the solution set is $\{2\}$.

71. $\dfrac{1}{(x-2)^2} > 0$

Since the denominator in the inequality is a square, we know it cannot be negative. Additionally, because the numerator is 1, the fraction will never be negative. As a result, x can be any real number except one that makes the denominator zero. Since 2 is the only value that makes the denominator zero, the solution set is $\{x|x$ is a real number and $x \neq 2\}$ or $(-\infty,2)\cup(2,\infty)$.

73. Write an inequality stating that the radicand must be greater than or equal to zero.

$27-3x^2 \geq 0$

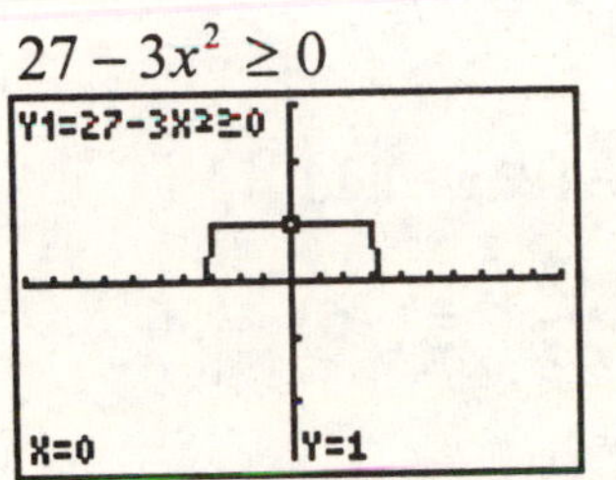

The inequality is true for values between 3 and –3. This means that the radicand is positive for values between 3 and –3, and the domain of the function is $\{x|-3 \leq x \leq 3\}$ or $[-3,3]$.

Review Exercises

74.

$$\left|\frac{x-5}{3}\right|<8$$

$$-8<\frac{x-5}{3}<8$$

$$-24<x-5<24$$

$$-19<x<29$$

The solution set is $\{x|-19<x<29\}$ or $(-19,29)$.

75.

$$\frac{2x+6}{x^2+8x+16} \div \frac{x^2-9}{x^2+3x-4}$$

$$=\frac{2x+6}{x^2+8x+16}\cdot\frac{x^2+3x-4}{x^2-9}$$

$$=\frac{2\cancel{(x+3)}}{\cancel{(x+4)}(x+4)}\cdot\frac{\cancel{(x+4)}(x-1)}{\cancel{(x+3)}(x-3)}$$

$$=\frac{2(x-1)}{(x+4)(x-3)}$$

76.

$$x^4-16y^4$$

$$=(x^2+4y^2)(x^2-4y^2)$$

$$=(x^2+4y^2)(x+2y)(x-2y)$$

Chapter 8 Review

8.1

1.

$$2x^2-3=125$$

$$2x^2=128$$

$$x^2=64$$

$$x=\pm 8$$

The solutions are –8 and 8 and the solution set is $\{-8,8\}$.

2.

$$3x^2-150=0$$

$$3x^2=150$$

$$x^2=50$$

$$x=\pm\sqrt{50}$$

$$x=\pm\sqrt{25\cdot 2}$$

$$x=\pm 5\sqrt{2}$$

The solutions are $-5\sqrt{2}$ and $5\sqrt{2}$ and the solution set is $\{-5\sqrt{2},5\sqrt{2}\}$.

3.

$$3x^2-2=0$$

$$3x^2=2$$

$x^2 = \frac{2}{3}$

$x = \pm\sqrt{\frac{2}{3}}$

Rationalize the denominator.

$x = \pm\frac{\sqrt{2}}{\sqrt{3}} \cdot \frac{\sqrt{3}}{\sqrt{3}} = \pm\frac{\sqrt{6}}{3}$

The solutions are $-\frac{\sqrt{6}}{3}$ and $\frac{\sqrt{6}}{3}$ and the solution set is $\left\{-\frac{\sqrt{6}}{3}, \frac{\sqrt{6}}{3}\right\}$.

4. $(x-4)^2 = 18$

$x - 4 = \pm\sqrt{18}$

$x = 4 \pm \sqrt{9 \cdot 2}$

$x = 4 \pm 3\sqrt{2}$

The solutions are $4 - 3\sqrt{2}$ and $4 + 3\sqrt{2}$ and the solution set is $\{4 - 3\sqrt{2}, 4 + 3\sqrt{2}\}$.

5. $(x+7)^2 = -36$

$x + 7 = \pm\sqrt{-36}$

$x = -7 \pm 6i$

The solutions are $-7 - 6i$ and $-7 + 6i$ and the solution set is $\{-7 - 6i, -7 + 6i\}$.

6. $x^2 + 20x +$ ______

Since $b = 20$, we add $\left(\frac{b}{2}\right)^2 = \left(\frac{20}{2}\right)^2 = (10)^2 = 100$.

$x^2 + 20x + 100 = (x+10)^2$

7. $x^2 - 3x +$ ______

Since $b = 3$, we add $\left(\frac{b}{2}\right)^2 = \left(\frac{3}{2}\right)^2 = \frac{9}{4}$.

$x^2 - 3x + \frac{9}{4} = \left(x - \frac{3}{2}\right)^2$

8. $x^2 - 12x + 27 = 0$

$x^2 - 12x \quad = -27$

Since $b = -12$, we add $\left(\frac{b}{2}\right)^2 = \left(\frac{-12}{2}\right)^2 = (-6)^2 = 36$.

$x^2 - 12x + 27 = 0$

$x^2 - 12x + 36 = -27 + 36$

$(x-6)^2 = 9$

Apply the square root property.

$x - 6 = 3 \qquad x - 6 = -3$

$x = 9 \qquad x = 3$

The solutions are 3 and 9 and the solution set is $\{3, 9\}$.

9. $x^2 - 7x - 1 = 0$

$x^2 - 7x \quad = 1$

Since $b = -7$, we add $\left(\frac{b}{2}\right)^2 = \left(\frac{-7}{2}\right)^2 = \frac{49}{4}$.

$x^2 - 7x + \frac{49}{4} = 1 + \frac{49}{4}$

$\left(x - \frac{7}{2}\right)^2 = \frac{4}{4} + \frac{49}{4}$

$\left(x - \frac{7}{2}\right)^2 = \frac{53}{4}$

Apply the square root property.

$$x-\frac{7}{2}=\pm\sqrt{\frac{53}{4}}$$

$$x=\frac{7}{2}\pm\frac{\sqrt{53}}{2}=\frac{7\pm\sqrt{53}}{2}$$

The solutions are $\frac{7\pm\sqrt{53}}{2}$ and the solution set is $\left\{\frac{7\pm\sqrt{53}}{2}\right\}$.

10. $2x^2+3x-4=0$

$$x^2+\frac{3}{2}x-2=0$$

$$x^2+\frac{3}{2}x \quad =2$$

Since $b=\frac{3}{2}$, we add

$$\left(\frac{b}{2}\right)^2=\left(\frac{\frac{3}{2}}{2}\right)^2=\left(\frac{3}{2}\div 2\right)^2$$

$$=\left(\frac{3}{2}\cdot\frac{1}{2}\right)^2=\left(\frac{3}{4}\right)^2=\frac{9}{16}.$$

$$x^2+\frac{3}{2}x+\frac{9}{16}=2+\frac{9}{16}$$

$$\left(x+\frac{3}{4}\right)^2=\frac{32}{16}+\frac{9}{16}$$

$$\left(x+\frac{3}{4}\right)^2=\frac{41}{16}$$

Apply the square root property.

$$x+\frac{3}{4}=\pm\sqrt{\frac{41}{16}}$$

$$x=-\frac{3}{4}\pm\frac{\sqrt{41}}{4}$$

$$x=\frac{-3\pm\sqrt{41}}{4}$$

The solutions are $\frac{-3\pm\sqrt{41}}{4}$ and the solution set is $\left\{\frac{-3\pm\sqrt{41}}{4}\right\}$.

11. $A=P(1+r)^t$

$$2916=2500(1+r)^2$$

$$\frac{2916}{2500}=(1+r)^2$$

Apply the square root property.

$$1+r=\pm\sqrt{\frac{2916}{2500}}$$

$$r=-1\pm\sqrt{1.1664}$$

$$r=-1\pm 1.08$$

The solutions are $-1-1.08=-2.08$ and $-1+1.08=0.08$. We disregard -2.08 since we cannot have a negative interest rate. The interest rate is 0.08 or 8%.

12. $W(t)=3t^2$

$$1200=3t^2$$

$$400=t^2$$

Apply the square root property.

$$t^2=400$$

$$t=\pm\sqrt{400}$$

$$t=\pm 20$$

The solutions are -20 and 20. We disregard -20, because we cannot have a negative time measurement. The fetus will weigh 1200 grams after 20 weeks.

13.

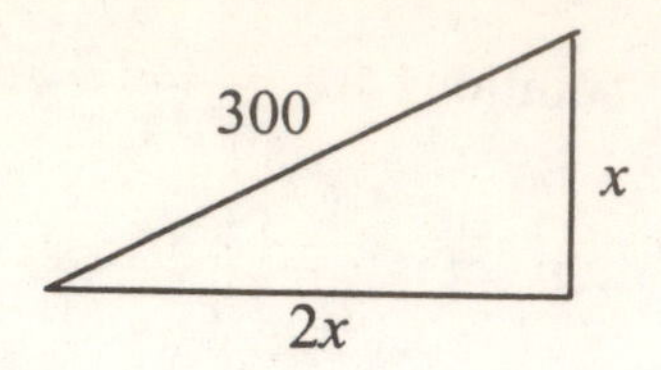

Use the Pythagorean Theorem.

$$(2x)^2 + x^2 = 300^2$$
$$4x^2 + x^2 = 90,000$$
$$5x^2 = 90,000$$
$$x^2 = 18,000$$
$$x = \pm\sqrt{18,000}$$
$$x = \pm\sqrt{3600 \cdot 5}$$
$$x = \pm 60\sqrt{5}$$

The solutions are $\pm 60\sqrt{5}$ meters. We disregard $-60\sqrt{5}$ meters, because we can't have a negative length measurement. Therefore, the building is $60\sqrt{5}$ or approximately 134.2 meters high.

8.2

14.

$$x^2 = 2x + 4$$
$$x^2 - 2x - 4 = 0$$
$$a = 1 \quad b = -2 \quad c = -4$$
$$x = \frac{-(-2) \pm \sqrt{(-2)^2 - 4(1)(-4)}}{2(1)}$$
$$= \frac{2 \pm \sqrt{4+16}}{2} = \frac{2 \pm \sqrt{20}}{2}$$
$$= \frac{2 \pm \sqrt{4 \cdot 5}}{2} = \frac{2 \pm 2\sqrt{5}}{2}$$
$$= \frac{2(1 \pm \sqrt{5})}{2} = 1 \pm \sqrt{5}$$

The solutions are $1 \pm \sqrt{5}$ and the solution set is $\{1 \pm \sqrt{5}\}$.

15.

$$x^2 - 2x + 19 = 0$$
$$a = 1 \quad b = -2 \quad c = 19$$
$$x = \frac{-(-2) \pm \sqrt{(-2)^2 - 4(1)(19)}}{2(1)}$$
$$= \frac{2 \pm \sqrt{4-76}}{2} = \frac{2 \pm \sqrt{-72}}{2}$$
$$= \frac{2 \pm \sqrt{-36 \cdot 2}}{2} = \frac{2 \pm 6\sqrt{2}i}{2}$$
$$= \frac{2(1 \pm 3\sqrt{2}i)}{2} = 1 \pm 3\sqrt{2}i$$

The solutions are $1 \pm 3\sqrt{2}i$ and the solution set is $\{1 \pm 3\sqrt{2}i\}$.

16.

$$2x^2 = 3 - 4x$$
$$2x^2 + 4x - 3 = 0$$
$$a = 2 \quad b = 4 \quad c = -3$$
$$x = \frac{-4 \pm \sqrt{4^2 - 4(2)(-3)}}{2(2)}$$
$$= \frac{-4 \pm \sqrt{16+24}}{4} = \frac{-4 \pm \sqrt{40}}{4}$$
$$= \frac{-4 \pm \sqrt{4 \cdot 10}}{4} = \frac{-4 \pm 2\sqrt{10}}{4}$$
$$= \frac{2(-2 \pm \sqrt{10})}{4} = \frac{-2 \pm \sqrt{10}}{2}$$

The solutions are $\frac{-2 \pm \sqrt{10}}{2}$ and the solution set is $\left\{\frac{-2 \pm \sqrt{10}}{2}\right\}$.

17.

$$x^2 - 4x + 13 = 0$$
$$a = 1 \quad b = -4 \quad c = 13$$

Find the discriminant.

$$b^2 - 4ac = (-4)^2 - 4(1)(13)$$
$$= 16 - 52 = -36$$

Since the discriminant is negative, there are two imaginary solutions which are complex conjugates.

18. $$9x^2 = 2 - 3x$$
$$9x^2 + 3x - 2 = 0$$
$a = 9 \quad b = 3 \quad c = -2$

Find the discriminant.

$$b^2 - 4ac = 3^2 - 4(9)(-2)$$
$$= 9 + 72 = 81$$

Since the discriminant is greater than zero, there are two unequal real solutions. Also, since the discriminant is a perfect square, the solutions are rational.

19. $$2x^2 + 4x = 3$$
$$2x^2 + 4x - 3 = 0$$
$a = 2 \quad b = 4 \quad c = -3$

Find the discriminant.

$$b^2 - 4ac = 4^2 - 4(2)(-3)$$
$$= 16 + 24 = 40$$

Since the discriminant is greater than zero, there are two unequal real solutions.

20. $$2x^2 - 11x + 5 = 0$$
$$(2x - 1)(x - 5) = 0$$

Apply the zero product principle.

$2x - 1 = 0$ and $x - 5 = 0$

$2x = 1 \qquad x = 5$

$x = \frac{1}{2}$

The solutions are $\frac{1}{2}$ and 5 and the solution set is $\left\{\frac{1}{2}, 5\right\}$.

21. $$(3x + 5)(x - 3) = 5$$
$$3x^2 - 9x + 5x - 15 = 5$$
$$3x^2 - 4x - 15 = 5$$
$$3x^2 - 4x - 20 = 0$$
$a = 3 \quad b = -4 \quad c = -20$

Use the quadratic formula.

$$x = \frac{-(-4) \pm \sqrt{(-4)^2 - 4(3)(-20)}}{2(3)}$$
$$= \frac{4 \pm \sqrt{16 + 240}}{6} = \frac{4 \pm \sqrt{256}}{6}$$
$$= \frac{4 \pm 16}{6}$$
$$= \frac{4 - 16}{6} \text{ or } \frac{4 + 16}{6}$$
$$= \frac{-12}{6} \text{ or } \frac{20}{6}$$
$$= -2 \text{ or } \frac{10}{3}$$

The solutions are -2 and $\frac{10}{3}$ and the solution set is $\left\{-2, \frac{10}{3}\right\}$.

22. $$3x^2 - 7x + 1 = 0$$
$a = 3 \quad b = -7 \quad c = 1$

Use the quadratic formula.

$$x = \frac{-(-7) \pm \sqrt{(-7)^2 - 4(3)(1)}}{2(3)}$$
$$= \frac{7 \pm \sqrt{49 - 12}}{6} = \frac{7 \pm \sqrt{37}}{6}$$

The solutions are $\frac{7 \pm \sqrt{37}}{6}$ and the solution set is $\left\{\frac{7 \pm \sqrt{37}}{6}\right\}$.

23. $x^2 - 9 = 0$

$x^2 = 9$

Apply the square root principle.

$x = \pm 3$

The solutions are –3 and 3 and the solution set is $\{-3, 3\}$.

24. $(x-3)^2 - 8 = 0$

$(x-3)^2 = 8$

Apply the square root principle.

$x - 3 = \pm\sqrt{8}$

$x = 3 \pm \sqrt{4 \cdot 2}$

$x = 3 \pm 2\sqrt{2}$

The solutions are $3 \pm 2\sqrt{2}$ and the solution set is $\{3 \pm 2\sqrt{2}\}$.

25. $3x^2 - x + 2 = 0$

$a = 3 \quad b = -1 \quad c = 2$

Use the quadratic formula.

$$x = \frac{-(-1) \pm \sqrt{(-1)^2 - 4(3)(2)}}{2(3)}$$

$$= \frac{1 \pm \sqrt{1 - 24}}{6} = \frac{1 \pm \sqrt{-23}}{6}$$

$$= \frac{1}{6} \pm \frac{\sqrt{23}}{6}i$$

The solutions are $\frac{1}{6} \pm \frac{\sqrt{23}}{6}i$ and the solution set is $\left\{\frac{1}{6} \pm \frac{\sqrt{23}}{6}i\right\}$.

26.

$$\frac{5}{x+1} + \frac{x-1}{4} = 2$$

$$4(x+1)\left(\frac{5}{x+1} + \frac{x-1}{4}\right) = 4(x+1)(2)$$

$$20 + (x+1)(x-1) = 8x + 8$$

$$20 + x^2 - 1 = 8x + 8$$

$$x^2 - 8x + 11 = 0$$

Use the quadratic formula.

$a = 1 \quad b = -8 \quad c = 11$

$$x = \frac{-(-8) \pm \sqrt{(-8)^2 - 4(1)(11)}}{2(1)}$$

$$= \frac{8 \pm \sqrt{64 - 44}}{2} = \frac{8 \pm \sqrt{20}}{2}$$

$$= \frac{8 \pm \sqrt{4 \cdot 5}}{2} = \frac{8 \pm 2\sqrt{5}}{2}$$

$$= \frac{2(4 \pm \sqrt{5})}{2} = 4 \pm \sqrt{5}$$

The solutions are $4 \pm \sqrt{5}$ and the solution set is $\{4 \pm \sqrt{5}\}$.

27. Because the solution set is $\left\{-\frac{1}{3}, \frac{3}{5}\right\}$, we have

$x = -\frac{1}{3}$ or $x = \frac{3}{5}$

$x + \frac{1}{3} = 0 \qquad x - \frac{3}{5} = 0$

$3x + 1 = 0 \qquad 5x - 3 = 0.$

Use the zero-product principle in reverse.

$$(3x+1)(5x-3) = 0$$

$$15x^2 - 9x + 5x - 3 = 0$$

$$15x^2 - 4x - 3 = 0$$

28. Because the solution set is $\{-9i, 9i\}$, we have

$x = -9i$ or $x = 9i$

$x + 9i = 0$ $x - 9i = 0.$

Use the zero-product principle in reverse.

$$(x+9i)(x-9i) = 0$$
$$x^2 - 81i^2 = 0$$
$$x^2 - 81(-1) = 0$$
$$x^2 + 81 = 0$$

29. Because the solution set is $\{-4\sqrt{3}, 4\sqrt{3}\}$, we have

$x = -4\sqrt{3}$ or $x = 4\sqrt{3}$

$x + 4\sqrt{3} = 0$ $x - 4\sqrt{3} = 0.$

Use the zero product principle in reverse.

$$\left(x+4\sqrt{3}\right)\left(x-4\sqrt{3}\right) = 0$$
$$x^2 - \left(4\sqrt{3}\right)^2 = 0$$
$$x^2 - 16 \cdot 3 = 0$$
$$x^2 - 48 = 0$$

30. $13 = 0.2x^2 - 1.2x + 2$

$0 = 0.2x^2 - 1.2x - 11$

Solve using the Pythagorean Theorem.

$a = 0.2$ $b = -1.2$ $c = -11$

$$x = \frac{-(-1.2) \pm \sqrt{(-1.2)^2 - 4(0.2)(-11)}}{2(0.2)}$$
$$= \frac{1.2 \pm \sqrt{1.44 + 8.8}}{0.4} = \frac{1.2 \pm \sqrt{10.24}}{0.4}$$
$$= \frac{1.2 \pm 3.2}{0.4}$$
$$= \frac{1.2 - 3.2}{0.4} \text{ or } \frac{1.2 + 3.2}{0.4}$$
$$= \frac{-2}{0.4} \text{ or } \frac{4.4}{0.4} = -5 \text{ or } 11$$

We disregard –5 because we cannot have a negative number of years. The solution is 11 and we conclude that the infection rate will be 11 PCs per month for every 1000 PCs in the year $1990 + 11 = 2001$.

31. $0 = -16t^2 + 140t + 3$

Solve using the Pythagorean Theorem.

$a = -16$ $b = 140$ $c = 3$

$$= \frac{-140 \pm \sqrt{19{,}600 + 192}}{-32}$$
$$= \frac{-140 \pm \sqrt{19{,}792}}{-32}$$
$$= \frac{-140 \pm 140.7}{-32}$$
$$= \frac{-140 - 140.7}{-32} \text{ or } \frac{-140 + 140.7}{-32}$$
$$= \frac{-280.7}{-32} \text{ or } \frac{0.7}{-32}$$
$$= 8.8 \text{ or } -0.02$$

We disregard –0.02 because we cannot have a negative time measurement. The solution is 8.8 and we conclude that the ball will hit the ground in 8.8 seconds.

8.3

32. $f(x) = -(x+1)^2 + 4$

Since $a = -1$ is negative, the parabola opens downward. The vertex of the parabola is $(h, k) = (-1, 4)$ and the axis of symmetry is $x = -1$. Replace $f(x)$ with 0 to find x–intercepts.

$0 = -(x+1)^2 + 4$

$-4 = -(x+1)^2$

$4 = (x+1)^2$

Apply the square root property.

$x+1 = \sqrt{4}$ or $x+1 = -\sqrt{4}$

$x+1 = 2$ $\quad x+1 = -2$

$x = 1$ $\quad x = -3$

The x–intercepts are 1 and –3. Set $x = 0$ and solve for y to obtain the y–intercept.

$y = -(0+1)^2 + 4$

$y = -(1)^2 + 4$

$y = -1 + 4 = 3$

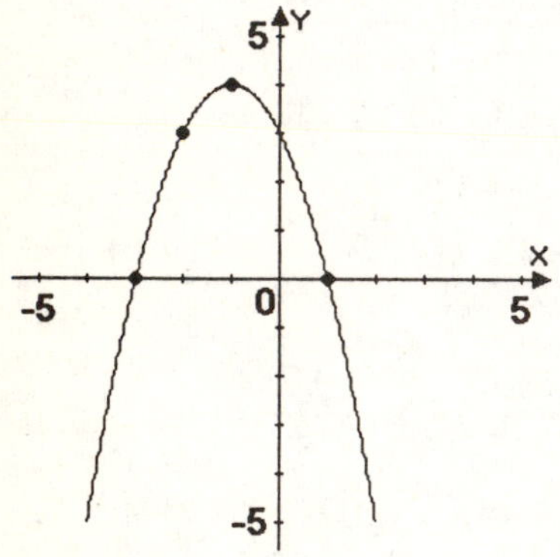

33. $f(x) = (x+4)^2 - 2$

Since $a = 1$ is positive, the parabola opens upward. The vertex of the parabola is $(h, k) = (-4, -2)$ and the axis of symmetry is $x = -4$. Replace $f(x)$ with 0 to find x–intercepts.

$0 = (x+4)^2 - 2$

$2 = (x+4)^2$

Apply the square root property.

$x+4 = \sqrt{2}$ or $x+4 = -\sqrt{2}$

$x = -4 + \sqrt{2}$ $\quad x = -4 - \sqrt{2}$

The x–intercepts are $-4 - \sqrt{2}$ and $-4 + \sqrt{2}$. Set $x = 0$ and solve for y to obtain the y–intercept.

$y = (0+4)^2 - 2$

$y = 4^2 - 2$

$y = 16 - 2$

$y = 14$

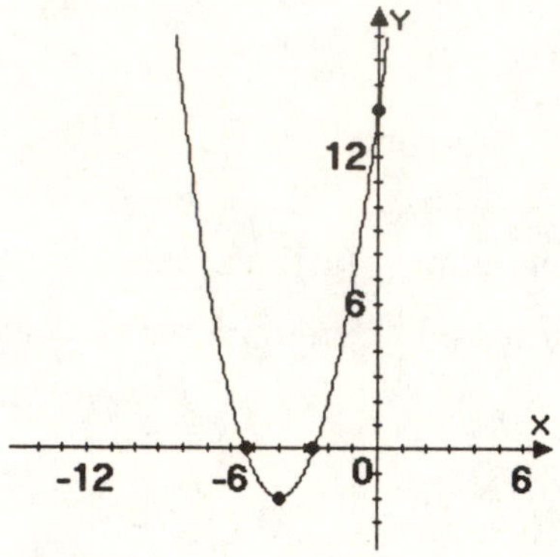

34. $f(x) = -x^2 + 2x + 3$

Since $a = -1$ is negative, the parabola opens downward. The x–coordinate of the vertex of the parabola is

$-\frac{b}{2a} = -\frac{2}{2(-1)} = -\frac{2}{-2} = 1$ and the

y–coordinate of the vertex of the parabola is

$$f\left(-\frac{b}{2a}\right) = f(1) = -1^2 + 2(1) + 3$$

$$= -1 + 2 + 3 = 4.$$

The vertex is (1, 4). Replace $f(x)$ with 0 to find x–intercepts.

$0 = -x^2 + 2x + 3$

$0 = x^2 - 2x - 3$

$0 = (x-3)(x+1)$

Apply the zero product principle.

$x-3=0$ or $x+1=0$

$x=3$ $\qquad$ $x=-1$

The x–intercepts are –1 and 3. Set $x=0$ and solve for y to obtain the y–intercept.

$y=-0^2+2(0)+3$

$y=0+0+3$

$y=3$

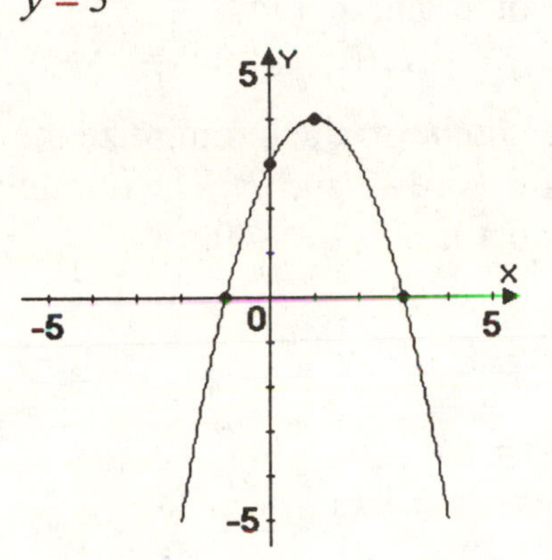

35. $f(x)=2x^2-4x-6$

Since $a=2$ is positive, the parabola opens upward. The x–coordinate of the vertex of the parabola is

$-\frac{b}{2a}=-\frac{-4}{2(2)}=-\frac{-4}{4}=1$ and the

y–coordinate of the vertex of the parabola is

$$f\left(-\frac{b}{2a}\right)=f(1)=2(1)^2-4(1)-6$$
$$=2(1)-4-6=2-4-6$$
$$=-8.$$

The vertex is $(1,-8)$. Replace $f(x)$ with 0 to find x–intercepts.

$0=2x^2-4x-6$

$0=x^2-2x-3$

$0=(x-3)(x+1)$

Apply the zero product principle.

$x-3=0$ or $x+1=0$

$x=3$ $\qquad$ $x=-1$

The x–intercepts are –1 and 3. Set $x=0$ and solve for y to obtain the y–intercept.

$y=2(0)^2-4(0)-6$

$y=2(0)-0-6$

$y=0-0-6=-6$

36. $f(x)=-0.02x^2+x+1$

Since $a=-0.02$ is negative, we know the function opens downward and has a maximum at

$x=-\frac{b}{2a}=-\frac{1}{2(-0.02)}=-\frac{1}{-0.04}=25.$

When 25 inches of rain falls, the maximum growth will occur. The maximum growth is

$$f(25)=-0.02(25)^2+25+1$$
$$=-0.02(625)+25+1$$
$$=-12.5+25+1=13.5.$$

A maximum yearly growth of 13.5 inches occurs when 25 inches of rain falls per year.

37. $s(t)=-16t^2+400t+40$

Since $a=-16$ is negative, we know the function opens downward and has a maximum at

$$x=-\frac{b}{2a}=-\frac{400}{2(-16)}=-\frac{400}{-32}=12.5.$$

At 12.5 seconds, the rocket reaches its maximum height. The maximum height is

$$s(12.5)=-16(12.5)^2+400(12.5)+40$$
$$=-16(156.25)+5000+40$$
$$=-2500+5000+40=2540.$$

The rocket reaches a maximum height of 2540 feet in 12.5 seconds.

38. According to the graph, the maximum is (20, 5.4). This means that the maximum divorce rate of 5.4 divorces per 1000 people occurred in the year 1960 + 20 = 1980.

39. Maximize the area using $A = lw$.

$$A(x)=x(100-2x)$$
$$A(x)=-2x^2+100x$$

Since $a=-2$ is negative, we know the function opens downward and has a maximum at

$$x=-\frac{b}{2a}=-\frac{100}{2(-2)}=-\frac{100}{-4}=25.$$

The maximum area is achieved when the width is 25 yards. The maximum area is

$$A(25)=25(100-2(25))$$
$$=25(100-50)$$
$$=25(50)=1250.$$

The area is maximized at 1250 square yards when the width is 25 yards.

40. Let x = one of the numbers
Let $14+x$ = the other number

We need to minimize the function

$$P(x)=x(14+x)$$
$$=14x+x^2$$
$$=x^2+14x.$$

The minimum is at

$$x=-\frac{b}{2a}=-\frac{14}{2(1)}=-\frac{14}{2}=-7.$$

The other number is
$14+x=14+(-7)=7.$

The numbers which minimize the product are –7 and 7. The minimum product is $-7\cdot 7=-49$.

8.4

41. Let $t=x^2$.

$$x^4-6x^2+8=0$$
$$(x^2)^2-6x^2+8=0$$
$$t^2-6t+8=0$$
$$(t-4)(t-2)=0$$

Apply the zero product principle.

$t-4=0$ or $t-2=0$

$t=4$ $\quad t=2$

Replace t by x^2.

$x^2=4$ or $x^2=2$

$x=\pm 2$ $\quad x=\pm\sqrt{2}$

The solutions are $\pm\sqrt{2}$ and ± 2 and the solution set is $\left\{-2,-\sqrt{2},\sqrt{2},2\right\}$.

42. Let $t=\sqrt{x}$.

$$x+7\sqrt{x}-8=0$$
$$\left(\sqrt{x}\right)^2+7\sqrt{x}-8=0$$
$$t^2+7t-8=0$$
$$(t+8)(t-1)=0$$

Apply the zero product principle.

$t+8=0 \quad \text{or} \quad t-1=0$

$t=-8 \qquad\qquad t=1$

Replace t by $\sqrt{x}$.

~~$\sqrt{x}=-8$~~ or $\sqrt{x}=1$

$x=1$

We disregard –8 because the square root of x cannot be a negative number. We need to check the solution, 1, because both sides of the equation were raised to an even power.

Check:

$$1+7\sqrt{1}-8=0$$
$$1+7(1)-8=0$$
$$1+7-8=0$$
$$8-8=0$$
$$0=0$$

The solution is 1 and the solution set is $\{1\}$.

43. Let $t=x^2+2x$.

$$\left(x^2+2x\right)^2-14\left(x^2+2x\right)=15$$
$$\left(x^2+2x\right)^2-14\left(x^2+2x\right)-15=0$$
$$t^2-14t-15=0$$
$$(t-15)(t+1)=0$$

Apply the zero product principle.

$t-15=0 \quad \text{or} \quad t+1=0$

$t=15 \qquad\qquad t=-1$

Replace t by x^2+2x.

First, consider $t=15$.

$$x^2+2x=15$$
$$x^2+2x-15=0$$
$$(x+5)(x-3)=0$$

Apply the zero product principle.

$x+5=0 \quad \text{or} \quad x-3=0$

$x=-5 \qquad\qquad x=3$

Next, consider $t=-1$.

$$x^2+2x=-1$$
$$x^2+2x+1=0$$
$$(x+1)^2=0$$

Apply the zero product principle to find the double root.

$$x+1=0$$
$$x=-1$$

The solutions are –5, –1, and 3 and the solution set is $\{-5,-1,3\}$.

44. Let $t=x^{-1}$.

$$x^{-2}+x^{-1}-56=0$$
$$\left(x^{-1}\right)^2+x^{-1}-56=0$$
$$t^2+t-56=0$$
$$(t+8)(t-7)=0$$

Apply the zero product principle.

$t+8=0 \quad \text{or} \quad t-7=0$

$t=-8 \qquad\qquad t=7$

Replace t by x^{-1}.

$x^{-1}=-8 \quad \text{or} \quad x^{-1}=7$

$\frac{1}{x}=-8 \qquad\qquad \frac{1}{x}=7$

$-8x=1 \qquad\qquad 7x=1$

$x=-\frac{1}{8} \qquad\qquad x=\frac{1}{7}$

The solutions are $-\frac{1}{8}$ and $\frac{1}{7}$, and the solution set is $\left\{-\frac{1}{8},\frac{1}{7}\right\}$.

45. Let $t = x^{\frac{1}{3}}$.

$$x^{\frac{2}{3}} - x^{\frac{1}{3}} - 12 = 0$$
$$\left(x^{\frac{1}{3}}\right)^2 - x^{\frac{1}{3}} - 12 = 0$$
$$t^2 - t - 12 = 0$$
$$(t-4)(t+3) = 0$$

Apply the zero product principle.

$t - 4 = 0$ or $t + 3 = 0$

$t = 4$ $\qquad t = -3$

Replace t by $x^{\frac{1}{3}}$.

$x^{\frac{1}{3}} = 4$ or $x^{\frac{1}{3}} = -3$

$\left(x^{\frac{1}{3}}\right)^3 = 4^3$ $\qquad \left(x^{\frac{1}{3}}\right)^3 = (-3)^3$

$x = 64$ $\qquad x = -27$

The solutions are -27 and 64, and the solution set is $\{-27, 64\}$.

46. Let $t = x^{\frac{1}{4}}$.

$$x^{\frac{1}{2}} + 3x^{\frac{1}{4}} - 10 = 0$$
$$\left(x^{\frac{1}{4}}\right)^2 + 3x^{\frac{1}{4}} - 10 = 0$$
$$t^2 + 3t - 10 = 0$$
$$(t+5)(t-2) = 0$$

Apply the zero product principle.

$t + 5 = 0$ or $t - 2 = 0$

$t = -5$ $\qquad t = 2$

Replace t by $x^{\frac{1}{4}}$.

$x^{\frac{1}{4}} = -5$ or $x^{\frac{1}{4}} = 2$

~~$\sqrt[4]{x} = -5$~~ $\qquad \left(x^{\frac{1}{4}}\right)^4 = 2^4$

$x = 16$

We disregard -5 because the fourth root of x cannot be a negative number. We need to check the solution, 16, because both sides of the equation were raised to an even power.

Check $x = 16$.

$$16^{\frac{1}{2}} + 3(16)^{\frac{1}{4}} - 10 = 0$$
$$4 + 3(2) - 10 = 0$$
$$4 + 6 - 10 = 0$$
$$10 - 10 = 0$$
$$0 = 0$$

The solution checks. The solution is 16 and the solution set is $\{16\}$.

8.5

47. $2x^2 + 5x - 3 < 0$

Solve the related quadratic equation.

$$2x^2 + 5x - 3 = 0$$
$$(2x-1)(x+3) = 0$$

Apply the zero product principle.

$2x - 1 = 0$ or $x + 3 = 0$

$2x = 1$ $\qquad x = -3$

$x = \frac{1}{2}$

The boundary points are -3 and $\frac{1}{2}$.

Test Interval	Test Number	Test	Conclusion
$(-\infty,-3)$	-4	$2(-4)^2+5(-4)-3<0$ $9<0$, false	$(-\infty,-3)$ does not belong to the solution set.
$\left(-3,\frac{1}{2}\right)$	0	$2(0)^2+5(0)-3<0$ $-3<0$, true	$\left(-3,\frac{1}{2}\right)$ belongs to the solution set.
$\left(\frac{1}{2},\infty\right)$	1	$2(1)^2+5(1)-3<0$ $4<0$, false	$\left(\frac{1}{2},\infty\right)$ does not belong to the solution set.

The solution set is $\left(-3,\frac{1}{2}\right)$ or $\left\{x\middle|-3<x<\frac{1}{2}\right\}$.

-3 1/2

48. $2x^2+9x+4\geq 0$

Solve the related quadratic equation.

$$2x^2+9x+4=0$$
$$(2x+1)(x+4)=0$$

Apply the zero product principle.

$2x+1=0$ or $x+4=0$

$2x=-1$ $\qquad x=-4$

$x=-\frac{1}{2}$

The boundary points are -4 and $-\frac{1}{2}$.

Test Interval	Test Number	Test	Conclusion
$(-\infty,-4]$	-5	$2(-5)^2+9(-5)+4\geq 0$ $9\geq 0$, true	$(-\infty,-4]$ belongs to the solution set.
$\left[-4,-\frac{1}{2}\right]$	-1	$2(-1)^2+9(-1)+4\geq 0$ $-3\geq 0$, false	$\left[-4,-\frac{1}{2}\right]$ does not belong to the solution set.
$\left[-\frac{1}{2},\infty\right)$	0	$2(0)^2+9(0)+4\geq 0$ $4\geq 0$, true	$\left[-\frac{1}{2},\infty\right)$ belongs to the solution set.

The solution set is $(-\infty,-4]\cup\left[-\frac{1}{2},\infty\right)$ or $\left\{x\middle|x\leq -4 \text{ or } x\geq -\frac{1}{2}\right\}$.

-4 -1/2

49. $\dfrac{x-6}{x+2}>0$

Find the values of x that make the numerator and denominator zero.

$x-6=0 \qquad x+2=0$

$x=6 \qquad x=-2$

The boundary points are –2 and 6.

Test Interval	Test Number	Test	Conclusion
$(-\infty,-2)$	-3	$\dfrac{-3-6}{-3+2}>0$ $9>0$, true	$(-\infty,-2)$ belongs to the solution set.
$(-2,6)$	0	$\dfrac{0-6}{0+2}>0$ $-3>0$, false	$(-2,6)$ does not belong to the solution set.
$(6,\infty)$	7	$\dfrac{7-6}{7+2}>0$ $\dfrac{1}{9}>0$, true	$(6,\infty)$ belongs to the solution set.

The solution set is $(-\infty,-2)\cup(6,\infty)$ or $\{x|x<-2 \text{ or } x>6\}$.

-2 6

50. $\dfrac{x+3}{x-4}\le 5$

Express the inequality so that one side is zero.

$$\frac{x+3}{x-4}-5\le 0$$

$$\frac{x+3}{x-4}-\frac{5(x-4)}{x-4}\le 0$$

$$\frac{x+3-5(x-4)}{x-4}\le 0$$

$$\frac{x+3-5x+20}{x-4}\le 0$$

$$\frac{-4x+23}{x-4}\le 0$$

Find the values of x that make the numerator and denominator zero.

$-4x+23=0 \quad \text{and} \quad x-4=0$

$-4x=-23 \qquad x=4$

$x=\dfrac{23}{4}$

The boundary points are 4 and $\frac{23}{4}$. We exclude 4 from the solution set, since this would make the denominator zero.

Test Interval	Test Number	Test	Conclusion
$(-\infty, 4)$	0	$\frac{0+3}{0-4} \le 5$ $\frac{3}{-4} \le 5$, true	$(-\infty, 4)$ belongs to the solution set.
$\left(4, \frac{23}{4}\right]$	5	$\frac{5+3}{5-4} \le 5$ $8 \le 5$, false	$\left(4, \frac{23}{4}\right]$ does not belong to the solution set.
$\left[\frac{23}{4}, \infty\right)$	6	$\frac{6+3}{6-4} \le 5$ $\frac{9}{2} \le 5$, true	$\left[\frac{23}{4}, \infty\right)$ belongs to the solution set.

The solution set is $(-\infty, 4) \cup \left[\frac{23}{4}, \infty\right)$ or $\left\{x \middle| x < 4 \text{ or } x \ge \frac{23}{4}\right\}$.

4 23/4

51. $s(t) = -16t^2 + 48t$

To find when the height is more than 32 feet above the ground, solve the inequality $-16t^2 + 48t > 32$.

Solve the related quadratic equation.

$$-16t^2 + 48t = 32$$

$$-16t^2 + 48t - 32 = 0$$

$$t^2 - 3t + 2 = 0$$

$$(t-2)(t-1) = 0$$

Apply the zero product principle.

$t - 2 = 0$ or $t - 1 = 0$

$t = 2$ $\quad$ $t = 1$

The boundary points are 1 and 2.

Test Interval	Test Number	Test	Conclusion
$(0,1)$	0.5	$-16(0.5)^2 + 48(0.5) > 32$ $20 > 32$, false	$(0,1)$ does not belong to the solution set.
$(1,2)$	1.5	$-16(1.5)^2 + 48(1.5) > 32$ $36 > 32$, true	$(1,2)$ belongs to the solution set.

$(2,\infty)$	3	$-16(3)^2+48(3)>32$ $0>32$, false	$(2,\infty)$ does not belong to the solution set.

The solution set is $(1,2)$. This means that the ball will be more than 32 feet above the graph between 1 and 2 seconds.

52. a. $H(0)=\frac{15}{8}(0)^2-30(0)+200=\frac{15}{8}(0)-0+200=0-0+200=200$

The heart rate is 200 beats per minute immediately following the workout.

b.

$$\frac{15}{8}x^2-30x+200>110$$

$$\frac{15}{8}x^2-30x+90>0$$

$$\frac{8}{15}\left(\frac{15}{8}x^2-30x+90\right)>\frac{8}{15}(0)$$

$$x^2-\frac{8}{15}(30x)+\frac{8}{15}(90)>0$$

$$x^2-16x+48>0$$

$$(x-12)(x-4)>0$$

Apply the zero product principle.

$x-12=0$ or $x-4=0$

$x=12$ $\quad$ $x=4$

The boundary points are 4 and 12.

Test Interval	Test Number	Test	Conclusion
$(0,4)$	1	$\frac{15}{8}(1)^2-30(1)+200>110$ $171\frac{7}{8}>110$, true	$(0,4)$ belongs to the solution set.
$(4,12)$	5	$\frac{15}{8}(5)^2-30(5)+200>110$ $96\frac{7}{8}>110$, false	$(4,\infty)$ does not belong to the solution set.
$(12,\infty)$	13	$\frac{15}{8}(13)^2-30(13)+200>110$ $126\frac{7}{8}>110$, false	$(12,\infty)$ does not belong to the solution set.

The solution set is $(0,4)\cup(12,\infty)$. This means that the heart rate exceeds 110 beats per minute between 0 and 4 minutes and more than 12 minutes after the workout. Between 0 and 4 minutes provides a more realistic answer since it is unlikely that the heart rate will begin to climb again without further exertion. Model breakdown occurs for the interval $(12,\infty)$.

Chapter 8 Test

1. $2x^2-5=0$

$2x^2=5$

$x^2=\frac{5}{2}$

$x=\pm\sqrt{\frac{5}{2}}$

Rationalize the denominators.

$x=\pm\frac{\sqrt{5}}{\sqrt{2}}\cdot\frac{\sqrt{2}}{\sqrt{2}}=\pm\frac{\sqrt{10}}{2}$

The solutions are $\pm\frac{\sqrt{10}}{2}$ and the solution set is $\left\{\pm\frac{\sqrt{10}}{2}\right\}$.

2. $(x-3)^2=20$

$x-3=\pm\sqrt{20}$

$x=3\pm\sqrt{4\cdot5}$

$x=3\pm2\sqrt{5}$

The solutions are $3\pm2\sqrt{5}$ and the solution set is $\{3\pm2\sqrt{5}\}$.

3. x^2-16x+ _____

Since $b=-16$, we add

$\left(\frac{b}{2}\right)^2=\left(\frac{-16}{2}\right)^2=(-8)^2=64$.

$x^2-16x+64=(x-8)^2$

4. $x^2+\frac{2}{5}x+$ _____

Since $b=\frac{2}{5}$, we add

$\left(\frac{b}{2}\right)^2=\left(\frac{\frac{2}{5}}{2}\right)^2=\left(\frac{2}{5}\div2\right)^2$

$=\left(\frac{2}{5}\cdot\frac{1}{2}\right)^2=\left(\frac{1}{5}\right)^2=\frac{1}{25}$

$x^2+\frac{2}{5}x+\frac{1}{25}=\left(x+\frac{1}{5}\right)^2$

5. $x^2-6x+7=0$

$x^2-6x \quad =-7$

Since $b=-6$, we add

$\left(\frac{b}{2}\right)^2=\left(\frac{-6}{2}\right)^2=(-3)^2=9$.

$x^2-6x+9=-7+9$

$(x-3)^2=2$

Apply the square root property.

$x-3=\pm\sqrt{2}$

$x=3\pm\sqrt{2}$

The solutions are $3\pm\sqrt{2}$ and the solution set is $\{3\pm\sqrt{2}\}$.

6. Use the Pythagorean Theorem.

$$50^2 + 50^2 = x^2$$
$$2500 + 2500 = x^2$$
$$5000 = x^2$$
$$\pm\sqrt{5000} = x$$
$$\pm\sqrt{2500 \cdot 2} = x$$
$$\pm 50\sqrt{2} = x$$

The solutions are $\pm 50\sqrt{2}$ feet. We disregard $-50\sqrt{2}$ feet because we can't have a negative length measurement. The width of the pond is $50\sqrt{2}$ feet.

7.
$$3x^2 + 4x - 2 = 0$$
$a = 3 \quad b = 4 \quad c = -2$

Find the discriminant.
$$b^2 - 4ac = 4^2 - 4(3)(-2)$$
$$= 16 + 24 = 40$$

Since the discriminant is greater than zero, there are two unequal real solutions.

8.
$$x^2 = 4x - 8$$
$$x^2 - 4x + 8 = 0$$
$a = 1 \quad b = -4 \quad c = 8$

Find the discriminant.
$$b^2 - 4ac = (-4)^2 - 4(1)(8)$$
$$= 16 - 32 = -16$$

Since the discriminant is negative, there are two imaginary solutions which are complex conjugates.

9.
$$2x^2 + 9x = 5$$
$$2x^2 + 9x - 5 = 0$$
$$(2x - 1)(x + 5) = 0$$

Apply the zero product principle.

$2x - 1 = 0$ and $x + 5 = 0$

$2x = 1 \qquad x = -5$

$x = \frac{1}{2}$

The solutions are $\frac{1}{2}$ and -5 and the solution set is $\left\{-5, \frac{1}{2}\right\}$.

10. $x^2 + 8x + 5 = 0$

Solve using the quadratic formula.

$a = 1 \quad b = 8 \quad c = 5$

$$x = \frac{-8 \pm \sqrt{8^2 - 4(1)(5)}}{2(1)}$$
$$= \frac{-8 \pm \sqrt{64 - 20}}{2} = \frac{-8 \pm \sqrt{44}}{2}$$
$$= \frac{-8 \pm \sqrt{4 \cdot 11}}{2} = \frac{-8 \pm 2\sqrt{11}}{2}$$
$$= \frac{2\left(-4 \pm \sqrt{11}\right)}{2} = -4 \pm \sqrt{11}$$

The solutions are $-4 \pm \sqrt{11}$ and the solution set is $\{-4 \pm \sqrt{11}\}$.

11. $(x + 2)^2 + 25 = 0$

$$(x + 2)^2 = -25$$

Apply the square root principle.

$$x + 2 = \pm\sqrt{-25}$$
$$x = -2 \pm 5i$$

The solutions are $-2 \pm 5i$ and the solution set is $\{-2 \pm 5i\}$.

12.
$$2x^2 - 6x + 5 = 0$$
$a = 2 \quad b = -6 \quad c = 5$

$$x = \frac{-(-6) \pm \sqrt{(-6)^2 - 4(2)(5)}}{2(2)}$$

$$= \frac{6 \pm \sqrt{36-40}}{4} = \frac{6 \pm \sqrt{-4}}{4}$$

$$= \frac{6 \pm 2i}{4} = \frac{2(3 \pm i)}{4}$$

$$= \frac{3 \pm i}{2} = \frac{3}{2} \pm \frac{1}{2}i$$

The solutions are $\frac{3}{2} \pm \frac{1}{2}i$ and the solution set is $\left\{\frac{3}{2} \pm \frac{1}{2}i\right\}$.

13. Because the solution set is $\{-3, 7\}$, we have

$x = -3$ or $x = 7$

$x + 3 = 0$ $\quad x - 7 = 0$

Use the zero-product principle in reverse.

$$(x+3)(x-7) = 0$$
$$x^2 - 7x + 3x - 21 = 0$$
$$x^2 - 4x - 21 = 0$$

14. Because the solution set is $\{-10i, 10i\}$, we have

$x = -10i$ or $x = 10i$

$x + 10i = 0$ $\quad x - 10i = 0$

Use the zero-product principle in reverse.

$$(x+10i)(x-10i) = 0$$
$$x^2 - 100i^2 = 0$$
$$x^2 - 100(-1) = 0$$
$$x^2 + 100 = 0$$

15. $f(x) = -0.5x^2 + 4x + 19$

$$20 = -0.5x^2 + 4x + 19$$
$$0 = -0.5x^2 + 4x - 1$$

Solve using the quadratic formula.

$a = -0.5 \quad b = 4 \quad c = -1$

$$x = \frac{-4 \pm \sqrt{4^2 - 4(-0.5)(-1)}}{2(-0.5)}$$

$$= \frac{-4 \pm \sqrt{16-2}}{-1} = \frac{-4 \pm \sqrt{14}}{-1}$$

$$= \frac{-4 - \sqrt{14}}{-1} \text{ or } \frac{-4 + \sqrt{14}}{-1}$$

$$= \frac{-4 - 3.7}{-1} \text{ or } \frac{-4 + 3.7}{-1}$$

$$= \frac{-7.7}{-1} \text{ or } \frac{-0.3}{-1}$$

$$= 7.7 \text{ or } 0.3$$

$$\approx 8 \text{ or } 0$$

In the years 1900 and 1998, 20 million people received food stamps.

16. $f(x) = (x+1)^2 + 4$

Since $a = 1$ is negative, the parabola opens upward. The vertex of the parabola is $(h, k) = (-1, 4)$ and the axis of symmetry is $x = -1$. Replace $f(x)$ with 0 to find x–intercepts.

$$0 = (x+1)^2 + 4$$
$$-4 = (x+1)^2$$

This will be result in complex solutions. As a result, there are no x–intercepts. Set $x = 0$ and solve for y to obtain the y–intercept.

$$y = (0+1)^2 + 4$$
$$y = (1)^2 + 4$$
$$y = 1 + 4$$
$$y = 5$$

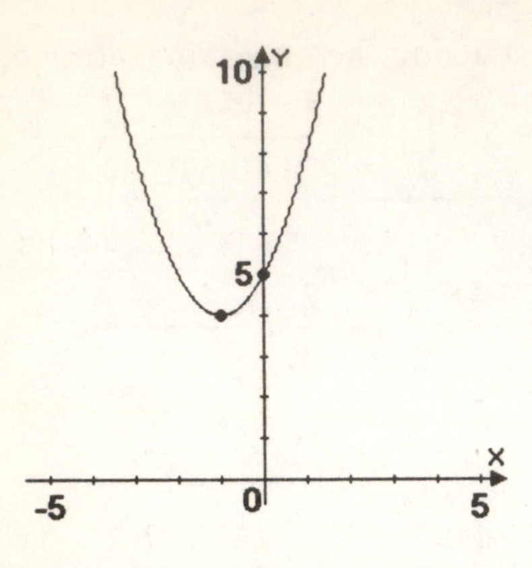

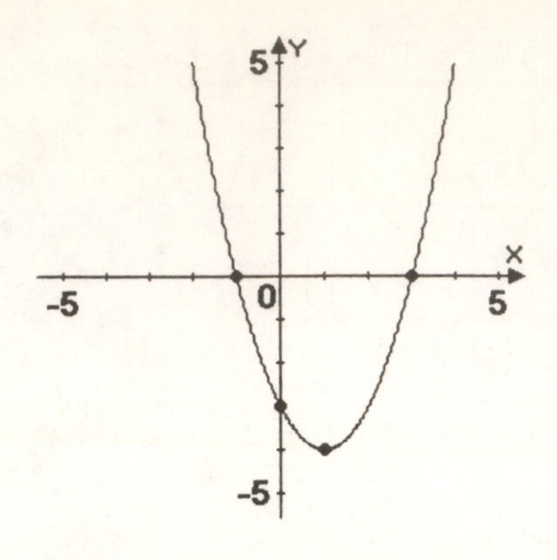

17. $f(x) = x^2 - 2x - 3$

Since $a = 1$ is positive, the parabola opens upward. The x–coordinate of the vertex of the parabola is

$$-\frac{b}{2a} = -\frac{-2}{2(1)} = -\frac{-2}{2} = 1$$ and the

y–coordinate of the vertex of the parabola is

$$f\left(-\frac{b}{2a}\right) = f(1) = 1^2 - 2(1) - 3$$
$$= 1 - 2 - 3 = -4.$$

The vertex is $(1,-4)$. Replace $f(x)$ with 0 to find x–intercepts.

$$0 = x^2 - 2x - 3$$
$$0 = (x-3)(x+1)$$

Apply the zero product principle.

$x - 3 = 0$ or $x + 1 = 0$

$x = 3$ $\quad$ $x = -1$

The x–intercepts are -1 and 3. Set $x = 0$ and solve for y to obtain the y–intercept.

$$y = 0^2 - 2(0) - 3$$
$$y = 0 - 0 - 3$$
$$y = -3$$

18. $s(t) = -16t^2 + 64t + 5$

Since $a = -16$ is negative, we know the function opens downward and has a maximum at

$$x = -\frac{b}{2a} = -\frac{64}{2(-16)} = -\frac{64}{-32} = 2.$$

The ball reaches its maximum height a two seconds. The maximum height is

$$s(2) = -16(2)^2 + 64(2) + 5$$
$$= -16(4) + 128 + 5$$
$$= -64 + 128 + 5 = 69.$$

The baseball reaches a maximum height of 69 feet after 2 seconds.

19. $0 = -16t^2 + 64t + 5$

Solve using the quadratic formula.

$a = -16 \quad b = 64 \quad c = 5$

$$x = \frac{-64 \pm \sqrt{64^2 - 4(-16)(5)}}{2(-16)}$$
$$= \frac{-64 \pm \sqrt{4096 + 320}}{-32}$$
$$= \frac{-64 \pm \sqrt{4416}}{-32}$$
$$= \frac{-64 - 66.5}{-32} \text{ or } \frac{-64 + 66.5}{-32}$$

$= \dfrac{-130.5}{-32}$ or $\dfrac{2.5}{-32}$

$= \dfrac{-130.5}{-32}$ or $\dfrac{2.5}{-32}$

≈ 4.1 or -0.1

We disregard –0.1 since we cannot have a negative time measurement. The solution is 4.1 and we conclude that the baseball hits the ground in approximately 4.1 seconds.

20. $f(x) = -x^2 + 46x - 360$

Since $a = -1$ is negative, we know the function opens downward and has a maximum at

$x = -\dfrac{b}{2a} = -\dfrac{46}{2(-1)} = -\dfrac{46}{-2} = 23.$

Profit is maximized when 23 computers are manufactured. The maximum profit is

$$\begin{aligned} f(23) &= -23^2 + 46(23) - 360 \\ &= -529 + 1058 - 360 \\ &= 169. \end{aligned}$$

23 computers should be manufactured each day. This produces a profit of 169 hundreds or $16,900.

21. Let $t = 2x - 5$.

$(2x-5)^2 + 4(2x-5) + 3 = 0$

$t^2 + 4t + 3 = 0$

$(t+3)(t+1) = 0$

Apply the zero product principle.

$t + 3 = 0$ or $t + 1 = 0$

$t = -3$ $\quad$ $t = -1$

Replace t by $2x - 5$.

First, consider $t = 15$.

$2x - 5 = -3$ or $2x - 5 = -1$

$2x = 2$ $\quad$ $2x = 4$

$x = 1$ $\quad$ $x = 2$

The solutions are 1 and 2 and the solution set is $\{1, 2\}$.

22. Let $t = x^2$.

$x^4 - 13x^2 + 36 = 0$

$(x^2)^2 - 13x^2 + 36 = 0$

$t^2 - 13t + 36 = 0$

$(t-9)(t-4) = 0$

Apply the zero product principle.

$t - 9 = 0$ or $t - 4 = 0$

$t = 9$ $\quad$ $t = 4$

Replace t by x^2.

$x^2 = 9$ or $x^2 = 4$

$x = \pm 3$ $\quad$ $x = \pm 2$

The solutions are ± 2 and ± 3 and the solution set is $\{-3, -2, 2, 3\}$.

23. Let $t = x^{\frac{1}{3}}$.

$x^{\frac{2}{3}} - 9x^{\frac{1}{3}} + 8 = 0$

$\left(x^{\frac{1}{3}}\right)^2 - 9x^{\frac{1}{3}} + 8 = 0$

$t^2 - 9t + 8 = 0$

$(t-8)(t-1) = 0$

Apply the zero product principle.

$t - 8 = 0$ or $t - 1 = 0$

$t = 8$ $\quad$ $t = 1$

Replace t by $x^{\frac{1}{3}}$.

$x^{\frac{1}{3}} = 8$ or $x^{\frac{1}{3}} = 1$

$\left(x^{\frac{1}{3}}\right)^3 = 8^3$ $\left(x^{\frac{1}{3}}\right)^3 = 1^3$

$x = 512$ $x = 1$

The solutions are 1 and 512 and the solution set is $\{1, 512\}$.

24. $x^2 - x - 12 < 0$

Solve the related quadratic equation.

$x^2 - x - 12 = 0$

$(x-4)(x+3) = 0$

Apply the zero product principle.

$x - 4 = 0$ or $x + 3 = 0$

$x = 4$ $x = -3$

The boundary points are -3 and 4.

Test Interval	Test Number	Test	Conclusion
$(-\infty, -3)$	-4	$(-4)^2 - (-4) - 12 < 0$ $8 < 0$, false	$(-\infty, -3)$ does not belong to the solution set.
$(-3, 4)$	0	$0^2 - 0 - 12 < 0$ $-12 < 0$, true	$(-3, 4)$ belongs to the solution set.
$(4, \infty)$	5	$5^2 - 5 - 12 < 0$ $8 < 0$, false	$(4, \infty)$ does not belong to the solution set.

The solution set is $(-3, 4)$ or $\{x \mid -3 < x < 4\}$.

-3 4

25. $\dfrac{2x+1}{x-3} \le 3$

Express the inequality so that one side is zero.

$$\frac{2x+1}{x-3} - 3 \le 0$$

$$\frac{2x+1}{x-3} - \frac{3(x-3)}{x-3} \le 0$$

$$\frac{2x+1-3(x-3)}{x-3} \le 0$$

$$\frac{2x+1-3x+9}{x-3} \le 0$$

$$\frac{-x+10}{x-3} \le 0$$

Find the values of x that make the numerator and denominator zero.

$-x+10=0$ and $x-3=0$

$-x=-10$ $\quad x=3$

$x=10$

The boundary points are 3 and 10. We exclude 3 from the solution set(s), since this would make the denominator zero.

Test Interval	Test Number	Test	Conclusion
$(-\infty,3)$	0	$\frac{2(0)+1}{0-3} \le 3$ $-\frac{1}{3} \le 3$, true	$(-\infty,3)$ belongs to the solution set.
$(3,10]$	4	$\frac{2(4)+1}{4-3} \le 3$ $9 \le 3$, false	$(3,10]$ does not belong to the solution set.
$[10,\infty)$	11	$\frac{2(10)+1}{10-3} \le 3$ $3 \le 3$, true	$[10,\infty)$ belongs to the solution set.

The solution set is $(-\infty,3)\cup[10,\infty)$ or $\{x \mid x<3 \text{ or } x \ge 10\}$.

3 10

Cumulative Review Exercises

Chapters 1-8

1. $8-(4x-5)=x-7$

$8-4x+5=x-7$

$13-4x=x-7$

$13=5x-7$

$20=5x$

$4=x$

The solution is 4 and the solution set is $\{4\}$.

2. $5x+4y=22$

$3x-8y=-18$

Multiply the first equation by 2 and solve by addition.

$10x+8y=44$

$3x-8y=-18$

$13x=26$

$x=2$

Back-substitute 2 for x to find y.

$5(2)+4y=22$

$10+4y=22$

$$4y = 12$$
$$y = 3$$

The solution is $(2,3)$ and the solution set is $\{(2,3)\}$.

3. $-3x + 2y + 4z = 6$

$7x - y + 3z = 23$

$2x + 3y + z = 7$

Multiply the second equation by 2 and add to the first equation to eliminate y.

$$\begin{aligned} -3x + 2y + 4z &= 6 \\ 14x - 2y + 6z &= 46 \\ \hline 11x + 10z &= 52 \end{aligned}$$

Multiply the second equation by 3 and add to the second equation to eliminate y.

$$\begin{aligned} 21x - 3y + 9z &= 69 \\ 2x + 3y + z &= 7 \\ \hline 23x + 10z &= 76 \end{aligned}$$

The system of two variables in two equations is:

$11x + 10z = 52$

$23x + 10z = 76$

Multiply the first equation by -1 and add to the second equation.

$$\begin{aligned} -11x - 10z &= -52 \\ 23x + 10z &= 76 \\ \hline 12x &= 24 \\ x &= 2 \end{aligned}$$

Back-substitute 2 for x to find z.

$$\begin{aligned} 11(2) + 10z &= 52 \\ 22 + 10z &= 52 \\ 10z &= 30 \\ z &= 3 \end{aligned}$$

Back-substitute 2 for x and 3 for z to find y.

$$\begin{aligned} -3(2) + 2y + 4(3) &= 6 \\ -6 + 2y + 12 &= 6 \\ 2y + 6 &= 6 \\ 2y &= 0 \\ y &= 0 \end{aligned}$$

The solution is $(2,0,3)$ and the solution set is $\{(2,0,3)\}$.

4. $|x - 1| > 3$

$x - 1 < -3$ or $x - 1 > 3$

$x < -2$ $\qquad$ $x > 4$

The solution set is $\{x | x < -2$ and $x > 4\}$ or $(-\infty, -2) \cup (4, \infty)$.

-2 4

5. $\sqrt{x+4} - \sqrt{x-4} = 2$

$$\begin{aligned} \sqrt{x+4} &= 2 + \sqrt{x-4} \\ \left(\sqrt{x+4}\right)^2 &= \left(2 + \sqrt{x-4}\right)^2 \\ x + 4 &= 4 + 4\sqrt{x-4} + x - 4 \\ \cancel{x} + 4 &= 4\sqrt{x-4} + \cancel{x} \\ 4 &= 4\sqrt{x-4} \\ 1 &= \sqrt{x-4} \\ 1^2 &= \left(\sqrt{x-4}\right)^2 \\ 1 &= x - 4 \\ 5 &= x \end{aligned}$$

The solution is 5 and the solution set is $\{5\}$.

6. $x-4\geq 0$ and $-3x\leq -6$

$x\geq 4 \qquad x\geq 2$

$x\geq 4$

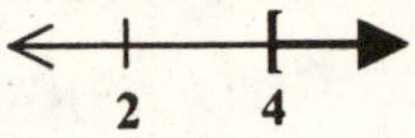

$x\geq 2$

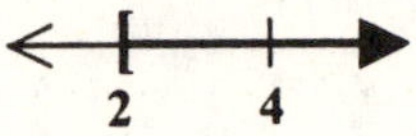

$x\geq 4$ and $x\geq 2$

The solution set is $\{x|x\geq 4\}$ or $[4,\infty)$.

7.
$$2x^2=3x-2$$
$$2x^2-3x+2=0$$

Solve using the quadratic formula.

$a=2 \qquad b=-3 \qquad c=2$

$$x=\frac{-(-3)\pm\sqrt{(-3)^2-4(2)(2)}}{2(2)}$$
$$=\frac{3\pm\sqrt{9-16}}{4}=\frac{3\pm\sqrt{-7}}{4}$$
$$=\frac{3\pm\sqrt{7}i}{4}=\frac{3}{4}\pm\frac{\sqrt{7}}{4}i$$

The solutions are $\frac{3}{4}-\frac{\sqrt{7}}{4}i$ and $\frac{3}{4}+\frac{\sqrt{7}}{4}i$ and the solution set is $\left\{\frac{3}{4}-\frac{\sqrt{7}}{4}i, \frac{3}{4}+\frac{\sqrt{7}}{4}i\right\}$.

8. $3x=15+5y$

Find the x–intercept by setting $y=0$ and solving.

$3x=15+5(0)$

$3x=15$

$x=5$

Find the y–intercept by setting $x=0$ and solving.

$3(0)=15+5y$

$0=15+5y$

$-15=5y$

$-3=y$

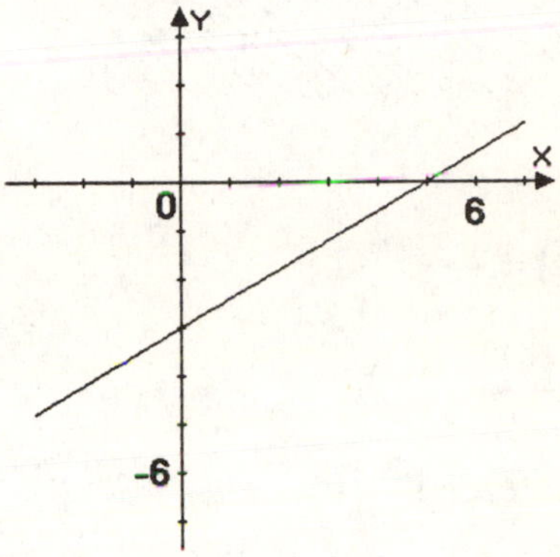

9. $2x-3y>6$

First, find the intercepts to the equation $2x-3y=6$.

Find the x–intercept by setting $y=0$ and solving.

$2x-3(0)=6$

$2x=6$

$x=3$

Find the y–intercept by setting $x=0$ and solving.

$2(0)-3y=6$

$-3y=6$

$y=-2$

Next, use the origin as a test point.

$2(0)-3(0)>6$

$0-0>6$

$0>6$

This is a false statement. This means that the origin will not fall in the shaded half-plane.

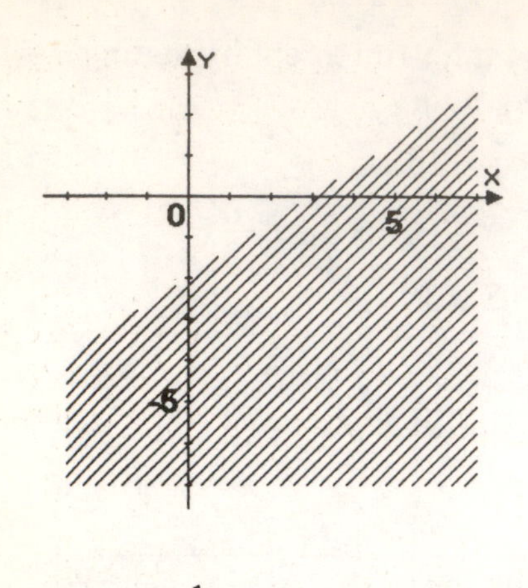

10. $f(x) = -\frac{1}{2}x + 1$

$m = -\frac{1}{2}$

$y-\text{intercept} = 1$

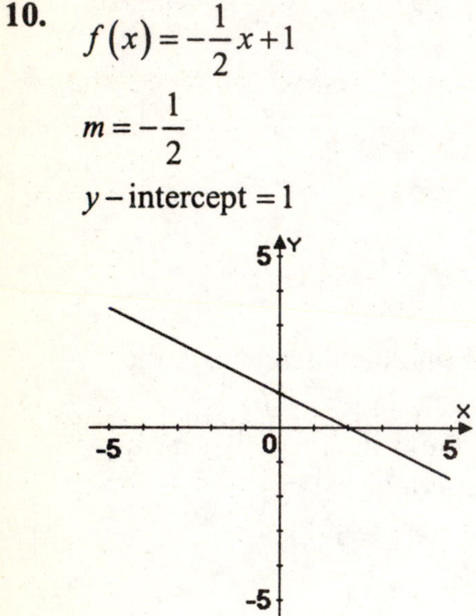

11. $f(x) = x^2 + 6x + 8$

Since $a = 1$ is positive, the parabola opens upward. The x–coordinate of the vertex of the parabola is

$-\frac{b}{2a} = -\frac{6}{2(1)} = -\frac{6}{2} = -3$ and the

y–coordinate of the vertex of the parabola is

$$f\left(-\frac{b}{2a}\right) = f(-3) = (-3)^2 + 6(-3) + 8$$
$$= 9 - 18 + 8 = -1.$$

The vertex is $(-3, -1)$. Replace $f(x)$ with 0 to find x–intercepts.

$0 = x^2 + 6x + 8$

$0 = (x+4)(x+2)$

Apply the zero product principle.

$x + 4 = 0$ or $x + 2 = 0$

$x = -4$ $\quad$ $x = -2$

The x–intercepts are –4 and –2. Set $x = 0$ and solve for y to obtain the y–intercept.

$y = 0^2 + 6(0) + 8$

$y = 0 + 0 + 8$

$y = 8$

12. $f(x) = (x-3)^2 - 4$

Since $a = 1$ is positive, the parabola opens upward. The vertex of the parabola is $(h, k) = (3, -4)$ and the axis of symmetry is $x = 3$. Replace $f(x)$ with 0 to find x–intercepts.

$0 = (x-3)^2 - 4$

$4 = (x-3)^2$

Apply the square root property.

$x - 3 = -2$ and $x - 3 = 2$

$x = 1$ $\quad$ $x = 5$

The x–intercepts are 1 and 5.

Set $x = 0$ and solve for y to obtain the y–intercept.

$$y = (0-3)^2 - 4$$
$$y = (-3)^2 - 4$$
$$y = 9 - 4$$
$$y = 5$$

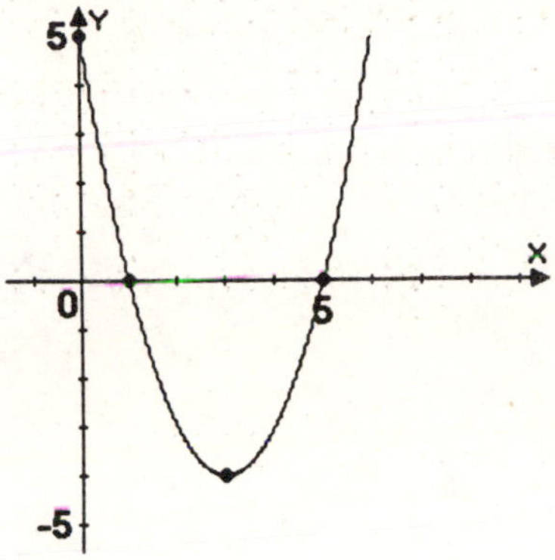

13. $\begin{vmatrix} 3 & 1 & 0 \\ 0 & 5 & -6 \\ -2 & -1 & 0 \end{vmatrix}$

$$= 3\begin{vmatrix} 5 & -6 \\ -1 & 0 \end{vmatrix} - 0\begin{vmatrix} 1 & 0 \\ -1 & 0 \end{vmatrix} + (-2)\begin{vmatrix} 1 & 0 \\ 5 & -6 \end{vmatrix}$$
$$= 3(5(0) - (-1)(-6)) + (-2)(1(-6) - 5(0))$$
$$= 3(-6) + (-2)(-6) = -18 + 12 = -6$$

14.

$$A = \frac{cd}{c+d}$$
$$A(c+d) = cd$$
$$Ac + Ad = cd$$
$$Ac - cd = -Ad$$
$$c(A-d) = -Ad$$
$$c = -\frac{Ad}{A-d} \text{ or } \frac{Ad}{d-A}$$

15. First, solve for y to obtain the slope of the line whose equation is $2x + y = 10.$

$$2x + y = 10$$
$$y = -2x + 10$$

The slope is –2. Since the line we are concerned with is perpendicular to this line, we know the slope will be $\frac{1}{2}$.

Using the point, $(-2, 4)$, and the slope, $\frac{1}{2}$, we can write the equation in point-slope form.

$$y - y_1 = m(x - x_1)$$
$$y - 4 = \frac{1}{2}(x - (-2))$$
$$y - 4 = \frac{1}{2}(x + 2)$$

Solve for y to obtain slope-intercept form.

$$y - 4 = \frac{1}{2}(x + 2)$$
$$y - 4 = \frac{1}{2}x + 1$$
$$y = \frac{1}{2}x + 5$$
$$f(x) = \frac{1}{2}x + 5$$

16. $\dfrac{-5x^3y^7}{15x^4y^{-2}} = \dfrac{-y^7y^2}{3x} = \dfrac{-y^9}{3x} = -\dfrac{y^9}{3x}$

17. $(4x^2 - 5y)^2$

$$= (4x^2)^2 + 2(4x^2)(-5y) + (-5y)^2$$
$$= 16x^4 - 40x^2y + 26y^2$$

18.

$$\begin{array}{r} x^2-5x+1 \\ 5x+1\overline{)5x^3-24x^2+0x+9} \\ \underline{5x^3+\ \ x^2} \qquad\qquad \\ -25x^2+0x \quad \\ \underline{-25x^2-5x} \quad \\ 5x+9 \\ \underline{5x+1} \\ 8 \end{array}$$

$$\frac{5x^3-24x^2+9}{5x+1}=x^2-5x+1+\frac{8}{5x+1}$$

19. $$\frac{\sqrt[3]{32xy^{10}}}{\sqrt[3]{2xy^2}}=\sqrt[3]{\frac{32xy^{10}}{2xy^2}}=\sqrt[3]{16y^8}$$
$$=\sqrt[3]{8\cdot 2y^6y^2}=2y^2\sqrt[3]{2y^2}$$

20. $$\frac{x+2}{x^2-6x+8}+\frac{3x-8}{x^2-5x+6}=\frac{x+2}{(x-4)(x-2)}+\frac{3x-8}{(x-2)(x-3)}$$
$$=\frac{(x+2)(x-3)}{(x-4)(x-2)(x-3)}+\frac{(3x-8)(x-4)}{(x-4)(x-2)(x-3)}$$
$$=\frac{(x+2)(x-3)+(3x-8)(x-4)}{(x-4)(x-2)(x-3)}$$
$$=\frac{x^2-3x+2x-6+3x^2-12x-8x+32}{(x-4)(x-2)(x-3)}$$
$$=\frac{4x^2-21x+26}{(x-4)(x-2)(x-3)}=\frac{(4x-13)\cancel{(x-2)}}{(x-4)\cancel{(x-2)}(x-3)}$$
$$=\frac{4x-13}{(x-4)(x-3)}$$

21. $x^4-4x^3+8x-32$
$=x^3(x-4)+8(x-4)$
$=(x-4)(x^3+8)$
$=(x-4)(x+2)(x^2-2x+4)$

22. $2x^2+12xy+18y^2$
$=2(x^2+6xy+9y^2)$
$=2(x+3y)^2$

23. Let x = the width of the carpet
Let $2x + 4$ = the length of the carpet

$$x(2x+4)=48$$
$$2x^2+4x=48$$
$$2x^2+4x-48=0$$
$$x^2+2x-24=0$$
$$(x+6)(x-4)=0$$

Apply the zero product principle.

$x+6=0$ and $x-4=0$

$x=-6$ $\quad x=4$

We disregard –6 because we can't have a negative length measurement. The width of the carpet is 4 feet and the length of the carpet is

$2x+4=2(4)+4=8+4=12$ feet.

24.

	Part Done in 1 Hour	Time Working Together	Part Done in x Hours
You	$\frac{1}{2}$	x	$\frac{x}{2}$
Your Sister	$\frac{1}{3}$	x	$\frac{x}{3}$

$$\frac{x}{2}+\frac{x}{3}=1$$
$$6\left(\frac{x}{2}+\frac{x}{3}\right)=6(1)$$
$$6\left(\frac{x}{2}\right)+6\left(\frac{x}{3}\right)=6$$
$$3x+2x=6$$
$$5x=6$$
$$x=\frac{6}{5}$$

If you and your sister work together, it will take $\frac{6}{5}$ hours or 1 hour and 12 minutes to clean the house.

25.

	d	r	$t=\frac{d}{r}$
Down Stream	20	$15+x$	$\frac{20}{15+x}$
Up Stream	10	$15-x$	$\frac{10}{15-x}$

$$\frac{20}{15+x}=\frac{10}{15-x}$$
$$20(15-x)=10(15+x)$$
$$300-20x=150+10x$$
$$300=150+30x$$
$$150=30x$$
$$5=x$$

The rate of the current is 5 miles per hour.

Chapter 9

Check Points 9.1

1. $f(60) = 13.49(0.967)^{60} - 1$
$= 13.49(0.1335329505) - 1$
≈ 0.8

Approximately one O-ring is expected to fail.

2. $f(x) = 3^x$

x	$f(x)$
-3	$3^{-3} = \frac{1}{3^3} = \frac{1}{27}$
-2	$3^{-2} = \frac{1}{3^2} = \frac{1}{9}$
-1	$3^{-1} = \frac{1}{3^1} = \frac{1}{3}$
0	$3^0 = 1$
1	$3^1 = 3$
2	$3^2 = 9$
3	$3^3 = 27$

3.

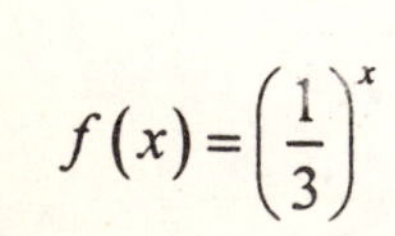

$f(x) = \left(\frac{1}{3}\right)^x$

x	$f(x) = \left(\frac{1}{3}\right)^x$ or 3^{-x}
-3	$3^{-(-3)} = 3^3 = 27$
-2	$3^{-(-2)} = 3^2 = 9$
-1	$3^{-(-1)} = 3^1 = 3$
0	$3^{-(0)} = 3^0 = 1$
1	$3^{-(1)} = 3^{-1} = \frac{1}{3}$
2	$3^{-(2)} = 3^{-2} = \frac{1}{3^2} = \frac{1}{9}$
3	$3^{-(3)} = 3^{-3} = \frac{1}{3^3} = \frac{1}{27}$

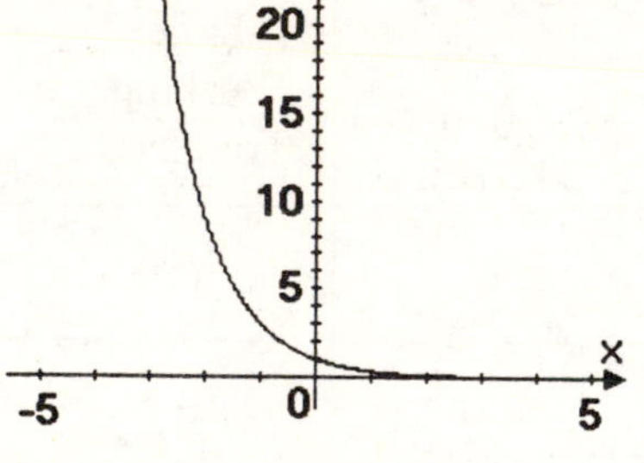

4. $f(x) = 3^x$ and $g(x) = 3^{x-1}$

x	$f(x)$	$g(x)$
-2	$\frac{1}{9}$	$\frac{1}{27}$
-1	$\frac{1}{3}$	$\frac{1}{9}$
0	1	$\frac{1}{3}$
1	3	1
2	9	3

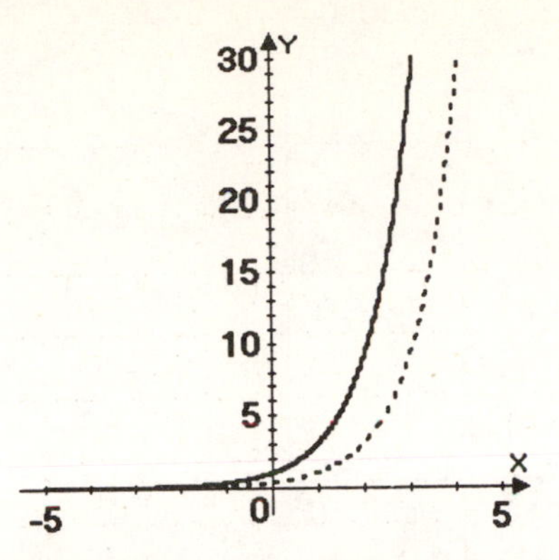

The graph of g is the graph of f shifted 1 unit to the right.

5. $f(x)=2^x$ and $g(x)=2^x+3$

x	$f(x)$	$g(x)$
-2	$\frac{1}{4}$	$3\frac{1}{4}$
-1	$\frac{1}{2}$	$3\frac{1}{2}$
0	1	4
1	2	5
2	4	7

The graph of g is the graph of f shifted up 3 units.

6. Since 2050 is 50 years after 2000, find $f(50)$.

$f(50)=6e^{0.013(50)}=6e^{0.65}\approx 11.49$

In 2050, the world population will be approximately 11.49 billion.

7. **a.**

$$A=10{,}000\left(1+\frac{0.08}{4}\right)^{4(5)}$$
$$=10{,}000(1+0.02)^{20}$$
$$=10{,}000(1.02)^{20}=14859.47$$

The balance in the account is \$14859.47 after 5 years of quarterly compounding.

b.

$$A=10{,}000e^{0.08(5)}$$
$$=10{,}000e^{0.4}=14918.25$$

The balance in the account is \$14918.25 after 5 years of continuous compounding.

Problem Set 9.1

Practice Exercises

1. $2^{3.4}=10.556$

3. $3^{\sqrt{5}}=11.665$

5. $4^{-1.5}=0.125$

7. $e^{2.3}=9.974$

9. $e^{-0.95}=0.387$

11. $f(x)=3^x$

x	$f(x)$
-2	$3^{-2}=\frac{1}{3^2}=\frac{1}{9}$
-1	$3^{-1}=\frac{1}{3^1}=\frac{1}{3}$
0	$3^0=1$
1	$3^1=3$
2	$3^2=9$

This functions matches graph (**d**).

13. $f(x) = 3^x - 1$

x	$f(x)$
-2	$3^{-2} - 1 = \frac{1}{3^2} - 1 = \frac{1}{9} - 1 = -\frac{8}{9}$
-1	$3^{-1} - 1 = \frac{1}{3^1} - 1 = \frac{1}{3} - 1 = -\frac{2}{3}$
0	$3^0 - 1 = 1 - 1 = 0$
1	$3^1 - 1 = 3 - 1 = 2$
2	$3^2 - 1 = 9 - 1 = 8$

This functions matches graph (**e**).

15. $f(x) = 3^{-x}$

x	$f(x)$
-2	$3^{-(-2)} = 3^2 = 9$
-1	$3^{-(-1)} = 3^1 = 3$
0	$3^{-(0)} = 3^0 = 1$
1	$3^{-(1)} = 3^{-1} = \frac{1}{3}$
2	$3^{-(2)} = 3^{-2} = \frac{1}{3^2} = \frac{1}{9}$

This functions matches graph (**f**).

17. $f(x) = 4^x$

x	$f(x)$
-2	$4^{-2} = \frac{1}{4^2} = \frac{1}{16}$
-1	$4^{-1} = \frac{1}{4^1} = \frac{1}{4}$
0	$4^0 = 1$
1	$4^1 = 4$
2	$4^2 = 16$

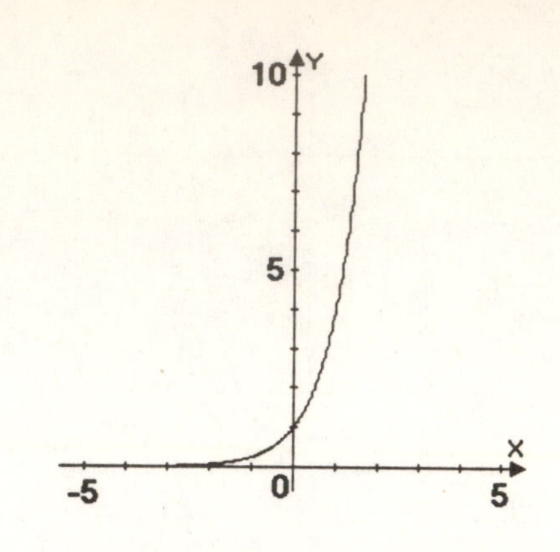

19. $g(x) = \left(\frac{3}{2}\right)^x$

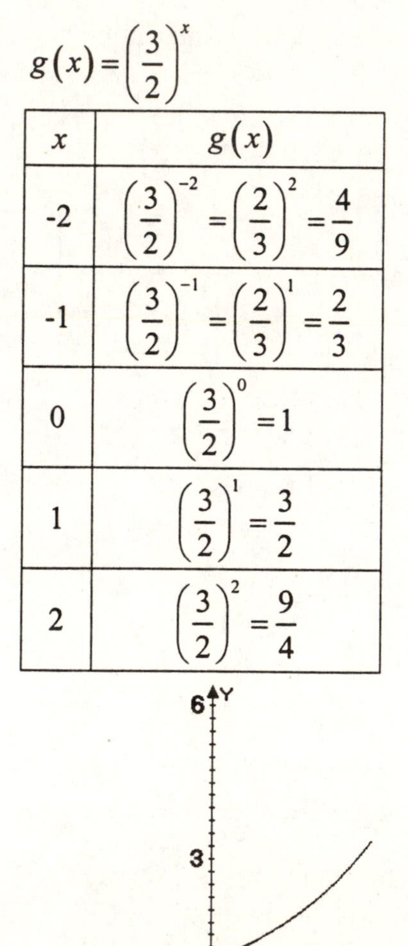

x	$g(x)$
-2	$\left(\frac{3}{2}\right)^{-2} = \left(\frac{2}{3}\right)^2 = \frac{4}{9}$
-1	$\left(\frac{3}{2}\right)^{-1} = \left(\frac{2}{3}\right)^1 = \frac{2}{3}$
0	$\left(\frac{3}{2}\right)^0 = 1$
1	$\left(\frac{3}{2}\right)^1 = \frac{3}{2}$
2	$\left(\frac{3}{2}\right)^2 = \frac{9}{4}$

21.

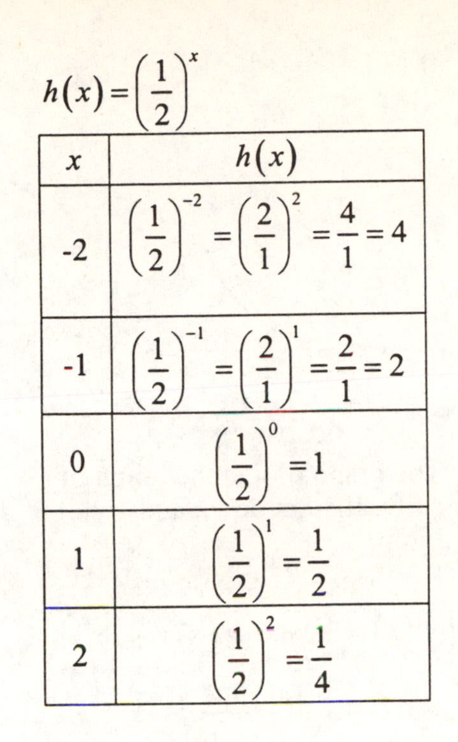

$h(x)=\left(\frac{1}{2}\right)^x$

x	$h(x)$
-2	$\left(\frac{1}{2}\right)^{-2}=\left(\frac{2}{1}\right)^2=\frac{4}{1}=4$
-1	$\left(\frac{1}{2}\right)^{-1}=\left(\frac{2}{1}\right)^1=\frac{2}{1}=2$
0	$\left(\frac{1}{2}\right)^0=1$
1	$\left(\frac{1}{2}\right)^1=\frac{1}{2}$
2	$\left(\frac{1}{2}\right)^2=\frac{1}{4}$

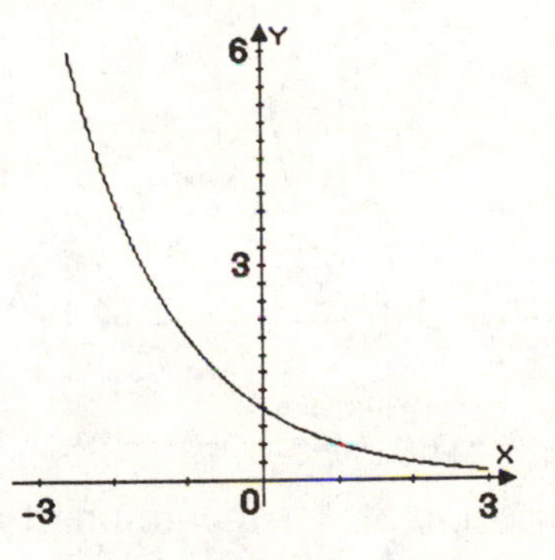

23.

$f(x)=(0.6)^x=\left(\frac{6}{10}\right)^x=\left(\frac{3}{5}\right)^x$

x	$f(x)$
-2	$\left(\frac{3}{5}\right)^{-2}=\left(\frac{5}{3}\right)^2=\frac{25}{9}$
-1	$\left(\frac{3}{5}\right)^{-1}=\left(\frac{5}{3}\right)^1=\frac{5}{3}$
0	$\left(\frac{3}{5}\right)^0=1$
1	$\left(\frac{3}{5}\right)^1=\frac{3}{5}$
2	$\left(\frac{3}{5}\right)^2=\frac{9}{25}$

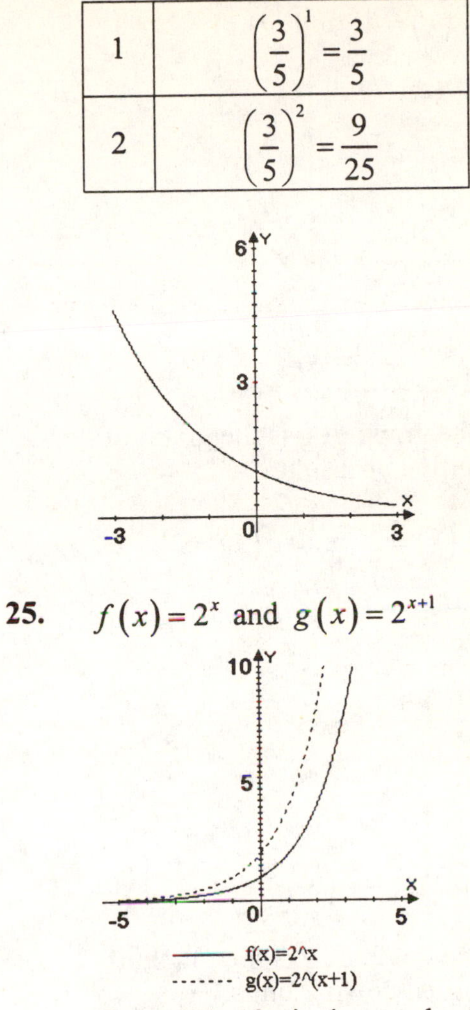

25. $f(x)=2^x$ and $g(x)=2^{x+1}$

The graph of g is the graph of f shifted 1 unit to the left.

27. $f(x)=2^x$ and $g(x)=2^{x-2}$

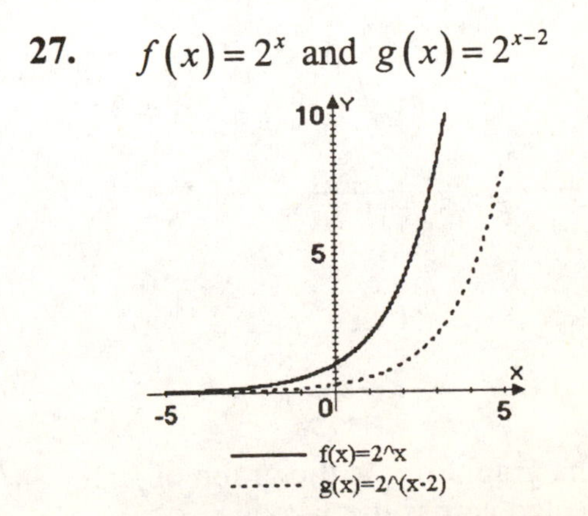

The graph of g is the graph of f shifted 2 units to the right.

29. $f(x)=2^x$ and $g(x)=2^x+1$

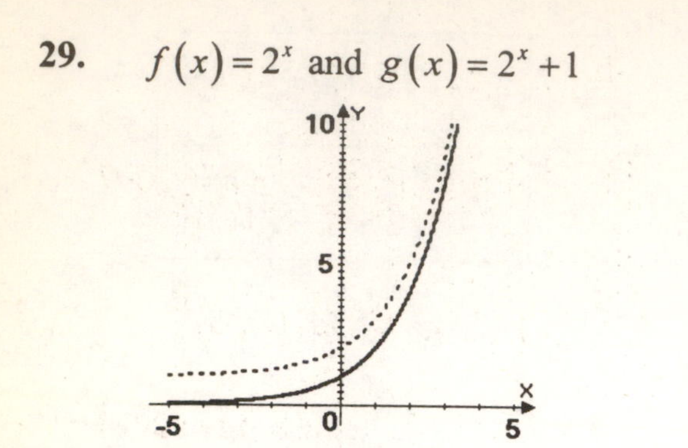

The graph of g is the graph of f shifted up 1 unit.

31. $f(x)=2^x$ and $g(x)=2^x-2$

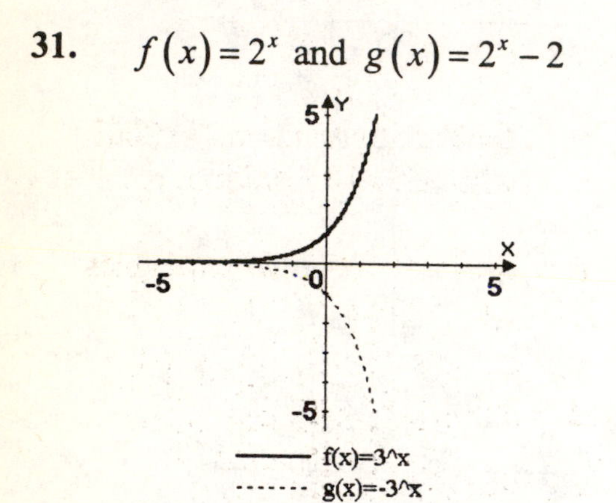

The graph of g is the graph of f shifted down 2 units.

33. $f(x)=3^x$ and $g(x)=-3^x$

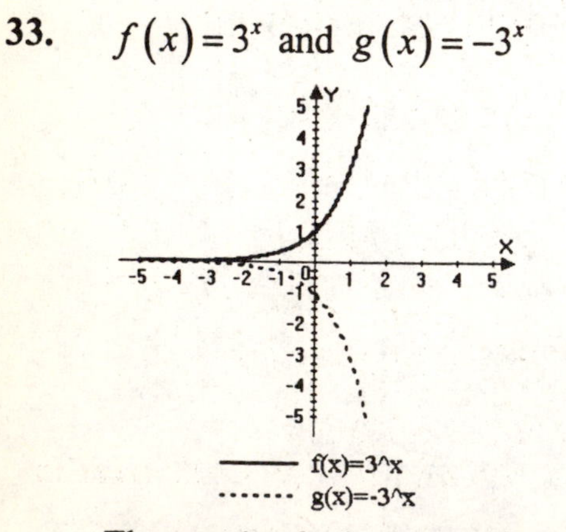

The graph of g is the graph of f reflected across the x–axis.

35. $f(x)=2^x$ and $g(x)=2^{x+1}-1$

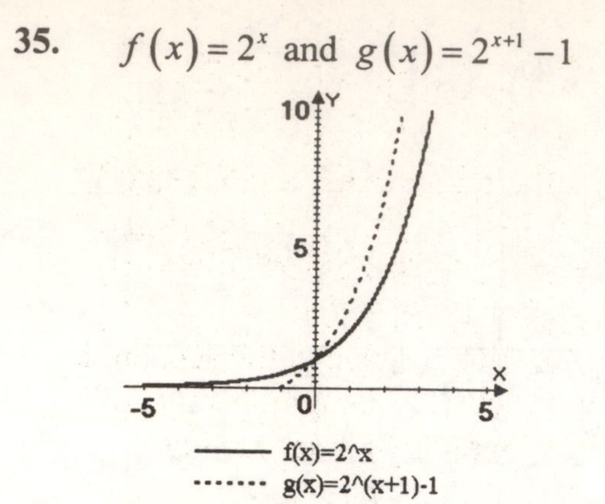

The graph of g is the graph of f shifted 1 unit down and 1 unit to the left.

37. $f(x)=3^x$ and $g(x)=\frac{1}{3}\cdot 3^x$

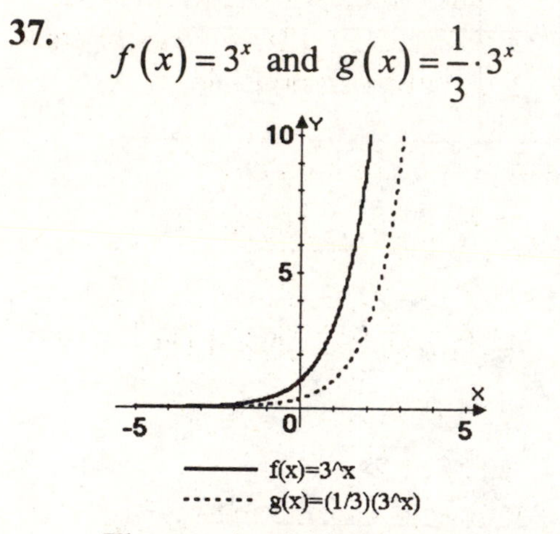

The graph of g is the graph of f compressed vertically by a factor of $\frac{1}{3}$.

39. **a.**

$$\begin{aligned}A&=10,000\left(1+\frac{0.055}{2}\right)^{2(5)}\\&=10,000(1+0.0275)^{10}\\&=10,000(1.0275)^{10}\\&=13116.51\end{aligned}$$

The balance in the account is \$13,116.51 after 5 years of semiannual compounding.

b.
$$A = 10,000\left(1+\frac{0.055}{12}\right)^{12(5)}$$
$$= 10,000(1+0.0045833)^{60}$$
$$= 10,000(1.0045833)^{60}$$
$$= 13157.04$$
The balance in the account is $13,157.04 after 5 years of monthly compounding.

c. $A = Pe^{rt} = 10,000e^{0.055(5)}$
$$= 10,000e^{0.275} = 13165.31$$
The balance in the account is $13,165.31 after 5 years of continuous compounding.

41. Monthly Compounding
$$A = 12,000\left(1+\frac{0.07}{12}\right)^{12(3)}$$
$$= 10,000(1+0.0058333)^{36}$$
$$= 10,000(1.0058333)^{36}$$
$$= 12329.24$$
Continuous Compounding
$$A = 12,000e^{0.0685(3)}$$
$$= 10,000e^{0.2055} = 12281.39$$
Monthly compounding at 7% yields the greatest return.

Application Exercises

43. **a.** $f(0) = 67.38(1.026)^0$
$$= 67.38(1) = 67.38$$
Mexico's population in 1980 was 67.38 million.

b. $f(27) = 67.38(1.026)^{27}$
$$\approx 134.74$$
Mexico's population in 2007 will be 134.74 million.

c. $f(54) = 67.38(1.026)^{54}$
$$\approx 269.46$$
Mexico's population in 2034 will be 269.46 million.

d. Since 2061 – 1980 is 81, find $f(81)$.
$$f(81) = 67.38(1.026)^{81}$$
$$\approx 538.85$$
Mexico's population in 2061 will be 538.85 million.

e. Mexico's population doubles every 27 years.

45. $S = 65,000(1+0.06)^{10}$
$$S = 65,000(1.06)^{10}$$
$$S = 116,405.10$$
In 10 years, the house will be worth $116,405.10.

47. $2^{1.7} = 3.249009585$
$$2^{1.73} = 3.317278183$$
$$2^{1.732} = 3.321880096$$
$$2^{1.73205} = 3.321995226$$
$$2^{1.7320508} = 3.321997068$$
$$2^{\sqrt{3}} = 3.321997085$$
As the number of decimal places in the approximation of $\sqrt{3}$ in the exponent increases, the value of the expression approaches $2^{\sqrt{3}}$.

49. $f(11) = 24000e^{0.21(11)} = 24000e^{2.31}$
$= 241786.1917$
Eleven years after 1989, in the year 2000, there were approximately 241,786 cases of AIDS in the US among intravenous drug users.

51. **a.** $f(0) = 80e^{-0.5(0)} + 20$
$= 80e^{0} + 20 = 80(1) + 20$
$= 80 + 20 = 100$
100% of information is remembered at the moment it is first learned.

b. $f(1) = 80e^{-0.5(1)} + 20$
$= 80e^{-0.5} + 20$
$= 48.522 + 20 = 68.522$
Approximately 68.5% of information is remembered after one week.

c. $f(4) = 80e^{-0.5(4)} + 20$
$= 80e^{-2} + 20$
$= 10.827 + 20 = 30.827$
Approximately 30.8% of information is remembered after four weeks.

d. $f(52) = 80e^{-0.5(52)} + 20$
$= 80e^{-26} + 20$
$= \left(4.087 \times 10^{-10}\right) + 20$
≈ 20
Approximately 20% of information is remembered after one year. (4.087×10^{-10} will be eliminated in rounding.)

53. $f(30) = \dfrac{90}{1 + 270e^{-0.122(30)}}$
$= \dfrac{90}{1 + 270e^{-3.66}}$
$= \dfrac{90}{1 + 6.948} = \dfrac{90}{7.948} \approx 11.3$
Approximately 11.3% of 30-year-olds have some coronary heart disease.

55. **a.** $N(0) = \dfrac{30,000}{1 + 20e^{-1.5(0)}} = \dfrac{30,000}{1 + 20e^{0}}$
$= \dfrac{30,000}{1 + 20(1)} = \dfrac{30,000}{1 + 20}$
$= \dfrac{30,000}{21} \approx 1428.6$
Approximately 1429 people became ill with the flu when the epidemic began.

b. $N(3) = \dfrac{30,000}{1 + 20e^{-1.5(3)}} = \dfrac{30,000}{1 + 20e^{-4.5}}$
$= \dfrac{30,000}{1 + .22218} = \dfrac{30,000}{1.22218}$
≈ 24546
Approximately 24,546 people became ill with the flu by the end of the third week.

c. The epidemic cannot grow indefinitely because there are a limited number of people that can become ill. Because there are 30,000 people in the town, the limit is 30,000.

Writing in Mathematics

57. Answers will vary.

59. Answers will vary.

61. Answers will vary.

Technology Exercises

63.

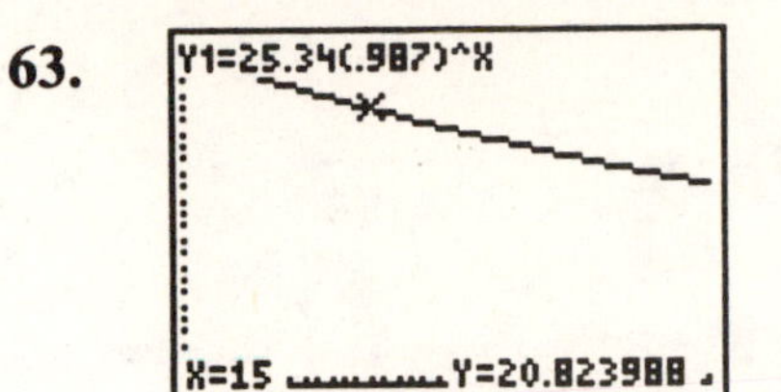

The student teacher ratio became less than 21 students per teacher 15 years after 1959 in the year 1974.

65.

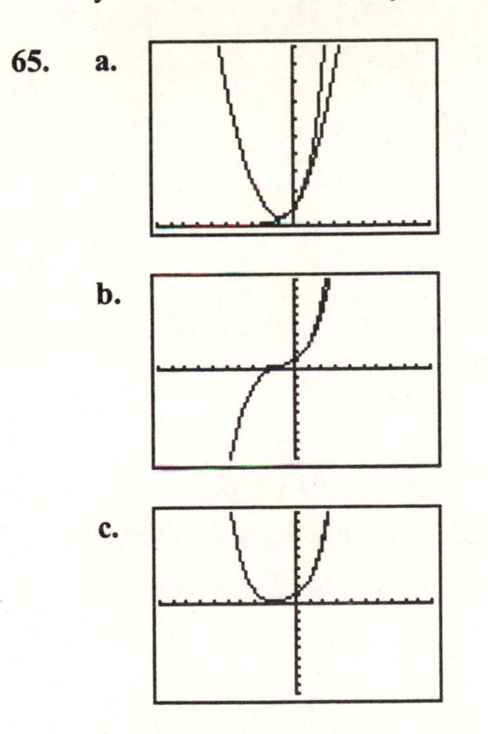

d. Answers will vary. The graph of $y = 1 + x + \frac{x^2}{2} + \ldots$ is approaching the graph of $y = e^x$.

Critical Thinking Exercises

67.

Graph **(a)** is $y = \left(\frac{1}{3}\right)^x$.

Graph **(b)** is $y = \left(\frac{1}{5}\right)^x$.

Graph **(c)** is $y = 5^x$.

Graph **(d)** is $y = 3^x$.

Answers will vary. A base between 0 and 1 will rise to the left and a base greater than 1 will rise to the right.

Review Exercises

69.

$$D = \frac{ab}{a+b}$$
$$D(a+b) = ab$$
$$Da + Db = ab$$
$$Da = ab - Db$$
$$Da = (a-D)b$$
$$b = \frac{Da}{a-D}$$

70.

$$\begin{vmatrix} 3 & -2 \\ 7 & -5 \end{vmatrix} = 3(-5) - 7(-2)$$
$$= -15 + 14 = -1$$

71.

$$x(x-3) = 10$$
$$x^2 - 3x = 10$$
$$x^2 - 3x - 10 = 0$$
$$(x-5)(x+2) = 0$$

Apply the zero product principle.

$x - 5 = 0$ or $x + 2 = 0$

$x = 5$ $\qquad$ $x = -2$

The solutions are 5 and –2 and the solution set is $\{-2,5\}$.

Check Points 9.2

1. **a.** $(f \circ g)(x)$
$= f(g(x)) = f(x^2-1)$
$= 5(x^2-1)+6 = 5x^2-5+6$
$= 5x^2+1$

b. $(g \circ f)(x) = g(f(x))$
$= g(5x+6)$
$= (5x+6)^2 - 1$
$= 25x^2 + 60x + 36 - 1$
$= 25x^2 + 60x + 35$

2.
$$f(g(x)) = f\left(\frac{x}{7}\right) = 7\left(\frac{x}{7}\right) = x$$
$$g(f(x)) = g(7x) = \frac{7x}{7} = x$$
The functions are inverses.

3.
$$f(g(x)) = f\left(\frac{x+7}{4}\right) = 4\left(\frac{x+7}{4}\right) - 7$$
$$= (x+7) - 7 = x + 7 - 7 = x$$
$$g(f(x)) = g(4x-7) = \frac{(4x-7)+7}{4}$$
$$= \frac{4x-7+7}{4} = \frac{4x}{4} = x$$
The functions are inverses.

4. $f(x) = 2x+7$
$y = 2x+7$
Interchange x and y and solve for y.
$x = 2y+7$
$x - 7 = 2y$
$\frac{x-7}{2} = y$
$f^{-1}(x) = \frac{x-7}{2}$

5. $f(x) = 4x^3 - 1$
$y = 4x^3 - 1$
Interchange x and y and solve for y.
$x = 4y^3 - 1$
$x + 1 = 4y^3$
$\frac{x+1}{4} = y^3$
$\sqrt[3]{\frac{x+1}{4}} = y$

6. Graphs (**b**) and (**c**) satisfy the horizontal line test and have inverse functions.

7. Since the points $(-2,1),(0,0)$ and $(1,-1)$ lie on the graph of the function, the points $(1,-2),(0,0)$ and $(-1,1)$ lie on the inverse function.

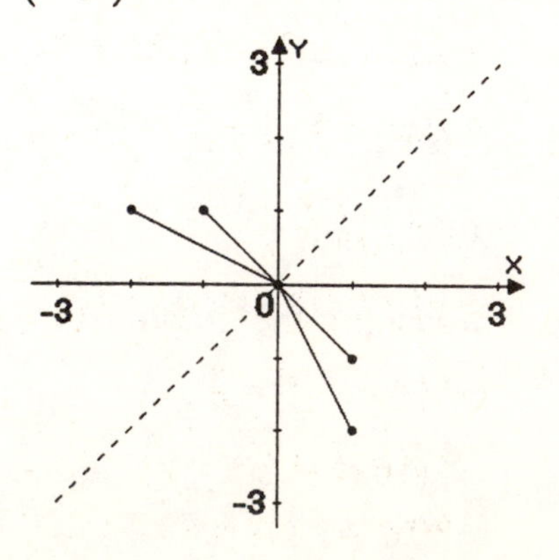

Problem Set 9.2

Practice Exercises

1. **a.** $(f \circ g)(x)$
$= f(g(x)) = f(x+7)$
$= 2(x+7) = 2x+14$

b. $(g \circ f)(x) = g(f(x)) = g(2x)$
$= 2x+7$

c. $(f \circ g)(2) = 2(2)+14$
$= 4+14 = 18$

3. **a.** $(f \circ g)(x)$
$= f(g(x)) = f(2x+1)$
$= (2x+1)+4 = 2x+5$

b. $(g \circ f)(x)$
$= g(f(x)) = g(x+4)$
$= 2(x+4)+1 = 2x+8+1 = 2x+9$

c. $(f \circ g)(2) = 2(2)+5 = 4+5 = 9$

5. **a.** $(f \circ g)(x) = f(g(x))$
$= f(5x^2-2) = 4(5x^2-2)-3$
$= 20x^2-8-3 = 20x^2-11$

b. $(g \circ f)(x)$
$= g(f(x)) = g(4x-3)$
$= 5(4x-3)^2-2$
$= 5(16x^2-24x+9)-2$
$= 80x^2-120x+45-2$
$= 80x^2-120x+43$

c. $(f \circ g)(2) = 20(2)^2-11$
$= 20(4)-11$
$= 80-11 = 69$

7. **a.** $(f \circ g)(x)$
$= f(g(x)) = f(x^2-2)$
$= (x^2-2)^2+2 = x^4-4x^2+4+2$
$= x^4-4x^2+6$

b. $(g \circ f)(x)$
$= g(f(x)) = g(x^2+2)$
$= (x^2+2)^2-2 = x^4+4x^2+4-2$
$= x^4+4x^2+2$

c. $(f \circ g)(2) = 2^4-4(2)^2+6$
$= 16-4(4)+6$
$= 16-16+6 = 6$

9. **a.** $(f \circ g)(x) = f(g(x))$
$= f(x-1) = \sqrt{x-1}$

b. $(g \circ f)(x) = g(f(x)) = g(\sqrt{x})$
$= \sqrt{x}-1$

c. $(f \circ g)(2) = \sqrt{2-1} = \sqrt{1} = 1$

11. **a.** $(f \circ g)(x)$
$= f(g(x)) = f\left(\frac{x+3}{2}\right)$
$= 2\left(\frac{x+3}{2}\right)-3 = x+3-3 = x$

b. $(g \circ f)(x) = g(f(x))$
$= g(2x-3)$
$= \dfrac{(2x-3)+3}{2}$
$= \dfrac{2x-3+3}{2}$
$= \dfrac{2x}{2} = x$

c. $(f \circ g)(2) = 2$

13. **a.** $(f \circ g)(x) = f(g(x)) = f\left(\dfrac{1}{x}\right)$
$= \dfrac{1}{\frac{1}{x}} = 1 \div \dfrac{1}{x} = 1 \cdot \dfrac{x}{1} = x$

b. $(g \circ f)(x) = g(f(x)) = g\left(\dfrac{1}{x}\right)$
$= \dfrac{1}{\frac{1}{x}} = 1 \div \dfrac{1}{x} = 1 \cdot \dfrac{x}{1} = x$

c. $(f \circ g)(2) = 2$

15. $f(g(x)) = f\left(\dfrac{x}{4}\right) = 4\left(\dfrac{x}{4}\right) = x$
$g(f(x)) = g(4x) = \dfrac{4x}{4} = x$
The functions are inverses.

17. $f(g(x)) = f\left(\dfrac{x-8}{3}\right) = 3\left(\dfrac{x-8}{3}\right) + 8$
$= x - 8 + 8 = x$

$g(f(x)) = g(3x+8) = \dfrac{(3x+8)-8}{3}$
$= \dfrac{3x+8-8}{3} = \dfrac{3x}{3} = x$
The functions are inverses.

19. $f(g(x)) = f\left(\dfrac{x+5}{9}\right) = 5\left(\dfrac{x+5}{9}\right) - 9$
$= \dfrac{5x+25}{9} - \dfrac{81}{9}$
$= \dfrac{5x+25-81}{9}$
$= \dfrac{5x-56}{9}$
Since $f(g(x)) \neq x$, we conclude the functions are not inverses.

21. $f(g(x)) = f\left(\dfrac{3}{x} + 4\right) = \dfrac{3}{\left(\frac{3}{x}+4\right) - 4}$
$= \dfrac{3}{\frac{3}{x}+4-4} = \dfrac{3}{\frac{3}{x}} = 3 \div \dfrac{3}{x}$
$= 3 \cdot \dfrac{x}{3} = x$
$g(f(x)) = g\left(\dfrac{3}{x-4}\right) = \dfrac{3}{\frac{3}{x-4}} + 4$
$= 3 \div \dfrac{3}{x-4} + 4 = 3 \cdot \dfrac{x-4}{3} + 4$
$= x - 4 + 4 = x$
The functions are inverses.

23. $f(g(x)) = f(-x) = -(-x) = x$
$g(f(x)) = g(-x) = -(-x) = x$
The functions are inverses.

25. **a.** $f(x) = x + 3$

$y = x + 3$

Interchange x and y and solve for y.

$x = y + 3$

$x - 3 = y$

$f^{-1}(x) = x - 3$

b. $f(f^{-1}(x)) = f(x-3)$

$= (x-3) + 3$

$= x - 3 + 3 = x$

$f^{-1}(f(x)) = f(x+3)$

$= (x+3) - 3$

$= x + 3 - 3 = x$

27. **a.** $f(x) = 2x$

$y = 2x$

Interchange x and y and solve for y.

$x = 2y$

$\frac{x}{2} = y$

$f^{-1}(x) = \frac{x}{2}$

b. $f(f^{-1}(x)) = f\left(\frac{x}{2}\right) = 2\left(\frac{x}{2}\right) = x$

$f^{-1}(f(x)) = f(2x) = \frac{2x}{2} = x$

29. **a.** $f(x) = 2x + 3$

$y = 2x + 3$

Interchange x and y and solve for y.

$x = 2y + 3$

$x - 3 = 2y$

$\frac{x-3}{2} = y$

$f^{-1}(x) = \frac{x-3}{2}$

b. $f(f^{-1}(x)) = f\left(\frac{x-3}{2}\right)$

$= 2\left(\frac{x-3}{2}\right) + 3$

$= x - 3 + 3 = x$

$f^{-1}(f(x)) = f^{-1}(2x+3)$

$= \frac{(2x+3)-3}{2}$

$= \frac{2x+3-3}{2} = \frac{2x}{2} = x$

31. **a.** $f(x) = x^3 + 2$

$y = x^3 + 2$

Interchange x and y and solve for y.

$x = y^3 + 2$

$x - 2 = y^3$

$\sqrt[3]{x-2} = y$

$f^{-1}(x) = \sqrt[3]{x-2}$

b. $f(f^{-1}(x)) = f\left(\sqrt[3]{x-2}\right)$

$= \left(\sqrt[3]{x-2}\right)^3 + 2$

$= x - 2 + 2 = x$

$f^{-1}(f(x)) = f^{-1}(x^3 + 2)$

$= \sqrt[3]{(x^3+2) - 2}$

$$= \sqrt[3]{x^3 + 2 - 2}$$
$$= \sqrt[3]{x^3} = x$$

33. **a.** $f(x) = (x+2)^3$
$$y = (x+2)^3$$
Interchange x and y and solve for y.
$$x = (y+2)^3$$
$$\sqrt[3]{x} = \sqrt[3]{(y+2)^3}$$
$$\sqrt[3]{x} = y+2$$
$$\sqrt[3]{x} - 2 = y$$
$$f^{-1}(x) = \sqrt[3]{x} - 2$$

b. $f(f^{-1}(x)) = f(\sqrt[3]{x} - 2)$
$$= ((\sqrt[3]{x} - 2) + 2)^3$$
$$= (\sqrt[3]{x} - 2 + 2)^3$$
$$= (\sqrt[3]{x})^3 = x$$
$$f^{-1}(f(x)) = f^{-1}((x+2)^3)$$
$$= \sqrt[3]{(x+2)^3} - 2$$
$$= x + 2 - 2 = x$$

35. **a.** $f(x) = \frac{1}{x}$
$$y = \frac{1}{x}$$
Interchange x and y and solve for y.
$$x = \frac{1}{y}$$
$$xy = 1$$
$$y = \frac{1}{x}$$
$$f^{-1}(x) = \frac{1}{x}$$

b. $f(f^{-1}(x)) = f\left(\frac{1}{x}\right) = \frac{1}{\frac{1}{x}}$
$$= 1 \div \frac{1}{x} = 1 \cdot \frac{x}{1} = x$$
$$f^{-1}(f(x)) = f^{-1}\left(\frac{1}{x}\right) = \frac{1}{\frac{1}{x}}$$
$$= 1 \div \frac{1}{x} = 1 \cdot \frac{x}{1} = x$$

37. **a.** $f(x) = \sqrt{x}$
$$y = \sqrt{x}$$
Interchange x and y and solve for y.
$$x = \sqrt{y}$$
$$x^2 = y$$
$$f^{-1}(x) = x^2$$

b. $f(f^{-1}(x)) = f(x^2) = \sqrt{x^2} = x$
$$f^{-1}(f(x)) = f^{-1}(\sqrt{x})$$
$$= (\sqrt{x})^2 = x$$

39. **a.** $f(x) = x^2 + 1$
$$y = x^2 + 1$$
Interchange x and y and solve for y.
$$x = y^2 + 1$$
$$x - 1 = y^2$$

$$\sqrt{x-1} = y$$
$$f^{-1}(x) = \sqrt{x-1}$$

b. $$f\left(f^{-1}(x)\right) = f\left(\sqrt{x-1}\right) = \left(\sqrt{x-1}\right)^2 + 1 = x - 1 + 1 = x$$

$$f^{-1}\left(f(x)\right) = f^{-1}\left(x^2+1\right) = \sqrt{\left(x^2+1\right)-1} = \sqrt{x^2+1-1} = \sqrt{x^2} = x$$

41. **a.** $$f(x) = \frac{2x+1}{x-3}$$
$$y = \frac{2x+1}{x-3}$$

Interchange x and y and solve for y.

$$x = \frac{2y+1}{y-3}$$
$$x(y-3) = 2y+1$$
$$xy - 3x = 2y + 1$$
$$xy - 2y = 3x + 1$$
$$(x-2)y = 3x+1$$
$$y = \frac{3x+1}{x-2}$$
$$f^{-1}(x) = \frac{3x+1}{x-2}$$

b. $$f\left(f^{-1}(x)\right)$$

$$= f\left(\frac{3x+1}{x-2}\right) = \frac{2\left(\frac{3x+1}{x-2}\right)+1}{\left(\frac{3x+1}{x-2}\right)-3}$$

$$= \frac{x-2}{x-2} \cdot \frac{2\left(\frac{3x+1}{x-2}\right)+1}{\left(\frac{3x+1}{x-2}\right)-3}$$

$$= \frac{2(3x+1)+1(x-2)}{(3x+1)-3(x-2)}$$

$$= \frac{6x+2+x-2}{3x+1-3x+6} = \frac{7x}{7} = x$$

$$f^{-1}\left(f(x)\right)$$

$$= f^{-1}\left(\frac{2x+1}{x-3}\right) = \frac{3\left(\frac{2x+1}{x-3}\right)+1}{\left(\frac{2x+1}{x-3}\right)-2}$$

$$= \frac{x-3}{x-3} \cdot \frac{3\left(\frac{2x+1}{x-3}\right)+1}{\left(\frac{2x+1}{x-3}\right)-2}$$

$$= \frac{3(2x+1)+1(x-3)}{(2x+1)-2(x-3)}$$

$$= \frac{6x+3+x-3}{2x+1-2x+6} = \frac{7x}{7} = x$$

43. **a.** $$f(x) = \sqrt[3]{x-4} + 3$$
$$y = \sqrt[3]{x-4} + 3$$

Interchange x and y and solve for y.

$$x = \sqrt[3]{y-4} + 3$$
$$x - 3 = \sqrt[3]{y-4}$$
$$(x-3)^3 = y - 4$$
$$(x-3)^3 + 4 = y$$
$$f^{-1}(x) = (x-3)^3 + 4$$

b. $f\left(f^{-1}(x)\right)$

$= f\left((x-3)^3+4\right)$

$= \sqrt[3]{\left((x-3)^3+4\right)-4}+3$

$= \sqrt[3]{(x-3)^3+4-4}+3$

$= \sqrt[3]{(x-3)^3}+3 = x-3+3 = x$

$f^{-1}\left(f(x)\right)$

$= f^{-1}\left(\sqrt[3]{x-4}+3\right)$

$= \left(\left(\sqrt[3]{x-4}+3\right)-3\right)^3+4$

$= \left(\sqrt[3]{x-4}+3-3\right)^3+4$

$= \left(\sqrt[3]{x-4}\right)^3+4 = x-4+4 = x$

45. The graph does not satisfy the horizontal line test so the function does not have an inverse.

47. The graph does not satisfy the horizontal line test so the function does not have an inverse.

49. The graph satisfies the horizontal line test so the function has an inverse.

51.

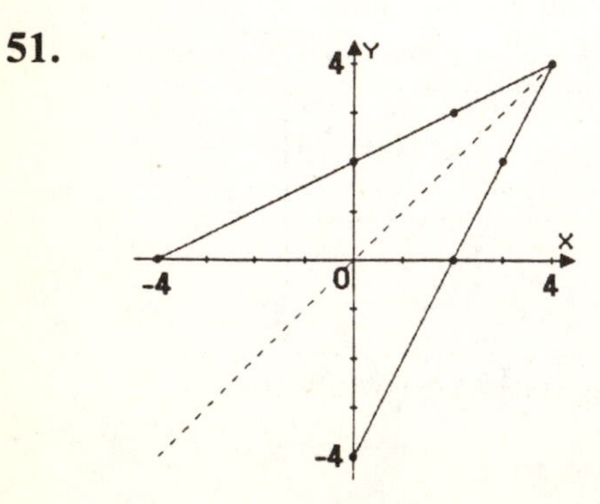

53.

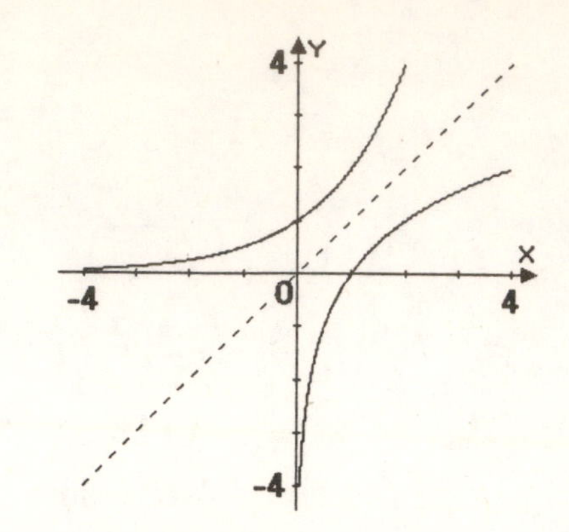

Application Exercises

55. **a.** f is the regular price of the computer less \$400.
g is 75% of the regular price of the computer.

b. $(f \circ g)(x) = f\left(g(x)\right)$
$= f(0.75x)$
$= 0.75x - 400$

$f \circ g$ represents 75% f the regular price of the computer less \$400.

c. $(g \circ f)(x) = g\left(f(x)\right)$
$= g(x-400)$
$= 0.75(x-400)$
$= 0.75x - 300$

$g \circ f$ represents 75% f the regular price of the computer less \$300.

d. $f \circ g$ models the greater discount. It is a savings of \$100 over $g \circ f$.

e. $f(x) = x - 400$

$y = x - 400$

Interchange x and y and solve for y.

$$x = y - 400$$
$$x + 400 = y$$
$$f^{-1}(x) = x + 400$$

f^{-1} models the regular price of the computer plus \$400.

57. **a.** We know that f has an inverse because it satisfies the horizontal line test.

b. $f^{-1}(0.25)$ or 15 represents the number of people who must be in a room so that the probability of 2 sharing a birthday would be 25%.
$f^{-1}(0.5)$ or 23 represents the number of people who must be in a room so that the probability of 2 sharing a birthday would be 50%. $f^{-1}(0.7)$ or 30 represents the number of people who must be in a room so that the probability of 2 sharing a birthday would be 70%.

59. Because the graph does not satisfy the vertical line test, it does not have an inverse function. This means that we cannot use age at first marriage to predict year.

Writing in Mathematics

61. Answers will vary

63. Answers will vary

65. Answers will vary

67. Answers will vary

Technology Exercises

69. $f(x) = \sqrt[3]{2 - x}$

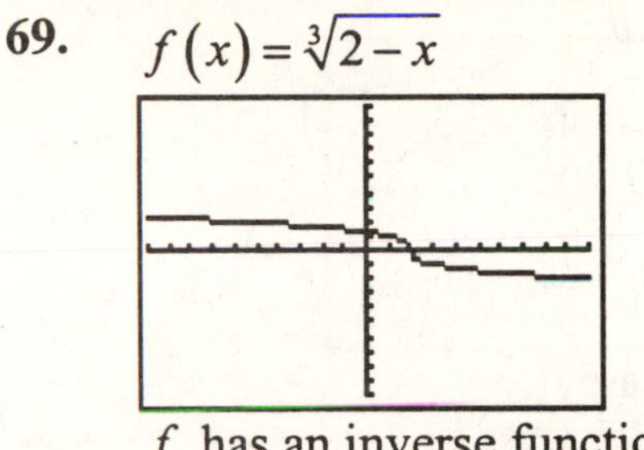

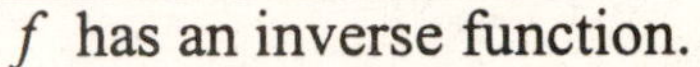

f has an inverse function.

71. $f(x) = \dfrac{x^4}{4}$

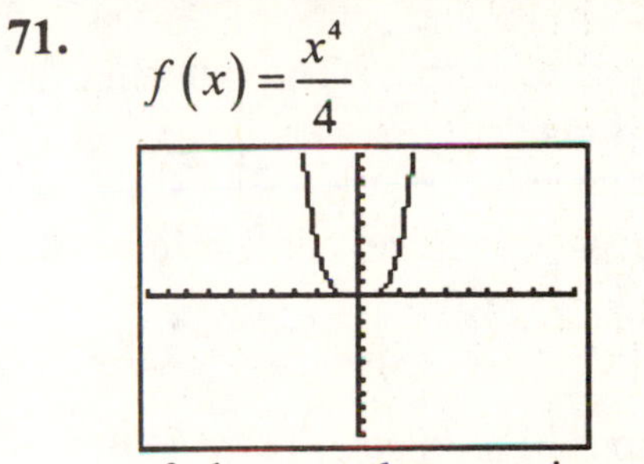

f does not have an inverse function.

73. $f(x) = (x - 1)^3$

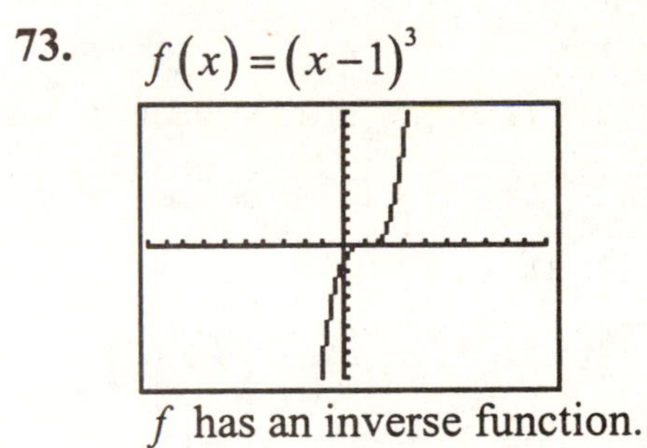

f has an inverse function.

75. $f(x) = x^3 + x + 1$

f has an inverse function.

77. $f(x)=\frac{1}{x}+2$

$g(x)=\frac{1}{x-2}$

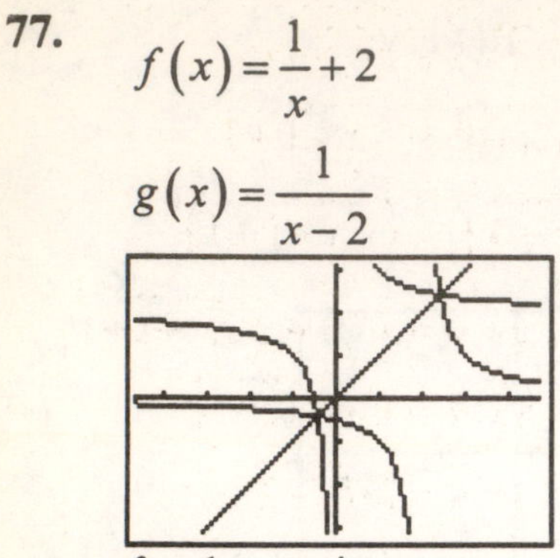

f and g are inverses.

Critical Thinking Exercises

79. Statement **d.** is true. The domain of f is the range of f^{-1}.

Statement **a.** is false. The inverse is $\{(4,1),(7,2)\}$.

Statement **b.** is false. $f(x) = 5$ does not satisfy the horizontal line test.

Statement **c.** is false. $f^{-1}=\frac{1}{3}x$

81. To find $(f\circ g)^{-1}(x)$, first find $(f\circ g)(x)$.

$$(f\circ g)(x)=f(g(x))=f(x+5)$$
$$=3(x+5)=3x+15$$

Now find $(f\circ g)^{-1}(x)$.

$$(f\circ g)(x)=3x+15$$
$$y=3x+15$$

Interchange x and y and solve for y.

$$x=3y+15$$
$$x-15=3y$$
$$\frac{x-15}{3}=y$$
$$(f\circ g)^{-1}(x)=\frac{x-15}{3}$$

To find $(g^{-1}\circ f^{-1})(x)$, find $g^{-1}(x)$ and $f^{-1}(x)$.

$$g(x)=x+5 \qquad f(x)=3x$$
$$y=x+5 \qquad y=3x$$

Interchange x and y and solve for y.

$$x=y+5 \qquad x=3y$$
$$x-5=y \qquad \frac{x}{3}=y$$
$$g^{-1}(x)=x-5 \qquad f^{-1}(x)=\frac{x}{3}$$

Now find $(g^{-1}\circ f^{-1})(x)$.

$$(g^{-1}\circ f^{-1})(x)=g^{-1}(f^{-1}(x))$$
$$=g^{-1}\left(\frac{x}{3}\right)=\frac{x}{3}-5$$
$$=\frac{x}{3}-\frac{15}{3}=\frac{x-15}{3}$$

Notice that $(f\circ g)^{-1}(x)=(g^{-1}\circ f^{-1})(x)$.

83. $f(x)=m_1x+b_1$

$g(x)=m_2x+b_2$

First find $(f\circ g)(x)$.

$$(f\circ g)(x)=f(g(x))=f(m_2x+b_2)$$
$$=m_1(m_2x+b_2)+b_1$$
$$=m_1m_2x+m_1b_2+b_1$$

The slope of the composite function is m_1m_2. The slope of f is m_1 and the slope of g is m_2, thus the product m_1m_2 is the same as the slope of the composite function.

Review Exercises

84. $\dfrac{4.3\times10^5}{8.6\times10^{-4}}=\dfrac{4.3}{8.6}\times\dfrac{10^5}{10^{-4}}=0.5\times10^9$
$=5\times10^{-1}\times10^9=5\times10^8$

85. $f(x)=x^2-4x+3$
Since $a=1$ is positive, the parabola opens upward. The x–coordinate of the vertex of the parabola is
$-\dfrac{b}{2a}=-\dfrac{-4}{2(1)}=-\dfrac{-4}{2}=2$ and the y–coordinate of the vertex of the parabola is
$f\left(-\dfrac{b}{2a}\right)=f(2)=2^2-4(2)+3$
$=4-8+3=-1.$
The vertex is at (2, – 1). Replace $f(x)$ with 0 to find x–intercepts.
$0=x^2-4x+3$
$0=(x-3)(x-1)$
Apply the zero product principle.
$x-3=0$ or $x-1=0$
$x=3$ $\quad$ $x=1$
The x–intercepts are 1 and 3. Set $x=0$ and solve for y to obtain the y–intercept.
$y=0^2-4(0)+3$
$y=0-0+3=3$

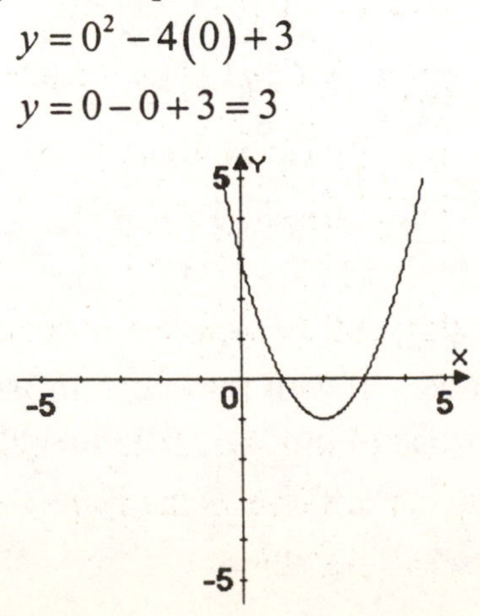

86. $\sqrt{x+4}-\sqrt{x-1}=1$
$\sqrt{x+4}=\sqrt{x-1}+1$
$\left(\sqrt{x+4}\right)^2=\left(\sqrt{x-1}+1\right)^2$
$x+4=x-1+2\sqrt{x-1}+1$
$x+4=x+2\sqrt{x-1}$
$4=2\sqrt{x-1}$
$2=\sqrt{x-1}$
$2^2=\left(\sqrt{x-1}\right)^2$
$4=x-1$
$5=x$
The solution is 5 and the solution set is $\{5\}$.

Check Points 9.3

1. **a.** $7^3=x$

b. $b^2=25$

c. $4^y=26$

2. **a.** $5=\log_2 x$

b. $3=\log_b 27$

c. $y=\log_e 33$

3. **a.** $\log_{10}100=2$ because $10^2=100$

b. $\log_3 3=1$ because $3^1=3$

c. $\log_{36}6=\dfrac{1}{2}$ because $36^{\frac{1}{2}}=6$

4. **a.** $\log_9 9 = 1$ because $\log_b b = 1$

b. $\log_8 1 = 0$ because $\log_b 1 = 0$

5. **a.** $\log_7 7^8 = 8$ because $\log_b b^x = x$

b. $3^{\log_3 17} = 17$ because $b^{\log_b x} = x$

6.

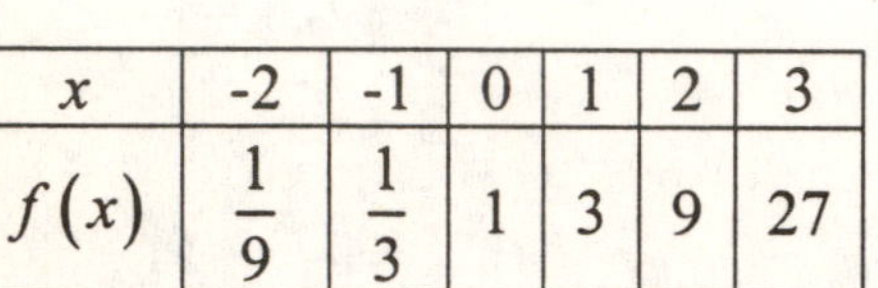

x	-2	-1	0	1	2	3
$f(x)$	$\frac{1}{9}$	$\frac{1}{3}$	1	3	9	27

x	$\frac{1}{9}$	$\frac{1}{3}$	1	3	9	27
$g(x)$	-2	-1	0	1	2	3

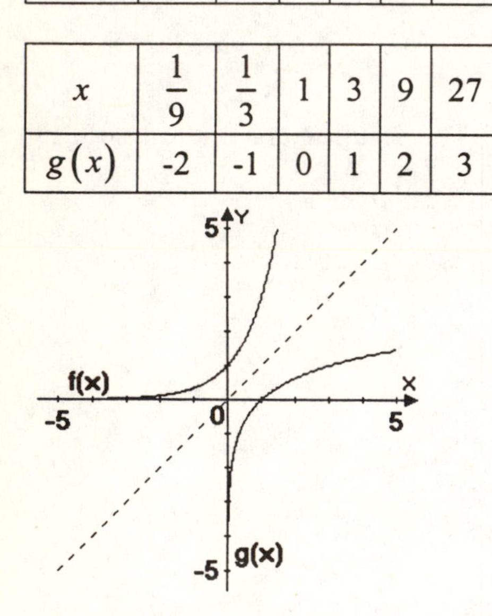

7. The domain of $h(x) = \log_4(x-5)$ is all values of x for which $x - 5 > 0$.

$$x - 5 > 0$$
$$x > 5$$

The domain of h is $\{x \mid x > 5\}$ or $(5, \infty)$.

8.
$$h(10) = 29 + 48.8\log(10+1)$$
$$= 29 + 48.8\log 11$$
$$\approx 79.8$$

A boy has reached approximately 80% of his adult height by age 10.

9.
$$R = \log\frac{I}{I_0} = \log\frac{10{,}000 I_0}{I_0}$$
$$= \log 10{,}000 = \log 10^4 = 4$$

The magnitude on the Richter scale is 4.

10. **a.** $f(x) = \ln(4-x)$

$$4 - x > 0$$
$$-x > -4$$
$$x < 4$$

The domain of f is $\{x \mid x < 4\}$ or $(-\infty, 4)$.

b. $g(x) = \ln x^2$

The domain of g is all real numbers for which $x^2 > 0$. The only number that must be excluded is 0. The domain of f is $\{x \mid x \neq 0\}$ or $(-\infty, 0) \cup (0, \infty)$.

11. **a.** Because $\ln e^x = x$, we conclude that $\ln e^{25x} = 25x$.

b. Because $e^{\ln x} = x$, we conclude that $e^{\ln\sqrt{x}} = \sqrt{x}$.

12.
$$W = 0.35\ln P + 2.74$$
$$W = 0.35\ln 197 + 2.74$$
$$W \approx 4.6$$

The average walking speed in Jackson, Mississippi is approximately 4.6 feet per second.

Problem Set 9.3

Practice Exercises

1. $4 = \log_2 16$

$2^4 = 16$

3. $2 = \log_3 x$

$3^2 = x$

5. $5 = \log_b 32$

$b^5 = 32$

7. $\log_6 216 = y$

$6^y = 216$

9. $2^3 = 8$

$\log_2 8 = 3$

11. $2^{-4} = \frac{1}{16}$

$\log_2 \frac{1}{16} = -4$

13. $\sqrt[3]{8} = 2$

$8^{\frac{1}{3}} = 2$

$\log_8 2 = \frac{1}{3}$

15. $13^2 = x$

$\log_{13} x = 2$

17. $b^3 = 1000$

$\log_b 1000 = 3$

19. $7^y = 200$

$\log_7 200 = y$

21. $\log_4 16 = y$

$4^y = 16$

$4^y = 4^2$

$y = 2$

23. $\log_2 64 = y$

$2^y = 64$

$2^y = 2^6$

$y = 6$

25. $\log_7 \sqrt{7} = y$

$7^y = \sqrt{7}$

$7^y = 7^{\frac{1}{2}}$

$y = \frac{1}{2}$

27. $\log_2 \frac{1}{8} = y$

$2^y = \frac{1}{8}$

$2^y = \frac{1}{2^3}$

$2^y = 2^{-3}$

$y = -3$

29. $\log_{64} 8 = y$

$64^y = 8$

$64^y = 64^{\frac{1}{2}}$

$y = \frac{1}{2}$

31. $\log_5 5 = y$

$5^y = 5^1$

$y = 1$

33. $\log_4 1 = y$

$4^y = 1$

$4^y = 4^0$

$y = 0$

35. $\log_5 5^7 = y$

$5^y = 5^7$

$y = 7$

37. Since $b^{\log_b x} = x$, $8^{\log_8 19} = 19$.

39. $f(x) = 4^x$

$g(x) = \log_4 x$

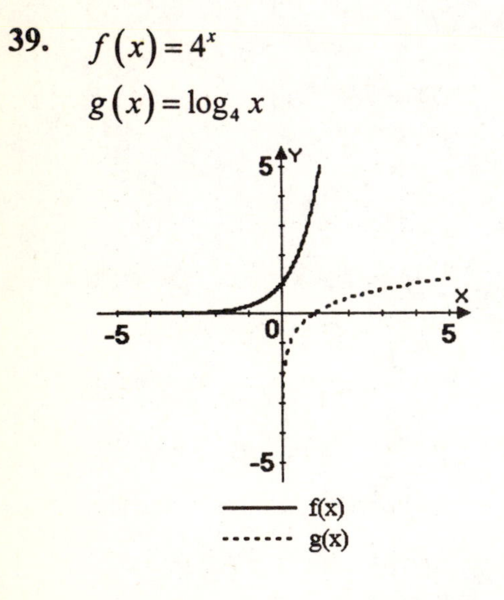

41. $f(x) = 5^x$

$g(x) = \log_5 x$

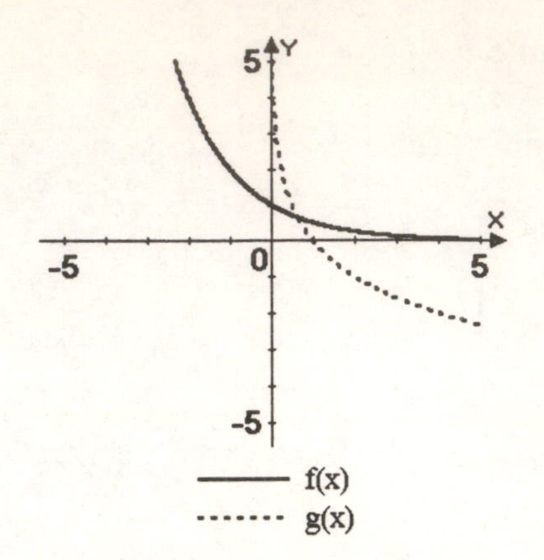

43. $f(x) = \log_5(x+4)$

$x + 4 > 0$

$x > -4$

The domain of f is $\{x | x > -4\}$ or $(-4, \infty)$.

45. $f(x) = \log_5(2-x)$

$2 - x > 0$

$-x > -2$

$x < 2$

The domain of f is $\{x | x < 2\}$ or $(-\infty, 2)$.

47. $f(x) = \ln(x-2)^2$

The domain of g is all real numbers for which $(x-2)^2 > 0$. The only number that must be excluded is 2. The domain of f is $\{x | x \neq 2\}$ or $(-\infty, 2) \cup (2, \infty)$.

49. $\log 100 = y$

$10^y = 100$

$$10^y = 10^2$$
$$y = 2$$

51. $\log 10^7 = y$

$$10^y = 10^7$$
$$y = 7$$

53. Since $10^{\log x} = x$, $10^{\log 33} = 33$.

55. $\ln 1 = y$

$$e^y = 1$$
$$e^y = e^0$$
$$y = 0$$

57. Since $\ln e^x = x$, $\ln e^6 = 6$.

59. $\ln \frac{1}{e^6} = \ln e^{-6}$

Since $\ln e^x = x$, $\ln e^{-6} = -6$.

61. Since $e^{\ln x} = x$, $e^{\ln 125} = 125$.

63. Since $\ln e^x = x$, $\ln e^{9x} = 9x$.

65. Since $e^{\ln x} = x$, $e^{\ln 5x^2} = 5x^2$.

67. Since $10^{\log x} = x$, $10^{\log \sqrt{x}} = \sqrt{x}$.

Application Exercises

69. $f(13) = 62 + 35\log(13-4)$

$$= 62 + 35\log(9) \approx 95.4$$

A 13-year-old girl is approximately 95.4% of her adult height.

71. Since $2000 - 1984 = 16$, find $f(16)$.

$$f(16) = 2.05 + 1.3\ln 16 \approx 5.7$$

In 2000, approximately $5.7 billion was spent on admission on spectator sports.

73. $D = 10\log\left(10^{12}\left(6.3\times 10^6\right)\right)$

$$= 10\log\left(6.3\times 10^{18}\right) \approx 188.0$$

The decibel level of a blue whale is approximately 188 decibels. At close range, the sound could rupture the human ear drum.

75. **a.** The original exam was at time, $t = 0$.

$$f(0) = 88 - 15\ln(0+1)$$
$$= 88 - 15\ln(1) \approx 88$$

The score on the original exam was 88%.

b. $f(2) = 88 - 15\ln(2+1)$

$$= 88 - 15\ln(3) \approx 71.5$$
$$f(4) = 88 - 15\ln(4+1)$$
$$= 88 - 15\ln(5) \approx 63.9$$
$$f(6) = 88 - 15\ln(6+1)$$
$$= 88 - 15\ln(7) \approx 58.8$$
$$f(8) = 88 - 15\ln(8+1)$$
$$= 88 - 15\ln(9) \approx 55.0$$
$$f(10) = 88 - 15\ln(10+1)$$
$$= 88 - 15\ln(11) \approx 52.0$$
$$f(12) = 88 - 15\ln(12+1)$$
$$= 88 - 15\ln(13) \approx 49.5$$

The average score for the tests is as follows.

2 months: 71.5
4 months: 63.9
6 months: 58.8
8 months: 55.0
10 months: 52.0
12 months: 49.5

c. 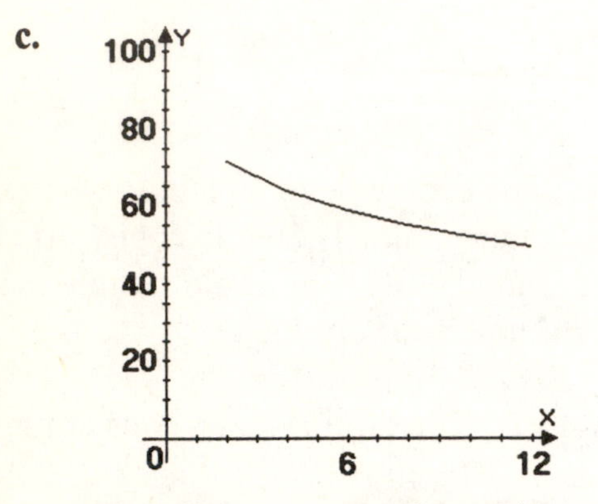

The students remembered less of the material over time.

Writing in Mathematics

77. Answers will vary.

79. Answers will vary.

81. Answers will vary.

83. Answers will vary.

Technology Exercises

85. $f(x)=\ln x \qquad g(x)=\ln(x+3)$

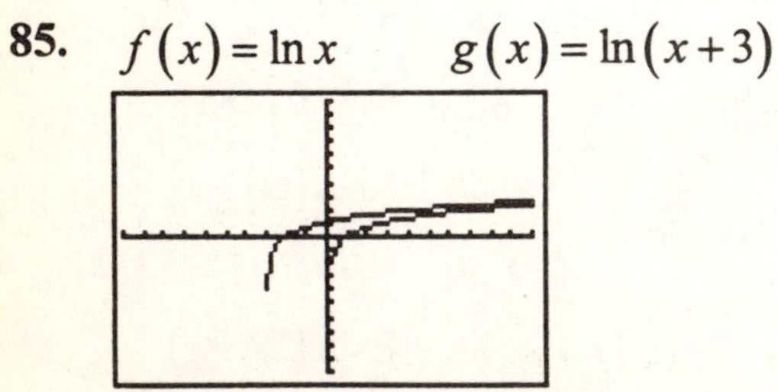

The graph of g is the graph of f shifted 3 units to the left.

87. $f(x)=\log x \qquad g(x)=-\log x$

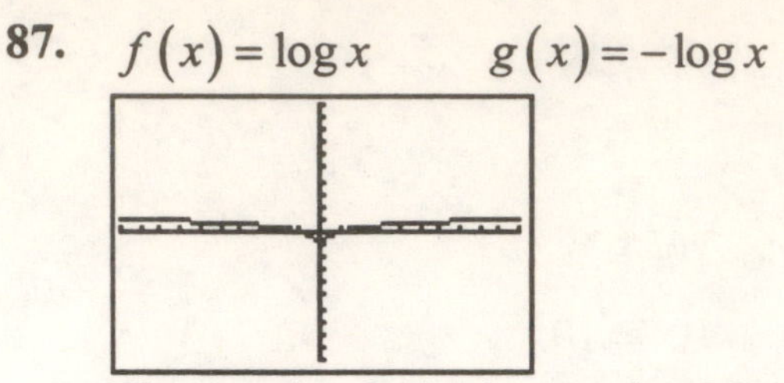

The graph of g is the graph of f reflected across the x–axis.

89. $f(t)=75-10\log(t+1)$

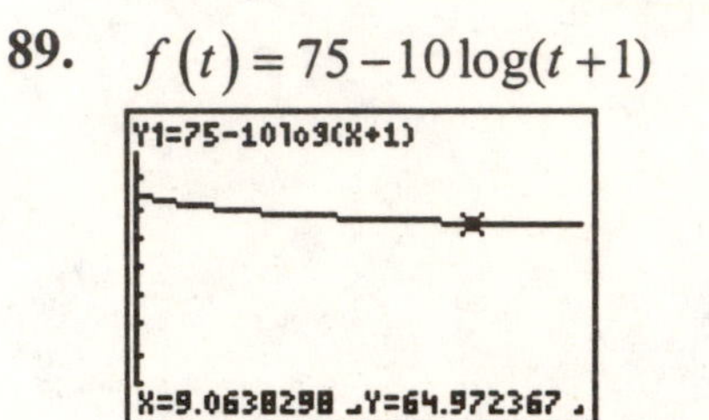

After approximately 9 months, the average score falls below 65.

91. 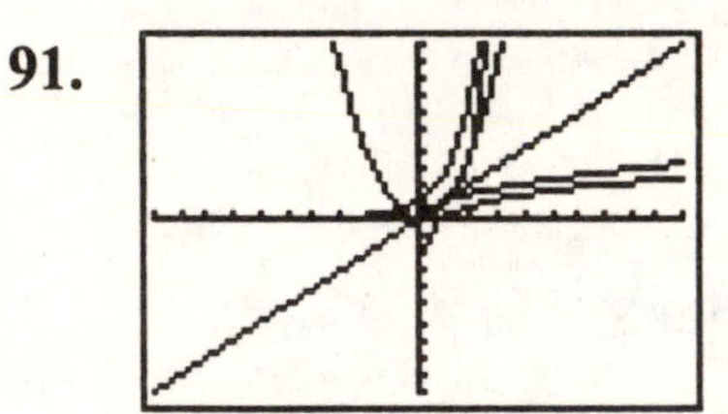

Use the trace function to compare how quickly the functions increase. In order from slowest to fastest, the functions are:

$y=\ln x,\ y=\sqrt{x},\ y=x,\ y=x^2,$

$y=e^x,$ and $y=x^x.$

Critical Thinking Exercises

93. To evaluate $\dfrac{\log_3 81-\log_\pi 1}{\log_{2\sqrt{2}} 8-\log 0.001}$, consider each of the terms independently.

$\log_3 81=y$	$\log_\pi 1=y$
$3^y=81$	$\pi^y=1$
$3^y=3^4$	$\pi^y=\pi^0$
$y=4$	$y=0$

$$\log_{2\sqrt{2}} 8 = y$$
$$\left(2\sqrt{2}\right)^y = 8$$
$$\left(2^1 2^{\frac{1}{2}}\right)^y = 2^3$$
$$\left(2^{\frac{3}{2}}\right)^y = 2^3$$
$$(2)^{\frac{3}{2}y} = 2^3$$
$$\frac{3}{2}y = 3$$
$$y = 3 \cdot \frac{2}{3} = 2$$

$$\log 0.001 = y$$
$$10^y = 0.001$$
$$10^y = 10^{-3}$$
$$y = -3$$

$$\frac{\log_3 81 - \log_\pi 1}{\log_{2\sqrt{2}} 8 - \log 0.001} = \frac{4-0}{2-(-3)} = \frac{4}{2+3} = \frac{4}{5}$$

95. To determine which expression represents a greater number, rewrite the expressions in exponential notation.

$$\log_4 60 = x \qquad \log_3 40 = y$$
$$4^x = 60 \qquad 3^y = 40$$

First consider $4^x = 60$. We know that $4^2 = 16$ and $4^3 = 64$, so x falls between 2 and 3 and is much closer to 3. Next consider $3^y = 40$. We know that $3^3 = 27$. This means that y is greater than 3, and therefore greater than x. This means that $\log_3 40$ represents the greater number.

Review Exercises

96.
$$2x = 11 - 5y$$
$$3x - 2y = -12$$

Rewrite the equations.

$$2x + 5y = 11$$
$$3x - 2y = -12$$

Multiply the first equation by 2 and the second equation by 5 and solve by addition.

$$4x + 10y = 22$$
$$15x - 10y = -60$$
$$19x = -38$$
$$x = -2$$

Back-substitute -2 for x to find y.

$$2(-2) + 5y = 11$$
$$-4 + 5y = 11$$
$$5y = 15$$
$$y = 3$$

The solution is $(-2,3)$ and the solution set is $\{(-2,3)\}$.

97.
$$6x^2 - 8xy + 2y^2 = 2\left(3x^2 - 4xy + y^2\right) = 2(3x - y)(x - y)$$

98.
$$x + 3 \le -4 \quad \text{or} \quad 2 - 7x \le 16$$
$$x \le -7 \qquad\qquad -7x \le 14$$
$$x \ge -2$$

The solution set is $\{x | x \le -7$ or $x \ge -2\}$ or $(-\infty, -7] \cup [-2, \infty)$.

Check Points 9.4

1. **a.** $\log_6(7 \cdot 11) = \log_6 7 + \log_6 11$

b. $\log(100x) = \log 100 + \log x = 2 + \log x$

2. **a.** $\log_8\left(\frac{23}{x}\right) = \log_8 23 - \log_8 x$

b. $\ln\left(\frac{e^5}{11}\right) = \ln e^5 - \ln 11$
$= 5 - \ln 11$

3. **a.** $\log_6 8^9 = 9\log_6 8$

b. $\ln\sqrt[3]{x} = \ln x^{\frac{1}{3}} = \frac{1}{3}\ln x$

4. **a.** $\log_b x^4\sqrt[3]{y} = \log_b x^4 + \log_b \sqrt[3]{y}$
$= 4\log_b x + \log_b y^{\frac{1}{3}}$
$= 4\log_b x + \frac{1}{3}\log_b y$

b. $\log_5\left(\frac{\sqrt{x}}{25y^3}\right)$
$= \log_5 \sqrt{x} - \log_5 25y^3$
$= \log_5 x^{\frac{1}{2}} - \left(\log_5 25 + \log_5 y^3\right)$
$= \frac{1}{2}\log_5 x - \log_5 5^2 - 3\log_5 y$
$= \frac{1}{2}\log_5 x - 2 - 3\log_5 y$

5. **a.** $\log 25 + \log 4 = \log(25\cdot 4)$
$= \log 100$
$= 2$

b. $\log(7x+6) - \log x = \log\left(\frac{7x+6}{x}\right)$

6. **a.** $2\ln x + \frac{1}{3}\ln(x+5)$
$= \ln x^2 + \ln(x+5)^{\frac{1}{3}}$
$= \ln x^2(x+5)^{\frac{1}{3}}$
$= \ln x^2\sqrt[3]{x+5}$

b. $2\log(x-3) - \log x$
$= \log(x-3)^2 - \log x$
$= \log\frac{(x-3)^2}{x}$

7. $\log_7 2506 = \frac{\log 2506}{\log 7} \approx 4.02$

8. $\ln_7 2506 = \frac{\ln 2506}{\ln 7} \approx 4.02$

Problem Set 9.4

Practice Exercises

1. $\log_5(7\cdot 3) = \log_5 7 + \log_5 3$

3. $\log_7(7x) = \log_7 7 + \log_7 x$
$= 1 + \log_7 x$

5. $\log(1000x) = \log 1000 + \log x$
$= 3 + \log x$

7. $\log_7\left(\frac{7}{x}\right) = \log_7 7 - \log_7 x$
$= 1 - \log_7 x$

9. $\log\left(\frac{x}{100}\right) = \log x - \log 100$
$= \log x - 2$

11. $\log_4\left(\frac{64}{y}\right) = \log_4 64 - \log_4 y$
$= 3 - \log_4 y$

13. $\ln\left(\frac{e^2}{5}\right) = \ln e^2 - \ln 5$
$= 2 - \ln 5$

15. $\log_b x^3 = 3\log_b x$

17. $\log N^{-6} = -6\log N$

19. $\ln \sqrt[5]{x} = \ln x^{\frac{1}{5}} = \frac{1}{5}\ln x$

21. $\log_b x^2 y = \log_b x^2 + \log_b y$
$= 2\log_b x + \log_b y$

23. $\log_4\left(\frac{\sqrt{x}}{64}\right) = \log_4 \sqrt{x} - \log_4 64$
$= \log_4 x^{\frac{1}{2}} - 3$
$= \frac{1}{2}\log_4 x \ - 3$

25. $\log_6\left(\frac{36}{\sqrt{x+1}}\right) = \log_6 36 - \log_6 \sqrt{x+1}$
$= 2 - \log_6 (x+1)^{\frac{1}{2}}$
$= 2 - \frac{1}{2}\log_6 (x+1)$

27. $\log_b\left(\frac{x^2 y}{z^2}\right)$
$= \log_b x^2 y - \log_b z^2$
$= \log_b x^2 + \log_b y - 2\log_b z$
$= 2\log_b x + \log_b y - 2\log_b z$

29. $\log\sqrt{100x} = \log(100x)^{\frac{1}{2}}$
$= \frac{1}{2}\log(100x)$
$= \frac{1}{2}(\log 100 + \log x)$
$= \frac{1}{2}(2 + \log x)$
$= 1 + \frac{1}{2}\log x$

31. $\log\sqrt[3]{\frac{x}{y}} = \log\left(\frac{x}{y}\right)^{\frac{1}{3}}$
$= \frac{1}{3}\log\left(\frac{x}{y}\right)$
$= \frac{1}{3}(\log x - \log y)$
$= \frac{1}{3}\log x - \frac{1}{3}\log y$

33. $\log 5 + \log 2 = \log(5\cdot 2) = \log 10 = 1$

35. $\ln x + \ln 7 = \ln(x\cdot 7) = \ln(7x)$

37. $\log_2 96 - \log_2 3 = \log_2 \frac{96}{3}$
$= \log_2 32 = 5$

39. $\log(2x+5) - \log x = \log\left(\frac{2x+5}{x}\right)$

41. $\log x + 3\log y = \log x + \log y^3$
$= \log xy^3$

43. $\frac{1}{2}\ln x + \ln y = \ln x^{\frac{1}{2}} + \ln y$
$= \ln x^{\frac{1}{2}} y = \ln y\sqrt{x}$

45. $2\log_b x + 3\log_b y = \log_b x^2 + \log_b y^3$
$= \log_b x^2 y^3$

47. $5\ln x - 2\ln y = \ln x^5 - \ln y^2$
$= \ln \frac{x^5}{y^2}$

49. $3\ln x - \frac{1}{3}\ln y = \ln x^3 - \ln y^{\frac{1}{3}}$
$= \ln \frac{x^3}{y^{\frac{1}{3}}} = \ln \frac{x^3}{\sqrt[3]{y}}$

51. $4\ln(x+6) - 3\ln x = \ln(x+6)^4 - \ln x^3$
$= \ln \frac{(x+6)^4}{x^3}$

53. $\log_5 13 = \frac{\log 13}{\log 5} \approx 1.5937$

55. $\log_{14} 87.5 = \frac{\log 87.5}{\log 14} \approx 1.6944$

57. $\log_{0.1} 17 = \frac{\log 17}{\log 0.1} \approx -1.2304$

59. $\log_\pi 63 = \frac{\log 63}{\log \pi} \approx 3.6193$

Application Exercises

61. **a.** $D = 10(\log I - \log I_0)$
$= 10\left(\log \frac{I}{I_0}\right)$

b. $D = 10\left(\log \frac{100}{1}\right) = 10(\log 100)$
$= 10(2) = 20$
The sound is 20 decibels louder on the decibel scale.

Writing in Mathematics

63. Answers will vary.

65. Answers will vary.

67. Answers will vary.

69. Answers will vary.

Technology Exercises

71. **a.** $y = \log_3 x = \frac{\log x}{\log 3}$

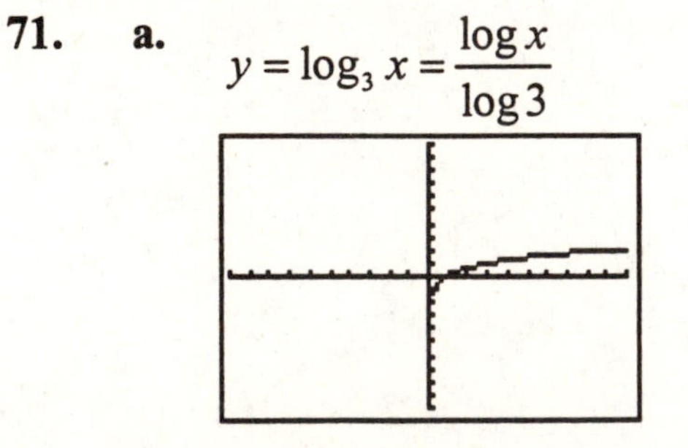

b. $y = 2 + \log_3 x$
$y = \log_3(x+2)$
$y = -\log_3 x$
$y = \log_3 x$

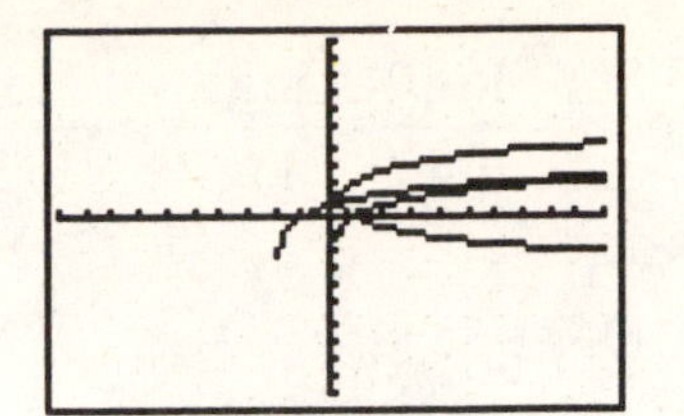

The graph of $y = 2 + \log_3 x$ is the graph of $y = \log_3 x$ shifted up two units.
The graph of $y = \log_3 (x+2)$ is the graph of $y = \log_3 x$ shifted 2 units to the left.
The graph of $y = -\log_3 x$ is the the graph of $y = \log_3 x$ reflected about the x–axis.

73. 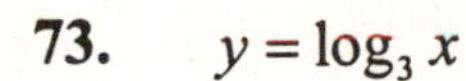

$y = \log_3 x$
$y = \log_{25} x$
$y = \log_{100} x$

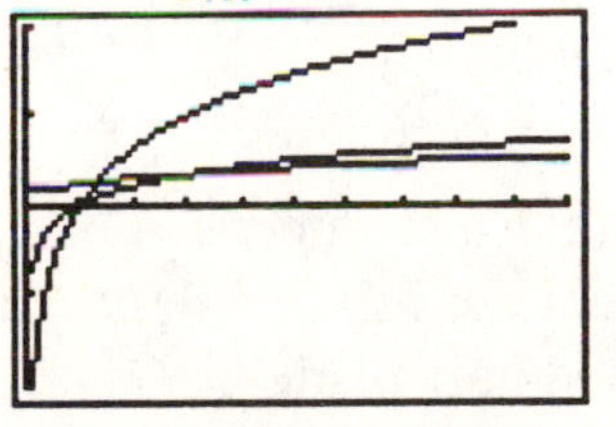

a. Change the window to focus on the (0, 1) interval.

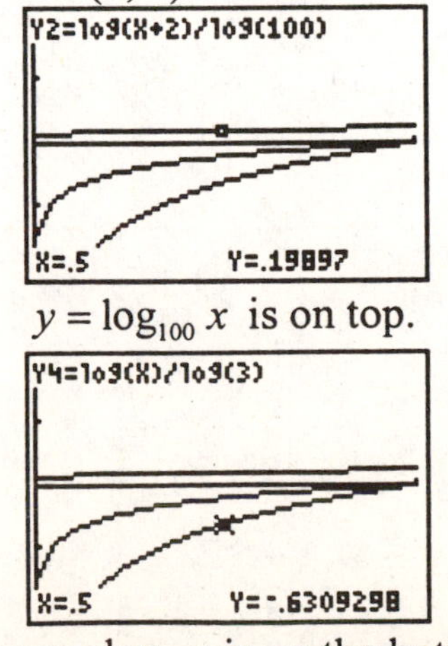

$y = \log_{100} x$ is on top.

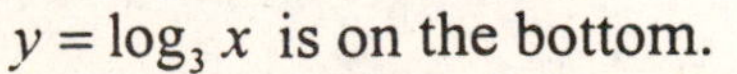

$y = \log_3 x$ is on the bottom.

b. Change the window to focus on the (1, 10) interval.

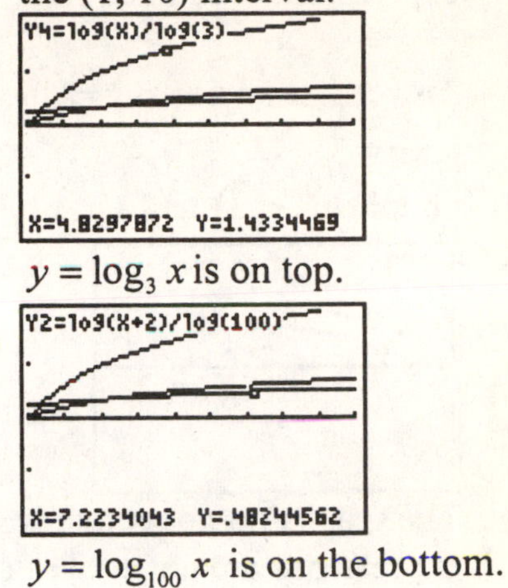

$y = \log_3 x$ is on top.

$y = \log_{100} x$ is on the bottom.

c. If $y = \log_b x$ is graphed for varying values of b, the graph of the one with the largest base will be on top in the interval (0, 1) and the one with the smallest base will be on top in the interval $(1, \infty)$. Conversely, If $y = \log_b x$ is graphed for varying values of b, the graph of the one with the smallest base will be on the bottom in the interval (0, 1) and the one with the largest base will be on the bottom in the interval $(1, \infty)$.

75. To verify that $\log \frac{x}{y} = \frac{\log x}{\log y}$, let $y =$ 3.

Graph $\log \frac{x}{3}$ and $\frac{\log x}{\log 3}$.

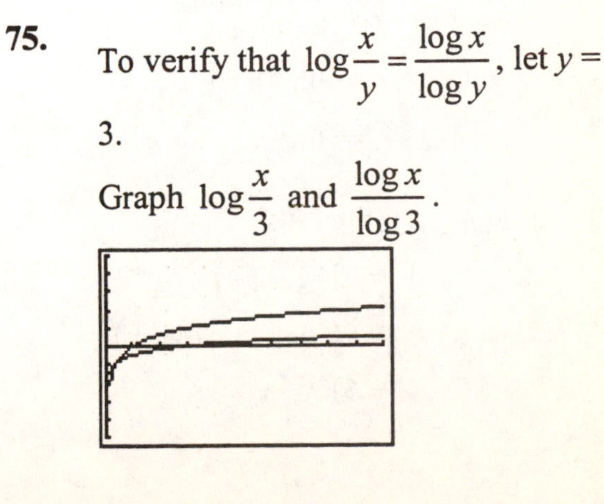

The graphs do not coincide, so the expressions are not equivalent.

77. To verify that $\ln(xy) = (\ln x)(\ln y)$, let $y = 3$.
Graph $\ln(x \cdot 3)$ and $(\ln x)(\ln 3)$.

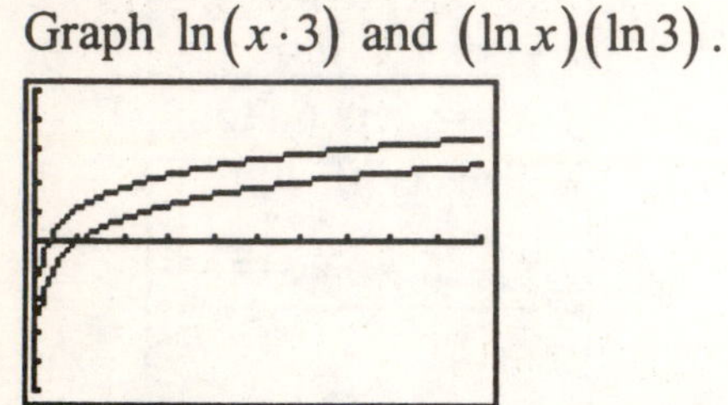

The graphs do not coincide, so the expressions are not equivalent.

Critical Thinking Exercises

79. Statement **d.** is true.

$$\ln\sqrt{2} = \ln 2^{\frac{1}{2}} = \frac{1}{2}\ln 2 = \frac{\ln 2}{2}$$

Statement **a.** is false.

$$\frac{\log_7 49}{\log_7 7} = \frac{\log_7 49}{1} = \log_7 49 = 2$$

Statement **b.** is false.
$\log_b(x^3 + y^3)$ cannot be simplified.
If we were taking the logarithm of a product and not a sum, we would have been able to simplify as follows.

$$\log_b(x^3y^3) = \log_b x^3 + \log_b y^3$$
$$= 3\log_b x + 3\log_b y$$

Statement **c.** is false.

$$\log_b(xy)^5 = 5\log_b(xy)$$
$$= 5(\log_b x + \log_b y)$$
$$= 5\log_b x + 5\log_b y$$

81. $$\log_7 9 = \frac{\log 9}{\log 7} = \frac{\log 3^2}{\log 7} = \frac{2\log 3}{\log 7} = \frac{2A}{B}$$

Review Exercises

83. $5x - 2y > 10$
First, find the intercepts to the equation $5x - 2y = 10$.
Find the x–intercept by setting y = 0.

$$5x - 2(0) = 10$$
$$5x = 10$$
$$x = 2$$

Find the y–intercept by setting $x = 0$.

$$5(0) - 2y = 10$$
$$-2y = 10$$
$$y = -5$$

Next, use the origin as a test point.

$$5(0) - 2(0) > 10$$
$$0 - 0 > 10$$
$$0 > 10$$

This is a false statement. This means that the origin will not fall in the shaded half-plane.

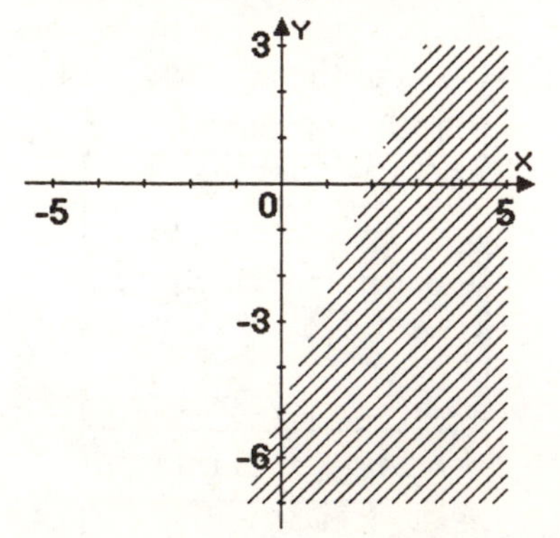

84. $$x - 2(3x - 2) > 2x - 3$$
$$x - 6x + 4 > 2x - 3$$
$$-5x + 4 > 2x - 3$$

$$-7x+4>-3$$
$$-7x>-7$$
$$x<1$$

The solution set is $\{x|x<1\}$ or $(-\infty,1)$.

85. $\dfrac{\sqrt[3]{40x^2y^6}}{\sqrt[3]{5xy}}=\sqrt[3]{\dfrac{40x^2y^6}{5xy}}=\sqrt[3]{8xy^5}$

$=\sqrt[3]{8xy^3y^2}=2y\sqrt[3]{xy^2}$

Check Points 9.5

1. a. $5^{3x-6}=125$

$$5^{3x-6}=5^3$$
$$3x-6=3$$
$$3x=9$$
$$x=3$$

The solution is 3 and the solution set is $\{3\}$.

b. $4^x=32$

$$2^{2x}=2^5$$
$$2x=5$$
$$x=\frac{5}{2}$$

The solution is $\frac{5}{2}$ and the solution set is $\left\{\frac{5}{2}\right\}$.

2. $5^x=134$

$$\ln 5^x=\ln 134$$
$$x\ln 5=\ln 134$$
$$x=\frac{\ln 134}{\ln 5}\approx 3.04$$

The solution is $\frac{\ln 134}{\ln 5}\approx 3.04$ and the solution set is $\left\{\frac{\ln 134}{\ln 5}\approx 3.04\right\}$.

3. $7e^{2x}=63$

$$e^{2x}=9$$
$$\ln e^{2x}=\ln 9$$
$$2x=\ln 9$$
$$x=\frac{\ln 9}{2}\approx 1.10$$

The solution is $\frac{\ln 9}{2}\approx 1.10$ and the solution set is $\left\{\frac{\ln 9}{2}\approx 1.10\right\}$.

4. $\log_2(x-4)=3$

$$2^3=x-4$$
$$8=x-4$$
$$12=x$$

The solution is 12 and the solution set is $\{12\}$.

5. $\log x+\log(x-3)=1$

$$\log(x(x-3))=1$$
$$10^1=(x(x-3))$$
$$10=x^2-3x$$
$$0=x^2-3x-10$$
$$0=(x-5)(x+2)$$

Apply the zero product principle.

$x-5=0$ and $x+3=0$

$x=5$ $\qquad x=-3$

We disregard –3 because it would result in taking the logarithm of a negative number in the original equation. The solution is 5 and the solution set is $\{5\}$.

6. $\log_5(x+1)-\log_5 x=2$

$$\log_5\frac{(x+1)}{x}=2$$

$$5^2=\frac{x+1}{x}$$

$$25=\frac{x+1}{x}$$

$$25x=x+1$$

$$24x=1$$

$$x=\frac{1}{24}$$

The solution is $\frac{1}{24}$ and the solution set is $\left\{\frac{1}{24}\right\}$.

7. $4\ln 3x=8$

$$\ln 3x=2$$

$$e^{\ln 3x}=e^2$$

$$3x=e^2$$

$$x=\frac{e^2}{3}$$

The solution is $\frac{e^2}{3}$ and the solution set is $\left\{\frac{e^2}{3}\right\}$.

8. $R=6e^{12.77x}$

$$6e^{12.77x}=7$$

$$\ln e^{12.77x}=\ln\frac{7}{6}$$

$$12.77x=\ln\frac{7}{6}$$

$$x=\frac{\ln\frac{7}{6}}{12.77}\approx 0.01$$

There is a 7% risk of a car accident when the blood alcohol concentration is 0.01.

9.

$$3600=1000\left(1+\frac{0.08}{4}\right)^{4t}$$

$$1000(1+0.02)^{4t}=3600$$

$$1000(1.02)^{4t}=3600$$

$$(1.02)^{4t}=\frac{3600}{1000}$$

$$\ln(1.02)^{4t}=\ln 3.6$$

$$4t\ln 1.02=\ln 3.6$$

$$\frac{4t\ln 1.02}{4\ln 1.02}=\frac{\ln 3.6}{4\ln 1.02}$$

$$t=\frac{\ln 3.6}{4\ln 1.02}\approx 16.2$$

It will take approximately 16.2 years for \$1000 to grow to \$3600.

10. $461.87+299.4\ln x=2000$

$$299.4\ln x=1538.13$$

$$\ln x=\frac{1538.13}{299.4}$$

$$e^{\ln x}=e^{\frac{1538.13}{299.4}}$$

$$x=e^{\frac{1538.13}{299.4}}\approx 170.2$$

Approximately 170 years after 1979, in the year 2149, there will be two million U.S. workers in the environmental industry.

Problem Set 9.5

Practice Exercises

1. $2^x = 64$

$2^x = 2^6$

$x = 6$

The solution is 6 and the solution set is $\{6\}$.

3. $5^x = 125$

$5^x = 5^3$

$x = 3$

The solution is 3 and the solution set is $\{3\}$.

5. $2^{2x-1} = 32$

$2^{2x-1} = 2^5$

$2x - 1 = 5$

$2x = 6$

$x = 3$

The solution is 3 and the solution set is $\{3\}$.

7. $4^{2x-1} = 64$

$4^{2x-1} = 4^3$

$2x - 1 = 3$

$2x = 4$

$x = 2$

The solution is 2 and the solution set is $\{2\}$.

9. $32^x = 8$

$(2^5)^x = 2^3$

$2^{5x} = 2^3$

$5x = 3$

$x = \frac{3}{5}$

The solution is $\frac{3}{5}$ and the solution set is $\left\{\frac{3}{5}\right\}$.

11. $9^x = 27$

$(3^2)^x = 3^3$

$3^{2x} = 3^3$

$2x = 3$

$x = \frac{3}{2}$

The solution is $\frac{3}{2}$ and the solution set is $\left\{\frac{3}{2}\right\}$.

13. $10^x = 3.91$

$\ln 10^x = \ln 3.91$

$x \ln 10 = \ln 3.91$

$x = \frac{\ln 3.91}{\ln 10} \approx 0.59$

The solution is $\frac{\ln 3.91}{\ln 10} \approx 0.59$ and the solution set is $\left\{\frac{\ln 3.91}{\ln 10} \approx 0.59\right\}$.

15. $e^x = 5.7$

$\ln e^x = \ln 5.7$

$x = \ln 5.7 \approx 1.74$

The solution is $\ln 5.7 \approx 1.74$ and the solution set is $\{\ln 5.7 \approx 1.74\}$.

17.
$$5^x = 17$$
$$\ln 5^x = \ln 17$$
$$x \ln 5 = \ln 17$$
$$x = \frac{\ln 17}{\ln 5} \approx 1.76$$

The solution is $\frac{\ln 17}{\ln 5} \approx 1.76$ and the solution set is $\left\{\frac{\ln 17}{\ln 5} \approx 1.76\right\}$.

19.
$$5e^x = 25$$
$$e^x = 5$$
$$\ln e^x = \ln 5$$
$$x = \ln 5 \approx 1.61$$

The solution is $\ln 5 \approx 1.61$ and the solution set is $\{\ln 5 \approx 1.61\}$.

21.
$$3e^{5x} = 1977$$
$$e^{5x} = 659$$
$$\ln e^{5x} = \ln 659$$
$$5x = \ln 659$$
$$x = \frac{\ln 659}{5} \approx 1.30$$

The solution is $\frac{\ln 659}{5} \approx 1.30$ and the solution set is $\left\{\frac{\ln 659}{5} \approx 1.30\right\}$.

23.
$$e^{0.7x} = 13$$
$$\ln e^{0.7x} = \ln 13$$
$$0.7x = \ln 13$$
$$x = \frac{\ln 13}{0.7} \approx 3.66$$

The solution is $\frac{\ln 13}{0.7} \approx 3.66$ and the solution set is $\left\{\frac{\ln 13}{0.7} \approx 3.66\right\}$.

25.
$$1250e^{0.055x} = 3750$$
$$e^{0.055x} = 3$$
$$\ln e^{0.055x} = \ln 3$$
$$0.055x = \ln 3$$
$$x = \frac{\ln 3}{0.055} \approx 19.97$$

The solution is $\frac{\ln 3}{0.055} \approx 19.97$ and the solution set is $\left\{\frac{\ln 3}{0.055} \approx 19.97\right\}$.

27.
$$30 - (1.4)^x = 0$$
$$-1.4^x = -30$$
$$1.4^x = 30$$
$$\ln 1.4^x = \ln 30$$
$$x \ln 1.4 = \ln 30$$
$$x = \frac{\ln 30}{\ln 1.4} \approx 10.11$$

The solution is $\frac{\ln 30}{\ln 1.4} \approx 10.11$ and the solution set is $\left\{\frac{\ln 30}{\ln 1.4} \approx 10.11\right\}$.

29.
$$e^{1-5x} = 793$$
$$\ln e^{1-5x} = \ln 793$$
$$1 - 5x = \ln 793$$
$$-5x = \ln 793 - 1$$
$$x = \frac{-(\ln 793 - 1)}{5}$$
$$x = \frac{1 - \ln 793}{5} \approx -1.14$$

The solution is $\frac{1 - \ln 793}{5} \approx -1.14$ and the solution set is $\left\{\frac{1 - \ln 793}{5} \approx -1.14\right\}$

31.
$$7^{x+2} = 410$$
$$\ln 7^{x+2} = \ln 410$$
$$(x+2)\ln 7 = \ln 410$$
$$x+2 = \frac{\ln 410}{\ln 7}$$
$$x = \frac{\ln 410}{\ln 7} - 2 \approx 1.09$$
The solution is $\frac{\ln 410}{\ln 7} - 2 \approx 1.09$ and the solution set is $\left\{\frac{\ln 410}{\ln 7} - 2 \approx 1.09\right\}$.

33.
$$2^{x+1} = 5^x$$
$$\ln 2^{x+1} = \ln 5^x$$
$$(x+1)\ln 2 = x\ln 5$$
$$x\ln 2 + \ln 2 = x\ln 5$$
$$x\ln 2 = x\ln 5 - \ln 2$$
$$x\ln 2 - x\ln 5 = -\ln 2$$
$$x(\ln 2 - \ln 5) = -\ln 2$$
$$x = \frac{-\ln 2}{\ln 2 - \ln 5}$$
$$x = \frac{\ln 2}{\ln 5 - \ln 2} \approx 0.76$$
The solution is $\frac{\ln 2}{\ln 5 - \ln 2} \approx 0.76$ and the solution set is $\left\{\frac{\ln 2}{\ln 5 - \ln 2} \approx 0.76\right\}$.

35. $\log_3 x = 4$
$$x = 3^4$$
$$x = 81$$
The solution is 81 and the solution set is $\{81\}$.

37. $\log_2 x = -4$
$$x = 2^{-4}$$
$$x = \frac{1}{2^4} = \frac{1}{16}$$
The solution is $\frac{1}{16}$ and the solution set is $\left\{\frac{1}{16}\right\}$.

39. $\log_9 x = \frac{1}{2}$
$$x = 9^{\frac{1}{2}}$$
$$x = \sqrt{9} = 3$$
The solution is 3 and the solution set is $\{3\}$.

41. $\log x = 2$
$$x = 10^2$$
$$x = 100$$
The solution is 100 and the solution set is $\{100\}$.

43. $\ln x = -3$
$$x = e^{-3}$$
$$x = \frac{1}{e^3}$$
The solution is $\frac{1}{e^3}$ and the solution set is $\left\{\frac{1}{e^3}\right\}$.

45. $\log_4(x+5) = 3$
$$x+5 = 4^3$$
$$x+5 = 64$$
$$x = 59$$

The solution is 59 and the solution set is $\{59\}$.

47. $\log_3(x-4)=-3$

$$x-4=3^{-3}$$
$$x-4=\frac{1}{3^3}$$
$$x-4=\frac{1}{27}$$
$$x=\frac{1}{27}+4$$
$$x=\frac{1}{27}+\frac{108}{27}=\frac{109}{27}$$

The solution is $\frac{109}{27}$ and the solution set is $\left\{\frac{109}{27}\right\}$.

49. $\log_4(3x+2)=3$

$$3x+2=4^3$$
$$3x+2=64$$
$$3x=62$$
$$x=\frac{62}{3}$$

The solution is $\frac{62}{3}$ and the solution set is $\left\{\frac{62}{3}\right\}$.

51. $\log_5 x+\log_5(4x-1)=1$

$$\log_5(x(4x-1))=1$$
$$x(4x-1)=5^1$$
$$4x^2-x=5$$
$$4x^2-x-5=0$$
$$(4x-5)(x+1)=0$$

Apply the zero product principle.

$4x-5=0$ and $x+1=0$

$4x=5$ $\quad x=-1$

$x=\frac{5}{4}$

We disregard -1 because it would result in taking the logarithm of a negative number in the original equation. The solution is $\frac{5}{4}$ and the solution set is $\left\{\frac{5}{4}\right\}$.

53. $\log_3(x-5)+\log_3(x+3)=2$

$$\log_3((x-5)(x+3))=2$$
$$(x-5)(x+3)=3^2$$
$$x^2-2x-15=9$$
$$x^2-2x-24=0$$
$$(x-6)(x+2)=0$$

Apply the zero product principle.

$x-6=0$ and $x+2=0$

$x=6$ $\quad x=-2$

We disregard -2 because it would result in taking the logarithm of a negative number in the original equation. The solution is 6 and the solution set is $\{6\}$.

55. $\log_2(x+2)-\log_2(x-5)=3$

$$\log_2\frac{x+2}{x-5}=3$$

$$\frac{x+2}{x-5}=2^3$$
$$\frac{x+2}{x-5}=8$$
$$x+2=8(x-5)$$
$$x+2=8x-40$$
$$-7x+2=-40$$
$$-7x=-42$$
$$x=6$$

The solution is 6 and the solution set is $\{6\}$.

57. $\log(3x-5)-\log(5x)=2$

$$\log\frac{3x-5}{5x}=2$$
$$\frac{3x-5}{5x}=10^2$$
$$\frac{3x-5}{5x}=100$$
$$3x-5=500x$$
$$-5=497x$$
$$-\frac{5}{497}=x$$

We disregard $-\frac{5}{497}$ because it would result in taking the logarithm of a negative number in the original equation. There is no solution. The solution set is $\varnothing$ or $\{\ \}$.

59. $\ln x=2$

$$e^{\ln x}=e^2$$
$$x=e^2\approx 7.39$$

The solution is $e^2\approx 7.39$ and the solution set is $\{e^2\approx 7.39\}$.

61. $5\ln 2x=20$

$$\ln 2x=4$$
$$e^{\ln 2x}=e^4$$
$$2x=e^4$$
$$x=\frac{e^4}{2}\approx 27.30$$

The solution is $\frac{e^4}{2}\approx 27.30$ and the solution set is $\left\{\frac{e^4}{2}\approx 27.30\right\}$.

63. $6+2\ln x=5$

$$2\ln x=-1$$
$$e^{\ln x}=e^{-\frac{1}{2}}$$
$$x=e^{-\frac{1}{2}}\approx 0.61$$

The solution is $e^{-\frac{1}{2}}\approx 0.61$ and the solution set is $\left\{e^{-\frac{1}{2}}\approx 0.61\right\}$.

65. $\ln\sqrt{x+3}=1$

$$\ln(x+3)^{\frac{1}{2}}=1$$
$$\frac{1}{2}\ln(x+3)=1$$
$$\ln(x+3)=2$$
$$e^{\ln(x+3)}=e^2$$
$$x+3=e^2$$
$$x=e^2-3\approx 4.39$$

The solution is $e^2-3\approx 4.39$ and the solution set is $\{e^2-3\approx 4.39\}$.

Application Exercises

67.

$$R = 6e^{12.77x}$$
$$6e^{12.77x} = 100$$
$$e^{12.77x} = \frac{100}{6}$$
$$\ln e^{12.77x} = \ln\frac{100}{6}$$
$$12.77x = \ln\frac{100}{6}$$
$$x = \frac{\ln\frac{100}{6}}{12.77} \approx 0.22$$

A blood alcohol concentration of 0.22 corresponds to a 100% risk.

69. **a.** Since 1994 is 0 years after 1994, find $f(0)$.

$$f(0) = 18.2e^{0.001(0)}$$
$$= 18.2e^{0} = 18.2(1) = 18.2$$

The population of New York was 18.2 million in 1984.

b.

$$18.5 = 18.2e^{0.001t}$$
$$e^{0.001t} = \frac{18.5}{18.2}$$
$$\ln e^{0.001t} = \ln\frac{18.5}{18.2}$$
$$0.001t = \ln\frac{18.5}{18.2}$$
$$t = \frac{\ln\frac{18.5}{18.2}}{0.001} \approx 16.3$$

The population of New York will reach 18.5 million approximately 16 years after 1994 in the year 2010.

71.

$$20000 = 12500\left(1+\frac{0.0575}{4}\right)^{4t}$$
$$20000 = 12500(1+0.014375)^{4t}$$
$$20000 = 12500(1.014375)^{4t}$$
$$\frac{20000}{12500} = (1.014375)^{4t}$$
$$1.6 = (1.014375)^{4t}$$
$$\ln 1.6 = \ln(1.014375)^{4t}$$
$$\ln 1.6 = 4t\ln 1.014375$$
$$\frac{4t\ln 1.014375}{4\ln 1.014375} = \frac{\ln 1.6}{4\ln 1.014375}$$
$$t = \frac{\ln 1.6}{4\ln 1.014375} \approx 8.2$$

It will take approximately 8.2 years.

73.

$$1400 = 1000\left(1+\frac{r}{360}\right)^{360(2)}$$
$$\frac{1400}{1000} = \left(1+\frac{r}{360}\right)^{720}$$
$$1.4 = \left(1+\frac{r}{360}\right)^{720}$$
$$\ln 1.4 = \ln\left(1+\frac{r}{360}\right)^{720}$$
$$\ln 1.4 = 720\ln\left(1+\frac{r}{360}\right)$$
$$\frac{\ln 1.4}{720} = \ln\left(1+\frac{r}{360}\right)$$
$$e^{\frac{\ln 1.4}{720}} = e^{\ln\left(1+\frac{r}{360}\right)}$$
$$e^{\frac{\ln 1.4}{720}} = 1+\frac{r}{360}$$

$$1+\frac{r}{360}=e^{\frac{\ln 1.4}{720}}$$

$$\frac{r}{360}=e^{\frac{\ln 1.4}{720}}-1$$

$$r=360\left(e^{\frac{\ln 1.4}{720}}-1\right)\approx 0.168$$

The annual interest rate is approximately 16.8%.

75. $16000=8000e^{0.08t}$

$$\frac{16000}{8000}=e^{0.08t}$$

$$2=e^{0.08t}$$

$$\ln 2=\ln e^{0.08t}$$

$$\ln 2=0.08t$$

$$t=\frac{\ln 2}{0.08}\approx 8.7$$

It will take approximately 8.7 years to double the money.

77. $7050=2350e^{r7}$

$$\frac{7050}{2350}=e^{7r}$$

$$3=e^{7r}$$

$$\ln 3=\ln e^{7r}$$

$$\ln 3=7r$$

$$r=\frac{\ln 3}{7}\approx 15.7$$

The annual interest rate would have to be 15.7% to triple the money.

79. $25000=15557+5259\ln x$

$$9443=5259\ln x$$

$$\ln x=\frac{9443}{5259}$$

$$e^{\ln x}=e^{\frac{9443}{5259}}$$

$$x=e^{\frac{9443}{5259}}\approx 6.0$$

The average price of a new car was \$25,000 6 years after 1989 in 1995.

81. $50=95-30\log_2 x$

$$-45=-30\log_2 x$$

$$\frac{-45}{-30}=\log_2 x$$

$$\log_2 x=\frac{3}{2}$$

$$x=2^{\frac{3}{2}}\approx 2.8$$

After approximately 2.8 days, only half the students recall the important features of the lecture. This is represented by the point (2.8, 50).

83. $\text{pH}=-\log x$

$$2.4=-\log x$$

$$-2.4=\log x$$

$$x=10^{-2.4}\approx .004$$

The hydrogen ion concentration is $10^{-2.4}$ or approximately 0.004 moles per liter.

Writing in Mathematics

85. Answers will vary.

87. Answers will vary.

89. Answers will vary.

91. Answers will vary.

Technology Exercises

93. $3^{x+1} = 9$

The solution is 1 and the solution set is $\{1\}$.

Verify the solution.

$3^{1+1} = 9$

$3^2 = 9$

$9 = 9$

95. $\log_3(3x-2) = 2$

Since $3\frac{2}{3} = \frac{11}{3}$, the solution is $\frac{11}{3}$ and the solution set is $\left\{\frac{11}{3}\right\}$.

Verify the solution.

$\log_3\left(3\left(\frac{11}{3}\right)-2\right) = 2$

$\log_3(11-2) = 2$

$\log_3(9) = 2$

$9 = 3^2$

$9 = 9$

97. $\log(x-15) + \log x = 2$

The solution is 20 and the solution set is $\{20\}$.

Verify the solution.

$\log(20-15) + \log 20 = 2$

$\log 5 + \log 20 = 2$

$\log(5 \cdot 20) = 2$

$\log(100) = 2$

$100 = 10^2$

$100 = 100$

99. $5^x = 3x + 4$

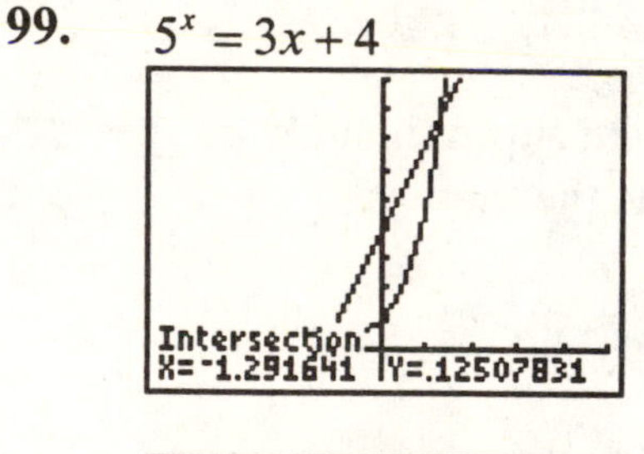

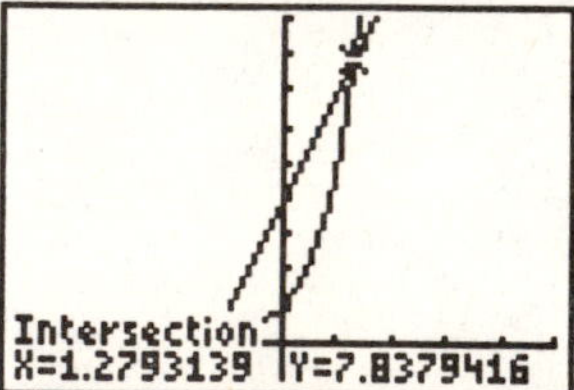

The solutions are -1.29 and 1.28 and the solution set is $\{-1.29, 1.28\}$.

Verify the solution, -1.29.

$5^{-1.29} = 3(-1.29) + 4$

$0.1254 = -3.87 + 4$

$0.1254 = 0.13$

The difference is due to rounding error.

Verify the solution, 1.28.

$$5^{1.28} = 3(1.28) + 4$$

$$7.847 = 3.84 + 4$$

$$7.847 = 7.84$$

The difference is due to rounding error.

101. $29 = 0.48\ln(x+1) + 27$

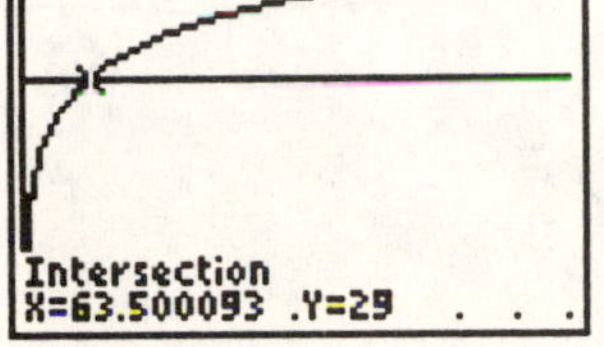

The barometric pressure is 29 inches of mercury approximately 63.5 miles from the eye of the hurricane.

103. $W(t) = 2600\left(1 - 0.51e^{-0.075t}\right)^3$

Y1=2600*(1-0.51*e^(-0.0_

X=19.680851 _Y=1792.7293

Y1=2600*(1-0.51*e^(-0.0_

X=20.212766 _Y=1820.6218

An adult elephant weighing 1800 kilograms is approximately 20 years old.

Critical Thinking Exercises

105. $A_{4000} = 4000\left(1 + \frac{0.03}{1}\right)^{1t}$

$$A_{2000} = 2000\left(1 + \frac{0.05}{1}\right)^{1t}$$

Set the right hand sides of the equations equal and solve for t.

$$4000\left(1 + \frac{0.03}{1}\right)^{1t} = 2000\left(1 + \frac{0.05}{1}\right)^{1t}$$

$$4000(1 + 0.03)^t = 2000(1 + 0.05)^t$$

$$2(1.03)^t = (1.05)^t$$

$$\ln 2(1.03)^t = \ln(1.05)^t$$

$$\ln 2 + \ln(1.03)^t = t\ln(1.05)$$

$$\ln 2 + t\ln(1.03) = t\ln(1.05)$$

$$\ln 2 = t\ln(1.05) - t\ln(1.03)$$

$$\ln 2 = t(\ln(1.05) - \ln(1.03))$$

$$t = \frac{\ln 2}{\ln(1.05) - \ln(1.03)}$$

$$t \approx 36.0$$

In approximately 36 years, the two accounts will have the same balance.

107. $(\log x)(2\log x + 1) = 6$

Let $t = \log x$.

$$(t)(2t + 1) = 6$$

$$2t^2 + t = 6$$

$$2t^2 + t - 6 = 0$$

$$(2t - 3)(t + 2) = 0$$

Apply the zero product principle.

$2t - 3 = 0$ and $t + 2 = 0$

$2t = 3$ $\quad$ $t = -2$

$t = \frac{3}{2}$

Substitute $\log x$ for t.

$$t = \frac{3}{2} \quad \text{and} \quad t = -2$$

$$\log x = \frac{3}{2} \qquad \log x = -2$$

$$x = 10^{\frac{3}{2}} \qquad x = 10^{-2}$$

The solutions are 10^{-2} and $10^{\frac{3}{2}}$ and the solution set is $\left\{10^{-2}, 10^{\frac{3}{2}}\right\}$.

Review Exercises

109. $\sqrt{x+4} - \sqrt{x-1} = 1$

$$\sqrt{x+4} = 1 + \sqrt{x-1}$$

$$\left(\sqrt{x+4}\right)^2 = \left(1+\sqrt{x-1}\right)^2$$

$$x + 4 = 1 + 2\sqrt{x-1} + x - 1$$

$$4 = 2\sqrt{x-1}$$

$$2 = \sqrt{x-1}$$

$$2^2 = \left(\sqrt{x-1}\right)^2$$

$$4 = x - 1$$

$$5 = x$$

The solution is 5 and the solution set is $\{5\}$.

110.

$$\frac{3}{x+1} - \frac{5}{x} = \frac{19}{x^2 + x}$$

$$\frac{3}{x+1} - \frac{5}{x} = \frac{19}{x(x+1)}$$

$$x(x+1)\left(\frac{3}{x+1} - \frac{5}{x}\right) = x(x+1)\left(\frac{19}{x(x+1)}\right)$$

$$x(3) - 5(x+1) = 19$$

$$3x - 5x - 5 = 19$$

$$-2x - 5 = 19$$

$$-2x = 24$$

$$x = -12$$

The solution is −12 and the solution set is $\{-12\}$.

111.

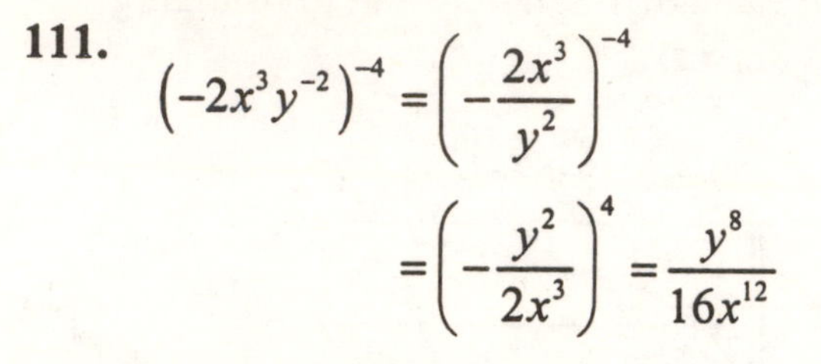

$$\left(-2x^3y^{-2}\right)^{-4} = \left(-\frac{2x^3}{y^2}\right)^{-4}$$

$$= \left(-\frac{y^2}{2x^3}\right)^4 = \frac{y^8}{16x^{12}}$$

Check Points 9.6

1. **a.**

$$643 = 491e^{k10}$$

$$\frac{643}{491} = e^{10k}$$

$$\ln\frac{643}{491} = \ln e^{10k}$$

$$\ln\frac{643}{491} = 10k$$

$$k = \frac{\ln\frac{643}{491}}{10} \approx 0.027$$

The exponential growth function is $A = A_0e^{0.027t}$.

b.

$$1000 = 491e^{0.027t}$$

$$\frac{1000}{491} = e^{0.027t}$$

$$\ln\frac{1000}{491} = \ln e^{0.027t}$$

$$\ln\frac{1000}{491} = 0.027t$$

$$t = \frac{\ln\frac{1000}{491}}{0.027} \approx 26.3$$

The population will reach one

billion approximately 26 years after 1980 in the year 2006.

2. a.
$$\frac{1}{2} = 1e^{k28}$$
$$\frac{1}{2} = e^{28k}$$
$$\ln\frac{1}{2} = \ln e^{28k}$$
$$\ln\frac{1}{2} = 28k$$
$$k = \frac{\ln\frac{1}{2}}{28} \approx -0.0248$$
The exponential growth function is $A = A_0e^{-0.0248t}$.

b.
$$10 = 60e^{-0.0248t}$$
$$\frac{10}{60} = e^{-0.0248t}$$
$$\frac{1}{6} = e^{-0.0248t}$$
$$\ln\frac{1}{6} = \ln e^{-0.0248t}$$
$$\ln\frac{1}{6} = -0.0248t$$
$$t = \frac{\ln\frac{1}{6}}{-0.0248} \approx 72.2$$
The strontium-90 will decay from 60 grams to 10 grams in 72.2 years.

Problem Set 9.6

Practice Exercises

1.
$$A = 208e^{0.008t}$$
$$A = 208e^{0.008(0)}$$
$$A = 208e^{0}$$
$$A = 208(1)$$
$$A = 208$$
In 1970, the population was 208 million.

3.
$$300 = 208e^{0.008t}$$
$$\frac{300}{208} = e^{0.008t}$$
$$\ln\frac{300}{208} = \ln e^{0.008t}$$
$$\ln\frac{300}{208} = 0.008t$$
$$t = \frac{\ln\frac{300}{208}}{0.008} \approx 45.8$$
The population will reach 300 million approximately 46 years after 1970 in the year 2016.

5. Since $k = 0.026$, each year the population will increase by 2.6%.

7.
$$A = 574e^{0.026t}$$
$$1624 = 574e^{0.026t}$$
$$\frac{1624}{574} = e^{0.026t}$$
$$\ln\frac{1624}{574} = \ln e^{0.026t}$$
$$\ln\frac{1624}{574} = 0.026t$$
$$t = \frac{\ln\frac{1624}{574}}{0.026} \approx 40.0$$
The population will be 1624 million 40 years after 1974 in the year 2014.

9. $V = 140e^{0.068t}$

$V = 140e^{0.068(0)}$

$V = 140e^{0}$

$V = 140(1)$

$V = 140$

You paid \$140 thousand, or \$140,000 for the house.

11. $200 = 140e^{0.068t}$

$\frac{200}{140} = e^{0.068t}$

$\ln\frac{10}{7} = \ln e^{0.068t}$

$\ln\frac{10}{7} = 0.068t$

$t = \frac{\ln\frac{10}{7}}{0.068} \approx 5.2$

The house will be worth \$200,000 approximately 5 years after 2000 in the year 2005.

13. $A = 200e^{kt}$

$680 = 200e^{k7}$

$\frac{680}{200} = e^{7k}$

$\ln\frac{17}{5} = \ln e^{7k}$

$\ln\frac{17}{5} = 7k$

$k = \frac{\ln\frac{17}{5}}{7} \approx 0.175$

The exponential growth model is $A = 200e^{0.175t}$. Since $k = 0.175$, the number of AIDS cases is increasing by 17.5% each year.

15. $A = 16e^{-0.000121t}$

$A = 16e^{-0.000121(5715)}$

$A = 16e^{-0.691515}$

$A \approx 8.01$

Approximately 8 grams of carbon-14 will be present in 5715 years.

17. After 10 seconds, there will be $16 \cdot \frac{1}{2} = 8$ grams present. After 20 seconds, there will be $8 \cdot \frac{1}{2} = 4$ grams present. After 30 seconds, there will be $4 \cdot \frac{1}{2} = 2$ grams present. After 40 seconds, there will be $2 \cdot \frac{1}{2} = 1$ grams present. After 50 seconds, there will be $1 \cdot \frac{1}{2} = \frac{1}{2}$ gram present.

19. $A = A_0 e^{-0.000121t}$

$15 = 100e^{-0.000121t}$

$\frac{15}{100} = e^{-0.000121t}$

$\ln 0.15 = \ln e^{-0.000121t}$

$\ln 0.15 = -0.000121t$

$t = \frac{\ln 0.15}{-0.000121} \approx 15,679$

The paintings are approximately 15,679 years old.

21. **a.** $\frac{1}{2} = 1e^{k1.31}$

$\ln\frac{1}{2} = \ln e^{1.31k}$

$$\ln\frac{1}{2} = 1.31k$$

$$k = \frac{\ln\frac{1}{2}}{1.31} \approx -0.52912$$

The exponential model is given by $A = A_0 e^{-0.52912t}$.

b.

$$A = A_0 e^{-0.52912t}$$

$$0.945A_0 = A_0 e^{-0.52912t}$$

$$0.945 = e^{-0.52912t}$$

$$\ln 0.945 = \ln e^{-0.52912t}$$

$$\ln 0.945 = -0.52912t$$

$$t = \frac{\ln 0.945}{-0.52912} \approx 0.1069$$

The age of the dinosaur ones is approximately 0.1069 billion or 106,900,000 years old.

23.

$$2A_0 = A_0 e^{kt}$$

$$2 = e^{kt}$$

$$\ln 2 = \ln e^{kt}$$

$$\ln 2 = kt$$

$$t = \frac{\ln 2}{k}$$

The population will double in $t = \frac{\ln 2}{k}$ years.

25.

$$t = \frac{\ln 2}{k}$$

$$t = \frac{\ln 2}{0.011} \approx 63.0$$

China's population will double in approximately 63 years.

Writing in Mathematics

27. Answers will vary.

29. Answers will vary.

31. Answers will vary.

33. Answers will vary.

Technology Exercises

35.

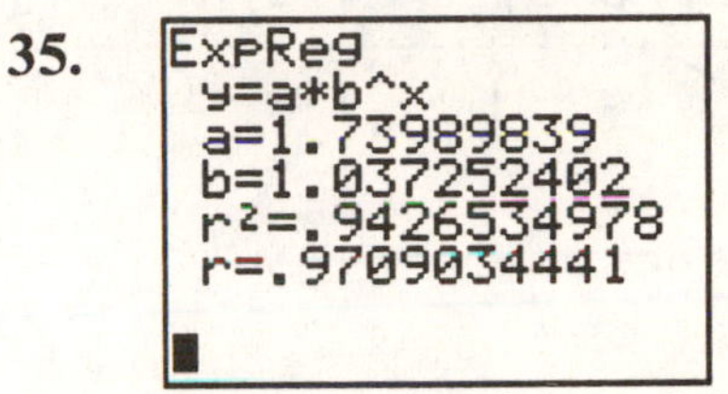

The exponential model is $y = 1.740(1.037)^x$. Since $r = 0.97$, the model fits the data very well.

37.

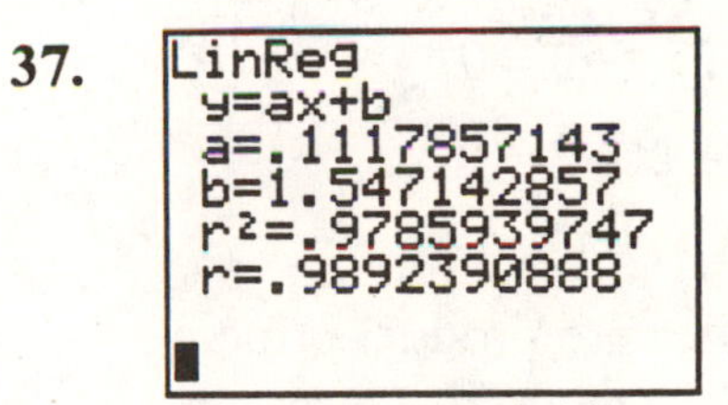

The linear model is $y = 0.112x + 1.547$. Since $r = 0.99$, the model fits the data very well.

39. Using r, the model of best fit is $y = 0.112x + 1.547$. Substitute 7.50 for y.

$$7.50 = 0.112x + 1.547$$

$$5.953 = 0.112x$$

$$x \approx 53$$

The minimum wage will be $7.50 53 years after 1970 in the year 2023. Answers will vary based on model chosen.

Critical Thinking Exercises

41. Answers will vary.

Review Exercises

43.
$$\frac{x^2-9}{2x^2+7x+3} \div \frac{x^2-3x}{2x^2+11x+5}$$
$$= \frac{x^2-9}{2x^2+7x+3} \cdot \frac{2x^2+11x+5}{x^2-3x}$$
$$= \frac{(x+3)(x-3)}{(2x+1)(x+3)} \cdot \frac{(2x+1)(x+5)}{x(x-3)}$$
$$= \frac{x+5}{x}$$

44.
$$x^{\frac{2}{3}} + 2x^{\frac{1}{3}} - 3 = 0$$
Let $t = x^{\frac{1}{3}}$.
$$\left(x^{\frac{1}{3}}\right)^2 + 2x^{\frac{1}{3}} - 3 = 0$$
$$t^2 + 2t - 3 = 0$$
$$(t+3)(t-1) = 0$$
Apply the zero product principle.
$t+3=0$ and $t-1=0$
$t=-3$ $\quad t=1$

Substitute $x^{\frac{1}{3}}$ for t.
$x^{\frac{1}{3}} = -3$ and $x^{\frac{1}{3}} = 1$
$\left(x^{\frac{1}{3}}\right)^3 = (-3)^3$ $\quad \left(x^{\frac{1}{3}}\right)^3 = (1)^3$
$x = -27$ $\quad x = 1$
The solutions are –27 and 1 and the solution set is $\{-27, 1\}$.

45.
$$6\sqrt{2} - 2\sqrt{50} + 3\sqrt{98}$$
$$= 6\sqrt{2} - 2\sqrt{25 \cdot 2} + 3\sqrt{49 \cdot 2}$$
$$= 6\sqrt{2} - 2 \cdot 5\sqrt{2} + 3 \cdot 7\sqrt{2}$$
$$= 6\sqrt{2} - 10\sqrt{2} + 21\sqrt{2} = 17\sqrt{2}$$

Chapter 9 Review

9.1

1. $f(x) = 4^x$

x	$f(x)$
-2	$4^{-2} = \frac{1}{4^2} = \frac{1}{16}$
-1	$4^{-1} = \frac{1}{4^1} = \frac{1}{4}$
0	$4^0 = 1$
1	$4^1 = 4$
2	$4^2 = 16$

The coordinates match graph **d**.

2. $f(x) = 4^{-x}$

x	$f(x)$
-2	$4^{-(-2)} = 4^2 = 16$
-1	$4^{-(-1)} = 4^1 = 4$
0	$4^{-0} = 4^0 = 1$
1	$4^{-1} = \frac{1}{4^1} = \frac{1}{4}$
2	$4^{-2} = \frac{1}{4^2} = \frac{1}{16}$

The coordinates match graph **a**.

3. $f(x) = -4^{-x}$

x	$f(x)$
-2	$-4^{-(-2)} = -4^2 = -16$
-1	$-4^{-(-1)} = -4^1 = -4$
0	$-4^{-0} = -4^0 = -1$

1	$-4^{-1} = -\frac{1}{4^1} = -\frac{1}{4}$
2	$-4^{-2} = -\frac{1}{4^2} = -\frac{1}{16}$

The coordinates match graph **b.**

4. $f(x) = -4^{-x} + 3$

x	$f(x)$
-2	$-4^{-(-2)} + 3 = -4^2 + 3$ $= -16 + 3 = -13$
-1	$-4^{-(-1)} + 3 = -4^1 + 3$ $= -4 + 3 = -1$
0	$-4^{-0} + 3 = -4^0 + 3$ $= -1 + 3 = 2$
1	$-4^{-1} + 3 = -\frac{1}{4^1} + 3$ $= -\frac{1}{4} + 3 = \frac{11}{4}$
2	$-4^{-2} + 3 = -\frac{1}{4^2} + 3$ $= -\frac{1}{16} + 3 = \frac{47}{16}$

The coordinates match graph **c.**

5. $f(x) = 2^x$ and $g(x) = 2^{x-1}$

x	$f(x)$	$g(x)$
-2	$\frac{1}{4}$	$\frac{1}{8}$
-1	$\frac{1}{2}$	$\frac{1}{4}$
0	1	$\frac{1}{2}$
1	2	1
2	4	2

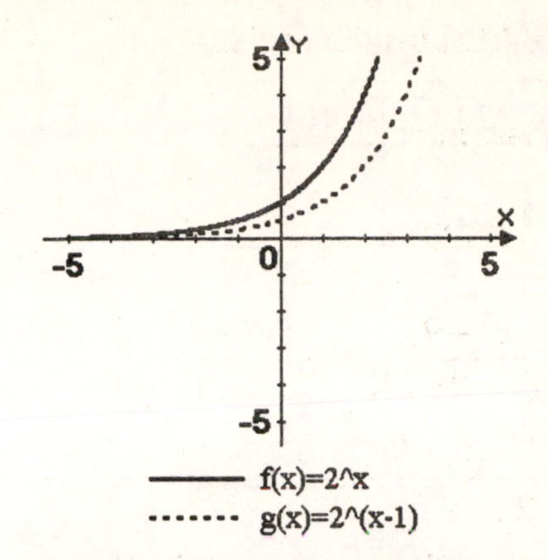

The graph of g is the graph of f shifted 1 unit to the right.

6. $f(x) = 2^x$ and $g(x) = \left(\frac{1}{2}\right)^x$

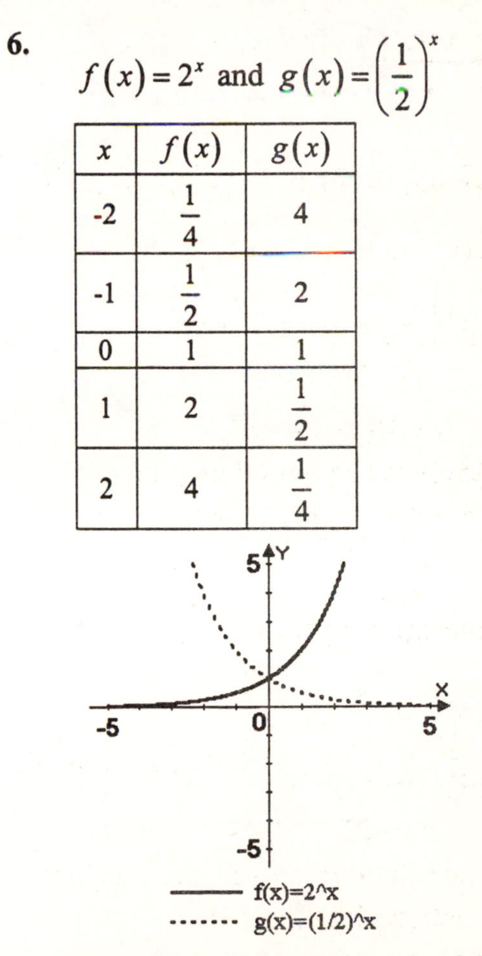

x	$f(x)$	$g(x)$
-2	$\frac{1}{4}$	4
-1	$\frac{1}{2}$	2
0	1	1
1	2	$\frac{1}{2}$
2	4	$\frac{1}{4}$

The graph of g is the graph of f reflected across the y–axis.

7. 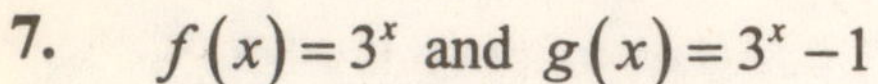
$f(x)=3^x$ and $g(x)=3^x-1$

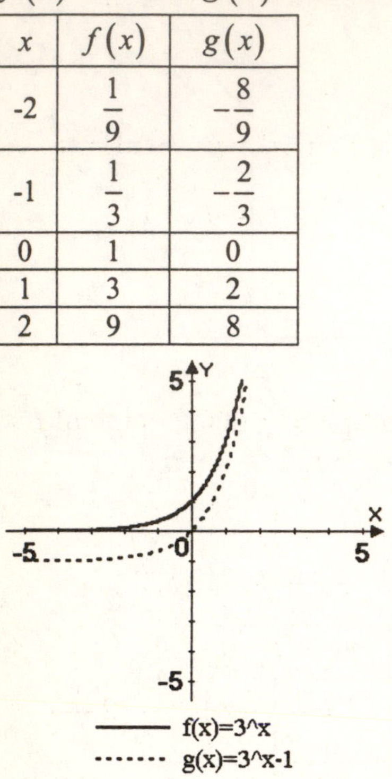

x	$f(x)$	$g(x)$
-2	$\frac{1}{9}$	$-\frac{8}{9}$
-1	$\frac{1}{3}$	$-\frac{2}{3}$
0	1	0
1	3	2
2	9	8

The graph of g is the graph of f shifted down 1 unit.

8. $f(x)=3^x$ and $g(x)=-3^x$

x	$f(x)$	$g(x)$
-2	$\frac{1}{9}$	$-\frac{1}{9}$
-1	$\frac{1}{3}$	$-\frac{1}{3}$
0	1	-1
1	3	-3
2	9	-9

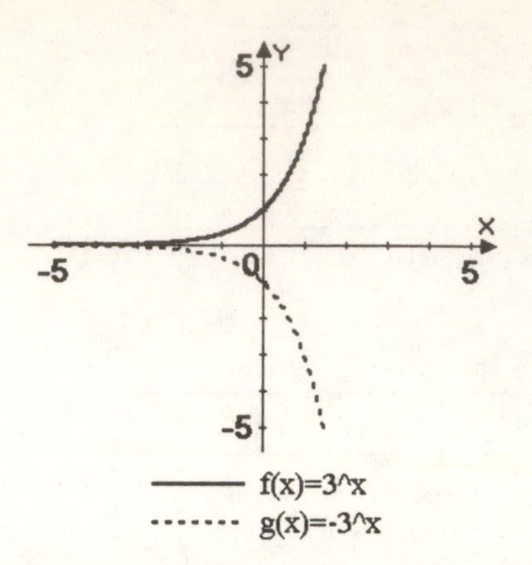

The graph of g is the graph of f reflected across the x–axis.

9. 5.5% Compounded Semiannually:

$$A=5000\left(1+\frac{0.055}{2}\right)^{2\cdot 5}$$
$$=5000(1+0.0275)^{10}$$
$$=5000(1.0275)^{10}\approx 6558.26$$

5.25% Compounded Monthly:

$$A=5000\left(1+\frac{0.0525}{12}\right)^{12\cdot 5}$$
$$=5000(1+0.004375)^{60}$$
$$=5000(1.004375)^{60}\approx 6497.16$$

5.5% compounded semiannually yields the greater return.

10. 7.0% Compounded Monthly:

$$A=14000\left(1+\frac{0.07}{12}\right)^{12\cdot 10}$$
$$=14000\left(1+\frac{7}{1200}\right)^{120}$$
$$=14000\left(\frac{1207}{1200}\right)^{120}\approx 28135.26$$

6.85% Compounded Continuously:

$A = 14000e^{0.0685 \cdot 10}$

$= 14000e^{0.685} \approx 27772.81$

7.0% compounded monthly yields the greater return.

11. a. The coffee was 200°F when it was first taken out of the microwave.

b. After 20 minutes, the temperature is approximately 119°F.

c. The coffee will cool to a low of 70°F. This means that the temperature of the room is 70°F.

9.2

12. a.

$$\begin{aligned}(f \circ g)(x) &= f(g(x)) \\ &= f(4x-1) \\ &= (4x-1)^2 + 3 \\ &= 16x^2 - 8x + 1 + 3 \\ &= 16x^2 - 8x + 4\end{aligned}$$

b.

$$\begin{aligned}(g \circ f)(x) &= g(f(x)) \\ &= g(x^2+3) \\ &= 4(x^2+3) - 1 \\ &= 4x^2 + 12 - 1 \\ &= 4x^2 + 11\end{aligned}$$

c.

$$\begin{aligned}(f \circ g)(3) &= 16(3)^2 - 8(3) + 4 \\ &= 16(9) - 24 + 4 \\ &= 144 - 24 + 4 \\ &= 124\end{aligned}$$

13. a.

$$\begin{aligned}(f \circ g)(x) = f(g(x)) &= f(x+1) \\ &= \sqrt{x+1}\end{aligned}$$

b.

$$\begin{aligned}(g \circ f)(x) = g(f(x)) &= g(\sqrt{x}) \\ &= \sqrt{x} + 1\end{aligned}$$

c. $(f \circ g)(3) = \sqrt{3+1} = \sqrt{4} = 2$

14. $f(x) = \frac{3}{5}x + \frac{1}{2}$ and $g(x) = \frac{5}{3}x - 2$

$$\begin{aligned}f(g(x)) &= f\left(\frac{5}{3}x - 2\right) \\ &= \frac{3}{5}\left(\frac{5}{3}x - 2\right) + \frac{1}{2} \\ &= \frac{3}{5}\left(\frac{5}{3}x\right) - \left(\frac{3}{5}\right)2 + \frac{1}{2} \\ &= x - \frac{6}{5} + \frac{1}{2} = x - \frac{7}{10}\end{aligned}$$

$$\begin{aligned}g(f(x)) &= g\left(\frac{3}{5}x + \frac{1}{2}\right) \\ &= \frac{5}{3}\left(\frac{3}{5}x + \frac{1}{2}\right) - 2 \\ &= \frac{5}{3}\left(\frac{3}{5}x\right) + \left(\frac{5}{3}\right)\frac{1}{2} - 2 \\ &= x + \frac{5}{6} - 2 = x - \frac{7}{6}\end{aligned}$$

The functions are not inverses.

15. $f(x) = 2 - 5x$ and $g(x) = \frac{2-x}{5}$

$$\begin{aligned}f(g(x)) &= f\left(\frac{2-x}{5}\right) \\ &= 2 - 5\left(\frac{2-x}{5}\right)\end{aligned}$$

$$=2-(2-x)$$
$$=2-2+x=x$$
$$g(f(x))=g(2-5x)$$
$$=\frac{2-(2-5x)}{5}$$
$$=\frac{2-2+5x}{5}=\frac{5x}{5}=x$$

The functions are inverses.

16. $f(x)=4x-3$

$y=4x-3$

Interchange x and y and solve for y.

$x=4y-3$

$x+3=4y$

$\frac{x+3}{4}=y$

$f^{-1}(x)=\frac{x+3}{4}$

16. **a.** $f(x)=4x-3$

$y=4x-3$

Interchange x and y and solve for y.

$x=4y-3$

$x+3=4y$

$\frac{x+3}{4}=y$

$f^{-1}(x)=\frac{x+3}{4}$

b.
$$f(f^{-1}(x))=f\left(\frac{x+3}{4}\right)$$
$$=4\left(\frac{x+3}{4}\right)-3$$
$$=x+3-3=x$$

$$f^{-1}(f(x))=f(4x-3)$$
$$=\frac{(4x-3)+3}{4}$$
$$=\frac{4x-3+3}{4}=\frac{4x}{4}=x$$

17. **a.** $f(x)=\sqrt{x+2}$

$y=\sqrt{x+2}$

Interchange x and y and solve for y.

$x=\sqrt{y+2}$

$x^2=y+2$

$x^2-2=y$

$f^{-1}(x)=x^2-2$ for $x\geq 0$

b.
$$f(f^{-1}(x))=f(\sqrt{x+2})$$
$$=(\sqrt{x+2})^2-2$$
$$=x+2-2=x$$
$$f^{-1}(f(x))=f(x^2-2)$$
$$=\sqrt{(x^2-2)+2}$$
$$=\sqrt{x^2-2+2}$$
$$=\sqrt{x^2}=x$$

18. **a.** $f(x)=8x^3+1$

$y=8x^3+1$

Interchange x and y and solve for y.

$x=8y^3+1$

$x-1=8y^3$

$\frac{x-1}{8}=y^3$

$$\sqrt[3]{\frac{x-1}{8}} = y$$
$$\frac{\sqrt[3]{x-1}}{2} = y$$
$$f^{-1}(x) = \frac{\sqrt[3]{x-1}}{2}$$

b.
$$f\left(f^{-1}(x)\right) = f\left(\frac{\sqrt[3]{x-1}}{2}\right)$$
$$= 8\left(\frac{\sqrt[3]{x-1}}{2}\right)^3 + 1$$
$$= 8\left(\frac{x-1}{8}\right) + 1$$
$$= x - 1 + 1 = x$$
$$f^{-1}\left(f(x)\right) = f\left(8x^3+1\right)$$
$$= \frac{\sqrt[3]{\left(8x^3+1\right)-1}}{2}$$
$$= \frac{\sqrt[3]{8x^3+1-1}}{2}$$
$$= \frac{\sqrt[3]{8x^3}}{2} = \frac{2x}{2} = x$$

19. Since the graph satisfies the horizontal line test, it has an inverse function.

20. Since the graph does not satisfy the horizontal line test, it does not have an inverse function.

21. Since the graph satisfies the horizontal line test, it has an inverse function.

22. Since the graph does not satisfy the horizontal line test, it does not have an inverse function.

23. Since the points $(-3,-1),(0,0)$ and $(2,4)$ lie on the graph of the function, the points $(-1,-3)$, $(0,0)$ and $(4,2)$ lie on the inverse function.

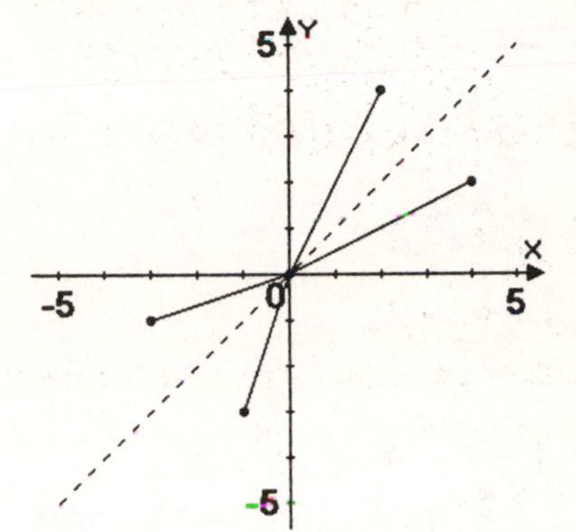

9.3

24.
$$\frac{1}{2} = \log_{49} 7$$
$$49^{\frac{1}{2}} = 7$$

25.
$$3 = \log_4 x$$
$$4^3 = x$$

26.
$$\log_3 81 = y$$
$$3^y = 81$$

27.
$$6^3 = 216$$
$$\log_6 216 = 3$$

28.
$$b^4 = 625$$
$$\log_b 625 = 4$$

29.
$$13^y = 874$$
$$\log_{13} 874 = y$$

30. $\log_4 64 = y$

$4^y = 64$

$4^y = 4^3$

$y = 3$

31. $\log_5 \dfrac{1}{25} = y$

$5^y = \dfrac{1}{25}$

$5^y = 5^{-2}$

$y = -2$

32. $\log_3(-9)$

This logarithm cannot be evaluated because -9 is not in the domain of $y = \log_3(-9)$.

33. $\log_{16} 4 = y$

$16^y = 4$

$16^y = 16^{\frac{1}{2}}$

$y = \dfrac{1}{2}$

34. $\log_{17} 17 = 1$ because $17^1 = 17$.

35. $\log_3 3^8 = 8$ because $\log_b b^x = x$.

36. Because $\ln e^x = x$, we conclude that $\ln e^5 = 5$.

37. Since $\log_8 8 = 1$, we have $\log_3(\log_8 8) = \log_3 1$. Now, evaluate $\log_3 1$.

$\log_3 1 = y$

$3^y = 1$

$3^y = 3^0$

$y = 0$

Therefore $\log_3(\log_8 8) = 0$.

38. $f(x) = 2^x$

$g(x) = \log_2 x$

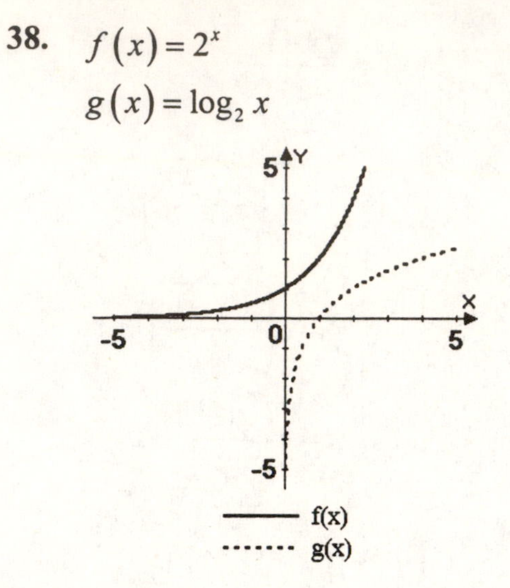

39. $f(x) = 2^x$

$g(x) = \log_2 x$

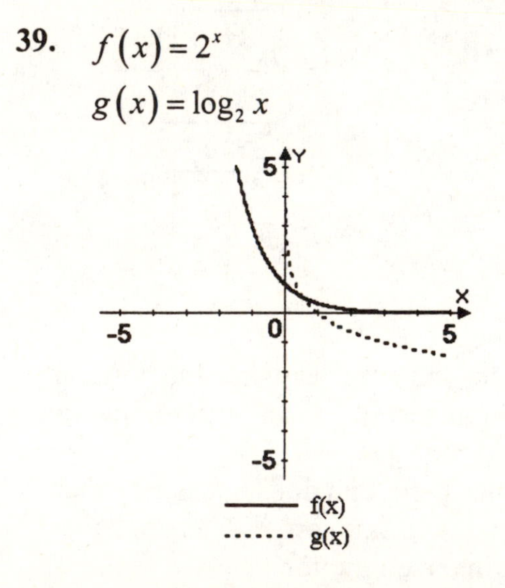

40. $f(x) = \log_8(x+5)$

$x + 5 > 0$

$x > -5$

The domain of f is $\{x | x > -5\}$ or $(-5, \infty)$.

41. $f(x) = \log(3-x)$

$3-x>0$

$-x>-3$

$x<3$

The domain of f is $\{x|x<3\}$ or $(-\infty,3)$.

42. $f(x) = \ln(x-1)^2$

The domain of g is all real numbers for which $(x-1)^2 > 0$. The only number that must be excluded is 1. The domain of f is $\{x|x \neq 1\}$ or $(-\infty,1)\cup(1,\infty)$.

43. Since $\ln e^x = x$, $\ln e^{6x} = 6x$.

44. Since $e^{\ln x} = x$, $e^{\ln\sqrt{x}} = \sqrt{x}$.

45. Since $10^{\log x} = x$, $10^{\log 4x^2} = 4x^2$.

46.
$$R = \log\frac{I}{I_0}$$
$$R = \log\frac{1000I_0}{I_0}$$
$$R = \log 1000$$
$$10^R = 1000$$
$$10^R = 10^3$$
$$R = 3$$

The magnitude on the Richter scale is 3.

47. **a.** $f(0) = 76 - 18\log(0+1)$

$= 76 - 18\log(1)$

$= 76 - 18(0) = 76 - 0 = 76$

The average score when the exam was first given was 76.

b. $f(2) = 76 - 18\log(2+1)$

$= 76 - 18\log(3) \approx 67.4$

$f(4) = 76 - 18\log(4+1)$

$= 76 - 18\log(5) \approx 63.4$

$f(6) = 76 - 18\log(6+1)$

$= 76 - 18\log(7) \approx 60.8$

$f(8) = 76 - 18\log(8+1)$

$= 76 - 18\log(9) \approx 58.8$

$f(12) = 76 - 18\log(12+1)$

$= 76 - 18\log(13) \approx 55.9$

The average scores were as follows:

2 months	67.4
4 months	63.4
6 months	60.8
8 months	58.8
12 months	55.9.

c.

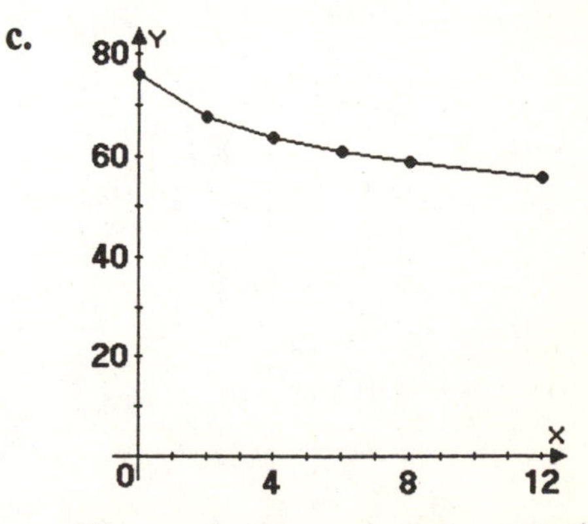

The students retain less material over time.

48. $t = \frac{1}{0.06}\ln\left(\frac{12}{12-5}\right)$
$= \frac{1}{0.06}\ln\left(\frac{12}{7}\right) \approx 9.0$

It will take approximately 9 weeks for the man to run 5 miles per hour.

9.4

49. $\log_6(36x^3) = \log_6 36 + \log_6 x^3$
$= 2 + 3\log_6 x$

50. $\log_4 \frac{\sqrt{x}}{64} = \log_4 \sqrt{x} - \log_4 64$
$= \log_4 x^{\frac{1}{2}} - 3$
$= \frac{1}{2}\log_4 x - 3$

51. $\log_2\left(\frac{xy^2}{64}\right) = \log_2 xy^2 - \log_2 64$
$= \log_2 x + \log_2 y^2 - 6$
$= \log_2 x + 2\log_2 y - 6$

52. $\ln\sqrt[3]{\frac{x}{e}} = \ln\left(\frac{x}{e}\right)^{\frac{1}{3}} = \frac{1}{3}\ln\left(\frac{x}{e}\right)$
$= \frac{1}{3}(\ln x - \ln e)$
$= \frac{1}{3}(\ln x - 1)$
$= \frac{1}{3}\ln x - \frac{1}{3}$

53. $\log_b 7 + \log_b 3 = \log_b(7\cdot 3)$
$= \log_b 21$

54. $\log 3 - 3\log x = \log 3 - \log x^3$
$= \log\frac{3}{x^3}$

55. $3\ln x + 4\ln y = \ln x^3 + \ln y^4$
$= \ln(x^3y^4)$

56. $\frac{1}{2}\ln x - \ln y = \ln x^{\frac{1}{2}} - \ln y$
$= \ln\sqrt{x} - \ln y$
$= \ln\frac{\sqrt{x}}{y}$

57. $\log_6 72{,}348 = \frac{\log 72{,}348}{\log 6} \approx 6.2448$

58. $\log_4 0.863 = \frac{\log 0.863}{\log 4} \approx -0.1063$

9.5

59. $2^{4x-2} = 64$
$2^{4x-2} = 2^6$
$4x - 2 = 6$
$4x = 8$
$x = 2$

The solution is 2 and the solution set is $\{2\}$.

60. $125^x = 25$
$(5^3)^x = 5^2$
$5^{3x} = 5^2$
$3x = 2$
$x = \frac{2}{3}$

The solution is $\frac{2}{3}$ and the solution set is $\left\{\frac{2}{3}\right\}$.

61.

$$9^x = \frac{1}{27}$$
$$\left(3^2\right)^x = 3^{-3}$$
$$3^{2x} = 3^{-3}$$
$$2x = -3$$
$$x = -\frac{3}{2}$$

The solution is $-\frac{3}{2}$ and the solution set is $\left\{-\frac{3}{2}\right\}$.

62.

$$8^x = 12{,}143$$
$$\ln 8^x = \ln 12{,}143$$
$$x \ln 8 = \ln 12{,}143$$
$$x = \frac{\ln 12{,}143}{\ln 8} \approx 4.52$$

The solution is $\frac{\ln 12{,}143}{\ln 8} \approx 4.52$ and the solution set is $\left\{\frac{\ln 12{,}143}{\ln 8} \approx 4.52\right\}$.

63.

$$9e^{5x} = 1269$$
$$e^{5x} = \frac{1269}{9}$$
$$\ln e^{5x} = \ln 141$$
$$5x = \ln 141$$
$$x = \frac{\ln 141}{5} \approx 0.99$$

The solution is $\frac{\ln 141}{5} \approx 0.99$ and the solution set is $\left\{\frac{\ln 141}{5} \approx 0.99\right\}$.

64.

$$30e^{0.045x} = 90$$
$$e^{0.045x} = \frac{90}{30}$$
$$\ln e^{0.045x} = \ln 3$$
$$0.045x = \ln 3$$
$$x = \frac{\ln 3}{0.045} \approx 24.41$$

The solution is $\frac{\ln 3}{0.045} \approx 24.41$ and the solution set is $\left\{\frac{\ln 3}{0.045} \approx 24.41\right\}$.

65.

$$\log_5 x = -3$$
$$x = 5^{-3}$$
$$x = \frac{1}{125}$$

The solution is $\frac{1}{125}$ and the solution set is $\left\{\frac{1}{125}\right\}$.

66.

$$\log x = 2$$
$$x = 10^2$$
$$x = 100$$

The solution is 100 and the solution set is $\{100\}$.

67.

$$\log_4 (3x-5) = 3$$
$$3x-5 = 4^3$$
$$3x-5 = 64$$

$$3x = 69$$
$$x = 23$$

The solution is 23 and the solution set is $\{23\}$.

68. $\log_2(x+3) + \log_2(x-3) = 4$

$$\log_2((x+3)(x-3)) = 4$$
$$\log_2(x^2 - 9) = 4$$
$$x^2 - 9 = 2^4$$
$$x^2 - 9 = 16$$
$$x^2 = 25$$
$$x = \pm 5$$

We disregard –5 because it would result in taking the logarithm of a negative number in the original equation. The solution is 5 and the solution set is $\{5\}$.

69. $\log_3(x-1) - \log_3(x+2) = 2$

$$\log_3 \frac{x-1}{x+2} = 2$$
$$\frac{x-1}{x+2} = 3^2$$
$$\frac{x-1}{x+2} = 9$$
$$x - 1 = 9(x+2)$$
$$x - 1 = 9x + 18$$
$$-8x - 1 = 18$$
$$-8x = 19$$
$$x = -\frac{19}{8}$$

We disregard $-\frac{19}{8}$ because it would result in taking the logarithm of a negative number in the original equation. There is no solution. The solution set is $\varnothing$ or $\{\ \}$.

70. $\ln x = -1$

$$x = e^{-1}$$
$$x = \frac{1}{e}$$

The solutions are $\frac{1}{e}$ and the solution set is $\left\{\frac{1}{e}\right\}$.

71. $3 + 4\ln 2x = 15$

$$4\ln 2x = 12$$
$$\ln 2x = 3$$
$$2x = e^3$$
$$x = \frac{e^3}{2}$$

The solutions are $\frac{e^3}{2}$ and the solution set is $\left\{\frac{e^3}{2}\right\}$.

72. $13 = 10.1e^{0.005t}$

$$\frac{13}{10.1} = e^{0.005t}$$
$$\ln\frac{13}{10.1} = \ln e^{0.005t}$$
$$\ln\frac{13}{10.1} = 0.005t$$
$$t = \frac{\ln\frac{13}{10.1}}{0.005} \approx 50$$

If the growth rate continues, the population will reach13 million approximately 50 years after 1992 in the year 2042.

73. $560 = 364(1.005)^t$

$\frac{560}{364} = (1.005)^t$

$\ln\frac{560}{364} = \ln(1.005)^t$

$\ln\frac{560}{364} = t\ln 1.005$

$t = \frac{\ln\frac{560}{364}}{\ln 1.005} \approx 86.4$

The carbon dioxide concentration will double the pre-industrial level approximately 86 years after the year 2000 in the year 2086.

74. $30,000 = 15,557 + 5259\ln x$

$14,443 = 5259\ln x$

$\frac{14,443}{5259} = \ln x$

$e^{\frac{14,443}{5259}} = e^{\ln x}$

$x = e^{\frac{14,443}{5259}} \approx 15.6$

The average cost of a new car will be $30,000 approximately 16 years after 1989 in the year 2005.

75.
$20,000 = 12,500\left(1+\frac{0.065}{4}\right)^{4t}$

$20,000 = 12,500(1+0.01625)^{4t}$

$20,000 = 12,500(1.01625)^{4t}$

$\frac{20,000}{12,500} = (1.01625)^{4t}$

$1.6 = (1.01625)^{4t}$

$\ln 1.6 = \ln(1.01625)^{4t}$

$\ln 1.6 = 4t\ln 1.01625$

$\frac{\ln 1.6}{4\ln 1.01625} = \frac{4t\ln 1.01625}{4\ln 1.01625}$

$t = \frac{\ln 1.6}{4\ln 1.01625} \approx 7.3$

It will take approximately 7.3 years.

76. $3(50,000) = 50,000e^{0.075t}$

$\frac{3(50,000)}{50,000} = e^{0.075t}$

$3 = e^{0.075t}$

$\ln 3 = \ln e^{0.075t}$

$\ln 3 = 0.075t$

$t = \frac{\ln 3}{0.075} \approx 14.6$

The money will triple in approximately 14.6 years.

77. $3 = e^{r5}$

$\ln 3 = \ln e^{5r}$

$\ln 3 = 5r$

$r = \frac{\ln 3}{5} \approx 0.220$

The money will triple in 5 years if the interest rate is approximately 22%.

78. **a.** $A = 14.6e^{kt}$

$29.3 = 14.6e^{k17}$

$\frac{29.3}{14.6} = e^{17k}$

$\ln\frac{29.3}{14.6} = \ln e^{17k}$

$\ln\frac{29.3}{14.6} = 17k$

$k = \frac{\ln\frac{29.3}{14.6}}{17} \approx 0.041$

b. Since 2005 is 25 years after 1980, find A for $t = 25$.

$A = 14.6e^{0.041(25)}$

$= 14.6e^{1.025} \approx 40.7$

The population will reach approximately 40.7 million in the year 2005.

c. $50 = 14.6e^{0.041t}$

$\frac{50}{14.6} = e^{0.041t}$

$\ln\frac{50}{14.6} = \ln e^{0.041t}$

$\ln\frac{50}{14.6} = 0.041t$

$t = \frac{\ln\frac{50}{14.6}}{0.041} \approx 30.0$

The population will reach 50 million approximately 30 years after 1980 in the year 2010.

79. $A = A_0 e^{-0.000121t}$

$15 = 100e^{-0.000121t}$

$\frac{15}{100} = e^{-0.000121t}$

$\ln\frac{3}{20} = \ln e^{-0.000121t}$

$\ln\frac{3}{20} = -0.000121t$

$t = \frac{\ln\frac{3}{20}}{-0.000121} \approx 15{,}679$

The paintings are approximately 15,679 years old.

80. Answers will vary.

Chapter 9 Test

1. $f(x) = 2^x$

$g(x) = 2^{x+1}$

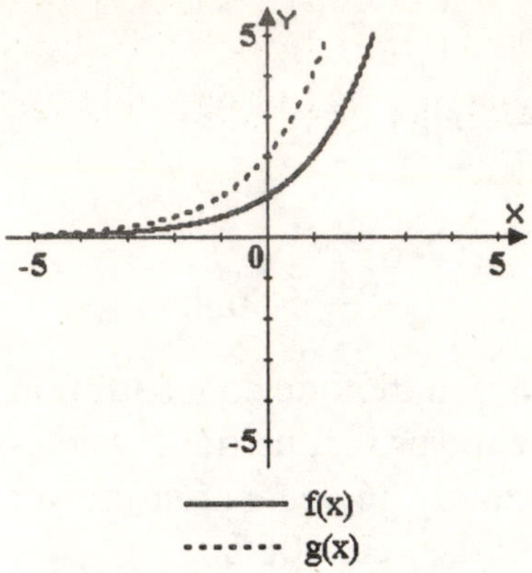

2. Semiannual Compounding:

$A = 3000\left(1 + \frac{0.065}{2}\right)^{2(10)}$

$A = 3000(1 + 0.0325)^{20}$

$A = 3000(1.0325)^{20} \approx 5687.51$

Continuous Compounding:

$A = 3000e^{0.06(10)}$

$A = 3000e^{0.6} \approx 5466.36$

Semiannual compounding at 6.5% yields a greater return. The difference in the yields is \$221.

3. $f(x) = x^2 + x$ and $g(x) = 3x - 1$

$(f \circ g)(x) = f(g(x)) = f(3x-1)$

$= (3x-1)^2 + (3x-1)$

$= 9x^2 - 6x + 1 + 3x - 1$

$= 9x^2 - 3x$

$(g \circ f)(x) = g(f(x))$

$= g(x^2 + x)$

$= 3(x^2 + x) - 1$

$= 3x^2 + 3x - 1$

4. $f(x) = 5x - 7$

$y = 5x - 7$

Interchange x and y and solve for y.

$x = 5y - 7$

$x + 7 = 5y$

$\frac{x+7}{5} = y$

$f^{-1}(x) = \frac{x+7}{5}$

5. **a.** Because the line satisfies the vertical line test, we know its inverse is a function.

b. $f(80) = 2000$

c. $f^{-1}(2000)$ represents the income, \$80,000, of a family that gives \$2000 to charity.

6. $\log_5 125 = 3$

$5^3 = 125$

7. $\sqrt{36} = 6$

$36^{\frac{1}{2}} = 6$

$\log_{36} 6 = \frac{1}{2}$

8. $f(x) = 3^x$

$g(x) = \log_3 x$

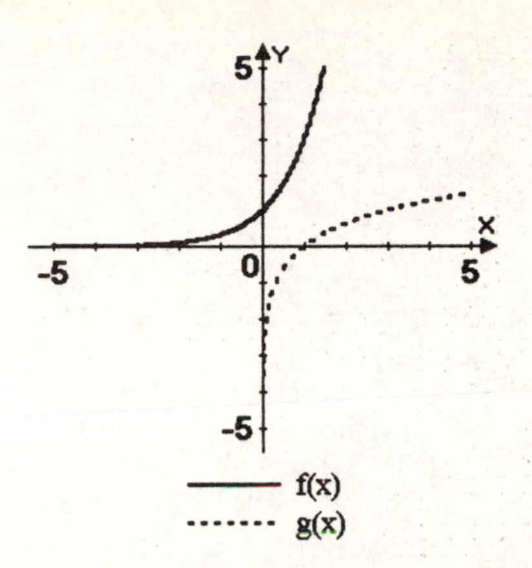

9. Since $\ln e^x = x$, $\ln e^{5x} = 5x$.

10. $\log_b b = 1$ because $b^1 = b$.

11. $\log_6 1 = 0$ because $6^0 = 1$.

12. $f(x) = \log_5 (x - 7)$

$x - 7 > 0$

$x > 7$

The domain of f is $\{x | x > 7\}$ or $(7, \infty)$.

13. $D = 10 \log \frac{I}{I_0}$

$D = 10 \log \frac{10^{12} I_0}{I_0}$

$= 10 \log 10^{12}$

$= 10(12) = 120$

The sound has a loudness of 120 decibels.

14. $\log_4 (64x^5) = \log_4 64 + \log_4 x^5$

$= 3 + 5 \log_4 x$

15.
$$\log_3 \frac{\sqrt[3]{x}}{81} = \log_3 \sqrt[3]{x} - \log_3 81$$
$$= \log_3 x^{\frac{1}{3}} - 4$$
$$= \frac{1}{3}\log_3 x - 4$$

16. $6\log x + 2\log y = \log x^6 + \log y^2$
$$= \log x^6 y^2$$

17. $\ln 7 - 3\ln x = \ln 7 - \ln x^3 = \ln \frac{7}{x^3}$

18. $\log_{15} 71 = \frac{\log 71}{\log 15} \approx 1.5741$

19.
$$3^{x-2} = 81$$
$$3^{x-2} = 3^4$$
$$x - 2 = 4$$
$$x = 6$$
The solution is 6 and the solution set is $\{6\}$.

20.
$$5^x = 1.4$$
$$\ln 5^x = \ln 1.4$$
$$x \ln 5 = \ln 1.4$$
$$x = \frac{\ln 1.4}{\ln 5} \approx 0.21$$
The solution is $\frac{\ln 1.4}{\ln 5} \approx 0.21$ and the solution set is $\left\{\frac{\ln 1.4}{\ln 5} \approx 0.21\right\}$.

21.
$$400e^{0.005x} = 1600$$
$$e^{0.005x} = \frac{1600}{400}$$
$$\ln e^{0.005x} = \ln 4$$
$$0.005x = \ln 4$$
$$x = \frac{\ln 4}{0.005} \approx 277.26$$
The solution is $\frac{\ln 4}{0.005} \approx 277.26$ and the solution set is $\left\{\frac{\ln 4}{0.005} \approx 277.26\right\}$.

22.
$$\log_{25} x = \frac{1}{2}$$
$$x = 25^{\frac{1}{2}} = \sqrt{25} = 5$$
The solution is 5 and the solution set is $\{5\}$.

23.
$$\log_6(4x-1) = 3$$
$$4x - 1 = 6^3$$
$$4x - 1 = 216$$
$$4x = 217$$
$$x = \frac{217}{4}$$
The solution is $\frac{217}{4}$ and the solution set is $\left\{\frac{217}{4}\right\}$.

24.
$$\log x + \log(x+15) = 2$$
$$\log(x(x+15)) = 2$$
$$x(x+15) = 10^2$$
$$x^2 + 15 = 100$$
$$x^2 + 15 - 100 = 0$$
$$(x+20)(x-5) = 0$$
Apply the zero product principle.

$$x+20=0 \quad \text{and} \quad x-5=0$$
$$x=-20 \qquad\qquad x=5$$

We disregard -20 because it would result in taking the logarithm of a negative number in the original equation. The solution is 5 and the solution set is $\{5\}$.

25. $2\ln 3x = 8$

$$\ln 3x = \frac{8}{2}$$
$$e^{\ln 3x} = e^4$$
$$3x = e^4$$
$$x = \frac{e^4}{3}$$

The solution is $\frac{e^4}{3}$ and the solution set is $\left\{\frac{e^4}{3}\right\}$.

26. **a.** $P(0) = 89.18e^{-0.004(0)}$

$$= 89.18e^0 = 89.18(1)$$
$$= 89.18$$

In 1959, 89.18% of married men were employed.

b. The percentage of married men who are employed is decreasing. We know this since the growth rate, k, is negative.

c. $77 = 89.18e^{-0.004t}$

$$\frac{77}{89.18} = e^{-0.004t}$$
$$\ln\frac{77}{89.18} = \ln e^{-0.004t}$$
$$\ln\frac{77}{89.18} = -0.004t$$
$$t = \frac{\ln\frac{77}{89.18}}{-0.004} \approx 36.7$$

77% of married men were employed approximately 37 years after 1959 in the year 1996.

27.

$$8000 = 4000\left(1+\frac{0.05}{4}\right)^{4t}$$
$$\frac{8000}{4000} = (1+0.0125)^{4t}$$
$$2 = (1.0125)^{4t}$$
$$\ln 2 = \ln(1.0125)^{4t}$$
$$\ln 2 = 4t\ln(1.0125)$$
$$\frac{\ln 2}{4\ln(1.0125)} = \frac{4t\ln(1.0125)}{4\ln(1.0125)}$$
$$t = \frac{\ln 2}{4\ln(1.0125)} \approx 13.9$$

It will take approximately 13.9 years for the money to grow to \$8000.

28. $2 = 1e^{r10}$

$$2 = e^{10r}$$
$$\ln 2 = \ln e^{10r}$$
$$\ln 2 = 10r$$
$$r = \frac{\ln 2}{10} \approx 0.069$$

The money will double in 10 years with an interest rate of approximately 6.9%.

29. $509 = 484e^{k(10)}$

$$\frac{509}{484} = e^{10k}$$

$$\ln\frac{509}{484} = \ln e^{10k}$$

$$\ln\frac{509}{484} = 10k$$

$$k = \frac{\ln\frac{509}{484}}{10} \approx 0.005$$

The exponential growth function is $A = 484e^{0.005t}$.

30.

$$A = A_0 e^{-0.000121t}$$

$$5 = 100e^{-0.000121t}$$

$$\frac{5}{100} = e^{-0.000121t}$$

$$\ln 0.05 = \ln e^{-0.000121t}$$

$$\ln 0.05 = -0.000121t$$

$$t = \frac{\ln 0.05}{-0.000121} \approx 24758$$

The man died approximately 24,758 years ago.

Cumulative Review Exercises

Chapters 1-9

1.

$$\sqrt{2x+5} - \sqrt{x+3} = 2$$

$$\sqrt{2x+5} = 2 + \sqrt{x+3}$$

$$\left(\sqrt{2x+5}\right)^2 = \left(2 + \sqrt{x+3}\right)^2$$

$$2x+5 = 4 + 4\sqrt{x+3} + x + 3$$

$$2x+5 = 7 + 4\sqrt{x+3} + x$$

$$x - 2 = 4\sqrt{x+3}$$

$$(x-2)^2 = \left(4\sqrt{x+3}\right)^2$$

$$x^2 - 4x + 4 = 16(x+3)$$

$$x^2 - 4x + 4 = 16x + 48$$

$$x^2 - 20x - 44 = 0$$

$$(x-22)(x+2) = 0$$

Apply the zero product principle.

$x - 22 = 0$ and $x + 2 = 0$

$x = 22$ $\quad$ $x = -2$

The solution, -2, does not check. The solution is 22 and the solution set is $\{22\}$.

2.

$$(x-5)^2 = -49$$

$$\sqrt{(x-5)^2} = \pm\sqrt{-49}$$

$$x - 5 = \pm 7i$$

$$x = 5 \pm 7i$$

The solutions are $5 - 7i$ and $5 + 7i$ and the solution set is $\{5 - 7i, 5 + 7i\}$.

3.

$$x^2 + x > 6$$

$$x^2 + x - 6 > 0$$

Solve the related quadratic equation.

$$x^2 + x - 6 = 0$$

$$(x+3)(x-2) = 0$$

Apply the zero product principle.

$x + 3 = 0$ or $x - 2 = 0$

$x = -3$ $\quad$ $x = 2$

The boundary points are -3 and 2.

Test Interval	Test Number	Test	Conclusion
$(-\infty,-3)$	-4	$(-4)^2+(-4)>6$ $12>6$, true	$(-\infty,-3)$ belongs to the solution set.
$(-3,2)$	0	$0^2+0>6$ $0>6$, false	$(-3,2)$ does not belong to the solution set.
$(2,\infty)$	3	$3^2+3>6$ $12>6$, true	$(2,\infty)$ does not belong to the solution set.

The solution set is $(-\infty,-3)\cup(2,\infty)$ or $\{x|x<-3 \text{ and } x>2\}$.

-3 2

4.
$$6x-3(5x+2)=4(1-x)$$
$$6x-15x-6=4-4x$$
$$-9x-6=4-4x$$
$$-5x=10$$
$$x=-2$$

The solution is –2 and the solution set is $\{-2\}$.

5.
$$\frac{2}{x-3}-\frac{3}{x+3}=\frac{12}{x^2-9}$$
$$\frac{2}{x-3}-\frac{3}{x+3}=\frac{12}{(x+3)(x-3)}$$
$$(x+3)(x-3)\left(\frac{2}{x-3}-\frac{3}{x+3}\right)=(x+3)(x-3)\left(\frac{12}{(x+3)(x-3)}\right)$$
$$2(x+3)-3(x-3)=12$$
$$2x+6-3x+9=12$$
$$-x+15=12$$
$$-x=-3$$
$$x=3$$

We disregard 3 since it would make one or more of the denominators in the original equation zero. There is no solution. The solution set is $\varnothing$ or $\{\ \}$.

6. $3x+2<4$ and $4-x>1$

$3x<2 \qquad -x>-3$

$x<\frac{2}{3} \qquad x<3$

$x<3$

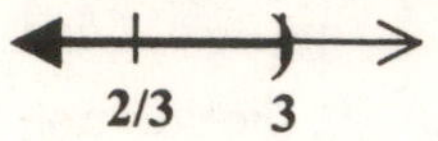

$x<\frac{2}{3}$

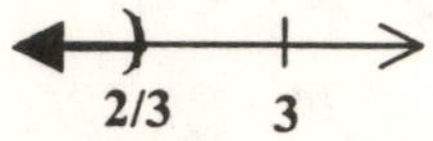

$x<3$ and $x<\frac{2}{3}$

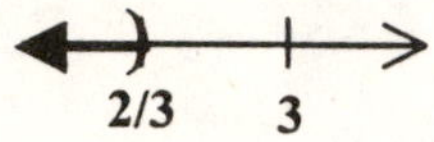

The solution set is $\left\{x \middle| x<\frac{2}{3}\right\}$ or $\left(-\infty,\frac{2}{3}\right)$.

7. $3x-2y+z=7$

$2x+3y-z=13$

$x-y+2z=-6$

Add the first two equations to eliminate z.

$3x-2y+z=7$

$2x+3y-z=13$

$5x+y=20$

Multiply the second equation by 2 and add to the third equation.

$4x+6y-2z=26$

$x-y+2z=-6$

$5x+5y=20$

We now have a system of two equations in two variables.

$5x+y=20$

$5x+5y=20$

Multiply the first equation by -1 and add to the second equation.

$-5x-y=-20$

$5x+5y=20$

$4y=0$

$y=0$

Back-substitute 0 for y to find x.

$5x+y=20$

$5x+0=20$

$5x=20$

$x=4$

Back-substitute 4 for x and 0 for y to find z.

$3x-2y+z=7$

$3(4)-2(0)+z=7$

$12+z=7$

$z=-5$

The solution is $(4,0,-5)$ and the solution set is $\{(4,0,-5)\}$.

8. $\log_9 x+\log_9(x-8)=1$

$\log_9(x(x-8))=1$

$(x(x-8))=9^1$

$x^2-8x=9$

$x^2-8x-9=0$

$(x-9)(x+1)=0$

Apply the zero product principle.

$x-9=0$ and $x+1=0$

$x=9 \qquad x=-1$

We disregard -1 because it would result in taking the logarithm of a negative number in the original equation. The solution is 9 and the solution set is $\{9\}$.

9. $f(x) = (x+2)^2 - 4$

Since $a = 1$ is positive, the parabola opens upward. The vertex of the parabola is $(h, k) = (-2, -4)$. Replace $f(x)$ with 0 to find x–intercepts.

$$0 = (x+2)^2 - 4$$
$$(x+2)^2 = 4$$
$$x + 2 = \pm 2$$
$$x = -2 \pm 2$$
$$x = -2 - 2 \text{ or } -2 + 2$$
$$x = -4 \text{ or } 0$$

The x–intercepts are –4 and 0. Set $x = 0$ and solve for y to obtain the y–intercept.

$$y = (0+2)^2 - 4$$
$$y = (2)^2 - 4$$
$$y = 4 - 4$$
$$y = 0$$

10. $y < -3x + 5$

First, find the intercepts to the equation $y = -3x + 5$. Find the x–intercept by setting $y = 0$.

$$0 = -3x + 5$$
$$-5 = -3x$$
$$\frac{5}{3} = x$$

Find the y–intercept by setting $x = 0$.

$$y = -3(0) + 5$$
$$y = 5$$

Next, use the origin as a test point.

$$0 < -3(0) + 5$$
$$0 < 0 + 5$$
$$0 < 5$$

This is a true statement. This means that the point, $(0, 0)$, will fall in the shaded half-plane.

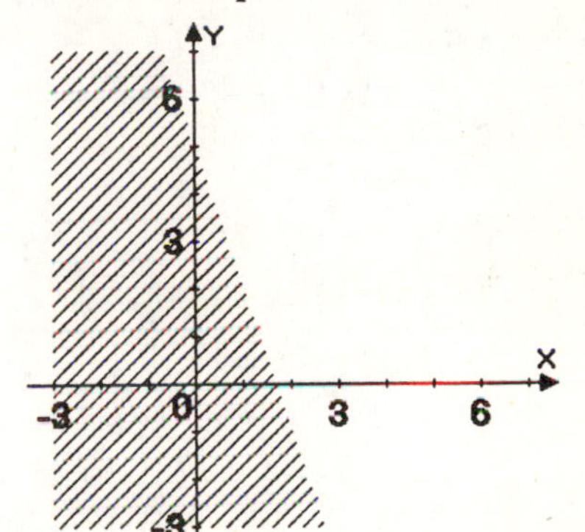

11. $f(x) = 3^{x-2}$

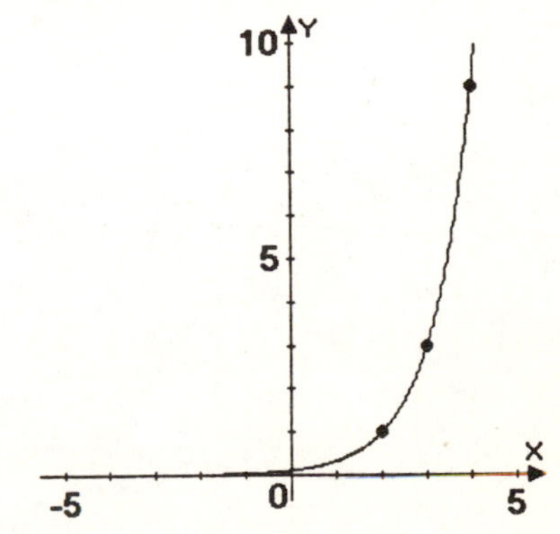

12. $\frac{2x+1}{x-5}-\frac{4}{x^2-3x-10}$
$=\frac{2x+1}{x-5}-\frac{4}{(x-5)(x+2)}$
$=\frac{2x+1}{x-5}\cdot\frac{(x+2)}{(x+2)}-\frac{4}{(x-5)(x+2)}$
$=\frac{(2x+1)(x+2)}{(x-5)(x+2)}-\frac{4}{(x-5)(x+2)}$
$=\frac{(2x+1)(x+2)-4}{(x-5)(x+2)}$
$=\frac{2x^2+5x+2-4}{(x-5)(x+2)}$
$=\frac{2x^2+5x-2}{(x-5)(x+2)}$

13. $\frac{\frac{1}{x-1}+1}{\frac{1}{x+1}-1}=\frac{(x-1)(x+1)}{(x-1)(x+1)}\cdot\frac{\frac{1}{x-1}+1}{\frac{1}{x+1}-1}$
$=\frac{(x+1)+(x-1)(x+1)}{(x-1)-(x-1)(x+1)}$
$=\frac{x+1+x^2-1}{x-1-(x^2-1)}$
$=\frac{x+x^2}{x-1-x^2+1}$
$=\frac{x^2+x}{x-x^2}=\frac{x(x+1)}{x(1-x)}$
$=\frac{x+1}{1-x}$ or $-\frac{x+1}{x-1}$

14. $\frac{6}{\sqrt{5}-\sqrt{2}}=\frac{6}{\sqrt{5}-\sqrt{2}}\cdot\frac{\sqrt{5}+\sqrt{2}}{\sqrt{5}+\sqrt{2}}$
$=\frac{6(\sqrt{5}+\sqrt{2})}{5-2}=\frac{6(\sqrt{5}+\sqrt{2})}{3}$
$=2(\sqrt{5}+\sqrt{2})=2\sqrt{5}+2\sqrt{2}$

15. $8\sqrt{45}+2\sqrt{5}-7\sqrt{20}$
$=8\sqrt{9\cdot 5}+2\sqrt{5}-7\sqrt{4\cdot 5}$
$=8\cdot 3\sqrt{5}+2\sqrt{5}-7\cdot 2\sqrt{5}$
$=24\sqrt{5}+2\sqrt{5}-14\sqrt{5}=12\sqrt{5}$

16. $\frac{5}{\sqrt[3]{2x^2y}}=\frac{5}{\sqrt[3]{2x^2y}}\cdot\frac{\sqrt[3]{2^2xy^2}}{\sqrt[3]{2^2xy^2}}$
$=\frac{5\sqrt[3]{4xy^2}}{\sqrt[3]{2^3x^3y^3}}=\frac{5\sqrt[3]{4xy^2}}{2xy}$

17. $5ax+5ay-4bx-4by$
$=5a(x+y)-4b(x+y)$
$=(x+y)(5a-4b)$

18. $5\log x-\frac{1}{2}\log y=\log x^5-\log y^{\frac{1}{2}}$
$=\log\frac{x^5}{y^{\frac{1}{2}}}=\log\frac{x^5}{\sqrt{y}}$

19. $\frac{1}{p}+\frac{1}{q}=\frac{1}{f}$
$pqf\left(\frac{1}{p}\right)+pqf\left(\frac{1}{q}\right)=pqf\left(\frac{1}{f}\right)$
$qf+pf=pq$
$qf=pq-pf$
$qf=p(q-f)$
$p=\frac{qf}{q-f}$

20. Since a denominator cannot equal zero, exclude from the domain all values which make $x^2+2x-15=0$.

$x^2+2x-15=0$

$(x+5)(x-3)=0$

Apply the zero product principle.

$x+5=0$ and $x-3=0$

$x=-5$ $x=3$

The domain is $\{x|x$ is a real number and $x \neq -5$ and $x \neq 3\}$ or $(-\infty,-5)\cup(-5,3)\cup(3,\infty)$.

21. To find the domain, find all values of x for which the radicand is greater than or equal to zero.

$2x-6 \geq 0$

$2x \geq 6$

$x \geq 3$

The domain is $\{x|x \geq 3\}$ or $[3,\infty)$.

22. To find the domain, find all values of x for which $1-x$ is greater than zero.

$1-x>0$

$-x>-1$

$x<1$

The domain is $\{x|x<1\}$ or $(-\infty,1)$.

23. Let x = the width of the garden
Let $2x + 2$ = the length of the garden

$2x+2(2x+2)=22$

$2x+4x+4=22$

$6x+4=22$

$6x=18$

$x=3$

The width of the garden is 3 feet and the length of the garden is $2x+2$ $=2(3)+2=6+2=8$ feet.

24. Let x = the salary before the raise

$x+0.06x=19,610$

$1.06x=19,610$

$x=\frac{19,610}{1.06}=18,500$

The salary before the raise is $18,500.

25. $F(t)=1-k\ln(t+1)$

$\frac{1}{2}=1-k\ln(3+1)$

$-\frac{1}{2}=-k\ln 4$

$k\ln 4=\frac{1}{2}$

$k=\frac{1}{2\ln 4}\approx 0.3607$

The equation becomes
$F(t)=1-0.3607\ln(t+1)$.

Next find $F(6)$.

$F(6)=1-0.3607\ln(6+1)$

$=1-0.3607\ln 7 \approx 0.298$

After 6 hours, 0.298 or $\frac{298}{1000}$ people will remember all the words.

Chapter 10

Check Points 10.1

1.
$$d = \sqrt{(x_2 - x_1)^2 + (y_2 - y_1)^2}$$
$$= \sqrt{(5-2)^2 + (2-(-2))^2}$$
$$= \sqrt{3^2 + 4^2} = \sqrt{9+16}$$
$$= \sqrt{25} = 5$$
The distance between the point is 5 units.

2.
$$\text{Midpoint} = \left(\frac{x_1 + x_2}{2}, \frac{y_1 + y_2}{2}\right)$$
$$= \left(\frac{1+7}{2}, \frac{2+(-3)}{2}\right)$$
$$= \left(\frac{8}{2}, \frac{-1}{2}\right) = \left(4, -\frac{1}{2}\right)$$
The midpoint is $\left(4, -\frac{1}{2}\right)$.

3.
$$(x-h)^2 + (y-k)^2 = r^2$$
$$(x-0)^2 + (y-0)^2 = 4^2$$
$$x^2 + y^2 = 16$$

4.
$$(x-h)^2 + (y-k)^2 = r^2$$
$$(x-5)^2 + (y-(-6))^2 = 10^2$$
$$(x-5)^2 + (y+6)^2 = 100$$

5.
$$(x+3)^2 + (y-1)^2 = 4$$
$$(x-(-3))^2 + (y-1)^2 = 2^2$$
The center is $(-3,1)$ and the radius is 2 units.

6.
$$x^2 + y^2 + 4x - 4y - 1 = 0$$
$$(x^2 + 4x \quad) + (y^2 - 4y \quad) = 1$$
Complete the squares.
$$\left(\frac{b}{2}\right)^2 = \left(\frac{4}{2}\right)^2 = (2)^2 = 4$$
$$\left(\frac{b}{2}\right)^2 = \left(\frac{-4}{2}\right)^2 = (-2)^2 = 4$$
$$(x^2 + 4x + 4) + (y^2 - 4y + 4) = 1 + 4 + 4$$
$$(x+2)^2 + (y-2)^2 = 9$$

Problem Set 10.1

Practice Exercises

1.
$$d = \sqrt{(14-2)^2 + (8-3)^2}$$
$$= \sqrt{12^2 + 5^2} = \sqrt{144+25}$$
$$= \sqrt{169} = 13$$
The distance is 13 units.

3.
$$d = \sqrt{(6-4)^2 + (3-1)^2}$$
$$= \sqrt{2^2 + 2^2} = \sqrt{4+4}$$
$$= \sqrt{8} = \sqrt{4 \cdot 2} = 2\sqrt{2} \approx 2.83$$
The distance is $2\sqrt{2}$ or 2.83 units.

5.
$$d = \sqrt{(-3-0)^2 + (4-0)^2}$$
$$= \sqrt{(-3)^2 + 4^2} = \sqrt{9+16}$$
$$= \sqrt{25} = 5$$
The distance is 5 units.

7. $d = \sqrt{(3-(-2))^2 + (-4-(-6))^2}$

$= \sqrt{5^2 + 2^2} = \sqrt{25+4}$

$= \sqrt{29} \approx 5.39$

The distance is $\sqrt{29}$ or 5.39 units.

9. $d = \sqrt{(4-0)^2 + (1-(-3))^2}$

$= \sqrt{4^2 + 4^2} = \sqrt{16+16}$

$= \sqrt{32} = \sqrt{16 \cdot 2} = 4\sqrt{2} \approx 5.66$

The distance is $4\sqrt{2}$ or 5.66 units.

11. $d = \sqrt{(3.5-(-0.5))^2 + (8.2-6.2)^2}$

$= \sqrt{4^2 + 2^2} = \sqrt{16+4}$

$= \sqrt{20} = \sqrt{4 \cdot 5} = 2\sqrt{5} \approx 4.47$

The distance is $2\sqrt{5}$ or 4.47 units.

13. $d = \sqrt{\left(\sqrt{5}-0\right)^2 + \left(0-\left(-\sqrt{3}\right)\right)^2}$

$= \sqrt{\left(\sqrt{5}\right)^2 + \left(\sqrt{3}\right)^2} = \sqrt{5+3}$

$= \sqrt{8} = \sqrt{4 \cdot 2} = 2\sqrt{2} \approx 2.83$

The distance is $2\sqrt{2}$ or 2.83 units.

15. $d = \sqrt{\left(3\sqrt{3}-\left(-\sqrt{3}\right)\right)^2 + \left(\sqrt{5}-4\sqrt{5}\right)^2}$

$= \sqrt{\left(4\sqrt{3}\right)^2 + \left(-3\sqrt{5}\right)^2}$

$= \sqrt{16 \cdot 3 + 9 \cdot 5} = \sqrt{48+45}$

$= \sqrt{93} \approx 9.64$

The distance is $\sqrt{93}$ or 9.64 units.

17. $d = \sqrt{\left(\frac{7}{3}-\frac{1}{3}\right)^2 + \left(\frac{1}{5}-\frac{6}{5}\right)^2}$

$= \sqrt{\left(\frac{6}{3}\right)^2 + \left(-\frac{5}{5}\right)^2}$

$= \sqrt{2^2 + (-1)^2} = \sqrt{4+1}$

$= \sqrt{5} \approx 2.24$

The distance is $\sqrt{5}$ or 2.24 units.

19. $\text{Midpoint} = \left(\frac{6+2}{2}, \frac{8+4}{2}\right)$

$= \left(\frac{8}{2}, \frac{12}{2}\right) = (4, 6)$

The midpoint is $(4, 6)$.

21. $\text{Midpoint} = \left(\frac{-2+(-6)}{2}, \frac{-8+(-2)}{2}\right)$

$= \left(\frac{-8}{2}, \frac{-10}{2}\right) = (-4, -5)$

The midpoint is $(-4, -5)$.

23. $\text{Midpoint} = \left(\frac{-3+6}{2}, \frac{-4+(-8)}{2}\right)$

$= \left(\frac{3}{2}, \frac{-12}{2}\right) = \left(\frac{3}{2}, -6\right)$

The midpoint is $\left(\frac{3}{2}, -6\right)$.

25. $\text{Midpoint} = \left(\frac{-\frac{7}{2}+\left(-\frac{5}{2}\right)}{2}, \frac{\frac{3}{2}+\left(-\frac{11}{2}\right)}{2}\right)$

$= \left(\frac{-\frac{12}{2}}{2}, \frac{-\frac{8}{2}}{2}\right)$

$= \left(-\frac{12}{2} \cdot \frac{1}{2}, -\frac{8}{2} \cdot \frac{1}{2}\right)$

$$=\left(-\frac{12}{4},-\frac{8}{4}\right)=(-3,-2)$$

The midpoint is $(-3,-2)$.

27.

$$\text{Midpoint}=\left(\frac{8+(-6)}{2},\frac{3\sqrt{5}+7\sqrt{5}}{2}\right)$$

$$=\left(\frac{2}{2},\frac{10\sqrt{5}}{2}\right)=\left(1,5\sqrt{5}\right)$$

The midpoint is $\left(1,5\sqrt{5}\right)$.

29.

$$\text{Midpoint}=\left(\frac{\sqrt{18}+\sqrt{2}}{2},\frac{-4+4}{2}\right)$$

$$=\left(\frac{\sqrt{9\cdot 2}+\sqrt{2}}{2},\frac{0}{2}\right)$$

$$=\left(\frac{3\sqrt{2}+\sqrt{2}}{2},0\right)$$

$$=\left(\frac{4\sqrt{2}}{2},0\right)=\left(2\sqrt{2},0\right)$$

The midpoint is $\left(2\sqrt{2},0\right)$.

31. $(x-h)^2+(y-k)^2=r^2$

$(x-0)^2+(y-0)^2=7^2$

$x^2+y^2=49$

33. $(x-h)^2+(y-k)^2=r^2$

$(x-3)^2+(y-2)^2=5^2$

$(x-3)^2+(y-2)^2=25$

35. $(x-h)^2+(y-k)^2=r^2$

$(x-(-1))^2+(y-4)^2=2^2$

$(x+1)^2+(y-4)^2=4$

37. $(x-h)^2+(y-k)^2=r^2$

$(x-(-3))^2+(y-(-1))^2=\left(\sqrt{3}\right)^2$

$(x+3)^2+(y+1)^2=3$

39. $(x-h)^2+(y-k)^2=r^2$

$(x-(-4))^2+(y-0)^2=10^2$

$(x+4)^2+y^2=100$

41. $x^2+y^2=16$

$(x-0)^2+(y-0)^2=4^2$

The center is $(0,0)$ and the radius is 4 units.

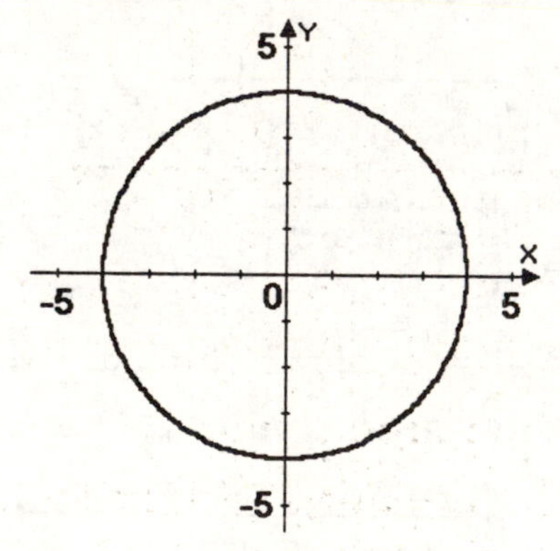

43. $(x-3)^2+(y-1)^2=36$

$(x-3)^2+(y-1)^2=6^2$

The center is $(3,1)$ and the radius is 6 units.

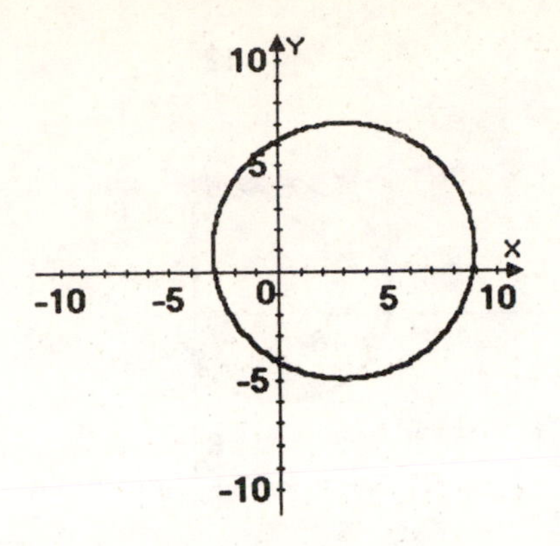

45. $$(x+3)^2+(y-2)^2=4$$

$$(x-(-3))^2+(y-2)^2=2^2$$

The center is $(-3,2)$ and the radius is 2 units.

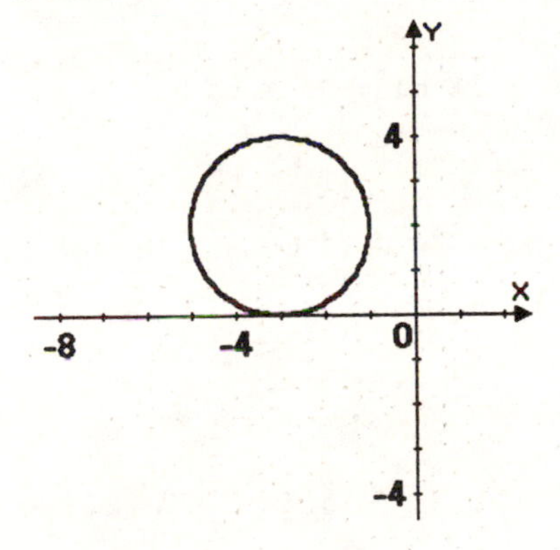

47. $$(x+2)^2+(y+2)^2=4$$

$$(x-(-2))^2+(y-(-2))^2=2^2$$

The center is $(-2,-2)$ and the radius is 2 units.

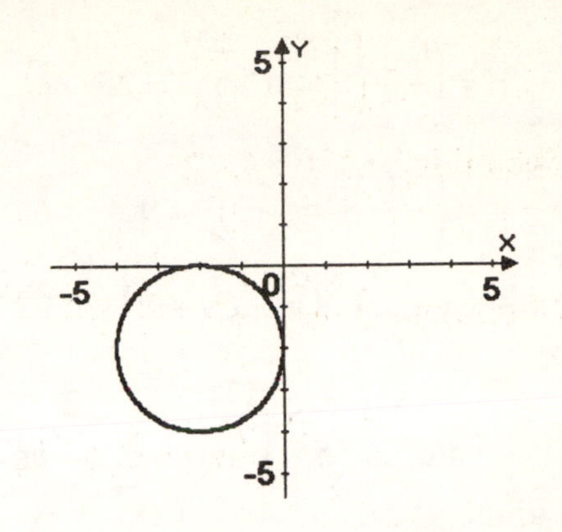

49. $$x^2+y^2+6x+2y+6=0$$

$$(x^2+6x\quad)+(y^2+2y\quad)=-6$$

Complete the squares.

$$\left(\frac{b}{2}\right)^2=\left(\frac{6}{2}\right)^2=(3)^2=9$$

$$\left(\frac{b}{2}\right)^2=\left(\frac{2}{2}\right)^2=(1)^2=1$$

$$(x^2+6x+9)+(y^2+2y+1)=-6+9+1$$

$$(x+3)^2+(y+1)^2=4$$

The center is $(-3,-1)$ and the radius is 2 units.

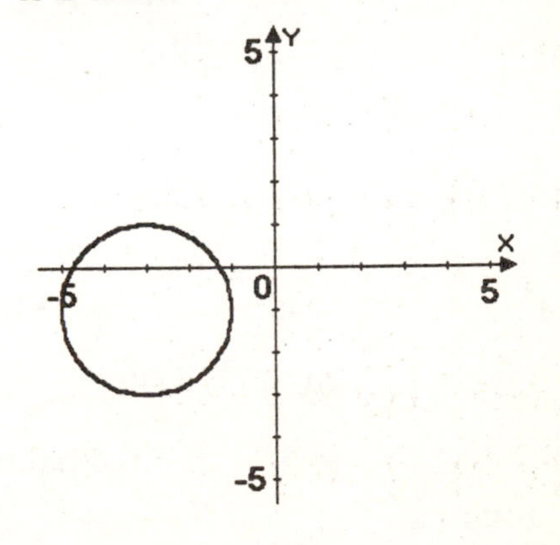

51. $$x^2+y^2-10x-6y-30=0$$

$$(x^2-10x\quad)+(y^2-6y\quad)=30$$

Complete the squares.

$$\left(\frac{b}{2}\right)^2 = \left(\frac{-10}{2}\right)^2 = (-5)^2 = 25$$

$$\left(\frac{b}{2}\right)^2 = \left(\frac{-6}{2}\right)^2 = (-3)^2 = 9$$

$$(x^2 - 10x + 25) + (y^2 - 6y + 9) = 30 + 25 + 9$$

$$(x-5)^2 + (y-3)^2 = 64$$

The center is $(5,3)$ and the radius is 8 units.

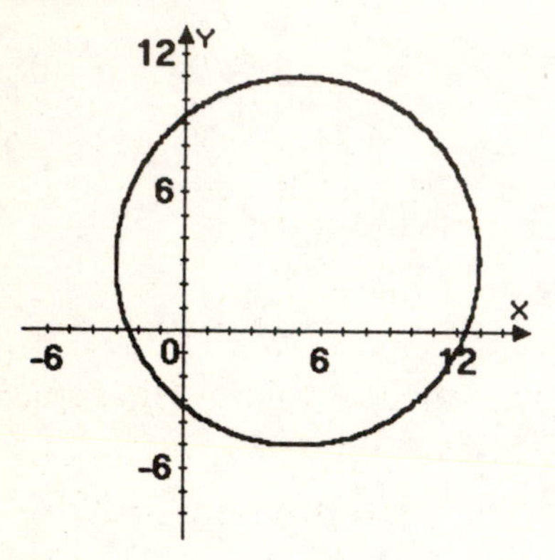

53. $$x^2 + y^2 + 8x - 2y - 8 = 0$$

$$(x^2 + 8x \quad) + (y^2 - 2y \quad) = 8$$

Complete the squares.

$$\left(\frac{b}{2}\right)^2 = \left(\frac{8}{2}\right)^2 = (4)^2 = 16$$

$$\left(\frac{b}{2}\right)^2 = \left(\frac{-2}{2}\right)^2 = (-1)^2 = 4$$

$$(x^2 + 8x + 16) + (y^2 - 2y + 1) = 8 + 16 + 1$$

$$(x+4)^2 + (y-1)^2 = 25$$

The center is $(-4,1)$ and the radius is 5 units.

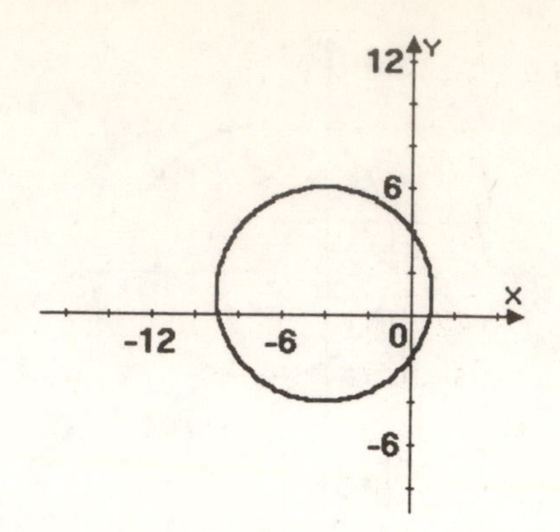

55. $$x^2 - 2x + y^2 - 15 = 0$$

$$(x^2 - 2x \quad) + y^2 = 15$$

Complete the squares.

$$\left(\frac{b}{2}\right)^2 = \left(\frac{-2}{2}\right)^2 = (-1)^2 = 1$$

$$(x^2 - 2x + 1) + y^2 = 15 + 1$$

$$(x-1)^2 + y^2 = 16$$

The center is $(1,0)$ and the radius is 4 units.

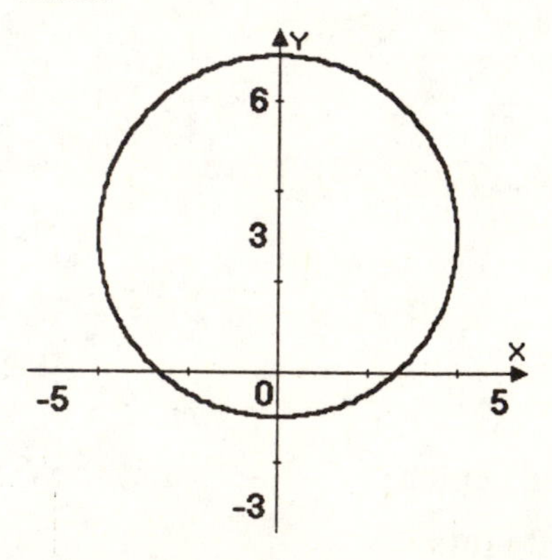

Application Exercises

57. First find the distance from Bangkok to Phnom Penh.

$$d=\sqrt{(65-(-115))^2+(70-170)^2}$$
$$=\sqrt{180^2+(-100)^2}$$
$$=\sqrt{32400+10000}$$
$$=\sqrt{42400}\approx 205.9$$

The distance is approximately 205.9 miles.

$$t=\frac{d}{r}=\frac{205.9}{400}\approx 0.5$$

It will take approximately 0.5 hours or 30 minutes to make the flight.

59. The center of the circle is (0, 68 +14) or (0, 82). The radius of the Ferris wheel is 68 feet. The equation of the circular wheel is:

$$x^2+(y-82)^2=68^2$$
$$x^2+(y-82)^2=4624.$$

Writing in Mathematics

61. Answers will vary.

63. Answers will vary.

65. Answers will vary.

67. Answers will vary.

Technology Exercises

69. $(y+1)^2=36-(x-3)^2$

$$y+1=\pm\sqrt{36-(x-3)^2}$$
$$y=-1\pm\sqrt{36-(x-3)^2}$$

Critical Thinking Exercises

71. Statement **d.** is true.

Statement **a.** is false. The equation of the circle is $x^2+y^2=16^2$ or $x^2+y^2=256$.

Statement **b.** is false. The circle is centered at $(3,-5)$.

Statement **c.** is false. The graph is not a circle. The x– and y–terms are not positive squares.

73. **a.** $d_{(x_1,y_1) \text{ and } \left(\frac{x_1+x_2}{2},\frac{y_1+y_2}{2}\right)}$

$$= \sqrt{\left(x_1 - \frac{x_1+x_2}{2}\right)^2 + \left(y_1 - \frac{y_1+y_2}{2}\right)^2} = \sqrt{\left(\frac{2x_1}{2} - \frac{x_1+x_2}{2}\right)^2 + \left(\frac{2y_1}{2} - \frac{y_1+y_2}{2}\right)^2}$$

$$= \sqrt{\left(\frac{2x_1 - x_1 - x_2}{2}\right)^2 + \left(\frac{2y_1 - y_1 - y_2}{2}\right)^2} = \sqrt{\left(\frac{x_1 - x_2}{2}\right)^2 + \left(\frac{y_1 - y_2}{2}\right)^2}$$

$$= \sqrt{\frac{(x_1 - x_2)^2}{4} + \frac{(y_1 - y_2)^2}{4}} = \sqrt{\frac{1}{4}\left((x_1 - x_2)^2 + (y_1 - y_2)^2\right)}$$

$$= \frac{1}{2}\sqrt{(x_1 - x_2)^2 + (y_1 - y_2)^2}$$

$d_{\left(\frac{x_1+x_2}{2},\frac{y_1+y_2}{2}\right),(x_2,y_2)}$

$$= \sqrt{\left(\frac{x_1+x_2}{2} - x_2\right)^2 + \left(\frac{y_1+y_2}{2} - y_2\right)^2} = \sqrt{\left(\frac{x_1+x_2}{2} - \frac{2x_2}{2}\right)^2 + \left(\frac{y_1+y_2}{2} - \frac{2y_2}{2}\right)^2}$$

$$= \sqrt{\left(\frac{x_1 + x_2 - 2x_2}{2}\right)^2 + \left(\frac{y_1 + y_2 - 2y_2}{2}\right)^2} = \sqrt{\left(\frac{x_1 - x_2}{2}\right)^2 + \left(\frac{y_1 - y_2}{2}\right)^2}$$

$$= \sqrt{\frac{(x_1 - x_2)^2}{4} + \frac{(y_1 - y_2)^2}{4}} = \sqrt{\frac{1}{4}\left((x_1 - x_2)^2 + (y_1 - y_2)^2\right)}$$

$$= \frac{1}{2}\sqrt{(x_1 - x_2)^2 + (y_1 - y_2)^2}$$

b. If the points are collinear,

$$d_{(x_1,y_1),\left(\frac{x_1+x_2}{2},\frac{y_1+y_2}{2}\right)} + d_{\left(\frac{x_1+x_2}{2},\frac{y_1+y_2}{2}\right),(x_2,y_2)} = d_{(x_1,y_1),(x_2,y_2)}.$$

$$d_{(x_1,y_1),(x_2,y_2)} = \sqrt{(x_1 - x_2)^2 + (y_1 - y_2)^2}$$

$$d_{(x_1,y_1),\left(\frac{x_1+x_2}{2},\frac{y_1+y_2}{2}\right)} + d_{\left(\frac{x_1+x_2}{2},\frac{y_1+y_2}{2}\right),(x_2,y_2)}$$

$$= \frac{1}{2}\sqrt{(x_1 - x_2)^2 + (y_1 - y_2)^2} + \frac{1}{2}\sqrt{(x_1 - x_2)^2 + (y_1 - y_2)^2} = \sqrt{(x_1 - x_2)^2 + (y_1 - y_2)^2}$$

Since the distances are equal, we conclude that the points are collinear.

75. Find the center of the circle by finding the midpoint of the line segment.

$$\text{Midpoint} = \left(\frac{x_1+x_2}{2}, \frac{y_1+y_2}{2}\right)$$
$$= \left(\frac{-7+1}{2}, \frac{2+2}{2}\right)$$
$$= \left(\frac{-6}{2}, \frac{4}{2}\right) = (-3, 2)$$

Find the diameter of the circle by finding the distance between the points.

$$d = \sqrt{(1-(-7))^2 + (2-2)^2}$$
$$= \sqrt{(8)^2 + (0)^2} = \sqrt{64} = 8$$

The radius is half the diameter, so $r = \frac{d}{2} = \frac{8}{2} = 4.$

The equation of the circle in standard form is:

$$(x-h)^2 + (y-k)^2 = r^2$$
$$(x-(-3))^2 + (y-2)^2 = 4^2$$
$$(x+3)^2 + (y-2)^2 = 16.$$

The equation of the circle in general form is:

$$(x+3)^2 + (y-2)^2 = 16$$
$$x^2 + 6x + 9 + y^2 - 4y + 4 = 16$$
$$x^2 + y^2 + 6x - 4y + 13 = 16$$
$$x^2 + y^2 + 6x - 4y - 3 = 0.$$

77. The center of the circle with equation, $x^2 + y^2 = 25$, is the point (0,0). First, find the slope of the line going through the center and the point, (3, −4).

$$m = \frac{y_2 - y_1}{x_2 - x_1} = \frac{-4-0}{3-0} = -\frac{4}{3}$$

Since the tangent line is perpendicular to the line going through the center and the point, (3, −4), we know that its slope will be $\frac{3}{4}$. We can now write the point-slope equation of the line.

$$y-(-4) = \frac{3}{4}(x-3)$$
$$y+4 = \frac{3}{4}(x-3)$$

Review Exercises

78.
$$f(g(x)) = f(3x+4) = (3x+4)^2 - 2$$
$$= 9x^2 + 24x + 16 - 2$$
$$= 9x^2 + 24x + 14$$
$$g(f(x)) = g(x^2-2) = 3(x^2-2) + 4$$
$$= 3x^2 - 6 + 4 = 3x^2 - 2$$

79.
$$2x = \sqrt{7x-3} + 3$$
$$2x - 3 = \sqrt{7x-3}$$
$$(2x-3)^2 = 7x - 3$$
$$4x^2 - 12x + 9 = 7x - 3$$
$$4x^2 - 19x + 12 = 0$$
$$(4x-3)(x-4) = 0$$

Apply the zero product principle.

$$4x - 3 = 0 \qquad x - 4 = 0$$
$$4x = 3 \qquad x = 4$$
$$x = \frac{3}{4}$$

The solution $\frac{3}{4}$ does not check. The

solution is 4 and the solution set is $\{4\}$.

80.
$$|2x-5|<10$$
$$-10<2x-5<10$$
$$-10+5<2x-5+5<10+5$$
$$-5<2x<15$$
$$-\frac{5}{2}<x<\frac{15}{2}$$

The solution set is $\left\{x \middle| -\frac{5}{2}<x<\frac{15}{2}\right\}$ or $\left(-\frac{5}{2},\frac{15}{2}\right)$.

Check Points 10.2

1. $$\frac{x^2}{36}+\frac{y^2}{9}=1$$

Because the denominator of the x^2 – term is greater than the denominator of the y^2 – term, the major axis is horizontal. Since $a^2=36$, $a=6$ and the vertices are $(-6,0)$ and $(6,0)$. Since $b^2=9$, $b=3$ and endpoints of the minor axis are $(0,-3)$ and $(0,3)$.

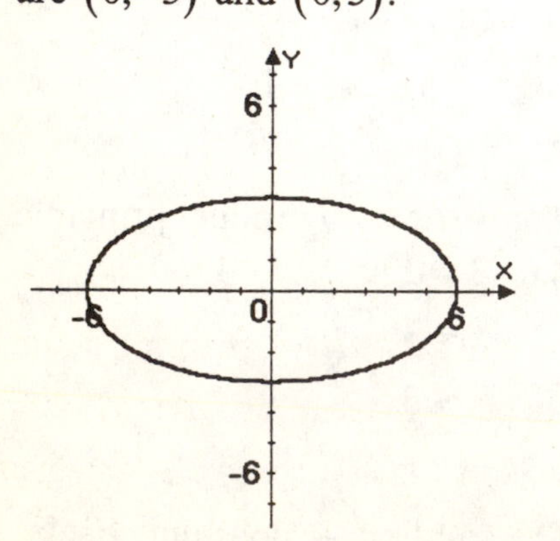

2.
$$16x^2+9y^2=144$$
$$\frac{16x^2}{144}+\frac{9y^2}{144}=\frac{144}{144}$$
$$\frac{x^2}{9}+\frac{y^2}{16}=1$$

Because the denominator of the y^2 – term is greater than the denominator of the x^2 – term, the major axis is vertical. Since $a^2=16$, $a=4$ and the vertices are $(0,-4)$ and $(0,4)$. Since $b^2=9$, $b=3$ and endpoints of the minor axis are $(-3,0)$ and $(3,0)$.

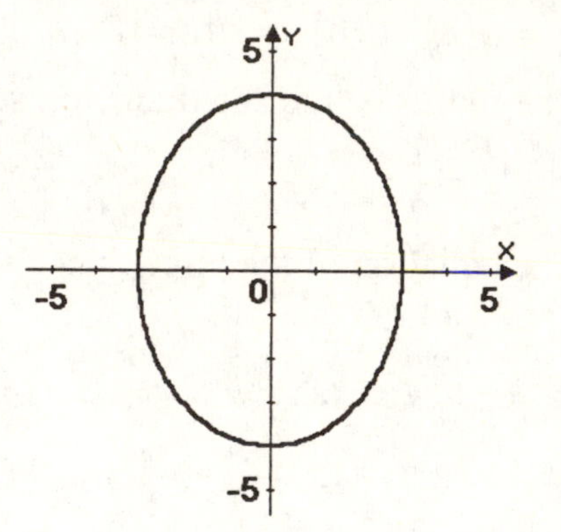

3. $$\frac{(x+1)^2}{9}+\frac{(y-2)^2}{4}=1$$

The center of the ellipse is $(-1,2)$. Because the denominator of the x^2 – term is greater than the denominator of the y^2 – term, the major axis is horizontal. Since $a^2=9$, $a=3$ and the vertices lie 3 units to the left and right of the center. Since $b^2=4$, $b=2$ and endpoints of the minor axis lie two units above and below the center.

Center	Vertices	Endpoints of Minor Axis
$(-1,2)$	$(-1-3,2)$ $=(-4,2)$	$(-1,2-2)$ $=(-1,0)$
	$(-1+3,2)$ $=(2,2)$	$(-1,2+2)$ $=(-1,4)$

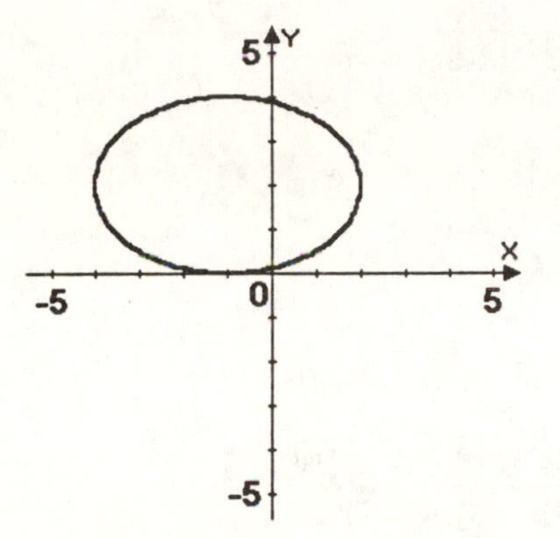

4. Since the truck is 12 feet wide, we need to determine the height of the archway at $\frac{12}{2}=6$ feet from the center.

$$\frac{x^2}{400}+\frac{y^2}{100}=1$$
$$\frac{6^2}{400}+\frac{y^2}{100}=1$$
$$\frac{36}{400}+\frac{y^2}{100}=1$$
$$\frac{9}{100}+\frac{y^2}{100}=1$$
$$100\left(\frac{9}{100}+\frac{y^2}{100}\right)=100(1)$$
$$9+y^2=100$$
$$y^2=91$$
$$y=\sqrt{91}\approx 9.54$$

The height of the archway 6 feet from the center is approximately 9.54 feet. Since the truck is 9 feet high, the truck will clear the archway.

Problem Set 10.2

Practice Exercises

1. $\frac{x^2}{16}+\frac{y^2}{4}=1$

Because the denominator of the x^2 – term is greater than the denominator of the y^2 – term, the major axis is horizontal. Since $a^2=16$, $a=4$ and the vertices are $(-4,0)$ and $(4,0)$. Since $b^2=4$, $b=2$ and endpoints of the minor axis are $(0,-2)$ and $(0,2)$.

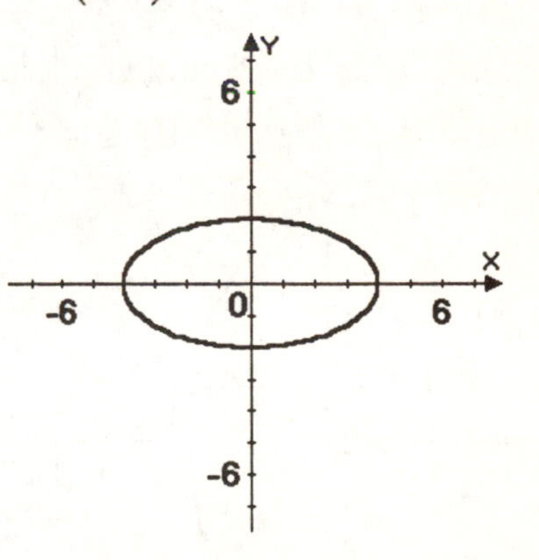

3. $\frac{x^2}{9}+\frac{y^2}{36}=1$

Because the denominator of the y^2 – term is greater than the denominator of the x^2 – term, the major axis is vertical. Since $a^2=36$, $a=6$ and the vertices are $(0,-6)$ and $(0,6)$.

Since $b^2 = 9$, $b = 3$ and endpoints of the minor axis are $(-3,0)$ and $(3,0)$.

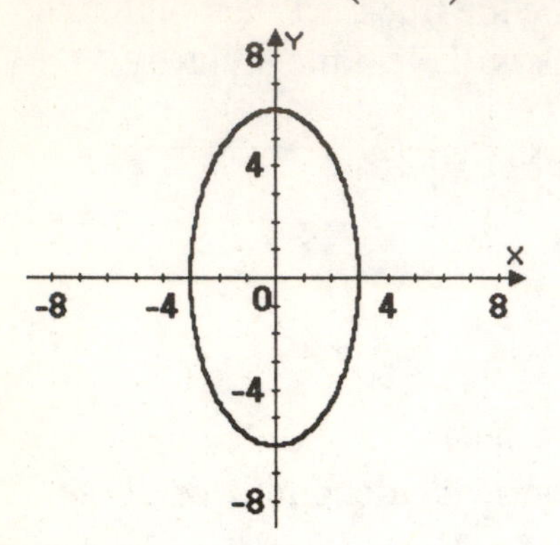

5. $\dfrac{x^2}{25}+\dfrac{y^2}{64}=1$

Because the denominator of the y^2 – term is greater than the denominator of the x^2 – term, the major axis is vertical. Since $a^2 = 64$, $a = 8$ and the vertices are $(0,-8)$ and $(0,8)$. Since $b^2 = 25$, $b = 5$ and endpoints of the minor axis are $(-5,0)$ and $(5,0)$.

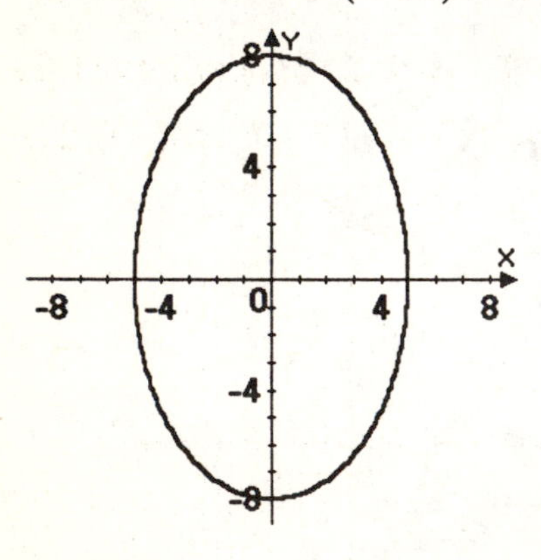

7. $\dfrac{x^2}{49}+\dfrac{y^2}{81}=1$

Because the denominator of the y^2 – term is greater than the denominator of the x^2 – term, the major axis is vertical. Since $a^2 = 81$, $a = 9$ and the vertices are $(0,-9)$ and $(0,9)$. Since $b^2 = 49$, $b = 7$ and endpoints of the minor axis are $(-7,0)$ and $(7,0)$.

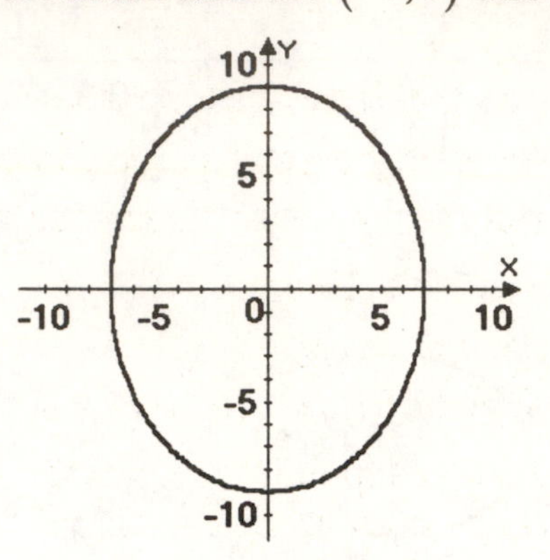

9. $25x^2 + 4y^2 = 100$

$$\frac{25x^2}{100}+\frac{4y^2}{100}=\frac{100}{100}$$

$$\frac{x^2}{4}+\frac{y^2}{25}=1$$

Because the denominator of the y^2 – term is greater than the denominator of the x^2 – term, the major axis is vertical. Since $a^2 = 25$, $a = 5$ and the vertices are $(0,-5)$ and $(0,5)$. Since $b^2 = 4$, $b = 2$ and endpoints of the minor axis are $(-2,0)$ and $(2,0)$.

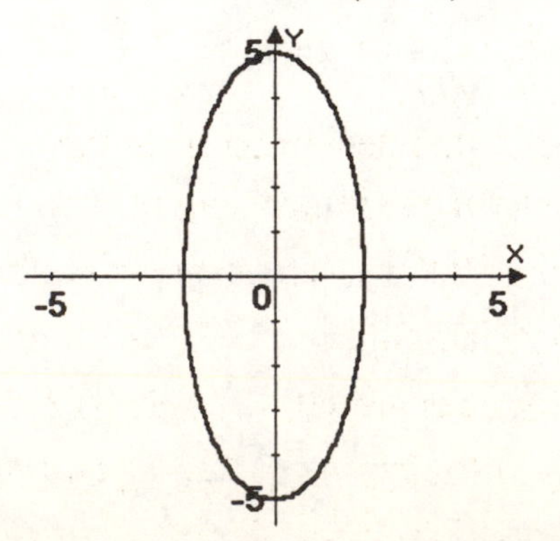

11. $4x^2+16y^2=64$

$$\frac{4x^2}{64}+\frac{16y^2}{64}=\frac{64}{64}$$

$$\frac{x^2}{16}+\frac{y^2}{4}=1$$

Because the denominator of the x^2 – term is greater than the denominator of the y^2 – term, the major axis is horizontal. Since $a^2=16$, $a=4$ and the vertices are $(-4,0)$ and $(4,0)$. Since $b^2=4$, $b=2$ and endpoints of the minor axis are $(0,-2)$ and $(0,2)$.

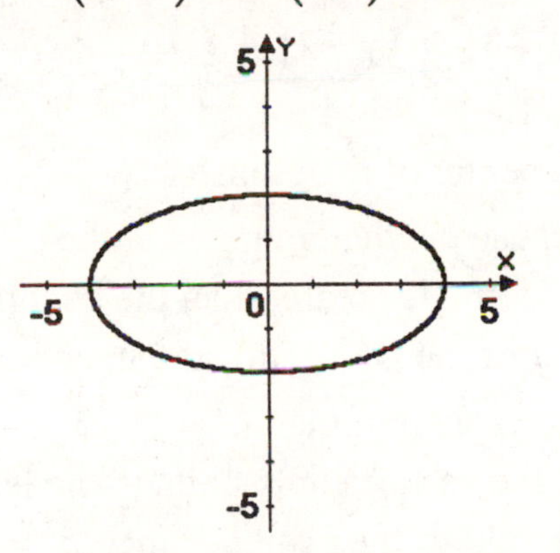

13. $25x^2+9y^2=225$

$$\frac{25x^2}{225}+\frac{9y^2}{225}=\frac{225}{225}$$

$$\frac{x^2}{9}+\frac{y^2}{25}=1$$

Because the denominator of the y^2 – term is greater than the denominator of the x^2 – term, the major axis is vertical. Since $a^2=9$, $a=3$ and the vertices are $(0,-3)$ and $(0,3)$. Since $b^2=25$, $b=5$ and endpoints of the minor axis are $(-5,0)$ and $(5,0)$.

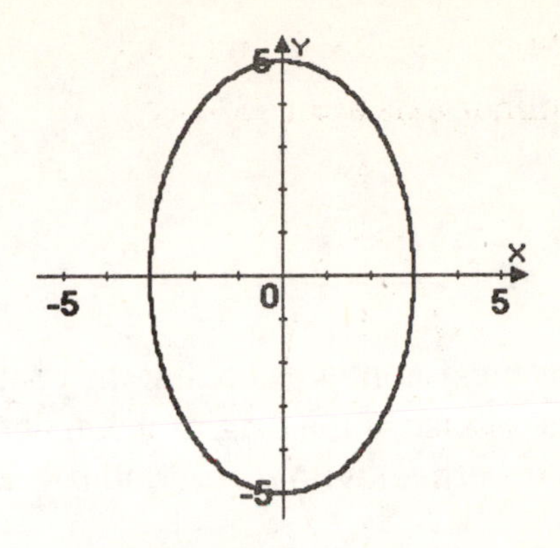

15. $x^2+2y^2=8$

$$\frac{x^2}{8}+\frac{2y^2}{8}=\frac{8}{8}$$

$$\frac{x^2}{8}+\frac{y^2}{4}=1$$

Because the denominator of the x^2 – term is greater than the denominator of the y^2 – term, the major axis is horizontal. Since $a^2=8$, $a=\sqrt{8}=2\sqrt{2}$ and the vertices are $(-2\sqrt{2},0)$ and $(2\sqrt{2},0)$. Since $b^2=4$, $b=2$ and endpoints of the minor axis are $(0,-2)$ and $(0,2)$.

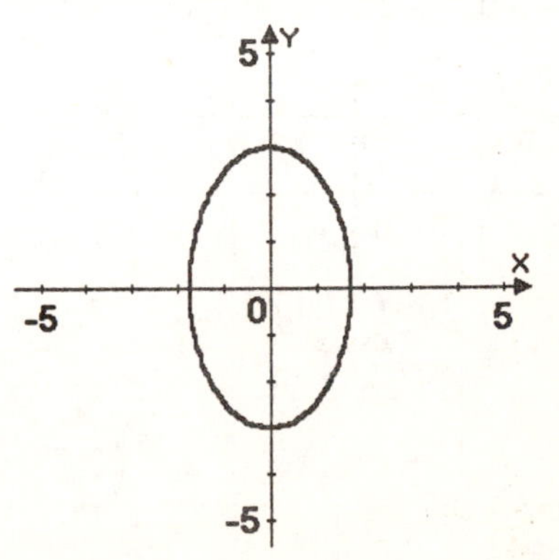

17. From the graph, we see that the center of the ellipse is the origin, the major axis is horizontal with $a=2$, and

$b = 1.$

$$\frac{x^2}{2^2} + \frac{y^2}{1^2} = 1$$

$$\frac{x^2}{4} + \frac{y^2}{1} = 1$$

19. From the graph, we see that the center of the ellipse is the origin, the major axis is vertical with $a = 2$, and $b = 1$.

$$\frac{x^2}{1^2} + \frac{y^2}{2^2} = 1$$

$$\frac{x^2}{1} + \frac{y^2}{4} = 1$$

21. $$\frac{(x-2)^2}{9} + \frac{(y-1)^2}{4} = 1$$

The center of the ellipse is $(2,1)$. Because the denominator of the x^2 – term is greater than the denominator of the y^2 – term, the major axis is horizontal. Since $a^2 = 9$, $a = 3$ and the vertices lie 3 units to the left and right of the center. Since $b^2 = 4$, $b = 2$ and endpoints of the minor axis lie two units above and below the center.

Center	Vertices	Endpoints of Minor Axis
$(2,1)$	$(2-3,1)$ $=(-1,1)$	$(2,1-2)$ $=(2,-1)$
	$(2+3,1)$ $=(5,1)$	$(2,1+2)$ $=(2,3)$

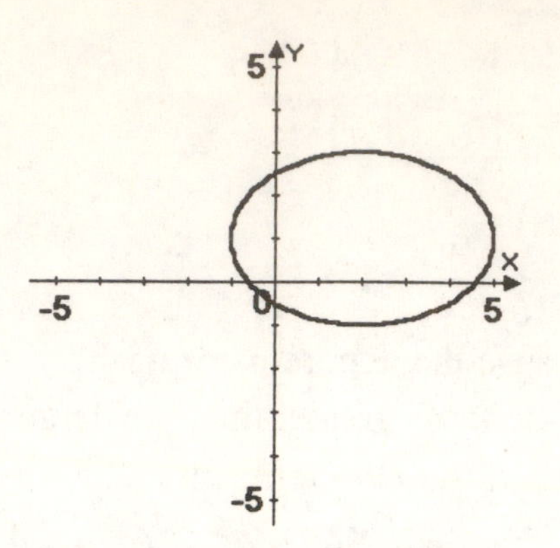

23. $$(x+3)^2 + 4(y-2)^2 = 16$$

$$\frac{(x+3)^2}{16} + \frac{4(y-2)^2}{16} = \frac{16}{16}$$

$$\frac{(x+3)^2}{16} + \frac{(y-2)^2}{4} = 1$$

The center of the ellipse is $(-3,2)$. Because the denominator of the x^2 – term is greater than the denominator of the y^2 – term, the major axis is horizontal. Since $a^2 = 16$, $a = 4$ and the vertices lie 4 units to the left and right of the center. Since $b^2 = 4$, $b = 2$ and endpoints of the minor axis lie two units above and below the center.

Center	Vertices	Endpoints of Minor Axis
$(-3,2)$	$(-3-4,2)$ $=(-7,2)$	$(-3,2-2)$ $=(-3,0)$
	$(-3+4,2)$ $=(1,2)$	$(-3,2+2)$ $=(-3,4)$

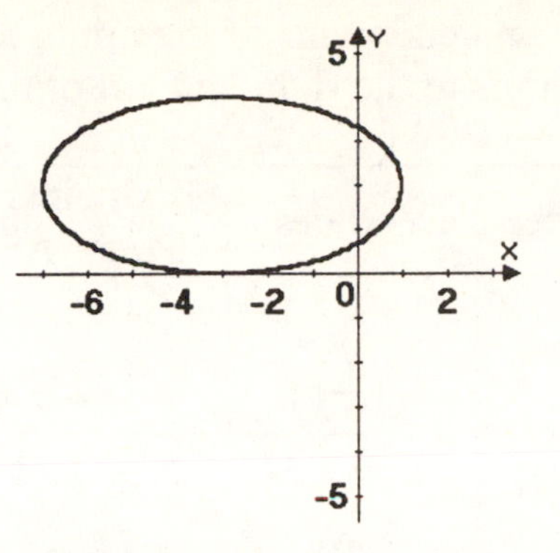

25. $\frac{(x-4)^2}{9}+\frac{(y+2)^2}{25}=1$

The center of the ellipse is $(4,-2)$. Because the denominator of the y^2 – term is greater than the denominator of the x^2 – term, the major axis is vertical. Since $a^2=25$, $a=5$ and the vertices lie 5 units to the above and below the center. Since $b^2=9$, $b=3$ and endpoints of the minor axis lie 3 units to the right and left of the center.

Center	Vertices	Endpoints Minor Axis
$(4,-2)$	$(4,-2-5)$ $=(4,-7)$	$(4-3,-2)$ $=(1,-2)$
	$(4,-2+5)$ $=(4,3)$	$(4+3,-2)$ $=(7,-2)$

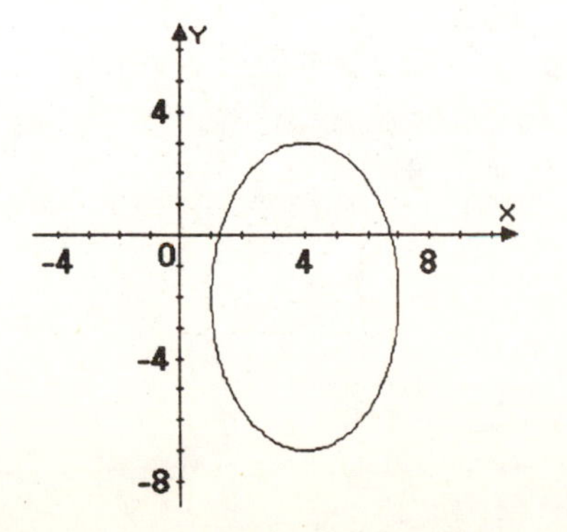

27. $\frac{x^2}{25}+\frac{(y-2)^2}{36}=1$

The center of the ellipse is $(0,2)$. Because the denominator of the y^2 – term is greater than the denominator of the x^2 – term, the major axis is vertical. Since $a^2=36$, $a=6$ and the vertices lie 6 units to the above and below the center. Since $b^2=25$, $b=5$ and endpoints of the minor axis lie 5 units to the left and right of the center.

Center	Vertices	Endpoint Minor Axis
$(0,2)$	$(0,2-6)$ $=(0,-4)$	$(0-5,2)$ $=(-5,2)$
	$(0,2+6)$ $=(0,8)$	$(0+5,2)$ $=(5,2)$

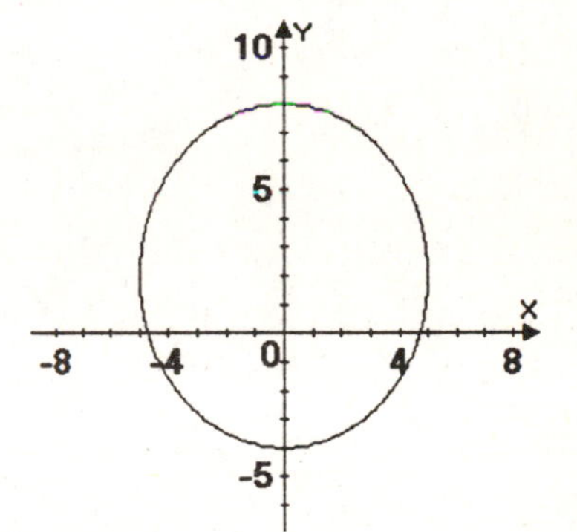

29. $\frac{(x+3)^2}{9}+(y-2)^2=1$

$\frac{(x+3)^2}{9}+\frac{(y-2)^2}{1}=1$

The center of the ellipse is $(-3,2)$. Because the denominator of the x^2 – term is greater than the denominator of the y^2 – term, the major axis is horizontal. Since $a^2 = 9$, $a = 3$ and the vertices lie 3 units to the left and right of the center. Since $b^2 = 1$, $b = 1$ and endpoints of the minor axis lie two units above and below the center.

Center	Vertices	Endpoints of Minor Axis
$(-3,2)$	$(-3+3,2) = (0,2)$	$(-3,2-1) = (-3,1)$
	$(-3-3,2) = (-6,2)$	$(-3,2+1) = (-3,3)$

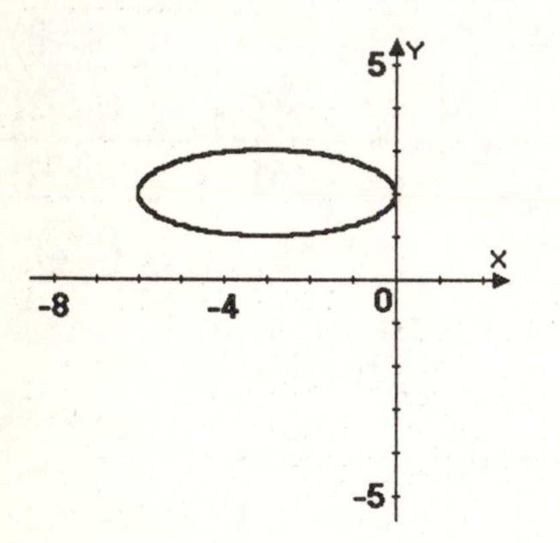

31. $$9(x-1)^2 + 4(y+3)^2 = 36$$

$$\frac{9(x-1)^2}{36} + \frac{4(y+3)^2}{36} = \frac{36}{36}$$

$$\frac{(x-1)^2}{4} + \frac{(y+3)^2}{9} = 1$$

The center of the ellipse is $(1,-3)$. Because the denominator of the y^2 – term is greater than the denominator of the x^2 – term, the major axis is vertical. Since $a^2 = 9$, $a = 3$ and the vertices lie 3 units to the above and below the center. Since $b^2 = 9$, $b = 3$ and endpoints of the minor axis lie 3 units to the right and left of the center.

Center	Vertices	Endpoints of Minor Axis
$(1,-3)$	$(1,-3-3) = (1,-6)$	$(1-3,-3) = (-2,-3)$
	$(1,-3+3) = (1,0)$	$(1+3,-3) = (4,-3)$

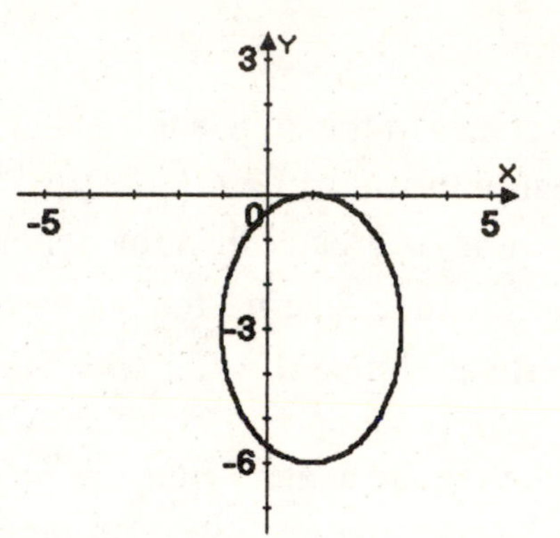

Application Exercises

33. From the figure, we see that the major axis is horizontal with $a = 15$, and $b = 10$.

$$\frac{x^2}{15^2} + \frac{y^2}{10^2} = 1$$

$$\frac{x^2}{225} + \frac{y^2}{100} = 1$$

Since the truck is 8 feet wide, we need to determine the height of the archway at $\frac{8}{2} = 4$ feet from the center.

$$\frac{4^2}{225} + \frac{y^2}{100} = 1$$

$$\frac{16}{225}+\frac{y^2}{100}=1$$
$$900\left(\frac{16}{225}+\frac{y^2}{100}\right)=900(1)$$
$$4(16)+9y^2=900$$
$$64+9y^2=900$$
$$9y^2=836$$
$$y^2=\frac{836}{9}$$
$$y=\sqrt{\frac{836}{9}}\approx 9.64$$

The height of the archway 4 feet from the center is approximately 9.64 feet. Since the truck is 7 feet high, the truck will clear the archway.

35. **a.**
$$\frac{x^2}{48^2}+\frac{y^2}{23^2}=1$$
$$\frac{x^2}{2304}+\frac{y^2}{529}=1$$

b.
$$c^2=a^2-b^2$$
$$c^2=48^2-23^2$$
$$c^2=2304-529$$
$$c^2=1775$$
$$c=\sqrt{1775}\approx 42.1$$

The desk was situated approximately 42 feet from the center of the ellipse.

Writing in Mathematics

37. Answers will vary.

39. Answers will vary.

41. Answers will vary.

Technology Exercises

43. Answers will vary. For example, consider Exercise 21.

$$\frac{(x-2)^2}{9}+\frac{(y-1)^2}{4}=1$$
$$\frac{(y-1)^2}{4}=1-\frac{(x-2)^2}{9}$$
$$(y-1)^2=4\left(1-\frac{(x-2)^2}{9}\right)$$
$$(y-1)^2=4-\frac{4(x-2)^2}{9}$$
$$y-1=\pm\sqrt{4-\frac{4(x-2)^2}{9}}$$
$$y=1\pm\sqrt{4-\frac{4(x-2)^2}{9}}$$

Critical Thinking Exercises

45. Graphing the points, we see that the center of the ellipse is at the origin and the major axis is vertical. We have $a=6$.

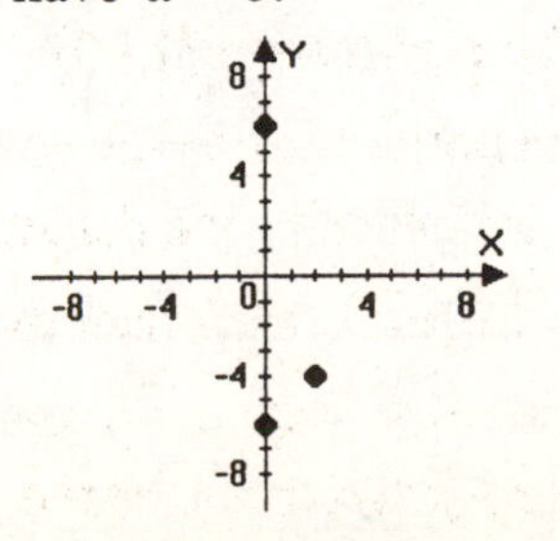

Using a and the given point, we can solve for b.

$$\frac{x^2}{b^2}+\frac{y^2}{a^2}=1$$

$$\frac{2^2}{b^2}+\frac{(-4)^2}{6^2}=1$$

$$\frac{4}{b^2}+\frac{16}{36}=1$$

$$36b^2\left(\frac{4}{b^2}+\frac{16}{36}\right)=36b^2(1)$$

$$36(4)+16b^2=36b^2$$

$$144+16b^2=36b^2$$

$$144=20b^2$$

$$\frac{144}{20}=b^2$$

$$\frac{36}{5}=b^2$$

The equation ellipse in standard form is

$$\frac{x^2}{\frac{36}{5}}+\frac{y^2}{36}=1.$$

47. $$4x^2+9y^2-32x+36y+64=0$$

$$\left(4x^2-32x\quad\right)+\left(9y^2+36y\quad\right)=-64$$

$$4\left(x^2-8x\quad\right)+9\left(y^2+4y\quad\right)=-64$$

Complete the squares.

$$\left(\frac{b}{2}\right)^2=\left(\frac{-8}{2}\right)^2=(-4)^2=16$$

$$\left(\frac{b}{2}\right)^2=\left(\frac{4}{2}\right)^2=(2)^2=4$$

$$4\left(x^2-8x+16\right)+9\left(y^2+4y+4\right)=-64+4(16)+9(4)$$

$$4(x-4)^2+9(y+2)^2=-64+64+36$$

$$4(x-4)^2+9(y+2)^2=36$$

$$\frac{4(x-4)^2}{36}+\frac{9(y+2)^2}{36}=\frac{36}{36}$$

$$\frac{(x-4)^2}{9}+\frac{(y+2)^2}{4}=1$$

$$\frac{(x-4)^2}{9}+\frac{(y+2)^2}{4}=1$$

49. The ellipse's vertices lie on the larger circle. This means that a is the radius of the circle. The equation of the larger circle is $x^2 + y^2 = 25$. The endpoints of the ellipse's minor axis lie on the smaller circle. This means that b is the radius of the smaller circle. The equation of the smaller circle is $x^2 + y^2 = 9$.

Review Exercises

51.
$$x^3 + 2x^2 - 4x - 8 = x^2(x+2) - 4(x+2) = (x+2)(x^2 - 4)$$
$$= (x+2)(x+2)(x-2) = (x+2)^2(x-2)$$

53. $\sqrt[3]{40x^4y^7} = \sqrt[3]{8 \cdot 5x^3xy^6y} = 2xy^2\sqrt[3]{5xy}$

55.
$$\frac{2}{x+2} + \frac{4}{x-2} = \frac{x-1}{x^2-4}$$
$$\frac{2}{x+2} + \frac{4}{x-2} = \frac{x-1}{(x+2)(x-2)}$$
$$(x+2)(x-2)\left(\frac{2}{x+2} + \frac{4}{x-2}\right) = (x+2)(x-2)\left(\frac{x-1}{(x+2)(x-2)}\right)$$
$$2(x-2) + 4(x+2) = x-1$$
$$2x - 4 + 4x + 8 = x - 1$$
$$6x + 4 = x - 1$$
$$5x = -5$$
$$x = -1$$

The solution is –1 and the solution set is $\{-1\}$.

Check Points 10.3

1. **a.** $\frac{x^2}{25} - \frac{y^2}{16} = 1$

Because the x^2–term is preceded by a plus sign, the transverse axis lies along the x–axis and the vertices lie a units to the left and right of the origin. Since $a^2 = 25$ and $a = 5$, the vertices are $(-5, 0)$ and $(5, 0)$.

b. $\frac{y^2}{25} - \frac{x^2}{16} = 1$

Because the y^2–term is preceded by a plus sign, the transverse axis lies along the y–axis and the vertices lie a units above and below the origin. Since $a^2 = 25$ and $a = 5$, the vertices are $(0,-5)$ and $(0,5)$.

2. $\frac{x^2}{36} - \frac{y^2}{9} = 1$

The equation is in the form $\frac{x^2}{a^2} - \frac{y^2}{b^2} = 1$ with $a^2 = 36$, and $b^2 = 9$.

We know the transverse axis lies on the x-axis and the vertices are $(-6,0)$ and $(6,0)$. Because $a^2 = 36$ and $b^2 = 9$, $a = 6$ and $b = 3$. Construct a rectangle using –6 and 6 on the x–axis, and –3 and 3 on the y–axis. Draw extended diagonals to obtain the asymptotes. Graph the hyperbola.

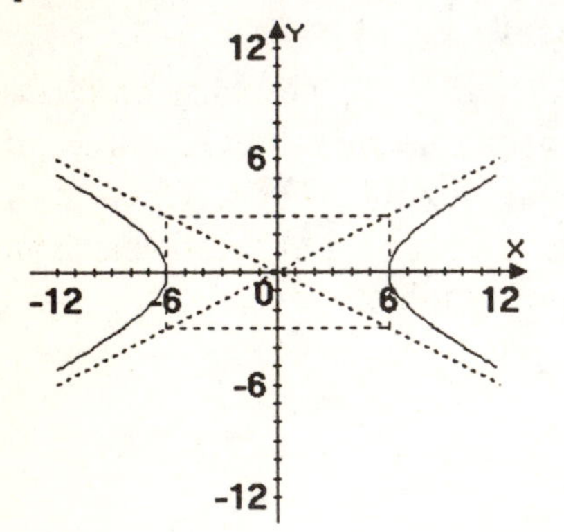

3. $y^2 - 4x^2 = 4$

$\frac{y^2}{4} - \frac{4x^2}{4} = \frac{4}{4}$

$\frac{y^2}{4} - \frac{x^2}{1} = 1$

The equation is in the form $\frac{y^2}{a^2} - \frac{x^2}{b^2} = 1$ with $a^2 = 4$, and $b^2 = 1$.

We know the transverse axis lies on the y-axis and the vertices are $(0,-2)$ and $(0,2)$. Because $a^2 = 4$ and $b^2 = 1$, $a = 2$ and $b = 1$. Construct a rectangle using –1 and 1 on the x–axis, and –2 and 2 on the y–axis. Draw extended diagonals to obtain the asymptotes. Graph the hyperbola.

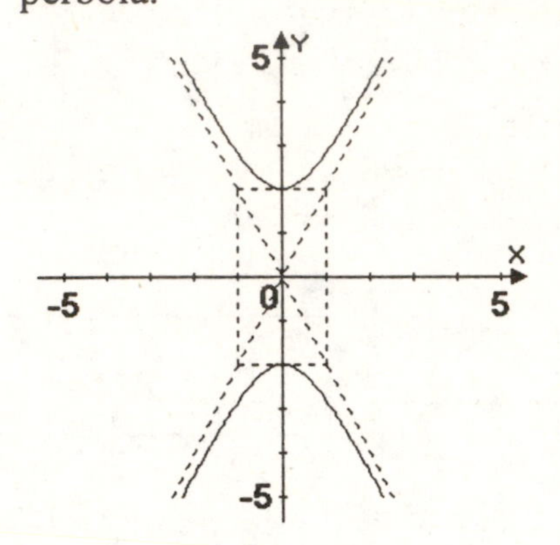

Problem Set 10.3

Practice Exercises

1. Since the x^2 – term is positive, the transverse axis lies along the x–axis. Also, since $a^2 = 4$ and $a = 2$, the vertices are $(-2,0)$ and $(2,0)$. This corresponds to graph (b).

3. Since the y^2 – term is positive, the transverse axis lies along the y–axis. Also, since $a^2 = 4$ and $a = 2$, the vertices are $(0,-2)$ and $(0,2)$. This corresponds to graph (a).

5. $\frac{x^2}{9} - \frac{y^2}{25} = 1$

The equation is in the form

$\frac{x^2}{a^2}-\frac{y^2}{b^2}=1$ with $a^2=9$, and $b^2=25$.
We know the transverse axis lies on the x-axis and the vertices are $(-3,0)$ and $(3,0)$. Because $a^2=9$ and $b^2=25$, $a=3$ and $b=5$. Construct a rectangle using –3 and 3 on the x–axis, and –5 and 5 on the y–axis. Draw extended diagonals to obtain the asymptotes. Graph the hyperbola.

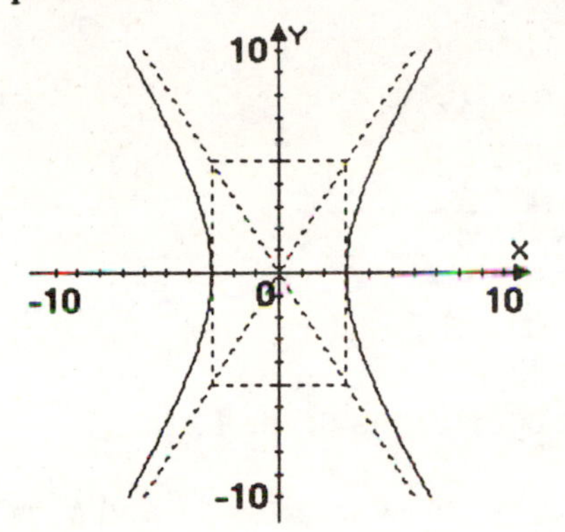

7. $\frac{x^2}{100}-\frac{y^2}{64}=1$
The equation is in the form
$\frac{x^2}{a^2}-\frac{y^2}{b^2}=1$ with
$a^2=100$, and $b^2=64$. We know the transverse axis lies on the x-axis and the vertices are $(-10,0)$ and $(10,0)$.
Because $a^2=100$ and $b^2=64$, $a=10$ and $b=8$. Construct a rectangle using –10 and 10 on the x–axis, and –8 and 8 on the y–axis. Draw extended diagonals to obtain the asymptotes. Graph the hyperbola.

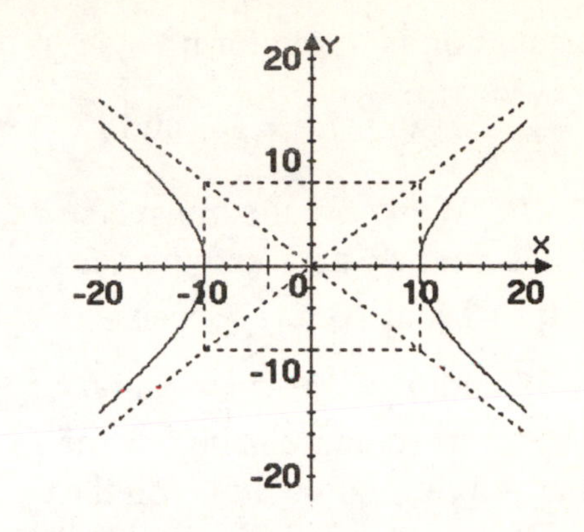

9. $\frac{y^2}{16}-\frac{x^2}{36}=1$
The equation is in the form
$\frac{y^2}{a^2}-\frac{x^2}{b^2}=1$ with
$a^2=16$, and $b^2=36$. We know the transverse axis lies on the y-axis and the vertices are $(0,-4)$ and $(0,4)$.
Because $a^2=16$ and $b^2=36$, $a=4$ and $b=6$. Construct a rectangle using –4 and 4 on the x–axis, and –6 and 6 on the y–axis. Draw extended diagonals to obtain the asymptotes. Graph the hyperbola.

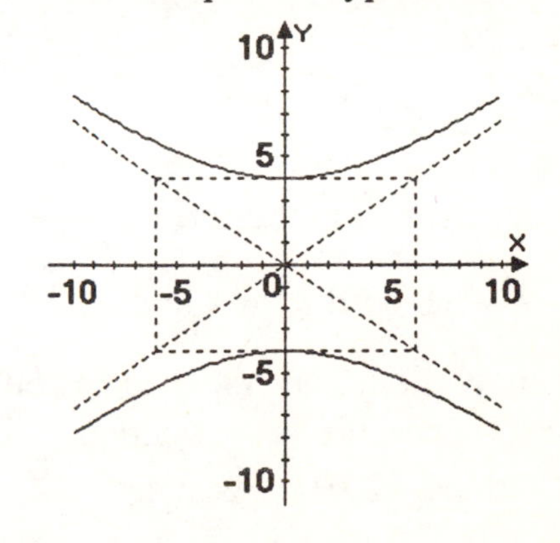

11. $\frac{y^2}{36}-\frac{x^2}{25}=1$

The equation is in the form

$\frac{y^2}{a^2}-\frac{x^2}{b^2}=1$ with $a^2=36$, and $b^2=25$. We know the transverse axis lies on the y-axis and the vertices are $(0,-6)$ and $(0,6)$. Because $a^2=36$ and $b^2=25$, $a=6$ and $b=5$. Construct a rectangle using –5 and 5 on the x–axis, and –6 and 6 on the y–axis. Draw extended diagonals to obtain the asymptotes. Graph the hyperbola.

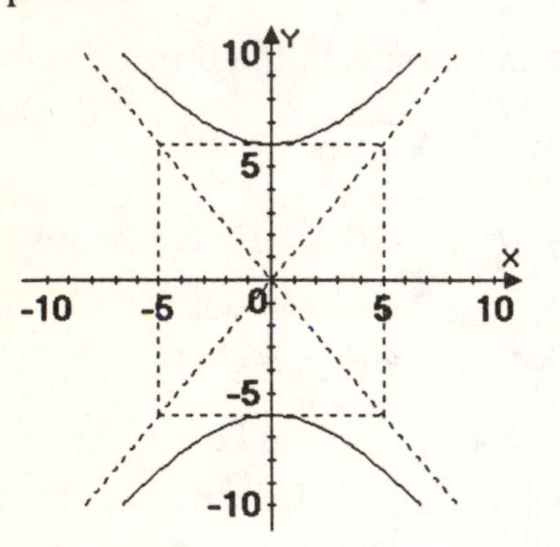

13. $9x^2-4y^2=36$

$$\frac{9x^2}{36}-\frac{4y^2}{36}=\frac{36}{36}$$

$$\frac{x^2}{4}-\frac{y^2}{9}=1$$

The equation is in the form

$\frac{x^2}{a^2}-\frac{y^2}{b^2}=1$ with $a^2=4$ and $b^2=9$.

We know the transverse axis lies on the x-axis and the vertices are $(-2,0)$ and $(2,0)$. Because $a^2=4$ and $b^2=9$, $a=2$ and $b=3$. Construct a rectangle using –2 and 2 on the x–axis, and –3 and 3 on the y–axis. Draw extended diagonals to obtain the asymptotes. Graph the hyperbola.

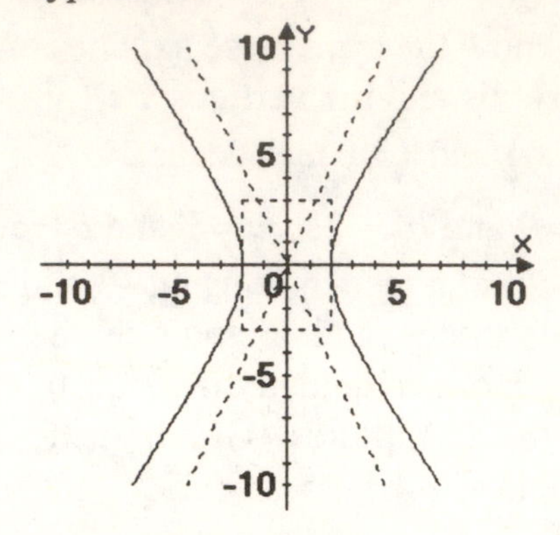

15. $9y^2-25x^2=225$

$$\frac{9y^2}{225}-\frac{25x^2}{225}=\frac{225}{225}$$

$$\frac{y^2}{25}-\frac{x^2}{9}=1$$

The equation is in the form

$\frac{y^2}{a^2}-\frac{x^2}{b^2}=1$ with $a^2=25$ and $b^2=9$.

We know the transverse axis lies on the y-axis and the vertices are $(0,-5)$ and $(0,5)$. Because $a^2=25$ and $b^2=9$, $a=5$ and $b=3$. Construct a rectangle using –3 and 3 on the x–axis, and –5 and 5 on the y–axis. Draw extended diagonals to obtain the asymptotes. Graph the hyperbola.

17.

$$4x^2 = 4 + y^2$$

$$4x^2 - y^2 = 4$$

$$\frac{4x^2}{4} - \frac{y^2}{4} = \frac{4}{4}$$

$$\frac{x^2}{1} - \frac{y^2}{4} = 1$$

The equation is in the form

$\frac{x^2}{a^2} - \frac{y^2}{b^2} = 1$ with $a^2 = 1$ and $b^2 = 4$.

We know the transverse axis lies on the x-axis and the vertices are $(-1, 0)$ and $(1, 0)$. Because $a^2 = 1$ and $b^2 = 4$, $a = 1$ and $b = 2$. Construct a rectangle using -1 and 1 on the x–axis, and -2 and 2 on the y–axis. Draw extended diagonals to obtain the asymptotes. Graph the hyperbola.

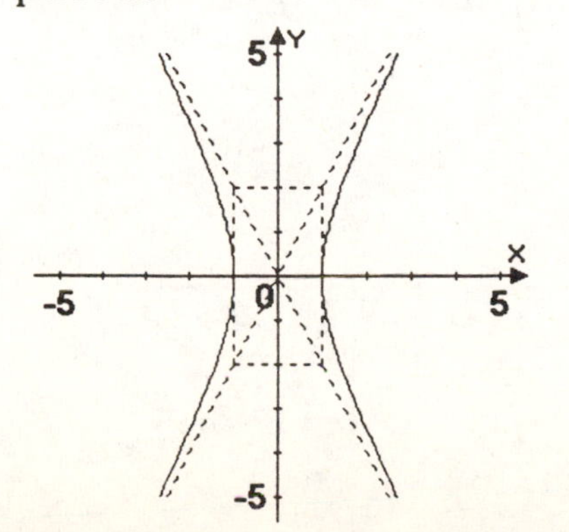

19. From the graph we see that the transverse axis lies along the x–axis and the vertices are $(-3, 0)$ and $(3, 0)$. This means that $a = 3$. We also see that $b = 5$.

$$\frac{x^2}{a^2} - \frac{y^2}{b^2} = 1$$

$$\frac{x^2}{3^2} - \frac{y^2}{5^2} = 1$$

$$\frac{x^2}{9} - \frac{y^2}{25} = 1$$

21. From the graph we see that the transverse axis lies along the y–axis and the vertices are $(0, -2)$ and $(0, 2)$. This means that $a = 2$. We also see that $b = 3$.

$$\frac{y^2}{a^2} - \frac{x^2}{b^2} = 1$$

$$\frac{y^2}{2^2} - \frac{x^2}{3^2} = 1$$

$$\frac{y^2}{4} - \frac{x^2}{9} = 1$$

Application Exercises

23.

$$625y^2 - 400x^2 = 250{,}000$$

$$\frac{625y^2}{250{,}000} - \frac{400x^2}{250{,}000} = \frac{250{,}000}{250{,}000}$$

$$\frac{y^2}{400} - \frac{x^2}{625} = 1$$

Since the houses at the vertices of the hyperbola will be closest, find the distance between the vertices. Since $a^2 = 400$, $a = 20$. The houses are $20 + 20 = 40$ yards apart.

Writing in Mathematics

25. Answers will vary.

27. Answers will vary.

29. Answers will vary.

Technology Exercises

31. $\frac{x^2}{4} - \frac{y^2}{9} = 0$

Solve the equation for y.

$$\frac{x^2}{4} = \frac{y^2}{9}$$
$$9x^2 = 4y^2$$
$$\frac{9}{4}x^2 = y^2$$
$$\pm\sqrt{\frac{9}{4}x^2} = y$$
$$\pm\frac{3}{2}x = y$$

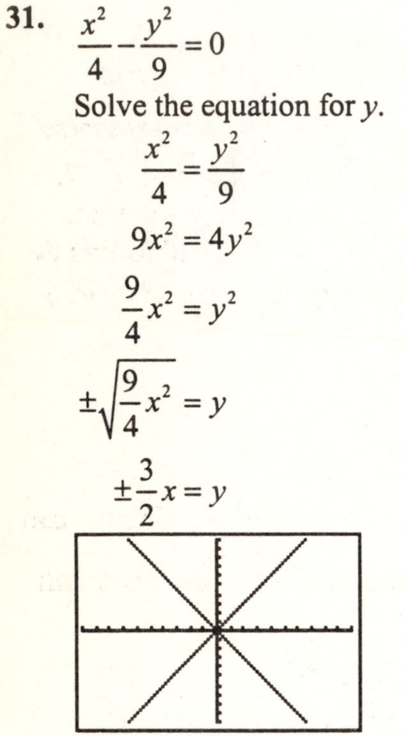

The graph is not a hyperbola. The graph is two lines.

33. Statement **c.** is true. Since $y = -\frac{2}{3}$ is an asymptote, the graph of the hyperbola does not intersect it.

Statement **a.** is false. If a hyperbola has a transverse axis along the y–axis and one of the branches is removed, the remaining branch does not define a function of x.

Statement **b.** is false. The points on the hyperbola's asymptotes do not satisfy the hyperbola's equation.

Statement **d.** is false. See Exercise 32 for two different hyperbolas that share the same asymptotes.

35. $\frac{(x+2)^2}{9} - \frac{(y-1)^2}{25} = 1$

This is the graph of a hyperbola with center $(-2,1)$. The equation is in the form $\frac{(x-h)^2}{a^2} - \frac{(y-k)^2}{b^2} = 1$ with $a^2 = 9$ and $b^2 = 25$. We know the transverse axis is horizontal and the vertices lie 3 units to the right and left of $(-2,1)$ at $(-2-3,1) = (-5,1)$ and $(-2+3,1) = (1,1)$. Because $a^2 = 9$ and $b^2 = 25$, $a = 3$ and $b = 5$. Construct two sides of a rectangle using -5 and 1 (the x–coordinates of the vertices) on the x–axis. The remaining two sides of the rectangle are constructed 5 units above and 5 units below the center, $(-2,1)$, at $1-5=-4$ and $1+5=6$. Draw extended diagonals to obtain the asymptotes. Graph the hyperbola.

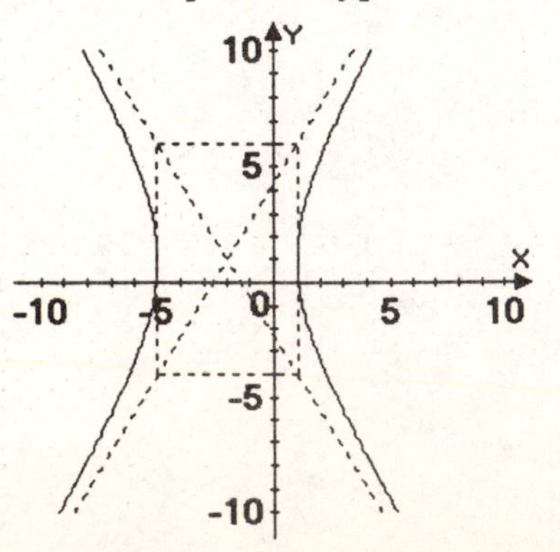

37. $x^2 - y^2 - 2x - 4y - 4 = 0$

Rearrange and complete the squares.

$$x^2 - 2x \quad - y^2 - 4y \quad = 4$$

$$\left(x^2 - 2x \quad\right) - \left(y^2 + 4y \quad\right) = 4$$

Complete the squares.

$$\left(\frac{b}{2}\right)^2 = \left(\frac{-2}{2}\right)^2 = (-1)^2 = 1$$

$$\left(\frac{b}{2}\right)^2 = \left(\frac{4}{2}\right)^2 = 2^2 = 4$$

$$\left(x^2 - 2x + 1\right) - \left(y^2 + 4y + 4\right) = 4 + 1 - 4$$

$$(x-1)^2 - (y+2)^2 = 1$$

$$\frac{(x-1)^2}{1} - \frac{(y+2)^2}{1} = 1$$

This is the graph of a hyperbola with center $(1, -2)$. The equation is in the form $\frac{(x-h)^2}{a^2} - \frac{(y-k)^2}{b^2} = 1$ with $a^2 = 1$ and $b^2 = 1$. We know the transverse axis is horizontal and the vertices lie 3 units to the right and left of $(1, -2)$ at $(1-1, -2) = (0, -2)$ and $(1+1, -2) = (2, -2)$. Because $a^2 = 1$ and $b^2 = 1$, $a = 1$ and $b = 1$. Construct two sides of a rectangle using 0 and 2 (the x–coordinates of the vertices) on the x–axis. The remaining two sides of the rectangle are constructed 1 unit above and 1 unit below the center, $(1, -2)$, at $-2 - 1 = -3$ and $-2 + 1 = -1$. Draw extended diagonals to obtain the asymptotes. Graph the hyperbola.

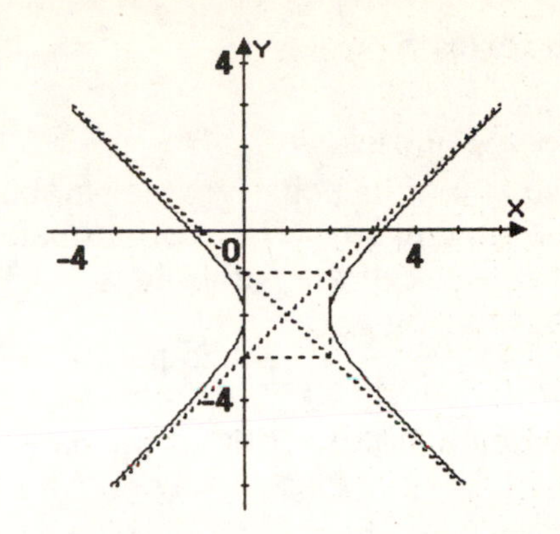

39. Since the vertices are $(0, 7)$ and $(0, -7)$, we know that the transverse axis lies along the y–axis and $a = 7$. Use the equation of the asymptote, $y = 5x$, to find b. We need to find the x–coordinate that corresponds with $y = 7$.

$$7 = 5x$$

$$\frac{7}{5} = x$$

This means that $b = \pm\frac{7}{5}$. Using a and b, write the equation of the hyperbola.

$$\frac{y^2}{7^2} - \frac{x^2}{\left(\frac{7}{5}\right)^2} = 1$$

$$\frac{y^2}{49} - \frac{x^2}{\frac{49}{25}} = 1$$

Review Exercises

40. $y=-x^2-4x+5$

Since $a=-1$ is negative, the parabola opens downward. The x–coordinate of the vertex of the parabola is

$-\frac{b}{2a}=-\frac{-4}{2(-1)}=-2$ and the y–coordinate of the vertex of the parabola is

$$f\left(-\frac{b}{2a}\right)=f(-2)$$
$$=-(-2)^2-4(-2)+5$$
$$=-4+8+5=9.$$

The vertex is at (–2, 9).
Replace y with 0 to find x–intercepts.

$0=-x^2-4x+5$

$0=x^2+4x-5$

$0=(x+5)(x-1)$

The vertex is at (–2, 9).
Replace y with 0 to find x–intercepts.

$0=-x^2-4x+5$

$0=x^2+4x-5$

$0=(x+5)(x-1)$

Apply the zero product principle.

$x+5=0$ or $x-1=0$

$x=-5$ $\qquad$ $x=1$

The x–intercepts are –5 and 1.
Set $x=0$ and solve for y to obtain the y–intercept.

$y=-0^2-4(0)+5=5$

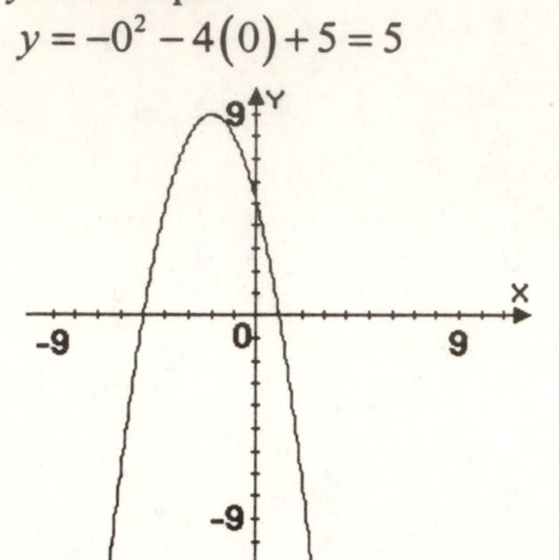

41. $3x^2-11x-4\ge 0$

Solve the related quadratic equation.

$3x^2-11x-4=0$

$(3x+1)(x-4)=0$

Use the zero product principle.

$3x+1=0$ or $x-4=0$

$3x=-1$ $\qquad$ $x=4$

$x=-\frac{1}{3}$

The boundary points are $-\frac{1}{3}$ and 4.

Test Interval	Test Number	Substitution	Conclusion
$\left(-\infty,-\frac{1}{3}\right]$	-1	$3(-1)^2-11(-1)-4\ge 0$ $10\ge 0$, true	$\left(-\infty,-\frac{1}{3}\right]$ belongs in the solution set
$\left[-\frac{1}{3},4\right]$	0	$3(0)^2-11(0)-4\ge 0$ $-4\ge 0$, false	$\left[-\frac{1}{3},4\right]$ does not belong in the solution set.
$[4,\infty)$	5	$3(5)^2-11(5)-4\ge 0$ $16\ge 0$, true	$[4,\infty)$ belongs in the solution set.

The solution set is $\left(-\infty,-\frac{1}{3}\right]\cup[4,\infty)$ or $\left\{x \middle| x \le -\frac{1}{3} \text{ or } x \ge 4\right\}$.

42. $\log_4(3x+1)=3$

$3x+1=4^3$

$3x+1=64$

$3x=63$

$x=21$

The solution is 21 and the solution set is $\{21\}$.

Check Points 10.4

1. $x=-(y-2)^2+1$

This is a parabola of the form $x=a(y-k)^2+h$. Since $a=-1$ is negative, the parabola opens to the left. The vertex of the parabola is $(1,2)$. Replace y with 0 to find the x–intercept.

$x=-(0-2)^2+1$

$=-(-2)^2+1$

$=-4+1=-3$

The x–intercept is –3. Replace x with 0 to find the y–intercepts.

$0=-(y-2)^2+1$

$0=-(y^2-4y+4)+1$

$0=-y^2+4y-4+1$

$0=-y^2+4y-3$

$0=y^2-4y+3$

$0=(y-3)(y-1)$

Apply the zero product principle.

$y-3=0$ and $y-1=0$

$y=3$ $\quad y=1$

The y–intercepts are 1 and 3.

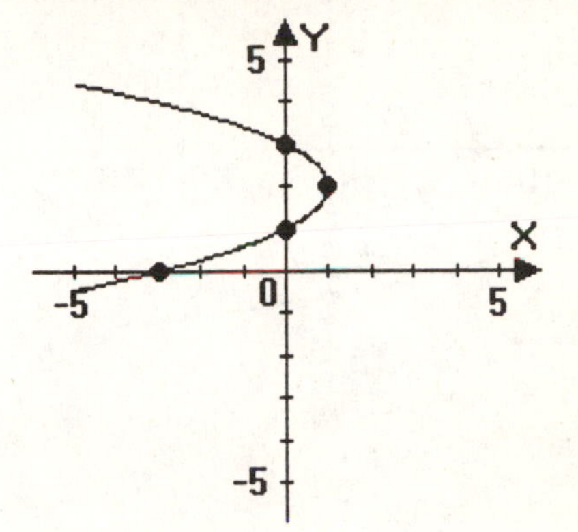

2. $x=y^2+8y+7$

This is a parabola of the form $x=ay^2+by+c$. Since $a=1$ is positive, the parabola opens to the right. The y–coordinate of the vertex is $-\frac{b}{2a}=-\frac{8}{2(1)}=-\frac{8}{2}=-4$. The x–coordinate of the vertex is

$x=(-4)^2+8(-4)+7$

$=16-32+7=-9.$

The vertex of the parabola is $(-9,-4)$.

Replace y with 0 to find the x–intercept.

$x=0^2+8(0)+7$

$=0+0+7=7$

The x–intercept is 7. Replace x with 0 to find the y–intercepts.

$0=y^2+8y+7$

$0=(y+7)(y+1)$

Apply the zero product principle.

$y+7=0$ and $y+1=0$

$y=-7$ $\quad y=-1$

The y–intercepts are -1 and -7.

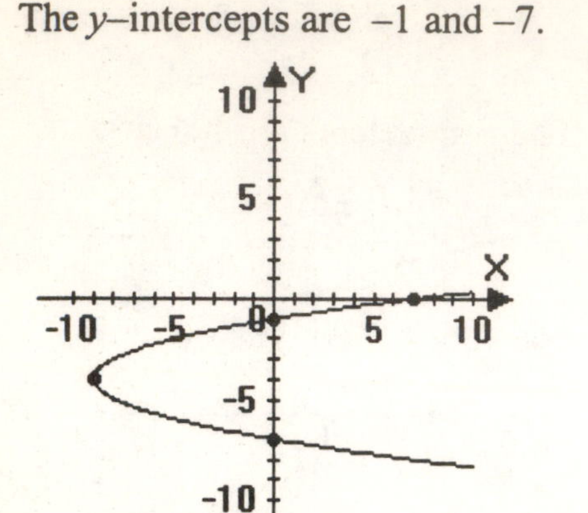

3. **a.** $x^2 = 4y^2 + 16$

Since both the x– and y–terms are squared, the graph of the equation cannot be a parabola. To determine whether the graph is a circle, ellipse, or hyperbola, move the x– and y–terms to the same side of the equation.

$x^2 - 4y^2 = 16$

Because x^2 and y^2 have opposite signs, the equation's graph is a hyperbola.

b. $x^2 = 16 - 4y^2$

Since both the x– and y–terms are squared, the graph of the equation cannot be a parabola. To determine whether the graph is a circle, ellipse, or hyperbola, move the x– and y–terms to the same side of the equation.

$x^2 + 4y^2 = 16$

Because x^2 and y^2 have different positive coefficients, the equation's graph is an ellipse.

c. $4x^2 = 16 - 4y^2$

Since both the x– and y–terms are squared, the graph of the equation cannot be a parabola. To determine whether the graph is a circle, ellipse, or hyperbola, move the x– and y–terms to the same side of the equation.

$4x^2 + 4y^2 = 16$

Because x^2 and y^2 have the same positive coefficient, the equation's graph is a circle.

d. $x = -4y^2 + 16y$

Since only one variable is squared, the graph of the equation is a parabola.

Problem Set 10.4

Practice Exercises

1. **a.** Since $a = 1$, the parabola opens to the right.

b. The vertex of the parabola is $(-1, 2)$.

c. Graph **b.** is the equation's graph.

3. **a.** Since $a = 1$, the parabola opens to the right.

b. The vertex of the parabola is $(1, -2)$.

c. Graph **f**.is the equation's graph.

5. **a.** Since $a = -1$, the parabola opens to the left.

b. The vertex of the parabola is $(1,2)$.

c. Graph **e**. is the equation's graph. Either graph a or graph e will match this. One will be changed to open to the left.

7. $x = 2y^2$

$x = 2(y-0)^2 + 0$

The vertex is the point $(0,0)$.

9. $x = (y-2)^2 + 3$

The vertex is the point $(3,2)$.

11. $x = -4(y+2)^2 - 1$

The vertex is the point $(-1,-2)$.

13. $x = 2(y-6)^2$

$x = 2(y-6)^2 + 0$

The vertex is the point $(0,6)$.

15. $x = y^2 - 6y + 6$

The y–coordinate of the vertex is

$$-\frac{b}{2a} = -\frac{-6}{2(1)} = -\frac{-6}{2} = 3.$$

The x–coordinate of the vertex is

$f(3) = 3^2 - 6(3) + 6 = 9 - 18 + 6 = -3.$

The vertex is the point $(-3,3)$.

17. $x = 3y^2 + 6y + 7$

The y–coordinate of the vertex is

$$-\frac{b}{2a} = -\frac{6}{2(3)} = -\frac{6}{6} = -1.$$

The x–coordinate of the vertex is

$$f(-1) = 3(-1)^2 + 6(-1) + 7$$
$$= 3(1) - 6 + 7 = 3 - 6 + 7 = 4.$$

The vertex is the point $(4,-1)$.

19. $x = (y-2)^2 - 4$

This is a parabola of the form $x = a(y-k)^2 + h$. Since $a = 1$ is positive, the parabola opens to the right. The vertex of the parabola is $(-4,2)$. The axis of symmetry is $y = 2$. Replace y with 0 to find the x–intercept.

$x = (0-2)^2 - 4 = 4 - 4 = 0$

The x–intercept is 0. Replace x with 0 to find the y–intercepts.

$0 = (y-2)^2 - 4$

$0 = y^2 - 4y + 4 - 4$

$0 = y^2 - 4y$

$0 = y(y-4)$

Apply the zero product principle.

$y = 0$ and $y - 4 = 0$

$y = 4$

The y–intercepts are 0 and 4.

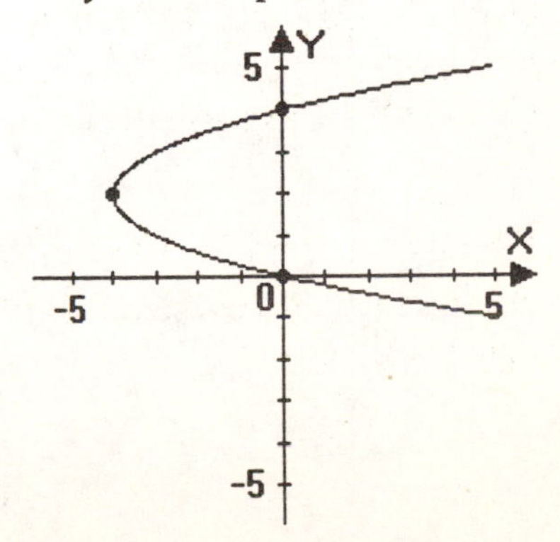

21. $x=(y-3)^2-5$

This is a parabola of the form $x=a(y-k)^2+h$. Since $a=1$ is positive, the parabola opens to the right. The vertex of the parabola is $(-5,3)$. The axis of symmetry is $y=3$. Replace y with 0 to find the x–intercept.

$x=(0-3)^2-5=(-3)^2-5=9-5=4$

The x–intercept is 0. Replace x with 0 to find the y–intercepts.

$0=(y-3)^2-5$

$0=y^2-6y+9-5$

$0=y^2-6y+4$

Solve using the quadratic formula.

$$x=\frac{-b\pm\sqrt{b^2-4ac}}{2a}$$

$$=\frac{-(-6)\pm\sqrt{(-6)^2-4(1)4}}{2(1)}$$

$$=\frac{6\pm\sqrt{36-16}}{2}=\frac{6\pm\sqrt{20}}{2}$$

$$=\frac{6\pm 2\sqrt{5}}{2}=3\pm\sqrt{5}$$

The y–intercepts are $3-\sqrt{5}$ and $3+\sqrt{5}$.

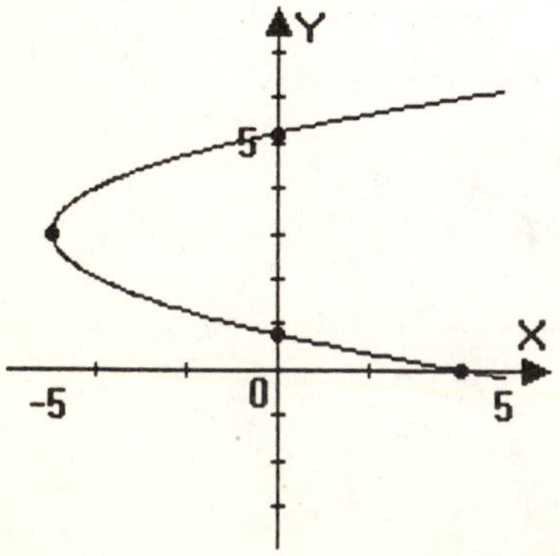

23. $x=-(y-5)^2+4$

This is a parabola of the form $x=a(y-k)^2+h$. Since $a=-1$ is negative, the parabola opens to the left. The vertex of the parabola is $(4,5)$. The axis of symmetry is $y=5$. Replace y with 0 to find the x–intercept.

$x=-(0-5)^2+4=-(-5)^2+4$

$=-25+4=-21$

The x–intercept is 0. Replace x with 0 to find the y–intercepts.

$0=-(y-5)^2+4$

$0=-(y^2-10y+25)+4$

$0=-y^2+10y-25+4$

$0=-y^2+10y-21$

$0=y^2-10y+21$

$0=(y-7)(y-3)$

Apply the zero product principle.

$y-7=0$ and $y-3=0$

$y=7$ $\qquad$ $y=3$

The y–intercepts are 3 and 7.

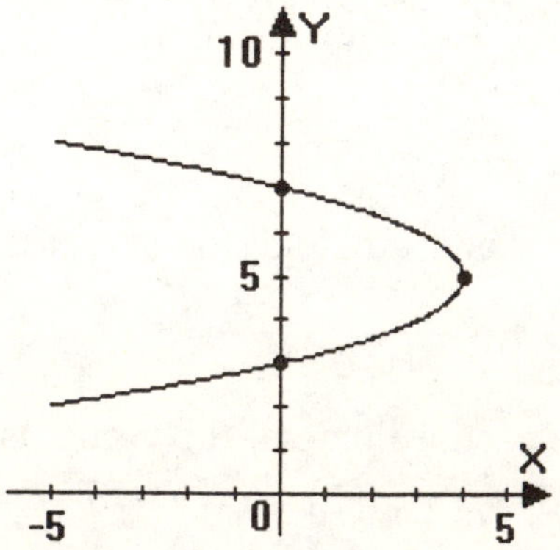

25. $x=(y-4)^2+1$

This is a parabola of the form

$x = a(y-k)^2 + h$. Since $a = 1$ is positive, the parabola opens to the right. The vertex of the parabola is $(1,4)$. The axis of symmetry is $y = 4$.

Replace y with 0 to find the x–intercept.

$x = (0-4)^2 + 1 = (-4)^2 + 1 = 16 + 1 = 17$

The x–intercept is 0. Replace x with 0 to find the y–intercepts.

$0 = (y-4)^2 + 1$

$0 = y^2 - 8y + 16 + 1$

$0 = y^2 - 8y + 17$

Solve using the quadratic formula.

$$y = \frac{-b \pm \sqrt{b^2 - 4ac}}{2a}$$

$$= \frac{-(-8) \pm \sqrt{(-8)^2 - 4(1)(17)}}{2(1)}$$

$$= \frac{8 \pm \sqrt{64 - 68}}{2} = \frac{8 \pm \sqrt{-4}}{2}$$

$$= \frac{8 \pm 2i}{2} = 4 \pm i$$

The solutions are complex, so there are no y–intercepts.

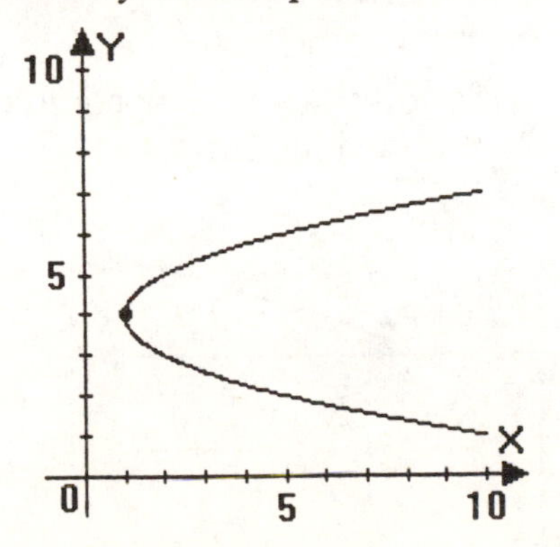

27. $x = -3(y-5)^2 + 3$

This is a parabola of the form $x = a(y-k)^2 + h$. Since $a = -3$ is negative, the parabola opens to the left. The vertex of the parabola is $(3,5)$. The axis of symmetry is $y = 5$.

Replace y with 0 to find the x–intercept.

$x = -3(0-5)^2 + 3 = -3(-5)^2 + 3$

$= -3(25) + 3 = -75 + 3 = -72$

The x–intercept is 0. Replace x with 0 to find the y–intercepts.

$0 = -3(y-5)^2 + 3$

$0 = -3(y^2 - 10y + 25) + 3$

$0 = -3y^2 + 30y - 75 + 3$

$0 = -3y^2 + 30y - 72$

$0 = y^2 - 10y + 24$

$0 = (y-6)(y-4)$

Apply the zero product principle.

$y - 6 = 0$ and $y - 4 = 0$

$y = 6$ $\qquad$ $y = 4$

The y–intercepts are 4 and 6.

29. $x = -2(y+3)^2 - 1$

This is a parabola of the form $x = a(y-k)^2 + h$. Since $a = -2$ is negative, the parabola opens to the left. The vertex of the parabola is

$(-1,-3)$. The axis of symmetry is $y=-3$. Replace y with 0 to find the x–intercept.

$x=-2(0+3)^2-1=-2(3)^2-1$

$=-2(9)-1=-18-1=-19$

The x–intercept is 0. Replace x with 0 to find the y–intercepts.

$0=-2(y+3)^2-1$

$0=-2(y^2+6x+9)-1$

$0=-2y^2-12x-18-1$

$0=-2y^2-12x-19$

$0=2y^2+12x+19$

Solve using the quadratic formula.

$$y=\frac{-12\pm\sqrt{12^2-4(2)(19)}}{2(2)}$$

$$=\frac{-12\pm\sqrt{144-152}}{4}$$

$$=\frac{-12\pm\sqrt{-8}}{4}$$

Since the solutions will be complex, there are no y–intercepts.

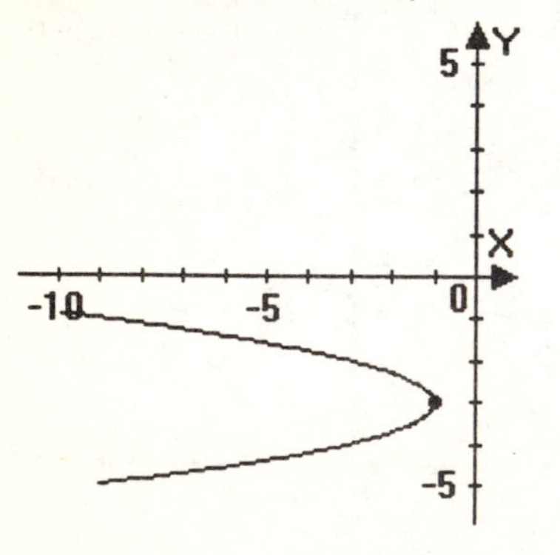

31. $x=\frac{1}{2}(y+2)^2+1$

This is a parabola of the form $x=a(y-k)^2+h$. Since $a=\frac{1}{2}$ is positive, the parabola opens to the right. The vertex of the parabola is $(1,-2)$. The axis of symmetry is $y=-2$. Replace y with 0 to find the x–intercept.

$x=\frac{1}{2}(0+2)^2+1$

$=\frac{1}{2}(4)+1=2+1=3$

The x–intercept is 0. Replace x with 0 to find the y–intercepts.

$0=\frac{1}{2}(y+2)^2+1$

$0=\frac{1}{2}(y^2+2y+4)+1$

$0=\frac{1}{2}y^2+y+2+1$

$0=\frac{1}{2}y^2+y+3$

$0=y^2+2y+6$

Solve using the quadratic formula.

$$y=\frac{-2\pm\sqrt{2^2-4(1)(6)}}{2(1)}$$

$$=\frac{-2\pm\sqrt{4-24}}{2}$$

$$=\frac{-2\pm\sqrt{-20}}{2}$$

Since the solutions will be complex, there are no y–intercepts.

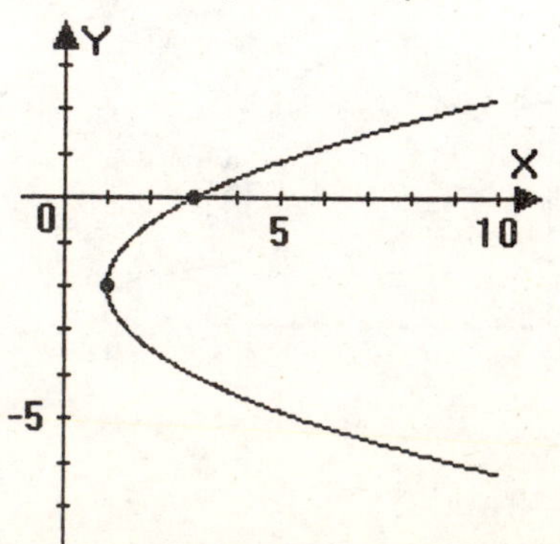

33. $x = y^2 + 2y - 3$

This is a parabola of the form $x = ay^2 + by + c$. Since $a = 1$ is positive, the parabola opens to the right. The y–coordinate of the vertex is $-\frac{b}{2a} = -\frac{2}{2(1)} = -\frac{2}{2} = -1$. The x–coordinate of the vertex is $x = (-1)^2 + 2(-1) - 3 = 1 - 2 - 3 = -4$. The vertex of the parabola is $(-4, -1)$. The axis of symmetry is $y = -1$. Replace y with 0 to find the x–intercept.

$x = 0^2 + 2(0) - 3 = 0 + 0 - 3 = -3$

The x–intercept is –3. Replace x with 0 to find the y–intercepts.

$0 = y^2 + 2y - 3$

$0 = (y+3)(y-1)$

Apply the zero product principle.

$y + 3 = 0$ and $y - 1 = 0$

$y = -3$ $\quad y = 1$

The y–intercepts are –3 and 1.

35. $x = -y^2 - 4y + 5$

This is a parabola of the form $x = ay^2 + by + c$. Since $a = -1$ is negative, the parabola opens to the left. The y–coordinate of the vertex is $-\frac{b}{2a} = -\frac{-4}{2(-1)} = -\frac{-4}{-2} = -2$. The x–coordinate of the vertex is $x = -(-2)^2 - 4(-2) + 5 = -4 + 8 + 5 = 9$.

The vertex of the parabola is $(9, -2)$. The axis of symmetry is $y = -2$. Replace y with 0 to find the x–intercept.

$x = -0^2 - 4(0) + 5 = 0 - 0 + 5 = 5$

The x–intercept is 5. Replace x with 0 to find the y–intercepts.

$0 = -y^2 - 4y + 5$

$0 = y^2 + 4y - 5$

$0 = (y+5)(y-1)$

Apply the zero product principle.

$y + 5 = 0$ and $y - 1 = 0$

$y = -5$ $\quad y = 1$

The y–intercepts are –5 and 1.

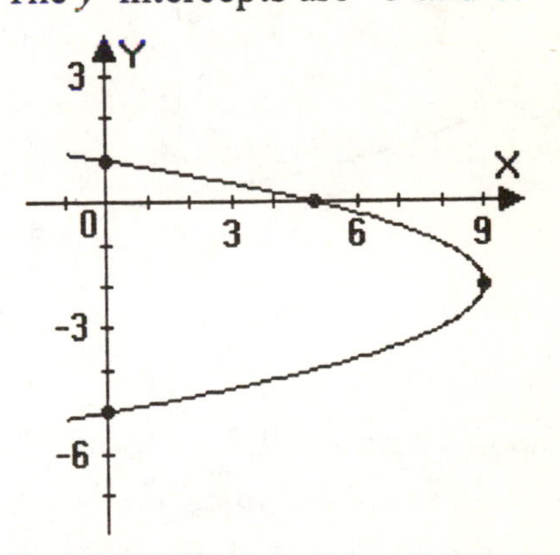

37. $x = y^2 + 6y$

This is a parabola of the form $x = ay^2 + by + c$. Since $a = 1$ is positive, the parabola opens to the right. The y–coordinate of the vertex is $-\frac{b}{2a} = -\frac{6}{2(1)} = -\frac{6}{2} = -3$.

The x–coordinate of the vertex is
$x=(-3)^2+6(-3)=9-18=-9.$
The vertex of the parabola is $(-9,-3)$. The axis of symmetry is $y=-3$. Replace y with 0 to find the x–intercept.
$x=0^2+6(0)=0$
The x–intercept is 0. Replace x with 0 to find the y–intercepts.
$0=y^2+6y$
$0=y(y+6)$
Apply the zero product principle.
$y=0$ and $y+6=0$
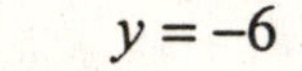
$y=-6$
The y–intercepts are –6 and 0.

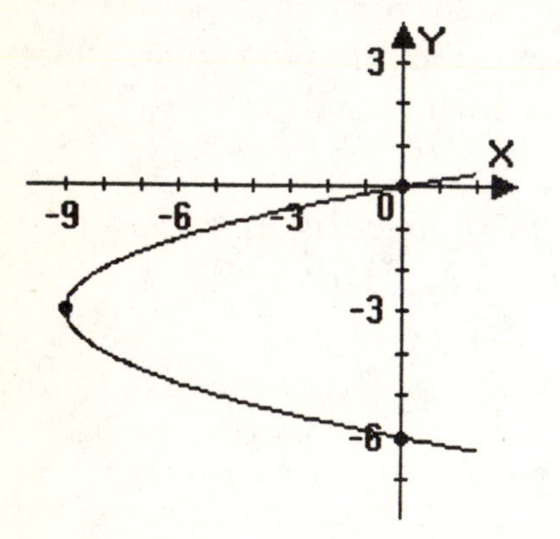

39. $x=-2y^2-4y$
This is a parabola of the form $x=ay^2+by+c$. Since $a=-2$ is negative, the parabola opens to the left. The y–coordinate of the vertex is
$-\frac{b}{2a}=-\frac{-4}{2(-2)}=-\frac{-4}{-4}=-1.$ The x–coordinate of the vertex is
$x=-2(-1)^2-4(-1)=-2(1)+4$
$=-2+4=2$
The vertex of the parabola is $(2,-1)$.

The axis of symmetry is $y=-1$. Replace y with 0 to find the x–intercept.
$x=-2(0)^2-4(0)=-2(0)-0=0$
The x–intercept is 0. Replace x with 0 to find the y–intercepts.
$0=-2y^2-4y$
$0=y^2+2y$
$0=y(y+2)$
Apply the zero product principle.
$y=0$ and $y+2=0$
$y=-2$
The y–intercepts are –2 and 0.

41. $x=-2y^2-4y+1$
This is a parabola of the form $x=ay^2+by+c$. Since $a=-2$ is negative, the parabola opens to the left. The y–coordinate of the vertex is
$-\frac{b}{2a}=-\frac{-4}{2(-2)}=-\frac{-4}{-4}=-1.$
The x–coordinate of the vertex is
$x=-2(-1)^2-4(-1)+1=-2(1)+4+1$
$=-2+4+1=3$

The vertex of the parabola is $(3,-1)$.
The axis of symmetry is $y=-1$. Replace y with 0 to find the x–intercept.

$x = -2(0)^2 - 4(0) + 1 = -2(0) - 0 + 1$
$= 0 - 0 + 1 = 1$
The x–intercept is 0. Replace x with 0 to find the y–intercepts.
$0 = -2y^2 - 4y + 1$
Solve using the quadratic formula.

$$y = \frac{-b \pm \sqrt{b^2 - 4ac}}{2a}$$
$$= \frac{-(-4) \pm \sqrt{(-4)^2 - 4(-2)(1)}}{2(-2)}$$
$$= \frac{4 \pm \sqrt{16+8}}{-4} = \frac{4 \pm \sqrt{24}}{-4} = \frac{4 \pm 2\sqrt{6}}{-4}$$
$$= \frac{2(2 \pm \sqrt{6})}{-4} = \frac{2 \pm \sqrt{6}}{-2} = \frac{-(2 \pm \sqrt{6})}{2}$$
$$= \frac{-2 \pm \sqrt{6}}{2}$$

The y–intercepts are $\frac{-2 \pm \sqrt{6}}{2}$.

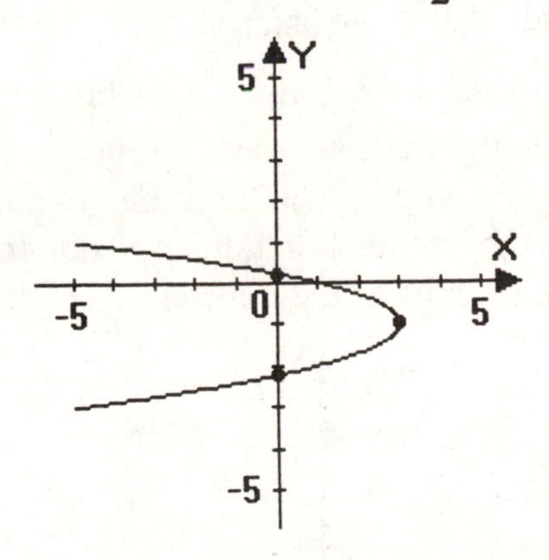

43. **a.** Since the squared term is y, the parabola is horizontal.

b. Since $a = 2$ is positive, the parabola opens to the right.

c. The vertex is the point $(2, 1)$.

45. **a.** Since the squared term is x, the parabola is vertical.

b. Since $a = 2$ is positive, the parabola opens up.

c. The vertex is the point $(1, 2)$.

47. **a.** Since the squared term is x, the parabola is vertical.

b. Since $a = -1$ is negative, the parabola opens down.

c. The vertex is the point $(-3, 4)$.

49. **a.** Since the squared term is y, the parabola is horizontal.

b. Since $a = -1$ is negative, the parabola opens to the left.

c. The vertex is the point $(4, -3)$.

51. **a.** Since the squared term is x, the parabola is vertical.

b. Since $a = 1$ is positive, the parabola opens up.

c. The x–coordinate of the vertex is
$-\frac{b}{2a} = -\frac{-4}{2(1)} = -\frac{-4}{2} = 2.$
The y–coordinate of the vertex is
$f(2) = 2^2 - 4(2) - 1$
$= 4 - 8 - 1 = -5.$
The vertex is the point $(2, -5)$.

53. **a.** Since the squared term is y, the parabola is horizontal.

b. Since $a = -1$ is negative, the parabola opens to the left.

c. The y–coordinate of the vertex is $-\frac{b}{2a} = -\frac{4}{2(-1)} = -\frac{4}{-2} = 2.$
The x–coordinate of the vertex is
$$f(2) = -(2)^2 + 4(2) + 1$$
$$= -4 + 8 + 1 = 5.$$
The vertex is the point $(5,2)$.

55. $x - 7 - 8y = y^2$
Since only one variable is squared, the graph of the equation is a parabola.

57.
$$4x^2 = 36 - y^2$$
$$4x^2 + y^2 = 36$$
Because x^2 and y^2 have different positive coefficients, the equation's graph is an ellipse.

59.
$$x^2 = 36 + 4y^2$$
$$x^2 - 4y^2 = 36$$
Because x^2 and y^2 have opposite signs, the equation's graph is a hyperbola.

61.
$$3x^2 = 12 - 3y^2$$
$$3x^2 + 3y^2 = 12$$
Because x^2 and y^2 have the same positive coefficient, the equation's graph is a circle.

63.
$$3x^2 = 12 + 3y^2$$
$$3x^2 - 3y^2 = 12$$
Because x^2 and y^2 have opposite signs, the equation's graph is a hyperbola.

65. $x^2 - 4y^2 = 16$
Because x^2 and y^2 have opposite signs, the equation's graph is a hyperbola.
$$\frac{x^2}{16} - \frac{4y^2}{16} = \frac{16}{16}$$
$$\frac{x^2}{16} - \frac{y^2}{4} = 1$$
The equation is in the form $\frac{x^2}{a^2} - \frac{y^2}{b^2} = 1$ with $a^2 = 16$, and $b^2 = 4$.
We know the transverse axis lies on the x-axis and the vertices are $(-4,0)$ and $(4,0)$. Because $a^2 = 16$ and $b^2 = 4$, $a = 4$ and $b = 2$. Construct a rectangle using –4 and 4 on the x–axis, and –2 and 2 on the y–axis. Draw extended diagonals to obtain the asymptotes. Graph the hyperbola.

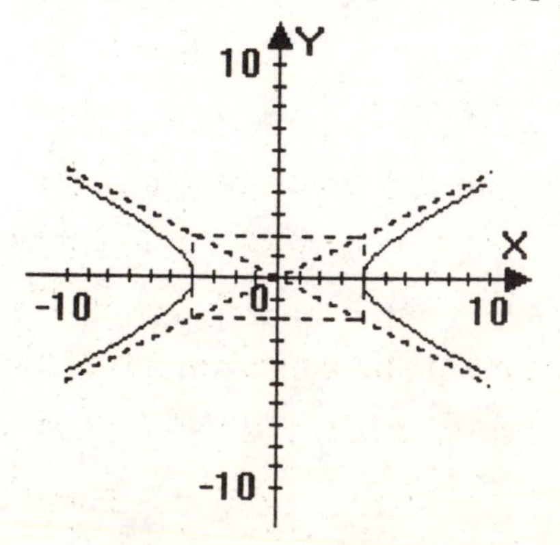

67. $4x^2 + 4y^2 = 16$

Because x^2 and y^2 have the same positive coefficient, the equation's graph is a circle.

$$\frac{4x^2}{4} + \frac{4y^2}{4} = \frac{16}{4}$$

$$x^2 + y^2 = 4$$

The center is $(0,0)$ and the radius is 2 units.

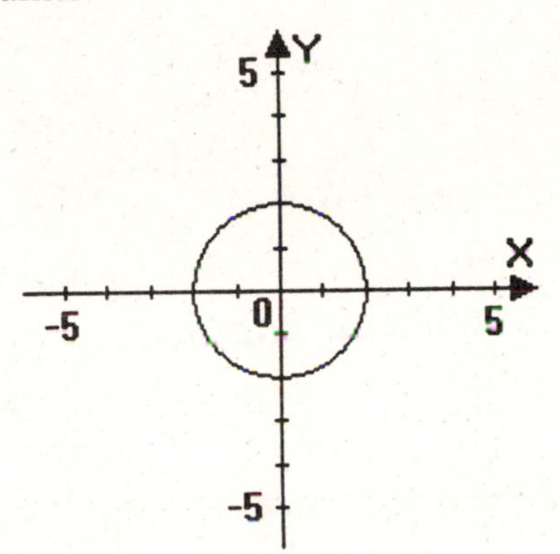

69. $x^2 + 4y^2 = 16$

Because x^2 and y^2 have different positive coefficients, the equation's graph is an ellipse.

$$\frac{x^2}{16} + \frac{4y^2}{16} = \frac{16}{16}$$

$$\frac{x^2}{16} + \frac{y^2}{4} = 1$$

Because the denominator of the x^2 – term is greater than the denominator of the y^2 – term, the major axis is horizontal. Since $a^2 = 16$, $a = 4$ and the vertices are $(-4,0)$ and $(4,0)$. Since $b^2 = 4$, $b = 2$ and endpoints of the minor axis are $(0,-2)$ and $(0,2)$.

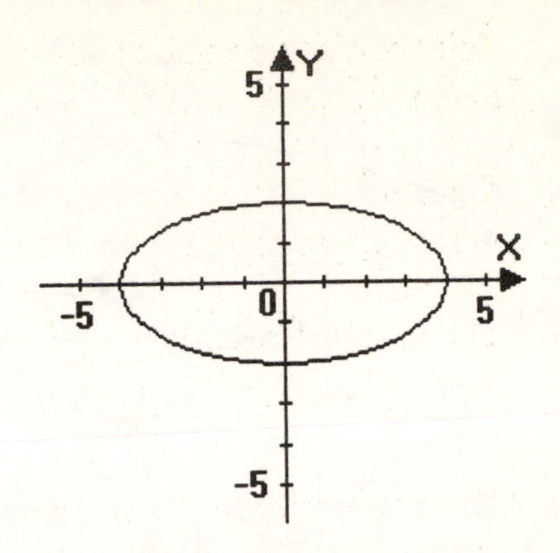

71. $x = (y-1)^2 - 4$

Since only one variable is squared, the graph of the equation is a parabola.

This is a parabola of the form $x = a(y-k)^2 + h$. Since $a = 1$ is positive, the parabola opens to the right. The vertex of the parabola is $(-4,1)$. The axis of symmetry is $y = 1$. Replace y with 0 to find the x–intercept.

$$x = (0-1)^2 - 4 = (-1)^2 - 4 = 1 - 4 = -3$$

The x–intercept is 0. Replace x with 0 to find the y–intercepts.

$$0 = (y-1)^2 - 4$$
$$0 = y^2 - 2y + 1 - 4$$
$$0 = y^2 - 2y - 3$$
$$0 = (y-3)(y+1)$$

Apply the zero product principle.

$y - 3 = 0$ and $y + 1 = 0$

$y = 3$ $\qquad$ $y = -1$

The y–intercepts are –1 and 3.

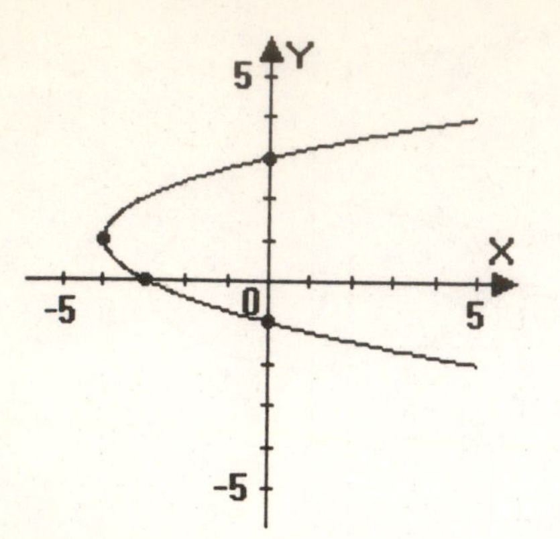

73. $(x-2)^2+(y+1)^2=16$

Because x^2 and y^2 have the same positive coefficient, the equation's graph is a circle.
The center is $(2,-1)$ and the radius is 4 units.

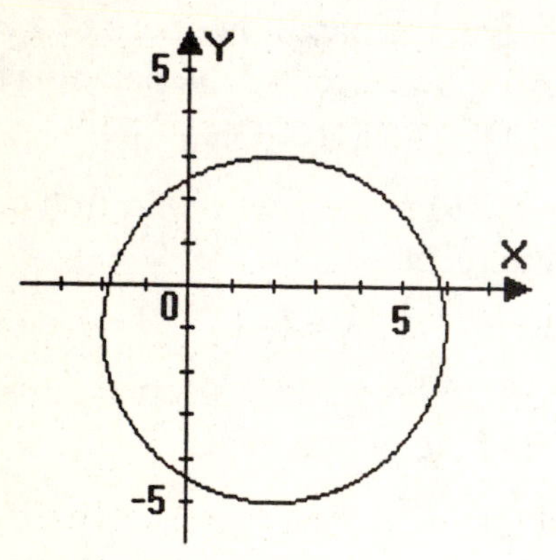

Application Exercises

75. a.

$$y=ax^2$$
$$316=a(1750)^2$$
$$316=a(3062500)$$
$$\frac{316}{3062500}=a$$
$$0.0001032=a$$

The equation is
$y=0.0001032x^2$.

b. To find the height of the cable 1000 feet from the tower, find y when $x=1750-1000=750$.

$$y=0.0001032(750)^2$$
$$=0.0001032(562,500)=58.05$$

The height of the cable is 58.05 feet.

77. a.

$$y=ax^2$$
$$2=a(6)^2$$
$$2=a(36)$$
$$\frac{2}{36}=a$$
$$\frac{1}{18}=a$$

The equation is $y=\frac{1}{18}x^2$.

b.

$$a=\frac{1}{4p}$$
$$\frac{1}{18}=\frac{1}{4p}$$
$$4p=18$$
$$p=\frac{18}{4}=4.5$$

The receiver should be placed 4.5 feet from the base of the dish.

Writing in Mathematics

79. Answers will vary.

81. Answers will vary.

83. Answers will vary.

85. Answers will vary.

Technology Exercises

87.
$$y^2+2y-6x+13=0$$
$$y^2+2y+(-6x+13)=0$$
$$a=1 \quad b=2 \quad c=-6x+13$$
$$y=\frac{-2\pm\sqrt{2^2-4(1)(-6x+13)}}{2(1)}$$
$$=\frac{-2\pm\sqrt{4-4(-6x+13)}}{2}$$
$$=\frac{-2\pm\sqrt{4+24x-52}}{2}$$
$$=\frac{-2\pm\sqrt{24x-48}}{2}$$

89. Answers will vary. For example, consider Exercise 19.
$$x=(y-2)^2-4$$
$$x+4=(y-2)^2$$
$$\pm\sqrt{x+4}=y-2$$
$$2\pm\sqrt{x+4}=y$$

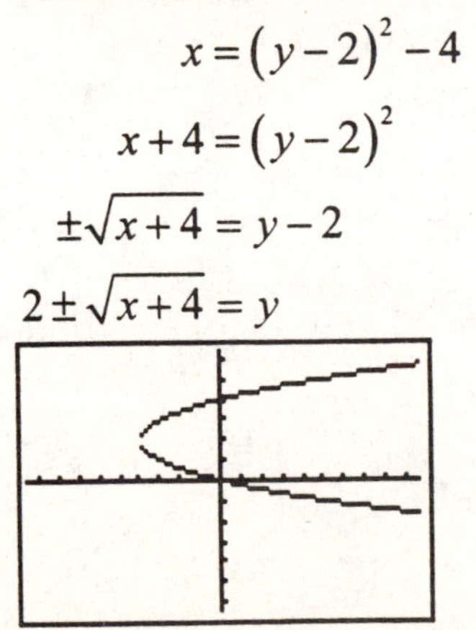

Critical Thinking Exercises

91. Answers will vary.

Review Exercises

93. $f(x)=2^{1-x}$

x	$f(x)$
-2	8
-1	4
0	2
1	1
2	$\frac{1}{2}$

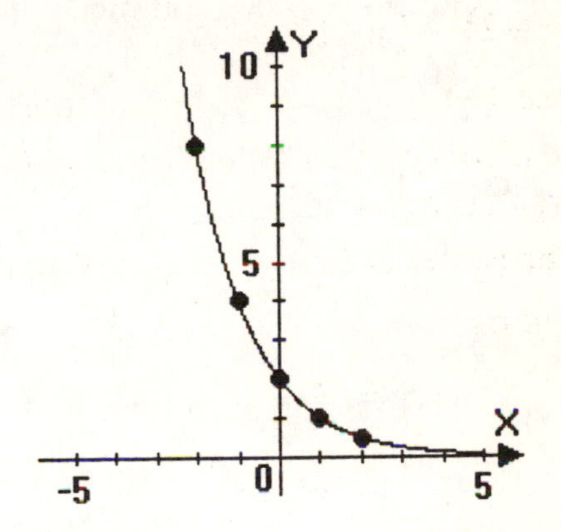

94.
$$f(x)=\frac{1}{3}x-5$$
$$y=\frac{1}{3}x-5$$
Interchange x and y and solve for y.
$$x=\frac{1}{3}y-5$$
$$x+5=\frac{1}{3}y$$
$$3x+15=y$$
$$f^{-1}(x)=3x+15$$

95.
$$4x-3y=12$$
$$3x-4y=2$$
Multiply the first equation by –3 and the second equation by 4.

$$\begin{aligned} -12x + 9y &= -36 \\ 12x - 16y &= 8 \\ \hline -7y &= -28 \\ y &= 4 \end{aligned}$$

Back-substitute 4 for y to find x.

$$4x - 3(4) = 12$$
$$4x - 12 = 12$$
$$4x = 24$$
$$x = 6$$

The solution is $(6,4)$ and the solution set is $\{(6,4)\}$.

Check Points 10.5

1. $x^2 = y - 1$

$4x - y = -1$

Solve the first equation for y.

$$x^2 = y - 1$$
$$x^2 + 1 = y$$

Substitute $x^2 + 1$ for y in the second equation and solve for x.

$$4x - y = -1$$
$$4x - (x^2 + 1) = -1$$
$$4x - x^2 - 1 = -1$$
$$4x - x^2 = 0$$
$$x(4 - x) = 0$$

Apply the zero product principle.

$x = 0$ or $4 - x = 0$

$4 = x$

Substitute 0 and 4 for x to find y.

$x = 0$ or $x = 4$

$y = x^2 + 1$ $\quad$ $y = x^2 + 1$

$y = 0^2 + 1$ $\quad$ $y = 4^2 + 1$

$y = 1$ $\quad$ $y = 16 + 1$

$y = 17$

The solutions are $(0,1)$ and $(4,17)$ and the solution set is $\{(0,1),(4,17)\}$.

2. $x + 2y = 0$

$(x-1)^2 + (y-1)^2 = 5$

Solve the first equation for x.

$$x + 2y = 0$$
$$x = -2y$$

Substitute $-2y$ for x in the second equation and solve for y.

$$(-2y-1)^2 + (y-1)^2 = 5$$
$$4y^2 + 4y + 1 + y^2 - 2y + 1 = 5$$
$$5y^2 + 2y + 2 = 5$$
$$5y^2 + 2y - 3 = 0$$
$$(5y-3)(y+1) = 0$$

Apply the zero product principle.

$5y - 3 = 0$ or $y + 1 = 0$

$5y = 3$ $\quad$ $y = -1$

$y = \frac{3}{5}$

Substitute -1 and $\frac{3}{5}$ for y to find x.

$y = -1$ or $y = \frac{3}{5}$

$x = -2(-1)$ $\quad$ $x = -2\left(\frac{3}{5}\right)$

$x = 2$ $\quad$ $x = -\frac{6}{5}$

The solutions are $(2,-1)$ and

$\left(-\frac{6}{5},\frac{3}{5}\right)$ and the solution set is

$\left\{(2,-1),\left(-\frac{6}{5},\frac{3}{5}\right)\right\}$.

3.
$$3x^2+2y^2=35$$
$$4x^2+3y^2=48$$
Multiply the first equation by –3 and the second equation by 2.
$$\begin{aligned}-9x^2-6y^2&=-105\\ 8x^2+6y^2&=96\\ \hline -x^2&=-9\\ x^2&=9\\ x&=\pm3\end{aligned}$$
Back substitute ±3 for x to find y.
$$x=\pm3$$
$$\begin{aligned}3(\pm3)^2+2y^2&=35\\ 3(9)+2y^2&=35\\ 27+2y^2&=35\\ 2y^2&=8\\ y^2&=4\\ y&=\pm2\end{aligned}$$
The solutions are $(3,2),(3,-2)$, $(-3,2)$ and $(-3,-2)$ and the solution set is $\{(-3,-2),(-3,2),(3,-2),(3,2)\}$.

4.
$$y=x^2+5$$
$$x^2+y^2=25$$
Rewrite the equations to solve by addition.
$$\begin{aligned}-x^2\quad+y&=5\\ x^2+y^2\quad&=25\\ \hline y^2+y&=30\end{aligned}$$
$$y^2+y-30=0$$
$$(y+6)(y-5)=0$$
Apply the zero product principle.

$y+6=0$ or $y-5=0$

$y=-6$ $\qquad y=5$

Substitute –6 and 5 for y to find x.

$y=-6$ or $y=5$

$-6=x^2+5$ $\qquad 5=x^2+5$

$-11=x^2$ $\qquad 0=x^2$

$0=x$

We disregard $y=-6$, there is no value of x for which x^2 is a negative number. The solution is $(0,5)$ and the solution set is $\{(0,5)\}$.

5. Let x = the length of the rectangle
Let y = the width of the rectangle

Perimeter: $2x+2y=20$

Area: $xy=21$

Solve the second equation for y.
$$xy=21$$
$$y=\frac{21}{x}$$
Substitute $\frac{21}{x}$ for y in the first equation and solve for x.
$$2x+2\left(\frac{21}{x}\right)=20$$
$$2x+\frac{42}{x}=20$$

$$x\left(2x+\frac{42}{x}\right)=x(20)$$
$$2x^2+42=20x$$
$$2x^2-20x+42=0$$
$$x^2-10x+21=0$$
$$(x-7)(x-3)=0$$

Apply the zero product principle.

$x-7=0$ or $x-3=0$

$x=7$ $\qquad x=3$

Substitute 3 and 7 for x to find y.

$x=7$ or $x=3$

$y=\frac{21}{7}$ $\qquad y=\frac{21}{3}$

$y=3$ $\qquad y=7$

The dimensions of the rectangle are 3 feet by 7 feet.

Problem Set 10.5

Practice Exercise

1. $x+y=2$

$y=x^2-4$

Substitute x^2+4 for y in the first equation and solve for x.

$$x+(x^2-4)=2$$
$$x+x^2-4=2$$
$$x^2+x-6=0$$
$$(x+3)(x-2)=0$$

Apply the zero product principle.

$x+3=0$ or $x-2=0$

$x=-3$ $\qquad x=2$

Substitute –3 and 2 for x in the second equation to find y.

$x=-3$ or $x=2$

$y=(-3)^2-4$ $\qquad y=2^2-4$

$y=9-4$ $\qquad y=4-4$

$y=5$ $\qquad y=0$

The solutions are $(-3,5)$ and $(2,0)$ and the solution set is $\{(-3,5),(2,0)\}$.

3. $x+y=2$

$y=x^2-4x+4$

Substitute x^2-4x+4 for y in the first equation and solve for x.

$$x+x^2-4x+4=2$$
$$x^2-3x+4=2$$
$$x^2-3x+2=0$$
$$(x-2)(x-1)=0$$

Apply the zero product principle.

$x-2=0$ or $x-1=0$

$x=2$ $\qquad x=1$

Substitute 1 and 2 for x to find y.

$x=2$ or $x=1$

$x+y=2$ $\qquad x+y=2$

$2+y=2$ $\qquad 1+y=2$

$y=0$ $\qquad y=1$

The solutions are $(2,0)$ and $(1,1)$ and the solution set is $\{(1,1),(2,0)\}$.

5. $y=x^2-4x-10$

$y=-x^2-2x+14$

Substitute $-x^2-2x+14$ for y in the first equation and solve for x.

$$-x^2-2x+14=x^2-4x-10$$
$$0=2x^2-2x-24$$

$0 = x^2 - x - 12$

$0 = (x-4)(x+3)$

Apply the zero product principle.

$x-4=0$ or $x+3=0$

$x=4$ $\quad$ $x=-3$

Substitute 3 and 4 for x to find y.

$x=4$

$y = 4^2 - 4(4) - 10$

$= 16 - 16 - 10 = -10$

$x=-3$

$y = (-3)^2 - 4(-3) - 10$

$= 9 + 12 - 10 = 11$

The solutions are $(4,-10)$ and $(-3,11)$ and the solution set is $\{(-3,11),(4,-10)\}$.

7. $x^2 + y^2 = 25$

$x - y = 1$

Solve the second equation for x.

$x - y = 1$

$x = y + 1$

Substitute $y+1$ for x to find y.

$x^2 + y^2 = 25$

$(y+1)^2 + y^2 = 25$

$y^2 + 2y + 1 + y^2 = 25$

$2y^2 + 2y + 1 = 25$

$2y^2 + 2y - 24 = 0$

$y^2 + y - 12 = 0$

$(y+4)(y-3) = 0$

Apply the zero product principle.

$y+4=0$ or $y-3=0$

$y=-4$ $\quad$ $y=3$

Substitute –4 and 3 for y to find x.

$y=-4$ $\quad$ $y=3$

$x=-4+1$ $\quad$ $x=3+1$

$x=-3$ $\quad$ $x=4$

The solutions are $(-3,-4)$ and $(4,3)$ and the solution set it $\{(-3,-4),(4,3)\}$.

9. $xy = 6$

$2x - y = 1$

Solve the first equation for y.

$xy = 6$

$y = \frac{6}{x}$

Substitute $\frac{6}{x}$ for y in the second equation and solve for x.

$2x - \frac{6}{x} = 1$

$x\left(2x - \frac{6}{x}\right) = x(1)$

$2x^2 - 6 = x$

$2x^2 - x - 6 = 0$

$(2x+3)(x-2) = 0$

Apply the zero product principle.

$x-2=0$ or $2x+3=0$

$x=2$ $\quad$ $2x=-3$

$x = -\frac{3}{2}$

Substitute 2 and $-\frac{3}{2}$ for x to find y.

$x = 2$ or $x = -\frac{3}{2}$

$2y = 6$

$y = 3$

$-\frac{3}{2}y = 6$

$-\frac{2}{3}\left(-\frac{3}{2}\right)y = \left(-\frac{2}{3}\right)6$

$y = -4$

The solutions are $(2,3)$ and $\left(-\frac{3}{2}, -4\right)$ and the solution set is $\left\{(2,3), \left(-\frac{3}{2}, -4\right)\right\}$.

11. $y^2 = x^2 - 9$

$2y = x - 3$

Solve the second equation for x.

$2y = x - 3$

$2y + 3 = x$

Substitute $2y + 3$ for x to find y.

$y^2 = (2y+3)^2 - 9$

$y^2 = 4y^2 + 12y + 9 - 9$

$y^2 = 4y^2 + 12y$

$0 = 3y^2 + 12y$

$0 = 3y(y+4)$

Apply the zero product principle.

$3y = 0$ or $y + 4 = 0$

$y = 0$ $\quad$ $y = -4$

Substitute –4 and 0 for y to find x.

$y = 0$ or $y = -4$

$2(0) + 3 = x$ $\quad$ $2(-4) + 3 = x$

$3 = x$ $\quad$ $-8 + 3 = x$

$-5 = x$

The solutions are $(3,0)$ and $(-5,-4)$ and the solution set is $\{(-5,-4), (3,0)\}$.

13. $xy = 3$

$x^2 + y^2 = 10$

Solve the first equation for y.

$xy = 3$

$y = \frac{3}{x}$

Substitute $\frac{3}{x}$ for y to find x.

$$x^2 + \left(\frac{3}{x}\right)^2 = 10$$

$$x^2 + \frac{9}{x^2} = 10$$

$$x^2\left(x^2 + \frac{9}{x^2}\right) = x^2(10)$$

$$x^4 + 9 = 10x^2$$

$$x^4 - 10x^2 + 9 = 0$$

$$(x^2 - 9)(x^2 - 1) = 0$$

$$(x+3)(x-3)(x+1)(x-1) = 0$$

Apply the zero product principle.

$x + 3 = 0$ $\quad$ $x - 3 = 0$

$x = -3$ $\quad$ $x = 3$

$x + 1 = 0$ $\quad$ $x - 1 = 0$

$x = -1$ $\quad$ $x = 1$

Substitute ± 1 and ± 3 for x to find y.

$x = -3$ $\quad$ $x = 3$

$y = \frac{3}{-3}$ $\quad$ $y = \frac{3}{3}$

$y = -1$ $\quad$ $y = 1$

$x = -1$ $\quad$ $x = 1$

$y = \frac{3}{-1}$ $\quad$ $y = \frac{3}{1}$

$y = -3$ $\quad$ $y = 3$

The solutions are $(-3,-1),(-1,-3),(1,3)$ and $(3,1)$ and the solution set is $\{(-3,-1),(-1,-3),(1,3),(3,1)\}$.

15. $x + y = 1$

$x^2 + xy - y^2 = -5$

Solve the first equation for y.

$x + y = 1$

$y = -x + 1$

Substitute $-x + 1$ for y and solve for x.

$$x^2 + x(-x+1) - (-x+1)^2 = -5$$
$$x^2 - x^2 + x - (x^2 - 2x + 1) = -5$$
$$x^2 - x^2 + x - x^2 + 2x - 1 = -5$$
$$-x^2 + 3x - 1 = -5$$
$$-x^2 + 3x + 4 = 0$$
$$x^2 - 3x - 4 = 0$$
$$(x-4)(x+1) = 0$$

Apply the zero product principle.

$x - 4 = 0$ or $x + 1 = 0$

$x = 4$ $\quad$ $x = -1$

Substitute -1 and 4 for x to find y.

$x = 4$ or $x = -1$

$y = -4 + 1$ $\quad$ $y = -(-1) + 1$

$y = -3$ $\quad$ $y = 1 + 1$

$y = 2$

The solutions are $(4,-3)$ and $(-1,2)$ and the solution set is $\{(4,-3),(-1,2)\}$.

17. $x + y = 1$

$(x-1)^2 + (y+2)^2 = 10$

Solve the first equation for y.

$x + y = 1$

$y = -x + 1$

Substitute $-x + 1$ for y to find x.

$$(x-1)^2 + ((-x+1)+2)^2 = 10$$
$$(x-1)^2 + (-x+1+2)^2 = 10$$
$$(x-1)^2 + (-x+3)^2 = 10$$
$$x^2 - 2x + 1 + x^2 - 6x + 9 = 10$$
$$2x^2 - 8x + 10 = 10$$
$$2x^2 - 8x = 0$$
$$2x(x-4) = 0$$

Apply the zero product principle.

$2x = 0$ or $x - 4 = 0$

$x = 0$ $\quad$ $x = 4$

Substitute 0 and 4 for x to find y.

$x = 0$ or $x = 4$

$y = -0 + 1$ $\quad$ $y = -4 + 1$

$y = 1$ $\quad$ $y = -3$

The solutions are $(0,1)$ and $(4,-3)$ and the solution set is $\{(0,1),(4,-3)\}$.

19. Solve the system by addition.

$$\begin{aligned} x^2+y^2&=13\\ x^2-y^2&=5\\ \hline 2x^2&=18\\ x^2&=9\\ x&=\pm 3 \end{aligned}$$

Substitute ± 3 for x to find y.

$$\begin{aligned} x&=\pm 3\\ (\pm 3)^2+y^2&=13\\ 9+y^2&=13\\ y^2&=4\\ y&=\pm 2 \end{aligned}$$

The solutions are $(-3,-2),(-3,2)$, $(3,-2)$ and $(3,2)$ and the solution set is $\{(-3,-2),(-3,2),(3,-2),(3,2)\}$.

21.

$$\begin{aligned} x^2-4y^2&=-7\\ 3x^2+y^2&=31 \end{aligned}$$

Multiply the first equation by –3 and add to the second equation.

$$\begin{aligned} -3x^2+12y^2&=21\\ 3x^2+y^2&=31\\ \hline 13y^2&=52\\ y^2&=4\\ y&=\pm 2 \end{aligned}$$

Substitute –2 and 2 for y to find x.

$$\begin{aligned} y&=\pm 2\\ 3x^2+(\pm 2)^2&=31\\ 3x^2+4&=31\\ 3x^2&=27\\ x^2&=9\\ x&=\pm 3 \end{aligned}$$

The solutions are $(-3,-2),(-3,2)$, $(3,-2)$ and $(3,2)$ and the solution set is $\{(-3,-2),(-3,2),(3,-2),(3,2)\}$.

23.

$$\begin{aligned} 3x^2+4y^2-16&=0\\ 2x^2-3y^2-5&=0 \end{aligned}$$

Multiply the first equation by 3 and the second equation by 4 and solve by addition.

$$\begin{aligned} 9x^2+12y^2-48&=0\\ 8x^2-12y^2-20&=0\\ \hline 17x^2-68&=0\\ 17x^2&=68\\ x^2&=4\\ x&=\pm 2 \end{aligned}$$

Substitute ± 2 for x to find y.

$$\begin{aligned} x&=\pm 2\\ 2(\pm 2)^2-3y^2-5&=0\\ 2(4)-3y^2-5&=0\\ 8-3y^2-5&=0\\ 3-3y^2&=0\\ 3&=3y^2\\ 1&=y^2\\ \pm 1&=y \end{aligned}$$

The solutions are $(-2,-1),(-2,1)$, $(2,-1)$ and $(2,1)$ and the solution set is $\{(-2,-1),(-2,1),(2,-1),(2,1)\}$.

25.

$$\begin{aligned} x^2+y^2&=25\\ (x-8)^2+y^2&=41 \end{aligned}$$

Multiply the first equation by –1 and solve by addition.

$$-x^2 \qquad - y^2 = -25$$
$$(x-8)^2 + y^2 = 41$$
$$-x^2 + (x-8)^2 = 16$$
$$-x^2 + x^2 - 16x + 64 = 16$$
$$-16x + 64 = 16$$
$$-16x = -48$$
$$x = 3$$

Substitute 3 for x to find y.

$$x = 3$$
$$3^2 + y^2 = 25$$
$$6 + y^2 = 25$$
$$y^2 = 16$$
$$y = \pm 4$$

The solutions are $(3,-4)$ and $(3,4)$ and the solution set is $\{(3,-4),(3,4)\}$.

27.

$$y^2 - x = 4$$
$$x^2 + y^2 = 4$$

Multiply the first equation by -1 and solve by addition.

$$-y^2 + x = -4$$
$$x^2 + y^2 = 4$$
$$x^2 + x = 0$$
$$x(x+1) = 0$$

Apply the zero product principle.

$x = 0$ or $x + 1 = 0$

$$x = -1$$

Substitute -1 and 0 for x to find y.

$x = 0$ or $x = -1$

$$y^2 - 0 = 4 \qquad y^2 - (-1) = 4$$
$$y^2 = 4 \qquad y^2 + 1 = 4$$
$$y = \pm 2 \qquad y^2 = 3$$
$$y = \pm\sqrt{3}$$

The solutions are $(0,-2),(0,2)$, $(-1,-\sqrt{3})$ and $(-1,\sqrt{3})$ and the solution set is $\{(-1,-\sqrt{3}),(-1,\sqrt{3}),(0,-2),(0,2)\}$.

29.

$$3x^2 + 4y^2 = 16$$
$$2x^2 - 3y^2 = 5$$

Multiply the first equation by -2 and the second equation by 3 and solve by addition.

$$-6x^2 - 8y^2 = -32$$
$$6x^2 - 9y^2 = 15$$
$$-17y^2 = -17$$
$$y^2 = 1$$
$$y = \pm 1$$

Substitute ± 1 for y to find x.

$$y = \pm 1$$
$$3x^2 + 4(\pm 1)^2 = 16$$
$$3x^2 + 4(1) = 16$$
$$3x^2 + 4 = 16$$
$$3x^2 = 12$$
$$x^2 = 4$$
$$x = \pm 2$$

The solutions are $(-2,1),(2,1)$, $(-2,-1)$ and $(2,-1)$ and the solution set is $\{(-2,-1),(-2,1),(2,-1),(2,1)\}$.

31. $2x^2 + y^2 = 18$

$xy = 4$

Solve the second equation for y.

$xy = 4$

$y = \frac{4}{x}$

Substitute $\frac{4}{x}$ for y in the second equation and solve for x.

$$2x^2 + \left(\frac{4}{x}\right)^2 = 18$$

$$2x^2 + \frac{16}{x^2} = 18$$

$$x^2\left(2x^2 + \frac{16}{x^2}\right) = x^2(18)$$

$$2x^4 + 16 = 18x^2$$

$$2x^4 - 18x^2 + 16 = 0$$

$$x^4 - 9x^2 + 8 = 0$$

$$(x^2 - 8)(x^2 - 1) = 0$$

$$(x^2 - 8)(x+1)(x-1) = 0$$

Apply the zero product principle.

$x^2 - 8 = 0$ $\quad$ $x + 1 = 0$ $\quad$ $x - 1 = 0$

$x^2 = 8$ $\quad$ $x = -1$ $\quad$ $x = 1$

$x = \pm\sqrt{8}$

$x = \pm 2\sqrt{2}$

Substitute $\pm 2\sqrt{2}$ and ± 1 for x to find y.

$x = 1$ $\quad$ $x = -1$

$y = \frac{4}{1}$ $\quad$ $y = \frac{4}{-1}$

$y = 4$ $\quad$ $y = -4$

$x = 2\sqrt{2}$ $\quad$ $x = -2\sqrt{2}$

$y = \frac{4}{2\sqrt{2}}$ $\quad$ $y = \frac{4}{-2\sqrt{2}}$

$y = \frac{2}{\sqrt{2}} \cdot \frac{\sqrt{2}}{\sqrt{2}}$ $\quad$ $y = -\frac{2}{\sqrt{2}} \cdot \frac{\sqrt{2}}{\sqrt{2}}$

$y = \frac{2\sqrt{2}}{2}$ $\quad$ $y = -\frac{2\sqrt{2}}{2}$

$y = \sqrt{2}$ $\quad$ $y = -\sqrt{2}$

The solutions are $(2\sqrt{2}, \sqrt{2})$, $(-2\sqrt{2}, -\sqrt{2})$, $(1, 4)$ and $(-1, -4)$ and the solution set is $\{(-2\sqrt{2}, -\sqrt{2}), (-1, -4), (1, 4), (2\sqrt{2}, \sqrt{2})\}$.

33. $x^2 + 4y^2 = 20$

$x + 2y = 6$

Solve the second equation for x.

$x + 2y = 6$

$x = 6 - 2y$

Substitute $6 - 2y$ for x to find y.

$$(6 - 2y)^2 + 4y^2 = 20$$

$$36 - 24y + 4y^2 + 4y^2 = 20$$

$$36 - 24y + 8y^2 = 20$$

$$8y^2 - 24y + 16 = 0$$

$$y^2 - 3y + 2 = 0$$

$$(y-2)(y-1) = 0$$

Apply the zero product principle.

$y - 2 = 0$ or $y - 1 = 0$

$y = 2$ $\quad$ $y = 1$

Substitute 1 and 2 for y to find x.

$y = 2$ or $y = 1$

$x = 6 - 2(2)$ $x = 6 - 2(1)$

$x = 6 - 4$ $x = 6 - 2$

$x = 2$ $x = 4$

The solutions are $(2,2)$ and $(4,1)$ and the solution set is $\{(2,2),(4,1)\}$.

35. Eliminate y by adding the two equations.

$$x^3 + y = 0$$
$$x^2 - y = 0$$
$$x^3 + x^2 = 0$$
$$x^2(x+1) = 0$$

Apply the zero product principle.

$x^2 = 0$ or $x + 1 = 0$

$x = 0$ $x = -1$

Substitute -1 and 0 for x to find y.

$x = 0$ or $x = -1$

$0^2 - y = 0$ $(-1)^2 - y = 0$

$-y = 0$ $1 - y = 0$

$y = 0$ $-y = -1$

$y = 1$

The solutions are $(0,0)$ and $(-1,1)$ and the solution set is $\{(-1,1), (0,0)\}$.

37. $x^2 + (y-2)^2 = 4$

$x^2 - 2y = 0$

Solve the second equation for x^2.

$$x^2 - 2y = 0$$
$$x^2 = 2y$$

Substitute $2y$ for x^2 in the first equation and solve for y.

$$2y + (y-2)^2 = 4$$
$$2y + y^2 - 4y + 4 = 4$$
$$y^2 - 2y + 4 = 4$$
$$y^2 - 2y = 0$$
$$y(y-2) = 0$$

Apply the zero product principle.

$y = 0$ or $y - 2 = 0$

$y = 2$

Substitute 0 and $\frac{4}{5}$ for y to find x.

$y = 0$ or $y = 2$

$x^2 = 2(0)$ $x^2 = 2(2)$

$x^2 = 0$ $x^2 = 4$

$x = 0$ $x = \pm 2$

The solutions are $(0,0), (-2,2)$ and $(2,2)$ and the solution set is $\{(0,0), (-2,2), (2,2)\}$.

39. $y = (x+3)^2$

$x + 2y = -2$

Substitute $(x+3)^2$ for y in the second equation.

$$x + 2(x+3)^2 = -2$$
$$x + 2(x^2 + 6x + 9) = -2$$
$$x + 2x^2 + 12x + 18 = -2$$
$$2x^2 + 13x + 18 = -2$$
$$2x^2 + 13x + 20 = 0$$
$$(2x+5)(x+4) = 0$$

Apply the zero product principle.

$$\begin{aligned} 2x+5&=0 \quad \text{or} \quad x+4=0 \\ 2x&=-5 \qquad\quad x=-4 \\ x&=-\frac{5}{2} \end{aligned}$$

Substitute $-\frac{5}{2}$ and -4 for x to find y.

$x=-\frac{5}{2}$ or $x=-4$

$$\begin{aligned} -\frac{5}{2}+2y&=-2 \\ -5+4y&=-4 \\ 4y&=1 \\ y&=\frac{1}{4} \end{aligned} \qquad \begin{aligned} -4+2y&=-2 \\ 2y&=2 \\ y&=1 \end{aligned}$$

The solutions are $\left(-\frac{5}{2},\frac{1}{4}\right)$ and $(-4,1)$ and the solution set is $\left\{(-4,1),\left(-\frac{5}{2},\frac{1}{4}\right)\right\}$.

$$\begin{aligned} 5x-12&=0 \quad \text{or} \quad x+2=0 \\ 5x&=12 \qquad\quad x=-2 \\ x&=\frac{12}{5} \end{aligned}$$

Substitute -2 and $\frac{12}{5}$ for x to find y.

$x=\frac{12}{5}$ or $x=-2$

$$\begin{aligned} y&=-2\left(\frac{12}{5}\right)-1 \\ y&=-\frac{24}{5}-\frac{5}{5} \\ y&=-\frac{29}{5} \end{aligned} \qquad \begin{aligned} y&=-2(-2)-1 \\ y&=4-1 \\ y&=3 \end{aligned}$$

The solutions are $\left(\frac{12}{5},-\frac{29}{5}\right)$ and $(-2,3)$ and the solution set is $\left\{\left(\frac{12}{5},-\frac{29}{5}\right),(-2,3)\right\}$.

41. $x^2+y^2+3y=22$

$2x+y=-1$

Solve the second equation for y.

$$\begin{aligned} 2x+y&=-1 \\ y&=-2x-1 \end{aligned}$$

Substitute $-2x-1$ for y to find x.

$$\begin{aligned} x^2+(-2x-1)^2+3(-2x-1)&=22 \\ x^2+4x^2+4x+1-6x-3&=22 \\ 5x^2-2x-2&=22 \\ 5x^2-2x-24&=0 \\ (5x-12)(x+2)&=0 \end{aligned}$$

Apply the zero product principle.

43. Let x = one of the numbers
Let y = the other number

$x+y=10$

$xy=24$

Solve the second equation for y.

$$\begin{aligned} xy&=24 \\ y&=\frac{24}{x} \end{aligned}$$

Substitute $\frac{24}{x}$ for y in the first equation and solve for x.

$$\begin{aligned} x+\frac{24}{x}&=10 \\ x\left(x+\frac{24}{x}\right)&=x(10) \end{aligned}$$

$x^2+24=10x$

$x^2-10x+24=0$

$(x-6)(x-4)=0$

Apply the zero product principle.

$x-6=0$ or $x-4=0$

$x=6$ $\quad$ $x=4$

Substitute 6 and 4 for x to find y.

$x=6$ $\quad$ $x=4$

$y=\frac{24}{6}$ or $y=\frac{24}{4}$

$y=4$ $\quad$ $y=6$

The numbers are 4 and 6.

45. Let x = one of the numbers
Let y = the other number

$$x^2-y^2=3$$
$$\underline{2x^2+y^2=9}$$
$$3x^2=12$$
$$x^2=4$$
$$x=\pm 2$$

Substitute ± 2 for x to find y.

$$x=\pm 2$$
$$(\pm 2)^2-y^2=3$$
$$4-y^2=3$$
$$-y^2=-1$$
$$y^2=1$$
$$y=\pm 1$$

The numbers are either 2 and –1, 2 and 1, –2 and –1, or –2 and 1.

Application Exercises

47. $16x^2+4y^2=64$

$y=x^2-4$

Solve the second equation for x^2.

$y=x^2-4$

$y+4=x^2$

Substitute x^2-4 for y in the first equation and solve for x.

$$16(y+4)+4y^2=64$$
$$16y+64+4y^2=64$$
$$16y+4y^2=0$$
$$4y(4+y)=0$$

Apply the zero product principle.

$4y=0$ or $4+y=0$

$y=0$ $\quad$ $y=-4$

Substitute 0 and 4 for y to find x.

$y=0$ or $y=-4$

$0=x^2-4$ $\quad$ $-4=x^2-4$

$4=x^2$ $\quad$ $0=x^2$

$\pm 2=x$ $\quad$ $0=x$

The comet intersects the planet's orbit at the points $(2,0),(-2,0)$ and $(0,-4)$.

49. Let x = the length of the rectangle
Let y = the width of the rectangle

Perimeter: $2x+2y=36$

Area: $xy=77$

Solve the second equation for y.

$xy=77$

$y=\frac{77}{x}$

Substitute $\frac{77}{x}$ for y in the first equation and solve for x.

$$2x+2\left(\frac{77}{x}\right)=36$$
$$2x+\frac{154}{x}=36$$

$$x\left(2x+\frac{154}{x}\right)=x(36)$$
$$2x^2+154=36x$$
$$2x^2-36x+154=0$$
$$x^2-18x+77=0$$
$$(x-7)(x-11)=0$$

Apply the zero product principle.

$x-7=0$ or $x-11=0$

$x=7$ $\quad x=11$

Substitute 7 and 11 for x to find y.

$x=7$ or $x=11$

$y=\frac{77}{7}$ $\quad y=\frac{77}{11}$

$y=11$ $\quad y=7$

The dimensions of the rectangle are 7 feet by 11 feet.

51. Let x = the length of the screen
Let y = the width of the screen

$$x^2+y^2=10^2$$
$$xy=48$$

Solve the second equation for y.

$$xy=48$$
$$y=\frac{48}{x}$$

Substitute $\frac{48}{x}$ for y to find x.

$$x^2+\left(\frac{48}{x}\right)^2=10^2$$
$$x^2+\frac{2304}{x^2}=100$$
$$x^2\left(x^2+\frac{2304}{x^2}\right)=x^2(100)$$
$$x^4+2304=100x^2$$
$$x^4-100x^2+2304=0$$
$$(x^2-64)(x^2-36)=0$$
$$(x+8)(x-8)(x+6)(x-6)=0$$

Apply the zero product principle.

$x+8=0$ $\quad x-8=0$

$x=-8$ $\quad x=8$

$x+6=0$ $\quad x+6=0$

$x=-6$ $\quad x=-6$

We disregard –8 and –6 because we cannot have a negative length. Substitute 8 and 6 for x to find y.

$x=8$ or $x=6$

$y=\frac{48}{8}$ $\quad y=\frac{48}{6}$

$y=6$ $\quad y=8$

The dimensions of the screen are 8 inches by 6 inches.

53.
$$x^2-y^2=21$$
$$4x+2y=24$$

Solve for y in the second equation.

$$4x+2y=24$$
$$2y=24-4x$$
$$y=12-2x$$

Substitute $12-2x$ for y and solve for x.

$$x^2-(12-2x)^2=21$$
$$x^2-(144-48x+4x^2)=21$$
$$x^2-144+48x-4x^2=21$$
$$-3x^2+48x-144=21$$
$$-3x^2+48x-165=0$$
$$x^2-16x+55=0$$
$$(x-5)(x-11)=0$$

Apply the zero product principle.

$x-5=0$ or $x-11=0$

$x=5$ $\quad$ $x=11$

Substitute 5 and 11 for x to find y.

$x=5$ or $x=11$

$y=12-2(5)$ $\quad$ $y=12-2(11)$

$y=12-10$ $\quad$ $y=12-22$

$y=2$ $\quad$ $y=-10$

We disregard -10 because we can't have a negative length measurement. The larger square is 5 meters by 5 meters and the smaller square to be cut out is 2 meters by 2 meters.

Writing in Mathematics

55. Answers will vary.

57. Answers will vary.

Technology Exercises

59. Answers will vary. For example, consider Exercise 1.

$x+y=2$ $\quad$ $y=x^2-4$

$y=-x+2$

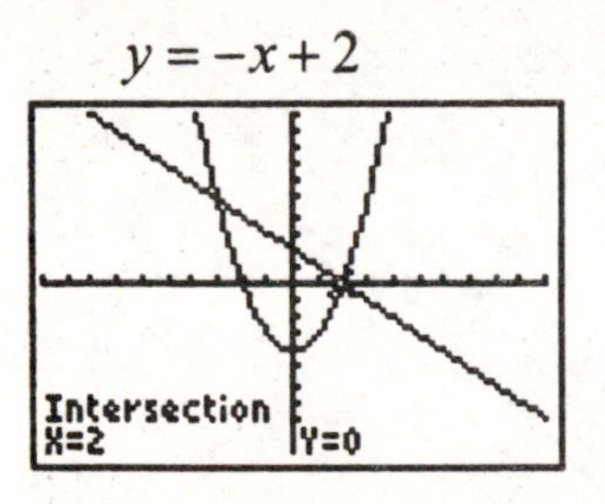

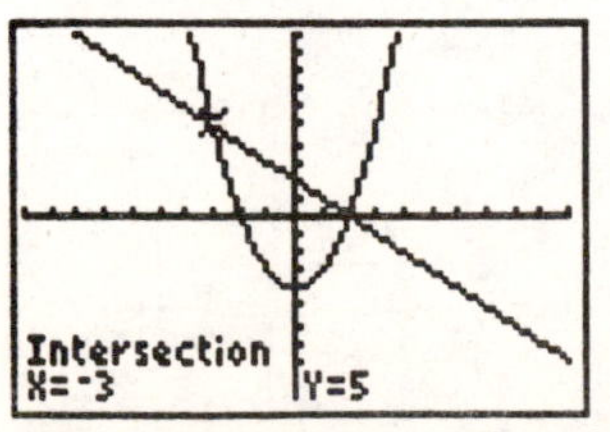

The solutions are $(2,0)$ and $(-3,5)$.

This is the same answer obtained in Exercise 1.

Critical Thinking Exercises

61. Statement **b.** is true. As shown in the graph below, a parabola and a circle can intersect in at most four points, and therefore, has at most four real solutions.

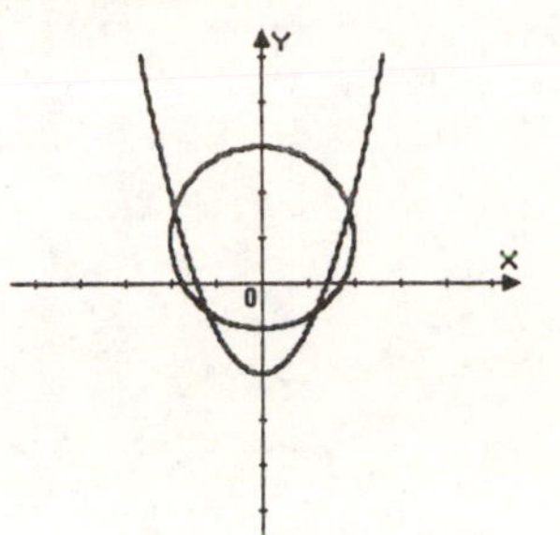

Statement **a.** is false. As shown in the graph below, a circle and a line can intersect in at most two points, and therefore has at most two real solutions.

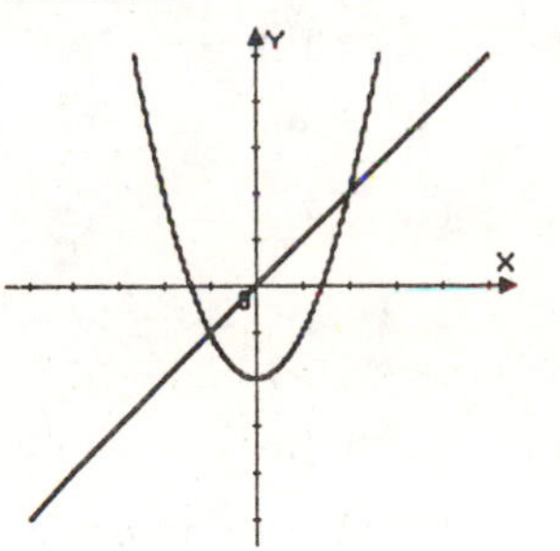

Statement **c.** is false. As shown in the graphs below, it is possible that a system of two equations in two variables whose graphs represent circles do not intersect, or intersect in a single point. This means that the system would have no solution, or a single solution, respectively.

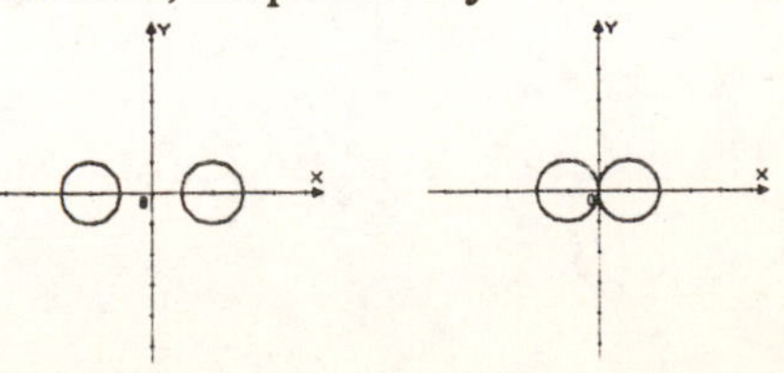

Statement **d.** is false. As shown in the graph below, a circle and a parabola can intersect in one point, and therefore have only one real solution.

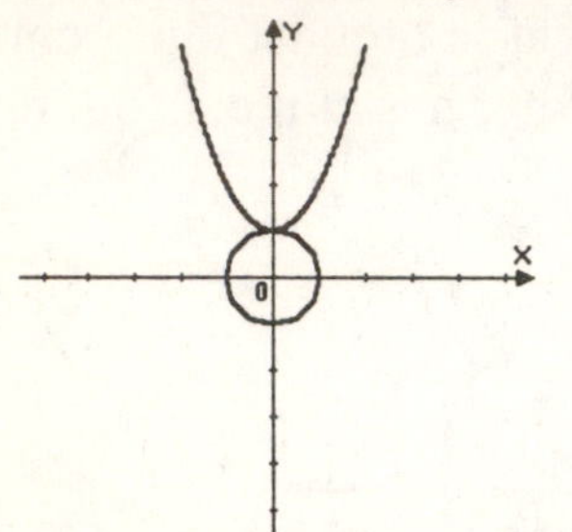

63. $\log_y x = 3$

$\log_y (4x) = 5$

Rewrite the equations.

$y^3 = x$

$y^5 = 4x$

Substitute y^3 for x in the second equation and solve for y.

$$y^5 = 4y^3$$
$$y^5 - 4y^3 = 0$$
$$y^3(y^2 - 4) = 0$$
$$y^3(y+2)(y-2) = 0$$

Apply the zero product principle.

$y^3 = 0 \qquad y+2=0 \qquad y-2=0$

$y = 0 \qquad y=-2 \qquad y=2$

We disregard 0 and -2 because the base of a logarithm must be greater than zero.

Substitute 2 for y to find x.

$y^3 = x$

$2^3 = x$

$8 = x$

The solution is $(8,2)$ and the solution set is $\{(8,2)\}$.

Review Exercises

65. $3x - 2y \le 6$

First, find the intercepts to the equation $3x - 2y = 6$. Find the x–intercept by setting y equal to zero.

$$3x - 2(0) = 6$$
$$3x = 6$$
$$x = 2$$

Find the y–intercept by setting x equal to zero.

$$3(0) - 2y = 6$$
$$-2y = 6$$
$$y = -3$$

Next, use the origin as a test point.

$$3(0) - 2(0) \le 6$$
$$0 - 0 \le 6$$
$$0 \le 6$$

This is a true statement. This means that the origin will fall in the shaded half-plane.

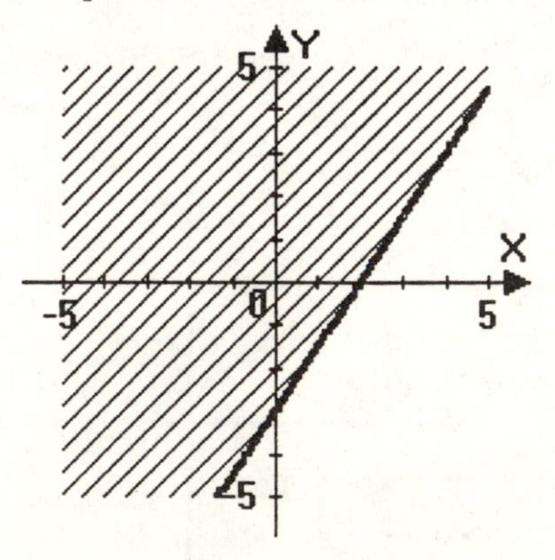

66.

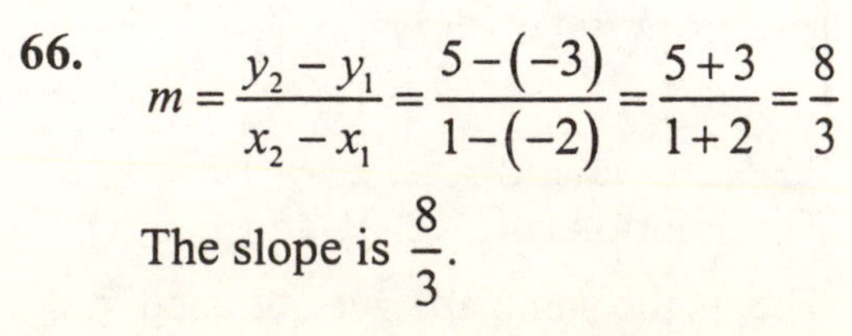

$$m = \frac{y_2 - y_1}{x_2 - x_1} = \frac{5-(-3)}{1-(-2)} = \frac{5+3}{1+2} = \frac{8}{3}$$

The slope is $\frac{8}{3}$.

Chapter 10 Review

10.1

1.
$$d=\sqrt{(3-(-2))^2+(9-(-3))^2}$$
$$=\sqrt{(3+2)^2+(9+3)^2}$$
$$=\sqrt{5^2+12^2}=\sqrt{25+144}$$
$$=\sqrt{169}=13$$
The distance between the point is 13 units.

2.
$$d=\sqrt{(-2-(-4))^2+(5-3)^2}$$
$$=\sqrt{(-2+4)^2+2^2}=\sqrt{2^2+4}$$
$$=\sqrt{4+4}=\sqrt{8}=\sqrt{4\cdot 2}=2\sqrt{2}$$
The distance between the point is $2\sqrt{2}$ or approximately 2.83 units.

3.
$$\text{Midpoint}=\left(\frac{2+(-12)}{2},\frac{6+4}{2}\right)$$
$$=\left(\frac{-10}{2},\frac{10}{2}\right)=(-5,5)$$
The midpoint is $(-5,5)$.

4.
$$\text{Midpoint}=\left(\frac{4+(-15)}{2},\frac{-6+2}{2}\right)$$
$$=\left(\frac{-11}{2},\frac{-4}{2}\right)=\left(-\frac{11}{2},-2\right)$$
The midpoint is $\left(-\frac{11}{2},-2\right)$.

5.
$$(x-0)^2+(y-0)^2=3^2$$
$$x^2+y^2=9$$

6.
$$(x-(-2))^2+(y-4)^2=6^2$$
$$(x+2)^2+(y-4)^2=36$$

7.
$$x^2+y^2=1$$
$$(x-0)^2+(y-0)^2=1^2$$
The center is $(0,0)$ and the radius is 1 units.

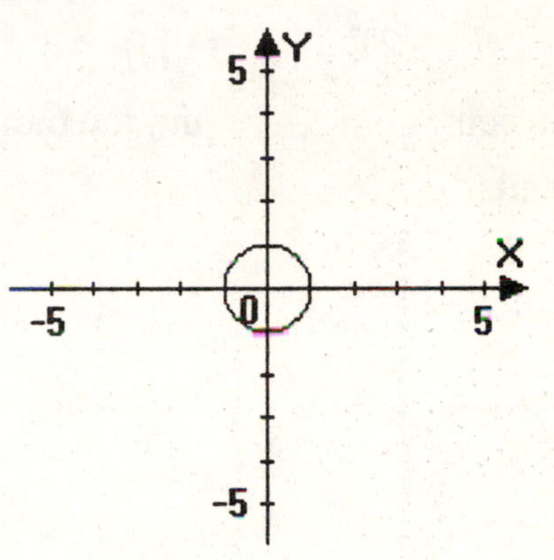

8.
$$(x+2)^2+(y-3)^2=9$$
$$(x-(-2))^2+(y-3)^2=3^2$$
The center is $(-2,3)$ and the radius is 3 units.

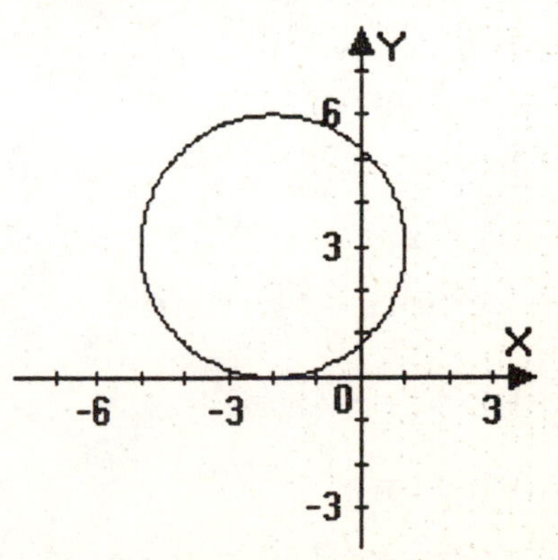

9.
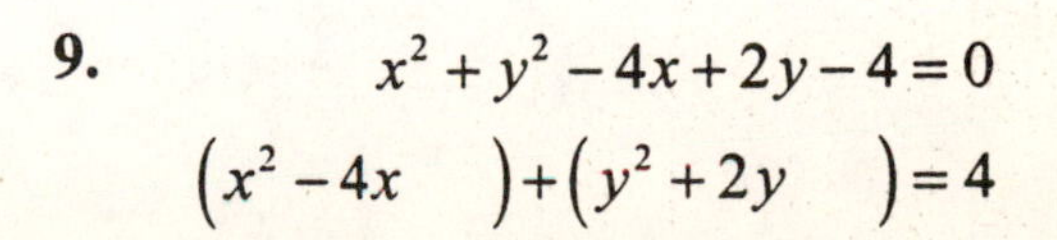
$$x^2+y^2-4x+2y-4=0$$
$$(x^2-4x\quad)+(y^2+2y\quad)=4$$

Complete the squares.

$$\left(\frac{b}{2}\right)^2 = \left(\frac{-4}{2}\right)^2 = (-4)^2 = 4$$

$$\left(\frac{b}{2}\right)^2 = \left(\frac{2}{2}\right)^2 = 1^2 = 1$$

$$(x^2 - 4x + 4) + (y^2 + 2y + 1) = 4 + 4 + 1$$

$$(x-2)^2 + (y+1)^2 = 9$$

$$(x-2)^2 + (y-(-1))^2 = 3^2$$

The center is $(2, -1)$ and the radius is 3 units.

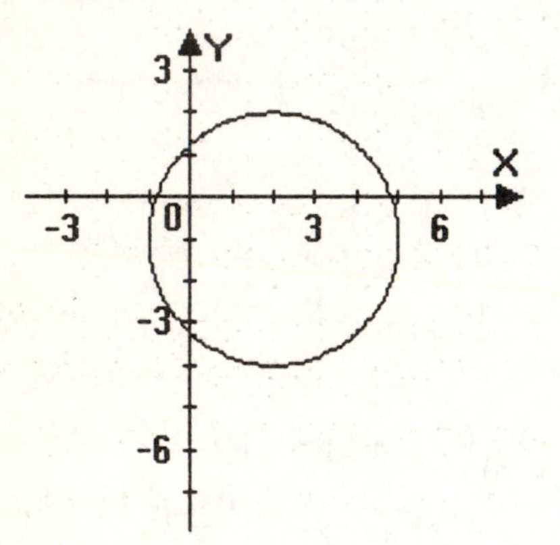

10. $x^2 + y^2 - 4y = 0$

$$x^2 + (y^2 - 4y \quad) = 0$$

Complete the square.

$$\left(\frac{b}{2}\right)^2 = \left(\frac{-4}{2}\right)^2 = (-4)^2 = 4$$

$$x^2 + (y^2 - 4y + 4) = 0 + 4$$

$$(x-0)^2 + (y-2)^2 = 4$$

$$(x-0)^2 + (y-2)^2 = 2^2$$

The center is $(0, 2)$ and the radius is 2 units.

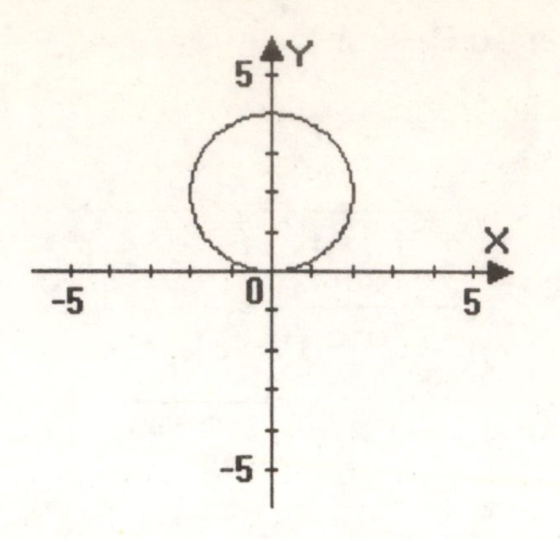

10.2

11. $\frac{x^2}{36} + \frac{y^2}{25} = 1$

Because the denominator of the x^2 – term is greater than the denominator of the y^2 – term, the major axis is horizontal. Since $a^2 = 36$, $a = 6$ and the vertices are $(-6, 0)$ and $(6, 0)$. Since $b^2 = 25$, $b = 5$ and endpoints of the minor axis are $(0, -5)$ and $(0, 5)$.

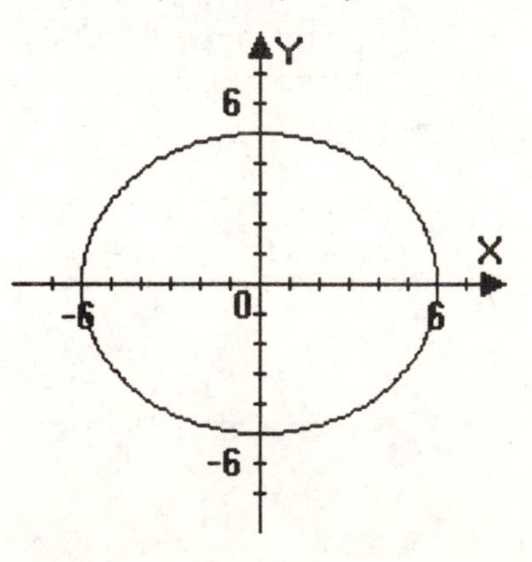

12. $\frac{x^2}{25} + \frac{y^2}{16} = 1$

Because the denominator of the x^2 – term is greater than the denominator of the y^2 – term, the major axis is horizontal. Since $a^2 = 25$, $a = 5$ and the vertices are $(-5,0)$ and $(5,0)$. Since $b^2 = 16$, $b = 4$ and endpoints of the minor axis are $(0,-4)$ and $(0,4)$.

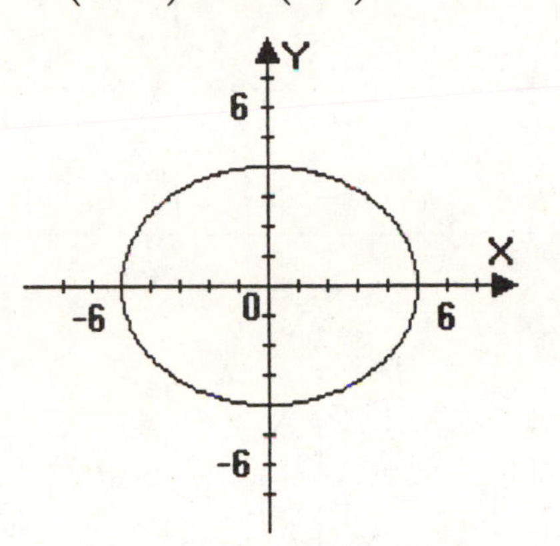

13. $4x^2 + y^2 = 16$

$$\frac{4x^2}{16} + \frac{y^2}{16} = \frac{16}{16}$$

$$\frac{x^2}{4} + \frac{y^2}{16} = 1$$

Because the denominator of the y^2 – term is greater than the denominator of the x^2 – term, the major axis is vertical. Since $a^2 = 16$, $a = 4$ and the vertices are $(0,-4)$ and $(0,4)$. Since $b^2 = 4$, $b = 2$ and endpoints of the minor axis are $(-2,0)$ and $(2,0)$.

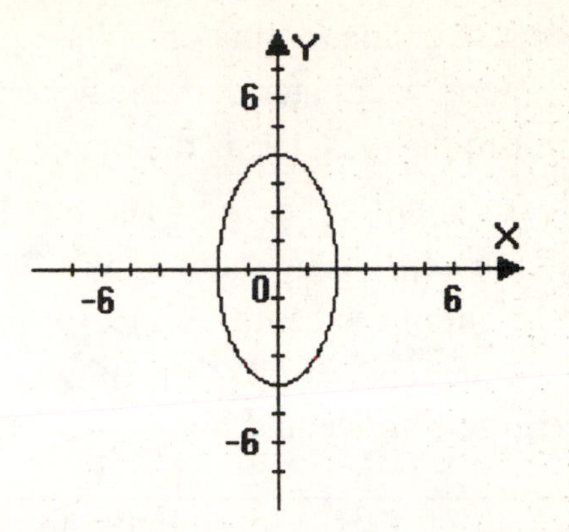

14. $4x^2 + 9y^2 = 36$

$$\frac{4x^2}{36} + \frac{9y^2}{36} = \frac{36}{36}$$

$$\frac{x^2}{9} + \frac{y^2}{4} = 1$$

Because the denominator of the x^2 – term is greater than the denominator of the y^2 – term, the major axis is horizontal. Since $a^2 = 9$, $a = 3$ and the vertices are $(-3,0)$ and $(3,0)$. Since $b^2 = 4$, $b = 2$ and endpoints of the minor axis are $(0,-2)$ and $(0,2)$.

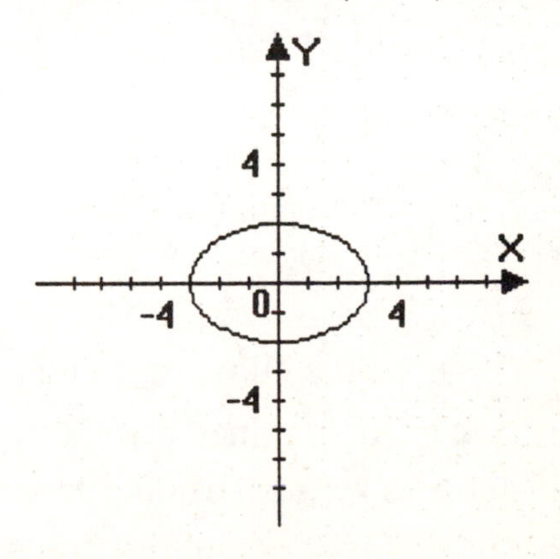

15.

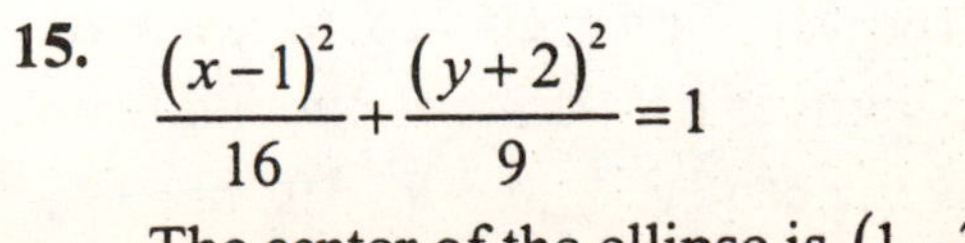

$$\frac{(x-1)^2}{16} + \frac{(y+2)^2}{9} = 1$$

The center of the ellipse is $(1,-2)$.

Because the denominator of the x^2 – term is greater than the denominator of the y^2 – term, the major axis is horizontal. Since $a^2 = 16$, $a = 4$ and the vertices lie 4 units to the left and right of the center. Since $b^2 = 9$, $b = 3$ and endpoints of the minor axis lie 3 units above and below the center.

Center	Vertices	Endpoints of Minor Axis
$(1,-2)$	$(1-4,-2)$ $=(-3,-2)$	$(1,-2-3)$ $=(1,-5)$
	$(1+4,-2)$ $=(5,-2)$	$(1,-2+3)$ $=(1,1)$

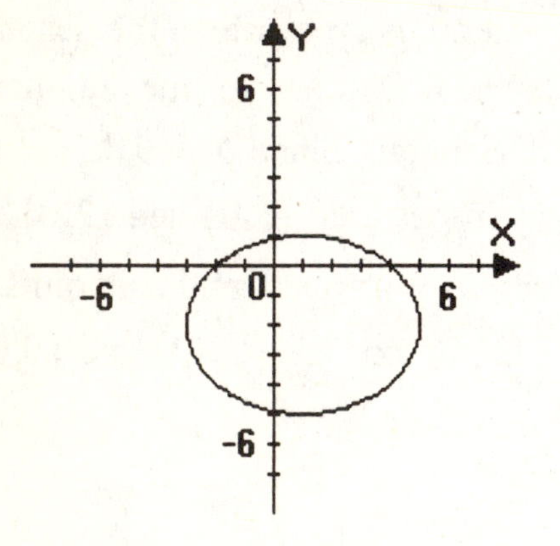

16. $$\frac{(x+1)^2}{9}+\frac{(y-2)^2}{16}=1$$

The center of the ellipse is $(-1,2)$. Because the denominator of the y^2 – term is greater than the denominator of the x^2 – term the major axis is vertical. Since $a^2 = 16$, $a = 4$ and the vertices lie 4 units above and below the center. Since $b^2 = 9$, $b = 3$ and endpoints of the minor axis lie 3 units to the left and right of the center.

Center	Vertices	Endpoints of Minor Axis
$(-1,2)$	$(-1,2-4)$ $=(-1,-2)$	$(-1-3,2)$ $=(-4,2)$
	$(-1,2+4)$ $=(-1,6)$	$(-1+3,2)$ $=(2,2)$

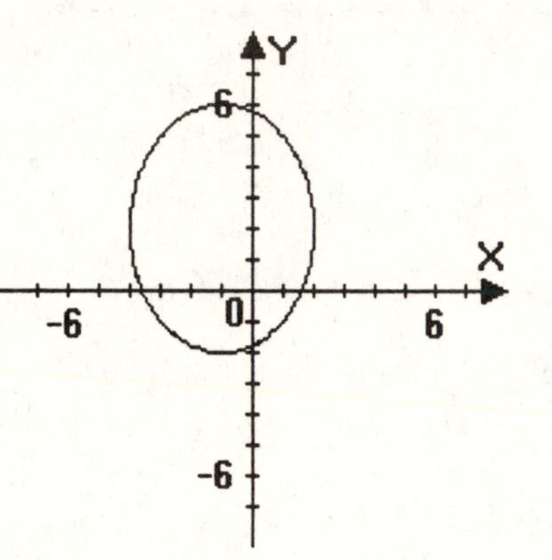

17. From the figure, we see that the major axis is horizontal with $a = 25$, and $b = 15$.

$$\frac{x^2}{25^2}+\frac{y^2}{15^2}=1$$

$$\frac{x^2}{625}+\frac{y^2}{225}=1$$

Since the truck is 14 feet wide, we need to determine the height of the archway at 14 feet to the right of center.

$$\frac{14^2}{625}+\frac{y^2}{225}=1$$

$$\frac{196}{625}+\frac{y^2}{225}=1$$

$$5625\left(\frac{196}{625}+\frac{y^2}{225}\right)=5625(1)$$
$$9(196)+25y^2=5625$$
$$1764+25y^2=5625$$
$$25y^2=3861$$
$$y^2=\frac{3861}{25}$$
$$y=\sqrt{\frac{3861}{25}}\approx 12.43$$

The height of the archway 14 feet from the center is approximately 12.43 feet. Since the truck is 12 feet high, the truck will clear the archway.

10.3

18. $\frac{x^2}{16}-y^2=1$

$\frac{x^2}{16}-\frac{y^2}{1}=1$

The equation is in the form

$\frac{x^2}{a^2}-\frac{y^2}{b^2}=1$ with $a^2=16$, and $b^2=1$.

We know the transverse axis lies on the x-axis and the vertices are $(-4,0)$ and $(4,0)$. Because $a^2=16$ and $b^2=1$, $a=4$ and $b=1$. Construct a rectangle using –4 and 4 on the x–axis, and –1 and 1 on the y–axis. Draw extended diagonals to obtain the asymptotes. Graph the hyperbola.

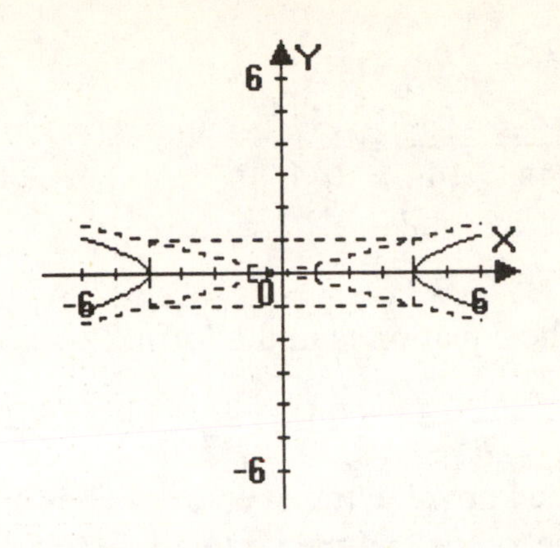

19. $\frac{y^2}{16}-x^2=1$

$\frac{y^2}{16}-\frac{x^2}{1}=1$

The equation is in the form

$\frac{y^2}{a^2}-\frac{x^2}{b^2}=1$ with $a^2=16$, and $b^2=1$.

We know the transverse axis lies on the y-axis and the vertices are $(0,-4)$ and $(0,4)$. Because $a^2=16$ and $b^2=1$, $a=4$ and $b=1$. Construct a rectangle using –1 and 1 on the x–axis, and –4 and 4 on the y–axis. Draw extended diagonals to obtain the asymptotes. Graph the hyperbola.

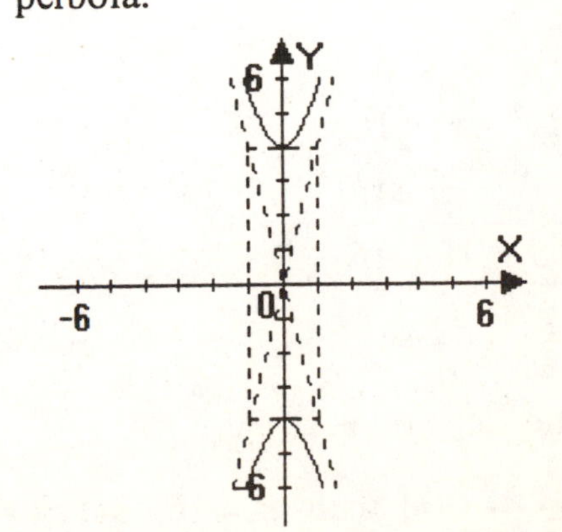

20. $9x^2 - 16y^2 = 144$

$$\frac{9x^2}{144} - \frac{16y^2}{144} = \frac{144}{144}$$

$$\frac{x^2}{16} - \frac{y^2}{9} = 1$$

The equation is in the form

$\frac{x^2}{a^2} - \frac{y^2}{b^2} = 1$ with $a^2 = 16$, and $b^2 = 9$.

We know the transverse axis lies on the x-axis and the vertices are $(-4,0)$ and $(4,0)$. Because $a^2 = 16$ and $b^2 = 9$, $a = 4$ and $b = 3$.

Construct a rectangle using –4 and 4 on the x–axis, and –3 and 3 on the y–axis. Draw extended diagonals to obtain the asymptotes. Graph the hyperbola.

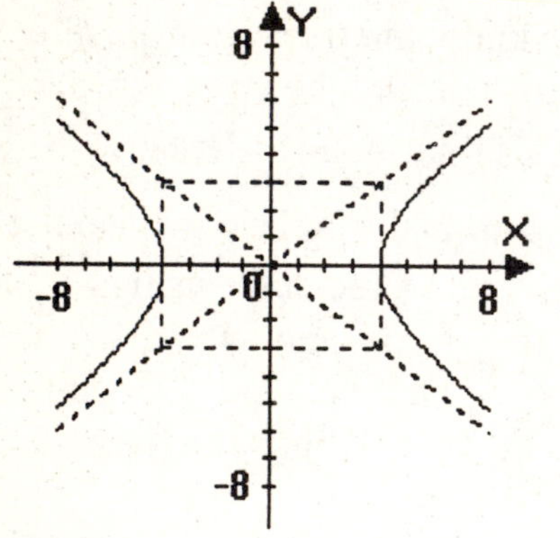

21. $4y^2 - x^2 = 16$

$$\frac{4y^2}{16} - \frac{x^2}{16} = \frac{16}{16}$$

$$\frac{y^2}{4} - \frac{x^2}{16} = 1$$

The equation is in the form

$\frac{y^2}{a^2} - \frac{x^2}{b^2} = 1$ with $a^2 = 16$, and $b^2 = 4$.

We know the transverse axis lies on the y-axis and the vertices are $(0,-4)$ and $(0,4)$. Because $a^2 = 16$ and $b^2 = 4$, $a = 4$ and $b = 2$.

Construct a rectangle using –2 and 2 on the x–axis, and –4 and 4 on the y–axis. Draw extended diagonals to obtain the asymptotes. Graph the hyperbola.

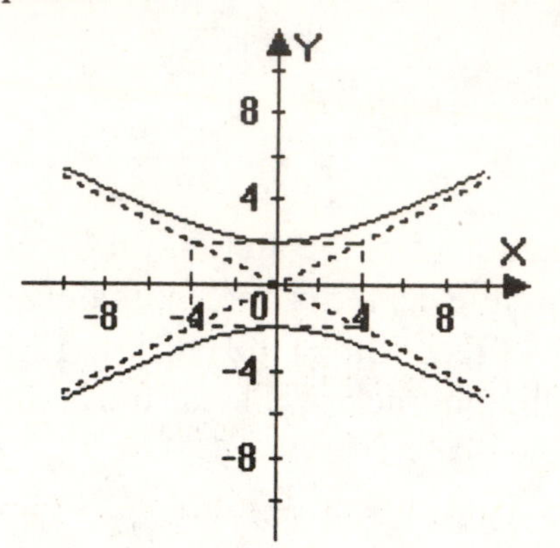

22. $x = (y-3)^2 - 4$

This is a parabola of the form $x = a(y-k)^2 + h$. Since $a = 1$ is positive, the parabola opens to the right. The vertex of the parabola is $(-4,3)$. The axis of symmetry is $y = 3$. Replace y with 0 to find the x–intercept.

$$x = (0-3)^2 - 4 = (-3)^2 - 4 = 9 - 4 = 5$$

The x–intercept is 5. Replace x with 0 to find the y–intercepts.

$$0 = (y-3)^2 - 4$$

$$0 = y^2 - 6y + 9 - 4$$

$$0 = y^2 - 6y + 5$$

$$0 = (y-5)(y-1)$$

Apply the zero product principle.

$y - 5 = 0$ and $y - 1 = 0$

$y = 5$ $\qquad$ $y = 1$

The y–intercepts are 1 and 5.

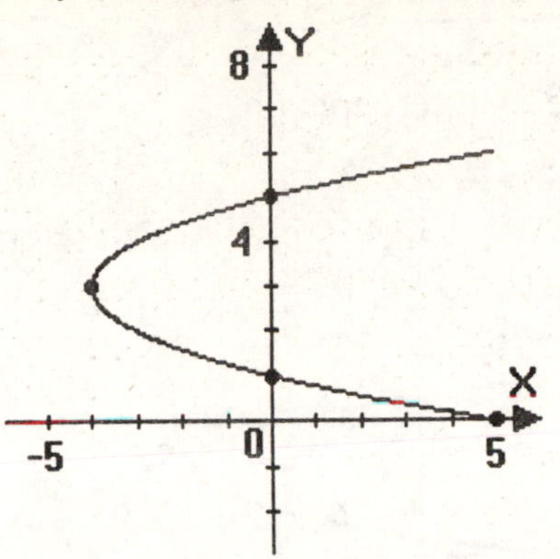

23. $x = -2(y+3)^2 + 2$

This is a parabola of the form $x = a(y-k)^2 + h$. Since $a = -2$ is negative, the parabola opens to the left. The vertex of the parabola is $(2,-3)$. The axis of symmetry is $y = -3$. Replace y with 0 to find the x–intercept.

$$x = -2(0+3)^2 + 2 = -2(3)^2 + 2$$
$$= -2(9) + 2 = -18 + 2 = -16$$

The x–intercept is -16. Replace x with 0 to find the y–intercepts.

$$0 = -2(y+3)^2 + 2$$
$$0 = -2(y^2 + 6y + 9) + 2$$
$$0 = -2y^2 - 12y - 18 + 2$$
$$0 = -2y^2 - 12y - 16$$
$$0 = y^2 + 6y + 8$$
$$0 = (y+4)(y+2)$$

Apply the zero product principle.

$y + 4 = 0$ and $y + 2 = 0$

$y = -4$ $\quad$ $y = -2$

The y–intercepts are –4 and –2.

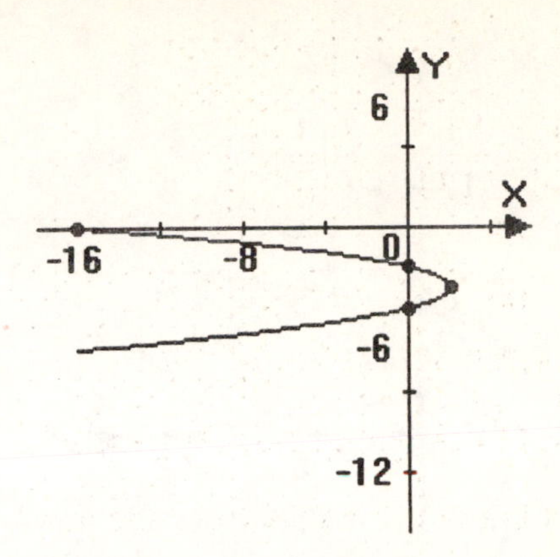

24. $x = y^2 - 8y + 12$

This is a parabola of the form $x = ay^2 + by + c$. Since $a = 1$ is positive, the parabola opens to the right. The y–coordinate of the vertex is $-\dfrac{b}{2a} = -\dfrac{-8}{2(1)} = -\dfrac{-8}{2} = 4$. The x–coordinate of the vertex is

$$x = 4^2 - 8(4) + 12 = 16 - 32 + 12$$
$$= 16 - 32 + 12 = -4.$$

The vertex of the parabola is $(-4, 4)$. The axis of symmetry is $y = 4$. Replace y with 0 to find the x–intercept.

$$x = 0^2 - 8(0) + 12 = 12$$

The x–intercept is 12. Replace x with 0 to find the y–intercepts.

$$0 = y^2 - 8y + 12$$
$$0 = (y-6)(y-2)$$

Apply the zero product principle.

$y - 6 = 0$ and $y - 2 = 0$

$y = 6$ $\quad$ $y = 2$

The y–intercepts are 2 and 6.

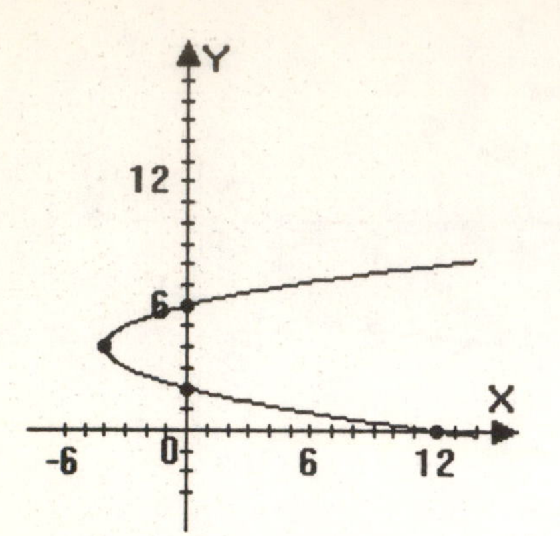

25. $x = -y^2 - 4y + 6$

This is a parabola of the form $x = ay^2 + by + c$. Since $a = -1$ is negative, the parabola opens to the left. The y–coordinate of the vertex is $-\frac{b}{2a} = -\frac{-4}{2(-1)} = -\frac{-4}{-2} = -2$. The x–coordinate of the vertex is

$x = -(-2)^2 - 4(-2) + 6$
$= -4 + 8 + 6 = 10.$

The vertex of the parabola is $(10, -2)$.

The axis of symmetry is $y = -2$.

Replace y with 0 to find the x–intercept.

$x = -0^2 - 4(0) + 6 = 0^2 - 0 + 6 = 6$

The x–intercept is 6. Replace x with 0 to find the y–intercepts.

$0 = -y^2 - 4y + 6$

Solve using the quadratic formula.

$$y = \frac{-b \pm \sqrt{b^2 - 4ac}}{2a}$$

$$= \frac{-(-4) \pm \sqrt{(-4)^2 - 4(-1)(6)}}{2(-1)}$$

$$= \frac{4 \pm \sqrt{16 + 24}}{-2} = \frac{4 \pm \sqrt{40}}{-2}$$

$$= \frac{4 \pm 2\sqrt{10}}{-2} = -2 \pm \sqrt{10}$$

The y–intercepts are $-2 \pm \sqrt{10}$.

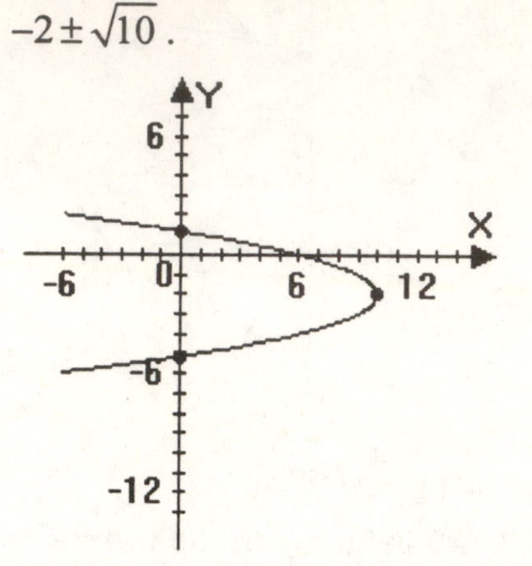

26. $x + 8y = y^2 + 10$

Since only one variable is squared, the graph of the equation is a parabola.

27. $16x^2 = 32 - y^2$

$16x^2 + y^2 = 32$

Because x^2 and y^2 have different positive coefficients, the equation's graph is an ellipse.

28. $x^2 = 25 + 25y^2$

$x^2 - 25y^2 = 25$

Because x^2 and y^2 have opposite signs, the equation's graph is a hyperbola.

29. $x^2 = 4 - y^2$

$x^2 + y^2 = 4$

Because x^2 and y^2 have the same positive coefficient, the equation's graph is a circle.

30. $36y^2 = 576 + 16x^2$

$36y^2 - 16x^2 = 576$

Because x^2 and y^2 have opposite signs, the equation's graph is a hyperbola.

31. $\frac{(x+3)^2}{9} + \frac{(y-4)^2}{25} = 1$

Because x^2 and y^2 have different positive coefficients, the equation's graph is an ellipse.

32. $y = x^2 + 6x + 9$

Since only one variable is squared, the graph of the equation is a parabola.

33. $5x^2 + 5y^2 = 180$

Because x^2 and y^2 have the same positive coefficient, the equation's graph is a circle.

Divide both sides of the equation by 5.

$x^2 + y^2 = 36$

The center is (0, 0) and the radius is 6 units.

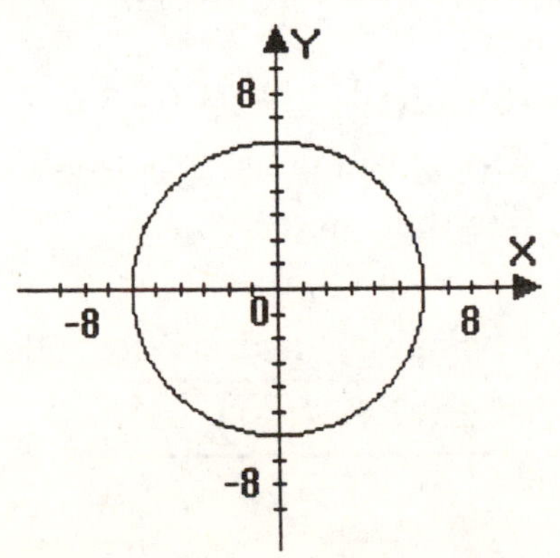

34. $4x^2 + 9y^2 = 36$

Because x^2 and y^2 have different positive coefficients, the equation's graph is an ellipse.

$$\frac{4x^2}{36} + \frac{9y^2}{36} = \frac{36}{36}$$

$$\frac{x^2}{9} + \frac{y^2}{4} = 1$$

Because the denominator of the x^2 – term is greater than the denominator of the y^2 – term, the major axis is horizontal. Since $a^2 = 9$, $a = 3$ and the vertices are $(-3, 0)$ and $(3, 0)$. Since $b^2 = 4$, $b = 2$ and endpoints of the minor axis are $(0, -2)$ and $(0, 2)$.

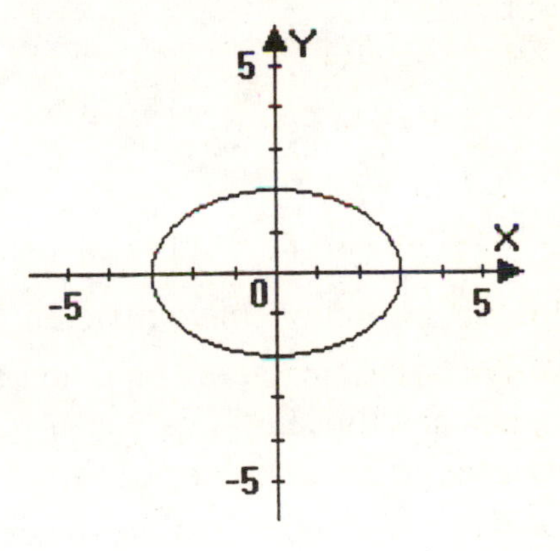

35. $4x^2 - 9y^2 = 36$

Because x^2 and y^2 have opposite signs, the equation's graph is a hyperbola.

$$\frac{4x^2}{36} - \frac{9y^2}{36} = \frac{36}{36}$$

$$\frac{x^2}{9} - \frac{y^2}{4} = 1$$

The equation is in the form

$\frac{x^2}{a^2} - \frac{y^2}{b^2} = 1$ with $a^2 = 9$, and $b^2 = 4$.

We know the transverse axis lies on the x-axis and the vertices are $(-3,0)$ and $(3,0)$. Because $a^2=9$ and $b^2=4$, $a=3$ and $b=2$. Construct a rectangle using –3 and 3 on the x–axis, and –2 and 2 on the y–axis. Draw extended diagonals to obtain the asymptotes. Graph the hyperbola.

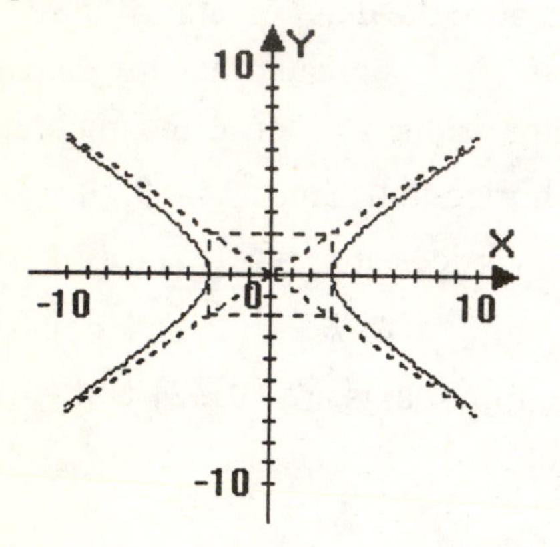

36. $\frac{x^2}{25}+\frac{y^2}{1}=1$

Because x^2 and y^2 have different positive coefficients, the equation's graph is an ellipse.
Because the denominator of the x^2 – term is greater than the denominator of the y^2 – term, the major axis is horizontal. Since $a^2=25$, $a=5$ and the vertices are $(-5,0)$ and $(5,0)$. Since $b^2=1$, $b=1$ and endpoints of the minor axis are $(0,-1)$ and $(0,1)$.

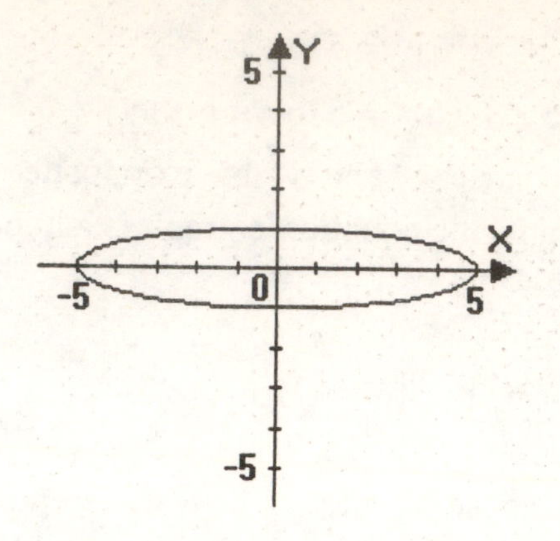

37. $x+3=-y^2+2y$

$$x=-y^2+2y-3$$

Since only one variable is squared, the graph of the equation is a parabola.
This is a parabola of the form $x=ay^2+by+c$. Since $a=-1$ is negative, the parabola opens to the left. The y–coordinate of the vertex is $-\frac{b}{2a}=-\frac{2}{2(-1)}=-\frac{2}{-2}=1$. The x–coordinate of the vertex is $x=-1^2+2(1)-3=-1+2-3=-2$.

The vertex of the parabola is $(-2,1)$.
Replace y with 0 to find the x–intercept.

$$x=-0^2+2(0)-3=0+0-3=-3$$

The x–intercept is –3. Replace x with 0 to find the y–intercepts.

$$0=-y^2+2y-3$$

Solve using the quadratic formula.

$$y=\frac{-2\pm\sqrt{2^2-4(-1)(-3)}}{2(-1)}$$

$$=\frac{-2\pm\sqrt{4-12}}{-2}=\frac{-2\pm\sqrt{-8}}{-2}$$

We do not need to simplify further.

The solutions are complex and there are no y–intercepts.

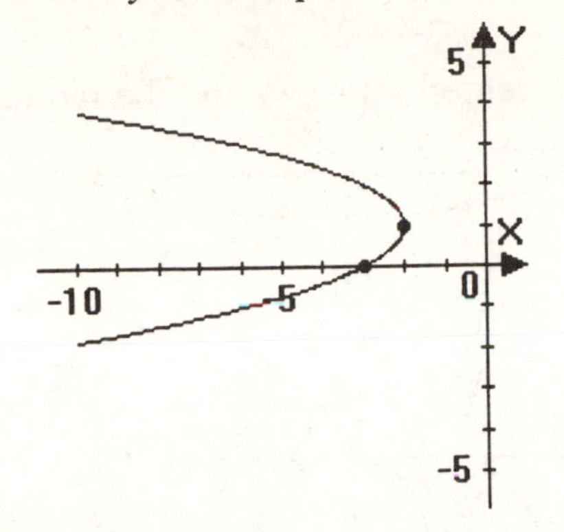

38. $y-3=x^2-2x$

$y=x^2-2x+3$

Since only one variable is squared, the graph of the equation is a parabola.

This is a parabola of the form $y=ax^2+bx+c$. Since $a=1$ is positive, the parabola opens to the right.

The x–coordinate of the vertex is

$-\frac{b}{2a}=-\frac{-2}{2(1)}=-\frac{-2}{2}=1$. The y–coordinate of the vertex is

$y=1^2-2(1)+3=1-2+3=2.$

The vertex of the parabola is $(1,2)$.

Replace x with 0 to find the y–intercept.

$y=0^2-2(0)+3=0-0+3=3$

The y–intercept is 3. Replace y with 0 to find the x–intercepts.

$0=x^2-2x+3$

Solve using the quadratic formula.

$$x=\frac{-2\pm\sqrt{2^2-4(-1)(-3)}}{2(-1)}$$

$$=\frac{-2\pm\sqrt{4-12}}{-2}=\frac{-2\pm\sqrt{-8}}{-2}$$

We do not need to simplify further. The solutions are complex and there are no x–intercepts.

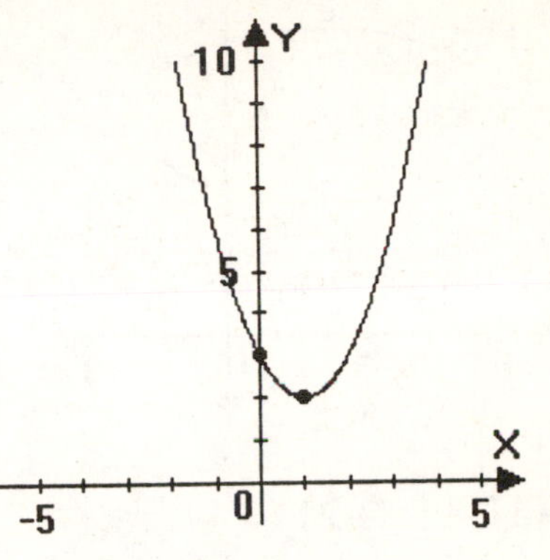

39. $\frac{(x+2)^2}{16}+\frac{(y-5)^2}{4}=1$

Because x^2 and y^2 have different positive coefficients, the equation's graph is an ellipse.

The center of the ellipse is $(-2,5)$.

Because the denominator of the x^2 – term is greater than the denominator of the y^2 – term, the major axis is horizontal. Since $a^2=16$, $a=4$ and the vertices lie 4 units to the left and right of the center. Since $b^2=4$, $b=2$ and endpoints of the minor axis lie two units above and below the center.

Center	Vertices	Endpoints of Minor Axis
$(-2,5)$	$(-2-4,5)$ $=(-6,5)$	$(-2,5-2)$ $=(-2,3)$
	$(-2+4,5)$ $=(2,5)$	$(-2,5+2)$ $=(-2,7)$

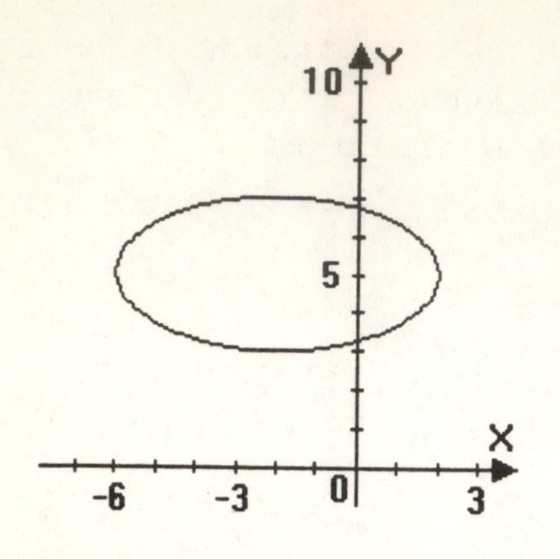

40. $(x-3)^2+(y+2)^2=4$

Because x^2 and y^2 have the same positive coefficient, the equation's graph is a circle.

$(x-3)^2+(y+2)^2=4$

The center is $(3,-2)$ and the radius is 2 units.

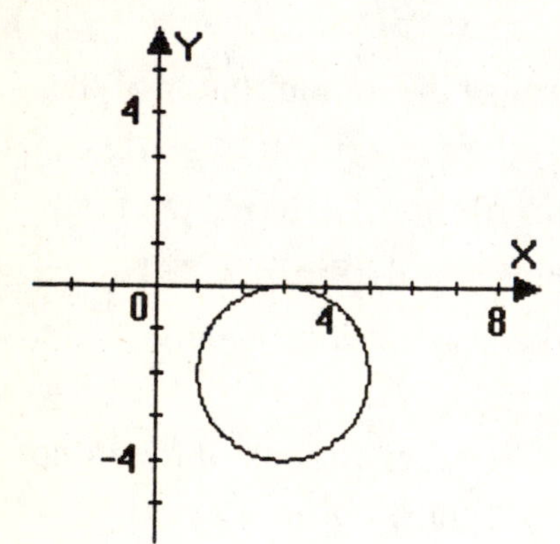

41. $$x^2+y^2+6x-2y+6=0$$

$$\left(x^2+6x\quad\right)+\left(y^2-2y\quad\right)=-6$$

Complete the squares.

$$\left(\frac{b}{2}\right)^2=\left(\frac{6}{2}\right)^2=(3)^2=9$$

$$\left(\frac{b}{2}\right)^2=\left(\frac{-2}{2}\right)^2=(-1)^2=1$$

$$\left(x^2+6x+9\right)+\left(y^2-2y+1\right)=-6+9+1$$

$$(x+3)^2+(y-1)^2=4$$

The center is $(-3,1)$ and the radius is 2 units.

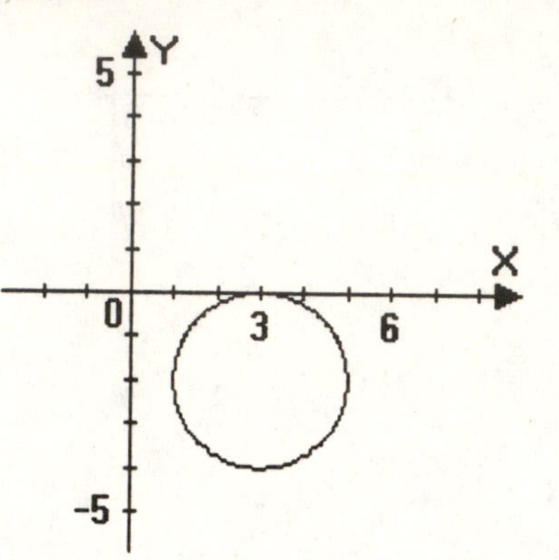

42. **a.** Using the point (6, 3), substitute for x and y to find a in $y=ax^2$.

$$3=a(6)^2$$

$$3=a(36)$$

$$a=\frac{3}{36}=\frac{1}{12}$$

The equation for the parabola is

$$y=\frac{1}{12}x^2.$$

b. $$a=\frac{1}{4p}$$

$$\frac{1}{12}=\frac{1}{4p}$$

$$4p=12$$

$$p=3$$

The light source should be placed at the point (0, 3). This is the point 3 inches above the vertex.

10.5

43. $5y = x^2 - 1$

$x - y = 1$

Solve the second equation for y.

$x - y = 1$

$-y = -x + 1$

$y = x - 1$

Substitute $x - 1$ for y in the first equation.

$5(x-1) = x^2 - 1$

$5x - 5 = x^2 - 1$

$0 = x^2 - 5x + 4$

$0 = (x-4)(x-1)$

Apply the zero product principle.

$x - 4 = 0$ and $x - 1 = 0$

$x = 4$ $\quad$ $x = 1$

Back-substitute 1 and 4 for x to find y.

$x = 4$ and $x = 1$

$y = x - 1$ $\quad$ $y = x - 1$

$y = 4 - 1$ $\quad$ $y = 1 - 1$

$y = 3$ $\quad$ $y = 0$

The solutions are $(1,0)$ and $(4,3)$ and the solution set is $\{(1,0),(4,3)\}$.

44. $y = x^2 + 2x + 1$

$x + y = 1$

Solve the second equation for y.

$x + y = 1$

$y = -x + 1$

Substitute $-x + 1$ for y in the first equation.

$-x + 1 = x^2 + 2x + 1$

$0 = x^2 + 3x$

$0 = x(x+3)$

Apply the zero product principle.

$x = 0$ and $x + 3 = 0$

$x = -3$

Back-substitute -3 and 0 for x to find y.

$x = 0$ and $x = -3$

$y = -x + 1$ $\quad$ $y = -x + 1$

$y = -0 + 1$ $\quad$ $y = -(-3) + 1$

$y = 1$ $\quad$ $y = 3 + 1$

$y = 4$

The solutions are $(-3,4)$ and $(0,1)$ and the solution set is $\{(-3,4),(0,1)\}$.

45. $x^2 + y^2 = 2$

$x + y = 0$

Solve the second equation for y.

$x + y = 0$

$y = -x$

Substitute $-x$ for y in the first equation.

$x^2 + (-x)^2 = 2$

$x^2 + x^2 = 2$

$2x^2 = 2$

$x^2 = 1$

$x = \pm 1$

Back-substitute -1 and 1 for x to find y.

$x = -1$ and $x = 1$

$y = -x$ $\quad$ $y = -x$

$y = -(-1)$ $\quad$ $y = -1$

$y = 1$

The solutions are $(-1,1)$ and $(1,-1)$ and the solution set is $\{(-1,1),(1,-1)\}$.

46. $2x^2 + y^2 = 24$

$x^2 + y^2 = 15$

Multiple the second equation by -1 and add to the first equation.

$$2x^2 + y^2 = 24$$
$$-x^2 - y^2 = -15$$
$$x^2 = 9$$
$$x = \pm 3$$

Back-substitute -3 and 3 for x to find y.

$$x = \pm 3$$
$$(\pm 3)^2 + y^2 = 15$$
$$9 + y^2 = 15$$
$$y^2 = 6$$
$$y = \pm\sqrt{6}$$

The solutions are $\left(-3, -\sqrt{6}\right)$, $\left(-3, \sqrt{6}\right)$, $\left(3, -\sqrt{6}\right)$ and $\left(3, \sqrt{6}\right)$ and the solution set is $\left\{\left(-3, -\sqrt{6}\right), \left(-3, \sqrt{6}\right), \left(3, -\sqrt{6}\right), \left(3, \sqrt{6}\right)\right\}$.

47. $xy - 4 = 0$

$y - x = 0$

Solve the second equation for y.

$$y - x = 0$$
$$y = x$$

Substitute x for y in the first equation and solve for x.

$$x(x) - 4 = 0$$
$$x^2 - 4 = 0$$
$$(x+2)(x-2) = 0$$

Apply the zero product principle.

$x + 2 = 0$ and $x - 2 = 0$

$x = -2$ $\quad$ $x = 2$

Back-substitute -2 and 2 for x to find y.

$x = -2$ and $x = 2$

$y = x$ $\quad$ $y = x$

$y = -2$ $\quad$ $y = 2$

The solutions are $(-2, -2)$ and $(2, 2)$ and the solution set is $\{(-2, -2), (2, 2)\}$.

48. $y^2 = 4x$

$x - 2y + 3 = 0$

Solve the second equation for x.

$$x - 2y + 3 = 0$$
$$x = 2y - 3$$

Substitute $2y - 3$ for x in the first equation and solve for y.

$$y^2 = 4(2y - 3)$$
$$y^2 = 8y - 12$$
$$y^2 - 8y + 12 = 0$$
$$(y - 6)(y - 2) = 0$$

Apply the zero product principle.

$y - 6 = 0$ and $y - 2 = 0$

$y = 6$ $\quad$ $y = 2$

Back-substitute 2 and 6 for y to find x.

$y = 6$ and $y = 2$

$x = 2y - 3$ $\quad$ $x = 2y - 3$

$x = 2(6) - 3$ $\quad$ $x = 2(2) - 3$

$x = 12 - 3$ $\quad$ $x = 4 - 3$

$x = 9$ $\quad$ $x = 1$

The solutions are $(1, 2)$ and $(9, 6)$ and the solution set is $\{(1, 2), (9, 6)\}$.

49. $x^2 + y^2 = 10$

$y = x + 2$

Substitute $x + 2$ for y in the first equation and solve for x.

$$x^2 + (x+2)^2 = 10$$
$$x^2 + x^2 + 4x + 4 = 10$$
$$2x^2 + 4x + 4 = 10$$
$$2x^2 + 4x - 6 = 0$$
$$x^2 + 2x - 3 = 0$$
$$(x+3)(x-1) = 0$$

Apply the zero product principle.

$x + 3 = 0$ and $x - 1 = 0$

$x = -3$ $\quad$ $x = 1$

Back-substitute -3 and 1 for x to find y.

$x = -3$ and $x = 1$

$y = x + 2$ $\quad$ $y = x + 2$

$y = -3 + 2$ $\quad$ $y = 1 + 2$

$y = -1$ $\quad$ $y = 3$

The solutions are $(-3, -1)$ and $(1, 3)$ and the solution set is $\{(-3, -1), (1, 3)\}$.

50. $xy = 1$

$y = 2x + 1$

Substitute $2x + 1$ for y in the first equation and solve for x.

$$x(2x+1) = 1$$
$$2x^2 + x = 1$$
$$2x^2 + x - 1 = 0$$
$$(2x-1)(x+1) = 0$$

Apply the zero product principle.

$2x - 1 = 0$ and $x + 1 = 0$

$2x = 1$ $\quad$ $x = -1$

$x = \frac{1}{2}$

Back-substitute -1 and $\frac{1}{2}$ for x to find y.

$x = -1$ and $x = \frac{1}{2}$

$y = 2x + 1$ $\quad$ $y = 2x + 1$

$y = 2(-1) + 1$ $\quad$ $y = 2\left(\frac{1}{2}\right) + 1$

$y = -2 + 1$ $\quad$ $y = 1 + 1$

$y = -1$ $\quad$ $y = 2$

The solutions are $(-1, -1)$ and $\left(\frac{1}{2}, 2\right)$ and the solution set is $\left\{(-1, -1), \left(\frac{1}{2}, 2\right)\right\}$.

51. $x + y + 1 = 0$

$x^2 + y^2 + 6y - x = -5$

Solve for y in the first equation.

$x + y + 1 = 0$

$y = -x - 1$

Substitute $-x - 1$ for y in the second equation and solve for x.

$$x^2 + (-x-1)^2 + 6(-x-1) - x = -5$$
$$x^2 + x^2 + 2x + 1 - 6x - 6 - x = -5$$
$$2x^2 - 5x - 5 = -5$$
$$2x^2 - 5x = 0$$
$$x(2x-5) = 0$$

Apply the zero product principle.

$x=0$ and $2x-5=0$

$2x=5$

$x=\frac{5}{2}$

Back-substitute 0 and $\frac{5}{2}$ for x to find y.

$x=0$ and $x=\frac{5}{2}$

$y=-x-1$ $\quad$ $y=-x-1$

$y=-0-1$ $\quad$ $y=-\frac{5}{2}-1$

$y=-1$ $\quad$ $y=-\frac{7}{2}$

The solutions are $(0,-1)$ and $\left(\frac{5}{2},-\frac{7}{2}\right)$ and the solution set is $\left\{(0,-1),\left(\frac{5}{2},-\frac{7}{2}\right)\right\}$.

52. $x^2+y^2=13$

$x^2-y=7$

Solve for x^2 in the second equation.

$x^2-y=7$

$x^2=y+7$

Substitute $y+7$ for x^2 in the first equation and solve for y.

$(y+7)+y^2=13$

$y^2+y+7=13$

$y^2+y-6=0$

$(y+3)(y-2)=0$

Apply the zero product principle.

$y+3=0$ and $y-2=0$

$y=-3$ $\quad$ $y=2$

Back-substitute -3 and 2 for y to find x.

$y=-3$ and $y=2$

$x^2=y+7$ $\quad$ $x^2=y+7$

$x^2=-3+7$ $\quad$ $x^2=2+7$

$x^2=4$ $\quad$ $x^2=9$

$x=\pm 2$ $\quad$ $x=\pm 3$

The solutions are $(-3,2),(-2,-3)$, $(2,-3)$ and $(3,2)$ and the solution set is $\{(-3,2),(-2,-3),(2,-3),(3,2)\}$.

53. $2x^2+3y^2=21$

$3x^2-4y^2=23$

Multiply the first equation by 4 and the second equation by 3.

$8x^2+12y^2=84$

$9x^2-12y^2=69$

$17x^2=153$

$x^2=9$

$x=\pm 3$

Back-substitute ± 3 for x to find y.

$x=\pm 3$

$2(\pm 3)^2+3y^2=21$

$2(9)+3y^2=21$

$18+3y^2=21$

$3y^2=3$

$y^2=1$

$y=\pm 1$

We have $x=\pm 3$ and $y=\pm 1$, the solutions are $(-3,-1),(-3,1),(3,-1)$ and $(3,1)$ and the solution set is $\{(-3,-1),(-3,1),(3,-1),(3,1)\}$.

54. Let x = the length of the rectangle
Let y = the width of the rectangle

$$2x+2y=26$$
$$xy=40$$

Solve the first equation for y.

$$2x+2y=26$$
$$x+y=13$$
$$y=13-x$$

Substitute $13-x$ for y in the second equation.

$$x(13-x)=40$$
$$13x-x^2=40$$
$$0=x^2-13x+40$$
$$0=(x-8)(x-5)$$

Apply the zero product principle.

$$x-8=0 \quad \text{and} \quad x-5=0$$
$$x=8 \qquad\qquad x=5$$

Back-substitute 5 and 8 for x to find y.

$x=8$ and $x=5$
$y=13-8$ $\quad$ $y=13-5$
$y=5$ $\quad$ $y=8$

The solutions are the same. The dimensions are 8 meters by 5 meters.

55. $2x+y=8$
$$xy=6$$

Solve the first equation for y.

$$2x+y=8$$
$$y=-2x+8$$

Substitute $-2x+8$ for y in the second equation.

$$x(-2x+8)=6$$
$$-2x^2+8x=6$$
$$-2x^2+8x-6=0$$
$$x^2-4x+3=0$$
$$(x-3)(x-1)=0$$

Apply the zero product principle.

$$x-3=0 \quad \text{and} \quad x-1=0$$
$$x=3 \qquad\qquad x=1$$

Back-substitute 1 and 3 for x to find y.

$x=3$ and $x=1$
$y=-2x+8$ $\quad$ $y=-2x+8$
$y=-2(3)+8$ $\quad$ $y=-2(1)+8$
$y=-6+8$ $\quad$ $y=-2+8$
$y=2$ $\quad$ $y=6$

The solutions are the points $(1,6)$ and $(3,2)$.

56. Using the formula for the area, we have $x^2+y^2=2900$. Since there are 240 feet of fencing available, we have:

$$x+(x+y)+y+y+(x-y)+x=240$$
$$x+x+y+y+y+x-y+x=240$$
$$4x+2y=240.$$

The system of two variables in two equations is as follows.

$$x^2+y^2=2900$$
$$4x+2y=240$$

Solve the second equation for y.

$$4x+2y=240$$
$$2y=-4x+240$$
$$y=-2x+120$$

Substitute $-2x+120$ for y to find x.

$$x^2+(-2x+120)^2=2900$$
$$x^2+4x^2-480x+14400=2900$$
$$5x^2-480x+11500=0$$
$$x^2-96x+2300=0$$
$$(x-50)(x-46)=0$$

Apply the zero product principle.

$x-50=0$ and $x-46=0$

$x=50$ $x=46$

Back-substitute 46 and 50 for x to find y.

$x=50$ $x=46$

$y=-2x+120$ $y=-2x+120$

$y=-2(50)+120$ $y=-2(46)+120$

$y=-100+120$ $y=-92+120$

$y=20$ $y=28$

The solutions are $x=50$ feet and $y=20$ feet or $x=46$ feet and $y=28$ feet.

Chapter 10 Test

1. $$d=\sqrt{(2-(-1))^2+(-3-5)^2}$$
$$=\sqrt{(3)^2+(-8)^2}$$
$$=\sqrt{9+64}=\sqrt{73}\approx 8.54$$
The distance between the point is $\sqrt{73}$ or 8.54 units.

2. $$\text{Midpoint}=\left(\frac{-5+12}{2},\frac{-2+(-6)}{2}\right)$$
$$=\left(\frac{7}{2},\frac{-8}{2}\right)=\left(\frac{7}{2},-4\right)$$
The midpoint is $\left(\frac{7}{2},-4\right)$.

3. $$(x-3)^2+(y-(-2))^2=5^2$$
$$(x-3)^2+(y+2)^2=25$$

4. $$(x-5)^2+(y+3)^2=49$$
$$(x-5)^2+(y-(-3))^2=7^2$$
The center is $(5,-3)$ and the radius is 7 units.

5. $$x^2+y^2+4x-6y-3=0$$
$$(x^2+4x\quad)+(y^2-6y\quad)=3$$
Complete the squares.
$$\left(\frac{b}{2}\right)^2=\left(\frac{4}{2}\right)^2=(2)^2=4$$
$$\left(\frac{b}{2}\right)^2=\left(\frac{-6}{2}\right)^2=(-3)^2=9$$
$$(x^2+4x+4)+(y^2-6y+9)=3+4+9$$
$$(x+2)^2+(y-3)^2=16$$
$$(x-(-2))^2+(y-3)^2=4^2$$
The center is $(-2,3)$ and the radius is 4 units.

6. $$x=-2(y+3)^2+7$$
$$x=-2(y-(-3))^2+7$$
The vertex of the parabola is $(7,-3)$.

7. $$x=y^2+10y+23$$
The y–coordinate of the vertex is
$$-\frac{b}{2a}=-\frac{10}{2(1)}=-\frac{10}{2}=-5.$$
The x–coordinate of the vertex is
$$x=(-5)^2+10(-5)+23$$
$$=25-50+23$$
$$=25-50+23=-2.$$
The vertex of the parabola is $(-2,-5)$.

8. $$\frac{x^2}{4}-\frac{y^2}{9}=1$$

Because x^2 and y^2 have opposite signs, the equation's graph is a hyperbola.
The equation is in the form
$\frac{x^2}{a^2}-\frac{y^2}{b^2}=1$ with $a^2=4$, and $b^2=9$.
We know the transverse axis lies on the x-axis and the vertices are $(-2,0)$ and $(2,0)$. Because $a^2=4$ and $b^2=9$, $a=2$ and $b=3$.
Construct a rectangle using –2 and 2 on the x–axis, and –3 and 3 on the y–axis. Draw extended diagonals to obtain the asymptotes. Graph the hyperbola.

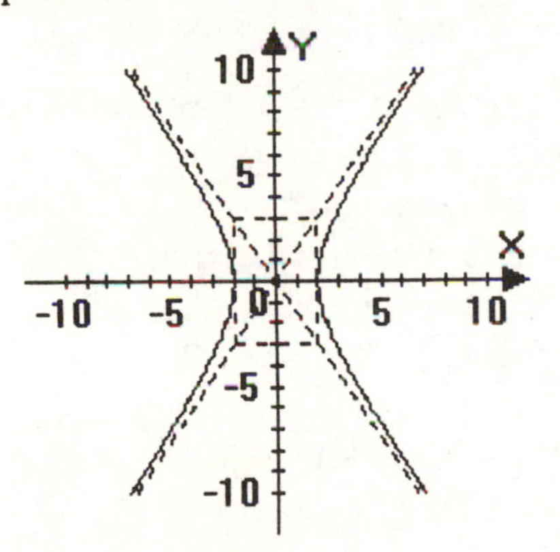

9. $4x^2+9y^2=36$
Because x^2 and y^2 have different positive coefficients, the equation's graph is an ellipse.
$$\frac{4x^2}{36}+\frac{9y^2}{36}=\frac{36}{36}$$
$$\frac{x^2}{9}+\frac{y^2}{4}=1$$
Because the denominator of the x^2 – term is greater than the denominator of the y^2 – term, the major axis is horizontal. Since $a^2=9$, $a=3$ and the vertices are $(-3,0)$ and $(3,0)$.
Since $b^2=4$, $b=2$ and endpoints of the minor axis are $(0,-2)$ and $(0,2)$.

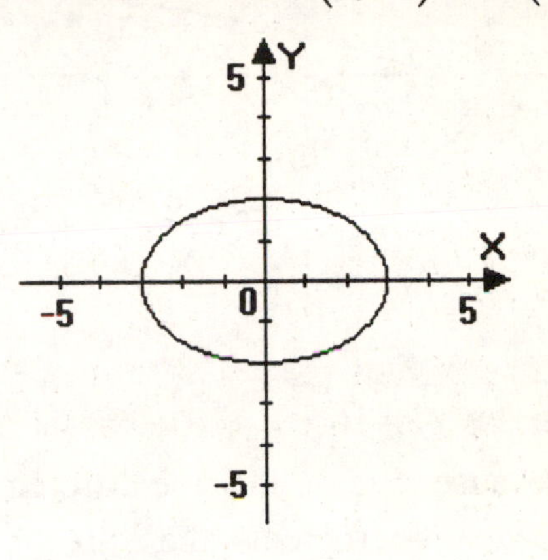

10. $x=(y+1)^2-4$
Since only one variable is squared, the graph of the equation is a parabola.
This is a parabola of the form $x=a(y-k)^2+h$. Since $a=1$ is positive, the parabola opens to the right. The vertex of the parabola is $(4,-1)$. Replace y with 0 to find the x–intercept.
$x=(0+1)^2-4=(1)^2-4=1-4=-3.$
The x–intercept is 0. Replace x with 0 to find the y–intercepts.
$$0=(y+1)^2-4$$
$$0=y^2+2y+1-4$$
$$0=y^2+2y-3$$
$$0=(y+3)(y-1)$$
Apply the zero product principle.
$y+3=0$ and $y+1=0$
$y=-3$ $\qquad$ $y=-1$
The y–intercepts are –3 and –1.

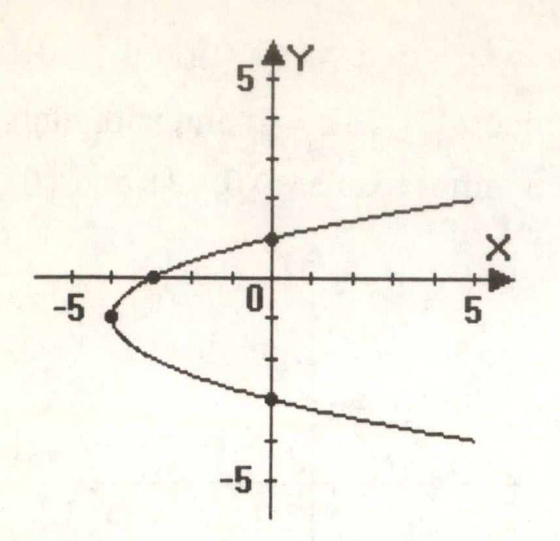

11. $16x^2 + y^2 = 16$

Because x^2 and y^2 have different positive coefficients, the equation's graph is an ellipse.

$$\frac{16x^2}{16} + \frac{y^2}{16} = \frac{16}{16}$$

$$\frac{x^2}{1} + \frac{y^2}{16} = 1$$

Because the denominator of the y^2 – term is greater than the denominator of the x^2 – term, the major axis is vertical. Since $a^2 = 16$, $a = 4$ and the vertices are $(0, -4)$ and $(0, 4)$. Since $b^2 = 1$, $b = 1$ and endpoints of the minor axis are $(-1, 0)$ and $(1, 0)$.

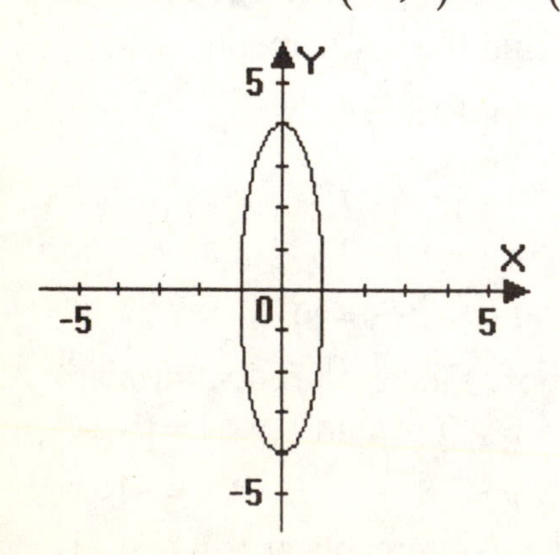

12. $25y^2 = 9x^2 + 225$

$25y^2 - 9x^2 = 225$

Because x^2 and y^2 have opposite signs, the equation's graph is a hyperbola.

$$\frac{25y^2}{225} - \frac{9x^2}{225} = \frac{225}{225}$$

$$\frac{y^2}{9} - \frac{x^2}{25} = 1$$

The equation is in the form

$\frac{y^2}{a^2} - \frac{x^2}{b^2} = 1$ with $a^2 = 9$, and $b^2 = 25$.

We know the transverse axis lies on the y-axis and the vertices are $(0, -3)$ and $(0, 3)$. Because $a^2 = 9$ and $b^2 = 25$, $a = 35$ and $b = 5$.

Construct a rectangle using –5 and 5 on the x–axis, and –3 and 3 on the y–axis. Draw extended diagonals to obtain the asymptotes. Graph the hyperbola.

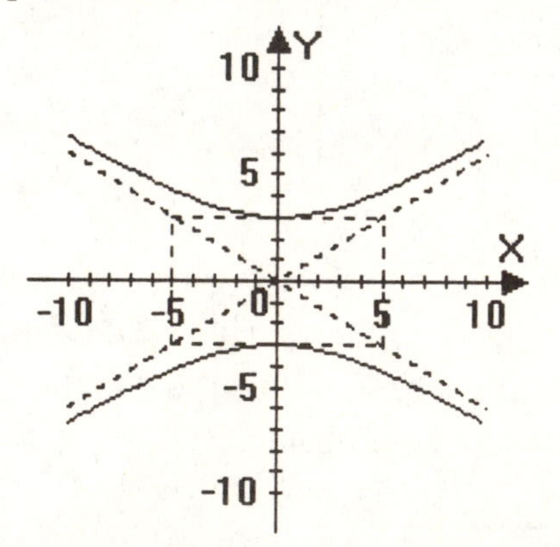

13. $x = -y^2 + 6y$

Since only one variable is squared, the graph of the equation is a parabola.

This is a parabola of the form $x = ay^2 + by + c$. Since $a = 1$ is positive, the parabola opens to the right. The y–coordinate of the vertex is $-\frac{b}{2a}=-\frac{6}{2(-1)}=-\frac{6}{-2}=3$. The x–coordinate of the vertex is $x=-3^2+6(3)=-9+18=9$.

The vertex of the parabola is $(9,3)$.

Replace y with 0 to find the x–intercept.

$x=-0^2+6(0)=0+0=0$

The x–intercept is 0. Replace x with 0 to find the y–intercepts.

$0=-y^2+6y$

$0=-y(y-6)$

Apply the zero product principle.

$-y=0$ and $y-6=0$

$y=0$ $\quad y=6$

The y–intercepts are 0 and 6.

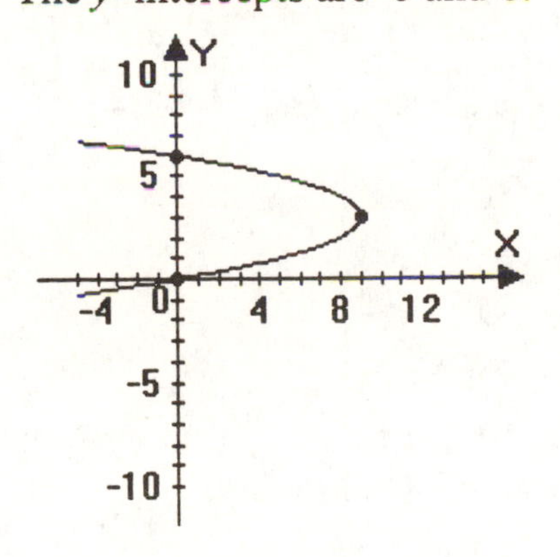

14. $\frac{(x-2)^2}{16}+\frac{(y+3)^2}{9}=1$

Because x^2 and y^2 have different positive coefficients, the equation's graph is an ellipse.

The center of the ellipse is $(2,-3)$.

Because the denominator of the x^2 – term is greater than the denominator of the y^2 – term, the major axis is horizontal. Since $a^2=16$, $a=4$ and the vertices lie 4 units to the left and right of the center. Since $b^2=9$, $b=3$ and endpoints of the minor axis lie 3 units above and below the center.

Center	Vertices	Endpoints of Minor Axis
$(2,-3)$	$(2-4,-3)$ $=(-2,-3)$	$(2,-3-3)$ $=(2,-6)$
	$(2+4,-3)$ $=(6,-3)$	$(2,-3+3)$ $=(2,0)$

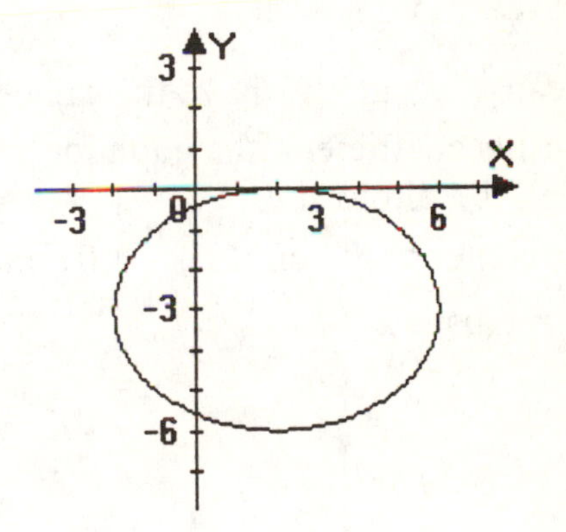

15. $(x+1)^2+(y+2)^2=9$

Because x^2 and y^2 have the same positive coefficient, the equation's graph is a circle.

The center of the circle is $(-1,-2)$ and the radius is 3.

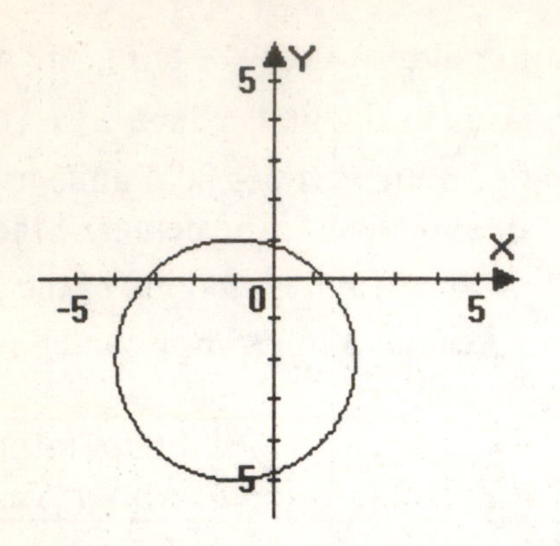

16.

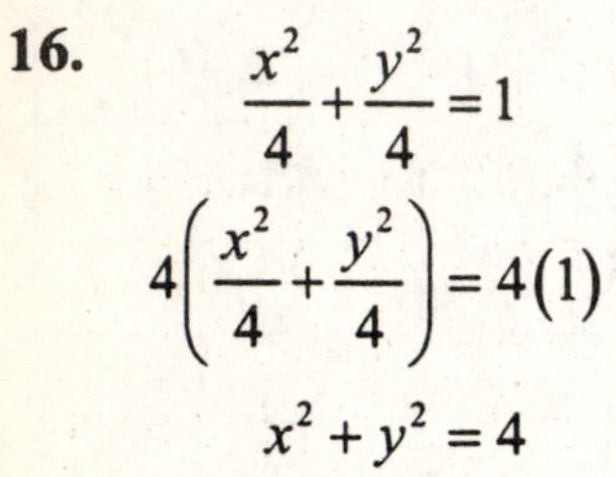

$$\frac{x^2}{4}+\frac{y^2}{4}=1$$

$$4\left(\frac{x^2}{4}+\frac{y^2}{4}\right)=4(1)$$

$$x^2+y^2=4$$

Because x^2 and y^2 have the same positive coefficient, the equation's graph is a circle.
The center of the circle is $(0,0)$ and the radius is 2.

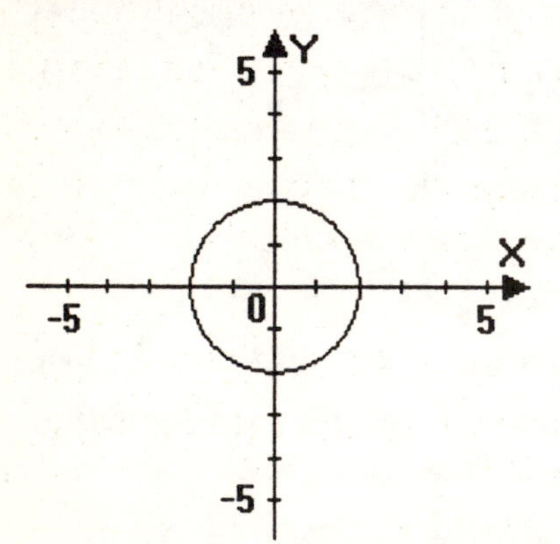

17. $x^2+y^2=25$

$x+y=1$

Solve the second equation for y.

$x+y=1$

$y=-x+1$

Substitute $-x+1$ for y to find x.

$$x^2+(-x+1)^2=25$$
$$x^2+x^2-2x+1=25$$
$$2x^2-2x+1=25$$
$$2x^2-2x-24=0$$
$$x^2-x-12=0$$
$$(x-4)(x+3)=0$$

Apply the zero product principle.

$x-4=0$ and $x+3=0$

$x=4$ $\qquad x=-3$

Back-substitute -3 and 4 for x to find y.

$x=4$	and	$x=-3$
$y=-x+1$		$y=-x+1$
$y=-4+1$		$y=-(-3)+1$
$y=-3$		$y=3+1$
		$y=4$

The solutions are $(-3,4)$ and $(4,-3)$ and the solution set is $\{(-3,4),(4,-3)\}$.

18. $2x^2-5y^2=-2$

$3x^2+2y^2=35$

Multiply the first equation by 2 and the second equation by 5.

$$4x^2-10y^2=-4$$
$$\underline{15x^2+10y^2=175}$$
$$19x^2=171$$
$$x^2=9$$
$$x=\pm 3$$

In this case, we can back-substitute 9 for x^2 to find y.

$$x^2 = 9$$
$$2x^2 - 5y^2 = -2$$
$$2(9) - 5y^2 = -2$$
$$18 - 5y^2 = -2$$
$$-5y^2 = -20$$
$$y^2 = 4$$
$$y = \pm 2$$

We have $x = \pm 3$ and $y = \pm 2$, the solutions are $(-3,-2),(-3,2),(3,-2)$ and $(3,2)$ and the solution set is $\{(-3,-2),(-3,2),(3,-2),(3,2)\}$.

19. $2x + y = 39$

$xy = 180$

Solve the first equation for y.

$$2x + y = 39$$
$$y = 39 - 2x$$

Substitute $39 - 2x$ for y to find x.

$$x(39 - 2x) = 180$$
$$39x - 2x^2 = 180$$
$$0 = 2x^2 - 39x + 180$$
$$0 = (2x - 15)(x - 12)$$

Apply the zero product principle.

$2x - 15 = 0$ and $x - 12 = 0$

$2x = 15$ $\quad x = 12$

$x = \frac{15}{2}$

Back-substitute $\frac{15}{2}$ and 12 for x to find y.

$x = \frac{15}{2}$ and $x = 12$

$y = 39 - 2x$ $\quad y = 39 - 2x$

$y = 39 - 2\left(\frac{15}{2}\right)$ $\quad y = 39 - 2(12)$

$\quad y = 39 - 24$

$y = 39 - 15$ $\quad y = 15$

$y = 24$

The dimensions are 15 feet by 12 feet or 24 feet by $\frac{15}{2}$ or 7.5 feet.

20. Let x = the length of the rectangle
Let y = the width of the rectangle

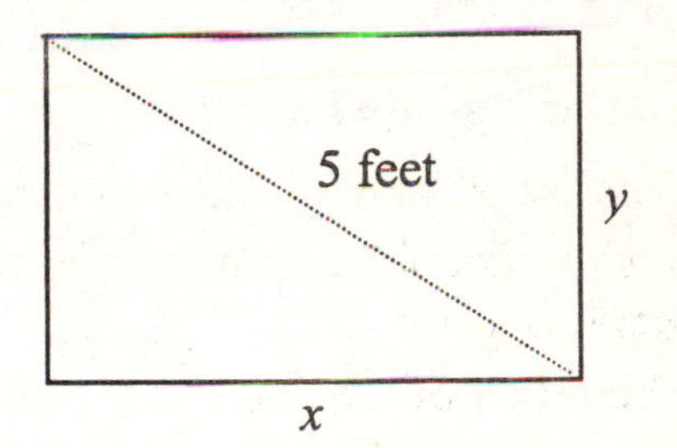

Using the Pythagorean Theorem, we obtain $x^2 + y^2 = 5^2$. Since the perimeter is 14 feet, we have $2x + 2y = 14$. The system of two equations in two variables is as follows.

$$x^2 + y^2 = 25$$
$$2x + 2y = 14$$

Solve the second equation for y.

$$2x + 2y = 14$$
$$2y = 14 - 2x$$
$$y = 7 - x$$

Substitute $7 - x$ for y to find x.

$$x^2 + (7 - x)^2 = 25$$
$$x^2 + 49 - 14x + x^2 = 25$$
$$2x^2 - 14x + 49 = 25$$

$$2x^2 - 14x + 24 = 0$$
$$x^2 - 7x + 12 = 0$$
$$(x-4)(x-3) = 0$$

Apply the zero product principle.

$x - 4 = 0$ and $x - 3 = 0$

$x = 4$ $\quad x = 3$

Back-substitute 3 and 4 for x to find y.

$x = 4$ and $x = 3$

$y = 7 - x$ $\quad y = 7 - x$

$y = 7 - 4$ $\quad y = 7 - 3$

$y = 3$ $\quad y = 4$

The solutions are the same. The dimensions are 4 feet by 3 feet.

Cumulative Review Exercises

Chapters 1-10

1. $3x + 7 > 4$ or $6 - x < 1$

$3x > -3$ $\quad -x < -5$

$x > -1$ $\quad x > 5$

$x > -1$

$x > 5$

$x > -1$ or $x > 5$

The solution set is $\{x | x > -1\}$ or $(-1, \infty)$.

2.
$$x(2x - 7) = 4$$
$$2x^2 - 7x = 4$$
$$2x^2 - 7x - 4 = 0$$
$$(2x+1)(x-4) = 0$$

Apply the zero product principle.

$2x + 1 = 0$ and $x - 4 = 0$

$2x = -1$ $\quad x = 4$

$x = -\frac{1}{2}$

The solutions are $-\frac{1}{2}$ and 4 and the solution set is $\left\{-\frac{1}{2}, 4\right\}$.

3.
$$\frac{5}{x-3} = 1 + \frac{30}{x^2 - 9}$$
$$\frac{5}{x-3} = 1 + \frac{30}{(x+3)(x-3)}$$

Multiply both sides of the equation by the LCD, $(x+3)(x-3)$.

$$(x+3)(x-3)\left(\frac{5}{x-3}\right) = (x+3)(x-3)\left(1 + \frac{30}{(x+3)(x-3)}\right)$$
$$(x+3)(5) = (x+3)(x-3) + 30$$
$$5x + 15 = x^2 - 9 + 30$$
$$15 = x^2 - 5x + 21$$

$0 = x^2 - 5x + 6$

$0 = (x-3)(x-2)$

Apply the zero product principle.

$x - 3 = 0$ and $x - 2 = 0$

$x = 3$ $\quad$ $x = 2$

We disregard 3 because it would make the denominator zero. The solution is 2 and the solution set is $\{2\}$.

4. $3x^2 + 8x + 5 < 0$

Solve the related quadratic equation.

$3x^2 + 8x + 5 = 0$

$(3x+5)(x+1) = 0$

Apply the zero product principle.

$3x + 5 = 0$ or $x + 1 = 0$

$3x = -5$ $\quad$ $x = -1$

$x = -\frac{5}{3}$

The boundary points are $-\frac{5}{3}$ and -1.

Test Interval	Test Number	Test	Conclusion
$\left(-\infty, -\frac{5}{3}\right)$	-2	$3(-2)^2 + 8(-2) + 5 < 0$ $1 < 0$, false	$\left(-\infty, -\frac{5}{3}\right)$ does not belong to the solution set.
$\left(-\frac{5}{3}, -1\right)$	$-\frac{4}{3}$	$3\left(-\frac{4}{3}\right)^2 + 8\left(-\frac{4}{3}\right) + 5 < 0$ $-11 < 0$, true	$\left(-\frac{5}{3}, -1\right)$ belongs to the solution set.
$(-1, \infty)$	0	$3(0)^2 + 8(0) + 5 < 0$ $5 < 0$, false	$(-1, \infty)$ does not belong to the solution set.

The solution set is $\left(-\frac{5}{3}, -1\right)$ or $\left\{x \middle| -\frac{5}{3} < x < -1\right\}$.

-5/3 -1

5. $3^{2x-1}=81$

$3^{2x-1}=3^4$

$2x-1=4$

$2x=5$

$x=\dfrac{5}{2}$

The solution is $\frac{5}{2}$ and the solution set is $\left\{\frac{5}{2}\right\}$.

6. $30e^{0.7x}=240$

$e^{0.7x}=80$

$\ln e^{0.7x}=\ln 8$

$0.7x=\ln 8$

$x=\dfrac{\ln 8}{0.7}=\dfrac{2.08}{0.7}\approx 2.97$

The solution is $\frac{2.08}{0.7}\approx 2.97$ and the solution set is $\left\{\frac{2.08}{0.7}\approx 2.97\right\}$.

7. $3x^2+4y^2=39$

$5x^2-2y^2=-13$

Multiply the second equation by 2 and add to the first equation.

$$\begin{aligned} 3x^2+4y^2 &= 39 \\ 10x^2-4y^2 &= -26 \\ \hline 13x^2 &= 13 \\ x^2 &= 1 \\ x &= \pm 1 \end{aligned}$$

In this case, we can back-substitute 9 for x^2 to find y.

$x^2=1$

$3x^2+4y^2=39$

$3(1)+4y^2=39$

$3+4y^2=39$

$4y^2=36$

$y^2=9$

$y=\pm 3$

We have $x=\pm 1$ and $y=\pm 3$, the solutions are $(-1,-3),(-1,3),(1,-3)$ and $(1,3)$ and the solution set is $\{(-1,-3),(-1,3),(1,-3),(1,3)\}$.

8. $f(x)=-\dfrac{2}{3}x+4$

$y=-\dfrac{2}{3}x+4$

The y–intercept is 4 and the slope is $-\frac{2}{3}$. We can write the slope as $m=\frac{-2}{3}=\frac{\text{rise}}{\text{run}}$ and use the intercept and the slope to graph the function.

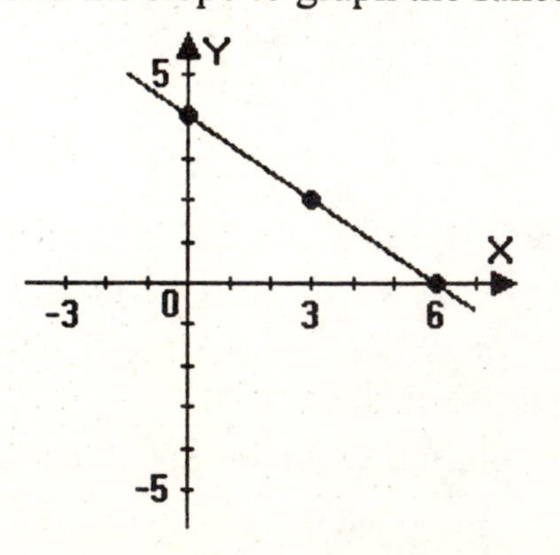

9. 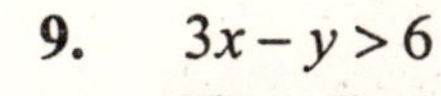
$3x-y>6$

First, find the intercepts to the equation $3x-y=6$.

Find the x–intercept by setting y equal to zero.

$3x-0=6$

$3x=6$

$x=2$

Find the y–intercept by setting x equal to zero.

$3(0)-y=6$

$-y=6$

$y=-6$

Next, use the origin as a test point.

$3(0)-0>6$

$0-0>6$

$0>6$

This is a false statement. This means that the origin will not fall in the shaded half-plane.

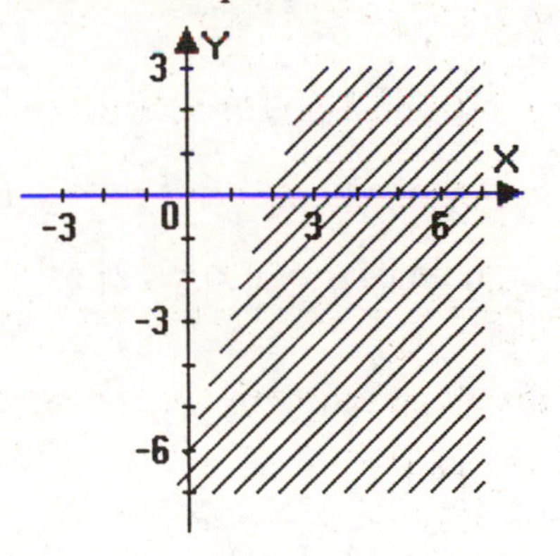

10. $x^2+y^2+4x-6y+9=0$

Because x^2 and y^2 have the same positive coefficient, the equation's graph is a circle.

$(x^2+4x \quad)+(y^2-6y \quad)=-9$

Complete the squares.

$\left(\frac{b}{2}\right)^2=\left(\frac{4}{2}\right)^2=(2)^2=4$

$\left(\frac{b}{2}\right)^2=\left(\frac{-6}{2}\right)^2=(-3)^2=9$

$(x^2+4x+4)+(y^2-6y+9)=-9+4+9$

$(x+2)^2+(y-3)^2=4$

The center of the circle is $(-2,3)$ and the radius is 2.

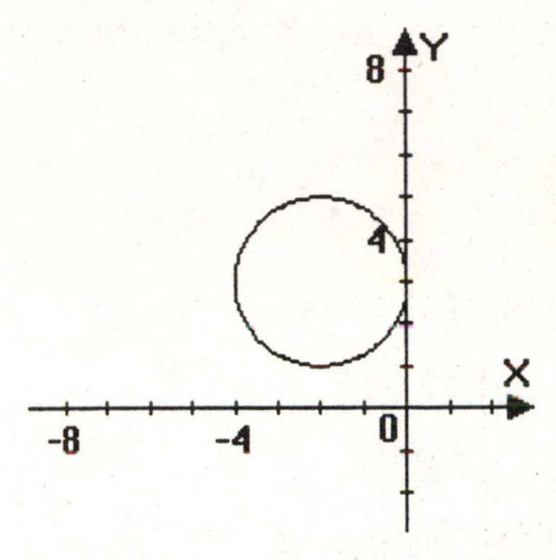

11. $9x^2-4y^2=36$

Because x^2 and y^2 have opposite signs, the equation's graph is a hyperbola.

$\frac{9x^2}{36}-\frac{4y^2}{36}=\frac{36}{36}$

$\frac{x^2}{4}-\frac{y^2}{9}=1$

The equation is in the form

$\frac{x^2}{a^2}-\frac{y^2}{b^2}=1$ with $a^2=4$, and $b^2=9$.

We know the transverse axis lies on the x-axis and the vertices are $(-2,0)$ and $(2,0)$. Because $a^2=4$ and $b^2=9$, $a=2$ and $b=3$.

Construct a rectangle using –2 and 2 on the x–axis, and –3 and 3 on the y–axis. Draw extended diagonals to obtain the asymptotes. Graph the hyperbola.

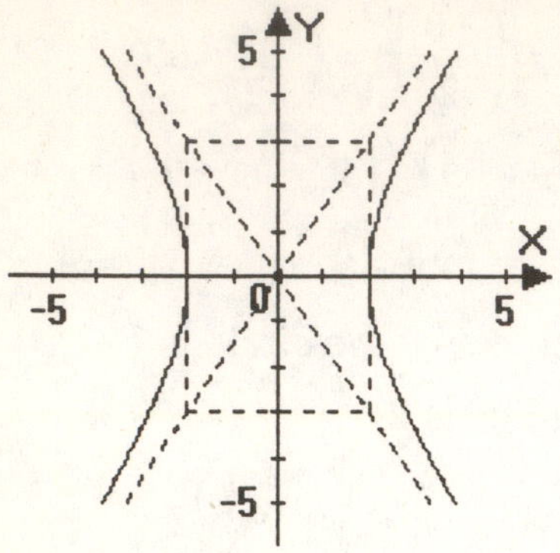

12. $-2\left(3^2-12\right)^3-45\div 9-3$

$=-2(9-12)^3-45\div 9-3$

$=-2(-3)^3-45\div 9-3$

$=-2(-27)-45\div 9-3$

$=54-5-3=46$

13. $\left(-9+7x-4x^2+4x^3\right)\div(2x-1)$

Rewrite the polynomials in descending order and divide.

$$\begin{array}{r} x^2-5x-1 \\ 3x-4\overline{)3x^3-19x^2+17x+4} \\ \underline{3x^3-\ 4x^2} \qquad\qquad \\ -15x^2+17x \qquad \\ \underline{-15x^2+20x} \qquad \\ -3x+4 \\ \underline{-3x+4} \\ 0 \end{array}$$

$$\frac{3x^3-19x^2+17x+4}{3x-4}=x^2-5x-1$$

14. $\sqrt[3]{4x^2y^5}\cdot\sqrt[3]{4xy^2}$

$=\sqrt[3]{4x^2y^5 4xy^2}=\sqrt[3]{16x^3y^7}$

$=\sqrt[3]{8\cdot 2x^3y^6y}=2xy^2\sqrt[3]{2y}$

15. $(2+3i)(4-i)$

$=8-2i+12i-3i^2=8+10i-3(-1)$

$=8+10i+3=11+10i$

16. $12x^3-36x^2+27x=3x\left(4x^2-12x+9\right)$

$=3x(2x-3)^2$

17. $x^3-2x^2-9x+18$

$=x^2(x-2)-9(x-2)$

$=(x-2)\left(x^2-9\right)$

$=(x-2)(x+3)(x-3)$

18. Since the radicand must be positive, the domain will exclude all values of x which make the radicand less than zero.

$6-3x\geq 0$

$-3x\geq -6$

$x\leq 2$

The domain of $f=\{x|x\leq 2\}$ or $(-\infty,2]$.

19. $\frac{1-\sqrt{x}}{1+\sqrt{x}}=\frac{1-\sqrt{x}}{1+\sqrt{x}}\cdot\frac{1-\sqrt{x}}{1-\sqrt{x}}$

$=\frac{\left(1-\sqrt{x}\right)^2}{1^2+\left(\sqrt{x}\right)^2}$

$=\frac{\left(1-\sqrt{x}\right)^2}{1+x}$ or $\frac{1-2\sqrt{x}+x}{1+x}$

20. $\frac{1}{3}\ln x+7\ln y=\ln x^{\frac{1}{3}}+\ln y^7$

$=\ln x^{\frac{1}{3}}y^7$

21. $\left(3x^3-5x^2+2x-1\right)\div(x-2)$

$$\begin{array}{r|rrrr} 2 & 3 & -5 & 2 & -1 \\ & & 6 & 2 & 8 \\ \hline & 3 & 1 & 4 & 7 \end{array}$$

$$\left(3x^3-5x^2+2x-1\right)\div(x-2)$$
$$=3x^2+x+4+\frac{7}{x-2}$$

22. $x=-2\sqrt{3}$ and $x=2\sqrt{3}$

$x+2\sqrt{3}=0$ $x-2\sqrt{3}=0$

Multiply the factors to obtain the polynomial.

$$\left(x+2\sqrt{3}\right)\left(x-2\sqrt{3}\right)=0$$
$$x^2-\left(2\sqrt{3}\right)^2=0$$
$$x^2-4\cdot3=0$$
$$x^2-12=0$$

23. Let x = the rate of the slower car

	r •	t	$= d$
Fast	$x+10$	2	$2(x+10)$
Slow	x	2	$2x$

$$2(x+10)+2x=180$$
$$2x+20+2x=180$$
$$4x+20=180$$
$$4x=160$$
$$x=40$$

The rate of the slower car is 40 miles per hour and the rate of the faster car is 40 + 10 = 50 miles per hour.

24. Let x = the number of miles driven in a day

$C_R=39+0.16x$

$C_A=25+0.24x$

Set the costs equal.

$$39+0.16x=25+0.24x$$
$$39=25+0.08x$$
$$14=0.08x$$
$$\frac{14}{0.08}=x$$
$$x=175$$

The cost is the same when renting from either company when 175 miles are driven in a day.

$C_R=39+0.16(175)=39+28=67$

When 175 miles are driven, the cost is $67.

25. Let x = the number of apples

Let y = the number of bananas

$3x+2y=354$

$2x+3y=381$

Multiply the first equation by –3 and the second equation by 2 and solve by addition.

$$\begin{array}{r} -9x-6y=-1062 \\ 4x+6y=\quad 762 \\ \hline -5x=-300 \\ x=60 \end{array}$$

Back-substitute 60 for x to find y.

$$3(60)+2y=354$$
$$180+2y=354$$
$$2y=174$$
$$y=87$$

There are 60 calories in an apple and 87 calories in a banana.

Chapter 11

Check Points 11.1

1. **a.** $a_n = 2n + 5$

$a_1 = 2(1) + 5 = 2 + 5 = 7$

$a_2 = 2(2) + 5 = 4 + 5 = 9$

$a_3 = 2(3) + 5 = 6 + 5 = 11$

$a_4 = 2(4) + 5 = 8 + 5 = 13$

The first four terms are 7, 9, 11, 13.

b. $a_n = \frac{(-1)^n}{2^n + 1}$

$a_1 = \frac{(-1)^1}{2^1 + 1} = \frac{-1}{2+1} = -\frac{1}{3}$

$a_2 = \frac{(-1)^2}{2^2 + 1} = \frac{1}{4+1} = \frac{1}{5}$

$a_3 = \frac{(-1)^3}{2^3 + 1} = \frac{-1}{8+1} = -\frac{1}{9}$

$a_4 = \frac{(-1)^4}{2^4 + 1} = \frac{1}{16+1} = \frac{1}{17}$

The first four terms are $-\frac{1}{3}, \frac{1}{5}, -\frac{1}{9}, \frac{1}{17}$.

2. $a_n = \frac{20}{(n+1)!}$

$a_1 = \frac{20}{(1+1)!} = \frac{20}{2!} = \frac{20}{2 \cdot 1} = 10$

$a_2 = \frac{20}{(2+1)!} = \frac{20}{3!} = \frac{20}{3 \cdot 2 \cdot 1} = \frac{10}{3}$

$a_3 = \frac{20}{(3+1)!} = \frac{20}{4!} = \frac{20}{4 \cdot 3 \cdot 2 \cdot 1} = \frac{5}{6}$

$a_4 = \frac{20}{(4+1)!} = \frac{20}{5!} = \frac{20}{5 \cdot 4 \cdot 3 \cdot 2 \cdot 1} = \frac{1}{6}$

The first four terms are $10, \frac{10}{3}, \frac{5}{6}, \frac{1}{6}$.

3. **a.** $\sum_{i=1}^{6} 2i^2 = 2(1)^2 + 2(2)^2 + 2(3)^2 + 2(4)^2 + 2(5)^2 + 2(6)^2$

$= 2(1) + 2(4) + 2(9) + 2(16) + 2(25) + 2(36)$

$= 2 + 8 + 18 + 32 + 50 + 72 = 182$

b. $\sum_{k=3}^{5} (2^k - 3) = (2^3 - 3) + (2^4 - 3) + (2^5 - 3) = (8-3) + (16-3) + (32-3)$

$= 5 + 13 + 29 = 47$

c. $\sum_{i=1}^{5} 4 = 4 + 4 + 4 + 4 + 4 = 20$

4. **a.** $1^2+2^2+3^2+...+9^2=\sum_{i=1}^{9} i^2$

b. $1+\frac{1}{2}+\frac{1}{4}+\frac{1}{8}+...+\frac{1}{2^{n-1}}=\sum_{i=1}^{n}\frac{1}{2^{n-1}}$

Problem Set 11.1

Practice Exercises

1. $a_n=3n+2$

$a_1=3(1)+2=3+2=5$

$a_2=3(2)+2=6+2=8$

$a_3=3(3)+2=9+2=11$

$a_4=3(4)+2=12+2=14$

The first four terms are 5, 8, 11, 14.

3. $a_n=3^n$

$a_1=3^1=3$

$a_2=3^2=9$

$a_3=3^3=27$

$a_4=3^4=81$

The first four terms are 3, 9, 27, 81.

5. $a_n=(-3)^n$

$a_1=(-3)^1=-3$

$a_2=(-3)^2=9$

$a_3=(-3)^3=-27$

$a_4=(-3)^4=81$

The first four terms are –3, 9, –27, 81.

7. $a_n=(-1)^n(n+3)$

$a_1=(-1)^1(1+3)=-1(4)=-4$

$a_2=(-1)^2(2+3)=1(5)=5$

$a_3=(-1)^3(3+3)=-1(6)=-6$

$a_4=(-1)^4(4+3)=1(7)=7$

The first four terms are –4, 5, –6, 7.

9. $a_n=\frac{2n}{n+4}$

$a_1=\frac{2(1)}{1+4}=\frac{2}{5}$

$a_2=\frac{2(2)}{2+4}=\frac{4}{6}=\frac{2}{3}$

$a_3=\frac{2(3)}{3+4}=\frac{6}{7}$

$a_4=\frac{2(4)}{4+4}=\frac{8}{8}=1$

The first four terms are $\frac{2}{5},\frac{2}{3},\frac{6}{7},1$.

11. $a_n=\frac{(-1)^{n+1}}{2^n-1}$

$a_1=\frac{(-1)^{1+1}}{2^1-1}=\frac{(-1)^2}{2-1}=\frac{1}{1}=1$

$a_2=\frac{(-1)^{2+1}}{2^2-1}=\frac{(-1)^3}{4-1}=\frac{-1}{3}=-\frac{1}{3}$

$a_3=\frac{(-1)^{3+1}}{2^3-1}=\frac{(-1)^4}{8-1}=\frac{1}{7}$

$a_4=\frac{(-1)^{4+1}}{2^4-1}=\frac{(-1)^5}{16-1}=\frac{-1}{15}=-\frac{1}{15}$

The first four terms are $1,-\frac{1}{3},\frac{1}{7},-\frac{1}{15}$.

13. $a_n=\frac{n^2}{n!}$

$a_1=\frac{1^2}{1!}=\frac{1}{1}=1$

$a_2 = \frac{2^2}{2!} = \frac{4}{2\cdot 1} = \frac{4}{2} = 2$

$a_3 = \frac{3^2}{3!} = \frac{9}{3\cdot 2\cdot 1} = \frac{3}{2}$

$a_4 = \frac{4^2}{4!} = \frac{16}{4\cdot 3\cdot 2\cdot 1} = \frac{2}{3}$

The first four terms are $1, 2, \frac{3}{2}, \frac{2}{3}$.

15. $a_n = 2(n+1)!$

$a_1 = 2(1+1)! = 2(2)!$
$= 2(2\cdot 1) = 2(2) = 4$

$a_2 = 2(2+1)! = 2(3)!$
$= 2(3\cdot 2\cdot 1) = 2(6) = 12$

$a_3 = 2(3+1)! = 2(4)!$
$= 2(4\cdot 3\cdot 2\cdot 1) = 2(24) = 48$

$a_4 = 2(4+1)! = 2(5)!$
$= 2(5\cdot 4\cdot 3\cdot 2\cdot 1) = 2(120) = 240$

The first four terms are 4, 12, 48, 240.

17. $\sum_{i=1}^{6} 5i = 5(1)+5(2)+5(3)+5(4)+5(5)+5(6) = 5+10+15+20+25+30 = 105$

19. $\sum_{i=1}^{4} 2i^2 = 2(1)^2 + 2(2)^2 + 2(3)^2 + 2(4)^2 = 2(1)+2(4)+2(9)+2(16)$

$= 2+8+18+32 = 60$

21. $\sum_{k=1}^{5} k(k+4) = 1(1+4)+2(2+4)+3(3+4)+4(4+4)+5(5+4)$

$= 1(5)+2(6)+3(7)+4(8)+5(9) = 5+12+21+32+45 = 115$

23. $\sum_{i=1}^{4}\left(-\frac{1}{2}\right)^i = \left(-\frac{1}{2}\right)^1 + \left(-\frac{1}{2}\right)^2 + \left(-\frac{1}{2}\right)^3 + \left(-\frac{1}{2}\right)^4 = -\frac{1}{2}+\frac{1}{4}+\left(-\frac{1}{8}\right)+\frac{1}{16}$

$= -\frac{1}{2}\cdot\frac{8}{8}+\frac{1}{4}\cdot\frac{4}{4}+\left(-\frac{1}{8}\right)\frac{2}{2}+\frac{1}{16} = -\frac{8}{16}+\frac{4}{16}-\frac{2}{16}+\frac{1}{16}$

$= \frac{-8+4-2+1}{16} = -\frac{5}{16}$

25. $\sum_{i=5}^{9} 11 = 11+11+11+11+11 = 55$

27. $$\sum_{i=0}^{4}\frac{(-1)^i}{i!}=\frac{(-1)^0}{0!}+\frac{(-1)^1}{1!}+\frac{(-1)^2}{2!}+\frac{(-1)^3}{3!}+\frac{(-1)^4}{4!}=\frac{1}{1}+\frac{-1}{1}+\frac{1}{2\cdot 1}+\frac{-1}{3\cdot 2\cdot 1}+\frac{1}{4\cdot 3\cdot 2\cdot 1}$$
$$=1-1+\frac{1}{2}-\frac{1}{6}+\frac{1}{24}=\frac{1}{2}\cdot\frac{12}{12}-\frac{1}{6}\cdot\frac{4}{4}+\frac{1}{24}=\frac{12}{24}-\frac{4}{24}+\frac{1}{24}=\frac{12-4+1}{24}=\frac{9}{24}=\frac{3}{8}$$

29. $$\sum_{i=1}^{5}\frac{i!}{(i-1)!}=\frac{1!}{(1-1)!}+\frac{2!}{(2-1)!}+\frac{3!}{(3-1)!}+\frac{4!}{(4-1)!}+\frac{5!}{(5-1)!}=\frac{1!}{0!}+\frac{2!}{1!}+\frac{3!}{2!}+\frac{4!}{3!}+\frac{5!}{4!}$$
$$=\frac{1}{1}+\frac{2\cdot \cancel{1!}}{\cancel{1!}}+\frac{3\cdot \cancel{2!}}{\cancel{2!}}+\frac{4\cdot \cancel{3!}}{\cancel{3!}}+\frac{5\cdot \cancel{4!}}{\cancel{4!}}=1+2+3+4+5=15$$

31. $1^2+2^2+3^2+...+15^2=\sum_{i=1}^{15}i^2$

33. $2+2^2+2^3+...+2^{11}=\sum_{i=1}^{11}2^i$

35. $1+2+3+...+30=\sum_{i=1}^{30}i$

37. $\frac{1}{2}+\frac{2}{3}+\frac{3}{4}+...+\frac{14}{14+1}=\sum_{i=1}^{14}\frac{i}{i+1}$

39. $4+\frac{4^2}{2}+\frac{4^3}{3}+...+\frac{4^n}{n}=\sum_{i=1}^{n}\frac{4^i}{i}$

41. $1+3+5+...+(2n-1)=\sum_{i=1}^{n}(2i-1)$

43. $5+7+9+11+...+31=\sum_{k=2}^{15}(2k+1)$

45. $a+ar+ar^2+...+ar^{12}=\sum_{k=0}^{12}ar^k$

47. $a+(a+d)+(a+2d)+...+a(a+nd)$
$=\sum_{k=0}^{n}(a+kd)$

Application Exercises

49. a. $\sum_{i=1}^{9}a_i=333.3+407.5+495.4+662.1+722.9+778.0+753.1+847.0+938.9=5939.1$

This means from 1991 to 1999, a total of 5939.1 million CDs were sold.

b. $\frac{1}{9}\sum_{i=1}^{9}a_i=\frac{1}{9}(5939.1)=659.9$

This means that from 1991 to 1999, an average of 659.9 million CDs sold each year.

51. $a_n = 0.16n^2 - 1.04n + 7.39$

$a_1 = 0.16(1)^2 - 1.04(1) + 7.39 = 0.16(1) - 1.04 + 7.39 = 0.16 - 1.04 + 7.39 = 6.51$

$a_2 = 0.16(2)^2 - 1.04(2) + 7.39 = 0.16(4) - 2.08 + 7.39 = 0.64 - 2.08 + 7.39 = 5.95$

$a_3 = 0.16(3)^2 - 1.04(3) + 7.39 = 0.16(9) - 3.12 + 7.39 = 1.44 - 3.12 + 7.39 = 5.71$

$a_4 = 0.16(4)^2 - 1.04(4) + 7.39 = 0.16(16) - 4.16 + 7.39 = 2.56 - 4.16 + 7.39 = 5.79$

$a_5 = 0.16(5)^2 - 1.04(5) + 7.39 = 0.16(25) - 5.20 + 7.39 = 4.00 - 5.20 + 7.39 = 6.19$

$\sum_{i=1}^{5} a_i = a_1 + a_2 + a_3 + a_4 + a_5 = 6.51 + 5.95 + 5.71 + 5.79 + 6.19 = 30.15$

From 1991 to 1995, Americans spent $30.15 billion on recreational boating.

53. $a_{20} = 6000\left(1 + \frac{0.06}{4}\right)^{20} = 6000(1 + 0.015)^{20} = 6000(1.015)^{20} = 8081.13$

The balance in the account after 5 years if $8081.13.

Writing in Mathematics

55. Answers will vary.

57. Answers will vary.

59. Answers will vary.

Technology Exercises

61. Answers will vary. For example, consider Exercise 1.

$a_n = 3n + 2$

```
seq(3N+2,N,1,4)
     {5 8 11 14}
```

This is the same result obtained in Exercise 1.

63.

```
10→N
                10
(1+1/N)^N
        2.59374246
```

```
100→N
               100
(1+1/N)^N
       2.704813829
```

```
1000→N
              1000
(1+1/N)^N
       2.716923932
```

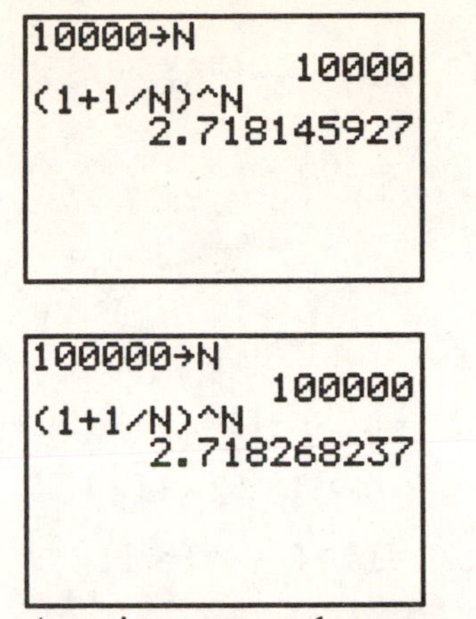

As n increases, the approximation gets closer to the decimal approximation of e, 2.7183.

65. $a_n = \dfrac{100}{n}$

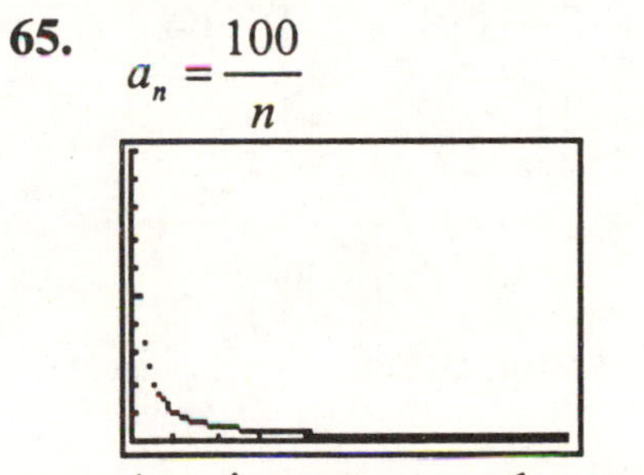

As n increases, a_n decreases.

67. $a_n = \dfrac{3n^4 + n - 1}{5n^4 + 2n^2 + 1}$

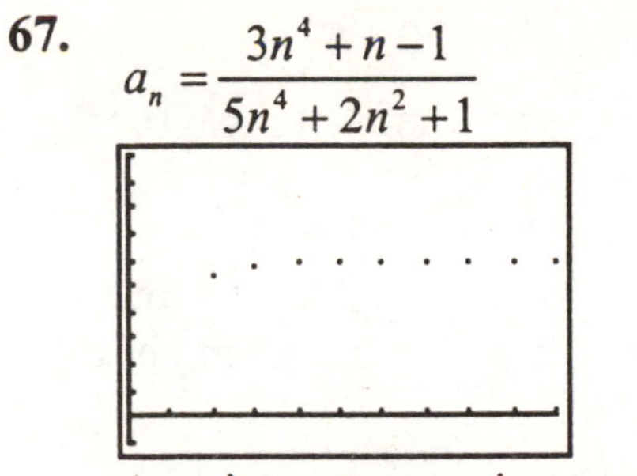

As n increases, a_n increases.

Critical Thinking Exercises

69. $\sum_{i=1}^{4} \log 2i$

$= \log 2(1) + \log 2(2) + \log 2(3) + \log 2(4)$

$= \log 2 + \log 4 + \log 6 + \log 8$

$= \log(2 \cdot 4 \cdot 6 \cdot 8) = \log 384$

71. $a_n = a_{n-1} + 5$

$a_1 = 7$

$a_2 = a_{2-1} + 5 = a_1 + 5 = 7 + 5 = 12$

$a_3 = a_{3-1} + 5 = a_2 + 5 = 12 + 5 = 17$

$a_4 = a_{4-1} + 5 = a_3 + 5 = 17 + 5 = 22$

The first four terms are 7, 12, 17, 22.

Review Exercises

73. $\sqrt[3]{40x^4y^7} = \sqrt[3]{8 \cdot 5x^3xy^6y} = 2xy^2\sqrt[3]{5xy}$

74. $27x^3 - 8 = (3x-2)(9x^2 + 6x + 4)$

75.

$$\frac{6}{x} + \frac{6}{x+2} = \frac{5}{2}$$

$$2x(x+2)\left(\frac{6}{x} + \frac{6}{x+2}\right) = 2x(x+2)\left(\frac{5}{2}\right)$$

$$2(x+2)(6) + 2x(6) = x(x+2)(5)$$

$$12(x+2) + 12x = 5x(x+2)$$

$$12x + 24 + 12x = 5x^2 + 10x$$

$$24x + 24 = 5x^2 + 10x$$

$$0 = 5x^2 - 14x - 24$$

$$0 = (5x+6)(x-4)$$

Apply the zero product principle.

$5x + 6 = 0$ or $x - 4 = 0$

$5x = -6$ $\quad x = 4$

$x = -\dfrac{6}{5}$

The solutions are $-\dfrac{6}{5}$ and 4 and the solution set is $\left\{-\dfrac{6}{5}, 4\right\}$.

Check Points 11.2

1. $a_1 = 100$
$a_2 = 100 - 30 = 70$
$a_3 = 70 - 30 = 40$
$a_4 = 40 - 30 = 10$
$a_5 = 10 - 30 = -20$
$a_6 = -20 - 30 = -50$
The first six terms are 100, 70, 40, 10, -20, -50.

2. $a_n = a_1 + (n-1)d$
$a_9 = 6 + (9-1)(-5)$
$= 6 + (8)(-5)$
$= 6 - 40 = -34$

3. **a.** $a_n = a_1 + (n-1)d$
$= 12808 + (n-1)2350$
$= 12808 + 2350n - 2350$
$= 2350n + 10458$

b. Since 2010 – 1983 = 27, find a_{27}.
$a_{27} = 2350(27) + 10458$
$= 63450 + 10458 = 73908$
U.S. travelers will spend $73908 million in other countries by 2010.

4. To find the sum, use $S_n = \frac{n}{2}(a_1 + a_n)$.
First find a_{15}.
$a_{15} = a_1 + (n-1)d = 3 + (15-1)3$
$= 3 + (14)3 = 45$
Now substitute.

$S_{15} = \frac{15}{2}(3 + 45) = \frac{15}{2}(48)$
$= 15(24) = 360$

5. $\sum_{i=1}^{30}(6i - 11) = 6(1) - 11 + 6(2) - 11$
$+6(3) - 11 + ... + 6(30) - 11$
$= 6 - 11 + 12 - 11 + 18$
$-11 + ... + 180 - 11$
$= -5 + 1 + 7 + ... + 169$

$S_n = \frac{n}{2}(a_1 + a_n)$

$S_{30} = \frac{30}{2}(-5 + 169) = 15(164) = 2460$

$\sum_{i=1}^{30}(6i - 11) = 2460$

6. To find the sum, use $S_n = \frac{n}{2}(a_1 + a_n)$.
First find a_{10}.
$a_{10} = 1800(10) + 49730$
$= 18000 + 49730 = 67730$
$S_{10} = \frac{10}{2}(51530 + 67730) = 5(119260)$
$= 596300$
It would cost $596,300 for nursing home care for a ten-year period beginning in 2001.

Problem Set 11.2

Practice Exercises

1. Since 6 – 2 = 4, $d = 4$.

3. Since $-2 - (-7) = 5$, $d = 5$.

5. Since $711-714=-3,\ d=-3.$

7. $a_1=200$
$a_2=200+20=220$
$a_3=220+20=240$
$a_4=240+20=260$
$a_5=260+20=280$
$a_6=280+20=300$
The first six terms are 200, 220, 240, 260, 280, and 300.

9. $a_1=-7$
$a_2=-7+4=-3$
$a_3=-3+4=1$
$a_4=1+4=5$
$a_5=5+4=9$
$a_6=9+4=13$
The first six terms are –7, -3, 1, 5, 9, 13.

11. $a_1=300$
$a_2=300-90=210$
$a_3=210-90=120$
$a_4=120-90=30$
$a_5=30-90=-60$
$a_6=-60-90=-150$
The first six terms are 300, 210, 120, 30, -60, -150.

13. $a_1=\frac{5}{2}$
$a_2=\frac{5}{2}-\frac{1}{2}=\frac{4}{2}=2$
$a_3=\frac{4}{2}-\frac{1}{2}=\frac{3}{2}$
$a_4=\frac{3}{2}-\frac{1}{2}=\frac{2}{2}=1$
$a_5=1-\frac{1}{2}=\frac{1}{2}$
$a_6=\frac{1}{2}-\frac{1}{2}=0$
The first six terms are $\frac{5}{2},2,\frac{3}{2},1,\frac{1}{2},0.$

15. $a_1=-0.4$
$a_2=-0.4-1.6=-2$
$a_3=-2-1.6=-3.6$
$a_4=-3.6-1.6=-5.2$
$a_5=-5.2-1.6=-6.8$
$a_6=-6.8-1.6=-8.4$
The first six terms are $-0.4,-2,-3.6,-5.2,-6.8,-8.4.$

17. $a_6=13+(6-1)4=13+(5)4$
$=13+20=33$

19. $a_{50}=7+(50-1)5=7+(49)5$
$=7+245=252$

21. $a_{200}=-40+(200-1)5=-40+(199)5$
$=-40+995=955$

23. $a_{60}=35+(60-1)(-3)=35+(59)(-3)$
$=35+(-177)=-142$

25. $a_n=a_1+(n-1)d=1+(n-1)4$
$=1+4n-4=4n-3$
$a_{20}=4(20)-3=80-3=77$

27. $a_n=a_1+(n-1)d=7+(n-1)(-4)$
$=7-4n+4=11-4n$

$a_{20} = 11 - 4(20) = 11 - 80 = -69$

29. $a_n = a_1 + (n-1)d = -20 + (n-1)(-4)$
$= -20 - 4n + 4 = -4n - 16$
$a_{20} = -4(20) - 16 = -80 - 16 = -96$

31. $a_n = a_1 + (n-1)d = -\frac{1}{3} + (n-1)\left(\frac{1}{3}\right)$
$= -\frac{1}{3} + \frac{1}{3}n - \frac{1}{3} = \frac{1}{3}n - \frac{2}{3}$
$a_{20} = \frac{1}{3}(20) - \frac{2}{3} = \frac{20}{3} - \frac{2}{3} = \frac{18}{3} = 6$

33. $a_n = a_1 + (n-1)d = 4 + (n-1)(-0.3)$
$= 4 - 0.3n + 0.3 = 4.3 - 0.3n$
$a_{20} = 4.3 - 0.3(20) = 4.3 - 6 = -1.7$

35. First find a_{20}.
$a_{20} = 4 + (20-1)6 = 4 + (19)6$
$= 4 + 114 = 118$
$S_{20} = \frac{20}{2}(4 + 118) = 10(122) = 1220$

37. First find a_{50}.
$a_{50} = -10 + (50-1)4 = -10 + (49)4$
$= -10 + 196 = 186$
$S_{50} = \frac{50}{2}(-10 + 186) = 25(176) = 4400$

39. First find a_{100}.
$a_{100} = 1 + (100-1)1 = 1 + (99)1$
$= 1 + 99 = 100$
$S_{100} = \frac{100}{2}(1 + 100) = 50(101) = 5050$

41. First find a_{60}.
$a_{60} = 2 + (60-1)2 = 2 + (59)2$
$= 2 + 118 = 120$
$S_{60} = \frac{60}{2}(2 + 120) = 30(122) = 3660$

43. Since there are 12 even integers between 21 and 45, find a_{12}.
$a_{12} = 22 + (12-1)2 = 22 + (11)2$
$= 22 + 22 = 44$
$S_{12} = \frac{12}{2}(22 + 44) = 6(66) = 396$

45. $\sum_{i=1}^{17}(5i+3) = (5(1)+3) + (5(2)+3) + (5(3)+3) + \ldots + (5(17)+3)$
$= (5+3) + (10+3) + (15+3) + \ldots + (85+3) = 8 + 13 + 18 + \ldots + 88$
$S_{17} = \frac{17}{2}(8 + 88) = \frac{17}{2}(96) = 17(48) = 816$

47. $\sum_{i=1}^{30}(-3i+5) = (-3(1)+5) + (-3(2)+5) + (-3(3)+5) + \ldots + (-3(30)+5)$
$= (-3+5) + (-6+5) + (-9+5) + \ldots + (-90+5) = 2 + (-1) + (-4) + \ldots + (-85)$

$$S_{30} = \frac{30}{2}(2 + (-85)) = 15(-83) = -1245$$

49. $$\sum_{i=1}^{100} 4i = 4(1) + 4(2) + 4(3) + \ldots + 4(100) = 4 + 8 + 12 + \ldots + 400$$

$$S_{100} = \frac{100}{2}(4 + 400) = 50(404) = 20,200$$

Application Exercises

51. **a.** $$\begin{aligned} a_n &= a_1 + (n-1)d \\ &= 126,424 + (n-1)(1265) \\ &= 126,424 + 1265n - 1265 \\ &= 1265n + 125159 \end{aligned}$$

b. $$\begin{aligned} a_{16} &= 1265(16) + 125159 \\ &= 20240 + 125159 = 145399 \end{aligned}$$

There will be 145,399 thousand, or 145,399,000 employees in the United States in 2005.

53. Company A

$$\begin{aligned} a_n &= 24000 + (n-1)1600 \\ &= 24000 + 1600n - 1600 \\ &= 1600n + 22400 \end{aligned}$$

$$\begin{aligned} a_{10} &= 1600(10) + 22400 \\ &= 16000 + 22400 = 38400 \end{aligned}$$

Company B

$$\begin{aligned} a_n &= 28000 + (n-1)1000 \\ &= 28000 + 1000n - 1000 \\ &= 1000n + 27000 \end{aligned}$$

$$\begin{aligned} a_{10} &= 1000(10) + 27000 \\ &= 10000 + 27000 = 37000 \end{aligned}$$

Company A will pay \$1400 more in year 10.

55. **a.** $$\begin{aligned} a_n &= 3.78 + (n-1)0.576 \\ &= 3.78 + 0.576n - 0.576 \\ &= 0.576n + 3.204 \end{aligned}$$

b. $$\begin{aligned} a_{41} &= 0.576(41) + 3.204 \\ &= 23.616 + 3.204 = 26.82 \end{aligned}$$

$$\begin{aligned} S_{40} &= \frac{41}{2}(3.78 + 26.82) \\ &= 20.5(30.6) = 627.3 \end{aligned}$$

The total amount of solid waste recovered from 1960 to 2000 is 627.3 million tons.

57. $$\begin{aligned} a_{10} &= 33000 + (10-1)2500 \\ &= 33000 + (9)2500 \\ &= 33000 + 22500 = 55500 \end{aligned}$$

$$\begin{aligned} S_{10} &= \frac{10}{2}(33000 + 55500) \\ &= 5(88500) = 442500 \end{aligned}$$

The total salary over a ten-year period is \$442,500.

59. $$\begin{aligned} a_{26} &= 30 + (26-1)2 = 30 + (25)2 \\ &= 30 + 50 = 80 \end{aligned}$$

$S_{26} = \frac{26}{2}(30+80) = 13(110)$
$= 1430$

There are 1430 seats in the theater.

Writing in Mathematics

61. Answers will vary.

63. Answers will vary.

65. Answers will vary.

Technology Exercises

67. Answers will vary. For example, consider Exercise 45.

$\sum_{i=1}^{17}(5i+3)$

```
sum(seq(5I+3,I,1
,17))
                816
```

This is the same result obtained in Exercise 45.

Critical Thinking Exercises

69. From the sequence, we see that
$a_1 = 21700$ and
$d = 23172 - 21700 = 1472.$
We know that $a_n = a_1 + (n-1)d$. We can substitute what we know to find n.

$$314628 = 21700 + (n-1)1472$$
$$292928 = (n-1)1472$$
$$\frac{292928}{1472} = \frac{(n-1)1472}{1472}$$
$$199 = n-1$$
$$200 = n$$

314,628 is the 200th term of the sequence.

71. $1+3+5+\ldots+(2n-1)$

$$S_n = \frac{n}{2}(a_1 + a_n) = \frac{n}{2}(1+(2n-1))$$
$$= \frac{n}{2}(1+2n-1) = \frac{n}{2}(2n)$$
$$= n(n) = n^2$$

Review Exercises

72. $\log(x^2-5) - \log(x+5) = 3$

$$\log\left(\frac{x^2-5}{x+5}\right) = 3$$
$$\frac{x^2-5}{x+5} = 10^3$$
$$\frac{x^2-5}{x+5} = 1000$$
$$x^2 - 5 = 1000(x+5)$$

$$x^2 - 5 = 1000x + 5000$$
$$x^2 - 1000x - 5005 = 0$$
$$a = 1 \quad b = -1000 \quad c = -5005$$

Solve using the quadratic equation.

$$x = \frac{-(-1000) \pm \sqrt{(-1000)^2 - 4(1)(-5005)}}{2(1)} = \frac{1000 \pm \sqrt{1000000 + 20020}}{2}$$

$$= \frac{1000 \pm \sqrt{1020020}}{2} = \frac{1000 \pm \sqrt{4 \cdot 255005}}{2} = \frac{1000 \pm 2\sqrt{255005}}{2} = 500 \pm \sqrt{255005}$$

We disregard $500 - \sqrt{255005}$ because we cannot take the log of a negative number. The solution is $500 + \sqrt{255005}$ or approximately 1005 and the solution set is $\{500 + \sqrt{255005}\}$.

73. $x^2 + 3x \le 10$

Solve the related quadratic equation.

$$x^2 + 3x - 10 = 0$$
$$(x+5)(x-2) = 0$$

Apply the zero product principle.

$x + 5 = 0$ or $x - 2 = 0$

$x = -5$ $\quad x = 2$

The boundary points are -5 and 2.

Test Interval	Test Number	Test	Conclusion
$(-\infty, -5]$	-6	$(-6)^2 + 3(-6) \le 10$ $18 \le 10$, false	$(-\infty, -5]$ does not belong to the solution set.
$[-5, 2]$	0	$0^2 + 3(0) \le 10$ $0 \le 10$, true	$[-5, 2]$ belongs to the solution set.
$[2, \infty)$	3	$3^2 + 3(3) \le 10$ $18 \le 10$, false	$[2, \infty)$ does not belong to the solution set.

The solution set is $[-5, 2]$ or $\{x \mid -5 \le x \le 2\}$.

-5 2

74.

$$A = \frac{Pt}{P+t}$$
$$(P+t)(A) = (P+t)\left(\frac{Pt}{P+t}\right)$$
$$AP + At = Pt$$
$$AP - Pt = -At$$
$$P(A-t) = -At$$
$$\frac{P(A-t)}{A-t} = -\frac{At}{A-t}$$
$$P = -\frac{At}{A-t} \text{ or } \frac{At}{t-A}$$

Check Points 11.3

1. The first term is 12. The second term is $12 \cdot \frac{1}{2} = 6$. The third term is $6 \cdot \frac{1}{2} = 3$. The fourth term is $3 \cdot \frac{1}{2} = \frac{3}{2}$. The fifth term is $\frac{3}{2} \cdot \frac{1}{2} = \frac{3}{4}$. The sixth term is $\frac{3}{4} \cdot \frac{1}{2} = \frac{3}{8}$.

2.
$$a_n = a_1 r^{n-1}$$
$$a_7 = 5(-3)^{7-1} = 5(-3)^6 = 5(729)$$
$$= 5(729)3645$$

3.
$$a_n = a_1 r^{n-1} = 3(2)^{n-1}$$
$$a_8 = 3(2)^{8-1} = 3(2)^7 = 3(128) = 384$$

4.
$$S_n = \frac{2(1-(-3)^n)}{1-(-3)} = \frac{2(1-(-3)^n)}{4}$$
$$= \frac{1-(-3)^n}{2}$$
$$S_9 = \frac{1-(-3)^9}{2} = \frac{1-(-19683)}{2}$$
$$= \frac{19684}{2} = 9842$$

5.
$$\sum_{i=1}^{8} 2 \cdot 3^i$$
$$S_n = \frac{a_1(1-r^n)}{1-r} = \frac{6(1-3^n)}{1-3}$$
$$= \frac{6(1-3^n)}{-2} = -3(1-3^n)$$
$$S_8 = -3(1-3^8) = -3(1-6561)$$
$$= -3(-6560) = -3(-6560) = 19680$$

6. We have $a_1 = 30,000$. We also know that over the next 29 years, the salary will increase by 6% each year. The first year, the salary is \$30,000. The second year, the salary is \$30,000 + 0.06(\$30,000) or 1.06(\$30,000). Extending this, we find that $r = 1.06$.

$$S_{30} = \frac{30000\left(1-(1.06)^{30}\right)}{1-1.06}$$
$$= \frac{30000(1-5.74349)}{-0.06}$$
$$= \frac{30000(-4.74349)}{-0.06}$$
$$= \frac{-142304.7}{-0.06} = 2,371,745$$

The total lifetime salary over the 30 years is \$2,371,745.

7.
$$A = P\frac{\left(1+\frac{r}{n}\right)^{nt} - 1}{\frac{r}{n}}$$

$$= 3000\frac{\left(1+\frac{0.10}{1}\right)^{1\cdot 40}-1}{\frac{.10}{1}}$$

$$= 3000\frac{(1.10)^{40}-1}{0.10} = 1,327,777.67$$

After 40 years, the value of the IRA is approximately $1,327,777.67.

8. $r = \frac{a_2}{a_1} = \frac{2}{3}$

$$S = \frac{a_1}{1-r} = \frac{3}{1-\frac{2}{3}} = \frac{3}{\frac{1}{3}} = 3 \div \frac{1}{3} = 3 \cdot \frac{3}{1} = 9$$

9. $0.\overline{9} = 0.9999....$

$$= \frac{9}{10} + \frac{9}{100} + \frac{9}{1000} + \frac{9}{10000} + ...$$

$$0.\overline{9} = \frac{a_1}{1-r} = \frac{\frac{9}{10}}{1-\frac{1}{10}} = \frac{\frac{9}{10}}{\frac{9}{10}} = 1$$

10. The common ratio is 80% or 0.8. We assume that each person will spend $0.8 \cdot 1000 = 800$.

$$S = \frac{a_1}{1-r} = \frac{800}{1-0.8} = \frac{800}{0.2} = 4000$$

Problem Set 11.3

Practice Exercises

1. $r = \frac{a_2}{a_1} = \frac{15}{5} = 3$

3. $r = \frac{a_2}{a_1} = \frac{30}{-15} = -2$

5. $r = \frac{a_2}{a_1} = \frac{\frac{9}{2}}{3} = \frac{9}{2} \div 3 = \frac{9}{2} \cdot \frac{1}{3} = \frac{3}{2}$

7. $r = \frac{a_2}{a_1} = \frac{-0.4}{4} = -0.1$

9. The first term is 2. The second term is $2 \cdot 3 = 6$. The third term is $6 \cdot 3 = 18$. The fourth term is $18 \cdot 3 = 54$. The fifth term is $54 \cdot 3 = 162$.

11. The first term is 20. The second term is $20 \cdot \frac{1}{2} = 10$. The third term is $10 \cdot \frac{1}{2} = 5$. The fourth term is $5 \cdot \frac{1}{2} = \frac{5}{2}$. The fifth term is $\frac{5}{2} \cdot \frac{1}{2} = \frac{5}{4}$.

13. The first term is –4. The second term is $-4(-10) = 40$. The third term is $40(-10) = -400$. The fourth term is $-400(-10) = 4000$. The fifth term is $4000(-10) = -40,000$.

15. The first term is $-\frac{1}{4}$. The second term is $-\frac{1}{4}(-2) = \frac{1}{2}$. The third term is $\frac{1}{2}(-2) = -1$. The fourth term is $-1(-2) = 2$. The fifth term is $2(-2) = -4$.

17. $a_8 = 6(2)^{8-1} = 6(2)^7 = 6(128) = 768$

19. $a_{12} = 5(-2)^{12-1} = 5(-2)^{11}$
$= 5(-2048) = -10240$

21. $a_6 = 6400\left(-\frac{1}{2}\right)^{6-1} = 6400\left(-\frac{1}{2}\right)^5$
$= -200$

23. $a_8 = 1,000,000(0.1)^{8-1}$
$= 1,000,000(0.1)^7$
$= 1,000,000(0.0000001) = 0.1$

25. $r = \frac{a_2}{a_1} = \frac{12}{3} = 4$
$a_n = a_1 r^{n-1} = 3(4)^{n-1}$
$a_7 = 3(4)^{7-1} = 3(4)^6$
$= 3(4096) = 12,288$

27. $r = \frac{a_2}{a_1} = \frac{6}{18} = \frac{1}{3}$

$a_n = a_1 r^{n-1} = 18\left(\frac{1}{3}\right)^{n-1}$

$a_7 = 18\left(\frac{1}{3}\right)^{7-1} = 18\left(\frac{1}{3}\right)^6$

$= 18\left(\frac{1}{729}\right) = \frac{18}{729} = \frac{2}{81}$

29. $r = \frac{a_2}{a_1} = \frac{-3}{1.5} = -2$
$a_n = a_1 r^{n-1} = 1.5(-2)^{n-1}$
$a_7 = 1.5(-2)^{7-1} = 1.5(-2)^6$
$= 1.5(64) = 96$

31. $r = \frac{a_2}{a_1} = \frac{-0.004}{0.0004} = -10$
$a_n = a_1 r^{n-1} = 0.0004(-10)^{n-1}$
$a_7 = 0.0004(-10)^{7-1} = 0.0004(-10)^6$
$= 0.0004(1000000) = 400$

33. $r = \frac{a_2}{a_1} = \frac{6}{2} = 3$

$S_{12} = \frac{2(1-3^{12})}{1-3} = \frac{2(1-531441)}{-2}$
$= \frac{2(-531440)}{-2} = \frac{-1,062,880}{-2}$
$= 531,441$

35. $r = \frac{a_2}{a_1} = \frac{-6}{3} = -2$

$S_{11} = \frac{a_1(1-r^n)}{1-r} = \frac{3(1-(-2)^{11})}{1-(-2)}$

$= \frac{\cancel{3}(1-(-2048))}{\cancel{3}} = 2049$

37. $r = \frac{a_2}{a_1} = \frac{3}{-\frac{3}{2}} = 3 \div \left(-\frac{3}{2}\right)$

$= 3 \cdot \left(-\frac{2}{3}\right) = -2$

$S_{14} = \frac{a_1(1-r^n)}{1-r} = \frac{-\frac{3}{2}(1-(-2)^{14})}{1-(-2)}$

$= \frac{-\frac{3}{2}(1-(-16384))}{3} = \frac{-\frac{3}{2}(16385)}{3}$

$$= -\frac{3}{2}(16385) \div 3 = -\frac{49155}{2} \cdot \frac{1}{3}$$
$$= -\frac{16385}{2}$$

39. $$\sum_{i=1}^{8} 3^i = \frac{3(1-3^8)}{1-3} = \frac{3(1-6561)}{-2}$$
$$= \frac{3(-6560)}{-2} = \frac{-19680}{-2} = 9840$$

41. $$\sum_{i=1}^{10} 5 \cdot 2^i = \frac{10(1-2^{10})}{1-2} = \frac{10(1-1024)}{-1}$$
$$= \frac{10(-1023)}{-1} = 10,230$$

43. $$\sum_{i=1}^{6} \left(\frac{1}{2}\right)^{i+1} = \frac{\frac{1}{4}\left(1-\left(\frac{1}{2}\right)^6\right)}{1-\frac{1}{2}} = \frac{\frac{1}{4}\left(1-\frac{1}{64}\right)}{\frac{1}{2}}$$
$$= \frac{\frac{1}{4}\left(\frac{64}{64}-\frac{1}{64}\right)}{\frac{1}{2}} = \frac{\frac{1}{4}\left(\frac{63}{64}\right)}{\frac{1}{2}}$$
$$= \frac{1}{4}\left(\frac{63}{64}\right) \div \frac{1}{2} = \frac{1}{4}\left(\frac{63}{64}\right) \cdot \frac{2}{1}$$
$$= \frac{63}{128}$$

45. $$r = \frac{a_2}{a_1} = \frac{\frac{1}{3}}{1} = \frac{1}{3}$$
$$S = \frac{a_1}{1-r} = \frac{1}{1-\frac{1}{3}} = \frac{1}{\frac{2}{3}} = 1 \div \frac{2}{3} = 1 \cdot \frac{3}{2} = \frac{3}{2}$$

47. $$r = \frac{a_2}{a_1} = \frac{\frac{3}{4}}{3} = \frac{3}{4} \div 3 = \frac{3}{4} \cdot \frac{1}{3} = \frac{1}{4}$$
$$S = \frac{a_1}{1-r} = \frac{3}{1-\frac{1}{4}} = \frac{3}{\frac{3}{4}} = 3 \div \frac{3}{4}$$
$$= 3 \cdot \frac{4}{3} = \frac{12}{3} = 4$$

49. $$r = \frac{a_2}{a_1} = \frac{-\frac{1}{2}}{1} = -\frac{1}{2}$$
$$S = \frac{a_1}{1-r} = \frac{1}{1-\left(-\frac{1}{2}\right)} = \frac{1}{\frac{3}{2}} = 1 \div \frac{3}{2}$$
$$= 1 \cdot \frac{2}{3} = \frac{2}{3}$$

51. $r = -0.3$
$$a_1 = 26(-0.3)^{1-1} = 26(-0.3)^0$$
$$= 26(1) = 26$$
$$S = \frac{26}{1-(-0.3)} = \frac{26}{1.3} = 20$$

53. $$0.\overline{5} = \frac{a_1}{1-r} = \frac{\frac{5}{10}}{1-\frac{1}{10}} = \frac{\frac{5}{10}}{\frac{9}{10}} = \frac{5}{10} \div \frac{9}{10}$$
$$= \frac{5}{10} \cdot \frac{10}{9} = \frac{5}{9}$$

55. $$0.\overline{47} = \frac{a_1}{1-r} = \frac{\frac{47}{100}}{1-\frac{1}{100}} = \frac{\frac{47}{100}}{\frac{99}{100}}$$
$$= \frac{47}{100} \div \frac{99}{100} = \frac{47}{100} \cdot \frac{100}{99} = \frac{47}{99}$$

57.

$$0.\overline{257} = \frac{a_1}{1-r} = \frac{\frac{257}{1000}}{1-\frac{1}{1000}} = \frac{\frac{257}{1000}}{\frac{999}{1000}}$$

$$= \frac{257}{1000} \div \frac{999}{1000} = \frac{257}{1000} \cdot \frac{1000}{999}$$

$$= \frac{257}{999}$$

59. The sequence is arithmetic with common difference $d = 5$.

61. The sequence is geometric with common ratio $r = 2$.

63. The sequence is neither arithmetic nor geometric.

Application Exercises

65.

$$r = \frac{a_2}{a_1} = \frac{2}{1} = 2$$

$$a_{15} = 1(2)^{15-1} = (2)^{14} = 16384$$

On the fifteenth day, you will put aside $16,384 for savings.

67. $r = 1.04$

$$a_7 = 3{,}000{,}000(1.04)^{7-1}$$
$$= 3{,}000{,}000(1.04)^6$$
$$= 3{,}000{,}000(1.265319)$$
$$= 3{,}795{,}957$$

The athlete's salary for year 7 will be $3,795,957.

69. **a.**

$$r_{1995 \text{ to } 1996} = \frac{21.36}{20.60} = 1.03689$$
$$\approx 1.04$$

$$r_{1996 \text{ to } 1997} = \frac{22.19}{21.36} = 1.03886$$
$$\approx 1.04$$

$$r_{1997 \text{ to } 1998} = \frac{23.02}{22.19} = 1.0374$$
$$\approx 1.04$$

b. $a_n = a_1 r^{n-1} = 20.60(1.04)^{n-1}$

c. Since 2005 – 1994 = 11, find a_{11}.

$$a_{11} = 20.60(1.04)^{11-1}$$
$$= 20.60(1.04)^{10} = 30.493$$

The population of Iraq will be 30.493 million in 2005.

71.

$$r = \frac{a_2}{a_1} = \frac{2}{1} = 2$$

$$S_{15} = \frac{a_1(1-r^n)}{1-r} = \frac{1(1-(2)^{15})}{1-2}$$
$$= \frac{(1-32768)}{-1} = \frac{(-32767)}{-1} = 32767$$

Your savings will be $32767 over the 15 days.

73. $r = 1.05$

$$S_{20} = \frac{a_1(1-r^n)}{1-r} = \frac{24000(1-(1.05)^{20})}{1-1.05}$$
$$= \frac{24000(1-2.6533)}{-0.05}$$
$$= \frac{24000(-1.6533)}{-0.05} = 793583$$

The total lifetime salary over the 20 years is $793,583.

75. $r = 0.9$

$$S_{10} = \frac{a_1(1-r^n)}{1-r} = \frac{20(1-(0.9)^{10})}{1-0.9}$$
$$= \frac{20(1-0.348678)}{0.1}$$
$$= \frac{20(0.651322)}{0.1} = 130.264$$

After 10 swings, the pendulum covers a distance of approximately 130 inches.

77.

$$A = P\frac{\left(1+\frac{r}{n}\right)^{nt} - 1}{\frac{r}{n}}$$
$$= 2500\frac{\left(1+\frac{0.09}{1}\right)^{1(40)} - 1}{\frac{0.09}{1}}$$
$$= 2500\frac{(1+0.09)^{40} - 1}{0.09}$$
$$= 2500\frac{(1.09)^{40} - 1}{0.09}$$
$$= 2500\frac{31.4094 - 1}{0.09}$$
$$= 2500\frac{30.4094}{0.09} = 844706$$

After 40 years, the value of the IRA will be \$844,706.

79.

$$A = P\frac{\left(1+\frac{r}{n}\right)^{nt} - 1}{\frac{r}{n}}$$
$$= 600\frac{\left(1+\frac{0.08}{4}\right)^{4(18)} - 1}{\frac{0.08}{4}}$$
$$= 600\frac{(1+0.02)^{72} - 1}{0.02}$$
$$= 600\frac{(1.02)^{72} - 1}{0.02}$$
$$= 600\frac{4.16114 - 1}{0.02} = 600\frac{3.16114}{0.02}$$
$$= 94834.2$$

The value of the TSA after 18 years will be \$94834.20.

81. $r = 60\% = 0.6$

$a_1 = 6(.6) = 3.6$

$$S = \frac{3.6}{1-0.6} = \frac{3.6}{0.4} = 9$$

The total economic impact of the factory will be \$9 million per year.

83. $r = \frac{1}{4}$

$$S = \frac{\frac{1}{4}}{1-\frac{1}{4}} = \frac{\frac{1}{4}}{\frac{3}{4}} = \frac{1}{4} \div \frac{3}{4} = \frac{1}{4} \cdot \frac{4}{3} = \frac{1}{3}$$

Eventually $\frac{1}{3}$ of the largest square will be shaded.

Writing in Mathematics

85. Answers will vary.

87. Answers will vary.

89. Answers will vary.

91. Answers will vary.

Technology Exercises

93. Answers will vary. For example, consider Exercise 25.

$a_n = 3(4)^{n-1}$

```
seq(3(4)^(N-1),N
,7,7)
                {12288}
```

This matches the result obtained in Exercise 25.

95.

$$f(x) = \frac{2\left[1-\left(\frac{1}{3}\right)^x\right]}{1-\frac{1}{3}}$$

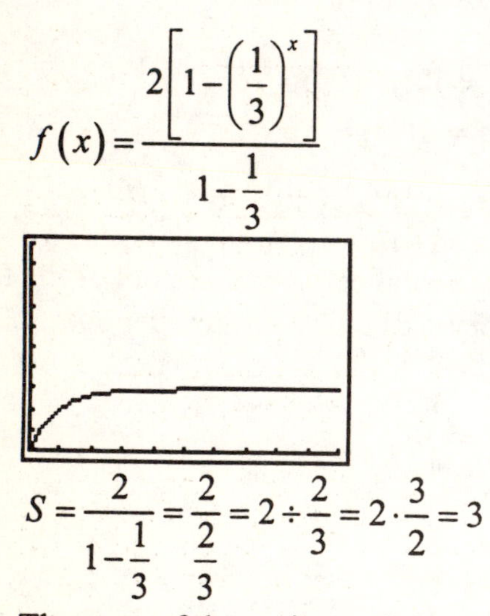

$$S = \frac{2}{1-\frac{1}{3}} = \frac{2}{\frac{2}{3}} = 2 \div \frac{2}{3} = 2 \cdot \frac{3}{2} = 3$$

The sum of the series and the asymptote of the function are both 3.

Critical Thinking Exercises

97. Statement **d.** is true. The common ratio is $0.5 = \frac{1}{2}$.

Statement **a.** is false. The sequence is not geometric. The fourth term would have to be $24 \cdot 4 = 96$ for the sequence to be geometric.

Statement **b.** is false. We do not need to know the terms between $\frac{1}{8}$ and $\frac{1}{512}$, but we do need to know how many terms there are between $\frac{1}{8}$ and $\frac{1}{512}$.

Statement **c.** is false. The sum of the sequence is $\frac{10}{1-\left(-\frac{1}{2}\right)}$.

99.

$$A = P\frac{\left(1+\frac{r}{n}\right)^{nt} - 1}{\frac{r}{n}}$$

$$1,000,000 = P\frac{\left(1+\frac{0.10}{12}\right)^{12(30)} - 1}{\frac{0.10}{12}}$$

$$1,000,000 = P\frac{\left(1+\frac{1}{120}\right)^{360} - 1}{\frac{1}{120}}$$

$$1,000,000 = P\frac{\left(1\frac{1}{120}\right)^{360} - 1}{\frac{1}{120}}$$

$$1,000,000 = P\frac{19.8374 - 1}{\frac{1}{120}}$$

$$\frac{1}{120}(1,000,000) = \frac{1}{120}\left(P\frac{18.8374}{\frac{1}{120}}\right)$$

$$\frac{25000}{3} = 18.8374P$$
$$\frac{25000}{3(18.8374)} = P$$
$$442.382 = P$$

You should deposit approximately \$442.38 per month.

Review Exercises

100. $\sqrt{28} - 3\sqrt{7} + \sqrt{63}$
$= \sqrt{4 \cdot 7} - 3\sqrt{7} + \sqrt{9 \cdot 7}$
$= 2\sqrt{7} - 3\sqrt{7} + 3\sqrt{7}$
$= 2\sqrt{7}$

101. $2x^2 = 4 - x$

$2x^2 + x - 4 = 0$

$a = 2 \quad b = 1 \quad c = -4$

Solve using the quadratic formula.

$$x = \frac{-1 \pm \sqrt{1^2 - 4(2)(-4)}}{2(2)}$$
$$= \frac{-1 \pm \sqrt{1+32}}{4} = \frac{-1 \pm \sqrt{33}}{4}$$

The solutions are $\frac{-1 \pm \sqrt{33}}{4}$ and the solution set is $\left\{\frac{-1 \pm \sqrt{33}}{4}\right\}$.

103.
$$\frac{6}{\sqrt{3}-\sqrt{5}} = \frac{6}{\sqrt{3}-\sqrt{5}} \cdot \frac{\sqrt{3}+\sqrt{5}}{\sqrt{3}+\sqrt{5}}$$
$$= \frac{6\left(\sqrt{3}+\sqrt{5}\right)}{3-5}$$
$$= \frac{6\left(\sqrt{3}+\sqrt{5}\right)}{-2}$$
$$= -3\left(\sqrt{3}+\sqrt{5}\right)$$

Check Points 11.4

1. a.
$$\binom{6}{3} = \frac{6!}{3!(6-3)!} = \frac{6!}{3!3!}$$
$$= \frac{\cancel{6} \cdot 5 \cdot 4 \cdot \cancel{3!}}{\cancel{3!}\,\cancel{3 \cdot 2} \cdot 1} = 20$$

b.
$$\binom{6}{0} = \frac{6!}{0!(6-0)!} = \frac{\cancel{6!}}{0!\cancel{6!}} = \frac{1}{1} = 1$$

c.
$$\binom{8}{2} = \frac{8!}{2!(8-2)!} = \frac{8!}{2!6!}$$
$$= \frac{8 \cdot 7 \cdot \cancel{6!}}{2 \cdot 1 \cdot \cancel{6!}} = 28$$

d.
$$\binom{3}{3} = \frac{3!}{3!(3-3)!} = \frac{\cancel{3!}}{\cancel{3!}0!} = \frac{1}{1}$$

2. Applying the Binomial Theorem to $(x+1)^4$, we have $a = x$, $b = 1$, and $n = 4$.

$$(x+1)^4 = \binom{4}{0}x^4 + \binom{4}{1}x^3 + \binom{4}{2}x^2 + \binom{4}{3}x + \binom{4}{4}$$
$$= \frac{4!}{0!(4-0)!}x^4 + \frac{4!}{1!(4-1)!}x^3 + \frac{4!}{2!(4-2)!}x^2 + \frac{4!}{3!(4-3)!}x + \frac{4!}{4!(4-4)!}$$

$$=\frac{\cancel{4!}}{0!\cancel{4!}}x^4+\frac{4!}{1!3!}x^3+\frac{4!}{2!2!}x^2+\frac{4!}{3!1!}x+\frac{\cancel{4!}}{\cancel{4!}0!}$$
$$=1x^4+\frac{4\cdot\cancel{3!}}{1\cdot\cancel{3!}}x^3+\frac{4\cdot3\cdot\cancel{2!}}{2\cdot1\cdot\cancel{2!}}x^2+\frac{4\cdot\cancel{3!}}{\cancel{3!}\cdot1}x+1$$
$$=x^4+4x^3+6x^2+4x+1$$

3. Applying the Binomial Theorem to $(2x-y)^5$, we have $a=2x,\ b=-y,$ and $n=5.$

$$(x-2y)^5$$
$$=\binom{5}{0}x^5+\binom{5}{1}x^4(-2y)+\binom{5}{2}x^3(-2y)^2+\binom{5}{3}x^2(-2y)^3+\binom{5}{4}x(-2y)^4+\binom{5}{5}(-2y)^5$$
$$=\frac{\cancel{5!}}{0!\cancel{5!}}x^5-\frac{5!}{1!(5-1)!}16x^4y+\frac{5!}{2!(5-2)!}8x^3y^2-\frac{5!}{3!(5-3)!}4x^2y^3+\frac{5!}{4!(5-4)!}2xy^4-\frac{5!}{5!(5-5)!}32y^5$$
$$=1x^5-\frac{5\cdot\cancel{4!}}{1\cdot\cancel{4!}}2x^4y+\frac{5\cdot4\cdot\cancel{3!}}{2\cdot1\cdot\cancel{3!}}4x^3y^2-\frac{5\cdot4\cdot\cancel{3!}}{\cancel{3!}2\cdot1}8x^2y^3+\frac{5\cdot\cancel{4!}}{\cancel{4!}\cdot1}16xy^4-\frac{\cancel{5!}}{\cancel{5!}0!}32y^5$$
$$=x^5-5(2x^4y)+10(4x^3y^2)-10(8x^2y^3)+5(16xy^4)-1(32y^5)$$
$$=x^5-10x^4y+40x^3y^2-80x^2y^3+80xy^4-32y^5$$

4. Applying the Binomial Theorem to find the fifth term of $(2x+y)^9$, we have $a=2x,\ b=y,$ and $n=9.$

$$\binom{n}{r-1}a^{n-r+1}b^{r-1}=\binom{9}{5-1}(2x)^{9-5+1}y^{5-1}=\binom{9}{4}(2x)^5y^4=\frac{9!}{4!(9-4)!}(2x)^5y^4$$
$$=\frac{9\cdot8\cdot7\cdot6\cdot\cancel{5!}}{4\cdot3\cdot2\cdot1\cdot\cancel{5!}}32x^5y^4=\frac{9\cdot8\cdot7\cdot\cancel{6}}{4\cdot\cancel{3}\cdot\cancel{2}\cdot1}32x^5y^4=126(32x^5y^4)=4032x^5y^4$$

Problem Set 11.4

Practice Exercises

1. $\binom{8}{3}=\frac{8!}{3!(8-3)!}$

$=\frac{8!}{3!5!}=\frac{8\cdot7\cdot\cancel{6}\cdot\cancel{5!}}{\cancel{3}\cdot\cancel{2}\cdot1\cdot\cancel{5!}}$

$=56$

3. $\binom{12}{1}=\frac{12!}{1!(12-1)!}=\frac{12\cdot\cancel{11!}}{1\cdot\cancel{11!}}=12$

5. $\binom{6}{6}=\frac{\cancel{6!}}{\cancel{6!}(6-6)!}=\frac{1}{0!}=\frac{1}{1}=1$

7. $\binom{100}{2}=\frac{100!}{2!(100-2)!}=\frac{100\cdot99\cdot\cancel{98!}}{2\cdot1\cdot\cancel{98!}}$

$=4950$

9. Applying the Binomial Theorem to $(x+2)^3$, we have $a = x$, $b = 2$, and $n = 3$.

$$(x+2)^3 = \binom{3}{0}x^3 + \binom{3}{1}x^2(2) + \binom{3}{2}x(2)^2 + \binom{3}{3}2^3$$

$$= \frac{3!}{0!(3-0)!}x^3 + \frac{3!}{1!(3-1)!}2x^2 + \frac{3!}{2!(3-2)!}4x + \frac{3!}{3!(3-3)!}8$$

$$= \frac{\cancel{3!}}{1\cdot\cancel{3!}}x^3 + \frac{3\cdot\cancel{2!}}{1\cdot\cancel{2!}}2x^2 + \frac{3\cdot\cancel{2!}}{\cancel{2!}1!}4x + \frac{\cancel{3!}}{\cancel{3!}0!}8 = x^3 + 3(2x^2) + 3(4x) + 1(8)$$

$$= x^3 + 6x^2 + 12x + 8$$

11. Applying the Binomial Theorem to $(3x+y)^3$, we have $a = 3x$, $b = y$, and $n = 3$.

$$(3x+y)^3 = \binom{3}{0}(3x)^3 + \binom{3}{1}(3x)^2 y + \binom{3}{2}(3x)y^2 + \binom{3}{3}y^3$$

$$= \frac{3!}{0!(3-0)!}27x^3 + \frac{3!}{1!(3-1)!}9x^2y + \frac{3!}{2!(3-2)!}3xy^2 + \frac{3!}{3!(3-3)!}y^3$$

$$= \frac{\cancel{3!}}{1\cdot\cancel{3!}}27x^3 + \frac{3\cdot\cancel{2!}}{1\cdot\cancel{2!}}9x^2y + \frac{3\cdot\cancel{2!}}{\cancel{2!}1!}3xy^2 + \frac{\cancel{3!}}{\cancel{3!}0!}y^3 = 27x^3 + 3(9x^2y) + 3(3xy^2) + 1(y^3)$$

$$= 27x^3 + 27x^2y + 9xy^2 + y^3$$

13. Applying the Binomial Theorem to $(5x-1)^3$, we have $a = 5x$, $b = -1$, and $n = 3$.

$$(5x-1)^3 = \binom{3}{0}(5x)^3 + \binom{3}{1}(5x)^2(-1) + \binom{3}{2}(5x)(-1)^2 + \binom{3}{3}(-1)^3$$

$$= \frac{3!}{0!(3-0)!}125x^3 - \frac{3!}{1!(3-1)!}25x^2 + \frac{3!}{2!(3-2)!}5x(1) - \frac{3!}{3!(3-3)!}$$

$$= \frac{\cancel{3!}}{1\cdot\cancel{3!}}125x^3 - \frac{3\cdot\cancel{2!}}{1\cdot\cancel{2!}}25x^2 + \frac{3\cdot\cancel{2!}}{\cancel{2!}1!}5x - \frac{\cancel{3!}}{\cancel{3!}0!} = 125x^3 - 3(25x^2) + 3(5x) - 1$$

$$= 125x^3 - 75x^2 + 15x - 1$$

15. Applying the Binomial Theorem to $(2x+1)^4$, we have $a = 2x$, $b = 1$, and $n = 4$.

$$(2x+1)^4 = \binom{4}{0}(2x)^4 + \binom{4}{1}(2x)^3 + \binom{4}{2}(2x)^2 + \binom{4}{3}2x + \binom{4}{4}$$

$$= \frac{4!}{0!(4-0)!}16x^4 + \frac{4!}{1!(4-1)!}8x^3\cdot 1 + \frac{4!}{2!(4-2)!}4x^2\cdot 1^2 + \frac{4!}{3!(4-3)!}2x\cdot 1^3 + \frac{4!}{4!(4-4)!}\cdot 1^4$$

$$=\frac{\cancel{4!}}{0!\cancel{4!}}16x^4+\frac{4!}{1!3!}8x^3\cdot 1+\frac{4!}{2!2!}4x^2\cdot 1+\frac{4!}{3!1!}2x\cdot 1+\frac{\cancel{4!}}{\cancel{4!}0!}\cdot 1$$

$$=1\left(16x^4\right)+\frac{4\cdot\cancel{3!}}{1\cdot\cancel{3!}}8x^3+\frac{4\cdot 3\cdot\cancel{2!}}{2\cdot 1\cdot\cancel{2!}}4x^2+\frac{4\cdot\cancel{3!}}{\cancel{3!}\cdot 1}2x+1=16x^4+4\left(8x^3\right)+6\left(4x^2\right)+4(2x)+1$$

$$=16x^4+32x^3+24x^2+8x+1$$

17. Applying the Binomial Theorem to $\left(x^2+2y\right)^4$, we have $a=x^2$, $b=2y$, and $n=4$.

$$\left(x^2+2y\right)^4=\binom{4}{0}\left(x^2\right)^4+\binom{4}{1}\left(x^2\right)^3(2y)+\binom{4}{2}\left(x^2\right)^2(2y)^2+\binom{4}{3}x^2(2y)^3+\binom{4}{4}(2y)^4$$

$$=\frac{4!}{0!(4-0)!}x^8+\frac{4!}{1!(4-1)!}2x^6y+\frac{4!}{2!(4-2)!}4x^4y^2+\frac{4!}{3!(4-3)!}8x^2y^3+\frac{4!}{4!(4-4)!}16y^4$$

$$=\frac{\cancel{4!}}{0!\cancel{4!}}x^8+\frac{4!}{1!3!}2x^6y+\frac{4!}{2!2!}4x^4y^2+\frac{4!}{3!1!}8x^2y^3+\frac{\cancel{4!}}{\cancel{4!}0!}16y^4$$

$$=1\left(x^8\right)+\frac{4\cdot\cancel{3!}}{1\cdot\cancel{3!}}2x^6y+\frac{4\cdot 3\cdot\cancel{2!}}{2\cdot 1\cdot\cancel{2!}}4x^4y^2+\frac{4\cdot\cancel{3!}}{\cancel{3!}\cdot 1}8x^2y^3+16y^4$$

$$=x^8+4\left(2x^6y\right)+6\left(4x^4y^2\right)+4\left(8x^2y^3\right)+16y^4=x^8+8x^6y+24x^4y^2+32x^2y^3+16y^4$$

19. Applying the Binomial Theorem to $(y-3)^4$, we have $a=y$, $b=-3$, and $n=4$.

$$(y-3)^4=\binom{4}{0}y^4+\binom{4}{1}y^3(-3)+\binom{4}{2}y^2(-3)^2+\binom{4}{3}y(-3)^3+\binom{4}{4}(-3)^4$$

$$=\frac{4!}{0!(4-0)!}y^4-\frac{4!}{1!(4-1)!}3y^3+\frac{4!}{2!(4-2)!}9y^2-\frac{4!}{3!(4-3)!}27y+\frac{4!}{4!(4-4)!}81$$

$$=\frac{\cancel{4!}}{0!\cancel{4!}}y^4-\frac{4!}{1!3!}3y^3+\frac{4!}{2!2!}9y^2-\frac{4!}{3!1!}27y+\frac{\cancel{4!}}{\cancel{4!}0!}81$$

$$=1\left(y^4\right)-\frac{4\cdot\cancel{3!}}{1\cdot\cancel{3!}}3y^3+\frac{4\cdot 3\cdot\cancel{2!}}{2\cdot 1\cdot\cancel{2!}}9y^2-\frac{4\cdot\cancel{3!}}{\cancel{3!}\cdot 1}27y+81$$

$$=y^4-4\left(3y^3\right)+6\left(9y^2\right)-4(27y)+81=y^4-12y^3+54y^2-108y+81$$

21. Applying the Binomial Theorem to $\left(2x^3-1\right)^4$, we have $a=2x^3$, $b=-1$, and $n=4$.

$$\left(2x^3-1\right)^4=\binom{4}{0}\left(2x^3\right)^4+\binom{4}{1}\left(2x^3\right)^3(-1)+\binom{4}{2}\left(2x^3\right)^2(-1)^2+\binom{4}{3}\left(2x^3\right)(-1)^3+\binom{4}{4}(-1)^4$$

$$=\frac{4!}{0!(4-0)!}16x^{12}-\frac{4!}{1!(4-1)!}8x^9+\frac{4!}{2!(4-2)!}4x^6-\frac{4!}{3!(4-3)!}2x^3+\frac{4!}{4!(4-4)!}$$

$$=\frac{\cancel{4!}}{0!\cancel{4!}}16x^{12}-\frac{4!}{1!3!}8x^9+\frac{4!}{2!2!}4x^6-\frac{4!}{3!1!}2x^3+\frac{\cancel{4!}}{\cancel{4!}0!}$$

$$=1\left(16x^{12}\right)-\frac{4\cdot\cancel{3!}}{1\cdot\cancel{3!}}8x^9+\frac{4\cdot 3\cdot\cancel{2!}}{2\cdot 1\cdot\cancel{2!}}4x^6-\frac{4\cdot\cancel{3!}}{\cancel{3!}\cdot 1}2x^3+1$$

$$=16x^{12}-4\left(8x^9\right)+6\left(4x^6\right)-4\left(2x^3\right)+1=16x^{12}-32x^9+24x^6-8x^3+1$$

23. Applying the Binomial Theorem to $(c+2)^5$, we have $a=c,\ b=2,$ and $n=5.$

$$(c+2)^5=\binom{5}{0}c^5+\binom{5}{1}c^4(2)+\binom{5}{2}c^3(2)^2+\binom{5}{3}c^2(2)^3+\binom{5}{4}c(2)^4+\binom{5}{5}2^5$$

$$=\frac{\cancel{5!}}{0!\cancel{5!}}c^5+\frac{5!}{1!(5-1)!}2c^4+\frac{5!}{2!(5-2)!}4c^3+\frac{5!}{3!(5-3)!}8c^2+\frac{5!}{4!(5-4)!}16c+\frac{5!}{5!(5-5)!}32$$

$$=1c^5+\frac{5\cdot\cancel{4!}}{1\cdot\cancel{4!}}2c^4+\frac{5\cdot 4\cdot\cancel{3!}}{2\cdot 1\cdot\cancel{3!}}4c^3+\frac{5\cdot 4\cdot\cancel{3!}}{\cancel{3!}2\cdot 1}8c^2+\frac{5\cdot\cancel{4!}}{\cancel{4!}\cdot 1}16c+\frac{\cancel{5!}}{\cancel{5!}0!}32$$

$$=c^5+5\left(2c^4\right)+10\left(4c^3\right)+10\left(8c^2\right)+5(16c)+1(32)=c^5+10c^4+40c^3+80c^2+80c+32$$

25. Applying the Binomial Theorem to $(x-1)^5$, we have $a=x,\ b=-1,$ and $n=5.$

$$(x-1)^5=\binom{5}{0}x^5+\binom{5}{1}x^4(-1)+\binom{5}{2}x^3(-1)^2+\binom{5}{3}x^2(-1)^3+\binom{5}{4}x(-1)^4+\binom{5}{5}(-1)^5$$

$$=\frac{\cancel{5!}}{0!\cancel{5!}}x^5-\frac{5!}{1!(5-1)!}x^4+\frac{5!}{2!(5-2)!}x^3-\frac{5!}{3!(5-3)!}x^2+\frac{5!}{4!(5-4)!}x-\frac{5!}{5!(5-5)!}$$

$$=1x^5-\frac{5\cdot\cancel{4!}}{1\cdot\cancel{4!}}x^4+\frac{5\cdot 4\cdot\cancel{3!}}{2\cdot 1\cdot\cancel{3!}}x^3-\frac{5\cdot 4\cdot\cancel{3!}}{\cancel{3!}2\cdot 1}x^2+\frac{5\cdot\cancel{4!}}{\cancel{4!}\cdot 1}x-\frac{\cancel{5!}}{\cancel{5!}0!}$$

$$=x^5-5x^4+10x^3-10x^2+5x-1$$

27. Applying the Binomial Theorem to $(x-2y)^5$, we have $a=x,\ b=-2y,$ and $n=5.$

$$(x-2y)^5$$

$$=\binom{5}{0}x^5+\binom{5}{1}x^4(-2y)+\binom{5}{2}x^3(-2y)^2+\binom{5}{3}x^2(-2y)^3+\binom{5}{4}x(-2y)^4+\binom{5}{5}(-2y)^5$$

$$= \frac{5!}{0!5!}x^5 - \frac{5!}{1!(5-1)!}2x^4y + \frac{5!}{2!(5-2)!}4x^3y^2$$

$$- \frac{5!}{3!(5-3)!}8x^2y^3 + \frac{5!}{4!(5-4)!}16xy^4 - \frac{5!}{5!(5-5)!}32y^5$$

$$= 1x^5 - \frac{5\cdot 4!}{1\cdot 4!}2x^4y + \frac{5\cdot 4\cdot 3!}{2\cdot 1\cdot 3!}4x^3y^2 - \frac{5\cdot 4\cdot 3!}{3!2\cdot 1}8x^2y^3 + \frac{5\cdot 4!}{4!\cdot 1}16xy^4 - \frac{5!}{5!0!}32y^5$$

$$= x^5 - 5\left(2x^4y\right) + 10\left(4x^3y^2\right) - 10\left(8x^2y^3\right) + 5\left(16xy^4\right) - 1\left(32y^5\right)$$

$$= x^5 - 10x^4y + 40x^3y^2 - 80x^2y^3 + 80xy^4 - 32y^5$$

29. Applying the Binomial Theorem to $(2a+b)^6$, we have $a = 2a$, $b = b$, and $n = 6$.

$$(2a+b)^6$$

$$= \binom{6}{0}(2a)^6 + \binom{6}{1}(2a)^5 b + \binom{6}{2}(2a)^4 b^2 + \binom{6}{3}(2a)^3 b^3 + \binom{6}{4}(2a)^2 b^4 + \binom{6}{5}2ab^5 + \binom{6}{6}b^6$$

$$= \frac{6!}{0!(6-0)!}64a^6 + \frac{6!}{1!(6-1)!}32a^5b + \frac{6!}{2!(6-2)!}16a^4b^2 + \frac{6!}{3!(6-3)!}8a^3b^3$$

$$+ \frac{6!}{4!(6-4)!}4a^2b^4 + \frac{6!}{5!(6-5)!}2ab^5 + \frac{6!}{6!(6-6)!}b^6$$

$$= \frac{6!}{1 6!}64a^6 + \frac{6\cdot 5!}{1\cdot 5!}32a^5b + \frac{6\cdot 5\cdot 4!}{2\cdot 1\cdot 4!}16a^4b^2 + \frac{6\cdot 5\cdot 4\cdot 3!}{3\cdot 2\cdot 1\cdot 3!}8a^3b^3$$

$$+ \frac{6\cdot 5\cdot 4!}{4!2\cdot 1}4a^2b^4 + \frac{6\cdot 5!}{5!1}2ab^5 + \frac{6!}{6!\cdot 1}b^6$$

$$= 64a^6 + 6\left(32a^5b\right) + 15\left(16a^4b^2\right) + 20\left(8a^3b^3\right) + 15\left(4a^2b^4\right) + 6\left(2ab^5\right) + 1b^6$$

$$= 64a^6 + 192a^5b + 240a^4b^2 + 160a^3b^3 + 60a^2b^4 + 12ab^5 + b^6$$

31. $(x+2)^8$

First Term $\binom{n}{r-1}a^{n-r+1}b^{r-1} = \binom{8}{1-1}x^{8-1+1}2^{1-1} = \binom{8}{0}x^8 2^0$

$$= \frac{8!}{0!(8-0)!}x^8 \cdot 1 = \frac{8!}{0!8!}x^8 = x^8$$

Second Term $\binom{n}{r-1}a^{n-r+1}b^{r-1} = \binom{8}{2-1}x^{8-2+1}2^{2-1} = \binom{8}{1}x^7 2^1 = \frac{8!}{1!(8-1)!}2x^7$

$$= \frac{8 \cdot \cancel{7!}}{1 \cdot \cancel{7!}}2x^7 = 8 \cdot 2x^7 = 16x^7$$

Third Term $\binom{n}{r-1}a^{n-r+1}b^{r-1} = \binom{8}{3-1}x^{8-3+1}2^{3-1} = \binom{8}{2}x^6 2^2 = \frac{8!}{2!(8-2)!}4x^6$

$$= \frac{8 \cdot 7 \cdot \cancel{6!}}{2 \cdot 1 \cdot \cancel{6!}}4x^6 = 28 \cdot 4x^6 = 112x^6$$

33. $(x-2y)^{10}$

First Term $\binom{n}{r-1}a^{n-r+1}b^{r-1} = \binom{10}{1-1}x^{10-1+1}(-2y)^{1-1} = \binom{10}{0}x^{10}(-2y)^0$

$$= \frac{10!}{0!(10-0)!}x^{10} \cdot 1 = \frac{\cancel{10!}}{0!\cancel{10!}}x^{10} = x^{10}$$

Second Term $\binom{n}{r-1}a^{n-r+1}b^{r-1} = \binom{10}{2-1}x^{10-2+1}(-2y)^{2-1} = \binom{10}{1}x^9(-2y)^1 = -\frac{10!}{1!(10-1)!}2x^9y$

$$= -\frac{10 \cdot \cancel{9!}}{1 \cdot \cancel{9!}}2x^9y = -10 \cdot 2x^9y = -20x^9y$$

Third Term $\binom{n}{r-1}a^{n-r+1}b^{r-1} = \binom{10}{3-1}x^{10-3+1}(-2y)^{3-1} = \binom{10}{2}x^8(-2y)^2 = \frac{10!}{2!(10-2)!}4x^8y^2$

$$= \frac{10 \cdot 9 \cdot \cancel{8!}}{2 \cdot 1 \cdot \cancel{8!}}4x^8y^2 = 45 \cdot 4x^8y^2 = 180x^8y^2$$

35. $(x^2+1)^{16}$

First Term $\binom{n}{r-1}a^{n-r+1}b^{r-1} = \binom{16}{1-1}(x^2)^{16-1+1}(1)^{1-1} = \binom{16}{0}(x^2)^{16}1^0$

$$= \frac{16!}{0!(16-0)!}x^{32} \cdot 1 = \frac{\cancel{16!}}{0!\cancel{16!}}x^{32} = x^{32}$$

Second Term $\binom{n}{r-1}a^{n-r+1}b^{r-1} = \binom{16}{2-1}(x^2)^{16-2+1}(1)^{2-1} = \binom{16}{1}(x^2)^{15}1^1$

$$= \frac{16!}{1!(16-1)!}x^{30} \cdot 1 = \frac{16 \cdot \cancel{15!}}{1 \cdot \cancel{15!}}x^{30} = 16x^{30}$$

Third Term $\binom{n}{r-1}a^{n-r+1}b^{r-1}=\binom{16}{3-1}\left(x^2\right)^{16-3+1}(1)^{3-1}=\binom{16}{2}\left(x^2\right)^{14}1^2$

$$=\frac{16!}{2!(16-2)!}x^{28}\cdot 1=\frac{16\cdot 15\cdot \cancel{14!}}{2\cdot 1\cdot \cancel{14!}}x^{28}=120x^{28}$$

37. $\left(y^3-1\right)^{20}$

First Term $\binom{n}{r-1}a^{n-r+1}b^{r-1}=\binom{20}{1-1}\left(y^3\right)^{20-1+1}(-1)^{1-1}=\binom{20}{0}\left(y^3\right)^{20}(-1)^0$

$$=\frac{20!}{0!(20-0)!}y^{60}\cdot 1=\frac{\cancel{20!}}{0!\cancel{20!}}y^{60}=x^{60}$$

Second Term $\binom{n}{r-1}a^{n-r+1}b^{r-1}=\binom{20}{2-1}\left(y^3\right)^{20-2+1}(-1)^{2-1}=\binom{20}{1}\left(y^3\right)^{19}(-1)^1$

$$=\frac{20!}{1!(20-1)!}y^{57}\cdot(-1)=-\frac{20\cdot \cancel{19!}}{1\cdot \cancel{19!}}y^{57}=-20y^{57}$$

Third Term $\binom{n}{r-1}a^{n-r+1}b^{r-1}=\binom{20}{3-1}\left(y^3\right)^{20-3+1}(-1)^{3-1}=\binom{20}{2}\left(y^3\right)^{18}(-1)^2$

$$=\frac{20!}{2!(20-2)!}y^{54}\cdot 1=\frac{20\cdot 19\cdot \cancel{18!}}{2\cdot 1\cdot \cancel{18!}}y^{54}=190y^{54}$$

39. $(2x+y)^6$

Third Term $\binom{n}{r-1}a^{n-r+1}b^{r-1}=\binom{6}{3-1}(2x)^{6-3+1}y^{3-1}=\binom{6}{2}(2x)^4y^2=\frac{6!}{2!(6-2)!}16x^4y^2$

$$=\frac{6\cdot 5\cdot \cancel{4!}}{2\cdot 1\cdot \cancel{4!}}16x^4y^2=15\left(16x^4y^2\right)=240x^4y^2$$

41. $(x-1)^9$

Fifth Term $\binom{n}{r-1}a^{n-r+1}b^{r-1}=\binom{9}{5-1}x^{9-5+1}(-1)^{5-1}=\binom{9}{4}x^5(-1)^4=\frac{9!}{4!(9-4)!}x^5\cdot 1$

$$=\frac{9\cdot 8\cdot 7\cdot \cancel{6}\cdot \cancel{5!}}{4\cdot \cancel{3}\cdot\cancel{2}\cdot 1\cdot \cancel{5!}}x^5=126x^5$$

43. $\left(x^2+y^3\right)^8$

Sixth Term $\binom{n}{r-1}a^{n-r+1}b^{r-1}=\binom{8}{6-1}\left(x^2\right)^{8-6+1}\left(y^3\right)^{6-1}=\binom{8}{5}\left(x^2\right)^3\left(y^3\right)^5=\frac{8!}{5!(8-5)!}x^6y^{15}$

$$=\frac{8\cdot 7\cdot \cancel{6}\cdot \cancel{5!}}{\cancel{5!}\cdot \cancel{3\cdot 2}\cdot 1}x^6y^{15}=56x^6y^{15}$$

45. $\left(x-\frac{1}{2}\right)^9$

Fourth Term $\binom{n}{r-1}a^{n-r+1}b^{r-1}=\binom{9}{4-1}x^{9-4+1}\left(-\frac{1}{2}\right)^{4-1}=\binom{9}{3}x^6\left(-\frac{1}{2}\right)^3=-\frac{9!}{3!(9-3)!}\cdot\frac{1}{8}x^6$

$$=-\frac{9\cdot \cancel{8}\cdot 7\cdot \cancel{6!}}{3\cdot 2\cdot 1\cdot \cancel{6!}}\cdot\frac{1}{\cancel{8}}x^6=-\frac{21}{2}x^6$$

Application Exercises

47. $g(t)=f(t+10)=0.002(t+10)^3-0.9(t+10)^2+1.27(t+10)+6.76$

Use the Binomial Theorem to expand $(t+10)^3$. (The Binomial Theorem could be used to expand $(t+10)^2$ also, but it can be multiplied more efficiently using the foil method.)

$$(t+10)^3=\binom{3}{0}t^3+\binom{3}{1}t^2(10)+\binom{3}{2}t(10)^2+\binom{3}{3}10^3$$

$$=\frac{3!}{0!(3-0)!}t^3+\frac{3!}{1!(3-1)!}10t^2+\frac{3!}{2!(3-2)!}100t+\frac{3!}{3!(3-3)!}1000$$

$$=\frac{\cancel{3!}}{1\cdot\cancel{3!}}t^3+\frac{3\cdot\cancel{2!}}{1\cdot\cancel{2!}}10t^2+\frac{3\cdot\cancel{2!}}{\cancel{2!}1!}100t+\frac{\cancel{3!}}{\cancel{3!}0!}1000$$

$$=t^3+3\left(10t^2\right)+3(100t)+1(1000)=t^3+30t^2+300t+1000$$

Now substitute the expanded expression.

$$g(t)=f(t+10)=0.002(t+10)^3-0.9(t+10)^2+1.27(t+10)+6.76$$

$$=0.002\left(t^3+30t^2+300t+1000\right)-0.9\left(t^2+20t+100\right)+1.27(t+10)+6.76$$

$$=0.002t^3+0.06t^2+0.6t+2-0.9t^2-18t-90+1.27t+12.7+6.76$$

$$=0.002t^3+0.06t^2-0.9t^2+0.6t-18t+1.27t+2-90+12.7+6.76$$

$$=0.002t^3-0.84t^2-16.13t-6854$$

Writing in Mathematics

49. Answers will vary.

51. Answers will vary.

53. Answers will vary.

55. Answers will vary.

57. Answers will vary.

Technology Exercises

59. $f_1(x) = (x+2)^3$

$f_2(x) = x^3$

$f_3(x) = x^3 + 6x^2$

$f_4(x) = x^3 + 6x^2 + 12x$

$f_5(x) = x^3 + 6x^2 + 12x + 8$

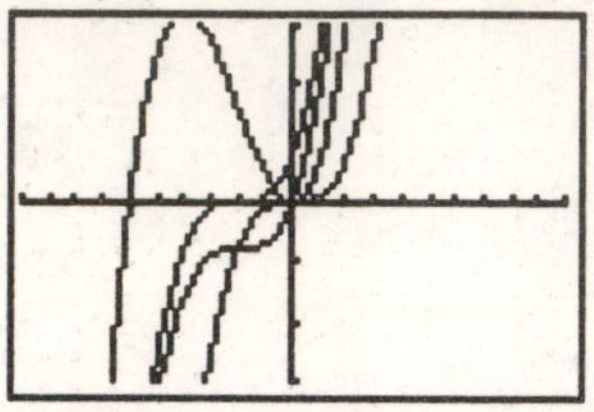

Graphs f_1 and f_5 are the same. This means that the functions are equivalent. Graphs f_2 through f_4 are increasingly similar to the graphs of f_1 and f_5.

61. Applying the Binomial Theorem to $(x-1)^3$, we have $a = x,\ b = -1,$ and $n = 3$.

$$(x-1)^3 = \binom{3}{0}x^3 + \binom{3}{1}x^2(-1) + \binom{3}{2}x(-1)^2 + \binom{3}{3}(-1)^3$$

$$= \frac{3!}{0!(3-0)!}x^3 - \frac{3!}{1!(3-1)!}x^2 + \frac{3!}{2!(3-2)!}x(1) - \frac{3!}{3!(3-3)!}$$

$$= \frac{\cancel{3!}}{1\cdot\cancel{3!}}x^3 - \frac{3\cdot\cancel{2!}}{1\cdot\cancel{2!}}x^2 + \frac{3\cdot\cancel{2!}}{\cancel{2!}1!}x - \frac{\cancel{3!}}{\cancel{3!}0!} = x^3 - 3x^2 + 3x - 1$$

Graph using the method from Exercises 59 and 60.

$f_1(x) = (x-1)^3$

$f_2(x) = x^3$

$f_3(x) = x^3 + 3x^2$

$f_4(x) = x^3 + 3x^2 + 3x$

$f_5(x) = x^3 - 3x^2 + 3x - 1$

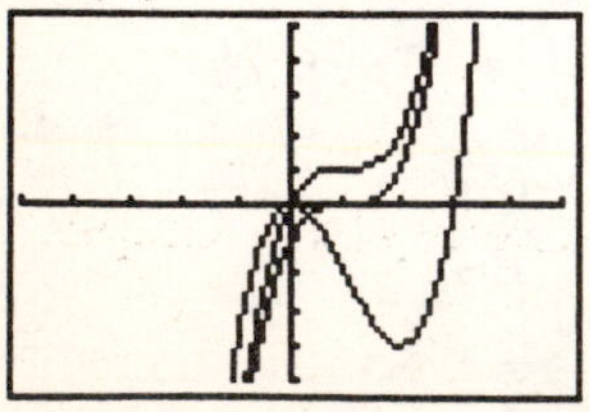

Graphs f_1 and f_5 are the same. This means that the functions are equivalent. Graphs f_2 through f_4 are increasingly similar to the graphs of f_1 and f_5.

63. Applying the Binomial Theorem to $(x+2)^6$, we have $a=x$, $b=2$, and $n=6$.

$$(x+2)^6=\binom{6}{0}x^6+\binom{6}{1}x^5 2+\binom{6}{2}x^4 2^2+\binom{6}{3}x^3 2^3+\binom{6}{4}x^2 2^4+\binom{6}{5}x 2^5+\binom{6}{6}2^6$$

$$=\frac{6!}{0!(6-0)!}x^6+\frac{6!}{1!(6-1)!}2x^5+\frac{6!}{2!(6-2)!}4x^4+\frac{6!}{3!(6-3)!}8x^3+\frac{6!}{4!(6-4)!}16x^2$$

$$+\frac{6!}{5!(6-5)!}32x+\frac{6!}{6!(6-6)!}64$$

$$=\frac{\cancel{6!}}{1\cancel{6!}}x^6+\frac{6\cdot\cancel{5!}}{1\cdot\cancel{5!}}2x^5+\frac{6\cdot 5\cdot\cancel{4!}}{2\cdot 1\cdot\cancel{4!}}4x^4+\frac{\cancel{6}\cdot 5\cdot 4\cdot\cancel{3!}}{\cancel{3\cdot 2}\cdot 1\cdot\cancel{3!}}8x^3+\frac{6\cdot 5\cdot\cancel{4!}}{\cancel{4!}2\cdot 1}16x^2+\frac{6\cdot\cancel{5!}}{\cancel{5!}1}32x+\frac{\cancel{6!}}{\cancel{6!}\cdot 1}64$$

$$=x^6+6\left(2x^5\right)+15\left(4x^4\right)+20\left(8x^3\right)+15\left(16x^2\right)+6(32x)+1(64)$$

$$=x^6+12x^5+60x^4+160x^3+240x^2+192x+64$$

Graph using the method from Exercises 59 and 60.

$f_1(x)=(x+2)^6$

$f_2(x)=x^6$

$f_3(x)=x^6+12x^5$

$f_4(x)=x^6+12x^5+60x^4$

$f_5(x)=x^6+12x^5+60x^4+160x^3$

$f_6(x)=x^6+12x^5+60x^4+160x^3+240x^2$

$f_7(x)=x^6+12x^5+60x^4+160x^3+240x^2+192x$

$f_8(x)=x^6+12x^5+60x^4+160x^3+240x^2+192x+64$

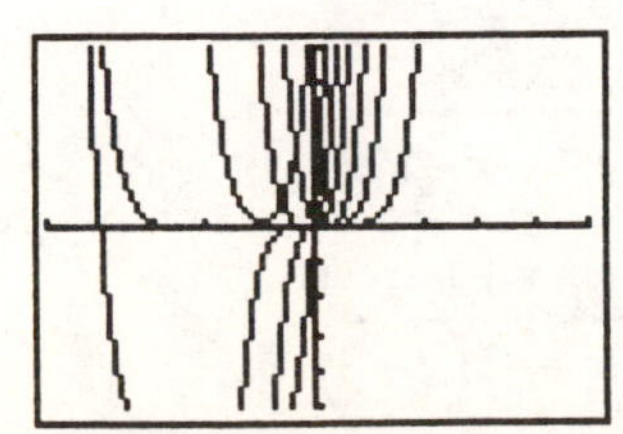

Graphs f_1 and f_8 are the same. This means that the functions are equivalent. Graphs f_2 through f_7 are increasingly similar to the graphs of f_1 and f_8.

Critical Thinking Exercises

65. Statement **b.** is true. The Binomial Theorem can be written in condensed form as $(a+b)^n = \sum_{r=0}^{n} \binom{n}{r} a^{n-r} b^r$

Statement **a.** is false. The binomial expansion for $(a+b)^n$ contains $n+1$ terms.

Statement **c.** is false. The sum of the binomial coefficients in $(a+b)^n$ is 2^n.

Statement **d.** is false. There are values of *a* and *b* for which $(a+b)^4 = a^4 + b^4$. Consider $a = 0$ and $b = 1$.

$$(0+1)^4 = 0^4 + 1^4$$
$$(1)^4 = 0+1$$
$$1 = 1$$

67. In $(x^2+y^2)^5$, the term containing x^4 is the term in which $a = x^2$ is squared. Applying the Binomial Theorem, the following pattern results. In the first term, x^2 is taken to the fifth power. In the second term, x^2 is taken to the fourth power. In the third term x^2 is taken to the third power. In the fourth term, x^2 is taken to the second power. This is the term we are looking for. Applying the Binomial Theorem to $(x^2+y^2)^5$, we have $a = x^2$, $b = y^2$, and $n = 5$. We are looking for the rth term where $r = 4$.

$$\binom{n}{r-1} a^{n-r+1} b^{r-1}$$
$$= \binom{5}{4-1} (x^2)^{5-4+1} (y^2)^{4-1}$$
$$= \binom{5}{3} (x^2)^2 (y^2)^3 = \frac{5!}{3!(5-3)!} x^4 y^6$$
$$= \frac{5!}{3!2!} x^4 y^6 = \frac{5 \cdot 4 \cdot \cancel{3!}}{\cancel{3!} 2 \cdot 1} x^4 y^6 = 10x^4 y^6$$

Review Exercises

68.
$$f(a+1) = (a+1)^2 + 2(a+1) + 3$$
$$= a^2 + 2a + 1 + 2a + 2 + 3$$
$$= a^2 + 4a + 6$$

69. $f(x) = x^2 + 5x \qquad g(x) = 2x - 3$

$$f(g(x)) = f(2x-3)$$
$$= (2x-3)^2 + 5(2x-3)$$
$$= 4x^2 - 12x + 9 + 10x - 15$$
$$= 4x^2 - 2x - 6$$
$$g(f(x)) = g(x^2 + 5x)$$
$$= 2(x^2 + 5x) - 3$$
$$= 2x^2 + 10x - 3$$

70.
$$\frac{x}{x+3} - \frac{x+1}{2x^2 - 2x - 24}$$
$$= \frac{x}{x+3} - \frac{x+1}{2(x^2 - x - 12)}$$
$$= \frac{x}{x+3} - \frac{x+1}{2(x-4)(x+3)}$$
$$= \frac{x}{x+3} \cdot \frac{2(x-4)}{2(x-4)} - \frac{x+1}{2(x-4)(x+3)}$$

$$=\frac{2x(x-4)}{2(x-4)(x+3)}-\frac{x+1}{2(x-4)(x+3)}$$
$$=\frac{2x(x-4)-(x+1)}{2(x-4)(x+3)}$$
$$=\frac{2x^2-8x-x-1}{2(x-4)(x+3)}$$
$$=\frac{2x^2-9x-1}{2(x-4)(x+3)}$$

Chapter 11 Review

11.1

1. $a_n = 7n-4$

$a_1 = 7(1)-4 = 7-4 = 3$

$a_2 = 7(2)-4 = 14-4 = 10$

$a_3 = 7(3)-4 = 21-4 = 17$

$a_4 = 7(4)-4 = 28-4 = 24$

The first four terms are 3, 10, 17, 24.

2. $a_n = (-1)^n \frac{n+2}{n+1}$

$a_1 = (-1)^1 \frac{1+2}{1+1} = -\frac{3}{2}$

$a_2 = (-1)^2 \frac{2+2}{2+1} = \frac{4}{3}$

$a_3 = (-1)^3 \frac{3+2}{3+1} = -\frac{5}{4}$

$a_4 = (-1)^4 \frac{4+2}{4+1} = \frac{6}{5}$

The first four terms are $-\frac{3}{2}, \frac{4}{3}, -\frac{5}{4}, \frac{6}{5}$.

3. $a_n = \frac{1}{(n-1)!}$

$a_1 = \frac{1}{(1-1)!} = \frac{1}{0!} = \frac{1}{1} = 1$

$a_2 = \frac{1}{(2-1)!} = \frac{1}{1!} = \frac{1}{1} = 1$

$a_3 = \frac{1}{(3-1)!} = \frac{1}{2!} = \frac{1}{2\cdot 1} = \frac{1}{2}$

$a_4 = \frac{1}{(4-1)!} = \frac{1}{3!} = \frac{1}{3\cdot 2\cdot 1} = \frac{1}{6}$

The first four terms are $1, 1, \frac{1}{2}, \frac{1}{6}$.

4. $a_n = \frac{(-1)^{n+1}}{2^n}$

$a_1 = \frac{(-1)^{1+1}}{2^1} = \frac{(-1)^2}{2} = \frac{1}{2}$

$a_2 = \frac{(-1)^{2+1}}{2^2} = \frac{(-1)^3}{4} = -\frac{1}{4}$

$a_3 = \frac{(-1)^{3+1}}{2^3} = \frac{(-1)^4}{8} = \frac{1}{8}$

$a_4 = \frac{(-1)^{4+1}}{2^4} = \frac{(-1)^5}{16} = -\frac{1}{16}$

The first four terms are $\frac{1}{2}, -\frac{1}{4}, \frac{1}{8}, -\frac{1}{16}$.

5. $\sum_{i=1}^{5}(2i^2-3)$

$=(2(1)^2-3)+(2(2)^2-3)+(2(3)^2-3)+(2(4)^2-3)+(2(5)^2-3)+(2(6)^2-3)$

$=(2(1)-3)+(2(4)-3)+(2(9)-3)+(2(16)-3)+(2(25)-3)$

$=(2-3)+(8-3)+(18-3)+(32-3)+(50-3)=-1+5+15+29+47=95$

6. $\sum_{i=0}^{4}(-1)^{i+1}i!=(-1)^{0+1}0!+(-1)^{1+1}1!+(-1)^{2+1}2!+(-1)^{3+1}3!+(-1)^{4+1}4!$

$=(-1)^1 1+(-1)^2 1+(-1)^3 2\cdot 1+(-1)^4 3\cdot 2\cdot 1+(-1)^5 4\cdot 3\cdot 2\cdot 1$

$=-1+1-2+6-24=-20$

7. $\frac{1}{3}+\frac{2}{4}+\frac{3}{5}+\ldots+\frac{15}{17}=\sum_{i=1}^{15}\frac{i}{i+2}$

8. $4^3+5^3+6^3+\ldots+13^3=\sum_{i=4}^{13}i^3$

11.2

9. $a_1=7$

$a_2=7+4=11$

$a_3=11+4=15$

$a_4=15+4=19$

$a_5=19+4=23$

$a_6=23+4=27$

The first six terms are $3,-1,-5,-9,-13.$

10. $a_1=-4$

$a_2=-4-5=-9$

$a_3=-9-5=-14$

$a_4=-14-5=-19$

$a_5=-19-5=-24$

$a_6=-24-5=-29$

The first six terms are $-4,-9,-14,-19,-24,-29.$

11. $a_1=\frac{3}{2}$

$a_2=\frac{3}{2}-\frac{1}{2}=\frac{2}{2}=1$

$a_3=1-\frac{1}{2}=\frac{1}{2}$

$a_4=\frac{1}{2}-\frac{1}{2}=0$

$a_5=0-\frac{1}{2}=-\frac{1}{2}$

$a_6=-\frac{1}{2}-\frac{1}{2}=-\frac{2}{2}=-1$

The first six terms are $\frac{3}{2},1,\frac{1}{2},0,-\frac{1}{2},-1.$

12. $a_6 = 5 + (6-1)3 = 5 + (5)3$
$= 5 + 15 = 20$

13. $a_{12} = -8 + (12-1)(-2) = -8 + 11(-2)$
$= -8 + (-22) = -30$

14. $a_{14} = 14 + (14-1)(-4) = 14 + 13(-4)$
$= 14 + (-52) = -38$

15. $d = -3 - (-7) = 4$
$a_n = -7 + (n-1)4 = -7 + 4n - 4$
$= 4n - 11$
$a_{20} = 4(20) - 11 = 80 - 11 = 69$

16. $a_n = 200 + (n-1)(-20)$
$= 200 - 20n + 20$
$= 220 - 20n$
$a_{20} = 220 - 20(20)$
$= 220 - 400 = -180$

17. $a_n = -12 + (n-1)\left(-\frac{1}{2}\right)$
$= -12 - \frac{1}{2}n + \frac{1}{2}$
$= -\frac{24}{2} - \frac{1}{2}n + \frac{1}{2}$
$= -\frac{1}{2}n - \frac{23}{2}$
$a_{20} = -\frac{1}{2}(20) - \frac{23}{2}$
$= -\frac{20}{2} - \frac{23}{2} = -\frac{43}{2}$

18. $d = 8 - 15 = -7$
$a_n = 15 + (n-1)(-7) = 15 - 7n + 7$
$= 22 - 7n$
$a_{20} = 22 - 7(20) = 22 - 140 = -118$

19. First, find d.
$d = 12 - 5 = 7$
Next, find a_{22}.
$a_{22} = 5 + (22-1)7 = 5 + (21)7$
$= 5 + 147 = 152$
Now, find the sum.
$S_{22} = \frac{22}{2}(5 + 152) = 11(157) = 1727$

20. First, find d.
$d = -3 - (-6) = 3$
Next, find a_{15}.
$a_{15} = -6 + (15-1)3 = -6 + (14)3$
$= -6 + 42 = 36$
Now, find the sum.
$S_{15} = \frac{15}{2}(-6 + 36) = \frac{15}{2}(30) = 225$

21. We are given that $a_1 = 300$.
$S_{100} = \frac{100}{2}(3 + 300)$
$= 50(303) = 15150$

22. $$\sum_{i=1}^{16}(3i+2)=(3(1)+2)+(3(2)+2)+(3(3)+2)+...+(3(16)+2)$$
$$=(3+2)+(6+2)+(9+2)+...+(48+2)$$
$$=5+8+11+...+50$$
$$S_{16}=\frac{16}{2}(5+50)=8(55)=440$$

23. $$\sum_{i=1}^{25}(-2i+6)=(-2(1)+6)+(-2(2)+6)+(-2(3)+6)+...+(-2(25)+6)$$
$$=(-2+6)+(-4+6)+(-6+6)+...+(-50+6)$$
$$=4+2+0+...+(-44)$$
$$S_{25}=\frac{25}{2}(4+(-44))=\frac{25}{2}(-40)=-500$$

24. $$\sum_{i=1}^{30}(-5i)=(-5(1))+(-5(2))+(-5(3))+...+(-5(30))$$
$$=-5+(-10)+(-15)+...+(-150)$$
$$S_{30}=\frac{30}{2}(-5+(-150))=15(-155)=-2325$$

25. **a.** $$a_n=1043.04+(n-1)(-0.4118)$$
$$=1043.04-0.4118n+0.4118$$
$$=1043.4518-0.4118n$$

b. $$a_{100}=1043.4518-0.4118(100)$$
$$=1043.4518-41.18$$
$$=1002.2718$$

26. $$a_{10}=31500+(10-1)2300$$
$$=31500+(9)2300$$
$$=31500+20700=52200$$
$$S_{10}=\frac{10}{2}(31500+52200)$$
$$=5(83700)=418500$$

The total salary over a ten-year period is $418,500.

27. $$a_{35}=25+(35-1)1=25+(34)1$$
$$=25+34=59$$
$$S_{35}=\frac{35}{2}(25+59)=\frac{35}{2}(84)=1470$$
There are 1470 seats in the theater.

11.3

28. The first term is 3. The second term is $3\cdot 2=6$. The third term is $6\cdot 2=12$. The fourth term is $12\cdot 2=24$. The fifth term is $24\cdot 2=48$.

29. The first term is $\frac{1}{2}$. The second term is $\frac{1}{2}\cdot\frac{1}{2}=\frac{1}{4}$. The third term is $\frac{1}{4}\cdot\frac{1}{2}=\frac{1}{8}$. The fourth term is $\frac{1}{8}\cdot\frac{1}{2}=\frac{1}{16}$. The fifth term is $\frac{1}{16}\cdot\frac{1}{2}=\frac{1}{32}$.

30. The first term is 16. The second term is $16\cdot-\frac{1}{4}=-4$. The third term is $-4\cdot-\frac{1}{4}=1$. The fourth term is $1\cdot-\frac{1}{4}=-\frac{1}{4}$. The fifth term is $-\frac{1}{4}\cdot-\frac{1}{4}=\frac{1}{16}$.

31. The first term is –5. The second term is $-5\cdot-1=5$. The third term is $5\cdot-1=-5$. The fourth term is $-5\cdot-1=5$. The fifth term is $5\cdot-1=-5$.

32. $a_7=2(3)^{7-1}=2(3)^6=2(729)=1458$

33.
$$a_6=16\left(\frac{1}{2}\right)^{6-1}=16\left(\frac{1}{2}\right)^5$$
$$=16\left(\frac{1}{32}\right)=\frac{1}{2}$$

34.
$$a_5=-3(2)^{5-1}=-3(2)^4$$
$$=-3(16)=-48$$

35.
$$a_n=a_1r^{n-1}=1(2)^{n-1}$$
$$a_8=1(2)^{8-1}=1(2)^7=1(128)=128$$

36.
$$a_n=a_1r^{n-1}=100\left(\frac{1}{10}\right)^{n-1}$$
$$a_8=100\left(\frac{1}{10}\right)^{8-1}=100\left(\frac{1}{10}\right)^7$$
$$=100\left(\frac{1}{10000000}\right)=\frac{1}{100000}$$

37.
$$d=\frac{-4}{12}=-\frac{1}{3}$$
$$a_n=a_1r^{n-1}=12\left(-\frac{1}{3}\right)^{n-1}$$
$$a_8=12\left(-\frac{1}{3}\right)^{8-1}=12\left(-\frac{1}{3}\right)^7$$
$$=12\left(-\frac{1}{2187}\right)=-\frac{12}{2187}=-\frac{4}{729}$$

38.
$$r=\frac{a_2}{a_1}=\frac{-15}{5}=-3$$
$$S_{15}=\frac{5\left(1-(-3)^{15}\right)}{1-(-3)}$$
$$=\frac{5(1-(-14348907))}{4}$$
$$=\frac{5(14348908)}{4}=\frac{71744540}{4}$$
$$=17{,}936{,}135$$

39.
$$r=\frac{a_2}{a_1}=\frac{4}{8}=\frac{1}{2}$$

$$S_7 = \frac{8\left(1-\left(\frac{1}{2}\right)^7\right)}{1-\frac{1}{2}} = \frac{8\left(1-\frac{1}{128}\right)}{\frac{1}{2}}$$

$$= \frac{8\left(\frac{128}{128}-\frac{1}{128}\right)}{\frac{1}{2}} = \frac{8\left(\frac{127}{128}\right)}{\frac{1}{2}}$$

$$= \frac{8}{1}\left(\frac{127}{128}\right) \div \frac{1}{2} = \frac{8}{1}\left(\frac{127}{128}\right) \cdot \frac{2}{1}$$

$$= \frac{2032}{128} = \frac{127}{8} = 15.875$$

40. $$\sum_{i=1}^{6} 5^i = \frac{5\left(1-5^6\right)}{1-5} = \frac{5(1-15625)}{-4}$$

$$= \frac{5(-15624)}{-4} = 5(3906)$$

$$= 19,530$$

41. $$\sum_{i=1}^{7} 3(-2)^i = \frac{-6\left(1-(-2)^7\right)}{1-(-2)}$$

$$= \frac{-6(1-(-128))}{3}$$

$$= \frac{-6(129)}{3}$$

$$= -2(129) = -258$$

42. $$\sum_{i=1}^{5} 2\left(\frac{1}{4}\right)^{i-1} = \frac{2\left(1-\left(\frac{1}{4}\right)^5\right)}{1-\frac{1}{4}}$$

$$= \frac{2\left(1-\frac{1}{1024}\right)}{\frac{3}{4}} = \frac{2\left(\frac{1024}{1024}-\frac{1}{1024}\right)}{\frac{3}{4}}$$

$$= \frac{2\left(\frac{1023}{1024}\right)}{\frac{3}{4}} = \frac{\frac{2046}{1024}}{\frac{3}{4}} = \frac{2046}{1024} \div \frac{3}{4}$$

$$= \frac{2046}{1024} \cdot \frac{4}{3} = \frac{682}{256} = \frac{341}{128}$$

43. $$r = \frac{a_2}{a_1} = \frac{3}{9} = \frac{1}{3}$$

$$S = \frac{9}{1-\frac{1}{3}} = \frac{9}{\frac{2}{3}} = 9 \div \frac{2}{3} = 9 \cdot \frac{3}{2} = \frac{27}{2}$$

44. $$r = \frac{a_2}{a_1} = \frac{-1}{2} = -\frac{1}{2}$$

$$S = \frac{2}{1-\left(-\frac{1}{2}\right)} = \frac{2}{\frac{3}{2}} = 2 \div \frac{3}{2} = 2 \cdot \frac{2}{3} = \frac{4}{3}$$

45. $$r = \frac{a_2}{a_1} = \frac{4}{-6} = -\frac{2}{3}$$

$$S = \frac{-6}{1-\left(-\frac{2}{3}\right)} = \frac{-6}{\frac{5}{3}} = -6 \div \frac{5}{3}$$

$$= -6 \cdot \frac{3}{5} = -\frac{18}{5}$$

46. $$\sum_{i=1}^{\infty} 5(0.8)^i = \frac{4}{1-0.8} = \frac{4}{0.2} = 20$$

47.

$$0.\overline{6} = \frac{a_1}{1-r} = \frac{\frac{6}{10}}{1-\frac{1}{10}} = \frac{\frac{6}{10}}{\frac{9}{10}} = \frac{6}{10} \div \frac{9}{10}$$
$$= \frac{6}{10} \cdot \frac{10}{9} = \frac{2}{3}$$

48.

$$0.\overline{47} = \frac{a_1}{1-r} = \frac{\frac{47}{100}}{1-\frac{1}{100}} = \frac{\frac{47}{100}}{\frac{99}{100}}$$
$$= \frac{47}{100} \div \frac{99}{100} = \frac{47}{100} \cdot \frac{100}{99} = \frac{47}{99}$$

49. $r = 1.06$

$$a_n = a_1 r^{n-1} = 32000(1.06)^{n-1}$$
$$a_6 = 32000(1.06)^{6-1} = 32000(1.06)^5$$
$$= 32000(1.338226) = 42823.22$$

The salary in the sixth year is approximately \$42,823.22.

$$S_6 = \frac{a_1(1-r^n)}{1-r} = \frac{32000\left(1-(1.06)^6\right)}{1-1.06}$$
$$= \frac{32000(1-1.418519)}{-0.06}$$
$$= \frac{32000(-0.418519)}{-0.06} = 223210.13$$

The total salary over the six years is approximately \$223,210.13.

50.

$$A = P\frac{\left(1+\frac{r}{n}\right)^{nt} - 1}{\frac{r}{n}}$$

$$= 200\frac{\left(1+\frac{0.10}{12}\right)^{12(18)} - 1}{\frac{0.10}{12}}$$

$$= 200\frac{\left(1+\frac{1}{120}\right)^{216} - 1}{\frac{1}{120}}$$

$$= 200\frac{\left(\frac{121}{120}\right)^{216} - 1}{\frac{1}{120}} = 200\frac{6.004693-1}{\frac{1}{120}}$$

$$= 200\frac{5.004693}{\frac{1}{120}} = 120112.63$$

After 18 years, the value of the account will be approximately \$120,112.63.

51. $r = 70\% = 0.7$

$$a_1 = 4(.7) = 2.8$$
$$S = \frac{2.8}{1-0.7} = \frac{2.8}{0.3} = 9.\overline{3}$$

The total spending in the town will be approximately \$9.3 million each year.

52.

$$\binom{11}{8} = \frac{11!}{8!(11-8)!}$$
$$= \frac{11 \cdot 10 \cdot 9 \cdot \cancel{8!}}{\cancel{8!} \cdot 3 \cdot 2 \cdot 1} = 165$$

53.

$$\binom{90}{2} = \frac{90!}{2!(90-2)!} = \frac{90 \cdot 89 \cdot \cancel{88!}}{2 \cdot 1 \cdot \cancel{88!}}$$
$$= 4005$$

54. Applying the Binomial Theorem to $(2x+1)^3$, we have $a=2x$, $b=1$, and $n=3$.

$$(2x+1)^3=\binom{3}{0}(2x)^3+\binom{3}{1}(2x)^2\cdot 1+\binom{3}{2}(2x)\cdot 1^2+\binom{3}{3}1^3$$

$$=\frac{3!}{0!(3-0)!}8x^3+\frac{3!}{1!(3-1)!}4x^2\cdot 1+\frac{3!}{2!(3-2)!}2x\cdot 1+\frac{3!}{3!(3-3)!}1$$

$$=\frac{\cancel{3!}}{1\cdot\cancel{3!}}8x^3+\frac{3\cdot\cancel{2!}}{1\cdot\cancel{2!}}4x^2+\frac{3\cdot\cancel{2!}}{\cancel{2!}1!}2x+\frac{\cancel{3!}}{\cancel{3!}0!}$$

$$=8x^3+3\left(4x^2\right)+3(2x)+1=8x^3+12x^2+6x+1$$

55. Applying the Binomial Theorem to $\left(x^2-1\right)^4$, we have $a=x^2$, $b=-1$, and $n=4$.

$$\left(x^2-1\right)^4=\binom{4}{0}\left(x^2\right)^4+\binom{4}{1}\left(x^2\right)^3(-1)+\binom{4}{2}\left(x^2\right)^2(-1)^2+\binom{4}{3}x^2(-1)^3+\binom{4}{4}(-1)^4$$

$$=\frac{4!}{0!(4-0)!}x^8-\frac{4!}{1!(4-1)!}x^6+\frac{4!}{2!(4-2)!}x^4-\frac{4!}{3!(4-3)!}x^2+\frac{4!}{4!(4-4)!}1$$

$$=\frac{\cancel{4!}}{0!\cancel{4!}}x^8-\frac{4!}{1!3!}x^6+\frac{4!}{2!2!}x^4-\frac{4!}{3!1!}x^2+\frac{\cancel{4!}}{\cancel{4!}0!}$$

$$=1\left(x^8\right)-\frac{4\cdot\cancel{3!}}{1\cdot\cancel{3!}}x^6+\frac{4\cdot 3\cdot\cancel{2!}}{2\cdot 1\cdot\cancel{2!}}x^4-\frac{4\cdot\cancel{3!}}{\cancel{3!}\cdot 1}x^2+1$$

$$=x^8-4x^6+6x^4-4x^2+1$$

56. Applying the Binomial Theorem to $(x+2y)^5$, we have $a=x$, $b=2y$, and $n=5$.

$$(x+2y)^5=\binom{5}{0}x^5+\binom{5}{1}x^4(2y)+\binom{5}{2}x^3(2y)^2+\binom{5}{3}x^2(2y)^3+\binom{5}{4}x(2y)^4+\binom{5}{5}(2y)^5$$

$$=\frac{\cancel{5!}}{0!\cancel{5!}}x^5+\frac{5!}{1!(5-1)!}2x^4y+\frac{5!}{2!(5-2)!}4x^3y^2+\frac{5!}{3!(5-3)!}8x^2y^3$$

$$+\frac{5!}{4!(5-4)!}16xy^4+\frac{5!}{5!(5-5)!}32y^5$$

$$=1x^5+\frac{5\cdot\cancel{4!}}{1\cdot\cancel{4!}}2x^4y+\frac{5\cdot 4\cdot\cancel{3!}}{2\cdot 1\cdot\cancel{3!}}4x^3y^2+\frac{5\cdot 4\cdot\cancel{3!}}{\cancel{3!}2\cdot 1}8x^2y^3+\frac{5\cdot\cancel{4!}}{\cancel{4!}\cdot 1}16xy^4+\frac{\cancel{5!}}{\cancel{5!}0!}32y^5$$

$$=x^5+5\left(2x^4y\right)+10\left(4x^3y^2\right)+10\left(8x^2y^3\right)+5\left(16xy^4\right)+1\left(32y^5\right)$$

$$=x^5+10x^4y+40x^3y^2+80x^2y^3+80xy^4+32y^5$$

57. Applying the Binomial Theorem to $(x-2)^6$, we have $a = x,\ b = -2,$ and $n = 6$.

$$
\begin{aligned}
&(x-2)^6 \\
&= \binom{6}{0}x^6 + \binom{6}{1}x^5(-2) + \binom{6}{2}x^4(-2)^2 + \binom{6}{3}x^3(-2)^3 + \binom{6}{4}x^2(-2)^4 + \binom{6}{5}x(-2)^5 + \binom{6}{6}(-2)^6 \\
&= \frac{6!}{0!(6-0)!}x^6 + \frac{6!}{1!(6-1)!}x^5(-2) + \frac{6!}{2!(6-2)!}x^4(-2)^2 + \frac{6!}{3!(6-3)!}x^3(-2)^3 \\
&\qquad + \frac{6!}{4!(6-4)!}x^2(-2)^4 + \frac{6!}{5!(6-5)!}x(-2)^5 + \frac{6!}{6!(6-6)!}(-2)^6 \\
&= \frac{\cancel{6!}}{1\cancel{6!}}x^6 - \frac{6\cdot\cancel{5!}}{1\cdot\cancel{5!}}2x^5 + \frac{6\cdot5\cdot\cancel{4!}}{2\cdot1\cdot\cancel{4!}}4x^4 - \frac{\cancel{6}\cdot5\cdot4\cdot\cancel{3!}}{\cancel{3\cdot2}\cdot1\cdot\cancel{3!}}8x^3 \\
&\qquad + \frac{6\cdot5\cdot\cancel{4!}}{\cancel{4!}2\cdot1}16x^2 - \frac{6\cdot\cancel{5!}}{\cancel{5!}1}32x + \frac{\cancel{6!}}{\cancel{6!}\cdot1}64 \\
&= x^6 - 6\left(2x^5\right) + 15\left(4x^4\right) - 20\left(8x^3\right) + 15\left(16x^2\right) - 6(32x) + 1\cdot64 \\
&= x^6 - 12x^5 + 60x^4 - 160x^3 + 240x^2 - 192x + 64
\end{aligned}
$$

58. $\left(x^2+3\right)^8$

First Term
$$
\begin{aligned}
\binom{n}{r-1}a^{n-r+1}b^{r-1} &= \binom{8}{1-1}\left(x^2\right)^{8-1+1}3^{1-1} = \binom{8}{0}\left(x^2\right)^8 3^0 \\
&= \frac{8!}{0!(8-0)!}x^{16}\cdot1 = \frac{\cancel{8!}}{0!\cancel{8!}}x^{16} = x^{16}
\end{aligned}
$$

Second Term
$$
\begin{aligned}
\binom{n}{r-1}a^{n-r+1}b^{r-1} &= \binom{8}{2-1}\left(x^2\right)^{8-2+1}3^{2-1} = \binom{8}{1}\left(x^2\right)^7 3^1 = \frac{8!}{1!(8-1)!}3x^{14} \\
&= \frac{8\cdot\cancel{7!}}{1\cdot\cancel{7!}}3x^{14} = 8\cdot3x^{14} = 24x^{14}
\end{aligned}
$$

Third Term
$$
\begin{aligned}
\binom{n}{r-1}a^{n-r+1}b^{r-1} &= \binom{8}{3-1}\left(x^2\right)^{8-3+1}3^{3-1} = \binom{8}{2}\left(x^2\right)^6 3^2 = \frac{8!}{2!(8-2)!}9x^{12} \\
&= \frac{8\cdot7\cdot\cancel{6!}}{2\cdot1\cdot\cancel{6!}}9x^{12} = 28\cdot9x^{12} = 252x^{12}
\end{aligned}
$$

59. $(x-3)^9$

First Term
$$
\binom{n}{r-1}a^{n-r+1}b^{r-1} = \binom{9}{1-1}x^{9-1+1}(-3)^{1-1} = \binom{9}{0}x^9(-3)^0
$$

$$= \frac{9!}{0!(9-0)!}x^9 \cdot 1 = \frac{\cancel{9!}}{0!\cancel{9!}}x^9 = x^9$$

Second Term $\binom{n}{r-1}a^{n-r+1}b^{r-1} = \binom{9}{2-1}x^{9-2+1}(-3)^{2-1} = \binom{9}{1}x^8(-3)^1 = -\frac{9!}{1!(9-1)!}3x^8$

$$= -\frac{9 \cdot \cancel{8!}}{1 \cdot \cancel{8!}}3x^8 = -9 \cdot 3x^8 = -27x^8$$

Third Term $\binom{n}{r-1}a^{n-r+1}b^{r-1} = \binom{9}{3-1}x^{9-3+1}(-3)^{3-1} = \binom{9}{2}x^7(-3)^2 = \frac{9!}{2!(9-2)!}9x^7$

$$= \frac{9 \cdot 8 \cdot \cancel{7!}}{2 \cdot 1 \cdot \cancel{7!}}9x^7 = 36 \cdot 9x^7 = 324x^7$$

60. $(x+2)^5$

Fourth Term $\binom{n}{r-1}a^{n-r+1}b^{r-1} = \binom{5}{4-1}x^{5-4+1}(2)^{4-1} = \binom{5}{3}x^2(2)^3 = \frac{5!}{3!(5-3)!}8x^2$

$$= \frac{5!}{3!2!}8x^2 = \frac{5 \cdot 4 \cdot \cancel{3!}}{\cancel{3!} \cdot 2 \cdot 1}8x^2 = (10)8x^2 = 80x^2$$

61. $(2x-3)^6$

Fifth Term $\binom{n}{r-1}a^{n-r+1}b^{r-1} = \binom{6}{5-1}(2x)^{6-5+1}(-3)^{5-1} = \binom{6}{4}(2x)^2(-3)^4$

$$= \frac{6!}{4!(6-4)!}4x^2(81) = \frac{6!}{4!2!}324x^2 = \frac{6 \cdot 5 \cdot \cancel{4!}}{\cancel{4!} \cdot 2 \cdot 1}324x^2$$

$$= (15)324x^2 = 4860x^2$$

Chapter 11 Test

1.

$$a_n = \frac{(-1)^{n+1}}{n^2}$$

$$a_1 = \frac{(-1)^{1+1}}{1^2} = \frac{(-1)^2}{1} = \frac{1}{1} = 1$$

$$a_2 = \frac{(-1)^{2+1}}{2^2} = \frac{(-1)^3}{4} = \frac{-1}{4} = -\frac{1}{4}$$

2.
$$\sum_{i=1}^{5}\left(i^2+10\right)$$
$$=\left(1^2+10\right)+\left(2^2+10\right)+\left(3^2+10\right)+\left(4^2+10\right)+\left(5^2+10\right)$$
$$=(1+10)+(4+10)+(9+10)+(16+10)+(25+10)$$
$$=11+14+19+26+35=105$$

3.
$$\frac{2}{3}+\frac{3}{4}+\frac{4}{5}+\ldots+\frac{21}{22}=\sum_{i=2}^{21}\frac{i}{i+1}$$

4.
$$d=9-4=5$$
$$a_n=4+(n-1)5=4+5n-5$$
$$=5n-1$$
$$a_{12}=5(12)-1=60-1=59$$

5.
$$d=\frac{a_2}{a_1}=\frac{4}{16}=\frac{1}{4}$$
$$a_n=a_1r^{n-1}=16\left(\frac{1}{4}\right)^{n-1}$$
$$a_{12}=16\left(\frac{1}{4}\right)^{12-1}=16\left(\frac{1}{4}\right)^{11}$$
$$=16\left(\frac{1}{4194304}\right)=\frac{16}{4194304}$$
$$=\frac{1}{262144}$$

6.
$$d=-14-(-7)=-7$$
$$a_{10}=-7+(10-1)(-7)$$
$$=-7+(9)(-7)$$
$$=-7+(-63)=-70$$
$$S_{10}=\frac{10}{2}(-7+(-70))$$
$$=5(-77)=-385$$

7.
$$\sum_{i=1}^{30}(3i-4)$$
$$=(3(1)-4)+(3(2)-4)+(3(3)-4)+\ldots+(3(20)-4)$$
$$=(3-4)+(6-4)+(9-4)+\ldots+(60-4)$$
$$=-1+2+5+\ldots+56$$
$$S_{20}=\frac{20}{2}(-1+56)=10(55)=550$$

8.
$$r=\frac{a_2}{a_1}=\frac{-14}{7}=-2$$
$$S_{10}=\frac{7\left(1-(-2)^{10}\right)}{1-(-2)}=\frac{7(1-1024)}{3}$$
$$=\frac{7(-1023)}{3}=-2387$$

9.
$$\sum_{i=1}^{15}(-2)^i=\frac{-2\left(1-(-2)^{15}\right)}{1-(-2)}$$
$$=\frac{-2(1-(-32768))}{3}$$
$$=\frac{-2(32769)}{3}=-21,846$$

10.
$$r=\frac{1}{2}$$
$$S=\frac{4}{1-\frac{1}{2}}=\frac{4}{\frac{1}{2}}=4\div\frac{1}{2}=4\cdot\frac{2}{1}=8$$

11.

$$0.\overline{73} = \frac{a_1}{1-r} = \frac{\frac{73}{100}}{1-\frac{1}{100}} = \frac{\frac{73}{100}}{\frac{99}{100}}$$

$$= \frac{73}{100} \div \frac{99}{100} = \frac{73}{100} \cdot \frac{100}{99} = \frac{73}{99}$$

12. $r = 1.04$

$$S_8 = \frac{a_1(1-r^n)}{1-r} = \frac{30000\left(1-(1.04)^8\right)}{1-1.04}$$

$$= \frac{30000(1-1.368569)}{-0.04}$$

$$= \frac{30000(-0.368569)}{-0.04}$$

$$= 276,426.75$$

The total salary over the eight years is approximately $276,426.75

13.

$$\binom{9}{2} = \frac{9!}{2!(9-2)!} = \frac{9!}{2!7!}$$

$$= \frac{9 \cdot 8 \cdot \cancel{7!}}{2 \cdot 1 \cdot \cancel{7!}} = 36$$

14. Applying the Binomial Theorem to $(x^2-1)^5$, we have $a = x^2$, $b = -1$, and $n = 5$.

$$(x^2-1)^5$$

$$= \binom{5}{0}(x^2)^5 + \binom{5}{1}(x^2)^4(-1) + \binom{5}{2}(x^2)^3(-1)^2 + \binom{5}{3}(x^2)^2(-1)^3 + \binom{5}{4}(x^2)(-1)^4 + \binom{5}{5}(-1)^5$$

$$= \frac{\cancel{5!}}{0!\cancel{5!}}x^{10} - \frac{5!}{1!(5-1)!}x^8 + \frac{5!}{2!(5-2)!}x^6 - \frac{5!}{3!(5-3)!}x^4 + \frac{5!}{4!(5-4)!}x^2 - \frac{5!}{5!(5-5)!}$$

$$= 1x^{10} - \frac{5 \cdot \cancel{4!}}{1 \cdot \cancel{4!}}x^8 + \frac{5 \cdot 4 \cdot \cancel{3!}}{2 \cdot 1 \cdot \cancel{3!}}x^6 - \frac{5 \cdot 4 \cdot \cancel{3!}}{\cancel{3!}2 \cdot 1}x^4 + \frac{5 \cdot \cancel{4!}}{\cancel{4!} \cdot 1}x^2 - \frac{\cancel{5!}}{\cancel{5!}0!}$$

$$= x^{10} - 5x^8 + 10x^6 - 10x^4 + 5x^2 - 1$$

15. $(x+y^2)^8$

First Term
$$\binom{n}{r-1}a^{n-r+1}b^{r-1} = \binom{8}{1-1}x^{8-1+1}(y^2)^{1-1} = \binom{8}{0}x^8(y^2)^0$$

$$= \frac{8!}{0!(8-0)!}x^8 \cdot 1 = \frac{\cancel{8!}}{0!\cancel{8!}}x^8 = x^8$$

Second Term
$$\binom{n}{r-1}a^{n-r+1}b^{r-1} = \binom{8}{2-1}x^{8-2+1}(y^2)^{2-1} = \binom{8}{1}x^7(y^2)^1 = \frac{8!}{1!(8-1)!}x^7y^2$$

$$= \frac{8 \cdot \cancel{7!}}{1 \cdot \cancel{7!}}x^7y^2 = 8x^7y^2$$

Third Term $\binom{n}{r-1}a^{n-r+1}b^{r-1}=\binom{8}{3-1}x^{8-3+1}\left(y^2\right)^{3-1}=\binom{8}{2}x^6\left(y^2\right)^2=\frac{8!}{2!(8-2)!}x^6y^4$

$$=\frac{8\cdot 7\cdot \cancel{6!}}{2\cdot 1\cdot \cancel{6!}}x^6y^4=28x^6y^4$$

Cumulative Review Exercises

Chapters 1-11

1. $9(x-1)=1+3(x-2)$

$9x-9=1+3x-6$

$9x-9=3x-5$

$6x-9=-5$

$6x=4$

$x=\frac{4}{6}=\frac{2}{3}$

The solution is $\frac{2}{3}$ and the solution set is $\left\{\frac{2}{3}\right\}$.

2. $3x+4y=-7$

$x-2y=-9$

Solve the second equation for x.

$x-2y=-9$

$x=2y-9$

Substitute $2y-9$ for x in the first equation and solve for y.

$3(2y-9)+4y=-7$

$6y-27+4y=-7$

$10y-27=-7$

$10y=20$

$y=2$

Back-substitute 2 for y to find x.

$x=2(2)-9$

$x=4-9$

$x=-5$

The solution is $(-5,2)$ and the solution set is $\{(-5,2)\}$.

3. $x-\ y+3z=-9$

$2x+3y-\ z=16$

$5x+2y-\ z=15$

Multiply the second equation by 3 and add to the first equation.

$x-\ y+3z=-9$

$\underline{6x+9y-3z=48}$

$7x+8y=39$

Multiply the second equation by -1 and add to the third equation.

$-2x-3y+\ z=-16$

$\underline{5x+2y-\ z=\ 15}$

$3x-y=-1$

We now have a system of two equations in two variables.

$7x+8y=39$

$3x-\ y=-1$

Multiply the second equation by 8 and add to the first equation.

$7x+8y=39$

$\underline{24x-8y=-8}$

$31x=31$

$x=1$

Back-substitute 1 for x to find y.

$$\begin{aligned} 7x+8y&=39\\ 7(1)+8y&=39\\ 7+8y&=39\\ 8y&=32\\ y&=4 \end{aligned}$$

Back-substitute 1 for x and 4 for y to find z.

$$\begin{aligned} x-y+3z&=-9\\ 1-4+3z&=-9\\ -3+3z&=-9\\ 3z&=-6\\ z&=-2 \end{aligned}$$

The solution is $(1,4,-2)$ and the solution set is $\{(1,4,-2)\}$.

4. $$\begin{aligned} 7x+18&\le 9x-2\\ -2x+18&\le -2\\ -2x&\le -20\\ x&\ge 10 \end{aligned}$$

The solution set is $\{x|x\ge 10\}$. In interval notation, the solution set is $[10,\infty)$.

5. $4x-3<13$ and $-3x-4\ge 8$

$4x<16$ $\quad -3x\ge 12$

$x<4$ $\quad x\le -4$

$x<4$

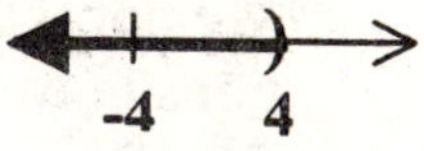

$x\le -4$

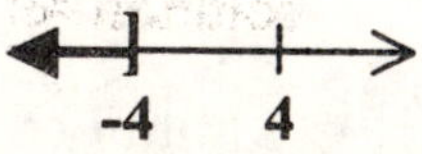

$x<4$ and $x\le -4$

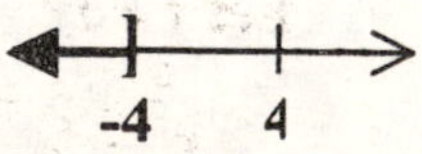

The solution set is $\{x|x\le -4\}$ or $(-\infty,-4]$.

6. $2x+4>8$ or $x-7\le 3$

$2x>4$ $\quad x\le 10$

$x>2$

$x>2$

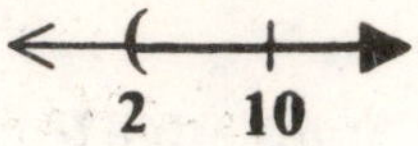

$x\le 10$

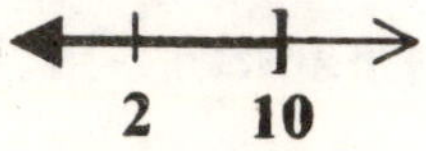

$x>2$ or $x\le 10$

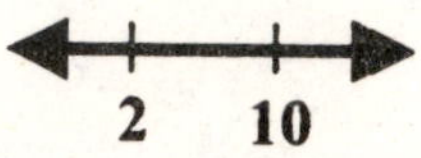

The solution set is $\mathbb{R}$, $(-\infty,\infty)$ or $\{x|x \text{ is a real number}\}$.

7. $$\begin{aligned} |2x-1|&<5\\ -5<2x&-1<5\\ -4<2&x<6\\ -2<&x<3 \end{aligned}$$

The solution set is $\{x|-2<x<3\}$. In interval notation the solution set is $(-2,3)$.

8. $$\left|\frac{2}{3}x-4\right|=2$$

$\frac{2}{3}x-4=2$ or $\frac{2}{3}x-4=-2$

$\frac{2}{3}x=6$ $\quad \frac{2}{3}x=2$

$\frac{3}{2}\cdot\frac{2}{3}x=\frac{3}{2}\cdot 6$ $\quad \frac{3}{2}\cdot\frac{2}{3}x=\frac{3}{2}\cdot 2$

$x=9$ $\quad x=3$

The solutions are 3 and 9 and the solution set is $\{3,9\}$.

9.

$$\frac{4}{x-3}-\frac{6}{x+3}=\frac{24}{x^2-9}$$

$$\frac{4}{x-3}-\frac{6}{x+3}=\frac{24}{(x-3)(x+3)}$$

$$(x-3)(x+3)\left(\frac{4}{x-3}-\frac{6}{x+3}\right)=(x-3)(x+3)\left(\frac{24}{(x-3)(x+3)}\right)$$

$$4(x+3)-6(x-3)=24$$

$$4x+12-6x+18=24$$

$$-2x+30=24$$

$$-2x=-6$$

$$x=3$$

Since 3 would make one or more of the denominators in the original equation zero, we disregard it and conclude that there is no solution. The solution set is $\varnothing$ or $\{\ \}$.

10. $\sqrt{x+4}-\sqrt{x-3}=1$

$$\sqrt{x+4}=1+\sqrt{x-3}$$

$$x+4=\left(1+\sqrt{x-3}\right)^2$$

$$x+4=1+2\sqrt{x-3}+x-3$$

$$x+4=2\sqrt{x-3}+x-2$$

$$6=2\sqrt{x-3}$$

$$3=\sqrt{x-3}$$

$$3^2=x-3$$

$$9=x-3$$

$$12=x$$

The solution is 12 and the solution set is $\{12\}$.

11.

$$2x^2=5-4x$$

$$2x^2+4x-5=0$$

$a=2 \quad b=4 \quad c=-5$

Solve using the quadratic formula.

$$x=\frac{-4\pm\sqrt{4^2-4(2)(-5)}}{2(2)}$$

$$=\frac{-4\pm\sqrt{16+40}}{4}=\frac{-4\pm\sqrt{56}}{4}$$

$$=\frac{-4\pm2\sqrt{14}}{4}=\frac{2\left(-2\pm\sqrt{14}\right)}{4}$$

$$=\frac{-2\pm\sqrt{14}}{2}$$

The solutions are $\frac{-2\pm\sqrt{14}}{2}$ and the solution set is $\left\{\frac{-2\pm\sqrt{14}}{2}\right\}$.

12. Let $t=x^{\frac{1}{3}}$

$$x^{\frac{2}{3}}-5x^{\frac{1}{3}}+6=0$$

$$\left(x^{\frac{1}{3}}\right)^2-5x^{\frac{1}{3}}+6=0$$

$t^2 - 5t + 6 = 0$

$(t-3)(t-2) = 0$

Apply the zero product principle.

$t - 3 = 0$ or $t - 2 = 0$

$t = 3$ $\quad$ $t = 2$

$x^{\frac{1}{3}} = 3$ $\quad$ $x^{\frac{1}{3}} = 2$

$x = 3^3$ $\quad$ $x = 2^3$

$x = 27$ $\quad$ $x = 8$

The solutions are 8 and 27 and the solution set is $\{8, 27\}$.

13. $2x^2 + x - 6 \le 0$

Solve the related quadratic equation.

$2x^2 + x - 6 = 0$

$(2x-3)(x+2) = 0$

Apply the zero product principle.

$2x - 3 = 0$ or $x + 2 = 0$

$2x = 3$ $\quad$ $x = -2$

$x = \frac{3}{2}$

The boundary points are -2 and $\frac{3}{2}$.

Test Interval	Test Number	Test	Conclusion
$(-\infty, -2]$	-3	$2(-3)^2 + (-3) - 6 \le 0$ $9 \le 0$, false	$(-\infty, -2]$ does not belong to the solution set.
$\left[-2, \frac{3}{2}\right]$	0	$2(0)^2 + 0 - 6 \le 0$ $-6 \le 0$, true	$\left[-2, \frac{3}{2}\right]$ belongs to the solution set.
$\left[\frac{3}{2}, \infty\right)$	2	$2(2)^2 + 2 - 6 \le 0$ $4 \le 0$, false	$\left[\frac{3}{2}, \infty\right)$ does not belong to the solution set.

The solution set is $\left[-2, \frac{3}{2}\right]$ or $\left\{x \middle| -2 \le x \le \frac{3}{2}\right\}$.

-2 3/2

14. $\log_8 x + \log_8 (x+2) = 1$

$$\log_8 (x(x+2)) = 1$$
$$x(x+2) = 8^1$$
$$x^2 + 2x = 8$$
$$x^2 + 2x - 8 = 0$$
$$(x+4)(x-2) = 0$$

Apply the zero product principle.

$x+4=0$ or $x-2=0$

$x=-4$ $\quad$ $x=2$

Since we cannot take a log of a negative number, we disregard –4 and conclude that the solution is 2 and the solution set is $\{2\}$.

15. $5^{2x+3} = 125$

$$5^{2x+3} = 5^3$$
$$2x+3 = 3$$
$$2x = 0$$
$$x = 0$$

The solution is 0 and the solution set is $\{0\}$.

16. $2x^2 - 3y^2 = 5$

$3x^2 - 4y^2 = 16$

Multiply the first equation by –3 and the second equation by 2 and solve by addition.

$$-6x^2 + 9y^2 = -15$$
$$6x^2 + 8y^2 = 32$$
$$17y^2 = 17$$
$$y^2 = 1$$
$$y = \pm 1$$

Back-substitute ± 1 for y to find x.

$$y = \pm 1$$
$$2x^2 - 3(\pm 1)^2 = 5$$
$$2x^2 - 3(1) = 5$$
$$2x^2 - 3 = 5$$
$$2x^2 = 8$$
$$x^2 = 4$$
$$x = \pm 2$$

The solutions are $(-2,-1), (-2,1)$, $(2,-1)$ and $(2,1)$ and the solution set is $\{(-2,-1),(-2,1),(2,-1),(2,1)\}$.

17. $2x^2 - y^2 = -8$

$x - y = 6$

Solve the second equation for x.

$x - y = 6$

$x = y + 6$

Substitute $y+6$ for x.

$$2(y+6)^2 - y^2 = -8$$
$$2(y^2 + 12x + 36) - y^2 = -8$$
$$2y^2 + 24x + 72 - y^2 = -8$$
$$y^2 + 24x + 72 = -8$$
$$y^2 + 24x + 80 = 0$$
$$(y+20)(y+4) = 0$$

Apply the zero product principle.

$y+20=0$ or $y+4=0$

$y=-20$ $\quad$ $y=-4$

Back-substitute –4 and –20 for y to find x.

$y=-20$ or $y=-4$

$x=-20+6$ $\quad$ $x=-4+6$

$x=-14$ $\quad$ $x=2$

The solutions are $(-14,-20)$ and $(2,-4)$ and the solution set is $\{(-14,-20),(2,-4)\}$.

18. $x-3y=6$

Find the x–intercept by setting $y=0$.

$$x-3(0)=6$$
$$x-0=6$$
$$x=6$$

Find the y–intercept by setting $x=0$.

$$0-3y=6$$
$$-3y=6$$
$$y=-2$$

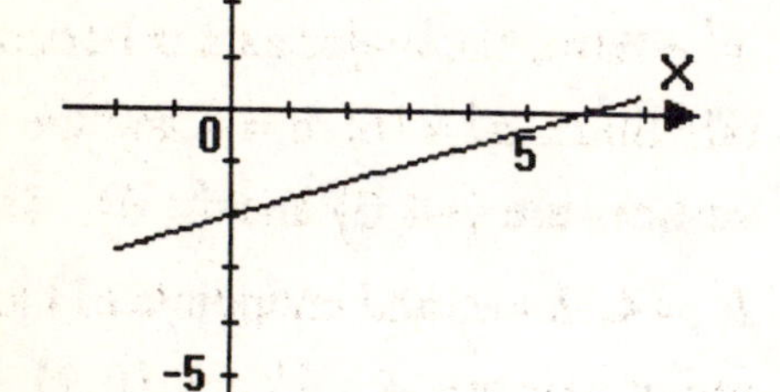

19.

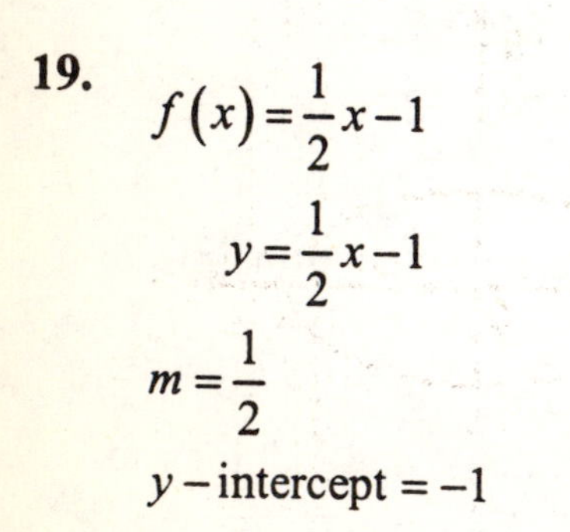

$$f(x)=\frac{1}{2}x-1$$
$$y=\frac{1}{2}x-1$$
$$m=\frac{1}{2}$$

y – intercept $=-1$

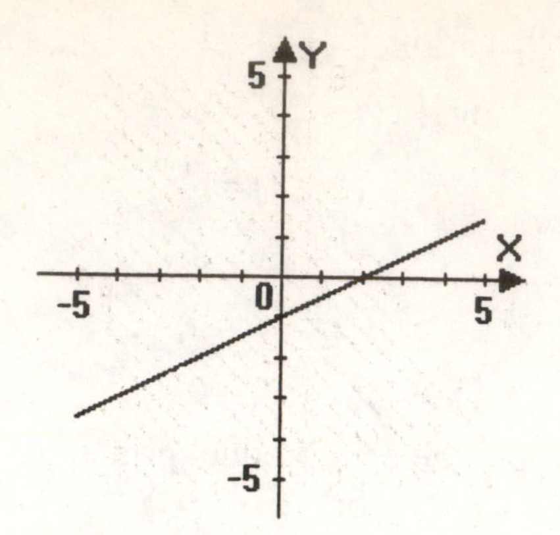

20. $3x-2y>-6$

First, find the intercepts to the equation $3x-2y=-6$.

find x – intercept, set $y=0$

$$3x-2y=-6$$
$$3x-2(0)=-6$$
$$3x=-6$$
$$x=-2$$

find y – intercept, set $x=0$

$$3x-2y=-6$$
$$3(0)-2y=-6$$
$$-2y=-6$$
$$y=3$$

Next, use the origin as a test point.

$$3x-2y>-6$$
$$3(0)-2(0)>-6$$
$$0>-6$$

This is a true statement. This means that the point, $(0,0)$, will fall in the shaded half-plane.

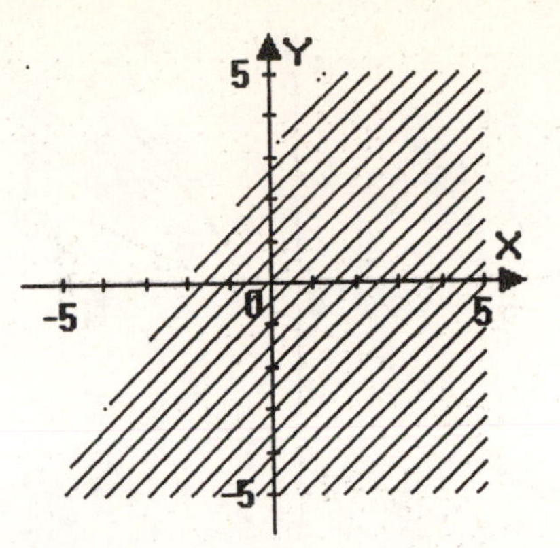

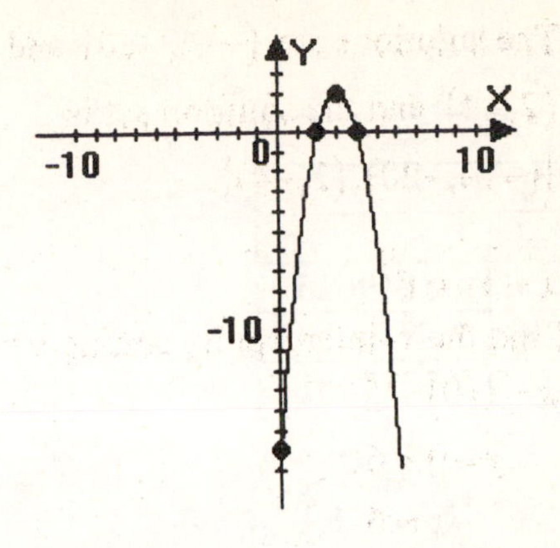

21. $f(x) = -2(x-3)^2 + 2$

This is a parabola of the form $y = a(x-h)^2 + k$. Since $a = -2$ is negative, the parabola opens downward. The vertex of the parabola is $(3, 2)$. The axis of symmetry is $x = 3$. Replace x with 0 to find the y–intercept.

$$y = -2(0-3)^2 + 2 = -2(-3)^2 + 2$$
$$= -2(9) + 2 = -18 + 2 = -16$$

The y–intercept is –16. Replace y with 0 to find the x–intercepts.

$$0 = -2(x-3)^2 + 2$$
$$0 = -2(x^2 - 6x + 9) + 2$$
$$0 = -2x^2 + 12x - 18 + 2$$
$$0 = -2x^2 + 12x - 16$$
$$0 = x^2 - 6x + 8$$
$$0 = (x-4)(x-2)$$

Apply the zero product principle.

$x - 4 = 0$ and $x - 2 = 0$

$x = 4$ $\quad$ $x = 2$

The x–intercepts are 2 and 4.

22. $\dfrac{x^2}{16} + \dfrac{y^2}{4} = 1$

Because x^2 and y^2 have different positive coefficients, the equation's graph is an ellipse. Since the denominator of the x^2 – term is greater than the denominator of the y^2 – term, the major axis is horizontal. Since $a^2 = 16$, $a = 4$ and the vertices are $(-4, 0)$ and $(4, 0)$. Since $b^2 = 4$, $b = 2$ and endpoints of the minor axis are $(0, -2)$ and $(0, 2)$.

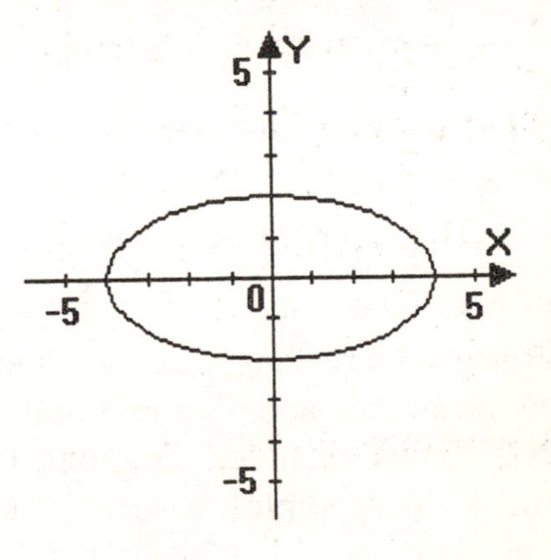

23. $y = \log_2 x$

x	y
1	0
2	1
3	1.5849625
4	2
5	2.3219281

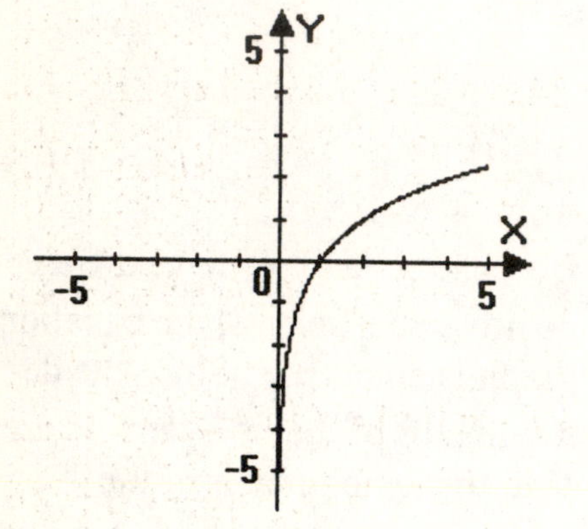

24. $x^2 - y^2 = 9$

$\frac{x^2}{9} - \frac{y^2}{9} = 1$

The equation is in the form

$\frac{x^2}{a^2} - \frac{y^2}{b^2} = 1$ with $a^2 = 9$, and $b^2 = 9$.

We know the transverse axis lies on the x-axis and the vertices are $(-3,0)$ and $(3,0)$. Because $a^2 = 3$ and $b^2 = 3$, $a = 3$ and $b = 3$. Construct a rectangle using –3 and 3 on the x–axis, and –3 and 3 on the y–axis. Draw extended diagonals to obtain the asymptotes. Graph the hyperbola.

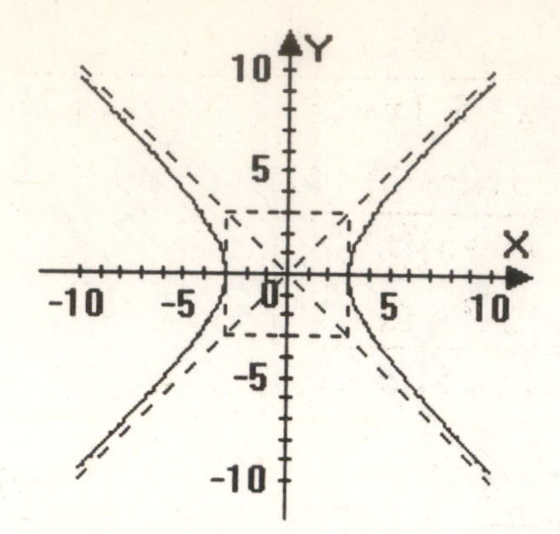

25. $4[2x - 6(x - y)] = 4[2x - 6x - 6y]$

$= 4[-4x - 6y]$

$= -16x - 24y$

26. $(-5x^3y^2)(4x^4y^{-6}) = -5(4)x^3x^4y^2y^{-6}$

$= -20x^7y^{-4}$

$= \frac{-20x^7}{y^4}$

27. $(8x^2 - 9xy - 11y^2) - (7x^2 - 4xy + 5y^2)$

$= 8x^2 - 9xy - 11y^2 - 7x^2 + 4xy - 5y^2$

$= x^2 - 5xy - 16y^2$

28. $(3x - 1)(2x + 5) = 6x^2 + 15x - 2x - 5$

$= 6x^2 + 13x - 5$

29. $(3x^2 - 4y)^2 = 9x^4 - 24x^2y + 16y^2$

30. $\frac{3x}{x+5} - \frac{2}{x^2 + 7x + 10}$

$= \frac{3x}{x+5} - \frac{2}{(x+2)(x+5)}$

$$= \frac{3x}{x+5} \cdot \frac{(x+2)}{(x+2)} - \frac{2}{(x+5)(x+2)}$$
$$= \frac{3x(x+2)-2}{(x+5)(x+2)}$$
$$= \frac{3x^2+6x-2}{(x+5)(x+2)}$$

31. $$\frac{1-\frac{9}{x^2}}{1+\frac{3}{x}} = \frac{x^2}{x^2} \cdot \frac{1-\frac{9}{x^2}}{1+\frac{3}{x}} = \frac{x^2 \cdot 1 - x^2 \cdot \frac{9}{x^2}}{x^2 \cdot 1 + x^2 \cdot \frac{3}{x}}$$
$$= \frac{x^2-9}{x^2+3x} = \frac{(x+3)(x-3)}{x(x+3)}$$
$$= \frac{x-3}{x}$$

32. $$\frac{x^2-6x+8}{3x+9} \div \frac{x^2-4}{x+3}$$
$$= \frac{x^2-6x+8}{3x+9} \cdot \frac{x+3}{x^2-4}$$
$$= \frac{(x-4)(x-2)}{3(x+3)} \cdot \frac{x+3}{(x+2)(x-2)}$$
$$= \frac{x-4}{3(x+2)} = \frac{x-4}{3x+6}$$

33. $$\sqrt{5xy} \cdot \sqrt{10x^2y} = \sqrt{50x^3y^2}$$
$$= \sqrt{25 \cdot 2x^2xy^2}$$
$$= 5xy\sqrt{2x}$$

34. $$4\sqrt{72} - 3\sqrt{50} = 4\sqrt{36 \cdot 2} - 3\sqrt{25 \cdot 2}$$
$$= 4 \cdot 6\sqrt{2} - 3 \cdot 5\sqrt{2}$$
$$= 24\sqrt{2} - 15\sqrt{2}$$
$$= 9\sqrt{2}$$

35. $$(5+3i)(7-3i) = 35 - 15i + 21i - 9i^2$$
$$= 35 + 6i - 9(-1)$$
$$= 35 + 6i + 9$$
$$= 44 + 6i$$

36. $$81x^4 - 1 = (9x^2+1)(9x^2-1)$$
$$= (9x^2+1)(3x+1)(3x-1)$$

37. $$24x^3 - 22x^2 + 4x = 2x(12x^2 - 11x + 2)$$
$$= 2x(4x-1)(3x-2)$$

38. $$x^3 + 27y^3 = (x+3y)(x^2 - 3xy + 9y^2)$$

39. $$(f-g)(x) = f(x) - g(x)$$
$$= (x^2+3x-15) - (x-2)$$
$$= x^2 + 3x - 15 - x + 2$$
$$= x^2 + 2x - 13$$
$$(f-g)(5) = 5^2 + 2(5) - 13$$
$$= 25 + 10 - 13 = 22$$

40. $$\left(\frac{f}{g}\right)(x) = \frac{f(x)}{g(x)} = \frac{x^2+3x-15}{x-2}$$
$$= \frac{(x+5)(x-2)}{x-2} = x+5$$

Domain of $\frac{f}{g}$ must exclude all numbers that will make the denominator of the function zero. Domain of $\frac{f}{g}$ is $\{x | x \text{ is a real number and } x \neq 2\}$.

41. $f(g(x)) = f(x-2)$

$= (x-2)^2 + 3(x-2) - 15$

$= x^2 - 4x + 4 + 3x - 6 - 15$

$= x^2 - x - 17$

42. $g(f(x)) = f(x^2 + 3x - 15)$

$= (x^2 + 3x - 15) - 2$

$= x^2 + 3x - 15 - 2$

$= x^2 + 3x - 17$

43. $f(x) = 7x - 3$

$y = 7x - 3$

Interchange x and y and solve for and solve for y.

$x = 7y - 3$

$x + 3 = 7y$

$\frac{x+3}{7} = y$

$f^{-1}(x) = \frac{x+3}{7}$

44. $d = \sqrt{(6-(-3))^2 + (-1-(-4))^2}$

$= \sqrt{9^2 + 3^2} = \sqrt{81+9} = \sqrt{90}$

$= \sqrt{9 \cdot 10} = 3\sqrt{10}$

The distance is $3\sqrt{10}$ units.

45. $(3x^3 - x^2 + 4x + 8) \div (x + 2)$

-2	3	-1	4	8
		-6	14	-36
	3	-7	18	-28

$(3x^3 - x^2 + 4x + 8) \div (x + 2)$

$= 3x^2 - 7x + 18 - \frac{28}{x+2}$

46. $I = \frac{R}{R+r}$

$I(R + r) = R$

$IR + Ir = R$

$IR - R = -Ir$

$R(I - 1) = -Ir$

$R = -\frac{Ir}{I-1}$

47. Since the line is parallel to the line whose equation is $3x + y = 9$, we know the lines will have the same slope. Solve for y to obtain the slope.

$3x + y = 9$

$y = -3x + 9$

The slope of the lines is -3.

Write the equation of the line in point-slope form and solve for y to obtain slope-intercept form.

$y - y_1 = m(x - x_1)$

$y - 5 = -3(x - (-2))$

$y - 5 = -3(x + 2)$

$y - 5 = -3x - 6$

$y = -3x - 1$

48. $\begin{vmatrix} -2 & -4 \\ 5 & 7 \end{vmatrix} = -2(7) - 5(-4)$

$= -14 + 20 = 6$

49.
$$2\ln x-\frac{1}{2}\ln y=\ln x^2-\ln y^{\frac{1}{2}}$$
$$=\ln\frac{x^2}{y^{\frac{1}{2}}}=\ln\frac{x^2}{\sqrt{y}}$$

50.
$$\sum_{i=2}^{5}\left(i^3-4\right)$$
$$=\left(2^3-4\right)+\left(3^3-4\right)+\left(4^3-4\right)+\left(5^3-4\right)$$
$$=(8-4)+(27-4)+(64-4)+(125-4)$$
$$=4+23+60+121=208$$

51. First, find d.
$d=6-2=4$
Next, find a_{30}.
$$a_{30}=2+(30-1)4=2+(29)4$$
$$=2+116=118$$
Now, find the sum.
$$S_{30}=\frac{30}{2}(2+118)=15(120)=1800$$

52.
$$0.\overline{6}=\frac{a_1}{1-r}=\frac{\frac{3}{10}}{1-\frac{1}{10}}=\frac{\frac{3}{10}}{\frac{9}{10}}=\frac{3}{10}\div\frac{9}{10}$$
$$=\frac{3}{10}\cdot\frac{10}{9}=\frac{1}{3}$$

53. Applying the Binomial Theorem to $\left(2x-y^3\right)^4$, we have $a=2x,\ b=-y^3$, and $n=4$.
$$\left(2x-y^3\right)^4$$
$$=\binom{4}{0}(2x)^4+\binom{4}{1}(2x)^3\left(-y^3\right)+\binom{4}{2}(2x)^2\left(-y^3\right)^2+\binom{4}{3}2x\left(-y^3\right)^3+\binom{4}{4}\left(-y^3\right)^4$$
$$=\frac{4!}{0!(4-0)!}16x^4-\frac{4!}{1!(4-1)!}8x^3y^3+\frac{4!}{2!(4-2)!}4x^2y^6-\frac{4!}{3!(4-3)!}2xy^9+\frac{4!}{4!(4-4)!}y^{12}$$
$$=\frac{\cancel{4!}}{0!\cancel{4!}}16x^4-\frac{4!}{1!3!}8x^3y^3+\frac{4!}{2!2!}4x^2y^6-\frac{4!}{3!1!}2xy^9+\frac{\cancel{4!}}{\cancel{4!}0!}y^{12}$$
$$=1\left(16x^4\right)-\frac{4\cdot\cancel{3!}}{1\cdot\cancel{3!}}8x^3y^3+\frac{4\cdot3\cdot\cancel{2!}}{2\cdot1\cdot\cancel{2!}}4x^2y^6-\frac{4\cdot\cancel{3!}}{\cancel{3!}\cdot1}2xy^9+y^{12}$$
$$=16x^4-4\left(8x^3y^3\right)+6\left(4x^2y^6\right)-4\left(2xy^9\right)+y^{12}$$
$$=16x^4-32x^3y^3+24x^2y^6-8xy^9+y^{12}$$

54.
$$f(x)=\frac{x-2}{x^2-3x+2}=\frac{x-2}{(x-2)(x-1)}$$
The domain of f is $\{x|x \text{ is a real number and } x\neq1 \text{ and } x\neq2\}$.

55. $f(x) = \ln(2x-8)$

$2x-8>0$

$2x>8$

$x>4$

The domain of f is $\{x|x>4\}$ or $(4,\infty)$.

56. Let x = the original price of the computer

$x-0.30x=434$

$0.70x=434$

$x=\dfrac{434}{0.70}$

$x=620$

The original price of the computer is \$620.

57. Let x = the width of the rectangle
Let $3x + 1$ = the length of the rectangle

$x(3x+1)=52$

$3x^2+x=52$

$3x^2+x-52=0$

$(3x+13)(x-4)=0$

Apply the zero product principle.

$3x+13=0$ or $x-4=0$

$3x=-13$ $\quad x=4$

$x=-\dfrac{13}{3}$

We disregard $-\dfrac{13}{3}$ because we cannot have a negative length measurement. We conclude that the width is 4 yards and the length is

$3x+1=3(4)+1=12+1=13$ yards.

The rectangle's dimensions are 4 yards by 13 yards.

58. Let x = the amount invested at 12%
Let y = the amount invested at 14%

$x+y=4000$

$0.12x+0.14y=508$

Solve for y in the first equation.

$x+y=4000$

$y=4000-x$

Substitute $4000-x$ for y in the second equation and solve for x.

$0.12x+0.14(4000-x)=508$

$0.12x+560-0.14x=508$

$560-0.02x=508$

$-0.02x=-52$

$x=2600$

Back-substitute 2600 for x to find y.

$y=4000-x$

$y=4000-2600$

$y=1400$

\$2600 was invested at 12% and \$1400 was invested at 14%.

59. $A=Pe^{rt}$

$18000=6000e^{r(10)}$

$\dfrac{18000}{6000}=e^{10r}$

$3=e^{10r}$

$\ln 3=\ln e^{10r}$

$1.0986=10r$

$0.110\approx r$

An interest rate of approximately 11% is required.

60. Since current varies inversely as resistance, we have $i = \frac{k}{R}$.

Use $R = 22$ and $i = 5$ to find k.

$$5 = \frac{k}{22}$$

$$110 = k$$

The equation becomes $i = \frac{110}{R}$. Find i when $R = 10$.

$$i = \frac{110}{10} = 11$$

When the resistance is 10 ohms, 11 amperes of current is needed.